T0189092

Lecture Notes in Computer Science 10636

Commenced Publication in 1973
Founding and Former Series Editors:
Gerhard Goos, Juris Hartmanis, and Jan van Leeuwen

More information about this series at http://www.springer.com/series/7407

Derong Liu · Shengli Xie
Yuanqing Li · Dongbin Zhao
El-Sayed M. El-Alfy (Eds.)

Neural Information Processing

24th International Conference, ICONIP 2017
Guangzhou, China, November 14–18, 2017
Proceedings, Part III

Editors
Derong Liu
Guangdong University of Technology
Guangzhou
China

Shengli Xie
Guangdong University of Technology
Guangzhou
China

Yuanqing Li
South China University of Technology
Guangzhou
China

Dongbin Zhao
Institute of Automation
Chinese Academy of Sciences
Beijing
China

El-Sayed M. El-Alfy
King Fahd University of Petroleum
and Minerals
Dhahran
Saudi Arabia

ISSN 0302-9743 ISSN 1611-3349 (electronic)
Lecture Notes in Computer Science
ISBN 978-3-319-70089-2 ISBN 978-3-319-70090-8 (eBook)
https://doi.org/10.1007/978-3-319-70090-8

Library of Congress Control Number: 2017957558

LNCS Sublibrary: SL1 – Theoretical Computer Science and General Issues

Printed on acid-free paper

This Springer imprint is published by Springer Nature
The registered company is Springer International Publishing AG
The registered company address is: Gewerbestrasse 11, 6330 Cham, Switzerland

Preface

ICONIP 2017 – the 24th International Conference on Neural Information Processing – was held in Guangzhou, China, continuing the ICONIP conference series, which started in 1994 in Seoul, South Korea. Over the past 24 years, ICONIP has been held in Australia, China, India, Japan, Korea, Malaysia, New Zealand, Qatar, Singapore, Thailand, and Turkey. ICONIP has now become a well-established, popular and high-quality conference series on neural information processing in the region and around the world. With the growing popularity of neural networks in recent years, we have witnessed an increase in the number of submissions and in the quality of papers. Guangzhou, Romanized as Canton in the past, is the capital and largest city of southern China's Guangdong Province. It is also one of the five National Central Cities at the core of the Pearl River Delta. It is a key national transportation hub and trading port. November is the best month in the year to visit Guangzhou with comfortable weather. All participants of ICONIP 2017 had a technically rewarding experience as well as a memorable stay in this great city.

A neural network is an information processing structure inspired by biological nervous systems, such as the brain. It consists of a large number of highly interconnected processing elements, called neurons. It has the capability of learning from example. The field of neural networks has evolved rapidly in recent years. It has become a fusion of a number of research areas in engineering, computer science, mathematics, artificial intelligence, operations research, systems theory, biology, and neuroscience. Neural networks have been widely applied for control, optimization, pattern recognition, image processing, signal processing, etc.

ICONIP 2017 aimed to provide a high-level international forum for scientists, researchers, educators, industrial professionals, and students worldwide to present state-of-the-art research results, address new challenges, and discuss trends in neural information processing and applications. ICONIP 2017 invited scholars in all areas of neural network theory and applications, computational neuroscience, machine learning, and others.

The conference received 856 submissions from 3,255 authors in 56 countries and regions across all six continents. Based on rigorous reviews by the Program Committee members and reviewers, 563 high-quality papers were selected for publication in the conference proceedings. We would like to express our sincere gratitude to all the reviewers for the time and effort they generously gave to the conference. We are very grateful to the Institute of Automation of the Chinese Academy of Sciences, Guangdong University of Technology, South China University of Technology, Springer's *Lecture Notes in Computer Science* (LNCS), IEEE/CAA *Journal of Automatica Sinica* (JAS), and the Asia Pacific Neural Network Society (APNNS) for their financial support. We would also like to thank the publisher, Springer, for their cooperation in publishing the proceedings in the prestigious LNCS series and for sponsoring the best paper awards at ICONIP 2017.

September 2017

Derong Liu
Shengli Xie
Yuanqing Li
Dongbin Zhao
El-Sayed M. El-Alfy

ICONIP 2017 Organization

Asia Pacific Neural Network Society

General Chair

Derong Liu	Chinese Academy of Sciences and Guangdong University of Technology, China

Advisory Committee

Sabri Arik	Istanbul University, Turkey
Tamer Basar	University of Illinois, USA
Dimitri Bertsekas	Massachusetts Institute of Technology, USA
Jonathan Chan	King Mongkut's University of Technology, Thailand
C.L. Philip Chen	The University of Macau, SAR China
Kenji Doya	Okinawa Institute of Science and Technology, Japan
Minyue Fu	The University of Newcastle, Australia
Tom Gedeon	Australian National University, Australia
Akira Hirose	The University of Tokyo, Japan
Zeng-Guang Hou	Chinese Academy of Sciences, China
Nikola Kasabov	Auckland University of Technology, New Zealand
Irwin King	Chinese University of Hong Kong, SAR China
Robert Kozma	University of Memphis, USA
Soo-Young Lee	Korea Advanced Institute of Science and Technology, South Korea
Frank L. Lewis	University of Texas at Arlington, USA
Chu Kiong Loo	University of Malaya, Malaysia
Baoliang Lu	Shanghai Jiao Tong University, China
Seiichi Ozawa	Kobe University, Japan
Marios Polycarpou	University of Cyprus, Cyprus
Danil Prokhorov	Toyota Technical Center, USA
DeLiang Wang	The Ohio State University, USA
Jun Wang	City University of Hong Kong, SAR China
Jin Xu	Peking University, China
Gary G. Yen	Oklahoma State University, USA
Paul J. Werbos	Retired from the National Science Foundation, USA

Program Chairs

Shengli Xie	Guangdong University of Technology, China
Yuanqing Li	South China University of Technology, China
Dongbin Zhao	Chinese Academy of Sciences, China
El-Sayed M. El-Alfy	King Fahd University of Petroleum and Minerals, Saudi Arabia

Program Co-chairs

Shukai Duan	Southwest University, China
Kazushi Ikeda	Nara Institute of Science and Technology, Japan
Weng Kin Lai	Tunku Abdul Rahman University College, Malaysia
Shiliang Sun	East China Normal University, China
Qinglai Wei	Chinese Academy of Sciences, China
Wei Xing Zheng	University of Western Sydney, Australia

Regional Chairs

Cesare Alippi	Politecnico di Milano, Italy
Tingwen Huang	Texas A&M University at Qatar, Qatar
Dianhui Wang	La Trobe University, Australia

Invited Session Chairs

Wei He	University of Science and Technology Beijing, China
Dianwei Qian	North China Electric Power University, China
Manuel Roveri	Politecnico di Milano, Italy
Dong Yue	Nanjing University of Posts and Telecommunications, China

Poster Session Chairs

Sung Bae Cho	Yonsei University, South Korea
Ping Guo	Beijing Normal University, China
Yifei Pu	Sichuan University, China
Bin Xu	Northwestern Polytechnical University, China
Zhigang Zeng	Huazhong University of Science and Technology, China

Tutorial and Workshop Chairs

Long Cheng	Chinese Academy of Sciences, China
Kaizhu Huang	Xi'an Jiaotong-Liverpool University, China
Amir Hussain	University of Stirling, UK

James Kwok	Hong Kong University of Science and Technology, SAR China
Huajin Tang	Sichuan University, China

Panel Discussion Chairs

Lei Guo	Beihang University, China
Hongyi Li	Bohai University, China
Hye Young Park	Kyungpook National University, South Korea
Lipo Wang	Nanyang Technological University, Singapore

Award Committee Chairs

Haibo He	University of Rhode Island, USA
Zhong-Ping Jiang	New York University, USA
Minho Lee	Kyungpook National University, South Korea
Andrew Leung	City University of Hong Kong, SAR China
Tieshan Li	Dalian Maritime University, China
Lidan Wang	Southwest University, China
Jun Zhang	South China University of Technology, China

Publicity Chairs

Jun Fu	Northeastern University, China
Min Han	Dalian University of Technology, China
Yanjun Liu	Liaoning University of Technology, China
Stefano Squartini	Università Politecnica delle Marche, Italy
Kay Chen Tan	National University of Singapore, Singapore
Kevin Wong	Murdoch University, Australia
Simon X. Yang	University of Guelph, Canada

Local Arrangements Chair

Renquan Lu	Guangdong University of Technology, China

Publication Chairs

Ding Wang	Chinese Academy of Sciences, China
Jian Wang	China University of Petroleum, China

Finance Chair

Xinping Guan	Shanghai Jiao Tong University, China

Registration Chair

Qinmin Yang	Zhejiang University, China

Conference Secretariat

Biao Luo	Chinese Academy of Sciences, China
Bo Zhao	Chinese Academy of Sciences, China

Contents

Computer Vision

Computer Vision

Scanpath Prediction Based on High-Level Features and Memory Bias

Xuan Shao[1], Ye Luo[1](✉), Dandan Zhu[1], Shuqin Li[1], Laurent Itti[2], and Jianwei Lu[1,3](✉)

[1] School of Software Engineering, Tongji University, Shanghai, China
yeluo@tongji.edu.cn
[2] Department of Computer Science, University of Southern California, Los Angeles, USA
[3] Institute of Translational Medicine, Tongji University, Shanghai, China
jwlu33@hotmail.com

Abstract. Human scanpath prediction aims to use computational models to mimic human gaze shifts under free view conditions. Previous works utilizing low-level features, hand-crafted high-level features, saccadic amplitude, memory bias cannot fully explain the mechanism of visual attention. In this paper, we propose a comprehensive method to predict scanpath from four aspects: low-level features, saccadic amplitude, semantic features learned via deep convolutional neural network, memory bias including short-term and long-term memory. By calculating the probabilities for all candidate regions in an image, the position of next fixation point can be selected via picking the one with the largest probability product. Moreover, fixation duration as a key factor is first used to model memory effect on scanpath prediction. Experiments on two public datasets demonstrate the effectiveness of the proposed method, and comparisons with state-of-the-art methods further validate the superiority of our method.

Keywords: Scanpath prediction · Fixation duration · Memory bias · Semantic features

1 Introduction

Visual attention as a fundamental process in our visual system enables us to allocate our limited processing resources to the most informative part of the visual scene. Most research activities relevant to visual attention are actually dealing with overt attention which involves eye movements. Eye movements are composed of fixations and saccades. Fixations show the interesting regions human watches and while saccades describe the changes of eye positions from one place to another. A sequence of fixations or a scanpath records the trajectory of our attention when we look a scene. Different from saliency map estimation which got well development during the past two decades, the scanpath prediction models can record not only the attended positions but also the order among fixation points. Thus, research on the scanpath prediction can better mimic the

D. Liu et al. (Eds.): ICONIP 2017, Part III, LNCS 10636, pp. 3–13, 2017.
https://doi.org/10.1007/978-3-319-70090-8_1

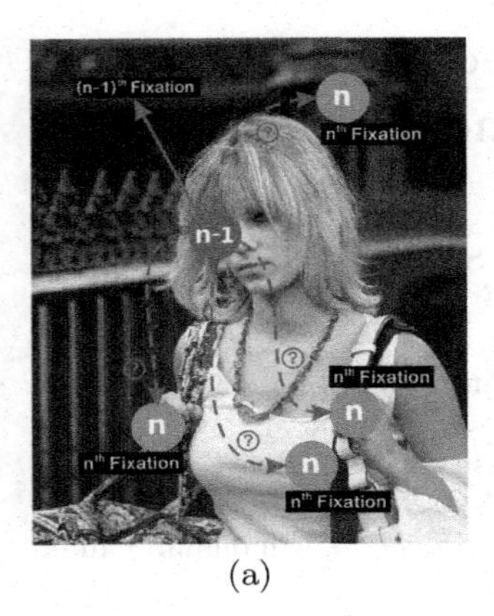

(a)

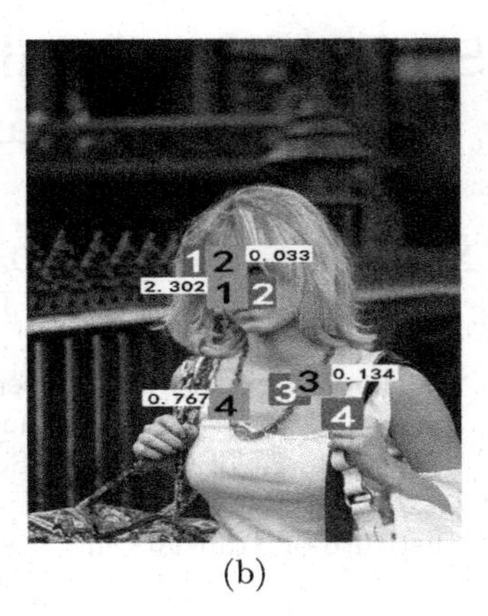

(b)

Fig. 1. (a) Given previous $n-1$ fixation points, our purpose in this paper is to select the best position for the n_{th} fixation point from all candidates in the image. (b) An illustration of the groundtruth and the estimated scanpath. The green scanpath with ordered numbers is groundtruth. The red one is our estimated result, and the float number next to fixation point is fixation duration (e.g. the duration for the first fixation point is 2.302 s). (Color figure online)

mechanism of human vision system and becomes an important research topic in computer vision nowadays. Moreover, various applications for the scanpath prediction are proposed such as medical diagnosis [1,2], advertising design [3,4] and automatic image cropping [5], etc.

To predict scanpath, most of the methods start from using the appearance based features. The influence of various low-level features on scanpath prediction is discussed in [6] but high-level feature is not mentioned. Recently, the amplitude and the orientation of the saccade are used for scanpath prediction [7]. However, high-level features are not considered in their model either. Considering the effects of high-level features, Liu et al. train a Hidden Markov Model (HMM) for scanpath prediction, and the high-level features were implicitly included via the learned hidden states [8]. However, the number of hidden states in HMM is limited and empirically set for each dataset. Therefore, a better way is needed to utilize both low-level and high-level features for scanpath prediction on images. Another key factor influencing gaze shift is memory bias (including short-term memory and long-term memory). Keech et al. point out that gaze shift is simultaneously controlled by the spatial attention and human memory [9], but how to utilize the memory effect is not clearly stated in the paper. According to [10], as the increasing of watching time, long-term memory is involved and affect the change of human gaze. Besides, it is short-term memory that prevents previously attended objects being watched again in the near future. All the above papers reach a consensus that either long-term memory or short-term memory did play an important role in scanpath prediction, but how to model and incorporate them is still an open problem.

Fixation duration, which is the time duration people focus on the fixation point, has close relationship to memory and human's behavior when watch an image. According to [11], the time duration an observer spends on a fixation point will change the position of the next one. This is in accordance with the memory effect: long fixation duration signifies strong attractiveness of a region. Besides, experiments in [12] point out that with fixation duration increasing, more objects are added to memory, and object impression left in memory (i.e. the content of memory) will be changed consequently. Hence it is a good choice to model memory effect by fixation duration.

In order to address above problems in prediction of scanpath, we propose a comprehensive method that combines low-level features, semantic features, saccadic amplitude, long-term memory and short-term memory for scanpath prediction. As is shown in Fig. 1(a), given previous $n-1$ fixation points, we aim to select the suitable position for the n_{th} fixation point from all the candidate regions. To select the position, we estimate the probability for each kind of feature and then select the candidate with the largest probability product as the position of the n_{th} fixation point. Examples of an estimated scanpath and the groundtruth of one observer can be found in Fig. 1(b). Fixation sequence with ordered number in green is groudtruth, while the red one is the result estimated by our method. From this figure we can see that the two sequences show high similarity thus manifest the effectiveness of our proposed method. To summarize, in this paper, the main contributions reside in two-fold:

1. We propose a comprehensive method to integrate various features and decide the positions of fixations during the scanpath prediction on images. These factors including low-level features, semantic features obtained via deep neural network, saccadic amplitude and memory bias are more compact compared to previous methods.
2. Different from other methods, we use fixation duration to measure the time effect when model long-term and short-term memory. Moreover, in order to thoroughly analyze the features influenced the computation of fixation durations, a regression model is built and used to predict the fixation durations.

2 Proposed Method

Given previous $n-1$ fixation points, our method targets to predict the most possible position for the n_{th} fixation point as shown in Fig. 2. At first, instead of image pixels, we employ superpixels as the basic processing unit, and adopt the method in [13] for superpixel segmentation. Then, the position of the n_{th} fixation point is decided through measuring the influence of four factors: low-level features, high-level features, saccadic amplitude and memory bias for each superpixel candidate. Finally, we multiply all the probabilities together and select the one with the largest product as the n_{th} fixation point. In other words, we are attracted by the superpixel with large response to low-level feature, strong semantic meaning, appropriate amplitude of saccade and deep impression in mind.

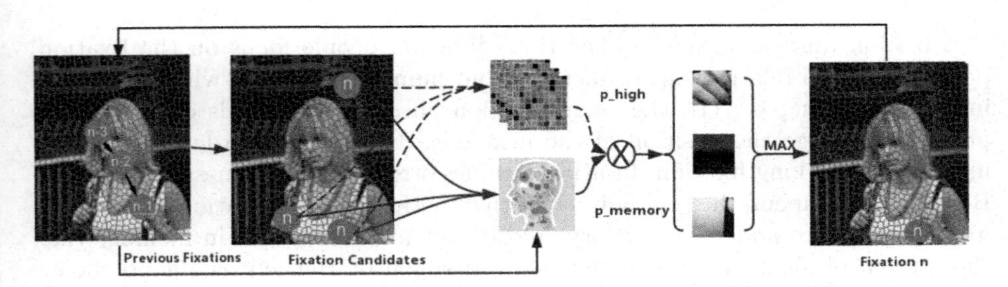

Fig. 2. Flow chart of our proposed method. For simplicity, we omit low-level features and explicitly show effects of memory bias and high-level features. Given previous $n-1$ fixation points, for each superpixel, we first estimate the probability of each feature and then multiply them together. The superpixel with the largest product is considered as the n_{th} fixation point.

Specifically, given an image I, we first segment it into M superpixels, and then the objective function to select the n_{th} fixation point can be formulated as:

$$R^{(n)} = \arg\max_{R}(\{p_l^n(R) \times p_h^n(R) \times p_s^n(R) \times p_m^n(R, d_1, \cdots, d_n)\}_{R \subset I}), \quad (1)$$

where $p_l^n(R)$, $p_h^n(R)$, $p_s^n(R)$ and $p_m^n(R, d_1, \ldots, d_n)$ are probabilities of the low-level features, high-level features, saccadic amplitude and memory bias, respectively, and d_n is the fixation duration for the n_{th} fixation. Note that R could be any superpixel including previous attended regions, and the effect of inhibition-of-return is included in the memory term p_m^n.

The way to obtain the probability of each kind of features is the key ste. For low-level features, we concatenate YUV color values and Gabor features to be a feature vector and then calculate $p_l^n(R)$ for R. For saccadic amplitude, we follow [14] to obtain $p_s^n(R)$. In this paper, we focus on calculations of $p_h^n(R)$ for high-level features and $p_m^n(R, d_1, \ldots, d_n)$ for memory bias. It is worthy noting that fixation durations $\{d_n\}$ are estimated and used to model memory bias.

2.1 High-Level Features

We use a multi-layer convolutional neural network (CNN) to extract high-level features for each superpixel. A superpixel is first wrapped into a bounding box and then sent to CNN for high-level feature extraction as did in [15]. The details of the network architecture can be found in Fig. 3.

Once obtained the learned features, followed by the two fully connected layers, a linear system is learned and used to show the likelihood that the learned high-level features attract human attention, and we approximate the probability of high-level features with the likelihood as shown in Eq. 2:

$$p_h^n(R) = \frac{1}{M}(\mathbf{w}_i^T \phi_c(R) + b_i), i = 1, 2, ..., l, \quad (2)$$

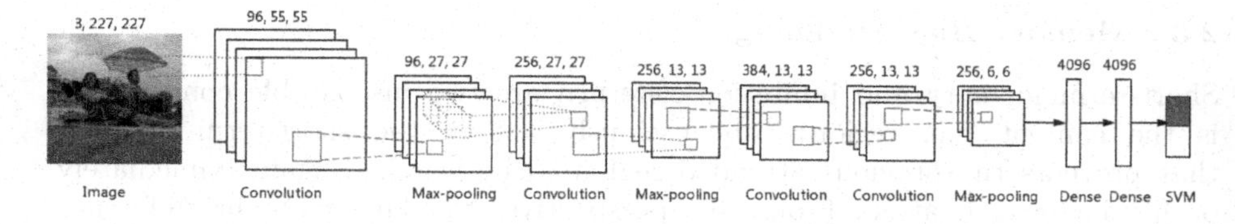

Fig. 3. Architecture of the convolutional neural network used.

where $\phi_c(.)$ is the extracted feature of R and l is the feature dimension. M is total number of superpixels. Parameters of the linear system $\mathbf{w}_i$ and b_i are obtained together with feature learning.

2.2 Fixation Duration Estimation

Fixation duration as an important information can be collected via recording eye movement with eye trackers. However, not all of the datasets provide this kind of information. In order to obtain a model which can estimate the fixation duration for any dataset, we build a regression model on OSIE dataset [16] in which twenty kinds of features influencing the fixation duration together with the ground-truth are provided. Details of all these features can be found in [17].

We employ Support Vector Regression (SVR) to describe the relationship between fixation duration and all used 20 features. In SVR, we first concatenate all the features into one vector $x \in R^{20}$ and then map the features into an m-dimensional feature space using Radial Basis Kernel Function (RBF). The reason we choose RBF is that it generally gives good results without much tuning and bias-term is dropped since we preprocess our fixation duration data to have zero means. Then a linear regression model $f(x, w)$ is written as:

$$f(x, w) = \sum_{i=1}^{m} w_i g_i(x), \tag{3}$$

where $g_i(x), (i = 1, ..., m)$ denotes a set of RBF functions, and w_i is the corresponding weight to the feature in i_{th} dimension. Denote y the groundtruth of a fixation duration and $d = f(x, w)$ the estimated duration. During the training stage, SVR tries to reduce model complexity by minimizing $||w||^2$. Thus SVR is formulated as minimization of the following function:

$$min \frac{1}{2} ||w||^2, \tag{4}$$

$$\text{s.t.} \begin{cases} y_i - f(x_i, w) <= \epsilon + \xi_i \\ f(x_i, w) - y_i >= \epsilon + \xi_i^* \end{cases}$$

where $\xi_i \geq 0$ and $\xi_i^* \geq 0$ are two slack variables to measure the deviation of training samples outside ϵ-insensitive zone. After training, the learned regression model can be used to estimate fixation duration d for any fixation point as long as its corresponded feature x are extracted. In the next section, we will use the predicted fixation duration d for memory modeling.

2.3 Memory Bias Modeling

Short-term memory and inhibition-of-return are two inseparable concepts. It is the content (i.e. region) subconsciously left in the short-term memory that prevents the previous attended region to be watched again immediately or in the next fixation. From this perspective, the large content difference between two successive fixation points will make the next fixation point easy to memorize. Besides, a region which has been fixated for a long time tends to have deep impression in our mind, thus fixation duration as another important factor should be included to model the short-term memory. Therefore, the short-term memory can be modeled as:

$$s_n(R, R^{(n-1)}, d_{n-1}) = G_\sigma\left(\frac{T - d_{n-1}}{||\phi_c(R) - \phi_c(R^{(n-1)})||_2 + \epsilon}\right), n = 2, \cdots, N \quad (5)$$

where $R^{(n-1)}$ and d_{n-1} are the superpixel and estimated fixation duration corresponded to the $(n-1)_{th}$ fixation point, and N is predefined length of the scanpath to be predicted. T is the maximum time that people focus on one point. $G_\sigma(.)$ is a Gaussian function to normalize data into 0 to 1. In our experiments, we set $T = 3$ s, $\sigma = 0.004$, $s_1 = 0$, $N = 5$ and $\epsilon = 0.5$ to make the denominator remain not being zero. It is worth mentioning that only one previous fixation point is taken into consideration to model the effect of short-term memory to the next fixation point. Two more points can be included but are out of scope in this paper.

Nevertheless, to select the next fixation point, the influence caused by all previous attended fixation points is called long-term memory. In other words, all previous attended fixation points have the different level of impact on the selection of the next fixation point, and the impact can be measured as an accumulative effect by short-term memory. Moreover, the role that each short-term memory plays in the long-term memory is quantized by the normalized fixation duration $\frac{d_n}{T}$ with the assumption that long fixation duration tends to make the region much more memorable. Thus, long-term memory can be modeled as:

$$l_n(R, d_1, \ldots, d_n) = l_{n-1}(R, d_1, ..., d_{n-1}) + \frac{s_n(R, R^{(n-1)}, d_{n-1})}{N}\frac{d_n}{T}, \quad (6)$$

where $l_{n-1}(R, d_1, ..., d_{n-1})$ is the long-term memory corresponded to the $(n-1)_{th}$ fixation point. We initialize $l_1(R, d_0, d_1) = 1$.

Generally, both long-term memory and short-term memory contribute equally to the next fixation point selection. The difference lies onto that short-term memory prevents people from looking back while long-term memory selects the fixations according to the impression stored in mind. By this way, the probability of the memory bias $p_m^n(R, d_1, \cdots, d_n)$ can be approximated as:

$$p_m^n(R, d_1, \ldots, d_n) = \frac{l_n(R, d_1, \ldots, d_n) - s_n(R, R^{(n-1)}, d_{n-1})}{Z}, \quad (7)$$

where Z is a normalization factor to make p_m^n a probability.

3 Experimental Results

3.1 Datasets and Evaluation Metric

To evaluate our method, we use two benchmark datasets: NUSEF [18] and JUDD [19]. We use Smith-Waterman metric to evaluate estimated scanpath.

NUSEF DATASET. NUSEF dataset typically consists of a pool of 758 portrait images of different sizes, and each was observed by 15 subjects in a free view environment. For fair comparison, we split the dataset into two subsets: portrait subset and face subset.

JUDD DATASET. Judd dataset consists of 1003 images, including scenery images as well as certain portrait images. In this database, for each image, each viewer fixated on faces, people and text. Other fixations were mainly on body parts(e.g. eyes, hands, etc.), cars and animals.

Evaluation Metric. To evaluate the accuracy of our estimated scanpath, we use Smith-Waterman metric to evaluate its similarity compared with groundtruth scanpath. The large similarity score means the predicted scanpath is closely matched to the groundtruth.

3.2 Comparison with Other Methods

In order to show the superiority of our proposed scanpath prediction method, we compare it to: [20] (Itti for short), [14] (WW for short) and [8] (Liu for short).

Qualitative Comparison. As is shown in Fig. 4, we visualize fixation sequences of three images from NUSEF and JUDD using four aforementioned scanpath prediction methods. From this figure, we can see our method generated favorably similar sequence as the groundtruth. Specifically, with regard to the first fixation point, our method performs better than other methods. Besides, the subsequent sequence generated by our method roughly matches the groudtruth. Nevertheless, our method is less effective in prediction of longer fixation point like the 5_{th} fixation point generated by our method in the first row. This is due to inconsistency among different people who adopt different strategies when watching an image.

Quantitative Comparison. We compare our method with aforementioned methods, using Smith-Waterman metric. From Fig. 5(a), we can see that our method outperforms the three methods on all the employed datasets, and we get extremely better result on JUDD dataset in that the content of this dataset is more complex than NUSEF which is dominant by human portrait, and our method takes more factors that influence gaze shift into consideration thus can get good performance. The reason that our method achieves comparative result with Liu's method [8] on NUSEF is that their method is dataset-oriented and parameters are specifically trained for this dataset.

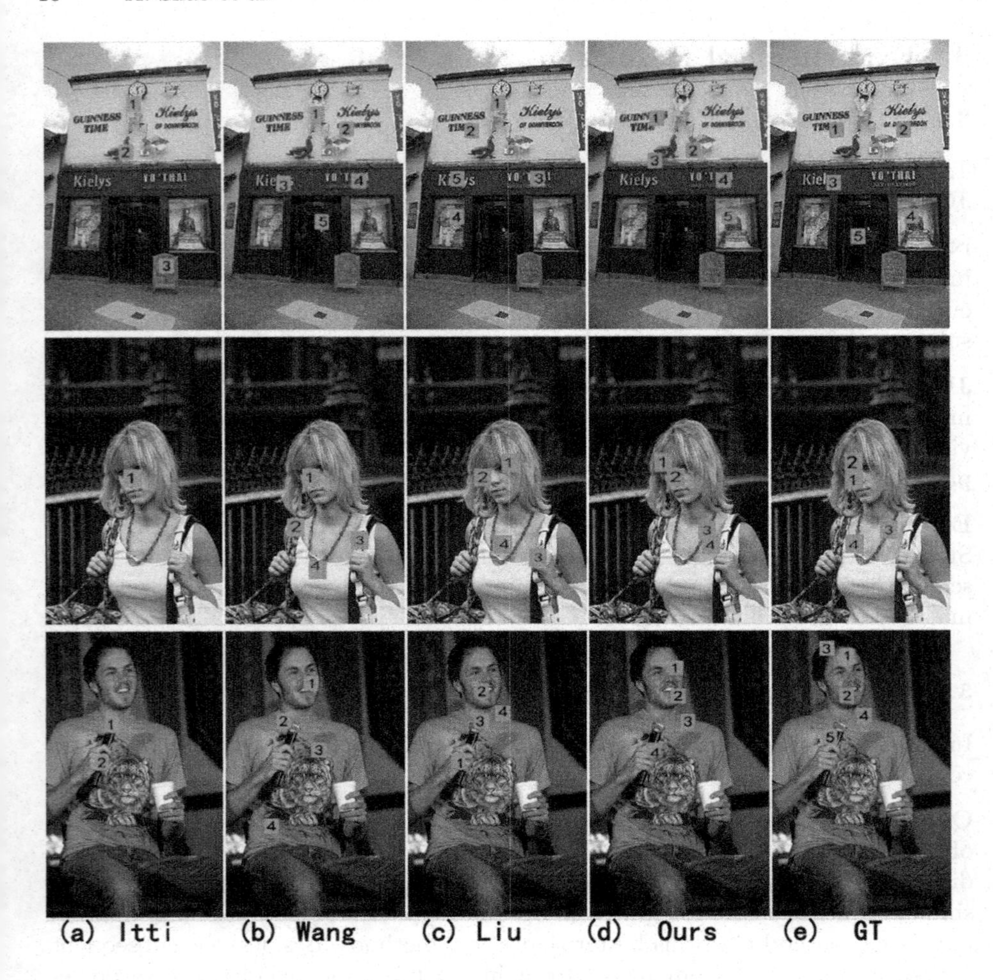

Fig. 4. Qualitative results of different methods compared with ours. The best aligned scanpaths are shown for each method. Each row (from left to right) is the estimated scanpath by methods ranging from Itti's method [20], Wang's method [14], Liu's method [8], Ours' and the groundtruth.

3.3 Effect of Memory Bias

In order to show the effect of memory bias on predicting scanpath, we conduct the experiments using our method without the effect of memory (N−M−D for short), and our method with memory (N+M−D for short) on the employed two datasets. Comparisons results can be seen in Fig. 5(b). From this figure, we can see that our method with memory bias achieves better results than the one without. It is about 8.6% and 11.4% improvement by adding the factor of memory on NUSEF and JUDD datasets respectively.

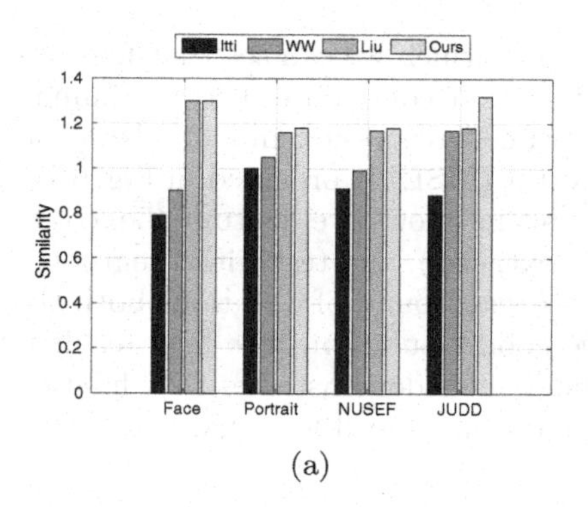

(a)

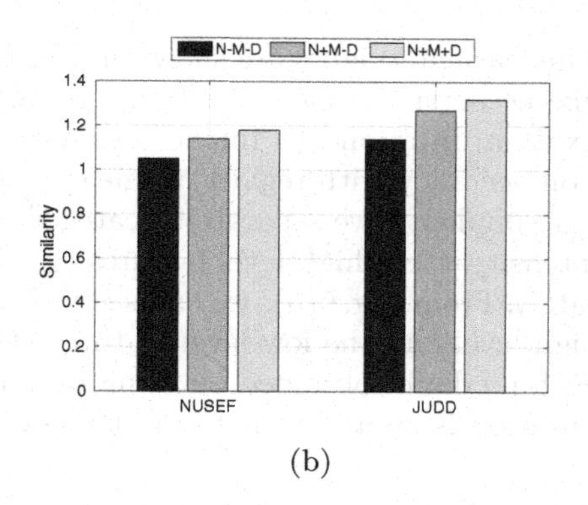

(b)

Fig. 5. (a) Comparison results on Face, Portrait, NUSEF and JUDD dataset with Itti's method [20], WW's method [14] and Liu's method [8]. (b) Comparison results of similarity with and without memory {N−M−D: without memory; N+M−D: memory without duration; N+M+D: memory with duration.}. Better viewed in color.

3.4 Fixation Duration Discussion

Effect of Fixation Duration. In order to show the contribution of fixation duration to scanpath prediction, we compare the performances of our method with and without the effect of fixation duration as shown in Fig. 5(b). From Fig. 5(b) we can see that our method with memory and fixation duration (N+M+D for short) achieves the best result on both NUSEF and Judd datasets, and the result comparisons between our method with memory only (N+M−D for short) and with memory & duration (N+M+D for short) can further validate that the duration of fixations is helpful to scanpath prediction.

Contribution of Features to Fixation Duration Estimation. In order to validate the effectiveness of the trained regression model for fixation duration estimation, we split the OSIE dataset into 80% for training and 20% for testing.

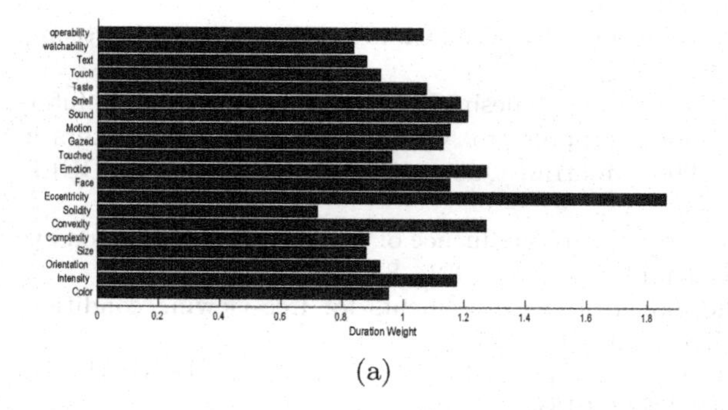

(a)

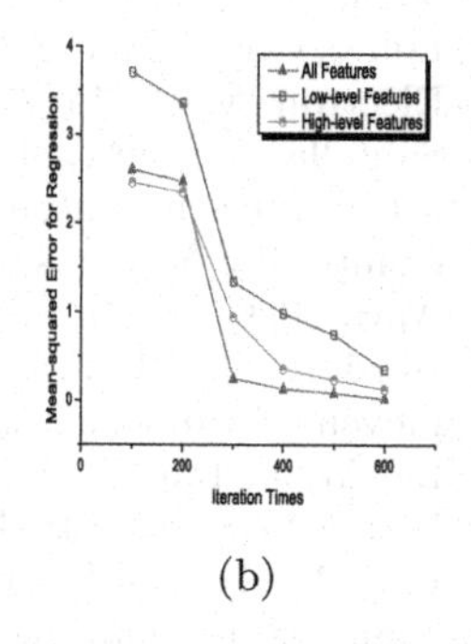

(b)

Fig. 6. (a) Learned weights of twenty features for the regression model on fixation durations. (b) Prediction results measured by mean-squared error per iteration are shown on the test dataset.

The learned weights are shown in Fig. 6(a), from which we can see that features like eccentricity, face, emotion, gazed, and motion contribute a lot to human fixation duration. To further evaluate the trained regression model, classification accuracy with regards to mean-square error (MSE) is provided in Fig. 6(b). For simplicity, we separate 20 category features into low-level features(i.e. color, intensity, etc.), high-level features (i.e. face, text, etc.), and test with them separately. From Fig. 6(b), we can see that after several round of iterations both the high-level and the low-level features achieve good results, but the best result is with no doubt obtained by using both of them. This demonstrates that fixation duration is related to not only low-level but also high-level features.

4 Conclusion

This paper proposes a novel method of predicting scanpath. Compared with other methods, our method combines semantic features learned via deep convolutional neural network and memory bias including short-term and long-term memory. Experiments show our method outperforms state-of-art methods. What's more, we explore the relationship between different features and fixation duration. We find that incorporating fixation duration into memory is beneficial to scanpath prediction. Our future work will focus on establishing an end-to-end framework to predict scanpath based on Recurrent Neural Network.

References

1. He, K., Zhang, X., Ren, S., Sun, J.: Spatial pyramid pooling in deep convolutional networks for visual recognition. IEEE Trans. Pattern Anal. Mach. Intell. **37**(9), 1904 (2014)
2. Galgani, F., Sun, Y., Lanzi, P.L., Leigh, J.: Automatic analysis of eye tracking data for medical diagnosis. In: IEEE Symposium on Computational Intelligence and Data Mining, 2009, CIDM 2009, pp. 195–202. IEEE (2009)
3. Higgins, E., Leinenger, M., Rayner, K.: Eye movements when viewing advertisements. Front. Psychol. **5**(5), 210 (2014)
4. Lohse, G.L., Wu, D.J.: Eye movement patterns on Chinese yellow pages advertising. Electron. Mark. **11**(2), 87–96 (2001)
5. Santella, A., Agrawala, M., Decarlo, D., Salesin, D., Cohen, M.: Gaze-based interaction for semi-automatic photo cropping. In: Conference on Human Factors in Computing Systems, CHI 2006, Montral, Qubec, Canada, pp. 771–780. DBLP, April 2006
6. Harding, G., Bloj, M.: Real and predicted influence of image manipulations on eye movements during scene recognition. J. Vis. **10**(2), 8.1 (2010)
7. Le, M.O., Liu, Z.: Saccadic model of eye movements for free-viewing condition. Vision Research **116**(Pt B), 152 (2015)
8. Liu, H., Xu, D., Huang, Q., Li, W., Xu, M., Lin, S.: Semantically-Based Human Scanpath Estimation with HMMs (2013)
9. Keech, T.D., Resca, L.: Eye movement trajectories in active visual search: contributions of attention, memory, and scene boundaries to pattern formation. Atten. Percept. Psychophys. **72**(1), 114–41 (2010)

10. Becker, M.W., Rasmussen, I.P.: Guidance of attention to objects and locations by long-term memory of natural scenes. J. Exp. Psychol. Learn. Mem. Cogn. **34**(6), 1325 (2008)
11. Kliegl, R., Nuthmann, A., Engbert, R.: Tracking the mind during reading: the influence of past, present, and future words on fixation durations. J. Exp. Psychol. Gen. **135**(1), 12 (2006)
12. Alvarez, G.A., Cavanagh, P.: The capacity of visual short term memory is set both by visual information load and by number of objects. Psychol. Sci. **15**(2), 106–111 (2004)
13. Liu, M.Y., Tuzel, O., Ramalingam, S., Chellappa, R.: Entropy rate superpixel segmentation. In: Computer Vision and Pattern Recognition, vol. 32, pp. 2097–2104. IEEE (2011)
14. Wang, W., Chen, C., Wang, Y., Jiang, T., Fang, F., Yao, Y.: Simulating human saccadic scanpaths on natural images. In: Computer Vision and Pattern Recognition, vol. 42, pp. 441–448. IEEE (2011)
15. Lee, G., Tai, Y.W., Kim, J.: Deep Saliency with Encoded Low Level Distance Map and High Level Features, pp. 660–668 (2016)
16. Xu, J., Jiang, M., Wang, S., Kankanhalli, M.S., Zhao, Q.: Predicting human gaze beyond pixels. J. Vis. **14**(1), 97–97 (2014)
17. Jiang, M., Boix, X., Roig, G., Xu, J., Gool, L.V., Zhao, Q.: Learning to predict sequences of human visual fixations. IEEE Trans. Neural Netw. Learn. Syst. **27**(6), 1241 (2016)
18. Ramanathan, S., Katti, H., Sebe, N., Kankanhalli, M., Chua, T.-S.: An eye fixation database for saliency detection in images. In: Daniilidis, K., Maragos, P., Paragios, N. (eds.) ECCV 2010. LNCS, vol. 6314, pp. 30–43. Springer, Heidelberg (2010). doi:10.1007/978-3-642-15561-1_3
19. Judd, T., Ehinger, K., Durand, F., Torralba, A.: Learning to predict where humans look. In: IEEE International Conference on Computer Vision, vol. 30, pp. 2106–2113. IEEE (2010)
20. Itti, L., Koch, C., Niebur, E.: A model of saliency-based visual attention for rapid scene analysis. IEEE Trans. Pattern Anal. Mach. Intell. **20**(11), 1254–1259 (1998)

Revisiting Faster R-CNN: A Deeper Look at Region Proposal Network

Guangxing Han, Xuan Zhang(✉), and Chongrong Li

Tsinghua National Laboratory for Information Science and Technology (TNList), Institute for Network Sciences and Cyberspace (INSC), Tsinghua University, Beijing 100084, China
hgx14@mails.tsinghua.edu.cn, {zhangx,licr}@cernet.edu.cn

Abstract. Currently, state-of-the-art object detectors are based on Faster R-CNN. We firstly revisit Faster R-CNN and explore problems in it, e.g., coarseness of feature maps for accurate localization, fixed-window feature extraction in RPN and insensitivity for small scale objects. Then a novel object detection network is proposed to address these problems. Specifically, we utilize a two-stage cascade multi-scale proposal generation network to get high accurate proposals: an original RPN is adopted to initially generate coarse proposals, then another network with multi-layer features and RoI pooling layer are introduced to refine these proposals. We also generate small scale proposals in the second stage simultaneously. After that, a detection network with multi-layer features further classifies and refines proposals. A novel 3-step joint training algorithm is introduced to optimize our model. Experiments on PASCAL VOC 2007 and 2012 demonstrate the effectiveness of our network.

Keywords: Faster R-CNN · General object detection · Multi-scale object proposal · Multi-layer feature aggregation

1 Introduction

Object detection is one of the most fundamental problems in computer vision. Currently top leading results on PASCAL VOC [1], MS COCO [2] object detection challenges all utilize Faster R-CNN [3] framework. The pioneering R-CNN [4] decomposes object detection into two primary tasks. Firstly thousands of candidate object locations are generated by traditional object proposal methods (e.g., Selective Search [5]). Then these candidates are classified and further refined by deep convolutional neural network (DCNN [6,7]). Based on R-CNN, Faster R-CNN introduces region proposal network (RPN [3]) to generate high quality proposals and adopts Fast R-CNN [8] to perform RoI-wise classification and refinement. Both shared convolutional layers and end-to-end joint training push object detection to real-time and state-of-art accuracy.

In Faster R-CNN, RPN is built on top of high-level convolutional feature layer (e.g., layer "conv5_3" in VGG-16 [7]) which is shared with Fast R-CNN.

D. Liu et al. (Eds.): ICONIP 2017, Part III, LNCS 10636, pp. 14–24, 2017.
https://doi.org/10.1007/978-3-319-70090-8_2

Specifically, a small network is slided over the feature layer. Each sliding window takes as input an 3×3 spatial window of the convolutional feature maps. Then it is mapped to a lower-dimensional feature (e.g., 512-d for VGG-16). After that, this feature is fed into two sibling fully-connected layers: a box-regression layer (reg) and a box-classification layer (cls) to generate object proposals. Using multiple reg and cls layers, RPN can simultaneously predict multiple object proposals centered at each sliding window with different scales and aspect ratios. Then top scored proposals are selected. Finally, Fast R-CNN, also built on top of high-level convolutional features (e.g., "conv5_3" in VGG16), are followed to further classify and refine the proposals.

However, three issues emerge in such an approach. Firstly, as we know, higher layer convolutional features capture high-level semantic knowledge. Lower layer features, on the other hand, contain rich low-level visual information. [3] only focuses on high-level features, which is too coarse for accurate object detection. Secondly, multiple proposals produced by one sliding window in RPN share the same features which is calculated by the fixed 3×3 spatial window of the DCNN feature. However, these proposals differ much in scales and aspect ratios. We expect RoI pooling features would improve localization accuracy. Thirdly, the receptive field of higher layers is relatively large, and therefore not suitable for small scale object detection. For example in VGG-16, the receptive fields of layer "conv3_3", "conv4_3" and "conv5_3" are 40, 92 and 192 respectively [9]. Higher layers with large receptive field are more suitable for large object detection. In contrast, low-level layers such as "conv4_3" are better matched to small scale object. We can detect different scale objects on different layers.

For the first problem, [10,11] adopt skip-layer connections to directly aggregate multi-layer convolutional features. We follow this with a top-down feature aggregation. Then we propose a novel parallel top-down cascade to refine proposals generated in higher layers. This is different from [12,13] which only use higher layer features for multi-stage refining. For the second problem, HyperNet [10] extracts features for candidate boxes by RoI pooling layer. But approximately 70% of the forward time is consumed by the proposal generation network. HyperNet-SP simplifies this network with detection accuracy loss. We expect careful cascade design can achieve better speed/accuracy trade-offs. As for the third problem, [14,15] employ RPNs on multiple layers to compute multi-scale proposals. Each layer focuses on objects within certain scale ranges. However, lower layers alone do not have strong semantic knowledge for detection [16]. We expect top-down feature aggregation would alleviate the problem.

To summarise, we propose an object detection network with region proposals generated in a top-down cascade manner. Instead of treating DCNNs as black-box feature extractors in [12,13], our approach explores the inherent hierarchies of DCNNs. Concretely, we first generate coarse object proposals on high-level convolutional layer using original RPN. Ablation experiments demonstrate excellent performance of RPN. Then another network, denoted as R-RPN, uses multi-layer features and RoI pooling layer to refine the top scored 2k proposals. Small scale object proposals are generated simultaneously in this stage. Using the two-stage cascade multi-scale proposal generation network, we achieve high recall with only a few proposals. Meanwhile we also introduce a 3-step method to optimize the whole network. Our main contributions are in two folds:

- We propose a novel two-stage cascade multi-scale proposal generation network. It achieves 98.2% recall at IoU = 0.5 with 300 proposals on PASCAL VOC 2007 test which is comparable to HyperNet, and outperforms RPN by a large margin (needs 2k proposals). Moreover, our method is 12× faster than HyperNet on proposal generation due to cascade design.
- Using the 3-step joint training algorithm, we achieve 76.8% and 73.2% mAP on PASCAL VOC 2007 and 2012 test respectively. Moreover, our method runs at 3.3 fps on TITAN X GPU. Both accuracy and running speed are competitive compared to other object detection networks.

2 Object Detection with Two-Stage Cascade Multi-scale Proposal Generation

2.1 Basic Proposal Generation Networks

In this section, we investigate into three basic proposal generation networks RPN, R-RPN and FRCN. In RPN, fixed-window convolutional feature extraction is adopted for fast computation. In FRCN, RoI pooling and two successive fully connected layers (fc) is adopted to extract accurate features for proposals. While R-RPN is a compromise of FRCN and RPN. In R-RPN, the ultimate 256-d feature for each box is obtained through RoI pooling layer and a single fc layer. The difference between R-RPN and RPN lies in that feature is extracted using a fixed window or RoI pooling layer. While R-RPN differs with FRCN on whether or not using more discriminative but computationally expensive fc layers.

2.2 Our Approach

The detailed architecture of our object detection network is shown in Fig. 1. It mainly consists of three parts. Firstly, we generate hierarchical feature maps through convolutional layers. Then a two-stage proposal generation network is followed to produce high quality proposals. Finally, object proposals are further classified and refined by a detection network. In the rest of our paper we use conv3, conv4 and conv5 to represent for layer conv3_3, conv4_3 and conv5_3 respectively. Conv45 and conv345 are the combination of corresponding layers.

Hierarchical convolutional layers feedforward. We use VGG16 as our backbone network. Initially, we resize the input image such that its shorter side is 600 pixels and keep the aspect ratio. Then we feed forward the re-scaled image through hierarchical convolutional layers. Specifically, we mainly focus on features from layers conv3, conv4 and conv5.

Two-stage cascade multi-scale proposal generation. Multi-scale proposals are generated in a two-stage cascade manner. Firstly, we directly build a RPN

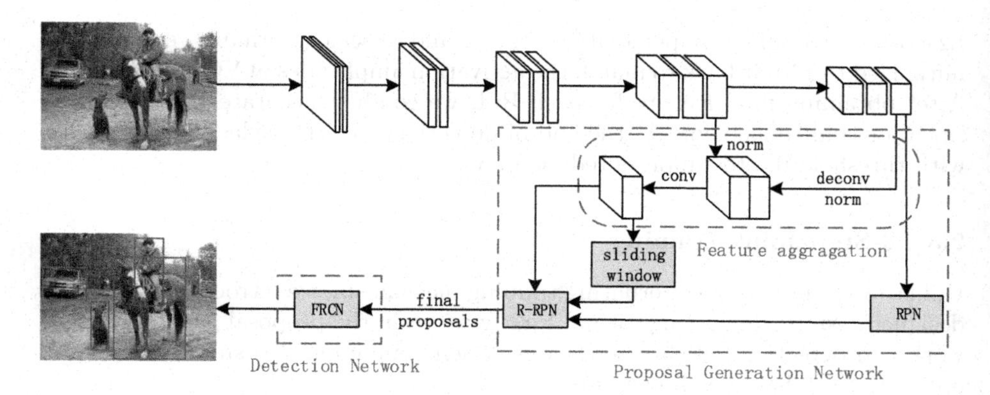

Fig. 1. Proposed object detection architecture. We firstly generate hierarchical feature maps which are shared by proposal generation network and detection network. Then a two-stage cascade multi-scale proposal generation network is introduced to produce proposals. Finally, proposals are further classified and refined by the detection network. Our model is jointly trained in a 3-step manner.

module on top of layer conv5. Roughly 20k anchors are generated as in [3]. We also adopt NMS with threshold 0.7 to suppress most of the redundancy. After this, top 2k ranked proposals are selected and they achieve very high recall, e.g., approximately 98% at IoU = 0.5 on PASCAL VOC 2007 test dataset.

Then a R-RPN module follows. In R-RPN, we combine features from layer conv4 and conv5 to serve as the input feature maps. To concatenate multi-layer features, we firstly use a deconvolutional layer upsampling conv5 to achieve the same resolution with conv4, then L2-normalise each layer per spatial location, rescale it with a learnable "scale layer" initialized to 20 and concat them together. At last, we reduce the dimension of combined feature to 256 channels with 1×1 convolutional layer. The input boxes can be divided into two parts. One comes from the generated 2k proposals in RPN, the other is the small scale boxes generated by sliding window on conv45 resolution (stride: 2) with single scale of 64^2 pixels and aspect ratio 1:1. We generate small scale object proposals on conv45 rather than conv5 to better fit the small object size and utilize fine-grained features. Only one aspect ratio is adopted for no obvious drop in recall when it comes to small scale object. Finally, top 300 proposals are selected after the same NMS operation in RPN before the final detection network.

Final object detection. The final detection network employs a core FRCN module. The input RoIs come from the proposal generation network. Input feature maps are derived from layer conv3, conv4 and conv5. Different from the multi-layer combination strategy used in R-RPN, we combine multi-layer features for each proposal individually. Firstly, we extract fixed feature descriptors from multiple layers simultaneously for each proposal, then perform L2-normalization, concatenation, and feature reduction in turn to produce fixed representation of size $7 \times 7 \times 512$. Different from method in [17], we again

L2-normalise the feature per spatial location and re-scale (learnable "scale layer" initialized to 130) it back to match the activation amplitudes in VGG16 fc6 layer. After obtaining final feature for each RoI, we conduct accurate RoI-wise classification and refinement. At the end of detection, we adopt class specific NMS with threshold 0.3 to suppress redundancy.

2.3 3-Step Joint Training

Unlike [3,10] use 4-step alternating training mechanism, we introduce a carefully designed 3-step training algorithm. Firstly we train the proposal generation network and detection network separately. Then combining two sub-networks, we optimize the whole pipeline end-to-end.

In the first step, we train the two-stage proposal generation network alone. As for RPN, We mainly follow the hyper-parameters in [3]. For R-RPN, we firstly generate boxes by sliding window on conv4 resolution (stride: 2) with 4 scales (64^2, 128^2, 256^2 and 512^2 pixels) and 3 aspect ratios (1:1, 1:2 and 2:1). The positive and negative labels are determined as in [3]. Moreover, we keep the ratios of positive and negative boxes at most 1:2. This small trick both improves the training speed and model accuracy. Moreover, unsampled boxes are neglected in R-RPN while RPN still forwards them due to its implementation by convolution. We jointly optimize two tasks with the loss function defined as:

$$L = l(c_1, c_1^*, b_1, b_1^*) + l(c_2, c_2^*, b_2, b_2^*) \tag{1}$$

$$l(c, c^*, b, b^*) = L_{cls}(c, c^*) + \lambda L_{reg}(b, b^*) \tag{2}$$

where c_1 and c_2 are predicted labels for RPN and R-RPN, c_1^* and c_2^* are corresponding true labels. Similarly, b_1, b_2 and b_1^*, b_2^* are predicted and true boxes for RPN and R-RPN respectively. The L_{cls} and L_{reg} losses are calculated as in [8]. Our model is initialized by VGG16 model. Newly added layers adopt "Xavier" initialization. By default $\lambda = 1$. We train the network with learning rate 10^{-3} on the first 50k iterations and 10^{-4} for the next 30k iterations.

Then in the second step, we train the FRCN together with a simple RPN. Specifically, we adopt the approximate joint training and mainly follow the hyper-parameters in [3]. The loss function is defined as:

$$L = l(c_1, c_1^*, b_1, b_1^*) + l(c_3, c_3^*, b_3, b_3^*) \tag{3}$$

where c_3 and c_3^* are predicted and true labels for FRCN, b_3 and b_3^* are predicted and true boxes for FRCN.

Thus so far proposal generation network and detection network are separately trained. In the third step, we jointly train two networks with shared convolutional layers. Firstly, parameters of the shared convolutional layers and FRCN are inherited from the trained model in step two, while parameters in RPN and R-RPN comes from the model in step one. Different from step one, the input boxes of R-RPN come from top 2k proposals in RPN together with ~2k small scale proposals generated by sliding window. We optimize the whole pipeline with approximate joint training and the loss function is defined as:

$$L = l(c_1, c_1^*, b_1, b_1^*) + l(c_2, c_2^*, b_2, b_2^*) + l(c_3, c_3^*, b_3, b_3^*) \quad (4)$$

We use relatively low learning rate in this step, 10^{-4} for the first 40k iterations, then 10^{-5} and 10^{-6} for the next 40k and 20k iterations.

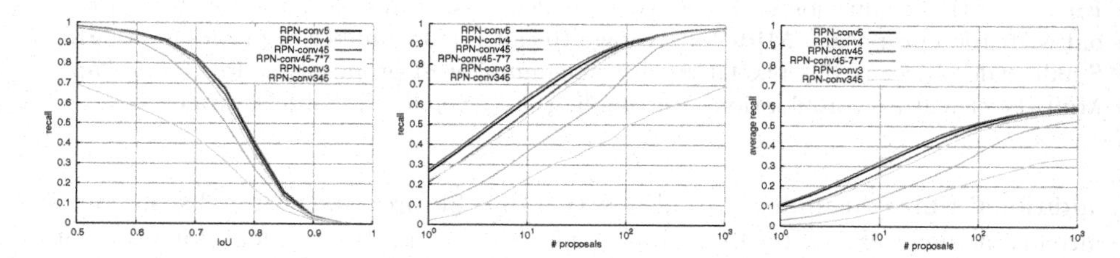

Fig. 2. RPN evaluation on PASCAL VOC 2007 test. **Left**: Recall versus IoU threshold with 1000 proposals. **Middle**: Recall versus number of proposals at IoU = 0.5. **Right**: Average Recall versus number of proposals. "7 * 7": Spatial window size for extracting features in RPN.

3 Experimental Results

3.1 Ablation Experiments

We conduct several ablation experiments to evaluate our design on PASCAL VOC 2007 test. The training set combines VOC 2007 trainval and VOC 2012 trainval dataset. There are about 16.5k trainval images and 5k test images over 20 categories. By default, all networks share the same VGG16 backbone network.

Is RPN with conv5 Feature Good Enough?
We evaluate RPN on a variety of convolutional layers including multi-layer combination. We mainly follow the hyper-parameters in [3]. By default, we use 3×3 spatial window to extract feature maps for each anchor.

The evaluation result is illustrated in Fig. 2. It shows that RPN with conv5 feature maps outperforms methods either using conv4 or conv3 feature maps by a large margin. When using multi-layer features such as conv45 and conv345, the performance significantly improves with the help of conv5, thus indicating that conv5 contains more semantics which is suit for proposal generation. However the combined features are still inferior to conv5 feature alone. The reason is that on conv3 or conv4 resolution, a 3×3 spatial window only has an receptive field of 48 and 108 respectively. It may not generate the exact feature description covering most of the object. If we enlarge the feature window to 7×7 on conv4 resolution, we achieve slightly better proposal recall than conv5 due to the help of fine-grained details of lower layers. But the improvement is minor considering

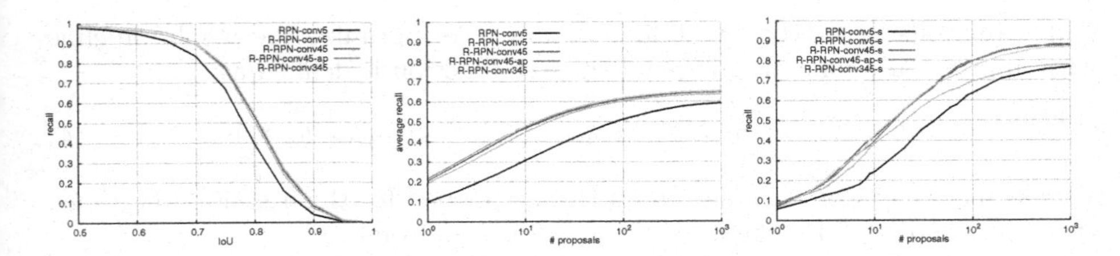

Fig. 3. R-RPN evaluation on PASCAL VOC 2007 test. **Left**: Recall versus IoU threshold with 1000 proposals. **Middle**: Average Recall versus number of proposals. **Right**: Small scale object ($<32^2$ pixels) recall versus number of proposals at IoU = 0.5. "ap": Multi-layer feature combination after pooling. "s": Small scale object proposals.

additional time consuming. We also notice that 2k proposals for RPN on conv5 already achieve 98.5% recall at IoU = 0.5. So it is pretty efficient for RPN on conv5 alone to generate coarse region proposals in the first-stage.

How Much Does the 2nd Stage Proposal Network Help?
We comprehensively evaluate the design choices of the 2nd stage proposal generation network. In practice, we use the training scheme of the first step in Sect. 2.3. By default, multi-layer features are concatenated before R-RPN.

The evaluation result is illustrated in Fig. 3. We can find that all two-stage cascade methods significantly outperform RPN alone. RPN gets 59.2% average recall (AR) with 1000 proposals, while RPN cascaded with R-RPN on conv5, conv45 and conv345 achieve 63.6%, 64.7% and 65.3% AR respectively. For higher IoU (e.g., 0.8), cascaded methods outperform RPN by more than 10 points. This indicates that RoI pooling extracts fine-grained feature which obtains more accurate object localization.

However, it seems R-RPN on conv5, conv45 or conv345 achieves similar performance. Now let's take a deeper look at small scale object (area $< 32^2$ pixels) in Fig. 3. For small scale object, R-RPN on conv5 only gets 77.7% recall with 1000 proposals at IoU = 0.5, which is similar to RPN alone. But R-RPN on conv45 achieves 87.8% recall, similar to conv345, which is substantially higher than conv5. This implies that features on conv4/conv3 resolution are more suitable for small scale objects due to its suitable receptive field and large resolution feature maps. In addition, R-RPN on conv45 is more competitive compared to conv345 for less time-consuming and similar performance.

Furthermore, we also compare different feature combination strategies. We explore on whether to combine multi-layer features before RoI pooling or individually for each box. Results in Fig. 3 indicate that the two methods achieve almost the same performance. However in practice, the later method consumes nearly twice as much time as the former. The reason is that R-RPN needs to process ∼4k boxes and a large amount of boxes are highly overlapped. If we manipulate each box individually, large computation resources will be wasted. So we directly combine conv4 and conv5 before RoI pooling in R-RPN.

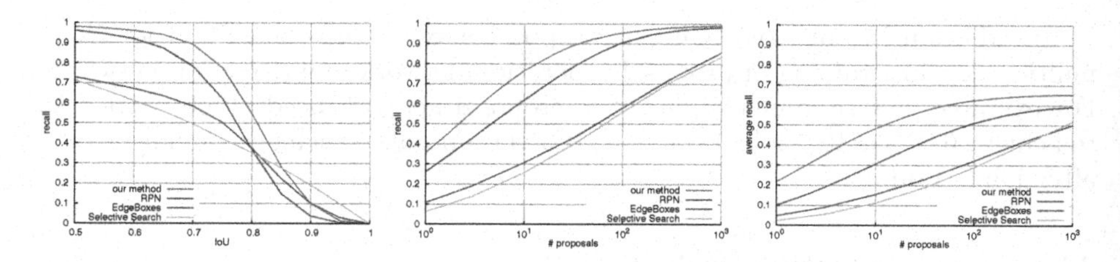

Fig. 4. Proposal evaluation on PASCAL VOC 2007 test. **Left**: Recall versus IoU threshold with 300 proposals. **Middle**: Recall versus number of proposals at IoU=0.5. **Right**: Average Recall versus number of proposals.

Table 1. Object detection mAP on VOC 2007 and 2012 for different features.

Feature	mAP on VOC 2007	mAP on VOC 2012
conv5	75.7	72.2
conv45	76.5	72.9
conv345	**76.8**	**73.2**

How Important Are Multi-layer Features?
We evaluate the effect of multi-layer features on detection network. By default, proposals come from two-stage proposal generation network and multi-layer features are combined individually for each proposal with skip-pooling.

The evaluation result is shown in Table 1. We achieve similar results as [17]. Features with conv345 achieve the best result both in VOC2007 and VOC2012 test dataset. We adopt skip-pooling here for two reasons. Firstly, normalization just before layer fc6 match the activation amplitudes of the original VGG16 fc6 layer, which helps for training. Secondly, only few proposals (e.g., 300 proposals) are evaluated in the final detection network. In practice we observe lower time consumption compared to layer combination before RoI pooling.

3.2 Proposal Evaluation

In this section, we evaluate the proposal generation network on PASCAL VOC 2007 test dataset. The training data consists of VOC 2007 trainval and VOC 2012 trainval. Evaluation result is shown in Fig. 4. Using only 300 proposals, our method achieves 64.7% AR, outperforming RPN by 5.5 points and also 25 points higher than Selective Search as well as EdgeBoxes. To compare with HyperNet and DeepProposal, [10] shows that HyperNet and HyperNet-SP needs 20 and 30 proposals respectively to achieve 75% recall at IoU = 0.7. DeepProposal [18] needs 540 proposals. Our method requires 25 proposals. However, HyperNet takes nearly 810 ms for proposal generation owing to the enormous number of input RoIs, while our method only costs 70 ms to achieve comparable results due to the efficient cascade design.

In addition, we observe that our proposal network also benefits a lot from multi-task joint training of proposal generation network and detection network. For example, before joint training, our proposal network needs 1000 proposals to achieve 64.7% AR, while it only needs 300 proposals to achieve similar results after joint training.

3.3 Object Detection Evaluation

We comprehensively compare our method with other VGG16 based object detection networks on PASCAL VOC 2007 and 2012 test dataset.

Table 2. PASCAL VOC 2007 & 2012 test detection results. For VOC2007 test, the training data is 2007trainval + 2012trainval. For VOC2012 test, the training data is 2007trainvaltest + 2012trainval. "#100": 100 proposals. "#300": 300 proposals.

Approach	FPS	mAP on VOC 2007	mAP on VOC 2012
Faster R-CNN	7	73.2	70.4
CRAFT	-	75.7	71.3
HyperNet-SP	5	74.8	71.3
HyperNet	0.9	76.3	71.4
Ours-#100	4	76.3	72.3[a]
Ours-#300	3.3	**76.8**	**73.2**[b]

[a] http://host.robots.ox.ac.uk:8080/anonymous/NMMVBD.html.
[b] http://host.robots.ox.ac.uk:8080/anonymous/ZMWN3P.html

PASCAL VOC 2007 results. The comparison result is shown in Table 2. Using 100 proposals, our method achieves 76.3% mean average precision (mAP) on VOC 2007 similar to HyperNet. While using 300 proposals, our method achieves 76.8% mAP, 0.5 points higher than HyperNet and 2 points higher than HyperNet-SP. Moreover, our method has a frame rate of 3.3 fps which is comparable with HyperNet-SP and 3.7× faster than HyperNet, which demonstrates the effectiveness of our cascade design. Our method also achieves 1.1 points higher than CRAFT, which proves the effectiveness of the top-down cascade multi-scale proposal generation and our 3-step joint training. Moreover, our method is especially good at small scale object detection. For example our method achieves 66.9% mAP on "bottle", which is 15 points higher than Faster R-CNN and 4.5 points higher than HyperNet.

PASCAL VOC 2012 results. We observe similar results on PASCAL VOC 2012 test dataset in Table 2. Using 100 proposals, our method achieves 72.3% mAP on VOC 2012 test, 0.9 points higher than HyperNet and CRAFT. Using 300 proposals, our method achieves 73.2% mAP, 1.8 points higher than HyperNet and CRAFT. We still outperforms other methods significantly on small scale objects such as "bottle" and "plant".

Our implementation adopt the publicly available code of [3] in python version. All of our networks are trained and tested on a single TITAN X GPU.

4 Conclusion

We propose an object detection network with a novel proposal generation subnetwork attempting to solve problems in Faster R-CNN. Two-stage cascade multi-scale proposal generation network is carefully designed for fast and accurate proposals generation. Specifically, initial multi-scale proposals are generated in RPN, then R-RPN is adopted to further refine those proposals. We also conduct several ablation experiments to evaluate our design. Experiments on PASCAL VOC 2007 and 2012 demonstrate the effectiveness of our method both on speed and accuracy compared to other object detection networks.

References

1. Everingham, M., Eslami, S.A., Van Gool, L., et al.: The Pascal visual object classes challenge: a retrospective. Int. J. Comput. Vis. **111**, 98–136 (2015). LNCS. Springer
2. Lin, T.-Y., Maire, M., Belongie, S., Hays, J., Perona, P., Ramanan, D., Dollár, P., Zitnick, C.L.: Microsoft COCO: Common Objects in Context. In: Fleet, D., Pajdla, T., Schiele, B., Tuytelaars, T. (eds.) ECCV 2014. LNCS, vol. 8693, pp. 740–755. Springer, Cham (2014). doi:10.1007/978-3-319-10602-1_48
3. Ren, S., He, K., Girshick, R., et al.: Faster R-CNN: towards real-time object detection with region proposal networks. In: Advances in Neural Information Processing Systems 28, pp. 91–99. Curran Associates, Montréal (2015)
4. Girshick, R., Donahue, J., Darrell, T., et al.: Region-based convolutional networks for accurate object detection and segmentation. In: IEEE Computer Vision and Pattern Recognition, pp. 580–587. IEEE Press, Columbus (2014)
5. Uijlings, J.R., Van De Sande, K.E., Gevers, T., et al.: Selective search for object recognition. Int. J. Comput. Vis. **104**, 154–171 (2013)
6. Krizhevsky, A., Sutskever, I., Hinton, G.E.: ImageNet classification with deep convolutional neural networks. In: Advances in Neural Information Processing Systems 25, pp. 1097–1105. Curran Associates, South Lake Tahoe (2012)
7. Simonyan, K., Zisserman, A.: Very deep convolutional networks for large-scale image recognition. arXiv preprint arXiv:1409.1556 (2014)
8. Girshick, R.: Fast R-CNN. In: IEEE International Conference on Computer Vision, pp. 1440–1448. IEEE Press, Santiago (2015)
9. Yu, W., Yang, K., Bai, Y., et al.: Visualizing and comparing convolutional neural networks. arXiv preprint arXiv:1412.6631 (2014)
10. Kong, T., Yao, A., Chen, Y., et al.: HyperNet: towards accurate region proposal generation and joint object detection. In: IEEE Computer Vision and Pattern Recognition, pp. 845–853. IEEE Press, Las Vegas (2016)
11. Zhang, L., Lin, L., Liang, X., He, K.: Is Faster R-CNN doing well for pedestrian detection? In: Leibe, B., Matas, J., Sebe, N., Welling, M. (eds.) ECCV 2016. LNCS, vol. 9906, pp. 443–457. Springer, Cham (2016). doi:10.1007/978-3-319-46475-6_28
12. Yang, B., Yan, J., Lei, Z., et al.: Craft objects from images. In: IEEE Computer Vision and Pattern Recognition, pp. 6043–6051. IEEE Press, Las Vegas (2016)

13. Gidaris, S., Komodakis, N.: Attend refine repeat: active box proposal generation via in-out localization. arXiv preprint arXiv:1606.04446 (2016)
14. Cai, Z., Fan, Q., Feris, R.S., Vasconcelos, N.: A unified multi-scale deep convolutional neural network for fast object detection. In: Leibe, B., Matas, J., Sebe, N., Welling, M. (eds.) ECCV 2016. LNCS, vol. 9908, pp. 354–370. Springer, Cham (2016). doi:10.1007/978-3-319-46493-0_22
15. Liu, W., Anguelov, D., Erhan, D., Szegedy, C., Reed, S., Fu, C.-Y., Berg, A.C.: SSD: single shot multibox detector. In: Leibe, B., Matas, J., Sebe, N., Welling, M. (eds.) ECCV 2016. LNCS, vol. 9905, pp. 21–37. Springer, Cham (2016). doi:10.1007/978-3-319-46448-0_2
16. Lin, T.Y., Dollár, P., Girshick, R., et al.: Feature pyramid networks for object detection. arXiv preprint arXiv:1612.03144 (2016)
17. Bell, S., Lawrence Zitnick, C., Bala, K., et al.: Inside-outside net: detecting objects in context with skip pooling and recurrent neural networks. In: IEEE Computer Vision and Pattern Recognition, pp. 2874–2883. IEEE Press, Las Vegas (2016)
18. Ghodrati, A., Diba, A., Pedersoli, M., et al.: DeepProposal: hunting objects by cascading deep convolutional layers. In: IEEE International Conference on Computer Vision, pp. 2578–2586. IEEE Press, Santiago (2015)

A Novel Ant Colony Detection Using Multi-Region Histogram for Object Tracking

Seid Miad Zandavi[1], Feng Sha[1(✉)], Vera Chung[1], Zhicheng Lu[1], and Weiming Zhi[2]

[1] School of Information Technologies, University of Sydney, Sydney, NSW 2006, Australia
{miad.zandavi,feng.sha,vera.chung}@sydney.edu.au
[2] Department of Engineering Science, University of Auckland, Auckland 1010, New Zealand
wzhi262@aucklanduni.ac.nz

Abstract. Efficient object tracking become more popular in video processing domain. In recent years, many researchers have developed excellent models and methods for complicated tracking problems in real environment. Among those approaches, object feature definition is one of the most important component to obtain better accuracy in tracking. In this paper, we propose a novel multi-region feature selection method which defines histogram values of basic areas and random areas (MRH) and combined with continuous ant colony filter detection to represent the original target. The proposed approach also achieves smooth tracking on different video sequences, especially with Motion Blur problem. This approach is designed and tested in MATLAB 2016b environment. The experiment result demonstrates better and faster tracking performance and shows continuous tracking trajectory and competitive outcomes regarding to traditional methods.

Keywords: Multi-Region Histogram · Ant colony filter · Histogram

1 Introduction

Video tracking plays an important role in computer vision applications. Researchers and companies seek to find accurate and precise object detection and tracking estimation for different classes of objects in different kinds of videos. During the tracking process, good feature definition with swarm intelligence estimation can help increasing the whole tracking performance in complicated environment and handle vary object change challenges.

Zhang and van der Maaten state three components to form a tracking process [1]: observation model, motion model and search strategy. In this paper, we propose a novel feature definition base on selecting specific regions inside target instead of the whole domain in order to distinguish the key areas and less important areas. This proposed observation model can map key regions to estimated solution and increase accuracy when significant change occurs. Our approach uses continuous ant colony filter [2, 3] as

D. Liu et al. (Eds.): ICONIP 2017, Part III, LNCS 10636, pp. 25–33, 2017.
https://doi.org/10.1007/978-3-319-70090-8_3

search strategy to find possible solution in next frame. We also optimize the parameters [4, 5] in continuous ant colony filter to give proper setting for the estimation.

This paper is organized in five sections. Section 1 introduces brief explanation about the proposed approach. Section 2 details our feature selection method and histogram representation process of object definition, it includes the structure of feature selection and mapping process from previous frame to next frame to overcome drawbacks about illumination, motion blur, etc. Section 3 explains the continuous ant colony filter and our approach for better parameter selection algorithm. In Sect. 4, we present experiment results and compare our MRH-ACD to traditional approaches and benchmark trackers in plural video sequences [6]. Finally, we conclude our approach and further work in Sect. 5.

2 Multi-Region Histogram Feature

In order to make observation model more efficient and to recognize the target object (like human, car, etc.) in first place, an excellent feature definition is important. Many researchers have already designed local invariant feature methods. They contain direct raw pixel definition [7], key points detection like SIFT [8] and SURF [9], and mathematical representation HOG [11]. These approaches performed good outcome to specific images and videos but could not handle complicated target and environment such as Occlusion, Deformation, Illumination Variation, etc. [6]

In this paper, our approach is to select fixed pixel areas (basic area) combined with several random pixel areas (supplementary area) as preliminary object descriptor. These selected areas are called as feature elements, then we assign different weight value for each element to determine their importance. By matching integrated histograms of original and possible targets with weight and elasticity parameters, the proposed method would find true solution much easier. The procedures of Multi-Region Histogram (MRH) calculation are as follows:

1. Load frame 0 and its gray version picture, then transfer image to HSV format;
2. Select pixel areas, then calculate and normalize each histogram value for feature description according to Fig. 1: suppose we have a target object with center point (x, y) and the width and height value are x_1 and y_1 respectively; red block is set as 1/4 of total height x_1 and width y_1 of the target, green blocks are set as 1/8 of x_1 and y_1; blue blocks are set random center point with target object and the size of each block cannot be larger than 1/4 of the original target size and their width and height cannot be larger than x_1 and y_1;
3. Compare element histograms use OpenCV method Bhattacharyya distance [7] in Eq. (1), where H_{k1} and H_{k2} represent the two histogram values in matching element area k (k = *basic area numbers* + *random are numbers*); I represent the length of histogram, N represents the bin numbers.

$$d(H_{k_1}, H_{k_2}) = \sqrt{1 - \frac{1}{\sqrt{\overline{H_{K_1}}\,\overline{H_{K_2}}N^2}} \sum_I \sqrt{H_{k_1}(I) \cdot H_{k_2}(I)}} \tag{1}$$

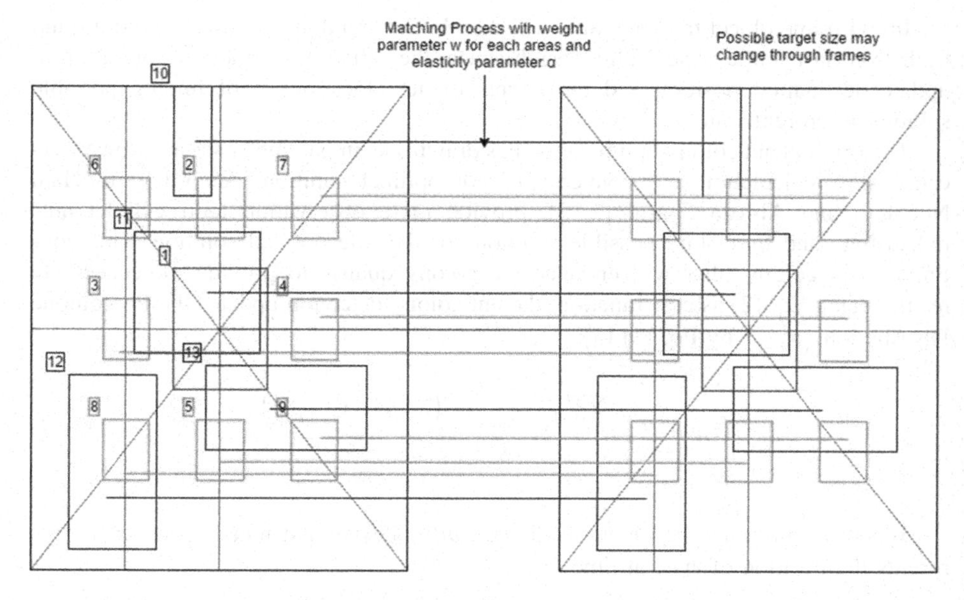

Fig. 1. Multi-Region Histogram (MRH) selection and mapping. (Color figure online)

The likelihood value depends on the similarity of two objects' histogram properties in Eq. (2), where α is the proportion of desired object and possible solution, and w depends on the selected elements' importance. The basic elements that have more contributions to the original target definition will be assigned larger weight value while random elements will have smaller weight value. Histogram bins b in k element area within original target *H1* and possible solution *H2* will compute the sum of all bins *hk-1* (*hk1*) and *hk-2* (*hk2*).

$$fitness = \alpha \cdot \sum\nolimits_{k} \left(w \cdot \sum\nolimits_{bk_2}^{hk_1} Hk_1 \times \sum\nolimits_{bk_2}^{hk_2} Hk_2 \right) \tag{2}$$

3 Ant Colony Detection

3.1 Continuous Ant Colony Filter

After study on natural ant colony behavior, researchers developed powerful swarm algorithms to solve optimization and estimation problems. Base on the ant movement to search and retrieve food source, the ant colony optimization has applied to many real-industry applications like scheduling, vehicle route and image processing [15].

The basic idea for ant colony intelligence is individual ant distribution and information interaction within the swarm. Each ant will search for possible food source within a proper range, once possible food sources identified, other ants will converge to the source destination follow pheromone information until food retrieved [15].

In 2D frame object tracking, ants will distribute to random location around ground truth from last frame. And within search space, they share pheromone information to update destination position and move ants to new locations until finding possible solution in current frame.

The original ant colony optimization has drawbacks in parameter setting which may cause slow calculation and converge to non-optimal solution. According to Hadi Nobahari and Alireza Sharifi [2, 3], provide more observation loop will let ants redistribute and find other possible solutions or the true optimal solution. This continuous ant colony filter will enhance the colony quality to provide more accurate result. Using Eq. (2) as cost function, the ant colony detection updates the Pheromone Distribution $\left(\sigma_{k-1}^{i}\right)$ by Eq. (3) [3]:

$$\left(\sigma_{k-1}^{i}\right)^{2}=\frac{\sum_{j=1}^{m}\frac{1}{f_{k-1}^{i,j}-f_{k-1,min}^{i,j}}\left(x_{k-1}^{i,j}-x_{k-1,min}^{i,j}\right)}{\sum_{i=1}^{m}\frac{1}{f_{k-1}^{i,j}-f_{k-1,min}^{i,j}}} \tag{3}$$

Where i represents the iteration number, m represents ant number, and $x_{k-1}^{i,j}$ represents the position of ant j at time $k-1$.

3.2 Parameter Selection Algorithm

Different parameter setting will affect the performance of search algorithm [4, 5], especially for ant colony algorithm. In order to determine the suitable parameters in continuous ant colony filter to improve the search performance, the proposed multi-region histogram with ant colony detection (MRH-ACD) conducts random sampling [4, 5] to tune the parameters of ant colony detection.

The intervals of parameters space are given in Table 1. These spaces are taken widely enough to be utilized for different benchmarks prevalently. The performance of ACD is evaluated for 500 sets of parameters, generated randomly using uniform distribution [4, 5]. For each set, precision is calculated using *Dag1*. Then the mean precision which shows the performance of ACD is defined as cost function for each parameter set in tracking scheme. Logically, mean precision is smaller than 1, however, closer to 1 determines the better performance.

Table 1. Search domain of ACD parameters

Parameters	min	max
Ant No.	10	100
Top Ant Fraction (TAF)	0.1	0.8
Expansion Factor (EF)	1.0	5
Iteration	5	100

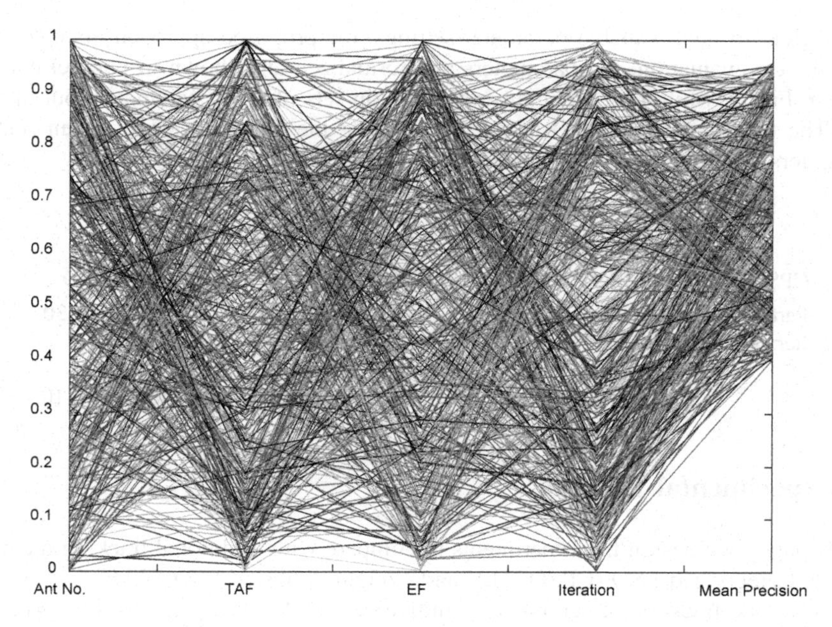

Fig. 2. Parallel visualization of 500 parameter sets

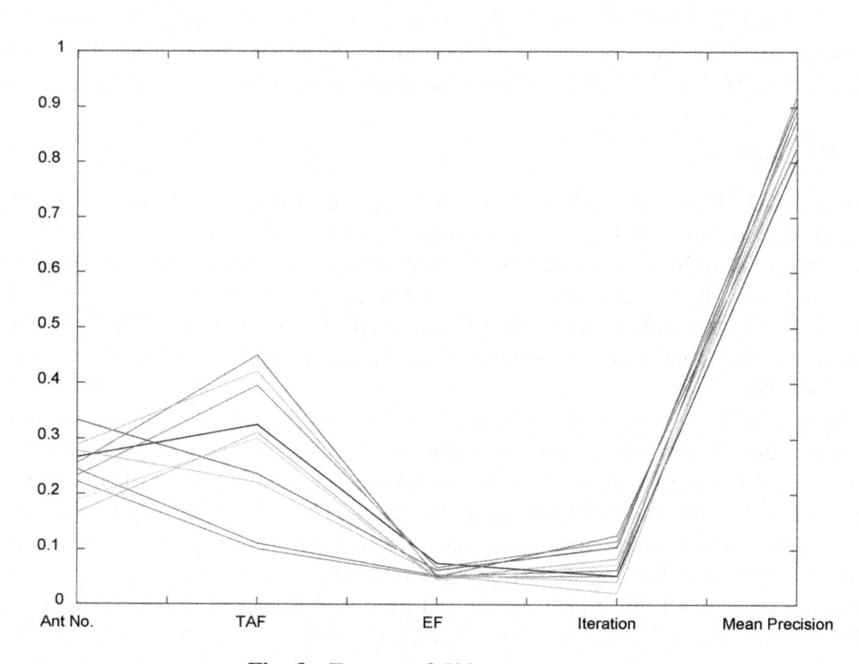

Fig. 3. Top ten of 500 parameter sets

In order to fetch suitable parameter settings, the proposed approach obtained 500 random sets evaluation. All parameters are normalized and shown in parallel coordinates in Fig. 2, while Fig. 3 demonstrates ten top sets of ACD parameters out of 500 sets. The tuned value of ACD is calculated using the weight average of ten sets of parameters. Finally, the best suitable tuned parameters are listed in Table 2.

Table 2. Parameter setting among test methods

CPSO	WAPSO	SCT	MRH-ACD
Particle No. = 30 Iteration = 20	Particle No. = 30 Iteration = 20	Degree of Influence = 1.5 Dependency = 3	Ant No. = 30 TAF = 0.3 EF = 1.2 Iteration = 10

4 Experimental Results

In this paper, we design the experiments to compare with traditional tracking methods such as Categorized PSO (CPSO) [12] and Weight Adjusted PSO(WAPSO) [13] and deep network trackers Structuralist cognitive model for tracking (SCT) [14]. The proposed MRH-ACD provides better outcomes in multiple sequences in OTB-50 [6] and OTB-100 datasets [6].

All experiments were performed in same system configuration as follows: Windows 10, Intel i7-6700 CPU@3.4 GHz, x64-based processor and 8 GB RAM. The parameters of SCT were set to the values presented according to [14]. Two major parameters of SCT and other PSO parameter settings are listed in Table 2.

4.1 Evaluation

Table 3 listed 10 frames each in 5 video sequences for the proposed MRH-ACD experiment results from OTB-50 and OTB-100 [6]. The testing videos include different target objects in different environments like car blur and car in night road, man face detection, human body movement with interference and size change of a dog doll. The tracking challenges expand to Scale Variation, Motion Blur, Fast Motion, In-Plane Rotation, Out-of-Plane Rotation, Illumination Variation, Background Clutters, Occlusion and Deformation.

The results show that CPSO [12] and WAPSO [13] which use color histogram to represent whole target cannot handle similar color environment and light reflection in BlurCar2 and CarDark videos; the same problem also occurs in complicated woman video which contains multiple tracking problems; they lost target in early frames. They also cannot provide stable track trajectory about blur human face when In-Plane Rotation occurs, and the dog doll scale up and down.

On the other hand, SCT [14] model and proposed MRH-ACD demonstrate good quality of tracking and focus on target trajectory during the whole tracking process, while MRH-ACD performs better outcomes according to general performance

Table 3. Experiments' screenshots.

Sequences	Tracking screenshots
BlurCar2 (Frame 540-549)	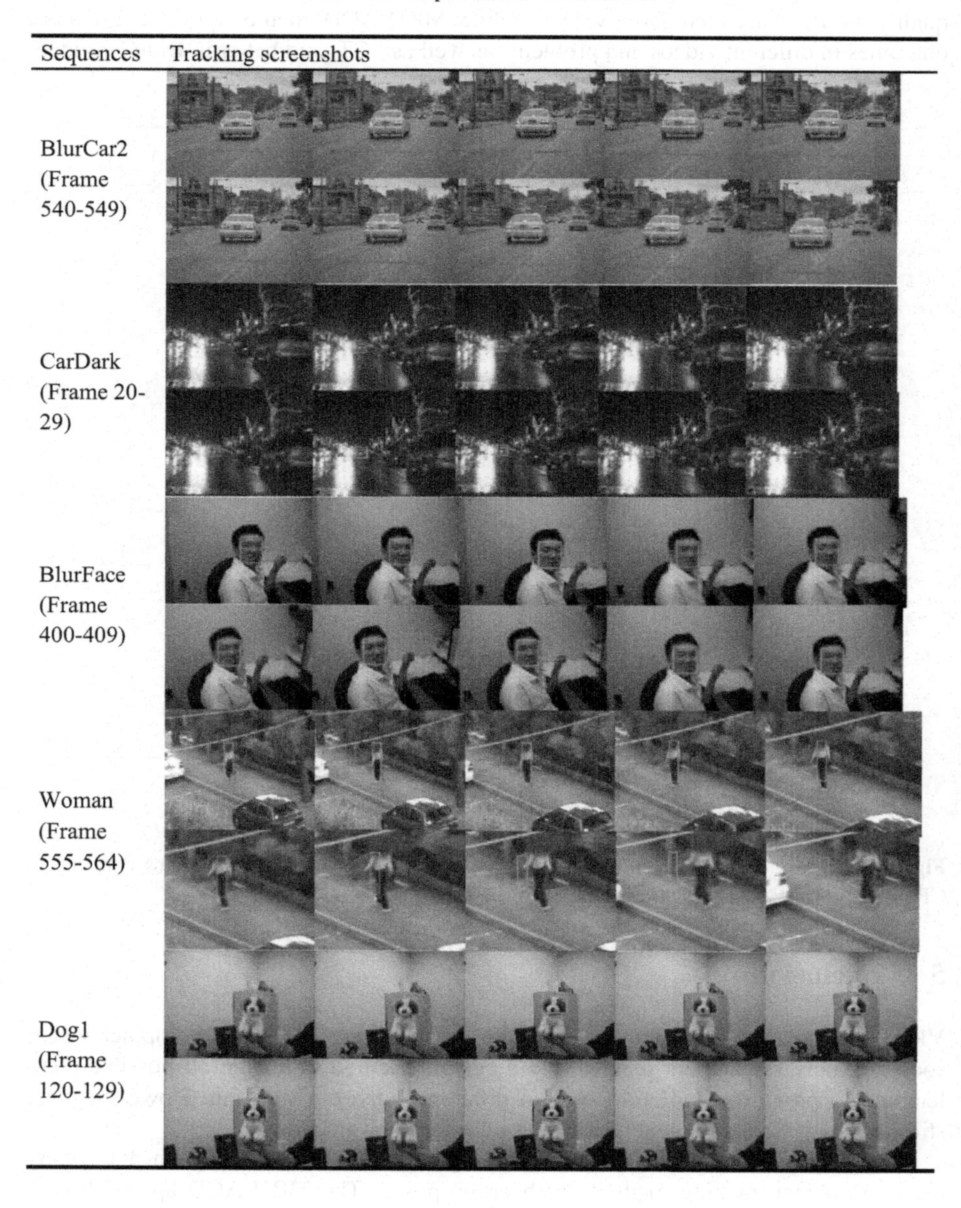
CarDark (Frame 20-29)	
BlurFace (Frame 400-409)	
Woman (Frame 555-564)	
Dog1 (Frame 120-129)	

evaluation function in benchmark (precision plots) [6] and only lost frames in woman video at frame 561 when target half-covered by trees.

Figure 4 shows 4 examples in OTB-50 and OTB-100 dataset [6] testing base on one-pass evaluation (OPE) of precision plots diagrams. The results show CPSO [12] and WAPSO [13] performs low quality in BlurCar2 and Woman videos and average quality in BlurFace and Doll videos, while MRH-ACD method provide excellent outcomes in different videos and problems as well as SCT, and better in BlurCar video.

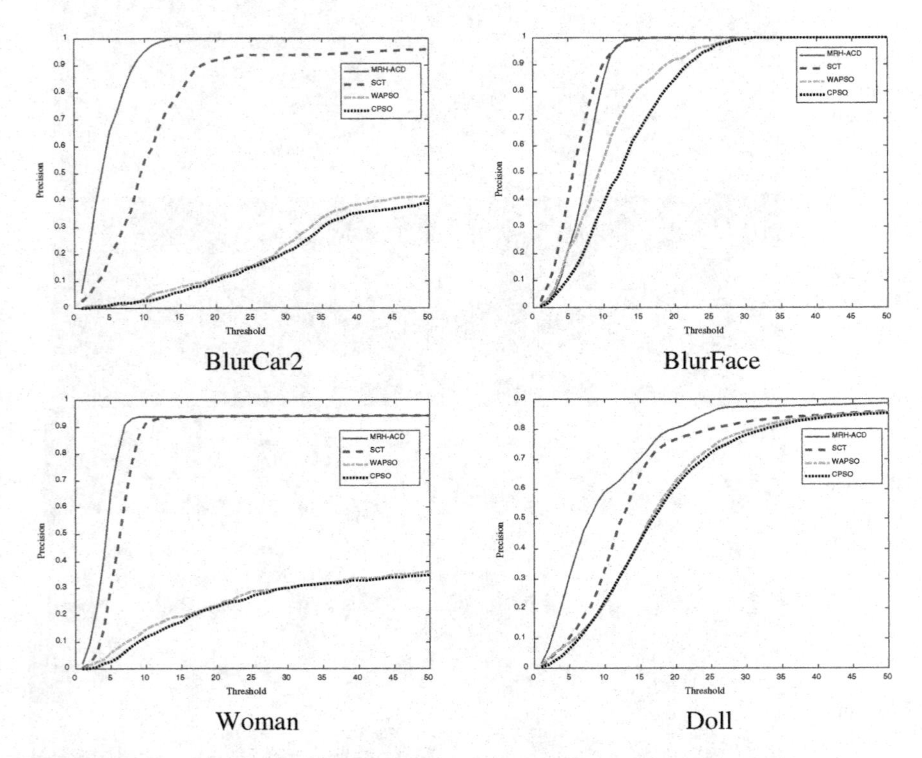

Fig. 4. One-pass evaluation (OPE) of precision plots base on threshold diagrams compare to CPSO [12], WAPSO [13] and SCT [14].

5 Conclusion

Video processing and object tracking gain great interest in today's computer vision society. Many researchers seek to find a general method to handle all possible challenges and provide reliable outcome in different video environment. However, there still have higher requirements for object detection and tracking.

In this paper, a novel multi-region histogram feature definition with ant colony detection object tracking method has been proposed. The MRH-ACD approach uses the advantages of multi-region histogram features and swarm intelligence to overcome multiple problems defined in object tracking area. This new approach separates target into different representative regions and compute overall histogram differences with possible solutions located by ant colony detection. According to our experiment results, MRH-ACD has demonstrated a better tracking progress for motion blur, occlusion, scale variation and other problem combination cases compare to traditional PSO methods and SCT.

In future research, more efficient multiple region selection, other swarm intelligence algorithm that help estimate solution and introduce deep network model to train base on MRH feature will be proposed; more test will be implemented about average performance in OTB-50, OTB-100 and VOT2014 datasets. In addition, reduce redundancy and increase robustness will be improved to meet higher quality.

References

1. Zhang, L., van der Maaten, L.: Structure preserving object tracking. In: Proceedings of the IEEE conference on computer vision and pattern recognition, pp. 1838–1845 (2013)
2. Nobahari, H., Sharifi, A.: Continuous ant colony filter applied to online estimation and compensation of ground effect in automatic landing of quadrotor. Eng. Appl. Artif. Intell. **32**, 100–111 (2014)
3. Nobahari, H., Sharifi, A.: A novel heuristic filter based on ant colony optimization for non-linear systems state estimation. In: Li, Z., Li, X., Liu, Y., Cai, Z. (eds.) ISICA 2012. CCIS, pp. 20–29. Springer, Heidelberg (2012). doi:10.1007/978-3-642-34289-9_3
4. Nobahari, H., Zandavi, S.M., Mohammadkarimi, H.: Simplex filter: a novel heuristic filter for nonlinear systems state estimation. Appl. Soft Comput. **49**, 474–484 (2016)
5. Franken, N.: Visual exploration of algorithm parameter space. In: IEEE Congress on Evolutionary Computation, CEC 2009, pp. 389–398. IEEE (2009)
6. Wu, Y., Lim, J., Yang, M.-H.: Online object tracking: a benchmark. In: Proceedings of the IEEE Conference on Computer Vision and Pattern Recognition, pp. 2411–2418 (2013)
7. OpenCV 2.4.9.0 documentation. http://docs.opencv.org/2.4.9/
8. Lowe, D.G.: Object recognition from local scale-invariant features. In: The Proceedings of the Seventh IEEE International Conference on Computer Vision, pp. 1150–1157. IEEE (1999)
9. Bay, H., Ess, A., Tuytelaars, T., Van Gool, L.: Speeded-up robust features (SURF). Comput. Vis. Image Underst. **110**, 346–359 (2008)
10. Rublee, E., Rabaud, V., Konolige, K., Bradski, G.: ORB: an efficient alternative to SIFT or SURF. In: 2011 IEEE international conference on Computer Vision (ICCV), pp. 2564–2571. IEEE (2011)
11. Dalal, N., Triggs, B.: Histograms of oriented gradients for human detection. In: IEEE Computer Society Conference on Computer Vision and Pattern Recognition, CVPR 2005, pp. 886–893. IEEE (2005)
12. Sha, F., Bae, C., Liu, G., Zhao, X., Chung, Y.Y., Yeh, W.: A categorized particle swarm optimization for object tracking. In: IEEE Congress on Evolutionary Computation (CEC), pp. 2737–2744. IEEE (2015)
13. Liu, G., Chen, Z., Yeung, H.W.F., Chung, Y.Y., Yeh, W.-C.: A new weight adjusted particle swarm optimization for real-time multiple object tracking. In: Hirose, A., Ozawa, S., Doya, K., Ikeda, K., Lee, M., Liu, D. (eds.) ICONIP 2016. LNCS, vol. 9948, pp. 643–651. Springer, Cham (2016). doi:10.1007/978-3-319-46672-9_72
14. Choi, J., Jin Chang, H., Jeong, J., Demiris, Y., Young Choi, J.: Visual tracking using attention-modulated disintegration and integration. In: Proceedings of the IEEE Conference on Computer Vision and Pattern Recognition, pp. 4321–4330 (2016)
15. Dorigo, M.: Optimization, learning and natural algorithms. Ph.D. thesis, Politecnico di Milano, Italy (1992)

Spatial Quality Aware Network for Video-Based Person Re-identification

Yujie Wang[1], Biao Leng[2(✉)], and Guanglu Song[1]

[1] School of Computer Science and Engineering, Beihang University, Beijing, China
[2] State Key Laboratory of Software Development Environment, School of Computer Science and Engineering, Beihang University, Beijing, China
lengbiao@buaa.edu.cn

Abstract. Person re-identification in video is challenging in computer vision. Most methods adopt feature aggregation to get a video-level representation. However, almost all of them do it on the final feature embedding, which neglects the spatial difference among feature maps. To address this problem, we proposed an effective approach, named Spatial Quality Aware Network (SQAN) for video-based person re-identification. SQAN distributes a score for each pixel in a feature map. Then scores are normalized across all frames and the weighted sum is used to aggregate them. To deal with overfitting, we also proposed a semantic dropout strategy. Experiments show that our proposed method is competitive with state-of-the-art methods in performance.

Keywords: Person re-identification · Deep learning · Feature aggregation

1 Introduction

Person re-identification(re-id), which is widely applicated in smart video surveillance, aims to identify a probe person from a gallery person set via visual information. Most previous works [1–5] focus on image-based re-id, i.e. given a probe person's image, the system should return the most similar person across the gallery person set. Impressive progress has been achieved in the image-based person re-id area. However, in video surveillance scenario, one person's information is encoded not only in individual frames but the correspondence among frames. Empirical evidences [6] confirm that the video-based re-id is superior to the others. However, many challenges still exist.

Due to the length of a video is variable, the feature representation of video is not fixed, which makes the comparison between videos hard. Most methods resort to feature aggregation to build a fix length representation of the video. The direct way to aggregate features is to fetch the max or average value among frame-level features [5], i.e. max/average pool. The max pool only maintains the most salient part of features, while the average pool neglects the importance differences among frame features. These information's loss degrades the robustness of the algorithm.

D. Liu et al. (Eds.): ICONIP 2017, Part III, LNCS 10636, pp. 34–43, 2017.
https://doi.org/10.1007/978-3-319-70090-8_4

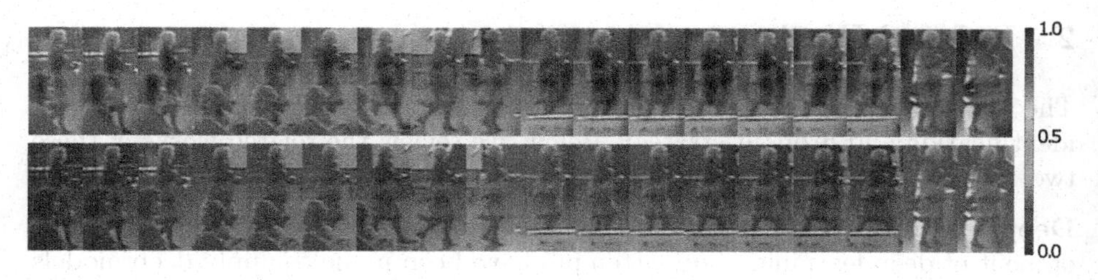

Fig. 1. The color represents the quality score, and warmer is higher quality score. The high quality score means this part should contribute more information for final video representation. The first row illustrates the spatial quality scores generated by our proposed method. The second row illustrates frame-level scores of QAN [7]. This figure shows that our proposed method can fully exploit spatial information, even the QAN determines a frame should have a low score. Best view in color. (Color figure online)

To overcome these weaknesses, Y. Liu et al. [7] proposed Quality Aware Network(QAN) to estimate the frames' quality. It generates a score for each frame and uses weighted sum of scores and corresponding feature embeddings to aggregate frame-level features. However, their method assumes all pixels in a feature map have the same score. This assumption losses the spatial differences of frames, as Fig. 1 shows to us.

In this paper, we mainly focus on aggregating features in spatial across frames. We consider the impact of different pixels in feature maps to improve person re-id performance. To achieve our goal, we proposed a network, named Spatial Quality Aware Network (SQAN). The SQAN has two branches and supports end-to-end training. The first branch is to learn a representation in frame-level. The second branch is to learn quality scores for different pixels of a feature map. Then the outputs of two branches will be aggregated to form a compact video-level representation. Note that in the second branch, we adopt an unsupervised like manner to learn scores, which means it does not depend on the human-made score label. What's more, to overcome the overfitting problem, we proposed an effective dropout strategy.

We evaluate our method in two datasets, iLIDS-VID and MARS. Experiments indicate that the proposed method is effective and is competitive with state-of-the-art methods.

In a word, the main contributions of the paper are as follows:

- The major contribution is that we proposed a Spatial Quality Aware Network (SQAN), which fully exploits spatial information of frames in a video. It shows big improvement in person re-id task than former method.
- The minor contribution is the proposed semantic dropout strategy that is used to effectively regularize spatial information.
- Experiments show that our proposed method reaches competitive performance compared with state-of-the-art methods.

2 Related Work

The proposed SQAN mainly builds upon deep learning based person re-identification and dropout strategy. Below, we review the related works in these two aspects.

Deep learning based person re-identification. Along with the rapid development of deep learning, many attempts have been made to apply deep models into person re-id. Wu et al. [8] proposed that hand-crafted histogram feature is complementary to Convolutional Neural Network(CNN) feature. Liu et al. [7] designed a quality generate unit to distribute different weights to frames, then use the weighted sum of them to represent a video. What's more, some methods adopt Recurrent Neural Network(RNN) and its variants to learn video-level feature for video based re-id task. McLaughlin et al. [5] use CNN to extract frame-level features from the frame and optical flow, then RNN is used to aggregate features across frames. Yan et al. [9] use Long-Short Term Memory network [10] to aggregate frame-level features into video-level feature.

To fully exploit frames' information, we proposed spatial quality aware network (SQAN). It can be seen as an extension of QAN proposed by [7]. Our SQAN fully exploits the spatial differences across frames, which is omitted by QAN.

Dropout strategy. Dropout [11] is a widely used method in deep learning to relief overfitting problem, which is mostly severe when training data is not enough. Due to the small scale in most existing person re-id datasets, this method should be useful in person re-id task. The traditional dropout [11] randomly set some values to zero for the given inputs. Geng et al. [12] proposed pairwise-consistent dropout, which is used for dropping the values in same positions among multiple input feature vectors. Tompson et al. [13] proposed a method to regularize for convolution layers, which sets all the values across the randomly selected channels of the feature map into zero.

However, [11,12] don't consider the spatial correlation and semantic structure of feature maps. [13] only consider the spatial correlation in randomly selected channels. Thus, we propose a semantic dropout strategy, which drops values in a feature map and all the values in the same position across channels will be dropped too. See details in Sect. 3.3

3 Proposed Method

3.1 Architecture Overview

Recent work [7] shows great improvement on person re-id by granting a score to each frame of a video. However, it considers every part of a frame owns the same weight. It ignores the useful information in some parts of a frame with a low score. To make the best use of useful information from all frames, we designed a network, named Spatial Quality Aware Network (SQAN). The core part of it is spatial quality generate module. It gives a score for each pixel of a frame's feature map. Note that a pixel in high level representation feature map corresponds to

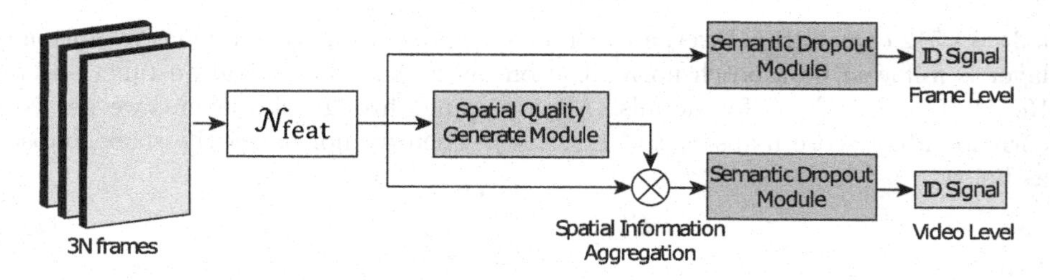

Fig. 2. The proposed Spatial Quality Aware Network (SQAN). The network's input is 3N frames, where N is the sampled frame number of one person's video. There are 2N frames belong to the same identity, while N frames belong to another identity. $\mathcal{N}_{feat}$ is the feature extraction network. The Spatial Quality Generate Module and Semantic Dropout Module are introduced at Sects. 3.2 and 3.3. The final representation is the feature after spatial information aggregation.

a specific part in original frame. And this operation can be seen as an quality evaluation to a specific part of original frame. Then the scores are normalized across feature maps in a video. Finally, these feature maps are aggregated to represent a video. Besides, we design a semantic dropout strategy to overcome overfitting. See Fig. 2 for details.

3.2 Spatial Quality Generate Module

Given the input video V with N frames of a person. Let $I_i(i = 1, \dots, N)$ to represent its frames. The module's target is to output scores for each pixel of feature maps. A deep neural network $\mathcal{N}_{feat}$ is used to extract frame-level feature. In this paper, we use GoogLeNet [14] as $\mathcal{N}_{feat}$. To encode spatial information, the last 7×7 feature maps is used to learn the score, formulated as $f_{7\times 7} = \mathcal{N}_{feat}(I)$. We use f to represent $f_{7\times 7}$ below for simplicity. The Spatial Quality Generate Module(SQGM) includes three layers: a $1 \times 1 \times 512$ convolution layer,

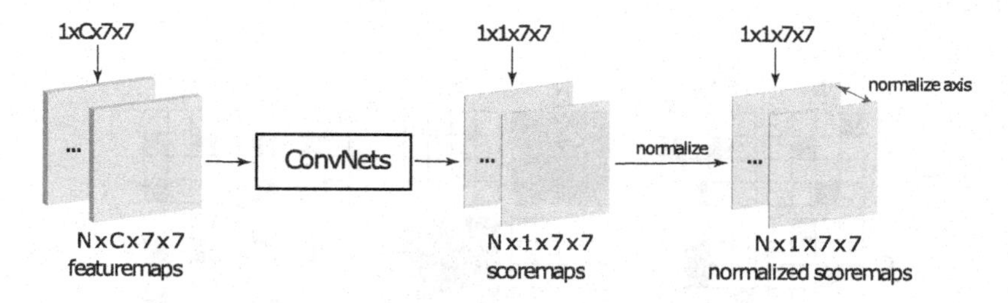

Fig. 3. Spatial Quality Generate Module. The input of this module is the spatial feature maps of a video, which contains N feature maps and C channels for each. Both height and width of feature maps are 7 for example. After passing through a ConvNets, N score maps are produced. Then the score maps are normalized across N feature maps, to be the final output of this module.

a $3\times3\times512$ convolution layer, and a $1\times1\times1$ convolution layer. Each convolution layer is followed by a batch normalization layer. And the activation function is ReLU [15]. See Fig. 3 for details. After passing through the layers, we get N corresponding score maps $S_i(i=1,\ldots,N)$. Then we normalize the score maps as below:

$$S^{x,y}_{norm_i} = \frac{e^{S^{x,y}_i}}{\sum_j e^{S^{x,y}_j}} \tag{1}$$

Note that $S^{x,y}_{norm_i}$ is the normalized score at position (x,y) of frame i's feature map.

Then we calculate the weighted sum of f and normalized score maps as the final video representation F:

$$F^{x,y,c} = \sum_{i=1}^{N} S^{x,y}_{norm_i} \cdot f_i^{x,y,c} \tag{2}$$

Note that $F^{x,y,c}$ and $f_i^{x,y,c}$ are values at position (x, y) in channel c of final feature map and input feature map separately.

3.3 Semantic Dropout Module

Overfitting is a severe problem in model optimization, especially in small datasets. Dropout is an effective method to relief this problem. The most common dropout strategy drops values randomly and is mostly applied to the feature vector. [13] proposed a convolution dropout strategy and it drops values across some randomly selected channels. However, in SQAN, the corresponding vector at each pixel in f is highly semantic. Our intuition is to make the representation more robust via dropping some pixel-wise vectors and letting the remains can also represent a person. Thus, we propose a dropout strategy, named Semantic Dropout. It drops randomly selected pixel's values in a feature map. And all the values in the same position across all channels will be dropped too. See Fig. 4 for details.

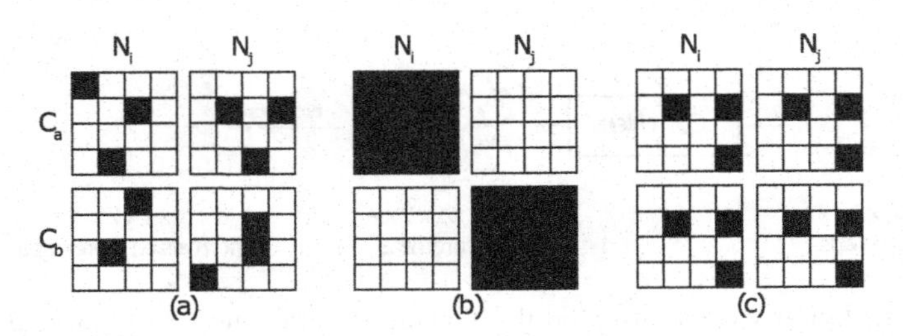

Fig. 4. The difference among three kinds of dropout strategies. (a) is the normal dropout. (b) is spatial dropout [13]. (c) is our proposed semantic dropout. C_a and C_b means two channels. N_i and N_j means two feature maps from different frames of a video.

Note that it is important to drop f and F in the same dropout pattern, i.e. the dropped pixels should be same. Because F is aggregated from f, if they adopt different dropout pattern, the optimized target will be inconsistent.

3.4 Multi-Loss Supervised Training

We hope the aggregated feature can not only classify identities, but the distance between different identities is distant. So we use triplet loss [16] to deal with this issue. The overall loss of SQAN can be formulated as follows:

$$\mathcal{L} = \mathcal{L}_{softmax_1} + \mathcal{L}_{softmax_2} + \mathcal{L}_{softmax_3} + \mathcal{L}_{trp} \tag{3}$$

$$\mathcal{L}_{trp} = \sum_{V_a, V_p, V_n} [||\mathcal{N}(V_a) - \mathcal{N}(V_p)||_2^2 - ||\mathcal{N}(V_a) - \mathcal{N}(V_n)||_2^2 + margin]_+ \tag{4}$$

Note that $\mathcal{L}_{softmax_1}$,$\mathcal{L}_{softmax_2}$,$\mathcal{L}_{softmax_3}$ are the original GoogLeNet softmax loss. $\mathcal{L}_{trp}$ represent triplet loss. $\mathcal{N}(V)$ is the representation for video V. V_a and V_p are same identity's video, while V_n is one different identity's video. What's more, the function $[z]_+ = max(z, 0)$ and $margin$ is a hyper-parameter which is set to 1.2 in all experiments.

4 Experiment

4.1 Datasets and Evaluation Protocol

iLIDS-VID. The iLIDS-VID [17] dataset contains 600 videos of 300 randomly sampled people. Each person has one pair of video from two camera views. Each video is comprised of 23 to 192 image frames, with an average length of 73 for each. The challenges of this dataset largely lie in clothing similarities, lighting and viewpoint changes across camera views, complicated background, and occlusions.

The evaluation on iLIDS-VID is the same as previous methods [9]. The dataset is randomly divided into training set and testing set by half, with no overlap between them. During testing, the sequence of the first camera is regarded as the query, while sequences from the second camera as the gallery set. The widely used cumulative matching characteristic (CMC) curve is employed for measuring the performance of methods on this dataset. To ensure statistically reliable evaluation, we repeat the procedure 10 times and use the average performance as the result.

MARS. MARS [6] is a recently released large scale video-based re-id dataset. It contains 1,261 identities and around 20,000 video sequences. The dataset has 1,191,003 images in total from six different cameras and each identity has 13.2 sequences on average. Different from the iLIDS-VID dataset, it has no manually annotated bounding boxes. Each sequence is automatically obtained by pedestrian detector and tracker. Besides, the dataset also contains 3,248 distractor sequences.

For the sake of large scale MARS dataset, the train/test split is fixed with 631 and 630 identities respectively. We use mean average precision score (mAP) and cumulative matching characteristic (CMC) to evaluate methods, which are recommended in [6]. The evaluation mode is video-to-video, single query.

4.2 Implementation Details

Our implementation is based on the open source deep learning framework Caffe [18]. All our experiments were carried on a NVIDIA TITAN X GPU with 12GB of onboard memory. The network is trained with stochastic gradient descent (SGD) end-to-end. The learning rate is set to 1e-3. The total iterations are 15,000 for iLIDS-VID and 250,000 for MARS. The weight decay is set to 0.002. The batch size is fixed to 24, and 8 frames are randomly sampled for anchor, positive and negative classes in triplet loss. As for SDM, the dropout ratio is set to 0.3, which means a pixel vector will be selected to drop in a probability of 30%.

4.3 Ablation Study on iLIDS-VID

Table 1 compares the results of different variants of SQAN. We remark that in this table all results are obtained in the same experiment settings, except (a). So the differences are contributed by the method itself.

Method (a) is the original GoogLeNet with Batch Normalization. It only has image level softmax supervision and uses average pool to aggregate features. What's more, it only has 4,000 iterations for it converge rapidly. It reaches 61.3% CMC1 performance, which is similar to previous works.

Method (b) is QAN proposed by [7]. The "QGM" means the frame-level quality generate module, which distribute a score to each frame. It improves 11% in CMC1 and we think the improvement is from two aspects: one is the video-level supervision and the other is the frame-level quality score.

Method (c) is our proposed method with SQGM. It brings about 14% improvement comparing with QAN. This shows to us that spatial information can not be omitted. Parts in a frame with low score may be important and vice versa.

Method (d) is the final version of SQAN. It introduces SDM based on (c) and further gains 1.3% in CMC1. All these results show the effectiveness of our proposed methods.

4.4 Comparison with State-of-the-art Methods

To further judge the effectiveness, we also compare our methods with other state-of-the-art methods. We evaluate our methods both in iLIDS-VID and MARS.

Table 1. Ablation Study on iLIDS-VID

	(a)	(b)	(c)	(d)
+QGM		✓		
+SQGM			✓	✓
+SDM				✓
CMC1	61.3	68.0	76.4	**77.7**
CMC5	83.3	86.8	93.4	**94.3**
CMC10	88.7	89.8	97.3	**97.4**
CMC20	92.4	97.4	99.1	**99.8**

Table 2 shows the results on iLIDS-VID. SQAN achieves higher CMC than most of the other methods, only a little bit lower than current state-of-the-art method PAM-LOMO+KISSME.

Table 2. Comparison of SQAN and other state-of-the-art methods on iLIDS-VID

	CMC1	CMC5	CMC10	CMC20
SQAN	77.7	94.3	97.4	99.8
QAN	68.0	86.8	89.8	97.4
CNN+RNN [5]	58.0	84.0	91.0	96.0
TDL [19]	56.3	87.6	95.6	98.3
FrameExtraction+CNN [20]	60.2	85.1	-	94.2
PAM-LOMO+KISSME [21]	79.5	95.1	97.6	99.1

Table 3 shows the results on MARS. Comparing with the results on iLIDS-VID, SQAN has a big gap below the state-of-the-art method on MARS. We can compensate it by adding XQDA and re-ranking [22]. The we can get 75.8% CMC1 and 67.4% mAP. This performance is close to the state-of-the-art TriNet [4] and better than those methods with similar additions. But we argue that the intrinsical reason of bad performance is the properties of attention like schema. The attention schema for feature aggregation has the assume that semantic part in each frame should be aligned. For a more realistic and not cropped dataset like MARS, the misalignment problem is more frequent and severe than it on iLIDS-VID. This problem will be left for our future work.

Table 3. Comparison of SQAN and other state-of-the-art methods on MARS.

	CMC1	CMC5	mAP
SQAN	67.5	80.3	41.0
SQAN+XQDA+re-ranking[a]	75.8	85.5	67.4
CNN+XQDA [6]	65.3	82.0	47.6
FrameExtraction+CNN [20]	55.5	70.2	-
CaffeNet+XQDA+re-ranking [22]	67.8	-	58.0
ResNet50+XQDA+re-ranking [22]	73.9	-	68.5
TriNet [4]	79.8	91.4	67.7

[a]We use the released code by [22] for XQDA and re-ranking

5 Conclusion and Future Work

In this paper, we propose a Spatial Quality Aware Network (SQAN) for person re-identification. The proposed method can distribute a quality score to each pixel of a frame's feature map, then the weighted feature maps are aggregated across frames to represent the video of a person. What's more, we also propose a dropout strategy, named semantic dropout, which effectively reduces the impact of overfitting. Experiments show the effectiveness of our method and our method is competitive with state-of-the-art methods in performance.

SQAN is a fine-grained spatial information aggregation model. It may suffer from the severe misalignment problem in a video. Thus, how to integrate alignment method into SQAN will be explored in our future work.

Acknowledgments. This work is supported by the National Natural Science Foundation of China (No. 61472023) and the State Key Laboratory of Software Development Environment (No. SKLSDE-2016ZX-24).

References

1. Chen, D., Yuan, Z., Hua, G., Zheng, N., Wang, J.: Similarity learning on an explicit polynomial kernel feature map for person re-identification. In: Proceedings of the IEEE Conference on Computer Vision and Pattern Recognition, pp. 1565–1573 (2015)
2. Liao, S., Hu, Y., Zhu, X., Li, S.Z.: Person re-identification by local maximal occurrence representation and metric learning. In: Proceedings of the IEEE Conference on Computer Vision and Pattern Recognition, pp. 2197–2206 (2015)
3. Su, C., Yang, F., Zhang, S., Tian, Q., Davis, L.S., Gao, W.: Multi-task learning with low rank attribute embedding for person re-identification. In: Proceedings of the IEEE International Conference on Computer Vision, pp. 3739–3747 (2015)
4. Hermans, A., Beyer, L., Leibe, B.: In defense of the triplet loss for person re-identification. arXiv preprint arXiv:1703.07737 (2017)
5. McLaughlin, N., Martinez del Rincon, J., Miller, P.: Recurrent convolutional network for video-based person re-identification. In: Proceedings of the IEEE Conference on Computer Vision and Pattern Recognition, pp. 1325–1334 (2016)

6. Zheng, L., Bie, Z., Sun, Y., Wang, J., Su, C., Wang, S., Tian, Q.: MARS: a video benchmark for large-scale person re-identification. In: Leibe, B., Matas, J., Sebe, N., Welling, M. (eds.) ECCV 2016. LNCS, vol. 9910, pp. 868–884. Springer, Cham (2016). doi:10.1007/978-3-319-46466-4_52
7. Liu, Y., Yan, J., Ouyang, W.: Quality aware network for set to set recognition. arXiv preprint arXiv:1704.03373 (2017)
8. Wu, S., Chen, Y.C., Li, X., Wu, A.C., You, J.J., Zheng, W.S.: An enhanced deep feature representation for person re-identification. In: 2016 IEEE Winter Conference on Applications of Computer Vision (WACV), pp. 1–8. IEEE (2016)
9. Yan, Y., Ni, B., Song, Z., Ma, C., Yan, Y., Yang, X.: Person re-identification via recurrent feature aggregation. In: Leibe, B., Matas, J., Sebe, N., Welling, M. (eds.) ECCV 2016. LNCS, vol. 9910, pp. 701–716. Springer, Cham (2016). doi:10.1007/978-3-319-46466-4_42
10. Hochreiter, S., Schmidhuber, J.: Long short-term memory. Neural Comput. **9**(8), 1735–1780 (1997)
11. Srivastava, N., Hinton, G.E., Krizhevsky, A., Sutskever, I., Salakhutdinov, R.: Dropout: a simple way to prevent neural networks from overfitting. J. Mach. Learn. Res. **15**(1), 1929–1958 (2014)
12. Geng, M., Wang, Y., Xiang, T., Tian, Y.: Deep transfer learning for person re-identification. arXiv preprint arXiv:1611.05244 (2016)
13. Tompson, J., Goroshin, R., Jain, A., LeCun, Y., Bregler, C.: Efficient object localization using convolutional networks. In: Proceedings of the IEEE Conference on Computer Vision and Pattern Recognition, pp. 648–656 (2015)
14. Szegedy, C., Liu, W., Jia, Y., Sermanet, P., Reed, S., Anguelov, D., Erhan, D., Vanhoucke, V., Rabinovich, A.: Going deeper with convolutions. In: Proceedings of the IEEE Conference on Computer Vision and Pattern Recognition, pp. 1–9 (2015)
15. Glorot, X., Bordes, A., Bengio, Y.: Deep sparse rectifier neural networks. In: Aistats, vol. 15, p. 275 (2011)
16. Schroff, F., Kalenichenko, D., Philbin, J.: Facenet: a unified embedding for face recognition and clustering. In: Proceedings of the IEEE Conference on Computer Vision and Pattern Recognition, pp. 815–823 (2015)
17. Wang, T., Gong, S., Zhu, X., Wang, S.: Person re-identification by video ranking. In: Fleet, D., Pajdla, T., Schiele, B., Tuytelaars, T. (eds.) ECCV 2014. LNCS, vol. 8692, pp. 688–703. Springer, Cham (2014). doi:10.1007/978-3-319-10593-2_45
18. Jia, Y., Shelhamer, E., Donahue, J., Karayev, S., Long, J., Girshick, R., Guadarrama, S., Darrell, T.: Caffe: Convolutional architecture for fast feature embedding. arXiv preprint arXiv:1408.5093 (2014)
19. You, J., Wu, A., Li, X., Zheng, W.S.: Top-push video-based person re-identification. In: Proceedings of the IEEE Conference on Computer Vision and Pattern Recognition, pp. 1345–1353 (2016)
20. Zhang, W., Hu, S., Liu, K.: Learning compact appearance representation for video-based person re-identification. arXiv preprint arXiv:1702.06294 (2017)
21. Khan, F.M., Brèmond, F.: Multi-shot person re-identification using part appearance mixture. In: 2017 IEEE Winter Conference on Applications of Computer Vision (WACV), pp. 605–614. IEEE (2017)
22. Zhong, Z., Zheng, L., Cao, D., Li, S.: Re-ranking person re-identification with k-reciprocal encoding. arXiv preprint arXiv:1701.08398 (2017)

Robust Visual Tracking via Occlusion Detection Based on Depth-Layer Information

Xiaoguang Niu, Zhipeng Cui, Shijie Geng, Jie Yang, and Yu Qiao(✉)

Institute of Image Processing and Pattern Recognition, Department of Automation, Shanghai Jiao Tong University, Shanghai, China
qiaoyu@sjtu.edu.cn

Abstract. In this paper, we propose a novel occlusion detection algorithm based on depth-layer information for robust visual tracking. The scene can be classified into the near, the target and the far layer. We find that when occlusion happens, some background patches in the near layer will move into the target region and hence occlude the target. Based on this feature of occlusion, we propose an algorithm which exploits both temporal and spatial context information to discriminate occlusion from target appearance variation. Using the framework of particle filter, our algorithm divides the background region around the target into multiple patches and tracks each of them. The background patch that occludes the target is identified collaboratively by the tracking results of both background and target trackers. Then the occlusion is evaluated with the target visibility function. If occlusion is detected, the target template stops updating. Comprehensive experiments in OTB-2013 and VOT-2015 show that our tracker achieves comparable performance with other state-of-art trackers.

Keywords: Visual tracking · Occlusion detection · Template update · Correlation filter

1 Introduction

Visual object tracking is a fundamental problem in computer vision. Despite significant progress in recent years, several factors including partial occlusion, target deformation and scale variations still remain unsolved for robust tracking. In this paper, we mainly consider the problem of partial occlusion and develop an effective approach for occlusion detection based on depth-layer information.

A common strategy to alleviate the challenge of partial occlusion is adaptive decontamination of the training set [1]. This requires the tracker to distinguish occlusion from target appearance variation such as deformation and illumination variation. In cases of occlusion, the target template should keep unchanged so that the training set will not be contaminated by the background information, while target appearance variation is the opposite situation. To balance the robustness and adaptiveness of a tracking algorithm, accurate occlusion reasoning is of great importance.

D. Liu et al. (Eds.): ICONIP 2017, Part III, LNCS 10636, pp. 44–53, 2017.
https://doi.org/10.1007/978-3-319-70090-8_5

To catch the essential characteristic of occlusion, consider the occlusion in terms of depth. The scene can be classified into be 3 layers according to their depth: the near layer, the target layer and the far layer. When occlusion happens, there are points in the near layer that move into the bounding box of the object, i.e., they are in the region of target in the 2-D scene but have different depth. However, instead of estimating the depth map, which can be computational inefficient, we divide the background region around the target into patches and deal with each of them independently to estimate their depth and position. This can be formulated by the framework of particle filter, aiming to estimate the state of the background patch that occludes the target, i.e., the patch that comes into the bounding box of the target.

Based these observations, we propose a novel occlusion detection algorithm for robust visual tracking. The tracking module of our algorithm consists of the tracker for target and trackers for background. The background around the target is divided into multiple patches, each with a tracker. The background trackers are initialized, tracked and updated individually, then the patches are classified according to their depth. We introduce the concept of target visibility which measures how well the target can be captured by the viewer. The near-layer background patches that overlap the target will reduce the visibility of the target. The template updating strategy is conducted by the template updater based on target visibility. By applying appropriate target template updating strategy in case of occlusion, our tracker is robust against occlusion.

2 Related Work

Correlation Filter-Based Trackers. In recent years, correlation filter (CF) based tracking algorithms have achieved amazing performance. D.S. Bolme et al. [2] propose MOSSE, where correlation filter is trained with the appearance of target and by correlating the trained filter with candidates of target, the position of target in the current frame is the position with the maximum response. Based on MOSSE, J.F. Henriques et al. [3] formulate the Kernelized Correlation Filter (KCF) tracker, which introduces circulant matrix, kernel method and ridge regression to improve the performance. Further improvements include extracting better features [4–6], handling scale variations of target [7,8], building more reliable memories [9,10], and so on.

Occlusion Detection Methods. Various methods for detecting occlusion have been proposed and can be divided roughly into two categories. The first one is to learn indicators that are sensitive to occlusion. The different states of the indicator show whether there is occlusion. A. Yilmaz et al. [11] handle the occlusion based on both the distance between the objects and the change of the object size. M. Mathias et al. [12] detect partially occluded pedestrians by using many occlusion-specific classifiers, including left/right/bottom occlusion cases. The second way is part-based tracking methods [13–16], which have favorable property of robustness against partial occlusion. The appearance of the object is modeled by multiple parts. The confidence scores of individual parts are exploited to construct a robust tracker.

3 Our Algorithm

3.1 Overview

Our tracking algorithm is based on depth-layer clue and takes into account both spatial and temporal context information for occlusion detection. The framework of our algorithm is shown in Algorithm 1 (Fig. 1). There are two types of trackers: the target tracker and the background trackers. First they estimate the positions of the target and background patches respectively. Then the occlusion detector calculates the probability of occluding the target for each background patch. The template updater applies appropriate target template updating strategy according to the result of occlusion detector, which can prevent the target template from false updating due to occlusion. Therefore our tracking algorithm is robust against occlusion. It is worth noting that our algorithm provides a general framework and the specific choice of tracking methods of the target and background patches can vary according to different applications.

Algorithm 1: Depth-based Occlusion Detection Tracking

Parameter : α: threshold for the tracking confidence of background trackers.
β: threshold for the overlap ratio of background patch and target.
γ: threshold for the target visibility function.
Input : I_t, the current frame.
Output : $bbox_t$: bounding box of target in I_t.

1 Estimate the target's $bbox_t$ in I_t, as described in Sec. 3.2.
2 Estimate the positions of background patches in the I_t, as described in Sec. 3.3.
3 Calculate the observation likelihood of background patches, $p(z_t^i|x_t^i)$, as described in Sec. 3.4
4 Calculate the visibility v of the target. If $v < \gamma$, do not update the target template; Otherwise, update the target template, as described in Sec. 3.5.
5 Output tracking result $bbox_t$.

Fig. 1. Algorithm: depth-layer based occlusion detection tracking

3.2 Target Tracker

The target tracker outputs the bounding box of target in current frame as the tracking result and will be updated by the target template updater according to the result of occlusion detector. We use a variant of Gaussian KCF [3], together with a scaling pool to handle the scale variation of target, as the target tracker, which is based on DSST [8]. (For more details of KCF and DSST, please refer to [3,8].) However, any other choice of tracking methods will work too.

3.3 Background Trackers

Most of the existing part-based tracking methods divide the *target* into several parts and observe whether these parts can be tracked with high confidence. However, if there is appearance variation, the corresponding tracking confidence will also drop, so the tracker may falsely detect non-existent occlusion and miss the template update, which may lead to the failure of future tracking. According to our observation of occlusion, when occlusion happens, there must be something in the near layer that moves into the bounding box of the object, i.e., they are in the same position in the 2-D scene but have different depth. Therefore, we utilize the *background* information for occlusion detection.

The background surrounding the target is divided into a number of patches, each with a tracker. The background trackers output the locations and response maps of each individual background patch in current frame. We apply linear KCF [3] tracker as the background tracker since it is more efficient than Gaussian KCF. Having these information, our algorithm will then check whether there are background patches overlapping the target. In this way, occlusion and target appearance variation can be accurately distinguished.

3.4 Occlusion Detector

The occlusion detector works by utilizing the location and response map information about the target and background patches obtained from the target tracker and background trackers respectively. All background patches are first examined based on their location relationship with the target. If a patch does not overlap the target bounding box, it does not influence the occlusion detection. If it overlaps the target bounding box, this patch may be occluded by the target (in the far layer) *or* may occlude the target (in the near layer). It is necessary to identify these two situations. Based on our observation, if the tracking confidence of a patch is lower than a threshold α, it may be blocked by the target. Here we make some simplifications, ignoring the fact that the low confidence may be attributed to appearance change or other situations. Based on the peak-to-sidelobe ratio (PSR) [2]

$$PSR = \frac{max(R) - avg(R)}{\sigma(R)}, \tag{1}$$

we define the tracking confidence (denoted as TC):

$$TC(x_t) = \mathbb{u}(PSR - \alpha) * PSR, \tag{2}$$

where R denotes the response map in KCF [3], σ, max and avg are operators to get the standard deviation, maximum and average value respectively, and $\mathbb{u}(x)$ is the unit step function whose value is zero for $x \leq 0$ and one for $x > 0$. It indicates that the tracking confidence is positive only when the value of PSR is higher than α.

The overlap ratio (denoted as OR) between the patch and the target is calculated as

$$OR(x_t) = \mathrm{u}(\frac{bbox(x_t) \cap bbox(target)}{bbox(x_t)} - \beta) * \frac{bbox(x_t) \cap bbox(target)}{bbox(x_t)}, \quad (3)$$

denoting the ratio of the bounding box x_t that comes into the target region. It shows that the two bounding boxes of patch and target are not regarded as overlapped if their overlap ratio is lower than a predefined threshold β.

We formulate our occlusion detector scheme by the framework of particle filter. The particle filter is a sequential Bayesian estimation technique through importance sampling. For simplicity and tractability, we assume that the target is occluded by one rectangular background region. Different from the previous works [14,17,18], we exploit particle filter to estimate the state of occluding background patch instead of the target. Let x_t denote the bounding box parameters of the background patch in frame t, z_t the observed feature of the corresponding patch. Hence the observation likelihood $p(z_t|x_t)$ describes the probability of the target's being occluded by the background patch cropped from x_t, defined as

$$p(z_t|x_t) = \frac{1}{Y} * TC(x_t) * OR(x_t), \quad (4)$$

where Y is the normalizing factor.

Given all the available observations $z_{1:t-1} = \{z_1, z_2, ..., z_{t-1}\}$ up to frame $t-1$, the distribution of x_t is predicted by

$$p(x_t|z_{1:t-1}) = \int p(x_t|x_{t-1})p(x_{t-1}|z_{1:t-1})dx_{t-1}. \quad (5)$$

At frame t, the observation z_t is available and the distribution is updated by

$$p(x_t|z_{1:t}) = \frac{p(z_t|x_t)p(x_t|z_{1:t-1})}{p(z_t|z_{1:t-1})} \quad (6)$$

according to Bayesian rule. To estimate the true value of x_t, a finite set of N samples $\{x_t^i\}_{i=1,...,N}$ are drawn with weights $\{w_t^i\}_{i=1,...,N}$. Since the Sequential Importance Re-sampling (SIS) is adopted here, the weights are updated by

$$w_t^i = w_{t-1}^i p(z_t|x_t). \quad (7)$$

Because there is no prior knowledge about where the occlusion may occur, we select the patches' locations, i.e., the samples $\{x_t^i\}_{i=1,...,N}$, uniformly and randomly along all four sides of the target bounding box.

Note that the weight of x_t calculated by Eq. 7 may be zero, which means that x_t should be discarded because it has no contribution to occlusion. Then a new patch will be generated around the new target. The other patches having nonzero weights are preserved, updated in a standard interpolation way and keep on tracking.

In summary, the occlusion detector classifies the background patches into 3 classes with the classification criteria as follows:

1. If a patch does not overlap the target bounding box, then we ignore it.
2. If a patch overlaps the target bounding box with low tracking confidence, it belongs to the far layer and may be occluded by the target.
3. If a patch overlaps the target bounding box with high tracking confidence, it belongs to the near layer and is very likely occluding the target.

3.5 Target Template Updater

Based to the information from occlusion detector, we define the visibility function of the target as follows:

$$v = 1 - \frac{1}{N}\sum_{i=1}^{N} \mathrm{u}(p(z_t^i|x_t^i)), \tag{8}$$

where N is the number of background patches. The visibility function measures how well the target can be captured by the camera, and hence can be used to evaluate target occlusion. Recall that the occluding patches have nonzero $p(z_t|x_t)$, so the smaller v is, the more occluding background patches there are. Let γ be the predefined threshold for the target visibility function. The updating strategy is as follows:

1. If $v < \gamma$, then the occlusion happens and the target template stops updating.
2. Otherwise, if $v \geq \gamma$, the occlusion is not observed and the target template will update as the standard DSST [8].

Our template updater is robust against occlusion because it applies a template updating strategy according to the information from occlusion detector. By detecting the occurrence of occlusion explicitly, the erroneous target template updating can be effectively avoided, which improves the quality of the training set and makes the model more discriminative.

4 Experimental Results

4.1 Implementation Details

To evaluate our occlusion detection tracking scheme, we implement a prototype. The target tracker is based on DSST [8] and most the parameters are the same as DSST except that the learning rate is changed to 0.001. There are 50 background patches and their size is 14×14. The background patch trackers is based on linear kernel KCF [3]. The parameters in occlusion detection scheme are set as $\alpha = 80$, $\beta = 0.3$, $\gamma = 6$. We evaluated this prototype on two widely used benchmarks, OTB-2013 [19] and VOT-2015 [20].

4.2 Evaluation on OTB-2013

OTB-2013 [19] contains 51 different tracking sequences with fully annotated attributes. Both quantitative and qualitative experiments are conducted on it.

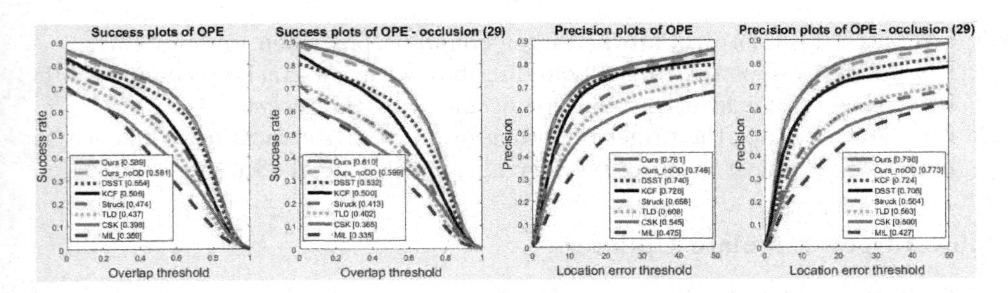

Fig. 2. Quantitative evaluation. From left to right: success plot of trackers on OTB-2013, success plot of trackers on *OCC* sequences, precision plot of trackers on OTB-2013, precision plot of trackers on *OCC* sequences.

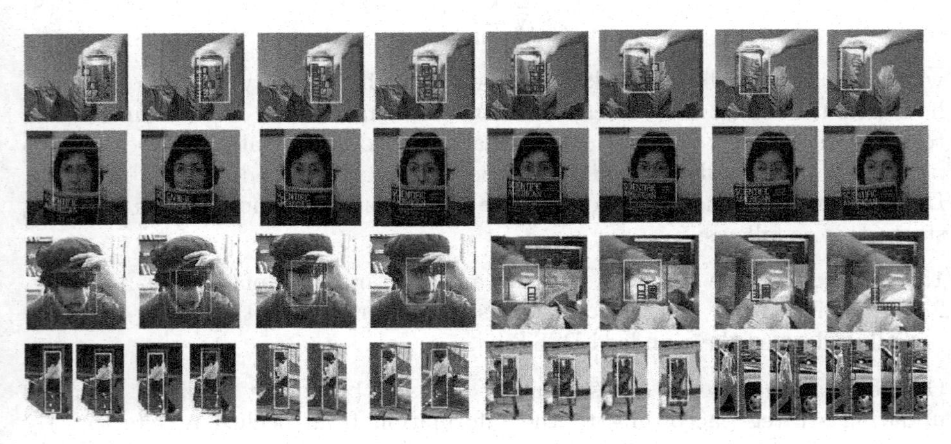

Fig. 3. Qualitative results of our tracker in sequences with partial occlusion. The red bounding boxes denote ground truth. The yellow bounding boxes denote tracking results. The thin blue squares show the background patches that occlude the target. (Color figure online)

Quantitative Evaluation. We choose KCF [3], DSST [8], IVT [17], Struck [21], SCM [22], TLD [23], CSK [24] and MIL [25] for comparison. Figure 2 shows the results of one-pass evaluation (OPE), using Success Plots and Precision Plots, both for all 51 sequences and the sequences with attribute "partial occlusion" (*OCC*). In Fig. 2, "Ours" denotes our tracker, and "Ours_noOD" denotes our tracker with the occlusion detector disabled. "Ours_noOD" method will not conduct occlusion detection and update the template of target in every frame, so it can be seen as a baseline to our tracker. The comparison between "Ours" and "Ours_noOD" shows that a great improvement is attained through the explicit incorporation of occlusion detection. In Precision Plot of *OCC* sequences, "Ours" (79.6%) outperforms its baseline "Ours_noOD" (77.3%) by 2.3%, and in Success Plot the gap is 0.8%.

Trackers	Accuracy	Robustness
RAJSSC	1.87	3.77
SME	2.22	4.82
SRDCF	2.22	2.90
OACF	2.33	5.07
Ours	2.42	4.97
NSAMF	3.07	2.78
SAMF	3.32	5.13
MUSTer	3.35	5.20
DSST	3.57	8.10
MvCFT	3.62	4.77
MKCF_+	3.98	4.67
KCF_MTSA	4.97	6.75
sKCF	5.02	7.10
KCF2	5.10	6.33
KCFv2	5.87	6.75
STC	10.48	11.35
LOFT_lite	12.17	13.27

(a)

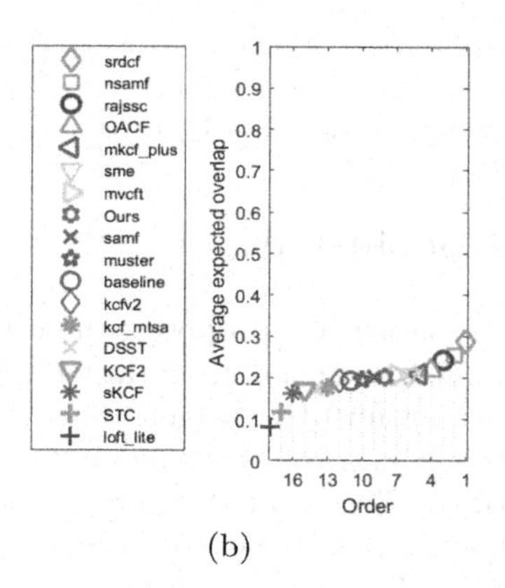

(b)

Fig. 4. (a) Accuracy and robustness ranking results. Our tracker provides comparable performance with other correlation filter based trackers. (b) The expected average overlap (EAO) graph. The trackers are ranked from left to right.

Qualitative Evaluation. We choose some sequences with partial occlusion in OTB-2013 [19] to show that our tracker will discern the presence of occlusion when facing with it, as shown in Fig. 3. The first row is sequence *Coke*. The coke can moves from right to left, being occluded by the plant. In these frames, the number of blue boxes is proportional to the severity of the occlusion, which shows that our tracker is aware of the occlusion and prevents itself from pollution by the background. The second row is sequence *FaceOcc1*, where the woman's face is undergoing a serious occlusion by the book. The following rows show more examples. In sequences *FaceOCC2*, the man's face is occluded by his finger and the hat; in *Tiger2*, the tiger doll moves behind the plant and is overlapped. The last row shows pedestrian examples. In *Woman*, the woman is occluded by a white car and then a black car; in *Basketball*, the player in green is occluded by the player in white; in *David3*, a tree covers the walking man. All the examples prove that our method takes effect.

4.3 Evaluation on VOT-2015

The VOT-2015 [20] benchmark contains 60 sequences and is per-frame annotated with several visual attributes. Accuracy and robustness are used as performance measures. The accuracy measures the overlap ratio between the prediction of trackers and the ground truth. The robustness measures the failure frequency of trackers, where the failure is defined as the zero overlap ratio. We evaluated our tracker on VOT-2015 and performed a comprehensive comparison with all other correlation filter based trackers that are included in the data set.

Figure 4(a) shows the results of accuracy and robustness (AR) ranking. It can be seen that our tracker achieves comparable performance with the state-of-art correlation filter trackers.

The expected average overlap (EAO) is a criterion that combines the raw values of per-frame accuracy and failures and is introduced in VOT-2015 [20]. Figure 4(b) shows the EAO scores.

5 Conclusions

In this paper, we propose a new framework of occlusion detection which can distinguish occlusion from target appearance variation. We track the background patches around the target. Then by examining their spatial relationship with the target and the tracking confidence, the patches that occlude the target are identified. Based on this information, we calculate the visibility of the target. If the visibility is smaller than a threshold, the template of target will stop updating. Extensive experiments demonstrate the effectiveness of our framework on occlusion detection and practical value in constructing a robust tracking algorithm.

Acknowledgments. This research is partly supported by NSFC (No: 61375048), USCAST2015-13, USCAST2016-23, SAST2016008.

References

1. Danelljan, M., Hager, G., Shahbaz Khan, F., Felsberg, M.: Adaptive decontamination of the training set: a unified formulation for discriminative visual tracking. In: Proceedings of the IEEE Conference on Computer Vision and Pattern Recognition, pp. 1430–1438 (2016)
2. Bolme, D.S., Beveridge, J.R., Draper, B.A., Lui, Y.M.: Visual object tracking using adaptive correlation filters. In: Proceedings of the IEEE Conference on Computer Vision and Pattern Recognition, pp. 2544–2550. IEEE (2010)
3. Henriques, J.F., Caseiro, R., Martins, P., Batista, J.: High-speed tracking with kernelized correlation filters. IEEE Trans. Pattern Anal. Mach. Intell. **37**(3), 583–596 (2015)
4. Danelljan, M., Shahbaz Khan, F., Felsberg, M., Van de Weijer, J.: Adaptive color attributes for real-time visual tracking. In: Proceedings of the IEEE Conference on Computer Vision and Pattern Recognition, pp. 1090–1097 (2014)
5. Ma, C., Huang, J.B., Yang, X., Yang, M.H.: Hierarchical convolutional features for visual tracking. In: Proceedings of the IEEE International Conference on Computer Vision, pp. 3074–3082 (2015)
6. Danelljan, M., Robinson, A., Shahbaz Khan, F., Felsberg, M.: Beyond correlation filters: learning continuous convolution operators for visual tracking. In: Leibe, B., Matas, J., Sebe, N., Welling, M. (eds.) ECCV 2016. LNCS, vol. 9909, pp. 472–488. Springer, Cham (2016). doi:10.1007/978-3-319-46454-1_29
7. Li, Y., Zhu, J.: A scale adaptive kernel correlation filter tracker with feature integration. In: Agapito, L., Bronstein, M.M., Rother, C. (eds.) ECCV 2014. LNCS, vol. 8926, pp. 254–265. Springer, Cham (2015). doi:10.1007/978-3-319-16181-5_18
8. Danelljan, M., Häger, G., Khan, F., Felsberg, M.: Accurate scale estimation for robust visual tracking. In: British Machine Vision Conference, Nottingham, 1–5 September 2014. BMVA Press (2014)

9. Hong, Z., Chen, Z., Wang, C., Mei, X., Prokhorov, D., Tao, D.: Multi-store tracker (muster): a cognitive psychology inspired approach to object tracking. In: Proceedings of the IEEE Conference on Computer Vision and Pattern Recognition, pp. 749–758 (2015)
10. Wang, S., Zhang, S., Liu, W., Metaxas, D.N.: Visual tracking with reliable memories. In: International Joint Conference on Artificial Intelligence, New York, United States, pp. 9–15 (2016)
11. Yilmaz, A., Li, X., Shah, M.N.: Contour-based object tracking with occlusion handling in video acquired using mobile cameras. IEEE Trans. Pattern Anal. Mach. Intell. **26**(11), 1531–1536 (2004)
12. Mathias, M., Benenson, R., Timofte, R., Van Gool, L.: Handling occlusions with Franken-classifiers. In: Proceedings of the IEEE International Conference on Computer Vision, pp. 1505–1512 (2013)
13. Zhang, T., Jia, K., Xu, C., Ma, Y., Ahuja, N.: Partial occlusion handling for visual tracking via robust part matching. In: Proceedings of the IEEE Conference on Computer Vision and Pattern Recognition. pp. 1258–1265. IEEE (2014)
14. Liu, T., Wang, G., Yang, Q.: Real-time part-based visual tracking via adaptive correlation filters. In: Proceedings of the IEEE Conference on Computer Vision and Pattern Recognition, pp. 4902–4912 (2015)
15. Li, Y., Zhu, J., Hoi, S.C.: Reliable patch trackers: robust visual tracking by exploiting reliable patches. In: Proceedings of the IEEE Conference on Computer Vision and Pattern Recognition, pp. 353–361 (2015)
16. Liu, S., Zhang, T., Cao, X., Xu, C.: Structural correlation filter for robust visual tracking. In: Proceedings of the IEEE Conference on Computer Vision and Pattern Recognition, June 2016
17. Ross, D.A., Lim, J., Lin, R.S., Yang, M.H.: Incremental learning for robust visual tracking. Int. J. Comput. Vis. **77**(1–3), 125–141 (2008)
18. Mei, X., Ling, H.: Robust visual tracking using 1 minimization. In: Proceedings of the IEEE International Conference on Computer Vision, pp. 1436–1443. IEEE (2009)
19. Wu, Y., Lim, J., Yang, M.H.: Online object tracking: a benchmark. In: Proceedings of the IEEE Conference on Computer Vision and Pattern Recognition, pp. 2411–2418 (2013)
20. Kristan, M., Matas, J., Leonardis, A., Felsberg, M., Čehovin, L., Fernandez, G., Vojir, T., Häger, G.: The visual object tracking VOT2015 challenge results. In: Visual Object Tracking Workshop 2015 at Proceedings of the IEEE International Conference on Computer Vision, December 2015
21. Hare, S., Saffari, A., Torr, P.H.: Struck: Structured output tracking with kernels. In: Proceedings of the IEEE International Conference on Computer Vision, pp. 263–270. IEEE (2011)
22. Zhong, W., Lu, H., Yang, M.H.: Robust object tracking via sparsity-based collaborative model. In: Proceedings of the IEEE Conference on Computer Vision and Pattern Recognition, pp. 1838–1845. IEEE (2012)
23. Kalal, Z., Mikolajczyk, K., Matas, J.: Tracking-learning-detection. IEEE Trans. Pattern Anal. Mach. Intell. **34**(7), 1409–1422 (2012)
24. Henriques, J.F., Caseiro, R., Martins, P., Batista, J.: Exploiting the circulant structure of tracking-by-detection with kernels. In: Fitzgibbon, A., Lazebnik, S., Perona, P., Sato, Y., Schmid, C. (eds.) ECCV 2012. LNCS, vol. 7575, pp. 702–715. Springer, Heidelberg (2012). doi:10.1007/978-3-642-33765-9_50
25. Babenko, B., Yang, M.H., Belongie, S.: Robust object tracking with online multiple instance learning. IEEE Trans. Pattern Anal. Mach. Intell. **33**(8), 1619–1632 (2011)

Object Tracking Based on Mean Shift Algorithm and Kernelized Correlation Filter Algorithm

Huazheng Zhou[1,2], Xiaohu Ma[1,2](✉), and Lina Bian[1,2]

[1] School of Computer Science and Technology, Soochow University, Suzhou 215006, China
xhma@suda.edu.cn

[2] Collaborative Innovation Center of Novel Software Technology and Industrialization, Nanjing 210023, China

Abstract. In order to solve the problems of motion blur and fast motion, a new robust object tracking algorithm using the Kernelized Correlation Filters (KCF) and the Mean Shift (MS) algorithm, called KCFMS is presented in this paper. The object tracking process can be described as: First, we give the initial position and size of the object and use the Mean Shift algorithm to obtain the position of the object. Second, the Kernelized Correlation Filtering algorithm is used to obtain the position of the object in the same frame. Third, we use the cross update strategy to update the object models. In order to improve the tracking speed as much as possible, our object tracking algorithm works only over one layer. This hybrid algorithm has a good tracking effect on the target fast motion and motion blur. We present extensive experimental results on a number of challenging sequences in terms of efficiency, accuracy and robustness.

Keywords: Kernelized Correlation Filters · Mean shift · Motion blur · Fast motion

1 Introduction

Visual tracking is a fundamental problem in computer vision, which finds a wide range of application areas [1]. Recently, hybrid discriminative generative methods have opened a promising direction to benefit from both types of methods. Several hybrid methods [2–6] have been proposed in many application domains. These methods train a model by optimizing a convex combination of the generative and discriminative log likelihood functions. The improper hybrid of discriminative generative model generates even worse performance than pure generative or discriminative methods. In this paper, our tracking algorithm only gives the initial position and size of the target in the first frame of the validation set GT (represented by a rectangular area). We use a tracking algorithm based on the discriminant and generated model to learn the appearance model. Discriminant methods focus on finding a decision boundary to distinguish between background and goals. Discriminant methods not only focus on the target but also focus on its background. The tracker based on the generation method focuses only

D. Liu et al. (Eds.): ICONIP 2017, Part III, LNCS 10636, pp. 54–64, 2017.
https://doi.org/10.1007/978-3-319-70090-8_6

on the target itself. For example, based on the histogram method, these methods are simple, but the tracking effect is very good.

Our hybrid algorithm uses a two-step tracking method. First, we use the mean shift algorithm to predict the position of the target in the current frame, and then use this position to sample. Second, we use the kernel correlation filtering algorithm to determine the position of the target in the frame. As shown in Fig. 1, we use cross update strategy to update the appearance model in the current frame. We use the position obtained by the MS method to update the KCF appearance model. Meanwhile, we use the position obtained by the KCF method to update the MS appearance model. This step occurs at the stage of the model updating.

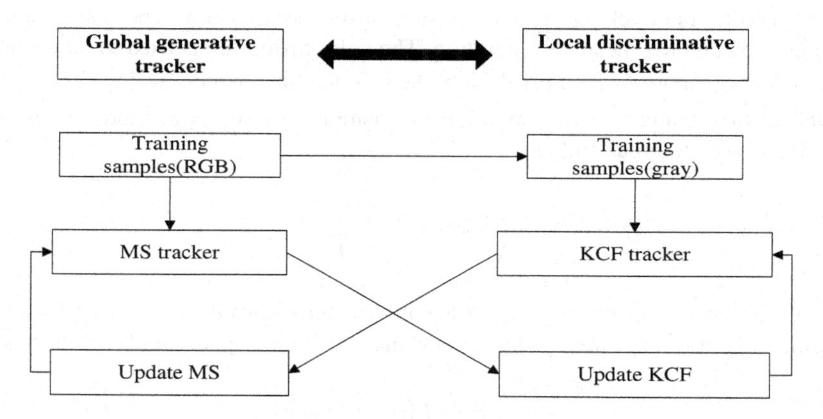

Fig. 1. The appearance models update strategy

In Sect. 2, the related work is briefly reviewed. Our method is proposed in Sect. 3. Section 4 shows the experimental results and analysis. Finally Sect. 5 draws the conclusion.

2 Related Work

A typical failure of visual tracking is drift, which means that tracking performance gradually degrades over time, and eventually the tracker lose the object. There exist several reasons to explain the drift issue. One explanation is that self-learning reinforces previous errors and causes the drift. Tang et al. [7] proposed to use co-training to online train two SVM trackers with color histogram features and HOG features. This method uses an SVM solver [8] to focus on recent appearance variations without representing the global object appearance. Babenko et al. [9] argued that the drift can be also caused by the ambiguities when using online labeled data to update the model. For example, the way of selecting positive and negative samples inevitably contain ambiguities, and these ambiguities lead to an offset of the online trained model. A multiple instance learning approach was introduced to avoid ambiguities by using a bag of samples instead of an individual sample. More recently, Kalal et al. [10], in another way, proposed a very efficient method using a combination of detection, tracking, and modeling modules. This tracker is robust against drifting by bootstrapping itself using positive and negative constraints.

In this paper, we propose to use the two stage methods to combine generative and discriminative models. That is to say, first we use the mean shift algorithm to predict the position of the target in the current frame, and then use this position to sample, using the kernel correlation filtering algorithm to determine the position of the target in the current frame.

2.1 The Mean Shift Algorithm

The mean shift [11] tracking process consists of two components. These are the target representation and the mean shift iteration. The color histogram is used to represent the target. The target region has n pixels, and the i-th pixel is donated by $\{x_{ms}^i\}_{i=1,\cdots,n}$. The probability of a feature u is an m-color-bin histogram. The target model $q_u(u = 1, 2, \cdots, m)$ is computed as follows:

$$q_u = \mathrm{C}\sum_{i=1}^{n} k\left(\left\|\frac{x_{ms}^i - x_{ms}^*}{h}\right\|^2\right)\delta\left[b(x_{ms}^i) - u\right] \tag{1}$$

where x_{ms}^* is the target center, $k(x)$ is the isotropic kernel profile. $\delta(x)$ is the Kronecker delta function, b(x) maps the pixel of a coordinate to feature space and C is the constant:

$$\mathrm{C} = 1/\sum_{i=1}^{n} k\left(\left\|\frac{x_{ms}^i - x_{ms}^*}{h}\right\|^2\right) \tag{2}$$

Similarly, the target candidate model from the target candidate region centered at position y is given by $\{p_u(y)\}_{u=1,2,\cdots,m}$.

A key issue in the Mean Shift algorithm is the computation of an offset from the current location to a new location. We use the Bhattacharyya coefficients to compute the similarity between the target histogram and the candidate histogram.

$$\rho[p(y), q] = \sum_{u=1}^{m} \sqrt{p_u(y)q_u} \tag{3}$$

2.2 The Kernelized Correlation Filters Algorithm

The KCF algorithm [12] uses the cyclic shifts to obtain large samples and then uses these samples to train the Classifier. The training process of the classifier can be described by the following formula:

$$\min_{w} \sum_{i} (f(X_i) - y_i)^2 + \lambda\|w\|^2 \tag{4}$$

where we need find the best w. We can refer to the paper [12].

$$w = \sum \alpha_i \varphi(X_i) \tag{5}$$

3 Overview of the Proposed Approach

KCF algorithm is difficult to deal with motion blur and fast motion. In order to resolve the drawback of KFC, a target tracking algorithm is proposed to joint kernel correlation filtering and mean shift. In each frame of the video, the hybrid algorithm first uses the mean shift algorithm to predict the target position in the current frame, and then uses this position as the input of the kernel correlation filtering algorithm to detect the target position, and finally uses the cross update strategy to update the target model. In addition, in order to maximize the speed of target tracking, the hybrid tracking algorithm has only one layer.

Our discriminant model uses the kernel correlation filter (KCF) algorithm to generate the model using the mean shift (MS) algorithm. Since our hybrid method is based on these two algorithms, we called our own algorithm as KCFMS. We first read RGB image at the t-th frame, then use MS algorithm to determine the target position l_{ms}. Next, we convert the image to gray image, and use l_{ms} and gray image as the input of kernel correlation filtering algorithm at the t frame. The output of the KCF is the position l_{kcf}. Then we use l_{kcf} as the t frame location l_t. This algorithm is parallel tracking algorithm. The specific process of the hybrid algorithm KCFMS proposed in this chapter is described as follows:

The KCFMS Algorithm.

Input: the number of the frames and the current image
Output: The target position of the t frame

(1) Read the first frame image and the initial position and size of the target
(2) Obtain the generation model by Eq.(1)
(3) Training the classifier by Eq.(4)
(4) Read a new frame image
(5) Determines the target position l_{ms} with the generating method MS,
(6) Use the location l_{ms} as the input of the KCF method, then Calculate the position of the prediction target is l_{kcf}
(7) Make $l_t = l_{kcf}$, as the estimate of the target position at the t frame.
(8) Updating the KCF classifier.
(9) If it is not the last frame, go to step (4), otherwise go to step (10).
(10) End the target tracking procedure.

We also give a flow chart of the algorithm, as shown in Fig. 2. In this flow chart, we have detailed the implementation of the algorithm in this chapter. The left dashed box indicates the work done by the algorithm on the first frame image, and the right side is the tracking flow of the algorithm on the remaining image sequence.

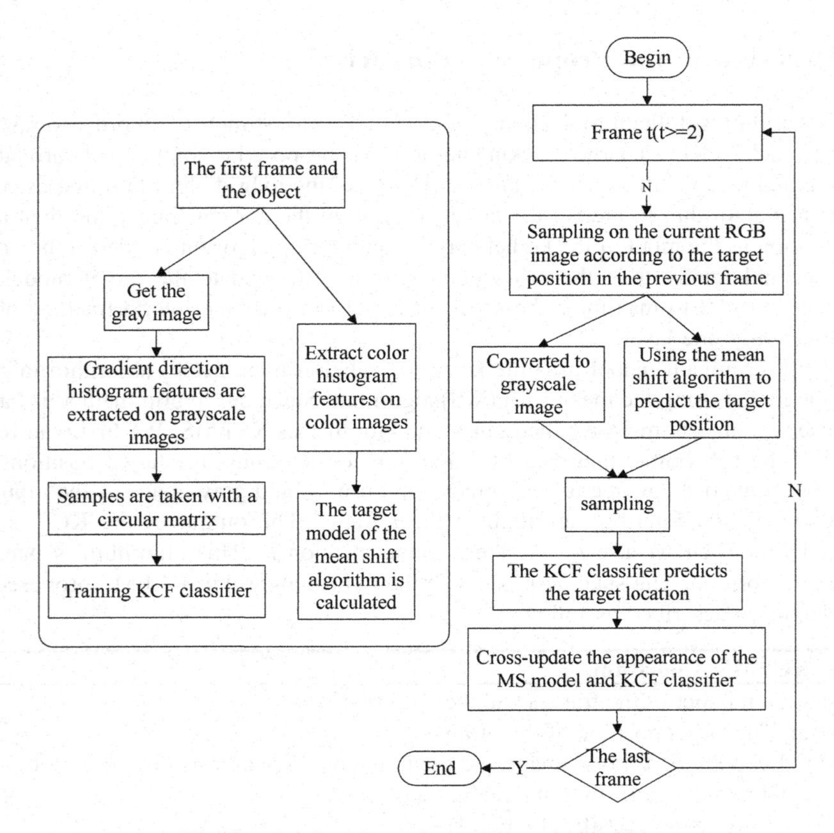

Fig. 2. The flowchart of our proposed algorithm

4 Experiments and Discussion

The experiment inputs the position and the size of the target in the first frame, and then tracks the target in the remaining image sequence. In order to fully analyze the performance of the algorithm in this paper, our experiment is carried out on 50 data sets [13], and the algorithm is analyzed comprehensively from qualitative analysis and quantitative analysis. In addition, the running environment is Windows10 operating system, and programming platform is Matlab2013a.

Our algorithm compares with the other four main algorithms. These algorithms are FCT [14], MIL [9], STC [15], and KCF [12], respectively. It is very likely that the target tracking algorithm based on the hybrid model have no better tracking effect than the single algorithm, while the kernel correlation filtering algorithm is closely related to the algorithm of our paper. Therefore, our algorithm must be compared with the KCF algorithm. For the mean shift algorithm MS is quite poor, so the article is not necessary to compared with this algorithm. In addition, our algorithm has only one layer. In this paper, we use the average center location error and the overall the average center location error to measure the performance of the algorithm.

4.1 Qualitative Comparison

Qualitative comparison gives the most intuitive description of the tracking results, and we can see the tracking result of each algorithm on the same data set. Qualitative comparison shows two data sets (David3 and Basketball) in the 50 datasets [13], which are used to test the ability of the algorithm, and thus improving the understanding of our algorithm. We will analyze the characteristics of the five data sets of Couple, David3, Boy, DragonBaby, and Girl2, as shown in Table 1. Then we analyze the intuitive tracking effect of each algorithm in the two data sets.

Table 1. Partial data set and characteristics

Sequence	Frames	Data characteristics
Couple	140	Scale Variation, Deformation, Fast Motion, Background Clutters
David3	252	Occlusion, Deformation, Rotation, Background Clutters
Boy	602	Scale Variation, Motion Blur, Fast Motion, Rotation
DragonBaby	113	Scale Variation, Occlusion, Motion Blur, Fast Motion, Rotation, Out-of-View
Girl2	1500	Scale Variation, Occlusion, Deformation, Motion Blur, Deformation

On the David3 dataset, the five algorithms tracked David. The dataset shows that David's journey is back and forth in the outdoor scene, with occlusion and rotation on the way. In general, our algorithm tracks the target throughout the process. Especially in the 83rd and the 187th frame when the object is occluded heavily by the tree, the KCFMS algorithm also can track the target. In the 139th frame around the target begin to go back, that is to say, the target appearance change, but our algorithm can track the object successfully. Part of the details shown in Fig. 3.

Fig. 3. Partial results of the five algorithms on the David3 dataset

Basketball is a complex data set with fast motion, target deformation, occlusion from similar object, illumination variation, and so on. Therefore, it is a data set favored by researchers of many target tracking algorithms. The result indicates that most of the algorithm's tracking results were good before the 230th frame, and only the MIL algorithm drifted slightly in the 290th frame due to the approximation of similar object. In the same time, the STC algorithm is drift. This data set also has a noticeable change in illumination variation, and we give the light changes in the 650th frame. In order to see the changes in illumination variation, we give the frame before and after the two tracking effect. We can see that our algorithm's tracking effect is very good in this case. In addition, the data set itself has low pixel. The details are shown in Fig. 4.

Fig. 4. Partial results of the five algorithms on the Basketball dataset

4.2 Quantitative Comparison

We analyze our algorithm for quantitative comparison from two aspects, the first is the average center error, and the other is the sum of error and mean error. Table 2 shows the overall mean center error values for each algorithm over 50 datasets, which is a measure of the target tracking algorithm. In this paper, the best value of the mean center error is marked in boldface, and the second mean center error is marked by the italicized slash. It can be seen from the Table 2 in which the tracking result of our algorithm is optimal.

Table 2. All mean center error values for each algorithm on 50 datasets (unit: pixels)

Algorithm	FCT [14]	MIL [9]	KCFMS	STC [15]	KCF [12]
Mean value	45.16	52.62	**26.53**	71.35	*35.98*

Table 2 shows the Mean Value of the algorithm, but it is not possible to gain a better understanding of the tracking effect of the algorithm in different datasets. Therefore, we will analyze the tracking result of each algorithm from the three aspects of illumination change (IV), motion blur (MB), fast moving (FM). For the analysis of these three characteristics, this paper uses the center error of each data set, the sum of center error and the average center error of these three aspects to analyze the tracking effect of the algorithm.

Table 3 gives the center errors for the eight data sets with the characteristics of the illumination change IV. In addition, we give the sum of the errors and calculate the mean value of the error. From Table 3, although the tracking results of the algorithm in each data set are not always good, the optimal tracking result is less than the KCF algorithm and the overall central error is the smallest which is 22.36 Pixels. Therefore, our algorithm is robustness to the illumination change.

Table 3. Average center location error with IV feature data set (unit: pixels)

Algorithm	FCT [14]	MIL [9]	KCFMS	STC [15]	KCF [12]
Basketball	88.52	103.80	**7.18**	78.66	*7.89*
David	29.35	44.83	**6.28**	36.35	*8.06*
Doll	27.15	22.41	*12.19*	140.15	**8.36**
FaceOcc2	22.39	17.62	*7.87*	18.57	**7.67**
Fish	8.91	12.68	7.61	*4.50*	**4.08**
Human7	**18.19**	32.98	30.89	*24.93*	48.20
Shaking	*27.61*	164.38	93.65	**16.14**	112.50
Tiger1	24.89	103.15	*13.21*	63.08	**8.05**
the Sum of Error	247.00	501.85	**178.88**	382.39	*204.80*
Mean Error	30.86	62.73	**22.36**	47.80	*25.60*

Table 4 shows the tracking results for each algorithm on a motion set with motion fuzzy MB. First, we can see from the algorithm that the average center error on each data set is optimal. Because in this 11 data set on our algorithm has four best (bold font representation) and three times (italic underlined) of the tracking results. Second, from the last two lines of data in Table 4, we can see that our algorithm error is 37.23 pixels, which is far better than the other four algorithms. We can also see that the STC algorithm has the second tracking effect, so we can deduce that the algorithm is very good for the target tracking effect in the case of occlusion.

Table 4. Average center location error with MB feature data set (unit: pixels)

Algorithm	FCT [14]	MIL [9]	KCFMS	STC [15]	KCF [12]
Biker	*27.99*	**16.63**	44.23	82.87	77.18
BlurCar3	180.53	90.46	*18.44*	52.09	**4.14**
BlurOwl	*173.18*	190.20	**14.34**	240.64	183.43
Box	104.03	**28.94**	118.35	100.56	*89.13*
Boy	8.99	15.97	**2.70**	18.30	*2.87*
DragonBaby	*37.89*	74.70	**16.08**	174.20	50.40
Girl2	*108.37*	172.37	**51.12**	267.49	264.58
Human7	**18.19**	32.98	30.89	*24.93*	48.20
Jumping	46.71	**12.85**	*18.78*	93.70	26.12
Tiger1	24.89	103.15	*13.21*	63.08	**8.05**
Tiger2	**34.94**	*46.83*	81.32	201.75	47.44
the Sum of Error	*765.71*	785.08	**409.48**	1319.61	801.52
Mean Error	*69.61*	71.37	**37.23**	119.96	72.87

Fast motion is also a hot topic in object tracking. Table 6 gives the tracking results of the algorithm on 15 datasets with fast motion. First, we can see that the optimal tracking result of KCF algorithm is the best from the average center error of the algorithm in each data set. The tracking result of STC algorithm is the worst, and the optimal tracking result of the remaining three algorithms is basically same. Then, looking at the overall tracking results, that is the sum of the errors in Table 5 and the mean of the errors. From this we can see that our algorithm is superior to KCF algorithm, KCF algorithm is ranked second, FCT and MIL algorithm are similar, STC algorithm tracking results are still the worst. That is to say, the KCFMS algorithm has better tracking performance than other algorithms for fast motion.

Table 5. Average center location error with FM feature data set (unit: pixels)

Algorithm	FCT [14]	MIL [9]	KCFMS	STC [15]	KCF [12]
Biker	*27.99*	**16.63**	44.23	82.87	77.18
Bird2	80.94	**15.71**	*16.02*	18.44	21.37
BlurCar3	180.53	90.46	*18.44*	52.09	**4.14**
BlurOwl	*173.18*	190.20	**14.34**	240.64	183.43
CarScale	26.63	40.04	*25.51*	96.18	**16.14**
ClifBar	**24.98**	*34.89*	35.10	36.96	36.72
Coke	43.81	75.84	97.08	**17.14**	*18.65*
Couple	*34.93*	38.39	**15.32**	826.12	47.56
DragonBaby	*37.89*	74.70	**16.08**	174.20	50.40
Human7	**18.19**	32.98	30.89	*24.93*	48.20
Jumping	46.71	**12.85**	*18.78*	93.70	26.12
Surfer	49.20	23.43	*10.56*	49.74	**8.74**
Tiger1	24.89	103.15	*13.21*	63.08	**8.05**
Tiger2	**34.94**	*46.83*	81.32	201.75	47.44
Vase	19.84	22.40	25.18	*15.01*	**12.43**
the Sum of Error	824.63	818.51	**462.07**	1992.85	*606.56*
Mean Error	54.98	54.58	**30.8**	132.86	*40.44*

In summary, the results of Tables 3, 4 and 5 show that the KCFMS algorithm is better than other algorithms in the three aspects of illumination, motion blur, fast motion. It can be seen that the tracking results of the hybrid tracking algorithm is perfect in this paper.

5 Conclusion

This paper presents a hybrid tracking algorithm. We use the MS tracker to track the RGB image, and then the RGB image is converted into the gray image. In the gray image we use the KCF method to track the object. When we obtain the object position, we use the cross update method to update the object appearance model. We do a lot of experiments and give the relevant experimental results in the public data sets.

Comparing with the mainstream of the algorithms is to analyze the tracking result. Experiments show that the proposed algorithm is more stable and has good tracking result on the changes of light, fast motion, motion blur, occlusion, background clutter and scale change. In this paper, the algorithm does not achieve the scale adaptability, but the scale adaptive is also a very important research direction of the object tracking, so our next goal is to achieve the scale of the algorithm.

Acknowledgment. This work is partially supported by the National Natural Science Foundation of China (61402310). Natural Science Foundation of Jiangsu Province of China (BK20141195).

References

1. Yilmaz, A., Javed, O., Shah, M.: Object tracking: a survey. ACM Comput. Surv. **38**(4), 1–17 (2006)
2. Lasserre, J.A., Bishop, C.M., Minka, T.P.: Principled hybrids of generative and discriminative models. In: 19th IEEE Computer Society Conference on Computer Vision and Pattern Recognition, vol. 1, pp. 87–94. IEEE Computer Society, New York (2006)
3. Ng, A., Jordan, M.I.: On discriminative vs. generative classifiers: a comparison of logistic regression and Naive Bayes. In: Proceedings of Advances in Neural Information Processing, vol. 28, no. 3, pp. 169–187 (2001)
4. Lin, R.S., Ross, D.A., Lim, J., et al.: Adaptive discriminative generative model and its applications. In: Neural Information Processing Systems, pp. 801–808 (2004)
5. Yang, M., Wu, Y.: Tracking non-stationary appearances and dynamic feature selection. In: 18th IEEE Computer Society Conference on Computer Vision and Pattern Recognition, pp. 1059–1066. IEEE Computer Society, San Diego (2005)
6. Yu, Q., Dinh, T.B., Medioni, G.: Online tracking and reacquisition using co-trained generative and discriminative trackers. In: Forsyth, D., Torr, P., Zisserman, A. (eds.) ECCV 2008. LNCS, vol. 5303, pp. 678–691. Springer, Heidelberg (2008). doi:10.1007/978-3-540-88688-4_50
7. Tang, F., Brennan, S., Zhao, Q., et al.: Co-tracking using semi-supervised support vector machines. In: 9th IEEE International Conference on Computer Vision, pp. 1–8. IEEE (2003)
8. Cauwenberghs, G., Poggio, T.: Incremental and decremental support vector machine learning. In: 13th International Conference on Neural Information Processing Systems, vol. 1, pp. 388–394. MIT Press, Denver (2000)
9. Babenko, B., Yang, M.H., Belongie, S.: Robust object tracking with online multiple instance learning. IEEE Trans. Pattern Anal. Mach. Intell. **33**(8), 1619–1632 (2011)
10. Kalal, Z., Matas, J., Mikolajczyk, K.: P-N learning: bootstrapping binary classifiers by structural constraints. In: 23rd IEEE Conference on Computer Vision and Pattern Recognition, vol. 238, pp. 49–56. IEEE Computer Society, San Francisco (2010)
11. Comaniciu, D., Menber, V.R., Meer, P.: Kernel-based object tracking. IEEE Trans. Pattern Anal. Mach. Intell. **25**(5), 564–575 (2003)
12. Henriques, J.F., Rui, C., Martins, P., et al.: High-speed tracking with Kernelized Correlation Filters. IEEE Trans. Pattern Anal. Mach. Intell. **37**(3), 583–596 (2014)
13. Wu, Y., Lim, J., Yang, M.H.: Online object tracking: a benchmark. IEEE Trans. Comput. Vis. Pattern Recogn. **37**(9), 1834–1848 (2015)

14. Zhang, K., Zhang, L., Yang, M.H.: Fast compressive tracking. IEEE Trans. Pattern Anal. Mach. Intell. **36**(10), 2002–2015 (2014)
15. Zhang, K., Zhang, L., Liu, Q., Zhang, D., Yang, M.-H.: Fast visual tracking via dense spatio-temporal context learning. In: Fleet, D., Pajdla, T., Schiele, B., Tuytelaars, T. (eds.) ECCV 2014. LNCS, vol. 8693, pp. 127–141. Springer, Cham (2014). doi:10.1007/978-3-319-10602-1_9

Deep Encoding Features for Instance Retrieval

Zhiming Ding, Zhengzhong Zhou, and Liqing Zhang(✉)

Key Laboratory of Shanghai Education Commission for Intelligent Interaction and Cognitive Engineering, Department of Computer Science and Engineering, Shanghai Jiao Tong University, Shanghai, China
{roromoade,tczhouzz,lqzhang}@sjtu.edu.cn

Abstract. In this paper, we propose a novel approach for instance retrieval. Compared with traditional retrieval pipeline, we first locate several candidate regions of target object with a region proposal network (RPN), instead of exhausting sliding window method. The candidate regions are detected through the trained RPN. Then we obtain the region-wise convolutional feature maps (CFMs) by forwarding them through a ROI pooling layer. Our feature encoding representation builds on the common sense that similar patterns have similar activations on feature maps. The target object is regarded as a combination of several meaningful patterns. In this way, we represent an image with the combination of encoded descriptors corresponding to the subsets of the proposed region. We also implement reranking algorithm to refine the proposed region in local retrieval. Through extensive experiments, we demonstrate the suitability of our feature encoding representation for instance retrieval, achieving comparable performance on both Oxford and Paris buildings benchmarks.

Keywords: Deep learning · Encoding scheme · Instance retrieval

1 Introduction

Image retrieval is divided into two classes. The traditional one focuses on category level which aims to find out the images in the same or similar class as the query image. The other one focuses on instance level. The instance-level retrieval is much harder because local and detailed image representation is necessarily to capture to distinguish two images. In this paper, we focus on instance-level image retrieval.

Traditional method dealing with instance retrieval is based on BoF [16] (bag of features) architecture using the local feature descriptors such as SIFT [9]. More recent research proposes a compact descriptor VLAD [7] which aggregates SIFT descriptors and outperforms the state-of-the-art result at that time. Combined with post-preprocessing techniques, such as query expansion [3] and spatial verification [12], slightly promotion is obtained.

In the recent years, Convolution Neural Networks (CNNs) have achieved the state-of-the-art performance in several computer vision tasks such as image

D. Liu et al. (Eds.): ICONIP 2017, Part III, LNCS 10636, pp. 65–74, 2017.
https://doi.org/10.1007/978-3-319-70090-8_7

classification [6], object detection [14]. Since CNNs have been shown to be good at feature representation after a large amount of data training. Moreover, the fine-tuning technique helps these representations eventually be reused to solve other tasks. Besides, features can be accessed conveniently and straightforward. Inspired by this, researches focus on feature extraction from intermediate convolutional layers and achieve better performance than before. R-MAC [17] is a kind of convolutional feature maps (CFM) based feature which achieves state-of-the-art performance on several public datasets [12,13]. The main idea of R-MAC is to extract representation from the maximum activation of feature map in each channel within a valid region. In order to reduce redundant calculation, the author proposes an approximate approach with integral sum-pooling.

Our feature extraction also relies on CFM and we make several improvements while dealing the retrieval. There exists two main problems in the previous method. First, they enumerate the proposal regions one by one with the sliding window method. The size of the sliding window changes when the scale of the image changes. They scan the image for many times until the satisfied result is obtained. Second, they adopt the same descriptor in both global and local retrieval stages. It is inefficient to calculate the descriptor for every image in the datasets. To deal with these problems, our method is made up of three parts, proposal network, global filtering and local retrieval. The proposal network is trained to learn candidates regions in the image. We design a binary encoding feature to filter unrelated image from the dataset. We consider maximum activation of feature maps at the same pixel along all channels, not channel wise anymore. For each strip-like pixel unit, we call it hyper-column. In fact, each hyper-column occurs a small part of the image, which representing a recognised pattern. In this way, the CFM of a given image can be regarded as a group of hyper-columns. Based on this idea, we propose a sparse local descriptor, which is called as Bag-of-Hyper-column (BoH), to represent a given region. In fact, each hyper-column is represented by a 1024D vector with only 50 non-zero elements. Our contributions can be summarised as follows:

1. We propose to use a pre-trained RPN to extract CFM-based feature of regions with a single forward pass.
2. We explore a binary feature encoding scheme for fast global initial retrieval.
3. We propose a novel descriptor and region-refined re-ranking algorithm for local retrieval. The experimental results of the proposed algorithm achieves the state-of-the-art performance.

2 Method

We first revisit RPN which locates candidate regions instead of rigid grid method in Sect. 2.1. Next, we provide the details of encoding schemes for the learned deep features in Sect. 2.2. Finally, our region refined algorithm based on proposed encoding schemes is discussed in Sect. 2.3. Figure 1 illustrates the pipeline of our proposed retrieval method.

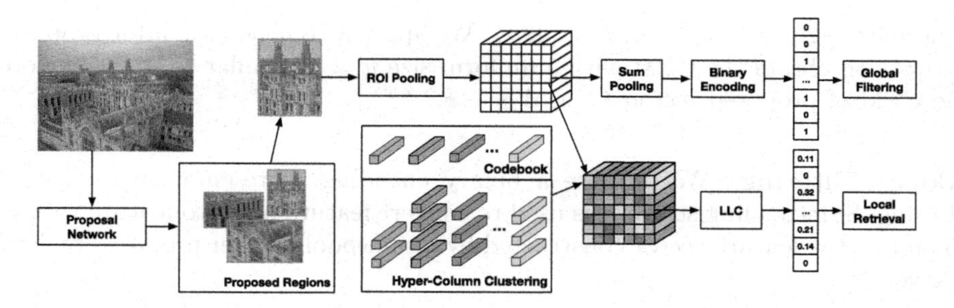

Fig. 1. Pipeline of our proposed retrieval method.

2.1 Proposal Network

Different from most previous works [11,17] on instance retrieval, we propose a pre-trained RPN to search several candidate regions from the whole image. Inspired by successful approach [14] on object detection task, we model this process with a fully convolutional network. The network architecture (see Fig. 2) of RPN is similar to CNN used in image classification tasks with additional two branches of convolutional layers, instead of fully connected layers. One branch regresses the coordinates of the proposal bounding box while the other predicts the score of containing an object rather than background. Candidate regions are conveniently achieved only with a forward pass of RPN. Since each regressed location is optimised when training the network, the resulted candidate regions occupy the object of interest more tightly than the fixed grid.

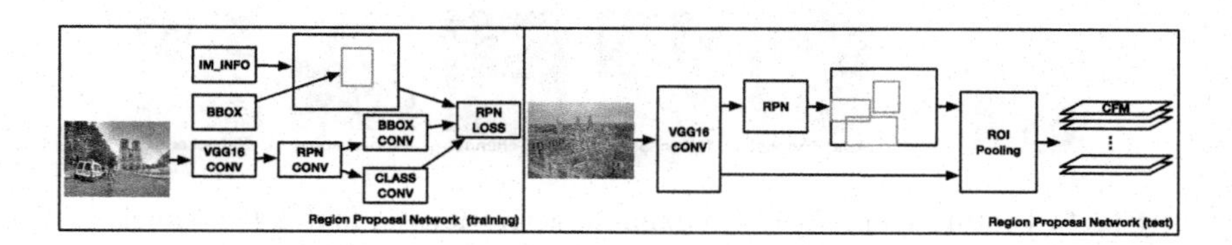

Fig. 2. The architecture of region proposal network.

We train our RPN on the cleaned version [5] of public Landmarks dataset [2]. For each bounding box, we assign a binary class label according to the IOU between the box and ground-truth region of interest. Then we apply SGD (stochastic gradient descent) and back-propagation algorithms to optimise a multi-task loss which consists of classification loss (log loss over object *vs* background classes) and a regression loss with smooth-$l1$ [4].

2.2 CFM-Based Encoding Features

Given an input image I of size $W \times H$ and the CFM of I is a 3D tensor, $\mathcal{X} \in \mathbb{R}^{K \times W \times H}$, where K is the number of channels. Our RPN tells us a set of candidate regions $\Omega = \{\Omega_1, \Omega_2, ..., \Omega_R\}$. We apply a ROI pooling after proposal network to obtain the CFM with a uniform size $w \times h$. Similar to $\mathcal{X}$, we denote the CFM of proposed region Ω_i as $\mathcal{X}_{\Omega_i} \in \mathbb{R}^{K \times w \times h}$.

Global Filtering. We propose a binary encoding representation for global filtering. Sum-pooling strategy is used to convert feature maps to feature vector. Therefore the feature vector constructed by a sum-pooling over proposed regions is given by

$$\mathbf{f}_{\Omega_i} = [f_{\Omega_i}^{(1)}, f_{\Omega_i}^{(2)}, ..., f_{\Omega_i}^{(K)}], \text{with } f_{\Omega_i}^{(k)} = \sum_{p \in \Omega_i} \mathcal{X}_{\Omega_i}^{(k)}(p). \tag{1}$$

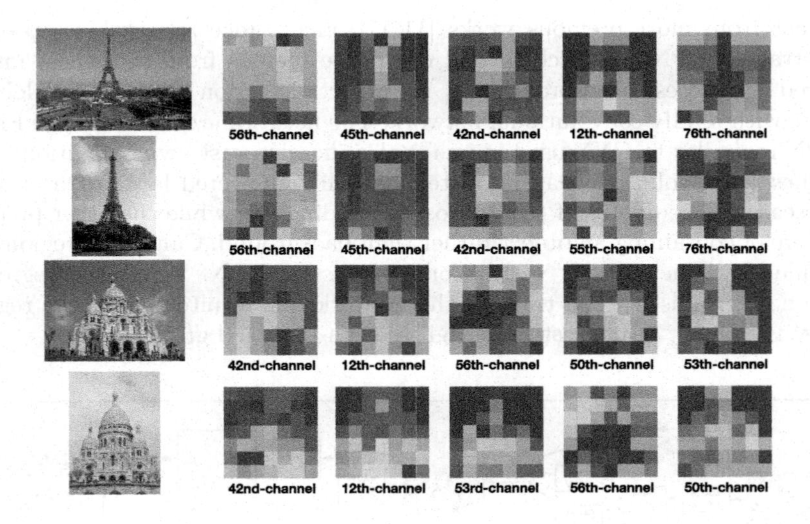

Fig. 3. Feature map of top 5 maximum activations in first 100 channels

Our global filtering feature vector considers the first N_{GF}th maximum activations in $\mathbf{f}_{\Omega}$. Figure 3 is a heat map visualisation for the feature maps corresponding to the top-5 stronger activations among the first 100 channels. Images at the left side in Fig. 3 are the famous Paris landmarks Eiffel Tower and Sacre Coeur respectively. The feature maps with much stronger response are listed in each row. For the same or similar object in different images, convolutional filters with stronger activation are also relative. Thus we propose a binary vector to replace the previous pooling feature vector. Denote the set of channels with the top N_{GF}th maximum activations to $ma(\Omega_i)$.

$$\mathbf{b}_{\Omega_i} = [b_{\Omega_i}^{(1)}, b_{\Omega_i}^{(2)}, ..., b_{\Omega_i}^{(K)}], \text{with } b_{\Omega_i}^{(k)} = \mathbb{1}[k \text{ in } ma(\Omega_i)] \tag{2}$$

Therefore we obtain our binary encoding feature. Compared with previous numeric feature vector, our global filtering feature records a part of whole channels with 1 and others with 0. Such design brings some convenience on both similarity calculation and memory storage.

Local Retrieval. Global filtering helps us reserve the images which have higher relevance to the query image. We continue our local retrieval on these images. We call our encoded feature in local retrieval as the Bag of Hyper-column (BoH) descriptor. In our problem, the CFM's activation of the candidate region is a 3D tensor, $\mathbf{X} = [\mathbf{x}_1, \cdots, \mathbf{x}_{w\times h}] \in \mathbb{R}^{K\times w\times h}$. Each hyper-column in CFM represents a small patch with $\frac{W}{w} \times \frac{H}{h}$ pixels in the input image. Thus the CFM can be regarded as the combination of $w \times h$ hyper-column patches.

We fit the codebook of hyper-column $\mathbf{B} = \{\mathbf{b}_1, \mathbf{b}_2, \cdots, \mathbf{b}_M\} \in \mathbb{R}^{K\times M}$ with K-Means algorithm. Inspired by the Locality-constrained Linear Coding (LLC)[18], we convert learned CFM into a M-dimensional code to generate the final image representation. It comes to a special case when we calculate the code for a hyper-column. We use this trick several times in Sect. 2.3 to refine the proposed region. In order to obtain the BoH descriptor $\mathbf{C}$, we deal with the following criteria:

$$\min_{\mathbf{C}} \sum_{i=1}^{w\times h} ||\mathbf{x}_i - \mathbf{B}\mathbf{c}_i||^2 + \lambda||\mathbf{d}_i \odot \mathbf{c}_i||^2, s.t.\ \mathbb{1}^\top \mathbf{c}_i = 1,\ \forall i \tag{3}$$

where $\odot$ denotes the element-wise multiplication, and we take

$$\mathbf{d}_i = \exp(\frac{\text{dist}(\mathbf{x}_i, \mathbf{B})}{\sigma}), \tag{4}$$

where $\text{dist}(\mathbf{x}_i, \mathbf{B}) = [\text{dist}(\mathbf{x}_i, \mathbf{b}_1), \cdots, \text{dist}(\mathbf{x}_i, \mathbf{b}_M)]^\top$, and $\text{dist}(\mathbf{x}_i, \mathbf{b}_j)$ is the Euclidean distance between $\mathbf{x}_i$ and $\mathbf{b}_j$. σ is used for adjusting the weight decay speed for the locality adaptor.

Given a reserved image, there are several proposed regions $\Omega = \{\Omega_1, \cdots, \Omega_R\}$. According to the encoding scheme discussed above, we calculate the BoH descriptor over Ω and select the region which has the minimum cosine distance with the query item as the best proposed region. The distance is regarded as the similarity metric between the reserved image and the query image.

2.3 Region Adaptive Reranking

The main idea of our reranking scheme is to refine the candidate region of matched top N images. Since our proposed region is detected from a pre-trained proposal network. In fact, the bounding box given for each region is a regular rectangle with prior knowledge. Generally, the main part of object is in the middle of its candidate region and unrelated background is at the edge of the bounding box. We refine the selected area in the bounding box when reranking the results. The representation is designed as the sum of hyper-columns' feature vector, we apply a variant breadth-first-depth algorithm to select the proper hyper-columns. Algorithm 1 shows the details of the reranking algorithm.

Algorithm 1: Region refining algorithm for reranking.

Input:
The BoH descriptor of the query item F_q
The CFM of the proposed region M
Output: The BoH descriptor of the proposed region F_p

```
1  Choose the middle part of the region as the base region R;
2  Maintain a list V to record whether one is visited for each pixel in region;
3  Maintain a queue Q to record whether one is selected for each pixel in region;
4  for pix in R do
5      Calculate the BoH descriptor of pix, F_pix;
6      Update the BoH descriptor, F_p = F_p + F_pix;
7      Append pix into V and put pix into Q;
8  Calculate the distance between F_q and F_p, D;
9  while Q is not empty do
10     Pop the front element pix in Q;
11     for nxt is a neighbour of pix and nxt is not in V do
12         Calculate the BoH descriptor of nxt, F_nxt;
13         Add nxt into V;
14         Update the BoH descriptor after adding nxt, F' = F_p + F_nxt;
15         Calculate the distance between F' and F_q, D';
16         if D' ≤ D then
17             Put nxt into Q;
18             Update the BoH descriptor and distance, (F_q, D) = (F', D');
```

3 Experiments

In this section, we first introduce the datasets that we used in our experiments. Then we present the implementation details of our algorithms. Finally, a comparison with the state-of-the-art method is given.

3.1 Datasets

Oxford5k. The Oxford dataset contains 5,062 images corresponding to 11 Oxford landmarks. There are five queries for each landmark and thus 55 queries in total. The dataset can be augmented by combining with another 100,071 distractors from Flickr100k for large-scale retrieval evaluation. We denote the augmented dataset as Oxford105k. The performance is measured by mean average precision (mAP) over all queries, as in [12].

Paris6k. Similar to Oxford5k, the Paris6k dataset contains 6,412 images corresponding to 12 Paris landmarks with 55 queries in total. For large-scale retrieval evaluation, we combine it with the Flickr100k, namely Paris106k. mAP over all queries is reported as in [13].

3.2 Implementation Details

In our experiments, we fine-tune our pre-trained RPN network under two based network, VGG16 [15] and ResNet101 [6]. We choose Landmarks dataset as our training data. The training procedure finishes after 10 epochs. From the trained RPN network, we extract the ROI pooling map for every proposed region of an image. We obtain a size of $512 \times 7 \times 7$ feature map with VGG16-based network and $1024 \times 14 \times 14$ with ResNet101-based network. After obtaining the ROI pooling map of a proposed region, we extract the global filtering feature according to the method introduced in Sect. 2.2. In practice, we choose $N_{GF} = 200$. In local retrieval stage, we construct a codebook with $M = 1024$ centroids. Every region is represented by its $K = 50$ nearest neighbours centroids.

Figure 4 shows the top retrieval results of several query items. The first image at each row is the query image. The images followed from left to right in each row is the retrieval results according to the similarity. The black number below represents the rank and the red number represents the rank before reranking strategy. Our reranking strategy improves the rank of the object that occurs small proportion of the whole image and thus bring the retrieval precision with a slight improvement.

Fig. 4. Top retrieval results of several query items.

3.3 Comparison with the State-of-the-art Algorithms

In this section, we compare our results with several instance search works in the literature. Table 1 is divided into two parts. The above part reports the retrieval results using different compact representations without post-processing, e.g. reranking, spatial verification or QE. The other part reports results of corresponding method with post-processing. From the first part, aggregated CNN based representations have better performance than naive CNN representations. Among these aggregated features, the R-MAC descriptor achieves the best mAP result. Because generating R-MAC descriptor considers both multi-scale feature map and local spatial information. The second part of Table 1 reports methods with post-processing. Our method achieves the best performance on Paris datasets. Our proposed BoH descriptor obtains comparable performance with the well developed BoW model [10] on Oxford datasets. Our method has a stable performance on both small and large scale datasets. When the scale of the retrieval dataset increases from 5k to 100k, our method still has a stable performance, the mAP metric slightly decreases. Thus we believe our BoH descriptor is more robust.

Table 1. Comparison with existing methods.

Method	Dim.	Datasets			
		Oxford5k	Paris6k	Oxford105k	Paris106k
Babenko and Lemptisky et al. [1]	512	68.5	79.8	62.2	70.4
Kalantidis et al. [8]	512	68.2	79.7	63.3	71.0
Tolias et al. [17] (VGG16)	512	66.9	83.0	61.6	75.7
Tolias et al. [17] (ResNet101)	2048	69.4	85.2	63.7	77.8
Kalantidis et al. [8]	512	72.2	85.5	67.8	79.7
Mikulik et al. [10]	-	**84.9**	82.4	**79.5**	77.3
Tolias et al. [17] (VGG16)	512	77.3	86.5	73.2	79.8
Tolias et al. [17] (ResNet101)	2048	78.9	89.7	75.5	85.3
Ours (VGG16)	1024	78.7	87.7	75.1	83.4
Ours (ResNet101)	1024	82.9	**89.8**	78.6	**86.1**

4 Conclusion

In this paper, we propose a novel approach for instance retrieval based on encoding deep features. We obtain proposed regions from an image with the popular RPN. Different from previous merely deep feature method, we introduce different encoding features to specific tasks. The binary encoding feature is suitable to global filtering while the LLC-based feature is fit for local retrieval. Because our BoH feature has a property of accumulation, we present a reranking algorithm on region refining. We conduct experiment on Oxford and Paris datasets and report a better result compared with the current state-of-the-art methods.

Acknowledgements. The work was supported by the National Natural Science Foundation of China (Grant No. 91420302) and the National Basic Research Program of China (Grant No. 2015CB856004), and the Key Basic Research Program of Shanghai, China (15JC1400103).

References

1. Babenko, A., Lempitsky, V.: Aggregating local deep features for image retrieval. In: Proceedings of the IEEE International Conference on Computer Vision, pp. 1269–1277 (2015)
2. Babenko, A., Slesarev, A., Chigorin, A., Lempitsky, V.: Neural codes for image retrieval. In: Fleet, D., Pajdla, T., Schiele, B., Tuytelaars, T. (eds.) ECCV 2014. LNCS, vol. 8689, pp. 584–599. Springer, Cham (2014). doi:10.1007/978-3-319-10590-1_38
3. Chum, O., Philbin, J., Sivic, J., Isard, M., Zisserman, A.: Total recall: automatic query expansion with a generative feature model for object retrieval. In: IEEE 11th International Conference on Computer Vision, ICCV 2007, pp. 1–8. IEEE (2007)
4. Girshick, R.: Fast R-CNN. In: Proceedings of the IEEE International Conference on Computer Vision, pp. 1440–1448 (2015)
5. Gordo, A., Almazán, J., Revaud, J., Larlus, D.: Deep image retrieval: learning global representations for image search. In: Leibe, B., Matas, J., Sebe, N., Welling, M. (eds.) ECCV 2016. LNCS, vol. 9910, pp. 241–257. Springer, Cham (2016). doi:10.1007/978-3-319-46466-4_15
6. He, K., Zhang, X., Ren, S., Sun, J.: Deep residual learning for image recognition. In: Proceedings of the IEEE Conference on CVPR, pp. 770–778 (2016)
7. Jégou, H., Douze, M., Schmid, C., Pérez, P.: Aggregating local descriptors into a compact image representation. In: 2010 IEEE Conference on Computer Vision and Pattern Recognition (CVPR), pp. 3304–3311. IEEE (2010)
8. Kalantidis, Y., Mellina, C., Osindero, S.: Cross-dimensional weighting for aggregated deep convolutional features. In: Hua, G., Jégou, H. (eds.) ECCV 2016. LNCS, vol. 9913, pp. 685–701. Springer, Cham (2016). doi:10.1007/978-3-319-46604-0_48
9. Lowe, D.G.: Distinctive image features from scale-invariant keypoints. Int. J. Comput. Vis. **60**(2), 91–110 (2004)
10. Mikulik, A., Perdoch, M., Chum, O., Matas, J.: Learning vocabularies over a fine quantization. Int. J. Comput. Vis. **103**(1), 163–175 (2013)
11. Mohedano, E., McGuinness, K., O'Connor, N.E., Salvador, A., Marqués, F., Giró-i Nieto, X.: Bags of local convolutional features for scalable instance search. In: Proceedings of the 2016 ACM on ICMR, pp. 327–331. ACM (2016)
12. Philbin, J., Chum, O., Isard, M., Sivic, J., Zisserman, A.: Object retrieval with large vocabularies and fast spatial matching. In: IEEE Conference on Computer Vision and Pattern Recognition, CVPR 2007, pp. 1–8. IEEE (2007)
13. Philbin, J., Chum, O., Isard, M., Sivic, J., Zisserman, A.: Lost in quantization: improving particular object retrieval in large scale image databases. In: IEEE Conference on CVPR 2008, pp. 1–8. IEEE (2008)

14. Ren, S., He, K., Girshick, R., Sun, J.: Faster R-CNN: towards real-time object detection with region proposal networks. In: Advances in Neural Information Processing Systems, pp. 91–99 (2015)
15. Simonyan, K., Zisserman, A.: Very deep convolutional networks for large-scale image recognition. arXiv preprint arXiv:1409.1556 (2014)
16. Sivic, J., Zisserman, A., et al.: Video google: a text retrieval approach to object matching in videos. In: ICCV, pp. 1470–1477 (2003)
17. Tolias, G., Sicre, R., Jégou, H.: Particular object retrieval with integral max-pooling of CNN activations. arXiv preprint arXiv:1511.05879 (2015)
18. Wang, J., Yang, J., Yu, K., Lv, F., Huang, T., Gong, Y.: Locality-constrained linear coding for image classification. In: 2010 IEEE Conference on CVPR, pp. 3360–3367. IEEE (2010)

Discriminative Semi-supervised Learning Based on Visual Concept-Like Features

Fang Liu and Xiaofeng Wu(✉)

Department of Electronic Engineering, Fudan University, Shanghai 200433, China
xiaofengwu@fudan.edu.cn

Abstract. A discriminative semi-supervised learning method based on visual concept-like high-level features is proposed in this paper. Previous semi-supervised learning methods usually use unlabeled data to augment the training set or regularize the decision boundary of classifiers. The classification results rely on the precision on unlabeled data using supervised classifiers trained with limited labeled samples. When a small number of labeled samples are provided, these methods are likely to get bad results. Differently, the proposed method directly uses the distribution information of all available data in the feature space to learn a new representation which is achieved by computing the similarities of a chosen image and some discriminative data exemplars sampled from the feature space. A semi-supervised distance metric learning method by learning a projection matrix under the equivalence constraints of similar pairs and dissimilar pairs is introduced to measure these similarities, and a pseudo-mahalanobis distance is thus obtained to represent the similarities between data samples instead of Euclidean distance. Experiments showed the effectiveness of this learned distance. The new representation can be fed into standard classifiers for image classification task. The training data of our system can either be original image data or handcrafted features or image features learned by deep architectures. Therefore, the proposed method can be applied in both feature extraction and feature enhancement. In the semi-supervised classification task on eight standard datasets, the proposed method achieves improved performance over many of the previous existing methods.

Keywords: Semi-supervised image classification · Discriminative feature learning · Metric learning

1 Introduction

Recent years have witnessed great progress in image classification with less or no annotations, since annotating data by human efforts is both time-consuming and expensive while unlabeled data is numerous and easy to achieve. Semi-supervised methods try to reveal the information carried by unlabeled data to improve performance. A large number of methods regularize the decision boundary by forcing it to pass through the region with lower density of unlabeled data. Another widely-used scheme is the self-training scheme [1]. It first annotates the unlabeled data by training a supervised classifier on labeled data. Then the training set is augmented by adding the

D. Liu et al. (Eds.): ICONIP 2017, Part III, LNCS 10636, pp. 75–83, 2017.
https://doi.org/10.1007/978-3-319-70090-8_8

most confident unlabeled data with their predicted labels. The system iterates between training models and augmenting the training set until some termination condition is reached. However, self-training relies on the predicted labels on unlabeled data for training a new classifier. It can probably make an error when a small number of labeled data is available. Differently, some ensemble algorithms assign a pseudo-label to unlabeled data, and then sample them for training a new classifier. They iterate to construct the ensemble classifier under the restriction of a cost function. The precision relies on the prediction of pseudo-labels using the constructed classifier at each iteration. In contrast, the proposed method compares the learned distance between data samples to obtain pseudo-labels. Then ensemble supervised classifiers are trained and used to extract a new visual concept-like representation of the input data.

Different object classes carry many discriminative visual concept-like features that can help us distinguish the classes, such as colors, shapes and textures. We call them visual concepts for simplicity in this paper. These concepts are lower-dimensional compared with the features or images. A new category can be learned by comparing the new object with the existing categories from the perspective of visual concepts. For instance, a volleyball has similar shape to a basketball but has different texture and color. This learning procedure is called learning by comparison. It is a part of Eleanor Rosch's prototype theory [2] which states that an object's class is determined by its similarities to prototypes that represent object classes. This theory has been used successfully in transfer learning, where labeled data from different classes are available.

Since the visual concepts exist in images regardless of labels and can provide discriminative information, we consider to learn these visual concepts from all available data. In particular, we aim to learn some data samples that contain several typical visual concepts. The typical concepts are called concept exemplars such as "spherical", "red" and "brown". And a group of data samples containing several similar concept exemplars are named a "subset". For instance, the class of apples can be viewed as a subset of "red" and "smooth" at least. These subsets are sampled from all available data in an unsupervised way based on the assumption that neighboring samples in the feature space share similar concepts exemplars. To be discriminative, samples of the same subset should be close to each other and samples from different subsets should be as far as possible. In other words, the subsets are expected to be inter-distinct and intra-compact. We combine several subsets to form a cluster. Discriminative information can be learned by concatenated the similarities between the chosen image and these subsets. However, a cluster is formed in one sampling trial and can be noisy because the concepts we learned in one trial are limited, so a rich set of clusters is necessary for our learning procedure to cover enough concepts.

As stated above, the learning-by-comparison procedure has a critical demand on similarity measurement. Euclidean distance is widely used to represent the similarity. However, it is not enough when features have high dimensionality. We introduce a semi-supervised distance metric learning method which learns a pseudo-mahalanobis distance by learning a projection matrix under the equivalence constraints. The pseudo-mahalanobis distance can measure the similarity in a better way, which means similar samples are closer and dissimilar samples are as far as possible.

The rest of this paper is organized as follows. Section 2 briefly reviews related work in the field of both metric learning and high-level feature learning. Section 3 introduces the proposed method in details, followed by experiments for performance evaluation in Sect. 4. Section 5 concludes the paper.

2 Related Work

Our work is generally relevant to image classification with metric learning under equivalence constraints and high-level feature learning based on Eleanor Rosch's prototype theory. In this section, some current related works in these two research fields are simply reviewed, and the similarities and differences among our method and these works are discussed.

2.1 Metric Learning Under Equivalence Constraints

There are many widely-used metric learning methods using equivalence constraints. Relevant components analysis [3] (RCA) learns an embedding which allocates larger weights to the most relevant dimensions of the features and lower weights to less relevant ones. But it does not incorporate dissimilarity constraints, and is limited in the original input space to learn linear transformations. Discriminative components analysis [4] (DCA) incorporates dissimilarity constraints into RCA and Kernel RCA. The semi-supervised discriminative common vector method [5] (SS-DCV) introduced in our method is similar in spirit to DCA, but it overcomes a serious shortcoming of DCA – the criterion of maximizing the classical LDA (Linear Discriminant Analysis) function does not have a unique solution while the dimensionality of the sample space is much larger than the number of similar sample pairs, which leads to miss the optimal projection direction. SS-DCV method projects the data onto the subspace orthogonal to the linear span of the difference vectors of similar sample pairs first, in which similar pairs have identical projections. Then it learns a linear embedding that maximizes the scatter of the dissimilar sample pairs. This corresponds to a pseudo-metric characterized by a positive semi-definite matrix in the original input space. The integrate derivation of SS-DCV can be found in [5].

2.2 High-Level Feature Learning Using Eleanor Rosch's Prototype Theory

A method that is closely related to ours is Ensemble Projection [6]. It is also based on the Eleanor Rosch's prototype theory. It samples an ensemble of prototype sets that present different classes. In return, an ensemble of diverse projection functions are learned based on these prototype sets. The prototype is similar to our visual concept exemplar. But their prototypes indicate the classes directly and our visual concept exemplar is expected to be a subset of several visual concepts shared among different object classes. Projection values of an individual data sample through these projection functions are stacked together to form a new feature representation. However, different from our approach, Ensemble Projection is purely unsupervised and do not leverage label information for a specific task. In particular, Ensemble Projection samples the prototypes in the original feature space. In our method, visual concept exemplars are learned in a lower-dimensional subspace. In the projected space, similar pairs have identical projections which indicate that they have similar subsets of visual concepts. In the experiments on different datasets for image classification, we will compare our algorithm with the Ensemble Projection method.

3 Discriminative Semi-supervised Learning Based on Visual Concept-Like Features

Following the notations that are widely used in semi-supervised feature learning, the input of our method includes labeled data $\mathbf{x}_{1:L}$, $y_{1:L}$ and unlabeled data $\mathbf{x}_{L+1:L+U}$, where $\mathbf{x}_i$ denotes the feature vector of image i, $y_i \in \{1,\ldots,C\}$ indicates its label, and C is the number of classes, L is the length of labeled data, U is the length of unlabeled data. A new image representation $\mathbf{f}$ is learned using both the unlabeled data and the labeled data.

3.1 Projection onto a Subspace Under Equivalence Constraints

Given a small amount of labeled data, we aim to learn a class-discriminative subspace and achieve a better judgement of similarity using SS-DCV [5].

Let $\mathbf{x}_i \in \mathbb{R}^d, i = 1,\ldots,N$ denote the samples of the training set. A set of equivalence constraints in the form of similar and dissimilar pairs are given and we aim to learn a pseudo-mahalanobis distance of the form

$$d_A(\mathbf{x}_i, \mathbf{x}_j) = ||\mathbf{x}_i - \mathbf{x}_j||_{\mathbf{A}} = \sqrt{(\mathbf{x}_i - \mathbf{x}_j)^T \mathbf{A} (\mathbf{x}_i - \mathbf{x}_j)}, \tag{1}$$

where $\mathbf{A} \geq 0$ is a symmetric positive semi-definite matrix reflecting the underlying relationships between different dimensions of the feature. If $\mathrm{q} = \mathrm{Rank}(\mathbf{A}) \leq d$, $\mathbf{A}$ can be written in the form $\mathbf{A} = \boldsymbol{W}\boldsymbol{W}^T$ where $\boldsymbol{W}$ is a full-rank rectangular matrix of size $d \times q$, so that

$$||\mathbf{x}_i - \mathbf{x}_j||_{\mathbf{A}}^2 = ||\mathbf{W}^T\mathbf{x}_i - \mathbf{W}^T\mathbf{x}_j||^2 \tag{2}$$

i.e. the pseudo-mahalanobis distance between samples are equivalent to Euclidean distances on their linear projections by $\mathbf{W}^T$.

$\mathbf{X}_s$ and $\mathbf{X}_D$ denote the matrixes whose columns are the difference vectors of the given similar and dissimilar pairs

$$\mathbf{X}_s = \left[\mathbf{x}_{s1,1} - \mathbf{x}_{s1,2},\ \mathbf{x}_{s2,1} - \mathbf{x}_{s2,2,\ldots}\mathbf{x}_{sn,1} - \mathbf{x}_{sn,2}\right], \tag{3}$$

$$\mathbf{X}_D = \left[\mathbf{x}_{d1,1} - \mathbf{x}_{d1,2},\ \mathbf{x}_{d2,1} - \mathbf{x}_{d2,2,\ldots}\mathbf{x}_{dm,1} - \mathbf{x}_{dm,2}\right], \tag{4}$$

where $\mathbf{x}_{si,1}$ and $\mathbf{x}_{si,2}$ respectively represent the first and second samples of the i-th similar sample pair; $\mathbf{x}_{di,1}$ and $\mathbf{x}_{di,2}$ respectively represent the first and second samples of the i-th dissimilar sample pair.

The full method of semi-supervised discriminative common vector can be presented as follows:

1. Compute $\mathbf{X}_s$ and its orthonormal basis matrix $\mathbf{U}$.
2. Project the dissimilar sample pairs to the null space of $\mathbf{X}_s$ using $\widetilde{\mathbf{X}}_D = (\mathbf{I} - \mathbf{U}\mathbf{U}^T)\mathbf{X}_D$.
3. Compute $\widetilde{\mathbf{S}}_D = \widetilde{\mathbf{X}}_D\widetilde{\mathbf{X}}_D^T$ and its leading eigenvectors $\boldsymbol{W}$, then output the final distance metric $\mathbf{A} = \boldsymbol{W}\boldsymbol{W}^T$.

3.2 Visual Concept Clusters Learning and Feature Extraction

In the projected lower-dimensional subspace, we learn S cluster sets $\mathcal{C}_{1:S}$. Each cluster $\mathcal{C}_s$ contains $e_{1:m \times n}, \mathrm{l}_{1:m \times n}$, where e_i denotes the i-th exemplar and $\mathrm{l}_i \in \{1, \ldots, n\}$ denotes the pseudo label. There are n subsets of exemplars in a cluster and the pseudo label indicates which subsets the exemplar belongs to. In each subset, we have m exemplars so the total number of exemplars in a cluster is $m \times n$.

Firstly, a group of feature samples that are far apart from each other are chosen as the skeleton of the subsets. Then their k-nearest neighbors are found to enrich the subsets. The pseudo label is shared in the same subset. The method of learning the cluster set can be summarized as follows:

1. Randomly choose n feature samples as a skeleton for t times and choose the furthest one.
2. Calculate m nearest neighbors of every feature sample in the skeleton.
3. Repeat step 1 to step 2 for S times to get the cluster sets $\mathcal{C}_{1:S}$.

After the cluster set is created, we train logistic regression on each cluster. The classification scores of a chosen feature are then concatenated to form the new representation. Classification scores indicate the similarities between the chosen feature and the visual concepts exemplars we created, so they can represent the correlation of the current feature and the shared visual concepts.

4 Experiments

The primary performance evaluation experiments are executed for semi-supervised image classification on eight standard datasets as follows:

1. Texture-25 [7]: texture images divided into 25 categories, with 40 images per class.
2. Caltech-101 [8]: 8677 images from 101 object classes, with 31 to 800 samples per class.
3. STL-10 [9]: 100000 unlabeled images and 13000 labeled images from 10 object classes with 500 training images and 800 test images per class.
4. Scene-15 [10]: 4485 scene images of indoor and outdoor environments divided into 15 classes, with 200 to 400 samples per class.
5. Indoor-67 [11]: 15620 images from 67 indoor classes, with at least 100 images per class.

6. Event-8 [12]: 1574 images from 8 sports event categories.
7. Building-25 [13]: 4794 images from 25 architectural styles, such as American craftsman, Baroque, and Gothic.
8. LandUse-21 [14]: 2100 satellite images divided into 21 classes, with 100 samples per class.

The inputs of our algorithm are CNN features with the dimensionality of 4096 obtained from an off-the-shelf CNN [15] pre-trained on the ImageNet dataset. For comparison, we use the same datasets and CNN features used by [6].

4.1 Experiment Settings

Three baselines are used in our primary performance evaluation experiments: k-nearest neighbor (k-NN) algorithm, Logistic Regression (LR) and support vector machines (SVMs) with radial basis function (RBF) kernels for semi-supervised classification. The original convolutional neural network (CNN) features, the features learned by the Ensemble Projection method and features learned by our algorithm were fed into these classifiers to evaluate the method. For different datasets, the parameters in the experiments were fixed to the following values: $\mathcal{S}$ = 100, m = 6, n = 30, and t = 50. Different numbers of training images per class were tested. In keeping with most existing systems for semi-supervised classification [16–19], we evaluate the method in the transductive manner, where we take the training and test samples as a whole, and randomly choose labeled samples from the whole dataset to learn and infer labels of other samples whose labels are held back as the unlabeled samples. The reported results are the average performance over 5 runs with random labeled-unlabeled splits.

4.2 Classification Results

Table 1 lists the precision of the methods using 5 labeled training examples per class. Three kinds of classifiers are used: k-NN, Logistic Regression and SVMs with RBF kernel. They worked with three feature inputs: the original CNN features, features learned by ensemble projection (indicated by "+EP") and features learned by our visual concept method (indicated by "+VC"). The best performance for each dataset is indicated in bold, and the second best is in bold italic. It is easy to observe that

Table 1. Precision (%) of image classification on the eight datasets, with 5 labeled training examples per class

Methods	Scene-15	LandUse-21	Texture-25	Building-25	Event-8	Caltech-101	Indoor-67	STL-10
k-NN	61.27	69.04	81.50	30.57	75.51	69.75	20.86	54.16
k-NN+EP	74.86	75.73	83.97	34.64	86.35	70.49	24.64	64.75
k-NN+VC	76.03	79.94	89.51	35.82	86.67	61.86	26.75	65.71
LR	73.07	77.70	86.96	36.61	84.71	**81.38**	31.02	65.66
LR+EP	80.06	80.56	87.52	***40.36***	***89.47***	79.28	***34.50***	***74.00***
LR+VC	***80.78***	**84.12**	**91.72**	**41.42**	**90.01**	***80.21***	**36.30**	**75.29**
SVMs	72.54	73.78	82.00	38.75	83.12	76.92	32.05	65.62
SVMs+EP	79.66	77.23	83.10	39.38	87.31	74.77	31.76	70.99
SVMs+VC	**81.39**	***81.91***	***90.35***	40.34	88.39	68.21	34.26	73.57

classifiers consistently have better performance when working with our features. Logistic regression performs best among these three classifiers. Working with our features, logistic regression gets the highest precision on six of the eight datasets followed by SVMs, which get higher precision than k-NN.

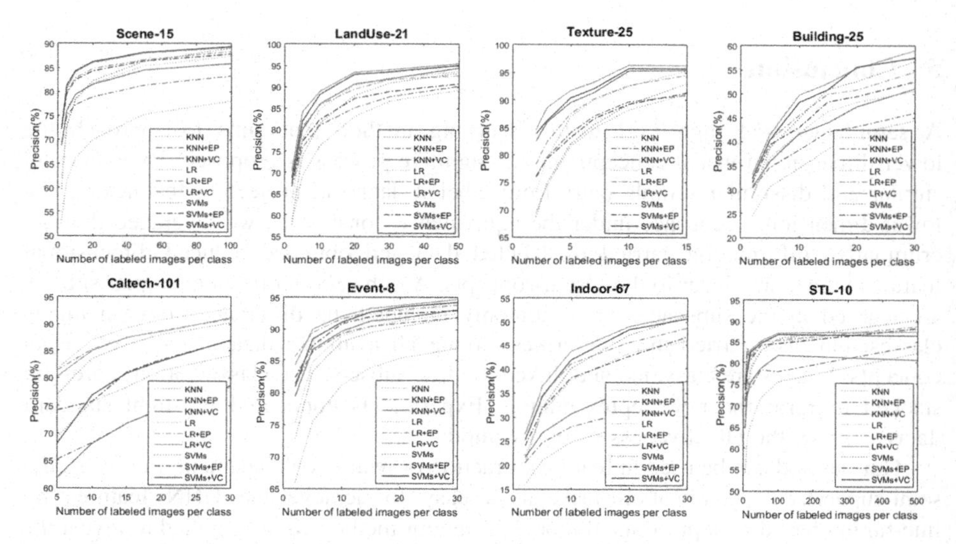

Fig. 1. Precision (%) of image classification on the eight datasets using different number of labeled images per class. Classifiers and features used are as same as Table 1.

Figure 1 shows the classification results on the eight datasets using these three kinds of classifiers and three feature inputs when different number of labeled training examples are provided. Numbers of labeled images per class are chosen according to the different structures of the datasets: different number of classes and different number of images per class. Results achieved by the same kind of classifier are shown in the same color and results generated by three classifiers working with same features are shown with the same line type. The figure shows the advantages of our features over the original CNN features and the features learned by Ensemble Projection across different datasets and classifiers. For example, when given 5 labeled samples per class, we obtain a 4.2% improvement over the Ensemble Projection on the Texture-25 dataset using LR and 3.56% on the LandUse-21 dataset.

The enhancement of precision is attributed to our discriminative common vector learning for optimal similarity measure metrics, and projecting all available data to the lower-dimensional subspace. This projection applies the supervisory information carried by the labeled sample pairs and learns a lower-dimensional visual concept in the subspace. When given a small number of labeled data, our method can be used as an effective feature enhancing method.

To clarify the influence and contributions of the parameters in our method on various image datasets, we tested a variety of values. In details, the number of clusters and the number of subsets per cluster and the number of exemplars per subset are adjusted. It shows that learning more exemplars or more clusters can slightly improve the accuracy. But in a large range of values, results are not sensitive to the parameters. Values fixed in the experiments can achieve a promising result. The proposed method is also efficient because the training of logistic regression is efficient.

5 Conclusion

A semi-supervised high-level feature learning method that aims to approach the lower-dimensional visual concepts in our cognitive process is proposed. By extracting similar and dissimilar sample pairs from labeled data and projecting features onto a lower-dimensional subspace under the equivalence constraints, we leveraged the discriminative information carried by labeled data and obtained reduced dimensional features which are closer to the visual concepts. A rich set of concept exemplar subsets are learned in the subspace. They not only included the difference between image classes, but also carried the similarities among all available data in terms of visual concepts. Images are classified and linked to these subsets. The classification scores are stacked to form the new representation. Experiments conducted on eight standard datasets show the effectiveness of our method.

Our method can be used in feature extraction or feature enhancement. For a specific semi-supervised classification task, it is easy to achieve the CNN features by fine-tuning based on a pre-trained model. Then our method can be applied to reveal the information carried by the labeled data and enhance the feature using both labeled and unlabeled data to improve the classification results. To prepare more comprehensive equivalence constraints and apply the proposed method to practical tasks would be our future work.

References

1. Rosenberg, C., Hebert, M., Schneiderman, H.: Semi-supervised self-training of object detection models. In: IEEE Workshops on Application of Computer Vision, vol. 1, pp. 29–36. IEEE Computer Society (2005)
2. Rosch, E.: Principles of categorization. In: Concepts: Core Readings, pp. 189–206 (1999)
3. Shental, N., Hertz, T., Weinshall, D., Pavel, M.: Adjustment learning and relevant component analysis. In: Heyden, A., Sparr, G., Nielsen, M., Johansen, P. (eds.) ECCV 2002. LNCS, vol. 2353, pp. 776–790. Springer, Heidelberg (2002). doi:10.1007/3-540-47979-1_52
4. Hoi, S.C., Liu, W., Lyu, M.R., Ma, W.-Y.: Learning distance metrics with contextual constraints for image retrieval. In: 2006 IEEE Computer Society Conference on Computer Vision and Pattern Recognition, pp. 2072–2078. IEEE (2006)
5. Cevikalp, H.: Semi-supervised discriminative common vector method for computer vision applications. Neurocomputing **129**, 289–297 (2014)
6. Dai, D., Van Gool, L.: Unsupervised high-level feature learning by ensemble projection for semi-supervised image classification and image clustering. arXiv preprint arXiv:1602.00955 (2016)

7. Lazebnik, S., Schmid, C., Ponce, J.: A sparse texture representation using local affine regions. IEEE Trans. Pattern Anal. Mach. Intell. **27**, 1265–1278 (2005)
8. Fei-Fei, L., Fergus, R., Perona, P.: Learning generative visual models from few training examples: an incremental bayesian approach tested on 101 object categories. Comput. Vis. Image Underst. **106**, 59–70 (2007)
9. Coates, A., Lee, H., Ng, A.Y.: An analysis of single-layer networks in unsupervised feature learning. In: AISTATS 2011, Ann Arbor, vol. 1001, no. 2 (2010)
10. Lazebnik, S., Schmid, C., Ponce, J.: Beyond bags of features: spatial pyramid matching for recognizing natural scene categories. In: 2006 IEEE Computer Society Conference on Computer Vision and Pattern Recognition, pp. 2169–2178. IEEE (2006)
11. Quattoni, A., Torralba, A.: Recognizing indoor scenes. In: IEEE Conference on Computer Vision and Pattern Recognition, CVPR 2009, pp. 413–420. IEEE (2009)
12. Li, L.-J., Fei-Fei, L.: What, where and who? Classifying events by scene and object recognition. In: IEEE 11th International Conference on Computer Vision, ICCV 2007, pp. 1–8. IEEE (2007)
13. Xu, Z., Tao, D., Zhang, Ya., Wu, J., Tsoi, A.C.: Architectural style classification using multinomial latent logistic regression. In: Fleet, D., Pajdla, T., Schiele, B., Tuytelaars, T. (eds.) ECCV 2014. LNCS, vol. 8689, pp. 600–615. Springer, Cham (2014). doi:10.1007/978-3-319-10590-1_39
14. Yang, Y., Newsam, S.: Bag-of-visual-words and spatial extensions for land-use classification. In: Proceedings of the 18th SIGSPATIAL International Conference on Advances in Geographic Information Systems, pp. 270–279. ACM (2010)
15. Chatfield, K., Simonyan, K., Vedaldi, A., Zisserman, A.: Return of the devil in the details: delving deep into convolutional nets. arXiv preprint arXiv:1405.3531 (2014)
16. Ebert, S., Larlus, D., Schiele, B.: Extracting structures in image collections for object recognition. In: Daniilidis, K., Maragos, P., Paragios, N. (eds.) ECCV 2010. LNCS, vol. 6311, pp. 720–733. Springer, Heidelberg (2010). doi:10.1007/978-3-642-15549-9_52
17. Fergus, R., Weiss, Y., Torralba, A.: Semi-supervised learning in gigantic image collections. In: Advances in Neural Information Processing Systems, pp. 522–530 (2009)
18. Liu, W., He, J., Chang, S.-F.: Large graph construction for scalable semi-supervised learning. In: Proceedings of the 27th International Conference on Machine Learning (ICML 2010), pp. 679–686 (2010)
19. Pitelis, N., Russell, C., Agapito, L.: Semi-supervised learning using an unsupervised atlas. In: Calders, T., Esposito, F., Hüllermeier, E., Meo, R. (eds.) ECML PKDD 2014. LNCS, vol. 8725, pp. 565–580. Springer, Heidelberg (2014). doi:10.1007/978-3-662-44851-9_36

Pedestrian Counting System Based on Multiple Object Detection and Tracking

Xiang Li, Haohua Zhao, and Liqing Zhang(✉)

Key Laboratory of Shanghai Education Commission for Intelligent Interaction and Cognitive Engineering, Department of Computer Science and Engineering, Shanghai Jiao Tong University, Shanghai, China
{lucianlee,haoh.zhao,lqzhang}@sjtu.edu.cn

Abstract. With the increasing demands on video surveillance and business promotion, effective pedestrian counting in surveillance environments has become a hot research topic in computer vision. In this paper, we implement a pedestrian counting system based on multiple object detection and tracking. Region proposal network (RPN) and Real Adaboost classifier are employed to train a head-shoulder detector with high accuracy. We utilize the DSST algorithm to track the position transformations and the size changes of pedestrians. By combining human detection with object tracking together and using detection results to optimize the tracking algorithm, the pedestrian counting system is developed with high robustness against occlusions. We evaluated the system on the videos recorded in the subway station. The results showed that our system achieves a high accuracy and can be used for pedestrian counting in crowded public places.

Keywords: Pedestrian counting · Human detection · Object tracking

1 Introduction

Effective pedestrian counting in surveillance environments has been more and more important in order to ensure various security demands of public places. The goal of pedestrian counting is to estimate the number of people passing through a given area during a certain period of time. Pedestrian counting has broad applications, such as avoiding the occurrence of stampede accidents, controlling the number of people in the subway. However, there still exist a number of difficulties for pedestrian counting in overcrowded public places. First of all, the resolution and the frame rate of surveillance cameras are relatively low. It is very hard to obtain sufficient detailed information in the surveillance videos. The position and posture of the same pedestrian also changes fast, which causes troubles to object tracking. Differences in humans' heights, clothes and motions also make the training much more difficult. Some other aspects will also affect the detection accuracy, including the high pedestrian density, occlusions and complex background.

D. Liu et al. (Eds.): ICONIP 2017, Part III, LNCS 10636, pp. 84–94, 2017.
https://doi.org/10.1007/978-3-319-70090-8_9

In this paper, we propose a pedestrian counting method based on multiple object detection and tracking. Figure 1 shows the framework of the proposed pedestrian counting system. This counting system chooses the head and shoulder of the pedestrian as the detection target. We use Region proposal network (RPN) and Real Adaboost to train this head-shoulder detector, as shown in Fig. 2. DSST algorithm is employed as a tracking tool. By uniting detection with tracking, we propose a pedestrian counting system. Experimental results show that our system is able to count pedestrians in public places.

The rest of this paper is organized as follows: Sect. 2 discusses the related works. Section 3 introduces the proposed counting system, including human detection method, human tracking method and pedestrian counting algorithm. Experimental results are given in Sect. 4, followed by conclusions in Sect. 5.

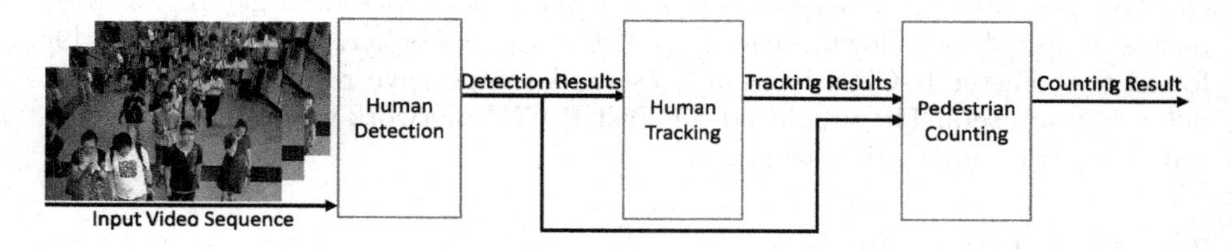

Fig. 1. Framework of the proposed pedestrian counting system

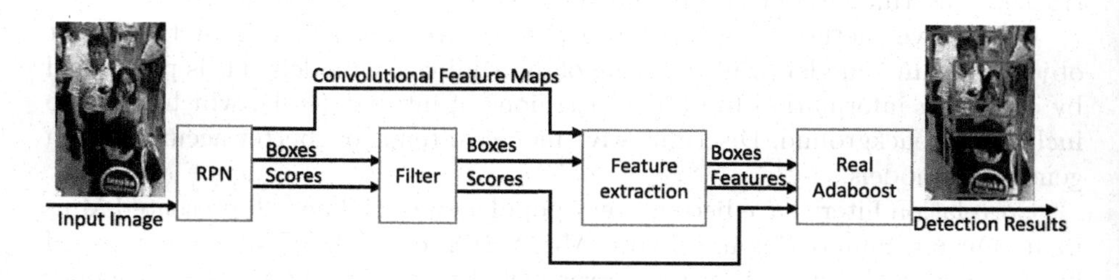

Fig. 2. Framework of the human detection method

2 Related Works

The work of pedestrian counting can be divided into 3 main parts: human detection, human tracking and pedestrian counting. Both detection and tracking in cluttering environments are challenging topics in computer vision.

2.1 Pedestrian Detection

Human detection methods can be divided into three families [1]: decision forests, deep networks and DPM (Deformable Part Detectors) variants.

Human detectors based on decision forests are often trained in the following way: (1) several channel images are generated; (2) then, features for each training example are extracted from these channels; and (3) finally, those features are used to train a decision forest via AdaBoost. Sliding window and image pyramid are also used for detection. Dollar [2,3] combined HOG (Histograms of Oriented Gradients) [4] features with LUV features, and proposed aggregated channel features (ACF). Fast feature pyramids is also implemented in order to accelerate the computation of channel images with different scales. Following the success of ACF, Locally Decorrelated Channel Features (LDCF) [5] and Checkerboards [6] were proposed and showed significant improvement. This two algorithms use the same channels as ACF.

Other state-of-art human detection methods employ neural networks to improve the detection accuracy. Ren [7] proposed Faster R-CNN, which is an effective general object detection method. RPN is proposed to get region proposals, followed by a downstream Fast R-CNN [8] classifier. However, Zhang [9] found that Faster R-CNN has not presented competitive results on pedestrian detection datasets. They replaced the Fast R-CNN network with boosted forests (BF) and got competitive accuracy.

2.2 Object Tracking

In the last years, great efforts have been made to improve the performance of object tracking. There are mainly three kinds of trackers, including generative trackers, discriminative trackers and correlation filter based trackers.

Generative methods [10] only use the features extracted from the target object to train a model, while training of discriminative models [11] is performed by collecting information from lots of regions of interest (RoI), which will also include the background. Discriminative models often show better accuracy than generative models.

Correlation filters have become very popular since Bolme [12] presented Minimum Output Sum of Squared Error (MOSSE) filter. MOSSE filter is computed by minimizing the sum of squared error between the desired gaussian response and the correlation of the object and the filter. Later, Danelljan [13] introduced Discriminative Scale Space Tracking (DSST) method, which achieves the goal of scale estimation. Because of the use of fast Fourier transform (FFT), correlation filter based trackers can run at a relatively high speed.

3 Approach

Our system consists of 3 components (illustrated in Fig. 1): a human detection method, a human tracking method and the final pedestrian counting algorithm.

3.1 Detection Target

For the sake of getting better detection results, we should choose the detection target wisely. Figure 3 shows some frames shot in a subway station, from which we can find that legs of pedestrians can be hardly captured by the monitors and the whole body is obviously a bad choice for pedestrian counting. Face detection algorithms are also not suitable for our system either, because people in those surveillance videos tend to look down and turn their faces away from cameras. The head and the shoulder of the pedestrian can be identified easily in surveillance videos. Besides, compared with human faces, heads and shoulders often contain more information in such a surveillance video, which makes them a better choice.

Fig. 3. Frames from a video recorded in a subway station

3.2 Human Detection

In the pedestrian counting system, we have a few special requirements for the human detection algorithm. First of all, we want the precision of this algorithm to be as higher as possible since all the false positive boxes will be considered as humans and lead to an excessive counting result. Besides, we don't need the detection method to reach a relatively high recall rate. That's because almost every pedestrian will be captured by the monitor for several consecutive frames. As long as the detection method can detect a person in one of those frames, being able to detect this person in other frames or not will make no difference to the final counting result.

Inspired by the RPN+BF method [9], we implement a human detection method which can detect pedestrians in the scenario of crowded public places and can also meet our needs of reaching high precision. Figure 2 shows the structure of our detection method. We will introduce all the components of our detection detailedly in the rest of this subsection.

RPN. RPN was first introduced in [7] as a part of Faster R-CNN, which can be used for multi-category object detection. In our system, RPN is only used for pedestrian detection. After taking in an input image, RPN will propose some regions, each with four coordinates and a score that means the probability of that region to be a human.

The architecture of RPN can be divided into two parts. Layers from conv1_1 to conv5_3 are the same as VGG-16. A simple network is connected to conv5_3 layer in order to get region proposals, including a convolutional layer and two sibling fully-connected layers. At each point of the convolutional layer mentioned above, RPN will generate k (k = 9) region proposals with different scales.

We analyze the sizes of humans' heads and shoulders, and then set the aspect ratio of our anchors (reference boxes) [7] to 1:1. We adopt anchors of 9 different scales. The side lengths of those anchors start from 41.6 pixels and increase with a scaling stride of 1.3×. Using anchors of different sizes helps our detection method to detect multi-scale pedestrians without the usage of image pyramids.

Feature Extraction. The proposals produced by RPN have different sizes, but the following Real Adaboost classifier need to use features with the same size as the inputs. Therefore, a RoI pooling layer is built on a convolution feature map, which will map the coordinates of the regions to the feature map, and then the max pooling will be used to get features with a fixed resolution of 7×7. All the convolution feature maps produced by RPN can be used for feature extraction, but using feature maps with large scaling factors (i.e. low resolution) will cause a serious problem: If a RoI's size is smaller than 7×7 after mapping, the pooling bins will collapse and pooling results are no longer independent of each other. Besides, due to the low resolution of most surveillance cameras, the heads and shoulders they captured are relatively small. As a result, feature maps with large stride (i.e. 8 or 16) won't be used for feature extraction. Higher-resolution feature map Conv3_3 is chosen to be our feature map.

However, the information provided by Conv3_3 may be too little for the following Real Adaboost classifier. A feature extraction network is used to compute several extra convolution feature maps. This feature extraction network includes a 1-stride pooling layer and three convolutional layers whose outputs are feature maps with a scaling factor of 4. These convolutional layers are named of Conv4_1_new, Conv4_2_new and Conv4_3_new. For each proposal, two features will be extracted from Conv3_3 layer and Conv4_3_new layer respectively. After using a fully-connected layer to combine this two features, we will get a fixed-length (i.e. 37632) feature.

Real Adaboost Classifier. The downstream Real Adaboost classifier can reduce the number of false positive boxes. We use decision trees of depth 2 as the weak classifiers. A "bootstrap" way is used to train our classifier: The training process will be divided into 4 stages, and the result of each stage is a decision forest with {64; 128; 256; 512} trees. The forest with 512 trees will be our final classifier. After each of the first three stages, additional hard negative examples will be mined and added into the training set. The scores generated by RPN can also be used in the Real Adaboost classifier. They will be regarded as the results of the 0th weak classifier.

3.3 Human Tracking

Our system still needs to track the detected humans, in order to achieve the goal of pedestrian counting. In the surveillance videos, almost every human heading for the monitor cameras changes its size over time, which means we should use a scale-adaptive tracking algorithm. We choose DSST method [13] to track humans, which utilizes two correlation filers [12], including a 2-dimension translation filter to track the new position of the object, and a 1-dimension scale filer to detect the size transformation of the target.

3.4 Pedestrian Counting

We further implement a pedestrian counting system based on multiple object detection and tracking. We will maintain a real-time set. The elements of this set will be the current trackers, each of which is used to track a human that can still be captured by the monitor camera. Note that this set will be called "the tracker set" for short.

At each frame of the surveillance video, we will firstly use the trackers in the tracker set to track the humans that have been detected before. Then, the detection method introduced in Sect. 3.2 is used to detect the humans that appear in the current frame. Figure 4 shows some detection and tracking results, which are labeled by red and blue respectively. After that, Overlaps (i.e. intersection-over-union, IoU) between detection boxes and tracking boxes will be computed in order to match the results of detection and tracking. Figure 5 shows some matching or failed-to-matching examples. For a pair of matching boxes, we assume that the detection box and the tracking box have circled a same person. For a detection box that fails to match with any tracking boxes, we assume that the detection algorithm has detected a new pedestrian, in which case our system will create a new tracker for this detection box and add 1 to the current number of pedestrians. There may exist some tracking boxes that fail to match with any detection boxes. That's caused by miss-detection. In this situation, our counting system will use the tracking results to update the corresponding trackers.

Fig. 4. Some detection and tracking examples (Color figure online)

We note that the low frame rates of the monitor cameras often result in the fast change of the motions and positions of a same pedestrian between two continuous frames. Simply using tracking results to update the trackers may

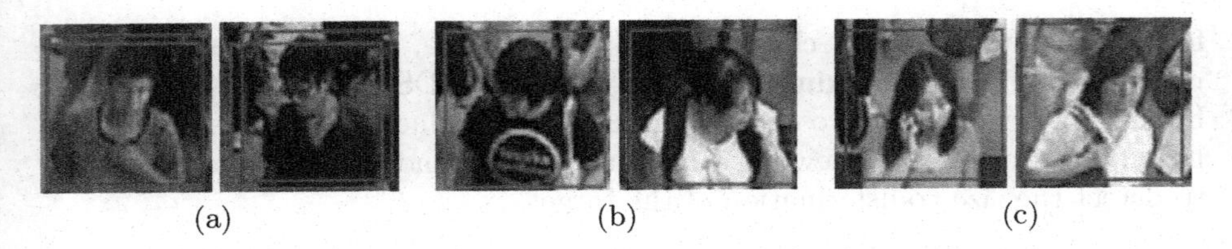

Fig. 5. Examples of matching and failed-to-matching boxes. (a) Pairs of matching boxes. (b) Detection boxes that fail to match with any tracking boxes. (c) Tracking boxes that fail to match with any detection boxes.

cause excursion during tracking, and may even lose some humans. In order to avoid such situations, detection results have been used to optimize the DSST algorithm: For a pair of matching boxes, we will use the detection result to create a new tracker. Then, this tracker will be updated by the tracking result and replace the previous tracker in the tracker set.

4 Experiments and Analysis

4.1 Datasets

To meet real-world needs, a video recorded in the subway station[1] is used for training and testing. This video has 1489 frames, with a resolution of 330 px × 205 px and a frame rate of 10 FPS. Figure 3 shows some frames of this video, in which pedestrians will get into the detection area from the top of the images and leave from the bottom, the left side or the right side. Dozens of pedestrians may show up at the same time with different ages, clothes, motions, genders and walking routes, which makes this video a suitable data set that closes to the reality of crowded public places. Note that this video will be called "the subway video" for short in the following subsections.

Besides, INRIA person dataset [14] are also used for training. Our training set consists of two parts, including the INRIA dataset and the first 400 frames of the subway video. 2630 positive examples are labeled in these two sub-datasets. Our detection algorithm will also extract some negative examples from these two sub-datasets during the training. Negative examples from INRIA are often backgrounds. Negative examples extracted from the subway video are often the other parts of pedestrians. These negative examples can help our detection method tell the differences between pedestrians and various backgrounds, as well as the differences between head-shoulders and the other parts of humans.

We will evaluate our system on the other frames of the subway video. Only the bottom 3/4 part of the subway video (~330 px × 150 px) is used for further testing due to the low resolution of the surveillance camera.

[1] This dataset can be downloaded at https://jbox.sjtu.edu.cn/l/eHE9Zh.

4.2 Experiments on Human Detection

We mentioned in Subsect. 3.2 that we need a detection method with high precision. In other words, we regard the false positives per image (FPPI) very important, but don't need the miss rate to be as lower as possible. In this subsection, several experiments were conducted on a testing dataset that is made up of 1640 ground truth boxes labeled by us.

The Effect of Real Adaboost. First of all, we evaluated the effect of the Real Adaboost classifier by comparing the performance of two methods. The first method only uses RPN for detection, while our method uses both RPN and Real Adaboost to detect humans. The ROC curves of these two methods are shown in Fig. 6(a). From the curves, we can find that when the miss rates of these two methods are same, RPN+Real Adaboost method can reach a lower FPPI than RPN method.

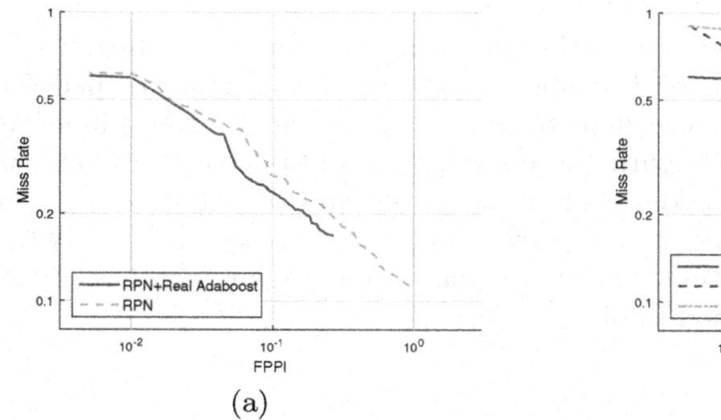

(a)

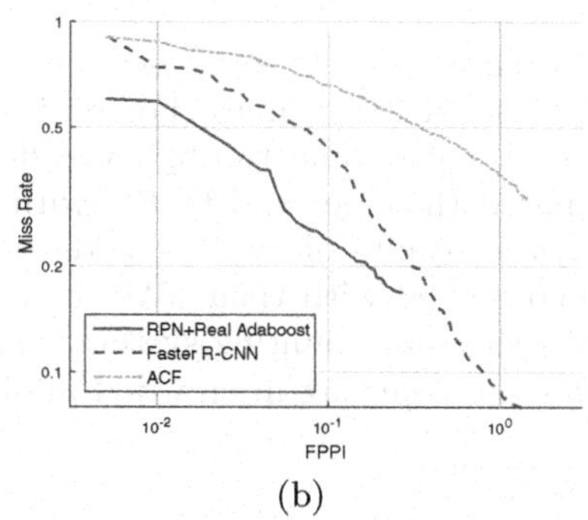

(b)

Fig. 6. Comparison of RPN+Real Adaboost and some existing methods. (a) RPN+Real Adaboost compared with RPN only. (b) RPN+Real Adaboost compared with Faster R-CNN and ACF.

The Detection Results. A threshold is chosen to get relatively high recall rate (i.e. 82.50%) and low FPPI (i.e. 0.22). To show the recall rate of 82.50% won't affect the final counting result, we also tested how many different people can our detection method detect. There exist 174 pedestrians in 200 frames of the test video, while 168 of them can be detected for at least once, which means an accuracy of 96.55% is reached and the detection method is suitable for pedestrian counting.

Comparisons with State-of-the-Art Methods. Faster R-CNN [7] and ACF [2,3] are compared with our detection method. Figure 6(b) shows the comparative results on our testing data set. The results show that our detection method can achieve higher accuracy when FPPI is relatively low. We also investigated why Faster R-CNN lost the competition with RPN+Real Adaboost: The use of convolutional feature maps with large scaling factors (i.e. 16) may lead to some false positive boxes with small sizes. However, RPN+Real Adaboost only uses the convolutional feature maps with a stride of 4, which makes it a better detector for small-size objects, such as the heads and shoulders of pedestrians.

4.3 Experiments on Human Tracking

We mentioned in Subsect. 3.4 that part of the detection results have been used to optimize the DSST algorithm in order to reduce the number of false tracking results caused by the low frame rates of the surveillance videos. We evaluated the effect of this optimization on 168 different pedestrians that can be detected by our detection method. The experimental results show that the number of incorrectly-tracked pedestrians decreases from 31 to 3 after this optimization.

4.4 Experiments on Pedestrian Counting

The accuracy of the proposed pedestrian counting system is also tested on a test video of 200 frames. For each frame, we labeled the number of pedestrians that have come through the detection area up to the current frame. We compared those ground truths with the counting results produced by our system. The results of another pedestrian counting method reported in [15] were also compared with them. We further drew a line chart shown in Fig. 7, which proves that our counting system can reach a high accuracy and can be used for pedestrian counting in crowded public places.

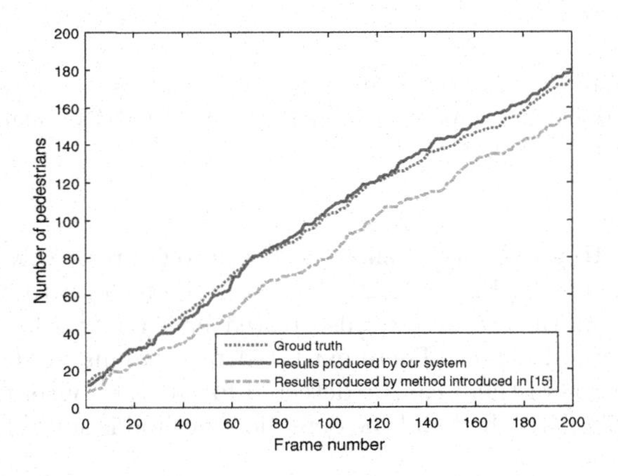

Fig. 7. Comparison of the counting results

5 Conclusions

In this paper, we proposed a new method for pedestrian counting. RPN and Real Adaboost are employed to reach high detection recall rate and lower FPPI in overcrowded public places. Instead of only using DSST method for human tracking, part of detection results are used for optimization. The characteristics of the surveillance videos and public places are taken into consideration to meet real-world needs. The experimental results show that the proposed system can achieve a high accuracy on detection, tracking and counting, which makes it a reliable pedestrian counting system for surveillance cameras.

Acknowledgement. The work was supported by the National Natural Science Foundation of China (Grant No. 91420302), the National Basic Research Program of China (Grant No. 2015CB856004) and the Key Basic Research Program of Shanghai, China (15JC1400103).

References

1. Benenson, R., Omran, M., Hosang, J., Schiele, B.: Ten years of pedestrian detection, what have we learned? In: Agapito, L., Bronstein, M.M., Rother, C. (eds.) ECCV 2014. LNCS, vol. 8926, pp. 613–627. Springer, Cham (2015). doi:10.1007/978-3-319-16181-5_47
2. Dollár, P., Tu, Z., Perona, P., Belongie, S.: Integral channel features. In: BMVC 2009. BMVA Press (2009)
3. Dollár, P., Appel, R., Belongie, S., Perona, P.: Fast feature pyramids for object detection. IEEE PAMI **36**(8), 1532–1545 (2014)
4. Dalal, N., Triggs, B.: Histograms of oriented gradients for human detection. In: CVPR 2005. vol. 1, pp. 886–893. IEEE (2005)
5. Nam, W., Dollár, P., Han, J.H.: Local decorrelation for improved pedestrian detection. In: NIPS 2014, pp. 424–432. MIT Press (2014)
6. Zhang, S., Benenson, R., Schiele, B.: Filtered channel features for pedestrian detection. In: CVPR 2015, pp. 1751–1760. IEEE (2015)
7. Ren, S., He, K., Girshick, R., Sun, J.: Faster R-CNN: towards real-time object detection with region proposal networks. IEEE PAMI **39**(6), 1137–1149 (2017)
8. Girshick, R.: Fast R-CNN. In: ICCV 2015, pp. 1440–1448. IEEE Computer Society (2015)
9. Zhang, L., Lin, L., Liang, X., He, K.: Is faster R-CNN doing well for pedestrian detection? In: Leibe, B., Matas, J., Sebe, N., Welling, M. (eds.) ECCV 2016. LNCS, vol. 9906, pp. 443–457. Springer, Cham (2016). doi:10.1007/978-3-319-46475-6_28
10. Comaniciu, D., Ramesh, V., Meer, P.: Real-time tracking of non-rigid objects using mean shift. In: CVPR 2000, vol. 2, pp. 142–149. IEEE (2000)
11. Wang, L., Ouyang, W., Wang, X., Lu, H.: Visual tracking with fully convolutional networks. In: CVPR 2015, pp. 3119–3127. IEEE Computer Society (2015)
12. Bolme, D.S., Beveridge, J.R., Draper, B.A., Lui, Y.M.: Visual object tracking using adaptive correlation filters. In: CVPR 2010, pp. 2544–2550. IEEE (2010)
13. Danelljan, M., Häger, G., Khan, F., Felsberg, M.: Accurate scale estimation for robust visual tracking. In: BMVC 2014. BMVA Press (2014)

14. INRIA Person Dataset. http://pascal.inrialpes.fr/data/human/
15. Zhang, X., Zhang, L.: Real time crowd counting with human detection and human tracking. In: Loo, C.K., Yap, K.S., Wong, K.W., Beng Jin, A.T., Huang, K. (eds.) ICONIP 2014. LNCS, vol. 8836, pp. 1–8. Springer, Cham (2014). doi:10.1007/978-3-319-12643-2_1

Mixture of Matrix Normal Distributions for Color Image Inpainting

Xiuling Zhou[1(✉)], Jing Wang[2], Ping Guo[3], and C.L. Philip Chen[4]

[1] The Department of Technology and Industry Development,
Beijing City University, Beijing, China
zxlmouse@bcu.edu.cn

[2] Peking University Health Science Center, Peking University, Beijing, China

[3] The Laboratory of Image Processing and Pattern Recognition,
Beijing Normal University, Beijing, China
pguo@bnu.edu.cn

[4] The Faculty of Science and Technology, University of Macau,
Macau SAR, China

Abstract. Gaussian mixture model is commonly used as image prior model to solve image restoration problem. However, vector representation leads to lose the inherent spatial relevant information and cause unstable estimation. In this paper, a mixture of matrix normal distributions (MMND) based image restoration algorithm is proposed, which incorporates the hidden structural information into prior image modeling. MMND is used as the prior image model and expectation maximization algorithm is used to optimize the maximum posterior criterion. Experiments conducted on color images indicate that MMND can achieve better peak signal to noise ratio (PSNR) as compared to other state-of-the-art methods.

Keywords: Mixture of matrix normal distributions · Gaussian mixture model · Color image inpainting

1 Introduction

For the degraded image Y, the image restoration problem can be expressed as follows:

$$\mathrm{vec}(Y) = U\mathrm{vec}(X) + \mathrm{vec}(W), \tag{1}$$

where vec() is a vector operator, U is an irreversible linear degradation operator, X is the original image, and W is an additive noise [1]. Here, we only focus on image inpainting problem.

Typically, image restoration is an ill-posed problem. Regularization methods are used to solve the aforementioned problem [2]. It requires some prior information of an image or an image model to determine the regularization term. Therefore, to find a better image prior model is the key issue to solve the image restoration problem.

Tailoring a prior for real-world images is a nontrivial and subjective matter. Many prior models have been proposed, such as Tikhonov regularization [3], Markov Random Field (MRF) prior [4], and heavy-tailed densities on the wavelet domain [5, 6].

D. Liu et al. (Eds.): ICONIP 2017, Part III, LNCS 10636, pp. 95–104, 2017.
https://doi.org/10.1007/978-3-319-70090-8_10

Sparse model is an emerging and powerful method to describe signals based on the sparsity and redundancy of their representations [7, 8]. It has shown promising results in image restoration problem with local patch-based sparse representations calculated with dictionaries learned from natural images [9–12]. However, dictionary learning requires high computational cost, since it is a large-scale and highly non-convex problem [8, 13].

Recently, Gaussian mixture model (GMM) used as image priors has shown surprisingly strong performance in solving image restoration problem at lower computational cost [14–16]. For GMM-based image restoration algorithm [14, 15], piecewise linear estimation (PLE) is proposed to estimate original image patches and model parameters. Meanwhile, a dual mathematical interpretation of the proposed framework is given by [15]: based on the PCA basis, the PLE can be interpreted as a structured sparse estimation.

In color image, each mode of the 3rd order tensor has certain physical meanings: such as mode-3 represents different color bands, joints of mode-1 and mode-2 represents the images of the same object with different bands. They contain useful spatial information, and there is a strong statistical relationship between different bands of an image, such as the co-occurrence of an image edges, texture patterns, shadings etc. 3rd order tensor (stacked as a vector) is used in GMM-based image restoration algorithm, which leads to lose the inherent spatial relevant information, as well as arise the problem of inaccuracy. For color image processing, such a process produces false colors and artifacts [10]. Moreover, in GMM, a 3rd order tensor $\mathrm{X} \in \mathrm{R}^{r \times p \times 3}$ has the variable dimension like $3rp$, and the number of parameters for covariance matrix is $3rp(3rp + 1)/2$. For small sample size, the high dimension makes the statistical inference procedure unstable.

Matrix normal models have important applications in various fields [17, 18]. In matrix normal distributions, the random variable is represented by matrix, which can maintain the inherent spatial structure of matrix data and explore the relationship between rows and columns of the matrix. Furthermore, the matrix representation reduces the estimated parameters, which is suitable for the small sample size problem.

In this paper, a mixture of matrix normal distributions based image restoration algorithm is proposed to overcome the problems of GMM for color image inpainting, which incorporates the hidden structural information of X into prior image modeling. Based on the criterion of MAP (maximum a posterior) estimation, the original image and the parameters of MMND are estimated by EM (expectation maximization) and flip-flop algorithm. The feasibility and efficiency of MMND-based image restoration algorithm are shown by the experiments on color images.

2 Finite Mixture of Matrix Normal Distributions

Given the observed matrix set $D = \{\mathrm{X}_i\}_{i=1}^{I}$ where $\mathrm{X}_i \in \mathrm{R}^{r \times p}$. Assume that the matrix in D is sampled from a mixture of matrix normal distributions (MMND) with K components [17]:

$$p(\mathrm{X}, \Theta) = \sum_{j=1}^{K} \pi_j G(\mathrm{X}, \mathrm{M}_j, \Omega_j, \mathrm{S}_j), \tag{2}$$

with $\pi_j \geq 0$ and $\sum_{j=1}^{K} \pi_j = 1$, where $G(\mathrm{X}, \mathrm{M}_j, \Omega_j, \mathrm{S}_j)$ is a general density function of matrix normal distribution:

$$G(\mathrm{X}, \mathrm{M}_j, \Omega_j, \mathrm{S}_j) = (2\pi)^{-\frac{rp}{2}} |\Omega_j|^{-\frac{p}{2}} |\mathrm{S}_j|^{-\frac{r}{2}} \exp\left\{ -\frac{1}{2} \mathrm{tr}\left[\mathrm{S}_j^{-1} (\mathrm{X} - \mathrm{M}_j)^{\mathrm{T}} \Omega_j^{-1} (\mathrm{X} - \mathrm{M}_j) \right] \right\}, \tag{3}$$

$\mathrm{X} \in \mathrm{R}^{r \times p}$ is a random matrix; $\Theta = \{\pi_j, \mathrm{M}_j, \Omega_j, \mathrm{S}_j\}_{j=1}^{K}$ denotes the parameter set of mixture of matrix normal distributions. S_j is a $p \times p$ symmetric and positive definite matrix, denoted as column covariance of the jth component, Ω_j is a $r \times r$ symmetric and positive definite matrix, denoted as row covariance of the jth component, M_j is the mean matrix of the jth component, and the weights π_j with $j = 1, \cdots, K$ represent the prior probabilities of belonging to each sub-population corresponding to a mixture component.

The posterior probability $p(j \mid \mathrm{X}, \Theta)$, the observed matrix X belongs to the jth component of the mixture can be expressed by Bayes theorem as follows:

$$p(j \mid \mathrm{X}, \Theta) = \frac{\pi_j G(\mathrm{X}, \mathrm{M}_j, \Omega_j, \mathrm{S}_j)}{p(\mathrm{X}, \Theta)}, j = 1, \cdots, K \tag{4}$$

In the case of MMND, the Bayesian decision rule $j^* = \mathrm{argmax}_j p(j \mid \mathrm{X}, \hat{\Theta})$ is adopted to classify the matrix X into class j^* with the largest posterior probability $p(j \mid \mathrm{X}, \hat{\Theta})$. And the classification rule becomes:

$$j^* = \mathrm{argmin}_j d_j(\mathrm{X}), j = 1, \cdots, K \tag{5}$$

where

$$d_j(\mathrm{X}) = p \ln |\Omega_j| + r \ln |\mathrm{S}_j| + \mathrm{tr}\left[\mathrm{S}_j^{-1} (\mathrm{X} - \mathrm{M}_j)^{\mathrm{T}} \Omega_j^{-1} (\mathrm{X} - \mathrm{M}_j) \right] - 2 \ln \pi_j \tag{6}$$

3 Mixture of Matrix Normal Distributions for Image Restoration

3.1 Regularized Image Restoration Model

Assume a color image $\mathrm{A} \in \mathrm{R}^{n_x \times n_y \times 3}$ is considered, which is seen as a 3rd-order tensor, where n_x and n_y represent the number of pixels in the two spatial dimensions. Since the following model will be represented by matrix or vector, the 3rd-order tensor A is transferred into a matrix $X \in R^{3\times(n_x n_y)}$ by mode–3 matrix unfolding. Detailed tensor concepts can refer to the reference [19].

Let X be the original image and Y be the degraded image. The degraded image is then formulated as Eq. (1). The image restoration problem is to estimate the original image X from the degraded image Y under some conditions.

For the simplicity of model, an image is decomposed into overlapping $r \times p$ local patches, each degraded image patch is formulated as in Eq. (7):

$$\mathrm{vec}(\mathrm{Y}_i) = \mathrm{U}_i\mathrm{vec}(\mathrm{X}_i) + \mathrm{vec}(\mathrm{W}_i),\ 1 \leq i \leq I, \tag{7}$$

where I is the total number of patches. Each original image patch is estimated and then the whole original image is obtained by combining and averaging the estimated patches.

Assume that the image patches as matrix variables $\{\mathrm{X}_i\}_{i=1}^{I}$ satisfy the mixture of matrix normal distributions, the probability density function is defined as Eq. (2). Each image patch X_i is independently sampled from one of the components with an unknown index j, whose probability density function is given as:

$$\begin{aligned} p(\mathrm{X}_i) &= G(\mathrm{X}_i, \mathrm{M}_{j_i}, \Omega_{j_i}, \mathrm{S}_{j_i}) \\ &= (2\pi)^{-\frac{rp}{2}}|\Omega_{j_i}|^{-\frac{p}{2}}|\mathrm{S}_{j_i}|^{-\frac{r}{2}}\exp\left\{-\tfrac{1}{2}\mathrm{tr}\left[\mathrm{S}_{j_i}^{-1}(\mathrm{X}_i - \mathrm{M}_{j_i})^{\mathrm{T}}\Omega_{j_i}^{-1}(\mathrm{X}_i - \mathrm{M}_{j_i})\right]\right\} \end{aligned} \tag{8}$$

Next, the optimization objective function to estimate the image patches is deduced by maximizing the log posterior probability.

Assume $\mathrm{vec}(\mathrm{W}_i) \sim \mathrm{N}(0, \sigma^2\mathrm{I}_N)$, $N = rp$. By Eq. (7):

$$p(\mathrm{vec}(\mathrm{Y}_i) \mid \mathrm{vec}(\mathrm{X}_i)) = \frac{1}{(2\pi)^{N/2}\sigma^N}\exp\left(-\frac{1}{2\sigma^2}\|\mathrm{vec}(\mathrm{Y}_i) - \mathrm{U}_i\mathrm{vec}(\mathrm{X}_i)\|^2\right) \tag{9}$$

There exists a relationship between trace and vec operator:

$$\mathrm{tr}\left[\mathrm{S}_j^{-1}(\mathrm{X} - \mathrm{M}_j)^{\mathrm{T}}\Omega_j^{-1}(\mathrm{X} - \mathrm{M}_j)\right] = \mathrm{vec}(\mathrm{X} - \mathrm{M}_j)^{\mathrm{T}}(\mathrm{S}_j \otimes \Omega_j)^{-1}\mathrm{vec}(\mathrm{X} - \mathrm{M}_j) \tag{10}$$

By using Bayes theorem and Eq. (10), the original image patch X_i can be estimated by maximizing the log posterior probability $\ln p(\mathrm{vec}(\mathrm{X}_i) \mid \mathrm{vec}(\mathrm{Y}_i), \mathrm{M}_j, \Omega_j, \mathrm{S}_j)$:

$$\left(\widehat{\mathrm{X}}_i, j_i\right) = \arg\max_{\mathrm{X}_i, j} \ln p(\mathrm{vec}(\mathrm{X}_i)|\mathrm{vec}(\mathrm{Y}_i), \mathrm{M}_j, \Omega_j, \mathrm{S}_j) = \arg\min_{\mathrm{X}_i, j} L, \tag{11}$$

where

$$\begin{aligned} L = &\|\mathrm{vec}(\mathrm{Y}_i) - \mathrm{U}_i\mathrm{vec}(\mathrm{X}_i)\|^2 + p\sigma^2\ln|\Omega_j| + r\sigma^2\ln|\mathrm{S}_j| \\ &+ \sigma^2\mathrm{vec}(\mathrm{X} - \mathrm{M}_j)^{\mathrm{T}}(\mathrm{S}_j \otimes \Omega_j)^{-1}\mathrm{vec}(\mathrm{X} - \mathrm{M}_j) \end{aligned}. \tag{12}$$

The first term in Eq. (12) introduces the likelihood force that demands a proximity between $\text{vec}(\text{Y}_i)$ and $\text{U}_i\text{vec}(\text{X}_i)$. The remaining terms represent the image prior. This regularization term assumes that image patches obey MMND, and each image patch is independently sampled from one of the components of MMND.

From above analysis and the objective function of Eq. (12), the estimation of the original image patches $\{\text{X}_i\}_{i=1}^{I}$ from degraded image patches $\{\text{Y}_i\}_{i=1}^{I}$ can be decomposed into the following sub-problems:

(1) Estimate the component parameters $\{\text{M}_j, \Omega_j, \text{S}_j\}_{j=1}^{K}$ of MMND from the original image patches $\{\text{X}_i\}_{i=1}^{I}$.

(2) Identify the matrix normal component j_i that generates the image patch X_i, $1 \le i \le I$.

(3) Estimate the patch X_i from its corresponding matrix normal distribution $\{\text{M}_{j_i}, \Omega_{j_i}, \text{S}_{j_i}\}$ and the degraded patch Y_i, $1 \le i \le I$.

3.2 Optimization Method

In order to estimate the original image patch, the objective function in Eq. (12) can be optimized by EM algorithm, which can obtain a local minimum solution [15].

(1) E step: estimation of the original image patches.

In the E step, the estimates of the matrix normal parameters $\left\{\hat{M}_j, \hat{\Omega}_j, \hat{\text{S}}_j\right\}_{j=1}^{K}$ are assumed to be known. For each patch, the estimates $\hat{X}_i^j$ with all the matrix normal components are calculated by minimum the objective function L in Eq. (12), and the best matrix normal distribution j_i is selected to obtain the estimate of the patch $\hat{X}_i = \hat{X}_i^{j_i}$.

For fixed component j, by minimizing Eq. (12) with respect to $\text{vec}(\text{X}_i)$, the following patch estimation formula can be obtained:

$$\text{vec}\left(\hat{\text{X}}_i^j\right) = \left(\text{U}_i^{\text{T}}\text{U}_i + \sigma^2\left(\hat{S}_j \otimes \hat{\Omega}_j\right)^{-1}\right)^{-1}\left(\text{U}_i^{\text{T}}\text{vec}(\text{Y}_i) + \sigma^2\left(\hat{S}_j \otimes \hat{\Omega}_j\right)^{-1}\text{vec}\left(\hat{M}_j\right)\right) \quad (13)$$

According to the estimated $\hat{\text{X}}_i^j$, by minimizing Eq. (12) with respect to j, the best matrix normal component j_i that generates the maximum MAP among all the component distributions is identified:

$$j_i = \arg\min_j \left(\begin{array}{c} \left\|\text{vec}(\text{Y}_i) - \text{U}_i\text{vec}\left(\hat{\text{X}}_i^j\right)\right\|^2 + p\sigma^2 \ln|\hat{\Omega}_j| + r\sigma^2 \ln|\hat{\text{S}}_j| \\ +\sigma^2\text{tr}\left[\hat{\text{S}}_j^{-1}\left(\hat{\text{X}}_i^j - \hat{\text{M}}_j\right)^{\text{T}}\hat{\Omega}_j^{-1}\left(\hat{\text{X}}_i^j - \hat{\text{M}}_j\right)\right] \end{array}\right) \quad (14)$$

The final image patch estimate is obtained by substituting the best model j_i in Eq. (13)

$$\hat{\text{X}}_i = \hat{\text{X}}_i^{j_i} \quad (15)$$

(2) M step: estimation of the matrix normal parameters.

In the M-step, the patch estimate $\hat{X}_i$ and the corresponding matrix normal component j_i which generates the patch are assumed to be known for each patch. The matrix normal parameters $\left\{\hat{M}_j, \hat{\Omega}_j, \hat{S}_j\right\}_{j=1}^{K}$ will be estimated.

Let ℓ_j be the ensemble of the patch indices i that are assigned to the j-th matrix normal component, i.e., $\ell_j = \{i \mid j_i = j\}$, and let $|\ell_j|$ be its cardinality. The parameters of each matrix normal component are estimated with the maximum likelihood estimate (MLE) by using all patches assigned to that matrix normal cluster,

$$\left(\hat{M}_j, \hat{\Omega}_j, \hat{S}_j\right) = \arg \max_{M_j, \Omega_j, S_j} \ln p\left(\left\{\hat{X}_i\right\}_{i \in \ell_j} | M_j, \Omega_j, S_j\right) \tag{16}$$

So the MLE of parameters of matrix normal distribution is shown as follows [20]:

$$\hat{M}_j = \frac{1}{|\ell_j|} \sum_{i \in \ell_j} \hat{X}_i, \tag{17}$$

$$\hat{\Omega}_j = \frac{1}{|\ell_j| p} \sum_{i \in \ell_j} \left[(\hat{X}_i - \hat{M}_j) \hat{S}_j^{-1} [(\hat{X}_i - \hat{M}_j)^T\right], \tag{18}$$

$$\hat{S}_j = \frac{1}{|\ell_j| r} \sum_{i \in \ell_j} [(\hat{X}_i - \hat{M}_j)^T \hat{\Omega}_j^{-1} (\hat{X} - \hat{M}_j)] \tag{19}$$

It is efficient to iteratively solve Eqs. (18) and (19) until convergence, known as the “flip-flop” algorithm [20]. For small sample size problem, the empirical row and column covariance Ω_j, S_j estimates may be singular. Regularization is a solution for this problem [21, 22]. The theoretical analysis of regularization term and parameter will be studied in the future.

4 Experiments

4.1 Experiment Setup

The MMND-based image restoration algorithm is compared with three representative methods for image inpainting problem, including FOE (fields of experts) [4], BP (beta process) [12] and GMM-based image restoration algorithm on berkeley segmentation database [23].The results of FOE are generated by the author’s original implementation, and those of GMM and MMND are calculated with our own implementation. The PSNRs of BP are cited from the corresponding papers.

In the case of inpainting, the original image X is masked with a random mask, vec(Y) = Uvec(X), where U is a diagonal matrix whose diagonal entries are randomly either 1 or 0, keeping or killing the corresponding pixels. The size of images in the Berkeley segmentation database is $481 \times 321 \times 3$.Uniform masks that retain 80%, 50%,

30% and 20% of the pixels are used. In the following experiments, the number of K components is set to 19.

4.2 Results and Analysis

The experimental results are shown in Table 1 and Fig. 1.

Table 1. PSNR results on berkeley segmentation database

Image	Data ratio	FOE	BP	GMM	MMND
Castle	80%	36.44	41.51	47.03	49.66
	50%	30.43	36.45	37.18	38.07
	30%	27.32	32.02	32.34	32.53
	20%	25.41	29.12	29.11	28.79
Mushroom	80%	40.69	42.56	47.97	52.19
	50%	33.32	38.88	39.29	41.34
	30%	29.63	34.63	34.55	35.52
	20%	27.59	31.56	31.28	30.72
Horse	80%	36.73	41.97	47.88	50.22
	50%	30.63	37.27	37.48	38.29
	30%	27.76	32.52	32.41	32.74
	20%	26.39	29.99	29.45	28.92
Train	80%	30.55	40.73	44.01	43.97
	50%	24.79	32.00	32.59	32.69
	30%	22.29	27.00	27.05	27.15
	20%	21.07	24.59	24.05	24.01
Kangaroo	80%	36.32	42.74	46.97	49.48
	50%	30.05	37.34	36.89	38.19
	30%	27.32	32.21	31.73	32.39
	20%	26.04	29.59	28.94	28.49
Average	80%	36.15	41.90	46.77	**49.11**
	50%	29.84	36.39	36.69	**37.72**
	30%	26.86	31.68	31.62	**32.07**
	20%	25.30	**28.97**	28.57	28.19

In most of the cases, MMND-based image restoration algorithm gives the best average PSNR. Compared with GMM and BP, MMND did not perform well for the available data ratio is less than 20%. When the available data ratio is higher than 20%, the average gain between MMND and GMM is between 0.4 dB and 2 dB. In the same context, the average gain between MMND and BP is between 0.4 dB and 7 dB. Compared with FOE, MMND outperforms them in all cases. The average gain between MMND and FOE is between 2.5 dB and 12.5 dB.

The visual quality comparisons on castle with only 30% available data are provided in Fig. 1. It is shown that all the methods produce good inpainting results on the

Fig. 1. Image inpainting results on castle with 30% available data. (a) Original castle image. (b) Castle image degraded with 30% available data. (c) FOE (PSNR = 27.32). (d) BP (PSNR = 32.02). (e) GMM (PSNR = 32.34). (f) MMND (PSNR = 32.53)

smooth regions. FOE fails in recovering edges and produce blurred effects. BP is able to recover some edges and to generate some blurred effects simultaneously. GMM and MMND can provide better restoration on both edges and details than other competing methods. MMND produces the best visual quality with the highest PSNR.

5 Conclusions

In this paper, mixture of matrix normal distributions as the prior image model is used for image restoration, where EM algorithm is used to optimize the MAP criterion. Experiments conducted on color images indicate that MMND can achieve better PSNR as compared to other state-of-the-art methods. In the future work, the proposed algorithm combined with more advanced regularization methods will be investigated. And it will be applied for other image restoration problem, such as image denoising and image deblurring.

Acknowledgment. This work is fully supported by the grants from Beijing Natural Science Foundation (Project No. 4162027), the National Natural Science Foundation of China (61375045) and the Joint Research Fund in Astronomy (U1531242) under cooperative agreement between the National Natural Science Foundation of China (NSFC) and Chinese Academy of Sciences (CAS). Prof. Ping Guo and Xiuling Zhou are the authors to whom all the correspondence should be addressed.

References

1. Banham, M.R.: Wavlet-based image restoration techniques. Ph.D. dissertation, Evanston, Northwestern Universtity (1994)
2. Tikhonov, A.N., Arsenin, V.Y.: Solutions of Ill-Posed Problems. Winston, New York (1977)
3. Haber, E., Tenorio, L.: Learning regularization functionals. Inv. Probl. **19**, 611–626 (2003)
4. Roth, S., Black, M.J.: Fields of experts. Int. J. Comput. Vis. **82**(2), 205–229 (2009)
5. Bioucas-Dias, J.M.: Bayesian wavelet-based image deconvolution: a GEM algorithm exploiting a class of heavy-tailed priors. IEEE Trans. Image Proc. **15**(4), 937–951 (2006)
6. Chantas, G., Galatsanos, N., Likas, A., Saunders, M.: Variational Bayesian image restoration based on a product of t-distributions image prior. IEEE Trans. Image Proc. **17**(10), 1795–1805 (2008)
7. Wright, J., Ma, Y., Mairal, J., Spairo, G., Huang, T., Yan, S.: Sparse representation for computer vision and pattern recognition. Proc. IEEE **98**, 1031–1044 (2010)
8. Aharon, M., Elad, M., Bruckstein, A.: K-SVD: an algorithm for designing overcomplete dictionaries for sparse representation. IEEE Trans. Signal Proc. **54**(11), 4311–4322 (2006)
9. Elad, M., Aharon, M.: Image denoising via sparse and redundant representations over learned dictionaries. IEEE Trans. Image Proc. **15**(12), 3736–3745 (2006)
10. Mairal, J., Elad, M., Sapiro, G.: Sparse representation for color image restoration. IEEE Trans. Image Proc. **17**(1), 53–69 (2008)
11. Mairal, J., Bach, F., Ponce, J., Sapiro, G., Zisserman, A.: Non-local sparse models for image restoration. In: International Conference on Computer Vision, pp. 2272–2279 (2009)
12. Zhou, M., Chen, H., Paisley, J., Ren, L., Li, L., Xing, Z., Dunson, D., Sapiro, G., Carin, L.: Nonparametric Bayesian dictionary learning for analysis of noisy and incomplete images. IEEE Trans. Image Proc. **2**(1), 130–144 (2012)
13. Engan, K., Aase, S., Husoy, J.: Multi-frame compression: theory and design. Signal Proc. **80** (10), 2121–2140 (2000)
14. Yu, G., Sapiro, G.: Statistical compressed sensing of Gaussian mixture models. IEEE Trans. Signal Proc. **59**(12), 5842–5858 (2011)
15. Yu, G., Sapiro, G., Mallat, S.: Solving inverse problems with piecewise linear estimators: from Gaussian mixture models to structured sparsity. IEEE Trans. Image Proc. **21**, 2481–2499 (2012)
16. Zoran, D., Weiss, Y.: Natural images, Gaussian mixtures and dead leaves. In: NIPS (2012)
17. Viroli, C.: Finite mixtures of matrix normal distributions for classifying three-way data. Stat. Comput. **21**, 511–522 (2011)
18. Viroli, C.: On matrix-variate regression analysis. J. Multivar. Anal. (2012). doi:10.1016/j.jmva.2012.04.005
19. Tao, D., Song, M., Li, X., et al.: Bayesian tensor approach for 3-D face modeling. IEEE Trans. Circ. Syst. Video Technol. (T-CSVT) **18**(10), 1397–1410 (2008)
20. Dutilleul, P.: The MLE algorithm for the matrix normal distribution. J. Stat. Comput. Simul. **64**(2), 105–123 (1999)

21. Zhou, X.L., Guo, P., Philip Chen, C.L.: Regularized covariance matrix estimation based on MDL principle. Frontiers Intell. Control Inf. Proc. **4**, 407–430 (2014)
22. Zhou, X.L., Guo, P., Philip Chen, C.L.: Covariance matrix estimation with multi-regularization parameters based on MDL principle. Neural Process. Lett. **38**, 227–238 (2013)
23. Martin, D., Fowlkes, C., Tal, D., Malik, J.: A database of human segmented natural images and its application to evaluating segmentation algorithms and measuring ecological statistics. In: Proceedings of IEEE International Conference on Computer Vision, vol. 2, pp. 416–423 (2001)

RGB-D Tracking Based on Kernelized Correlation Filter with Deep Features

Shuang Gu[1], Yao Lu[1(✉)], Lin Zhang[1], and Jian Zhang[2]

[1] Beijing Laboratory of Intelligent Information Technology, School of Computer Science, Beijing Institute of Technology, Beijing 100081, China
{cs_gs,vis_yl}@bit.edu.cn

[2] Advanced Analytics Institute, University of Technology Sydney, Sydney, NSW 2007, Australia

Abstract. This paper proposes a new RGB-D tracker which is upon Kernelized Correlation Filter(KCF) with deep features. KCF is a high-speed target tracker. However, the HOG feature used in KCF shows some weaknesses, such as not robust to noise. Therefore, we consider using RGB-D deep features in KCF, which refer to deep features of RGB and depth images and the deep features contain abundant and discriminated information for tracking. The mixture of deep features highly improves the performance of the tracker. Besides, KCF is sensitive to scale variations while depth images benefit for handling this problem. According to the principle of similar triangle, the ratio of scale variation can be observed simply. Tested over Princeton RGB-D Tracking Benchmark, Our RGB-D tracker achieves the highest accuracy when no occlusion happens. Meanwhile, we keep the high-speed tracking even if deep features are calculated during tracking and the average speed is 10 FPS.

Keywords: RGB-D · KCF · Deep features · Scale estimation

1 Introduction

Target tracking is an important research branch of computer vision, which generally takes use of RGB images. Nowadays, depth cameras which can capture distances from targets to the camera lens, such as Kinect, RealSence, are more available to form depth images. Considering that depth images provide spatial information which totally distinguish from traditional RGB images, tracking on RGB-D datasets takes on more potential and challenges.

KCF [6] tracker has achieved the promising performance with high speed up to 40 FPS at the same time. It proves that the description capability of HOG features is out-standing. In [4], it adds color space features to enhance the performance of KCF [6]. However, features extracted from Convolution Neural Network(CNN) have shown a huge superiority in many Computer Vision areas,

D. Liu et al. (Eds.): ICONIP 2017, Part III, LNCS 10636, pp. 105–113, 2017.
https://doi.org/10.1007/978-3-319-70090-8_11

like detection, segmentation and the others. MDNet [10] and TCNN [7] design Convolution Neural Networks for visual tracking and access higher accuracy. All of these are done in RGB images without any depth information. S.Song proposes RGBD-OF [12] that makes use of depth image to solve the problem of occlusion but tracking speed is too slow only around 0.26 FPS.

Besides, KCF [6] tracker is sensitive to scale variation, so DSST [3] and SAMF [8] are proposed to fix this problem. DSST [3] creates another scale filter with 33 scales at the cost of time because the speed is 0.94 FPS in MATLAB. SAMF [8] searches the best position and scale at the same time but its scale estimation at 7 levels is coarse.

Due to the huge advantages of deep features from CNN, we select RGB-D deep features in our tracker to improve the precision of visual tracking. Additionally, another filter to estimate the scale of the target is no need because the ratio of scale variation can be easily obtained according to the principle of similar triangle.

2 Proposed Method

2.1 Our RGB-D Tracker

KCF tracker proposes to use circulant sampling to extend correlation filters to speed up target tracking. Each sample is shifted in KCF to get a circulant matrix, like Eq. (1)

$$X = C(x) = \begin{pmatrix} x_1 & x_2 & x_3 & \dots & x_n \\ x_n & x_1 & x_2 & \dots & x_{n-1} \\ \dots & \dots & \dots & \dots & \dots \\ x_2 & x_3 & x_4 & \dots & x_1 \end{pmatrix}. \tag{1}$$

All circulant matrices are made diagonal by the Discrete Fourier Transform (DFT), regardless of the generating vector x. So, it can be expressed as Eq. (2)

$$X = C(x) = F diag(\hat{x}) F^H. \tag{2}$$

in which, F is a constant matrix that is independent on x, and $\hat{x} = \mathcal{F}(x)$. F^H means Hermitian transpose.

For the constant matrix F, it has $F^H F = I$. Equation (2) can be transformed to Eq. (3)

$$X^H X = F diag(\hat{x}^* \odot \hat{x}) F^H. \tag{3}$$

$\odot$ is defined as element-wise product. Applying them recursively to the full expression for linear regression,

$$\hat{w} = \frac{\hat{x}^* \odot \hat{y}}{\hat{x}^* \odot \hat{x} + \lambda}. \tag{4}$$

w can be easily recovered in the spatial domain with Inverse DFT. By Eq. (4), a response map can be achieved where the maximum response represents the new position of the target.

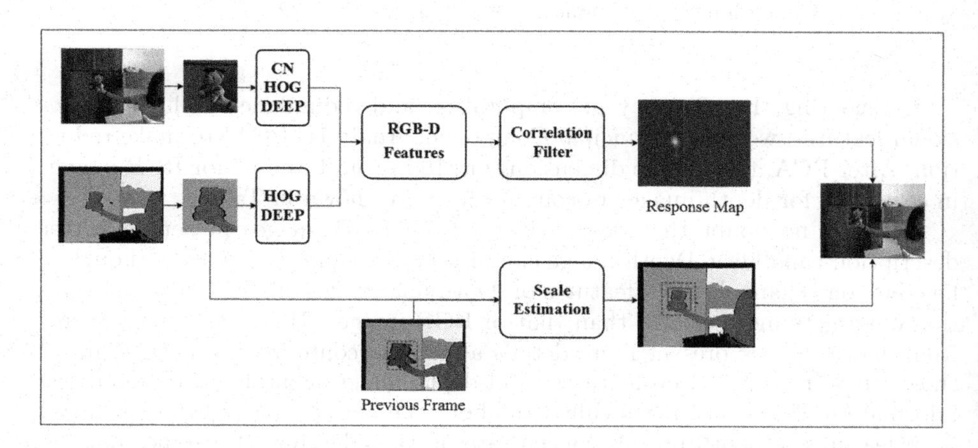

Fig. 1. In the n-th frame, we extract a RGB-D patch to obtain different features called RGB-D features and send these features to one KCF tracker. Then, the best score in the response map is the most possible translation of the target. After that, estimate the scale of the target according to the previous target size. Finally, get the position and size of a target box.

The framework of our method is shown in Fig. 1. The input is a pair of RGB-D image which means we need two trackers for RGB and depth images respectively or single tracker by fusing RGB and depth features. Experiments show that such single tracker outperforms multiple trackers using single feature. We reserve an interface for choosing different feature combinations in which RGB image has three optional features, HOG, Color Name(CN) [4], deep features from CNN, and depth image has two, HOG and deep features. KCF gives the translation of the tracked target from a response map. The new position is the place of maximum response. Scale estimation is dependent on the variation of the target's distance to the camera. The depth of the initial target is a basis for the next frames. If the depth of the target is farther, the scale of it will be smaller and vice versa.

2.2 Deep Feature

Deep features are extracted from VGG-19 [11] network which is pre-trained in ImageNet[1]. In this paper, features in convolution layers are more useful for us. Conv1 and Conv4 layers are chosen. The dimensions of each layer are 96 and 512.

[1] http://www.image-net.org/.

Table 1. Feature DimensionsFeature Dimensions

Feature	CN [4]	HOG	Deep-conv1	Deep-conv2
Dimension(RGB)	10	31	16	64
Dimension(Depth)	none	31	16	32

Considering the efficiency of computation and redundancy of high dimensional features, we adopt Principal Component Analysis (PCA) to make reduction. After PCA, the feature dimensions are 16 in conv1 and 64 for RGB images in conv4, 32 for depth images in conv4, which are shown in Table 1.

There is no doubt that deep features from RGB images present a better description capability. Depth image can also be a meaningful image although it has only one channel. HOG features of depth images describe the edge information of targets more clearly than that of RGB images. However, deep features from depth images present more details about the contour of targets. Figure 2 shows these features, deep features from depth images contain more about the information of contour, meanwhile less about the background. In the low layer, deep features of depth images concentrate on the edge but also some contents inside of the target, like the depth of the target itself. In the high layer, edge features are weakened but semantic features spring up which can be supplementaries. The amount of information in depth images could not be as various as RGB images because of single channel. So 32-dimension features in conv4 are enough.

(a) RGB-conv1-1 (b) RGB-conv4-1 (c) Depth-hog (d) Depth-conv1-1 (e) Depth-conv4-1

Fig. 2. (a) is deep feature of an RGB image in conv1-1, (b) is that in conv4-1, (c) is a hog image generated from a depth image, (d) and (e) are deep feature of a depth image in conv1-1 and conv4-1.

2.3 Scale Estimation

The size of the target is linear to the distance from the target to the camera. Since given the depth of targets, it will be easier to estimate the size of targets. The change ratio of the target size is similar to that of the depth value. According to the principle of camera imaging, we have:

$$f = \frac{s * h}{S} . \tag{5}$$

where f is focal length, s is the size of target, S is the actual size of target, h is the distance from the target to the camera lens. In different scenes, as Fig. 3 shows, the target captured by the one camera has different imaging size, so it can be expressed as:

$$scalefactor = \frac{s_2}{s_1} = \frac{h_2}{h_2}. \tag{6}$$

scalefactor indicates the scale variation in each frame.

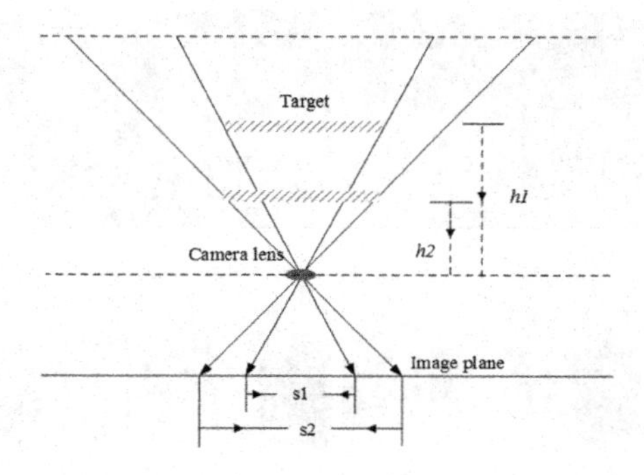

Fig. 3. Due to motions of the target, the size in an image plane changes from s_1 to s_2 following with the depth value from h_1 to h_2.

Depth Value Estimation. A given target box also contains some background pixels what we do not want. To get rid of background pixels, we use *K-means* or *Gaussian distribution* to handle it.

K-means: for a given target box, K is set to 2 to separate foreground and background pixels. Depth value is the average value of all foreground pixels.

Gaussian distribution: depth histogram can be fitted to a Gaussian model and the mean value μ of Gaussian distribution is treated as target's depth value.

3 Experiments

In the Princeton RGB-D benchmark[2], 95 Evaluation datasets and 5 Validation datasets are available, which include situations of scale variation, occlusion, active motion and so on. Our experiments are done in Ubuntu 14.04 with Inter Core i7-4790 and GPU GeForce GTX TITAN X. VGG network uses a matlab wrapper called *MatConvNet*[3].

The evaluation criteria comes from Princeton RGB-D benchmark which is using overlap ratio to measure the precision of tracker answers.

[2] http://tracking.cs.princeton.edu/.
[3] http://www.vlfeat.org/matconvnet/.

3.1 Results

Figure 4 shows some tracking results. It contains the situations of scale variation and small target tracking. Table 2 is a comparison between different feature selections in our RGB-D tracker. This ranking table is evaluated on Princeton RGB-D Tracking Benchmark. Table 3 compares our method with other RGB-D tracking methods.

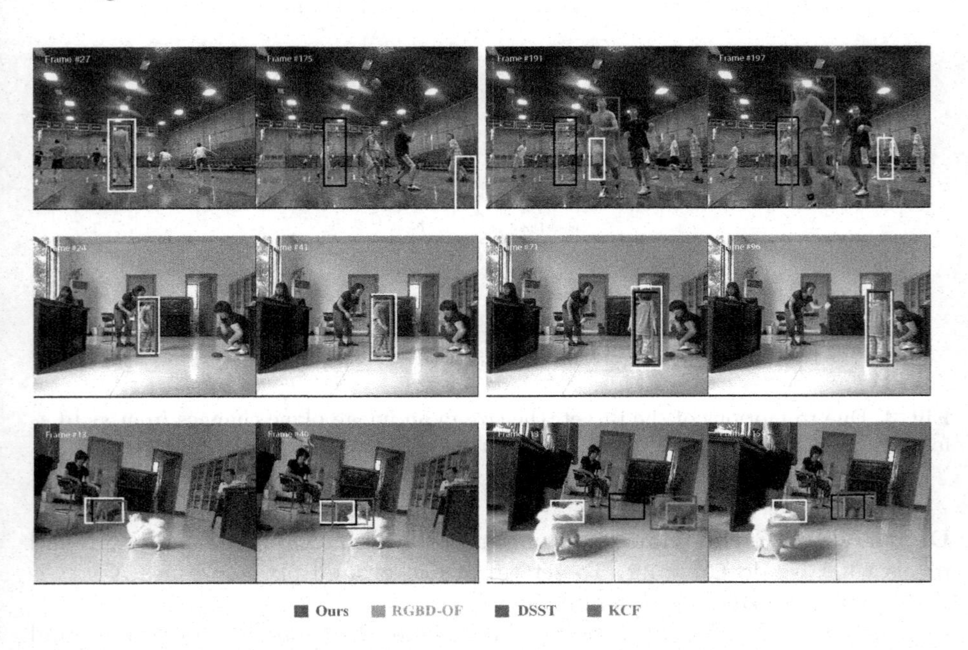

Fig. 4. It shows some tracking results on the videos, "basketball1", "child_no1", "bdog_occ2".

3.2 Analysis

Table 2 gives the tracking evaluation of different feature combinations. Firstly, after adding depth images, the performance of KCF is totally improved on this dataset. Secondly, when depth images only uses HOG features, it is hard for our tracker to enhance its tracking capability and its average ranking raises from 11.09 (RGB(HOG)+Depth(HOG)) to 10.64 (RGB(HOG+DEEP)+Depth(HOG)). However, when HOG of depth images replaced by deep features, our tracker arrives at a new rank which is around 9.36. Thirdly, we find that CN features do not work with deep features because color information has been contained in deep features. At last, That we use deep features of RGB-D images and add scale estimation achieves the average rank about 8.55.

Table 2. Comparsion between different feature combinations

RGB Feat.	Depth Feat.	Avg. Rank	Target type			Target size	
			Human	Animal	Rigid	Large	Small
HOG	none	14.73	0.42(17)	0.50(15)	0.65(13)	0.48(16)	0.55(14)
HOG	HOG	11.09	0.49(11)	0.58(11)	0.66(12)	0.52(14)	0.61(9)
DEEP	HOG	10.91	0.45(15)	0.62(8)	0.71(11)	0.52(14)	0.63(9)
CN+DEEP	HOG	10.91	0.45(15)	0.62(8)	0.71(11)	0.52(14)	0.63(9)
HOG+DEEP	HOG	10.64	0.46(13)	0.62(8)	0.71(11)	0.52(14)	0.64(9)
DEEP	DEEP	9.36	0.55(8)	0.63(8)	0.69(11)	0.60(10)	0.63(9)
CN+DEEP	DEEP	9.55	0.55(8)	0.63(8)	0.69(11)	0.59(11)	0.63(9)
HOG+DEEP	DEEP	8.82	**0.57(8)**	0.64(5)	0.70(11)	**0.62(9)**	0.64(9)
HOG+DEEP	HOG+DEEP	8.82	0.54(8)	0.64(5)	0.71(11)	0.61(10)	0.63(9)
DEEP	DEEP+Scale	**8.55**	0.45(15)	**0.72(2)**	**0.75(5)**	0.55(12)	**0.67(8)**
RGB Feat.	Depth Feat.	Movement		Occlusion		Motion	
		Low	Fast	Yes	No	Passive	Active
HOG	none	0.65(12)	0.47(16)	0.41(16)	0.68(13)	0.65(13)	0.47(17)
HOG	HOG	0.66(11)	0.54(10)	0.44(13)	0.75(9)	0.65(12)	0.54(10)
DEEP	HOG	0.67(10)	0.54(10)	0.44(13)	0.78(6)	0.63(15)	0.56(9)
CN+DEEP	HOG	0.67(10)	0.55(10)	0.44(13)	0.78(6)	0.63(15)	0.56(9)
HOG+DEEP	HOG	0.68(10)	0.55(10)	0.44(13)	0.78(6)	0.64(14)	0.57(9)
DEEP	DEEP	0.67(10)	0.60(9)	0.48(11)	0.81(3)	0.63(15)	0.61(9)
CN+DEEP	DEEP	0.67(10)	0.59(9)	0.47(12)	0.81(3)	0.63(15)	0.61(9)
HOG+DEEP	DEEP	0.69(10)	**0.61(9)**	**0.50(11)**	0.81(3)	0.64(14)	**0.63(8)**
HOG+DEEP	HOG+DEEP	0.70(10)	0.59(9)	0.48(11)	0.82(3)	0.65(12)	0.61(9)
DEEP	Depth+Scale	**0.73(9)**	0.57(9)	0.43(13)	**0.87(1)**	**0.69(11)**	0.59(9)

Table 3 compares our method with other RGB-D trackers. First of all, occlusion is a big problem in this datasets. But without occlusion, our tracking precision is around 0.87, which is the highest one in Princeton RGB-D tracking benchmark. In this item, OAPF [9] also has a high score around 0.85, but its speed is 0.9 FPS which is too slow. Because of deep features calculation, our RGB-D tracker is slower than DS-KCF [5]. However, its score on no occlusion datasets is 0.78 which is lower than us. DS-KCF, also based on KCF, collects occlusion detection, scale adjustment, segmentation to solve different problems of tracking. It indicates that DS-KCF might be hard to make improvement, but if occlusion problem is figured out, we can step forward.

It is impossible to neglect occlusion. Our method can handle short-time occlusion, but for long-time occlusion and multi-times occlusions, it will be invalid. So our future work is to add Depth-Constraint to solve this problem.

Table 3. Comparsion between different RGB-D trackers

Algorithm	Avg. Rank	Target type			Target size		Movement	
		Human	Animal	Rigid	Large	Small	Low	Fast
3D-T [2]	2.45	0.81(1)	0.64(4)	0.73(6)	0.80(1)	0.71(2)	0.75(3)	0.75(1)
OAPF [9]	2.45	0.64(5)	0.85(1)	0.77(2)	0.73(3)	0.73(1)	0.85(1)	0.68(3)
RGBDOcc-OF [12]	2.64	0.74(2)	0.63(5)	0.78(1)	0.78(2)	0.70(3)	0.76(2)	0.72(2)
DS-KCF [5]	4.09	0.67(3)	0.61(6)	0.76(3)	0.69(5)	0.70(4)	0.75(4)	0.67(4)
RGBD-OF [12]	4.64	0.64(4)	0.65(3)	0.75(5)	0.72(4)	0.65(6)	0.73(6)	0.66(5)
Ours	5.82	0.45(10)	0.72(2)	0.75(4)	0.55(7)	0.67(5)	0.73(7)	0.57(6)
SAMF+Depth [8]	9.45	0.45(11)	0.50(10)	0.67(8)	0.52(10)	0.55(9)	0.65(8)	0.49(11)
LDP-SVT [1]	9.73	0.47(8)	0.59(7)	0.56(12)	0.53(9)	0.52(11)	0.56(12)	0.51(9)

Algorithm	Avg. Rank	Occlusion		Motion		Avg. FPS
		Yes	No	Passive	Active	
3D-T [2]	2.45	0.73(1)	0.78(5)	0.79(2)	0.73(1)	
OAPF [9]	2.45	0.64(3)	**0.85(2)**	0.78(4)	0.71(2)	0.9
RGBDOcc-OF [12]	2.64	0.72(2)	0.75(6)	0.82(1)	0.70(3)	0.26
DS-KCF [5]	4.09	0.63(4)	**0.78(4)**	0.79(3)	0.66(5)	39
RGBD-OF [12]	4.64	0.60(5)	0.79(3)	0.74(6)	0.66(4)	0.26
Ours-Deep	5.82	0.43(9)	0.87(1)	0.69(7)	0.59(6)	10.49
SAMF+Depth [8]	9.45	0.41(11)	0.72(7)	0.66(8)	0.49(11)	
LDP-SVT [1]	9.73	0.41(10)	0.68(9)	0.58(11)	0.50(9)	

4 Conclusion

In this paper, we propose a new RGB-D tracker using deep features that can improve the accuracy of tracking. The deep features extracted from depth images are also meaningful. It is a common method that build up a scale pyramid to predict the scale of targets, but in our paper, we use the principle of camera imaging to deal with scale variation problem dependent on depth images to avoid lots of computation.

Acknowledgments. This work is supported by the National Natural Science Foundation of China (No. 61273273) and by Research Fund for the Doctoral Program of Higher Education of China (No. 20121101110034).

References

1. Awwad, S., Piccardi, M.: Prototype-based budget maintenance for tracking in depth videos. Multimedia Tools Appl. 1–16 (2016)
2. Bibi, A., Zhang, T., Ghanem, B.: 3D part-based sparse tracker with automatic synchronization and registration. In: Proceedings of the IEEE Conference on Computer Vision and Pattern Recognition, pp. 1439–1448 (2016)
3. Danelljan, M., Hager, G., Khan, F.S., Felsberg, M.: Discriminative scale space tracking. IEEE Trans. Pattern Anal. Mach. Intell. **39**(8), 1561–1575 (2016)

4. Danelljan, M., Shahbaz Khan, F., Felsberg, M., Van de Weijer, J.: Adaptive color attributes for real-time visual tracking. In: Proceedings of the IEEE Conference on Computer Vision and Pattern Recognition, pp. 1090–1097 (2014)
5. Hannuna, S., Camplani, M., Hall, J., Mirmehdi, M., Damen, D., Burghardt, T., Paiement, A., Tao, L.: DS-KCF: a real-time tracker for RGB-D data. J. Real-Time Image Proc. 1–20 (2016)
6. Henriques, J.F., Caseiro, R., Martins, P., Batista, J.: High-speed tracking with kernelized correlation filters. IEEE Trans. Pattern Anal. Mach. Intell. **37**(3), 583–596 (2015)
7. Kang, K., Li, H., Yan, J., Zeng, X., Yang, B., Xiao, T., Zhang, C., Wang, Z., Wang, R., Wang, X., et al.: T-cnn: Tubelets with convolutional neural networks for object detection from videos. arXiv preprint arXiv:1604.02532 (2016)
8. Li, Y., Zhu, J.: A Scale Adaptive Kernel Correlation Filter Tracker with Feature Integration. In: Agapito, L., Bronstein, M.M., Rother, C. (eds.) ECCV 2014. LNCS, vol. 8926, pp. 254–265. Springer, Cham (2015). doi:10.1007/978-3-319-16181-5_18
9. Meshgi, K., Maeda, S.I., Oba, S., Skibbe, H., Li, Y.Z., Ishii, S.: An occlusion-aware particle filter tracker to handle complex and persistent occlusions. Comput. Vis. Image Underst. **150**, 81–94 (2016)
10. Nam, H., Han, B.: Learning multi-domain convolutional neural networks for visual tracking. In: Proceedings of the IEEE Conference on Computer Vision and Pattern Recognition, pp. 4293–4302 (2016)
11. Simonyan, K., Zisserman, A.: Very deep convolutional networks for large-scale image recognition. arXiv preprint arXiv:1409.1556 (2014)
12. Song, S., Xiao, J.: Tracking revisited using rgbd camera: Baseline and benchmark. arXiv preprint arXiv:1212.2823 (2012)

Convolutional Gated Recurrent Units Fusion for Video Action Recognition

Bo Huang, Hualong Huang, and Hongtao Lu(✉)

Key Laboratory of Shanghai Education Commission for Intelligent Interaction and Cognitive Engineering, Department of Computer Science and Engineering, Shanghai Jiao Tong University, Shanghai, People's Republic of China
{948465700,hlhuang,htlu}@sjtu.edu.cn

Abstract. Two-stream Convolutional Networks (ConvNets) have achieved great success in video action recognition. Research also shows that early fusion of the two-stream ConvNets can further boost the performance. Existing fusion methods focus on short snippets thus fails to learn global representations for videos. We introduce a Convolutional Gated Recurrent Units (ConvGRU) fusion method to model long-term dependency inside actions. This fusion method takes advantage of both Recurrent Neural Networks (RNN) models which have strong capacity to handle long-term dependency for sequence modeling and early fusion architecture which learns the evolution of appearance feature and motion feature. We further propose an end-to-end architecture according to this fusion method and evaluate our approach using a widely used action recognition dataset named UCF101. We investigate different input lengths and fusion layers and find that fusing at the last convolutional layer with an input length of 10 entries yields best performance (93.0%) which is comparable to the state-of-the-art.

Keywords: Two-stream ConvNets · ConvGRU · Fusion

1 Introduction

Human action recognition in videos attracts extensive research interests nowadays due to its wide applications in video retrieval, surveillance, human-computer interaction and so on. Early work [1–3] focused on using hand-crafted spatial-temporal local descriptors for video representation and classification. As Convolutional Neural Networks (ConvNets) achieved dramatic success in many areas of computer vison such as image classification [21], object detection [22] etc., recent work in action recognition has concentrated on applying ConvNets to this task. However, early attempts [4–6] to applying ConvNets to action recognition did not yield comparable performance to the state-of-the-art hand-crafted descriptors [3] until the proposal of two-stream ConvNets [7] which decomposes video into spatial component as RGB and temporal component as optical flow. Based on the two-stream ConvNets, several improved methods [8–12] have been proposed recently and finally ConvNets achieved the state-of-the-art performance in action recognition. Among these improved methods, [12] achieved 93.4% classification accuracy with an end-to-end deep architecture in the widely used UCF101 dataset.

D. Liu et al. (Eds.): ICONIP 2017, Part III, LNCS 10636, pp. 114–123, 2017.
https://doi.org/10.1007/978-3-319-70090-8_12

Our network architecture is inspired by the two-stream fusing architecture in [8]. [8] proposed two kinds of fusion methods, namely, spatial fusion and temporal fusion, to improve the two-stream architecture. In their proposed architecture, the convolutional feature maps of both streams were concatenated and a 3D convolutional layer was added after the concatenated layer to learn the pixel-wise correspondences between appearance stream and motion stream. Moreover, they also changed the pooling layer to a 3D max pooling layer to extend the receptive field in the temporal domain. However, we find a main drawback of these two fusion methods: they tend to distinguish actions by short video clips instead of digging the long-term dependences inside actions. We came to this conclusion mainly by two reasons: (i) 3D convolutional layers are sensitive to short clips but cannot handle relatively long-term dependency, and (ii) pooling in the temporal domain results in bias towards snippets with high responses in the 3D convolutional layer. So if two actions resemble in short snippets, though appear to be distinct in the long term, the fusion methods by [8] may fail due to the above two reasons we mentioned. A natural thought to overcome this drawback is to utilize Recurrent Neural Networks (RNN) to handle the long-term dependency inside actions. [11, 12] attempted to model long-term dependency inside actions by Long Short Term Memory (LSTM) and Gated Recurrent Unit (GRU) separately. [11] yielded a rather disappointing result with no improvement over temporal pooling and [12] yielded a minor increase over two-stream architecture with a highly increase of computation cost. We argue that the main reason these two approaches failed is that they did not make full use of the complementation between appearance information and motion information. Therefore, we introduce a Convolutional Gated Recurrent Unit (ConvGRU) fusion method to model long-term dependency inside actions. This approach use ConvGRU architecture to fuse the two-stream ConvNets both spatially and temporally so as to learn the evolution of appearance feature and motion feature in the long term.

2 Related Work

Several deep learning approaches have been used to learn video representations from both frame-wise appearance information and temporal information. [4] studied a natural extension of ConvNets to learn spatiotemporal features, which stacked consecutive video frames and extended 2D ConvNets into time. [5] tested ConvNets with deep structures and several temporal sampling methods on a large dataset, called Sports-1 M. However, these temporal sampling methods did not learn strong spatiotemporal features, with similar levels of performance achieved by a purely spatial network. The recently proposed C3D ConvNets [6] got better performance than [5] by using 3 * 3 * 3 filter kernels and a limited temporal support of 16 consecutive frames.

However, these deep models achieved lower performance compared with shallow hand-crafted representations [3]. This might be caused by relatively small datasets available and complex motion patterns. Recent work on deep approaches that matched performance of the state-of-the-art hand-crafted representations is the two-stream ConvNets architecture proposed in [7]. They decomposed a video into spatial component as RGB and temporal component as optical flow and then fed the two components into separate deep ConvNets architectures to learn both appearance and motion representations for the video. Each stream was trained separately and a late fusion of softmax scores was used to combine the two streams and make the final decision. Most recent state-of-the-art deep approaches were based on the two-stream ConvNets architecture.

One branch has investigated learning motion patterns in network architectures across longer time period. [11] compared several temporal pooling architectures and suggested that temporal pooling of convolutional layers outperforms slow, local, or late pooling, as well as temporal convolution. They also attempted to feed ConvNets features into a recurrent network with Long Short-Term Memory (LSTM) [15] cells to model ordered sequential information but yields a rather disappointing result with no improvement over temporal pooling. [12] used Gated Recurrent Units (GRU) [16] instead of LSTM and replaced the inner product operations inside the gates with 2D convolution operations, thus taking advantage of the local spatial similarity in images. This approach yielded a minor increase over two-stream ConvNets architectures. [10] utilized deep architectures to learn discriminative convolutional feature maps and conducted trajectory-constrained pooling to aggregate these convolutional features into effective descriptors. This approach shares the merits of both hand-crafted features and deep-learned features.

Another branch has investigated learning pixel-wise correspondences between appearance and motion by fusion the two-stream networks at an early stage rather than the softmax layer. [8] investigated how and where to fuse the two-stream networks. They found that fusion the temporal network into the spatial network at the last convolutional layer performs better than fusion earlier. They also found that spatial-temporal neighborhoods pooling of abstract convolutional features can further boost performance. Moreover, they proposed a new ConvNets architecture for spatiotemporal fusion of video snippets which yielded a considerable increase of 4.5% over two-stream ConvNets architectures. [9] replaced the VGG-M-2048 models [13] used in both streams with Residual Networks (ResNets) [14] to investigate deeper models in two-stream network. They injected residual connections between the appearance and motion pathways to allow spatiotemporal interaction between the two streams. They also transform temporal residual connections into learnable convolutional filters that operate on adjacent feature maps in time, thus increasing the spatial-temporal receptive field as the depth of model increases.

3 Our Approach

3.1 GRU

In this section, we review Gated Recurrent Units (GRU) networks which are the prototype of our ConvGRU fusion networks. GRU networks are a particular type of RNN model. An RNN model defines a recurrent hidden state whose activation at each time is dependent on that of the previous time, so it can handle a sequence of inputs which may have variable lengths. To be more concrete, given a input sequence $X = (x_1, x_2, \ldots, x_T)$, the hidden state of a RNN model is defined as $h_t = \varphi(h_{t-1}, x_t)$, where φ is a nonlinear activation function. It is difficult to train a RNN model due to the exploding or vanishing gradient effect [18]. Therefore, variants of RNNs such as Long Short Term Memory (LSTM) [15] and Gated Recurrent Units (GRU) [16] were proposed to model long-term temporal dependency. In this paper, we will mainly focus on GRU networks as they have shown similar performance to LSTM networks but with a lower memory requirement and computation cost [17].

The activation function h_t of GRU is defined by the following equations:

$$r_t = \sigma(W^r x_t + U^r h_{t-1}), \tag{1}$$

$$z_t = \sigma(W^z x_t + U^z h_{t-1}), \tag{2}$$

$$\tilde{h}_t = g(W x_t + U(r_t \odot h_{t-1})), \tag{3}$$

$$h_t = (1 - z_t) \odot h_{t-1} + z_t \odot \tilde{h}_t. \tag{4}$$

Here $\odot$ represents an element-wise multiplication, σ represents a sigmoid function and g represents a nonlinear activation function. GRU networks define two gates, namely, reset gate denoted by r_t and update gate denoted by z_t, and a candidate hidden layer $\tilde{h}_t$ to allow each recurrent unit to adaptively capture dependencies of different time scales. The reset gate r_t decides how much information of h_{t-1} the candidate hidden layer should discard, and the update gate z_t decides the degree to which the unit updates its activation.

3.2 ConvGRU Fusion

This section delves the main contribution of this work. We introduce a ConvGRU fusion architecture to model long-term dependency inside actions by the evolution of appearance feature and motion feature.

Suppose $x_t^a \in \mathbb{R}^{H \times W \times D}$ and $x_t^b \in \mathbb{R}^{H \times W \times D}$ are two feature maps from spatial streams and temporal streams at time t. A recurrent fusion function can be defined as $f: x_t^a, x_t^b, h_{t-1} \rightarrow h_t$, where $h_t \in \mathbb{R}^{H \times W \times D}$ is the output feature map at time t. Here we assume the feature maps of x^a, x^b and h_t have the same size for simplicity. The ConvGRU fusion function can be denoted by the following equations.

$$x_t^{cat} = f^{cat}\left(x_t^a, x_t^b\right), \tag{5}$$

$$r_t = \sigma\left(W^r * x_t^{cat} + U^r * h_{t-1}\right), \tag{6}$$

$$z_t = \sigma\left(W^z * x_t^{cat} + U^z * h_{t-1}\right), \tag{7}$$

$$\tilde{h}_t = g\left(W * x_t^{cat} + U * (r_t \odot h_{t-1})\right), \tag{8}$$

$$h_t = (1 - z_t) \odot h_{t-1} + z_t \odot \tilde{h}_t. \tag{9}$$

The first equation stacks the two feature maps at the same spatial locations across feature channels. To be more concrete, suppose $x_t^a(i, j, d)$ denotes the pixel of x_t^a at position (i, j) and channel d and $x_t^b(i, j, d)$ denotes the pixel of x_t^b at position (i, j) and channel d, then x_t^{cat} can be denoted as:

$$x_t^{cat}(i, j, 2d - 1) = x_t^a(i, j, d) \quad x_t^{cat}(i, j, 2d) = x_t^b(i, j, d), \tag{10}$$

where $x_t^{cat}(i, j, d)$ denotes the pixel of x_t^{cat} at position (i, j) and channel d. The next four equations are similar to GRU equations, but we make three modifications. First, the input to the GRU is the concatenation of spatial stream and temporal stream, thus making full use of the complementation between appearance information and motion information. We address that this can be more helpful to learn spatial-temporal features than a late fusion at softmax scores. Second, we replace all the matrix multiplication by convolutional filters because the original design of GRU leads to two main problems: (i) Applying a GRU directly can lead to a drastic increase of parameters because all the convolutional feature map inputs are 3D tensors. (ii) The original GRU does not take advantage of the underlying structure of feature maps that they have strong local correlations. Convolutional filters are well adapted to solve these two problems when applying GRU to 3D convolutional feature maps. The third modifications is the activation function g applied on the candidate hidden state $\tilde{h}_t$. In the original implementation of GRU, function g is the hyperbolic tangent function, but the activation function of all the other activation layers of a deep convolutional networks is the rectified linear units function. So here we define function g as the rectified linear units function for consistency.

As we can see, the channels of output feature maps are doubled since we concatenate two streams before we feed them into convGRU architecture, which leads to double parameters of the next layer. This may cause a drastic increase of parameters if the next layer is a full-connected layer which holds a major part of the whole parameters in the network. Therefore, we add a convolutional layer after the convGRU fusion layer to apply channel reduction on the convGRU output feature maps.

3.3 Proposed Architecture

Based on the ConvGRU fusion approach, we propose an end-to-end architecture shown in Fig. 1 which applies ConvGRU fusion at the last convolutional layer of VGG-16 ConvNets. This architecture recurrently takes frames and optical flows along time as input and extract feature maps by the 5 blocks of convolutional layers. Then the ConvGRU cells output a new hidden state according to the previous hidden state and the concatenated feature maps from both streams. In the end, the output hidden state of the last time step undergoes a channel reduction layer and then full-connected layers to produce classification scores. Parameters are shared for each time step.

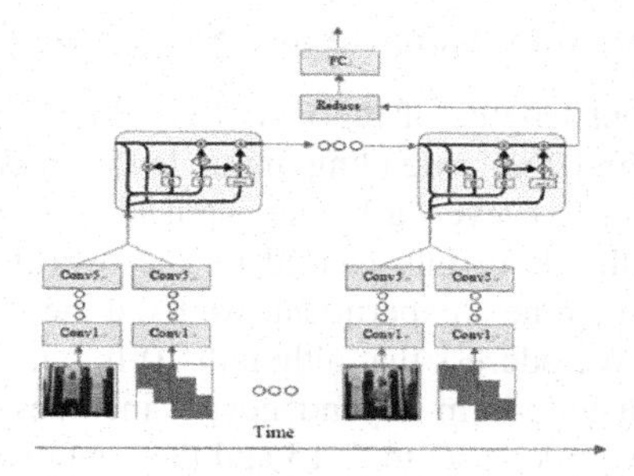

Fig. 1. Proposed architecture

3.4 Implementation Details

Two-Stream Architecture. We employ the very deep VGG-16 model [20] which has been pre-trained on ImageNet ILSVRC-2012 for both streams for sake of comparison to the fusion approaches in [8]. We use the same training strategy as [8]: In spatial training, we first rescale a selected frame so that the size of the smallest side is equal to 256, then a 224×224 sub-image is randomly cropped from the rescaled image, the sub-image then undergoes horizontal flipping. In temporal training, an optical flow volume was computed for the selected frame and then a $224 \times 224 \times 2L$ input is randomly cropped and flipped. L is set to 10 the same as [7, 8]. The dropout is set to 0.85 for spatial stream and 0.9 for temporal stream for consistency with [8]. The batch size is set to 256 and we do not use batch normalization.

ConvGRU Fusion. We apply additional augmentation similar to [8]. For each video, we sample T frames by randomly sampling a start frames and a temporal stride $\tau \in [5, 15]$. We evaluate different values of T by experiment. Instead of directly cropping a 224×224 sub-image, we first randomly jitter its width and height by $\pm$ 25% and rescale it to 224×224, then we randomly crop the rescaled image at a maximum of 25% distance from the image boarders. The cropped sub-image is then rescaled to 224×224 again.

Testing. For each video, we randomly sampling 50/T inputs to the ConvGRU fusion networks and average their classification scores. Each input to the ConvGRU fusion networks contains T frames. This ensures that for different value of T, the frame numbers of each video used for testing are the same.

4 Evaluation

4.1 Dataset and Experimental Protocols

We evaluate our approach on UCF101 [19], which is one of the most popular datasets for action recognition. The UCF101 dataset consists of 13320 action videos in 101 categories. We use the provided evaluation protocol and report the mean average accuracy.

4.2 How Many Frames to Be Sampled in a Single ConvNets Input

We evaluate different input lengths of the ConvNets through experiment with result shown in Table 1. We find that extending input length from 5 to 10 results in a performance gain of 0.5%. However, a longer input length of 25 drops the accuracy sharply, we believe that this is mainly caused by the overfitting problem due to the relatively small dataset. In the next experiment, we fixed the input length to 10 since it yields best performance. We address that although 10 frames input length seem to be rather small for modeling long-term dependency, it includes enough temporal information because the frame stride is set to 5–15 and the temporal input at each time step contains a stack of optical flows from 10 frames.

Table 1. Accuracy of different input lengths to convGRU

Fusion method	Dataset	Fusion layer	Input lengths	Acc.
ConvGRU	UCF101(split1)	ReLU5	5	90.4%
ConvGRU	UCF101(split1)	ReLU5	10	90.9%
ConvGRU	UCF101(split1)	ReLU5	25	87.6%

4.3 Where to Fuse the Two-Stream ConvNets

In the following experiment, we evaluate how different fusion layers can effect on the performance. The result can be seen in Table 2. As we can see, fusing at the ReLU5 layer achieves the best accuracy of 90.9% and fusing at an early layer drops the accuracy sharply. We do not list the classification accuracy of fusing at ReLU2 or ReLU1 layer because It is difficult to train and it converges very slowly. Actually fusing at ReLU4 layer or ReLU3 layer also takes longer time to converge than that at ReLU5 layer. We think this is mainly because early fusing does not make full use of the pre-trained ConvNets. As we know, the convolutional layers play a role of feature extraction to the original input images. When we fusing the two-stream ConvNets at ReLU5 layer, the features extracted by the previous 5 convolutional blocks are highly related clues for the final classification, but early fusing results in such drastic changes of ConvNets response that they drop the patterns after the fusing layer learned by being pre-trained on large images datasets. And current action recognition datasets may not big enough to learn spatial-temporal features from scratch.

Table 2. Accuracy of different fusion layer

Fusion method	Dataset	Fusion layer	Acc.
ConvGRU	UCF101(split1)	ReLU3	83.4%
ConvGRU	UCF101(split1)	ReLU4	84.2%
ConvGRU	UCF101(split1)	ReLU5	90.9%

4.4 Whether to Fuse

In this experiment, we investigate how using the concatenation of two-stream feature maps as GRU inputs contributes to learning long-term dependency inside actions with results shown in Table 3. We carry out a contrast experiment that use ConvGRU separately for each stream and average the softmax scores to get the final prediction. This approach results in 1.1% drop of accuracy with much more parameters required, which proves the effectiveness of ConvGRU fusion approach.

Table 3. Accuracy of whether to fuse before ConvGRU

Fusion method	Dataset	Fusion layer	#parameters	Acc.
Average	UCF101(split1)	Softmax	297.63 M	89.8%
ConvGRU	UCF101(split1)	ReLU5	210.71 M	90.9%

4.5 Compared to Different Fusion Methods

We use VGG-16 networks [20] for both streams in the following experiment. The fusion layer is set to ReLU5 which yields the best performance in [8] for fair comparison. After the fusion layer, we only use a single process stream to get the final prediction.

We compare our approach with the original Two-Stream ConvNets approach [7] and Convolutional Fusion method proposed by [8]. The result is shown in Table 4. We first observe that the accuracies of spatial stream and temporal stream are 83.8% and 86.4%, which are better than 73.0% and 83.7% reported in [7] owing to using deeper ConvNet network VGG-16 instead of VGG-M-2048 [13]. Second we can observe that both Conv fusion method [8] and ConvGRU fusion method can boost accuracy with only one stream remained after fusion. Moreover, ConvGRU fusion exceeds Conv Fusion by 2.2%, which is a considerable improvement.

Table 4. Accuracy of different fusion methods

Fusion method	Dataset	Fusion layer	#parameters	Acc.
Spatial	UCF101(split1)	ReLU5	134.66 M	83.8%
Temporal	UCF101(split1)	ReLU5	134.67 M	86.4%
Conv	UCF101(split1)	ReLU5	154.09 M	88.7%
ConvGRU	UCF101(split1)	ReLU5	210.71 M	90.9%

4.6 Compared to the State-of-the-Art

Inspired by [8] which boosts accuracy by further average the classification scores of their Conv fusion approach with the original temporal stream classification scores, we achieve our best classification accuracy of 93.0% by further average the classification scores with both spatial stream and temporal stream. Table 5 shows the comparison to the state-of-the-art approaches. We can see that the accuracy of our approach exceeds the base-line accuracy of two-stream ConvNets [7] by a large margin of 5%.

Our approach also performs better than the Conv fusion approach [8] which indicates ConvGRU architecture has stronger capacity to handle long-term dependency inside actions. And compared to other approaches which directly apply RNN architecture on single stream [11, 12], our approach also yields a considerable increase of accuracy, which addresses the importance of the evolution of appearance feature and motion feature. We also notice that the ST-ResNet [9] approach yields a higher accuracy of 93.4%, partially owing to deeper ConvNets architecture: ResNet-50 [14].

Table 5. Mean classification accuracy of the state-of-art on UCF101 over 3 train/test splits

Method	UCF101
Spatiotemporal ConvNet [5]	65.4%
C3D [6]	85.2%
Two-Stream ConvNet [7]	88.0%
Two-Stream Conv Pooling [11]	88.2%
GRU-RCN [12]	90.8%
Conv Fusion [8] (average score with temporal stream)	92.5%
ST-ResNet [9]	93.4%
ConvGRU Fusion (single stream)	90.9%
ConvGRU Fusion (average score with both streams)	92.8%
ConvGRU Fusion (weighted average score with both streams)	93.0%

5 Conclusion

We introduce a Convolutional Gated Recurrent Units (ConvGRU) fusion method to model long-term dependency inside actions. This approach takes advantage of RNN models which show strong capacity to handle long-term dependency and early fusion architecture which learns the evolution of appearance feature and motion feature. According to this fusion method, we propose an end-to-end architecture and achieve the state-of-the-art performance on a popular action recognition dataset named UCF101.

Acknowledgement. This paper is supported by NSFC (No. 61772330, 61272247, 61533012, 61472075), the 863 National High Technology Research and Development Program of China (SS2015AA020501), the Basic Research Project of Innovation Action Plan (16JC1402800) and the Major Basic Research Program (15JC1400103) of Shanghai Science and Technology Committee.

References

1. Laptev, I.: On space-time interest points. Int. J. Comput. Vis. **64**(2–3), 107–123 (2005)
2. Wang, H., Klaser, A., Schmid, C., Liu, C.L.: Dense trajectories and motion boundary descriptors for action recognition. Int. J. Comput. Vis. **103**(1), 60–79 (2013)
3. Wang, H., Schmid, C.: Action recognition with improved trajectories. In: Proceedings of ICCV, pp. 3551–3558 (2013)

4. Ji, S., Xu, W., Yang, M., Yu, K.: 3D convolutional neural networks for human action recognition. IEEE Trans. Pattern Anal. Mach. Intell. **35**(1), 221–231 (2013)
5. Karpathy, A., Toderici, G., Shetty, S., Leung, T., Sukthankar, R., Li, F.F.: Large-scale video classification with convolutional neural networks. In: Proceedings of CVPR, pp. 1725–1732 (2014)
6. Tran, D., Bourdev, L., Fergus, R., Torresani, L., Paluri, M.: Learning spatiotemporal features with 3D convolutional networks. In: Proceedings of ICCV, pp. 4489–4497 (2015)
7. Simonyan, K., Zisserman, A.: Two-stream convolutional networks for action recognition in videos. In: Proceedings of NIPS, pp. 568–576 (2014)
8. Feichtenhofer, C., Pinz, A., Zisserman, A.: Convolutional two-stream network fusion for video action recognition. In: Proceedings of CVPR, pp. 1933–1941 (2016)
9. Feichtenhofer, C., Pinz, A., Wildes., R.P.: Spatiotemporal residual networks for video action recognition. In: Proceedings of NIPS, pp. 3468–3476 (2016)
10. Wang, L., Qiao, Y., Tang, X.: Action recognition with trajectory-pooled deep-convolutional descriptors. In: Proceedings of CVPR, pp. 4305–4314 (2015)
11. Yue-Hei Ng, J., Hausknecht, M., Vijayanarasimhan, S., Vinyals, O., Monga, R., Toderici, G.: Beyond short snippets: deep networks for video classification. In: Proceedings of CVPR, pp. 4694–4702 (2015)
12. Ballas, N., Yao, L., Pal, C., Courville, A.: Delving deeper into convolutional networks for learning video representations. In: Proceedings of ICLR (2016)
13. Chatfield, K., Simonyan, K., Vedaldi, A., Zisserman, A.: Return of the devil in the details: delving deep into convolutional nets. In: Proceedings of BMVC (2014)
14. He, K., Zhang, X., Ren, S., Sun, J.: Deep residual learning for image recognition. arXiv preprint arXiv:1512.03385 (2015)
15. Hochreiter, S., Schmidhuber, J.: Long short-term memory. Neural Comput. **9**(8), 1735–1780 (1997)
16. Cho, K., Van Merriënboer, B., Gulcehre, C., Bahdanau, D., Bougares, F., Schwenk, H., Bengio, Y.: Learning phrase representations using RNN encoder-decoder for statistical machine translation. arXiv preprint arXiv:1406.1078 (2014)
17. Chung, J., Gulcehre, C., Cho, K., Bengio, Y.: Empirical evaluation of gated recurrent neural networks on sequence modeling. arXiv preprint arXiv:1412.3555 (2014)
18. Bengio, Y., Simard, P., Frasconi, P.: Learning long-term dependencies with gradient descent is difficult. IEEE Trans. Neural Netw. **5**(2), 157–166 (1994)
19. Soomro, K., Zamir, A.R., Shah, M.: UCF101: A dataset of 101 human actions classes from videos in the wild. Technical report CRCV-TR-12-01, UCF Center for Research in Computer Vision (2012)
20. Simonyan, K., Zisserman, A.: Very deep convolutional networks for large-scale image recognition. In: Proceedings of ICLR (2014)
21. Krizhevsky, A., Sutskever, I., Hinton, G.E.: ImageNet classification with deep convolutional neural networks. In: Proceedings of NIPS, pp. 1106–1114 (2012)
22. Redmon, J., Divvala, S., Girshick, R., Farhadi, A.: You only look once: unified, real-time object detection. In: Proceedings of CVPR, pp. 779–788 (2016)

Stereo Matching Using Conditional Adversarial Networks

Hualong Huang, Bo Huang, Hongtao Lu(✉), and Huiyu Weng

Key Laboratory of Shanghai Education Commission for Intelligent Interaction and Cognitive Engineering, Department of Computer Science and Engineering, Shanghai Jiao Tong University, Shanghai, People's Republic of China
{hlhuang,948465700}@sjtu.edu.cn, {lu-ht,weng-hy}@cs.sjtu.edu.cn

Abstract. Recently, adversarial networks have attracted increasing attentions for the promising results of generative tasks. In this paper we present the first application of conditional adversarial networks to stereo matching task. Our approach performs a conditional adversarial training process on two networks: a generator that learns the mapping from a pair of RGB images to a dense disparity map, and a discriminator that distinguishes whether the disparity map comes from the ground truth or from the generator. Here, both the generator and the discriminator take the same RGB image pair as an input condition. During this conditional adversarial training process, our discriminator gradually captures high-level contextual features to detect inconsistencies between the ground truth and the generated disparity maps. These high-level contextual features are incorporated into loss function in order to further help the generator to correct predicted disparity maps. We evaluate our model on the Scene Flow dataset and an improvement is achieved compared with the most related work *pix2pix*.

Keywords: Stereo matching · Conditional adversarial networks

1 Introduction

Accurately estimating depth from RGB images is a significant topic in many computer vision applications, including robotics and autonomous vehicles. One of the most promising solution to this problem is stereo matching, which tries to mimic the human vision system. A stereo matching algorithm takes a pair of RGB images token by two cameras, just like our two eyes, as input and tries to predict a dense disparity map. Here, disparity refers to a measure of depth and can be converted to depth by simple formulations [21]. To compute a disparity map, the core task is computing the correspondence of each pixel pair between two rectified RGB images [21].

Over the past decades, a vast amount of stereo matching algorithms have sprung up. In this paper, we only focus on recent methods, which can be roughly categorized into two main classes of approaches.

D. Liu et al. (Eds.): ICONIP 2017, Part III, LNCS 10636, pp. 124–132, 2017.
https://doi.org/10.1007/978-3-319-70090-8_13

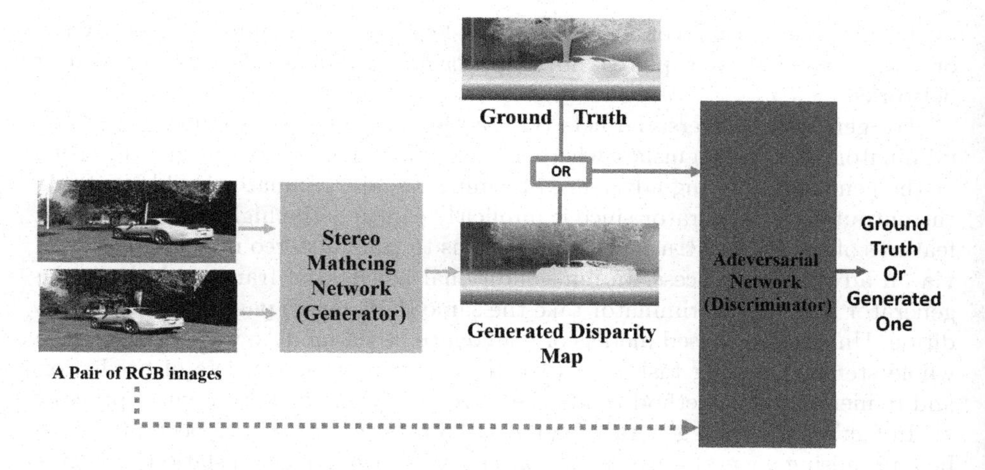

Fig. 1. Overview of the proposed approach. The left generator, stereo matching network, takes a pair of RGB images as input to produce a per-pixel disparity map. Then this generated disparity map or the corresponding ground truth is randomly put into the right discriminator to make a decision whether the input disparity map is the ground truth or a generated one. Both the generator and the discriminator take the same RGB image pair as an input condition.

The first class is local methods which estimate the disparity of each pixel via local information, such as image patches. While early methods rely on simple hand-crafted features, such as gradients, to compute the matching cost of image pathes, recent methods turn to convolutional neural networks (CNNs) for more powerful features [14,16,23]. However, it is worth noting that local stereo matching methods actually predict each pixel's disparity independently and do not take the global semantic contextual information into account. To overcome the drawback of local methods, various traditional post-processing steps have been explored, such as left right consistency checks [3] and semi-global matching [7]. Unfortunately, these hand-engineered steps are susceptible due to their limited ability to improve output disparity maps.

In contrasted to local methods, the second class, global methods, considers more high-level contextual information. One of them uses conditional random field (CRF) to model the dependencies between pixels, such as the pairwise smoothness assumption, and then converts the stereo matching task into a maximum a posteriori (MAP) problem [20]. However, a major shortcoming of most CRF based approaches is the difficulty of designing energy items that not only capture pixel dependencies but also are tractable to perform inference. Another promising global method, confidence based models, focuses on the refinement of disparity maps produced by a given method. Most of the confidence based models need to first estimate a confidence map that measures the correctness of predicted disparity at each pixel [18]. Unsurprisingly, CNNs can be introduced to help estimating confidence map [22] as well as refining the disparity predictions [4].

In this paper, instead of integrating high-level contextual information by CRF or confidence map, we explore a novel approach based on conditional adversarial networks.

For generative adversarial nets (GANs) [6,9,15], the core module is the discriminator that distinguishes whether a sample comes from the ground truth or the generator. During adversarial training, the discriminator is able to guide the output of the generator since it implicitly captures the high-level contextual features of the ground truth. This inspires us to train a stereo matching network via an adversarial process. In our conditional adversarial framework, both the generator and the discriminator take the same RGB image pair as an input condition. Unlike CRF based models or confidence based models, which address the whole stereo matching task as a two-steps pipeline that consists of prediction and refinement, our method treats the stereo matching task as a whole process.

In fact, conditional generative adversarial nets (cGANs) have been proven to be a promising general-purpose solution to image-to-image translation problems [9]. These image translation problems include graphics tasks, like synthesizing photos from label maps, and vision tasks, like semantic segmentation. However, such framework has still not been applied in stereo matching task.

Our contributions are as follows:

1. We present, to the best of our knowledge, the first application of conditional adversarial networks to stereo matching task.
2. We achieve an improvement on Scene Flow dataset [16] compared with the most related work *pix2pix* [9].

The remainder of the paper is organized as follows. We present our adversarial training approach and network architectures in Sect. 2. After that, experimental results are shown in Sect. 3 and finally we conclude the paper in Sect. 4.

2 Conditional Adversarial Networks for Stereo Matching

We first describe our adversarial training framework for stereo matching task in Sect. 2.1 and then detail the network architectures in Sect. 2.2.

2.1 Adversarial Training Framework

Original GANs [6] can be directly extended to conditional GANs (cGANs) [15] by feeding extra information for both the generator and the discriminator as an additional input. Thus, the loss function of cGAN consisting of a generator G and a discriminator D would be

$$\mathcal{L}_{cGAN}(G, D) = \mathbb{E}_{x \sim p_{data}(x)} \left[\log D(x|y)\right] + \mathbb{E}_{z \sim p_z(z)} \left[\log(1 - D(G(z|y))\right]. \quad (1)$$

where x is the ground truth data, z is random noise, and y represents extra input conditions. In our work, x represents the ground truth disparity map and y represents a pair of observed RGB images. During the adversarial training process, G tries to minimize this objective function while D tries to maximize it, i.e.

$$G^* = \arg \min_G \max_D \mathcal{L}_{cGAN}(G, D). \tag{2}$$

Like [13], we train a generator by optimizing a hybrid loss function, which is a weighted sum of two loss terms. The first loss term is a traditional L1 loss which captures low-level information effectively, such as the pixel intensity. The second loss term is the adversarial loss term in Eq. 2, which is introduced to capture high-level global contextual information. Therefore, our final loss function is as follows

$$G^* = \arg \min_G \max_D \mathcal{L}_{cGAN}(G, D) + \lambda \mathcal{L}_{L1}(G). \tag{3}$$

where λ is a hyper-parameter to balance the L1 loss term and the adversarial loss term. The adversarial loss term becomes large once the discriminator detects the generated disparity map is 'fake', which will guide the generator's output space become more close to the ground truth space.

2.2 Network Architecture

The overview of our architecture is illustrated in Fig. 1. With the guidance of the discriminator, the generator tries to produce disparity maps that cannot be distinguished from the ground truth during the adversarial training process.

Generative Network. The architecture of the generator is illustrated in Fig. 2(a). The left and right RGB images are first passed through a Siamese network [2]. We choose Siamese network for two reasons. Firstly, before fusing information from the left and right RGB images, we hope to extract more powerful features rather than raw pixels. Secondly, these two branches of Siamese

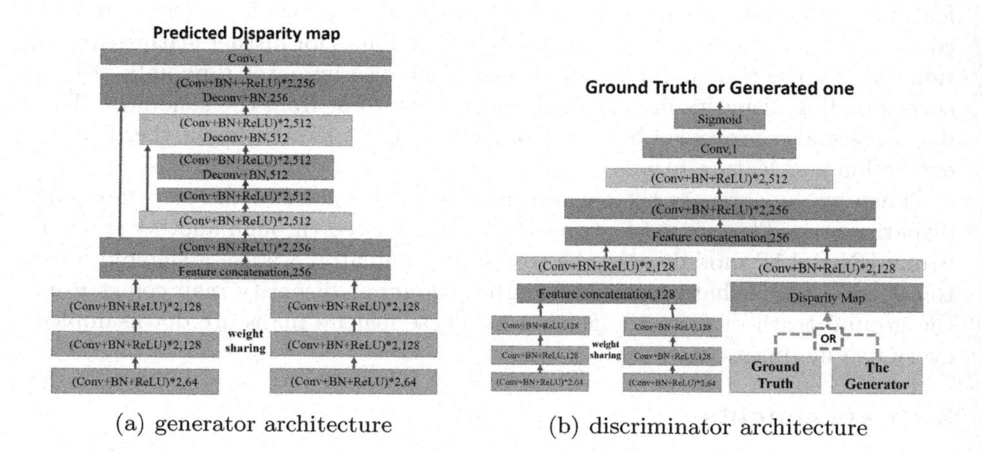

(a) generator architecture (b) discriminator architecture

Fig. 2. Network architectures

network share the identical CNN architecture as well as weights, which take the similarity of the left and right RGB images into consideration. These two branches of Siamese network consist of 6 'Conv-BN-ReLU' modules, which are made of the following rule: each convolutional layer is followed by a batch normalization layer [8] and a rectified non-linearity. All the kernel size of convolutional layer is 3 and the numbers of outputs of each layer are 64, 128, 128 sequentially and the feature maps are downsampled between each two modules.

These feature maps extracted by Siamese network are then simply concatenated for further processing by a U-net architecture [19]. U-net architecture can be seen as an improved version of the traditional encoder-decoder architecture. Traditional encoder-decoder architectures usually involve a number of pooling and deconvolution layers [17]. Therefore, the encoder increases the receptive field size while loses high-frequent details. To recover the high-frequent details, we use a U-net architecture [19] with skip connections to introduce low-level features from the encoder layers into the corresponding decoder layers. The U-net consists of an encoder of three '(Conv-BN-ReLU)$\times$2' modules and a decoder of three 'Deconv-BN and (Conv-BN-ReLU)$\times$2' modules. The features of the ith layer of the encoder is passed into the $(7 - i)$th layer of the decoder as an additional input. The numbers of outputs of the encoder are 256, 256, 512, 512, 512, 512 sequentially and the feature maps are downsampled every one modules. The numbers of outputs of the decoder are 512, 512, 512, 512, 256, 256 sequentially and the feature maps are upsampled every one modules.

Discriminative Network. The architecture of the discriminator is illustrated in Fig. 2(b). The discriminator takes a pair of RGB images and the corresponding disparity map as input.

There are two different branches to process the RGB images and the ground truth disparity map separately. Similar to the generator, the left RGB image and right RGB image are passed through a Siamese network first for high-level features extraction. Then the output of the Siamese network is processed by a concatenation layer and two 'Conv-BN-ReLU' modules for further feature extraction and feature fusion. On the other hand, the discriminator randomly picks a corresponding disparity map from the ground truth or from the generator. This disparity is also processed by two 'Conv-BN-ReLU' modules for further feature extraction and feature fusion.

Then we concatenate the features from RGB images and the corresponding disparity map. These fused features are further passed through another stack of 'Conv-BN-ReLU' modules. At the top of discriminator, we use a sigmoid layer to output a value that represents weather the input disparity map comes from the ground truth or from the generator. These feature maps are downsampled every two modules.

3 Experiments

In this section we present qualitative and quantitative results on the Scene Flow dataset [16]. In Sect. 3.1 we present the overall experiment settings including dataset and optimization details. Then in Sect. 3.2 we experiment different variants of our model and justify the design choices.

3.1 Experimental Settings

We train our architecture using Scene Flow dataset [16] for two main reasons. Firstly, this synthesized dataset contains rectified RGB image pairs as well as corresponding perfect ground truth disparity maps. These RGB image pairs are real and complex enough to simulate real scenes. Secondly, the Scene Flow dataset is large enough to train models without over-fitting.

We choose the *driving* subset from Scene Flow dataset, which is very similar to the real scenes of autonomous driving. The *driving* subset contains 4400 images pairs including a training set with 3520 image pairs and a validation set with 880 image pairs. For evaluation we use the percentage of disparity whose absolute difference from the ground truth disparity is more than x pixels.

We implement all the architectures in this section using TensorFlow [1]. Before training we normalize all the images such that the pixel intensities range from -1 to 1. To compare with the most related work *pix2pix* [9], our adversarial network is trained with the same hyper-parameters: a weight hyper-parameter $\lambda = 100$, a batch size of 1 and a iterative process with 2000 epochs. We optimize our adversarial models using Adam [11] with a initial learning rate of 10^{-5}. All our result is resized to 256×256 for comparing with the most related work *pix2pix* [9]. All our architectures can run on single NVIDIA 1080 GPU.

3.2 Qualitative and Quantitative Results

As Table 1 shows, we compare a number of different model variants. Here, G represents the generator and D represents the discriminator. These ablation experiments justify several key architecture choices in this paper: Siamese generator, Siamese discriminator and the adversarial loss term.

The 'single image as input' model is the baseline model from the most related GAN based model *pix2pix* [9], which take one RGB image as input and output a corresponding disparity map. We improve this model by taking the left and right RGB images as input and then concatenating them for further processing. However this model causes notable errors. To improve this network, we first explore the architecture of the generator. A Siamese network is utilized to extract features from the left and right RGB images first, then a 'U-Net' with optimized network hyper-parameter does the following processing. This structural change reduces the mean error from 2.16 px to 1.87 px. To further improve the structure, instead of simply concatenating a pair of RGB images and the corresponding disparity map as an input of the discriminator, we address the RGB image pair first. Here we adopt the same Siamese network as the generator for its powerful feature extraction ability. In Fig. 3, We give illustrations of the qualitative results on the Scene Flow dataset.

To demonstrate the regularization property of the adversarial term, we also compare the architecture where the generator's loss function only has $L1$ term

Fig. 3. Qualitative results on Scene Flow dataset. From left to right: left RGB image, disparity prediction and the ground truth.

Fig. 4. Results without and with adversarial term. From left to right: left RGB image, result without the discriminator, result with the discriminator.

with the architecture where the generator's loss function consists of both $L1$ term and the adversarial term. In Fig. 4, we give illustrations of the disparity maps generated with and without adversarial training, which shows the adversarial term leads to more sharp and more precise results.

Table 1. Results on Scene Flow dataset.

Model	>2px	>3px	4px	>5px	Mean Error (px)
Comparison of the generator's architectures					
Input single RGB image	74.12	66.39	2.82	2.14	2.91
Input two RGB images + concatenated G	57.09	48.44	1.67	1.49	2.16
Input two RGB images + Siamese G	**40.26**	**35.04**	**1.10**	**0.93**	**1.87**
Comparison of the discriminator's architectures					
Siamese G + concatenated D	40.26	35.04	1.10	0.93	1.87
Siamese G + Siamese D	**23.84**	**20.58**	**0.40**	**0.29**	**1.05**
Comparison of the loss functions of the generator					
G with L1 loss	27.15	23.2	0.73	0.55	1.35
G with L1 loss and adversarial loss	**23.84**	**20.58**	**0.40**	**0.29**	**1.05**
Final Siamese G + Siamese D	**23.84**	**20.58**	**0.40**	**0.29**	**1.05**

4 Conclusion

In this paper, we have investigate a novel conditional adversarial architecture for stereo matching task. An conditional adversarial loss term based on the discriminator is incorporated into the loss function of the generator to further guide the disparity prediction task. Such conditional adversarial architecture works due to the discriminator's ability to capture the high-level inconsistencies between the ground truth and the generated disparity maps. It is worth noting that our model use GAN at a condition setting and both the generator and the discriminator should take the same pair of RGB images as an extra input. To demonstrate the regularization property of our conditional adversarial networks, a series of experiments on the Scene Flow dataset are conducted. Our results show that the conditional adversarial architecture helps to improve the results of stereo matching task.

Acknowledgments. This paper is supported by NSFC(No.61772330, 61272247, 61533012, 61472075), the 863 National High Technology Re-search and Development Program of China (SS2015AA020501), the Basic Research Project of Innovation Action Plan (16JC1402800) and the Major Basic Research Program (15JC1400103) of Shanghai Science and Technology Committee.

References

1. Abadi, M., Agarwal, A., Barham, P., et al.: TensorFlow: Large-scale machine learning on heterogeneous distributed systems. arXiv preprint arXiv:1603.04467 (2016)
2. Bromley, J., Guyon, I., LeCun, Y., et al.: Signature verification using a šiamesetime delay neural network. In: Advances in Neural Information Processing Systems, pp. 737–744 (1994)
3. Fua, P.: A parallel stereo algorithm that produces dense depth maps and preserves image features. Mach. Vis. Appl. **6**, 35–49 (1993)
4. Gidaris, S., Komodakis, N.: Detect, Replace, Refine: Deep Structured Prediction for Pixel Wise Labeling. arXiv preprint arXiv:1612.04770 (2016)

5. Girshick, R., Donahue, J., Darrell, T., et al.: Rich feature hierarchies for accurate object detection and semantic segmentation. In: Proceedings of the IEEE Conference on Computer Vision and Pattern Recognition, pp. 580–587 (2014)
6. Goodfellow, I., et al.: Generative adversarial nets. In: Advances in Neural Information Processing Systems (2014)
7. Hirschmuller, H.: Accurate and efficient stereo processing by semiglobal matching and mutual information. In: IEEE Computer Society Conference on Computer Vision and Pattern Recognition, vol. 2, pp. 807–814. IEEE (2005)
8. Ioffe, S., Szegedy, C.: Batch normalization: accelerating deep network training by reducing internal covariate shift. arXiv preprint arXiv:1502.03167 (2015)
9. Isola, P., Zhu, J.Y., Zhou, T., et al.: Image-to-image translation with conditional adversarial networks. arXiv preprint arXiv:1611.07004 (2016)
10. Krizhevsky, A., Sutskever, I., Hinton, G.E.: Imagenet classification with deep convolutional neural networks. In: Advances in Neural Information Processing Systems, pp. 1097–1105 (2012)
11. Kingma, D., Ba, J.: Adam: A method for stochastic optimization. arXiv preprint arXiv:1412.6980 (2014)
12. Long, J., Shelhamer, E., Darrell, T.: Fully convolutional networks for semantic segmentation. In: Proceedings of the IEEE Conference on Computer Vision and Pattern Recognition, pp. 3431–3440 (2015)
13. Luc, P., Couprie, C., Chintala, S., et al.: Semantic Segmentation using Adversarial Networks. arXiv preprint arXiv:1611.08408 (2016)
14. Luo, W., Schwing, A.G., Urtasun, R.: Efficient deep learning for stereo matching. In: Proceedings of the IEEE Conference on Computer Vision and Pattern Recognition, pp. 5695–5703 (2016)
15. Mirza, M., Osindero, S.: Conditional generative adversarial nets. arXiv preprint arXiv:1411.1784 (2014)
16. Mayer, N., Ilg, E., Hausser, P., et al.: A large dataset to train convolutional networks for disparity, optical flow, and scene flow estimation. In: IEEE International Conference on Computer Vision and Pattern Recognition (CVPR), arXiv:1512.02134 (2016)
17. Noh, H., Hong, S., Han, B.: Learning deconvolution network for semantic segmentation. In: Proceedings of the IEEE International Conference on Computer Vision, pp. 1520–1528 (2015)
18. Park, M.G., Yoon, K.J.: Leveraging stereo matching with learning-based confidence measures. In: Proceedings of the IEEE Computer Society Conference on Computer Vision and Pattern Recognition, 101–109 (2015)
19. Ronneberger, O., Fischer, P., Brox, T.: U-Net: convolutional networks for biomedical image segmentation. In: Navab, N., Hornegger, J., Wells, W.M., Frangi, A.F. (eds.) MICCAI 2015. LNCS, vol. 9351, pp. 234–241. Springer, Cham (2015). doi:10.1007/978-3-319-24574-4_28
20. Scharstein, D., Pal, C.: Learning conditional random fields for stereo. In: Proceedings of the IEEE Computer Society Conference on Computer Vision and Pattern Recognition (2007)
21. Scharstein, D., Szeliski, R.: A taxonomy and evaluation of dense two-frame stereo correspondence algorithms. Int. J. Comput. Vis. **47**(1), 7–42 (2002)
22. Seki, A., Pollefeys, M.: Patch based confidence prediction for dense disparity map. In: British Machine Vision Conference, 10 September 2016
23. Zbontar, J., LeCun, Y.: Computing the stereo matching cost with a convolutional neural network. In: Proceedings of the IEEE Conference on Computer Vision and Pattern Recognition, pp. 1592–1599 (2015)

A Point and Line Features Based Method for Disturbed Surface Motion Estimation

Xiang Li and Yue Zhou(✉)

Institute of Image Processing and Pattern Recognition,
Shanghai Jiao Tong University, No. 800, Dongchuan Road, Shanghai 200240, China
{lostxine,zhouyue}@sjtu.edu.cn

Abstract. Calculating the motion of disturbed surface such as a reflective monochromatic one is often a difficult part, especially when using single feature based method. The error introduced from the feature extraction and matching will gradually accumulate into a larger final error. For a texture-less surface, the number of features makes the situation even more challenging. In this paper, point and line features from stereo sequences are combined to estimate 3D motion of disturbed surfaces. Taking the advantage of feature combination by two-stage iterative optimization and multiple filtering, the motion of surfaces can be estimated accurately, even under little motion blur. This paper also explored the relationship between measurement accuracy and object motion mode. This may provide a reference for the design of a vision based motion measuring system.

Keywords: Motion estimation · Combined features · Iterative optimization

1 Introduction

Estimating the motion of objects is a classic subject in computer vision. It has many applications such as human-computer interaction, robot navigation. The motion estimation problem shares most mathematical theorems [4] with Simultaneous Localization Mapping (SLAM) or Structure from Motion (SFM). That is, the motion of camera can be estimated by SLAM or SFM algorithms. [3,8,11,13] are several classic and reliable solutions. However, most of solutions depend on single fundamental feature and may not work in texture-less or blur scenes. Reflective objects will also affect the performance. Therefore, motion estimation of disturbed objects, which contains reflective and texture-less surfaces, is still a challenging problem that many authors have started to focus on over the last few years.

The work of this paper is inspired by [3,9]. In order to obtain accurate measurement, the most basic features such as points and lines which are supported and proved by mathematic should not be ignored. Tips like filtering and optimization are introduced to overcome disturbed surfaces and unstable features.

D. Liu et al. (Eds.): ICONIP 2017, Part III, LNCS 10636, pp. 133–140, 2017.
https://doi.org/10.1007/978-3-319-70090-8_14

In this paper, we proposed a novel motion estimation method for stereo system: converting traditional 3D-2D problem into 3D-3D and combining points with lines in 3D space. With the help of constraint from local color and 3D position relationship, an iterative optimization method like 4D ICP algorithm [7], handing both points and lines, are proposed. Bundle adjustment [12] and Random Sample Consensus (RANSAC) filtering were also involved. Our main contributions include: (i) tried to solve estimation problem with the help of 3D feature pairs; and (ii) combined point and line features.

Here comes the outline of this paper. Background, motivation and main idea are proposed first in this section. Then we will introduce the pipeline of proposed method. Detail discussions on point and line constraints come after the pipeline. Post processing containing filtering goes later. Finally, experiments are presented in Sect. 3, followed by analysis and conclusion.

2 Proposed Method

In this section, we present the proposed motion estimation algorithm which using both point and line features from stereo camera. Figure 1 shows the pipeline. Firstly, image sequence passed through the preprocessing, lines and points in 3D space were extracted and reserved for motion estimation. Then, we applied two-stage iterative optimization, in order to calculate the rotation and translation separately. After that, post processing, including bundle adjustment and RANSAC filtering, gathered several adjacent frames and gave final prediction of 6-dof motion. Finally, 'inertia' kept the result and helped estimation at the next frame.

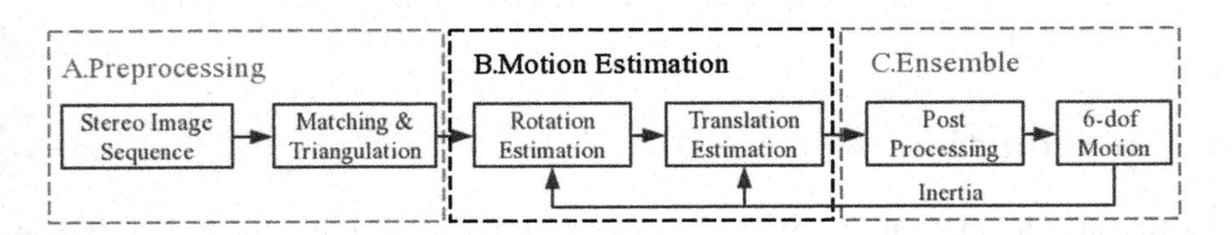

Fig. 1. Pipeline of proposed method

2.1 Preprocessing

The purpose of preprocessing is to extract the features in image sequence and to calculate the specific positions of the features in 3D space. As for stereo camera, with the help of offline calibration, SGBM [5] could generate reliable disparity map, but features needed to be extracted in region that disparity map had covered. The extraction of 3D feature is shown in green zone of Fig. 2. After extraction, matching and filtering began with features from different moments. That is the orange zone in Fig. 2.

Feature points were located by FAST [10] and described by SURF [1] to lower time complexity. Feature lines were generated from Canny [2]. Matching algorithm of lines came from Jia et al. [6].

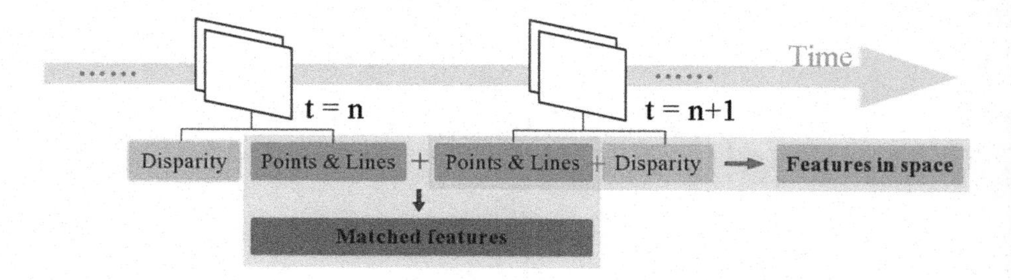

Fig. 2. Flow of preprocessing. (Color figure online)

2.2 Motion Estimation

In our solution, motion estimation module was designed handle 3D-3D matched points and lines. A new cost function with physical meaning, was introduce to cover both point and line features. In order to reduce the difficulty of solving and achieve convergence faster, the cost function, which includes 6 degrees of freedom, was split into two little ones: rotation cost function and translation cost function. Solving these two sub-problems was what we called two-stage iterative optimization.

Rotation Estimation. At this stage, the desired solution is rotation vector $\vec{\mathbf{r}}$ containing 3-dof. Figure 3 shows several variables will be used. Suppose N point and M line features from $t = n$ are marked in blue and corresponding features in $t = n + 1$ are marked in red. The green line between j^{th} point pair is called $d_j(\vec{\mathbf{r}})$. It is a vector from a point in n but after $\vec{\mathbf{r}}$ transformation, pointing to corresponding point in $n+1$. Rotation cost function was made of four parts like Eq. (1).

$$\arg\min_{\vec{\mathbf{r}}} \left(c_p\theta_p(\vec{\mathbf{r}}) + c_l\theta_l(\vec{\mathbf{r}}) + c_i\theta_{(\vec{\mathbf{r}}-\mathbf{r}')} + c_n\|\vec{\mathbf{r}}\|\right) \tag{1}$$

In Eq. (1), c_p, c_l, c_i, c_n are four constant weights. $\theta_p(\vec{\mathbf{r}})$ is the cost from point features and $\theta_l(\vec{\mathbf{r}})$ comes from line features, which will be introduced later. $\theta_{(\vec{\mathbf{r}}-\mathbf{r}')}$ represent the angle between this rotation vector $\vec{\mathbf{r}}$ and last one $\mathbf{r}'$. That is the representation of 'inertia' part. $\|\vec{\mathbf{r}}\|$ is designed to constrain the range of $\vec{\mathbf{r}}$, otherwise $\|\vec{\mathbf{r}}\|$ may be larger than 2π or smaller than 0.

The cost from point features $\theta_p(\vec{\mathbf{r}})$ can be divided into 2 parts like Eq. (2). c_1, c_2 are both constants. The first part represents the average angle between d_j and their mean. Because when two point clouds towards the same direction, all the connections between corresponding points should be parallel to each other. c_2 is followed by the standard deviation of d_j's length. This part prevents degradation of the first part due to symmetry.

$$\theta_p(\vec{\mathbf{r}}) = c_1\frac{1}{N}\sum_{j=1}^{N}\|\theta_{(d_j-\bar{d})}\| + c_2\sqrt{\frac{1}{N}\sum_{j=1}^{N}\|d_j\|} \tag{2}$$

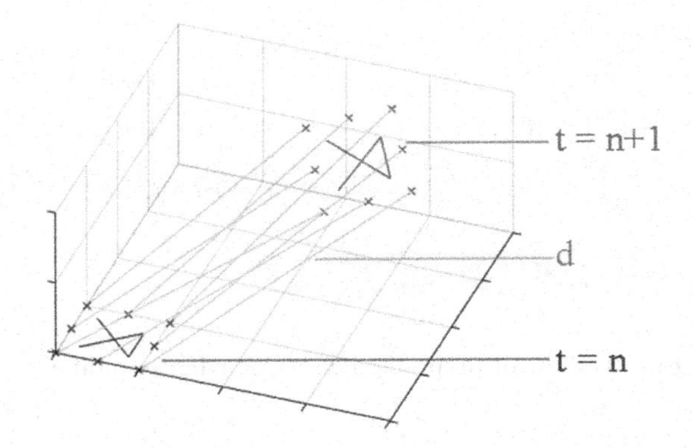

Fig. 3. Variables in the cost function. (Color figure online)

Suppose there were matched lines l_m, l'_m at n and $n+1$. They would contribute their angle $\theta_{(l,l')} = \theta_m$ to the cost from line features $\theta_l(\vec{\mathbf{r}})$. So:

$$\theta_l(\vec{\mathbf{r}}) = c_3 \frac{1}{M} \sum_{m=1}^{M} \|\theta_m\| \tag{3}$$

Translation Estimation. After the rotation estimation and transformation, features from two moments would share the same orientation. It would be easy to estimate translation vector $\vec{\mathbf{t}}$ using Eq. (4).

$$\arg\min_{\vec{\mathbf{t}}} \sum_{j=1}^{N} \|d_j(\vec{\mathbf{t}})\| \tag{4}$$

The code of motion estimation are available online[1].

2.3 Ensemble

The ensemble module contains two main parts in order to achieve higher precision. Bundle adjustment [12] helped localize points; RANSAC filtering picked out the frames which held an obvious error. At the same time, number of inner frames reflected the robustness of motion estimation algorithm.

3 Experiment

To evaluate our algorithm, we built a indoor rotating platform with cube model on it and acquired stereo image sequences. Figure 4 showed design of the experimental device. For every sequence, the cube rotated around a fixed axis at a

[1] https://github.com/LostXine/motionEstimation.

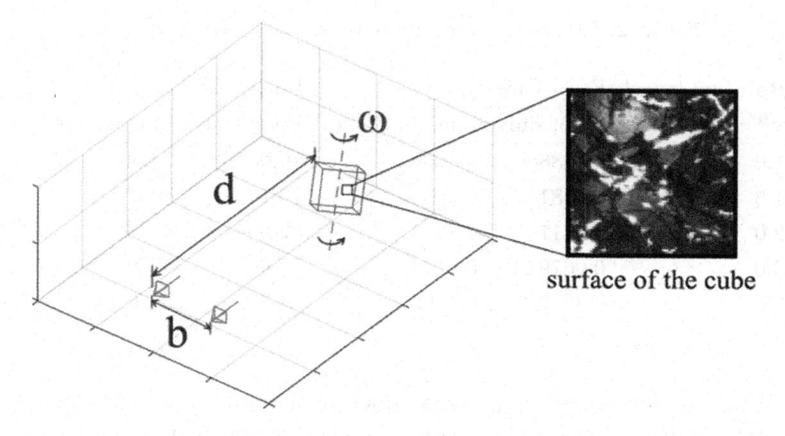

Fig. 4. Design of experimental device

constant speed. We obtained several different sequences by changing the combination of rotation speed and sample interval. In Fig. 4, baseline of stereo rig $b = 18\,\text{cm}$; distance from the rig to rotating cube $d \approx 200\,\text{cm}$; rotation speed $\omega \in [1, 14]\,\text{deg/s}$ and surface of the cube was made of reflective monochrome metallic material. Each camera produces a color picture size of $1280 \times 1024 \times 3$ pixels, and the side length of the cube is approximately 440 pixels in the image.

The experiment used five unique sequences, and the error of rotation speed was the highest priority. The basic information for each sequence is shown in Table 1.

Table 1. Information of test sequences

Index	ω (deg/s)	Sample interval (s)	Rotation speed (deg/frame)	Length (frame)
1	1	1.00	1.0	402
2	6	0.25	1.5	276
3	1	2.00	2.0	201
4	6	0.50	3.0	138
5	14	0.25	3.5	206

For the same pipeline, we had implemented two methods using different features: one used only points and the other used both points and lines. Comparison between two methods are shown in Table 2. The error reflects accuracy of the algorithm, the smaller the better. Inner is the proportion of reserved frames after filtering, reflecting the robustness of the algorithm, the higher the better.

In general, both methods achieved the motion estimation task for the disturbed surface and prove the feasibility of our pipeline. In this experiment, there were about 100–700 available feature point pairs between every two frames.

Table 2. Comparison of different feature based methods

Rotation speed (deg/frame)	Error (deg/frame)		Inner (%)	
	Point only	Line and point	Point only	Line and point
1.0	0.4688	0.4066	79.05	79.30
1.5	0.2881	0.2631	73.19	80.07
2.0	0.3417	0.2993	82.00	85.00
3.0	0.0679	0.0144	57.66	57.66
3.5	0.3200	0.1896	44.39	56.10

In addition, there were 0–25 pairs of lines, although they were few, but for better performance played a very critical role: (i) absolute error of rotation estimation was reduced. Especially when the motion between frames went larger, this reduction effect is more pronounced. (ii) combined lines helped increase the percentage of inner frames after RANSAC filtering, that means a greater probability of success and better robustness because of the increased signal to noise ratio.

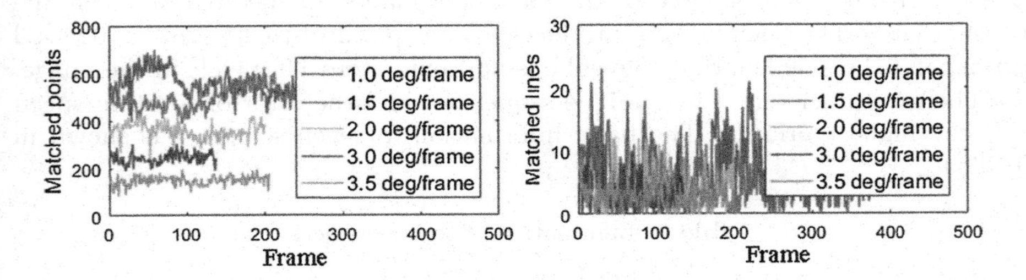

Fig. 5. Number of matched features between frames

An intuitive conclusion is that, with the increase of rotation speed, the reliability (Inner) of algorithm was declining. Because with the increase in angle, reflective surfaces begun to change in brightness even shape. This would increase the difficulty of matching, especially without valid texture information. Figure 5 supports this view, that degradation of algorithm reliability is actually the decline of matching reliability. But an unexpected phenomenon is that error went up when the rotation between frames was closer to 1.0 deg/frame. Because the angle was tiny, the real distance of the same feature point between two frames was often less than one pixel. At the same time, it would difficult to locate the feature points precisely to superpixel level. Resulting in a larger error could also be understood.

As for the time complexity, our method runs at 0.2–0.5 fps on a PC with Intel i7-6700 CPU @ 3.40 GHz. Feature extraction and optimization process took most of the time.

4 Conclusion

In this paper, we proposed a combined line and point feature based method for motion estimation, especially when the surface was disturbed by reflection and other noises. Provided image sequences from stereo camera, all the features would be converted into 3D space after matching and filtering. Then, a brand new cost function including points and lines was optimized. Post processing filtered out invalid frame and gave final output. Experiments showed that our solution could estimate the 6-dof motion stably and accurately without losing much computational efficiency.

However, this method still depends on the fundamental features of the objects. Although we had overcome some of the disturbed reflective surfaces by combining several types of features, for texture-less objects that have a small number of features, this algorithm still has limitations. A method that combines features and shading model may perform better on disturbed objects, and will be done in the future.

Acknowledgments. This work is supported by the National High Technology Research and Development Program of China (863 Program) under Grant 2015AA016402.

References

1. Bay, H., Tuytelaars, T., Van Gool, L.: SURF: speeded up robust features. In: Leonardis, A., Bischof, H., Pinz, A. (eds.) ECCV 2006. LNCS, vol. 3951, pp. 404–417. Springer, Heidelberg (2006). doi:10.1007/11744023_32
2. Canny, J.: A computational approach to edge detection. IEEE Trans. Pattern Anal. Mach. Intell. **6**, 679–698 (1986)
3. Feng, Y., Wu, Y., Fan, L.: On-line object reconstruction and tracking for 3D interaction. In: 2012 IEEE International Conference on Multimedia and Expo (ICME), pp. 711–716. IEEE (2012)
4. Hartley, R., Zisserman, A.: Multiple View Geometry in Computer Vision. Cambridge University Press, New York (2003)
5. Hirschmüller, H.: Accurate and Efficient Stereo Processing by Semi-global Matching and Mutual Information, vol. 2, pp. 807–814 (2005)
6. Jia, Q., Gao, X., Fan, X., Luo, Z., Li, H., Chen, Z.: Novel coplanar line-points invariants for robust line matching across views. In: Leibe, B., Matas, J., Sebe, N., Welling, M. (eds.) ECCV 2016. LNCS, vol. 9912, pp. 599–611. Springer, Cham (2016). doi:10.1007/978-3-319-46484-8_36
7. Men, H., Gebre, B., Pochiraju, K.: Color point cloud registration with 4D ICP algorithm. In: 2011 IEEE International Conference on Robotics and Automation (ICRA), pp. 1511–1516. IEEE (2011)
8. Nistér, D.: Preemptive RANSAC for live structure and motion estimation. Mach. Vis. Appl. **16**(5), 321–329 (2005)
9. Pilz, F., Pugeault, N., Krüger, N.: Comparison of point and line features and their combination for rigid body motion estimation. In: Cremers, D., Rosenhahn, B., Yuille, A.L., Schmidt, F.R. (eds.) Statistical and Geometrical Approaches to Visual Motion Analysis. LNCS, vol. 5604, pp. 280–304. Springer, Heidelberg (2009). doi:10.1007/978-3-642-03061-1_14

10. Rosten, E., Drummond, T.: Machine learning for high-speed corner detection. In: Leonardis, A., Bischof, H., Pinz, A. (eds.) ECCV 2006. LNCS, vol. 3951, pp. 430–443. Springer, Heidelberg (2006). doi:10.1007/11744023_34
11. Taylor, C.J., Kriegman, D.J.: Structure and motion from line segments in multiple images. IEEE Trans. Pattern Anal. Mach. Intell. **17**(11), 1021–1032 (1995)
12. Triggs, B., McLauchlan, P.F., Hartley, R.I., Fitzgibbon, A.W.: Bundle adjustment — a modern synthesis. In: Triggs, B., Zisserman, A., Szeliski, R. (eds.) IWVA 1999. LNCS, vol. 1883, pp. 298–372. Springer, Heidelberg (2000). doi:10.1007/3-540-44480-7_21
13. Wu, C.: Towards linear-time incremental structure from motion. In: 2013 International Conference on 3DTV-Conference, pp. 127–134. IEEE (2013)

Improving Object Detection with Convolutional Neural Network via Iterative Mechanism

Xin Qiu and Chun Yuan(✉)

Graduate School at Shenzhen, Tsinghua University, Shenzhen, China
qx14@mails.tsinghua.edu.cn, yuanc@sz.tsinghua.edu.cn

Abstract. The iterative mechanism is prevalent and widely used in many fields, since iterations of simple functions can make complex behaviors. But this mechanism is often overlooked by the state-of-the-art convolutional neural network (CNN)-based object detection methods. In this paper, we propose to use the iterative mechanism to improve the object detection performance of the CNN algorithms. In order to show the benefits of using the iterative mechanism in object detection from more aspects, the main contributions of our work are two aspects: Firstly, we train an iterative version of Faster RCNN to show the application of the iterative mechanism in improving the localization accuracy; Secondly, we present a prototype CNN model that iteratively searches for objects on a very simple dataset to generate proposals. The thoughtful experiments on object detection benchmark datasets show that the proposed two iterative methods consistently improve the performance of the baseline methods, e.g. in PASCAL VOC2007 test set, our iterative version of Faster RCNN has 0.7115 mAP about 1.5 points higher than the baseline Faster RCNN (0.6959 mAP).

Keywords: Object detecion · Convolutional neural network · Iterative

1 Introduction

The iterative mechanism is widely used in many fields. For example, in mathematics, iterative methods are often used to produce approximate numerical solutions to some problems. The iterative mechanism is suitable for approximating the solution of nonlinear problems by means of stepwise approximation, since iterations of simple functions can also make complex behaviors [1].

Object detection is an important problem in computer vision because of its wide range of applications and challenges. And the problem of object detection has a large non-linear search space, see [2], as the goal of object detection is to identify and localize all the objects of predefined classes. Recently, significant advances have been made on object detection. To a large extent, these advances are driven by the leading detection methods based on CNN.

These state-of-the-art detection methods can be roughly divided into two types [3]. The first type uses a two-stage detection pipeline, known as a region-based CNN (RCNN) framework. In the first stage, the RCNN methods first

D. Liu et al. (Eds.): ICONIP 2017, Part III, LNCS 10636, pp. 141–150, 2017.
https://doi.org/10.1007/978-3-319-70090-8_15

use proposal methods (e.g. Selective Search) to extract candidate regions, called "proposals", which may contain objects. But with the advent of Faster RCNN [4], a Region Proposal Network (RPN) is usually used, instead of the proposal methods. Firstly, a set of reference boxes, called "anchors", are predefined overlaid on the image. Then the RPN generates proposals by outputting two predictions for each anchor: the "objectness"[1] score and the offsets of the anchor box from the ground-truth bounding box. In the second stage, since the proposals provide only rough localization, a regressor network is used to refine these proposals to achieve precise localization and classification.

Apart from the RCNN framework, there is another type of methods that adopts a one-stage pipeline, such as YOLO [5], SSD [6]. Compared with the RCNN methods, these methods essentially use the predicted object classes and offsets of each "anchor" (or "grid" in YOLO) as the final detection results without requiring a second stage refinement operation. Their detection speed is improved, but their detection accuracy is often reduced, especially when using the same backbone network and data augmentation schemes.

Based on this observation, we reason that since the anchor boxes are a set of pre-defined fixed-size boxes, it is difficult to learn how to regress these anchors to the ground-truth bounding boxes in only one step. And empirically humans usually determine the exact boundaries of an object in an iterative way.

Thus, we propose to use the iterative mechanism to improve the performance of CNN-based object detection methods. We introduce two separate models.

Firstly, we propose to improve the localization accuracy by iterating the bounding-box regression. To this end, we reimplement the standard Faster RCNN model in Tensorflow and train an iterative version using the PASCAL VOC dataset. Our experiments show that the iterative version of Faster RCNN (IF-RCNN) has 0.7115 mAP about 1.5 points higher than the baseline Faster RCNN that obtains 0.6959 mAP.

Secondly, in order to show the benefits of using the iterative mechanism in object detection from more aspects, we also propose an iterative CNN model that can iteratively generates object proposals. As humans usually find objects in an iterative way. Our experiments, on a simple dataset (Brainwash), show that the iterative proposal generation method (IPG-CNN) has 0.69 mAP about 2 points higher than the baseline methods. Note that though IPG-CNN is a toy model that works on the simple dataset, we argue that it can still show some of the application value of the iterative mechanism in proposal generation.

2 Related Work

The iterative mechanism has been used for different applications including attention-based detection methods and question-answering [7]. Our work on improving the localization accuracy is perhaps most similar to the Attentional Object Detection Network (AOD) [8]. AOD propose a model that firstly places

[1] "Objectness" measures the extent to which an image patch belongs to the foreground or background.

a sequence of glimpses at different locations in the image and then synthesizes these information to improve detection accuracy. But AOD mainly focuses on the attention mechanism and achieves a lower mAP than us.

To some extent, we think that the attention-based methods also use the iterative mechanism. [9] is one of the most influential models. [9] classifies digit numbers by sequentially selecting image patches. [10] trains an agent network to determine the most accurate bounding box by iteratively choosing an action to deform a bounding box.

While most of the recent object detection explores two directions to improve performance (i.e. making the backbone CNN deeper [11] or incorporating additional training data or information), a few pure iterative mechanism based detection methods are proposed. For example, AttentionNet [12] detects objects inside an image by iteratively classifying quantized directions to approximate the bounding boxes. But it is a single category detector, which makes detection inefficient. Another recent attempt is G-CNN [2]. G-CNN is an object detection technique that works without proposal algorithms. It performs detection by moving and scaling grids (i.e. anchors) towards objects iteratively. But G-CNN uses Fast-RCNN as its base framework and achieves lower mAP than us, too.

3 Iterative CNN for Object Detection

We first introduce an iterative version of Faster RCNN model that iteratively refines bounding boxes and then introduce another CNN model that iteratively generates proposals. Note that, these two models are independent, and we introduce both of them in this paper to show the benefits of using iterative mechanisms in object detection from more aspects.

3.1 Iterative Bounding Box Refinement

Network Design. We propose a Faster RCNN based iterative bounding box refinement model (IF-RCNN), as shown in Fig. 1. We decompose the Faster RCNN into two parts: the global part and the regression part. IF-RCNN uses the regression part to iteratively refine the predicted bounding boxes towards the ground truth boxes. The backbone of the global part can be any CNN network (e.g. VGG, ResNet101 [11]). In the standard Faster RCNN, the global and the regression parts are only run once, while we recommend iterating the same regression part twice during both the training and testing phases, which can lead to the most accurate detection in our experiments.

At every iteration t, for each proposed region box, the regression network uses an ROI layer to extract a feature vector with fixed dimension (7×7) from the computed global feature maps that within the box. The extracted feature vector is then fed into a fully connected network. Then, two predictions are made: C^t and B^t, where C represents class probabilities and B represents the bounding box. We use the same weights for regression network at each time step.

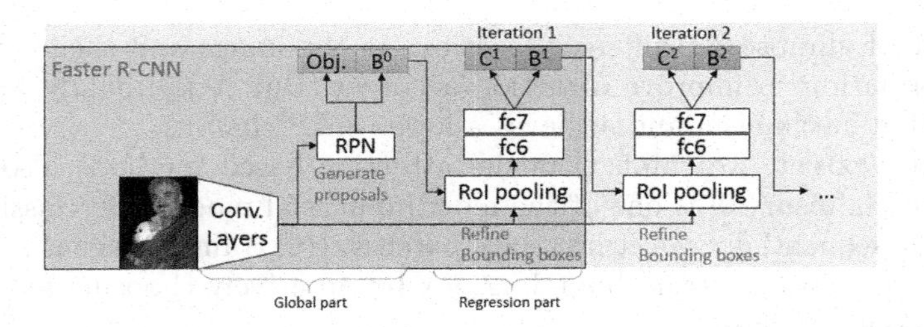

Fig. 1. Iterative bounding box refinement. $(Obj., B^0)$ represents the objectness score and the bounding box of each proposal. C represents the predicted object class.

In this way, a very simple strategy, i.e. we train a standard Faster RCNN model but test it with two iterative refinement steps, may provide a slight performance boost (to a maximum of 0.4 mAP). If further trained with our training methods, the iterative version of Faster RCNN can achieve 71.15% mAP, surpassing the standard Faster RCNN (69.59%) by 1.5% mAP.

Training Methods. In order to achieve better performance, we propose two essential improvements to the original training methods presented in Faster RCNN, as follows.

Firstly, during training, we define the loss function over all the iterative steps, since IF-RCNN iteratively refines bounding boxes towards ground-truths as accurate as possible.

$$Loss(C, B) = \sum_t \frac{1}{N_{cls}^t} \sum_i L_{cls}(c_i^t, C^*) + \sum_t \frac{1}{N_{box}^t} \sum_i L_{box}(b_i^t, B^*) \quad (1)$$

where i is the index of a proposal in a mini-batch, c the predicted class label, b the 4 parameterized coordinates of the predicted proposal bounding box, (C^*, B^*) the ground-truth and t the iteration step.

During training, each ground truth is assigned to at least one anchor according to the training method described in Faster RCNN and the assignment is not changed during the subsequent iterations of refinement.

Secondly, inspired by the ϵ-greedy policy which is widely used in reinforcement learning in continuous environments, we combine a training technique that at each time step we use *i.i.d.* Gaussian noises α to jitter the predicted bounding boxes where we progressively decrease the standard deviation of the noise.

$$B_{noised}^t = B_{pred}^t \cdot (\mathbf{1} + \alpha) = (x_{pred}, y_{pred}, w_{pred}, h_{pred}) \cdot (\mathbf{1} + \alpha) \quad (2)$$

We train IF-RCNN for 70k steps with a learning rate of 0.001 which is decreased by 10 at the 60k steps. And the standard deviation of the noise is set to be linearly decreased from 0.2 to 0 over the first 60k training steps. By using this technology, the regressor network will not only be trained by more diverse training samples, but will also gradually learn from its own successes or failures. Experiment results show that this improvement is essential.

3.2 Iterative Proposal Generation

In this section, we present a prototype iterative proposal generation model (IPG-CNN) that can generate proposals in a simple dataset (Brainwash). Nevertheless, we argue that this toy model can still show some of the application value of the iterative mechanisms in proposal generation.

Network Design. IPG-CNN is composed of three parts, as depicted in Fig. 3. To better explain the network structure, we firstly show the behavior of IPG-CNN during test time, Fig. 2.

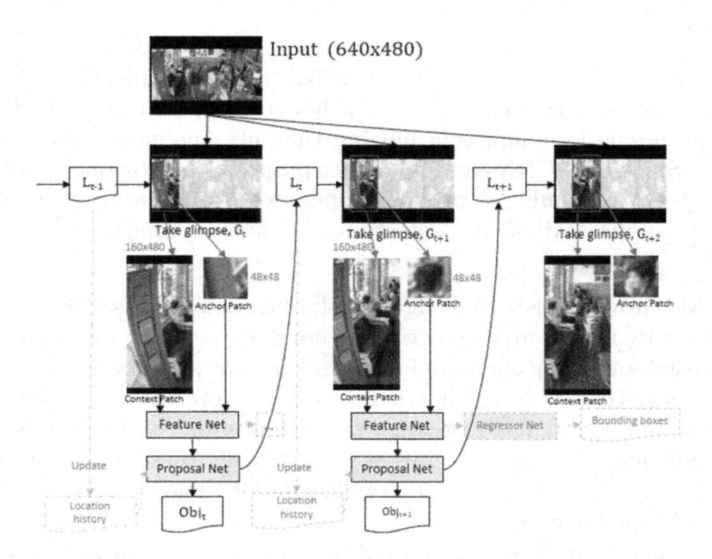

Fig. 2. In IPG-CNN, a Proposal Network is trained to search for locations L that may contain objects by adaptively "moving its fixation point" from left to right until the edge of the image, like attention mechanisms. At each time step t, given the location L_{t-1} of the last fixation point and the input image, IPG takes a "glimpse" G_t. Inspired by [9], each "glimpse" is a retina-like representation that consists of two resolution image patches: the anchor patch that also serves as an anchor and the context patch. Then, the Proposal Network outputs two predictions: the objectness score Obj_t of the anchor patch and the next glimpse location L_t.

Similar to Faster RCNN, when the predicted objectness score of an anchor patch is higher than 0.5, a Regressor Network is used to refine the anchor box to get an accurate bounding box. And we also marks the image with the predicted bounding box (see Fig. 2). These marks and the Location History vector act as an inhibition-or-return (IoR) mechanism, which have been widely applied in visual attention works, to prevent the model from endlessly attending to positions with strong stimulus. The Location History is a vector that records the last 10 locations.

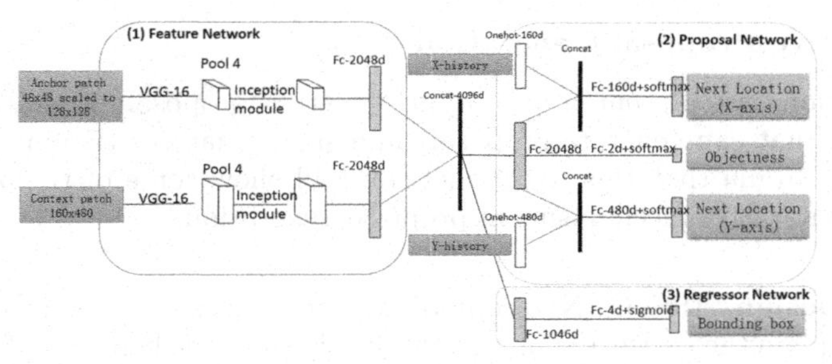

Fig. 3. The network architecture of IPG-CNN

That is, instead of using a set of fixed anchor boxes that are evenly distributed in the image and then evaluating every anchor to generate proposals, IPG-CNN generates proposals by adaptively moving the only one active anchor towards each object. Since IPG-CNN can find object proposals one by one which ensures that each object can only be proposed once, we do not need a normal post-processing to suppress duplicate detections, i.e. non-maximum suppression.

Training Methods. It is a challenging task to train a Proposal Network to find all the objects by using only one active anchor because there are any number of randomly distributed objects in an image and each object can be found through various search paths. We actually tried to use Reinforcement Learning to train the network, but found it to be very time consuming and unstable. Indeed, it is even difficult to train an iterative attention network in MNIST (28×28 pixels) to predict more than 25 discrete glimpse locations, see [9]. And we have $160 \times 480 = 76800$ locations.

To simplify this problem, we limit IPG-CNN to detect objects from left to right. By specifying the order in which objects are detected, the "fastest" path of "traversing" all objects is determined. Then the Proposal Network can be trained with supervised learning. Nevertheless, we believe that the IPG model can still show some application value of the iterative mechanism in proposal generation.

To prepare training samples, we use the idea of Experience Replay Buffer, which is also widely used in Reinforcement Learning to break correlations between successive samples. Firstly, we run the IPG-CNN model and store the predictions and the corresponding ground-truths as training samples in the Replay Buffer (of size 100k). Then training samples are randomly selected from the replay buffer. The oldest samples are discarded when the replay buffer is full.

We use the ADAM optimizer to train IPG-CNN with the total loss defined as a common multi-task loss: a summation of the regression loss of bounding boxes and two cross-entropy losses as depicted in Fig. 3. Note that, we use the softmax layers to predict locations for better accuracy.

4 Experiments

4.1 Iterative Bounding Box Refinement

For a fair comparison, we use the same experimental setup to train the iterative versions of Faster RCNN model and the standard (i.e. baseline) Faster RCNN model.

Experimental Setup. We trained these models on the trainval sets of PASCAL VOC 2007 and tested it on its testset. We re-implement the standard (i.e. baseline) Faster RCNN model in TensorFlow following the details presented in the Faster RCNN paper closely. But instead of using the 4-stage training procedure, we use the end-to-end training approach to train the model and we also use Tensorflow's 'crop_and_resize' operation to replace the original ROI Pooling layer. Conditioned on the backbone network, the mAP of the baseline Faster RCNN we reproduced is 0.6959 (VGG16) and 0.743 (ResNet101).

Experimental Results. We first show that, in our experiments, we can even improve the performance of the standard Faster RCNN by simply increasing the iterative steps to 2 during test time, as shown in Table 1.

Table 1. The standard Fasert R-CNN (VGG16), with different iterative steps during the test time.

Steps	mAP	aero	bike	bird	boat	bottle	bus	car	cat	chair	cow	table	dog	horse	mbike	person	plant	sheep	sofa	train	tv
1	69.59	66.9	77.7	**67.5**	**57.6**	**56.3**	80.2	80.1	**84.9**	50.8	74.8	66.8	**75.1**	80.4	72.0	**78.1**	45.8	**67.6**	66.0	70.9	**72.2**
2	**70.02**	69.3	**78.4**	67.5	56.9	56.1	**80.7**	80.1	84.2	**50.9**	73.7	**68.8**	75.1	**80.5**	**74.1**	78.0	**46.2**	67.4	**66.7**	75.7	70.0
3	69.52	**69.3**	77.4	66.7	57.6	55.7	80.3	**80.3**	79.7	50.0	**75.7**	67.9	74.8	79.2	72.7	77.6	45.6	67.3	65.8	**76.5**	70.4

We show the detection results of IF RCNN on PASCAL VOC 2007 test set using different setups. We use IF-RCNN (train = A, test = B) to represent an iterative Faster RCNN model that is trained with A refinement steps, but is tested with B steps. So, the standard Faster RCNN can also be called IF-RCNN (train = 1, test = 1) (Table 2).

Table 2. The mAP of the IF-RCNN (VGG16) models that are trained with the training method that defines loss function over all the iterative steps but does not combine the ϵ-greedy policy.

IF-RCNN mAP (%)	test = 1	test = 2	test = 3
train = 2	**70.03**	69.91	69.64
train = 3	68.55	68.86	68.28

The Number of the Iterative Steps. One may think that the more iterative steps the better performance. But according to the experiments, no improvement is achieved by training the network for a larger number of steps. One of the reason is that as the training progresses, the adjustments required for subsequent iterations are getting smaller and smaller which will lead to very imbalanced training samples. So, we limit the number of iterations to 2.

Table 3. The mAP of the IF-RCNN models that use different networks as their backbones. Our methods surpass both the two baseline models.

mAP (%)	test = 1	test = 2	test = 3	Faster RCNN baseline
IF-RCNN (VGG16) (train = 2)	70.01	71.15	69.52	69.59
IF-RCNN (ResNet101) (train = 2)	74.82	75.19	74.84	74.30

Table 3 shows the detection results of IF-RCNN models, which are trained by using both of the training methods mentioned in Sect. 3.1.

4.2 Iterative Proposal Generation

We evaluated the performance of the proposed model (IPG-CNN) using a people detection dataset (Brainwash). IPG-CNN was trained for 100K steps with a batch size of 32 and achieved a better mean average precision of 0.691 than the baselines, as shown in Table 4. And IPG-CNN does not need to suppress duplicate detection by applying non-maximum suppression, which is widely used by current object detection methods.

In this dataset, the target objects are usually small (a mean size of 23×23 pixels) while each image has a size of 640×448 pixels. Experiments showed that R-CNN based models did not appear suitable for this kind of dataset since the proposals generated by selective search achieved poor recall as argued in [13] and we use two OverFeat models presented in [13] as our baseline.

Our experiments show that most of the time IPG-CNN can detect an object in 4 steps. Figure 4 shows the process of localizing object.

The main difference between IPG-CNN and these baselines is that though IPG-CNN use only one anchor it learns where to place its anchor iteratively, based on different image content, while these baselines use a set of predefined fixed windows independent of the image content.

Table 4. mAP evaluation

Method	mAp (%)
Overfea-AlexNet	62.0
Overfea-GoogleNet	67.0
IPG-CNN	**69.1**

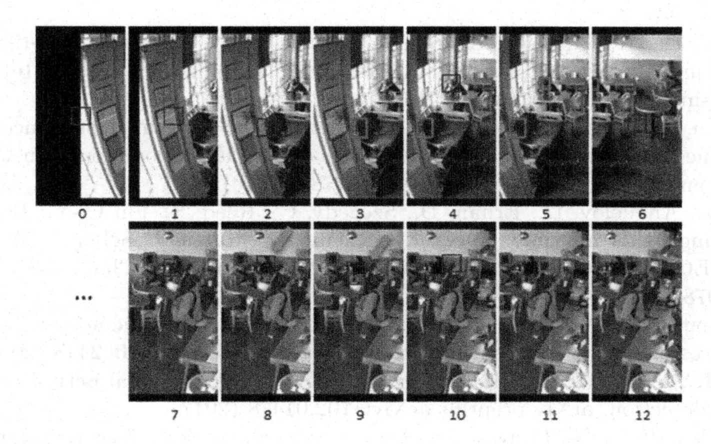

Fig. 4. Visualizations of serachs for object proposals. The red boxes represent the final detection results. Bottom row displays an example that the IPG-CNN "met" an object but failed to propose it. We think one of the reason is that the object is small and the IPG-CNN was "attracted" to a more salient object. (Color figure online)

5 Conclusion

We propose to improve object detection with convolutional neural network via the iterative mechanism. To this end, in order to show the benefits of using the iterative mechanism in object detection from more aspects, firstly, we proposed an iterative version of Faster RCNN to show the application of the iterative mechanism in improving the localization accuracy. Secondly, we presented a prototype CNN model that generates proposals on a very simple dataset by iteratively searches for objects. Our experiment results show that both the two iterative models are outperforming their baselines. This validates the usefulness of the iterative mechanism for object detection task.

Acknowledgments. This work is supported by the National High Technology Research and Development Plan (863 Plan) under Grant No.2015AA015800, the NSFC project under Grant No. U1433112, the Joint Research Center of Tencent & Tsinghua University.

References

1. Hoffman, J.D., Frankel, S.: Numerical Methods for Engineers and Scientists. CRC Press, Boca Raton (2001)
2. Najibi, M., Rastegari, M., Davis, L.S.: G-CNN: an iterative grid based object detector. In: Proceedings of the IEEE Conference on Computer Vision and Pattern Recognition, pp. 2369–2377 (2016)
3. Huang, J., Rathod, V., Sun, C., Zhu, M., Korattikara, A., Fathi, A., Fischer, I., Wojna, Z., Song, Y., Guadarrama, S., et al.: Speed/accuracy trade-offs for modern convolutional object detectors. arXiv preprint arXiv:1611.10012 (2016)

4. Ren, S., He, K., Girshick, R., Sun, J.: Faster R-CNN: towards real-time object detection with region proposal networks. In: Advances in Neural Information Processing Systems, pp. 91–99 (2015)
5. Redmon, J., Divvala, S., Girshick, R., Farhadi, A.: You only look once: unified, real-time object detection. In: Proceedings of the IEEE Conference on Computer Vision and Pattern Recognition, pp. 779–788 (2016)
6. Liu, W., Anguelov, D., Erhan, D., Szegedy, C., Reed, S., Fu, C.-Y., Berg, A.C.: SSD: single shot multibox detector. In: Leibe, B., Matas, J., Sebe, N., Welling, M. (eds.) ECCV 2016. LNCS, vol. 9905, pp. 21–37. Springer, Cham (2016). doi:10.1007/978-3-319-46448-0_2
7. Sukhbaatar, S., Weston, J., Fergus, R., et al.: End-to-end memory networks. In: Advances in Neural Information Processing Systems, pp. 2440–2448 (2015)
8. Hara, K., Liu, M.Y., Tuzel, O., Farahmand, A.M.: Attentional network for visual object detection. arXiv preprint arXiv:1702.01478 (2017)
9. Mnih, V., Heess, N., Graves, A., et al.: Recurrent models of visual attention. In: Advances in Neural Information Processing Systems, pp. 2204–2212 (2014)
10. Caicedo, J.C., Lazebnik, S.: Active object localization with deep reinforcement learning. In: Proceedings of the IEEE International Conference on Computer Vision, pp. 2488–2496 (2015)
11. He, K., Zhang, X., Ren, S., Sun, J.: Deep residual learning for image recognition. In: Proceedings of the IEEE Conference on Computer Vision and Pattern Recognition, pp. 770–778 (2016)
12. Yoo, D., Park, S., Lee, J.Y., Paek, A.S., So Kweon, I.: AttentionNet: aggregating weak directions for accurate object detection. In: Proceedings of the IEEE International Conference on Computer Vision, pp. 2659–2667 (2015)
13. Stewart, R., Andriluka, M., Ng, A.Y.: End-to-end people detection in crowded scenes. In: Proceedings of the IEEE Conference on Computer Vision and Pattern Recognition, pp. 2325–2333 (2016)

An Improved Random Walk Algorithm for Interactive Image Segmentation

Peitao Wang(✉), Zhaoshui He, and Shifeng Huang

School of Automation and Institute of Intelligent Signal Processing, Guangdong University of Technology, Guangzhou 510006, China
wapeitao@163.com

Abstract. Interactive image segmentation is an important issue in computer vision. Many algorithms have been proposed for this problem. Among them, random walk based algorithms have been proved to be efficient. However, a large number of seeds (i.e., pixels with user-specified labels) must be given in advance to achieve a desirable segmentation for such algorithms, which makes user interaction inconvenient. To solve this problem, we improve the random walk algorithm in two aspects: (1) label prior is taken into account when computing edge weights between adjacent pixels; (2) each unseeded pixel is assigned with the same label as the seed with maximum first arrival probability to reduce the bias effect of seed size. The improved algorithm can achieve a desirable segmentation with few seeds. Experiment results on natural images illustrate the accuracy of the proposed algorithm.

Keywords: Interactive image segmentation · Random walks · Label prior

1 Introduction

Interactive image segmentation, a crucial issue in computer vision, has gained considerable attention in the last decades. It allows a user to define the content to be extracted by simple user interaction. So far, many algorithms have been developed for this problem [1–17]. According to the forms of user interaction, they can be divided into three categories [1]. The first type of algorithms, such as active contour models [2] and level set method [3], need providing an initialized contour close to the true boundary. The second ones, such as intelligent scissors [4], need giving some pieces of the true boundary. The third ones, such as graph cut [5,6] and random walk based algorithms [7–11], need giving a labeling of some pixels. This type of algorithms have become popular recently due to the convenience that they can be extended to higher dimension without modification, while such an extension may be not available for the first two types [1].

Random walk based algorithms [8–10] formulate the image segmentation problem on a graph, where each pixel is represented as a node in the graph. Random walk (RW) algorithm was originally proposed for the general image

D. Liu et al. (Eds.): ICONIP 2017, Part III, LNCS 10636, pp. 151–159, 2017.
https://doi.org/10.1007/978-3-319-70090-8_16

segmentation by Leo Grady [8]. It assumes that a random walker starts from an unseeded pixel and then walks to one of its neighbor nodes randomly at each step. The first arrival probabilities that the random walker first reaches the foreground (background) seeds are computed in [8]. After that, the unseeded pixel takes a label with maximum probability. Later, Kim et al. [9] proposed a generative model based on random walks with restart (RWR) for interactive image segmentation. Differing from the RW [8], a random walker is supposed to start from each seed in RWR. Moreover, at each step, it will return to the starting seed with a restarting probability c, or walk to its neighborhood with probability $1 - c$. The steady-state probability that a random walker will finally stay at a pixel is utilized to construct a generative model for each label. Lazy random walk (LRW) was developed by Shen et al. for image segmentation and superpixel segmentation [10]. A lazy random walker will remain at the current position with a probability $1 - \alpha$ or transmit to its neighbor pixels with probability α at each step. This method simply classifies an unseeded pixel by comparing the commute time between the unseeded pixel and the seeds. The commute time is the expected time it takes a random walker to travel to another node and return to the original node.

Though good performances are obtained by the existing random walk based algorithms, a large number of seeds are required in order to achieve a desirable segmentation, which is inconvenient for users. In this paper, we propose an improved algorithm of RW [8]. It can achieve a desirable segmentation with few seeds. The rest of paper is organized as follows. The proposed improved algorithm of RW is presented in Sect. 2. The experiment results are given in Sect. 3. Finally, the conclusions are made in Sect. 4.

2 Improved Random Walk Algorithm

Consider the interactive image segmentation problem, where an image is expected to be divided into two segments: object and background. First, some important notations and their corresponding explanations are given as follows. An image is represented as a weighted undirected graph $G = (V, E)$ with nodes $v \in V$ and edges $e \in E \subseteq V \times V$. Each node v_i represents an image pixel. Each edge e_{ij}, connecting two adjacent nodes v_i and v_j, is assigned a weight w_{ij}. The degree of a node v_i is denoted by d_i and defined as the sum of weights $\sum w_{ij}$ of all edges e_{ij} incident on v_i. Next, we improve the RW algorithm [8] in two aspects: the improvement on edge weighting and the improvement on segmentation.

2.1 Improvement on Edge Weighting

In random walk based algorithms, the edge weight is used to measure the likelihood that a random walker will cross this edge. So far, many functions have been used to define edge weights, and the most popular one is the Gaussian weighting function represented as follows:

$$w_{ij} = exp(-\beta \|I_i - I_j\|^2), \tag{1}$$

where I_i and I_j are the pixel colors of nodes v_i and v_j in Lab color space, β is a constant. However, Gaussian weighting function only incorporates local information from adjacent nodes. In order to encode global and discriminant information into the edge weight, we define it as follows:

$$w_{ij} = exp(-\beta \|I_i - I_j\|^2) \times p_{ij}, \tag{2}$$

where p_{ij} denotes the probability that two adjacent nodes v_i and v_j have the same label.

We estimate p_{ij} in a very simple way. First, suppose that some seeds on foreground objects and background have been specified by users in advance. Consequently, all the nodes in a segmented image can be partitioned into two sets V_M (marked nodes, i.e., seeds) and V_U (unseeded nodes) such that $V_M \cup V_U = V$ and $V_M \cap V_U = \emptyset$. Moreover, $V_M = \{\mathcal{O}, \mathcal{B}\}$, where $\mathcal{O}$ denotes the seed set on foreground objects and $\mathcal{B}$ on background. Then, for an unseeded node $v_i \in V_U$ with pixel color I_i, we seek for the seeds $S(v_i)$ in V_M which are most similar to v_i in color, i.e.,

$$S(v_i) = \arg\min_{v_k \in V_M} \|I_k - I_i\|^2, \tag{3}$$

where I_k is the pixel color of node $v_k \in V_M$, and $\|\cdot\|$ denotes the Euclidean distance. After that, we set the label prior empirically. For a node $v_i \in V_U$, we set

$$Pr(obj|v_i) = \begin{cases} p & \text{if } S(v_i) \in \mathcal{O} \\ 1-p & \text{if } S(v_i) \in \mathcal{B} \\ 0.5 & \text{if } S(v_i) \cap \mathcal{O} \neq \emptyset \text{ } and \text{ } S(v_i) \cap \mathcal{B} \neq \emptyset \end{cases}, \tag{4}$$

$$Pr(bkg|v_i) = 1 - Pr(obj|v_i), \tag{5}$$

where $Pr(obj|v_i)$ ($Pr(bkg|v_i)$) denotes the normalized probability that node v_i fits the foreground objects (background) prior model. For any user-specified seed $v_i \in V_M$, we set

$$Pr(obj|v_i) = \begin{cases} 1 & \text{if } v_i \in \mathcal{O} \\ 0 & \text{if } v_i \in \mathcal{B} \end{cases}, \tag{6}$$

$$Pr(bkg|v_i) = 1 - Pr(obj|v_i). \tag{7}$$

To achieve a desirable segmentation, the parameter p ($0 \leq p \leq 1$) plays a critical role, which can be determined empirically according to following observations. If the foreground objects and background in a segmented image are well separable in the color space, p can be set a large value, e.g., $1-10^{-6}$. Because in this case a pixel having similar color with seeds on foreground (background) is very likely to be inside the region of foreground objects (background). Otherwise, p will be set a relatively small value if the foreground objects and background share similar color distributions. Finally, p_{ij} can be approximately estimated as follows:

$$p_{ij} = Pr(obj|v_i)Pr(obj|v_j) + Pr(bkg|v_i)Pr(bkg|v_j). \tag{8}$$

Although the estimation in Eq. (8) may be not very exact, but is good enough for the proposed algorithm.

2.2 Improvement on Segmentation

As in the traditional RW [8], image segmentation is formulated as the process of random walks. A random walker is supposed to start from an unseeded node and then walks to one of its neighbor nodes randomly at each step. However, instead of calculating the probabilities that the random walker first arrives at the foreground (background) seeds as in the traditional RW [8], the first arrival probabilities that the random walker first reaches each of the seeds are computed and a new decision rule of each pixel is used. The details including the formulae are presented in the following.

The Laplacian matrix $\boldsymbol{L}$ is defined as follows [8]:

$$L_{ij} = \begin{cases} d_i & \text{if } i = j \\ -w_{ij} & \text{if } v_i \text{ and } v_j \text{ are adjacent nodes} \\ 0 & \text{otherwise} \end{cases}, \tag{9}$$

where L_{ij} is indexed by nodes v_i and v_j. Without loss of generality, it may be assumed that the nodes in $\boldsymbol{L}$ are ordered such that the marked nodes are in first and the unseeded nodes are in second. Therefore, the Laplacian matrix $\boldsymbol{L}$ can be rewritten in block form as follows:

$$\boldsymbol{L} = \begin{bmatrix} \boldsymbol{L}_M & \boldsymbol{B} \\ \boldsymbol{B}^T & \boldsymbol{L}_U \end{bmatrix}. \tag{10}$$

Denote $\boldsymbol{x}_U^i \in \mathbb{R}_+^{|V_U| \times 1}$ as the first arrival probabilities that random walkers first arrive at the seed node $v_i \in V_M$ starting from nodes in V_U (where $|\cdot|$ denotes cardinality). Denote $\boldsymbol{x}_M^i \in \mathbb{R}_+^{|V_M| \times 1}$ as the indicating vector whose components are all zero except for the ith component which is equal to one. Then the first arrival probabilities $\boldsymbol{x}_U^i$ can be obtained by solving the following equation:

$$\boldsymbol{L}_U \boldsymbol{x}_U^i = -\boldsymbol{B} \boldsymbol{x}_M^i. \tag{11}$$

After that, each unseeded pixel is assigned with the same label as the seed with maximum first arrival probability. The proposed algorithm is summarized in Table 1.

For comparison, we briefly review the decision rule of pixels in the traditional RW [8]. Denote $\boldsymbol{x}^s \in \mathbb{R}_+^{|V_U| \times 1}$ as the probabilities that random walkers starting from nodes in V_U first reach one of the seeds with label s and $\boldsymbol{m}^s = [m_i^s]_{|V_M| \times 1}$ as the indicating vector with $m_i^s = 1$ if the label of seed v_i is s and $m_i^s = 0$ otherwise. For each label s,

$$\boldsymbol{L}_U \boldsymbol{x}^s = -\boldsymbol{B} \boldsymbol{m}^s. \tag{12}$$

Then each unseeded pixel takes a label with maximum probability in the traditional RW [8]. It can be seen that $\boldsymbol{x}^s$ is equal to the sum of $\boldsymbol{x}_U^i$ of the seeds with label s. Compared to the decision rule in [8], the proposed decision rule can reduce the bias effect of seed size. An example is illustrated in Fig. 1, which is obtained by random walks with different definitions of edge weight and different decision rules of pixels. From Fig. 1, we can see that when there exists a large difference between the number of green seeds and that of blue seeds, the proposed decision rule can obtain better segmentation results than that in [8], which demonstrates that the proposed decision rule is less dependent on the number of seeds.

Table 1. Algorithm

0: For a given image, specify seed sets $\mathcal{O}/\mathcal{B}$ for objects/background, respectively. Then $V_M = \{\mathcal{O}, \mathcal{B}\}$, $V_U = V \setminus V_M$, where V denotes the set of all pixels. Set the values of β and p, respectively.
1: **For** $v_i \in V_U$
2: $S(v_i) = \arg\min_{v_k \in V_M} ||I_k - I_i||^2$ by (3)
3: **If** $S(v_i) \in \mathcal{O}$
4: $Pr(obj|v_i) = p$, $Pr(bkg|v_i) = 1 - p$
5: **End**
6: **If** $S(v_i) \in \mathcal{B}$
7: $Pr(obj|v_i) = 1 - p$, $Pr(bkg|v_i) = p$
8: **End**
9: **If** $S(v_i) \cap \mathcal{O} \neq \emptyset$ and $S(v_i) \cap \mathcal{B} \neq \emptyset$
10: $Pr(obj|v_i) = Pr(bkg|v_i) = 0.5$
11: **End**
12: **End**
13: **For** $v_i \in V_M$
14: **If** $v_i \in \mathcal{O}$
15: $Pr(obj|v_i) = 1$, $Pr(bkg|v_i) = 0$
16: **End**
17: **If** $v_i \in \mathcal{B}$
18: $Pr(obj|v_i) = 0$, $Pr(bkg|v_i) = 1$
19: **End**
20: **End**
21: Compute the edge weights in the lattice by (2) and (8).
22: Compute the Laplacian matrix by (9).
23: Compute the first arriving probabilities by (11).
24: Assign each unseeded pixel with the same label as the seed with maximum first arrival probability.

3 Experiments and Results Analysis

In this section, we compared the improved RW algorithm with three well-known algorithms including the traditional RW [8], LRW [10] and RWR [9]. The parameters of algorithms were set as follows: the label prior probability in the proposed algorithm $p = 1 - 10^{-6}$, the leaving probability in LRW $\alpha = 1 - 10^{-4}$, the restarting probability in RWR $c = 10^{-4}$. Note that all images have been converted into the Lab color space in advance. The Lab color space is perceptually uniform,

Fig. 1. Comparison of the proposed decision rule with that in [8]. (a) the segmented images, where green/blue pixels indicate the foreground/background seeds, respectively. (b) the segmentation results is obtained by random walks with Gassuian weighting function as edge weight and the decision rule in [8]. (c) the segmentation results with Gassuian weighting function as edge weight and the proposed decision rule. (d) the segmentation results with the proposed edge weighting function and the decision rule in [8]. (e) the segmentation results with the proposed edge weighting function and the proposed decision rule. (Color figure online)

i.e., the Euclidean distance between two different colors corresponds approximately to the color difference perceived by the human eye [18]. All the experiments were performed on a computer with a 1.6-GHz Intel Xeon CPU E5-2603 v3 and 16-GB memory running on 64-bit MATLAB2015a in Windows 7.

Example 1. Weak boundary Problem
We evaluated the performance of the proposed algorithm for the weak boundary problem on synthetic images. We set $\beta = 90$. The segmentation results of algorithms are shown in Fig. 2. We can see that though LRW algorithm is slight better than ours, our algorithm is comparable to RWR and better than the traditional RW algorithm.

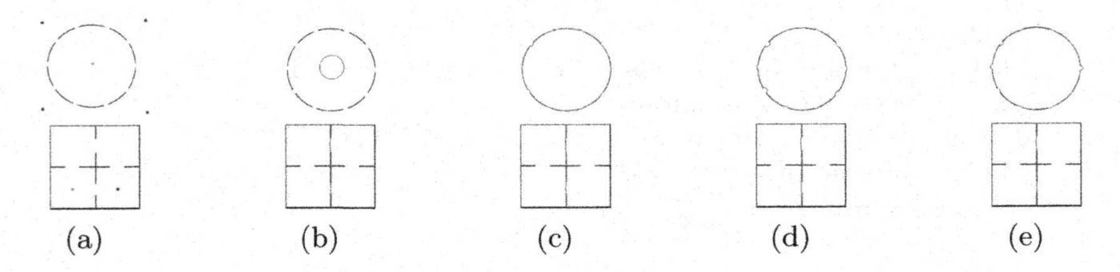

Fig. 2. Comparison of the proposed algorithm with three RW-based algorithms. (a) the segmented images, where green/blue pixels indicate the foreground/background seeds, respectively. (b)–(e) the segmentation boundaries by RW [8], LRW [10] with $\alpha = 1 - 10^{-4}$, RWR [9] with $c = 10^{-4}$ and the proposed algorithm with $p = 1 - 10^{-6}$, respectively. (Color figure online)

Example 2. Natural Image Segmentation
We evaluated the performance of the proposed algorithm on natural image segmentation. All the images, including the manual labeled ground-truth mask, are

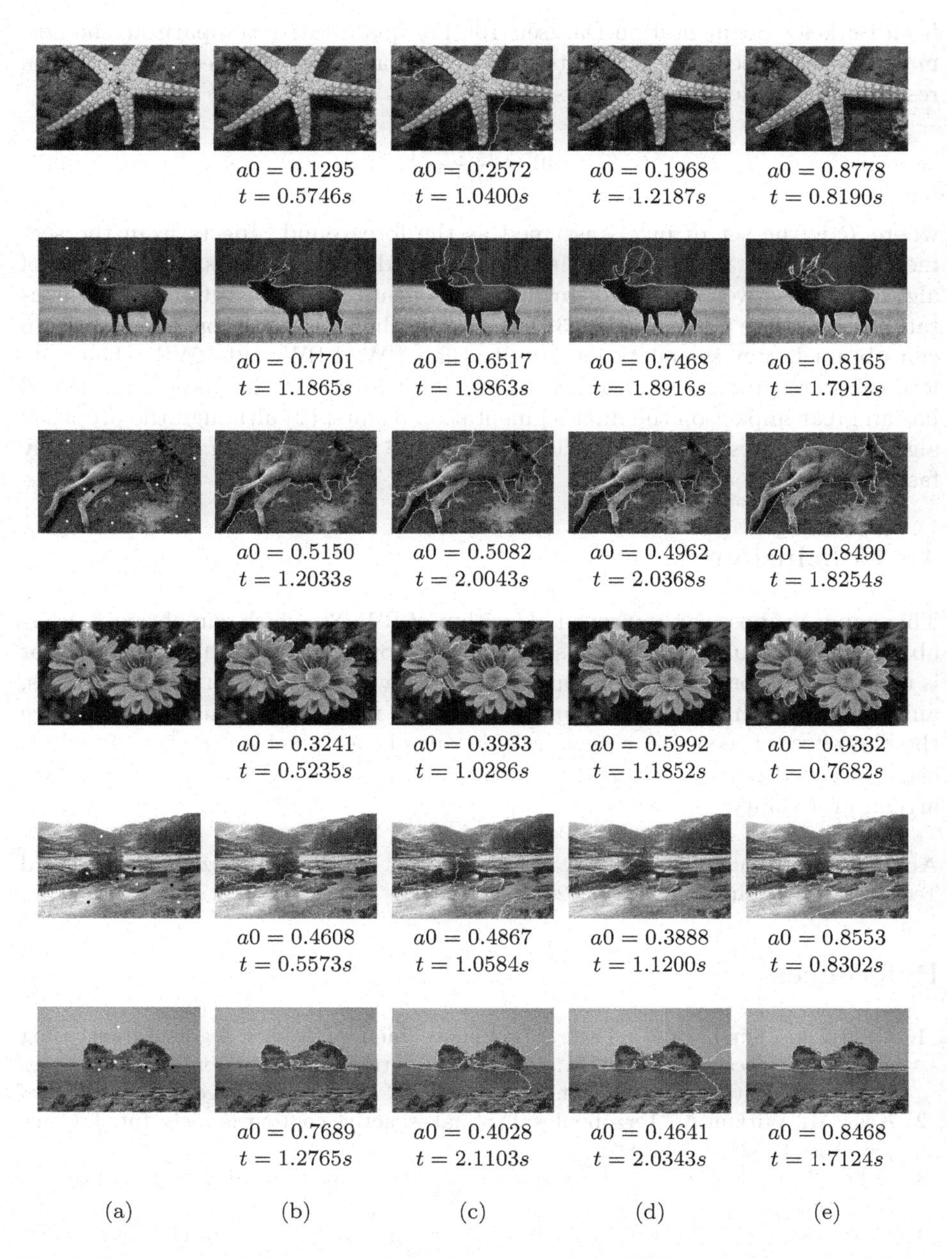

Fig. 3. Comparison of the proposed algorithm with three RW-based algorithms. (a) the segmented images, where green/blue pixels indicate the foreground/background seeds, respectively. (b)–(e) the segmentation results by RW [8], LRW [10] with $\alpha = 1 - 10^{-4}$, RWR [9] with $c = 10^{-4}$ and the proposed algorithm with $p = 1 - 10^{-6}$, respectively. (Color figure online)

from Berkeley Segmentation Dataset [19]. For quantitative comparison, the normalized overlap score $a0$ is used to measure the accuracy of object segmentation result, which is defined as follows:

$$a0 = \frac{|R \cap G|}{|R \cup G|}, \tag{13}$$

where R is the set of pixels assigned as the foreground objects from the segmentation result and G is that from the ground-truth. Besides, the runtime of algorithms also were recorded. In the experiments, we set $\beta = 60$. The segmentation results are shown in Fig. 3. We can see that: (1) the proposed algorithm can obtain better segmentation results than RW, LRW and RWR when only few seeds are provided by user, which demonstrates that the label prior indeed has an great impact on the final segmentation results; (2) although the proposed algorithm requires a little extra time to compute the label prior, it is still slightly faster than LRW and RWR.

4 Conclusions

This paper proposes an improved algorithm of RW [8], which can obtain a desirable segmentation only with few seeds. In the proposed algorithm, the label prior is taken into account when computing the edge weights between adjacent pixels, and it can be estimated in a simple, yet efficient way. Besides, in order to reduce the bias effect of seed size, a new decision rule of each pixel is proposed, in which each unseeded pixel is assigned the same label as the seed with maximum first arrival probability.

Acknowledgments. The work was supported in part by Guangzhou Science and Technology Program under Grant 201508010007.

References

1. Grady, L., Sinop, A.K.: Fast approximate random walker segmentation using eigenvector precomputation. In: 21st IEEE International Conference on Computer Vision and Pattern Recognition, pp. 1–8. IEEE Press, Anchorage (2008)
2. Kass, M., Witkin, A., Terzopoulos, D.: Snakes: active contour models. Int. J. Comput. Vis. **1**, 321–331 (1988)
3. Osher, S., Fedkiw, R.: Level Set Methods and Dynamic Implicit Surfaces. Springer, New York (2003)
4. Mortensen, E.N., Barrett, W.A.: Interactive segmentation with intelligent scissors. Graph. Models Image Process. **60**, 349–384 (1998)
5. Boykov, Y.Y., Jolly, M.-P.: Interactive graph cuts for optimal boundary and region segmentation of objects in N-D images. In: 8th IEEE International Conference on Computer Vision, pp. 105–112. IEEE Press, Vancouver (2001)
6. Boykov, Y., Funka-Lea, G.: Graph cuts and efficient ND image segmentation. Int. J. Comput. Vis. **70**, 109–131 (2006)

7. Zhou, D., Schölkopf, B.: Learning from labeled and unlabeled data using random walks. In: Rasmussen, C.E., Bülthoff, H.H., Schölkopf, B., Giese, M.A. (eds.) DAGM 2004. LNCS, vol. 3175, pp. 237–244. Springer, Heidelberg (2004). doi:10.1007/978-3-540-28649-3_29
8. Grady, L.: Random walks for image segmentation. IEEE Trans. Pattern Anal. Mach. Intell. **28**, 1768–1783 (2006)
9. Kim, T.H., Lee, K.M., Lee, S.U.: Generative image segmentation using random walks with restart. In: 10th European Conference on Computer Vision, pp. 264–275. Springer Press, Marseille (2008)
10. Shen, J., Du, Y., Wang, W., Li, X.: Lazy random walks for superpixel segmentation. IEEE Trans. Image Process. **23**, 1451–1462 (2014)
11. Dong, X., Shen, J., Shao, L., Van Gool, L.: Sub-Markov random walk for image segmentation. IEEE Trans. Image Process. **25**, 516–527 (2016)
12. Zhou, G., Xie, S., Yang, Z., Yang, J.-M., He, Z.: Minimum-volume-constrained nonnegative matrix factorization: enhanced ability of learning parts. IEEE Trans. Neural Netw. **20**, 1626–1637 (2011)
13. Yang, Z., Xiang, Y., Rong, Y., Xie, S.: Projection-pursuit-based method for blind separation of nonnegative sources. IEEE Trans. Neural Netw. Learn. Syst. **24**, 47–57 (2013)
14. Zhou, G., Yang, Z., Xie, S., Yang, J.-M.: Online blind source separation using incremental nonnegative matrix factorization with volume constraint. IEEE Trans. Neural Netw. **22**, 550–560 (2011)
15. Yang, Z., Zhou, G., Xie, S., Ding, S.: Blind spectral unmixing based on sparse nonnegative matrix factorization. IEEE Trans. Image Process. **20**, 1112–1125 (2011)
16. He, Z., Xie, S., Zdunek, R., Zhou, G., Cichocki, A.: Symmetric nonnegative matrix factorization: algorithms and applications to probabilistic clustering. IEEE Trans. Neural Netw. **22**, 2117–2131 (2011)
17. Zhou, G., Cichocki, A., Xie, S.: Fast nonnegative matrix/tensor factorization based on low-rank approximation. IEEE Trans. Signal Process. **60**, 2928–2940 (2012)
18. Leon, K., Mery, D., Pedreschi, F., Leon, J.: Color measurement in L * a * b units from RGB digital images. Food Res. Int. **39**, 1084–1091 (2006)
19. Martin, D., Fowlkes, C., Tal, D., Malik, J.: A database of human segmented natural images and its application to evaluating segmentation algorithms and measuring ecological statistics. In: 8th IEEE International Conference on Computer Vision, pp. 416–423. IEEE Press, Vancouver (2001)

Scene Recognition Based on Multi-feature Fusion for Indoor Robot

Xiaocheng Liu(✉), Wei Hong(✉), and Huiqiu Lu

College of Communication Engineering, Jilin University, Changchun 130022, China
2506977895@qq.com, hongweijilin@163.com

Abstract. In this paper, a method of scene recognition based on multi feature fusion is proposed to solve the problems of poor accuracy in scene recognition of intelligent home robot. Firstly, the H/I color model is used to extract the color feature from the scene. Secondly, the characteristics of the uniform background are extracted by the DS descriptors, and the scene of great difference is extracted using the SURF descriptors. The extracted feature descriptors are quantized using the "visual bag of words", and The SURF-DS-BOW model is generated by weighted fusion of the two feature descriptors. Finally, the multi kernel learning support vector machine (MKL-SVM) is used to fuse the color feature and the SURF-DS-BOW model to improve the accuracy of scene recognition. The experimental results show that the recognition rate of the method in indoor scene recognition is 86.4%, which is better than the relevant literature algorithm.

Keywords: Scene recognition · MKL-SVM · SURF-DS-BOW · H/I color model

1 Introduction

Scene recognition is one of the hallmark tasks of computer vision [1]. In order to realize intelligent behaviors autonomously, such as task scheduling, robot planning and task cooperation among multi robots, indoor robots must have the function of self-location. And the scene recognition is the core part for robots to complete localization. The scene recognition of mobile robot requires that the robot have the ability to make use of the visual information for environment or scene category judge, such as living room, bedroom, corridor and so on.

The existing methods of scene recognition mainly include three types: (1) based on low-level features. (2) based on the mid-level features. (3) based on the visual bag of words. In previous research, the methods based on low-level features were widely used, such as color, texture. However, with the increase requirement of scene recognition accuracy, the low-level of poor generalization ability is not suitable for today's research. Oliva [2] first used and improved the global feature Gist. Spectrum analysis is used to capture the spectral spatial properties of the image and the image features are extracted in terms of openness, roughness, naturalness, expansion, roughness. However, this feature is not prominent in the indoor scene recognition. In recent years, the "visual bag of words" model has become a research hotspot [3]. The "visual bag of

D. Liu et al. (Eds.): ICONIP 2017, Part III, LNCS 10636, pp. 160–169, 2017.
https://doi.org/10.1007/978-3-319-70090-8_17

words" model (BoW model) can be applied to image classification, by treating image features as words. In document classification, a bag of word is a sparse vector of occurrence counts of words; that is, a sparse histogram over the vocabulary. In computer vision, a bag of visual word is a vector of occurrence counts of a vocabulary of local image features. Wang [4] propose a kind of scene classification method combining visual attention and BOW model, although the accuracy of recognition is higher, but it still relies on the manual operator. Espinace [5] used image segmentation methods to separate the typical objects to infer the scenario category, but the image segmentation technique itself is a difficult technical. So this method is not suitable for indoor scene with complex environment. Yu [6] used visual saliency to extract regions of interest, and generate visual bag of words through SURF features, but it shows bad performance in real-time situation.

According to the complex and disorderly arrangement of indoor environment, this paper propose an indoor scene recognition method of multi feature fusion that com-bine low-level features with bag of words model, to solve the single method of the high rate of error recognition. First, using the H/I (Hue/Intensity) model to exacted the color feature. Then, the background of the indoor scenes was extracted by the DS (dense-sift) descriptors, the foreground is extracted by SURF descriptors, then DS-BOW model and SURF-BOW model is formed by establishing a visual bag of words. combine the two models to describe the scene information. The SURF-DS-BOW model is fused by changing the descriptor weight relationship. Finally, the H/I color histogram and the SURF-DS-BOW model are sent into the multi kernel learning support vector machine (MKL-SVM) for fusion recognition.

2 H/I Color Feature Model

Compared with other visual features, color feature has higher robustness due to its less dependence on the image size, shooting angle. So color is firstly selected as a feature channel for scene recognition.

HSV (Hue, Saturation and Value) color space that is more consistent with human eyes than RGB is selected to describe color feature [7]. Hue (H) has an angular dimension, starting at the red at 0°, passing through the green at 120° and the blue at 240°, and then wrapping back to red at 360°. Saturation (S) value range is from 0.0 to 1.0, and the larger the value, the more saturated the color. Value (V) is also often called brightness, ranging from black at value 0 to white at value 255. In order to improve real time performance, a new one-dimensional feature, called H/I color feature, is presented instead of HSV. From Cylindrical coordinate in HSV color space, the colors can be clearly distinguished with H when S and V are the maximum value. As S and V value decrease, and all colors are gradually collected to one point. As a result, it is more and more difficult to identify colors with H. To solve the problem, colors can be represented with its gray level Intensity (I) converted by coordinate transformation of

RGB model. Therefore, all pixel points in the HSV color space are divided into two subsets by the Formula (1). If the formula is satisfied, uses H, otherwise uses I.

$$s + \frac{(1 - 0.2)V}{255} > 1.0 \tag{1}$$

Figure 1(a)–(c) showed that the feature is extracted from a bedroom image with different color model. Figure 1(a) is the feature extraction in RGB model, only by changing the brightness can separate two kinds of visual color, Fig. 1(b) is the feature extraction in HSV model, which can separate color and brightness information, and HSV is more suitable for color perception than the RGB model. Figure 1(c) is the feature extraction in H/I model, As the shown from the edge of the bed, through sharpening its boundary and retaining the color of each pixel information, it determines the strength and shadow changes near the edges. As the shown from the wall and photo frame, the H/I model is further smoother than the HSV model.

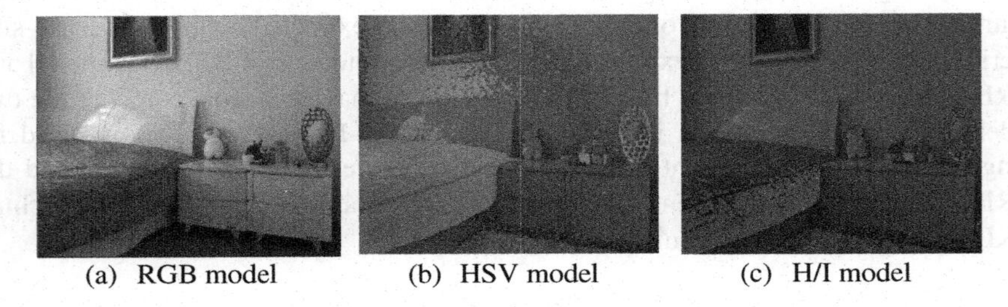

Fig. 1. Local feature is extracted from a bedroom image with different color model (Color figure online)

3 SURF-DS-BOW Model

3.1 BOW Model

In a BOW model, a scene is treated as a document and a visual dictionary is created corresponding to a scene by four-step process shown in Fig. 2(a)–(d). Firstly, images are collected for the scene, only an image is shown in Fig. 2(a) to reduce display complexity. Secondly, features are extracted by one descriptor as shown in Fig. 2(b). and then features are clustered by K-means to get a set of cluster centers which correspond to the words in the document [8]. The set of all words is called the visual dictionary which is shown in Fig. 2(c). Finally, the histogram of visual words is generated to construct the BOW model is shown in Fig. 2(d).

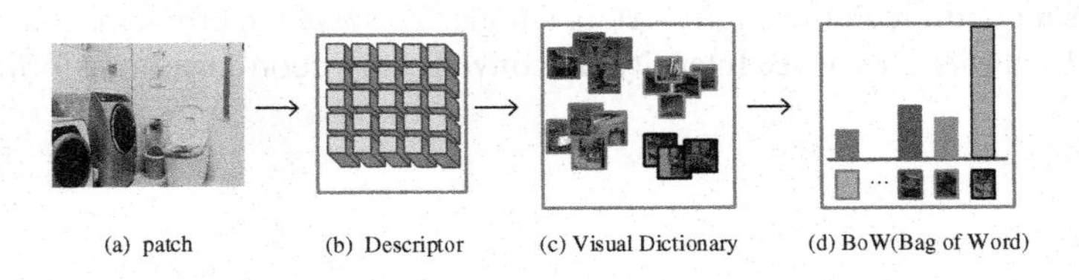

Fig. 2. Generate BOW model flow chart

In order to solve the contradiction between the large number of image features and the quick matching time, an improved K-means algorithm, called hierarchical K-means, is used to generate tree-like visual dictionary. Here K represents branch factor, that is different from the traditional K-means. The whole training image features is divided into K branches by using K-means recursively until the maximum level is reached. If the depth of tree is L, the number of nodes required can be estimated by the Formula (2). Through hierarchical K-means clustering, K cluster centers are obtained as word lists.

$$\sum\nolimits_{i=1}^{L} K^{i} = \frac{K^{L+1} - K}{K - 1} \approx K^{L} \tag{2}$$

3.2 DS and SURF Descriptor

Both DS descriptor and SURF descriptor are adopted to extract features in BOW model. The former is based on mesh segmentation image and then suitable for representing background which has similar style in large area. The latter is good at describing objects against background.

The DS descriptor has evolved from sift descriptor, in which an interest point is still described with a 128-dimension vector. But the big difference between them is how to select interest points. In the DS, an image is divided into m * n grids, and the center of each grid is chosen as the interest point. The experimental results in reference [9] show that the performance of DS is superior to that of SIFT.

SURF is an improved version of SIFT proposed by Bay [10] in order to decrease computation cost. It is divided into the following steps: Constructing the Gaussian pyramid scale space based on Hessian matrix and integral image, finding candidate points via non-maximal suppression, interpolated locating the extreme points known as the interest points accurately, extracting descriptor vectors of length 64 with Haar wavelet relative to their dominant direction. As a result, the SURF descriptor is invariant to rotation, scale, brightness, after reduction to unit length, contrast.

DS and SURF descriptors are applied to the images of kitchen respectively shown in Fig. 3. It is obvious that DS and SURF descriptors are complementary.

(a) original image

(b) DS features

(c) SURF features

Fig. 3. Features extraction with DS and SURF descriptors

3.3 BOW Based on SURF and DS Descriptor

According to the process mentioned above, two BOW models can be created by using DS or SURF descriptors respectively, recorded as WH_S and WH_d. Obviously, it is necessary to fuse them to improve performance of scene recognition. The fusion function is

$$WH_{Sd} = (\mathrm{w}_s \times WH_S) + (w_d \times WH_d) \tag{3}$$

where w_s is the weight of SURF, w_d is the weight of DS, w_s and w_d are relative to $\mathrm{w}_s + w_d = 1$. In order to highlight the role of foreground and weaken the effect of the background, w_s and w_d can be adjusted dynamically by $r = \frac{N_S}{N_S + N_d}$, $\mathrm{w}_s = c \times r^{\eta}$, $\mathrm{w}_d = 1 - \mathrm{w}_s$, where N_s, N_d are the number of SURF and DS descriptors, c is the scaling factor, η is the adjusting factor. An example of fusing process of WH_{Sd} is shown in Fig. 4, in which the image is the same with Fig. 3.

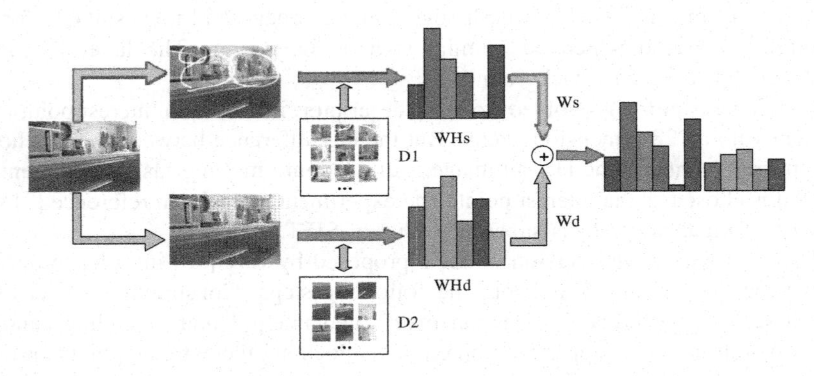

Fig. 4. SURF-DB-BOW feature generation process

4 Multiple Kernel Learning Support Vector Machine

The classification algorithm is a key element of the scene recognition. SVM often perform well, but in this paper, two types of features correspond to the best kernel function is not the same, one problem with simply adding kernels is that using uniform weights is possibly not optimal. For instance, if one kernel is not correlated with the labels at all, then giving it a positive weight will just add noise and degrade the performance. Therefore, in the present paper, a multiple kernel learning SVM, MKL-SVM, is proposed to improve scene recognition rate [11].

4.1 A Support Vector Machines

SVM uses an optimal separating hyperplane between classes by focusing on the named the support vectors which are training samples lying at the edge of the class distributions. The hyperplane H is described as $\omega^T x + b = 0$. Where ω is the normal vector of the optimal class, determined by the known sample. The left other training samples can be effectively discarded by this way. The principle of the SVM classifier is that only the training samples lying on the class boundaries are considered for discrimination. The original optimization problem is expressed as the Formulas (4) and (5), where c is the penalty term, ξ_i is the slack variable, $\xi_i \geq 0 \forall i$.

$$\Gamma = \frac{1}{2}\|\omega\|^2 + \mathrm{c}\sum\nolimits_{i=1}^{l} \xi_i \tag{4}$$

$$y_i(\omega^T x_i + \mathrm{b}) \geq 1 - \xi_i \tag{5}$$

In SVM, a feature map $\mathrm{k}(x_i, x_j) = <\phi(x_i), \phi(x_j)>$, named kernel function K, implicitly maps samples x to a feature space ϕ. For kernel algorithms, the solution of the learning problem is of the form

$$f(x) = \sum\nolimits_{i=1}^{l} \omega\,\mathrm{k}(x_i, x) + \mathrm{b} \tag{6}$$

Since there are different kernel types such as linear kernel, polynomial kernel, RBF kernel, and Sigmoid kernel are as follow.

(1) The linear kernel $\mathrm{K}(x, y) = x^T y + c$, is the simplest kernel function, it is given by the inner product $<x, y>$, plus an optional constant c.
(2) Polynomial kernel function $\mathrm{K}(x, y) = (\alpha x^T y + c)^{\mathrm{d}}$, adjustable parameters are the slope alpha α, the constant term c, and the polynomial degree d.
(3) RBF kernel function $\mathrm{K}(x, y) = \exp\left(-\frac{\|x-y\|^2}{2\sigma^2}\right)$, $\sigma > 0$, σ is the width of the kernel;
(4) Sigmoid kernel function $\mathrm{K}(x, y) = \tanh(\alpha x^T y + c)$, adjustable parameters are the slope alpha α and the intercept constant c. A common value for alpha is 1/N, where N is the data dimension;

4.2 Multiple Kernel Learning

In this paper, the H/I color model and SURF-DS-BOW model kernel functions are heterogeneous. A useful way of optimizing kernel weights optimally is MKL [12]. In the MKL approach, a convex formation of M kernels is used as follows:

$$\mathrm{K}(x_i, x_j) = \sum\nolimits_{m=1}^{M} \beta_m K_m(x_i, x_j) \tag{7}$$

Where $\beta \geq 0$, $\sum_{m=1}^{M} \beta_m = 1$ and M is the total number of kernels used. K_m is the basic kernel function, and β_m is the kernel function weight parameter. When using different features, the weight β_m represents the size of the contribution of the different characteristics to the classification. multiple kernel learning can automatically determine the weighting factor of the kernel function.

The MKL-SVM after the kernel function weighting factor can be expressed as follow, $\xi_i \geq 0 \forall i$, $\sum_{m=1}^{M} d_m = 1$, $d_m \geq 0 \forall m$. Solving the minimum Γ under the above constraint can get the solution of the problem.

$$\Gamma = \frac{1}{2}\sum\nolimits_{m=1}^{M} \frac{1}{d_m}\|\omega\|^2 + \mathrm{c}\sum\nolimits_{i=1}^{l} \xi_i \tag{8}$$

$$y_i\left[\sum\nolimits_{m=1}^{M} f_m(x_i) + b\right] \geq 1 - \xi_i \tag{9}$$

5 Experimental Results and Analysis

Experiments were carried out in an apartment shown in Fig. 5, which included five rooms: living room, bathroom, bedroom, kitchen and dining room. The robot was controlled to move to collect 120 scene images for each room under various lighting conditions, and 600 images in total. Among them, the first 80 images were used for training, and the last 40 are for recognition.

Fig. 5. Some scene images

5.1 Comparison of Recognition Rates of Different BOW Models

In this paper, DS, SURF and SURF-DS descriptor were used to construct the visual bag of words for the extracted features. The five scenarios for indoor home has been recognized by the visual bag of words. The recognition rate is shown in Fig. 6. On the whole, SURF-DS-BOW is obviously better than DS-BOW and SURF-BOW. In addition, SURF-DS-BOW can solve the problem that, the image smoothing region can't be described by SURF and the local salient regions can't be described by DS-BOW very well. By using fewer training samples in the experiment, the accuracy can be maintained above 83%, which shows the superiority of the proposed method. Dense-sift sampling step length is 10.

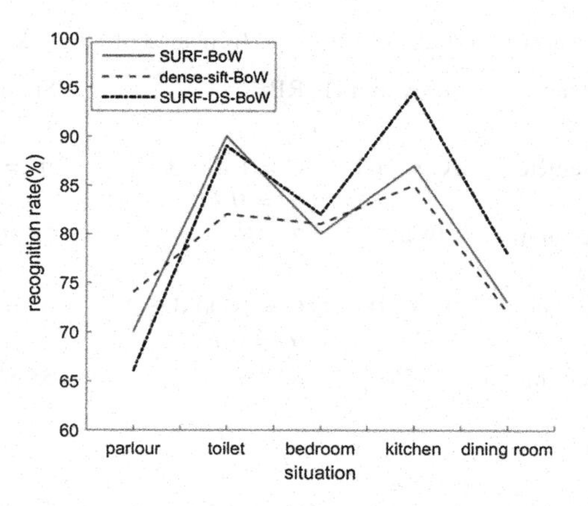

Fig. 6. Comparison of scene recognition rate

5.2 Feature Fusion Experiment of Multiple Kernel Learning

The fusion recognition is mainly divided into training stage and recognition stage. The training stage is mainly to generate MKL-SVM through multi kernel learning, the recognition stage is mainly to deals with the feature extraction of classified images and uses MKL-SVM to recognize. Experimental framework is shown in Fig. 7:

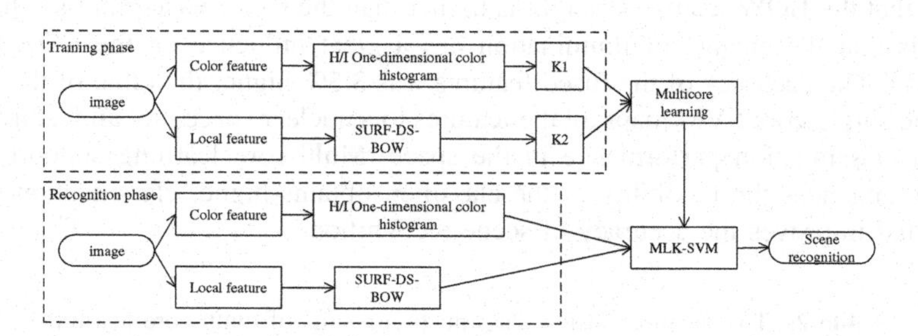

Fig. 7. Multi-feature fusion scene recognition framework

5.3 Selection of Kernel with Single Feature

Each feature was tested under four common kernel functions by SVM. The experimental results were shown in Table 1:

Table 1. Comparison on classification performance of single kernel function

Kernel / Feature	Linear	Polynomial	RBF	Sigmoid	
H/I color histogram	Parameter	**C = 100**	C = 1 d = 3 α = 0.4	σ = 0.5	C = 1000 α = 0.1
	Recognition rate	**79.1%**	76.4%	78.3%	71.4%
SURF-DS-BOW	Parameter	C = 100	C = 1000 d = 2 α = 0.7	**σ = 0.5**	C = 100 α = 0.4
	Recognition rate	80.2%	82.1%	**83.3%**	81.8%

As shown from Table 1, when Color Histogram used Linear kernel and SURF-DS-BOW used RBF kernel function, the performance was the best. This experiment has determined the optimal kernel function of each feature.

5.4 Feature Fusion Based on Multiple Kernel Learning

As the results show that the recognition performance of the two features has been improved by comparing with the methods in Table 1. It can be seen from Table 2 that, integrate the incorporation of H/I one-dimensional color information into local features can improve the recognition performance. From the weight of the kernel function, we can see that the BOW feature's weight is higher than the Color Histogram's. The main reason is that the change of illumination has a great influence on the color feature extraction. The accuracy of the fused features was 3.5% higher than that of the single. Since the single-core SVM maps all the features to a nuclear space, not all features have the best classification performance in the space. Multi-core learning support vector machine enhances the flexibility of the classifier, obtains higher classification performance and improves the accuracy of scene recognition.

Table 2. Two features fusion classification results of multi-core learning

Features fusion	SVM	MKL-SVM		
H/I histogram+ SURF-DS-BOW	Average accuracy	Kernel function	Weight	Average accuracy
	82.9%	Linear kernel+RBF	0.43:0.57	**86.4%**

6 Conclusion

This paper proposed a scene recognition method based on multi feature fusion which is combined the low-level visual feature with BOW model. The method improves the BOW model. SURF operator can describe the changing foreground information well, and DS operator can describe the local background area with varying flatness. Then the color feature is extracted by using the drop-dimensional H/I model. Finally, the color feature and SURF-DS-BOW feature are fused to recognize by MKL-SVM. Experimental results report the improvements of the accuracy in scene recognition.

References

1. Zhou, B., Lapedriza, A., Xiao, J.: Learning deep features for scene recognition using places database. In: 21st International Conference on Neural Information Processing Systems, pp. 487–495. MIT Press, Sarawak (2014)
2. Oliva, A., Torralba, A.: Building the gist of a scene: the role of global image features in recognition. J. Prog. Brain Res. **155**(2), 23–36 (2006)
3. Sivic, J., Zisserman, A.: Video Google: a text retrieval approach to object matching in videos. In: 9th IEEE International Conference on Computer Vision, p. 1470. IEEE Computer Society, France (2003)
4. Wang, R., Wang, Z., Ma, X.: Indoor scene classification based on the bag-of-words model of local feature information gain. J. IEICE Trans. Inf. Syst. **E96.D**(4), 984–987 (2013)
5. Espinace, P., Kollar, T., Soto, A.: Indoor scene recognition through object detection. In: IEEE International Conference on Robotics and Automation, pp. 1406–1413. IEEE Xplore, Anchorage, Alaska, (2010)
6. Yu, J., Hong, C., Tao, D.: Semantic embedding for indoor scene recognition by weighted hypergraph learning. J. Signal Process. **112**, 129–136 (2015)
7. Sural, S., Qian, G., Pramanik, S.: Segmentation and histogram generation using the HSV color space for image retrieval. In: International Conference on Image Processing, Proceedings, vol. 2, pp. II-589–II-592. IEEE, New York (2002)
8. Nister, D., Stewenius, H.: Robust scalable recognition with a vocabulary tree. J. **2**(10), 2161–2168 (2006)
9. Quan, Z., Rehman, S.U., Yu, Z.: Face recognition using dense sift feature alignment. J. Chin. J. Electron. **25**(6), 1034–1039 (2016)
10. Bay, H., Ess, A., Tuytelaars, T.: Speeded-up robust features. J. Comput. Vis. Image Underst. **110**(3), 404–417 (2008)
11. Lu, X., Li, X., Mou, L.: Semi-supervised multitask learning for scene recognition. J. IEEE Trans. Cybern. **45**(9), 1967–1976 (2015)
12. Lee, J., Lim, J.H., Choi, H., Kim, D.-S.: Multiple kernel learning with hierarchical feature representations. In: Lee, M., Hirose, A., Hou, Z.-G., Kil, R.M. (eds.) ICONIP 2013. LNCS, vol. 8228, pp. 517–524. Springer, Heidelberg (2013). doi:10.1007/978-3-642-42051-1_64

Correlation Filters with Adaptive Memories and Fusion for Visual Tracking

Cheng Peng[1], Fanghui Liu[1], Haiyan Yang[1], Jie Yang[1(✉)], and Nikola Kasabov[2]

[1] Institute of Image Processing and Pattern Recognition,
Shanghai Jiao Tong University, Shanghai, China
pynchon1899@gmail.com, {lfhsgre,umiiwa.y}@outlook.com,
jieyang@sjtu.edu.cn

[2] Knowledge Engineering and Discovery Research Institute,
Auckland University of Technology, Auckland, New Zealand
nkasabov@aut.ac.nz

Abstract. Correlation filter-based trackers (CFTs) with multiple features have recently achieved competitive performance. However, such conventional CFTs simply combine these features via a fixed weight. Likewise, these trackers also utilize a fixed learning rate to update their models, which makes CFTs easily drift especially when the target suffers heavy occlusions. To tackle these issues, we propose a dynamic decision fusion strategy to automatically learn the weight from the corresponding response map, and accordingly, models are adaptively updated based on a reliability metric. Moreover, a novel kernelized scale estimation scheme is proposed by exploiting the nonlinear relationship over targets of different sizes. Qualitative and quantitative comparisons on the benchmark have demonstrated that the proposed approach significantly outperforms other state-of-the-art trackers.

Keywords: Visual tracking · Decision fusion · Adaptive memories · Scale estimation

1 Introduction

Visual tracking is one of the most important research topics in computer vision. It aims to locate a given target in each frame of a video sequence. Although significant progress has been made over the decades, many challenging problems still remain due to several factors, such as illumination variations, background clutter and occlusions.

In general, tracking methods can be either generative or discriminative. Generative methods tackle the tracking problem as finding the best image candidate with minimal reconstruction error [1,14,17]. Comparably, discriminative methods train a classifier to distinguish the target from the background. The representative discriminative methods are the Correlation Filter-based Trackers (CFTs) [2,4,7,10,11]. They aim to learn a correlation filter from a set of training image patches, and then efficiently locate the target position in a new frame by

D. Liu et al. (Eds.): ICONIP 2017, Part III, LNCS 10636, pp. 170–179, 2017.
https://doi.org/10.1007/978-3-319-70090-8_18

utilizing the Discrete Fourier Transform (DFT). Bolme et al. [4] firstly introduce the correlation filter method into visual tracking. Henriques et al. [11] propose a Kernelized Correlation Filter (KCF) approach by exploiting the circulant structure of training samples. Based on [7], Bertinetto et al. [2] simply combine the scores of template and color distribution, which has achieved excellent results.

Despite their promising tracking performance, there still exist several problems within the CFTs. First, since the scale space consists of patches of different sizes, such nonlinear classification issue cannot be tackled well by linear correlation filters in [2,7]. It not only results in inaccurate target position, but also contaminates the training set during updating, which easily leads the tracker to drift. Second, although the fusion strategy of multiple models is incorporated in [2], estimations are roughly combined by a fixed fusion weight regardless of each reliability. Such fusion scheme does not make full use of each feature, and is not robust to drastic appearance variations either. Third, these former CFTs [2,4,7,10,11] utilize a fixed learning rate to update the filter coefficients and appearance models, which makes trackers unable to recourse to samples accurately tracked long before to help resist accumulated drift error, especially when the target suffers heavy occlusions.

To tackle the above issues, we attempt to build a novel correlation filter-based tracker, which investigates a kernelized scale estimation strategy, learns adaptive memories, and incorporates a dynamic fusion scheme into the tracking framework. First, we propose a novel scale estimation scheme with kernel trick to search in the scale space by exploiting the nonlinear relationship over targets of different sizes. Second, we design a dynamic decision fusion scheme based on Peak-to-Sidelobe Ratio (PSR) [12], which takes color distributions of the target into account to remedy the weaknesses of the Histogram of Oriented Gradients (HOG) [9] features. Third, different from existing CFTs above, an adaptive online learning scheme with PSR as the criterion is proposed to update the models, which helps to decontaminate the training set.

Experiments on Object Tracking Benchmark (OTB) [15] with 51 sequences demonstrate that the proposed tracker achieves superior performance when compared with other state-of-the-art trackers, especially on illumination variation, scale variation, and rotation attributes.

2 Related Work

The CFTs share the similar framework [6]. In this section, we briefly introduce the KCF method [11], which is closely related to the proposed approach.

Given the training samples $\{\mathbf{x}_i\}_{i=1}^{m}$, the KCF tracker aims to train a classifier $f(\mathbf{z}) = \mathbf{w}^\top \mathbf{z}$ that minimizes the squared error between $f(\mathbf{x}_i)$ and the corresponding function label y_i, *i.e.*,

$$\mathbf{w} = \arg\min_{\mathbf{w}} \sum_{i=1}^{m} \|f(\mathbf{x}_i) - y_i\|^2 + \lambda \|\mathbf{w}\|^2, \tag{1}$$

where $\lambda > 0$ is a regularization parameter. According to the Representer Theorem [13], the optimal weight vector $\mathbf{w}$ can be represented as $\mathbf{w} = \sum_{i=1}^{m} \alpha_i \varphi(\mathbf{x}_i)$, where φ is the mapping to a high-dimensional space with the kernel trick, and $\boldsymbol{\alpha}$ is the dual conjugate of $\mathbf{w}$ in the dual space, which can be calculated as

$$\boldsymbol{\alpha} = \mathcal{F}^{-1}\Big(\frac{\mathcal{F}(\mathbf{y})}{\mathcal{F}(\langle\varphi(\mathbf{x}), \varphi(\mathbf{x})\rangle) + \lambda}\Big), \tag{2}$$

where $\mathcal{F}$ and $\mathcal{F}^{-1}$ denote the DFT and its inverse, respectively. Given a new patch $\mathbf{z}$ from the search window in the next frame, the target location can be obtained by searching for the maximal value in response $\bar{\mathbf{y}}$, namely,

$$\bar{\mathbf{y}} = \mathcal{F}^{-1}\Big(\mathcal{F}(\boldsymbol{\alpha}) \odot \mathcal{F}(\langle\varphi(\mathbf{z}), \varphi(\bar{\mathbf{x}})\rangle)\Big), \tag{3}$$

where $\odot$ denotes element-wise product, and $\bar{\mathbf{x}}$ is the learned target appearance model.

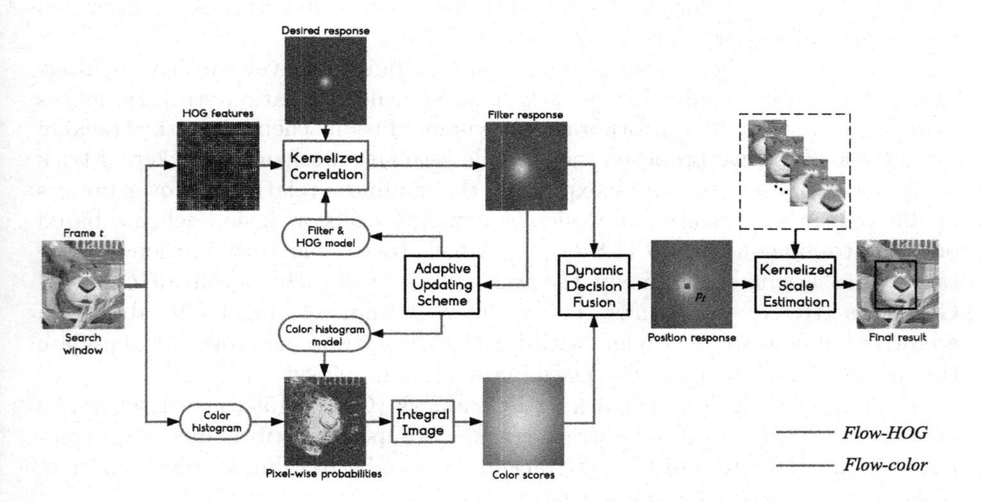

Fig. 1. The flowchart of the proposed method. The position response is incorporated with the filter response from *Flow − HOG* and the color scores from *Flow − color*. After that, the current scale is then obtained by the proposed scale estimation with kernel trick to achieve the final result. (Color figure online)

3 Proposed Approach

In this following section, we detail the proposed approach. The flowchart of the overall estimation and learning procedure is shown in Fig. 1. The proposed method is incorporated with the color information and the HOG features to comprehensively depict the target appearance. In frame t, we first obtain the filter response $\bar{\mathbf{y}}_{pf}^{t}$ based on the learned HOG models and filter coefficient, as well as the color scores $\bar{\mathbf{y}}_{ph}^t$ by utilizing the learned color histogram models respectively. A dynamic decision fusion is then proposed to merge these two estimations to derive the final position response $\bar{\mathbf{y}}_p^t$. Meanwhile, all models are updated adaptively with PSR as a criterion. The final result is obtained by the translation and the scale estimation.

3.1 Translation Estimation

Filter Response via HOG Features. The filter response yielded by the proposed approach in *Flow-HOG* process is similar to that of KCF. The filter coefficient $\boldsymbol{\alpha}_p^t$ can be derived based on 2-dimensional Gaussian-shaped function labels $\mathbf{y}_p$ using Eq. (2). Given the learned target appearance model $\bar{\mathbf{x}}_p^{t-1}$, the filter response $\bar{\mathbf{y}}_{pf}^t$ of the image patch $\mathbf{z}_p^t$ in the new frame is then calculated as in Eq. (3).

Color Scores via RGB Histograms. In *Flow-color* process, the search window is partitioned into the object region $\mathcal{O}$ and the background region $\mathcal{B}$. Similar to [3], we learn two models $P(\boldsymbol{\Omega}|M_{\mathcal{O}}^{t-1})$ and $P(\boldsymbol{\Omega}|M_{\mathcal{B}}^{t-1})$ represented by RGB histograms with 32 bins in frame $t-1$. They correspond to the color distributions of $\mathcal{O}$ and $\mathcal{B}$ respectively, and $\boldsymbol{\Omega}$ denote pixel values.

Subsequently, the pixel-wise posteriors of the object $P(M_{\mathcal{O}}^t|\boldsymbol{\Omega})$ in frame t can be calculated using the Bayes' theorem as

$$P(M_{\mathcal{O}}^t|\boldsymbol{\Omega}) = \frac{P(\boldsymbol{\Omega}|M_{\mathcal{O}}^{t-1})P(M_{\mathcal{O}}^t)}{P(\boldsymbol{\Omega}|M_{\mathcal{O}}^{t-1})P(M_{\mathcal{O}}^t) + P(\boldsymbol{\Omega}|M_{\mathcal{B}}^{t-1})P(M_{\mathcal{B}}^t)}. \tag{4}$$

Note that the model priors $P(M_{\mathcal{O}}^t)$ and $P(M_{\mathcal{B}}^t)$ are given by $P(M_{\mathcal{O}}^t) = \frac{N_{\mathcal{O}}^t}{N^t}$ and $P(M_{\mathcal{B}}^t) = \frac{N_{\mathcal{B}}^t}{N^t}$, where $N_{\mathcal{O}}^t$, $N_{\mathcal{B}}^t$ and N^t denote the pixel numbers of object region, background region and search window, respectively. Finally, the color scores $\bar{\mathbf{y}}_{ph}^t$ in frame t can be calculated using an integral image based on the pixel-wise probabilities $P(M_{\mathcal{O}}^t|\boldsymbol{\Omega})$.

Dynamic Decision Fusion. We propose the final position response as a linear combination of the kernelized correlation response and the color scores, which arrives at

$$\bar{\mathbf{y}}_p^t = (1-\xi^t)\bar{\mathbf{y}}_{ph}^t + \xi^t\bar{\mathbf{y}}_{pf}^t, \tag{5}$$

where ξ^t is a fusion weight associated with the current PSR value of the filter response. The PSR value is defined as $\text{psr}(\mathbf{R}) = \frac{\max(\mathbf{R})-\mu_\Theta(\mathbf{R})}{\sigma_\Theta(\mathbf{R})}$, where $\mathbf{R}$ is a response map, and Θ denotes the part of $\mathbf{R}$ except the sidelobe area around the peak, the mean value and standard deviation of which are μ_Θ and σ_Θ, respectively. Since the color information resists on the target appearance variations, we associate the fusion weight ξ^t in each frame with the PSR value of current response $\bar{\mathbf{y}}_{pf}^t$ as $\xi^t = \Psi(\frac{psr(\bar{\mathbf{y}}_{pf}^t)}{psr(\bar{\mathbf{y}}_{pf}^1)})$. Herein, the function Ψ which ensures both smoothness and boundness is defined as

$$\Psi(x) = \frac{\kappa}{1+e^{-\tau x}}, \tag{6}$$

where κ and τ are tuning parameters of the logistic function.

3.2 Robust Scale Estimation with Kernel Trick

In our approach, a novel robust scale estimation is proposed by learning an extra 1-dimensional scale KCF to search in the scale space. Suppose that the current target size is $M \times N$. We extract S patches centered at the position obtained in Eq. (5) and at sizes of $\rho^n M \times \rho^n N, n \in \{\lfloor -\frac{S-1}{2} \rfloor, \lfloor -\frac{S-3}{2} \rfloor, \ldots, \lfloor \frac{S-3}{2} \rfloor, \lfloor \frac{S-1}{2} \rfloor\}$, where ρ is a scale factor. These patches are resized to the same size and vectorized to make up the scale sample $\mathbf{x}_s$. The coefficient of the proposed scale KCF is then trained as

$$\boldsymbol{\alpha}_s^t = \mathcal{F}^{-1}\Big(\frac{\mathcal{F}(\mathbf{y}_s)}{\mathcal{F}(\langle \varphi(\mathbf{x}_s), \varphi(\mathbf{x}_s) \rangle) + \lambda}\Big), \tag{7}$$

where $\mathbf{y}_s$ are $1 \times S$ Gaussian-shaped function labels. The test sample $\mathbf{z}_s^t$ is obtained in the similar way of the training sample. The corresponding scale responses $\bar{\mathbf{y}}_s^t$ can be calculated as

$$\bar{\mathbf{y}}_s^t = \mathcal{F}^{-1}\Big(\mathcal{F}(\boldsymbol{\alpha}_s^{t-1}) \odot \mathcal{F}(\langle \varphi(\mathbf{z}_s^t), \varphi(\bar{\mathbf{x}}_s^{t-1}) \rangle)\Big), \tag{8}$$

where $\bar{\mathbf{x}}_s^{t-1}$ is the learned target scale model, and the current scale is the one which maximizes $\bar{\mathbf{y}}_s^t$.

3.3 Adaptive Updating Scheme

This section introduces an adaptive model learning strategy, which aims to consider different reliable degrees of models via PSR [12] in each frame. A high PSR value of the response map usually implies that the corresponding sample suffers casual occlusion or illumination variations with a low probability. Similar to Sect. 3.1, the updating weight η_i^t in each frame can be calculated as $\eta_i^t = \Phi_i\big(\frac{psr(\bar{\mathbf{y}}_{pf}^t)}{psr(\bar{\mathbf{y}}_{pf}^1)}\big), i \in \{f, h\}$. The function Φ_i is defined as

$$\Phi_i(x) = \frac{\sigma_i}{1 + e^{-\upsilon_i x}}, \tag{9}$$

where σ_i and υ_i are tuning parameters. Accordingly, we update the filter coefficient $\boldsymbol{\alpha}_p^t$ and target appearance model $\bar{\mathbf{x}}_p^t$ as

$$\boldsymbol{\alpha}_p^t = (1 - \eta_f^t)\boldsymbol{\alpha}_p^{t-1} + \eta_f^t \boldsymbol{\alpha}_p', \tag{10a}$$

$$\bar{\mathbf{x}}_p^t = (1 - \eta_f^t)\bar{\mathbf{x}}_p^{t-1} + \eta_f^t \bar{\mathbf{x}}_p', \tag{10b}$$

where the apostrophe denotes that the model is estimated from frame t alone. The histogram models can be updated similarly as

$$P(\boldsymbol{\Omega}|M_{\mathcal{O}}^t) = (1 - \eta_h^t)P(\boldsymbol{\Omega}|M_{\mathcal{O}}^{t-1}) + \eta_h^t P(\boldsymbol{\Omega}|M_{\mathcal{O}}'), \tag{11a}$$

$$P(\boldsymbol{\Omega}|M_{\mathcal{B}}^t) = (1 - \eta_h^t)P(\boldsymbol{\Omega}|M_{\mathcal{B}}^{t-1}) + \eta_h^t P(\boldsymbol{\Omega}|M_{\mathcal{B}}'). \tag{11b}$$

Algorithm 1. Proposed tracking approach

Input: Target position p_1 and scale s_1 in the first frame.
Output: Target position p_t and scale s_t in each frame ($t \geqslant 2$).
Train $\boldsymbol{\alpha}_p^1$ using Eq. (2) and $\boldsymbol{\alpha}_s^1$ using Eq. (7).
Obtain color models $P(\boldsymbol{\Omega}|M_O^1)$ and $P(\boldsymbol{\Omega}|M_B^1)$ via RGB histograms.
while frame t exists **do**
Extract a translation sample $\mathbf{z}_p^t$ at p_{t-1} and s_{t-1}.
Calculate $\bar{\mathbf{y}}_p^t$ using Eq. (5) and get p_t.
Extract S patches at p_t and $\rho^n s_{t-1}$ to make up the scale sample $\mathbf{z}_s^t$.
Calculate $\bar{\mathbf{y}}_s^t$ using Eq. (8) and get s_t.
Update models using Eq. (10), Eq. (11), and Eq. (12).
end while

Considering the fact that the scale difference is usually quite small between two adjacent frames, we use a constant learning weight η_s $(0 < \eta_s < 1)$ to update the scale KCF model as

$$\boldsymbol{\alpha}_s^t = (1 - \eta_s)\boldsymbol{\alpha}_s^{t-1} + \eta_s \boldsymbol{\alpha}'_s, \tag{12a}$$

$$\bar{\mathbf{x}}_s^t = (1 - \eta_s)\bar{\mathbf{x}}_s^{t-1} + \eta_s \bar{\mathbf{x}}'_s, \tag{12b}$$

which also helps to cut down the computational cost. Finally, the process of the proposed method is summarized in Algorithm 1.

4 Implementation and Experiments

4.1 Experimental Setup

The proposed tracker was implemented in MATLAB on a PC with Intel Xeon E5506 CPU (2.13 GHz) and 16 GB RAM. The parameters in our implementation were set as follows. The regularization parameter $\lambda = 0.0001$; the Gaussian kernel was chosen to calculate the kernelized correlation in Eqs. (3) and (8), with the kernel width = 0.5; the size S of the proposed scale KCF was set to 33 and the scale factor $\rho = 1.02$; the parameters of the logistic function were set as $\kappa = 0.3$, $\sigma_f = 0.01$, $\sigma_h = 0.04$, and $\tau = \upsilon_f = \upsilon_h = 6$; the scale learning weight $\eta_s = 0.025$. All these parameters were fixed for all sequences.

4.2 Visual Benchmark

We test the proposed approach on OTB [15], compared with 35 state-of-the-art trackers such as Staple [2], BIT [5], CNT [16], RPT [12], KCF [11], DSST [7], CSK [10], and CN [8]. The datasets of OTB contain 51 sequences annotated with different attributes including fast motion (FM), background clutter (BC), motion blur (MB), deformation (DEF), illumination variation (IV), in-plane rotation (IPR), occlusion (OCC), out-of-plane rotation (OPR), out of view (OV), and scale variation (SV).

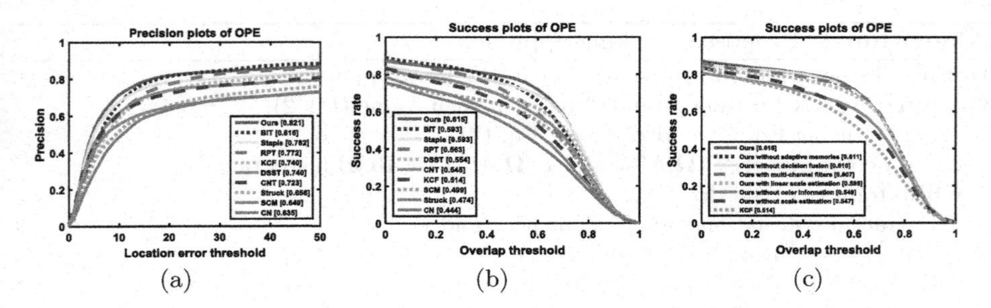

Fig. 2. (a, b) The precision and success rate plots of OPE with other state-of-the-art trackers on OTB [15]; (c) The success rate plots of OPE for the proposed approach with different key components. Specifically, we take the KCF tracker [11] as a baseline.

Fig. 3. A comparison of our proposed method with some other state-of-the-art trackers including KCF [11], DSST [7], Staple [2], and CN [8]. From left to right and top to bottom, the representative frames are from the *soccer*, *shaking*, *freeman4*, *tiger1*, *walking2*, and *girl* sequences respectively.

The precision to evaluate a tracker on a sequence is associated with the Center Location Error (CLE). It denotes the average Euclidean distance between the center locations of r_T and r_G, where r_T denotes the tracked bounding box and r_G denotes the ground truth. The final precision score is chosen from the precision plot for the threshold = 20 pixels. The accuracy metric of a tracker is the overlap score, which is defined as $\frac{|r_T \cap r_G|}{|r_T \cup r_G|}$, where $|\cdot|$ represents the number of pixels in this region. The success rate denotes the ratio of the number of successful frames (whose overlap scores are larger than a given threshold ε_0) to the total frame number. Trackers are ranked by the Area Under Curve (AUC) of each success plot.

4.3 Qualitative Comparisons

Overall Performance. The overall performance of One Pass Evaluation (OPE) is shown in Fig. 2(a) and (b). Note that only the top 10 trackers are listed.

The proposed method ranks first in both precision and success plots. It achieves an 8.1% improvement in mean CLE and 10.1% improvement in success rate over KCF [11], and achieves a 3.9% improvement in mean CLE and 2.2% improvement in success rate over a recent state-of-the-art tracker Staple [2]. Besides, the proposed approach is able to run up to 40 frames per second.

Attribute-Based Comparison. Table 1 shows the attribute-based comparison between different methods, which demonstrates that our tracker outperforms others especially in case of illumination variations, scale variations, and rotations. The tracking results of the five trackers are illustrated directly in Fig. 3. Specifically, the DSST and Staple methods do not perform well in *soccer* sequence, of which the challenging issues are illumination variations and rotations. The proposed tracker accurately tracks the woman while the CN and KCF methods drift in *walking2* sequence with the target undergoing significant scale variations. Since the proposed method updates the models adaptively, our tracker outperforms others in *girl* sequence when the girl's face is severely occluded by the man's.

Table 1. Ranked AUC scores (%) of the success rates on eleven attributes. The number of sequences associated with the corresponding attribute is shown in parenthesis. The best, second and third performance are indicated by colors.

Tracker	Attribute										
	FM (17)	BC (21)	MB (12)	DEF (19)	IV (25)	IPR (31)	OCC (29)	OPR (39)	OV (6)	SV (28)	Ave
Ours	50.8	61.2	54.0	61.7	60.1	60.7	60.2	60.0	52.8	58.6	58.0
BIT (2016) [5]	50.4	56.6	52.1	61.0	56.0	55.7	62.8	59.5	55.2	56.6	56.5
Staple (2016) [2]	50.1	55.7	52.6	60.7	56.1	57.6	58.5	56.9	51.8	54.5	55.4
CNT (2016) [16]	40.4	48.8	41.7	52.4	45.6	49.5	50.3	50.1	43.9	50.8	47.3
RPT (2015) [12]	52.5	59.6	53.2	51.4	53.4	54.8	51.1	53.7	52.5	52.3	53.4
KCF (2015) [11]	45.9	53.5	49.7	53.4	49.3	49.7	51.4	49.5	55.0	42.7	50.0
DSST (2014) [7]	42.8	51.7	45.5	50.6	56.1	56.3	53.2	53.6	46.2	54.6	51.0
CN (2014) [8]	37.3	45.3	41.0	43.8	41.7	46.9	42.8	44.3	41.0	38.4	42.2

Ablation Analysis. In order to validate the effectiveness of the key components of our approach, the comparison between the following variants is shown in Fig. 2(c): (i) the proposed method; (ii) the setting without scale estimation scheme like that in [4,11]; (iii) the setting using the linear scale estimation scheme which is applied in [2,7]; (iv) the setting without color information, similar to [4,7,11]; (v) the setting without dynamic decision fusion, similar to [2]; (vi) the setting using multi-channel filters, which is applied in [2,7]; (vii) the setting without adaptive memories, which is similar to [2,4,7,11]. The result demonstrates that each key component of the proposed approach is significantly conducive to the improvement of final tracking performance, especially the kernelized scale estimation scheme.

5 Conclusion

In this paper, we propose a novel scale estimation scheme with kernel trick by exploiting the nonlinear relationship over targets of different sizes. Besides, a dynamic decision fusion strategy is proposed to combine the color information with the HOG features, which makes the translation estimation more robust to both deformations and color changes. Moreover, we design an adaptive online learning scheme in order to decontaminate the training set, which efficiently prevents the tracker from drifting. The proposed tracker has achieved encouraging empirical performance in comparison to other state-of-the-art trackers on OTB, especially on conditions of scale variations, illumination variations, and rotations of the target.

Acknowledgments. This work was supported in part by the National Natural Science Foundation of China under Grant 61572315, Grant 6151101179, in part by 863 Plan of China under Grant 2015AA042308.

References

1. Bao, C., Wu, Y., Ling, H., Ji, H.: Real time robust L1 tracker using accelerated proximal gradient approach. In: Computer Vision and Pattern Recognition (CVPR), pp. 1830–1837 (2012)
2. Bertinetto, L., Valmadre, J., Golodetz, S., Miksik, O., Torr, P.H.: Staple: complementary learners for real-time tracking. In: Computer Vision and Pattern Recognition (CVPR), pp. 1401–1409 (2016)
3. Bibby, C., Reid, I.: Robust real-time visual tracking using pixel-wise posteriors. In: Forsyth, D., Torr, P., Zisserman, A. (eds.) ECCV 2008. LNCS, vol. 5303, pp. 831–844. Springer, Heidelberg (2008). doi:10.1007/978-3-540-88688-4_61
4. Bolme, D.S., Beveridge, J.R., Draper, B.A., Lui, Y.M.: Visual object tracking using adaptive correlation filters. In: Computer Vision and Pattern Recognition (CVPR), pp. 2544–2550 (2010)
5. Cai, B., Xu, X., Xing, X., Jia, K.: BIT: biologically inspired tracker. Trans. Image Process. (TIP) **25**(3), 1327–1339 (2016)
6. Chen, Z., Hong, Z., Tao, D.: An experimental survey on correlation filter-based tracking. Comput. Sci. **53**(6025), 68–83 (2015)
7. Danelljan, M., Häger, G., Khan, F.S., Felsberg, M.: Accurate scale estimation for robust visual tracking. In: British Machine Vision Conference (BMVC), pp. 65.1–65.11 (2014)
8. Danelljan, M., Khan, F.S., Felsberg, M., Weijer, J.V.D.: Adaptive color attributes for real-time visual tracking. In: Computer Vision and Pattern Recognition (CVPR), pp. 1090–1097 (2014)
9. Felzenszwalb, P.F., Girshick, R.B., McAllester, D., Ramanan, D.: Object detection with discriminatively trained part-based models. Trans. Pattern Anal. Mach. Intell. (TPAMI) **32**(9), 1627–1645 (2010)
10. Henriques, J.F., Caseiro, R., Martins, P., Batista, J.: Exploiting the circulant structure of tracking-by-detection with kernels. In: Fitzgibbon, A., Lazebnik, S., Perona, P., Sato, Y., Schmid, C. (eds.) ECCV 2012. LNCS, vol. 7575, pp. 702–715. Springer, Heidelberg (2012). doi:10.1007/978-3-642-33765-9_50

11. Henriques, J.F., Caseiro, R., Martins, P., Batista, J.: High-speed tracking with kernelized correlation filters. Trans. Pattern Anal. Mach. Intell. (TPAMI) **37**(3), 583–596 (2015)
12. Li, Y., Zhu, J., Hoi, S.C.H.: Reliable patch trackers: robust visual tracking by exploiting reliable patches. In: Computer Vision and Pattern Recognition (CVPR), pp. 353–361 (2015)
13. Schölkopf, B., Smola, A.J.: Learning with Kernels: Support Vector Machines, Regularization, Optimization, and Beyond. MIT press, Cambridge (2002)
14. Wang, N., Wang, J., Yeung, D.Y.: Online robust non-negative dictionary learning for visual tracking. In: International Conference on Computer Vision (ICCV), pp. 657–664 (2013)
15. Wu, Y., Lim, J., Yang, M.H.: Online object tracking: a benchmark. In: Computer Vision and Pattern Recognition (CVPR), pp. 2411–2418 (2013)
16. Zhang, K., Liu, Q., Wu, Y., Yang, M.H.: Robust visual tracking via convolutional networks without training. Trans. Image Process. (TIP) **25**(4), 1779 (2016)
17. Zhang, T., Liu, S., Ahuja, N., Yang, M.H., Ghanem, B.: Robust visual tracking via consistent low-rank sparse learning. Int. J. Comput. Vis. (IJCV) **111**(2), 171–190 (2015)

End-to-End Chinese Image Text Recognition with Attention Model

Fenfen Sheng[1,2](✉), Chuanlei Zhai[1,2], Zhineng Chen[1], and Bo Xu[1]

[1] Institute of Automation, Chinese Academy of Sciences, Beijing 100190, China
{shengfenfen2015,zhaichuanlei2014,zhineng.chen,xubo}@ia.ac.cn
[2] University of Chinese Academy of Sciences, Beijing 100190, China

Abstract. This paper presents an attention-based model for end-to-end Chinese image text recognition. The proposed model includes an encoder and a decoder. For each input text image, the encoder part firstly combines deep convolutional layers with bidirectional Recurrent Neural Network to generate an ordered, high-level feature sequence, which could avoid the complicated text segmentation pre-processing. Then in the decoder, a recurrent network with attention mechanism is developed to generate text line output, enabling the model to selectively exploit image features from the encoder correspondingly. The whole segmentation-free model allows end-to-end training within a standard backpropagation algorithm. Extensive experiments demonstrate significant performance improvements comparing to baseline systems. Furthermore, qualitative analysis reveals that the proposed model could learn the alignment between input and output in accordance with the intuition.

Keywords: Chinese images text recognition · End-to-end · Attention · Segmentation-free

1 Introduction

Chinese Image Text Recognition (ChnITR), which aims to read Chinese text in natural images, is an essential step for various commercial applications, such as geolocation, caption reading, and image-based machine translation. Despite the maturity of researches on Optical Character Recognition (OCR) [1], text recognition in natural images instead of scanned documents still remains challenging. Difficulties mainly come from the diversity of text patterns (e.g. low resolution, low contrast, and blurring) and cluttered background.

There are many ChnITR systems [2–6] focus on developing powerful word-level classifiers to improve performance. Most of them follow a three-step pipeline: first segmenting text into words, then recognizing each individual word, and finally employing post-processing to combine words results back into text. There is a trend to adopt Deep Neural Networks (DNN) for word representation learning in step 2. However, ChnITR accuracy is still confined by the word-level recognition, which lacks meaningful context-dependent information of

D. Liu et al. (Eds.): ICONIP 2017, Part III, LNCS 10636, pp. 180–189, 2017.
https://doi.org/10.1007/978-3-319-70090-8_19

whole text and further influences the following text combination step. Besides, the pre-segmentation step is quite tricky, which could bring in unavoidable troubles such as getting error text cutting points.

To deal with these problems, some methods regard ChnITR as a sequence labelling task. They are mainly based on Recurrent Neural Networks (RNN) [7]. Zhai et al. [8] combine bidirectional Long Short-Term Memory Recurrent Neural Networks (BiLSTM) with Connectionist Temporal Classification (CTC) for ChnITR without segmentation pre-processing. However, this system can not be trained end-to-end because it uses manually extracted HOG features as input, which also limit the learning of high-level features.

Inspired by recent advances on machine translation [9], this paper adopts the encoder-decoder model with attention mechanism, and proposes several meaningful modifications to accommodate ChnITR. For each input image, the encoder, i.e., deep convolutional layers (CNN) with BiRNN, is employed to generate a discriminative feature sequence. And the decoder, an attention-based network, recurrently outputs recognition result by decoding relevant contents from the feature sequence, which is determined by its attention mechanism at each step. Because of these, the whole attention-based architecture can be jointly trained end-to-end to maximize correction of the recognized result.

The contributions of this paper are as below: First, an attention-based architecture is adopted for ChnITR, which could automatically learn the alignment between input and output, and therefore dispense with the segmentation preprocessing. Second, a CNN-BiRNN structure is applied in encoder to extract both high-level and context-dependent features from the input. The whole system can be trained end-to-end without any pre or post processing thus could better explores information contained in the original text image. Experiments show that a significantly performance improvement has been achieved comparing to baseline systems. There are also a detailed ablation study to examine the effectiveness of each proposed components. Furthermore, qualitative analysis reveals that the proposed model finds the alignment between the source image and destination text as mentioned above.

2 Related Work

The ChnITR problem has been discussed for a long time but still remains unperfectly solved. By using segmentation methods, most of ChnITR systems capture individual words from text as the first step in the recognition pipeline. Mishra et al. [10] and Bai et al. [11] apply different binarization algorithms separately to distinguish text pixels from non-text's, then use existing OCR system to segment text into words and do word-level recognition work. A method from Bai et al. [4] is based on over-segmentation process. It firstly finds all potential cutting points from input text, then combines beam search with a language model to get the proper cutting points dynamically. After these steps, a word-level classifier is exploited to recognize each word. We use method [4] as a baseline to compare performances between our system and traditional ChnITR approach.

The advances of DNN for image representation encourage the development of more powerful word classifiers. A multi-layer DNN system is proposed for ChnITR in [12]. SHL-CNN system [13] employs a shared-hidden-layer deep convolutional neural network to extract word-level features. Similarly, Zhong et al. [14] put forward a multi-pooling layer on top of final convolutional layer; which is robust to spatial layout variations and deformations in multi-font printed Chinese words. And a 7-layer CNN based ChnITR architecture with synthetic data is presented by X Ren et al. [15].

Approaches above do not utilize whole-text context information because isolated word classification and subsequent words combination are treated separately. Usually, they have to design complicated optimization algorithm to train their systems and post-processing steps to refine output results. In order to leverage context-dependent information of the whole text, some researches regard text reading as a sequence labelling task avoiding explicitly segmentation. Ronaldo et al. [16] come up with a Chinese handwritten text recognition system based on Multi-Dimensional Long Short-Term Memory Recurrent Neural Network (MDLSTM-RNNs) trained with CTC. In [8], a novel BLSTM-CTC method is proposed to solve ChnITR problem. However, the use of HOG feature in [8] restricts its performance and end-to-end optimization. We use this system as another baseline to validate the effectiveness of our end-to-end system.

The proposed system is motivated by the recent advances of "attention-based" model, which has been proven successful for machine translation [9], image caption generation [17,18], and speech recognition [19]. Attention mechanism has been widely applied to the recognition of symbols sequences [20–23]. Baoguang et al. [22] extend the STN framework [24] with an attention-based sequence recognizer to train the whole model end-to-end. In [21], the authors come up with a new lexicon-free photo OCR framework that incorporates recursive CNNs for image encoding, RNNs for language modeling, and attention-based mechanism for better image feature usage. At the same time, Theodore et al. [20] present an attention-based model to transcribe complete paragraphs of text without an explicit line segmentation with a multi-dimensional LSTM network. Similar to English text recognition task, this paper attempts to solve ChnITR problem end-to-end by extending attention-based model without segmentation pre-processing.

3 An Attention-Based Model for End-to-End ChnITR

This section describes the proposed method for ChnITR, including an encoder and a decoder. The whole architecture is depicted in Fig. 1.

3.1 Encoder: Combination of Deep CNN and BiRNN

As illustrated in Fig. 1, there are several convolutional layers at the bottom of the encoder. It should be noted that, each image I has been resized into a fixed height before being fed to the encoder, while keeping its original aspect

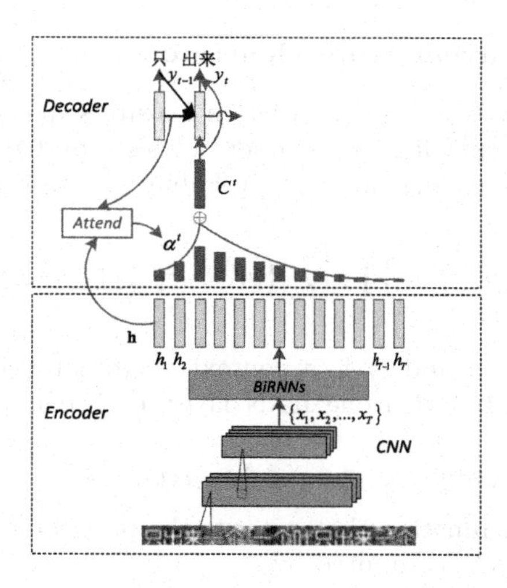

Fig. 1. The end-to-end architecture for ChnITR. The encoder combines deep CNN with BiRNN to extract feature sequence (**h**) from each input image. The decoder then generates the text result based on **h**.

ratio unchanged. After applying deep CNN to the resized image, we get a CNN sequence $\{x_1, x_2, \ldots, x_T\}$, each component of which contains high-level semantic information due to powerful feature extraction ability of deep CNN. What's more, the translation invariance property of deep CNN enables each x_i have a corresponding local image region, i.e. receptive fields.

Restricted by the size of receptive fields, the generated CNN sequence leverages limited context-dependent information. To get long-distance dependency, BiRNN is adopted on top of deep CNN. The BiRNN has also shown strong capability in learning meaningful structure from an ordered sequence. The BiRNN consists of a forward and a backward RNN. The forward RNN $\overrightarrow{f}$ reads the input sequence $\{x_1, x_2, \ldots, x_T\}$ as it is ordered (from x_1 to x_T) and calculates a sequence of forward hidden states $\left\{\overrightarrow{h_1}, \ldots, \overrightarrow{h_T}\right\}$. The backward RNN $\overleftarrow{f}$ reads the sequence in reverse order (from x_T to x_1), resulting in a sequence of backward hidden states $\left\{\overrightarrow{h_T}, \ldots, \overrightarrow{h_1}\right\}$. We obtain a feature sequence $\mathbf{h} = \{h_1, \ldots, h_T\}$ from $\{x_1, x_2, \ldots, x_T\}$ as the encoder output, each of which h_j concatenates the forward hidden state $\overrightarrow{h_j}$ and the backward one $\overleftarrow{h_j}$, i.e. $h_j = \left[\overrightarrow{h_j}; \overleftarrow{h_j}\right]$. In this way, the encoder vector h_j contains the summary of both the preceding words and the following words. As the BiRNN tends to focus on recent input x_j, the encoder vector h_j performs a better representation for the word around x_j.

3.2 Decoder: Recurrent Word Generator

The decoder generates text result, based on feature sequence $\mathbf{h} = \{h_1, \ldots, h_T\}$ from the encoder. Specifically, the decoder defines a probability over the recognition text $\mathbf{y}$ by decomposing the joint probability into ordered conditionals:

$$p(\mathbf{y}) = \prod_{t=1}^{T} p(y_t | \{y_1, \ldots, y_{t-1}\}, c_t). \tag{1}$$

where $\mathbf{y} = (y_1, \ldots, y_T)$, and c_t is a context vector at time t generated from the encoder sequence $\mathbf{h}$. With one RNN layer, each conditional probability is modeled as

$$p(y_t | y_1, \ldots, y_{t-1}, c_t) = g(y_{t-1}, s_t, c_t). \tag{2}$$

where g is a nonlinear function that outputs the probability of y_t, and s_t is the hidden state of the RNN, computed by

$$s_t = f(s_{t-1}, y_{t-1}, c_t). \tag{3}$$

It should be noted that, the probability for each target word y_t is conditioned on a distinct context vector c_t. The context vector c_t depends on the feature sequence $\mathbf{h}$:

$$c_t = \sum_{j=1}^{T} \alpha_{tj} h_j. \tag{4}$$

The weight α_{tj} of each feature vector h_j is computed by:

$$\alpha_{tj} = \frac{\exp(e_{tj})}{\sum_{k=1}^{T} \exp(e_{tk})}, \tag{5}$$

where

$$e_{tj} = a(s_{t-1}, h_j). \tag{6}$$

is an alignment model which scores how well the inputs around position j matches the output at position t. The score is based on the RNN hidden state s_{t-1} and the j-th feature vector h_j of the input sentence. The attention model a is parameterized as a feedforward neural network which is trained jointly with all the other components of the proposed system. With this approach, the information can be spread throughout the encoder feature sequence, which can be selectively retrieved by the decoder accordingly.

4 Experiments

4.1 Dataset

This paper carries out experiments on the dataset described in [4]. The dataset consists of 13 TV channels with various text patterns and backgrounds.

It includes 6633 real labeled text images (i.e., frames extracted from TV channels) and 1617636 artificial text images generated automatically by using the method in [4], since the number of labeled text images is far from enough for model training. Similarly to [4], we use all artificial text images for training, 4846 labeled text images for fine-tuning, and the rest 1787 for testing. For more details about dataset, please refer to [4].

4.2 Implementation Details

The proposed architecture is shown in Fig. 1. For the encoder, the CNN part has 7 convolutional layers, whose filter size, number of filters, stride and padding size are respectively $\{3, 64, 1, 1\}$, $\{3, 128, 1, 1\}$, $\{3, 256, 1, 1\}$, $\{3, 256, 1, 1\}$, $\{3, 512, 1, 1\}$, $\{3, 512, 1, 1\}$, $\{2, 512, 1, 0\}$. 2×2 max pooling follows the first, second, fourth, and sixth convolutional layers. The third, fifth and seventh convolutional layers are followed by a batch normalization operation separately. The input of the CNN is a resized $W \times 32$ gray scale image, and the output is a T length CNN sequence (different text image has different T because of different W). A BiRNN is on the top of the CNN, each of which has 1024 hidden units. For the decoder, the system uses a one-layer RNN network with 1024 cells and a softmax layer with 40000 output units. The number of the softmax function is equal to the class number of Chinese words in the dataset. The whole network is trained with the adadelta algorithm [25]. After that, beam search is employed to determine the recognition result that approximately maximizes the conditional probability.

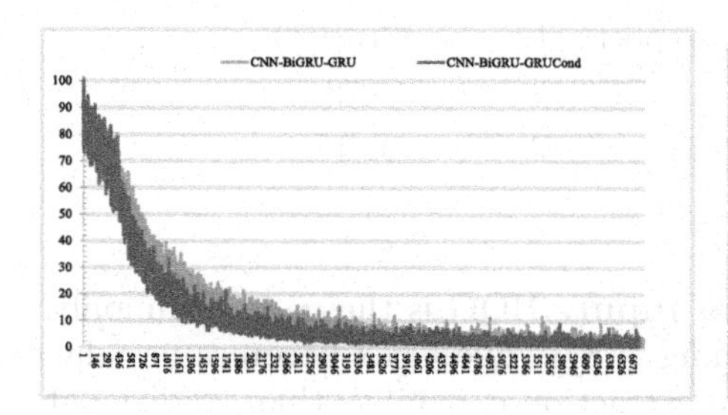

Fig. 2. Convergence process of models trained with different RNN units.

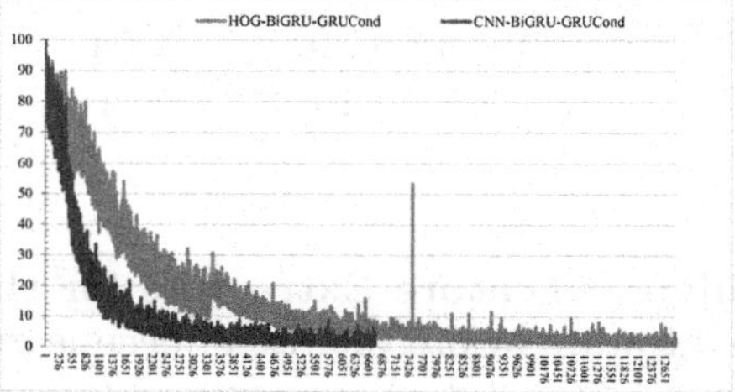

Fig. 3. Convergence process of models trained with different feature extractors.

4.3 Ablation Study

In this section, we empirically investigate the contributions made by different components in the proposed model, namely: different RNN units for model performance and different feature extractors for the input text image. After that, we cite methods [4,8] as baselines, and use character error rate (CER) as evaluation metric to compare their recognition performances.

Table 1. *Performance of models trained with different RNN units.*

Models	CNN−BiGRU−GRU	CNN−BiGRU−GRUCond
CER(%)	10.17	**9.06**

Different RNN Units for Model Performance. Gated Recurrent Unit (GRU) [26], one of RNN units, could overcome the vanishing gradient problem in standard RNN and transmit the gradient information consistently over long time. Compared to LSTM, GRU is simpler because it only has two gates but no memory cell or peephole connection. Recently, a novel conditional GRU with attention (GRUCond) is presented in [27]. One GRUCond layer with attention mechanism consists of three components: two GRU state transition blocks and an attention mechanism in between. The way of combining RNN blocks is different from normal GRU. Readers are referred to [27] for more details.

In this paper, we build two encoder-decoder models: CNN-BiGRU-GRU and CNN-BiGRU-GRUCond to validate the influence of different RNN units on the system's performance. Table 1 shows that CNN-BiGRU-GRUCond has lower character error than CNN-BiGRU-GRUCond, and converges faster as Fig. 2 indicated. As mentioned above, one GRUCond layer can be regarded as two GRU layers, which implicitly deepen and widen the network and thus enable our model to learn more abstract information in a certain extent.

Table 2. *Performance of models trained with different feature extractors.*

Feature extractors	Time (hour/epoch)	Epochs	CER(%)
HOG-BiGRU	**18.4**	6	10.78
CNN-BiGRU	29.3	**2**	**9.06**

Different Feature Extractors for the Input. HOG is the most commonly used feature before the rise of deep learning [4,8], while CNN becomes popular now because of its more powerful feature extraction capability than traditional features.

In this section, we combine HOG and CNN respectively with BiGRU as differnent feature extractors in the encoder: CNN-BiGRU and HOG-BiGRU. Results are presented in Table 2. Compared to HOG-BiGRU, CNN-BiGRU achieves better recognition performance, which agree well with the our intuition. It should also be noted that, though with more training time for each epoch, CNN-BiGRU conveges after 2 epoches, faster than HOG-BiGRU which needs 6 epoches, indicated in Fig. 3. Based on what we described above, we apply CNN-BiGRU as the feature extractor in the following experiments.

Table 3. *Performance of different methods on the 13 TV channels.*

Channels	Images	Bai's	Zhai's	Proposed	Channels	Images	Bai's	Zhai's	Proposed
AHTV	109	18.14	10.47	**10.20**	HuNtv	218	24.53	16.26	**13.10**
BJTV	159	35.69	15.56	**7.99**	JXTV	54	34.86	18.35	**9.54**
CCTV1	134	13.73	10.26	**7.31**	SDTV	109	18.63	12.08	**5.92**
CCTV4	133	14.44	**7.47**	8.74	SZTV	54	24.38	15.30	**8.72**
CQTV	155	20.13	10.07	**5.5**	XJYV	128	29.33	20.56	**13.89**
DFTV	306	14.74	12.23	**9.92**	YNTV	87	25.70	18.09	**9.21**
HeNTV	141	10.66	5.84	**3.58**	ALL	1787	21.99	13.27	**9.06**

Compared to the Baselines. Two typical systems are established as baselines in our experiments. The system proposed by [4] performs ChnITR following the three-step pipeline, i.e., over-segmentation, word classification and beam search, while the system proposed by [4] is a BiLSTM-CTC based segmentation-free method. Compared to [8] whose system can not be optimized at once because of the use of HOG features, our system achieves end-to-end in the real sense.

To quantitatively analyze our proposed pipeline, we test the text images one-by-one using our method and baseline systems respectively. The CERs(%) are listed in Table 3, which are grouped channel-by-channel. Our method gets the best results for ChnITR, where 58.80% and 31.73% relative CER reductions are observed when compared with [4, 8].

4.4 Attention Weight Visualization

Attention-based model can automatically learn the alignment information between input and output. For CNN-BiGRU-GRU and CNN-BiGRU-GRUCond, Figs. 4 and 5 provide an intuitive way to inspect this alignment by visualizing the attention weights α_{tj} in Eq. 5 respectively. Each row of a matrix in each plot indicates the weights associated with the feature sequences. From this we could see the source image and destination text align well.

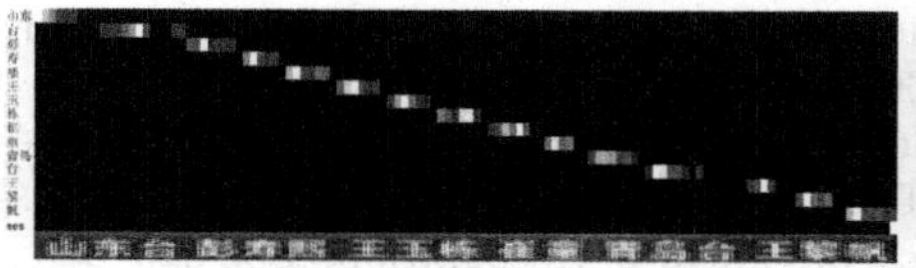

Fig. 4. One alignment exemplar obtained from CNN-BiGRU-GRU.

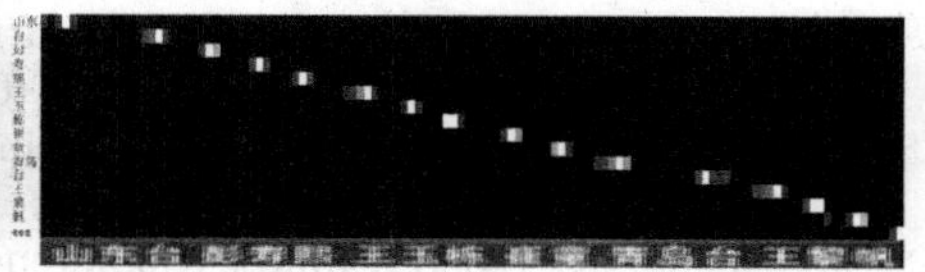

Fig. 5. One alignment exemplar obtained from CNN-BiGRU-GRUCond.

5 Conclusion

Most conventional approaches on ChnITR are based on costly pre-segmentation process, where unsatisfactory segmentation may harm the whole recognition.

This paper proposes an encoder-decoder model with attention mechanism to address this problem. Given an text image, the whole system can be trained end-to-end to maximize the probability of correct recognition. Extensive experiments show that the proposed ChnITR system significantly improve the recognition performance. Furthermore, from the qualitative analysis where we investigate the alignment generated by the proposed system, we conclude that the model can correctly align each target word with the relevant source word in the text image. In the future, we plan to investigate the end-to-end Chinese Image Text Spotting problem by combining text recognition with text detection.

Acknowledgments. This work was supported by National Key Technology R&D Program of China under No. 2015BAH53F02.

References

1. Nagy, G.: Twenty years of document image analysis in PAMI. IEEE Trans. Pattern Anal. Mach. Intell. **22**(1), 38–62 (2000)
2. Xu, L., Yin, F., Wang, Q.F., Liu, C.L.: An over-segmentation method for single-touching Chinese handwriting with learning-based filtering. Int. J. Doc. Anal. Recogn. (IJDAR) **17**(1), 91–104 (2014)
3. Saidane, Z., Garcia, C., Dugelay, J.L.: The image text recognition graph (iTRG). In: IEEE International Conference on Multimedia and Expo, ICME 2009, pp. 266–269. IEEE (2009)
4. Bai, J., Chen, Z., Feng, B., Xu, B.: Chinese image text recognition on grayscale pixels. In: 2014 IEEE International Conference on Acoustics, Speech and Signal Processing (ICASSP), pp. 1380–1384. IEEE (2014)
5. Song, Y., Chen, J., Xie, H., Chen, Z., Gao, X., Chen, X.: Robust and parallel Uyghur text localization in complex background images. Machine Vision and Applications, vol. 28, pp. 755–769. Springer, Heidelberg (2017). doi:10.1007/s00138-017-0837-3
6. Fang, S., Xie, H., Chen, Z., Zhu, S., Gu, X., Gao, X.: Detecting Uyghur text in complex background images with convolutional neural network. Multimedia Tools Appl. **76**(13), 15083–15103 (2017)
7. Graves, A., Mohamed, A.R., Hinton, G.: Speech recognition with deep recurrent neural networks. In: 2013 IEEE International Conference on Acoustics, Speech and Signal Processing (ICASSP), pp. 6645–6649. IEEE (2013)
8. Zhai, C., Chen, Z., Li, J., Xu, B.: Chinese image text recognition with BLSTM-CTC: a segmentation-free method. In: Tan, T., Li, X., Chen, X., Zhou, J., Yang, J., Cheng, H. (eds.) Chinese Conference on Pattern Recognition, pp. 525–536. Springer, Singapore (2016). doi:10.1007/978-981-10-3005-5_43
9. Bahdanau, D., Cho, K., Bengio, Y.: Neural machine translation by jointly learning to align and translate. arXiv preprint arXiv:1409.0473 (2014)
10. Mishra, A., Alahari, K., Jawahar, C.: An MRF model for binarization of natural scene text. In: 2011 International Conference on Document Analysis and Recognition (ICDAR), pp. 11–16. IEEE (2011)
11. Bai, J., Feng, B., Xu, B.: Binarization of natural scene text based on L1-Norm PCA. In: 2013 IEEE International Conference on Multimedia and Expo Workshops (ICMEW), pp. 1–4. IEEE (2013)

12. Bai, J., Chen, Z., Feng, B., Xu, B.: Chinese image character recognition using DNN and machine simulated training samples. In: Wermter, S., Weber, C., Duch, W., Honkela, T., Koprinkova-Hristova, P., Magg, S., Palm, G., Villa, A.E.P. (eds.) ICANN 2014. LNCS, vol. 8681, pp. 209–216. Springer, Cham (2014). doi:10.1007/978-3-319-11179-7_27
13. Bai, J., Chen, Z., Feng, B., Xu, B.: Image character recognition using deep convolutional neural network learned from different languages. In: 2014 IEEE International Conference on Image Processing (ICIP), pp. 2560–2564. IEEE (2014)
14. Zhong, Z., Jin, L., Feng, Z.: Multi-font printed Chinese character recognition using multi-pooling convolutional neural network. In: 2015 13th International Conference on Document Analysis and Recognition (ICDAR), pp. 96–100. IEEE (2015)
15. Ren, X., Chen, K., Sun, J.: A CNN based scene Chinese text recognition algorithm with synthetic data engine. arXiv preprint arXiv:1604.01891 (2016)
16. Messina, R., Louradour, J.: Segmentation-free handwritten Chinese text recognition with LSTM-RNN. In: 2015 13th International Conference on Document Analysis and Recognition (ICDAR), pp. 171–175. IEEE (2015)
17. Cho, K., Courville, A., Bengio, Y.: Describing multimedia content using attention-based encoder-decoder networks. IEEE Trans. Multimedia **17**(11), 1875–1886 (2015)
18. Xu, K., Ba, J., Kiros, R., Cho, K., Courville, A., Salakhudinov, R., Zemel, R., Bengio, Y.: Show, attend and tell: neural image caption generation with visual attention. In: International Conference on Machine Learning, pp. 2048–2057 (2015)
19. Chorowski, J.K., Bahdanau, D., Serdyuk, D., Cho, K., Bengio, Y.: Attention-based models for speech recognition. In: Advances in Neural Information Processing Systems, pp. 577–585 (2015)
20. Bluche, T., Louradour, J., Messina, R.: Scan, attend and read: end-to-end handwritten paragraph recognition with MDLSTM attention. arXiv preprint arXiv:1604.03286 (2016)
21. Lee, C.Y., Osindero, S.: Recursive recurrent nets with attention modeling for OCR in the wild. In: Proceedings of the IEEE Conference on Computer Vision and Pattern Recognition, pp. 2231–2239 (2016)
22. Shi, B., Wang, X., Lyu, P., Yao, C., Bai, X.: Robust scene text recognition with automatic rectification. In: Proceedings of the IEEE Conference on Computer Vision and Pattern Recognition, pp. 4168–4176 (2016)
23. Ba, J., Mnih, V., Kavukcuoglu, K.: Multiple object recognition with visual attention. arXiv preprint arXiv:1412.7755 (2014)
24. Jaderberg, M., Simonyan, K., Zisserman, A., et al.: Spatial transformer networks. In: Advances in Neural Information Processing Systems, pp. 2017–2025 (2015)
25. Zeiler, M.D.: Adadelta: an adaptive learning rate method. arXiv preprint arXiv:1212.5701 (2012)
26. Cho, K., Van Merriënboer, B., Bahdanau, D., Bengio, Y.: On the properties of neural machine translation: encoder-decoder approaches. arXiv preprint arXiv:1409.1259 (2014)
27. Sennrich, R., Firat, O., Cho, K., Birch, A., Haddow, B., Hitschler, J., Junczys-Dowmunt, M., Läubli, S., Barone, A.V.M., Mokry, J., et al.: Nematus: a toolkit for neural machine translation. arXiv preprint arXiv:1703.04357 (2017)

Application of Data Augmentation Methods to Unmanned Aerial Vehicle Monitoring System for Facial Camouflage Recognition

Yanyang Li, Sanqing Hu, Wenhao Huang, and Jianhai Zhang(✉)

College of Computer Science, Hangzhou Dianzi University, Hangzhou 310018, Zhejiang, China
lyyzzly@gmail.com, hwhzzly@gmail.com, {sqhu,jhzhang}@hdu.edu.cn

Abstract. Recently, the Unmanned Aerial Vehicle (UAV) monitoring system based on face recognition technology has attracted much attention. However, partly because of human hair changes, glasses wearing and other camouflage behavior, the accuracy of UAV face recognition system is still not high enough. In this paper, two kinds of data augmentation methods (the hairstyle hypothesis and eyeglass hypothesis) are used to expand the face dataset to make up the shortage of the original face data. In addition, the UAV locates human's face in the air from special distance and elevation, the collected face characteristics are vastly different from those in the public face library. Considering the peculiarity of UAV face localization, the data augmentation program is implemented to improve the accuracy of UAV identification of camouflage face to be 97.5%. The results show that our approach is effective and feasible.

Keywords: Data augmentation · UAV · Face camouflage · Face recognition

1 Introduction

Applying target recognition technology to UAV platform is the research hotspot over the past decades. Carnie [1] proposed a method to use target flight detection and identification for airborne drills to reduce aircraft collision accidents. Vladimir [2] tried to achieve independent tracking of dynamic targets in the UAV platform. Recently, the UAV monitoring system based on face recognition technology has received increasing attention. Davis [3] performed facial face recognition on a small face database on a UAV platform and achieved high accuracy with the AR face database. But they ignore the characteristics of UAV. UAV positioning human's face in the air with different distances and angles, so it makes facial characteristics different with public face library.

The scale of the face sample database is critical in face recognition. Currently, face-aligned data sets are extensively used for face recognition data training set, however these face-aligned data sets are often limited. In order to expand

D. Liu et al. (Eds.): ICONIP 2017, Part III, LNCS 10636, pp. 190–197, 2017.
https://doi.org/10.1007/978-3-319-70090-8_20

the data sets, many researchers put forward the concept of data augmentation. For example, cropping the human face to augment the faces by simple translation, rotation and scaling [4,5], these methods have played a role in increasing the size of the face data. However, they did not consider that the face picture had its own unique feature. The methods above increased the number of face images, but did not play the real role of augmentation. Lv [6] proposed five kinds of data augmentation methods: Landmark perturbation, hair augmentation, glasses augmentation, posture, illumination. These five kinds of methods were used to augment the scale of data for deep neural network training. Leng [7] used data augmentation on the feature level, trained by the CNN, and improved the accuracy of the classification. In this paper, two kinds of data augmentation methods (the hairstyle hypothesis and eyeglass hypothesis) were used to expand the face dataset to obtain higher accuracy even if the original face data were insufficient. In addition, when we implemented the data augmentation program, we also considered the characteristics of UAV face localization.

2 Materials and Methods

2.1 Experimental Scenarios

In this paper, the experimental face data were collected by UAV. 10 subjects participated in the experiments (5 males and 5 females, 20–25 years old). A total of six experimental scenarios were designed to collect face videos of subjects at different angles and distances. Six experimental scenarios are shown in Fig. 1.

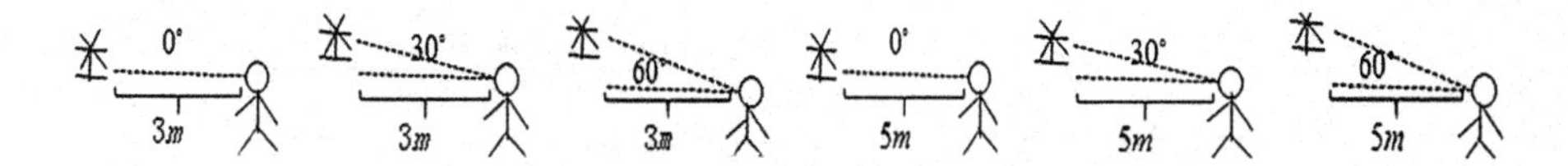

Fig. 1. Six experimental scenarios

2.2 Data Acquisition and Pre-processing

We collect data through the DJI matrice 100 platform. The system is stable and endurable. The high-definition map of DJI Matrice 100 can generate high quality video streams, and these contribute to investigate camouflage face recognition. The development platform is shown in Fig. 2.

Fig. 2. DJI matrice 100 development platform

Because of outdoor environments influence data obtainment, so we take multi-threads data acquisition method to deal with video stream, shown in Fig. 3, to reduce external disturbance. We collect data twice from each scene, take 2s video stream each time. The first 2s video stream is collected in the case of each subject without camouflage, and the second 2s video stream is collected in the case of each subject with camouflage.

We use the OpenCV toolbar [8] that Joe compiled to calculate and analyze, one can see that about 17 face pictures can be taken per second, so each subject in 2s can produce 34 pictures. Applying the multi-threads preprocessing for 2s video stream, the purpose is that removes the jitter, exposure, fisheye photos and reserves clear pictures.

In the six scenes, clear and positive face picture images were selected as training samples, which obtained by pre-processing the sum of the first pictures without camouflage in all the scenes. Then we put 10 subjects training samples into the training set, and each subject had three. Twelve images were randomly selected from the disguised images in all the scenarios. The test set consists of the same 10 subjects, with whom had 12 pictures for each.

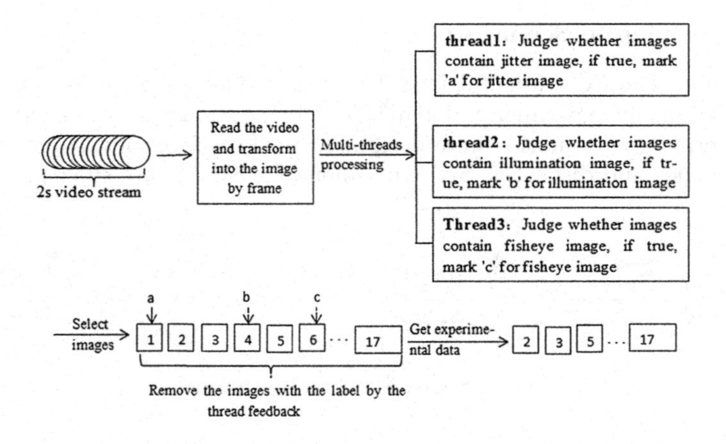

Fig. 3. Multi-threads collecting images

2.3 Data Augmentation

The first thing for UAV camouflage face recognition is to positioning key points and extracting features through face detection technology. Haar classifier [6] is one of the face detection methods to effectively extract facial features, detect contours of multiple people from one image, and implement face segmentation. After applying Haar classifier to face detection and segmentation, we locate the key points of the face by the split up snow classifier algorithm proposed by Nisson [9]. It can locate the human eyes, mouth and other key parts, and lay the foundation for the piecewise linear affine transformation.

The piecewise linear affine transformation [6,10,11] is used to augment the training set of hair camouflage. Aligning the hair template and face image simply is the first step. Then the coordinates of the key points in the hair template are transformed into the vector $[X, y]$, the key points which correspond with hair template in the face image are transformed into the vector $[x, y]$, and the piecewise linear affine transformation equations as follow:

$$\begin{array}{l} X = f(x, y) \\ Y = g(x, y) \end{array} \tag{1}$$

For $X = f(x, y)$, we use the linear interpolation method to achieve affine transformation, firstly, selecting three points from key points $[X, Y]$ in the face image forms a triangle, next selecting three points from the key points of hair template which correspond with the three points of face image. According to the following equation to achieve affine transformation:

$$Ax + By + CX + D = 0 \tag{2}$$

where

$$A = \begin{vmatrix} y_1 & X_1 & 1 \\ y_2 & X_2 & 1 \\ y_3 & X_3 & 1 \end{vmatrix}; B = -\begin{vmatrix} x_1 & X_1 & 1 \\ x_2 & X_2 & 1 \\ x_3 & X_3 & 1 \end{vmatrix}; C = \begin{vmatrix} x_1 & y_1 & 1 \\ x_2 & y_2 & 1 \\ x_3 & y_3 & 1 \end{vmatrix}; D = -\begin{vmatrix} x_1 & y_1 & X_1 \\ x_2 & y_2 & X_2 \\ x_3 & y_3 & X_3 \end{vmatrix}$$

Similarly, for $Y = g(x, y)$ is the same operating steps.

Using the affine transformation to achieve the local alignment of two images repeatedly. At last we use the alpha blending method [11] to achieve the superposition of the two images, which follows the formula:

$$S = \alpha I + (1 - \alpha)T \tag{3}$$

Here, α is a weight parameter, obtained from the hair template, I represents the data vector of hair template, T represents the data vector of face image, S is a new face after alpha blending. An example of demonstration by this step is shown in Fig. 4:

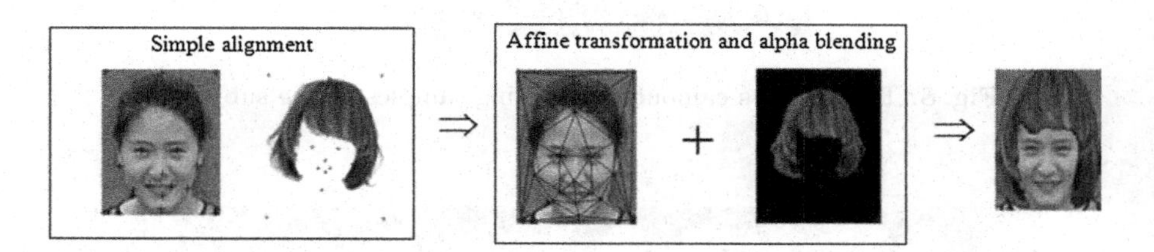

Fig. 4. Hairstyle camouflage

For eyeglass camouflage phenomenon, we provide increasing the features of wearing glasses in the training samples, the specific steps are: (1) Making two images aligned, which needs simple translation, Scaling and rotation for eyeglass template and face template; (2) Using alpha blending method to achieve the superposition of two templates. The overlay process is shown in the Fig. 5.

Fig. 5. Eyeglass camouflage

2.4 Application of Data Augmentation in Face Training Set

We design four training sets by data augmentation. The original face database contains three main faces of each subject, the original training samples of one subject are shown in Fig. 6.

Fig. 6. The original training samples of one subject

Fig. 7. The hairstyle camouflage training samples of one subject

Fig. 8. The eyeglass camouflage training samples of one subject

Fig. 9. Eyeglass and hairstyle camouflage training samples of one subject

Based on the original face database, we construct the hairstyle camouflage training set with the hairstyle hypothesis method, the hairstyle camouflage training samples of one subject are shown in Fig. 7.

Similarly, based on the original face database, we also construct the glass camouflage training set with the glass hypothesis method, the eyeglass camouflage training samples of one subject are shown in shown in Fig. 8.

Finally, based on the method of hairstyle and eyeglass hypothesis, a comprehensive training set is established, the comprehensive training samples of one subject are shown in Fig. 9.

2.5 Face Classification

Firstly, we carried out preprocessing work, grayed each pixel of image and normalized them to [0, 1]. Then chose the logical regression algorithm as multi-classifier. Because it could classify faces in a short time and reduce the cost of UAV. Here, we assumed the logical regression model was:

$$h_\theta(x) = \frac{1}{1+e^{\theta^{\mathrm{T}}x}} \tag{4}$$

where, h represents Sigmoid function, x represents the feature vector.

On the hypothetical model, established cost function was:

$$J(\theta) = -\frac{1}{m}[\sum_{i=1}^{m} y^{(i)} \log h_\theta(x^i) + (1-y^i)\log(1-\log h_\theta(x^{(i)}))] \tag{5}$$

Finally, we used the gradient descent algorithm to optimize θ, and found that when learning rate $\lambda = 0.003$, iterations $= 40$, the recognition results was the best.

3 Result

For comparison, we designed four different training sets, including original training set, hairstyle camouflage training set, eyeglass camouflage training set, hairstyle and eyeglass training set. The original training set has 3 training samples for each subject and a total of 30 training samples. The hairstyle camouflage training set has a total of 60 training samples and 6 training samples for each subject. The eyeglass training set has a total of 60 training samples and 6 training samples for each subject. The hair and glass training set has a total of 120 training samples an 12 training samples for each subject. The test set includes ten subjects' camouflage images, and each subject has 12 images, so there are a total of 120 test samples in the test set. From Table 1 one can see that: (1) In order to identify the test set of face, we use the parameters obtained by training the original data set to test the test training set, the recognition accuracy is 93.33%, which fully shows the classification algorithm suits multi-classification, and learning speed is fast. So apply it on the UAV is very suitable. (2) Using the parameters obtained by only the hairstyle camouflage or only eyeglass camouflage training set to identify the facial test set, the accuracy rate has improved, but the increase is not much, because the training focused on the camouflage face samples not enough, however the final test results are still better. (3) The recognition result is 97.5% when the parameters obtained from both hairstyle and glasses camouflage training set. This shows that when the face samples are relatively less in the training set, using face augmentation to increase facial features can improve the accuracy of UAV identification camouflage face. The four different training sets were trained to test same test set, and the identification results were shown as follows:

Table 1. Accuracy of results based on different data training sets

Different data training sets	Recognition accuracy
Original face training set	93.33%
Hairstyle camouflage training set	95.80%
Glass camouflage training set	94.20%
Hairstyle and glass camouflage training set	97.50%

4 Conclusions

In this paper, in the case of few normal training samples, we use two methods both hairstyle and eyeglass hypothesis to augment face training sets, face camouflage recognition is still relatively accurate. In the 4 kinds of experiments, the accuracy rate of UAV recognizes face camouflage increased by 4.17% after data augmentation. Our future works include: (1) applying this method to online UAV face camouflage recognition system; (2) further improving the recognition performance; (3) decreasing the calculation time; (4) finding suitable application fields like using small UAVs to identify facial camouflage suspects and maintaining social security.

Acknowledgments. This work was funded by National Natural Science Foundation of China under Grants (No. 61473110, No. 61633010), International Science and Technology Cooperation Program of China, Grant No. 2014DFG12570, Key Lab of Complex Systems Modeling and Simulation, Ministry of Education, China.

References

1. Ryan, C., Rodney, W., Reter, C.: Image processing algorithms for UAV "Sense and Avoid". In: IEEE International Conference on Robotics and Automation, pp. 2848–2853 (2006)
2. Vladimir, D., Isaac, K., Kevin, J., Reza, G.: Vision-based tracking and motion estimation for moving targets using small UAVs. In: American Control Conference, pp. 1428–1433 (2006)
3. Nicholas, D., Francesco, P., Karen, P.: Facial recognition using human visual system algorithms for robotic and UAV platforms. In: IEEE International Conference on Technologies for Practical Robot Applications (2013)

4. Krizhevsky, A., Sutskever, I., Hinton, G.: ImageNet classification with deep convolutional neural networks. In: Proceedings of the Advances in Neural information Processing Systems, pp. 1097–1105 (2012)
5. Howard, A.: Some improvements on deep convolutional neural network based image classification. arXiv:hepth/1501.02876
6. Jiangjing, L., Xiaohu, S., Jiashui, H., Xiangdong, Z., Xi, Z.: Data augmentation for face recognition. Neurocomputing **230**, 184–196 (2017)
7. Biao, L., Kai, Y., Jingyan, Q.: Data augmentation for unbalanced face recognition training sets. Neurocomputing **23**, 10–14 (2017)
8. Joe, M., Joseph, H.: Learning OpenCV 3 Computer Vision with Python, 2nd edn. China Machine Press, Beijing (2016)
9. Mikael, N., Jörgen, N., Ingvar, C.: Face detection using local SMQT features and split up snow classifier. In: IEEE International Conference on Acoustics, pp. 589–592 (2009)
10. Ardeshir, G.: Piecewise linear mapping functions for image registration. Pattern Recogn. **19**, 459–466 (1986)
11. Luoqi, L., Junliang, X., Si, L.: Wow! You are so beautiful today. ACM Trans. Multimedia Comput. Commun. 11–33 (2014)

3D Reconstruction with Multi-view Texture Mapping

Xiaodan Ye[1,2(✉)], Lianghao Wang[1,2,3(✉)], Dongxiao Li[1,2], and Ming Zhang[1,2]

[1] College of Information Science and Electronic Engineering, Zhejiang University, Hangzhou, China
{yexiaodan,wanglianghao,lidx,zhangm}@zju.edu.cn

[2] Zhejiang Provincial Key Laboratory of Information Processing, Communication and Networking, Hangzhou, China

[3] State Key Lab for Novel Software Technology, Nanjing University, Nanjing, China

Abstract. In this paper, a novel 3D reconstruction with multi-view texture mapping method based on Kinect 2 is proposed. Camera poses of all chosen key frames are optimized according to photometric consistency. Optimized camera poses can make the projected point from vertices to different views get closer. A small range of translations with limited calculation is added in this method. A new form of data term and smoothness term in Markov Random Field (MRF) objective function is presented. The outlier images are rejected before view selection and Poisson blending are applied in the end. Experimental results show that our method achieves a high-quality 3D model with high fidelity texture.

Keywords: Camera poses optimization · Markov Random Field (MRF) · Texture mapping

1 Introduction

Generating 3D models with high quality texture plays an important role in many fields, e.g., virtual reality (VR) and medical image processing. Nowadays, 3D reconstruction has made tremendous progress. The focus of 3D reconstruction is on the accuracy of the reconstructed shape, rather than the texture of surface on the reconstructed objects. However, a texture map is an important component of reconstructed model, and the quality of texture and resolution have a key impact on the fidelity of model.

Some household cameras have been applied to 3D reconstruction. Take Kinect as an example, it is a low-cost RGB-D camera. KinectFusion [1] is a real-time 3D reconstruction algorithm. However, KinectFusion tends to create a 3D model with high-quality geometry but low-quality texture. Actually, texture of Kinectfusion is per-vertex color fused from a color volume. In order to create high-quality texture, Song [2] rigidly attached an external high definition (HD) RGB camera to Kinect. Then Song textured the model with high-resolution RGB

D. Liu et al. (Eds.): ICONIP 2017, Part III, LNCS 10636, pp. 198–207, 2017.
https://doi.org/10.1007/978-3-319-70090-8_21

images, but the quality of texture mapping depended on the accuracy of mapping relationship between the HD RGB camera and the depth camera of Kinect. With the release of Kinect 2, the resolution of RGB-D images is improved and texture map can be generated with RGB images captured by Kinect 2 directly.

Zhou [3] utilized HD RGB images to render model as well, but texture was still per-vertex. In order to achieve a high fidelity rendering, a large number of vertices were required, which brought a high demands for storage and computation. Different from that [3] used voxel or vertex colors for rendering, Junho [4] used a recovered texture map instead. Multiple sub-textures for each face were estimated and blended for the global texture map. But if the reconstructed geometry or camera parameters are slightly inaccurate, the blended texture map may suffer from ghosting. Therefore, multi-view texture mapping with some pretreatment is a good trade-off between model fidelity and computation.

Multi-view texture mapping. Assigning one texture for each face demands less computational resource. The recovered model and multiple registered HD RGB views are regarded as input. Each RGB image can only represent a part of the model, so a complete texture needs multiple views. Texture mapping is performed in two steps: view selection and seam levelling.

As for view selection, multiple texture views were blended for each face [4, 5]. Lempisky and Ivanov [6] posed texture mapping as an image stitching problem, whereas the optimal stitching was sought within MRF energy optimization framework. They aimed to choose the best texture patch for each face and proposed a gradient-domain method to level seam. Based on this, Gal [7] added the component of local image translations. But if the local image translations are large, a large translational range is needed for search. It slows down the texture mapping procedure. Seam levelling is the other step of texture mapping. After view selection, adjacent texture patches may have strong color discontinuities. Poisson blending is good for seam levelling.

In this paper, a novel multi-view texture mapping method based on Kinect 2 is proposed. 3D model is generated by KinectFusion. Camera poses of all chosen key frames are optimized according to photometric consistency. Optimized camera poses can make projected points from vertices to different views get closer. A small range of translations with limited calculation can be added. MRF objective function is adopted for view selection where a new form of data term and smoothness term is presented. We reject the outlier images before view selection and apply Poisson blending to the seam levelling procedure. In the experimental results, this method achieves a 3D model with high fidelity texture.

In summary, the main contributions of this paper are two folds: (1) A camera pose optimization is proposed according to photometric consistency. (2) A new form of data term and smoothness term is proposed in MRF objective function for view selection with limited increase of calculation.

2 Overview

Figure 1 shows the framework of this 3D reconstruction with multi-view texture mapping method.

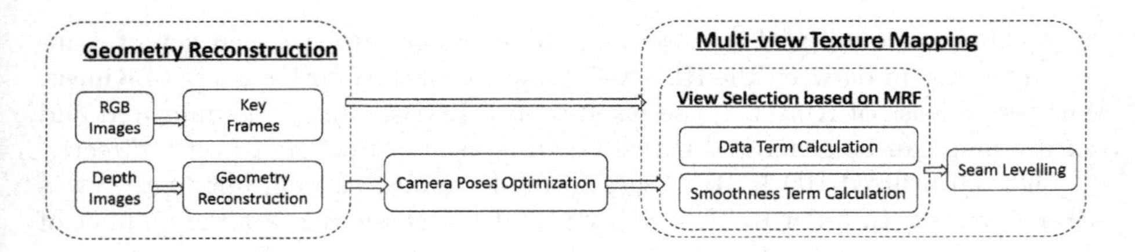

Fig. 1. Our 3D reconstruction with multi-view texture mapping framework.

Input. The input of the proposed method are a stream of depth images and RGB images captured by Kinect 2. The resolution of RGB image is 1920 × 1080 and that of depth image is 512 × 424. In order to align depth image with RGB image, we map the depth image to the RGB image. In this way, a depth image of 1920 × 1080 is obtained and the points that do not exist are set to zero.

Geometric reconstruction. KinectFusion is adopted to generate a geometric model with a triangular mesh representation. After geometric reconstruction, an initial mesh M and a set of camera poses registered with M is obtained.

Key frames. Since Kinect 2 is handheld, many captured images suffer from motion blur. In order to get high-quality images for texture mapping, we quantity the blurriness of all RGB images and select the clearest images as key frames using the method introduced in [8]. The selected key frames cover the entire model with the least blurriness. The set of key frames is denoted by $\{\mathrm{I}_i\}$.

Overall Process. The focus is on texturing the mesh M with the key frames $\{\mathrm{I}_i\}$. Each face triangle is projected onto the images where the face triangle is visible. Since each image covers only part of the object, the final texture will be a combination of texture patches. Our goal is to select the best texture for each face triangle, and penalize mismatches across triangle boundaries. This was represented as a Markov Random Field (MRF) problem in [6]. Since reconstructed geometry and the camera poses are not perfectly accurate, it will certainly result in mismatches across triangle boundaries. Such mismatches result from the inconsistency of color, intensity and illumination. Therefore before building a MRF function, the camera poses of key frames should be optimized according to photometric consistency.

3 Camera Poses Optimization

The reconstructed mesh M consists of triangular faces $\{F_k\}$, i.e., a set of vertices P. Each vertex $p \in P$ is projected onto key frames using the corresponding camera poses and intrinsic parameters. p is a 3 × 1 vector. If the difference between depth value of vertex p and corresponding value in depth image is within a certain threshold, p is considered to be visible in this key frame. Each vertex is projected onto the key frame where it is visible, and the colors of the projected points that are from the same vertex should be the same. This assumption is known as photo-consistency and holds as long as the sensor is noiseless, the scene is static, and the illumination is constant. Camera poses of all key frames can be optimized using this principle.

Objective. As described above, a set of key frames $\{\mathrm{I}_i\}$ and vertices P can be regarded as the input. For each key frame I_i, the corresponding camera pose H_i maps vertex p from the world coordinate to the local camera coordinate. H_i comprises a rotation matrix and a translation vector:

$$\mathrm{H}_i = \begin{bmatrix} R_i \; t_i \\ 0 \;\; 1 \end{bmatrix}, \tag{1}$$

where R_i and t_i are 3×3 rotational matrix and 3×1 translational vector, respectively. We aim to maximize the agreement of the colors of p in all key frames where p is visible. Color of the first frame is set as reference color. The reference color of p is denoted by $C(p)$. The following objective is to be minimized:

$$e_{rgb}(\mathrm{H}_i) = \sum_i \sum_{p \in P} \left(C(p) - \Gamma_i(p, \mathrm{H}_i)\right)^2 \tag{2}$$

Note that $\Gamma_i(p, \mathrm{H}_i)$ is the color at the image coordinate of the projection of p onto the image I_i, using the extrinsic matrix H_i. In this work, the internal matrix is known. In detail, $\Gamma_i(p, \mathrm{H}_i)$ can be expressed as $\Gamma_i(u(g(p, \mathrm{H}_i)))$. p is the point in the world coordinate, and g is the transformation function that transforms p from world coordinate to corresponding camera coordinate. u is the projection function that projects the point in the camera coordinate to the image plane.

$e_{rgb}(\mathrm{H}_i)$ is a nonlinear least-squares objective and can be minimized using Gaussian-Newton method. $\mathrm{H}_i \triangleq \exp(\xi)$ is a matrix of 4×4 and there is a representation ξ given by the Lie algebra $se(3)$ associated with the Lie Group $SE(3)$. ξ is a six-vector and H_i is represented as: $\mathrm{H}_i \triangleq \exp(\xi)$ [9]. $e_{rgb}(\xi)$ is minimized by iterative optimization. Then optimized ξ and H_i are obtained.

4 Texture Mapping

4.1 MRF Based View Selection

Rendering method adopted in this paper is selecting an image for each triangular face of reconstructed mesh M. Lempitsky [6] posed texturing as an image stitching problem. The goal is to select the best texture and penalize mismatches across triangle boundaries. Mesh M has a set of faces: $\{F_k\} = \{F_1, F_2, ..., F_K\}$. Assume there are N key frames and each key frame I_i corresponds to a texture fragment V^i. Then texture mapping is defined by a labeling vector $L = \{l_1, l_2, ..., l_K\}$, where $l_k \in \{0, 1, ..., N\}$ indicates that face F_k is rendered by texture fragment V^{l_k}. Each face label is expanded to include an image-space transformation [7].

Since the camera poses of all selected key frames have been optimized above, the range of translational vectors can be much smaller than that of Gal [7]. The decrease of translation range will speed up the calculation of labels. Specifically, the label can be represented as $l_k = (s_k, t_k)$, where s_k is the source key frame and t_k is a 2×1 translational vector. The texture fragment V^{l_k} is generated by translating the corresponding texture patch s_k in the key frame by t_k. Figure 2 shows the multi-view texture mapping with additional translational label component.

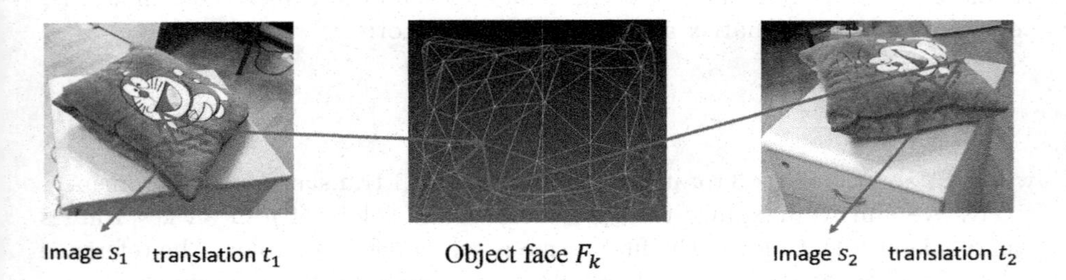

Fig. 2. For each face F_k, there are some texture patch candidates, e.g., the patches from s_1 and s_2. The translational components t_1 and t_2 can correct the remaining projection inaccuracy after camera poses optimization.

Finding a good label assignment is the goal. The MRF objective function is:

$$E(l) = \sum_{F_k} E_{data}(F_k, l_k) + \sum_{(F_k, F_j) \in Edges} E_{smooth}(F_k, F_j, l_k, l_j) \tag{3}$$

Note that the data term E_{data} tends to select good views for texturing and the smoothness term E_{smooth} minimizes the seam. F_k and F_j share an edge.

As for the data term E_{data}, Lempitsky and Ivanov [6] computed the angle θ between the local viewing direction of the corresponding view and the face normal. This method took into account that a better image should be got with a smaller viewing angle. But it is insufficient to select good texture fragment. Gal et al. [7] utilized the gradient magnitude of the image integrated over the projection of the face as the data term. This term took account for the out-of-focus blur and projection area. This term is larger if the projection area is larger or the gradient magnitude is larger. Our method combines these two approaches:

$$E_{data}(F_k, l_k) = \int_{\phi_{l_k}(F_k)} \left\| \nabla \left(I_{s_k}(p) \right) \sin \theta \right\|^2 dp, \tag{4}$$

where we compute the gradient magnitude $\nabla(I_{s_k}(p))$ of the image s_k into which face F_k is projected. $\phi_{l_k}(F_k)$ is the projection of F_k.

Since kinect 2 is handheld, there are some images suffering from blurriness. Images with the least blurriness are selected, but there are still some blurred images remaining. While calculating the data term, the frames whose average gradient magnitude is too far away from the mean value are rejected. In this case, the bad-quality texture fragment is avoided.

As for the smoothness term E_{smooth}, Michael [10] adopted Potts model, which is extremely fast to compute. This term preferred that texture fragments of adjacent faces are selected from the same image view. However, when only part of the image view is of good quality, this term will bring in unsatisfactory texture fragments. In this paper, a smoothness term corresponds to the discrepancy between texture fragments of adjacent faces. The mean color of each edge of the triangle face is calculated. Then the mean colors from different texture fragments of the common edge are subtracted to obtain the smoothness value of the edge:

$$E_{smooth}\left(F_k, F_j, l_k, l_j\right) = \delta_{k,j} \int_{F_k \cap F_j} \left\| I_{s_k}\left(\phi_{l_k}\left(p\right)\right) - I_{s_j}\left(\phi_{l_j}\left(p\right)\right) \right\|^2 dp. \quad (5)$$

Note that $\delta_{k,j}$ is the ratio between the actual length of the edge in the unit of millimeter and numbers of pixels of the edge. Then the value of smoothness term is expanded reasonably, but the relative size of smoothness value of all edges is preserved. $\phi_{l_k}(p)$ is the projection of p. Since the label l_k equals (s_k, t_k), discrete pixel values are adopted for the dislocation of adjacent texture patches. We allow the range of translational offset ± 4 pixels, which is much faster compared to ± 64 pixels of [7]. Because camera poses have been optimized before, a small range of translational offsets is also sufficient. Alpha-expansion Graph Cuts is adopted to minimize the energy of Eq. (3).

4.2 Seam Levelling

After view selection, every face is assigned a texture fragment from a certain key frame. The optimization of camera poses and additional component t_k of labels can deal with large errors of seams between adjacent texture patches. There are still some visible seams remaining. Poisson editing [11] is leveraged to resolve this problem. A boundary around each texture patch is identified and the color of the boundary is adjusted using the pixel values of the other side of the texture patch boundary. Taking into account the problem of computational complexity, the color of the boundary is adjusted only by a 15-pixel-wide border strip of the other side. Almost the same results can be obtained using the 15-pixel-wide border strip instead of the entire texture patch of the other side.

5 Experimental Results

In the experiments, datasets collected by Kinect 2 are used to evaluate the performance of the proposed method. Since there are not RGB-D datesets collected by Kinect 2 for 3D modeling available, all datasets are collected by ourselves using Kinect 2. Table 1 shows the statistics of the datasets.

Figure 3 shows the 3D models of the dataset "bag" with final texture generated by the proposed method. Note that a 3D model with high fidelity texture is achieved in Fig. 3. Some details are marked with red rectangles and enlarged. These details include stripes, fine lines, highly repetitive texture, etc. Minor errors are unnoticeable in flat-texture area, but very obvious in detail-intensive area. If there are mismatches between adjacent texture patches or projection errors caused by inaccurate camera poses, these errors will be exposed in the

Table 1. Statistics and the camera poses optimization information of the test datasets in our method.

Dataset	bag	backpack	doll
Number of faces	15701	21553	60887
Number of key frames	30	27	32
Cost before camera poses optimization	92672	160089	454415
Cost after camera poses optimization	11822.9	23967.1	61368.8
Optimization time/s	1249.0394	1089.4367	2375.9626

detailed areas marked in Fig. 3. As shown in Fig. 3, the proposed method solves most of the reconstruction and multi-view texture mapping errors.

The achievements of our two main contributions will be evaluated, i.e., the camera poses optimization and our novel texture mapping method, respectively.

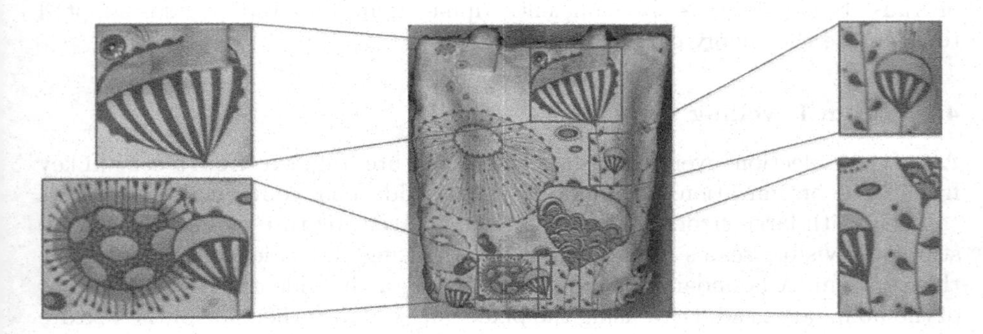

Fig. 3. The reconstructed 3D model with texture generated by proposed method. Some details are marked with red rectangles and enlarged. (Color figure online)

5.1 Evaluation on Camera Poses Optimization

Since the datasets utilized in this paper are manually collected using Kinect 2, there is not groundtruth of camera poses. Hence the performance of camera poses optimization cannot be evaluated with groundtruth. The camera poses are used to project the point from the world coordinate to the corresponding camera coordinate, and the subsequent multi-view texture mapping is performed based on this. In this case, the accuracy of the camera poses can be evaluated by the quality of the texture mapping or MRF energy. Figure 4 shows the final energy after MRF optimization before and after camera poses optimization.

Camera poses are optimized iteratively, so the energy changes as the number of iterations increases. As is shown in Fig. 4, the MRF energy decreases quickly when the number of iterations increases, and converges when the number of iterations is greater than 20. Hence we optimize camera poses with 20 iterations in

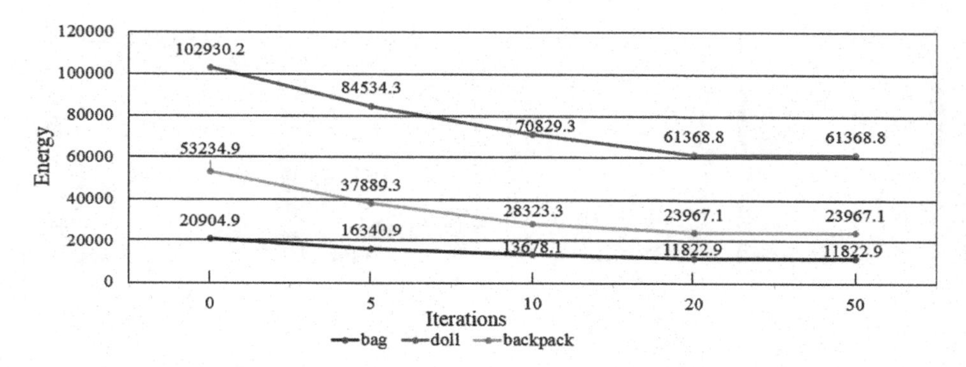

Fig. 4. The camera poses are optimized with 0, 5, 10, 20, 50 iterations. The energy after running MRF model decreases as the number of iterations increases. The camera poses optimization converges at about 20 iterations.

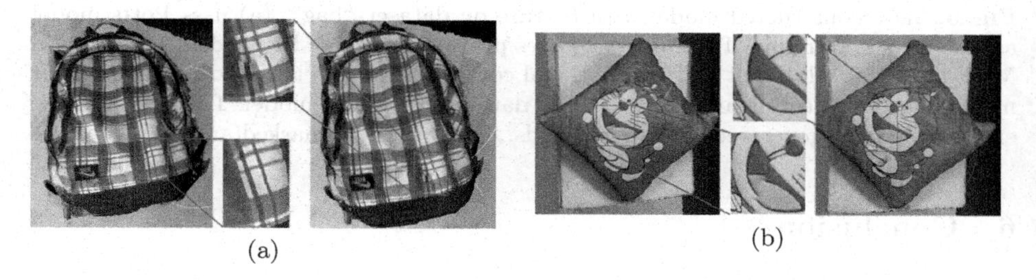

(a) (b)

Fig. 5. (a) shows the difference of the reconstructed model with texture before and after camera poses optimization. The right one is without camera poses optimization and the left one is optimized, (b) demonstrates the difference of the reconstructed model with and without translational component. The right one considers the translational component and the left one does not.

this work. Figure 5(a) shows the difference of the reconstructed model with texture before and after camera poses optimization, using the dataset "backpack".

5.2 Evaluation on Texture Mapping

MRF energy optimization is applied to texture mapping. A novel data term, smoothness term and additional label component are presented. Figure 5(b) shows the difference of the reconstructed model with and without translational component, using the dataset "doll". The range of translational offset is set as ± 4 pixels and the step is 1 pixel. The deviation of the doll is corrected after applying translational component, which is shown in Fig. 5(b).

Figure 6 shows the 3D reconstructed model with texture on dataset "bag", using different methods. The detail is marked and enlarged. These four results can be compared to that of Fig. 3. The rectangular areas marked in these four models are also marked and enlarged in Fig. 3. Obviously, the model in Fig. 3, i.e., the model generated using the proposed method outperforms other four.

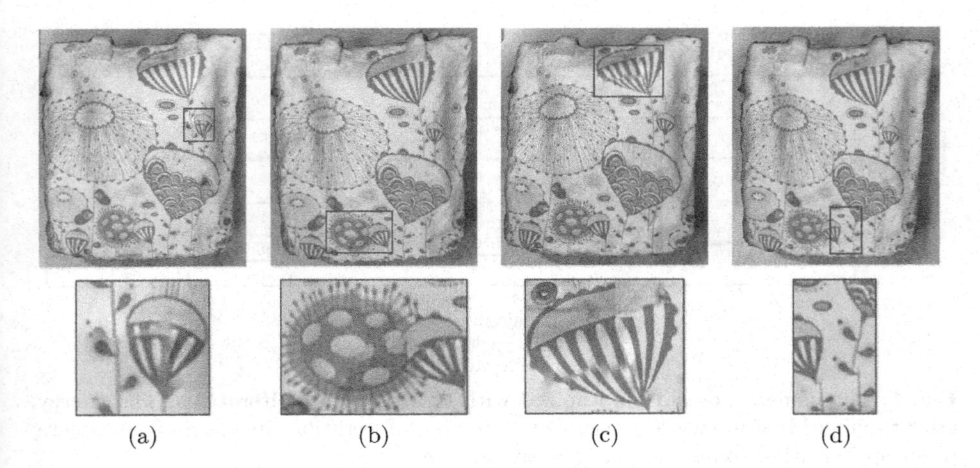

Fig. 6. 3D reconstructed model with texture on dataset "bag". (a) uses Potts model as smoothness term in MRF model, (b) uses projection area of the face as data term in MRF model, (c) does not add translational component to the label, (d) sums gradient magnitude of projection area of the face as data term in MRF model. There are texture errors in models generated by four methods. The errors are marked and enlarged.

6 Conclusion

A novel 3D reconstruction with multi-view texture mapping based on Kinect 2 is proposed in this paper. Camera poses after geometric model reconstruction are optimized according to photometric consistency. A novel data term, smoothness term and additional translational component are applied in MRF model. Experimental results show that novel format of MRF energy function and camera poses optimization improve the quality of the reconstructed model with texture.

Acknowledgments. This work is supported in part by the National Natural Science Foundation of China (Grant No. 61401390).

References

1. Newcombe, R.A., et al.: KinectFusion: real-time dense surface mapping and tracking. In: 2011 10th IEEE International Symposium on Mixed and Augmented Reality, Basel, pp. 127–136 (2011). doi:10.1109/ISMAR.2011.6092378
2. Liu, S., Li, W., Ogunbona, P., Chow, Y.W.: Creating simplified 3D models with high quality textures. In: 2015 International Conference on Digital Image Computing: Techniques and Applications (DICTA), Adelaide, SA, pp. 1–8 (2015). doi:10.1109/DICTA.2015.7371249
3. Zhou, Q.Y., Koltun, V.: Color map optimization for 3D reconstruction with consumer depth cameras. ACM Trans. Graph. **33**(4), 1–10 (2014)
4. Jeon, J., Jung, Y., Kim, H., Lee, S.: Texture map generation for 3D reconstructed scenes. Vis. Comput. **32**(6), 955–965 (2016)

5. Wang, L., Kang, S.B., Szeliski, R., Shum, H.Y.: Optimal texture map reconstruction from multiple views. In: Proceedings of the 2001 IEEE Computer Society Conference on Computer Vision and Pattern Recognition, CVPR 2001, vol. 1, pp. I-347–I-354 (2001)
6. Lempitsky, V., Ivanov, D.: Seamless mosaicing of image-based texture maps. In: IEEE Conference on Computer Vision and Pattern Recognition, pp. 1–6 (2007)
7. Gal, R., Wexler, Y., Ofek, E., Hoppe, H., Cohen-Or, D.: Seamless montage for texturing models. Comput. Graph. Forum **29**, 479–486 (2010)
8. Crete, F., Dolmiere, T., Ladret, P., Nicolas, M.: The blur effect: perception and estimation with a new no-reference perceptual blur metric. In: Electronic Imaging, pp. 64920I–64920I-11 (2007)
9. Kerl, C., Sturm, J., Cremers, D.: Robust odometry estimation for RGB-D cameras. In: 2013 IEEE International Conference on Robotics and Automation, Karlsruhe, pp. 3748–3754 (2013)
10. Waechter, M., Moehrle, N., Goesele, M.: Let there be color! Large-scale texturing of 3d reconstructions. In: Fleet, D., Pajdla, T., Schiele, B., Tuytelaars, T. (eds.) ECCV 2014. LNCS, vol. 8693, pp. 836–850. Springer, Cham (2014). doi:10.1007/978-3-319-10602-1_54
11. Rez, P., Gangnet, M., Blake, A.: Poisson image editing. ACM Trans. Graph. **22**(3), 313–318 (2003)

Online Tracking with Convolutional Neural Networks

Xiaodong Liu and Yue Zhou(✉)

Institute of Image Processing and Pattern Recognition,
Shanghai Jiao Tong University, Shanghai, China
{lxd7714059,zhouyue}@sjtu.edu.cn

Abstract. Convolutional neural networks (CNNs) have recently been widely applied to visual applications, but there are still not much attempts to employ CNNs for object tracking. In this paper, we propose a novel visual tracking method which utilizes the powerful representations of CNNs. We regard the visual tracking as a traditional binary classification task along with an online model update. The binary classification network is pre-trained on ImageNet dataset and fine-tuned on visual tracking benchmark dataset by sequentially training to avoid overfitting. In the tracking process, we conduct a short-term and long-term model update mechanism for adaptiveness and robustness, respectively. Extensive experiments on two visual tracking datasets demonstrate that our algorithm is comparable to state-of-art methods in terms of accuracy and robustness.

Keywords: CNNs · Visual tracking · Binary classification · Short-term update · Long-term update

1 Introduction

Visual tracking has been an important research topic over the past decades due to its numerous applications, such as surveillance, driverless car and human-computer interaction. Given the initial position of arbitrary object in the first frame, the tracker is expected to locate the target in the following frames. Despite many efforts put into visual tracking, it is still a challenging problem due to many complicated factors like abrupt object motion, non-rigid object deformation, occlusion, etc.

In the past decades, many tracking algorithms have been focusing on how to represent the target appearance robustly, which plays an important role on visual tracking. These algorithms can be divided into two categories: generative [1,19] and discriminative methods [7,8,11]. Generative algorithms model target appearance as a series of templates and search for the most similar region to the templates in the tracking process. However, discriminative methods consider visual tracking as a binary classification problem, which can be solved by training a classifier to distinguish the target from its surrounding background. These

D. Liu et al. (Eds.): ICONIP 2017, Part III, LNCS 10636, pp. 208–216, 2017.
https://doi.org/10.1007/978-3-319-70090-8_22

methods usually use hand-crafted features to model the target appearance, such as HOG, which maybe not so effective in handling tracking challenges.

Convolutional neural networks (CNNs) have drawn more and more attentions since a neural network named AlexNet [10] was proposed in 2012 which significantly improves the performance on ImageNet classification challenge [5]. More and more researchers apply CNNs in various computer vision applications [2,10,15]. Despite the success of CNNs in these filed, there are still not much attempts to transfer the CNNs into visual tracking. [9] takes outputs from hidden layer of the CNN as feature descriptors and uses an online SVM to train a classifier. [14] proposes an ensemble method which hedges outputs from different layers into a stronger one. [13] adopts a multi-domain training method to get a generic neural network which avoids overfitting and shows outstanding performance. However, there still exists troubles on exploiting CNN in visual tracking, such as how to train a generic network and what kind of CNN architecture is proper to visual tracking.

Inspired by discriminative tracking methods, visual tracking is similar to object classification in distinguishing the target from background. Nevertheless, tracking aims to locating arbitrary target while object classification needs to give the class labels of the objects. Considering the consistency and inconsistency between visual tracking and object classification, we propose a novel method that transfer CNN into visual tracking by regarding it as a binary classification task along with online model update. In addition, we adopt a sequentially training method to avoid overfitting when fine-tuning the pre-trained network on visual tracking dataset. Furthermore, we train several CNNs with different number of convolutional layers to explore the influence of different architectures to tracking performance in our framework.

The main contributions of this paper can be summarized as below: (i) we propose a novel method which transfers pre-trained CNNs to visual tracking by sequentially training a binary neural network with public tracking dataset; (ii) we conduct a short-term and long-term online update mechanism for adaptiveness and robustness in the tracking process; (iii) extensive experiments on two public benchmark demonstrate that our algorithm is comparable to state-of-art methods.

2 Training Neural Networks

Since we regard visual tracking as a binary classification task along with online update, our network architecture is similar to those used for multi-object detection and classification. Although, the performance of the networks on public classification benchmarks becomes better and better as these networks go deeper and deeper, we still think deeper CNNs are not appropriate for visual tracking due to the following considerations. First, visual tracking is only a binary classification problem which does not need complicated structure compared to multi-object classification task. Second, the computational complexity goes high as network goes deeper which is not desired by visual tracking.

Based on above reasons, our network (named CNN-c3) is comparatively shallow which only consists of three convolutional layers and two fully connected layers except input and output layer. Moreover, we train another two networks (named CNN-c2 and CNN-c5) that have a different number of convolutional layers (2 convolutional layers and 5 convolutional layers) for simple comparison. The convolutional layers of these networks are almost the same as the corresponding convolutional layers of VGG-M network proposed in [2] but with some minor adjustments. Table 1 shows the architectures of the three networks, and the activation function of all hidden layers is the Rectification Linear Unit (RELU).

Due to lack of training samples, fine-tuning pre-trained CNNs on tracking dataset is prone to overfitting. Meanwhile, it is hard to get a generic representation model since tracking scenarios are significantly different from each other. Considering that each sequence should contribute to the final model equally to avoid network dominated by a special training sequence, we adopt a sequentially training method. That is to say, we randomly initialize the output layer for each training sequence, and keep all the hidden layer shared while just replace the output layer of the network with sequence-specific output layer through training. The network is trained by Stochastic Gradient Descent (SGD), and the training is conducted sequence by sequence and repeated dozens of times to learning a generic model.

Table 1. CNN architectures. The convolutional layers are described in three sub-rows: the first row indicates the mumble and size of filters; the second indicates the stride (st.) and spatial padding (pad); in the third row, LRN donates Local Response Normalization, and pool means max pooling. The fc1 and fc2 are applied with dropout while fc3 still use soft-max for classification.

Arch.	conv1	conv2	conv3	conv4	conv5	fc1	fc2	fc3
CNN-c2	$96 \times 7 \times 7$ st. 2, pad 0 LRN, ×2 pool	$256 \times 5 \times 5$ st. 2, pad 0 LRN, ×2 pool	-	-	-	512 drop-out	512 drop-out	2 soft-max
CNN-c3	$96 \times 7 \times 7$ st. 2, pad 0 LRN, ×2 pool	$256 \times 5 \times 5$ st. 2, pad 0 LRN, ×2 pool	$512 \times 3 \times 3$ st. 1, pad 0 -	-	-	512 drop-out	512 drop-out	2 soft-max
CNN-c5	$96 \times 7 \times 7$ st. 2, pad 0 LRN, ×2 pool	$256 \times 5 \times 5$ st. 2, pad 0 LRN, ×2 pool	$512 \times 3 \times 3$ st. 1, pad 0 -	$512 \times 3 \times 3$ st. 1, pad 0 -	$512 \times 3 \times 3$ st. 1, pad 0 ×2 pool	512 drop-out	512 drop-out	2 soft-max

3 Tracking Algorithm

In this section, we first introduce the initialization of our network when running on new sequence, then describe the online tracking procedure and finally present the model update scheme. The overall tracking procedure is presented in Algorithm 1.

Algorithm 1. Proposed tracking algorithm.

Input: Initial target bounding box x_1 and fine-tuned CNN model.
Output: Estimated object state x_t in the following frames.

1: Generate training samples from first frame and initialize the CNN model.
2: **for** $t = 2 : m$ (m is the number of frames of a video) **do**
3: Generate the candidate regions.
4: Find the region with highest score S_t^* as tracking result.
5: **if** $S_t^* > 0.5$ **then**
6: Add current frame to successful frames.
7: Perform short-term update.
8: **else**
9: Perform long-term update.
10: **end if**
11: **end for**

3.1 Initialization

When new sequence comes, the sequentially trained network is utilized and the output fully connected layer is randomly initialized again. When the initial target is given in the first frame, we obtain training data by conducting a Gaussian distribution whose mean is the position and scale of initial target, where positive samples have $\geq$0.7 and negative samples have $\leq$0.5 IoU overlap ratios with the bounding box. We keep the parameters of convolutional layers fixed and only learn the fully connected layers, the initial training is conducted 50 times to get a sequence-specific model.

3.2 Online Tracking

Based on the tracking result of the previous frame, N candidate regions are generated by conducting a Gaussian distribution as done in model initialization. These candidate regions go through the network and the region with highest score are selected as the new target location. The tracking can be summarized as the following mathematical problem:

$$x_f = \underset{x_i}{argmax}\, F(x_i) \tag{1}$$

where x_i, $i \in \{1, ..., N\}$ is the candidates, $F(\cdot)$ donates the network computation and x_f is the final tracking result.

3.3 Model Update

For the sake of adaptiveness and robustness, the short-term and long-term update scheme are conducted. We record the frames with tracking scores high than 0.5 as successful frames that are used for generating training examples for model update. The recent 5 successful frames are used for adaptively short-term update while recent 60 frames are gathered for long-term online update.

4 Experimental Results

In this section, we first introduce the implementation details of the proposed method, referred to as CNNfT (using CNN-c3), then represent the evaluation results of the algorithm on two public benchmark datasets, including the comparison results of the three networks mentioned in Sect. 2.

4.1 Implementation Details

The proposed algorithm is implemented in MATLAB with MatConvNet toolbox [16], and runs at around 2 fps with a 2.10 GHz Intel Xeon CPU and a TITAN X GPU. The network is trained by SGD with momentum and weight decay set to 0.9 and 0.0001, respectively. The learning rate of convolutional layers and fully connected layers are set to 0.00004 and 0.0004, respectively. It takes 50 iterations to initialize the CNN model, 2 iterations for short-term update and 10 iterations for long-term update. 200 candidate regions are generated to satisfy various situations through tacking. The parameters are fixed throughout the whole experiments.

4.2 Evaluation on OTB Dataset

OTB100 [17] is a popular tracking benchmark which now consists of 100 videos with 11 attributes, including fast motion, background cluttered, etc. The proposed algorithm is compared with DeepSRDCF [4], HDT [14], CF2 [12], CNN-SVM [9], MEEM [18], KCF [8], DSST [3], TGPR [6] and Struck [7]. The experiments are conducted with one-pass evaluation (OPE) based on two metrics: (i) precision plot, which shows the percentage of frames whose estimated location is within the given threshold distance of the ground truth; (ii) success plot, which indicates the percentage of frames whose bounding box overlap is large than a threshold t_o.

To make experiments more convincing, the data used for training CNNs is obtained from VOT2015 benchmark, excluding the videos in the OTB dataset. Figure 1 presents the precision and success plots of 10 trackers. It indicates that our method achieves high performance on both metrics and outperforms the second tracker with a considerable margin which means our tracker can find location of target precisely with less failures.

Furthermore, we report the performance (success score) of top four trackers on different attributes in Fig. 2. It illustrates that our algorithm can handle various situations well and outperform the other trackers in all attributes.

Additionally, Fig. 4 shows the qualitative evaluation of the proposed method on some challenging sequences.

Also, to compare performance of CNNs with different number of convolutional layers, we run CNN-c2 and CNN-c5 on the OTB benchmark and present the performance on Fig. 3. It shows that smaller network architectures (CNN-c2, CNN-c3) are more proper to visual tracking compared to bigger network (CNN-c5) in our tracking framework.

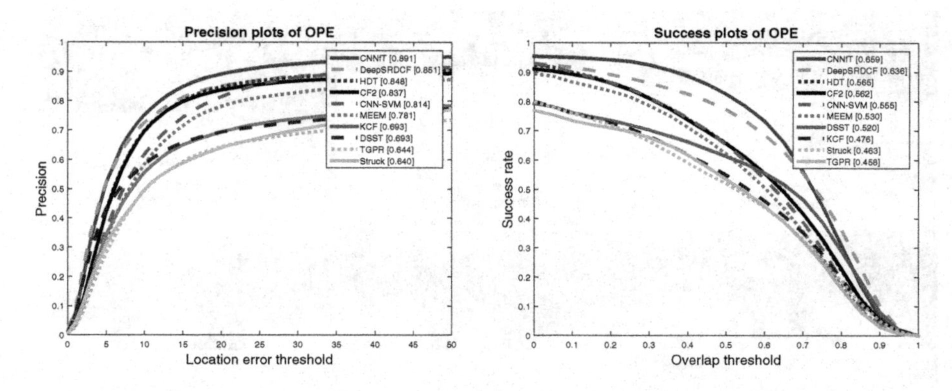

Fig. 1. Overall precision and success plots of 10 trackers on OTB10 benchmark.

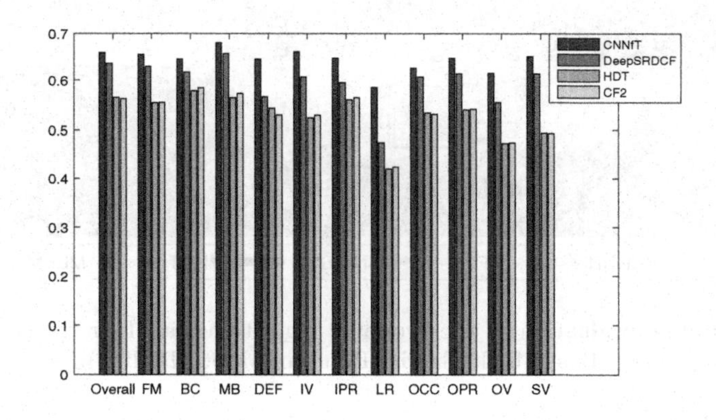

Fig. 2. The detailed OTB100 performance (success plots) of top four trackers on various attributes. Note that the other three trackers are all based on CNNs.

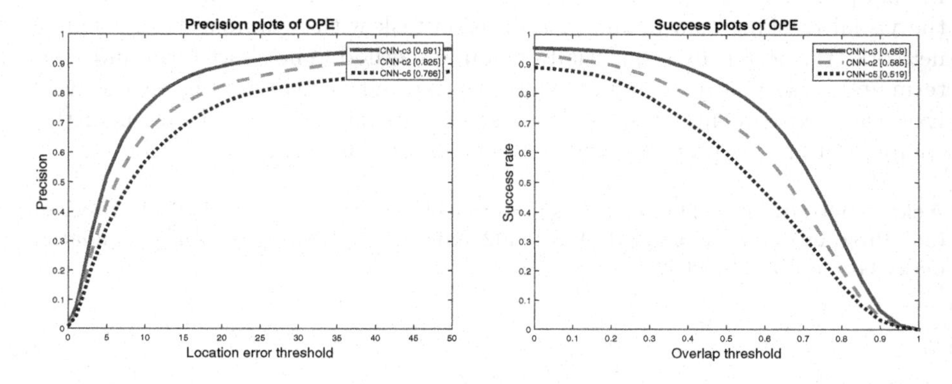

Fig. 3. Internal comparison on OTB100. The ranking order is same as the overall performance whereas in LR situation, the CNN-c2 network performs best of all.

Fig. 4. Qualitative evaluation of the proposed algorithm with four trackers on some challenging sequences (Basketball, Matrix, Bolt1 and MotorRolling).

5 Conclusion

In this paper, we propose a novel CNN based tracking algorithm that regards the visual tracking as a binary classification task with a comparatively shallow network transferred by sequentially training method. The short-term and long-term update scheme are conducted for adaptive and robust tracking, respectively. Extensive experimental results demonstrate the effectiveness of our algorithm compared to state-of-art methods on benchmark dataset.

Acknowledgments. The work is supported by National High-Tech R&D Program (863 Program) under Grant 2015AA016402 and Shanghai Natural Science Foundation under Grant 14Z111050022.

References

1. Avidan, S.: Ensemble tracking. IEEE Trans. Pattern Anal. Mach. Intell. **29**(2), 261–271 (2007)
2. Chatfield, K., Simonyan, K., Vedaldi, A., Zisserman, A.: Return of the devil in the details: delving deep into convolutional nets. arXiv preprint arXiv:1405.3531 (2014)
3. Danelljan, M., Häger, G., Khan, F., Felsberg, M.: Accurate scale estimation for robust visual tracking. In: British Machine Vision Conference, Nottingham, 1–5 September 2014. BMVA Press (2014)
4. Danelljan, M., Hager, G., Shahbaz Khan, F., Felsberg, M.: Convolutional features for correlation filter based visual tracking. In: Proceedings of the IEEE International Conference on Computer Vision Workshops, pp. 58–66 (2015)
5. Deng, J., Dong, W., Socher, R., Li, L.J., Li, K., Fei-Fei, L.: Imagenet: A large-scale hierarchical image database. In: IEEE Conference on Computer Vision and Pattern Recognition, 2009, CVPR 2009, pp. 248–255. IEEE (2009)
6. Gao, J., Ling, H., Hu, W., Xing, J.: Transfer learning based visual tracking with gaussian processes regression. In: Fleet, D., Pajdla, T., Schiele, B., Tuytelaars, T. (eds.) ECCV 2014. LNCS, vol. 8691, pp. 188–203. Springer, Cham (2014). doi:10.1007/978-3-319-10578-9_13
7. Hare, S., Saffari, A., Torr, P.H.: Struck: structured output tracking with kernels. In: 2011 IEEE International Conference on Computer Vision (ICCV), pp. 263–270. IEEE (2011)
8. Henriques, J.F., Caseiro, R., Martins, P., Batista, J.: High-speed tracking with kernelized correlation filters. IEEE Trans. Pattern Anal. Mach. Intell. **37**(3), 583–596 (2015)
9. Hong, S., You, T., Kwak, S., Han, B.: Online tracking by learning discriminative saliency map with convolutional neural network. In: ICML, pp. 597–606 (2015)
10. Krizhevsky, A., Sutskever, I., Hinton, G.E.: ImageNet classification with deep convolutional neural networks. In: Advances in Neural Information Processing Systems, pp. 1097–1105 (2012)
11. Liu, X., Zhou, Y.: Robust part-based correlation tracking. In: Hirose, A., Ozawa, S., Doya, K., Ikeda, K., Lee, M., Liu, D. (eds.) ICONIP 2016. LNCS, vol. 9948, pp. 635–642. Springer, Cham (2016). doi:10.1007/978-3-319-46672-9_71
12. Ma, C., Huang, J.B., Yang, X., Yang, M.H.: Hierarchical convolutional features for visual tracking. In: Proceedings of the IEEE International Conference on Computer Vision, pp. 3074–3082 (2015)
13. Nam, H., Han, B.: Learning multi-domain convolutional neural networks for visual tracking. In: Proceedings of the IEEE Conference on Computer Vision and Pattern Recognition, pp. 4293–4302 (2016)
14. Qi, Y., Zhang, S., Qin, L., Yao, H., Huang, Q., Lim, J., Yang, M.H.: Hedged deep tracking. In: Proceedings of the IEEE Conference on Computer Vision and Pattern Recognition, pp. 4303–4311 (2016)
15. Ren, S., He, K., Girshick, R., Sun, J.: Faster R-CNN: towards real-time object detection with region proposal networks. In: Advances in Neural Information Processing Systems, pp. 91–99 (2015)
16. Vedaldi, A., Lenc, K.: MatConvNet: convolutional neural networks for matlab. In: Proceedings of the 23rd ACM International Conference on Multimedia, pp. 689–692. ACM (2015)

17. Wu, Y., Lim, J., Yang, M.H.: Object tracking benchmark. IEEE Trans. Pattern Anal. Mach. Intell. **37**(9), 1834–1848 (2015)
18. Zhang, J., Ma, S., Sclaroff, S.: MEEM: robust tracking via multiple experts using entropy minimization. In: Fleet, D., Pajdla, T., Schiele, B., Tuytelaars, T. (eds.) ECCV 2014. LNCS, vol. 8694, pp. 188–203. Springer, Cham (2014). doi:10.1007/978-3-319-10599-4_13
19. Zhang, K., Zhang, L., Yang, M.-H.: Real-Time Compressive Tracking. In: Fitzgibbon, A., Lazebnik, S., Perona, P., Sato, Y., Schmid, C. (eds.) ECCV 2012. LNCS, vol. 7574, pp. 864–877. Springer, Heidelberg (2012). doi:10.1007/978-3-642-33712-3_62

Sharp and Real Image Super-Resolution Using Generative Adversarial Network

Dongyang Zhang, Jie Shao(✉), Gang Hu, and Lianli Gao

School of Computer Science and Engineering, Center for Future Media, University of Electronic Science and Technology of China, Chengdu 611731, China
{dyzhang,hugang}@std.uestc.edu.cn, {shaojie,lianli.gao}@uestc.edu.cn

Abstract. Recent studies have achieved great progress on accuracy and speed of single image super-resolution (SISR) based on neural networks. Most current SISR methods use mean squared error (MSE) loss as objective function. As a result, they can get high peak signal-to-noise ratios (PSNR) which are however not in full agreement with the visual qualities by experiments, and thus the output from these methods could be prone to blurry and over-smoothed. Especially at large upscaling factors, the output images are perceptually unsatisfactory in general. In this paper, we firstly propose a novel residual network architecture based on generative adversarial network (GAN) for image super-resolution (SR), which is capable of inferring photo-realistic images for 4× upscaling factors. Perceptual loss is applied as the objective function to make output image sharper and more real. In addition, we adopt some tricks to preprocess the input dataset and use improved techniques to train the generator and discriminator separately, which are proved to be effective for the result. We validate our GAN-based approach on CelebA dataset with mean opinion score (MOS) as performance measure. The results demonstrate that the proposed approach performs better than previous methods.

Keywords: Super-resolution · Generative adversarial network · Residual network

1 Introduction

Image super-resolution (SR) is a hot research problem in computer vision because of its practicability in many application scenarios. For example, the recovery from low resolution (LR) image to high correspondence can reduce network bandwidth and storage requirements in Internet environment. SR technique also helps doctors to diagnose illness according to medical image [1]. Nevertheless, SR is an inherently underdetermined inverse problem. Given a low resolution input, there are multiplicity of plausible high resolution (HR) outputs. Therefore, how to constrain the transformation space by prior information is critical to the quality of HR image.

To this end, the example-based strategy [2] adopted by recent advanced methods gives the prior information. One representative work for the example-based

D. Liu et al. (Eds.): ICONIP 2017, Part III, LNCS 10636, pp. 217–226, 2017.
https://doi.org/10.1007/978-3-319-70090-8_23

strategy is sparse-coding-based method [3], such as SRCNN [4] and ESPCN [5]. The original image is cropped into overlapping patches by fixed stride. Then, these patches are processed to produce corresponding LR and HR pairs fed to learn mapping function. In this work, a novel residual network architecture [6] based on generative adversarial network (GAN) [7] is proposed to super-resolve images for constraining the transformation space. We also use the pair of low and high resolution examples to train our network to address the challenge.

Reconstruction of a high resolution image can be divided into two categories: image super-resolution from a single input image, and image super-resolution from multiple images [8]. In this work, we focus on single image super-resolution (SISR). The SISR methods such as [4,5,9], use pixel-wise mean squared error (MSE) loss as objection function to make empirical risk minimization. MSE loss calculates the mean squared sum on pixel-wise differences between two images. In practice, it is observed that MSE loss has shortcomings when the dataset distribution is multimodal and nontrivial [10] because of the properties of convex function. Moreover, MSE as object function easily leads to higher peak signal-to-noise ratio (PSNR), which is one of the image evaluation criteria. However, the high PSNR may result in flaws in visual perception. Note that, structural similarity (SSIM) index [11] is another image evaluation criterion which measures image similarity from three aspects of brightness, contrast and structure respectively, but it is also weak in reflecting the perceptually better SR result. In this work, we adopt mean opinion score (MOS) to measure image SR performance. MOS is the most representative subjective measurement of image quality because it judges the quality of images by the observers.

This paper proposes a novel residual network architecture based on GAN for image SR. Different from existing methods [4,5] which only use MSE as the optimization target, we combine the adversarial loss together with the convolution loss which is the L2 penalty between the LR input and the down-sampled output. In implementation, originally we wanted to use sub-pixel convolution layer [5] to upscale the feature map. However, we find that the sub-pixel convolution layer may increase the instability of the model, particularly in GAN during training phase. We also apply some tricks to train the network, such as dataset preprocessing by adding some noise or changing the image with flipping/saturation/brightness. These tricks increase the sample diversity and help the generator behave better. Figure 1 shows the effect of the preprocessing step with those tricks, which confirms the benefit of them.

The main contributions of our work can be summarized as follows:

- We propose a novel residual architecture based on GAN, which is different from existing methods adopting conventional convolution structure. The framework is capable of inferring photo-realistic images for $4\times$ upscaling factors.
- We introduce the weighted adversarial loss and the convolution loss as the perceptual loss instead of conventional MSE loss. This change makes SR image sharper and more real.

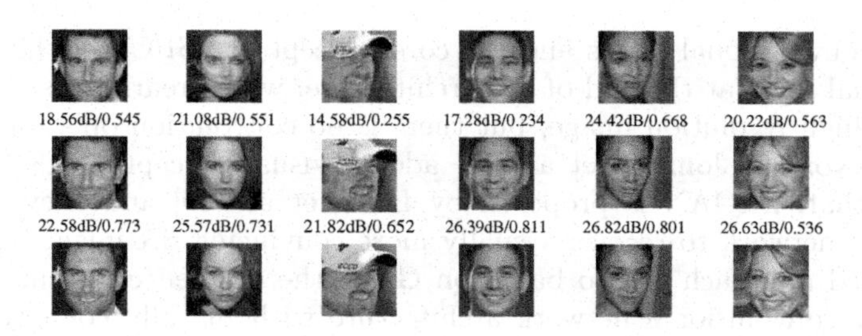

Fig. 1. First line shows the output images without preprocessing step. Second line shows with preprocessing step. Third line shows the original images to reference. Corresponding PSNR and SSIM are shown under each image.

- We adopt some tricks to preprocess input dataset and use several improved techniques to train the generator and discriminator separately, which are both proved to be effective for the result.

The rest of the paper is organized as follows. In Sect. 2, we briefly discuss the related studies. Section 3 describes our approach in detail. We report our experimental results in Sect. 4, and finally Sect. 5 concludes the paper.

2 Related Work

SISR algorithms can be categorized into four types: prediction models, edge-based methods [12], image statistical methods [13] and example-based methods [4,5,14], and example-based methods perform best among them. Prediction methods based on interpolation (linear or bicubic) are the first attempts to solve the problem. Although they are fast and simple, they easily lead to overly smoothed textures. Edge-based methods learn priors from edge features which include the depth and width of an edge for reconstructing HR images. Image statistical methods utilize various image properties such as gradient distribution for reconstruction, and example-based methods have been the most popular recently. More details about SISR can be found in [2].

Recently, convolutional neural network (CNN) achieves great success in computer version because of its strong data fitting capability and has been widely used in SR problem [4,5,14,15]. CNN learns nonlinear mappings between LR space and HR space employing the back-propagation algorithm with large image dataset to train. In [16], a three-layer convolutional network named SRCNN is proposed, and it shows that sparse-coding-based SR methods [3,17] can be viewed as a CNN. In [9], Kim et al. present a highly accurate SISR method using a very deep convolutional network with depth of 20. They also show a significant improvement in accuracy when the depth increases. Kim et al. [18] also introduce deeply-recursive convolutional network (DRCN), and the model has smaller number of parameters but better performance than SRCNN.

To achieve real-time SR, in [5] an efficient method named ESPCN is proposed for obtaining high resolution images directly from low resolution images. ESPCN has three convolutional layers and the core concept of ESPCN is the sub-pixel convolutional layer at the end of the architecture, which rearranges the feature maps into high resolution image, but there is no convolution operation for the layer. The work of Johnson et al. [15] adopts visual perceptual loss function. Based on that, SRGAN is proposed by Ledig et al. [19] and they use GAN to train the network to recover visually more convincing SR images. Although our proposed approach is also based on GAN, the difference is that we adopt a new fully convolutional network architecture without fully connected layers and propose a convolution content loss. Fully connected layers and the very deep VGG network in loss function used in SRGAN bring a large number of parameters and computational burden, so it is difficult to train the network well.

3 Our Approach

The goal of SISR is to super-resolve an image from low resolution to high resolution. The original LR image with C color channels can be represented by the tensor of size $W \times H \times C$. After the SISR with r upscaling factors, the HR image has the shape of $rW \times rH \times C$. In order to train a good generator which can estimate a visual friendly SR image, we use a new network with residual architecture based on GAN. We also adopt a new loss function to constrain the blurry and over smooth that happen in the conventional MSE loss frequently.

3.1 GAN-Based Residual Network Architecture

Inspired by generative adversarial network (GAN) [7], we define a generator network G_{θ_G} to super-resolve the images from low resolution to high resolution. In training phase, the discriminator network D_{θ_D} distinguishes whether the images are real or not. The discriminator D provides the gradient information for the generator G to produce the image that fools the discriminator during training. The training phase will achieve global optimum until network convergence which indicates that it is difficult for the discriminator to distinguish super-resolved images from real images. Completely different from MSE, the network solves the adversarial min-max problem as Eq. 1.

$$\min_{\theta_G} \max_{\theta_D} \mathbb{E}_{I^{HR} \sim p_{train}(I^{HR})} \left[\log D_{\theta_D}(I^{HR})\right] + \mathbb{E}_{I^{LR} \sim p_G(I^{LR})} \left[\log(1 - D_{\theta_D}(G_{\theta_G}(I^{LR})))\right]. \tag{1}$$

Our proposed architecture is illustrated in Fig. 2. Inspired by the deep convolutional generative adversarial networks (DCGAN) [20], we replace the D and G in classical GAN with two full convolutional neural networks and use batch-normalization (BN) [21] layer to increase the stability of the model. For the generator, we increase the resolution of the input image with deconvolution layer [22] which is a popular choice for recovering resolution from max-pooling and other image down-sampling layers. In the front of the generator network, we adopt two residual blocks [6] with identical layout which help converge faster.

In every residual block, the inputs pass a convolution layer to ensure the units are matched. Specifically, the convolutional layers have small 3×3 kernels and 1 stride. Except for the deconvolution layer, other layers do not change the feature map size. As Fig. 2 shows, we apply two deconvolution layers at the end of every residual block and every deconvolution layer broadens the feature maps two times. After residual blocks, the feature maps flow with convolution layer and ReLU but without batch-normalization. Obviously, the process that the input image upscales with nearest neighbor interpolation has to go through all layers until it reaches the output layer. We add the two parts multiplying with weight factors, and then use sigmoid to activate it. Actually, we propose the network structure that learns the residual images inspired by [9]. Adopting the residual architecture makes the network converge much faster and performance better.

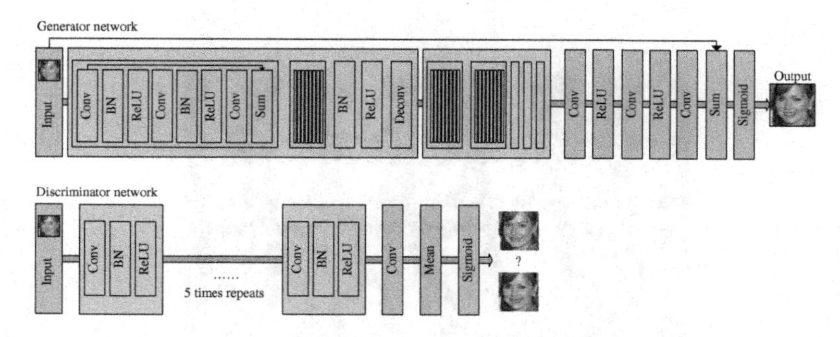

Fig. 2. Architecture of generator and discriminator network. Fed with an LR image, the generator gives the corresponding SR output directly. For every image, the discriminator generates one number to indicate the fake or real.

For discriminator network, we use ReLU as activation and avoid max-pooling in the network. For every image, the dimensionality of the output from the discriminator is only one to represent the level of truth. Instead of using fully connected layer, we use the mean layer to reduce feature maps to only one dimension, and then pass it to a sigmoid activation function at the end of the network. Thus, it supports any size image as input and end-to-end training. We design the discriminator network with seven repeated blocks consisting of convolution layer, ReLU and batch-normalization. As Fig. 2 shows, BN is not used at the generator output layer and the discriminator input layer. This is because applying BN directly to all layers will lead to sample turbulence and model instability [20].

3.2 Loss Function

The loss function is important for the performance of the generator network. During training phase, if the loss function is complex, it will affect the speed of convergence. For SR problem, most state-of-the-art models optimize MSE loss which leads to high PSNR [4,5]. However, its lack of high frequency content will lead to perceptually poor result (c.f. Fig. 3). Actually, GAN-based methods guide the reconstruction towards the real image dataset distribution, producing perceptually more convincing outputs.

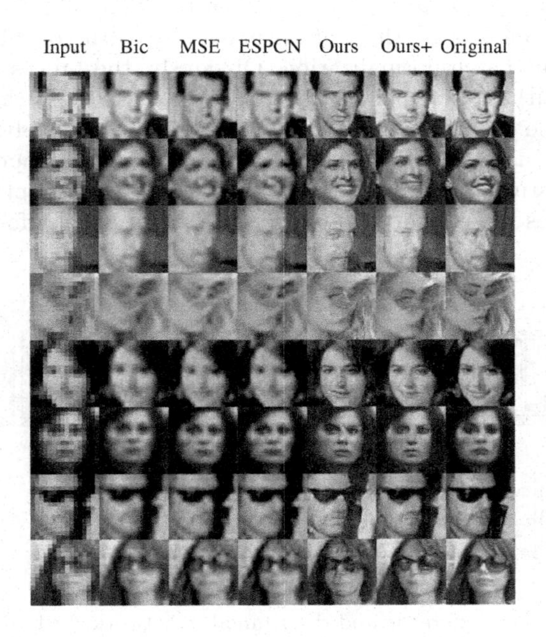

Fig. 3. From left to right, each column corresponds to input, Bic method, MSE method, ESPCN method, Ours method, Ours+ method and original image. Ours+ denotes applying some tricks based on Ours method, such as data preprocessing and enhancing the way of training.

In this work, we propose a convolution content loss as follow:

$$l_X^{LR} = \frac{1}{r^2WH}\sum_{x=1}^{rW}\sum_{y=1}^{rH}(I_{x,y}^{LR} - Conv(G_{\theta_G}(I^{LR})_{x,y}))^2. \quad (2)$$

We feed the output HR image $G_{\theta_G}(I^{LR})$ through multiple convolution layers ($Conv$ in Eq. 2). After down-sampling, the size of output is the same to the input entirely upside-down. Then, the squared Euclidean distance between them is computed as the convolution content loss. Besides the convolution content loss, we add cross entropy loss to complete the generator loss as:

$$l_G = factor \times \underbrace{l_X^{LR}}_{conv\ loss} + (1 - factor) \times \underbrace{l_{Gen}(D_{\theta_D}(G_{\theta_G}(I^{LR})),\ ones)}_{cross\ entropy\ loss}. \quad (3)$$

As Eq. 3 shows, we set a weight factor to control the influence of the two losses. In our experiments, we assign 0.9 to the factor. The discriminator loss is defined as:

$$l_D = \underbrace{l_{Dis}(D_{\theta_D}(I^{HR}),\ ones)}_{cross\ entropy\ loss} + \underbrace{l_{Dis}(D_{\theta_D}(G_{\theta_G}(I^{LR})),\ zeros)}_{cross\ entropy\ loss}. \tag{4}$$

Here, I^{HR} is images from the real dataset, and cross entropy function with the label which is a vector full of ones is used. The cross entropy of generated images is computed with the label of zeros. Finally, we sum the two losses directly as the discriminator loss.

4 Experiments

4.1 Experimental Dataset and Measurement

We conduct our experiments on CelebFaces Attributes dataset (CelebA) which is a large-scale dataset of face attributes with more than 200K celebrity images. All experiments are performed with a scale factor of $4\times$ between low and high resolution images. The input size is 16×16, while the output size increases to 64×64.

Both PSNR and SSIM are used as the performance measures in the evaluation. It is noteworthy that PSNR and SSIM are calculated on gray space. At the same time, we also evaluate the ability of different methods to reconstruct perceptually convincing images by carrying on MOS test. We invite 20 raters to mark 1 (worst) to 5 (best) to represent the visual perception quality. For the compared methods, as Fig. 3 shows, Bic represents the bicubic interpolation and MSE represents that we only optimize the pixel-wise error based three deconvolution layers architecture. ESPCN is proposed in [5]. Moreover, based on Ours method, we apply some tricks on data preprocessing as mentioned above and use novel training method to train. It is represented by Ours+ in Fig. 3. For each compared method, 30 images are chosen randomly from the generated dataset. Thus, each rater rated 180 instances which were presented in a random fashion.

4.2 Training Details and Parameters

Now, we will introduce the training details. All networks are trained on a NVIDIA Titan X GPU. ESPCN takes the advantage of sub-pixel convolutional layer which rearranges the feature maps into high resolution image directly. We wanted to import the layer to our GAN architecture originally. However, we find that it will increase the risk of model collapse during training. The output images come with checkerboard pattern [23] when adding the sub-pixel convolutional layer. As Fig. 4 shows, the PSNR and SSIM scores decrease heavily after several epochs but recover for a while. At last, we figure out that activation function plays a critical role for sub-pixel convolutional layer. In the experiments, we use the sigmoid activation function. However, we replace the sigmoid activation function with tanh behind that layer, and the small change will improve the collapse situation. The blue line in Fig. 4 is a non-adversarial network and the others are based on GAN. As Fig. 4 shows, the blue line is more smooth and stable than the other lines. It also explains that training GAN is not an easy job.

In the training phase, Ours+ method updates discriminator five times before updating the generator one time every batch for the first 10 epochs, which is different from Ours method that updates the generator and discriminator alternately. After 10 epochs, we update generator and discriminator equally. Meanwhile, we optimize the pixel-wise error between ground truth and SR images for generator for every 10 batches. For optimization, we assign the β value of Adam as 0.9 and learning rate as 0.0001. For generator network, the input image after up-sampling is added to output layer directly through skip connection. We set the weight factor to 0.5 for the two parts separately.

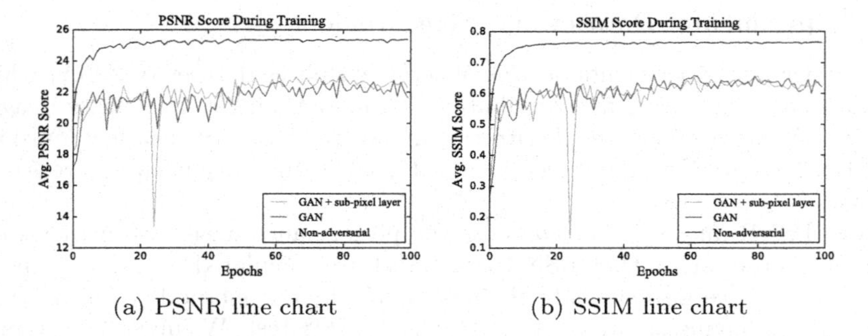

(a) PSNR line chart (b) SSIM line chart

Fig. 4. The sub-pixel convolutional layer in GAN-based architecture will increase the risk of model collapse. The non-adversarial line stands for the generator architecture we proposed which optimizes the pixel-wise error. Comparing the non-adversarial line with others, we can find that non-adversarial network is more stable than GAN-based network. (Color figure online)

4.3 Result and Analysis

In this experiment, we verify that the GAN-based model can reconstruct high resolution images sharper and more real. From Table 1, we show that the MSE trained model can achieve better scores at PSNR and SSIM as expected. However, Ours method performs better on visual perception in terms of MOS. Applying some tricks in Ours+ method (including preprocessing the dataset with saturation, brightness, contrast ratio and direction, training more times discriminator than generator in the beginning stage and optimizing the pixel-wise error

Table 1. PSNR, SSIM and MOS test results of different methods.

	Bic	MSE	ESPCN	Ours	Ours+	Original
PSNR	21.722	24.855	25.492	23.281	24.577	∞
SSIM	0.580	0.715	0.741	0.688	0.731	1
MOS	1.85	2.14	2.44	3.14	3.34	4.24

between ground truth and SR images for some intervals) can further improve the quality of the SR image.

5 Conclusion

In this paper, we introduce a GAN-based residual network which super-resolve images for large 4× upscaling factors. We also validate that the MSE loss can lead to higher PSNR but result in perceptually poor output. The GAN-based architecture which produces perceptually more convincing SR image by narrowing the difference between the real image dataset and reconstructed images is very promising. We not only combine both the convolution content loss and adversarial loss, but also use some tricks on dataset preprocessing and network training. These changes can make the generator perform better. As for MOS testing, our enhanced method obtains sharper and more real SR images than other methods.

Acknowledgments. This work is supported by the National Natural Science Foundation of China (grants No. 61672133, No. 61502080 and No. 61632007) and the Fundamental Research Funds for the Central Universities (grants No. ZYGX2015J058 and No. ZYGX2014Z007).

References

1. Nasrollahi, K., Moeslund, T.B.: Super-resolution: a comprehensive survey. Mach. Vis. Appl. **25**(6), 1423–1468 (2014)
2. Yang, C.-Y., Ma, C., Yang, M.-H.: Single-image super-resolution: a benchmark. In: Fleet, D., Pajdla, T., Schiele, B., Tuytelaars, T. (eds.) ECCV 2014. LNCS, vol. 8692, pp. 372–386. Springer, Cham (2014). doi:10.1007/978-3-319-10593-2_25
3. Yang, J., Wright, J., Huang, T.S., Ma, Y.: Image super-resolution as sparse representation of raw image patches. In: 2008 IEEE Computer Society Conference on Computer Vision and Pattern Recognition (CVPR 2008), 24–26 June 2008, Anchorage, Alaska, USA (2008)
4. Dong, C., Loy, C.C., He, K., Tang, X.: Image super-resolution using deep convolutional networks. IEEE Trans. Pattern Anal. Mach. Intell. **38**(2), 295–307 (2016)
5. Shi, W., Caballero, J., Huszar, F., Totz, J., Aitken, A.P., Bishop, R., Rueckert, D., Wang, Z.: Real-time single image and video super-resolution using an efficient sub-pixel convolutional neural network. In: 2016 IEEE Conference on Computer Vision and Pattern Recognition, CVPR 2016, Las Vegas, NV, USA, 27–30 June 2016, pp. 1874–1883 (2016)
6. He, K., Zhang, X., Ren, S., Sun, J.: Identity mappings in deep residual networks. In: Leibe, B., Matas, J., Sebe, N., Welling, M. (eds.) ECCV 2016. LNCS, vol. 9908, pp. 630–645. Springer, Cham (2016). doi:10.1007/978-3-319-46493-0_38
7. Goodfellow, I.J., Pouget-Abadie, J., Mirza, M., Xu, B., Warde-Farley, D., Ozair, S., Courville, A.C., Bengio, Y.: Generative adversarial nets. In: Advances in Neural Information Processing Systems 27: Annual Conference on Neural Information Processing Systems 2014, 8–13 December 2014, Montreal, Quebec, Canada, pp. 2672–2680 (2014)

8. Farsiu, S., Robinson, M.D., Elad, M., Milanfar, P.: Fast and robust multiframe super resolution. IEEE Trans. Image Process. **13**(10), 1327–1344 (2004)
9. Kim, J., Lee, J.K., Lee, K.M.: Accurate image super-resolution using very deep convolutional networks. In: 2016 IEEE Conference on Computer Vision and Pattern Recognition, CVPR 2016, Las Vegas, NV, USA, 27–30 June, pp. 1646–1654 (2016)
10. Sønderby, C.K., Caballero, J., Theis, L., Shi, W., Huszár, F.: Amortised MAP inference for image super-resolution. CoRR abs/1610.04490 (2016)
11. Wang, Z., Bovik, A.C., Sheikh, H.R., Simoncelli, E.P.: Image quality assessment: from error visibility to structural similarity. IEEE Trans. Image Process. **13**(4), 600–612 (2004)
12. Sun, J., Sun, J., Xu, Z., Shum, H.: Gradient profile prior and its applications in image super-resolution and enhancement. IEEE Trans. Image Process. **20**(6), 1529–1542 (2011)
13. Yang, J., Lin, Z., Cohen, S.: Fast image super-resolution based on in-place example regression. In: 2013 IEEE Conference on Computer Vision and Pattern Recognition, Portland, OR, USA, 23–28 June 2013, pp. 1059–1066 (2013)
14. Freeman, W.T., Jones, T.R., Pasztor, E.C.: Example-based super-resolution. IEEE Comput. Graph. Appl. **22**(2), 56–65 (2002)
15. Johnson, J., Alahi, A., Fei-Fei, L.: Perceptual losses for real-time style transfer and super-resolution. In: Leibe, B., Matas, J., Sebe, N., Welling, M. (eds.) ECCV 2016. LNCS, vol. 9906, pp. 694–711. Springer, Cham (2016). doi:10.1007/978-3-319-46475-6_43
16. Dong, C., Loy, C.C., He, K., Tang, X.: Learning a deep convolutional network for image super-resolution. In: Fleet, D., Pajdla, T., Schiele, B., Tuytelaars, T. (eds.) ECCV 2014. LNCS, vol. 8692, pp. 184–199. Springer, Cham (2014). doi:10.1007/978-3-319-10593-2_13
17. Yang, J., Wright, J., Huang, T.S., Ma, Y.: Image super-resolution via sparse representation. IEEE Trans. Image Process. **19**(11), 2861–2873 (2010)
18. Kim, J., Lee, J.K., Lee, K.M.: Deeply-recursive convolutional network for image super-resolution. In: 2016 IEEE Conference on Computer Vision and Pattern Recognition, CVPR 2016, Las Vegas, NV, USA, 27–30 June 2016, pp. 1637–1645 (2016)
19. Ledig, C., Theis, L., Huszar, F., Caballero, J., Aitken, A.P., Tejani, A., Totz, J., Wang, Z., Shi, W.: Photo-realistic single image super-resolution using a generative adversarial network. CoRR abs/1609.04802 (2016)
20. Radford, A., Metz, L., Chintala, S.: Unsupervised representation learning with deep convolutional generative adversarial networks. CoRR abs/1511.06434 (2015)
21. Ioffe, S., Szegedy, C.: Batch normalization: accelerating deep network training by reducing internal covariate shift. In: Proceedings of the 32nd International Conference on Machine Learning, ICML 2015, Lille, France, 6–11 July 2015, pp. 448–456 (2015)
22. Zeiler, M.D., Taylor, G.W., Fergus, R.: Adaptive deconvolutional networks for mid and high level feature learning. In: IEEE International Conference on Computer Vision, ICCV 2011, Barcelona, Spain, 6–13 November 2011, pp. 2018–2025 (2011)
23. Odena, A., Dumoulin, V., Olah, C.: Deconvolution and checkerboard artifacts. Distill **1**(10), e3 (2016)

An ELU Network with Total Variation for Image Denoising

Tianyang Wang[1(✉)], Zhengrui Qin[1], and Michelle Zhu[2]

[1] Northwest Missouri State University, Maryville, MO 64468, USA
toseattle@siu.edu, zqin@nwmissouri.edu
[2] Montclair State University, Montclair, NJ 07043, USA
zhumi@montclair.edu

Abstract. In this paper, we propose a novel convolutional neural network (CNN) for image denoising, which uses exponential linear unit (ELU) as the activation function. We investigate the suitability by analyzing ELU's connection with trainable nonlinear reaction diffusion model (TNRD) and residual denoising. On the other hand, batch normalization (BN) is indispensable for residual denoising and convergence purpose. However, direct stacking of BN and ELU degrades the performance of CNN. To mitigate this issue, we design an innovative combination of activation layer and normalization layer to exploit and leverage the ELU network, and discuss the corresponding rationale. Moreover, inspired by the fact that minimizing total variation (TV) can be applied to image denoising, we propose a TV regularized L2 loss to evaluate the training effect during the iterations. Finally, we conduct extensive experiments, showing that our model outperforms some recent and popular approaches on Gaussian denoising with specific or randomized noise levels for both gray and color images.

Keywords: Image denoising · Convolutional neural network · ELU · Total variation · Deep learning · Image processing

1 Introduction

Image denoising has been a long-time open and challenging research topic in computer vision, aiming to restore the latent clean image from a noisy observation. Generally, a noisy image can be modeled as $y = x + v$, where x is the latent clean image and v is the additive Gaussian white noise. To restore the clean mapping x from a noisy observation y, there are two main categories of methods, namely image prior modeling based and discriminative learning based. Traditional methods, such as BM3D [1], LSSC [2], EPLL [3], and WNNM [4], lie in the first category. And the second category, pioneered by Jain et al. [5], includes MLP [6], CSF [7], DGCRF [8], NLNet [9], and TNRD [10]. Until recently, Zhang et al. [11] discovered a deep residual denoising method to learn the noisy mapping

D. Liu et al. (Eds.): ICONIP 2017, Part III, LNCS 10636, pp. 227–237, 2017.
https://doi.org/10.1007/978-3-319-70090-8_24

with excellent results. However, there is still leeway to boost the denoising performance by reconsidering the activation and the loss function in convolutional neural network (CNN).

In this paper, we propose a deep CNN with exponential linear unit (ELU) [12] as the activation function and total variation (TV) as the regularizer of L2 loss function for image denoising, which achieves noticeable improvement compared to the state-of-the art work [11] in which the rectified linear unit (ReLU) [13] was used as the activation function. By analyzing the traits of ELU and its connection with trainable nonlinear reaction diffusion (TNRD) [10] and residual denoising [11], we show that ELU is more suitable for image denoising applications. Specifically, our method is based on residual learning, and the noisy mapping learned with ELU has a higher probability to obtain a desired 'energy' value than that learned with ReLU. It indicates that more noise can be removed from the original noisy observation, hence the denoising performance can be improved. On the other hand, batch normalization (BN) [14] is also applied in the model for the purpose of training convergence. However, Clevert et al. [12] pointed out that the direct combination of BN and ELU would degrade the network performance. Instead, we construct a new combination of layers by incorporating 1×1 convolutional layers, which can better integrate the BN and ELU layers. In our model, we set 'Conv-ELU-Conv-BN' as the fundamental block, where the second 'Conv' denotes the 1×1 convolutional layer. Furthermore, we utilize TV, which is a powerful regularizer in traditional denoising methods [15–17], to regularize L2 loss to further improve the network training performance. Without considering the dual formulation, the TV regularizer can still be solved by stochastic gradient decent (SGD) algorithm during the network training. Finally, we conduct extensive experiments to validate the effectiveness of our proposed approach.

The main contributions of this work can be generalized in three-folds. First, we have analyzed the suitability of ELU to denoising task. Second, we have proposed a novel combination of layers to better accommodate ELU and BN. Third, we have applied total variation to regularize L2 loss function. The rest of paper is organized as follows. The proposed network with ELU and TV is presented in Sect. 2 with the analysis of rationale. Extensive experiments and evaluation results can be found in Sect. 3. Section 4 concludes our work.

2 The Proposed Network

In our approach, a noisy mapping, rather than a clean mapping, is learned since residual learning had been proven successful for image denoising [11]. Besides, residual learning had been validated effective for scatter correction in medical image processing [18] which requires higher reliability. Before presenting our network architecture, we first discuss the ELU and its intrinsic property for denoising task, followed by how to regularize L2 loss with total variation. Our analysis on both ELU and TV are mainly derived from the energy perspective as denoising is closely relevant to energy reduction.

2.1 Exponential Linear Unit

The primary contribution of an activation function is to incorporate nonlinearity into a stack of linear convolutional layers to increase the network ability of capturing discriminative image features. As one of the activation functions, ELU [12] is defined as:

$$f(x) = \begin{cases} x & \text{if } x > 0 \\ \alpha(e^x - 1) & \text{if } x \leq 0 \end{cases} \tag{1}$$

where parameter α is used to control the level of ELU's saturation for negative inputs and a pre-determined value can be used for the entire training procedure. Unlike ReLU, the most frequently used activation function, ELU does not force the negative input to be zero, which can make the mean unit activation approach zero value since both positive and negative values can counteract each other in the resulted matrix. The near zero mean unit activation not only speeds up learning with a faster convergence but also enhances system robustness to noise. Although ELU has higher time complexity than other activation functions due to the exponential calculation, it can be tolerated if better domain performance is desired.

2.2 Motivation of Using ELU

For ELU network, Clevert et al. [12] reported a significant improvement on CIFAR-100 classification over the ReLU network with batch normalization. On ImageNet, ELU network also obtained a competitive performance with faster convergence compared to ReLU network. To the best of our knowledge, there is no existing work exploring the connection of ELU with Gaussian image denoising. In our work, we note that using different activation functions can generate residual mappings with different 'energy', which can be interpreted as angular second moment (ASM) and computed as follows

$$ASM = \sum_{i,j=0}^{N-1} P_{i,j}^2 \tag{2}$$

In practice, $P_{i,j}$ is an element of the gray-level co-occurrence matrix (GLCM) of a noisy mapping: $P_{i,j} \in \text{GLCM}(v)$. Since noisy image has lower ASM compared to a clean one, learning a noisy mapping with lower ASM can be expected. For better clarification, we study the connection between the residual denoising and TNRD [10] which was initially analyzed by Zhang et al. in [11]. According to their work, such a relation can be described by

$$v = y - x = \lambda \sum_{k=1}^{K} (\bar{f}_k * \phi_k(f_k * y)) \tag{3}$$

where v is the estimated residual of the latent clean image x with respect to the noisy observation y. f_k is a convolutional filter used in a typical CNN, and $\bar{f}_k$

is the filter obtained by rotating the filter f_k by 180°. We ignore the constant parameter λ since it only weights the right side term in Eq. (3). The influence function ϕ can be an activation function applied to feature maps or the original input. For residual denoising problem, the noisy mapping v should contain as much noise as possible. Therefore, the ASM is expected to be low. According to Eqs. (2) and (3), our goal is to choose the right activation function ϕ to have $ASM(v)_{\phi} < ASM(v)_{ReLU}$. To choose an appropriate ϕ, we conduct a simple experiment on three benchmark datasets, namely Pascal VOC2011, Caltech101, and 400 images of size 180×180 from BSD500 dataset that we use to train our network in Sect. 3. For each clean image, Gaussian white noise ($\sigma = 25$) is added to obtain the noisy observation denoted by y. We generate a randomized 3×3 filter as f_k, and take ELU as the function ϕ. The parameter α in Eq. (1) is set to 0.1 for ELU. The comparison of $ASM(v)_{ELU}$ and $ASM(v)_{ReLU}$ is given in Table 1.

Table 1. The comparison of $ASM(v)_{ELU}$ and $ASM(v)_{ReLU}$

	VOC 2011	Caltech 101	BSD 400
$ASM(v)_{ELU} > ASM(v)_{ReLU}$	5310	3275	130
$ASM(v)_{ELU} < ASM(v)_{ReLU}$	9651	5868	270
Percentage of $ASM(v)_{ELU} < ASM(v)_{ReLU}$	65%	64%	68%

It can be observed that there is a higher probability to get a lower ASM value when ELU is utilized as the activation function. As mentioned above, a low ASM corresponds to high noisy image. In residual denoising, higher noisy mapping means that more noise can be removed from the original noisy input, resulting in a better denoising effect. In other words, $ASM(v)$ should be small. Therefore, based on Table 1, ELU is preferred over ReLU as the activation function for higher noisy residual mapping.

2.3 TV Regularizer

In Sect. 2.2, we discuss activation selection to reduce ASM energy of a noisy mapping, and we know that the ASM for a noisy image is smaller than that of a clean counterpart. Unlike the ASM, total variation (TV) evaluates the energy directly from the original input signal. A noisy image has larger TV value than that of a clean one, and image denoising can be performed by minimizing the TV value [15]. Similarly, in residual denoising, the original L2 loss which measures the distance between the residual mapping and the ground truth noise also needs to be minimized. We thus use TV to regularize L2 loss function which is to be minimized by CNN, and the new loss function is defined as:

$$L = \frac{1}{2N} \sum_{i=1}^{N} ||R - (y_i - x_i)||^2 + \beta TV(y_i - R) \tag{4}$$

and according to [15], the TV value can be computed by

$$TV(u) \approx \sum_{i,j} \sqrt{(\nabla_x u)_{i,j}^2 + (\nabla_y u)_{i,j}^2} \tag{5}$$

where we take R as the learned noisy mapping of the latent clean image x_i with respect to the noisy observation y_i, and ∇_x, ∇_y are discretizations of the horizontal and vertical derivatives, respectively. Here, $\{(y_i, x_i)\}_{i=1}^N$ represents the noisy-clean image patch for training. β is used to weigh the total variation term. Though β can be a fixed value during training, our experiments show that updating its value with the change of training epochs could achieve better results. In general, solving a TV regularizer usually requires the dual formulation, however, it can be solved by stochastic gradient decent (SGD) algorithm during training without considering the dual formulation in our work. In Eq. (4), the minimization of the first term (L2 loss) will learn the noisy mapping, and the second term (TV) can be regarded as further denoising the obtained clean mapping.

2.4 Network Architecture

Our model is derived from the vgg-verydeep-19 pre-trained network [19], and includes a total of 15 convolutional layer blocks and 2 separate convolutional layers. There is no fully connected layer. The network architecture is shown in Fig. 1. The first convolutional layer is connected to an ELU layer to add nonlinearity, and the output of the last convolutional layer is fed into the loss layer. Between the two ends, the network is composed of 15 convolutional layer blocks with 'Conv-ELU-Conv-BN' pattern.

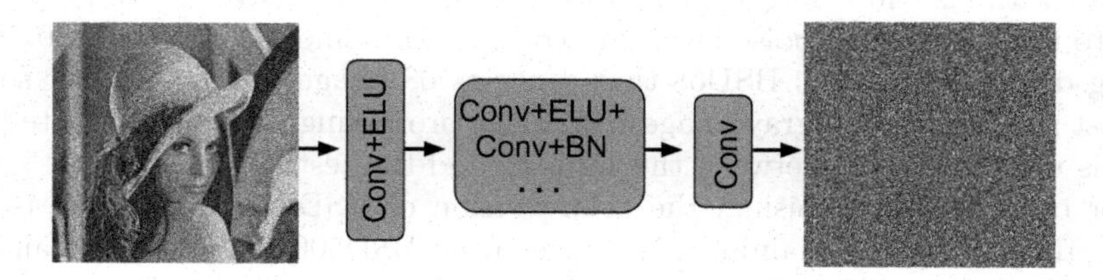

Fig. 1. The network architecture with pipe-lined components.

It has been shown that ELU can replace ReLU as the activation function in Sect. 2.2. Therefore, 'Conv-ELU' is built in each convolutional block. Batch normalization (BN) is necessary for residual denoising as reported by Zhang et al. [11]. However, direct combination of BN and ELU will adversely affect the network performance [12]. Fortunately, it is known that the pixel-wise coefficient transformation can be achieved by a 1×1 convolutional layer, which can also increase the non-linearity of the decision function [19, 20]. We thus utilize a 1×1 convolutional layer between ELU and BN layer. Every second 'Conv' in each block holds 1×1 filters, and other filters are all in the size of 3×3. Such configuration not only exerts the advantages of 1×1 convolutional layer, but also avoids direct connection of BN and ELU.

Note that our model does not contain any pooling layer since the final output must have the same size as the original input. One may argue that fully convolutional networks (FCN) [21] can also restore the output size, however it cannot be used in our case because it contains a pooling layer and thus needs up-sampling operation, which is not desirable for image denoising. Furthermore, FCN was originally designed for pixel-level classification without fully considering the relationships between pixels.

3 Experiments

Our experiments are conducted in Matlab using MatConvNet framework [22], which provides convenient interface to design network structure by adding or removing predefined layers. One NVidia Geforce TITAN X GPU is used to accelerate the mini-batch processing. To validate the efficacy of our method, we train three networks. The first network is for gray image Gaussian denoising with specific noise levels; the second and the third one are for color image Gaussian denoising with specific and randomized noise levels, respectively.

3.1 Data Sets

We choose the experiment datasets similar to the work from [11]. For gray image denoising with a specific noise level, 400 images of size 180×180 from Berkeley segmentation dataset (BSD500) are used for training and 128×1600 patches are cropped with size 40×40 for each. All color images are converted to gray ones prior to training. Three noise levels are considered, namely $\sigma = 15, 25, 50$. Two testing datasets are used: BSD68 that contains 68 images, and the other set of 12 most frequently used gray images[1] in image processing community. Note that there is no overlapping between the training and the testing datasets.

For color image denoising, the color version of BSD68 is employed as the testing data and the remaining 432 images from BSD500 are used for training. $\sigma = 15, 25, 50$ are still used as the specific noise levels, and 128×3000 patches with size 50×50 are cropped. However, for blind denoising, the noise levels are randomly selected from range [0, 55].

3.2 Compared Methods

Besides the well-known methods such as BM3D [1], LSSC [2], WNNM [4], EPLL [3], MLP [6], CSF [7], we also consider another four similar neural network based methods, namely DGCRF [8], NLNet [9], TNRD [10] and DnCNN [11], since these methods have reported promising results.

[1] https://github.com/cszn/DnCNN/tree/master/testsets/Set12.

3.3 Network Training

As explained in Sect. 2.4, our network has 15 convolutional blocks and 2 separate convolutional layers. We use the same depth for both gray and color image denoising. We initialize the weights using MSRA as He et al. [23] did for image classification. The TV regularizer is incorporated into the L2 loss function, and the entire network is trained by SGD with a momentum of 0.9. The initial learning rate is set to be 0.001, and changed to 0.0001 after 30 out of 50 epochs. The initial value of β in Eq. (4) is set to 0.0001, and increased to 0.0005 after 30 epochs. The weight decay is set to 0.0001. It is worth noting that weight decay regularizes the filter weights, whereas total variation regularizes the L2 loss.

3.4 Results Analysis

In our work, peak signal-to-noise ratio (PSNR) is utilized to evaluate the denoising effect. We first compare our method with other well-known methods on BSD68 gray images. The results are given in Table 2, where the best ones are highlighted in bold. It can be seen that our model shows the best average PSNR for all the three specific noise levels. When $\sigma = 50$, our method outperforms BM3D by 0.7 dB, which reaches the estimated upper bound over BM3D in [1]. We further validate our method on the 12 commonly used test images for image processing task, and the average PSNR is compared in Table 3. Our method outperforms DnCNN by around 0.1 dB, which gives similar increments as in Table 2.

Table 2. The average PSNR of different methods on the gray version of BSD68 dataset.

Methods	BM3D	MLP	EPLL	LSSC	CSF	WNNM	DGCRF	TNRD	NLNet	DnCNN	Ours
$\sigma = 15$	31.08	-	31.21	31.27	31.24	31.37	31.43	31.42	31.52	31.73	**31.82**
$\sigma = 25$	28.57	28.96	28.68	28.71	28.74	28.83	28.89	28.92	29.03	29.23	**29.34**
$\sigma = 50$	25.62	26.03	25.67	25.72	-	25.87	-	25.96	26.07	26.23	**26.32**

Table 3. The average PSNR of different methods on the 12 most commonly used gray images in image processing community.

Methods	BM3D	WNNM	EPLL	MLP	CSF	TNRD	DnCNN	Ours
$\sigma = 15$	32.37	32.70	32.14	-	32.32	32.50	32.86	**32.96**
$\sigma = 25$	29.97	30.26	29.69	30.03	29.84	30.06	30.44	**30.55**
$\sigma = 50$	26.72	27.05	26.47	26.78	-	26.81	27.21	**27.29**

Besides gray image denoising, we also train our model with specific and randomized noise levels for color image denoising. Table 4 depicts the competency of our model trained with specific noise levels. Similar to gray image case, our method increases the PSNR by about 0.1 dB compared to DnCNN, which is trained with specific noise levels as well. Note that training with randomized noise levels also generates satisfied results, which, however, are inferior to the results achieved by the models trained with specific noise levels.

Table 4. The average PSNR of different methods on the color version of BSD68 dataset.

Methods	CBM3D	MLP	TNRD	NLNet	DnCNN	Ours
$\sigma = 15$	33.50	-	31.37	33.69	33.99	**34.10**
$\sigma = 25$	30.69	28.92	28.88	30.96	31.31	**31.41**
$\sigma = 50$	27.37	26.01	25.96	27.64	28.01	**28.11**

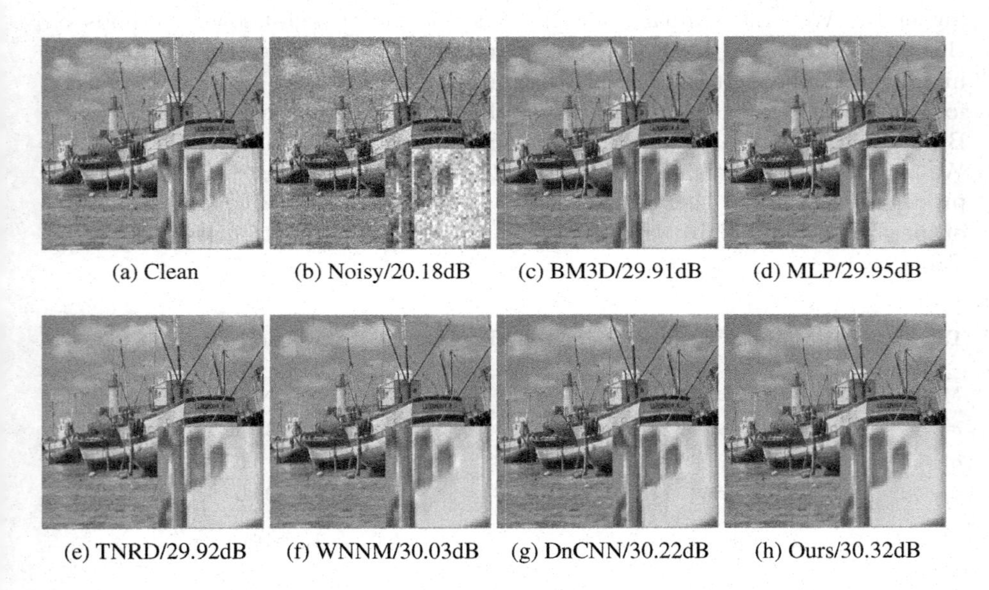

(a) Clean (b) Noisy/20.18dB (c) BM3D/29.91dB (d) MLP/29.95dB

(e) TNRD/29.92dB (f) WNNM/30.03dB (g) DnCNN/30.22dB (h) Ours/30.32dB

Fig. 2. Visual comparison of gray image denoising between our method and other methods. Our model is trained with specific noise level ($\sigma = 25$). The clean image is polluted by noise ($\sigma = 25$).

The visual comparison between our method and other well-known methods are given in Figs. 2, 3 and 4. We add noise ($\sigma = 25$) for one gray image, and our model is trained with a specific noise level ($\sigma = 25$). The denoising effect is shown in Fig. 2. While in Figs. 3 and 4, to validate randomized level (blind) denoising effect, we add two different noise levels ($\sigma = 35, 50$) for each color image, respectively. Note that color denoising visual comparison is carried only between our method and DnCNN, since DnCNN, to our best knowledge, is the state-of-the-art denoising method. Moreover, DnCNN also supports blind denoising. We compare our model with the version of DnCNN which was trained with randomized noise levels in the range of [0, 55]. To achieve fair comparison, our model is also trained with randomized noise levels within the same range. Results show that our model preserves more image details. Moreover, the over-smooth issue of the background scene is also alleviated. Hence, the utilization of total variation does not over-smooth the image.

Fig. 3. Visual comparison of color image denoising between our method and DnCNN. Both models are trained with randomized noise level from range [0, 55]. The noise ($\sigma = 35$) is added to the clean image.

Fig. 4. Visual comparison of color image denoising between our method and DnCNN. Both models are trained with randomized noise level from range [0, 55]. The noise ($\sigma = 50$) is added to the clean image.

4 Conclusion

In this paper, we propose a deep convolutional neural network with exponential linear unit as the activation function and total variation as the regularizer of L2 loss for Gaussian image denoising. By analyzing the advantages of ELU and the connection with residual denoising and trainable nonlinear reaction diffusion model, we have validated that ELU is more suitable for image denoising problem. To better accommodate ELU and BN layer, we design a novel structure by incorporating 1×1 convolutional layer. By studying the traits of total variation, we have shown the feasibility of regularizing L2 loss with TV in convolutional nets. Extensive experiments show promising quantitative and visual results compared with other reputed denoising methods which are based on image prior modeling or discriminative learning.

Acknowledgments. This project was partially supported by the new faculty start-up research grant at Montclair State University.

References

1. Dabov, K., Foi, A., Katkovnik, V., Egiazarian, K.: Image denoising by sparse 3-D transform-domain collaborative filtering. IEEE Trans. Image Process. **16**(8), 2080–2095 (2007)
2. Mairal, J., Bach, F., Ponce, J., Sapiro, G., Zisserman, A.: Non-local sparse models for image restoration. In: 2009 IEEE 12th International Conference on Computer Vision, pp. 2272–2279. IEEE (2009)
3. Zoran, D., Weiss, Y.: From learning models of natural image patches to whole image restoration. In: 2011 IEEE International Conference on Computer Vision (ICCV), pp. 479–486. IEEE (2011)
4. Gu, S., Zhang, L., Zuo, W., Feng, X.: Weighted nuclear norm minimization with application to image denoising. In: Proceedings of the IEEE Conference on Computer Vision and Pattern Recognition, pp. 2862–2869 (2014)
5. Jain, V., Seung, S.: Natural image denoising with convolutional networks. In: Advances in Neural Information Processing Systems, pp. 769–776 (2009)
6. Burger, H.C., Schuler, C.J., Harmeling, S.: Image denoising: can plain neural networks compete with BM3D? In: 2012 IEEE Conference on Computer Vision and Pattern Recognition (CVPR), pp. 2392–2399. IEEE (2012)
7. Schmidt, U., Roth, S.: Shrinkage fields for effective image restoration. In: Proceedings of the IEEE Conference on Computer Vision and Pattern Recognition, pp. 2774–2781 (2014)
8. Vemulapalli, R., Tuzel, O., Liu, M.Y.: Deep Gaussian conditional random field network: a model-based deep network for discriminative denoising. In: Proceedings of the IEEE Conference on Computer Vision and Pattern Recognition, pp. 4801–4809 (2016)
9. Lefkimmiatis, S.: Non-local color image denoising with convolutional neural networks. arXiv Preprint arXiv:1611.06757 (2016)
10. Chen, Y., Pock, T.: Trainable nonlinear reaction diffusion: a flexible framework for fast and effective image restoration. IEEE Trans. Pattern Anal. Mach. Intell. **39**(6), 1256–1272 (2017)

11. Zhang, K., Zuo, W., Chen, Y., Meng, D., Zhang, L.: Beyond a Gaussian denoiser: residual learning of deep CNN for image denoising. IEEE Trans. on Image Process. **26**(7), 3142–3155 (2017)
12. Clevert, D.A., Unterthiner, T., Hochreiter, S.: Fast and accurate deep network learning by Exponential Linear Units (ELUs). arXiv Preprint arXiv:1511.07289 (2015)
13. Krizhevsky, A., Sutskever, I., Hinton, G.E.: ImageNet classification with deep convolutional neural networks. In: Advances in Neural Information Processing Systems, pp. 1097–1105 (2012)
14. Ioffe, S., Szegedy, C.: Batch normalization: accelerating deep network training by reducing internal covariate shift. arXiv Preprint arXiv:1502.03167 (2015)
15. Chan, T., Esedoglu, S., Park, F., Yip, A.: Recent developments in total variation image restoration. Math. Models Comput. Vis. **17**(2) (2005)
16. Goldluecke, B., Cremers, D.: An approach to vectorial total variation based on geometric measure theory. In: 2010 IEEE Conference on Computer Vision and Pattern Recognition (CVPR), pp. 327–333. IEEE (2010)
17. Wang, Y., Chen, W., Zhou, S., Yu, T., Zhang, Y.: MTV: Modified Total Variation model for image noise removal. Electron. Lett. **47**(10), 592–594 (2011)
18. Xu, S., Prinsen, P., Wiegert, J., Manjeshwar, R.: Deep residual learning in CT physics: scatter correction for spectral CT. arXiv Preprint arXiv:1708.04151 (2017)
19. Simonyan, K., Zisserman, A.: Very deep convolutional networks for large-scale image recognition. arXiv Preprint arXiv:1409.1556 (2014)
20. He, K., Zhang, X., Ren, S., Sun, J.: Deep residual learning for image recognition. In: Proceedings of the IEEE Conference on Computer Vision and Pattern Recognition, pp. 770–778 (2016)
21. Long, J., Shelhamer, E., Darrell, T.: Fully convolutional networks for semantic segmentation. In: Proceedings of the IEEE Conference on Computer Vision and Pattern Recognition, pp. 3431–3440 (2015)
22. Vedaldi, A., Lenc, K.: MatConvNet: convolutional neural networks for matlab. In: Proceedings of the 23rd ACM International Conference on Multimedia, pp. 689–692. ACM (2015)
23. He, K., Zhang, X., Ren, S., Sun, J.: Delving deep into rectifiers: surpassing human-level performance on ImageNet classification. In: Proceedings of the IEEE International Conference on Computer Vision, pp. 1026–1034 (2015)

End-to-End Disparity Estimation with Multi-granularity Fully Convolutional Network

Guorun Yang and Zhidong Deng(✉)

State Key Laboratory of Intelligent Technology and Systems,
Tsinghua National Laboratory for Information Science and Technology,
Department of Computer Science, Tsinghua University, Beijing 100084, China
ygr13@mails.tsinghua.edu.cn, michael@mail.tsinghua.edu.cn

Abstract. Disparity estimation is a challenging task in the field of computer stereo vision. In this paper, we propose a multi-granularity fully convolutional network architecture for end-to-end dense disparity estimation. First, we use single well-pretrained residual network for extraction of multi-granularity and multi-layer features. Second, correlation layers at three different granularities are used to gain hierarchical matching cues between left and right feature maps. Third, we conduct concatenation-deconvolution operations to output disparity maps. Finally, the experimental results show that our method achieves state of the art results, taking the second place on the KITTI Stereo 2012 task.

Keywords: Multi-granularity · Correlation · Concatenation-deconvolution · Disparity estimation

1 Introduction

Disparity estimation is a classical problem in the field of stereo vision. It has been extensively applied to many areas such as view synthesis, object detection, and robot navigation. The main goal of disparity estimation is to calculate the displacement of corresponding pixels between left and right images, where corresponding pixels result from identical 3D points projected onto the two image planes. Displacement values at each location forms so-called disparity map.

It is challenging to perform disparity estimation accurately, particularly predict dense disparity map. The majority of stereo algorithms treat such task as a matching problem, measuring similarity between two corresponding patches of left and right images. From this point of view, the main idea of those algorithms is to develop powerful feature representation for image patches. Then the resulting feature vectors can be employed to compute match cost and then pick the best matching pixel between left and right images. In recent years, deep convolutional neural network (CNN) has demonstrated remarkable performance in many fields including computer vision, speech recognition, natural language processing, self-driving, and big data analysis through representation learning of

D. Liu et al. (Eds.): ICONIP 2017, Part III, LNCS 10636, pp. 238–248, 2017.
https://doi.org/10.1007/978-3-319-70090-8_25

hierarchical features based on a large scale labeled data. With utilization of CNN features, such patch-based matching methods can be significantly improved in terms of accuracy of disparity prediction.

Except for disparity calculation based on similarity of image patches, dense disparity map estimation problem could be considered as a pixel-wise labeling task, where each pixel would be assigned a real-value disparity. Lately, inspired by the success of fully-convolutional network (FCN) [1] for semantic segmentation task, such an end-to-end learning structure was introduced to predict disparity map [2]. The combination of encoder (top-down) and decoder (bottom-up) architecture can effectively link the global scene information with local disparity estimation, which leads to further improvements in both accuracy and speed.

In general, FCN models for disparity estimation contain a correlation module to extract matching information from left and right feature maps. Several approaches like [2,3] deploy correlation operations on low-level feature maps. In our opinion, matching cues not only exist in low-level features, but also occur in high-level features. Furthermore, the category information in high-level feature maps could be utilized to compensate matching cues lost in low-level features. For example, in an urban scene, adjacent road and sidewalk are difficult to distinguish from low-level features due to similar colors and textures. However, it would be convenient to differentiate in high-level semantic features. As a result, this paper attempts to extract matching cues from multiple granularity feature maps and aggregate them together. In the proposed method, we first exploit well-pretrained ResNet-50 [4] to obtain different granular hierarchical features. Second, different granular correlation layers are presented to produce feature maps that embed a diversity of matching cues. Finally, we design a concatenation-deconvolution sub-structure to aggregate all the matching information from different granularities and carry out regression of pixel-level disparity values.

The main contributions of this paper are summarized below:

- We learn to represent three different granularities of matching information.
- Those matching cues are aggregated to enhance capabilities of stereo disparity regression.
- On the KITTI Stereo 2012 task [5], the proposed multi-granularity FCN achieves state-of-the-art performance.

2 Related Work

There has been a large amount of work on stereo disparity estimation. In [6] proposed by Scharstein et al., stereo algorithms are regarded to generally include the following four steps: matching cost computation, cost aggregation, disparity computation, and disparity refinement. Several local descriptors based on gradient or binary patterns are designed to compute local matching cost [7,8], accompanying by some global optimization methods to improve results [9].

Zbontar and LeCun [10] used CNN for matching cost computation. Luo et al. [11] proposed a siamese network that extracted marginal distributions over all possible disparities for each pixel. Chen et al. [12] presented a multi-scale deep embedding model that fused features vectors learned within different scale-spaces. Shaked and Wolf [13] proposed a highway network architecture with a hybrid loss that conducted multi-level comparison of image patches.

Inspired by other pixel-wise labeling tasks such as semantic segmentation [1,14,15], the FCN is introduced for the end-to-end learning of disparity map. In 2016, Mayer et al. [2] proposed DispNet for real time disparity estimation. There is a structure similar to their previous work called FlowNet [3], which directly inspires us to use correlation layers for encoding matching cues.

Lately, several researchers extended FCN architecture to make further improvement for disparity or depth estimation. Kendall et al. [16] proposed an architecture called GC-Net that incorporates contextual information by means of 3D convolutions over a cost volume. Gidaris and Komodakis [17] presented a cascade network that had a pipeline to detect, replace, and refine the predicted errors. Kuznietsov et al. [18] proposed a semi-supervised approach for monocular depth map prediction. During training phase, they not only use ground-truth depth for supervised learning, but also define an alignment loss based on photo consistency. In this paper, we first employ well-pretrained ResNet-50 to have extraction of multi-scale and multi-layer feature maps and then adopt three different granularitie of correlation layers to get a diversity of matching information. Those matching cues are further aggregated to improve performance of stereo disparity regression. Finally, on the KITTI Stereo 2012 task [5], the proposed multi-granularity FCN achieves state-of-the-art results, ranking second compared to the other 94 competitors.

3 Model Architecture

Our multi-granularity FCN (MG-FCN) architecture is shown in Fig. 1. This is a data-driven model that enables end-to-end disparity learning. It is observed from Fig. 1 that it could be divided into three sub-structures: representation of multi-granularity features, correlation layers and concatenation-deconvolution.

3.1 Representation of Multi-granularity Features

Unlike computing matching cost on the pair of original images, this paper extracts three granularities of hierarchical features of left and right raw images. For this sake, we exploit a ResNet-50 that was well pretrained on a large scale benchmark of ImageNet. ResNet [4] is currently believed as one of the best CNN model due to allowing the network to have much deeper layers. As shown in Fig. 1, such a ResNet-50 model comprises three blocks that output three different granularities of features, respectively, which implies that the feature maps of ${M_1}^L$, ${M_1}^R$, ${M_2}^L$, ${M_2}^R$, ${M_3}^L$, ${M_3}^R$ will be used as inputs of incoming correlation layers.

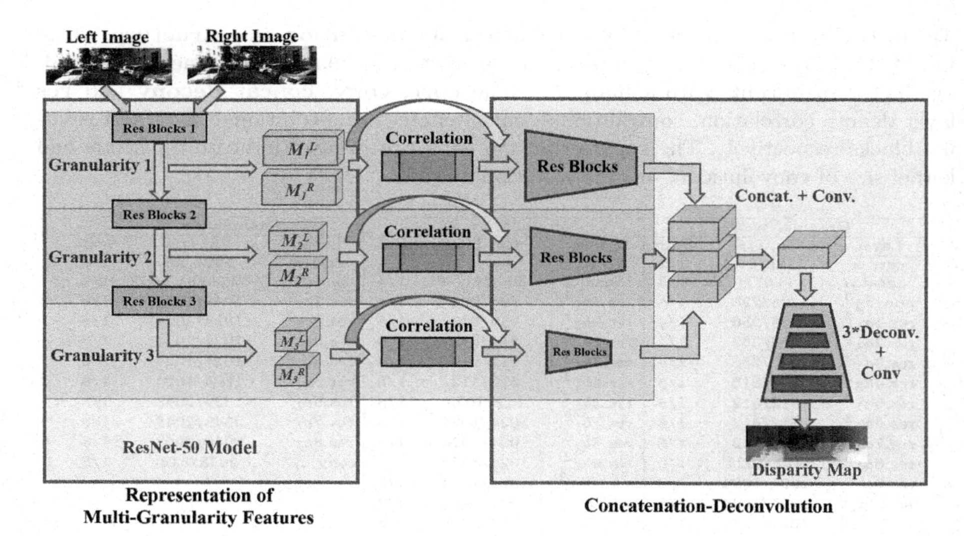

Fig. 1. Our MG-FCN architecture. *Res Blocks* indicate components in residual network, which comprises convolutional, batch normalization, and ReLU layers with split-transform-merge strategy, and the blue cubes represent feature maps. (Color figure online)

3.2 Correlation Layers

The three granularities of correlation layers, which involves the description of matching cost between corresponding patches, are critical in the MG-FCN architecture. Fischer et al. [3] defined a correlation layer in the FlowNet for optical flow estimation. This paper presents three different granularities of correlation layers for a diversity of matching cues. Given one displacement value, correlation layers are used to convolve left and right feature maps of ${M_i}^L$, ${M_i}^R$ $(i = 1, 2, 3)$ and further make summation of resulting multi-channel maps to generate one final matching feature map. The correlation of two patches centered at x_1 in M_i^L and x_2 in M_i^R is defined as

$$c(x_1, x_2) = \sum_{o \epsilon [-k,k] \times [-k,k]} \langle {M_i}^L(x_1 + o), {M_i}^R(x_2 + o) \rangle \quad (1)$$

where $K = 2k + 1$ is the size of patch. We set the maximum displacement β_i $(i = 1, 2, 3)$ to restrict search of possible patch-pairs. The correlation $c(x_1, x_2)$ is only calculated in the neighborhood of size $s_i = \beta_i + 1$, which implies uni-direction searching on ${M_i}^R$. Finally, the size of resulting feature maps for each of three granularities of correlation layers is $(s_i \times w \times h)$, where w indicates the width and h the height.

Table 1. The layers in our Concatenation-deconvolution sub-structure, where **Ch. I/O** denotes channels of input and output feature maps, **Scale** means the scaling factor of output feature map size. The **corr**, **conv**, **concat deconv** and **res** layer denote correlation, convolutional, concatenate, deconvolutional layer and residual blocks respectively. The superscript and subscript of layer indicate the stride and kernel size of convolutional or deconvolutional layer.

Granularity #1			Granularity #2			Granularity #3		
Layer	Ch. I/O	Scale	Layer	Ch. I/O	Scale	Layer	Ch. I/O	Scale
corr_a	128/97	1/2	corr_b	256/49	1/4	corr_c	1024/25	1/8
concat_a	(128+97)/225	1/2	concat_b	(256+49)/305	1/4	concat_c	(1024+25)/1049	1/8
poola_2^3	225/225	1/4	res_1b_1^3	305/256	1/4	res_1c_1^3	1049/1024	1/8
res_1a_1^3	225/256	1/4	res_2b_2^3	256/512	1/8	res_2c_1^3	1024/1024	1/8
res_2a_1^3	256/256	1/4	res_3b_1^3	512/512	1/8	res_3c_1^3	1024/1024	1/8
res_3a_1^3	256/256	1/4	res_4b_1^3	512/512	1/8	res_4c_1^3	1024/1024	1/8
res_4a_2^3	256/512	1/8	res_5b_1^3	512/512	1/8	res_5c_1^3	1024/1024	1/8
res_5a_1^3	512/512	1/8	res_6b_1^3	512/1024	1/8	res_6c_1^3	1024/2048	1/8
res_6a_1^3	512/512	1/8	res_7b_1^3	1024/1024	1/8	res_7c_1^3	2048/2048	1/8
res_7a_1^3	512/512	1/8	res_8b_1^3	1024/1024	1/8	res_8c_1^3	2048/2048	1/8
res_8a_1^3	512/1024	1/8	res_9b_1^3	1024/1024	1/8	conv_c_1^3	2048/512	1/8
res_9a_1^3	1024/1024	1/8	res_10b_1^3	1024/1024	1/8			
res_10a_1^3	1024/1024	1/8	res_11b_1^3	1024/1024	1/8			
res_11a_1^3	1024/1024	1/8	res_12b_1^3	1024/2048	1/8			
res_12a_1^3	1024/1024	1/8	res_13b_1^3	2048/2048	1/8			
res_13a_1^3	1024/1024	1/8	res_14b_1^3	2048/2048	1/8			
res_14a_1^3	1024/2048	1/8	conv_b_1^3	2048/512	1/8			
res_15a_1^3	2048/2048	1/8						
res_16a_1^3	2048/2048	1/8						
conv_a_1^3	2048/512	1/8						

Layer	Channels I/O	Scale	Inputs
concat	(512+512+512)/1536	1/8	conv_a, conv_b, conv_c
conv_2_1^1	1536/512	1/8	concat
deconv_1_1^3	512/256	1/4	conv_2
deconv_2_1^3	256/128	1/2	deconv_1
deconv_3_1^3	128/64	1	deconv_2
conv_3_1^1	64/1	1	deconv_3

3.3 Concatenation-Deconvolution

The concatenation-deconvolution sub-structure is designed to conduct feature aggregation and regression of stereo disparity values based on preceding feature maps that contain three different granularities of matching information from the correlation layers. As shown in Fig. 1 and Table 1, three residual blocks are used to further encode corresponding matching features before concatenation. In order to reduce the number of feature channels, we employ one $1*1$ convolutional layer to merge the concatenated feature maps. Finally, three deconvolutional layers and an extra convolutional layers are adopted to generate stereo deparity values.

The last convolutional layer outputs the predicted disparity maps. For end-to-end learning, it is required to define a loss function to measure the errors between the predicted disparity maps and the ground truths. This paper directly computes the absolute errors (L1-norm) between the predicted values d_i and the ground-truths $\hat{d}_i$ for each valid disparity pixels. Compared to other norms used for loss functions, we believe that the L1-norm function is more intuitive to describe the deviation between predicted disparities and the ground truths.

$$Loss(I_l, I_r, D) = \frac{1}{N_{\Omega_D}} \sum_{i \epsilon \Omega_D} ||d_i - \hat{d}_i||_1 \tag{2}$$

where Ω_D denotes the set of valid pixels that have the ground truths, N_{Ω_D} the number of valid pixels, I_l the left image, I_r the right image and D the ground truth of disparity map.

4 Experimental Results

We evaluated our method on CityScapes [19] and KITTI Stereo 2012 [5] datasets. Both of the two datasets provide stereo images and disparit ground truths. On CityScapes benchmark, the disparity maps are pre-computed by the SGM algorithm [9]. We use the "*gtFine*" subset that contains 5,000 images. The official split on this subset is that 2,975 images are exploited for training and 500 images are used for validation. On KITTI Stereo 2012 dataset, there are 194 training images with sparse disparity ground truth and 195 test images. To facilitate the comparison among different architectures, we split the training dataset like that of Luo et al. [11], in which 160 images are randomly selected for training and the remaining 34 images are adopted for validation.

In order to verify the performance of the MG-FCN, we compare it with SG-FCNs on both CityScapes and KITTI Stereo 2012. In each of three SG-FCNs, a single-granularity feature is extracted, followed by one correlation layer, residual blocks, and three deconvolution layers to learn disparity.

4.1 Implementation Details

We initially pre-trained the ResNet-50 [4] on a large scale ImageNet dataset. The three feature maps at *conv*1, *pool*1 and *res*4*a* layer of well-pretrained ResNet-50 are used for the three granularities of correlation operations. In three granularities of correlation layers, we set the maximum displacement $d = 96, 48, 24$, respectively. In the concatenation-deconvolution sub-structure, we adopted the same initialization procedure as He et al. [20] in the convolution and deconvolution layers involved. Meanwhile, we used Caffe framework and stochastic gradient descent (SGD) with momentum of 0.9 to train the MG-FCN. To avoid overfittings, we employed L2 regularization on the weights with decay of $w_d = 0.0001$.

Considering that the KITTI dataset only contains small and sparse labeled samples, we first trained the MG-FCN on CityScapes dataset with the initial learning rate $lr = 0.01$ and then fine-tuned it on KITTI dataset with the initial learning rate $lr = 0.001$. We exploited the polynomial learning rate policy with 90k iterations. Moreover, we took a random resize factor of $\alpha \in [0.5, 2.0]$ and the crop size of $513 * 321$ for data augmentation.

4.2 Results

The experimental results in Table 2 show the test error of three single-granularity FCN (SG-FCN) models and MG-FCN model on the validation dataset of CityScapes and KITTI Stereo. The SG-FCN #1 means that we concatenate correlation layer on *conv*1 feature maps of ResNet-50. The SG-FCN #2 and SG-FCN #3 are linked to *pool*1 and *conv*4*a* feature maps, respectively. The items $> i$ pixels ($i = 1, 2, 3, 4$) indicate different thresholds adopted to decide whether an estimated disparity value is correct. Numerical results in Table 2 measure the proportion of mistaken disparity pixels. The above comparison demonstrates that the MG-FCN model performs significantly better than the three single-granularity FCN (SG-FCN) models through the aggregation of matching cues on multiple granularities. Figure 2 shows the qualitative results on CityScapes, KITTI validation and test datasets respectively.

Table 2. The test error of SG-FCNs and MG-FCN across different error thresholds on the CityScapes and KITTI 2012 dataset

	CityScapes				KITTI			
	>2 px	>3 px	>4 px	>5 px	>2 px	>3 px	>4 px	>5 px
SG-FCN#1	5.38	3.16	2.28	1.81	4.61	2.86	2.06	1.60
SG-FCN#2	5.92	3.33	2.37	1.88	5.24	3.12	2.23	1.73
SG-FCN#3	7.20	4.01	2.75	2.12	8.64	4.67	2.89	2.00
MG-FCN	**4.35**	**2.60**	**1.90**	**1.55**	**4.31**	**2.67**	**1.94**	**1.52**

In Table 3, we evaluated our method on KITTI 2012 benchmark [5]. The item "Noc" refers to evaluation on non-occluded regions, i.e., regions for which the matching correspondence is inside the image domain, while "All" refers to evaluation on all image regions for which ground truth could be measured. "End-Point" denote the average end-point deviation between predicted disparity values and ground truth. Our MG-FCN achieves state-of-the-art results, which outperforms most patch-based methods [10,11,21] on both accuracy and runtime. Among FCN methods [2], our model is also competitive, just behind the GC-Net [16], ranking second on the ratings (http://www.cvlibs.net/datasets/kitti/eval_stereo_flow.php?benchmark=stereo).

(a) CityScapes data qualitative results on validation dataset. From left: left stereo input image, ground truth, disparity prediction

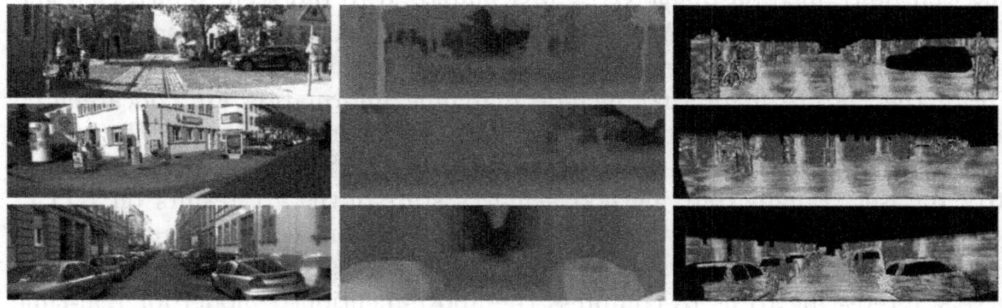

(b) KITTI data qualitative results on validation dataset. From left: left stereo input image, disparity prediction, error map.

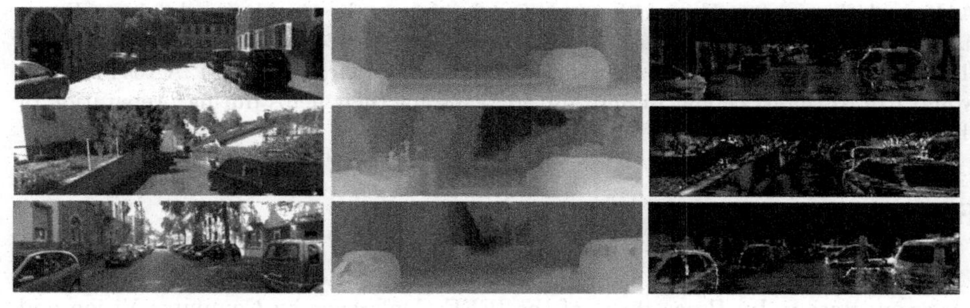

(c) KITTI data qualitative results on test dataset. From left: left stereo input image, disparity prediction, error map.

Fig. 2. Qualitative results. By learning to aggregate multi-granularity matching cues, our method could perform accurate disparity estimation on challenging scenarios.

Table 3. Comparision to state-of-art results on the KITTI 2012 benchmark

	>2 pixels		>3 pixels		>4 pixels		>5 pixels		End-Point		Runtime (s)
	Noc	All	Noc	All	Noc	All	Noc	All	Noc	All	
GC-NET [16]	2.71	3.46	1.77	2.30	1.36	1.77	1.12	1.46	0.6 px	0.7 px	0.9
L-ResMatch [13]	3.64	5.06	2.27	3.40	1.76	2.67	1.50	2.26	0.7 px	1.0 px	48
PBCP [21]	3.62	5.01	2.36	3.45	1.88	2.74	1.62	2.32	0.7 px	0.9 px	68
Displets v2 [22]	3.43	4.46	2.37	3.09	1.97	2.52	1.72	2.17	0.7 px	0.8 px	265
MC-CNN-arct [10]	3.90	5.45	2.43	3.63	1.90	2.85	1.64	2.39	0.7 px	0.9 px	67
Content-CNN [11]	4.98	6.51	3.07	4.29	2.39	3.36	2.03	2.82	0.8 px	1.0 px	0.7
Deep Embed [12]	5.05	6.47	3.10	4.24	1.73	2.32	1.92	2.68	0.9 px	1.1 px	3
DispNetC [2]	7.38	8.11	4.11	4.65	2.77	3.30	2.05	2.39	0.9 px	1.0 px	0.06
MG-FCN (Ours)	**3.73**	**4.41**	**2.17**	**2.68**	**1.56**	**1.97**	**1.22**	**1.56**	**0.8 px**	**0.8 px**	**0.6**

5 Conclusions

In this paper, we propose a MG-FCN model for end-to-end disparity estimation. In such a new pixel-level disparity prediction method, one ResNet-50 that was well pretrained on ImageNet is first employed to represent multi-scale and multi-layer features of raw left and right images. Second, we present three different granularities of correlation layers to seek a diversity of matching cues. Thrid, the feature maps that include matching information are concatenated and merged so as to perform stereo disparity regression. We evaluate the performance of the proposed MG-FCN model on both CityScapes and KITTI Stereo 2012 dataset. Finally, our method achieves state-of-the-art results on KITTI Stereo 2012 benchmark. In the future, we will focus on semi-supervised or even unsupervised learning methods for such a challenging problem. Meanwhile, it is very interesting to us to compress the above-mentioned model for real-time applications such as self-driving car.

Acknowledgments. This work was supported in part by the National Science Foundation of China (NSFC) under Grant Nos. 91420106, 90820305, and 60775040, and by the National High-Tech R&D Program of China under Grant No. 2012AA041402. We would like to thank Zeping Li and Shiyao Wang for their helps during preparation of this paper.

References

1. Long, J., Shelhamer, E., Darrell, T.: Fully convolutional networks for semantic segmentation. In: Proceedings of the IEEE Conference on Computer Vision and Pattern Recognition, pp. 3431–3440 (2015)
2. Mayer, N., Ilg, E., Häusser, P., Fischer, P., Cremers, D., Dosovitskiy, A., Brox, T.: A large dataset to train convolutional networks for disparity, optical flow, and scene flow estimation. In: IEEE International Conference on Computer Vision and Pattern Recognition (CVPR) arXiv:1512.02134 (2016)

3. Dosovitskiy, A., Fischer, P., Ilg, E., Häusser, P., Hazırbaş, C., Golkov, V., van der Smagt, P., Cremers, D., Brox, T.: FlowNet: learning optical flow with convolutional networks. In: IEEE International Conference on Computer Vision (ICCV) (2015)
4. He, K., Zhang, X., Ren, S., Sun, J.: Deep residual learning for image recognition. In: Proceedings of the IEEE Conference on Computer Vision and Pattern Recognition, pp. 770–778 (2016)
5. Geiger, A., Lenz, P., Urtasun, R.: Are we ready for autonomous driving? The KITTI vision benchmark suite. In: Conference on Computer Vision and Pattern Recognition (CVPR) (2012)
6. Scharstein, D., Szeliski, R., Zabih, R.: A taxonomy and evaluation of dense two-frame stereo correspondence algorithms. In: Proceedings of IEEE Workshop on Stereo and Multi-Baseline Vision, 2001, (SMBV 2001), pp. 131–140. IEEE (2001)
7. Geiger, A., Roser, M., Urtasun, R.: Efficient large-scale stereo matching. In: Kimmel, R., Klette, R., Sugimoto, A. (eds.) ACCV 2010. LNCS, vol. 6492, pp. 25–38. Springer, Heidelberg (2011). doi:10.1007/978-3-642-19315-6_3
8. Heise, P., Jensen, B., Klose, S., Knoll, A.: Fast dense stereo correspondences by binary locality sensitive hashing. In: 2015 IEEE International Conference on Robotics and Automation (ICRA), pp. 105–110. IEEE (2015)
9. Hirschmuller, H.: Stereo processing by semiglobal matching and mutual information. IEEE Trans. Pattern Anal. Mach. Intell. **30**(2), 328–341 (2008)
10. Zbontar, J., LeCun, Y.: Stereo matching by training a convolutional neural network to compare image patches. J. Mach. Learn. Res. **17**, 1–32 (2016)
11. Luo, W., Schwing, A.G., Urtasun, R.: Efficient deep learning for stereo matching. In: Proceedings of the IEEE Conference on Computer Vision and Pattern Recognition, pp. 5695–5703 (2016)
12. Chen, Z., Sun, X., Wang, L., Yu, Y., Huang, C.: A deep visual correspondence embedding model for stereo matching costs. In: Proceedings of the IEEE International Conference on Computer Vision, pp. 972–980 (2015)
13. Shaked, A., Wolf, L.: Improved stereo matching with constant highway networks and reflective confidence learning. arXiv preprint arXiv:1701.00165 (2016)
14. Chen, L.C., Papandreou, G., Kokkinos, I., Murphy, K., Yuille, A.L.: Semantic image segmentation with deep convolutional nets and fully connected CRFs. In: ICLR (2015)
15. Papandreou, G., Chen, L.C., Murphy, K., Yuille, A.L.: Weakly- and semi-supervised learning of a DCNN for semantic image segmentation arXiv:1502.02734 (2015)
16. Kendall, A., Martirosyan, H., Dasgupta, S., Henry, P., Kennedy, R., Bachrach, A., Bry, A.: End-to-end learning of geometry and context for deep stereo regression. arXiv preprint arXiv:1703.04309 (2017)
17. Gidaris, S., Komodakis, N.: Detect, replace, refine: deep structured prediction for pixel wise labeling. arXiv preprint arXiv:1612.04770 (2016)
18. Kuznietsov, Y., Stückler, J., Leibe, B.: Semi-supervised deep learning for monocular depth map prediction. arXiv preprint arXiv:1702.02706 (2017)
19. Cordts, M., Omran, M., Ramos, S., Rehfeld, T., Enzweiler, M., Benenson, R., Franke, U., Roth, S., Schiele, B.: The cityscapes dataset for semantic urban scene understanding. In: Proceedings of the IEEE Conference on Computer Vision and Pattern Recognition, pp. 3213–3223 (2016)
20. He, K., Zhang, X., Ren, S., Sun, J.: Delving deep into rectifiers: surpassing human-level performance on ImageNet classification. In: Proceedings of the IEEE International Conference on Computer Vision, pp. 1026–1034 (2015)

21. Seki, A., Pollefeys, M.: Patch based confidence prediction for dense disparity map. In: British Machine Vision Conference (BMVC), vol. 10 (2016)
22. Guney, F., Geiger, A.: Displets: resolving stereo ambiguities using object knowledge. In: Proceedings of the IEEE Conference on Computer Vision and Pattern Recognition, pp. 4165–4175 (2015)

MC-DCNN: Dilated Convolutional Neural Network for Computing Stereo Matching Cost

Xiao Liu[1], Ye Luo[1], Yu Ye[2], and Jianwei Lu[1(✉)]

[1] School of Software Engineering, Tongji University, Shanghai, China
{1532787,yeluo,jwlu33}@tongji.edu.cn

[2] College of Architecture and Urban Planning, Tongji University, Shanghai, China
yye@tongji.edu.cn

Abstract. Designing a model for computing better matching cost is a fundamental problem in stereo method. In this paper, we propose a novel convolutional neural network (CNN) architecture, which is called MC-DCNN, for computing matching cost of two image patches. By adding dilated convolution, our model gains a larger receptive field without adding parameters and losing resolution. We also concatenate the features of last three convolutional layers as a better descriptor that contains information of different image levels. The experimental results on Middlebury datasets validate that the proposed method outperforms the baseline CNN network on stereo matching problem, and especially performs well on weakly-textured areas, which is a shortcoming of traditional methods.

Keywords: Stereo method · Matching cost · CNN

1 Introduction

In recent years, binocular vision has been widely used in areas such as robots, smart cars, and remote sensing. Most binocular vision systems are based on stereo matching methods. Given a point in three-dimensional space and two images that meet the epipolar constraint, the point will be imaged to two pixels on each image at a same vertical coordinate and different horizontal coordinates. The difference between the horizontal coordinates of matched pixels is called disparity. Using disparity, we can recover the 3D position of matched pixels conversely. The goal of a stereo matching method is to find matched pixels on two images and calculate the disparity of each pixel pair. A typical stereo matching method comprises two steps [1]: firstly, designing a model to compute the matching cost of different disparity of left and right image, then optimizing the matching cost and calculating disparity between two images by using specific prior knowledge. A lot of prior researches have focused on the second step and designed kinds of optimization algorithms to reach a better result from a pre-calculated matching cost [2,3]. However, researchers didn't pay so much attention to computing a better matching cost, which is the basis of subsequent stereo

D. Liu et al. (Eds.): ICONIP 2017, Part III, LNCS 10636, pp. 249–259, 2017.
https://doi.org/10.1007/978-3-319-70090-8_26

methods. Although some traditional stereo correspondence algorithms have been proposed over the past few decades, these algorithms are not intelligent enough to handle situations such as the target pixel is lack of context information.

As we all know, during recent years, convolutional neural networks (CNN) have made great progress and became the mainstream of computer vision. Especially, CNN shows state-of-the-art performance in high level vision tasks including classification, object detection and semantic image segmentation. CNN has high comprehension of image patterns by learning features with more invariance and descriptive power, which are very appropriate for computing matching cost.

Thus this paper focuses on computing matching cost with CNN. We propose a novel CNN architecture, which is named Matching Cost Dilated Convolutional Neural Network (MC-DCNN). Following the work of MC-CNN [4,5], we modify the convolutional layers of MC-CNN from two aspects: using dilated convolution, and concatenating convolutional features of different scales. The former expands receptive field without adding CNN layers and reducing the feature map resolution, while the latter merges features in different scales around the target pixel.

1.1 Related Works

In this subsection, related studies in the field of matching cost are reviewed.

Pixel-wise matching cost is a simple but widely used method [6]. It works pretty good in preserving the structures near the disparity discontinuities. However, the algorithm failed in low-texture and repeated texture area, and is not robust to noise.

Common window-based matching cost, including the sum of absolute or squared difference (SAD/SSD) [7] and normalized cross correlation (NCC) [8], were introduced for providing a more reliable result by using image patches around target pixels. However, outliers frequently occurs near object boundaries when calculating window-based matching cost.

Nonparametric matching cost such as Rank and Census methods solved the above problem but not robust to orientation and distortion because they only rely on the relative ordering of pixel values [10].

A related problem to computing matching cost is learning local image descriptors. Several methods have been suggested for solving the problem of learning local image descriptors, such as boosting [15], convex optimization [16], and convolutional neural networks [17]. However, these methods compared image patches with larger variation, ignored some details which are significant for stereo matching. Moreover, the inclusion of pooling and subsampling to account for larger patch sizes [17] leads to the reduction of resolution.

Ẑbontar and LeCun [4,5] firstly proposed MC-CNN to compute matching cost of two image patches, which quickly becomes the most popular front-end for stereo matching methods. Their proposed CNN architecture takes a small 11×11 window without the use of pooling, that restricts the receptive field increasing. The method is also post-processed by using cross-based cost aggregation (CBCA) [14], semi-global matching (SGM) [2] and additional refinement procedures.

1.2 Our Motivations and Contributions

Having investigated the literature, we find that convolutional neural network (CNN) is becoming a developing trend to compute the matching cost. However, there is still a lot of room for improvement. MC-CNN [4,5], which is state-of-the-art method and proposed in 2016 to compute the matching cost, still has a 22.81% error rate without post processing on Middlebury datasets [1,11]. Further more, due to the simple architecture of MC-CNN, the limitations of MC-CNN on solving the image patch matching problem are obvious:

(1) MC-CNN has a relatively small respective field (9×9 in KITTI and 11×11 in Middlebury). Consider when people matches two images, the viewer would observe a wide area around the target object to gain context information. However it is hard to enlarge the respective field of it if restricted to original CNN architecture.
(2) The matching cost of the two image patches to be compared is purely determined via a single scale high-level features. However, only using high-level CNN feature leads to the loss of image details. Therefore multi-scale feature, which can represent the target pixel precisely, is needed.

To tackle the aforementioned problems, we propose a novel deep neural network architecture based on MC-CNN. By keeping other parts of MC-CNN unaltered, we made two improvements:

(1) We replace the traditional convolutional operation with dilated convolution to enlarge the receptive field. The dilated convolution operation can expand receptive field exponentially to achieve a better image patch based matching cost computation result. It also avoids the degradation of feature map resolution problem caused by using traditional convolution-pooling operations.
(2) We concatenate features from various convolution layers to incorporate multiple scale context information. The concatenated features increase the ability to describe the target pixel.

Experimental results have shown that our proposed model reaches a better performance than MC-CNN and works well on low texture areas.

2 Architecture of MC-DCNN

In this section, we propose a novel CNN architecture MC-DCNN (Matching Cost Dilated Convolutional Neural Network) for computing the similarity score of two image patches. As is mentioned in Sect. 1.2, we make modifications on the architecture of MC-CNN. MC-CNN is composed of two main parts:

(1) The first part is a pair of siamese networks. The whole network is composed of five convolutional layers (i.e. conv1, conv2, ..., conv5), and each layer has a kernel size of 3×3. Two image patches to be compared are fed into the network as the inputs, and the features extracted from each siamese network are concatenated as the final output of the CNN network.

(2) The second part of the MC-CNN consists of three fully connected (i.e. fc) layers with 384 neurons and a final layer activated by sigmoid function. In order to avoid repeated calculation and reuse the model on the whole image, we replace the fc layers with the convolutional layers of kernel size 1×1. The output of the network is the similarity score of the input patches.

As shown in Fig. 1, in our proposed model MC-DCNN, we made two improvements on the first part of MC-CNN, and keep the original structure of the second part.

(1) We replace traditional convolutional layers with dilated convolution layers. We set dilation rates (1, 1, 2, 4, 8) on the 3×3 convolutional layers successively, increasing the final receptive field from 11×11 to 33×33.
(2) Instead of only using the conv5 output as image feature, we concatenate the output of conv3, conv4 and conv5 as a descriptor that merges multi-scale image information.

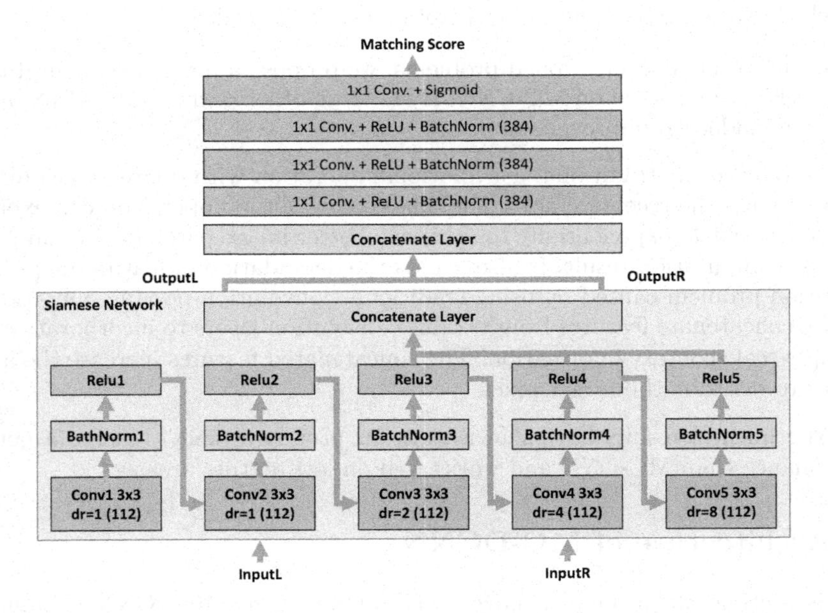

Fig. 1. Architecture of MC-DCNN. Dilation rates (1, 1, 2, 4, 8) were added on the 3×3 convolutional layers. We also concatenate outputs of the last three layers as a final descriptor of input image patch.

In the following subsection, we will firstly introduce the dilated convolution, and then multi-scale feature fusion method.

2.1 Dilated Convolution

It is obviously that enlarging the receptive field of CNN can effectively improve the result of matching cost computing. There are three common methods to enlarge the receptive field: (1) using larger convolution kernels; (2) adding convolution layers; (3) including a few strided pooling layers.

However, the methods are not appropriate for matching cost computing, because: (1) the parameters increase quadratically with the size of convolutional layers, that will reduce operation efficiency; (2) more convolution layers also increase the number of parameters and make the network difficult to train; (3) strided pooling layers can multiply the receptive field by downsampling the feature maps but the target matching cost matrix needs a pixel-level resolution. Even though the resolution can be recovered by fractional strided convolution, small image details filtered by pooling layers are difficult to recover.

In order to enlarge the receptive field efficiently without losing calculation efficiency and feature details, we draw lessons from [12,13], which obtained very good results in semantic segmentation. We also introduce their core idea dilated convolution to our network.

Given a 2-dimensional matrix $\mathbf{I}$ and a convolution kernel $\mathbf{k}$ of size $(2r+1)^2$, the convolution operation at position (p,q) in $\mathbf{I}$ can be defined as:

$$C(p,q|\mathbf{I},\mathbf{k}) = \sum_{i=-r}^{r}\sum_{j=-r}^{r} \mathbf{I}_{p+i,q+j}\mathbf{k}_{i+r,j+r} \tag{1}$$

As shown in Fig. 2, the receptive fields are enlarged linearly. Three convolutional layers with 3×3 kernel have receptive fields of size 3×3, 5×5 and 7×7 respectively. Given a few convolutional layers of same kernel size $(2r+1) \times (2r+1)$, the receptive field size of $n_{th} (n = 1, 2, ...)$ convolutional layers is:

$$\mathbf{S}_n = 2rn + 1 \tag{2}$$

that is, the receptive field is linearly increasing size.

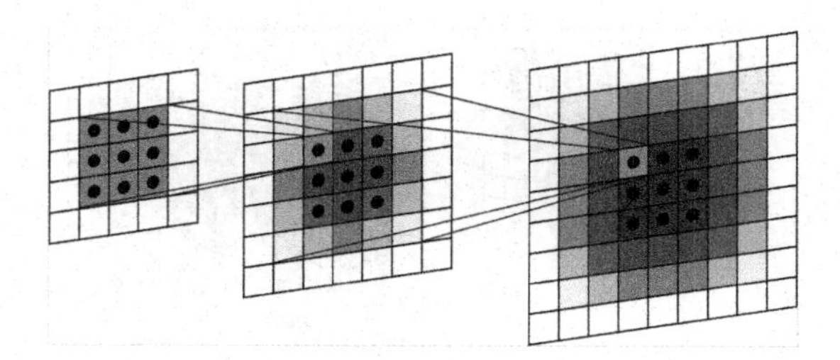

Fig. 2. Three 3×3 convolutional layers without dilation rate. The receptive field sizes are 3×3, 5×5 and 7×7, with linearly increased size.

Now we extend convolution operation. Let d be the dilation rate. A dilated convolution can be defined as:

$$C_d(p, q|\mathbf{I}, \mathbf{k}) = \sum_{i=-r}^{r} \sum_{j=-r}^{r} \mathbf{I}_{p+di,q+dj} \mathbf{k}_{i+r,j+r} \tag{3}$$

As shown in Fig. 3, there are three dilated convolutional layers with same 3×3 kernels, and dilation rates of 1, 2 and 4 respectively. These layers have also successive receptive field, which have larger size of 3×3, 7×7 and 15×15. When using same kernel size $(2r + 1) \times (2r + 1)$, the receptive field size of $n_{th} (n \geq 2)$ convolutional layers with dilation rate d_n is:

$$\mathbf{S}_n = \mathbf{S}_{n-1} + 2r \cdot d_n \tag{4}$$

which is a recursion formula, meanwhile

$$\mathbf{S}_1 = 2r \cdot d_1 + 1 \tag{5}$$

Simplify the Eqs. (4) and (5), the final form of $\mathbf{S}_n$ is:

$$\mathbf{S}_n = 2r \cdot \sum_{i=1}^{n} d_i + 1 \tag{6}$$

If we set $d_n = 2^{n-1}$, similar to the model shown in Fig. 3 and our proposed model, the receptive field size will be:

$$\mathbf{S}_n = 2r(2^n - 1) + 1 \tag{7}$$

which is exponentially increasing size.

It is noteworthy that none kernel weight is added when using dilated convolution. That means this improvement will not influence the performance efficiency theoretically.

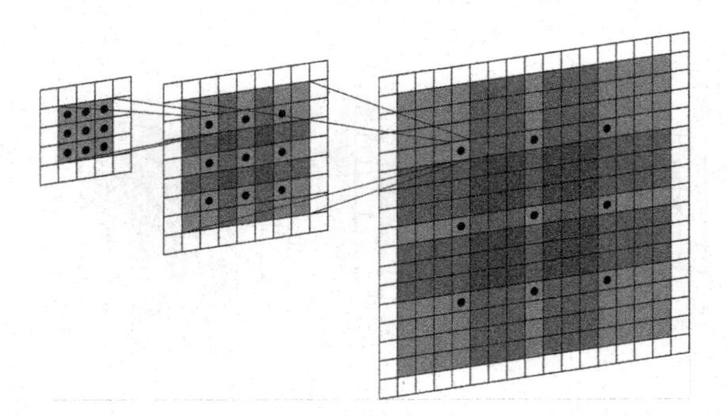

Fig. 3. Three 3×3 convolutional layers with dilation rates 1, 2 and 4 respectively. The receptive field sizes are 3×3, 7×7 and 15×15, with exponentially increased size.

2.2 Multi-scale Feature Fusion

As we know, high-level convolutional features have good invariance, but also lose low level details, which plays an important role in localized matching. To combine features of different level, we concatenate ($\uplus$) the output of conv3, conv4 and conv5 as the final descriptor of target pixel. Let $\mathbf{f}$ be a descriptor symbol, the final output of the siamese network is:

$$\mathbf{f}_{cnn} = \mathbf{f}_{conv3} \uplus \mathbf{f}_{conv4} \uplus \mathbf{f}_{conv5} \tag{8}$$

According to formula (4), the receptive field size of each convolutional layer are 3×3, 5×5, 9×9, 17×17, and 33×33. Features extracted by the first two layers have quite small receptive field, and are low level descriptors that lack of invariance. So we only concatenate the last three convolutional layers as the final descriptor.

3 Experimental Results and Discussion

In this section, we firstly introduce the Middlebury stereo datasets, and then some preparation schemes of the experiments are provided. At last, we conduct the comparison and discussion.

3.1 Middlebury Stereo Datasets

Middlebury stereo datasets [1,11] provide image pairs of indoor scenes with ground truth disparity. The datasets were published in five separates works in the years 2001, 2003, 2005, 2006, and 2014. They also provide an online leaderboard to display a ranked list of all submitted methods. A training set and a test set were provided for training and evaluation with 15 image pairs each, mainly taken from the 2014 dataset. The images are available in three resolutions, full (F), half (H), and quarter (Q). The error is computed at full resolution. If submitted outputs of the method is half or quarter resolution disparity maps, they will be upsampled before the error is computed. We chose to train and evaluate our model on the training dense set with 15 image pairs with half resolution because of limitation of hardware. The dataset were splitted into three parts, each part has 5 image pairs. We compute the error rates of our method by using two parts for training and the other for validation.

Table 1. Splitted training dense set

Name					
Set1	PlaytableP	ArtL	Playtable	Playroom	Recycle
Set2	Teddy	Piano	PianoL	Jadeplant	Pipes
Set3	Adirondack	Vintage	Motorcycle	MotorcycleE	Shelves

3.2 Training Set Preparation

Because our model has a larger receptive field than MC-CNN, we build our own training samples with image patches of size 33 × 33 segmented from left and right image. Each sample is a pair of image patches centered at position p in left image and position q in right image. For each pixel $p = (x, y)$ in left image, given the ground truth disparity d of p, we generate one positive sample at position q_{pos} and one negative sample at position q_{neg} in right image:

$$q_{pos} = (\langle x - d \rangle, y) \tag{9}$$

$$q_{neg} = (\langle x - d + o_{neg} \rangle, y) \tag{10}$$

where $\langle x - d \rangle$ denotes rounding of x, and o_{neg} is a random number chose from $(-18, -2)$ and $(2, 18)$. We produce samples pixel by pixel in each image pairs in accordance with the above rules, and finally obtain 28 million samples in total.

3.3 Comparison and Discussion

In order to evaluate our method, we compare the result of MC-DCNN with the baseline MC-CNN and other published methods. As is mentioned in Subsect. 3.2, we train and evaluate our model with the divided training dense set for three times. Each time we use samples extracted from two subsets listed in Table 1, about 20 million samples for training, and compute the disparity of the other five image pairs on the remaining subset. We also perform the post-processing pipeline on our raw-disparity map. The pipeline consists of a series of stereo method, including cross-based cost aggregation [14] and semi-global matching [2]. A median filter and a bilateral filter are also performed to smooth the disparity map.

Table 2 shows the results of our method and MC-CNN. Our method outperforms MC-CNN in both raw results and results after post-processing. Quantitatively, the accuracy is improved by 7.2% and 0.61%, respectively. We also compare our method with published methods in Table 3.

Table 2. Comparison of MC-DCNN and MC-CNN. Results on the training dense set of Middlebury datasets. The avg. error represents the percentage of bad pixels with threshold 2.0.

Methods		Avg. error
MC-DCNN	Raw-disparity	**15.71**
	After post-processing	**9.65**
MC-CNN-acrt	Raw-disparity	22.91
	After post-processing	10.26

Table 3. Comparison of our method and published methods.

Methods	Author	Resolution	Avg. error
MC-DCNN	After post-processing	Half	**9.65**
NTDE [19]	Kim et al. (2016)	Half	9.94
MC-CNN+TDSR [20]	Drouyer et al. (2017)	Full	10.2
MC-CNN-acrt [4]	Zbontar et al. (2015)	Half	10.26
MC-CNN+RBS [18]	Barron et al. (2016)	Half	10.8
MC-CNN-fst [4]	Zbontar et al. (2015)	Half	11.7

Moreover, according to [5], as the number of training sample increasing, the error rate of the matching will be decreased. Since we only use the half number of MC-CNN's training samples, better performance is expected when more training samples are used.

Figure 4 shows comparison of disparity maps between our method and MC-CNN. From this figure we can see that our proposed network works well on weakly-textured areas such as floors and walls. This further validates that features extracted via our proposed new network show strong capability on image matching problem.

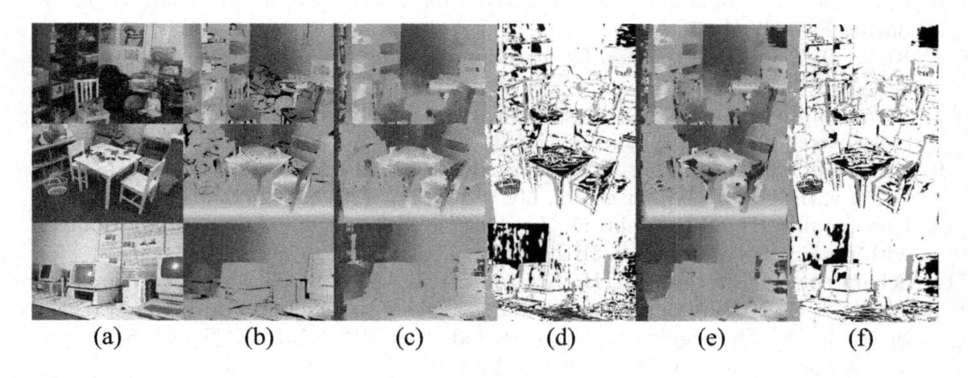

Fig. 4. Comparison of our method and MC-CNN. From left to right: (a) left image of image pair, (b) ground-truth disparity map, (c) result of MC-CNN, (d) error map of MC-CNN, (e) result of our method, (f) error map of our method. Black pixels on error maps represent bad disparity results with threshold 2.0.

4 Conclusions and Future Work

In this paper, we focus on computing matching cost of two image patches by CNN. A novel CNN architecture is proposed to learn features with more invariance and descriptive power for computing matching cost. We improve our model by two main thoughts: (1) enlarge the receptive field of CNN, and (2) merge multi-scale features. Firstly, we introduce dilated convolution to gain a large receptive field without adding parameters and losing resolution. Secondly, we concatenate the features of last three convolutional layers as a better descriptor that contains information of different image levels. The experiment results prove that the proposed model performs well on weakly-textured areas, which is a shortcoming of previous methods. In the future, we will explore hyper-parameters of our model, and improve the execution efficiency on some other frameworks. We will also transplant the components of our model in other dense image tasks.

References

1. Scharstein, D., Szeliski, R., Zabih, R.: A taxonomy and evaluation of dense two-frame stereo correspondence algorithms. In: IEEE SMBV, pp. 131–140 (2001)
2. Hirschmuller, H.: Stereo processing by semiglobal matching and mutual information. IEEE Trans. PAMI **30**(2), 328–341 (2008)
3. Woodford, O., Torr, P., Reid, I.: Global stereo reconstruction under second-order smoothness priors. IEEE Trans. PAMI **31**(12), 2115–2128 (2009)
4. Zbontar, J., LeCun, Y.: Computing the stereo matching cost with a convolutional neural network. In: IEEE CVPR, pp. 1592–1599 (2015)
5. Zbontar, J., LeCun, Y.: Stereo matching by training a convolutional neural network to compare image patches. JMLR **17**(1), 2287–2318 (2016)
6. Birchfield, S., Tomasi, C.: Depth discontinuities by pixel-to-pixel stereo. IJCV **35**(3), 269–293 (1999)
7. Kong, D., Tao, H.: A method for learning matching errors for stereo computation. BMVC **1**, 2–11 (2004)
8. Heo, Y.S., Lee, K.M., Lee, S.U.: Robust stereo matching using adaptive normalized cross-correlation. IEEE Trans. PAMI **33**(4), 807–822 (2011)
9. Hirschmuller, H., Innocent, P.R., Garibaldi, J.: Real-time correlation-based stereo vision with reduced border errors. IJCV **47**(1–3), 229–246 (2002)
10. Hirschmuller, H., Scharstein, D.: Evaluation of stereo matching costs on images with radiometric differences. IEEE Trans. PAMI **31**(9), 1582–1599 (2009)
11. Scharstein, D., Hirschmüller, H., Kitajima, Y., Krathwohl, G., Nešić, N., Wang, X., Westling, P.: High-resolution stereo datasets with subpixel-accurate ground truth. In: Jiang, X., Hornegger, J., Koch, R. (eds.) GCPR 2014. LNCS, vol. 8753, pp. 31–42. Springer, Cham (2014). doi:10.1007/978-3-319-11752-2_3
12. Chen, L.C., Papandreou, G., Kokkinos, I.: Semantic image segmentation with deep convolutional nets and fully connected CRFs. arXiv preprint arXiv:1412.7062 (2014)
13. Yu, F., Koltun, V.: Multi-scale context aggregation by dilated convolutions. arXiv preprint arXiv:1511.07122 (2015)
14. Zhang, K., Lu, J., Lafruit, G.: Cross-based local stereo matching using orthogonal integral images. IEEE Trans. CSVT **19**(7), 1073–1079 (2009)
15. Trzcinski, T., Christoudias, M., Lepetit, V.: Learning image descriptors with boosting. IEEE Trans. PAMI **37**(3), 597–610 (2013)
16. Simonyan, K., Vedaldi, A., Zisserman, A.: Learning local feature descriptors using convex optimisation. IEEE Trans. PAMI **36**(8), 1573–1585 (2014)
17. Zagoruyko, S., Komodakis, N.: Learning to compare image patches via convolutional neural networks. In: IEEE CVPR, pp. 4353–4361 (2015)

18. Barron, J.T., Poole, B.: The fast bilateral solver. In: ECCV, pp. 617–632 (2016)
19. Kim, K.R., Kim, C.S.: Adaptive smoothness constraints for efficient stereo matching using texture and edge information. In: IEEE ICIP, pp. 3429–3433 (2016)
20. Drouyer, S., Beucher, S., Bilodeau, M., Moreaud, M.: Sparse stereo disparity map densification using hierarchical image segmentation. In: ISMM, pp. 172–184 (2017)
21. Luo, W., Schwing, A.G., Urtasun, R.: Efficient deep learning for stereo matching. In: IEEE CVPR, pp. 5695–5703 (2016)

Improving Deep Crowd Density Estimation via Pre-classification of Density

Shunzhou Wang[1(✉)], Huailin Zhao[1], Weiren Wang[2], Huijun Di[3], and Xueming Shu[4]

[1] School of Electrical and Electronic Engineering, Shanghai Institute of Technology, Shanghai 201418, China
albertwangsz@gmail.com
[2] Department of Computer Science and Technology, Tsinghua University, Beijing 100084, China
[3] School of Computer Science and Technology, Beijing Institute of Technology, Beijing 100081, China
[4] Department of Engineering Physics, Tsinghua University, Beijing 100084, China

Abstract. Previous works about deep crowd density estimation usually chose one unified neural network to learn different densities. However, it is hard to train a compact neural network when the crowd density distribution is not uniform in the image. In order to get a compact network, a new method of pre-classification of density to improve the compactness of counting network is proposed in this paper. The method includes two networks: classification neural network and counting neural network. The classification neural network is used to classify crowd density into different classes and each class is fed to its corresponding counting neural networks for training and estimating. To evaluate our method effectively, the experiments are conducted on UCF_CC_50 dataset and Shanghaitech dataset. Comparing with other works, our method achieves a good performance.

Keywords: Crowd counting · Density classification · Deep learning

1 Introduction

Crowd counting is a computer vision problem, which estimates the total number of people in images or surveillance videos. Accurate crowd counting method has a significant importance in crowd control, crowd analysis, video surveillance and public safety management.

In recent years, deep learning technology has been made a big progress in computer vision field and there are a lot of works about convolutional neural network for crowd counting. These works can be divided into two categories: convolutional neural network for directly learning crowd counts [4,8] and convolutional neural network for learning crowd density map [1,2,4–7]. For directly learning count number, Shang et al. [8] used GoogLeNet [9] to learn the crowd feature and LSTM was used to decode these feature into crowd counts.

D. Liu et al. (Eds.): ICONIP 2017, Part III, LNCS 10636, pp. 260–269, 2017.
https://doi.org/10.1007/978-3-319-70090-8_27

Zhang et al. [4] proposed a cross-scene crowd counting convolutional neural network and used switch learning method to train the network alternatively with two learning objectives: crowd density map and crowd counts. They argued that switch learning method can learn a better local optimum and increase the counting accuracy.

Fig. 1. Some representative images of dense crowd from (a) UCF_CC_50 dataset, (b) Shanghaitech Part_A, and (c) Shanghaitech Part_B.

Comparing with directly learning crowd counts, learning crowd density map can help neural network acquire spatial information of the crowd and more and more works select crowd density map as their learning objective. Zhang et al. [5] designed a multi column convolutional neural network to learn the crowd density map. In multi column network, each column filters had different kernel size so that these filters could learn different scale size of crowd head. In [6], a counting neural network Hydra CNN was proposed and each column in Hydra CNN had the same architecture. In order to learning different scale head representations, they processed input image into different scale patches which are fed to each column for training. Mark et al. [1] only designed a column convolutional neural network to estimate the crowd density map and also achieved a comparable result. Zeng et al. [2] designed a multi-scale blob with refer to [3] and proposed a multi-scale convolutional neural network to learn crowd density. Lokesh et al. [7] designed a shallow-deep neural network whose shallow column learned the low-level feature and deep column learned the high-level feature. The learning objective of the shallow-deep network is also the crowd density map.

From these works above, selecting crowd density map as the learning objective actually improved counting result. However, we see from Fig. 1 that in many scenarios, the crowd density distribution is not uniform. This makes it hard to learn density maps for these inputs only with different filter kernel sizes [5]. Also, it is difficult to obtain accurate counting results by only using an unified multi column network [2,5–7]. A more compact counting network should be trained to learn different densities. Therefore, a method of pre-classification of density to improve the compactness of counting network is proposed. For each input image, a pyramid of patches are extracted and these patches are classified into different classes. Each class will be fed to corresponding counting network for training so that each counting network can learn a specific density. The experiments are conducted on UCF_CC_50 dataset and Shanghaitech dataset. The results show that our method can increase the counting accuracy no matter how many people and how scene change greatly in image.

2 Proposed Method

The pipeline of our proposed method is shown in Fig. 2. It consists of the classification network and the counting network. For an input image, we crop a pyramid of patches like [6]. The classification network accepts these patches as input and predicts the class for each patch. Then, each patch is further fed to a specific counting network that are trained for the corresponding class. The counting network generates a density map for each patch. Finally, we obtain the density map of the input image by integrating the density maps of all patches. The red box in Fig. 2 shows an example to explain how the network works. Given the red patch, the classification network, *i.e.*, GoogLeNet [9], predicts its class as **C1**. Hence, the patch is only fed to the counting network corresponding to **C1**, while other counting networks are left to be inactivated. The chosen network produces a density map for the patch which will be used to form the final density map. In this paper, we use GoogLeNet [9] as the classification network and Hydra3 [6] as the counting network. The details of the networks will be introduced below.

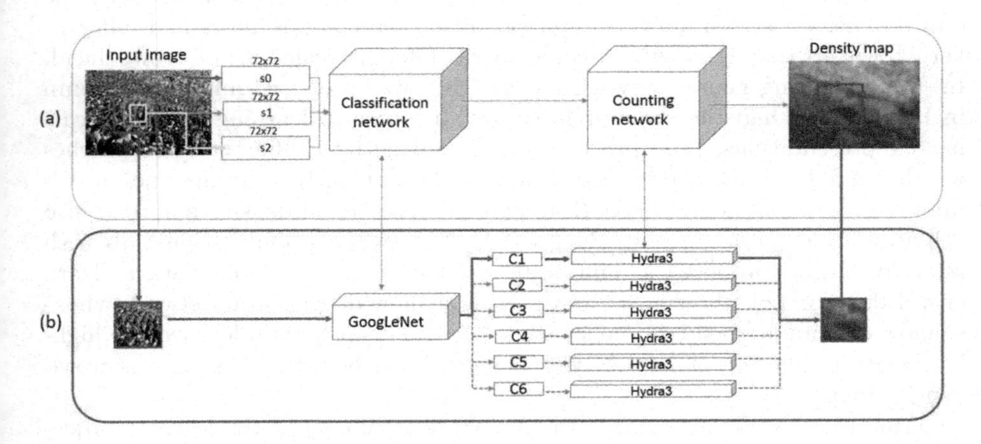

Fig. 2. The pipeline of our method. (a) blue box gives the whole framework and (b) red box gives an example of how the red patch be classified into **C1** class and be estimated by the counting network corresponding to **C1**. (Color figure online)

2.1 The Classification Neural Network

In order to solve the impact of different crowd density to counting result, we classify crowd density into different classes. The classification label is generated by K-means firstly. To avoid the problem of unbalance amounts of each class, the data augmentation strategies such as scales and flipping are used to enlarge patches in the smallest class. Each patch has only one class label and these patches are fed to GoogLeNet for training.

The input of the network is modified and the details structure is shown in Fig. 3. With the reference [6], the GoogLeNet input layers are changed from one column to three columns. The input dimension of each column is set as $1 \times 1 \times 72 \times 72$, the output dimensions depends on number of class our presetting. The Softmax and Cross Entropy Loss is selected as the loss function. After the stage of training, the classification network has the ability to judge the class of each patch and each class will be fed to its corresponding counting network for training and density estimation.

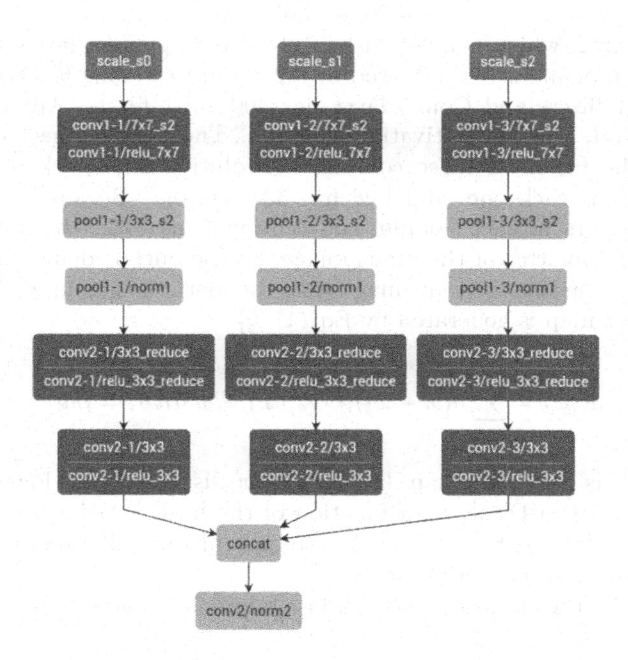

Fig. 3. The input of GoogLeNet changes from one column input layer to three columns input layer.

2.2 The Counting Neural Network

The input patches of each counting network are from the result of classification network so that each counting network is trained for learning a specific crowd density representations. The Hydra3 model [6] is chosen as the counting network and its structure is shown in Fig. 4. Each Hydra3 model has three columns convolutional network before fully connected layer. In Hydra3, the structure parameters of each column are the same and each column neural network is

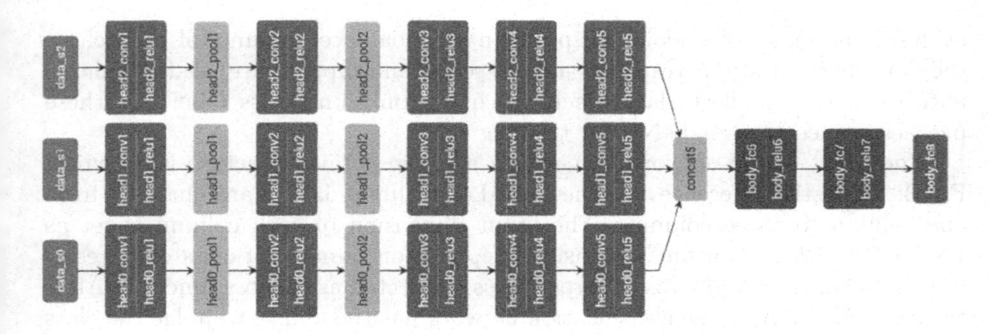

Fig. 4. The structure of Hydra3 model.

followed by the order Conv-Pool-Conv-Pool-Conv. There are 32 7×7 filters in Conv1 and Conv2, which is followed by a 2×2 max-pooling layer. Conv3 layer has 64 5×5 filters and is also followed by a 2×2 max-pooling layer. Conv4 layer has 1000 1×1 filters and Conv5 layer has 400 1×1 filters. All convolutional layers select ReLU as the activation function. The fully connected layers are followed by the fusion of three columns convolutional network. Fc6 and Fc7 have 512 neurons each one, and Fc8 has 324 neurons. Because there are two max-pooling layers in each column convolutional network, the output size of the network is a quarter of the input image. So the output density map will be resized to the same size of input image with the method used in [6]. The ground truth of density map is generated by Eq. (1) [5].

$$F(x) = \sum_{i=1}^{N} \delta(\boldsymbol{x} - \boldsymbol{x}_i) * G_{\sigma_i}(\boldsymbol{x}), \quad with\, \sigma_i = \beta \bar{d^i} \tag{1}$$

where $G_{\sigma_i}(\boldsymbol{x})$ is the Gaussian kernel [5], $\boldsymbol{x}_i$ is the head location in the image, $\delta(\boldsymbol{x} - \boldsymbol{x}_i)$ is a Dirac delta function of the head location, N is the number of people, $d^i = \frac{1}{m}\sum_{j=1}^{m} d_j^i$ is the average distance between $\boldsymbol{x}_i$ and the nearest m head. β is set as the same as [5].

Following [5], the counting network loss function is defined as

$$L(\Theta) = \frac{1}{2N} \sum_{i=1}^{N} \|F(X_i; \Theta) - F_i\|_2^2 \tag{2}$$

In the equation, Θ represents the optimization parameters of the network, N is the number of samples in one iteration, X_i is the input image samples, and F_i is the corresponding ground truth density map.

2.3 Evaluation Metric

Different methods are evaluated with the absolute error (MAE) and the mean squared error (MSE). The two evaluations are defined as follows.

$$MAE = \frac{1}{N}\sum_{1}^{N}|z_i - \hat{z}_i| \tag{3}$$

$$MSE = \sqrt{\frac{1}{N}\sum_{1}^{N}(z_i - \hat{z}_i)^2} \tag{4}$$

where N stands for total number of test set images, z_i is the actual number and $\hat{z}_i$ is the prediction result.

3 Experiments

To evaluate our propose method effectively, we conduct our experiments on two representative crowd dataset: UCF_CC_50 dataset and Shanghaitech dataset. The networks are trained based on Caffe [12] framework. The operating system is Ubuntu 14.04, processor is Intel Core i5-4430 and GPU is Nvidia GTX Titan. Two different optimization strategies are used to train the classification network and the counting network. For classification network, Mini-batch SGD and back propagate algorithm are used. The initial learning rate is set as 0.01, and after each 3200 iterations the learning rate multiplies 0.96. Further, the SGD momentum is set as 0.9. For counting network, the RMSProp [13] and back propagate algorithm are selected. The initial learning rate is 0.0001, rms_decay is set as 0.98 and 'inv' learning policy is chosen to decrease the learning rate with iterations. L1 regularization is chosen to prevent the counting network from over-fitting.

3.1 The UCF_CC_50 Dataset

The UCF _CC_50 dataset is first introduced by [10], which contains 50 images collected from Internet with the number of people from 94 to 4543. The dataset is randomly divided into five subsets and each subset contains 10 images. We conduct five-fold cross-validation on this dataset. Each time four subsets are chosen as the training set and the remaining subset is as the testing set. Following [6] steps, the longest side of training images is resized to 800 pixel, and then 150×150 pixel patches are randomly cropped from each training image. Data augmentation strategies such as flipping and scales are used to augment training set. A total of 1600 patches from each image are got and all training set includes 64,000 patches. In order to generate the class label, these patches are clustered into 6 classes by K-means according to crowd density firstly. Then, the classification network is trained to judge the class of each patch and each class is fed to the corresponding counting network for training and estimation. At the time of testing, the longest side of test images is also resized to 800 pixel, and 150×150 pixel patches are cropped from the image in sequential scan with a step of 10 pixel.

When the classification accuracy is one hundred percent, an incredible result is got that the MAE is 80.41 and MSE is 61.95. Actually, following [6] dataset

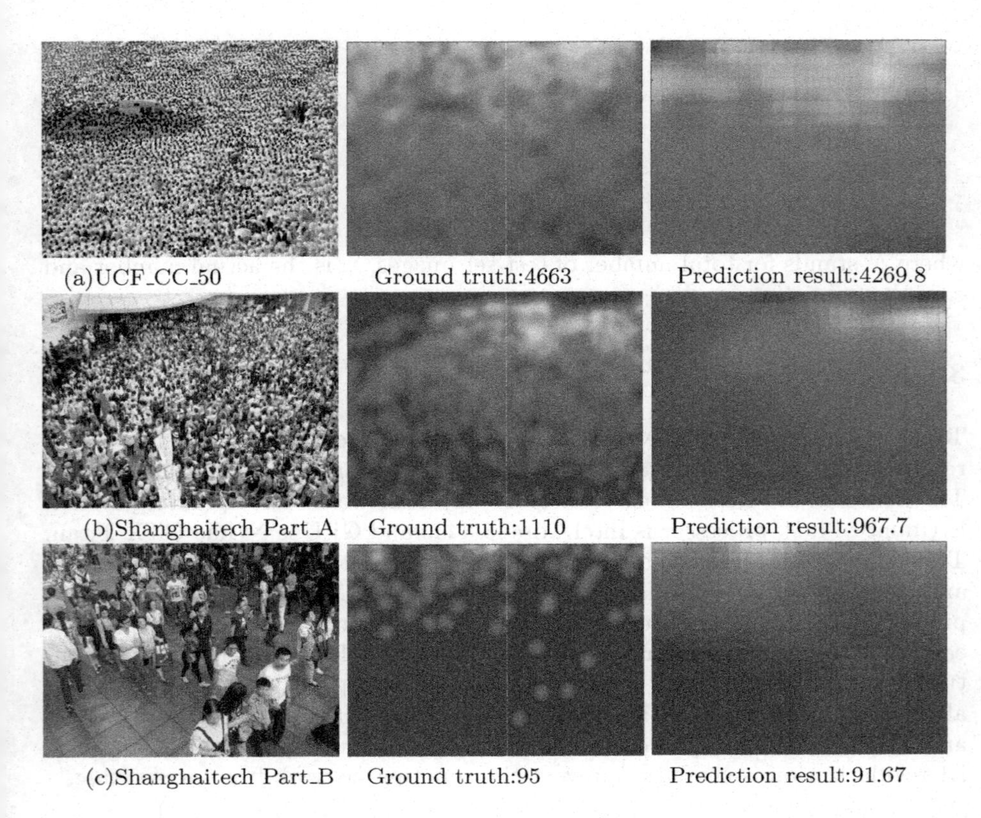

Fig. 5. Some representative results on (a) UCF_CC_50 dataset, (b) Shanghaitech Part_A and (c) Shanghaitech Part_B.

Table 1. Comparison different methods on UCF_CC_50 Dataset

Method	MAE	MSE
Lempitsky et al. [11]	493.4	487.1
Zhang et al. [4]	467.0	498.5
Idrees et al. [10]	419.5	541.6
Shang et al. [8]	270.3	-
Mark et al. [1]	338.6	424.5
MCNN [5]	377.6	509.1
MSCNN [2]	363.7	468.4
CrowdNet [7]	452.5	-
Hydra2 [6]	333.73	425.26
Hydra3 [6]	465.73	371.84
Ours - 53.67%	463.65	659.85
Ours - 91.35%	**148.07**	**212.83**

setting, a 53.67% accuracy classification network can be trained and the counting result with this classification accuracy is not ideal. Therefore, classifying crowd density correctly actually improve the counting results. However, due to the limitation of the UCF_CC_50 dataset size, a ideal accuracy classification network can't be got.

In order to increase the classification accuracy, the training set setting is changed when we train the classification network. Specially, at each fold cross-validation, the test images are added into the training set in order to enlarge the training samples and a classification network with 91.35% classification accuracy is got. Then, the classification network processes the original training set of each fold cross-validation and the corresponding counting network of each class is trained. The performance of our method is shown in Table 1 and some representative result is shown in Fig. 5(a). Comparing with other methods, it achieves the state of the art result on UCF_CC_50 dataset.

3.2 Shanghaitech Dataset

The Shanghaitech dataset is presented for the first time by [5], which is divided into Part_A and Part_B. The training set of Part_A contains 300 images, and the remaining 182 images are for testing. The training set of Part_B contains 400 images, and the remaining 316 images are for testing. The image preprocessing method and the data augmentation strategies are the same as UCF_CC_50. A total of 480,000 patches in Part_A and a total of 640,000 patches in Part_B will be got. In order to generate the class label, these patches are clustered into 6 classes by K-means according to crowd density firstly. Then, the classification network is trained to judge the class of each patch and each class is fed to the corresponding counting network for training and estimation.

The classification accuracy is 68.58% on Shanghaitech Part_A and is 75.64% on Shanghaitech Part_B. Comparing the performance of our method with other algorithms, we get the best MSE on Part_B and other comparable results with existing methods. The performance is shown in Table 2 and some experiment results are shown in Fig. 5(b) and (c).

Table 2. Comparison different methods on Shanghaitech Dataset

Method	PartA_MAE	PartA_MSE	PartB_MAE	PartB_MSE
Zhang et al. [4]	181.8	277.7	32.0	49.8
MCNN-CCR [5]	245.0	336.1	70.9	95.9
MCNN [5]	110.2	173.2	26.4	41.3
MSCNN [2]	**83.8**	**127.4**	**17.7**	30.2
Mark et al. [1]	-	-	22.1	31.5
Ours	119.37	147.48	20.83	**27.56**

3.3 Evaluation on Transfer Learning

Since the Shanghaitech Part_A and the UCF_CC_50 dataset are all with large numbers of people, we use the network trained on the UCF_CC_50 dataset to do transfer learning on the Shanghaitech Part_A. Two experiments are conducted for evaluating the transfer learning ability of our method. Firstly, the whole neural network weight trained on all images from UCF_CC_50 is directly used to estimate the test images of Shanghaitech Part_A. Then, the classification network weight with 91.35% classification accuracy trained on UCF_CC_50 dataset will be load and fixed, and the counting network will be fine-tuning on the Shanghaitech Part_A. The performance is performed in Table 3 and the result of fine-tuning counting network achieves a better result on MSE. In some sense, our method has a generalizability.

Table 3. Transfer learning from UCF_CC_50 to Shanghaitech Part_A

Method	MAE	MSE
MCNN [5]	**110.2**	173.2
Our method without transfer learning	119.37	147.48
Fine-tuning the counting network	143.81	**140.18**
The whole network trained on UCF_CC_50	189.1	259.46

4 Conclusion

In this paper, we propose a method via pre-classifying crowd density to improve crowd estimation result. Some experiments are conducted on UCF_CC_50 dataset and Shanghaitech dataset, and transfer learning experiments are also included. These results show that classifying crowd density correctly actually improves the crowd counting result and our proposed method also has a generalization ability among different crowd datasets.

Acknowledgments. This work is supported by the National Natural Science Foundation of China (No. 9142020013), the National Natural Science Foundation of China (No. 71774094) and the National Science and Technology Pillar Program during the 12th Five-year Plan Period (No. 2015BAK12B03).

References

1. Marsden, M., McGuiness, K., Little, S., O'Connor, N.E.: Fully convolutional crowd counting on highly congested scenes. arXiv preprint arXiv:1612.00220 (2016)
2. Zeng, L., Xu, X., Cai, B., Qiu, S., Zhang, T.: Multi-scale convolutional neural networks for crowd counting. arXiv preprint arXiv:1702.02359 (2017)
3. Lin, M., Chen, Q., Yan, S.: Network in network. arXiv preprint arXiv:1312.4400 (2013)

4. Zhang, C., Li, H., Wang, X., Yang, X.: Cross-scene crowd counting via deep convolutional neural networks. In: Proceedings of the IEEE Conference on Computer Vision and Pattern Recognition (CVPR), pp. 833–841. IEEE (2015)
5. Zhang, Y., Zhou, D., Chen, S., Gao, S., Ma, Y.: Single-image crowd counting via multi-column convolutional neural network. In: Proceedings of the IEEE Conference on Computer Vision and Pattern Recognition (CVPR), pp. 589–597. IEEE (2016)
6. Oñoro-Rubio, D., López-Sastre, R.J.: Towards perspective-free object counting with deep learning. In: Leibe, B., Matas, J., Sebe, N., Welling, M. (eds.) ECCV 2016. LNCS, vol. 9911, pp. 615–629. Springer, Cham (2016). doi:10.1007/978-3-319-46478-7_38
7. Boominathan, L., Kruthiventi, S.S., Babu, R.V.: CrowdNet: a deep convolutional network for dense crowd counting. In: Proceedings of the 2016 ACM on Multimedia Conference, pp. 640–644. ACM (2016)
8. Shang, C., Ai, H., Bai, B.: End-to-end crowd counting via joint learning local and global count. In: 2016 IEEE International Conference on Image Processing (ICIP), pp. 1215–1219. IEEE (2016)
9. Szegedy, C., Liu, W., Jia, Y., Sermanet, P., Reed, S., Anguelov, D., Erhan, D., Vanhoucke, V., Rabinovich, A.: Going deeper with convolutions. In: Proceedings of the IEEE Conference on Computer Vision and Pattern Recognition (CVPR), pp. 1–9. IEEE (2015)
10. Idrees, H., Saleemi, I., Seibert, C., Shah, M.: Multi-source multi-scale counting in extremely dense crowd images. In: Proceedings of the IEEE Conference on Computer Vision and Pattern Recognition (CVPR), pp. 2547–2554. IEEE (2013)
11. Lempitsky, V., Zisserman, A.: Learning to count objects in images. In: Advances in Neural Information Processing Systems, pp. 1324–1332 (2010)
12. Jia, Y., Shelhamer, E., Donahue, J., Karayev, S., Long, J., Girshick, R., Guadarrama, S., Darrell, T.: Caffe: Convolutional architecture for fast feature embedding. In: Proceedings of the 22nd ACM International Conference on Multimedia, pp. 675–678. ACM (2014)
13. Tieleman, T., Hinton, G.: Lecture 6.5-rmsprop: divide the gradient by a running average of its recent magnitude. COURSERA: Neural networks for machine learning.4,2 (2012)

A Pixel-to-Pixel Convolutional Neural Network for Single Image Dehazing

Chengkai Zhu[1,2], Yucan Zhou[1], and Zongxia Xie[1(✉)]

[1] Tianjin University, Tianjin, China
{zhuchengkai,zhouyucan,zongxiaxie}@tju.edu.cn
[2] Hong Kong University of Science and Technology, Kowloon, Hong Kong

Abstract. Estimating transmission maps is the key to single image dehazing. Recently, Convolutional Neural Networks based methods (CNNs), which aim to minimize the difference between the predictions and the transmission maps, have achieved promising dehazing results and outperformed traditional feature-based algorithms. However, two transmission maps with the same estimation error can produce quite different dehazing results. Therefore, these models are incapable to directly affect the quality of the restorations. To address this issue, we propose a pixel-to-pixel dehazing convolutional neural network in this paper, which learns a map from the hazy images to the haze-free screens. Specifically, we intuitively maximize the visual similarity between the predicted images and the ground truth with some visual-relevant loss functions, e.g., the mean square error and the gradient difference loss. Experiments on synthetic dataset and real images demonstrate that our method is effective and outperforms the state-of-the-art dehazing methods.

Keywords: Single image dehazing · Image restoration · Convolutional neural network · Deep learning

1 Introduction

Haze is an atmospheric phenomenon which reduces the clarity of what we see. In recent decades, hazy weather has occurred more and more frequently, because of the excessive light scattering caused by the increasing fine polluted suspended particles in the atmosphere. This phenomenon brings serious side effects on the quality of photography. Accordingly, both the visual feelings and the performance of computer vision applications (e.g., autonomous cars) are unsatisfactory. To alleviate this impact, it is meaningful to remove the haze in images with computer vision methods.

According to the atmospheric scattering model [1], estimating transmission maps is the key step to haze removal in images. Although acquiring the depth map with additional measurements is helpful for transmission map estimation [2], it is not always feasible in practice. Therefore, most dehazing methods focus on predicting the transmission map using only one single image without any additional information.

D. Liu et al. (Eds.): ICONIP 2017, Part III, LNCS 10636, pp. 270–279, 2017.
https://doi.org/10.1007/978-3-319-70090-8_28

Recently, both feature-based and learning-based single image dehazing methods have achieved promising results. By observing a number of hazy images, He et al. propose the Dark Channel Prior (DCP) for transmission map estimation, and refine the result by soft matting and guided image filtering [3,4]. To improve the quality, Meng et al. regard dehazing as an optimization problem based on the boundary constraint and contextual regularization (BCCR) [5]; Zhu et al. propose a trainable linear model named Color Attenuation Prior (CAP) to regress depth maps [6]. In spite of the noticeable progress, these methods may fail when complex screens break the assumptions of priors and rules. To overcome this limitation, several Convolutional Neural Networks (CNNs) have been applied to restore transmission maps, such as Multi-Scale CNN (MSCNN) [7] and DehazeNet [8]. Trained with synthetic training set to regress transmission maps, these deep models outperform previous dehazing methods and obtain more refined restorations.

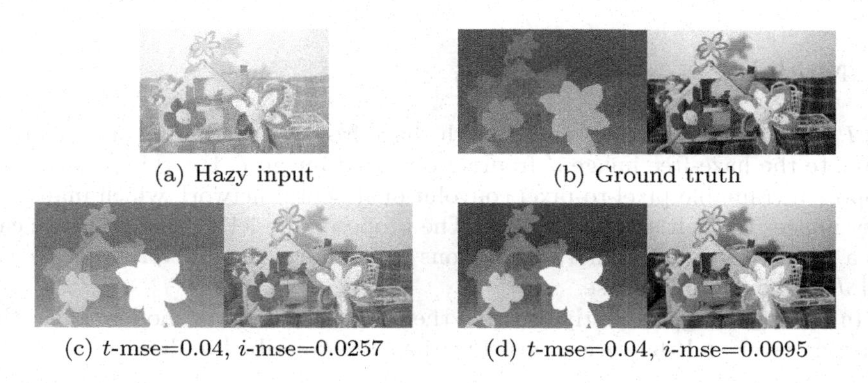

(a) Hazy input
(b) Ground truth
(c) t-mse=0.04, i-mse=0.0257
(d) t-mse=0.04, i-mse=0.0095

Fig. 1. The difference in mean square error metric between transmission maps (t-mse, on the left) and restored images (i-mse, on the right).

Although CNNs achieve a remarkable improvement, the dehazing quality of these works remains unconsidered in the training phase, which is important for recovering a visually satisfying dehazing image. As illustrated in Fig. 1(c) and (d) share the same error of transmission map estimation, but the dehazing results are quite different because the mean square error of transmission maps and that of restored screens assess distinct domains. Thus, even the error of the predicted transmission maps can be significantly reduced and a noticeable dehazing quality can be obtained, the further promotion remains uncertain. To this end, we aim to train a dehazing CNN by intuitively maximizing the visual similarity between the predicted and the haze-free images.

Different from previous transmission map estimation methods, we train a CNN utilizing the hazy images and the haze-free screens directly. To achieve this purpose, a pixel-to-pixel convolutional neural network is proposed. Specifically, a specially designed dehazing layer is attached to an Encoder-Decoder

Convolutional Neural Network (EDCNN) [9] with convolutional-deconvolutional architecture to predict haze-free images. Then, the image similarity between the restored and the haze-free images can be intuitively optimized in the training phase with suitable loss functions, such as mean square error (MSE) and gradient difference loss (GDL) [10].

The contributions of our work can be summarized as follows. Firstly, we propose a deep pixel-to-pixel trainable dehazing framework which learns a map from hazy images to haze-free images. And with this framework, a combination of some loss functions can be introduced to maximize the visual similarity between the restored and the haze-free images. Secondly, a novel dehazing layer is designed to restore the haze-free images, which makes it possible to extend existing CNNs to pixel-to-pixel dehazing learning. Thirdly, we conduct extensive experiments to demonstrate the effectiveness of the proposed method. The results show that the proposed model with a combination of MSE and GDL achieves the best performance on both synthetic and real world images.

2 Model

Let I denote the hazy image and J be the haze-free scene radiance, our goal is to estimate the haze-free image $\hat{J}$ from an observed image I. For this purpose, we propose a trainable pixel-to-pixel convolutional neural network which maps the hazy images to their scene radiances. The proposed model suggests that we can use a variety of visual similarity functions to minimize the difference between $\hat{J}$ and J in the training phase.

In the following, we firstly describe the architecture of our model. Then, the dehazing layer and the various loss functions are described in detail.

2.1 Architecture

The proposed pixel-to-pixel convolutional neural network aims to recover the haze-free image from its hazy observation. As illustrated in Fig. 2, it contains four operations: convolution, maxout, deconvolution and a specially designed dehazing operation. Given a hazy RGB image as input, this network firstly uses a convolutional maxout layer (ConvMaxout) for feature extraction. To restore a fine image, it uses a convolution-deconvolution unit to refine the content details. After that, a convolutional layer with sigmoid activation is used. Then, a 1-channel feature map is generated, which can be seen as a hidden transmission map. Finally, as the original hazy RGB image, the hidden transmission map and the global light have been acquired or estimated, the haze-free image is restored by the dehazing layer.

2.2 Dehazing Layer

The design of dehazing layer is inspired by the atmospheric scattering model [1]. It reveals that the relationship between the hazy image $I(x)$ and the screen radiance $J(x)$ is:

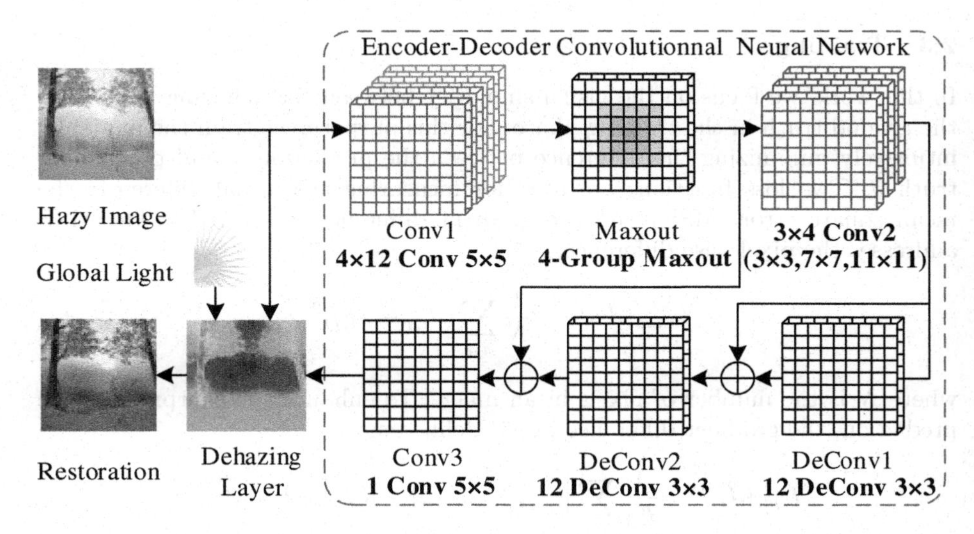

Fig. 2. The architecture of the proposed model. The $\oplus$ operator is a pixel-wise summation. (Color figure online)

$$I(x) = J(x)\,t(x) + A(1 - t(x)), \tag{1}$$

where $t(x)$ is the transmission map indicating the light portion which is not scattered, and A is the global light describing the intensity of the light source. Assuming that the atmosphere is evenly distributed, the transmission map $t(x)$ is expressed as $t(x) = e^{-\beta d(x)}$, where β is the atmospheric scattering coefficient, and $d(x)$ is the depth map. Obviously, we can recover the haze-free image $\hat{J}(x)$ when the hazy image $I(x)$, the transmission map $t(x)$ and the global light A are offered.

As we have mentioned above, the 1-channel feature map can be seen as a hidden transmission map t, which is generated from the sigmoid activation of the last convolutional layer. As for the global light A, we regard it as a prior known variable in the training phase, and estimate it by averaging the intensity values of the deepest 0.1% pixels in the predicted transmission map when applying to real-world scenes. Therefore, the haze-free image $\hat{J}(x)$ is easy to be calculated. In order to improve the numerical stability especially in the training phase, the scene radiance restored by this layer is defined as follows:

$$\hat{J}(x) = \frac{I(x) - A(1 - t(x))}{\max(t(x), 0.2)}. \tag{2}$$

In real-world applications, the last feature map can be further refined by guided image filter [4], leading to a better dehazing restoration in visual.

2.3 Training

In this work, we focus on the visual similarity between the dehazing image and the ground truth in the training phase. For this purpose, we train the model by intuitively minimizing the difference between the prediction $\hat{J}$ and the ground truth J. Two loss functions are used for measuring the visual difference: the mean square error (MSE) and the gradient difference loss (GDL). The MSE addresses the pixel-wise difference:

$$L_{mse}(\hat{J}, J) = \frac{1}{N} \sum_{i,j} (\hat{J}_{i,j} - J_{i,j})^2, \tag{3}$$

where N is the number of pixels in an image. To enhance the sharpness of the prediction, the gradient difference loss is defined as:

$$\begin{aligned} L_{gdl}(\hat{J}, J) =& \frac{1}{K_1} \sum_{i,j} \left| |\hat{J}_{i,j} - \hat{J}_{i-1,j}| - |J_{i,j} - J_{i-1,j}| \right|^{\alpha} \\ &+ \frac{1}{K_2} \sum_{i,j} \left| |\hat{J}_{i,j} - \hat{J}_{i,j-1}| - |J_{i,j} - J_{i,j-1}| \right|^{\alpha}, \end{aligned} \tag{4}$$

where K_1 and K_2 are the number of gradient differences, and α satisfies that $\alpha \geq 1$ ($\alpha = 2$ in this work).

We use Adam [11] algorithm to train the model. The learning rate is set to 10^{-4}. The parameters are initialized from a normal distribution $U(0, \sqrt{\frac{2}{fin}})$, where fin is the number of input units. A combined loss function $L = (1 - \lambda)L_{mse} + \lambda L_{gdl}$ is used in our model, where $0 \leq \lambda \leq 1$. In experiments, λ is set to 0 and 0.5, respectively.

3 Experiments

To verify the effectiveness of the proposed method, we compare with several state-of-the-art dehazing methods on both synthetic images and real-world pictures. We also illustrate what the dehazing layer has learned by visualizing the last feature map. The code is available at https://github.com/zck921031/Pixel-to-Pixel-Dehazing.

3.1 Experimental Settings

Different from the existing deep learning based dehazing methods [7,8], we employ hazy/haze-free image pairs to train our model without transmission maps. We use Keras [12] to implement the proposed model. In what follows, we will introduce the dataset, comparison methodologies and measurement metrics.

Dataset. It is hard to collect hazy and haze-free image pairs in practice. Therefore, we use NYU Depth Dataset V2 (NYUv2) [13] to generate synthetic hazy images as the benchmark dataset. There are 1449 pairs of RGB and depth images (640×480 pixels) in this dataset. We partition the dataset into three sets: the first 1350 pairs as the training set, the following 50 pairs as the validation set, and the last 49 pairs as the test set.

According to Eq. 1, the hazy observations can be generated with haze-free images and depth maps if the global light A and the scattering coefficient β are offered. In our experiments, A is set to $(1.0, 1.0, 1.0)$. To explore the robustness to haze density, β is randomly sampled from a uniform distribution $\beta \in U\,(0,15,0.225)$ in the training phase. The validation data and test data are generated with $\beta \in \{0.15, 0.175, 0.2, 0.225\}$. We do not use $\beta > 0.225$ because it leads to an over bright image in this dataset. On the contrary, we ignore $\beta < 0.15$ as the haze is too thin.

Methodology. We compare our model with 4 state-of-the-art methods: DCP [3], BCCR [5], MSCNN [7] and DehazeNet [8]. DCP and BCCR are two dehazing methods based on hand-crafted rules, while the others are trainable CNNs which learn a map between the hazy images and the transmission maps. In our experiments, the parameter setting of DCP and BCCR follows the original papers. For fairness, we re-implement MSCNN and DehazeNet, and train them in the same setting as that of the proposed model.

Evaluation Metric. To quantitatively evaluate the haze removal results, we adopt several criteria as evaluation metrics. Mean square error (MSE) aims to compute the pixel-by-pixel difference. Structural similarity (SSIM) index is a criterion to address perceived similarity with texture information [14]. CIEDE2000 criterion is to measure the color difference under the CIEDE 2000 standard [15]. And peak signal-to-noise ratio (PSNR) reflects the peak power ratio of a signal to its noise.

3.2 Quantitative Experiments on Synthetic Dataset

We conduct experiments on the synthetic dataset NYUv2. Our proposed model is trained by two loss functions (introduced in Sect. 2.3): a MSE ($\lambda = 0$) flagged with "M", and a combination of MSE and GDL ($\lambda = 0.5$) marked as "C". The architecture of "EDCNN" is illustrated in the red box of Fig. 2, aiming to regress the transmission map. For all trainable methods, the maximum iteration is set to 100. The best model is selected according to the PSNR of the validation set.

Table 1 summaries the average test results for 4 parameters on the synthetic dataset. The performance of EDCNN is better than MSCNN and DehazeNet. Using the pixel-to-pixel training framework (Ours (M) and Ours (C)), the dehazing images have notable promotions. Moreover, the combined loss (Ours (C)) achieves the best performance. As illustrated in Fig. 3, Ours (M) and Ours (C) are more robust to large β (thick haze) than EDCNN, but lose some accuracy

Table 1. The average experimental results on synthetic dataset.

Metric	Hazy	DCP	BCCR	MSCNN	DehazeNet	EDCNN	Ours (M)	Ours (C)
MSE (↓)	0.0799	0.0125	0.0244	0.0115	0.0108	0.0104	0.0090	**0.0084**
CIEDE (↓)	21.1256	7.9947	11.0124	6.1757	5.9295	5.7317	5.6899	**5.5994**
SSIM (↑)	0.7065	0.8802	0.8124	0.8834	0.8917	0.8980	0.8942	**0.9024**
PSNR (↑)	11.4134	19.8188	16.6833	20.8271	21.1478	21.3352	21.4060	**21.5475**

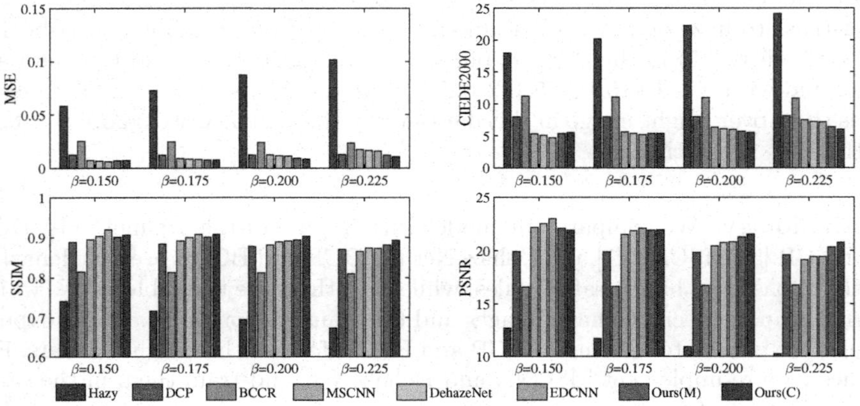

Fig. 3. Experimental results on synthetic dataset by different scattering coefficients.

when β is small (thin mist). But dealing with thick haze is more important in practice. Overall, we conclude that the pixel-to-pixel framework, which allows us to intuitively maximize the image similarity, can benefit the training of dehazing CNNs, and Ours (C) outperforms others.

3.3 Qualitative Evaluation on Real Images

Since the proposed model is trained and evaluated on synthetic dataset, the performance in real-world applications remains unknown. Hence, we demonstrate some dehazing results of natural hazy images, comparing with other state-of-the-art methods.

As the last feature map before the dehazing layer can be seen as a hidden transmission map, we firstly visualize it with some hazy images to illustrate what has been learned in our model. Figure 4 shows the dehazing results and the hidden transmission maps of several hazy images. We can see that the hazy area in the original images is mapped to low intensity in the hidden transmission maps (visualized as the blue area in the third row). It demonstrates that the transmission map prediction can be implicitly learned with the proposed pixel-to-pixel learning framework.

Fig. 4. A visualization of the dehazing results and the hidden transmission maps before the dehazing layer in our proposed model. The first row is the original image, the second row is the dehazing result, and the third row is the hidden transmission map. (Color figure online)

(a) Hazy (b) DCP (c) BCCR (d) DehazeNet (e) MSCNN (f) Ours(C)

Fig. 5. Dehazing results of real-world hazy images.

Figure 5 shows the dehazing results of real-world hazy images. For the first image, the restorations of DCP and BCCR still have some thin mist, while MSCNN and DehazeNet achieve darker results. Ours (C) gives out a brighter result without thin mist. For the second image, the results of DCP and BCCR look unnatural with unexpected coarse lumps and sharpness. MSCNN fails to remove the haze around the sun, while DehazeNet obtains a more clear output. In the dehazing image of Ours (C), the sun is clearly visible. The third row shows the restoration results of a heavy haze image. There exist many coarse lumps in DCP and BCCR. Compared with DehazeNet and MSCNN, Ours (C) gives out a more clear restoration. Although it is hard to assess the dehazing quality among these models, the illustrations demonstrate that our model can be applied to real-world applications even though it is trained on synthetic dataset by minimizing the difference between the predicted images and the haze-free screens.

4 Conclusions

In this paper, we have proposed a pixel-to-pixel convolutional neural network for single image dehazing. To minimize the visual difference, a novel dehazing layer has been adopted in the training phase. The model has been optimized with a combination of mean square error and gradient difference loss between the hazy observations and the haze-free images. Experimental results have demonstrated that our model outperforms other state-of-the-art dehazing methods, and the pixel-to-pixel training strategy has led to better dehazing restorations than transmission map estimation based methods.

Acknowledgments. This work is supported by the National Program on Key Basic Research Project under Grant 2013CB329304, and National Natural Science Foundation of China under Grants 61432011.

References

1. Narasimhan, S.G., Nayar, S.K.: Contrast restoration of weather degraded images. IEEE Trans. Pattern Anal. Mach. Intell. **25**(6), 713–724 (2003)
2. Kopf, J., Neubert, B., Chen, B., Cohen, M., Cohen-Or, D., Deussen, O., Uyttendaele, M., Lischinski, D.: Deep photo: model-based photograph enhancement and viewing. ACM Trans. Graph. (TOG) **27**, 116 (2008)
3. He, K., Sun, J., Tang, X.: Single image haze removal using dark channel prior. IEEE Trans. Pattern Anal. Mach. Intell. **33**(12), 2341–2353 (2011)
4. He, K., Sun, J., Tang, X.: Guided image filtering. IEEE Trans. Pattern Anal. Mach. Intell. **35**(6), 1397–1409 (2013)
5. Meng, G., Wang, Y., Duan, J., Xiang, S., Pan, C.: Efficient image dehazing with boundary constraint and contextual regularization. In: Proceedings of the IEEE International Conference on Computer Vision, pp. 617–624 (2013)
6. Zhu, Q., Mai, J., Shao, L.: A fast single image haze removal algorithm using color attenuation prior. IEEE Trans. Image Process. **24**(11), 3522–3533 (2015)
7. Ren, W., Liu, S., Zhang, H., Pan, J., Cao, X., Yang, M.-H.: Single image dehazing via multi-scale convolutional neural networks. In: Leibe, B., Matas, J., Sebe, N., Welling, M. (eds.) ECCV 2016. LNCS, vol. 9906, pp. 154–169. Springer, Cham (2016). doi:10.1007/978-3-319-46475-6_10
8. Cai, B., Xu, X., Jia, K., Qing, C., Tao, D.: DehazeNet: an end-to-end system for single image haze removal. IEEE Trans. Image Process. **25**(11), 5187–5198 (2016)
9. Mao, X., Shen, C., Yang, Y.B.: Image restoration using very deep convolutional encoder-decoder networks with symmetric skip connections. In: Advances in Neural Information Processing Systems, pp. 2802–2810 (2016)
10. Mathieu, M., Couprie, C., LeCun, Y.: Deep multi-scale video prediction beyond mean square error. arXiv preprint arXiv:1511.05440 (2015)
11. Kingma, D., Ba, J.: Adam: A method for stochastic optimization. In: International Conference on Learning Representations (2015)
12. Chollet, F.: Keras (2015). https://github.com/fchollet/keras
13. Silberman, N., Hoiem, D., Kohli, P., Fergus, R.: Indoor segmentation and support inference from RGBD images. In: Fitzgibbon, A., Lazebnik, S., Perona, P., Sato, Y., Schmid, C. (eds.) ECCV 2012. LNCS, vol. 7576, pp. 746–760. Springer, Heidelberg (2012). doi:10.1007/978-3-642-33715-4_54

14. Wang, Z., Bovik, A.C., Sheikh, H.R., Simoncelli, E.P.: Image quality assessment: from error visibility to structural similarity. IEEE Trans. Image Process. **13**(4), 600–612 (2004)
15. Sharma, G., Wu, W., Dalal, E.N.: The CIEDE2000 color-difference formula: Implementation notes, supplementary test data, and mathematical observations. Color Res. Appl. **30**(1), 21–30 (2005)

Level Set Based Online Visual Tracking via Convolutional Neural Network

Xiaodong Ning and Lixiong Liu(✉)

Beijing Laboratory of Intelligent Information Technology,
School of Computer Science and Technology,
Beijing Institute of Technology, Beijing 100081, China
{xdning, lxliu}@bit.edu.com

Abstract. In this paper, we propose a level set tracking algorithm, which integrates the information of the original frame and the confidence predicted by the deep feature based detector. First, we extract features from convolutional neural network and select part of them to avoid redundancy. Secondly, the features are used to generate a confidence map of the tracked object through the detector. And then the confidence along with the original frame is applied in level set model to acquire the segmentation result. We introduce an outlier rejection scheme to further improve the result. Finally, updating is employed to the detector to adapt to the changes in the video. One important contribution of our work is to use the deep features in confidence prediction, particularly the usage of low-level features in the neural network. Experimental results show that our model delivers a better performance than the state-of-the-art on a series of challenging videos.

Keywords: Object tracking · Level set · Convolutional neural network · Deep feature

1 Introduction

Object tracking is a fundamental task in the area of computer vision. It has a large variety of subfields, e.g. location tracking, trajectory tracking and contour tracking, which are classified by the characteristics or properties it tracks [1]. We are focused on contour tracking in this paper. A typical contour tracking algorithm aims to segment the object in the following each frame with an initialization in the first one. It is a challenging problem due to the complicated appearance changes in the videos.

Prior researches [2–4] mostly depend on hand-crafted features to build segmentation models. Those traditional features are limited in recognizing the sudden change of the object and not robust to complex scenes. Differing from conventional methods, deep feature based tracking algorithms [5, 6] demonstrate a remarkable performance. Deep features are extracted from convolutional neural networks (CNNs) and have strong capabilities of distinguishing objects.

While the performance is promising, limitations still remain, e.g. the deep features at a relatively low level are often ignored. Most current methods only adapt higher layer features such as the fully-connected layer features in R-CNN network [7],

D. Liu et al. (Eds.): ICONIP 2017, Part III, LNCS 10636, pp. 280–290, 2017.
https://doi.org/10.1007/978-3-319-70090-8_29

conv4-3 and conv5-3 layers in VGG network [8] and conv3 to conv5 in VGG [9]. Consequently, the intra class variations encoded by lower layers are discarded, which prevents accurate segmentation of the object contour. Therefore, our model includes some low-level features, and deploys an AdaBoost selection model [10] to remove unrelated feature maps. Hence we can both select best features and predict a confidence map efficiently. An outlier rejection scheme and a different updating scheme from [10] are introduced to further improve the effectiveness of our model.

Besides, current methods [7, 11] apply graph-cut segmentation and this leads to relatively slow efficiency [12]. Some attempts have been made to reduce the computational cost such as correlation filter [9] and probabilistic soft segmentation [13], whereas the segmentation of the object is not satisfactory. Instead, we develop a novel level set method. Level set methods are widely used in image segmentation problems, e.g. Chan-Vese model [14] and Liu et al.'s method [15], and show high efficiency in tracking problem [16]. Meanwhile, present level set models lack prior knowledge restrictions, which lead to unnecessary topological changes. Thus we construct a level set energy function which contains new length and weighted area terms. It balances the result of appearance model and the image information, and ensures the level set evolution can be restricted by the appearance model result.

In this paper, we propose an online visual tracking algorithm which consists of a deep feature based discriminative model and a level set method. The contributions of our work are as follows. First, we build a framework using a deep feature based model and level set method, where the CNN based detector can be updated online. Second, we improve the discriminative model which uses both low-level and relatively high-level deep features to detect the location of the object and obtains a coarse map for segmentation. Finally, we incorporate the result of detector with information of the original image in the energy function of the level set method, yielding a novel segmentation model.

2 Proposed Algorithm

The proposed tracker mainly contains two components: an appearance model to generate the confidence map of the object (i.e. an object detector) and a segmentation model to acquire the contour. An overview of our algorithm is shown in Fig. 1. For a given initial contour and its corresponding frame, a series of feature maps are first extracted from the region of interest (ROI) of the frame. The features include both hand-crafted features and deep features from VGG network. Then an AdaBoost feature selection process is performed on the extracted feature maps to reduce redundant ones. A discriminative model is also learned through the features and the contour in this procedure. For each subsequent frame, a ROI centered at the object contour in the last frame is cropped and the selected features are extracted. Finally, the confidence map is predicted by the discriminative model and a level set algorithm based on confidence is used to generate the object contour.

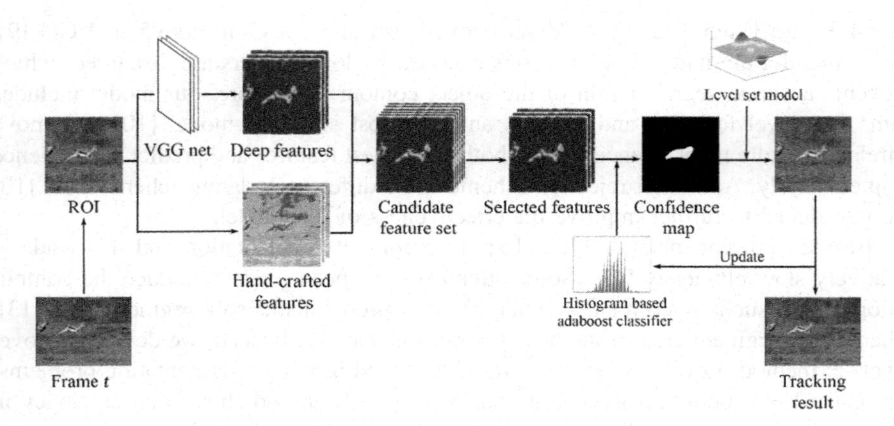

Fig. 1. Pipeline of the proposed algorithm. For Frame *t*, our tracker first extracts the deep features and handcrafted features. Then the features are selected according to the arrangement at the initialization stage of the tracking. Confidence map is predicted by using the selected features and an AdaBoost model. Finally, we use a level set method based on confidence to segment Frame *t*. Weak classifiers in the AdaBoost model are updated online while the arrangement of features is initialized at the beginning of tracking and will not be changed.

2.1 AdaBoost Feature Selection and Model Training

Our selection model is based on [10], an AdaBoost feature selection method. An extra outlier rejection scheme and updating are also introduced (see Sects. 2.3 and 2.4). This model can be used in feature selection at the initialization stage of our algorithm and confidence prediction during tracking.

In our model, Candidate feature set F contains hand-crafted features and VGG16 net [17] features. The hand-crafted features are extracted in both RGB and HSV color space in order to prevent the insufficiency of color information caused by a single space. Meanwhile, deep features including all the convolutional layers in VGG are shown in Fig. 2. In CNNs, the resolution of maps reduces along with the increasing of layer depth due to pooling operators. So we abandon higher layer features because they encode too small contour information of the object.

A detailed review of the extracted 1486-dim candidate features is shown in Table 1.

Table 1. Extracted candidate feature set.

Feature ID	Feature description	Feature ID	Feature description
f_1–f_3	RGB color	f_{79}–f_{206}	conv2-2 layer features
f_4–f_6	HSV color	f_{207}–f_{462}	conv3-3 layer features
f_7–f_{14}	8-oriented HOG descriptors calculated on a 5×5 window	f_{463}–f_{974}	conv4-3 layer features
f_{15}–f_{78}	conv1-2 layer features	f_{975}–f_{1486}	conv5-3 layer features

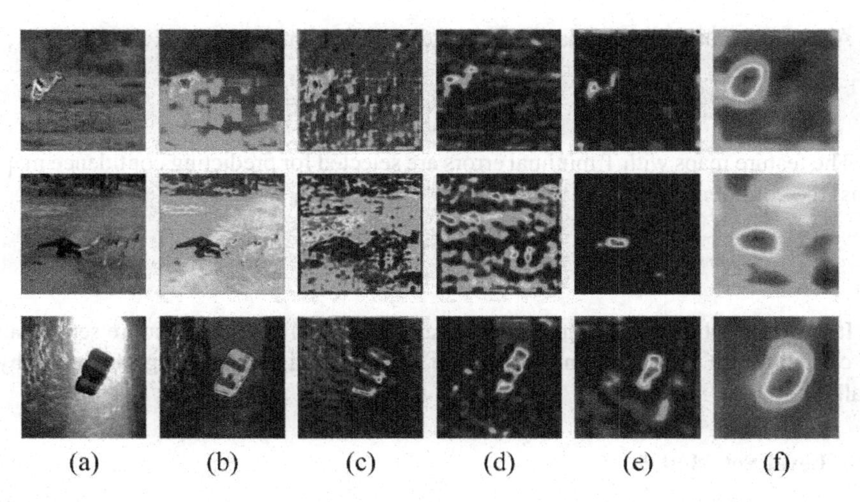

Fig. 2. Visualization of feature maps from different VGG convolutional layers. (a) ROIs of original frames. (b)–(f) features extracted from conv1-2, conv2-2, conv3-3, conv4-3 and conv5-3 layers respectively. It can be seen that the latter layer features have strengths at semantical discrimination of the object. However, the details of the object contour suffer loss with the network forwarding.

Given a single feature map $f \in \mathrm{F}$, the weak classifier is defined as:

$$\mathrm{L}(f) = \log\frac{\max\{\mathrm{H_{Obj}}(f), \delta\}}{\max\{\mathrm{H_{Bg}}(f), \delta\}} \tag{1}$$

where $\mathrm{H_{Obj}}$ and $\mathrm{H_{Bg}}$ are the histograms of object and background, and δ is a small value fixed to 0.001 that prevents dividing by zero. Feature map f is also tuned to have better discrimination by Eq. (1) [18].

The error of each feature is calculated with the weak classifier result $\mathrm{LI}_f = \mathrm{L}(f)$ and the sample weight $\mathrm{w_i^0}$. Initially, the sample weight $\mathrm{w_i^0} = 1/\mathrm{N}$ is equal for each feature where N is the number of samples. And at the tth iteration, error of the feature f_t is weighted by $\mathrm{W}^t = \{\mathrm{w_i^t} | 1 \leq \mathrm{i} \leq \mathrm{N}\}$:

$$\mathrm{Error}(\mathrm{W}^t; f_t) = \sum\nolimits_{\mathrm{i}=1}^{\mathrm{N}} \mathrm{w_i^t}\left(\mathrm{sgn}\left(\mathrm{LI}_{f_t}(\mathrm{i})\right) \neq \mathrm{y_i}\right) \tag{2}$$

$\mathrm{y_i} \in \{1, -1\}$ is the training label which represents the sample belong to the object ($\mathrm{y_i} = 1$) or background ($\mathrm{y_i} = -1$).

The weak classifier weight α_t for the feature f_t is given as follows:

$$\alpha_t = \frac{1}{2}\ln\left(\frac{1 - \mathrm{Error}(\mathrm{W}^t; f_t)}{\mathrm{Error}(\mathrm{W}^t; f_t)}\right) \tag{3}$$

At every iteration, we update W^t according to the error of weak classifier LI_{f_t},

$$w_i^{t+1} = w_i^t \exp(-\alpha_t \mathrm{sgn}(LI_{f_t}(i) \neq y_i)) \tag{4}$$

The feature maps with T minimal errors are selected for predicting confidence map S. Thus, the confidence is

$$S = \mathrm{sgn}\left(\sum_{t=0}^{T-1} \alpha_t LI_{f_t}\right) \tag{5}$$

It is worth noting that T is much less than the size of candidate feature set, because we can eliminate noise in confidence map and get a high predicting efficiency by a small T.

2.2 Level Set Model

We develop a level set method to further segment the object in the frame. The initial level set ϕ_t is determined by the segmentation result ϕ_{t-1} of the last frame (ϕ is defined by the initial contour C_0 for the first frame),

$$\phi_t(x) = \begin{cases} -1, & \text{if } \phi_{t-1}(x) \circ G_d < 0 \\ 1, & \text{otherwise} \end{cases} \tag{6}$$

where $\phi_{t-1} \circ G_d$ is a dilation operation with an element G_d adapted on the ϕ_{t-1}.

Like the traditional level set methods [16], we also used the edge indicator in our model. Let I be the frame represented in gray value, the conventional edge indicator can be defined as

$$g = \frac{1}{1 + |\nabla G_\sigma * I|^2} \tag{7}$$

where G_σ is Gaussian kernel with a standard deviation σ. This edge indicator is employed in the length and area terms respectively,

$$\mathcal{A}(\phi) = \int_\Omega g H_\varepsilon(-\phi) dx \tag{8}$$

$$\mathcal{L}(\phi) = \int_\Omega g \delta_\varepsilon(\phi) |\nabla \phi| dx \tag{9}$$

where Ω is the whole image domain, and H_ε is Heaviside function with parameter ε,

$$H_\varepsilon(x) = \begin{cases} \frac{1}{2}\left(1 + \frac{x}{\varepsilon} + \frac{1}{\pi}\sin\left(\frac{\pi x}{\varepsilon}\right)\right), & |x| \le \varepsilon \\ 1, & x > \varepsilon \\ 0, & x < -\varepsilon \end{cases} \tag{10}$$

The area term $\mathcal{A}(\phi)$ computes a weighted area the region inside the contour, while $\mathcal{L}(\phi)$ computes the energy along the length of the contour. In our model, the novel area and length terms adopt the confidence map after erosion and dilation operation $S_d = S \circ G_e \circ G_d$ to guide the edge indicator in both the area term and length term,

$$\mathcal{A}'(\phi) = \int_{\Omega} g_{\mathcal{A}} H_{\varepsilon}(-\phi)dx, \quad g_{\mathcal{A}} = \frac{1}{2}\left(\frac{1 - S_d}{2} + g\right) \tag{11}$$

$$\mathcal{L}'(\phi) = \int_{\Omega} g_{\mathcal{L}} \delta_{\varepsilon}(\phi)|\nabla\phi|dx, \quad g_{\mathcal{L}} = \frac{1}{2}\left(\frac{1}{1 + |\nabla S_d|^2} + g\right) \tag{12}$$

where $\frac{1-S_d}{2} \in \{0, 1\}$ can speed up the evolution process toward the predicted map S, and $\frac{1}{1+|\nabla S_d|^2}$ is boundary adjustment to the edge indicator g. The proposed indicators use the linear combination to compensate for the errors caused by g, and avoid the segmentation failure when S is inaccurate.

Hence the energy function $E(\phi)$ is obtained by three energy terms, improved area term and length term $\mathcal{A}'(\phi)$, $\mathcal{L}'(\phi)$, and penalty term $\mathcal{P}(\phi)$ respectively,

$$E(\phi) = \alpha\mathcal{A}'(\phi) + \nu\mathcal{L}'(\phi) + \mu\mathcal{P}(\phi) \tag{13}$$

where $\mathcal{P}(\phi)$ is

$$\mathcal{P}(\phi) = \frac{1}{2}\int_{\Omega} (|\nabla\phi| - 1)^2 dx \tag{14}$$

where $\delta_{\varepsilon} = H'_{\varepsilon}$ is the Dirac function,

$$\delta_{\varepsilon}(x) = \begin{cases} \frac{1}{2\varepsilon}\left(1 + \cos\left(\frac{\pi X}{\varepsilon}\right)\right), & |x| \le \varepsilon \\ 0, & |x| > \varepsilon \end{cases} \tag{15}$$

We can use the following gradient flow to minimize the energy function Eq. (13),

$$\frac{\partial\phi}{\partial t} = \mu\left(\Delta\phi - \operatorname{div}\left(\frac{\nabla\phi}{|\nabla\phi|}\right)\right) + \nu\delta_{\varepsilon}(\phi)\operatorname{div}\left(g_{\mathcal{L}}\frac{\nabla\phi}{|\nabla\phi|}\right) + \alpha g_{\mathcal{A}}\delta_{\varepsilon}(\phi) \tag{16}$$

The curve of zero level set can iteratively segment the object contour by minimizing $E(\phi)$ according to Eq. (16). Finally, we obtain the object contour $-\mathrm{sgn}(\phi)$.

2.3 Outlier Rejection

Since AdaBoost is sensitive to outliers [19], we introduce an outlier rejection scheme to compensate the discriminative model for wrong classification. Basically, the classification result after AdaBoost and segmentation can be written as

$$y_i(x) = \begin{cases} 1, & \text{if } \phi_t(x) < 0 \\ -1, & \text{otherwise} \end{cases} \tag{17}$$

The outlier rejection is defined as follows,

$$y_i'(x) = \begin{cases} 1, & \text{if } \phi_t(x) < 0 \wedge w_i < \Theta \wedge \|x - \text{center}(\phi_{t-1})\|_2 < D \\ -1, & \text{otherwise} \end{cases} \tag{18}$$

In Eq. (18), w_i is sample weight for the pixel x. The weight for each pixel is first initialized equally as $w_i = 1/N$. Then the selected feature f_t and the y_i in Eq. (17) are used to compute the error and α_t according to Eqs. (2) and (3). We update w_i by the error and α_t same as the Eq. (4). This scheme ensures the sample with a too large weight (larger than the predefined threshold $\Theta = 9/N$) is labeled negative. The samples that are too difficult to classify will be defined as background, which can lead to a cleaner result.

Moreover, we introduce a distance based rejection scheme. $\|\cdot\|_2$ means 2-norm, D is a threshold and $\text{center}(\phi_{t-1})$ is the center location of the object in the last frame in Eq. (18). The positive samples which are too far away from the object in the last frame will be removed because the object location in a video is usually continuous.

2.4 Model Updating

Once the segmentation and the outlier rejection are finished, the weak classifiers of AdaBoost model will be updated online. In Eq. (1), a weak classifier consists of two parts, object and background histograms. Thus we update the two histograms by linear combination of the current histograms and the ones in last frame,

$$P_t(c|H_i) = (1 - \gamma_i)P_{t-1}(c|H_i) + \gamma_i P_t(c|H_i), i = \{\text{Obj}, \text{Bg}\} \tag{19}$$

γ_{Obj} and γ_{Bg} are set to 0.08 and 0.1 respectively in all our experiments.

3 Experiments

Our model is implemented in C++ based on Caffee and OpenCV 2.4.10. All the tests run on a computer with a 3.3 GHz CPU, 4 GB RAM and a GK208 GPU. We select $T = 6$ best features to predict the confidence map in our experiments. The parameter setting for level set segmentation are shown in Table 2,

Table 2. Level set parameter setting.

n	Δt	ε	α	ν	μ
20	1.0	1.5	3.0	6.0	0.2

where n is the iteration number of evolution, Δt is the step of iteration, and ε, α, ν and μ are the same as parameters in Eqs. (10) and (13).

We evaluate our algorithm on SegTrack database [20] and compare it with other three top-performance non-rigid contour trackers [11, 13, 16]. The dataset consists of six sequences: birdfall2, cheetah, girl, monkeydog, parachute and penguin. Our tracker is evaluated by three metrics: success/precision plots of one-pass evaluation (OPE), error pixels and frame per second.

We first compare the success and precision plots of different trackers. For fair comparison, we convert the tracking contours of HT [11], SLSM [16] and our model into bounding boxes while the result of PT [13] is directly bounding box. The converted trackers are evaluated by success ratio, the ratio of frames whose tracked box has more overlap with ground truth box than the threshold; and precision, the ratio of frames whose tracking result is within the threshold from ground truth. Figure 3 shows the results of trackers. It is demonstrated that our model has better performance than other three trackers in both success and precision.

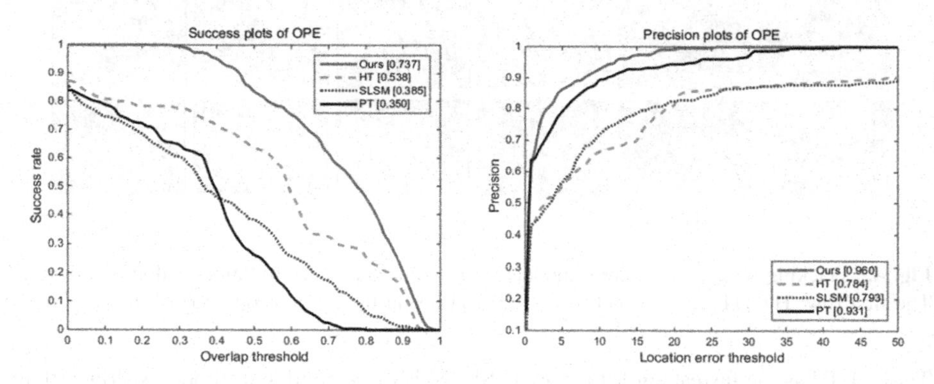

Fig. 3. The success plots and precision plots comparison of different trackers. The performance score of each tracker is presented in the legend. Values in the brackets are AUC (area under curve) and show the overall performance of trackers.

For the trackers whose results are in "contour" form, we further evaluate them by average number of error pixels per frame. As we discussed above, results of PT is in the form of bounding box and the metrics of error pixels is not fair for PT. Therefore, PT is not involved in this part. Table 3 lists the quantitative results of HT [11], SLSM [16] and our model. It can be seen that our model has smaller errors than HT and SLSM on the six videos. This advantage over SLSM, which also uses the level set method, may be due to more accurate confidence maps generated by deep features. Figure 4 shows the qualitative comparison in some selected frames from the dataset.

We measure the processing speed of all selected algorithms by frame per second in

Table 3. Average number of error pixels per frame of different trackers. The lowest errors are in boldface.

Sequence	HT [11]	SLSM [16]	Our model
girl	19995	10076	**3462**
parachute	587	2233	**321**
penguin	5597	5658	**2918**
monkeydog	1883	1080	**560**
cheetah	1193	1464	**787**
birdfall2	327	430	**261**
average	4930	3490	**1384**

(a) (b)

Fig. 4. Tracking results of different models for (a) "cheetah" and (b) "monkeydog" sequences. The first row: HT [11], the second row: SLSM [16] and the bottom row: our method.

Table 4. PT is the fastest algorithm while SLSM is the second. Our model is slower than SLSM, but faster than HT. Moreover, our model obtains the similar FPS values on all the videos because it processes the frame within a 224 × 224 ROI and the computational complexity is almost the same for different sequences. This inferior time consumption to the level set based tracker, SLSM, is due to the fixed size of segmentation region. SLSM segments the object in a window which is twice the size of the object and this strategy usually generates a smaller ROI than ours. In future work, we plan to introduce acceleration scheme [21] to enhance the efficiency of our model. Besides, it should be noted that the speed of our model is acceptable, when it has far better performance than others and still can compete with high-performance algorithm, HT.

Table 4. Comparison of running speed of methods (evaluated by frame per second). The best results are in boldface.

Sequence	HT [11]	PT [13]	SLSM [16]	Our model
girl	0.1140	**1.2494**	0.4769	0.4920
parachute	0.2559	**3.4570**	2.0394	0.4927
penguin	0.1962	**3.5360**	1.1009	0.4957
monkeydog	0.3922	**5.8909**	3.1212	0.4941
cheetah	0.4317	**6.2957**	4.5045	0.4921
birdfall2	0.6573	5.2128	**9.9931**	0.4885
average	0.3412	**4.2736**	3.5393	0.4925

4 Conclusion

We proposed a level set tracking framework which uses convolutional neural network features in this paper. We combine hand-crafted features and deep features, and then an AdaBoost selection procedure is employed on the features in order to select features and generate a coarse confidence map of the object. Afterwards, we segment the current frame based on the confidence along with the original image by a level set method. We also introduce an outlier rejection scheme and updating to further improve the result. Experimental results indicate that our method outperforms the state-of-the-art non-rigid contour tracking algorithms on SegTrack benchmark.

In future work, we will investigate how to improve the computational efficiency of our model, especially in the level set model. Other interests involve applying the superpixel to our model to generate more accurate confidence map, and exploring more effective segmentation methods.

Acknowledgments. This work is supported by National Natural Science Foundation of China under grant 61370133 and grant 61672095.

References

1. Li, X., Hu, W., Shen, C., Zhang, Z., Dick, A., Hengel, A.V.D.: A survey of appearance models in visual object tracking. ACM Trans. Intell. Syst. Technol. **4**(4), 478–488 (2013)
2. Hu, W., Zhou, X., Li, W., Luo, W., Zhang, X., Maybank, S.: Active contour-based visual tracking by integrating colors, shapes and motions. IEEE Trans. Image Process. **22**(5), 1778–1792 (2013)
3. Bibby, C., Reid, I.: Robust real-time visual tracking using pixel-wise posteriors. In: Forsyth, D., Torr, P., Zisserman, A. (eds.) ECCV 2008. LNCS, vol. 5303, pp. 831–844. Springer, Heidelberg (2008). doi:10.1007/978-3-540-88688-4_61
4. Barbu, T.: Template matching based video tracking system using a novel n-step search algorithm and HOG features. In: International Conference on Neural Information Processing (ICONIP), Doha, pp. 328–336 (2012)
5. Nam, H., Han, B.: Learning multi-domain convolutional neural networks for visual tracking. In: IEEE Conference on Computer Vision and Pattern Recognition, Las Vegas, pp. 4293–4302 (2016)

6. Zhang, K., Liu, Q., Wu, Y., Yang, M.H.: Robust visual tracking via convolutional networks without training. IEEE Trans. Image Process. **25**(4), 1779–1792 (2016)
7. Hong, S., You, T., Kwak, S., Han, B.: Online tracking by learning discriminative saliency map with convolutional neural network. In: International Conference on Machine Learning, Lille, pp. 597–606 (2015)
8. Wang, L., Ouyang, W., Wang, X., Lu, H.: Visual tracking with fully convolutional networks. In: IEEE Conference on Computer Vision and Pattern Recognition, Santiago, pp. 3119–3127 (2015)
9. Ma, C., Huang, J., Yang, X., Yang, M.: Hierarchical convolutional features for visual tracking. In: IEEE Conference on Computer Vision, Santiago, pp. 3074–3082 (2015)
10. Yeh, Y., Hsu, C.: Online selection of tracking features using AdaBoost. IEEE Trans. Circ. Syst. Video Technol. **19**(3), 442–446 (2009)
11. Godec, M., Roth, P.M., Bischof, H.: Hough-based tracking of non-rigid objects. In: IEEE International Conference on Computer Vision, Barcelona, vol. 117, no. 10, pp. 81–88 (2011)
12. Duffner, S., Garcia, C.: Fast pixelwise adaptive visual tracking of non-rigid objects. IEEE Trans. Image Process. **26**(5), 2368–2380 (2017)
13. Duffner, S., Garcia, C.: PixelTrack: a fast adaptive algorithm for tracking non-rigid objects. In: IEEE International Conference on Computer Vision, Sydney, pp. 2480–2487 (2013)
14. Chan, T.F., Sandberg, B.Y., Vese, L.A.: Active contours without edges for vector-valued images. J. Vis. Commun. Image Represent. **11**(2), 130–141 (2000)
15. Liu, L., Zhang, Q., Wu, M., Li, W., Shang, F.: Adaptive segmentation of magnetic resonance images with intensity inhomogeneity using level set method. Magn. Reson. Imaging **31**(4), 567–574 (2013)
16. Sun, X., Yao, H., Zhang, S., Li, D.: Non-rigid object contour tracking via a novel supervised level set model. IEEE Trans. Image Process. **24**(11), 3386–3399 (2015)
17. Simonyan, K., Zisserman, A.: Very deep convolutional networks for large-scale image recognition. arXiv:1409.1556 [cs.CV] (2014)
18. Collins, R.T., Liu, Y., Leordeanu, M.: Online selection of discriminative tracking features. IEEE Trans. Pattern Anal. Mach. Intell. **27**(10), 1631–1643 (2005)
19. Avidan, S.: Ensemble tracking. IEEE Trans. Pattern Anal. Mach. Intell. **29**(2), 261–271 (2007)
20. Tsai, D., Flagg, M., Rehg, J.M.: Motion coherent tracking with multi-label MRF optimization. Int. J. Comput. Vis. **100**(2), 1–11 (2010)
21. Liu, L., Fan, S., Ning, X., Liao, L.: An efficient level set model with self-similarity for texture segmentation. Neurocomputing **266**, 150–164 (2017)

End-to-End Scene Text Recognition with Character Centroid Prediction

Wei Zhao and Jinwen Ma(✉)

Department of Information Science, School of Mathematical Sciences And LMAM, Peking University, Beijing 100871, China
jwma@math.pku.edu.cn

Abstract. Scene text recognition tries to extract text information from natural images, being widely applied in computer vision and intelligent information processing. In this paper, we propose a novel end-to-end approach to scene text recognition with a specially trained fully convolutional network for predicting the centroid and pixel cluster of each character. With the help of this new information, we can solve the character instance segmentation problem effectively and then combine the recognized characters into words to accomplish the text recognition task. It is demonstrated by the experimental results on ICDAR2013 dataset that our proposed method with character centroid prediction can get a promising result on scene text recognition.

Keywords: Scene text recognition · Character centroid prediction · Fully convolutional networks · Character instance segmentation

1 Introduction

Text recognition has been investigated and applied for many years. Actually, Optical Character Recognition (OCR) is considered to be a powerful character recognition tool for the scanned images of papers or articles, but it is still rather difficult to detect and recognize text in natural scenes due to the complex environment, low image quality and other uncontrollable factors [1]. In recent years, deep convolutional neural networks have shown great capability on solving the computer vision problems and they also work well for scene text recognition. When using a CNN to undertake a text recognition task, it is usual to divide the whole problem into two subproblems: text localization and cropped word recognition, and we then train two different models to solve them separately. For the text localization problem, some specific characteristics of text are utilized to detect the fields of text in the image [3–5]. Apart from those methods, some general object detection techniques like the Faster-RCNN [2] can work as well. For the cropped word recognition problem, most recent methods are based on CRNN [6] that utilizes a CNN to extract the features and a RNN to deal with the sequence learning. Under this localization-and-recognition framework, we can get some good results in certain cases. However, it has certain innate drawbacks

D. Liu et al. (Eds.): ICONIP 2017, Part III, LNCS 10636, pp. 291–299, 2017.
https://doi.org/10.1007/978-3-319-70090-8_30

to train the two models separately. First, since each model only learns from a part of the training data, there will be certain loss in the final result. Second, since the both models use a DCNN to extract the features from an image, a lot of sharable computation are repeated, which leads to low efficiency.

In order to resolve these issues, we propose a novel end-to-end method for scene text recognition. Our main idea is to firstly make the instance segmentation for all the characters in an image and then combine the recognized characters into words. We consider the character instance segmentation as a clustering analysis problem where each character is corresponding to a cluster and our object is to divide the related pixels into a number of clusters which are corresponding to the characters, respectively. In order to do so, we train a fully convolutional neural network to predict the centroid and pixel cluster of each character. It is demonstrated by the experiments on ICDAR2013 dataset that our proposed method leads to a promising result on scene text recognition.

2 Related Works

2.1 Character Instance Segmentation

As shown in Fig. 1, character instance segmentation can be roughly considered as a combination of object detection and semantic segmentation. In fact, most general instance segmentation methods are either detection-based or segmentation-based methods. The detection-based methods [9] first detect all the possible instances in an image and then predicts the mask of each instance, while the segmentation-based methods [8] first put a label on each pixel and then group the pixels into some instances. Here, we adopt the segmentation-based method to make the character instance segmentation since the current object detection methods are not so good with small objects which is also the reason why we do not take text recognition with direct character detection.

2.2 Fully Convolutional Network

Fully convolutional network (FCN) only contains a number of convolutional layers. In comparison with the other deep learning neural network architectures, it has neither pooling layer nor fully connected layer. As a result, it can take an arbitrary-sized image as input and predict what we need for each pixel which makes them good at solving semantic segmentation problem. In a typical convolutional neural network, feature map will be down-sampled several times while it goes through the convolutional layers. As it goes deeper, more semantic information are extracted but a lot spatial information are abandoned. To deal with this effect, one common technique is to up-sample deep layer and concatenate it with shallow layer and do the final prediction base on this concatenated feature map, as proposed in [10]. With this modification, FCN can predict pixel label accurately. Of course, character instance segmentation cannot be done with only pixel label so that we train an FCN to predict character centroid as well as pixel label.

(a) Image classification

(b) Object localization

(c) Semantic segmentation

(d) Instance segmentation

Fig. 1. An example of image classification, object localization (also known as object detection), semantic segmentation and instance segmentation from [7]. (Color figure online)

2.3 Character Centroid Prediction

In fact, character centroid prediction is important for text recognition. Last year, Zhang et al. [5] utilized this idea to make text localization via a FCN-based model. In particular, a common FCN was trained to make the two-class semantic segmentation, with the segmentation result shown in Fig. 2. Actually, character centroid prediction acted as an auxiliary helper to select text line candidate so they didn't put too much effort on it. As we can see, their centroid prediction result is not very good, especially in some cases centroid regions even overlap with each other. There is one unnatural setting in their model that may be the reason to harm their result: they consider the whole problem as a classification

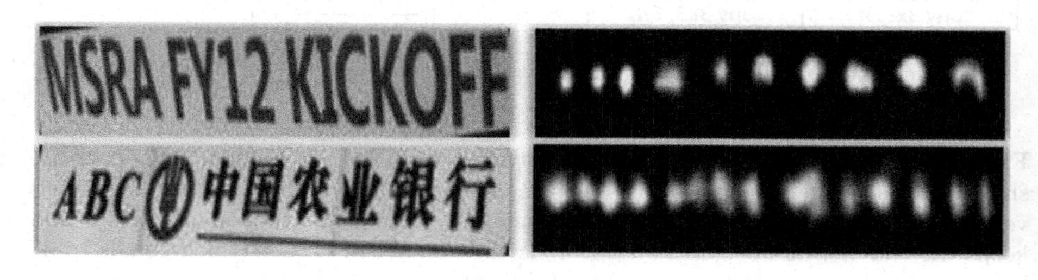

Fig. 2. The centroid prediction result given in [5].

problem and the label of a pixel is determined by setting a threshold on the distance between a pixel and the real centroid.

Now, we consider the character centroid prediction as a regression problem. Instead of predicting whether a pixel is near the real centroid, we train our FCN to learn pixel's relative position to the real centroid. As shown in Fig. 5, our FCN learns this target very well and we can easily accomplish character instance segmentation with our FCN's prediction.

3 Proposed Network with Character Centroid Prediction

3.1 Network Structure

As mentioned in the previous sections, our method's main component is an FCN. Figure 3 shows the network structure. We design our network based on two principles: 1. The size of feature map shouldn't shrink too much, in order to get precise character mask; 2. The size of receptive field must be large enough so that context information can be used. We adopt the residual learning structure proposed in [11] and use the consecutive downsample-upsample to enlarge receptive field while keeping the size of feature map. For each pixel, our FCN gives two predictions. The first one is an n_c-d vector representing pixel label probability. The other one is an $6n_c$-d vector, representing the pixel's relative position to its corresponding character centroid and neighbor character centroid given the pixel's label. Figure 4 shows an illustration of our network's prediction.

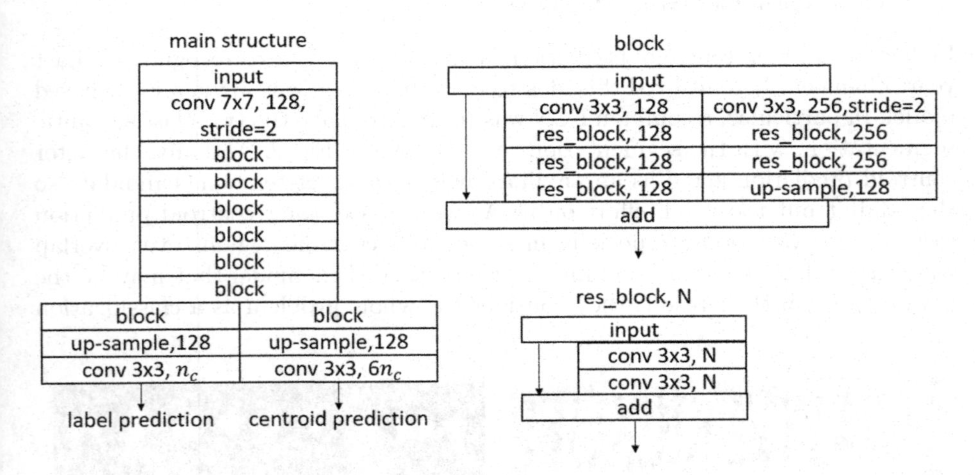

Fig. 3. Our network structure. Convolutional layer's kernel size and channels are showed in the figure and stride is 1 if not specified. Up-sampling is done by deconvolution (transposed convolution) which enlarges previous feature map 2 times and keeps the channel number unchanged. n_c is the number of classes which equals to 37 in our case (10 digits, 26 letters and background).

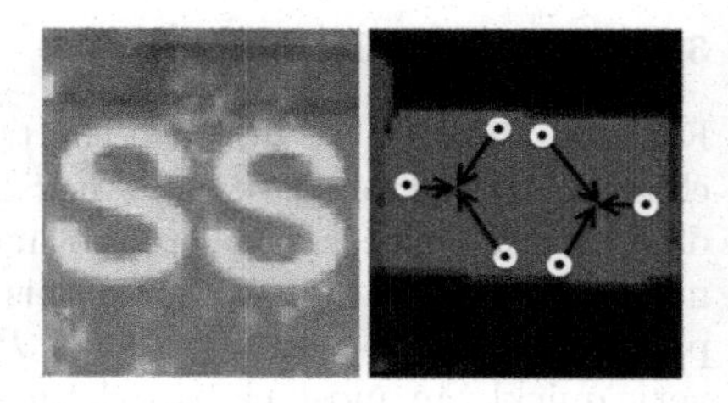

Fig. 4. Illustration of our network's prediction. The left two images show the label prediction result on test image. As we can see, the prediction is very accurate. However, without other information, we can't group pixels into characters, especially for those consecutive characters. The right two images show our goal of centroid prediction. With every pixel pointing to corresponding character's centroid, we can easily accomplish instance segmentation for character. We also let our network to predict neighbor character's centroid in order to combine characters into words. (Color figure online)

3.2 Training Process

During the training, we optimize the following two-part loss function:

$$Loss = \lambda \sum_{i,j=0}^{n-1} \mathbf{1}_{c_{ij}\neq 0} L1_{smooth}(\mathbf{pos}_{ijc_{ij}}^{pred}, \mathbf{pos}_{ij}^{truth}) - \sum_{i,j=0}^{n-1} log(p_{ij}(c_{ij})), \quad (1)$$

where

$$L1_{smooth}(f,g) = \begin{cases} 0.5(f-g)^2, & \text{if } |f-g| < 1; \\ |f-g| - 0.5, & \text{otherwise.} \end{cases} \quad (2)$$

In fact, the first part is just the smoothed L1 loss which guides relative centroid position regression, where n is the size of input image, $\mathbf{1}_{c_{ij}\neq 0}$ denotes if pixel $I(i,j)$ belongs to a character, $\mathbf{pos}_{ijc_{ij}}^{pred}$ denotes predicted relative position for pixel $I(i,j)$ if $I(i,j)$'s true label is c_{ij}, $\mathbf{pos}_{ij}^{truth}$ denotes the true relative position for pixel $I(i,j)$. Both $\mathbf{pos}_{ijc_{ij}}^{pred}$ and $\mathbf{pos}_{ij}^{truth}$ are 6-d vectors. If a character is the first or the last in a word, we let its left/right neighbor to be itself. The second part is the cross-entropy loss for pixel label prediction where c_{ij} is the true label of pixel $I(i,j)$ and $p_{ij}(c_{ij})$ is predicted probability that $I(i,j)$'s label is c_{ij}. We use a hyper-parameter λ to balance these two parts of loss and we set λ to 10^{-6} during in practice.

We train our network on synthetic data made by [12]. Our loss function is very simple and can be optimized by any gradient-based method. We use SGD with momentum to train our model. Our learning rate starts at 0.01, and cuts into half after [9600, 19200, 48000, 96000, 192000, 384000, 768000, 1152000] times parameter update.

3.3 Inference Principle

For a test image, we first use the trained FCN to predict the pixel's label and character's centroid. We can't just group together pixels which have same predicted centroid because there are small error in centroid prediction. We use non-maximum-suppression to solve this problem. In detail, for a non-background pixel $I(i,j)$, the sequence $I(i,j), PC(I(i,j)), PC(PC(I(i,j))), ...$ "converges" very quickly in most cases, where $PC(I(i,j))$ denotes $I(i,j)$'s predicted centroid's nearest pixel, and we use $PC^*(I(i,j))$ to denote the "limitation" of this sequence. We say $I(i,j)$ is $I(i',j')$'s "supporter" if $PC^*(I(i,j)) = I(i',j')$. We let pixel $I(i',j')$ to be a centroid candidate if it has enough supporters. For each pair of two centroid candidates $I(i_1,j_1)$ and $I(i_2,j_2)$, we change $I(i_1,j_1)$'s supporters' final predicted centroid to $I(i_2,j_2)$ and remove $I(i_1,j_1)$ from centroid candidates if the distance between these two candidates is too small and $I(i_1,j_1)$ has less supporters than $I(i_2,j_2)$. At last, for each character there is only one centroid candidate left, which we use as the final predicted character centroid and its supporters make up character's mask. For the probability of character's label, we let it be the average of its pixels' label probability. At this point, character instance segmentation is done and we now combine them into words. As described above, our network not only predicts pixel's corresponding character's

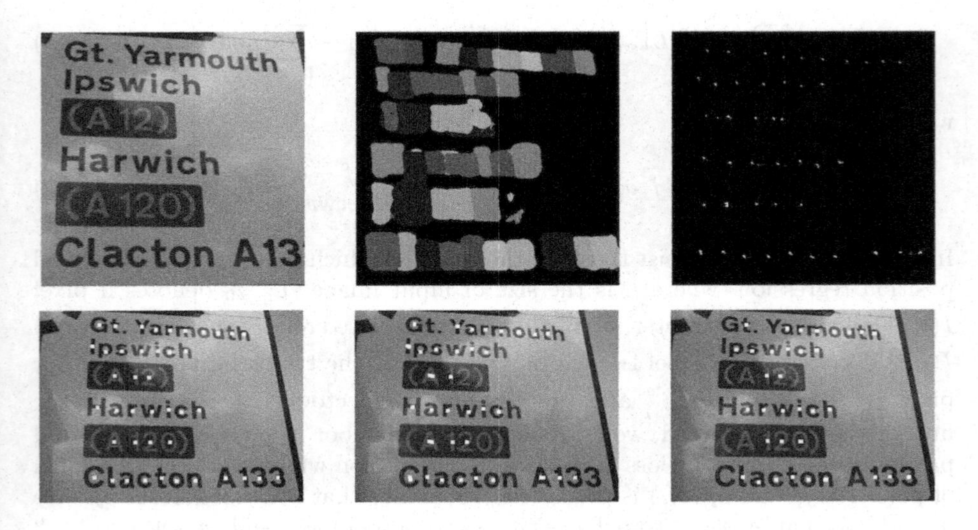

Fig. 5. Top row: test image, pixel label prediction, and heat map of pixel's supporter count(clipped to an appropriate interval). As we can see, the centroid prediction is very compact and consistent so we can easily get the final centroid prediction result. Bottom row: final centroid prediction for characters, characters' left neighbors, and characters' right neighbors(if a character is the start/end of a word, we set its left/right neighbor to itself). Our network not only predict character centroid accurately, the neighbor centroid prediction result is also good enough for us to combine characters into words. (Color figure online)

centroid, but also neighbor character's centroid. For a character, we use its final centroid candidate's prediction as neighbor character prediction result. Given a pair of characters, if one's centroid prediction is close enough to the other one's neighbor centroid prediction, then we determine that they belong to the same word. Figure 5 visualizes our network's prediction result and we can see that our method is valid.

4 Experimental Results

We test our proposed method on ICDAR2013 dataset [15]. Since this dataset contains images of many different scales, we just adopt the image pyramid technique to make multi-scale text recognition on task4. Our end-to-end recognition experimental results are listed in Table 1, in comparison with some state-of-the-art methods. We also test our proposed method on text localization task and give the experimental results in Table 2.

Table 1. The experimental results of text recognition on ICDAR2013 task 4.

Method	End-to-end			Word spotting		
	Recall	Precision	HMean	Recall	Precision	HMean
hust_mclab [6,16]	87.68%	95.83%	91.57%	90.77%	97.25%	93.90%
vggmaxbbnet [17,18]	82.12%	91.05%	86.35%	86.68%	94.64%	90.49%
Deep2Text II+[17,19]	72.08%	94.56%	81.81%	75.82%	96.29%	84.84%
Proposed method	84.62%	90.76%	87.58%	91.82%	93.24%	92.53%

In fact, ICDAR2013 task 4 was strongly contextualized. It can be seen from Table 1 that our proposed method can get a rather good result. Although it does not surpass the result of some state-of-the-art methods on the web, we just use the single-model end-to-end method while the other three methods all make text localization and cropped word recognition in two steps, respectively. So, our proposed method is more promising since our network can be further optimized

Table 2. The experimental results of text location on ICDAR2013 task1.

Method	Recall	Precision	HMean
CTPN [3]	82.98%	92.98%	87.69%
TextConv+WordGraph [4]	81.02%	93.38%	86.76%
MCLAB_FCN [5]	79.65%	88.40%	83.80%
IWRR2014 [13]	78.65%	85.89%	82.11%
HUST_MCLAB [14]	76.05%	87.96%	81.58%
Proposed method	87.16%	88.82%	87.98%

globally. Moreover, it can be seen from Table 2 that our proposed method is remarkably better than the other existing methods according to deteval criterion. It should be noted that our recall rate surpasses the other methods by a large margin.

5 Conclusion

We have proposed an end-to-end text recognition method with character centroid prediction. It is based on a specially trained fully convolutional network to predict the centroid and pixel cluster of each character so that the character instance segmentation problem can be solved effectively and then the recognized characters can be combined into words to accomplish the text recognition task. It is demonstrated by the experimental results on ICDAR2013 dataset that our proposed method can get a promising result on scene text recognition.

Acknowledgments. This work was supported by the Natural Science Foundation of China under Grant 61171138.

References

1. Ye, Q., Doermann, D.: Text detection and recognition in imagery: a survey. IEEE Trans. Pattern Anal. Mach. Intell. **37**(7), 1480–1500 (2015)
2. Ren, S., He, K., Girshick, R., et al.: Faster R-CNN: towards real-time object detection with region proposal networks. In: Advances in Neural Information Processing Systems, vol. 28, pp. 91–99 (2015)
3. Tian, Z., Huang, W., He, T., He, P., Qiao, Y.: Detecting text in natural image with connectionist text proposal network. In: Leibe, B., Matas, J., Sebe, N., Welling, M. (eds.) ECCV 2016. LNCS, vol. 9912, pp. 56–72. Springer, Cham (2016). doi:10.1007/978-3-319-46484-8_4
4. Zhu, S., Zanibbi, R.: A text detection system for natural scenes with convolutional feature learning and cascaded classification. In: Proceedings of 2016 IEEE Conference on Computer Vision and Pattern Recognition (CVPR 2015), pp. 625–632 (2015)
5. Zhang, Z., Zhang, C., Shen, W., et al.: Multi-oriented text detection with fully convolutional networks. In: Proceedings of 2016 IEEE Conference on Computer Vision and Pattern Recognition (CVPR 2016), pp. 4159–4167 (2016)
6. Shi, B., Bai, X., Yao, C.: An end-to-end trainable neural network for image-based sequence recognition and its application to scene text recognition. IEEE Trans. Pattern Anal. Mach. Intell. **PP**(99), 1 (2016)
7. Lin, T.-Y., Maire, M., Belongie, S., Hays, J., Perona, P., Ramanan, D., Dollár, P., Zitnick, C.L.: Microsoft COCO: common objects in context. In: Fleet, D., Pajdla, T., Schiele, B., Tuytelaars, T. (eds.) ECCV 2014. LNCS, vol. 8693, pp. 740–755. Springer, Cham (2014). doi:10.1007/978-3-319-10602-1_48
8. Zhang, Z., Fidler, S., Urtasun, R.: Instance-level segmentation for autonomous driving with deep densely connected mrfs. In: Proceedings of 2016 IEEE Conference on Computer Vision and Pattern Recognition (CVPR 2016), pp. 669–677 (2016)

9. Dai, J., He, K., Sun, J.: Instance-aware semantic segmentation via multi-task network cascades. In: Proceedings of 2016 IEEE Conference on Computer Vision and Pattern Recognition (CVPR 2016), pp. 3150–3158 (2016)
10. Long, J., Shelhamer, E., Darrell, T.: Fully convolutional networks for semantic segmentation. In: Proceedings of 2015 IEEE Conference on Computer Vision and Pattern Recognition (CVPR 2015), pp. 3431–3440 (2015)
11. He, K., Zhang, X., Ren, S., et al.: Deep residual learning for image recognition. In: Proceedings of 2016 IEEE conference on Computer Vision and Pattern Recognition (CVPR 2016), pp. 770–778 (2016)
12. Gupta, A., Vedaldi, A., Zisserman, A.: Synthetic data for text localisation in natural images. In: Proceedings of 2016 IEEE Conference on Computer Vision and Pattern Recognition (CVPR 2016), pp. 2315–2324 (2016)
13. Zamberletti, A., Noce, L., Gallo, I.: Text localization based on fast feature pyramids and multi-resolution maximally stable extremal regions. In: Jawahar, C.V., Shan, S. (eds.) ACCV 2014. LNCS, vol. 9009, pp. 91–105. Springer, Cham (2015). doi:10.1007/978-3-319-16631-5_7
14. Zhang, Z., Shen, W., Yao, C., et al.: Symmetry-based text line detection in natural scenes. In: Proceedings of 2015 IEEE Conference on Computer Vision and Pattern Recognition (CVPR 2015), pp. 2558–2567 (2015)
15. Karatzas, D., Shafait, F., Uchida, S., et al.: ICDAR 2013 robust reading competition. In: Proceedings of 2013 International Conference on Document Analysis and Recognition, pp. 1484–1493 (2013)
16. Liao, M., Shi, B., Bai, X., et al.: TextBoxes: a fast text detector with a single deep neural network. In: Prooceedings of AAAI 2017, pp. 4161–4167 (2017)
17. Jaderberg, M., Simonyan, K., Vedaldi, A., et al.: Reading text in the wild with convolutional neural networks. arXiv preprint arXiv:1412.1842 (2014)
18. Jaderberg, M., Simonyan, K., Vedaldi, A., et al.: Synthetic data and artificial neural networks for natural scene text recognition. arXiv preprint arXiv:1406.2227 (2014)
19. Yin, X.C., Yin, X., Huang, K., et al.: Robust text detection in natural scene images. IEEE Trans. Pattern Anal. Mach. Intell. **36**(5), 970–983 (2014)

Region-Based Face Alignment with Convolution Neural Network Cascade

Yu Zhang, Fei Jiang, and Ruimin Shen(✉)

Department of Computer Science and Engineering, Shanghai Jiao Tong University, No.800 Dongchuan Road, Minhang District, Shanghai, China
bellewanglu@163.com, {jiangf,rmshen}@sjtu.edu.cn

Abstract. Most face alignment approaches perform landmark detection over the entire face. However, it has been shown that the difficulty for landmark detection is unbalanced among different facial parts. Thus, in this paper, we propose a novel region-based facial landmark detection algorithm based on a two-level convolutional neural networks (CNNs). In the first level, we partition the whole face into four regions including three facial components (eyebrow-eyes, nose, and mouth) and the face contour. Regions are detected through an improved CNN model which is incorporated with a feature fusion scheme. To simultaneously detect three facial components and face contour landmarks, a novel weighted loss function combining bounding box regression with landmark localization is presented. In the second level, the landmarks are separately detected for three facial components. Experimental results on the public benchmarks demonstrate the superiority of the proposed algorithm over several state-of-the-art face alignment algorithms.

Keywords: Face alignment · Region-based · Convolution neural network · Feature fusion

1 Introduction

Face alignment is an appealing technology due to its close relevance to other face analysis tasks like emotion classification [1], face recognition [2], and face verification [3]. Face alignment aims at locating facial landmarks such as landmarks around eyes, mouth and nose. Recent years, impressive progresses have been made, but under unconstrained conditions, face alignment is still a challenge due to partial occlusions and large head pose variations.

Prevalent algorithms for the face alignment include template fitting algorithms [4,5] and regression-based algorithms [6–8]. Deep learning models also have been applied to the face alignment recently. For instance, Sun et. al. [9] propose a coarse-to-fine regression using a cascaded deep neural network architecture for face alignment, which is superior to pervious methods [10] and existing commercial systems. However, such cascaded deep neural network does not consider the property of unbalanced detection difficulties across different components. It has been shown that the detection of the contour landmarks is

D. Liu et al. (Eds.): ICONIP 2017, Part III, LNCS 10636, pp. 300–309, 2017.
https://doi.org/10.1007/978-3-319-70090-8_31

significantly more difficult than that of the inner landmarks [11]. Zhou et al. [11] designed a model to detect the face contour separately, which takes such unbalance into account. But the detection for inner landmarks is based on a coarse-to-fine process, which increases the computational complexity.

In this paper, we propose an efficient region-based face alignment algorithm based on a two-level deep convolutional neural networks(CNNs). To address the property of unbalanced detection difficulties across different landmarks, we firstly partition the whole face into four regions, including three facial components (eyebrow-eyes, nose, and mouth), and the face contour. Then, we detect the landmarks separately for each region which is less complex than that of the whole face. Specifically, in the first level of our proposed model, a novel convolutional neural network incorporated with a feature fusion scheme is proposed for the regions detection. Meanwhile, a novel weighted loss function combining bounding box regression and landmark localization is presented. In the second level, the landmarks are separately detected for three facial components using three different convolution neural networks without refine processing. Our contributions lie in

(1) A novel region-based face alignment algorithm is proposed. The algorithm divides the extensive landmark detections into several subproblems, which significantly simplify the task. Then, for each region, a deep convolutional neural network is adopted for the landmark detection, which dramatically reduces the network complexity and training burden compared with that for the whole face.

(2) Propose weighted loss function combining bounding box regression with landmark localization for precisely detecting regions and face contour landmarks at the same time. Moreover, by considering the fact that different regions contain different level features, a deep neural network incorporated with feature fusion process is designed for better region detection.

(3) Extensive results on the 300 W datasets show our algorithm outperforms several state-of-the-art algorithms.

2 Related Work

A large number of approaches have been proposed to tackle the challenge of face alignment. The proposed approaches can be grouped as regression-based methods and template-fitting methods. Template-fitting methods are also referred as generative methods. These methods fit generative models for both the shape and the appearance of a face. A representative model is AAM [12], which mapped shape model by coined landmark distribution model (PDM) [13], and [14] warped the training faces onto a common reference frame to obtain the appearance model.

Regression-based approach is more common in recent papers. Usually it starts with an initial shape, and then regressors are learned by extracting features around the shape landmarks. During this process, different features and regressors are used. In SDM [15], supervised descent method is proposed to handle the optimization of nonlinear least square problems, and nonlinear SIFT [16] is applied as the input feature for the linear regressor. ERT [6] utilized an shape index feature which is randomly sampled by exponential prior to train cascades of gradient boosting random forests. Since a initial shape is required for the regression, if the initialization is far from the ground truth, it is impossible to completely rectify the discrepancy by the subsequent iteration. CFSS [7] overcame the problem of initialization by proposing a method of shape search and estimate in a coarse-to-fine manner.

Deep convolutional network based regression methods also avoids the initialization problem. Sun et al. [9] designed a three-level cascaded deep convolutional networks. The first level predicts the rough position of five landmarks, and the refinement is done by the next two levels. Zhou et al. [11] proposed an extensive convolutional net work for detecting 68 landmarks. Zhou et al. [11] estimated bounding boxes for contour landmarks and inner landmarks in the first level. Then those landmarks are detected and refined by the subsequent level. Zhang et al. [8] proposed a convolutional network for multi-task learning which combined facial landmark localization with head pose estimation and smiling detection.

3 Our Method

The core idea of our method is that we simplify the extensive landmark detection task by dividing it into several subproblems. Specifically, we partition the whole face into four different regions and detect the landmarks in each region separately. Thus the network in our approach doesn't need to be very complicate to learn object features.

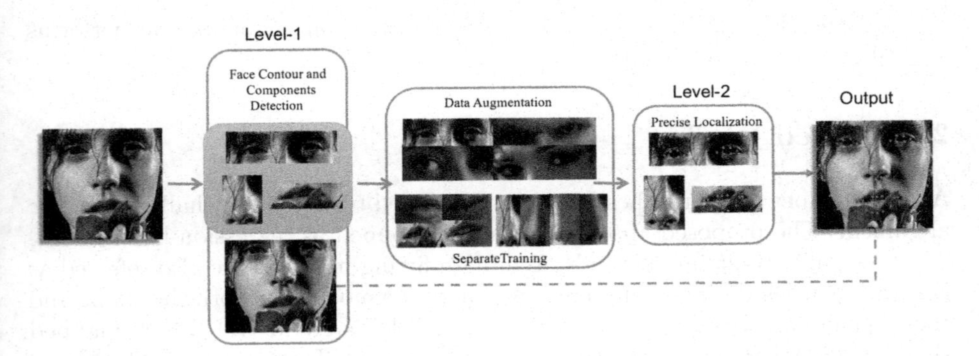

Fig. 1. Approach overview. The first level network predicts partial landmarks (face contour) and three bounding box for different face components. Then in the second level, landmarks are detected separately for three facial component. Lastly, we combine predictions of both level to locate all landmarks.

3.1 Architecture

Figure 1 gives a brief illustration of our region-based face alignment algorithms. A two-level deep convolutional neural network architecture is proposed. In the first level, we detect three facial components (eyebrow-eyes, nose, and mouth) and the face contour, which takes the whole face image as the input. After the first level, we get seventeen facial contour landmarks and the bounding boxes of three facial components. In the second level, three independent deep convolutional neural networks are proposed for the landmark detection of different facial components. The inputs of these deep models are the bounding boxes we acquired in the first level. More details will be disscussed in the following sections.

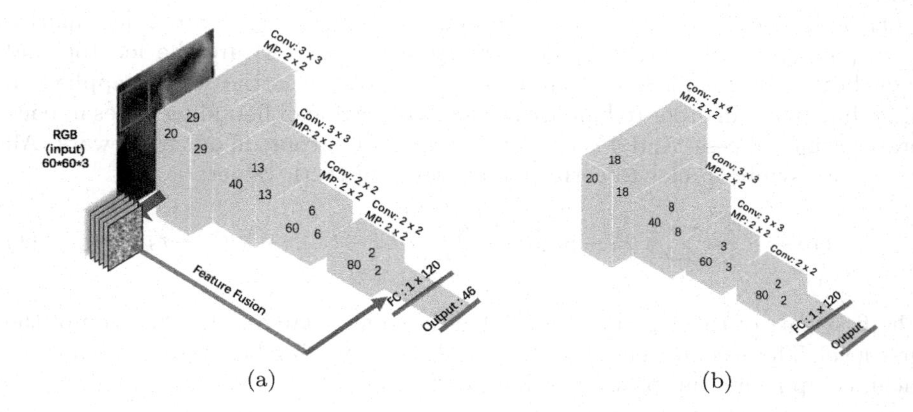

Fig. 2. Structure Details. (a) Deep structure of level 1. We use four convolutional layers followed by relu units and max pooling layers. This figure shows specific parameter for every layer. Here we fuse the feature map of the first convolutional layer (illustrated by a group of parallelograms) with that of the fourth convolutional layer.(b) Typical structure specification for level 2. Every component is trained separately through this architecture. Only the number of outputs is different.

3.2 Level 1: Region Detection

Localizing different landmarks require unbalanced efforts. Contour landmarks are shown to be significantly more difficult than the other landmarks [11] due to less local texture information and the irrelevant information from the background. To alleviate the aforementioned problem, we divide the whole face into face contour and three facial components (eyebrow-eyes, nose, and mouth) in the first level. The face contour region corresponds to seventeen contour landmarks. Since the landmarks detection and the facial components detection are both considered, we provide a new weighted loss function for better detection of landmarks and components. Meanwhile, we provide a feature fusion scheme for the facial components detection by considering the complementary information of features of different levels. Figure 2(a) illustrates detailed structure for level 1. We have four convolutional layers in our network. Figure 2(a) shows specific parameter for every layer. All convolutional layer is followed by relu unit. The group of parallelograms stands for the feature map of the first convolutional layer.

Weighted Loss Function. Since we need to localize seventeen landmarks over three bounding boxes in the first level, the loss of landmarks in back propagation process is more influential to that of the component bounding boxes. However, accurate detection in level 2 counts on precise bounding box regressions. To acquire accurate loaction of both contour landmarks and component bounding boxes, here we modify the L2 loss function to better meet our intention. Let $p_i(x_i, y_i)$ (where $i = 1, 2, \ldots 17$), represent seventeen face contour landmarks. $l_j = (lx_j, ly_j)$ and $r_j = (rx_j, ry_j)$ (where $j = 1, 2, 3, 4$) are the left-top and right-bottom coordinates of four bounding boxes. w is the weight applied to bounding boxes in order to balance the loss proportion of bounding boxes in back propagation process. And p is the total number of outputs in our framework. All uppercase symbols refer to corresponding ground truth values:

$$Loss = \frac{1}{2p}(\sum_i \parallel P_i - p_i \parallel^2 + w \sum_j (\parallel L_j - l_j \parallel^2 + \parallel R_j - r_j \parallel^2)) \quad (1)$$

The first part of this equation calculates the loss for landmark regression of the first level. The second weighted part stands for bounding box regression of three facial components (eyebrow-eyes, nose and mouth).

Feature Fusion. According to Sun et al. [9], the convolutional layers tend to extract same feature over the entire picture. However, different facial parts contain different kinds of features. For instance, eyebrows contain more low-level features (e.g. edges) than eyes. As the network forwards, low-level features are deprecated. So it's not possible to detect all components accurately. To maintain low-level feature while obtaining high-level feature, here we fuse the feature map acquired by the first convolutional layer and that of the fourth convolutional layer.

3.3 Level 2: Landmark Localization

Different components are trained separately in level 2. Bounding boxes calculated by landmark contour of each component is resized to 39×39 before training. Network structure is similar to level 1 except that we don't apply weighted loss and feature fusion in this level. Figure 2(b) illustrates detail for level 2 network. Apart from the number of the outputs, three components share the same network parameters.

4 Experiment

4.1 Datasets and Date Augmentation

Images are exacted according to face bounding box provided by 300 W [17] dataset. 300 W offers 3837 images with 68 annotations from four datasets: LFP- W [18], HELEN [19], AFW [20] and IBUG [21]. These datasets provide large quantities of valid images under unconstrained environments. 3148 training data are used as our training data. We test our method on the remaining images. The input image is resized to 60×60 for level 1, and 39×39 for level 2.

To reduce over fitting on the training data, we employed two artificial data augmentation. First we rotate the image by $\pm 30°$ and $\pm 15°$. Then we flip the image at random. These manipulations are implemented with opencv[1].

4.2 Training Details

Our approach is implemented with the open source Caffe [22] framework. The learning rate is initialized in range (0.003, 0.015). Besides, the network is trained by stochastic gradient descent with a weighted decay of 0.0005.

Level-1. We randomly choose 600 images out of 3148 training data as our validation set to estimate our weight loss. Precise bounding box prediction is the key for accurate localization of each component landmarks. Therefore, bounding box detection deserves priority over face contour detection. We have tried $w = 1, 2, \ldots 17$ and we got the best performance when $w = 2$. Figure 3 shows some visual results of the regions detection, including three facial components and the facial contour. From the detection result, we can see that the proposed deep architecture in the first level is robust for various poses.

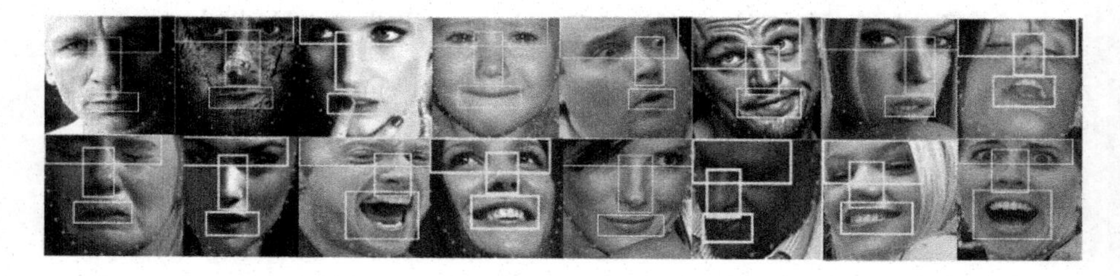

Fig. 3. Visual results of the regions detection. Green points mark the facial contour landmarks. And regions enclosed by yellow rectangle represent three components we defined (eyebrow-eyes, nose and mouth). (Color figure online)

[1] http://opencv.org/.

Level-2. We utilize the ground truth coordinates of each component to obtain our input for level 2. We randomly do shifting and padding (0.2 for each side: left, right, top, bottom) on these bounding boxes for model robustness.

4.3 Evaluation

We evaluate the model by two popular metrics, the mean error and the cumulative error distribution(CED). The mean error is measured by absolute distances between the prediction and the ground truth, normalized by the inner-pupil distance, which can be evaluated by

$$mean\ error = \frac{1}{n}\sum_{i=1}^{n}\frac{\sum_{j=1}^{p} \parallel G_{ij} - P_{ij} \parallel^2}{pD_i} \tag{2}$$

here P_{ij} is the predicted location and G_{ij} is the corresponding ground truth. D_i is the distance between two eyes. $i = 1, 2, \ldots n, j = 1, 2, \ldots p$ where p is the overall number of landmarks and n is the quantity of test image.

Cumulative error distribution (CED) is also reported. Cumulative error is defined as:

$$CED = \frac{N_{e \leqslant l}}{n}, \tag{3}$$

where e is the mean error. $N_{e \leqslant l}$ stands for total number of images which the error satisfy $e < l$.

4.4 Comparison with State-of-art Algorithms

We compare our model performance on the 300 W test image with ERT [6], CFSS [7], TCDCN [8]. Table 1 demonstrates the result.

Table 1. Mean Error Comparison between Different Methods.

Methods	Error	Ours	Improvement
ERT [6]	6.40	4.03	37.03%
CFSS [7]	5.76	4.03	30.03%
TCDCN [8]	5.54	4.03	27.26%

Table 1. shows that our method outperforms the compared methods on 300 W dataset. It can be observed that our method performs significantly better than ERT [6], CFSS [7] and TCDCN [8]. Figure 4 shows the CED curves for different error levels on 300 W dataset. It can be seen that our approach achieves the best performance among the compared methods. More visual results are shown in Fig. 5.

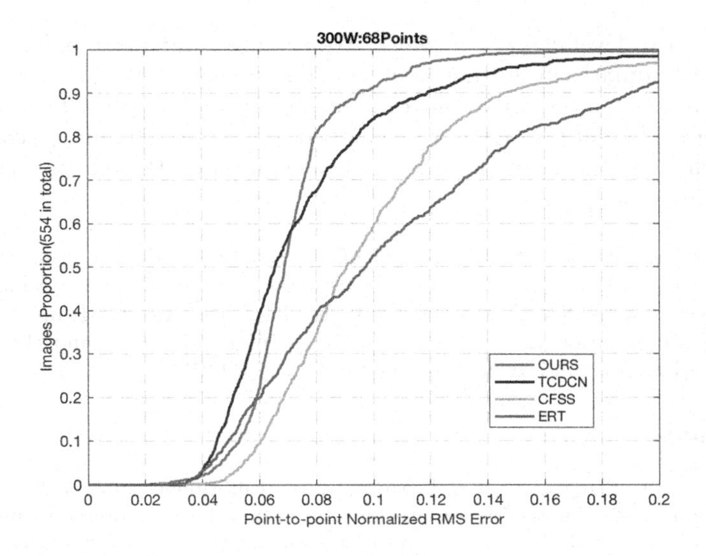

Fig. 4. Accumulative error curve for 68 landmarks detection over 300 W data

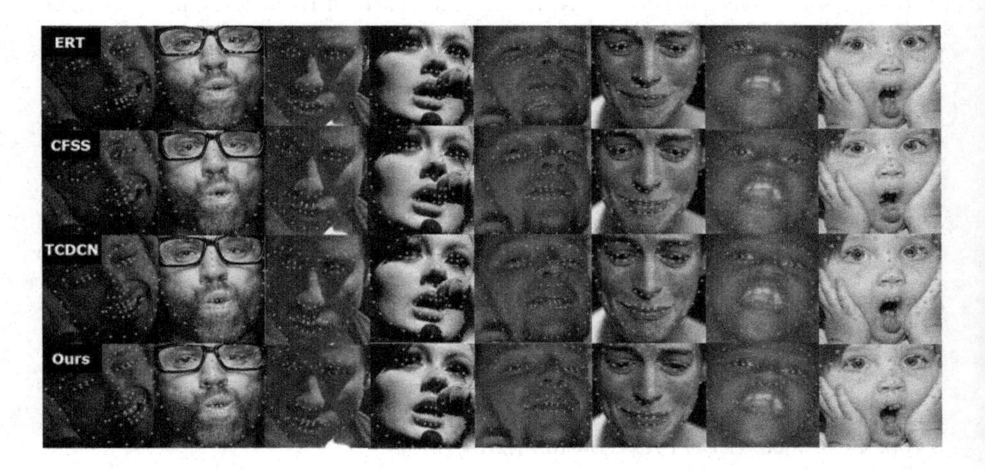

Fig. 5. Example images where our approach outperforms ERT [6], CFSS [7], and TCDCN [8]. The images are challenging because of large head poses and extreme illuminations.

5 Conclusion

In this paper, we proposed a novel region-based structure for face alignment detection. Different from most existing methods, we adopt a divide-and-conquer method, which, instead of detecting the landmarks for the entire face, predicts partial landmarks for facial components separately. Specifically, we solve this task on our two-level cascade network. We acquire component bounding boxes and partial landmarks in the first level. For simultaneous detection of regions and face contour landmarks, we applied feature fusion and rectified L2 loss. In the second level, different components are trained separately to maintain component independency. Experimental results have demonstrated the effectiveness of the propose algorithm.

Acknowledgments. The authors would like to thank the editor and all the anonymous reviewers of this paper for their constructive suggestions and comments. This work is supported by NSFC (No.61671290) in China, the Key Program for International S&T Cooperation Project of China (No.2016YFE0129500), and the Shanghai Committee of Science and Technology, China (No.17511101903).

References

1. Fabian, B.Q., Srinivasan, R., Martinez, A.M.: Emotionet: an accurate, real-time algorithm for the automatic annotation of a million facial expressions in the wild. In: 29th IEEE Conference on Computer Vision and Pattern Recognition, pp. 5562–5570. IEEE Press, Las Vegas (2016)
2. Chen, C., Dantcheva, A., Ross, A.: Automatic facial makeup detection with application in face recognition. In: 6th IEEE Conference on Biometrics, pp. 1–8. IEEE Press, Madrid (2013)
3. Lu, C., Tang, X.: Surpassing human-level face verification performance on LFW with GaussianFace. In: 29th AAAI Conference on Artifical Intelligence, pp. 3811–3819. AAAI Press, Austin Texas (2015)
4. Cootes, T.F., Taylor, C.J.: An algorithm for tuning an active appearance model to new data. In: 17th British Machine Vision Conference, pp. 919–928. DBLP, Edinburgh (2006)
5. Ashraf, A.B., Lucey, S., Cohn, J.F., Chen, T., Ambadar, Z., Prkachin, K.M.: The painful face - pain expression recognition using active appearance models. Image Vis. Comput. **27**(12), 1788–1796 (2009)
6. Kazemi, V., Sullivan, J.: One millisecond face alignment with an ensemble of regression trees. In: 27th IEEE Conference on Computer Vision and Pattern Recognition, pp. 1867–1874. IEEE Press, Columbus (2014)
7. Zhu, S., Li, C., Change Loy, C., Tang, X.: Face alignment by coarse-to-fine shape searching. In: 28th IEEE Conference on Computer Vision and Pattern Recognition, pp. 4998–5006. IEEE Press, Boston (2015)
8. Zhang, Z., Luo, P., Chen, C.L., Tang, X.: Learning deep representation for face alignment with auxiliary attributes. IEEE Trans. Pattern Anal. Mach. Intell. **38**(5), 918–930 (2016)
9. Sun, Y., Wang, X., Tang, X.: Deep convolutional network cascade for facial point detection. In: 26th IEEE Conference on Computer Vision and Pattern Recognition, pp. 3476–3483. IEEE Press, Portland (2013)
10. Sauer, P., Cootes, T., Taylor, C.: Accurate regression procedures for active appearance models. In: 18th British Machine Vision Conference, pp. 681–685. DBLP, Warwickshire (2007)
11. Zhou, E., Fan, H., Cao, Z., Jiang, Y., Yin, Q.: Extensive facial landmark localization with coarse-to-fine convolutional network cascade. In: 26th IEEE Conference on Computer Vision and Pattern Recognition Workshops, pp. 386–391. IEEE Press, Portland (2013)

12. Saragih, J., Goecke, R.: A nonlinear discriminative approach to AAM fitting. In: 11th International Conference on Computer Vision, pp. 1–8. IEEE Press, Rio de Janeiro (2007)
13. Cootes, T.F., Taylor, C.J.: Active Shape Models-'smart snakes'. In: 3th British Machine Vision Conference, pp. 266–275. DBLP, Oxford (1992)
14. Jin, X., Tan, X.: Face alignment in-the-wild: a survey. arXiv preprint. arXiv:1608.04188 (2016)
15. Xiong, X., Torre, F.D.L.: Supervised descent method and its applications to face alignment. In: 26th IEEE Conference on Computer Vision and Pattern Recognition, pp. 532–539. IEEE Press, Portland (2013)
16. Lowe, D.G.: Distinctive image features from scale-invariant keypoints. Int. J. Comput. Vis. **60**(2), 91–110 (2004)
17. Sagonas, C., Tzimiropoulos, G., Zafeiriou, S., Pantic, I.: 26th IEEE Conference on Computer Vision and Pattern Recognition Workshops, pp. 397–403. IEEE Press, Portland (2013)
18. Belhumeur, P.N., Jacobs, D.W., Kriegman, D.J., Kumar, N.: Localizing parts of faces using a consensus of exemplars. IEEE Trans. Pattern Anal. Mach. Intell. **35**(12), 2930–2940 (2013)
19. Le, V., Brandt, J., Lin, Z., Bourdev, L., Huang, T.S.: Interactive Facial Feature Localization. In: Fitzgibbon, A., Lazebnik, S., Perona, P., Sato, Y., Schmid, C. (eds.) ECCV 2012. LNCS, vol. 7574, pp. 679–692. Springer, Heidelberg (2012). doi:10.1007/978-3-642-33712-3_49
20. Zhu, X., Ramanan, D.: Face detection, pose estimation, and landmark localization in the wild. In: 25th IEEE Conference on Computer Vision and Pattern Recognition, pp. 2879–2886. IEEE Press, Rhode Island (2012)
21. Sagonas, C., Antonakos, E., Tzimiropoulos, G., Zafeiriou, S., Pantic, M.: 300 faces in-the-wild challenge: database and results. Image Vis. Comput. **47**, 3–18 (2016)
22. Jia, Y.Q.: Caffe: convolutional architecture for fast feature embedding. In: 22nd ACM international Conference on Multimedia, pp. 675–678. ACM, Netherlands (2014)

Generating Low-Rank Textures via Generative Adversarial Network

Shuyang Zhao and Jianwu Li(✉)

Beijing Key Laboratory of Intelligent Information Technology,
School of Computer Science and Technology, Beijing Institute of Technology,
Beijing 100081, China
{zsyprich,ljw}@bit.edu.cn

Abstract. Achieving structured low-rank representation from the original image is a challenging and significant task, owing to the capacity of the low-rank structure in expressing structured information from the real world. It is noteworthy that, most of the existing methods to obtain the low-rank textures, treat this issue as a "transformational problem", which lead to the poor quality of the images with complex backgrounds. In order to jump out of this interference, we try to explore this issue as a "generative problem" and propose the Low-rank texture Generative Adversarial Network (LR-GAN) using an unsupervised image-to-image network. Our method generates the high-quality low-rank texture gradually from the low-rank constraint after many iterations of training. Considering that the low-rank constraint is difficult to optimize (NP-hard problem) in the loss function, we introduce the layer of the low-rank gradient filter to approach the optimal low-rank solution. Experimental results demonstrate that the proposed method is effective on both synthetic and real world images.

Keywords: Generative adversarial network · Low-rank texture generative adversarial network · Structured low-rank representation · Low-rank constraint

1 Introduction

In the real world, the original images contain a wealth of interferences of transition, such as translation transformation, rotation transformation, scale transformation, which restrict the performance in many vision tasks. A large number of methods about invariant texture analysis have been extended for these tasks, including object recognition [1], image stitching [2], broadcast video analysis [3], and 3D reconstruction [4], etc.

In the past few years, low-rank representation has attracted significant attention in the field of computer vision and image processing. Zhang et al. reestablished the low-rank textures in the 3D scene by transforming invariant low-rank textures (TILT) [5].

D. Liu et al. (Eds.): ICONIP 2017, Part III, LNCS 10636, pp. 310–318, 2017.
https://doi.org/10.1007/978-3-319-70090-8_32

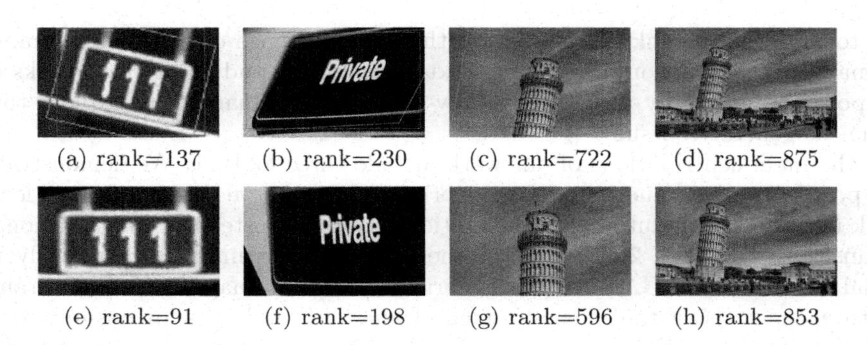

(a) rank=137 (b) rank=230 (c) rank=722 (d) rank=875

(e) rank=91 (f) rank=198 (g) rank=596 (h) rank=853

Fig. 1. (a)–(d) are four original images containing low-rank textures in the real world, and (e)–(h) denote the low-rank texture images after using the TILT transforms based on (a)–(d), respectively.

In TILT, a 2D texture is considered as a matrix $\chi(x, y) \in \mathbb{R}^2$,

$$\overline{\chi}(x, y) = \chi \circ \tau^{-1}(x, y) = \chi\left(\tau^{-1}(x, y)\right), \tag{1}$$

where $\overline{\chi}(x, y)$ is a low-rank texture image, and τ: $R^2 \rightarrow R^2$ belongs to a certain group of transforms. Because a lot of unknown noises are contained in the real-world images, the noise E is included to Eq. 1,

$$\overline{\chi} = (\chi + E) \circ \tau^{-1}. \tag{2}$$

The goal is to get the transformation function τ for recovering the low-rank texture χ and minimizing the influence of the noise. The process can be formulated mathematically as

$$\min_{\chi, E, \tau} rank(\chi) + \gamma ||E||_0 \text{ s.t. } \overline{\chi} \circ \tau = \chi + E, \tag{3}$$

where rank(∗) is the rank function, $|| * ||_0$ denotes the L_0 normalization, and $\gamma > 0$ is a weighting parameter to balance the rank and the sparsity of error.

Based on the TILT, Zhang et al. used no prior knowledge on the epipolar geometry [6] and remitted the conversion issue to ensure the matching in projective distortion images. Further, Zhang et al. proposed the approach of learning a structured low-rank representation [7] for image classification and low-rank images greatly improved the performance in classification tasks. Moreover, most of the existing solutions for the TILT issue are based on the alternating directional cost and are not theoretically guaranteed to converge to a correct solution. Therefore, Lin et al. proposed adaptive penalty (LADMAP) [8] using a linearized alternating direction method to solve the problem.

Note that the existing methods all treat the issue as a transformation problem. But as shown in Fig. 1(c), (d), (g) and (h), when the backgrounds become more complex, the quality of transformation is more ungraded. In order to jump out of this interference from the image backgrounds, we transform our ideas and try to make a low-rank integration on the original image directly. From transformation to generation, a low-rank texture generative adversarial networks is proposed to gradually generate the low-rank texture images and explore this issue as "generative issue".

The main contributions of our work include three folds: **1.** We demonstrate the possibility of a "generation" framework and first achieve the transformed low-rank textures by an automatic "generation program" instead of the traditional "transition program"; **2.** We propose the Low-Rank texture Generative Adversarial Networks (LR-GAN) that transform the original image into the low-rank texture image using the image-to-image framework and it is an unsupervised training algorithm; **3.** In order to ensure the low-rank constraint, we design the layer of the low-rank gradient filter for training to restrain the low-rank loss in a network framework.

2 Overview on Our Method

We propose a mechanism for transforming the original image into low-rank texture image. Figure 2(a) shows our pipeline of transformation for low-rank texture, the generative network Fig. 2(b) generates the low-rank texture image from the original image, and the discriminator network Fig. 2(c) compares the generation with the TILT image. The structure details of both the generative and the discriminative network will be explained in Sects. 3.1 and 3.2.

Assume that $\mathcal{X} = \{\chi^1, \chi^2, ...\chi^N\}$, $\chi \in \mathbb{R}^{m \times n \times c}$ be N original images with $m \times n$ scale and c channels. At the training phase, TILT is used to generate the low-rank texture image $\chi_{TILT} = TILT(\chi)$, $TILT(\chi) \in \mathbb{R}^{m \times n \times c}$. Let D denote

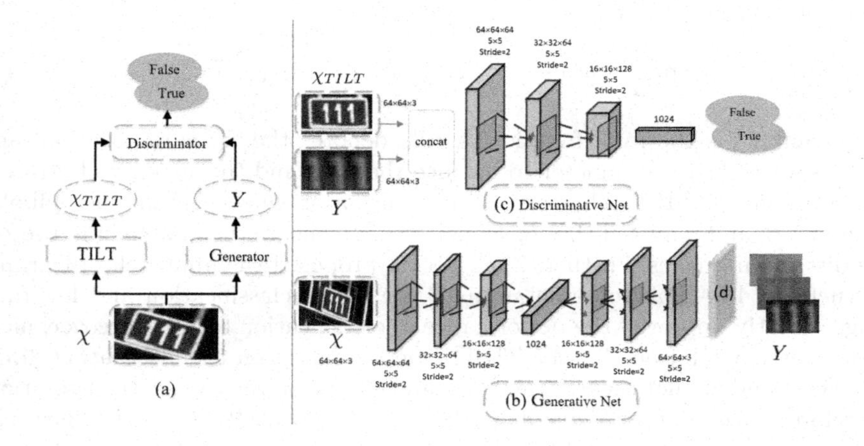

Fig. 2. (a) The general framework of LR-GAN; (b) the Generator generates the low-rank texture image from the original image; (c) the Discriminator distinguishes between the generative image and the TILT image; (d) the layer of the low-rank gradient filter for training.

the discriminative function and G denote the generative function. The objective of LR-GAN can be formalized as

$$\begin{aligned} &\min_{G}\max_{D} \mathbb{E}_{TILT(\chi)\sim TILT(\mathcal{X})}\left[\log p\left(y=1|TILT\left(\chi\right),D\right)\right] \\ &+\mathbb{E}_{\chi\sim\mathcal{X}}\left(\chi\right)\left[\log\left(1-p\left(y=1|G\left(\chi\right),D\right)\right)\right], \end{aligned} \tag{4}$$

where the function D and G play the two-player *min-max* game in the phase of training. Note that, with the TILT process, LR-GAN can easily converge to the optimal point, and we will expound it detailedly in Sect. 4.1.

After training, with the input $\chi \in \mathbb{R}^{m\times n\times c}$, the output $Y = G(\chi)$ will be the low-rank texture image in the phase of generation.

The rest of the paper is organized as follows. In Sect. 3, the architecture of LR-GAN network is described in detail. The training of the whole LR-GAN is presented in Sect. 4. Experimental results are given in Sect. 5, and the last section concludes this paper.

3 Low-Rank Texture Generative Networks (LR-GAN)

Based on Sect. 2, we introduce the LR-GAN using an unsupervised learning framework in this section. Deep Convolutional Generative Adversarial Network (DCGAN) [9] has more advantages over the other simple all-connected neural networks in image processing. Compared with other non-supervision methods, the discriminative network of it is more effective to distinguish the image features extracted from the network and it is more suitable for image recognition tasks. The generative network of it can not only keep the continuity between images but also learn a meaningful filter layer. And the training of it is stable in most cases. Inspired by DCGAN that using the convolution neural network to generate the output image from the input image, we combined the TILT with the DCGAN which has enhanced low-rank generation ability by the low-rank gradient filter and propose a generative adversarial network to generate the low-rank texture image from the original image. The LR-GAN is composed of two modules: the generator and discriminative network with TILT.

3.1 Formulation of the Generator Network

The general architecture of LR-GAN is shown in Fig. 2. LR-GAN employs the original image $\chi \in \mathbb{R}^{m\times n\times c}$, where $m \times n$ is the scale and c is the channel number (in this paper, $m = n = 64$, $c = 3$). The generative network can be seen as an encode-decode network. The encoder module stacks four convolution layers (G-conv1 to G-conv4). Rectified linear units (*ReLU*) [10] are used as the nonlinearity activation for every convolution layer and the batch normalization is adopted [11] after the G-conv4 layer. And the decoder employs three deconvolution layers (G-deconv1 to G-deconv3). Similar to the encoder, *ReLU* is used as our nonlinearity activation. Meanwhile, in order to ensure the connection between the Generative Network and Discriminator Network, *Tanh* is employed to associate the two nets in the phase of training.

3.2 Formulation of the Discriminator Network

Let $Data = \{\chi_i\}_{i=1}^{K} \in \mathcal{X}$ be the K training set images in $\mathcal{X}$, and TILT solves the following:

$$\min_{\chi_{TILT}, E, \tau} rank\,(\chi_{TILT}) + \gamma ||E||_0 \text{ s.t. } \chi \circ \tau = \chi_{TILT} + E, \tag{5}$$

where χ_{TILT} is the input to aid the Discriminator Network for the task of discriminating. Note that our method does not employ any label for training and it is an unsupervised method to get the low-rank texture.

Discriminator Network employs the concatenation of (χ_{TILT}, Y) as input by stacking three convolutions (D-conv1 to D-conv3), and uses rectified linear units (*LeakyReLU*) [12] activation in each layer of these convolutions as the nonlinearity activation function.

To ensure the robustness and non-saturating, the least squares loss function with a smoother and non-saturating gradient [13] is used in the discriminator,

$$\begin{aligned} \min_D L(D) = \frac{1}{2} E_{TILT(\chi) \sim TILT(\mathcal{X})} \left[\left(D\left(\chi_{TILT}\right) - 1 \right)^2 \right] \\ + \frac{1}{2} E_{\chi \sim \mathcal{X}} \left[D\left(Y\right)^2 \right], \end{aligned} \tag{6}$$

and the generator,

$$\min_G L(G) = \frac{1}{2} E_{\chi \sim \mathcal{X}} \left[\left(D(Y) - 1 \right)^2 \right]. \tag{7}$$

4 Training LR-GAN

It's our goal to generate the image that is highly similar to the $\mathcal{X}_{TILT}$ and has the low-rank. We introduce the unitary loss function with the low-rank constraint for the generator:

$$\min_G L(G) = \frac{1}{2} E_{\chi \sim \mathcal{X}} \left[\left(D(Y) - 1 \right)^2 \right] + \lambda \left(||\chi||_* \right), \tag{8}$$

where $|| * ||_*$ is the nuclear norm to ensure the low-rank constraint, and λ is a hyper-parameter. Considering that, it is extremely difficult to optimize the problem (general NP-hard problem), we design the layer of the low-rank gradient filter to approach the low-rank constraint.

4.1 The Layer of the Low-Rank Gradient Filter for Training

We assume that w_t be the weight of network and g_t be the gradient mapping back propagation at t-th propagation. We propose a strategy to filter the gradient to ensure the constraint on low-rank using the filtration function,

$$||rank\,(w_t \chi) - rank\,(w_{t+1} \chi)\,|| < \delta. \tag{9}$$

In order to get the goal of keeping a balance between the similarity and the low-rank, we train both G and D networks without the layer of the low-rank gradient filter at the beginning.

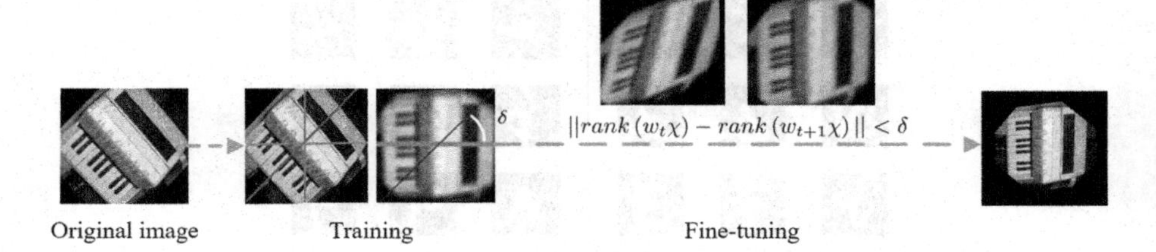

Fig. 3. Training and fine-tuning.

4.2 The Update of Weights

After many iterations, the generated image will have a high similarity to the χ_{TILT}. Now we fine-tune the network using the weight update method in Stochastic Gradient Descent (SGD) [14]:

$$w_{t+1} = \begin{cases} w_t - \alpha g_t \ , \ ||rank(w_t\chi) - rank(w_{t+1}\chi)|| < \delta \\ w_t - \beta\alpha g_t \ , \ ||rank(w_t\chi) - rank(w_{t+1}\chi)|| \geq \delta \end{cases}, \tag{10}$$

where α is the learning rate, β is the penalty factor to punish the g_t if the update with g_t has a larger fluctuation to the current weight as described in Eq. (9), and δ ensures the fluctuation in a steerable range.

As shown in Fig. 3, there are two directions to converge in the iterations for an original image. The TILT progress chooses a direction for our network training and makes the whole generative process be an unsupervised learning. After many iterations, the layer of the low-rank gradient filter is employed in the fine-tuning, and this layer will make the rank of the generative image waggling in the range of δ.

5 Experiments

We demonstrate the performance of our model on qualitative results. The first goal of our experiments is to investigate the effectiveness of the low-rank gradient filter layer. The second goal is to evaluate if LR-GAN can generate the high-quality low-rank image from the origin image. Our network was implemented on Tensorflow with NVIDIA GeForce GTX 1080 $\times$ 2. In all experiments, we empirically set $\beta = 0.1$, $\delta = 4$.

Distorted MNIST. In order to evaluate the effectiveness of our method, we distorted the images in the MNIST dataset like in [15] and used them as our baseline in the evaluative process. As shown in Fig. 5(a), the loss of Discriminator and Generator converges gradually. After 500 iterations, the low-rank gradient filter layer was added to the network, and the loss values varied very little. And as shown in Fig. 5(b), we evaluated the rank value in the iterative process, and the variation of the generative image converged in a local minimum value

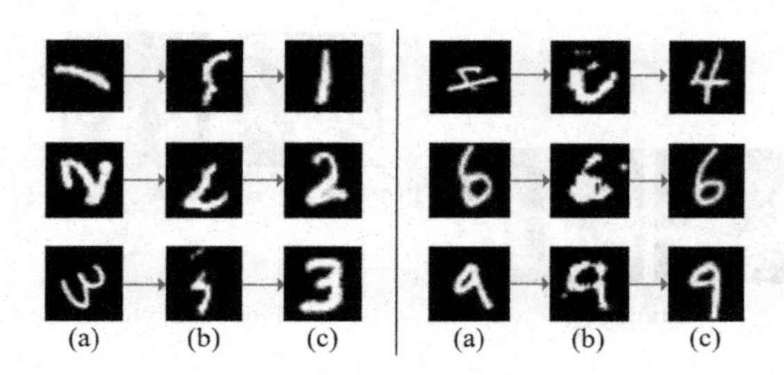

Fig. 4. The generative process on MNIST dataset. (a) the original images (b) the generative images in iteration process (c) the final generative images.

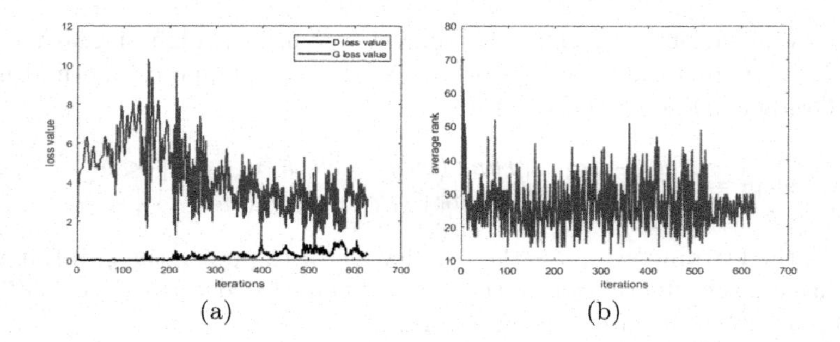

Fig. 5. Results on MNIST dataset: (a): the loss of both the Generator and the Discriminator during the iterations, (b): the changes of the rank during the generator iterations.

(the direction is guided by TILT) with the high similarity to the original image. From Fig. 4, the generated images are closer to the original images with lower ranks than the initial ones. Table 1 shows that LR-GAN achieves **0.916 (s/img)** and **27** average rank in the generative phase compared to the TILT on distorted MNIST dataset.

Table 1. The average rank and generative speed on MNIST dataset.

Method	Average rank	Generative speed (s/img)
TILT	31	2.456
LR-GAN	29	**0.916**
LR-GAN+Gradient filter	**27**	0.941

SVHN. We tested our LR-GAN networks on a challenging real-world dataset, Street View House Numbers (SVHN) [16]. This dataset contains around 200k real

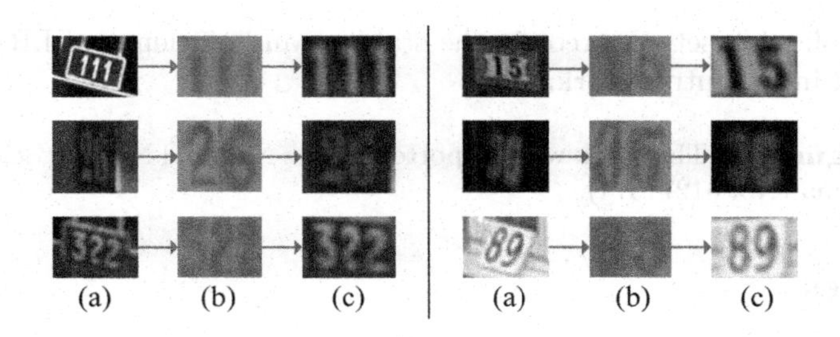

Fig. 6. The generative process on SVHN dataset. (a) the original images (b) the generative images in iteration process (c) the final generative images.

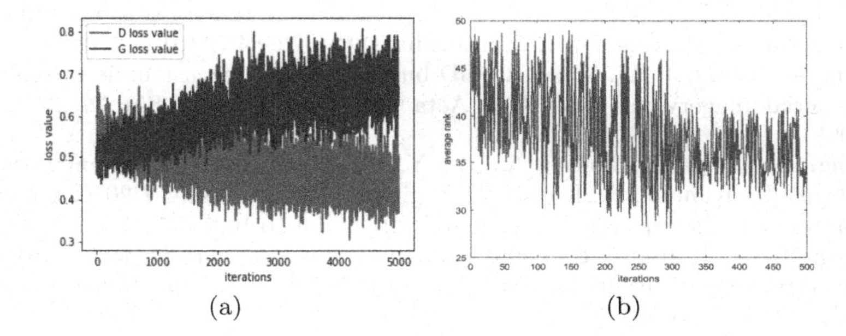

Fig. 7. Results on SVHN dataset: (a): the loss of both the Generator and the Discriminator during the iterations, (b): the changes of the rank during the generator iterations.

world images of house numbers. There are between 1 and 3 digits in each image, with the complex background and a large variability in spatial arrangement. The data were preprocessed by taking (32×32) crops around each digit sequence. We trained the LR-GAN to generate the digit at a particular position in the sequence. As shown in Fig. 7(a), the loss of Discriminator and Generator has a trend of convergence. We performed 5000 iterations and added the low-rank gradient filter layer in the 3500-th iteration on our network. From Fig. 7(b), the changes of the rank (generating 1 image per 100 iterations) show that the low-rank gradient filter layer constrains the low-rank. As shown in Fig. 6, our net gets a high performance under the complex background.

6 Conclusions

This paper presents a Low-Rank texture Generative Adversarial Networks (LR-GAN) for gradually generating the low-rank texture, based on a "generative strategy" using an unsupervised learning method. Furthermore, in order to ensure the low-rank constraint, the layer of low-rank gradient filter is proposed for this image-to-image framework. Experimental results demonstrate that LR-GAN achieves the low average rank and effective generative speed than TILT on the used public datasets. Moreover, the stability and efficiency of LR-GAN will be improved in the future work.

Acknowledgments. This work was supported by the National Natural Science Foundation of China (No. 61271374).

References

1. Yang, S., Wei, E., Guan, R., et al.: Triangle chain codes for image matching. Neurocomputing **120**(10), 268–276 (2013)
2. Brown, M., Lowe, D.G.: Automatic panoramic image stitching using invariant features. Int. J. Comput. Vis. **74**(1), 59–73 (2007)
3. Han, J., Farin, D., de With, P.H.N.: A mixed-reality system for broadcasting sports video to mobile devices. IEEE Multimedia **18**(2), 72–84 (2010)
4. Cheng, L., Gong, J., Li, M., et al.: 3D building model reconstruction from multi-view aerial imagery and Lidar data. Acta Geodaetica Cartogr. Sin. **77**(2), 125–139 (2009)
5. Zhang, Z., Liang, X., Ganesh, A., Ma, Y.: TILT: Transform Invariant Low-Rank Textures. In: Kimmel, R., Klette, R., Sugimoto, A. (eds.) ACCV 2010. LNCS, vol. 6494, pp. 314–328. Springer, Heidelberg (2011). doi:10.1007/978-3-642-19318-7_25
6. Zhang, Q., Li, Y., Blum, R.S., et al.: Matching of images with projective distortion using transform invariant low-rank textures. J. Vis. Commun. Image Represent. **38**(C), 602–613 (2016)
7. Zhang, Y., Jiang, Z., Davis, L.S.: Learning structured low-rank representations for image classification. In: Proceedings of the IEEE Conference on Computer Vision and Pattern Recognition (CVPR), pp. 676–683. IEEE (2013)
8. Lin, Z., Liu, R., Su, Z.: Linearized alternating direction method with adaptive penalty for low-rank representation. Neural Inf. Process. Syst. **24**, 612–620 (2011)
9. Radford, A., Metz, L., Chintala, S.: Unsupervised representation learning with deep convolutional generative adversarial networks. arXiv preprint arXiv: 1511.06434 (2015)
10. Nair, V., Hinton, G.E.: Rectified linear units improve restricted Boltzmann machines. In: International Conference on International Conference on Machine Learning (ICML), pp. 807–814 (2010)
11. Ioffe, S., Szegedy, C.: Batch normalization: accelerating deep network training by reducing internal covariate shift. In: International Conference on International Conference on Machine Learning (ICML), pp. 448–456 (2015)
12. Maas, A.L., Hannun, A.Y., Ng, A.Y.: Rectifier nonlinearities improve neural network acoust models. In: International Conference on International Conference on Machine Learning (ICML) (2013)
13. Mao, X., Li, Q., Xie, H., et al.: Least squares generative adversarial networks. arXiv preprint arXiv:1611.04076 (2016)
14. Zhao, S.Y., Li, W.J.: Fast asynchronous parallel stochastic gradient decent. arXiv preprint arXiv:1508.05711 (2015)
15. Jaderberg, M., Simonyan, K., Zisserman, A., et al.: Spatial transformer networks. In: Neural Information Processing Systems, pp. 2017–2025 (2015)
16. Netzer, Y., Wang, T., Coates, A., et al.: Reading digits in natural images with unsupervised feature learning. In: NIPS Workshop on Deep Learning Unsupervised Feature Learning (2012)

Deep Salient Object Detection via Hierarchical Network Learning

Dandan Zhu[1], Ye Luo[1(✉)], Lei Dai[2], Xuan Shao[1], Laurent Itti[3], and Jianwei Lu[1,4]

[1] School of Software Engineering, Tongji University, Shanghai, China
fendoualllife@163.com, yeluo@tongji.edu.cn, xaoshuan@gmail.com
[2] School of Automotive and Traffic Engineering, Jiangsu University, Zhenjiang, China
dl3140401034@gmail.com
[3] Department of Computer Science and Neuroscience Program, University of Southern California, Los Angeles, USA
itti@usc.edu
[4] Institute of Translational Medicine, Tongji University, Shanghai, China
jwlu33@tongji.edu.cn

Abstract. Salient object detection is a fundamental problem in both pattern recognition and image processing tasks. Previous salient object detection algorithms usually involve various features based on priors/assumptions about the properties of the objects. Inspired by the effectiveness of recently developed feature learning, we propose a novel deep salient object detection (DSOD) model using the deep residual network (ResNet 152-layers) for saliency computation. In particular, we model the image saliency from both local and global perspectives. In the local feature estimation stage, we detect local saliency by using a deep residual network (ResNet-L) which learns local region features to determine the saliency value of each pixel. In the global feature extraction stage, another deep residual network (ResNet-G) is trained to predict the saliency score of each image based on the global features. The final saliency map is generated by a conditional random field (CRF) to combining the local and global-level saliency map. Our DSOD model is capable of uniformly highlighting the objects-of-interest from complex background while well preserving object details. Quantitative and qualitative experiments on three benchmark datasets demonstrate that our DSOD method outperforms state-of-the-art methods in the salient object detection.

Keywords: Salient object detection · Deep residual network · Local and global perspectives

1 Introduction

Saliency detection attempts to identify the most important and conspicuous object regions in an image by the human visual and cognitive system. It is

D. Liu et al. (Eds.): ICONIP 2017, Part III, LNCS 10636, pp. 319–329, 2017.
https://doi.org/10.1007/978-3-319-70090-8_33

a fundamental problem in neural science, psychology and computer vision. Many computer vision researchers propose computational models to simulate the process of human visual attention or identify salient objects. Recently, salient object detection have drawn a large amount of attention in variety of computer vision tasks, such as object detection [1], person re-identification [2], object retargeting [3], image retrieval [4] and video summarization [5], etc.

Visual saliency can be viewed into different perspectives and contrast is one of them. Based on the observation that salient object is always distinguishing itself from its surroundings, contrast as a prior has been widely used to detect salient object. According to the range of the context that the contrast is computed to, it can be further categorized into local contrast and global contrast methods. The local contrast based methods usually compute center-surround difference to obtain the object-of-interest region standing out from their surroundings [6,7]. Due to the lack of the global information, methods of this category tend to highlight the boundaries of salient objects and neglect the interior content of the object. Meanwhile, the global contrast based methods take the entire image into consideration to estimate the saliency of every pixel or every image segment, thus the whole salient object is detected but the details of the object structure are always missing. There are also methods proposed to improve the performance of salient object detection via integrating both local and global cues [8].

The aforementioned methods may work well for low-level saliency, but they are neither sufficient nor necessary, especially in the cases when the saliency is also related to the human perception or is task-dependent. In addition, many of the mentioned methods perform saliency detection based on the segments of images (i.e. regions or superpixels), and it also requires a post-processing operation to smooth the saliency value of the pixels. However, the operation of the image segmentation may have negative effect on the saliency detection.

In this paper, we propose a novel image salient object detection model named Deep Salient Object Detection (i.e. DSOD) using deep residual network (i.e. ResNet) with 152-layers. Different from previous models [9,10], we use two ResNets to extract high-level semantic features from local and global perspectives respectively. Features extracted from a global ResNet (ResNet-G) roughly identify the global concepts of the salient object e.g. the location, the scale and the size of the salient object, while features from a local ResNet (ResNet-L) provide the details of the salient object such as the subtle structure of the object boundary.

In summary, the main contributions of our work are:

(1) A novel salient object detection model is proposed to integrate the global and local features extracted from ResNet-G and ResNet-L.
(2) A complete salient object detection framework is developed by further integrating our DSOD salient object detection model with multi-level image segmentation and a spatial coherence refinement method in the local feature extraction stage.
(3) Extensive experiments on the three benchmark datasets demonstrate that our proposed method significantly outperforms state-of-the-art methods.

2 Deep Salient Object Detection Framework

In this section, we present the proposed DSOD method in detail.The pipeline of DSOD is illustrated in Fig. 1. We model the complete framework of salient object detection with two deep ResNets: ResNet-G and ResNet-L. The ResNet-G takes the original image as input and generates semantic features globally. Since features extracted at the global level only can not accurately describe the detailed structure of the salient object, we consider using the ResNet-L to extract local features as a complement to the global information. Besides, in order to deal with the noise caused by the image segmentation, the spatial coherence method is integrated in the ResNet-L as smoothing step.

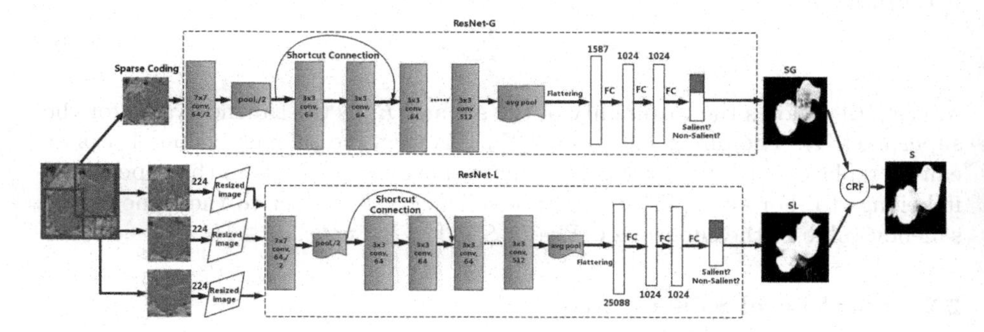

Fig. 1. Architecture overview of our proposed deep salient object detection model. The ResNet-G takes the entire image as input and generates global saliency map SG. The ResNet-L takes the image segments as input and produces the local-level saliency map SL. SL and SG are fused by CRF to obtain the aggregated saliency map S.

2.1 Global-Level Saliency Map

In order to extract the global level semantic features of the image, we takes the entire image as the input of the deep ResNet-G. Specifically, for each pixel in image I, a global-level feature is obtained via a feature extractor ϕ_g which is implemented by ResNet-G. Then, for each feature, we apply a linear transformation that assigns a corresponding saliency score to the pixel, which can be expressed as

$$SG(I_i) = w_{g,i}^T \phi_g(I_i) + b_{g,i}, i = 1, 2, ..., N, \tag{1}$$

where $w_{g,i}$ and $b_{g,i}$ represent the weights and biases of detector, the values of these parameters are obtained by performing a linear transformation. N is the pixel number of the original image I and I_i is the ith pixel. By this way, we can obtain a saliency map SG. Although the saliency map generated via the deep ResNet-G can highlight the entire salient regions, it suffers two problems: (1) the generated saliency map on global level may be difficult to distinguishes between the foreground and background; (2) when the background region is heavily cluttered, it can't preserve the subtle structure of salient objects. The reasons directly leading to the above two problems are the lack of consideration of the neighboring local information when extracting global-level features.

SLCI Smoothing. In order to solve these problems, we use the superpixel-based local context information (SLCI) method [11] to preserve the spatial structure of the salient object when generating a global-level saliency map. Given an image I, we first use the Simple Linear Iterative Clustering segmentation method (SLIC) [12] to segment I into M superpixels as $\{H_i\}_{i=1 \cdots M}$. Once obtaining the global-level saliency map SG, for each pixel, the saliency value is assigned the same as the ones with similar appearance in the local region. Denotes H_i the superpixel containing the pixel $p(x, y)$. Then the formula of SLCI smoothing for the pixel $p(x, y)$ is as follows:

$$R_i^g = \frac{1}{|H_i|} \sum_{p \in H_i} SG(p), \tag{2}$$

where $|H_i|$ denotes the cardinality of the set and R_i^g is the saliency value for the superpixel H_i. The direct meaning of Eq. 2 is that the saliency value of a pixel can be replaced with the average saliency value over all pixels in the superpixels it belongs to. For each pixel in H_i, the saliency value is set to the same as the saliency value of the superpixel. That's $S_g(p) = R_i^g, \forall p(x, y) \in H_i$.

2.2 Local-Level Saliency Map

In order to obtain the local-level saliency map, we first segment the image into multiple sub-regions, and then the multiple sub-regions are fed into the ResNet-L to generate the local-level semantic features. At last, saliency is estimated for each sub-region via the extracted local level features. We also introduce a method to refine the local level saliency map by considering the spatial coherence among multiple sub-regions.

Multi-level Image Segmentation. There are two steps to perform multi-level image segmentation: first, we employ the graph-based image segmentation method [13] to segment the image into multi-level sub-regions; secondly, in order to produce accurate segmentation and reduce the number of small meaningless segments, we apply region merging method [14]. Specifically, for each image, we divide it into $L = 10$ different levels. From the coarsest level (level 1) to the finest level (level 10), the number of regions is from 20 to 200, and the number of image regions in the intermediate levels follows a geometric series.

Multiscale Local Feature Extraction. We extract features for each image segment with a deep residual network which has been trained on the ImageNet datasets using the Caffe framework. Due to the segments obtained by the multi-level image segmentation having variant size, we first wrap each of them in a bounding box size of 200×200. Then, at each level, all the warped segments are fed into the ResNet-L and to obtain L saliency maps $\{S^{(1)}, S^{(2)}, ..., S^{(L)}\}$, $S^{(k)}(H_i) = w_{l,i}^T \phi_l(H_k) + b_{l,i}, k = 1, 2, ...L$. Here, $w_{l,i}$ and $b_{l,i}$ are obtained similarly as in Eq. 1. Once obtaining the saliency maps at all levels, we aim to further fuse these saliency map together to get an aggregated saliency map as $SL = \sum_{k=1}^{L} a_k S^{(k)}$, and the weights a_k are learned by performing a least-squares regression method on a training dataset.

$$\{a_k\}_{k=1}^{L} = \arg\min_{a_k} \sum_{i \in I_u} \|Gd - \sum_{k} a_k S_i^{(k)}\|^2, \tag{3}$$

where I_u represents the set of indices of the images in the training dataset. Gd represents the saliency map of groundtruth.

Spatial Coherence Refinement. Due to the fact that image segmentation method can not get the perfect segmentation results and our model assigns saliency values to each region individually, problems by lacking of the spatial coherence will inevitably appear in the resulting saliency map. Therefore, in order to enhance spatial coherence, we use a superpixel-based saliency refinement method. As we did before, the saliency value of the superpixel is set as the average saliency value of all pixels in this superpixel as $R_i = \frac{1}{|H_i|} \sum_{p \in H_i} SL(p)$ and $S_l(p) = R_i, \forall p \in H_i$, where R_i is the saliency value of the superpixel H_i at local saliency map SL. We enhance the saliency map by setting an objective function, which can be reduced to solving a linear system as:

$$\arg\min_{\{R_i^l\}} \sum_{i} (R_i^l - R_i)^2 + \sum_{i,j} \omega_{i,j} (R_i^l - R_j^l)^2, \tag{4}$$

where R_i^l is the saliency value of the superpixel H_i after refinement. The first term in the formula denotes the difference between the original saliency map and the refined saliency map, and the second term indicates the difference of spatial consistency between different superpixels. $\omega_{i,j}$ refers to the weight between superpixels R_i^l and R_j^l aiming to enhance the spatial consistency. The $\omega_{i,j}$ is thus defined as follows:

$$\omega_{i,j} = exp(-\frac{d^2(H_i, H_j)}{2\sigma_1^2}), \tag{5}$$

where $d(H_i, H_j)$ is the normalized Euclidean distance between the color histogram of H_i and H_j in $CIEL * a * b$. In our experiment, σ_1 is the standard deviation of the distance between pairwise superpixels and set to 0.1. After refinement, the refined local saliency map $S_l(p) = R_i^l, \forall p \in H_i$ is significantly improved.

2.3 Saliency Map Optimization

In order to get the better fusion result of S_l and S_g, we take conditional random field (CRF) approach on the two level saliency maps as:

$$S = \sum_{i=1}^{M} (w_i^g (R_i^g)^2 + w_i^l (R_i^l)^2) + \sum_{i,j} w_{i,j} \\ exp\left(-\frac{(R_i^g - R_j^g)^2 + (R_i^l - R_i^l)^2}{2\sigma_2^2}\right), \tag{6}$$

where R_i^g and R_i^l represent the saliency value assigned to superpixel H_i at global level and local level, respectively. w_i^g has a large accuracy derived from our ResNet-G detection, w_i^l denotes the ResNet-L detection weights associated with superpixel H_i, the two weights are the optimal value obtained by training. The last term is smooth term that keeps continuous saliency value. In our experiment, σ_2 is set to 0.15. Then, for any pair of adjacent superpixels, such as H_i and H_j, the weight $w_{i,j}$ can be defined as follows:

$$w_{i,j} = exp(-\frac{d^2(H_i, H_j)}{2\sigma_3^2}) + \lambda. \tag{7}$$

Here, $d(H_i, H_j)$ shares the same meaning as it is in Eq. 5. In order to minimize the influence of small noise generated by images with cluttered background, we set the parameters σ_3 to 0.1 and λ to 0.25 in our experiment. After optimization, the fused saliency map S is significantly improved.

3 Experiment

In this section, experimental results are reported to validate the proposed salient object detection model. Firstly, we compare our method with eight state-of-the-art methods on three public datasets qualitatively and quantitatively. In order to show the effectiveness of our DSOD method by using deep residual network, we perform comparisons by using different deep neural networks under our DSOD framework. Moreover, in order to show the effectiveness of salient object detection by ResNet-L and ResNet-G in our proposed DSOD method, we perform the performance analysis of them on ECSSD dataset.

3.1 Dataset

To evaluate our model, three public benchmark datasets are used: ECSSD, MSRA-B, PASCAL-S. All the experiments are run on the MATLAB R2011b platform on a workstation with Intel Xeon(R) CPU(2.60GHz) and 500 GB RAM.

3.2 Performance Comparisons with State-of-the-art Methods

In this subsection, we evaluate the proposed model on ECSSD, MSRA-B and PASCAL-S datasets. We trained our model using the 5000 images from the MSRA-B dataset and divide MSRA-B into two subsets, one subset of 4000 images for training and the other dataset of 1000 images for test. The results

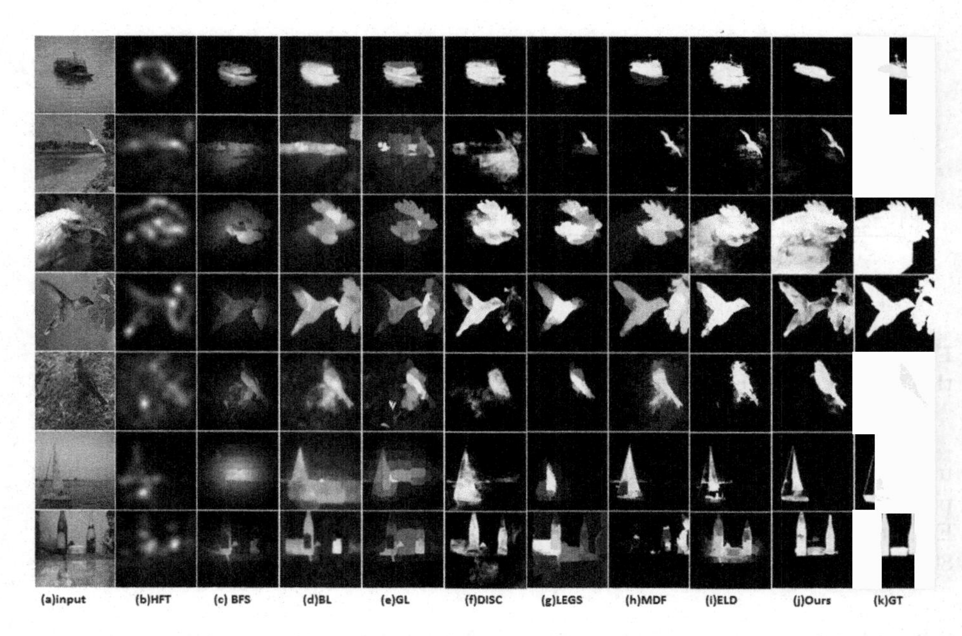

Fig. 2. Visual comparison of our results with the state-of-the-art methods on ECSSD, MSRA-B and PASCAL-S datasets. Top and bottom two rows are images from the ECSSD, MSRA-B datasets, middle three rows are images from the PASCAL-S dataset. The ground truth (GT) is shown in the last column. Our method produces saliency maps closest to the ground truth.

of our experiments with eight recent state-of-the-art methods: Visual Saliency Based on Scale-Space Analysis in the Frequency Domain (HFT [15]), Background and Foreground Seed (BFS [16]), Bootstrap Learning (BL [17]), Global and Local cues (GL [18]), Deep Image Saliency Computing via Progressive Representation Learning (DISC [11]), Local Estimation and Global Search (LEGS [19]), Multiscale Deep CNN Features (MDF [20]) and Encoded Low level Distance map (ELD [21]). We compare with them quantitatively and qualitatively.

Quantitative Comparison. Figure 2 shows the saliency maps generated by our method and eight other methods. Experimental results have shown that our method not only highlights entire salient objects, but also preserves the detail very well. In particular, our method can accurately detect the salient object and clear display the object. As can be seen, our method perform well in a variety of challenging case, such as the low contrast between salient object and background (row 1, row 3 and row 6), cluttered background (row 2), multiple salient objects (4-th and 7-th rows) and objects touching the image boundary (3-th and 4-th rows).

Qualitative Comparison. As shown in the Fig. 3, our method achieves the highest precision in the entire recall range in three datasets. The results of P-R curves demonstrate that the proposed method outperforms the state-of-the-art

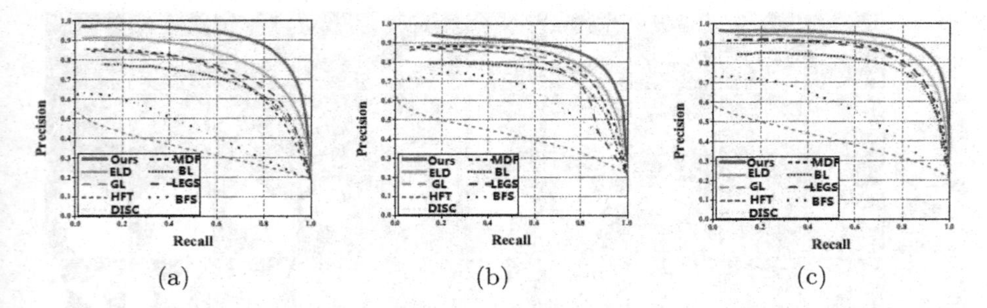

Fig. 3. Quantitative comparison the precision-recall curves of different methods on three datasets: (a) ECSSD dataset (b) MSRA-B dataset (c) PASCAL-S dataset.

methods. On the ECSSD dataset, the proposed method obtains the highest precision value of 98.5%, which is 7.5% higher than the best one (91% in the ELD method). On the other hand, the minimal recall value of Ours method is 80%, significantly higher than those of the other methods.

3.3 Comparison with Other Deep Neural Network

To further demonstrate the effectiveness of the proposed framework, we replace our DSOD architecture with CNN, VGG16 and AlexNet, and compare their performance, respectively. It is worth noting that both VGG16 and AlexNet networks are improved and enhanced on the CNN network structure, and are widely used in various computer vision tasks. In order to make a fair comparison, we put the DSOD framework of different networks on the same dataset (PASCAL-S). The experiments are all carried out on a workstation with NVIDIA GeForce GTX TITAN Black and 1 TB RAM. The results are shown in Fig. 4. The DSOD architecture with 152-layers ResNet perform consistently better than DSOD framework with VGG16 and AlexNet. As can be seen from Fig. 4, our DSOD framework with 152-layers Resnet is significantly superior to DSOD framework with VGG16 and AlexNet. Therefore, it can demonstrates the effectiveness of our DSOD framework with ResNet.

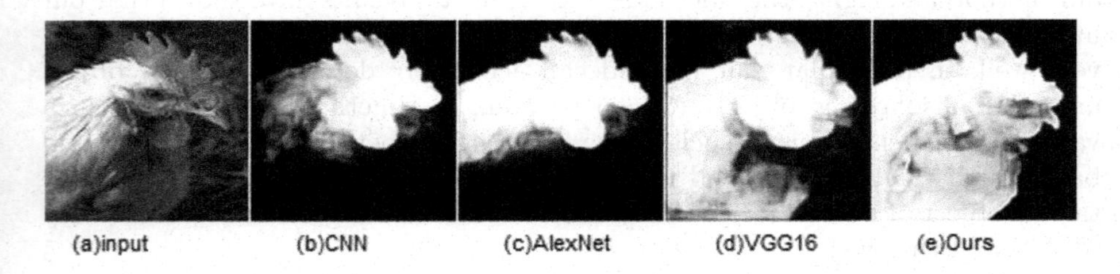

Fig. 4. Visual comparison of our results with other different types of deep neural networks. The samples are taken from the MSRA-B dataset.

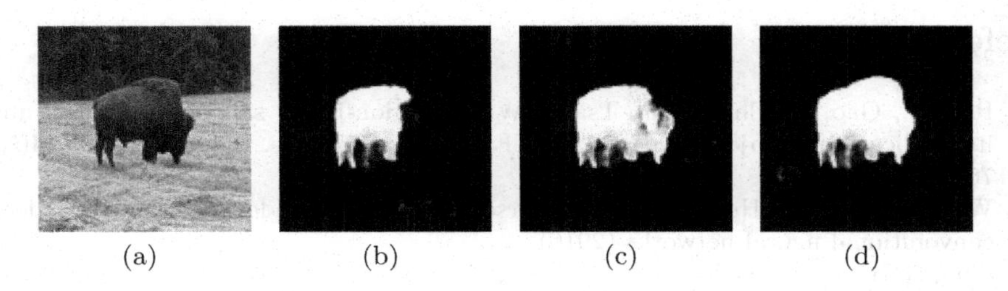

(a) (b) (c) (d)

Fig. 5. Visual comparison of the salient object detection result by ResNet-L and ResNet-G in our DSOD method. (a) Original image (b) Global saliency map by ResNet-G (c) Local saliency map by ResNet-L (d) the fused result.

3.4 ResNet-L v.s. ResNet-G on Salient Object Detection

In order to show the effectiveness of salient object detection by ResNet-L and ResNet-G in our proposed DSOD method, we perform the performance analysis of them on ECSSD dataset. The sampled detection results are shown in Fig. 5. From this figure, we can see that the global saliency map obtained via ResNet-G (i.e. Figure 5 (b)) cannot keep the complete object boundary (e.g. the head of the yak) and some of the background on the right bottom of the image are wrongly detected, while the local saliency map generated via ResNet-L (i.e. Figure 5 (c)) can obtain complete object boundary but the interior of the salient object cannot be highlighted homogeneously (e.g. the neck and the head). Only the fused result by our proposed DSOD method can not only highlight the overall object but preserve the object boundary and the details of the structure.

4 Conclusion

In this paper, we propose a novel salient object detection model. Compared with existing image saliency computing methods, our model achieves superior performance without relying on various features (priors/assumptions). We model the image saliency from both local and global observations. Specifically, our salient object detection model is built upon two deep ResNets. Firstly, ResNet-G generates a global level saliency map by takes the overall image as the input. Secondly, we decompose the input image into a series of regions, then put theses regions into the ResNet-L to produce local level saliency map while preserving object details. Our method outperforms eight state-of-the-art approaches in term of visual qualitative and quantitative results on three public datasets. As a future work, we are planning to combine our model with object location or object recognition to explore the efficient algorithms.

Acknowledgment. This work was supported by the General Program of National Natural Science Foundation of China (NSFC) under Grant No.61572362. This research was also partially supported by the General Program of National Natural Science Foundation of China (NSFC) under Grant No.81571347.

References

1. Ren, Z., Gao, S., Chia, L.T., Tsang, W.H.: Region-based saliency detection and its application in object recognition. IEEE Trans. Circ. Sys. Video Technol. **24**(5), 769–779 (2014)
2. Wu, L., Shen, C., Hengel, A.V.D.: Personnet: person re-identification with deep convolutional neural networks (2016)
3. Sung, Y.H., Tseng, W.Y., Lin, P.H., Kang, L.W., Lin, C.Y., Yeh, C.H.: Significance-preserving-guided content-aware image retargeting. In: Pan, J.S., Yang, C.N., Lin, C.C. (eds.) Advances in Intelligent Systems and Applications -, vol. 2. Springer, Heidelberg (2013). doi:10.1007/978-3-642-35473-1_34
4. Johnson, J., Krishna, R., Stark, M., Li, L., Shamma, D.A., Bernstein, M.S., et al.: Image retrieval using scene graphs (2015)
5. Marat, S., Guironnet, M., Pellerin, D.: Video summarization using a visual attention model. In: European Signal Processing Conference, pp. 1784–1788. IEEE (2015)
6. Itti, L., Koch, C.: A saliency-based search mechanism for overt and covert shifts of visual attention. Vis. Res. **40**(10), 1489–1506 (2000)
7. Zhou, Q.: Object-based attention: saliency detection using contrast via background prototypes. Electron. Lett. **50**(14), 997–999 (2014)
8. Goferman, S., Zelnik-Manor, L., Tal, A.: Context-aware saliency detection. IEEE Trans. Pattern Anal. Mach. Intell. **34**(10), 1915–1926 (2012)
9. Kruthiventi, S.S.S., Gudisa, V., Dholakiya, J.H., Babu, R.V.: Saliency unified: a deep architecture for simultaneous eye fixation prediction and salient object segmentation. In: IEEE Conference on Computer Vision and Pattern Recognition, pp. 5781–5790. IEEE Computer Society (2016)
10. Pan, J., Sayrol, E., Giro-I-Nieto, X., Mcguinness, K., OConnor, N.E.: Shallow and deep convolutional networks for saliency prediction. In: Computer Vision and Pattern Recognition, pp. 598–606. IEEE (2016)
11. Chen, T., Lin, L., Liu, L., Luo, X., Li, X.: Disc: deep image saliency computing via progressive representation learning. IEEE Trans. Neural Netw. Learn. Syst. **27**(6), 1135–1149 (2016)
12. Achanta, R., Shaji, A., Smith, K., Lucchi, A., Fua, P.: Slic superpixels compared to state-of-the-art superpixel methods. IEEE Trans. Pattern Anal. Mach. Intell. **34**(11), 2274 (2012)
13. Felzenszwalb, P.F., Huttenlocher, D.P.: Efficient graph-based image segmentation. Int. J. Comput. Vis. **59**(2), 167–181 (2004)
14. Zhou, C., Wu, D., Qin, W., Liu, C.: An efficient two-stage region merging method for interactive image segmentation. Comput. Electr. Eng. **54**(C), 220–229 (2016)
15. Li, J., Levine, M.D., An, X., Xu, X., He, H.: Visual saliency based on scale-space analysis in the frequency domain. IEEE Trans. Pattern Anal. Mach. Intell. **35**(4), 996–1010 (2013)
16. Wang, J., Lu, H., Li, X., Tong, N., Liu, W.: Saliency detection via background and foreground seed selection. Neurocomputing **152**(C), 359–368 (2015)
17. Tong, N., Lu, H., Xiang, R., Yang, M.H.: Salient object detection via bootstrap learning. In: Computer Vision and Pattern Recognition, pp. 1884–1892. IEEE (2015)
18. Tong, N., Lu, H., Zhang, Y., Xiang, R.: Salient object detection via global and local cues. Pattern Recogn. **48**(10), 3258–3267 (2015)

19. Wang, L., Lu, H., Xiang, R., Yang, M.H.: Deep networks for saliency detection via local estimation and global search. In: Computer Vision and Pattern Recognition, pp. 3183–3192. IEEE (2015)
20. Li, G., Yu, Y.: Visual saliency based on multiscale deep features, pp. 5455–5463 (2015)
21. Lee, G., Tai, Y.W., Kim, J.: Deep saliency with encoded low level distance map and high level features, pp. 660–668 (2016)

Multi-scale Region Proposal Network Trained by Multi-domain Learning for Visual Object Tracking

Yang Fang, Seunghyun Ko, and Geun-Sik Jo(✉)

School of Computer and Information Engineering, Inha University, Incheon 22212, South Korea
{fangyang,kosehy1}@inha.edu, gsjo@inha.ac.kr
http://ailab.inha.ac.kr

Abstract. This paper presents a multi-scale region proposal network (RPN) for visual object tracking, inspired by Faster R-CNN and Yolo detectors which adopt an RPN to significantly speed up the detection time and achieve state-of-the-art detection performance. We expand them to apply a multi-scale region proposal network for visual tracking. Our proposed network can utilize both fine-grained features from shallow convolutional layers and discriminative features from deep convolutional layers. The features of shallow layers are good at accurate objects localization, and the features of deep convolutional layers can efficiently distinguish between target objects and backgrounds. A multi-domain learning mechanism is applied to train our network in an end-to-end way. To predict a new target object and its location in a new frame, we propose an re-ranking algorithm to determine a true object by exploiting spatial modeling, scale variants and color attributes of object proposals. Our tracker is validated on the OTB-15 object tracking benchmark, and achieves 0.603 for the success rate and 0.760 for the precision rate of the one-pass evaluation. Additionally, our tracker can run at 22 frames per second, which is very close to real-time speed. Experiment results show its outstanding performance in both tracking accuracy and speed by comparing it with existing state-of-the-art methods.

Keywords: Color attributes · Multi-domain learning · Multi-scale RPN · Re-ranking algorithm · Visual tracking

1 Introduction

Single object tracking problem focuses on the task of predicting the trajectory of a moving object in videos. Given an initial location of the target object in the first frame, trackers should be capable of overcoming unpredictable variations in object appearance and backgrounds during tracking, and must find the target object in subsequent frames. Although object tracking has been studied for several decades, and much research progress has been made in recent years [1–3],

D. Liu et al. (Eds.): ICONIP 2017, Part III, LNCS 10636, pp. 330–339, 2017.
https://doi.org/10.1007/978-3-319-70090-8_34

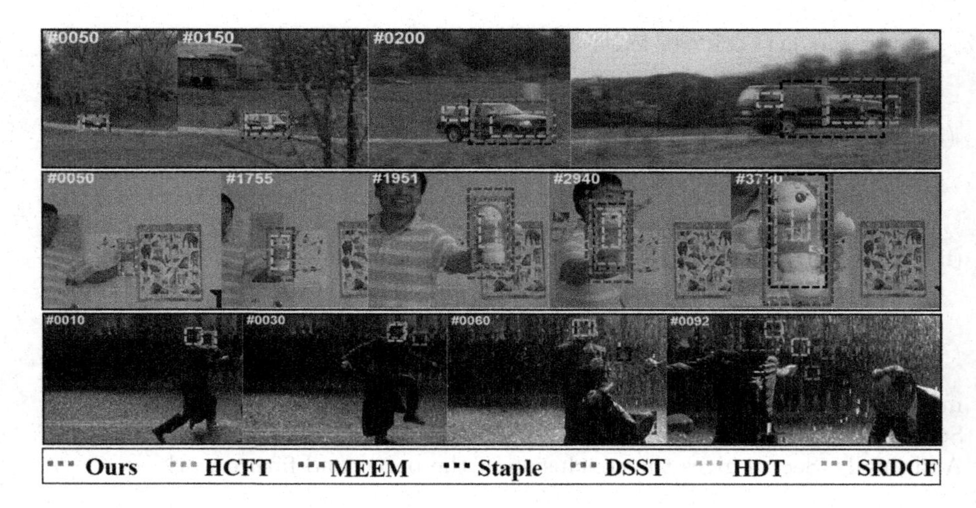

Fig. 1. Comparison of our approach with state-of-the-art trackers, including HCFT [6], MEEM [2], Staple [13], DSST [14], HDT [15] and SRDCF [7] on OTB-100

it remains a challenging problem in computer vision because of many factors [4], e.g. illumination variations, occlusions, background clutters, scale variation, deformation, and fast motion, and no single tracking algorithm can handle all scenarios. The long-term tracking task is explicitly broken down into tracking, learning, and detection in TLD [1]. Zhu [5] presented object tracker that is not limited to a local search window and has the ability to probe the entire frame. Recently, some methods [6,7] directly adopt the offline pre-trained models of vgg-16 [8] and its variations for features extraction, and they then feed convolutional neural network (CNN) features into independent classifiers, e.g. support vector machine (SVM) and correlation filters. And [9,10] train the CNN in an end-to-end way using labeled sequence datasets, and then conduct tracking without external classifiers, referred to as an *End-to-End Tracker*.

The region proposal network has been successfully applied in object detection tasks. Ren et al. [11] introduced a novel region proposal network that shares convolutional features with Fast R-CNN [12] to generate region proposals using hand-crafted anchor boxes. However, hand-crafted anchor boxes lack prior information for the network, so it cannot converge as soon as possible during the training stage. In this paper, we apply a multi-scale region proposal network by utilizing both fine-grained conv features and discriminative conv features for the tracking task. Figure 1 shows the tracking results on some of OTB-15 sequences, and it shows our tracker gets competitive results. We demonstrate that our tracking algorithm has three folds contributions, as follows.

(i) We first introduce a ***multi-scale RPN*** into tracking fields. Instead of using a *single* RPN, we add one additional region proposal network on top of passthrough layer, which follows the second convolutional layer. In that way, our network can learn both higher spatial resolution information and rich discriminative information to better localize the target object in a new frame.

(ii) To provide better ***prior knowledge*** for the network to easily learn the structural and scale information of the training data, we abandon the hand-crafted anchor boxes; instead, we use a *k-means algorithm* to cluster the ground-truth bounding boxes and select k cluster centers for anchor boxes.

(iii) We propose a novel ***re-ranking algorithm*** to rank object proposals, and predict new target objects by exploiting spatial modeling, scale variants, and color attributes between a true target and object proposals.

The rest of this paper is organized as follows. Details on multi-scale RPN, multi-domain learning mechanisms, and re-ranking algorithm are presented in Sect. 2. The experiments under OTB-100 benchmark are presented in Sect. 3. And finally, Sect. 4 gives the conclusion of the paper and future work.

2 Multi-scale Region Proposal Network and Tracking Algorithm

Section 2.1 first describes our proposed deep network architecture and Sect. 2.2 depicts how it is trained by a multi-domain learning mechanism. Then Sect. 2.3 proposes an efficient proposal re-ranking algorithm by exploiting the spatial, and structural relationship between the true target and object proposals and their color attributes.

2.1 Multi-scale Region Proposal Network Architecture

The architecture of our network is illustrated in Fig. 2. It consists of five convolutional layers, and one passthrough layer, which follows the activation of second conv layer, and two region proposal layers RPN_1 and RPN_2, where RPN_1 is built on top of a passthrough identity mapping layer, and RPN_2 is built on top of convolutional feature maps of the fifth convolutional layer. RPN_1 produces region proposals from fine-grained feature maps; fine-grained features are

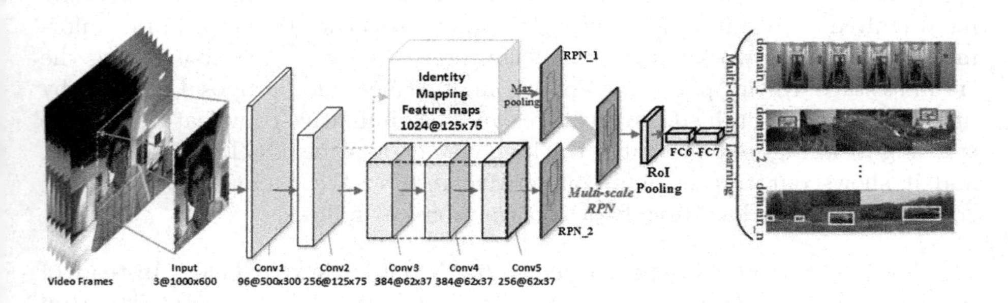

Fig. 2. Multi-scale region proposal network architecture

helpful in localizing small-scale objects. RPN_2 produces region proposals from coarse-grained feature maps, which provide sematic information. After RPN_1 and RPN_2, we apply region of interest (RoI) pooling layer(s) on top of each region of interest location. Then, we use two fully-connected layers as a finer bounding boxes regressor. Our input to the network is the entire frame of a single scale, and we re-scale the images such that $shorter\ size = 600\ pixels$ [11]. Figure 2 shows an input at a fixed size of 1000×600 pixels for an example.

Region proposal network produces region proposals regressed by anchor boxes [11]. To give anchor boxes better priors and make RPNs easier to learn and better at predicting tight bounding boxes surrounding objects, we first use a standard k-means clustering algorithm to cluster all ground-truth bounding boxes among the training data into m clusters, where m cluster centers out of all bounding boxes are picked as our anchor boxes. Specifically, the training set used in this paper contains $M(M = 19780)$ ground-truth bounding boxes, and these boxes have different scales and aspect ratios. Therefore, these bounding boxes must be clustered into N clusters in terms of scales and aspect ratio. Let b_i denotes the ith bounding box, and if c_j is the centroid of the jth cluster, then the distance metric is defined as $d(b_i, c_j) = 1 - IoU(b_i, c_j)$[1]. We initialize the N centroids by randomly selecting N boxes $\{c_1^{init}, c_2^{init}, ..., c_N^{init}\}$ from all ground-truth boxes with $\forall i, j,\ d(c_i^{init}, c_j^{init}) > threshold$. The anchor-boxes clustering algorithm is Algorithm 1, which generates anchor boxes with m scales and aspect ratios. In Fig. 3, the left image contains 19,780 object ground-truth bounding boxes from the training set, and the middle image shows nine anchor boxes produced by Algorithm 1, and their pixel sizes are in Table 1, and the right image represents the final object proposal generated by our multi-scale RPN.

Algorithm 1. Anchor box generating algorithm

Input: $B = \{b_1, b_2, ..., b_M\}$
Output: $C = \{c_1, c_2, ..., c_N\}$
Initialization: $\begin{cases} C^{init} = \{\mu_1, \mu_2, ..., \mu_N\} \\ J_{pre} = 0 \end{cases}$
while *1* **do**
  $g^{(i)} := \underset{j}{\operatorname{argmin}}\, d(b_i, \mu_j)$;
  $\mu_j := \underset{b_i}{\operatorname{argmin}} \frac{\sum_{k=1}^{M} 1\{g^{(i)}, g^{(k)} = j\} d(b_i, b_k)}{\sum_{k=1}^{M} 1\{g^{(i)} = j\}}$;
  $J = \frac{\sum_{j=1}^{M} d(b_i, \mu_{g^{(i)}})}{M}$; $J_{pre} = J$;
  if $|J_{pre} - J| < \sigma \;||\; iteration > iter_{max}$ **then**
    $c_i = \mu_i, i = 1, 2, ..., N$; break;
  end
end

[1] IoU donates Intersection over Union function.

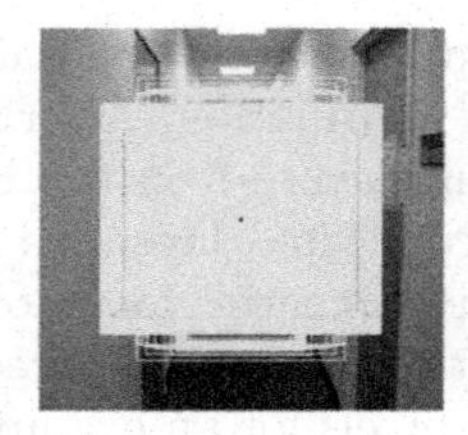
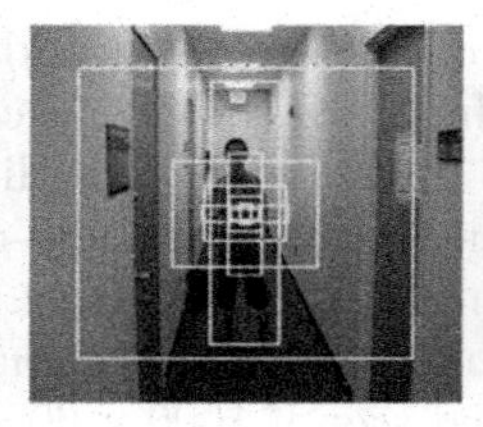

Fig. 3. From left to right: ground-truth bounding boxes, nine anchor boxes produced by Algorithm 1, and the final object region proposal regressed by anchor boxes

Table 1. Anchor boxes size in pixel value

Height (pixels)	13	34	40	100	40	94	168	80	390
Width (pixels)	28	15	48	22	163	73	140	350	385

2.2 Multi-domain Learning Mechanism

Inspired by earlier work [10], we designed K branches at the end of the network as the classifier layers. Each branch stands for one specific video domain, and each domain denotes one specific training sequence. On the network training stage, we applied a standard stochastic gradient descent (SGD) method, and when the training is on the kth iteration, only the kth branch layer of the classifier is utilized to update the network until the network converges or the predefined number of max iterations is reached. Specifically, for RPN training, we minimize the objective function with a multi-task loss [11]. Our multi-task loss function is defined as:

$$L(\{p_i\}, \{t_i\}) = \frac{1}{N_{cls}} \sum_i L_{cls}(p_i, p_i^*) + \lambda \frac{1}{N_{reg}} [p^* \geq 1] \sum_i L_{reg}(t_i, t_i^*) \quad (1)$$

where $\{p^*\}$ denotes the label of the anchor boxes containing an object or not, $p^* = 1$ when $IoU(anchor\ box, groundtruth\ box) \geq 0.7$[2], and $p^* = 0$ when $IoU \leq 0.3$. The regression loss is activated only for anchors with $p^* = 1$. t_i and t_i^* are predicted coordinates parameters and the ground-truth associated with a positive anchor, respectively.

For final bounding box refinement, we follow a previous method [12] and because it is not necessary to predict object class during tracking, the loss function only focuses on bounding box regression, which simplifies the learning process. The loss function is as follows:

$$L_{ref}(t, t^*) = \sum_{i \in \{x,y,w,h\}} smooth_{L_1}(t_i - t_i^*) \quad (2)$$

[2] Here, IoU means Interaction over Union between anchor boxes and ground-truth boxes. If $IoU \geq 0.7$, the anchor box is considered the true object location (positive), and if $IoU \leq 0.3$, it is considered as false location (negative).

where

$$smooth_{L_1}(x) = \begin{cases} 0.5x^2, & if\ |x| < 1 \\ |x| - 0.5, & otherwise \end{cases} \tag{3}$$

and the scale-invariant translation and log-space height/width shift of the predicted bounding box and the ground truth relative to the object proposal, t_i and t_i^* $(i = x, y, w, h)$ are defined as

$$\begin{aligned} t_x &= (x - x_p)/w_p,\ t_y = (y - y_p)/h_p, t_w = log(w/w_p),\ t_h = log(h/h_p) \\ t_x^* &= (x^* - x_p)/w_p,\ t_y^* = (y^* - y_p)/h_p, t_w^* = log(w^*/w_p),\ t_h^* = log(h^*/h_p) \end{aligned} \tag{4}$$

In the above, x, x^* and x_p represent predicted, ground-truth, and region proposal coordinations, respectively. Note that here, x_p is the output results of the region proposal networks.

In this paper, the maximum number of iterations is 100 K with a learning rate of 10^{-4} for convolutional layers and 10^{-3} for a fully connected layer, and the momentum and weight decay are set to 0.9 and 5.0×10^{-4}, respectively, during training. On the testing stage, domain-specific layers are replaced by a new classifier layer for test sequences.

2.3 Tracking and Update

A robust tracker should be computationally efficient and must also possess discriminative power. Our tracking system is modeled as a tracking-by-detection mechanism with the advantage of a high object recall rate from the proposed multi-scale RPN, and the tracker has to distinguish the target from distractors among all object candidates based on spatial information and unique patterns of the target object. Therefore, the tracking process is separated into two stages: objectness detection and proposals re-ranking.

Objectness detection: First, the multi-scale RPN network generates an object candidates pool $O_t = \{o_t^1, o_t^2, ..., o_t^n\}$ including n candidates in the current frame, t, and each candidate has a objectness detection score $\mathbf{f}_t = \{f_t^1, f_t^2, ..., f_t^n\}$. The number of candidates depends on the threshold of the detection score, θ, that we set. To get a higher object recall rate, θ is set to 0.1 in practice to get as many potential target proposals as possible.

Proposals re-ranking and model updating: To determine the target object o_t^* being tracked (while discarding other false-positive proposals), we utilize the spatial modeling and pattern relationship between target proposals O_t in the current frame, t, and tracked result o_{t-1}^* from the previous frame, $t-1$. Considering motion smoothness in real-world visual tracking, the expected target location in the current frame should not be far from the previous location, which means that the farther the proposal from the previous location, the less the probability that it is the new target in the current frame. Thus, we create a spatial constraint function, $S_{sc}(o_t^i, o_{t-1}^*) = \exp(-\frac{1}{2\sigma_1^2}\|p(o_t^i) - p(o_{t-1}^*)\|^2$, and we also take the smoothness of a scale variant of the object into account, and use the IoU function as a scale variant constraint to represent it as $S_{sv}(o_t^i, o_{t-1}^*) = \exp(-\frac{1}{2\sigma_2^2}\|IoU(o_t^i, o_{t-1}^*)\|^2$. Additionally, we adapt color attributes (*CN*) features [16] to build a robust color-based binary classifier. First, object proposals are rescaled to fixed size $H \times W$ in RGB channels, and then color features are extracted from the resized image patch to construct a color feature space, denoted by $\hat{o}_t^i$ for each object proposal o_t^i. The classifier is trained and updated by previous tracked result o_{t-1}^* and its cyclic shifts, and outputs the tracking score for each proposal as $S_c(\hat{o}_t^i) = \langle \phi(\hat{o}_t^i), \omega \rangle$. For more detail, please refer to earlier work [16]. Finally, the new tracked object in a new frame is estimated by maximizing the score function:

$$\mathbf{o_t^*} = \underset{i}{\operatorname{argmax}} \left[\gamma_{sc} S_{sc}(o_t^i, o_{t-1}^*) + \gamma_{sv} S_{sv}(o_t^i, o_{t-1}^*) + \gamma_c S_c(\hat{o}_t^i)\right] \tag{5}$$

in which, $\gamma_{sc} = \gamma_{sv} = 0.2, \gamma_c = 0.6$ and σ_1, σ_2 equal the diagonal length and the area size of the initialized bounding box, respectively.

3 Experiments

In this section, we present experiments that compare the proposed tracker with state-of-the-art trackers on a public online tracking benchmark OTB-15 [4], which contains a total of 100 sequences from different real-world scenarios. And the 100 sequences are manually tagged with 11 attributes, e.g. illumination variation (IV), scale variation (SV), occlusion (OCC), deformation (DEF), motion blur (MB), fast motion (FM), in-plane rotation (IPR), out-of-plane rotation (OPR), out-of-view (OV), background clutters (BC) and low resolution (LR). Our tracker is evaluated for each of them, as shown in Fig. 5.

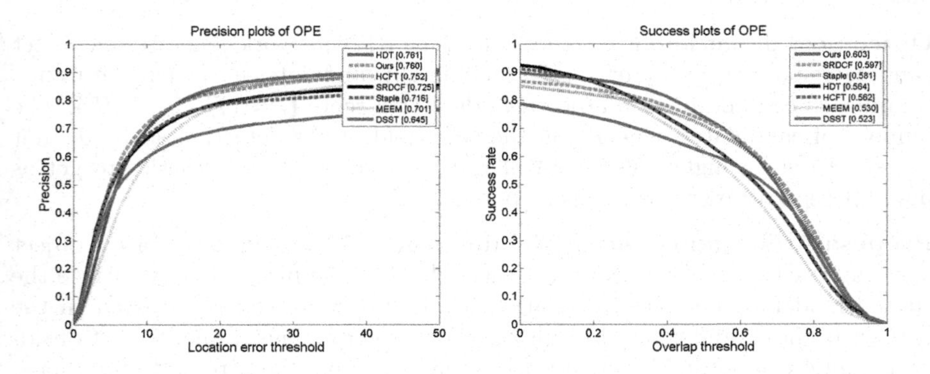

Fig. 4. Comparison of overall performance between the proposed tracker and other state-of-the-art trackers in terms of precision rate and success rate for one-pass evaluation (OPE)

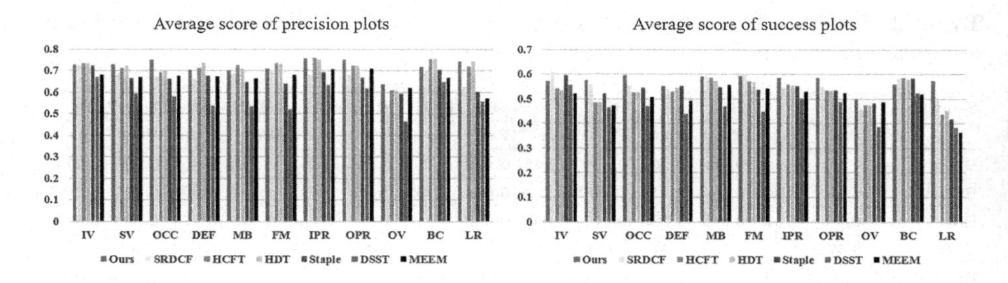

Fig. 5. Precision rate and success rate from the experiment results for 11 attributes

We conducted experiments on a server with an Intel Core i7-6700 CPU at 3.4 GHz $\times$ 8 cores, with a single NVIDIA GeForce GTX Titan X GPU. Our Matlab implementation runs at an average speed of around 22 fps. The proposed algorithm was compared with state-of-the-art trackers, and we chose the trackers based on their tracking speed with at least five frames per second or above, including SRDCF [7], Staple [13], HCFT [6], HDT [15], DSST [14] and MEEM [2]. Figure 4 shows the overall performance of our tracker and the compared trackers for one-pass evaluation with precision and success rate. The area under curve (AUC) of both precision and success plots is used to rank the overall performance. For an robustness evaluation, we performed a temporal robustness evaluation (SRE) and a spatial robustness evaluation (TRE), as well. Figure 6 demonstrates that our tracker provides the best tracking performance, compared to the others, in terms of spatial initialization perturbation. It proves that our multi-scale RPN can precisely re-initialize the object location in each frame. This kind of property can significantly mitigate the model drift problem and makes the tracker robust to spatial perturbation. We also evaluated our tracker for 11 attributes in the 100 sequences, and from Fig. 5, we can see that our tracking algorithm outperforms the others in IV, SV, OCC, IPR, OPR, OV, and LR, in seven of the 11 attributes. It further illustrates that multi-scale RPN can overcome many hard situations and retain a higher robustness ability.

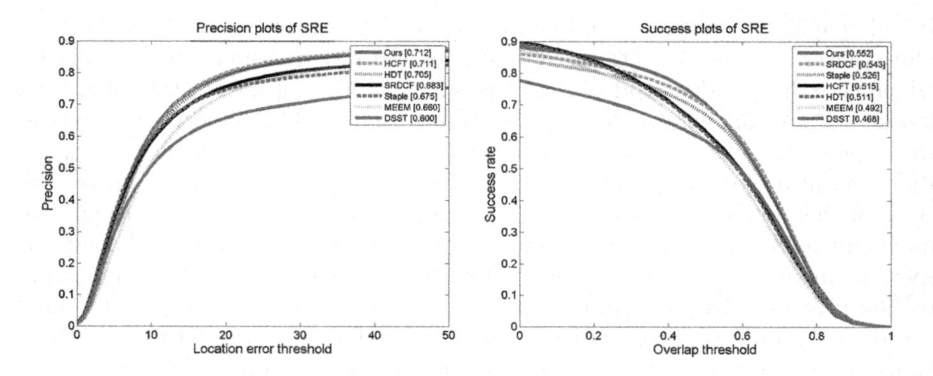

Fig. 6. Precision and success rates from the experiment for spatial robustness evaluation

Table 2. Experiments on 12 selected sequences of OTB-15. The best and second best results are in red and green colors, respectively

Sequences	Ours		HCFT [6]		HDT [15]		SRDCF [7]		Staple [13]		MEEM [2]		DSST [14]	
	Pre	Suc	Pre	Suc	Pre	Suc	Pre	Suc	Pre	Suc	Pre	Suc	Pre	Suc
Biker	0.505	0.369	0.489	0.252	0.489	0.245	0.488	0.353	0.488	0.256	0.508	0.265	0.496	0.274
Board	0.103	0.133	0.735	0.727	0.784	0.739	0.684	0.709	0.686	0.581	0.532	0.605	0.710	0.708
Box	0.808	0.579	0.332	0.283	0.071	0.098	0.366	0.355	0.367	0.354	0.329	0.274	0.353	0.340
ClifBar	0.572	0.307	0.783	0.477	0.696	0.439	0.779	0.547	0.609	0.458	0.811	0.534	0.884	0.664
DragonBaby	0.817	0.700	0.758	0.620	0.749	0.611	0.306	0.252	0.713	0.500	0.753	0.633	0.055	0.057
Dudek	0.815	0.817	0.781	0.726	0.805	0.727	0.745	0.801	0.743	0.706	0.704	0.706	0.731	0.765
Human6	0.825	0.690	0.369	0.219	0.443	0.231	0.825	0.731	0.885	0.800	0.600	0.265	0.411	0.372
Ironman	0.610	0.468	0.593	0.437	0.628	0.464	0.051	0.032	0.172	0.127	0.460	0.333	0.163	0.125
Lemming	0.652	0.581	0.246	0.228	0.255	0.225	0.330	0.269	0.256	0.236	0.812	0.664	0.404	0.327
Liquor	0.435	0.349	0.757	0.695	0.664	0.625	0.898	0.845	0.864	0.720	0.776	0.775	0.391	0.404
Panda	0.867	0.508	0.788	0.392	0.843	0.501	0.356	0.119	0.521	0.313	0.846	0.488	0.413	0.130
Tiger2	0.662	0.522	0.631	0.564	0.640	0.568	0.762	0.650	0.779	0.674	0.607	0.538	0.339	0.332

To analyze tracking performance in more detail, we chose 12 representative sequences from the OTB-15 benchmark and evaluated each of them independently. Table 2 presents the precision rate (Pre) and success rate (Suc), and tracking speed of the 12 sequences. We can see that our tracker outperforms the others in seven of the 12 sequences in terms of precision and success rate. The performance demonstrates that applying a multi-domain learning method to train our network is the key factor in handling both tracking of objects in multiple categories and in sequences of various properties. Additionally, the proposed tracker runs at 22 frames per second that is faster than the trackers which utilize deep convolutional features, e.g. HDT and HCFT, and even exceeds two of 4 hand-crafted feature-based tracker, e.g. MEEM and SRDCF. Therefore, we conclude that the proposed tracker provides outstanding performance in both tracking accuracy and tracking speed.

4 Conclusion

In this paper, we propose a novel tracking-by-detection-based tracking system. Our tracker achieves a 0.603 success rate and a 0.760 precision rate on OTB-15, which took first and second place, respectively, with respect to state-of-the-art trackers. Additionally, it can efficiently run at 22 fps, which is very close to real-time speed. Experiment results demonstrate that our tracker achieves competitive overall performance in terms of tracking precision and speed compared to state-of-the-art trackers. However, the proposed multi-scale RPN consists of two independent region proposal networks, so it bears high computational complexity to train and test two independent RPNs, thus slowing down both training and testing process. There should be room for improvement, increasing the tracking speed by exploring how to utilize a single RPN for multi-resolution CNN features. Moreover, our network is trained only on a small training data set, so it is possible to further improve tracking performance by using more training data. We consider these open research topics for future works.

Acknowledgments. This work was supported by the National Research Foundation of Korea (NRF) grant funded by the Korea government (MSIP) (No. 2015-R1A2A2A03006190) and also supported by Nvidia GPU Grant.

References

1. Kalal, Z., Mikolajczyk, K., Matas, J.: Tracking-learning-detection. IEEE Trans. Pattern Anal. Mach. Intell. **34**(1), 1409–1422 (2010)
2. Zhang, J., Ma, S., Sclaroff, S.: MEEM: robust tracking via multiple experts using entropy minimization. In: European Conference on Computer Vision, pp. 188–203 (2014)
3. Hare, S., Saffari, A., Torr, P.H.S.: Struck: structured output tracking with kernels. IEEE Trans. Pattern Anal. Mach. Intell. **38**(10), 2096–2109 (2016)
4. Wu, Y., Lim, J., Yang, M.H.: Online object tracking: a benchmark. In: IEEE Conference on Computer Vision and Pattern Recognition, pp. 2411–2418 (2013)
5. Zhu, G., Porikli, F., Li, H.: Beyond local search: tracking objects everywhere with instance-specific proposals. In: IEEE Computer Vision and Pattern Recognition, pp. 943–951 (2016)
6. Ma, C., Huang, J.B., Yang, X., Yang, M.H.: Hierarchical convolutional features for visual tracking. In: IEEE International Conference on Computer Vision, pp. 3074–3082 (2015)
7. Danelljan, M., Khan, F., Felsberg, M., van de Weijer, J.: Learning spatially regularized correlation filters for visual tracking. In: IEEE International Conference on Computer Vision, pp. 4310–4318 (2015)
8. Simonyan, K., Zisserman, A.: Very deep convolutional networks for large-scale image recognition. CoRR, abs/1409.1556 (2014)
9. Luca, B., Jack, V., Andrea, V., Philip, T.: Fully-convolutional Siamese networks for object tracking. In: European Conference on Computer Vision, pp. 850–865 (2016)
10. Nam, H., Han, B.: Learning multi-domain convolutional neural networks for visual tracking. In: IEEE Conference on Computer Vision and Pattern Recognition, pp. 4293–4302 (2016)
11. Ren, S., He, K., Girshick, R., Sun, J.: Faster R-CNN: towards real-time object detection with region proposal networks. In: Advances in Neural Information Processing Systems, pp. 91–99 (2015)
12. Girshick, R.: Fast R-CNN. In: IEEE International Conference on Computer Vision, pp. 1440–1448 (2015)
13. Bertinetto, L., Valmadre, J., Golodetz, S., Miksik, O., Torr, P.H.S.: Staple: complementary learners for real-time tracking. In: The IEEE Conference on Computer Vision and Pattern Recognition, pp. 1401–1409 (2016)
14. Danelljan, M., Hager, G., Shahbaz Khan, F., Felsberg, M.: Accurate scale estimation for robust visual tracking. In: British Machine Vision Conference (2014). doi:10.5244/C.28.65
15. Qi, Y., Zhang, S., Qin, L., Yao, H., Huang, Q., M.-H.Yang, J.L.: Hedged deep tracking. In: IEEE Conference on Computer Vision and Pattern Recognition, pp. 4303–4311 (2016)
16. Danelljan, M., Khan, F., Felsberg, M., van de Weijer, J.: Adaptive color attributes for real-time visual tracking. In: IEEE Conference on Computer Vision and Pattern Recognition, pp. 1090–1097 (2014)

Deep Part-Based Image Feature for Clothing Retrieval

Laiping Zhou, Zhengzhong Zhou, and Liqing Zhang(✉)

Key Laboratory of Shanghai Education Commission for Intelligent Interaction and Cognitive Engineering, Department of Computer Science and Engineering, Shanghai Jiao Tong University, Shanghai, China
{zlp,tczhouzz}@sjtu.edu.cn, zhang-lq@cs.sjtu.edu.cn

Abstract. In this paper, we propose a straightforward way to extract part-based features only with the supervision of part-based attributes. As we know, regions can be highlighted by labels through weakly-supervised segmentation algorithms, and deep features can be extracted from CNN convolutional layers. We develop a new approach to combine them, leading to simpler procedure with only one CNN forward pass and better interpretation. We apply this method to our database of over 100,000 clothing images, and achieve comparable results to the state of the art. Moreover, the part-based features support functionalities of tuning weights among the parts, and substituting visual part features from other clothes. Because of its simplicity, the method is promising to be transferred to other image retrieval domains.

Keywords: Convolutional neural networks · Class activation mapping · Locality-constrained linear coding · Clothing retrieval

1 Introduction

In recent years, Convolutional Neural Networks (CNNs) have achieved great success on image classification and detection tasks. In the meantime, researchers keep seeking effective ways to exploit features from inside CNNs for image retrieval. At first, activations of fully connected (FC) layers are considered to be a good representation [1]. But this encoding has a high dimension, and spatial information is distorted through the FC process. In recent years, convolutional activations have drawn more attention for preservation of spatial correlation. They are globally max-pooled or sum-pooled into a feature vector [2,3], or encoded with Bag of Visual Words (BoVW) and VLAD [4,5] for the convenience of retrieval.

However, these deep features only represent the discrimination of the *entirety* (or ROI). In many scenarios of image retrieval (especially for clothing retrieval), deep features should be divided spatially or semantically into several parts. This is necessary for the following two reasons:

D. Liu et al. (Eds.): ICONIP 2017, Part III, LNCS 10636, pp. 340–347, 2017.
https://doi.org/10.1007/978-3-319-70090-8_35

Interactivity. In an interactive image retrieval system, a user should be able to increase (or decrease) the importance of one or more components of an instance, and even substitute a part with the one from another instance. With a global deep feature, this can hardly be achieved.

Interpretability. Using separated features to search for a similar image resembles human mental activity, so it tends to provide retrieval results closer to those in customers' mind. Taking clothing retrieval as an example, when a customer uploads a query clothing image, he usually likes some of its spatial or semantic parts, for instance, collar, pocket, or style, pattern, etc. If global features are well separated into those perspectives, the system is more likely to give better results. In addition, the retrieval result will also be more interpretable for both users and researchers.

There have also been some researches on clothing retrieval, in which visual features are divided with attributes or semantically arranged. Some traditional methods first detect human faces, then guess the position of clothing components with prior knowledge of human body structure. Usually, such methods are not accurate enough, and not compatible with different datasets or image retrieval domains. Hybrid topic models are used to bridge semantic gaps by training visual words together with clothing attributes [6]. Some researchers achieve high precision with rich annotation and specific CNN structures [7]. Our method only requires part-related attributes, so is more resource-saving and also easier to be applied to other image retrieval tasks.

In this work, we first use Class Activation Mapping (CAM) [8], a weakly-supervised image segmentation method, to generate part heat-maps. Then we use a novel method to separate the convolutional activations with the heat-maps. The separated activations are then transformed to each part feature through LLC encoding. The contribution of our work can be summarized as followed:

1. We introduce CAM to separate the global visual feature, and bridge the semantic gap between visual features and image attributes.
2. We develop a novel algorithm of separating the CNN activations with CAM outputs, which simplifies the whole procedure.
3. We also present higher precision, better interactivity and interpretability of such part-based clothing features.

2 Method

In this section, we will introduce the summaries and details of our feature extraction procedure. The overview of our system will be introduced first. Then, we will get to explain the details of Class Activation Mapping, and its applications in our system. At last, we will discuss on the LLC encoding.

2.1 The Clothing Retrieval Procedure

The whole feature extraction and clothing retrieval procedure is summarized in Fig. 1. It is composed of two main steps, as is shown in the figure. Here we use VGG16 [9] as our CNN architecture, which is slightly modified to give out both CAM outputs and predicted attributes. The conv5_3 activations are sent to LLC encoder to generate main feature, as is described in [4,5], for coarse retrieval. On the other branch, they are separated by CAM, and sent to LLC encoder, too. This branch yields the part features for fine-grained retrieval. The CAM output serves as a visualization, to suggest whether our method works.

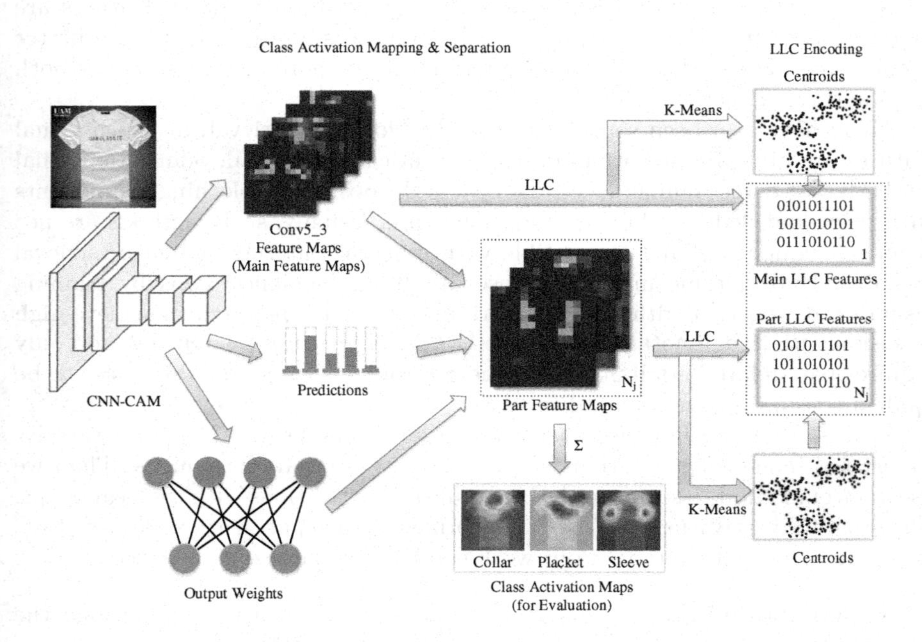

Fig. 1. The framework of our method.

2.2 Class Activation Mapping

Class Activation Mapping [8] is first proposed for CNN understanding or weakly-supervised object localization. See Fig. 2 for a brief introduction. First of all, the last convolutional layer should be connected to the output with only one fully-connected layer, in order to preserve spatial information. In this way, Global Average Pooling (GAP) replaces all extra FC layers, followed by the final one with a softmax output.

Now we have a modified CNN architecture, named as VGG16-CAM. Suppose the last convolutional layer (conv5_3) activations as A, where the value at coordinate (x, y) of kth channel is $A_k(x, y)$. Also suppose the output layer param is w (ignoring b here), where the value of connection of the kth channel to output node j is w_{kj}. Let y be the output of the final layer. As softmax function is incremental, we ignore it here. So we have

$$y_j = \sum_k w_{kj} \sum_{x,y} A_k(x, y) = \sum_{x,y} \sum_k w_{kj} A_k(x, y). \tag{1}$$

Equation (1) shows that the GAP process can be shifted to the last, so we can omit it to get the heat-map instead of a single value of this attribute's likelihood. As no spatial information is lost in VGG16-CAM, the map may well be resized to the size of the original image as the probability relationship between input pixels to output nodes. The heat-map before resizing can be described as

$$Y_j = \sum_k w_{kj} A_k. \tag{2}$$

CAM is a simple and elegant way to visualize CNN decisions and localize objects. However, its resolution is rather low, so using CAM to frame parts of input image directly is not practical.

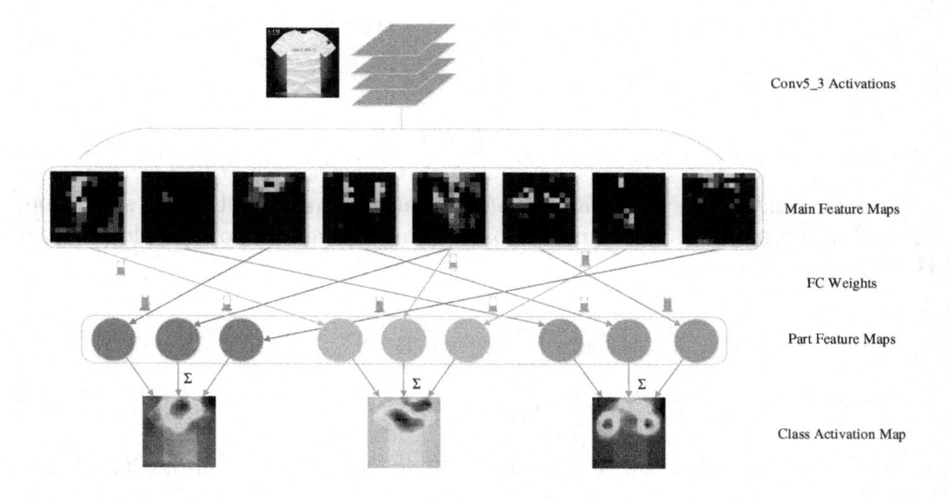

Fig. 2. CAM and CAM separation.

2.3 Separating Convolutional Activations with CAM

Actually, using CAM to frame parts directly is also time-consuming, as it leads to multiple forward computations of CNN. Here we introduce a trick just taking one forward pass, that is taking $A_k^j = w_{kj} A_k$ from Eq. (2) to serve as part-related convolutional activations.

CAM has the same resolution with conv5_3 activations. As Y_j is the likelihood heat-map of attribute j, it will highlight on the related part. Thus in any component $A_k^j = w_{kj} A_k$ of Y_j, if the convolutional activation A_k is irrelevant to attribute j, the param w_{kj} will suppress the component, and vice versa. In this way, CAM suppresses irrelevant channels and strengthens relevant channels, so that injects part information to the feature maps. See Fig. 2, after this step we obtain main feature maps and part feature maps.

2.4 LLC Encoding

After main feature maps and part feature maps being computed, they are sent to LLC encoding [10]. LLC is firstly proposed to encode SIFT features. Here we use it to encode convolutional activations.

At first, we must do some preparations. For main feature maps A and part feature maps $\{A^j\}_j$ of each image, we split them at spatial X and Y dimensions into vectors $\mathbf{a}_i$ and $\{\mathbf{a}_i^j\}_j$, with which we build up codebooks B and B^j by K-Means clustering. Then it comes to LLC encoding. It is a process of optimization of Eq. (3), where $\odot$ denotes element-wise multiplication, and $\mathbf{d}_i \in \mathbf{R}^M$ is a locality adaptor representing the similarity from the data point to centroids, as shown in Eq. (4).

$$\begin{aligned} &\min_{C} \sum_{i=1}^{N} ||\mathbf{a}_i - \mathbf{B}\mathbf{c}_i||^2 + \lambda||\mathbf{d}_i \odot \mathbf{c}_i||^2, \\ &s.t.\ \mathbf{1}^T\mathbf{c}_i = 1, \forall i. \end{aligned} \tag{3}$$

$$\mathbf{d}_i = \exp\left(\frac{dist(\mathbf{a}_i, \mathbf{B})}{\sigma}\right). \tag{4}$$

But in practice, we use a faster approximation version of LLC to save time. Instead of solving Eq. (3), we just choose k nearest neighbors of $\mathbf{a}_i$ as local bases $\mathbf{B}_i$ to reconstruct $\mathbf{a}_i$:

$$\begin{aligned} &\min_{\tilde{\mathbf{C}}} \sum_{i=1}^{N} ||\mathbf{a}_i - \tilde{\mathbf{c}}_i\mathbf{B}_i||^2, \\ &s.t.\ \mathbf{1}^T\tilde{\mathbf{c}}_i = 1, \forall i. \end{aligned} \tag{5}$$

This approximation can significantly reduce computation complexity when k is small, thus works well in practice. Finally, we sum up the LLC encodings $\tilde{\mathbf{c}}_i$ for each image to get the main or part encoding.

3 Experiments

In this section, we will evaluate our part-based features. At first, our method requires annotations on attributes, which are not common in public instance retrieval datasets. So we create our own clothing dataset, to test out our method. Then our implementation details will be introduced. Finally the experiment results are presented and analyzed.

3.1 Dataset

Our dataset is a clothing database of 116,196 upper garments with 11 attributes. The dataset is collected from e-commerce websites like Taobao.com, and their attributes are analyzed from tags and descriptions. The attributes are listed as followed: Shape, Pattern, Material, Collar, Sleeve, Placket, Style, Age, Category, Gender, and Season. Among them, collar, sleeve and placket are spatial part attributes, and the others are semantic part attributes. We select 100,000 of them as retrieval database, and leave the rest to be query items for evaluation. The retrieval database is then divided into training set and validation set in the proportion of 9:1.

3.2 Implementation Details

We first train a VGG16-CAM as shown in Sect. 2.2. The parameters of convolutional layers are transferred from VGG16 parameters pretrained with ImageNet dataset. The network is first finetuned with the attribute *Category* for a few epochs, then finetuned with all 11 attributes until validation loss plateaus, so as to steadily train the model. Those experiments are implemented on Tensorflow platform.

We use K-Means algorithm to generate centroids from feature maps. For main feature, we cluster them into 500 centroids. For part features, we cluster vectors from each part into $10 * N(y_j)$ centroids, where $N(y_j)$ is the number of outcomes of attribute j, to get 600 part centroids in total. In LLC encoding of Sect. 2.4, we choose $k = 5$ for main feature and $k = 3$ for part features.

3.3 Results

We evaluate our part features with two metrics: Normalized Discounted Cumulative Gain (NDCG) and mean Average Precision (mAP), while $NDCG@k = \sum_{j=1}^{k} \frac{2^{r(j)}-1}{\log(j+1)}$. The NDCG@10 and mAP@10 results are shown in Table 1. We can conclude that our deep part-based feature can achieve better retrieval precision than state-of-the-art deep features.

Table 1. NDCG and mAP of 10 results.

Methods	NDCG@10	mAP@10
R-MAC [2,3]	0.6203	0.4864
BoVW+VLAD [4,5]	0.8061	0.7647
Part Feature (Ours)	**0.8223**	**0.7743**

4 Applications

In this part, we discuss about the applications of our part-based deep feature. As we mentioned before, it can achieve better interactivity in image retrieval tasks. Here we show two examples of leveraging such advantages in clothing retrieval: Tuning weights among attributes and substituting visual part features.

Tuning Weights Among Attributes. When we want to pay more attention on some attribute, we can increase the weight of the corresponding feature. In Fig. 3 Line 3, 4, we want more results with the same *pattern* (dotted) to the query. By simply strengthening the *pattern* attribute, we get more dotted clothes in the return list.

Fig. 3. Examples of common search, weight tuning and feature substitution.

Substituting Part Features. When we like most parts of the query image, except for some part, we can simply substitute the corresponding feature with those from other clothes. In Fig. 3 Line 5, 6, we substitute the T-shirt's round *collar* with the one from a shirt, and get more results of T-shirts with shirt-collar.

5 Conclusion

In this paper, we propose a novel method to extract deep part-based features only based on part-based attributes with CAM and CAM separation, and adopt it to clothing retrieval problems. Such features can not only achieve higher accuracy than the state of the art, but also show better interpretability and interactivity during retrievals. As the feature extraction procedure is very straightforward, this method should be easily adapted to other scenarios of image retrieval.

Acknowledgments. The work was supported by the National Natural Science Foundation of China (Grant No. 91420302) and the National Basic Research Program of China (Grant No. 2015CB856004), and the Key Basic Research Program of Shanghai, China (15JC1400103).

References

1. Babenko, A., Slesarev, A., Chigorin, A., Lempitsky, V.: Neural codes for image retrieval. In: Fleet, D., Pajdla, T., Schiele, B., Tuytelaars, T. (eds.) ECCV 2014. LNCS, vol. 8689, pp. 584–599. Springer, Cham (2014). doi:10.1007/978-3-319-10590-1_38
2. Babenko, A., Lempitsky, V.: Aggregating local deep features for image retrieval. In: Proceedings of the IEEE International Conference on Computer Vision, pp. 1269–1277. IEEE (2015)
3. Tolias, G., Sicre, R., Jgou, H.: Particular object retrieval with integral max-pooling of CNN activations. arXiv preprint (2015). arXiv:1511.05879
4. Yue-Hei Ng, J., Yang, F., Davis, L.S.: Exploiting local features from deep networks for image retrieval. In: CVPR, pp. 53–61. IEEE (2015)
5. Mohedano, E., McGuinness, K., O'Connor, N.E., Salvador, A., Marqus, F., Gir-i-Nieto, X.: Bags of local convolutional features for scalable instance search. In: Proceedings of the 2016 ACM on International Conference on Multimedia Retrieval, pp. 327–331. ACM (2016)
6. Zhou, Z., Zhou, J., Zhang, L.: Demand-adaptive clothing image retrieval using hybrid topic model. In: Proceedings of the 2016 ACM on Multimedia Conference, pp. 496–500. ACM (2016)
7. Liu, Z., Luo, P., Qiu, S., Wang, X., Tang, X.: Deepfashion: powering robust clothes recognition and retrieval with rich annotations. In: CVPR, pp. 1096–1104. IEEE (2016)
8. Zhou, B., Khosla, A., Lapedriza, A., Oliva, A., Torralba, A.: Learning deep features for discriminative localization. In: CVPR, pp. 2921–2929. IEEE (2016)
9. Simonyan, K., Zisserman, A.: Very deep convolutional networks for large-scale image recognition. arXiv preprint (2014). arXiv:1409.1556
10. Wang, J., Yang, J., Yu, K., Lv, F., Huang, T., Gong, Y.: Locality-constrained linear coding for image classification. In: CVPR, pp. 3360–3367. IEEE (2010)

A Metric Learning Method Based on Damped Momentum with Threshold

Le Zhang, Lei Liu, and Zhiguo Shi(✉)

School of Computer and Communication Engineering,
University of Science and Technology Beijing, Beijing 100083, China
szg@ustb.edu.cn

Abstract. The convolutional neural networks in deep learning have become one of the mainstream algorithms of face recognition technology. Moreover, metric learning is also an important method to train deep learning models, as its ability of verification is very powerful, especially for the face images which are often used in CNNs. Recently, a new type method of metric learning named Center Loss has been proposed. It is simple to use and can enhance the model performance obviously. However, since the updating mechanism of Center Loss is simplistic, it can hardly process large-scale data when the categories are too much. This paper proposes an improved algorithm of Center Loss to accelerate the updating process of feature centers of original algorithm with a damped momentum, which urges deep learning models to have more rapid and steady convergence and better performance. Meanwhile, almost no additional computation cost is added since the new method has an optional threshold. The experimental results show that the improved Center Loss algorithm can further improve the recognition ability of the model, which is very helpful to enhancing the user experience of complex face recognition systems.

Keywords: Deep learning · Metric learning · Face recognition

1 Introduction

Recently, deep learning (DL) [1] has almost become one of the most popular machine learning techniques, and its performance, which is compared to the traditional algorithm, has greatly improved. In the field of computer vision, convolutional neural networks (CNNs) [2] have also undoubtedly become one of the most widely used algorithms.

With the distributed structure characteristics, the CNNs can easily process various input data that contains large amounts of information. In a large challenge called ILSVRC (ImageNet Large Scale Visual Recognition Challenge), various algorithms based on the CNNs have improved the top-1 accuracy from the initial 55% (AlexNet [3]) to 80% (Inception, v4 [4]). Among them, the famous one is GoogleNet [5] proposed by Szegedy et al. in 2014 and Deep Residual Net proposed by Kaiming He et al. in 2015 [6]. GoogleNet uses the idea of sparse learning to increase the size of network, so as to achieve a better classification result. Residual Net introduces the concept of residual, which makes the solver easier to converge and can lower the training error while deepening the network.

D. Liu et al. (Eds.): ICONIP 2017, Part III, LNCS 10636, pp. 348–356, 2017.
https://doi.org/10.1007/978-3-319-70090-8_36

However, most of the above CNNs algorithms only focus on the innovation of model structure, but do not take into account the differences of sample data between intra-class and inter-class. To avoid this, researchers in the field of deep learning have proposed many discriminative training methods for deep models, the most typical of which is Metric Learning (ML). Traditional metric learning algorithms include Contrastive Embedding [7], Triplet Embedding [8], Lifted Structured Embedding [9] and so on. In recent years, a new algorithm named Center Loss has been proposed [10]. Combining with Softmax Loss, Center Loss establishes the centers of each classes of training set in the feature space and adds a distance constraint between the sample feature and its appropriate feature center.

Although Center Loss algorithm is very simple and efficient, it is not sufficient for all types of training data. In this paper, we try to improve the convergence speed and the accuracy of model with no additional computational cost by modifying the Center Loss algorithm, so that it can be applicable for a variety of complex face recognition systems.

This paper is divided into 6 sections. The Sect. 2 describes the theoretical basis of metric learning and the origin of Center Loss. The Sect. 3 improves the Center Loss training algorithm according to the existing problems of it. In the Sect. 4, the experiment results show that the improved Center Loss is more effective for improving the accuracy of deep models than its origin version. In the Sect. 5, the related contents are summarized. Finally, the acknowledgement section is listed.

2 Related Work

Face Recognition Technology (FRT) is a biometric technique that uses computer to analyze the face images and extract effective identification information from those images to recognize the individual identities. In recent years, deep learning algorithm has become one of the most popular face recognition technologies, and most of the face recognition problems can be solved by using CNNs algorithm models.

To acquire a high recognition ability, metric learning is necessary for the training process of deep learning models. The concept of metric learning was originally proposed by Eric Xing in 2002 [11], he pointed out that the distance between feature vectors should not be measured simply by Euclidean distance or cosine distance, because the features of the same category do not exhibit linear distribution in a high-dimensional space. There should be a criterion satisfying the statistical properties of existing calibration data so as to reduce the distance between similar data and increase the distance between heterogeneous data, as shown in Fig. 1.

In fact, Center Loss is such a metric learning algorithm. Comparing with Softmax Loss, it constructs a new loss function:

$$\mathcal{L}_C = \frac{1}{2}\sum\nolimits_{i=1}^{m} \left\| x_i - c_{y_i} \right\|_2^2 \tag{1}$$

where x_i represents the ith feature in a batch which consists of m features (x_i is usually one dimensional), y_i represents the index of the feature center corresponds to x_i. This loss function can be combined with Softmax Loss to train a deep model together.

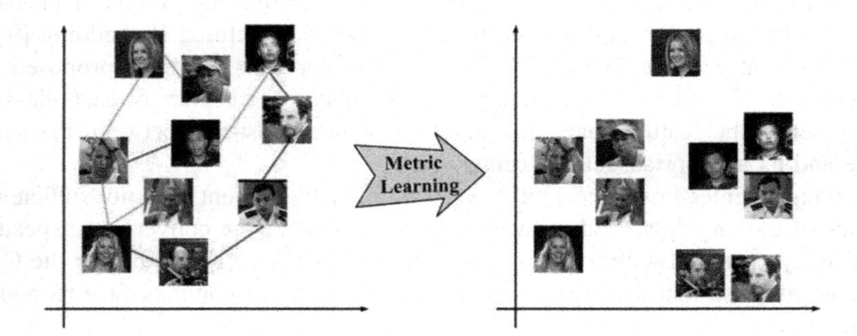

Fig. 1. Example of metric learning.

As a simple and effective metric learning algorithm, Center Loss avoids the vague selection process of the sample pairs, generates the feature centers and trains the CNNs models with the help of Softmax Loss. However, the updating process of the feature centers is too simple, often resulting in a very slow adjustment in high dimensional space. To solve this problem, this paper tries to accelerate the iterative process of the centers to prompt the deep models to converge more quickly.

3 Center Loss Algorithm with Damping and Threshold

3.1 General Framework of Metric Learning

As previously mentioned, since the models only trained by Softmax Loss have weaker cohesion, the feature vectors of each category overlap each other in high dimensional space, causing bad verification abilities in the end. If we construct a simple AlexNet model, train it by Softmax Loss and project its output features onto a two-dimensional plane, we could find that although the samples of different categories have been distinguished from each other, the inter-class distances among them are still small and the intra-class distances are still large, as shown in Fig. 2(left).

This indicates that the model has strong classification ability, while the identification ability is relatively low. When the samples are similar but belongs different categories, they may be erroneously divided into the same category.

If we combine the Center Loss algorithm which belongs to metric learning with Softmax Loss, the output feature distribution of AlexNet in Mnist will be shown as Fig. 2(right). It can be seen that the output features of Center Loss are more aggregated than those of Softmax Loss, and therefore they are easier to match the correct category. Of course, this is not just the characteristic of Center Loss. Most metric learning algorithms can achieve a similar result.

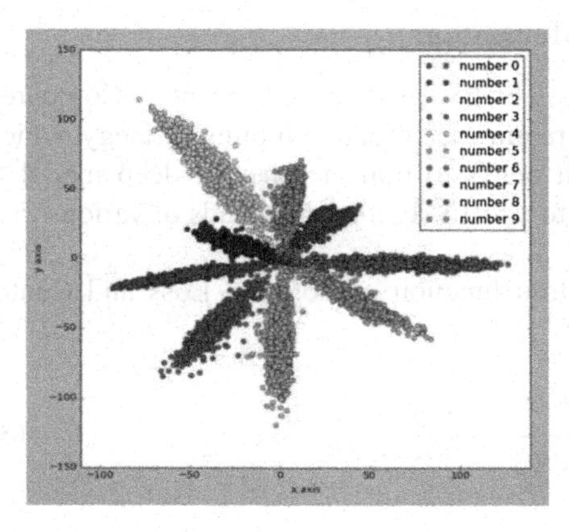

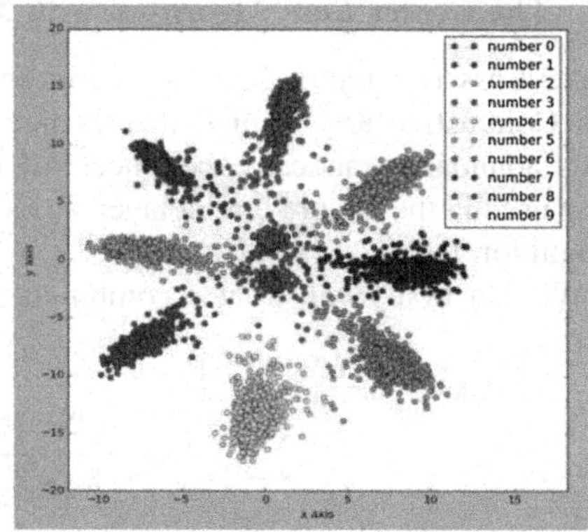

Fig. 2. The output feature distribution of Mnist. The left figure is trained only by Softmax Loss and the right figure is trained by Softmax Loss and Center Loss simultaneously.

3.2 Damped Momentum with Threshold

In physics, damping is the ability of a system to consume energy. From the mechanical vibration point of view, damping transforms the energy of mechanical vibration into heat energy or other energy that can be wasted, so as to achieve the purpose of vibration reduction. However, if the damping is always present, the speed of the whole system will become very slow, and the energy wasted will be too much. For this reason, we usually use damping with threshold instead of pure damping.

In classical physics, momentum is more abstract than damping, which refers to the increment of momentum of any object system (that is, the final momentum minus the initial momentum). It is produced by the resultant force of the objects. Actually, momentum is applied to deep learning widely, including the stochastic gradient descent algorithm, which uses a momentum factor to control the updating speed of the model parameters.

This paper introduces a damped momentum factor with threshold to improve the Center Loss algorithm. For convenience of understanding, given the variable x as the regulated one, the regulation formula can be written as:

$$\rho \circ x = \begin{cases} \rho(ln(x) + 1), & x > 1 \\ \rho x, & -1 \leq x \leq 1 \\ -\rho(ln(-x) + 1), & x < -1 \end{cases} \tag{2}$$

where ρ represents the damped momentum factor with threshold, it is usually a positive number less than 1.

Since the momentum simulates the moving inertia of the object, it can retain the updated direction of deep models to a certain extent, and uses the gradient of the current training batch to fine tune the final updated direction. These can increase the stability of training, speed up the learning process and also help to get rid of the local optimum.

3.3 The Center Loss Algorithm with Momentum

Center Loss is a metric learning algorithm which was proposed recently. Compared with Contrastive and Triplet, it does not require a special sampling strategy which needs abundant practical experience. All it needs to train a cohesive deep model is being used in the feature output layer, so as to effectively meet the needs of various face recognition tasks.

We can firstly construct a combining loss function of Softmax Loss and Center Loss:

$$\mathcal{L} = -\sum\nolimits_{i=1}^{m} \log \frac{e^{W_{y_i}^T x_i + b_{y_i}}}{\sum\nolimits_{j=1}^{n} e^{W_j^T x_i + b_j}} + \frac{1}{2} \sum\nolimits_{i=1}^{m} \left\| x_i - c_{y_i} \right\|_2^2 \tag{3}$$

This function is the basis of the whole Center Loss algorithm, and its partial derivatives can be used to update all the parameters of a deep model.

For the weight W of center loss layer, its updating formula is:

$$W^{t+1} = W^t - \mu^t \cdot \frac{\partial \mathcal{L}^t}{\partial W^t} = W^t - \mu^t \cdot \frac{\partial \mathcal{L}_S^t}{\partial W^t} \tag{4}$$

where t represents the time and μ represents the learning rate.

For other parameters of the deep model (weights and biases), their unified updating formula is:

$$\theta_C^{t+1} = \theta_C^t - \mu^t \sum\nolimits_i^m \frac{\partial \mathcal{L}^t}{\partial x_i^t} \cdot \frac{\partial x_i^t}{\partial \theta_C^t} \tag{5}$$

where $\frac{\partial \mathcal{L}^t}{\partial x_i^t}$ represents the back propagation error of the whole network.

For the feature center c_{y_i} of the Center Loss algorithm, since it is almost impossible to define the centers in the entire training set all at once, an iterative approach is needed to find the optimal center vectors in the training process. Therefore, the updating formulas of c_j are:

$$\Delta c_j^t = \frac{\sum\nolimits_{i=1}^{m} \delta(y_i = j) \cdot (c_j - x_i)}{1 + \sum\nolimits_{i=1}^{m} \delta(y_i = j)} \tag{6}$$

$$c_j^{t+1} = c_j^t - \alpha \cdot \Delta c_j^t \tag{7}$$

where the coefficient α is used to control the updating speed of feature centers to avoid jitter.

In this way, the Center Loss algorithm is combined with the Softmax Loss algorithm, which can be used simply and expediently with significant effect. However, although Center Loss can help deep models to distinguish the heterogeneous samples which are similar to each other, there may be a serious problem that the deep models are hard to converge with a large number of training samples.

To solve this, we introduced a damped momentum with threshold mentioned in Sect. 3.2 to improve the Center Loss algorithm. Here, we redefine the mean gradient of the feature centers in the original Center Loss algorithm as Eq. (8) to avoid the symbol confusion:

$$\Delta g_j^t = \frac{\sum_{i=1}^{m} \delta(y_i = j) \cdot (c_j - x_i)}{1 + \sum_{i=1}^{m} \delta(y_i = j)} \tag{8}$$

Then, by introducing a damped momentum factor with threshold, the iterative updating process of the feature centers is changed to:

$$\Delta c_j^t = \rho \circ \Delta c_j^{t-1} - \alpha \cdot \Delta g_j^t \tag{9}$$

$$c_j^{t+1} = c_j^t + \Delta c_j^t \tag{10}$$

Through this change, the iteration process of the feature centers of the Center Loss algorithm can be obviously and stably accelerated, and the convergence speed of deep models in training will be faster.

As a matter of fact, for a multi-classification problem, it is difficult to distinguish all of the categories strongly even in the high-dimensional space when the number of categories is very large. This is caused by the mathematical mechanism of the classification problem and can be partially solved by increasing the feature dimension and adjusting the learning rate of the final full-connected layer of the model. However, the original Center Loss algorithm also needs training iteration to determine the optimal location of each feature center, which results in a slow and unstable training process, like the stochastic gradient descent (SGD) algorithm [12]. To improve it, the Center Loss algorithm with momentum accelerates the iteration process of the feature centers. This can smooth the updating process so as to gain a deep model with better convergence.

4 Experiments

4.1 Verification in Small-Scale Dataset with a Few Categories

In order to verify the effectiveness of the improved algorithm, we still train an AlexNet model on the Mnist data set for experimental comparison, which is also easily accessible to visualization. Since the core of the Center Loss algorithm is the iterative process of its feature centers, we map all of them to a two-dimensional plane which is the same as before after each training iteration to observe their moving trajectory.

After 10K iterations, the updating trajectories of the feature centers of the original Center Loss and the improved Center Loss are respectively shown as Fig. 3(left) and (right).

It can be seen that the feature centers will gradually separate from each other at the beginning of the training process and find the right place for themselves regardless of whether or not the improvements were made to the Center Loss algorithm. However, the feature centers constrained by the Center Loss algorithm with momentum will

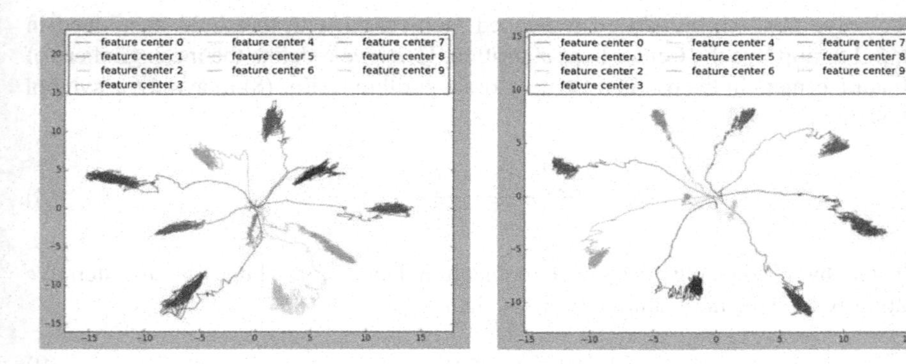

Fig. 3. Comparison of updating trajectory of feature centers in different Center Loss algorithms.

move more smoothly and more evenly than those constrained by the original algorithm obviously, so the categories can be distinguished more simply with each other.

4.2 Verification in Large-Scale Dataset with Many Categories

After verification on small-scale dataset, we should also need to validate the improved algorithm on a large-scale face dataset to observe its practical application effect in complex face recognition systems. Due to a significant rise in the amount of training data, the AlexNet model is unable to meet the needs of the simulation experiment in this section, so we choose to construct a residual network [6] to verify our algorithm.

The experiment of this section uses MsCeleb-1M [13] as the training set, uses LFW as the testing set and uses ROC curve to evaluate the performance of models. MsCeleb-1M is a huge collection of face images including about 1 million celebrities. Its open source version consists of nearly 100K celebrities, a total of about 8 million faces. LFW (Labeled Faces in the Wild) is a non restricted face recognition test set. It contains 5749 people and 13233 images with natural light, facial expression, pose and occlusion.

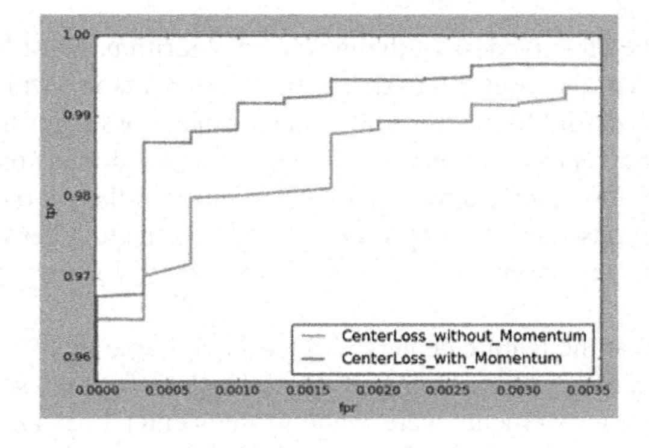

Fig. 4. Comparison of the ROC curves of different Center Loss algorithms.

In Sect. 4.1, the results of the experiments can be shown on a two-dimensional plane since the number of categories is very small. In contrast, MsCeleb-1M includes a great many categories, so we respectively test the models trained by two Center Loss algorithms on LFW dataset and draw their ROC curves in the same graph, as shown in Fig. 4.

It can be seen that the ROC curve of the model trained by the Center Loss algorithm with momentum is closer to the upper left of the coordinate system than another, and accordingly, the performance of this model will be better.

In order to quantitatively evaluate the performance of the models, we evaluate them by cutting the TPR values corresponding to the points at FPR = 0.01, FPR = 0.001 and FPR = 0 from the ROC curves. The comparison results of the models in this experiment with other well-known models are shown in Table 1.

Table 1. The comparison of the verification abilities of different deep models on LFW.

Model	TPR@FPR = 0.01	TPR@FPR = 0.001	TPR@FPR = 0
DeepFace [7]	94.73%	91.34%	89.44%
DeepID2 [14]	99.03%	98.14%	96.23%
FaceNet [8]	99.11%	98.47%	96.54%
Center Loss [10]	99.07%	98.03%	96.47%
Our method	**99.27%**	**98.73%**	**96.77%**

When FPR = 0, the TPR value can best reflect the verification abilities of the models, and the actual experience of users will be improved as the value increases. From the table it can be seen that the TPR0 value of the model trained by the Center Loss algorithm with momentum has reached 96.77%, which is significantly higher than the Center Loss algorithm without momentum and higher than other mainstream algorithms. This shows that the new Center Loss can really improve the verification abilities of the models.

Comparing with the original Center Loss algorithm, the improved algorithm has better performance in ROC results, which illustrates its validity clearly. The original Center Loss algorithm does not take into account the unstable process in the beginning of training and does not set a buffer mechanism for the feature centers near the optimal location, which affects the training process of the deep model negatively. In contrast, the improved Center Loss algorithm solves all these problems, so it can gain a better training result. Of course, comparing with other mainstream metric learning algorithms, the improved Center Loss algorithm still has no need to select the sample pairs from the training set like its original version, so its applications to complex face recognition tasks will be still easier to be realized.

5 Conclusions

The Center Loss algorithm eliminates the ambiguous sampling process of the traditional algorithm and uses a way that is similar to the data clustering to train the deep models. However, when the training samples have a large number of categories, this algorithm can hardly gain better performance since its updating process of the feature centers is very slow. This paper improves the Center Loss algorithm by introducing a damped momentum with threshold to speed up the updating of its centers. The experiment results show that the ROC curve of Center Loss with momentum is much better than the original algorithm, which means that the subjective experience of users will improve a lot in practical applications.

Acknowledgments. This work was funded by State's Key Project of Research and Development Plan (2016YFC0901303).

References

1. Lecun, Y., Bengio, Y., Hinton, G.E.: Deep learning. Nature **521**, 436–444 (2015)
2. Lecun, Y., Boser, B., Denker, J.: Handwritten digit recognition with a back-propagation network. Adv. Neural. Inf. Process. Syst. **2**, 396–404 (1997)
3. Krizhevsky, A., Sutskever, I., Hinton, G.E.: ImageNet classification with deep convolutional neural networks. In: International Conference on Neural Information Processing Systems, pp. 1097–1105. Curran Associates Inc. (2012)
4. Szegedy, C., Ioffe, S., Vanhoucke, V.: Inception-v4, Inception-ResNet and the Impact of Residual Connections on Learning (2016)
5. Szegedy, C., Liu, W., Jia, Y.: Going deeper with convolutions. In: IEEE Conference on Computer Vision and Pattern Recognition (CVPR), pp. 1–9 (2015)
6. He, K., Zhang, X., Ren, S.: Deep residual learning for image recognition. In: Computer Vision and Pattern Recognition, pp. 770–778. IEEE (2016)
7. Taigman, Y., Yang, M., Ranzato, M.: DeepFace: closing the gap to human-level performance in face verification. In: Conference on Computer Vision and Pattern Recognition, pp. 1701–1708 (2014)
8. Schroff, F., Kalenichenko, D., Philbin, J.: FaceNet: a unified embedding for face recognition and clustering. In: IEEE Conference on Computer Vision and Pattern Recognition (2015)
9. Song, H., Xiang, Y., Jegelka, S.: Deep metric learning via lifted structured feature embedding. Computer Science, pp. 4004–4012 (2015)
10. Wen, Y., Zhang, K., Li, Z.: A discriminative feature learning approach for deep face recognition. In: European Conference on Computer Vision, vol. 47, pp. 499–515. Springer, Cham (2016)
11. Xing, E., Ng, A., Jordan, M.: Distance metric learning, with application to clustering with side-information. Adv. Neural Inf. Process. Syst. **15**, 505–512 (2003)
12. Ruder, S.: An overview of gradient descent optimization algorithms (2016)
13. Guo, Y., Zhang, L., Hu, Y.: MS-Celeb-1M: challenge of recognizing one million celebrities in the real world. Electron. Imaging (2016)
14. Sun, Y., Chen, Y., Wang, X.: Deep learning face representation by joint identification-verification. In: Proceedings of Advances in Neural Information Processing Systems, vol. 27, pp. 1988–1996 (2014)

FCN and Unit-Linking PCNN Based Image Saliency Detection

Lecheng Zhou and Xiaodong Gu(✉)

Department of Electronic Engineering, Fudan University,
200433 Shanghai, China
xdgu@fudan.edu.cn

Abstract. Detecting salient regions of an image can significantly increase the efficiency of follow-up processing, and thus improve the performance of the whole system. In this paper, we proposed a novel model of image saliency detection, which combines pulse coupled neural network (PCNN) with a fully convolutional neural network (FCN). In our proposed model, an image is firstly fed into a unit-linking PCNN, and the segmentation result, providing topological properties of the objects, serves as an input channel of the FCN. Guided by the topological features, the deep neural network then provides a coarse saliency map. Finally, we use the segmentation result to refine the boundaries of the salient objects to generate a fine saliency map. Furthermore, in this model various techniques are introduced to the PCNN to refine the segmentation result, preserving structural integrity between different objects. Experimental results on several benchmarks show that our model outperforms other state-of-the-art approaches without retuning on different datasets.

Keywords: Image saliency detection · Fully convolutional neural network · Pulse coupled neural network

1 Introduction

Image saliency detection aims at discovering the most visually attractive region of an image automatically and accurately. With the rapid increase of image data to be processed, separating objects from the background can avoid the processing and storage of redundant information, thus has a wide range of applications such as object recognition, image compression, semantic segmentation, and so on. Since L. Itti proposed the visual attention model [1] mimicking the perspective of human perception, many studies [2–12] have worked on building computational models for salient object detection. These models present a saliency map showing the pixel-accurate saliency values for each input image, which is the aim of this paper. The saliency values represent the probability of being salient objects.

Inspired by the mechanism of human perception, many works focus on detecting image saliency through various priors, such as contrast priors [10], background priors [11], center priors [12] and so on. However, these priors may cause mistakes due to the inconformity in certain images. For example, it may be difficult to detect a salient

D. Liu et al. (Eds.): ICONIP 2017, Part III, LNCS 10636, pp. 357–366, 2017.
https://doi.org/10.1007/978-3-319-70090-8_37

object similar to its surrounding with contrast priors. Moreover, center priors may not work well with salient objects deviating from the image center.

To overcome the existing problems, we proposed a learning-based model for image saliency detection with raw images as input and saliency maps as output. There are two main components in the proposed model: a fully convolutional neural network (FCN) and a pulse coupled neural network (PCNN). These two different neural networks have their respective advantages in image processing. FCN has a powerful and adaptive function of extracting various features from images. Through adequate training, it can provide an optimal mapping between images and their saliency. PCNN can efficiently segment images without training, extracting the topological integrity of objects in the images. Such topological features can well preserve boundaries and details of salient objects. In our proposed model, PCNN segmentation results are used as an additional input channel of the FCN, and after that they are fused with the deep saliency maps produced by the FCN to get fine saliency maps.

The remainder of this paper is organized as followed: Sect. 2 presents a detail description of our proposed approach. The experimental results and corresponding analysis are given in Sect. 3. Finally, Sect. 4 gives a brief summary of this paper.

2 Proposed Approach

2.1 Overview

The architecture of our proposed image saliency detection approach is illustrated in Fig. 1. An image is first sent to a unit-linking PCNN, and the topological integrity of objects are extracted by highlighting the intensity of objects. The segmentation results, together with original input images, are then fed to the FCN to produce a deep saliency map. That means, the FCN in our model has four input channels. Finally, we perform a fusion of the deep saliency map and the segmentation result to refine the saliency maps, preserving clear boundaries and details of objects.

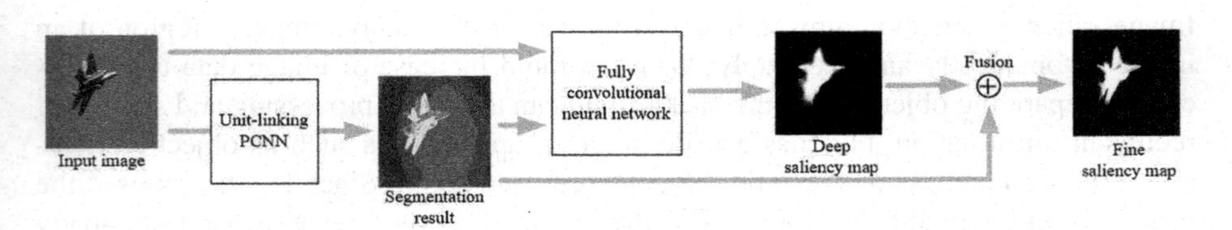

Fig. 1. The whole architecture of the proposed saliency detection model. The segmentation result created by unit-linking PCNN is fed into FCN, and as well used to refine the deep saliency map. Note that the segmentation result and deep saliency map both have the same size as the input image.

2.2 Image Segmentation with Unit-Linking PCNN

As a critical part of saliency detection model, whether the objects can be accurately segmented greatly influences the quality of the final saliency map. Unit-linking PCNN has proved to have impressive performance on image processing [13], and can be easily applied without any training. Besides, a series of techniques have been adapted to achieve better segmentation results from the unit-linking PCNN.

Unit-Linking PCNN. Pulse Coupled Neural Network (PCNN) is developed from Eckhorn's linking field model [14], and has a wide application in image processing due to its spatio-temporal linking feature. However, PCNN is too complex to compute and it is difficult to set all the network parameters. Thus a simplified model, named unit-linking PCNN [13], has been introduced in many researches in order to improve the efficiency in application.

The structure of a single neuron j in unit-linking PCNN is shown in Fig. 2. In image segmentation tasks, each pixel in the input grayscale image is considered as a neuron. The neuron is linked with its neighbor neurons through the L channel, and once any one of its neighbor neurons is fired, $L_j = 1$. $F_j = I_j$ represents the intensity of the pixel. After F and L channels are united as shown in Fig. 2, if U_j is greater than the threshold θ_j, the neuron will output a pulse. The neurons with higher intensity may fire earlier than those with lower intensity. Neuron sparks will pass to neighbor neurons, making them fire one by one until the sparks reach object boundaries, which may have obvious intensity difference. The segmentation result of an image can be observed through the firing image of its corresponding unit-linking PCNN.

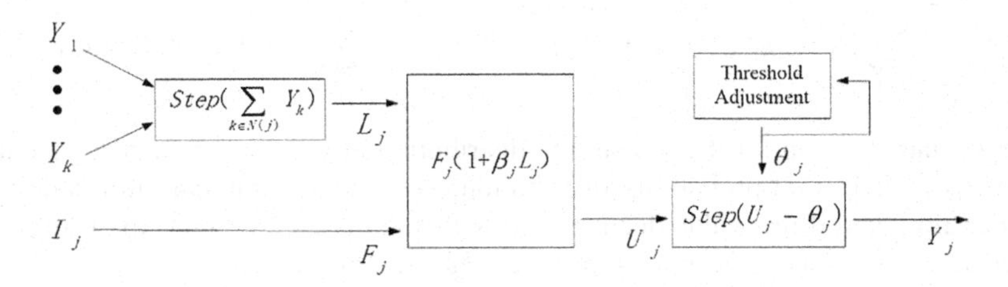

Fig. 2. Structure of a single neuron j in unit-linking PCNN [14]

Initial Threshold Computation. A problem in using unit-linking PCNN is that the initial threshold significantly affects the final result. If an inappropriate initial threshold is set, the object integrity may be destroyed after the segmentation. Besides, many images consist of various kinds of objects, and single initial threshold may not distinguish well between them, as shown in Fig. 3(b) and (c).

Different from the conventional segmentation approaches, we use multi threshold segmentation based on the gray histogram of the image. Through the intensity distribution shown in the histogram, the troughs represent the intensity boundaries of different objects. We smooth the histogram, and find the gray values corresponding to these troughs, selecting them as the initial thresholds. For each initial threshold, we do image segmentation separately, and then combine all the results together. After normalization, we will get the multi level segmentation results, which preserve more structural features (different objects can be discriminated from the background) as shown in Fig. 3(f).

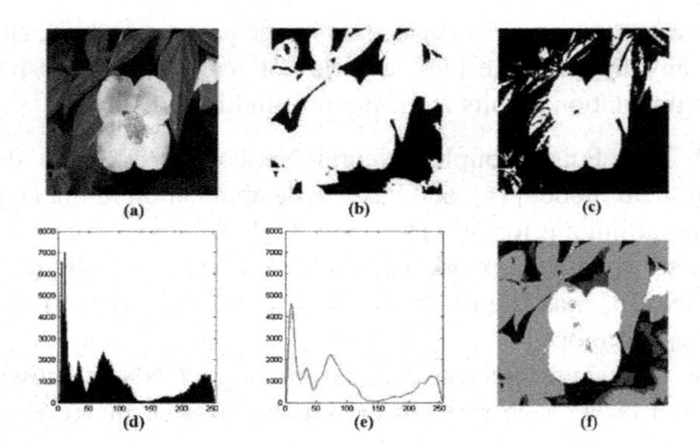

Fig. 3. Comparison between segmentation results using different initial threshold.

Iteration Times Determination. The firing image of PCNN changes periodically, so it is important to determine whether the unit-linking PCNN has obtained the best segmentation result and stop iteration. Quantitative indicators such as information entropy [15] has been introduced to determine the best iteration times. In our approach, we use mutual information (MI) [16] to measure the structural similarity between segmentation result and input image. The mutual information of two images can be calculated as follow:

$$MI = -\sum_{x \in I_1} P_1(x) lg P_1(x) - \sum_{x \in I_2} P_2(y) lg P_2(y) + \sum_{x,y \in I_1, I_2} P_{12}(x,y) lg P_{12}(x,y), \quad (1)$$

where P_1 and P_2 stand for the probability distribution of input image and segmentation result, P_{12} is their joint probability distribution. We calculate mutual information after each iteration, and if this value declines, we will stop iteration and use the firing image created in the previous iteration as the final segmentation result.

2.3 FCN with Topological Input Channel

Fully convolutional neural network (FCN) [17] was first designed for image semantic segmentation. A significant advantage of FCN is that it produces pixel-accurate prediction results, with the same size as input images. The original FCN has 16 convolutional layer followed by a deconvolution layer, and uses softmax loss function to produce a multi-classification results. As our target is saliency detection, the last convolutional layer is modified to make the network produce a saliency map with single dimension. A sigmoid function is used to normalize the output map to [0, 1], and the network ends up with a logistic (cross entropy) loss function. Moreover, the first convolutional layer is modified to suit the four input channels (RGB channels and topological channel). The additional channel provides topological integrity features to FCN, which may let the FCN focus more on objects-oriental features extraction, improving the quality of saliency maps. During training phase, images and corresponding segmentation results are fed into the FCN synchronously, and the stochastic gradient decent (SGD) algorithm is launched to minimize the loss of the network.

2.4 Saliency Map Refinement

Though FCN can produce relatively accurate saliency detection results, the object boundaries and details are usually fuzzy, which is caused by the large stride taken in the deconvolution layer. Since the PCNN segmentation results well preserve the boundaries and details of objects, we can reuse them to refine the saliency maps. In our approach, some pre-processing is done to the segmentation results before fusion. The deep saliency maps produced by FCN is slightly dilated, serving as a mask for the segmentation results. The gray level in the masked segmentation results is adjusted, and a hole filter is applied to fill the possible vacancies in the objects. The fusion of the deep saliency map (*DSM*) and segmentation results (*SR*) can be described as follow:

$$FSM = DSM \circ (SR^{\alpha} + \beta), \tag{2}$$

where α and β are the parameters controlling the fusion ratio, and $\circ$ represents elementwise product operator. In the experiment, we set $\alpha = 0.8$ and $\beta = 0.1$. The fine saliency maps (*FSM*) are then normalized to [0, 255], which become the final output of our proposed model. As illustrated in Fig. 1, the boundaries and details of the fine saliency map are much clearer, bringing about higher accuracy for saliency detection.

3 Experimental Results

3.1 Experiment Settings and Evaluation Metrics

Experiment Datasets. To evaluate our proposed approach, we conduct a series of experiments on five benchmark datasets: MSRA-10k [18], SED2 [19], ECSSD [12] and PASCAL-S [20]. For MSRA-10K, we randomly pick 1000 images from each of them. For three other datasets, which contain no more than 1000 images, we use all of their images to perform our test.

Implementation Details. The FCN used in our model is implemented on Caffe [21], and trained with 5000 images from MSRA-10k dataset (no overlap between training set and testing set). We resize each input image and its ground truth to 500 × 500, and do randomly sorting and flipping before training. Part of the net parameters are initialized with the FCN model used in semantic segmentation [17]. The training is conducted on a NVIDIA Tesla K20, with 100000 iterations in total.

Evaluation Metrics. We use precision and recall curve (P-R curve), F-measure and mean absolute error (MAE) in the experiments, to quantitatively evaluate the performance of saliency detection approaches. The P-R curve can be plotted by using different binary thresholds ranging from 0 to 255 and calculating corresponding precision and recall. It reflects a model's performance of discovering salient object. The F-measure combines precision and recall to provide a quantitative result as follow:

$$F_{\alpha} = \frac{(1+\alpha^2)Precision \times Recall}{\alpha^2 \times Precision + Recall}, \tag{3}$$

where α^2 is set to 0.3 in our experiment. We calculate maximum F-measure (F_{max}) from the P-R curve and average F-measure (F_{avg}) using the adaptive binary threshold (twice the average value of the saliency map). Finally, MAE refers to average pixel-wise error between the saliency map and ground truth.

3.2 Comparison with State-of-the-Art Approaches

Our approach is compared with 10 state-of-the-art methods in the experiment including DRFI [9], DSR [7], RBD [11], GMR [6], MC [8], GC [22], RC [2], HC [2], FT [5] and

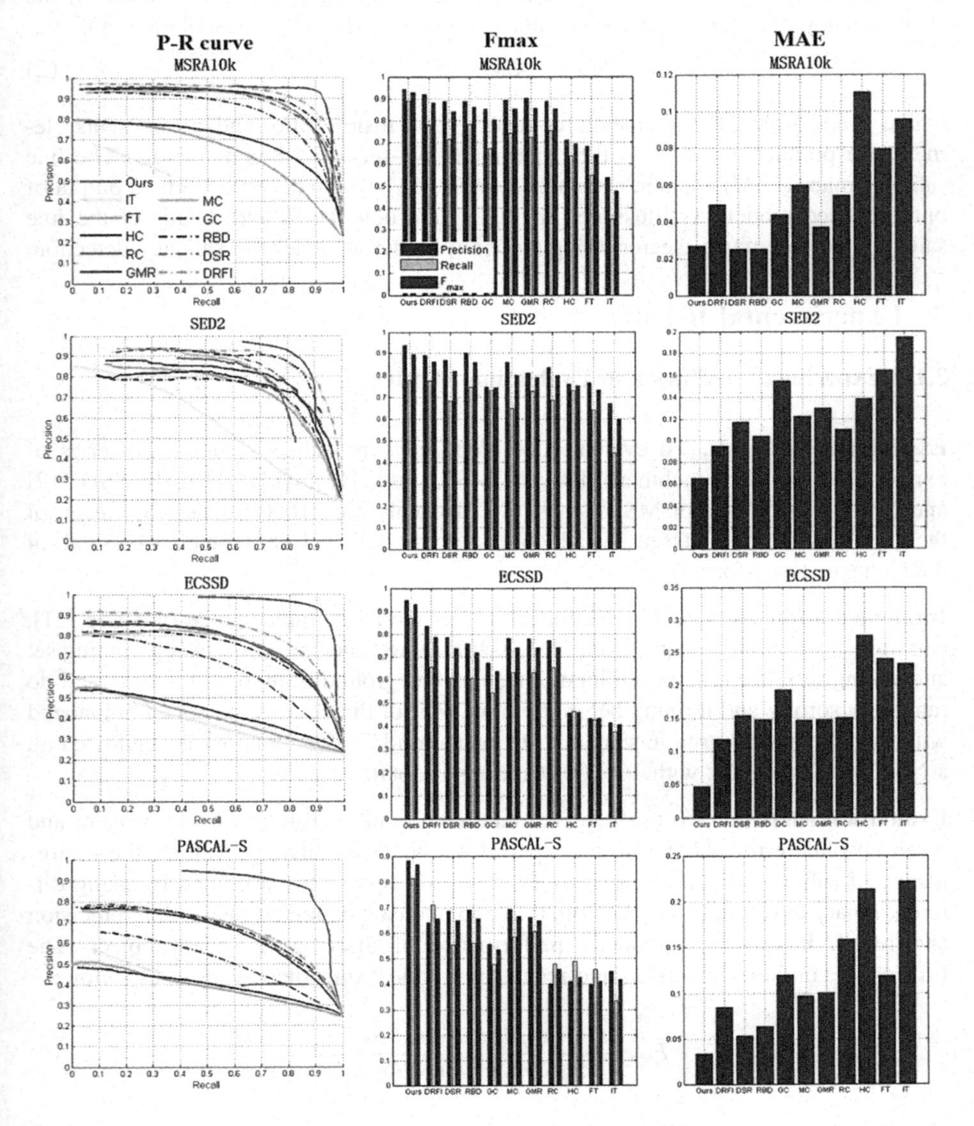

Fig. 4. Experimental results on four datasets. For P-R curves, red lines represent our approach, and for F measure and MAE chart, the results of our approach are shown in the leftmost bars.

IT [1]. The results of these approaches are generated by their open source code [23] and all the saliency maps are normalized to [0, 255].

The experimental results (P-R curve, F_{max}, MAE) of above-mentioned approaches and our approach on different datasets are displayed in Fig. 4. It can be seen clearly that in most cases, our approach have larger areas under P-R curves, larger F-measure and lower MAE, which indicates the superiority and efficiency of our approach. More specifically, quantitative results on maximum F-measure, average F-measure and MAE are shown in Table 2, and our approach achieves the best performances in these four datasets.

3.3 Contribution of Segmentation Results

To prove the contribution of PCNN image segmentation, we compare the results produced by our entire saliency detection model with a simple FCN without the topological channel and output refinement. The comparison results are demonstrated in Table 1. By comparing their performance, we can clearly see the contribution brought by the part of PCNN image segmentation.

Table 1. Comparison of the experimental results between model with PCNN image segmentation and without PCNN image segmentation on four datasets.

		F_{max}	F_{avg}	MAE
MSRA10k	With seg	**0.9293**	**0.8987**	**0.0268**
	Without seg	0.9007	0.8776	0.0290
SED2	With seg	**0.8948**	**0.8725**	**0.0646**
	Without seg	0.8753	0.8580	0.0657
ECSSD	With seg	**0.9294**	**0.9015**	**0.0474**
	Without seg	0.8923	0.8675	0.0581
PASCAL-S	With seg	**0.8685**	**0.8307**	**0.0339**
	Without seg	0.8175	0.7643	0.0483

Table 2. Quantitative comparison between our proposed model with other state-of-the-art approaches on four datasets, using three evaluation metrics (maximum F-measure, average F-measure and MAE).

		Ours	DRFI	DSR	RBD	GC
MSRA10k	F_{max}	**0.9293**	0.8825	0.8398	0.8612	0.8021
	F_{avg}	**0.8987**	0.7644	0.7767	0.7762	0.7398
	MAE	0.0268	0.0493	**0.0251**	0.0252	0.0436
		MC	GMR	RC	HC	FT
	F_{max}	0.8528	0.8546	0.8523	0.6908	0.6433
	F_{avg}	0.7434	0.7768	0.7699	0.6118	0.4307
	MAE	0.0600	0.0371	0.0544	0.1103	0.0796

(continued)

Table 2. (*continued*)

		Ours	DRFI	DSR	RBD	GC
SED2		Ours	DRFI	DSR	RBD	GC
	F_{max}	**0.8948**	0.8597	0.8173	0.8583	0.7433
	F_{avg}	**0.8725**	0.7759	0.7495	0.8040	0.6829
	MAE	**0.0646**	0.0945	0.1171	0.1039	0.1545
		MC	GMR	RC	HC	FT
	F_{max}	0.7986	0.7883	0.7949	0.7508	0.7317
	F_{avg}	0.6958	0.7324	0.7467	0.6626	0.6086
	MAE	0.1223	0.1298	0.1099	0.1384	0.1639
ECSSD		Ours	DRFI	DSR	RBD	GC
	F_{max}	**0.9294**	0.7860	0.7353	0.7163	0.6367
	F_{avg}	**0.9015**	0.6672	0.6453	0.6219	0.5756
	MAE	**0.0474**	0.1185	0.1545	0.1478	0.1928
		MC	GMR	RC	HC	FT
	F_{max}	0.7393	0.7375	0.7381	0.4564	0.4299
	F_{avg}	0.6104	0.6423	0.6771	0.3978	0.2842
	MAE	0.1466	0.1481	0.1506	0.2756	0.2395
PASCAL-S		Ours	DRFI	DSR	RBD	GC
	F_{max}	**0.8685**	0.6560	0.6505	0.6566	0.5370
	F_{avg}	**0.8307**	0.5942	0.5852	0.5974	0.4727
	MAE	**0.0339**	0.0848	0.0531	0.0635	0.1195
		MC	GMR	RC	HC	FT
	F_{max}	0.6657	0.6476	0.4589	0.4256	0.4112
	F_{avg}	0.6037	0.5897	0.3617	0.3618	0.3370
	MAE	0.0972	0.1002	0.1587	0.2133	0.1191

4 Conclusion

In this paper, we have proposed an effective image saliency detection model based on fully convolutional neural network, which combines the image segmentation results created by a unit-linking pulse coupled neural network. The segmentation results serve as the topological features for the FCN. Moreover, the output saliency maps can be refined by the segmentation results to have clearer boundaries and details. Experimental results on five benchmark datasets have proved that our approach performs better than other state-of-the-art methods in several evaluation metrics.

Acknowledgment. This work was supported in part by National Natural Science Foundation of China under grants 61371148 and 61771145.

References

1. Itti, L., Koch, C., Niebur, E.: A model of saliency-based visual attention for rapid scene analysis. IEEE Trans. Patt. Anal. Mach. Intell. **20**(11), 1254–1259 (1998)
2. Cheng, M.M., Zhang, G.X., Mitra, N.J., Huang, X., Hu, S.M.: Global contrast based salient region detection. In: Proceedings of the IEEE Conference on Computer Vision and Pattern Recognition, vol. 37, no. 3, pp. 409–416 (2011)
3. Li, X., Zhao, L., Wei, L., Yang, M.H.: DeepSaliency: multi-task deep neural network model for salient object detection. IEEE Trans. Image Process. **25**(8), 3919–3930 (2016)
4. Chen, T., Lin, L., Liu, L., Luo, X., Li, X.: DISC: deep image saliency computing via progressive representation learning. IEEE Trans. Neural Netw. Learn. Syst. **27**(6), 1135–1149 (2016)
5. Achanta, R., Hemami, S., Estrada, F., Susstrunk, S.: Frequency-tuned salient region detection. In: Proceedings of the IEEE Conference on Computer Vision and Pattern Recognition, vol. 22, no. 9–10, pp. 1597–1604 (2009)
6. Yang, C., Zhang, L., Lu, H., Ruan, X., Yang, M.H.: Saliency detection via graph-based manifold ranking. In: Proceedings of the IEEE Conference on Computer Vision and Pattern Recognition, vol. 9, no. 4, pp. 3166–3173 (2013)
7. Li, X., Lu, H., Zhang, L., Xiang, R., Yang, M.H.: Saliency detection via dense and sparse reconstruction. In: Proceedings of the IEEE Conference on Computer Vision and Pattern Recognition, pp. 2976–2983 (2013)
8. Jiang, B., Zhang, L., Lu, H., Yang, C., Yang, M.H.: Saliency detection via absorbing Markov chain. In: Proceedings of the 2013 IEEE International Conference on Computer Vision, pp. 1665–1672 (2013)
9. Jiang, H., Wang, J., Yuan, Z., Wu, Y., Zheng, N., Li, S.: Salient object detection: a discriminative regional feature integration approach. In: Proceedings of the IEEE Conference on Computer Vision and Pattern Recognition, pp. 2083–2090 (2013)
10. Klein, D.A., Frintrop, S.: Center-surround divergence of feature statistics for salient object detection. In: Proceedings of the 2011 IEEE International Conference on Computer Vision, vol. 50, no. 2, pp. 2214–2219 (2012)
11. Zhu, W., Liang, S., Wei, Y., Sun, J.: Saliency optimization from robust background detection. In: Proceedings of the IEEE Conference on Computer Vision and Pattern Recognition, vol. 1049–1050, no. 8, pp. 2814–2821 (2014)
12. Yan, Q., Xu, L., Shi, J., Jia, J.: Hierarchical saliency detection. In: Proceedings of the IEEE Conference on Computer Vision and Pattern Recognition, vol. 38, no. 4, pp. 1155–1162 (2013)
13. Gu, X.: Feature extraction using unit-linking pulse coupled neural network and its applications. Neural Process. Lett. **27**(1), 25–41 (2008)
14. Eckhorn, R., Reitboeck, H.J., Arndt, M., Dicke, P.: Feature linking via synchronization among distributed assemblies: simulations of results from cat visual cortex. Neural Comput. **2**(3), 293–307 (2014)
15. Gu, X., Guo, S., Yu, D.: A new approach for automated image segmentation based on unit-linking PCNN. Mach. Learn. Cybern. **1**, 175–178 (2002)
16. Wei, W., Li, Z.: Automated image segmentation based on modified PCNN and mutual information entropy. Comput. Eng. **36**(13), 199 (2010)
17. Long, J., Shelhamer, E., Darrell, T.: Fully convolutional networks for semantic segmentation. In: Proceedings of the IEEE Conference on Computer Vision and Pattern Recognition, vol. 79, no. 10, pp. 3431–3440 (2015)

18. Liu, T., Sun, J., Zheng, N., Tang, X., Shum, H.: Learning to detect a salient object. In: Proceedings of the IEEE Conference on Computer Vision and Pattern Recognition, vol. 33, no. 2, pp. 1–8 (2007)
19. Alpert, S., Galun, M., Basri, R., Brandt, A.: Image segmentation by probabilistic bottom-up aggregation and cue integration. In: Proceedings of the IEEE Conference on Computer Vision and Pattern Recognition, vol. 34, no. 2, pp. 1–8 (2007)
20. Li, Y., Hou, X., Koch, C., Rehg, J.M., Yuille, A.L.: The secrets of salient object segmentation. In: Proceedings of the IEEE Conference on Computer Vision and Pattern Recognition, pp. 280–287 (2014)
21. Jia, Y., Shelhamer, E., Donahue, J., et al.: Caffe: convolutional architecture for fast feature embedding. In: 22nd ACM International Conference on Multimedia, Orlando, FL, pp. 675–678 (2014)
22. Cheng, M.M., Warrell, J., Lin, W.Y., Zheng, S., Vineet, V., Crook, N.: Efficient salient region detection with soft image abstraction. In: Proceedings of the IEEE International Conference on Computer Vision, pp. 1529–1536 (2013)
23. Borji, A., Cheng, M.M., Jiang, H., Li, L.: Salient object detection: a benchmark. In: Proceedings of the IEEE Conference on Computer Vision and Pattern Recognition, pp. 5706–5722 (2015)

Disparity Estimation Using Convolutional Neural Networks with Multi-scale Correlation

Samer Jammal[1,2(✉)], Tammam Tillo[1,2], and Jimin Xiao[1,2]

[1] X'ian Jiaotong Liverpool University, Suzhou, China
{samer.jammal,jimin.xiao}@xjtlu.edu.cn
[2] University of Bozen-Bolzano, Bolzano, Italy
tammam.tillo@unibz.it

Abstract. Disparity estimation is a long-standing task in computer vision and multiple approaches have been proposed to solve this problem. A recent work based on convolutional neural networks, which uses a correlation layer to perform the matching process, has achieved state-of-the-art results for the disparity estimation task. This correlation layer employs a single kernel unit which is not suitable for low texture content and repeated patterns. In this paper we tackle this problem by using a multi-scale correlation layer with several correlation kernels and different scales. The major target is to integrate the information of the local matching process by combining the benefits of using both a small correlating scale for fine details and bigger scales for larger areas. Furthermore, we investigate the training approach using horizontally elongated patches that fits the disparity estimation task. The results obtained demonstrate the benefits of the proposed approach on both synthetic and real images.

Keywords: Disparity estimation · Convolutional neural networks · Multi-scale correlation · Stereo vision · Depth estimation

1 Introduction

Disparity estimation from left and right views of a particular scene has been an active research field for decades. Several approaches have been introduced to improve accuracy and reduce the computational cost [1]. The importance of a dense disparity map is apparent in the fields of autonomous driving, robotics, 3D modeling, 3D Television, Free Viewpoint Television (FTV) and Multi-View Coding (MVC).

Typical stereo-based methods for disparity estimation fall into two categories: local and global methods. Local methods [2] tend to be applied in real-time applications where computational efficiency is valued over accuracy. On the other hand, global approaches [3] focus on disparity estimation accuracy more than the computation efficiency. Between these two categories some methods provide a trade-off between accuracy and speed such as SGM algorithm [4]. In fact, for practical applications both disparity estimation accuracy and computational efficiency are important factors.

D. Liu et al. (Eds.): ICONIP 2017, Part III, LNCS 10636, pp. 367–376, 2017.
https://doi.org/10.1007/978-3-319-70090-8_38

Recently, deep learning has been successfully applied to many tasks in the computer vision area. New convolutional neural network architectures have been designed to provide per-pixel depth estimation from a single image [5] and disparity prediction from a pair of stereo images [6,8,9]. In [6], a correlation layer with single kernel that performs the matching process between the left and right CNN feature maps is employed. However, a single 1×1 kernel unit, as used in [6], is not suitable for large objects or repeated patterns, because a small number of pixels in the neighborhood involved is used in the matching process. Our method integrates the state-of-the-art disparity estimation network from [6] with a multi-scale correlation layer to improve the accuracy and solve the matching problem for large regions with low texture. We also propose to train the network using horizontally elongated images instead of the whole image, which eventually reduces training time and increases prediction accuracy.

The remainder of this paper is organized as follows. Related work is reviewed in Sect. 2 while, the proposed CNN with multi-scale correlation is described in Sect. 3. We present and discuss the training method and our results in Sect. 4. Finally, we present our conclusions in Sect. 5.

2 Related Work

Convolutional neural networks have been investigated as a possible solution to the stereo correspondence problem. Zagoruyko and Komodakis [7] exploited CNN architectures to learn directly from image patches a general similarity function. In particular, a two channels siamese models were used for stereo matching as specific case of image matching. Zbontar and LeCun [8,10] used a convolutional neural network to determine corresponding patches in left and right traditional stereo images. These reconstruction methods include semi-global block matching for the estimation of the depth of each pixel with a series of post-processing steps, which renders the model computationally intensive. To reduce the computation time [9] exploited a matching network with a product layer that computes the inner product between the two representations of a siamese architectures. However, the stereo matching between patches come up with incorrect correspondences due to various reasons such as occlusion, saturation, pixel intensity noise. As a solution to this problem, Seki and Pollefeys [11] exploited a two channels CNN to predict the correspondence confidence between disparity patches. Followed by a fusion method that employ the confidence.

In contrast, Mayer et al. [6] designed an end-to-end fully convolutional neural network, processing full-resolution of the left and right stereo images to predict a disparity map without any post-processing. This CNN includes a contracting part that progressively decreases the spatial size of the convolutional features, providing large receptive fields for higher-level convolutional layers, which in turn, enables the network to capture more global information. However, the pooling in the contracting part reduces the resolution of the predicted map. In an attempt to help the network in the matching process, a correlation layer with single kernel unit that performs the matching between the left and right feature maps is employed.

In order to improve the quality of the estimated disparity map, we proposed to use a convolutional neural network employing multi-scale correlation to learn the mapping between a pair of stereo images and its corresponding disparity map. It jointly exploits the strength of the highly efficient CNN and the benefits of performing the matching process for convolutional features with multi-scale correlation kernels. The main contributions of this paper are: First, a novel multi-scale correlation layer for a fully convolutional neural network architecture for disparity map prediction is proposed, the network takes a pair of RGB stereos images as its input, and outputs a single image representing the real disparity map without any pre or post-processing of the input images. Second, an efficient method for training the network based on horizontally elongated images is introduced. Moreover, an evaluation of the proposed model is provided.

3 Proposed Work

In this section a CNN with Multi-scale correlation for disparity estimation without pre and post-processing steps is proposed.

3.1 CNN Architecture

In our work, we introduce several modifications to both the network architecture and the training process compared to [6]. The network architecture is shown in Fig. 1. First, we propose to perform the matching process between the left and right convolutional features using multi-scale correlation layer. Second, a new strategy for training the network using horizontally elongated images is implemented.

3.2 Multi-Scale Correlation

In this section the concept of the multi-scale correlation layer is introduced. We start with the case of single-scale correlation unit for convolutional features. Although the single-scale correlation unit with a 1×1 kernel size is suitable for

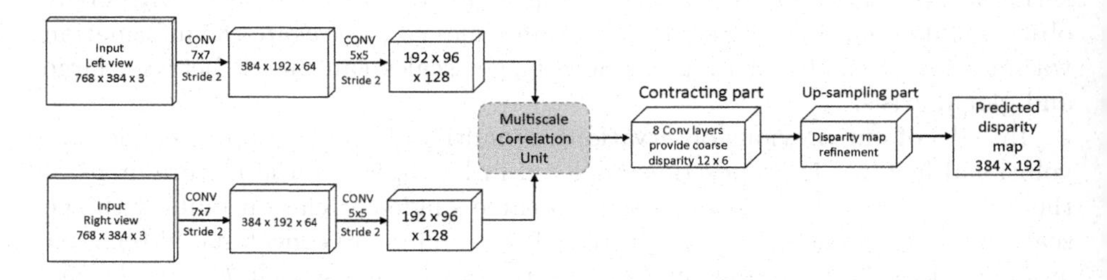

Fig. 1. The CNN architecture with the proposed multi-scale correlation layer. This layer perform the matching process between convolutional features using different correlation scales. The dashed block is presented in detail in Fig. 3.

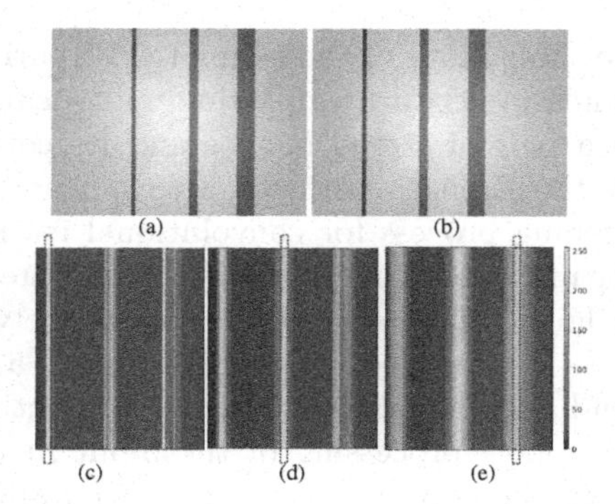

Fig. 2. The matching process using three single-scale correlation kernels with different sizes. In (a) and (b) the left and right views are illustrated. The figures (c), (d) and (e) show the matching results in the feature domain using the scales 1×1, 1×3 and 1×7.

finding the correspondence between small objects and details, it has limitations for big objects with uniform areas, and objects with low texture and/or repeated patterns. Indeed this problem can be minimized by employing a correlation unit with sizable scale, which performs the matching process between a set of pixels in the neighborhood and not just a single pixel in the case of small kernel.

To illustrate this point, a pair of stereo synthetic images generated using the software package Blender are shown Fig. 2(a). The scene consists of three vertical dark objects with different sizes (small, medium and large). The two images are input into two identical and separate CNN streams to generate features. Then, three correlation units with different scales $1 \times 1, 1 \times 3$ and 1×7 are applied to the obtained features. By comparing the three figures Fig. 2(c), (d) and (e) it is possible to observe that the maximum output for the small object is obtained in Fig. 2(c) which correspond to the output of the smallest correlation unit. Whereas, for the medium object, the largest correlation value is obtained using the scale 1×3. Similarly, the large object is best matched using a bigger scale, i.e., 1×7, as shown in Fig. 2(e). From this example we find that using a multi-scale is suitable for objects and regions with various sizes, whereas a single-scale kernel will produce sub-optimal results especially for a complex scene with many objects of different sizes. In summary, small scales can aggregate the information within a set of small local regions, and bigger scales can match between large and global areas.

In light of our previous observations, a multi-scale correlation layer for convolutional features is proposed as depicted in Fig. 3, where the feature maps of the left and right streams are passed to three single-correlation units with the scales $1 \times 1, 1 \times 3$ and 1×7. The output of these units are concatenated together and passed to the next convolutional layer. Let us suppose that F_L and F_R are the left and right feature maps of the second convolutional layer, respectively, $H \times W$ is the size of the feature and K is the total number of features for each branch. In addition let $N \times M$ be the size of the correlating block.

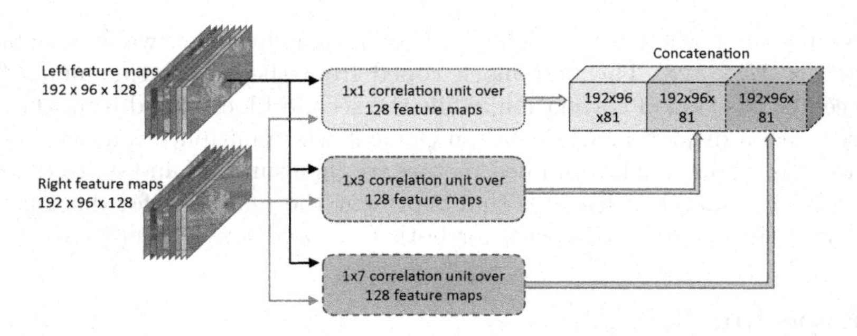

Fig. 3. The multi-scale correlation layer contains three kernels with different scales 1×1, 1×3, 1×7. The obtained features are concatenated together then passed again to the CNN.

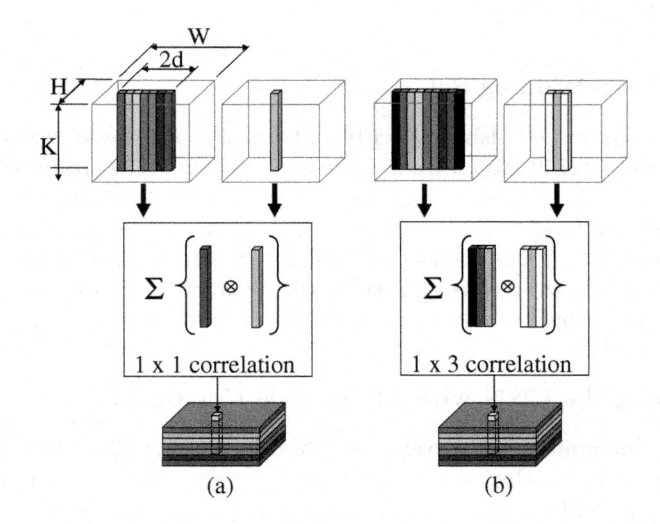

Fig. 4. The matching process between the left and right convolutional features. Due to the limited space only the correlation scale (a) 1×1 and (b) 1×3 are shown. Where black vectors indicate zero padding and $\otimes$ is the element-wise multiplication.

For parallel camera alignment, the most common setting for stereo cameras, the difference between the left and right views is mainly due to the objects' horizontal translation. As a result, the correlation process only covers the horizontal displacement. The correlation process between the left and right features will produce maps with the same size. The correlation map for d pixel displacement is:

$$F_m(x,y) = \sum_{n=1}^{K}\sum_{j=0}^{N}\sum_{i=0}^{M} Fl_n(x+i, y+j) \times Fr_n(x+i+d, y+j) \qquad (1)$$

where Fl_n and Fr_n represent channel n of the left and right feature, respectively, and $N \times M$ is the correlation kernel size. In our case, $N = 1$. The matching process in a single-scale correlation unit is performed between two feature blocks of size $1 \times M \times 128$. The first block coped from the left feature maps for a given center position of correlation, while the second block coped from the right features with a displacement d, and an element-wise multiplication between the two blocks is followed. The obtained vectors are then summed and set to the same position in the matched features that represent the matching for displacement d. In Fig. 4 the correlation process for both 1×1 and 1×3 is depicted.

4 Experiments

The model and architecture described in the previous section was trained on a large synthetic training set, and evaluated on four data-sets in comparison to the state-of-the-art work [6].

4.1 Synthetic Training Set

The network is trained using the FlyingThings3D synthetic stereo set. This dataset contains 22740 stereo frames for training, and 4760 frames for testing. Each frame contains between 5–20 everyday objects flying in the scene. The parameters of the objects (size, type, positions, texture, rotation) are randomly sampled, which makes this dataset suitable for training a deep CNN. An end-to-end learning approach is chosen to train the network to extract stereo features and predict the disparity map from the left and right views.

4.2 Training the CNN with Multi-scale Correlation

Following [6] the multi-scale correlation network contains 21 convolutional layers and a multi-scale correlation layer with three units. It takes a pair of stereo images as input, and outputs a single image representing the disparity map. Our network is trained on the FlyingThings3D training set. Initially, the training was performed using patches of size $768 \times 384 \times 3$ which were cropped from the original image of size $960 \times 540 \times 3$ from the FlyingThings3D training set, and a batch size of 4 per iteration.

During the experiments, we found that the training loss exhibits a considerable amount of noise as a result of the small batch size employed in the training process. This noise can be reduced by using a larger batch size. Increasing the batch size increases the demand on memory and hence increasing the training time. Training the network with a batch size of 8 images takes 20 days to reach the $660k$ iterations using one Titan X GPU.

In contrast, we propose the slicing of each image into four horizontally elongated patches, each with a size of $768 \times 128 \times 3$ to speed up the training process. The result is 91,040 pairs of stereo images for training the network and leads to use of a batch size of 12 per epoch or less. The memory requirement for training the network with both bach size of 12 with sliced images is the same as that for a batch size of 4 for the complete image. On the other hand, even thought the memory sizes used for training per iteration in both cases are the same, the use of sliced images provide more diversity and information than using one complete image, as the complete image may contains redundancy and repeated regions. Figure 5 shows The training loss using complete and sliced images, we observed that the training loss for sliced images is smoother and smaller.

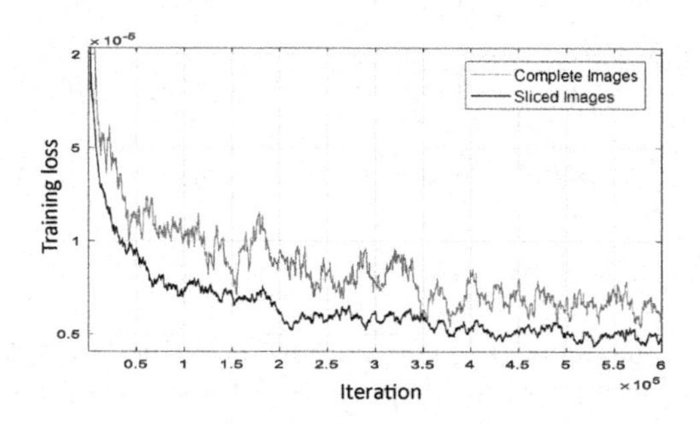

Fig. 5. Training loss analysis of multi-scale network with complete and sliced images on FlyingThings3D.

4.3 Results

The aim in this section is to test the performance of the disparity estimation using a single-scale correlation layer and the impact of using different scales in comparison with the multi-scale correlation unit. The different scales are 1×1, 1×3 and 1×7, where 1×1 is used in [6]. For each model, one training experiment is conducted, with each taking 10 days to train using one Titan X GPU. All of the models are trained on the same FlyingThings3D training set in order to provide a fair comparison, and the evaluation is performed on different datasets including a FlyingThings3D clean test, Driving clean, Monkaa clean [6], Sintel clean train, KITTI train and Middlebury. All the models are trained for 460k iterations. The feed forward run time in [6] is 0.06 s, whereas using the proposed model it is slightly increased and becomes 0.08 s.

Table 1 reports the mean absolute error (MAE) for each network on all testing sets, which is one of the standard error measures used for Middlebury stereo evaluation, except for the KITTI where we follow the error measurement method provided by the KITTI. For the purpose of ensuring a fair comparison between the different models, the evaluation is performed on models of the last five iterations on each training set, then the average is calculated. From Table 1 we noticed that the convolutional network with multi-scale correlation outperforms the other methods with single-scale kernels on KITTI 2012 Train, FlyingThings3D Clean, Monkaa Calean, Sintel Clean and Middlebury, which indicates the importance of using multi-scale correlation.

Meanwhile, we also notice that all the models have difficulties with large displacements as we can see in Table 1 on Middlebury 2014, for which the disparity

Table 1. Disparity map errors.

Method	KITTI 2012 Train	Driving clean	FlyingThings3D clean test	Monkaa clean	Sintel clean	Middlebury 2014
Multi-scale (Proposed)	**36.06**	**14.73**	**4.54**	**25.23**	**8.70**	**41.29**
Single-scale 1×1 [6]	40.64	15.42	4.59	25.24	9.82	44.71
Single-scale 1×3	36.76	14.75	4.66	25.26	10.32	46.33
Single-scale 1×7	40.63	18.84	4.98	26.60	10.85	42.09

reaches 600 pixels. This is because the network is trained on the FlyingThings3D training set which contains small disparities (of less than 160 pixels). The network learns to estimate disparities of around 160 pixels or less, so it is unable to effectively estimate large disparities. Nevertheless, for Middlebury 2014, the gain of the proposed method is large in comparison to other methods. The proposed Multi-scale correlation CNN shows better performance for flat regions with uniform areas, as shown in Fig. 6.

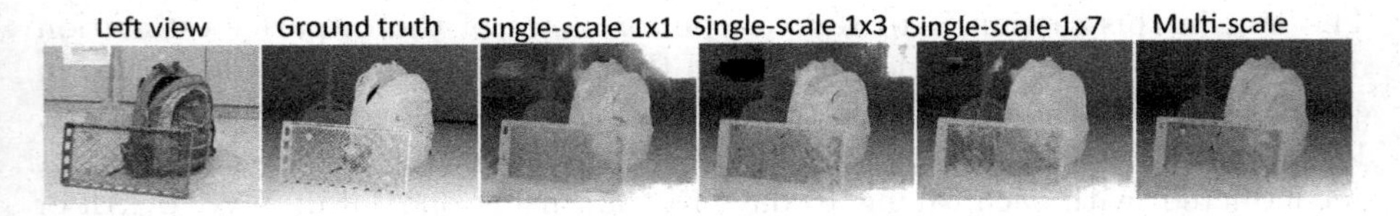

Fig. 6. Examples of disparity estimation for subjective comparison between multi-scale correlation and single scales on Middlebury 2014 dataset.

Figure 7 illustrates that proposed method learned to efficiently combine the information from different scales. In the first row (the image with chair) it could be observed that the proposed network uses the large scale to accurately estimate the disparity for the chair and from the small scales keep the edges. On the other hand, the network in [6] fails in estimating the disparity of the chairs back.

In fact each model with a different single correlation filter shows specific characteristics. The model using a 1×1 kernel size provides generally good performance with small objects and very rich texture, while using the 1×7 scale achieves a more accurate results for large objects and uniform areas. On the other hand, 1×3 shows good performance for objects with less complex texture and average size. The main advantage of this method is the ability to produce smooth predictions of the depth map as well as on texture-less regions.

In Table 2, we compare the performance of CNN with multi-scale correlation using new training method with horizontally spliced images with that use the

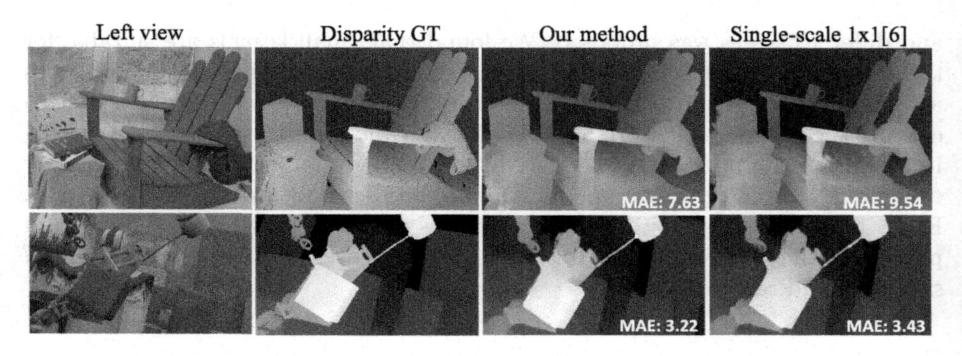

Fig. 7. Examples of disparity estimation using multi-scale correlation. The models trained on FlyingThings3D for 460k iterations. Rows from top to down: Middlebury 2014, FlyingThings3D (clean).

whole image for training as described in Sect.4. We can see that, in general, network trained with sliced images improves the performance as well as the training time.

Table 2. Disparity errors comparison for two types of training; namely: horizontally elongated patches and complete image.

Training multi-scale	KITTI 2012 Train	Driving clean	FlyingThings3D clean test	Monkaa clean	Sintel clean	Middlebury 2014
Sliced images	**36.06**	**15.18**	4.54	**25.23**	8.70	**41.29**
Complete images	36.30	17.99	**4.43**	25.66	**7.01**	45.98

Finally it is worth reporting that a multi-scale correlation layer with kernels sizes: 2, 4, 8, 16, 32 could have been chosen to form a base for different objects sizes, but practically odd-size kernels are more convenient, since destination feature maps are mapped directly onto the source neighborhood centers. Thus, two multi-scale networks were tested, namely: $\{1, 3, 7\}$ and $\{1, 3, 7, 15, 31\}$. It has been found that for the resolution we tested bigger, i.e. 15 and 31, kernels have very limited contribution, nevertheless, we believe that for images with larger resolution large kernels will have some major contribution. For the limited space we include the results only for the first network.

5 Conclusion

In this work a disparity estimation method based on convolutional neural network with a multi-scale correlation layer employing several correlation kernels and different scales was proposed. We found that small kernels are suitable for the disparity estimation of small objects with fine details while larger scales are suitable for larger objects with uniform areas. The proposed model is able to capture the key stereo features of the two views and generate an accurate disparity map. Furthermore, we investigated the training approach using horizontally elongated patches to speed up the training process and improve the accuracy. Experimental results showed the effectiveness of the proposed approach in comparison to the single kernel correlation, which achieves accurate results on both synthetic and real stereo images.

References

1. Scharstein, D., Szeliski, R.: A taxonomy and evaluation of dense two-frame stereo correspondence algorithms. Int. J. Comput. Vis. **47**, 7–42 (2002)
2. De-Maeztu, L., Mattoccia, S., Villanueva, A., Cabeza, R.: Linear stereo matching. In: Proceedings of the 2011 International Conference on Computer Vision, Barcelona, pp. 1708–1715 (2011)
3. Sun, J., Zheng, N.-N., Shum, H.-Y.: Stereo matching using belief propagation. In: IEEE Transactions on Pattern Analysis and Machine Intelligence, TPAMI, vol. 25, no. 7, pp. 787–800 (2003)
4. Hirschmuller, H.: Stereo processing by semiglobal matching and mutual information. IEEE Trans. Patt. Anal. Mach. Intell. **30**(2), 328–341 (2008)
5. Laina, I., Rupprecht, C., Belagiannis, V., Tombari, F., Navab, N.: Deeper depth prediction with fully convolutional residual networks. CoRR abs/1606.00373 (2016)
6. Mayer, N., Ilg, E., Hausser, P., Fischer, P., Cremers, D., Dosovitskiy, A., Brox, T.: A large dataset to train convolutional networks for disparity, optical flow, and scene flow estimation. In: IEEE International Conference on Computer Vision and Pattern Recognition, pp. 4040–4048 (2016)
7. Zagoruyko, S., Komodakis, N.: Learning to compare image patches via convolutional neural networks. In: Proceedings of the IEEE Conference on Computer Vision and Pattern Recognition, pp. 4353–4361 (2015)
8. Zbontar, J., LeCun, Y.: Computing the stereo matching cost with a convolutional neural network. In: Proceedings of the IEEE Conference on Computer Vision and Pattern Recognition, pp. 1592–1599 (2014)
9. Luo, W., Schwing, A.G., Urtasun, R.: Efficient deep learning for stereo matching. In: IEEE International Conference on Computer Vision and Pattern Recognition, pp. 5695–5703 (2016)
10. Zbontar, J., LeCun, Y.: Stereo matching by training a convolutional neural network to compare image patches. J. Mach. Learn. Res. **17**, 4 (2016)
11. Seki, A., Pollefeys, M.: Patch based confidence prediction for dense disparity map. In: Proceedings of the British Machine Vision Conference, BMVC, p. 23 (2016)

A Spatio-Temporal Convolutional Neural Network for Skeletal Action Recognition

Lizhang Hu and Jinhua Xu(✉)

Department of Computer Science and Technology,
Shanghai Key Laboratory of Multidimensional Information Processing, East China Normal University, 3663 North Zhongshan Road, Shanghai, China
51151201077@stu.ecnu.edu.cn, jhxu@cs.ecnu.edu.cn

Abstract. Human action recognition based on 3D skeleton data is a rapidly growing research area in computer vision. Convolutional Neural Networks (CNNs) have been proved to be the most effective representation learning in many vision tasks, but there is little work of CNNs for skeletal action recognition due to the variable-length of time sequences and lack of big skeleton datasets. In this paper, we propose a Spatio-Temporal CNN for skeleton based action recognition. A CNN architecture with two convolutional layers is used, in which the first layer is used to capture the spatial patterns and second layer for spatio-temporal patterns. Some techniques including data augmentation and segment pooling strategy are employed for long sequences. Experimental results on MSR Action3D, MSR DailyActivity3D and UT-Kinect show that our approach achieves comparable results with those of the state-of-the-art models.

Keywords: Convolutional neural networks · Skeletal action recognition · Deep learning · Action recognition

1 Introduction

Action recognition is an important and challenging research area in computer vision. It has a wide range of applications, such as video surveillance, human-computer interaction, health care and so on. Traditional action recognition studies are mainly based on video sequences of 2D frames with RGB channels [1], in which depth information is lost. Recently, with the emergence of the low-cost RGB-D sensors such as Microsoft Kinect [2], the information based on depth has been applied to action recognition [3,4]. Since the real-time skeleton estimation algorithms was proposed in [5], skeleton based action recognition has attracted a lot of attention due to their robustness to variations of viewpoint, human body scale and motion speed.

Traditional methods usually use hand-crafted features extracted from the skeleton sequences, such as histograms of 3D joint locations (HOJ3D) [6], Eigen-Joints [7], Moving Pose [8], sequence of the most informative joints (SMIJ) [9] and

D. Liu et al. (Eds.): ICONIP 2017, Part III, LNCS 10636, pp. 377–385, 2017.
https://doi.org/10.1007/978-3-319-70090-8_39

covariance descriptor [10, 11]. In recent years, automatic representation learning has been used for 3D skeletal based action recognition, and most work is recurrent neural network (RNN) based work. For example, in [12], Du *et al.* proposed an end-to-end hierarchical RNN for skeleton based action recognition. A Co-Occurrence long short-term memory (LSTM) was proposed in [13], which can select discriminative joints automatically. Although CNNs have been applied for RGB video sequences and gained some success, few work has used CNNs for skeletal action recognition. To our knowledge, the only work is [14], which proposed a Moving Poselet model for skeletal action recognition. The model has one convolutional layer and one pooling layer.

In this paper, we propose a Spatio-Temporal CNN for skeleton based action recognition. It has two convolutional layers. In the first layer, we use convolutional filters with kernel size one to extract spatial dependencies between the joints, and then we fed the spatial features to second temporal convolutional layer. In this way, more discriminative spatio-temporal patterns can be captured. Besides, for long time series dataset, such as DailyActivity3D [15], we use segment max pooling instead of global max pooling to maintain more temporal information. And we also use a data augmentation technique to increase the data for small size datasets. Experiments on three benchmark datasets demonstrate the effectiveness of our approach and the model achieves the state-of-the-art performance compared to those of recent models.

The rest of the paper is organized as follows. In Sect. 2 we introduce the details of the proposed method. Section 3 show the experimental results on three datasets and do some analysis of our methods. Finally we conclude the paper in Sect. 4.

2 The Proposed Spatio-Temporal CNN

The architecture of the proposed network is shown in Fig. 1. Different from the 2D convolutional filters of traditional CNNs for an image input, the convolutional layers of our model are one dimensional (1D) temporal convolution, whose channel number is equivalent to the dimension of input data, and kernel size represents the length of time slot. The kernel size of the filters in the first layer is one, therefore their inputs are joint data at one frame. In this way, the filter output can be regarded as the spatial descriptor of a frame. The second layer is temporal convolutional layer over spatial descriptors to obtain the spatio-temporal representation, and the kernel size is greater than one. Since different sequences have different length, the output size of the second layer for different sequences are not same. Therefore we use the next max pooling layer to obtain the fixed-length global representation for each sequence, which will be fed to a fully connected layer followed by a softmax layer for classification.

2.1 Part Based Spatio-Temporal CNN

Inspired by [12, 14, 16], we group the skeletal data into five parts, torso, left/right arm, left/right leg respectively. For the kth part, we use a vector $\mathbf{x}_t^k \in \mathbf{R}^{3P^k}$ to

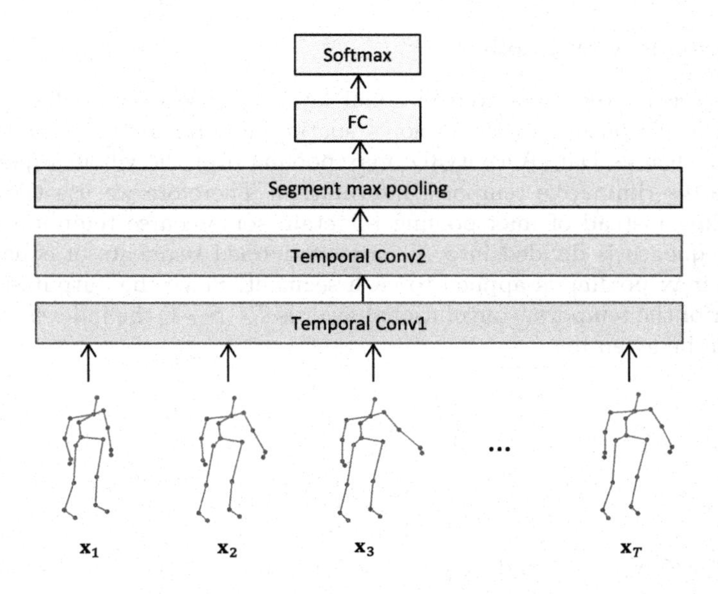

Fig. 1. The architecture of the proposed model.

represent the joints data at frame $t \in [1, \cdots, T]$, where P^k is the number of the joints for the part, and T is the number of frames. In the first layer, the kernel size is 1, and its output can be computed as:

$$h_{j,t}^k = \mathbf{w}_j^{1,k^T} \mathbf{x}_t^k + b_j^{1,k}, \quad j = 1, \cdots, J^k \tag{1}$$

where $h_{j,t}^k$ is the output of the jth filter at the tth frame, and $\mathbf{w}_j^{1,k}$ and $b_j^{1,k}$ are the weights and bias of the jth filter respectively. J^k is the number of filters for the kth part.

By stacking all parts' outputs, we obtain the output for the tth frame:

$$\mathbf{h}_t = [h_{1,t}^1, \cdots, h_{J^1,t}^1, \cdots, h_{1,t}^5, \cdots, h_{J^5,t}^5] \in \mathbf{R}^{J_1},$$

here $J_1 = \sum_k J^k$. The output of the second convolutional layer can be computed as:

$$o_{j,i} = \sum_{a=1}^{S} \mathbf{w}_{a,j}^{2\ T} \mathbf{h}_{a+i-1} + b_j^2, \quad j = 1, \cdots, J_2 \tag{2}$$

where $\mathbf{w}_{a,j}^2$ and b_j^2 are the weights and bias of the jth filter respectively, and $1 \leq i \leq T - S + 1$, S is the filter size of the second layer.

We also test a full body model in the experiments. The full body model can be regarded as a special case of the body parts model described above, in which there is only one part consisting of all joints.

2.2 Segment Max Pooling

After the second temporal convolutional layer, we use max pooling to generate fixed-length output as the action's spatio-temporal feature. For long time sequences, such as DailyActivity3D, max pooling over the whole sequence may lose some discriminative temporal information. Therefore we use the segment max pooling instead of max pooling to retain some coarse temporal information. A sequence is divided into N_s non-overlapped segments of equal length, and then max pooling is applied to each segment. Since the output size of the each filter of the temporal convolutional layer is $T-S+1$, the indexes of pooling region can be given by:

$$t^i_{start} = \left\lfloor \frac{T-S+1}{N_s} \cdot (i-1) \right\rfloor + 1 \tag{3}$$

$$t^i_{end} = \left\lceil \frac{T-S+1}{N_s} \cdot i \right\rceil \tag{4}$$

where $1 \leq i \leq N_s$, t^i_{start} and t^i_{end} are the start and end index for the ith pooling segment respectively.

Let $\mathbf{o}_j = [o_{j,1}, o_{j,2}, ..., o_{j,T-S+1}]$ be the output of the temporal convolutional layer, the jth output of the nth pooling segment is:

$$q^n_j = \max_{t^n_{start} \leq i < t^n_{end}} o_{j,i}, \tag{5}$$

3 Experiments

In this section, we conducted extensive experiments on three datasets, MSR Action3D dataset [17], MSR Daily Activity3D dataset [15] and UT-Kinect dataset [6], to demonstrate the effectiveness of our models.

3.1 Datasets and Settings

MSR Action3D Dataset. The MSR Action3D dataset are extracted from RGB-D videos by Kinect. It includes 567 action sequences of 20 action types performed by 10 subjects. There are 20 joints in each skeleton. The sequences are relatively short, with 30–50 frames. This dataset has two standard evaluation protocols [14,15]. In protocol 1, all sequences from subjects 1, 3, 5, 7 and 9 are used for training and the remaining are for testing. In protocol 2, the dataset is divided into three subsets, AS1, AS2 and AS3. For each subset, the training and testing sets are split in the same way as protocol 1. Besides, 10 skeleton sequences were not used in our experiment because of missing data or highly erroneous joint positions [15].

MSR DailyActivity3D Dataset. The MSR DailyActivity3D dataset contains 16 daily activity actions and has 320 sequences performed by 10 subjects. The sequences are longer, with 100–300 frames. The body is also represented by 20 joints. Each subject performed an action twice: once seated on a sofa and once standing, and it contains human-object interactions. Therefore it is more challenging than MSR Action3D. We also used the protocol in [15]. Due to the complexity of the actions in MSR DailyActivity3D dataset, we used the following techniques on this dataset. First, we normalized the skeleton data according to Algorithm 1 described in [8] since the positions are very noisy when it was performed near the sofa. Second, inspired by the data augment techniques for image recognition, we cropped each sample sequence into 7 sub-sequences because the dataset has longer sequences and less samples. Each subsequence accounted for 3/5 of the original sequence length (L). For the ith subsequence, the start index is $t_s^i = round((i-1)/6*2L/5)+1$, end index is $t_e^i = t_s^i + 3L/5 - 1$. Third, we used segment max pooling rather than global max pooling to maintain some coarse temporal information for the long sequences. Fourth, a ReLU layer was added between the max pooling layer and the fully connected layer.

UT-Kinect Dataset. The UT-Kinect dataset was captured by a stationary Kinect. It contains 200 sequences of 10 action classes performed by 10 subjects, and each action performed 2 times by each subject. The skeleton data also contains 3D positions of 20 joints. We used the experimental protocol in [18], in which half of the subjects were used for training and the remaining for testing.

Implementation Details In our experiments, we used Torch [19] toolbox as the deep learning platform. We used variable-length sequences as input by setting the batch size as one. The number of filters were set as $J_1 = 100$ (in full body model), $J^k = 100$(in body parts model), $J_2 = 200$ for all datasets. The temporal convolution filter size was set as $S = 6$ for MSR Action3D and MSR DailyActivity3D, and $S = 5$ for UT-Kinect because the minimum sequence length is 5 in this dataset. We set the segment max pooling's parameter $N_s = 3$ for MSR DailyActivity3D. We trained the network using stochastic gradient descent, and set learning rate, learning rate decay, momentum and weight decay as 1×10^{-2}, 1×10^{-4}, 1×10^{-3}, and 1×10^{-4} respectively. These parameters are same for all three datasets.

3.2 Results and Comparisons to the State-of-the-Art Models

We show the experimental results of our proposed model and comparisons with other methods in Tables 1, 2, 3 and 4 for three datasets. Our part-based model capturing spatio-temporal patterns with two convolutional layers is called as Part ST-CNN, and the full body model as FB ST-CNN. Compared FB ST-CNN with Part ST-CNN, we found that part-based model has slight improvement on most datasets except for MSR DailyActivity3D. From the tables it can be seen that our models outperform the Moving Poselet model [14] on MSR Action3D

Table 1. Experimental results and comparison on MSR Action3D Dataset (protocol 1).

Method	Accuracy(%)
Actionlet Ensemble [15]	88.20
Lie Group [20]	89.48
Pose Base Approach [16]	90.22
Moving Pose [8]	91.70
Moving Poselets [14]	**93.60**
FB ST-CNN	93.41
Part ST-CNN	93.41

Table 2. Experimental results and comparison on MSR Action3D Dataset (protocol 2).

Method	AS1	AS2	AS3	Avg
Bag of 3D Points [17]	72.90	71.90	79.20	74.70
HOD [21]	92.39	90.18	91.43	91.26
Lie Group [20]	95.29	83.87	98.22	92.46
Moving Poselets [14]	89.81	93.57	97.03	93.50
HURNN-L [12]	92.38	93.75	94.59	93.57
HBRNN-L [12]	93.33	94.64	95.50	94.49
FB ST-CNN	90.48	94.64	97.30	94.14
Part ST-CNN	90.48	96.43	97.30	**95.04**

Table 3. Experimental results and comparison on MSR DailyActivity3D dataset.

Method	Accuracy(%)
Only Joint Features [15]	68.0
Actionlet Ensemble [15]	74.0
Moving Pose [8]	73.8
Moving Poselets [14]	74.5
FB ST-CNN	**75.6**
Part ST-CNN	75.0

(protocol 2), MSR DailyActivity3D, but slightly worse than it on MSR Action3D (protocol 1), 93.41% vs. 93.60%. It should be pointed out that Moving Poselet model was not tested on UT-Kinect Dataset and we did not re-implement it, therefore no comparison was made with it on this dataset. The HBRNN model was only tested on MSR Action3D (protocol 2) in [12], and our Part ST-CNN is slightly better than it, 95.04% vs. 94.49%.

Table 4. Experimental results and comparison on UT-Kinect dataset.

Method	Accuracy(%)
Skeleton Joint Features [18]	87.9
Lie Group [20]	93.6
Elastic functional coding [22]	94.9
FB ST-CNN	98.0
Part ST-CNN	**100.0**

3.3 Effectiveness of the Proposed Model

In order to validate the effectiveness of our two-convolutional-layer designs, we conducted some contrast experiments with one temporal convolutional layer with kernel size greater than 1, denoted as T-CNN. And the Part T-CNN is similar to [14] except two differences. The first is that we used softmax layer for classification rather than SVM. The second is that we used 5 body parts rather than the 10 parts in [14]. We used 3 data modes as input, joints' position, velocity, position plus velocity. The experimental results for the three datasets are shown in Table 5.

Table 5. Results of the extensive experiments with different combination configurations on three datasets. And P represents the position, V represents the velocity.

Method	MSR Action3D (protocol 1)	MSR Action3D (protocol 2)	MSR DailyActivity3D	UT-Kinect
FB T-CNN P	85.35	87.77	73.75	91.92
FB T-CNN V	91.58	94.44	55.00	94.95
FB T-CNN P+V	90.11	92.89	73.13	95.96
Part T-CNN P	86.45	90.82	71.25	95.96
Part T-CNN V	89.38	92.63	54.38	94.95
Part T-CNN P+V	89.01	92.02	71.88	97.98
FB ST-CNN P	90.48	93.84	**75.63**	94.95
FB ST-CNN V	89.38	92.89	58.75	93.94
FB ST-CNN P+V	**93.41**	94.14	74.38	97.98
Part ST-CNN P	90.84	91.40	75.00	**100.00**
Part ST-CNN V	90.48	**95.04**	59.38	94.95
Part ST-CNN P+V	**93.41**	92.91	73.78	98.99

From the table, we have the following conclusions. First, for position data input, the FB(Part) ST-CNN with two convolutional layers are more effective than FB(Part) T-CNN with only one temporal convolutional layer. Second, for the velocity data, Part ST-CNN is better than Part T-CNN, but FB ST-CNN is worse than FB T-CNN on Action3D Dataset and UT-Kinect. Velocity is more suitable for FB T-CNN may be because it has lost some space information through the frame subtraction. Third, the combination of position and velocity or body part strategy can overcome the velocity's weakness in FB ST-CNN and has a slight accuracy improvement.

To evaluate the effect of segment max pooling layer for long sequences data, we did experiments with different N_s on MSR DailyActivity3D dataset, and all others parameters were kept unchanged. When $N_s = 1$, it is global max pooling. Through the experiments, we found that the accuracies of segment max pooling ($1 < N_s \leq 5$) were similar, about 75%, which are significantly better than that of $N_s = 1$ (about 70%). When $N_s > 5$, the accuracies degraded. Besides, data augmentation was not used for MSR Action3D Dataset because the sequences are short.

4 Conclusion

In this paper, we proposed a Spatio-Temporal CNN architecture for skeletal action recognition. Through the two-convolutional-layer architecture, discriminative spatio-temporal features can be learned. With the segment max pooling strategy, our model can also be applied to long sequence recognition. In the experiments on three datasets, it can be seen that our results were comparable or better than the state-of-the-art methods. In future work, we will explore more effective spatio-temporal representation learning, such as the combination of CNNs and long short-term memory (LSTM).

References

1. Aggarwal, J.K., Ryoo, M.S.: Human activity analysis: a review. ACM Comput. Surv. **43**(3), 1–43 (2011)
2. Han, J.: Enhanced computer vision with microsoft kinect sensor: a review. IEEE Trans. Cybern. **43**(5), 1318–1334 (2013)
3. Chen, L., Wei, H., Ferryman, J.M.: A survey of human motion analysis using depth imagery. Pattern Recogn. Lett. **34**, 1995–2006 (2013)
4. Ye, M., Zhang, Q., Wang, L., Zhu, J., Yang, R., Gall, J.: A survey on human motion analysis from depth data. In: Grzegorzek, M., Theobalt, C., Koch, R., Kolb, A. (eds.) Time-of-Flight and Depth Imaging. Sensors, Algorithms, and Applications. LNCS, vol. 8200, pp. 149–187. Springer, Heidelberg (2013). doi:10.1007/978-3-642-44964-2_8
5. Shotton, J., Fitzgibbon, A., Cook, M., Sharp, T., Finocchio, M., Moore, R., Kipman, A., Blake, A.: Real-time human pose recognition in parts from single depth images. IEEE Conf. Comput. Vis. Pattern Recogn. **411**(1), 1297–1304 (2011)
6. Xia, L., Chen, C.C., Aggarwal, J.K.: View invariant human action recognition using histograms of 3D joints. In: IEEE Computer Society Conference on Computer Vision and Pattern Recognition Workshops (CVPRW), pp. 20–27 (2012)

7. Yang, X., Tian, Y.L.: Effective 3D action recognition using eigenjoints. J. Vis. Commun. Image Represent. **25**(1), 2–11 (2014)
8. Zanfir, M., Leordeanu, M., Sminchisescu, C.: The moving pose: an efficient 3D kinematics descriptor for low-latency action recognition and detection. In: IEEE International Conference on Computer Vision, pp. 2752–2759 (2013)
9. Ofli, F., Chaudhry, R., Kurillo, G., Vidal, R., Bajcsy, R.: Sequence of the most informative joints (smij): a new representation for human skeletal action recognition. J. Vis. Commun. Image Represent. **25**(1), 24–38 (2014)
10. Hussein, M.E., Torki, M., Gowayyed, M.A., El-Saban, M.: Human action recognition using a temporal hierarchy of covariance descriptors on 3D joint locations. In: The 23rd International Joint Conference on Artificial Intelligence (2013)
11. Sivalingam, R., Somasundaram, G., Bhatawadekar, V., Morellas, V., Papanikolopoulos, N.: Sparse representation of point trajectories for action classification. In: IEEE International Conference on Robotics and Automation (ICRA), pp. 3601–3606 (2012)
12. Du, Y., Wang, W., Wang, L.: Hierarchical recurrent neural network for skeleton based action recognition. In: IEEE Conference on Computer Vision and Pattern Recognition, pp. 1110–1118 (2015)
13. Zhu, W., Lan, C., Xing, J., Zeng, W., Li, Y., Shen, L., Xie, X.: Co-occurrence feature learning for skeleton based action recognition using regularized deep LSTM networks. In: Thirtieth AAAI Conference on Artificial Intelligence (2016)
14. Tao, L., Vidal, R.: Moving poselets: a discriminative and interpretable skeletal motion representation for action recognition. In: IEEE Conference on Computer Vision Workshop (2015)
15. Wang, J., Liu, Z., Wu, Y., Yuan, J.: Mining actionlet ensemble for action recognition with depth cameras. In: IEEE Conference on Computer Vision and Pattern Recognition, pp. 1290–1297 (2012)
16. Wang, C., Wang, Y., Yuille, A.: An approach to pose-based action recognition. In: IEEE Conference on Computer Vision and Pattern Recognition (2013)
17. Li, W., Zhang, Z., Liu, Z.: Action recognition based on a bag of 3D points. In: Workshop on Human Activity Understanding from 3D Data, pp. 9–14 (2010)
18. Zhu, Y., Chen, W., Guo, G.: Fusing spatiotemporal features and joints for 3D action recognition. In: IEEE Conference on Computer Vision and Pattern Recognition Workshops, pp. 486–491 (2013)
19. Collobert, R., Kavukcuoglu, K., Farabet, C.: Torch7: a matlab-like environment for machine learning. In: BigLearn, NIPS Workshop (2011)
20. Vemulapalli, R., Arrate, F., Chellappa, R.: Human action recognition by representing 3D skeletons as points in a lie group. In: IEEE Conference on Computer Vision and Pattern Recognition (2014)
21. Gowayyed, M.A., Torki, M., Hussein, M.E., El-Saban, M.: Histogram of oriented displacements (HOD): describing trajectories of human joints for action recognition. In: International Joint Conference on Artificial Intelligence, pp. 1351–1357 (2013)
22. Anirudh, R., Turaga, P., Su, J., Srivastava, A.: Elastic functional coding of human actions: from vector-fields to latent variables. In: IEEE Conference on Computer Vision and Pattern Recognition, pp. 3147–3155 (2015)

Autonomous Perceptual Projection Correction Technique of Deep Heterogeneous Surface

Fan Yang[1,2], Baoxing Bai[1,2], Cheng Han[1,2](✉), Chao Zhang[1,2], and Yuying Du[1,2]

[1] School of Computer Science and Technology, Changchun University of Science and Technology, Changchun 130022, China
hancheng@cust.edu.cn

[2] Special Film Technology and Equipment National Local Joint Engineering Research Center, Changchun 130022, China

Abstract. This paper proposes a projection correction method which to improve the adaptive perception projection of the projection equipment in different environments. Firstly, in the process of photon signal transmission, projector-camera can cause the loss of photon signal due to the coupling of system channel. Therefore, this paper proposes a system coupling correction scheme, which effectively reduces the system coupling crosstalk. Secondly, in order to establish the feature mapping relationship between the projection image and the deep heterogeneous surface quickly, a projection feature image of color structured light mesh fringe is designed. Finally, due to the feature point of the heterogeneous surface is quite different in topological structure, it will lead to the problem of inconsistent geometric mapping relation. For this reason, a projective geometric correction algorithm for topological analysis is proposed, analyzing the spatial topological distribution of the depth heterogeneous surface and solving the homography matrix of each region in the heterogeneous surface, then the geometric correction of the projected distortion image is solved by using the homography matrix set. From the experimental analysis we can see that, in the deep heterogeneous surface environment, the average error, the maximum error and the root-mean-square error of the correction image respectively are 0.424 pixels, 0.862 pixels and 0.216 pixels. At the same time, the parallelism of the distortion correction image is kept 90° substantially. It can be seen that the geometric distortion correction accuracy of this method has reached the sub-pixel level and the imaging screen consistency level.

Keywords: Computer vision · Irregular surfaces · Geometric correction · Color structured light · Depth perception

1 Introduction

With the development of virtual reality and visualization technology, the traditional display device is unable to produce a strong visual impact to the user, so the projection display device has become an important component to improve the immersive virtual reality [1]. The current projection display technology is widely used in digital cinema, exhibition, military simulation and other fields to visual image display [2, 3]. In the

D. Liu et al. (Eds.): ICONIP 2017, Part III, LNCS 10636, pp. 386–396, 2017.
https://doi.org/10.1007/978-3-319-70090-8_40

study of geometric distortion correction of the projection screen, the papers [4, 5] perceive the corresponding relationship between projector display wall and the projective image are based on the multi-images structured light mode, which involve excessive sampling points, high time complexity and other issues. In the [6, 7], correcting the distortion projective image by using the way of projecting checkerboard pattern, that is the pixel mapping between the projector image and the camera image is established to correcting projective geometry by optimizing the detection of chessboard corners, this approach avoids the high temporal complexity of multi-frame structured light modes, but the corner of the checkerboard is susceptible to the occlusion problem of the projected curved surface. The [8] studies the flexible projection-surface correction by constructing the depth camera-projector system, it uses the depth camera to perceive the depth information of the flexible surface key points, and combined with surface fitting method to correcting the projection. However, this approach is limited by the low resolution of the depth camera.

In this work, we present autonomous perceptual projection correction method for Deep heterogeneous surface. Firstly, we propose to model and analyze the coupling of the system, which effectively reduces the system coupling crosstalk. Second, we designed a projection feature image of color structured light mesh fringe, that it to establish the feature mapping relationship between the projection image and the depth anisotropic surface. Finally, we propose a projective geometric correction algorithm for topological analysis, which achieve human visual consistency display.

2 Modeling and Analysis of Channel Coupling

In the intelligence projection system, a color image is projected by a projector, and a camera is used for image acquisition, in the process, because of the overlap interference of color spectrum curves between the color channel of the projector and the color channel of the camera, which results in energy superimposed interference loss of color photon signals in the channel [9, 10]. The camera focuses the captured color spectral signals on the CCD of the lens, but the optical signal captured by the camera is the optical signal projected from the projector in this paper, therefore, the color pixel information of the image is processed by the coloring mechanism of the projector, and be reflected by the color rendering surface, then be transmitted to the inductive surface by the physical device for color rendering of a camera, the signal processing model in the process is shown in formula (1), the simplified form is $I_C = M_{cross}KF\{I_P\} + A_o$, that $\mathrm{I_C}$ and $\mathrm{I_P}$ represent pixel information values in the camera and the projector, M_{cross} represents the light transmission crosstalk matrix of the system, K is the reflection matrix of object surface, $F\{I_P\}$ is the nonlinear transformation matrix, A_o is the information value of ambient light.

$$\underbrace{\begin{bmatrix} \mathrm{I_C^R} \\ \mathrm{I_C^G} \\ \mathrm{I_C^B} \end{bmatrix}}_{I_C} = \underbrace{\begin{bmatrix} m_{rr} & m_{rg} & m_{rb} \\ m_{gr} & m_{gg} & m_{gb} \\ m_{br} & m_{bg} & m_{bb} \end{bmatrix}}_{M_{Cross}} \underbrace{\begin{bmatrix} k_r & 0 & 0 \\ 0 & k_g & 0 \\ 0 & 0 & k_b \end{bmatrix}}_{K} F\left\{ \underbrace{\begin{bmatrix} \mathrm{I}_P^{\mathrm{R}} \\ \mathrm{I}_P^{\mathrm{R}} \\ \mathrm{I}_P^{\mathrm{R}} \end{bmatrix}}_{I_P} \right\} + \underbrace{\begin{bmatrix} \mathrm{I}_o^{\mathrm{R}} \\ \mathrm{I}_o^{G} \\ \mathrm{I}_o^{B} \end{bmatrix}}_{A_o} \tag{1}$$

The color coupling problem of the system is converted to the problem of solving parameters in formula (1), therefore, in order to solve the above parameters easily, firstly, set the calibration module with a reflection coefficient of 1, and a camera is used to capture the image of the calibration template placed in the natural environment, then A_o is calculated, next, single channel pure standard colors are projected by projectors respectively, the corresponding images are captured by the camera, the above parameter information values can be obtained by combining formula (1) and least square method for optimization. Finally, a correction matrix is constructed by using the parameter values obtained from the above solution to correct the color crosstalk.

3 Automatic Projection Correction

3.1 Design of Projection Feature Image

Since the projection feature image does not need to use each fringe pixel as the projection feature information, the feature information meshing for the generated color fringe image in order to quickly establish the feature mapping relationship of the projected image, so the meshed color grid intersection is used as the feature sample of the projection feature image, that the projected feature image is shown in Fig. 1.

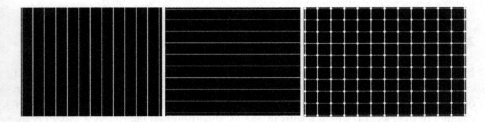

Fig. 1. Generation of projected feature image

3.2 Projective Geometric Correction of Topological Analysis

However, as the medium of projection is an anisotropic surface, because the topological structure of the feature points in the anisotropic surface is quite different, it leads to the inconsistency of the geometric mapping relations, therefore, it is necessary to establish the mapping relation between different regions in the anisotropic surface. By projecting the projection feature image to the anisotropic surface, a local of mapping topological analysis (LMTA) is used to solve the homography correspondence of each region, the specific process of LMTA algorithm is as follows:

Gathering information value $S_C^i = (x_i, y_i)$ of projection feature points by camera, due to the projection feature dot matrix has the character of window uniqueness, this character is used to establish the correspondence between projection feature and feature information of camera acquisition, $f : S_P^j(p_j, q_j) \rightarrow S_C^i(x_i, y_i)(i, j = 1, 2 \cdots N)$, in this formula, $S_P^j = (p_j, q_j)$ is the original dot matrix feature information value of the projector, and then use the parameter information of the projector-camera system obtained by calibration [11]. We can get the three-dimension topical information of feature points in space by formula (2), in formula (2), H_{cam} and H_{pro} are the homography matrix of the camera and the homography matrix of the projector respectively. This paper uses the Householder transform least squares method to calculate because the formula (2) is overdetermined equation set.

$$\begin{cases} x_i H_{cam}(3,4) = X_{Wi} H_{cam}(1,1) + Y_{Wi} H_{cam}(1,2) + Z_{Wi} H_{cam}(1,3) + H_{cam}(1,4) \\ \quad -x_i X_{Wi} H_{cam}(3,1) - x_i Y_{Wi} H_{cam}(3,2) - x_i Z_{Wi} H_{cam}(3,3) \\ y_i H_{cam}(3,4) = X_{Wi} H_{cam}(2,1) + Y_{Wi} H_{cam}(2,2) + Z_{Wi} H_{cam}(2,3) + H_{cam}(2,4) \\ \quad -y_i X_{Wi} H_{cam}(3,1) - y_i Y_{Wi} H_{cam}(3,2) - y_i Z_{Wi} H_{cam}(3,3) \\ p_j H_{pro}(3,4) = X_{Wi} H_{pro}(1,1) + Y_{Wi} H_{pro}(1,2) + Z_{Wi} H_{pro}(1,3) + H_{pro}(1,4) \\ \quad -p_j X_{Wi} H_{pro}(3,1) - p_j Y_{Wi} H_{pro}(3,2) - p_j Z_{Wi} H_{pro}(3,3) \\ q_j H_{pro}(3,4) = X_{Wi} H_{pro}(2,1) + Y_{Wi} H_{pro}(2,2) + Z_{Wi} H_{pro}(2,3) + H_{pro}(2,4) \\ \quad -q_j X_W H_{pro}(3,1) - q_j Y_W H_{pro}(3,2) - q_j Z_{Wi} H_{pro}(3,3) \end{cases} \tag{2}$$

And then save the feature points collected by the camera and the three-dimensional information values, then, divided and analyzed the feature points collected by the camera, the feature points of the local rectangular region in $S_C^i = (x_i, y_i)$ are divided into $\mathrm{Rect}Point^k(S_{c1}^{r1}, S_{c2}^{r2}, \cdots S_{cm}^{rn})$. We assume that the acquired matrix set is $\mathrm{Rect}Set\{\mathrm{Rect}Point^k\}$, at the same time, the remaining set of feature points is $Point\mathrm{Set}\{S_c^l\}$, however in $\mathrm{Rect}Set\{\mathrm{Rect}Point^k\}$, any two local rectangles area do not have intersection, $\mathrm{Rect}Point^i \cap \mathrm{Rect}Point^j = \emptyset$. Using the corresponding three-dimensional information value of the feature points, we can get the spatial 3D topological information of the feature points contained in the local rectangular region, and remember as $\mathrm{Rect}TDPoint^k(W_{c1}^{r1}, W_{c2}^{r2}, \cdots W_{cm}^{rn})$, and the corresponding matrix set is $\mathrm{Rect}TDSet\{\mathrm{Rect}TDPoint^k\}$.

We perform spatial topological surface fitting for each feature point in $\mathrm{Rect}TDPoint^k(W_{c1}^{r1}, W_{c2}^{r2}, \cdots W_{cm}^{rn})$, and remember as $\pi^k : Z_s^k = AX_s^k + BY_s^k + C$, and we get $SetSurface^k$. We can find the nearest spatial topological surface through the feature points in $PointSet\{S_c^l\}$. And then calculate the attribution difference to each topological surface through formula (3), where δ is the difference threshold. That is, using this threshold to further determine whether the feature points are best suited to be divided into a topological region. And the selection of the threshold is according to the anisotropic surface.

$$\zeta_{S_c^l}^s = \min(d_{S_c^l}^k) \leq \delta, \quad d_{S_c^l}^k = [Z_s^k - (A_k X_s^k + B_k Y_s^k + C_k)] \Big/ \sqrt{A_k^2 + B_k^2 + 1} \tag{3}$$

By analyzing the topological structure above, obtained N local topological space surfaces. In order to establish the mapping relationship between the pixels in each region of the projection image and the imaging pixels of the camera, a linear interpolation mapping method is adopted. Firstly, suppose that there are num pairs of feature points which collected by the camera in the Kth topological surface $I^{\mathrm{plane\_k}}$. Then

$X_a^k = (x_a^k, y_a^k)$ is used to represent the a-th feature dot pair which acquired by the camera in the k-th local area, and the corresponding feature points of the projector is $P_b^k = (p_b^k, q_b^k)$. Construct polynomial $X_{num}^k = A^k V_{num}^k$ using num feature dot pairs, which A^k represents the polynomial coefficient matrix, X_{num}^k means that all feature point pairs in the k-th local region collected by camera. The value of $X_{num}^k, A^k, V_{num}^k$, as shown in formula (4).

$$X_{num}^k = \begin{bmatrix} x_1^k & \cdots & x_{num}^k \\ y_1^k & \cdots & y_{num}^k \end{bmatrix}, A^k = \begin{bmatrix} a_{11}^k & a_{12}^k & a_{13}^k & a_{14}^k & a_{15}^k & a_{16}^k \\ a_{21}^k & a_{22}^k & a_{23}^k & a_{24}^k & a_{25}^k & a_{26}^k \end{bmatrix},$$
$$\hat{P}_b^k = \begin{bmatrix} (p_1^k)^2 & (q_1^k)^2 & p_1^k q_1^k & p_1^k & q_1^k & 1 \\ & & \cdots & & & \\ (p_{\text{num}}^k)^2 & (q_{num}^k)^2 & p_{num}^k q_{num}^k & p_{num}^k & q_{num}^k & 1 \end{bmatrix} \tag{4}$$

The polynomial coefficient matrix A^k between the k-th topological surface and the projected image is solved by $X_{num}^k = A^k V_{num}^k$, $\mathrm{A}_k = (V_{num}^k{}^T V_{num}^k)^{-1} V_{num}^k{}^T X_{num}^k$ can be obtained by the least square method. Calculate the set of polynomial coefficient matrices $\mathrm{A}_{\{1,2,\cdots N\}}$ between the original projection image and topological structure of each region. Then, the polynomial coefficient matrix set is composed of multiple polynomial coefficients matrices, and form a piecewise linear mapping relation, as shown in formula (5).

$$\mathrm{A}_{\{1,2,\cdots N\}} = \begin{cases} \mathrm{A}_1 = (V_1^T V_1)^{-1} V_1^T X_1 \\ \mathrm{A}_2 = (V_2^T V_2)^{-1} V_2^T X_2 \\ \vdots \\ \mathrm{A}_\mathrm{N} = (V_N^T V_N)^{-1} V_N^T X_N \end{cases} \tag{5}$$

After getting the mapping set of polynomial coefficient matrix that the projection original image corresponded, then, the projective primitive images that all local topological structure regions corresponded can be geometrically transformed, according to the polynomial coefficient matrix mapping set. After all the transformed images are spliced, a corrected pre-projected image is obtained.

4 Experimental Results

To verify the effectiveness of the adaptive projection correction algorithm for depth anisotropic surfaces proposed in this paper. Firstly, build a projector-camera projection correction system. The hardware of the system framework includes: the View Sonic projection device and the Daheng MER-U3x industrial camera, in the whole system development process, using C++, OpenCV, OpenGL, MATLAB and other development language and tools to test and analyze the experiment. The parameters of the projector - camera should be calibrated before the projection of the anisotropic surface projection is carried out. Then, obtain the topological structure distribution of the depth anisotropic surface through the projection feature image designed in this paper. Finally, through the computer to correct the projection image. In this paper, the anisotropic surface adaptive correction algorithm relates to the projector - camera equipment. Therefore, in order to avoid the error of the projection correction caused by the system hardware itself, firstly, analyze the coupling correction of the projector - camera system, and then for the method proposed in this paper, analyze the error of the projective correction-image precision of the deep anisotropic surface.

4.1 System Coupling Correction Analysis

In order to highlight the coupling between the different color channels, the projector projected an image which pure red, pure green and pure blue stripes alternated, as shown in Fig. 2, it can be found that the three channels interfere with each other. So when the projector project an image and the camera collected is completed, it will be found that the color of the collected image is significantly darker than the original projection image color, So when the projector project an image and the camera collected is completed, it will be found that the color of the collected image is significantly darker than the original projection image color, as shown in left figure in Fig. 3, the purity of each the color information can be improved through the color coupling correction. As shown in right figure in Fig. 3, the loss of the color signal collected by camera reduced.

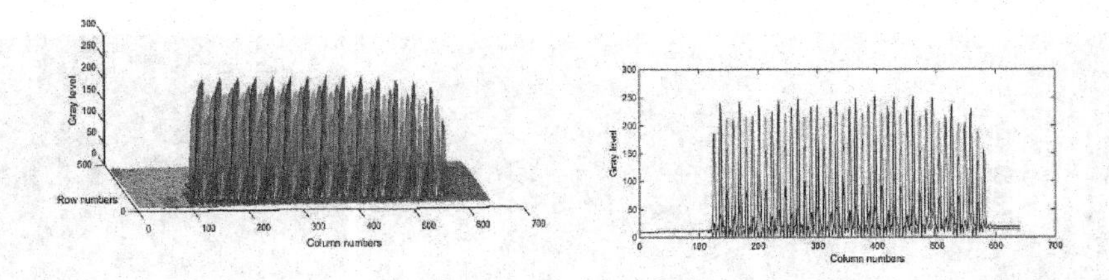

Fig. 2. Coupling signal for multi-channel of system

4.2 Projection Feature Image Processing

The projection feature image of the coding design is projected on the depth anisotropic surface, shoot the modulated image with the camera. In order to obtain the spatial topological distribution of the depth anisotropic surface, use the camera to collect the color grid coding pattern of the projection, the corresponding pre-decoding processing to the collection of images be taken, that is, extract the central stripe for the encoded horizontal color stripe, and clustering dispose the stripe color, specific treating processes can refer to the paper [12]. Then, gray processing and smooth processing to the collected images be taken, and then binarized the processed images in order to extract the circular cross point of the color grid coding pattern, extract the center position coordinates of the cross point using the ellipse fitting algorithm. However, the perspective transformation of the projector and the camera will lead to the extracted center

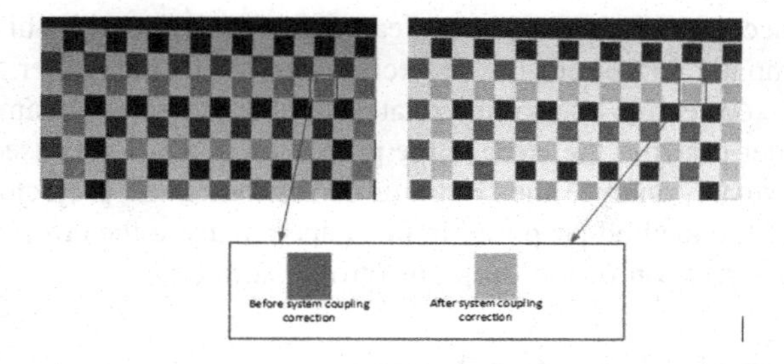

Fig. 3. System coupling correction

coordinates of the grid intersection is not accurate enough, so it is necessary to use the sub-pixel positioning method of the cross window to re-locate and extract. That is, for each grid intersection, extract the corresponding color information of horizontal center stripe. The central sub-pixel position of the grid cross-fringes can be extracted by constrained optimizing the fitting extracted center coordinates which combined with the horizontal circle center stripe and the longitudinal stripe. Figure 4(a) is the distorted image after modulated in the depth anisotropic surface that projective feature image shoot by the camera, Fig. 4(b) is a circular lattice space distribution map of the projective distorted image, and Fig. 4(c) is a distorted image after color fringe center extraction.

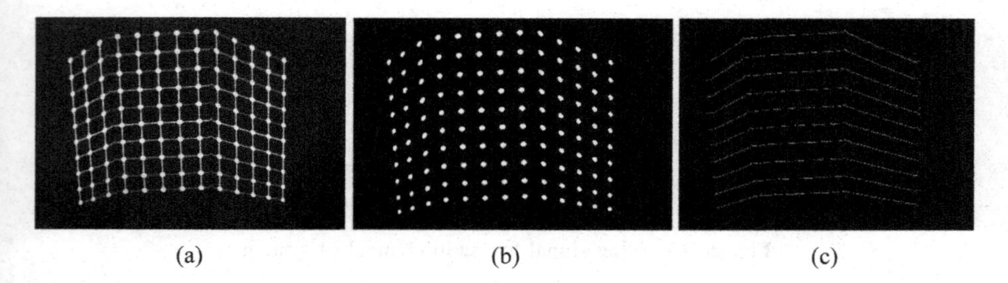

Fig. 4. Feature information processing

In order to match the projection feature information and the characteristic information captured by the camera easily, color cueing the extracted circle center sub-pixel position X_i, that is, identify the color pixel values of the horizontal stripes and vertical stripes of each circle center around using a window detection function, is $\mathrm{h}_k^{\mathrm{c}}, \mathrm{v}_k^c$, and then color cueing the circle center is $\left(X_i, \mathrm{h}_k^{\mathrm{c}}, \mathrm{v}_k^c\right)$. Color cueing the circle center of the characteristic image of projection distortion and the feature circle center of the projective original image, then, the relationship between the projector and the camera can be created using the uniqueness of the strip code (Fig. 5).

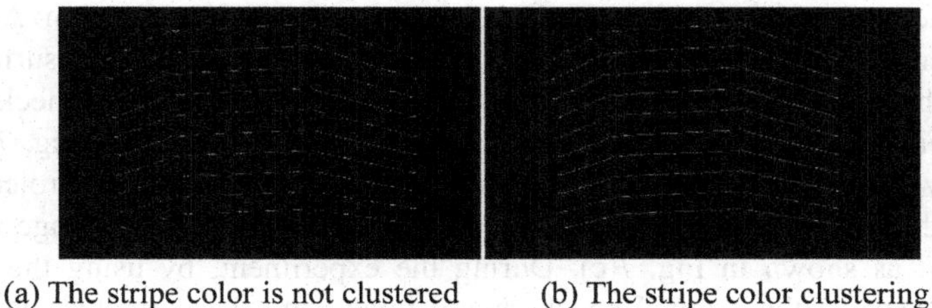

(a) The stripe color is not clustered (b) The stripe color clustering

Fig. 5. The clue of stripe color information

4.3 Accuracy Analysis of Adaptive Projection Correction

By using the projection feature image designed in this paper, the pixel mapping relationship between the projector image and the camera image is established. Due to the deformation of the projection surface, the projection screen produces a geometric distortion, so we can't use the plane projection correction method to solve the mapping relationship simply. The projection surface is subdivided by the topology analysis method proposed in this paper, and the projection display area is divided into areas with geometrical topology consistent characteristics. We perform geometric mapping for each topology area, solve the distortion and transformation problem of the projection screen effectively. In order to evaluate the proposed algorithm effectively in the paper, the projection correction analysis is performed on various depth anisotropic surfaces related in our daily life. First of all, we perform the projection experimental analysis to the corner and the cardboard, which have a very prominent depth anisotropic surface. As shown in Fig. 6, It can be seen that the projection geometry correction by this method is basically consistent with the normal picture viewed by the human eye.

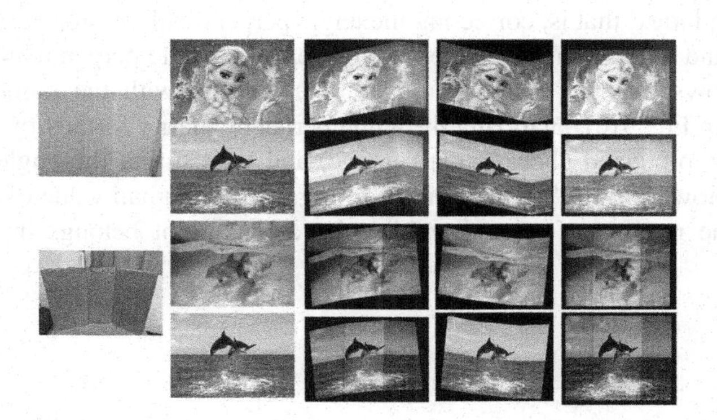

Fig. 6. Adaptive projection geometric correction effect

In order to verify the effectiveness of the adaptive projection correction algorithm, projecting a black and white checkerboard pattern on the depth anisotropic surface and performing the experimental analysis of projection geometry correction, and testing the effect and accuracy of the method for adaptive projection twist correction. As shown in Fig. 7, after projecting checkerboard image on the depth anisotropic surface with a projector, the camera captures the modulating distorted image of the checkerboard as shown in Fig. 7(a), the pre-projected image of projector as shown in Fig. 7(b) can be obtained by geometric distortion correction. And then, project the pre-projected image on the depth anisotropic surface, and we can get the updated output image with visual consistency, as shown in Fig. 7(c). During the experiment, by using the scaling of projection between the original checkerboard image and the corrected output image, and then, we put the corner of the checkerboard as a sampling point. In the depth anisotropic surface, by calculating the average error, the maximum error and the root mean square error of the distance between the ideal coordinate value and the actual coordinate value of the checkerboard distortion correction image, they are 0.424 pixels, 0.862 pixels, 0.216 pixels. It can be seen that the geometric distortion correction accuracy of this method has reached the sub-pixel level.

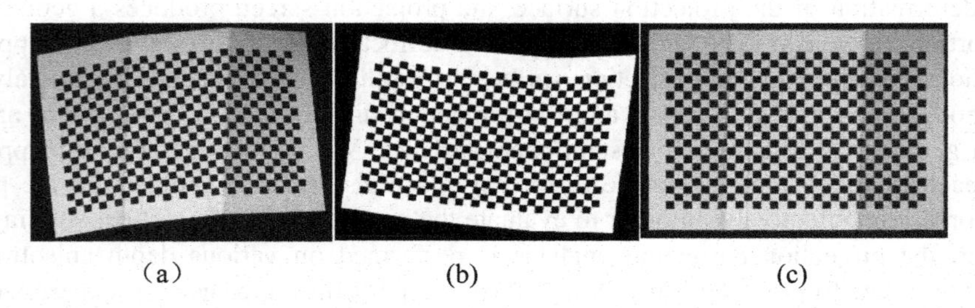

Fig. 7. (a) uncorrected image (b) pre-projection correction image (c) correction image

In order to further illustrate the distortion correction image projected by this method is conform to the visual consistency, the angle of the checkerboard is used to construct the evaluation index, that is, correcting the angle perpendicularity formed by the corner of each row and column in the image. Therefore, if the angle perpendicularity is closer to 90°, it shows that the corrected image is consistent with the human eye visual consistency. In Fig. 7(c), analyzing the angle perpendicularity formed by each row and column in the projected checkerboard image, and we can get the angle vertical distribution as shown in Fig. 8. The ordinate in Fig. 8 is the radian value of the angle, the abscissa is the number of rows to which the corner point belongs in the corrected

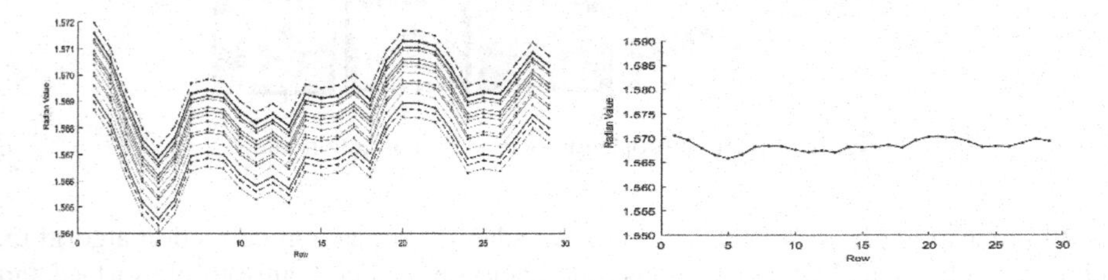

Fig. 8. Visual consistency analysis of geometry correction

image. It can be seen that the angle between each row, each column is 90°, which is consistent with the visual consistency of the human eye.

5 Conclusions

The projection display technology has some problems such as the geometric distortion of projection image and the interference of texture color information. It makes the observer can't perceive and understand the content of the projection image display visually. At the same time, in order to improve the adaptive perception projection of the projection equipment in different environments, we propose a projection correction method which can perceive deep heterogeneous surface autonomously. According to the experimental data, our method has reached the sub-pixel level and the imaging screen consistency level. In this way, we can perform adaptive projection display for deep heterogeneous surface, and the optimized projection screen conforms to the consistency of the human visual perception. At the same time, this method has strong anti-noise and robustness qualities.

Acknowledgement. This work was supported by the National Natural Science Foundation of China (Grant No. 61602058), Jilin province science and technology development plan item (20160101258JC), Jilin province science and technology development plan item (20150101015JC).

References

1. Majumder, A., Sajadi, B.: Large area displays: the changing face of visualization. Computer **46**(5), 26–33 (2013)
2. Park, J., Lee, B.U.: Defocus and geometric distortion correction for projected images on a curved surface. Appl. Optics **55**(4), 1–25 (2016)
3. Okatani, T., Deguchi, K.: Autocalibration of a projector-camera system. IEEE Trans. Patt. Anal. Mach. Intell. **27**(12), 1845–1855 (2015)
4. Zhu, B., Xie, L.J., Yang, T.J., Wang, Q.H., Zheng, Y.: An adaptive calibration algorithm for projected images in daily environment. J. Comput.-Aided Des. Comput. Graph. **24**(7), 941–948 (2012)
5. Xiao, C., Yang, H.Y., Liang, H.J., Ji, Y.L., Li, X.S.: Geometric calibration for multi-projector display system based on structured light. J. Comput.-Aided Des. Comput. Graph. **25**(6), 802–808 (2013)
6. Sun, W., Yang, X., Xiao, S., Hu, W.: Robust checkerboard recognition for efficient nonplanar geometry registration in projector-camera systems. In: Proceedings of the 5th ACM/IEEE International Workshop on Projector Camera Systems, pp. 504–542. ACM, New York (2008)
7. Xie C., Wang Q., Cheng W.: Simple auto-geometric correction for non-planar projection. In: International Conference on Automatic Control and Artificial Intelligence, pp. 1834–1837. IET, London (2012)
8. Steimle J., Jordt A., Maes P.: Flexpad: highly flexible bending interactions for projected handheld displays. In: Proceedings of the SIGCHI Conference on Human Factors in Computing Systems, pp. 237–246. ACM, New York (2013)

9. Madi, A., Ziou, D.: Color constancy for visual compensation of projector displayed image. Displays **35**(1), 6–17 (2014)
10. Xu, J., Wang, P., Yao, Y., Liu, S., Zhang, G.: 3D multi-directional sensor with pyramid mirror and structured light. Optics Lasers Eng. **93**, 156–163 (2017)
11. Boroomand, A., Sekkati, H., Lam, M., Clausi, D., Wong, A.: Saliency-guided projection geometric correction using a projector-camera system. In: 2016 IEEE International Conference on Image Processing (ICIP), pp. 2951–2955. IEEE (2016)
12. Fan, J.T., Han, C., Zhang, C., Li, M.X., Bai, B.X., Yang, H.M.: Study of a new decoding technology for de bruijn structured light. Acta Electronica Sinica **40**(3), 483–488 (2012)

Multi-camera Tracking Exploiting Person Re-ID Technique

Yiming Liang and Yue Zhou(✉)

Institute of Image Processing and Pattern Recognition,
Shanghai Jiao Tong University, Shanghai, China
{liangyiming,zhouyue}@sjtu.edu.cn

Abstract. Multi-target multi-camera tracking is an important issue in image processing. It is meaningful to improve matching performance across cameras with high computational efficiency. In this paper, we apply high performance feature representation LOMO and metric learning XQDA in person re-identification across cameras to improve tracking performance. We also exploit direction information of trajectories to handle viewpoint variation. Experiments on DukeMTMCT dataset show that the proposed method improves tracking performance and is also competitive in running time.

Keywords: Multi-camera tracking · Multi-target tracking · Person re-identification

1 Introduction

Pedestrian tracking is a fundamental topic in computer vision. In the past decade, multi-target tracking has attracted lots of attentions, and large amounts of algorithms have been proposed to solve it. State-of the-art methods obtain impressive performance, but multi-target problem still needs further study, especially when it comes to a multi-camera problem.

Compared to single-target tracking, multi-target tracking mainly focuses on data association rather than appearance model [1]. Multi-target tracking methods are supposed to tackle difficult problems such as occlusions, detection failures, appearance similarity among targets and improving computational efficiency. Numerous approaches [1–3] have made contribution to improvement in performance and computational efficiency. [4] casts the problem of tracking multiple people as a graph partitioning problem and proposed Correlation Clustering by Binary Integer Programming (BIPCC). Since solving a BIP is NP hard, the size of the problem is reduced by a multi-stage cascade in which data is clustered according to space-time and appearance criteria. BIPCC obtains significant accuracy and high computational efficiency and is also employed in multi-camera tracking [5].

When it comes to tracking pedestrians across cameras, there are no strong space-time constraints if the camera pairs have no overlapping areas or we lack

D. Liu et al. (Eds.): ICONIP 2017, Part III, LNCS 10636, pp. 397–404, 2017.
https://doi.org/10.1007/978-3-319-70090-8_41

for real world information. Disregarding the weak space-time constraints, tracking across cameras can be consider as a person re-identification problem. To this end, we exploit effective approaches in person re-identification to improve the performance of multi-camera tracking.

Due to illumination changes, viewpoint variations, pose variations and occlusions, appearance-based person re-identification encounters difficulty in the past decade [6]. Most existing methods [7–11] concentrate their efforts on feature representation and metric learning. A feature representation named Local Maximal Occurrence (LOMO) together with a subspace and a metric learning method named Cross-view Quadratic Discriminant Analysis (XQDA) have been proposed in [12]. Experiments show that the combination of LOMO and XQDA obtains impressive performance as well as high computational efficiency. We find that applying LOMO and XQDA is useful in tracking multiple pedestrians across cameras.

In this paper, we propose a multi-target multi-camera tracking (MTMCT) approach named LXB (LOMO and XQDA based on BIPCC). We take tracking results within single camera as input of our system. The single-camera trajectories we used in experiments are accomplished by BIPCC. When it comes to tracking across cameras, LOMO is used to extract appearance features and XQDA is used to determine the similarity among targets. After we obtain the similarity matrix, another BIPCC is performed on targets across cameras. We also analyze the viewpoint, or direction, information of the targets. When comparing similarity between targets, bounding boxes indicating same or near viewpoints are given extra weightings. The proposed method is tested on a multi-target multi-camera tracking dataset named DukeMTMCT [5]. Experiments show that the proposed method improves the state-of-the-art scores on the dataset and is also competitive in running time.

2 Related Work

Numerous approaches [1–3] have been proposed to solve multi-target tracking problem. However, better performance usually requires higher computational complexity [4]. [13–15] approximate the solution to reduce computational complexity. [16,17] relax constraints in the BIP. [4] decomposes the problem into tracklets phase and trajectories phase and solves the sub-problems exactly with a BIP solver. This method has also been extended into multi-camera style in [5].

In most existing appearance-based methods [8,9,12] for person re-identification, different kinds of appearance features, mainly color and texture histograms, are combined in order to obtain higher robustness and matching performance. The feature histograms are weighted globally [18] or object-specifically [8] according to their capacity in distinguishing an object from a gallery. State-of-the-art approaches [1,7,12] deal with viewpoint changes by extracting better features or learning good metrics. The LOMO method in [12] maximizes the local occurrence of each SILTP and HSV histogram at the same horizontal location. A practical computation method for Cross-view QDA is also proposed.

[7] presents a descriptor that models a region as a set of multiple Gaussian distributions, and each Gaussian represents the appearance of a local patch. Then The characteristics of the Gaussian set are also described by a Guassian distribution. [1] trains a 16-layer VGGNet to extract appearance features.

Person re-identification methods mentioned above exhibit high performance on viewpoint variations. In contrast to them, we research utilizing appearance of a pedestrian from each available viewpoints under a camera to improve re-identification performance across cameras. It is reasonable to exploit information from as many viewpoints as possible. For instance, a pedestrian with frontal view under one camera can appear with back view under another camera [12]. If the pedestrian's clothes look different from the front and the back, matching across cameras becomes difficult and error-prone.

There have been numerous methods [19–22] for pedestrian direction estimation, most of which are aimed at the improvement of driver assistance systems. For example, [19] introduces a simple method to estimate walking direction of a pedestrian. Haar wavelets are used to generate feature vectors and SVMs with linear kernel are used to classify 16 directions. [20] is also a SVM-based method and estimates the discrete probability distribution of the directions. It also use a Hidden Markov Model to handle direction changes. For simplicity, our approach is also a SVM-based method and HOG is used to generate feature vectors.

3 MTMCT Framework

Section 3.1 introduces direction estimation of pedestrians. Section 3.2 describes how to exploit direction information in comparison between two trajectories. Section 3.3 introduces the BIP method we used in data association. Section 3.4 discusses feature representation and metric learning for person re-identification across cameras.

3.1 Direction Estimation of Pedestrian

We use HOG feature to generate feature vectors of pedestrians and train 8 SVM classifiers corresponding to 8 directions. The intervals between adjacent directions are 45°. Each of the classifiers generates an output at run time, and the direction corresponding to highest output is chosen for the pedestrian.

There is also a constraint that directions of a pedestrian can not change rapidly during tracking. As for a pedestrian trajectory, the difference between the viewpoints of adjacent frames is smoothed.

HOG are used to generate feature vectors for direction estimation. SVMs are used to classify the HOG feature vectors. According to our experiments results and [19], we chose a linear kernel function for better overall performance.

3.2 Exploiting Direction Information

For a pedestrian trajectory, the directions corresponding to each frame are estimated first. Then we respectively compute appearance features for the available directions. For example, if a trajectory includes direction a, b and c, we extract 3 feature vectors respectively from the frames whose directions are a, b and c. Only appearance features with valid directions can be calculated when computing similarity between two trajectories.

When computing distance between features with directions of two trajectories, the directions are used to determine the weighting of the result. We denote the set of directions which appear in a trajectory as D. One D can contain at most 8 direction elements and at least 1 element. For trajectory A and B, their direction set are respectively D_A and D_B. Then the distance between A and B is defined as:

$$dir\_dist(A, B) = \frac{\sum_{i \in D_A,\ j \in D_B} w(i, j) \times dist(A_i, B_j)}{\sum_{i \in D_A,\ j \in D_B} w(i, j)} \tag{1}$$

where A_i is the feature vector of A corresponding to direction i, and B_j is the feature vector of B corresponding to direction j. $dist(\cdot, \cdot)$ can be any specific distance function between two feature vectors. $w(i, j)$ is the weighting for distance between two feature vectors with different directions. Since there are 8 directions in this system, the difference between two directions is at most 4. For direction i and j, the difference can be [0, 1, 2, 3, 4], and the corresponding $w(i, j)$ value are set as [1.0, 0.8, 0.6, 0.8, 1.0] according to performance in experiments. In Sect. 4.2 we illustrate the experiments and explain the choice among several groups of parameters.

3.3 Data Association

The system takes trajectories as input, and associate them into identities. We consider data association as a graph partitioning problem. Let V be a set of n trajectories. For $i, j \in V$, let $c_{ij} \in [-1, 1]$ represent the correlation between them. A higher correlation means they are more likely to belong to a same identity, and a lower value indicates that they are more unlikely to be the same person. Let the graph $G = (V, E, C)$ be a weighted graph on V. E is the set of edges connecting i and j in condition of c_{ij}. The system partitions V into sets in which trajectories belong to a same identity. The correlation clustering problem [23] is solved by a Binary Integer Program (BIP) on G:

$$\arg\max_X \sum_{(i,j) \in E} c_{ij} x_{ij} \tag{2}$$

subject to

$$x_{ij} \in \{0, 1\} \quad \forall (i, j) \in E \tag{3}$$

If x_{ij} equals to 1, i and j are considered to be a same person, and vice versa. X is the set of all possible combinations of x_{ij}.

Solving this BIP is NP hard and the approximation is also hard [23,24]. Keeping the problems small can help improve efficiency. Similar to [4,13,25], we employ a sliding temporal window. The window moves forward by half of its temporal length. Solutions from previous overlapping windows are also fed into the current window. Trajectories in a window are first divided into groups according to their appearance and, if available, space-time information. The division is to reduce the size of BIP and should be conservative to keep trajectories belonging to same person in the same group. Similar to [4], we employ simple k-means here. Then the solutions are computed in each group and in each temporal window.

3.4 Feature Representation and Metric Learning

Correlations in Sect. 3.3 are generated by appearance similarity and simple space-time criteria. As for appearance criteria, efficient and high-performance solutions in person re-identification can be employed here. LOMO and XQDA [12] show impressive performance as well as high computational efficiency in re-identifying person across cameras. LOMO maximizes the horizontal occurrence of local features and applies Retinex transform [26,27] to person images. XQDA learns a discriminant low dimensional subspace, and the metric is learned on the subspace. We estimate the direction information for each trajectory employing method in Sect. 3.1, and extract LOMO feature for each available direction. Then in each group and in each temporal window, features of all available trajectories are fed into a pre-trained XQDA model. Afterwards a distance matrix is generated. Elements in the matrix are transformed linearly into the range of $[-1, 1]$:

$$c_{ij} = -\frac{2}{d_{max} - d_{min}} \cdot \left(d_{ij} - \frac{d_{min} + d_{max}}{2}\right) \tag{4}$$

where c_{ij} and d_{ij} are the element in row i column j of correlation matrix C and distance matrix D respectively. d_{min} and d_{max} are the maximum and minimum value in D. Therefore, a longer distance is transformed into a lower correlation.

4 Experiments

We report the results on the DukeMTMCT [5] dataset. The DukeMTMCT dataset was captured from 8 synchronized cameras. It lasts for 85 min and contains more than 7,000 single camera trajectories and over 2,000 unique identities. Two camera pairs (2–8 and 3–5) have small overlapping areas, while the other cameras are disjoint. The running time was measured on a desktop PC with an Intel i7-6700 @ 3.40 GHz CPU.

4.1 Direction Estimation of Pedestrian

We randomly chose 1000 bounding boxes from ground truth training data of each camera in DukeMTMCT dataset [5], and got total 8000 samples. These bounding boxes are chosen from frame 49700 to 130000. All directions of the pedestrian samples are manual Annotated. 8 direction SVM classifiers are trained using 10-fold cross validation method. According to Table 1, This model is able to correctly classify most of the samples. In fact, most mis-classifications turn out to be adjacent directions of the correct ones. Therefore, the direction estimation model meets requirements.

Table 1. Mis-classification rates of direction estimation model

Direction (degree)	0	45	90	135	180	225	270	315
Mis-classification rate	0.088	0.063	0.057	0.097	0.087	0.066	0.051	0.089

4.2 Tracking Results on DukeMTMCT

Method in [5] is chosen to be the baseline for comparison, and we test our method from frame 130001 to 227540. The baseline method is also based on BIPCC and use striped color histograms together with simple temporal reasoning in matching targets across cameras. The input single camera trajectories are generated from [4]. All following experiments are based on these same input trajectories. The result is shown in Table 2. We also compare the performance of using different weighting for distance in Table 3. IDF1, IDP and IDR [5] are used to measure the performance.

Table 2. Performance comparison of our method

Tracker	IDF1 (%)	IDP (%)	IDR (%)	Time (seconds)
Baseline method	55.9	66.9	48.1	**2182.3**
Our method	**58.6**	**68.4**	**49.9**	2526.4
Baseline method+Direction	56.1	67.1	48.6	2362.9
Our method without direction	58.3	68.1	49.5	2345.5

As shown in Table 2, in comparison to the baseline method, our method yielded improved tracking score on the DukeMTMCT dataset. Both direction information and XQDA with LOMO help improve performance. However, improvement from direction information is not satisfactory considering the increased time consumption. In total, the increased time consumption of our method is acceptable.

Table 3. Performance comparison of different weighting for distance

Weighting	IDF1 (%)	IDP (%)	IDR (%)
Baseline method	55.9	66.9	48.1
$w = [1.0,\ 0.8,\ 0.6,\ 0.8,\ 1.0]$	**56.1**	**67.1**	**48.6**
$w = [1.0,\ 0.8,\ 0.6,\ 0.4,\ 0.2]$	56.0	67.0	48.4
$w = [1.0,\ 0.9,\ 0.8,\ 0.9,\ 1.0]$	56.0	67.0	48.3
$w = [1.0,\ 0.5,\ 0.01,\ 0.01,\ 0.01]$	55.3	66.1	47.5

As shown in Table 3, when the difference between two directions is as large as 3 or 4, higher weighting yield higher performance. It may be because the appearance of a person are usually more similar between opposite viewpoints than orthogonal viewpoints. Therefore, $w = [1.0,\ 0.8,\ 0.6,\ 0.8,\ 1.0]$ is finally chosen in our system.

5 Conclusion

In this paper, we innovatively import the LOMO and XQDA method and direction information to match person across cameras in MTMCT. Experiments prove that our approach outperforms the original baseline and is also competitive in running time.

Acknowledgments. This work is supported by National High-Tech R&D Program (863 Program) under Grant 2015AA016402.

References

1. Sadeghian, A., Alahi, A., Savarese, S.: Tracking the untrackable: learning to track multiple cues with long-term dependencies. arXiv preprint arXiv:1701.01909 (2017)
2. Yu, S.I., Meng, D., Zuo, W., Hauptmann, A.: The solution path algorithm for identity-aware multi-object tracking. In: Proceedings of the IEEE Conference on Computer Vision and Pattern Recognition, pp. 3871–3879 (2016)
3. Yoon, J.H., Lee, C.R., Yang, M.H., Yoon, K.J.: Online multi-object tracking via structural constraint event aggregation. In: Proceedings of the IEEE Conference on Computer Vision and Pattern Recognition, pp. 1392–1400 (2016)
4. Ristani, E., Tomasi, C.: Tracking multiple people online and in real time. In: Asian Conference on Computer Vision, pp. 444–459 (2014)
5. Ristani, E., Solera, F., Zou, R., Cucchiara, R., Tomasi, C.: Performance measures and a data set for multi-target, multi-camera tracking. In: European Conference on Computer Vision, pp. 17–35 (2016)
6. Doretto, G., Sebastian, T., Tu, P., Rittscher, J.: Appearance-based person reidentification in camera networks: problem overview and current approaches. J. Ambient Intell. Hum. Comput. **2**(2), 127–151 (2011)
7. Matsukawa, T., Okabe, T., Suzuki, E., Sato, Y.: Hierarchical Gaussian descriptor for person re-identification. In: Proceedings of the IEEE Conference on Computer Vision and Pattern Recognition, pp. 1363–1372 (2016)
8. Liu, C., Gong, S., Loy, C.C., Lin, X.: Person re-identification: what features are important? In: Fusiello, A., Murino, V., Cucchiara, R. (eds.) ECCV 2012. LNCS, vol. 7583, pp. 391–401. Springer, Heidelberg (2012). doi:10.1007/978-3-642-33863-2_39
9. Farenzena, M., Bazzani, L., Perina, A., Cristani, M., Murino, V.: Person reidentification by symmetry-driven accumulation of local features. In: Proceedings of the IEEE Conference on Computer Vision and Pattern Recognition, pp. 2360–2367 (2010)
10. Li, W., Zhao, R., Xiao, T., Wang, X.: DeepReID: deep filter pairing neural network for person re-identification. In: IEEE Conference on Computer Vision and Pattern Recognition, pp. 152–159 (2014)

11. Wang, T., Gong, S., Zhu, X., Wang, S.: Person re-identification by discriminative selection in video ranking. IEEE Trans. Pattern Anal. Mach. Intell. **38**(12), 2501–2514 (2016)
12. Liao, S., Hu, Y., Zhu, X., Li, S.Z.: Person re-identification by local maximal occurrence representation and metric learning. In: Proceedings of the IEEE Conference on Computer Vision and Pattern Recognition, pp. 2197–2206 (2015)
13. Shafique, K., Shah, M.: A noniterative greedy algorithm for multiframe point correspondence. IEEE Trans. Pattern Anal. Mach. Intell. **27**(1), 51–65 (2005)
14. Brendel, W., Amer, M., Todorovic, S.: Multiobject tracking as maximum weight independent set. In: Computer Vision and Pattern Recognition, pp. 1273–1280 (2011)
15. Roshan Zamir, A., Dehghan, A., Shah, M.: GMCP-tracker: global multi-object tracking using generalized minimum clique graphs. In: Computer Vision, CECCV, pp. 343–356 (2012)
16. Jiang, H., Fels, S., Little, J.J.: A linear programming approach for multiple object tracking. In: Computer Vision and Pattern Recognition, pp. 1–8 (2007)
17. Butt, A.A., Collins, R.T.: Multi-target tracking by Lagrangian relaxation to min-cost network flow. In: Proceedings of the IEEE Conference on Computer Vision and Pattern Recognition, pp. 1846–1853 (2013)
18. Mignon, A., Jurie, F.: PCCA: a new approach for distance learning from sparse pairwise constraints. In: Proceedings of the IEEE Conference on Computer Vision and Pattern Recognition, pp. 2666–2672 (2012)
19. Shimizu, H., Poggio, T.: Direction estimation of pedestrian from multiple still images. In: Intelligent Vehicles Symposium, pp. 596–600. IEEE (2004)
20. Gandhi, T., Trivedi, M.M.: Image based estimation of pedestrian orientation for improving path prediction. In: Intelligent Vehicles Symposium, pp. 506–511. IEEE (2008)
21. Flohr, F., Dumitru-Guzu, M., Kooij, J.F., Gavrila, D.M.: Joint probabilistic pedestrian head and body orientation estimation. In: Intelligent Vehicles Symposium Proceedings, pp. 617–622. IEEE (2014)
22. Tao, J., Klette, R.: Part-based RDF for direction classification of pedestrians, and a benchmark. In: Asian Conference on Computer Vision, pp. 418–432 (2014)
23. Bansal, N., Blum, A., Chawla, S.: Correlation clustering. In: Foundations of Computer, Science, pp. 238–247 (2002)
24. Tan, J.: A note on the inapproximability of correlation clustering. Inf. Process. Lett. **108**(5), 331–335 (2008)
25. Javed, O., Shafique, K., Rasheed, Z., Shah, M.: Modeling inter-camera space-time and appearance relationships for tracking across non-overlapping views. Comput. Vis. Image Underst. **109**(2), 146–162 (2008)
26. Jobson, D.J., Rahman, Z.U., Woodell, G.A.: Properties and performance of a center/surround retinex. IEEE Trans. Image Process. **6**(3), 451–462 (1997)
27. Jobson, D.J., Rahman, Z.U., Woodell, G.A.: A multiscale retinex for bridging the gap between color images and the human observation of scenes. IEEE Trans. Image Process. **6**(7), 965–976 (1997)

Towards Simulating Foggy and Hazy Images and Evaluating Their Authenticity

Ning Zhang, Lin Zhang(✉), and Zaixi Cheng

School of Software Engineering, Tongji University, Shanghai, China
cslinzhang@tongji.edu.cn

Abstract. To train and evaluate fog/haze removal models, it is highly desired but burdensome to collect a large-scale dataset comprising well-aligned foggy/hazy images with their fog-free/haze-free versions. In this paper, we propose a framework, namely *Foggy and Hazy Images Simulator* (FoHIS for short), to simulate more realistic fog and haze effects at any elevation in images. What's more, no former studies have introduced objective methods to evaluate the authenticity of synthetic foggy/hazy images. We innovatively design an *Authenticity Evaluator for Synthetic foggy/hazy Images* (AuthESI for short) to objectively measure which simulation algorithm could achieve more natural-looking results. We compare FoHIS with another two state-of-the-art methods, and the subjective results show that it outperforms those competitors. Besides, the prediction on simulated image's authenticity made by AuthESI is highly consistent with subjective judgements (Source codes are publicly available at https://github.com/noahzn/FoHIS).

Keywords: Fog · Haze · Simulation · Authenticity evaluation

1 Introduction

The decline in visibility due to fog and haze seriously threatens lives of drivers. Fortunately, research on defogging or haze removal has drawn attention during the past decade [1–3]. Accompanied by the rise of deep learning [4], one of the potential topics is to design a haze removal framework using deep architecture [5], while the difficulties of collecting a great number of pixel-wise aligned foggy/fog-free and hazy/haze-free training images hinder the idea. Thus, reliable algorithms of simulating natural-looking fog and haze are urgently needed.

Rendering fog and haze has been widely concerned in the field of computer graphics, and early studies mainly focused on fog and haze effects in virtual scenes. Kazufumi et al. displayed fog effects in outdoor 3D models [6]. Zhou et al. described an analytic approximation to the airlight integral from scattering media to render inhomogeneous fog effects [7]. Anthony and Venceslas modeled the fog function in a B-Spline function basis, and rendered fog in a navigable scene [8]. Although these virtual-scene-based methods can achieve pleasing results, they need to be implemented in pre-established virtual 3D scenes. Also, they require complex settings and massive computing resources.

D. Liu et al. (Eds.): ICONIP 2017, Part III, LNCS 10636, pp. 405–415, 2017.
https://doi.org/10.1007/978-3-319-70090-8_42

Recent research has shifted attention to image-based simulation. Compared with virtual-scene-based methods, simulating fog and haze in single images is not only convenient but also not dependent on expensive computing resources. In view of these advantages, image-based simulation method is also our focus.

1.1 Related Work

Since we focus on simulating foggy and hazy images in this paper, some recent representative studies which are image-based methods will be reviewed here.

Commercial image editing software such as Adobe Photoshop and CorelDRAW Graphics Suite can be used to generate lifelike fog and haze effects. Nevertheless, the process is very complicated and time-consuming. Whether it can produce realistic effects largely depends on the user's skill level. Zhao et al. [9] took advantage of single scattering model to render fog effects into both interior and exterior photos, and they achieved realistic results. However, their method is hard to be implemented because of requiring a great deal of user assistance. Guo et al. [10] estimated the transmission map by using Markov random field model and bilateral filter, then they rendered heterogeneous foggy scenes by Perlin noise. Due to the unreliability of estimating the transmission map, they are unable to achieve satisfactory results all the time. Since depth information plays a pivotal role in simulating fog and haze effects, Liu et al. [11] estimated the depth information by stereo matching. Whereas the different density of depth-aware fog effects are manually controlled by users, which could not guarantee the authenticity of output.

1.2 Our Motivations and Contributions

Having investigated the literature, we perceive that in the field of image-based fog/haze simulation, there's still large room for further research in at least two aspects. For one thing, in reality, the visibilities of two pixels having the same depth but different elevations are not the same. However, all the aforementioned image-based methods took the depth information as the distances from objects to the camera. Thereby, they cannot simulate fog/haze at particular elevations as shown in Fig. 1(b). What' more, when evaluating the authenticity of synthetic images, subjective assessment is usually involved in the experiments, which is effective, but not efficient. It's a great pity that how to measure the authenticity of simulated foggy/hazy images has not been investigated in the literature.

In this work, we attempt to fill the aforementioned research gaps, and the main contributions of our work are summarized as follows.

(1) We propose a framework based on atmospheric scattering model which can simulate both fog and haze effects at any elevation in image, namely *Foggy and Hazy Images Simulator* (FoHIS). Since the distances from objects through specific particle layers to camera are very important, instead of using the depth map as distance values, we estimate the elevation of each pixel in the image using perspective projection transformation and then compute

the distances interacted with the particle layer. Moreover, to simulate the heterogeneous fog effects, 3D Perlin noise is used to present more natural-looking results.

(2) As far as we know, we are the first to design an efficient *Authenticity Evaluator for Synthetic foggy/hazy Images* (AuthESI for short) to objectively measure the simulation results. We get inspiration from no-reference image quality assessment [12,13]. A collection of typical fog and haze features based on natural scene statistic (NSS) are selected and fit them to a multivariate Gaussian(MVG) model. Given a synthetic image, the authenticity is measured by computing a modified Bhattacharyya distance.

The remainder of this paper is organized as follows. Section 2 introduces the procedures of FoHIS. Section 3 presents our novel AuthESI. The experimental results and discussions are arranged in Sect. 4. Section 5 concludes the paper. Source codes are publicly available at https://github.com/noahzn/FoHIS.

2 FoHIS: Foggy and Hazy Images Simulator

2.1 Homogeneous Fog/Haze Simulation

According to [14], the fractional change of radiance I_o passing through thickness of $\mathrm{d}x$ in the direction ω due to absorption and scattering can be described as:

$$\frac{\mathrm{d}I_o(x,\omega)}{I_o(x,\omega)} = (-\beta_a(\omega) - \beta_s(\omega))\mathrm{d}x, \tag{1}$$

where β_a and β_s are absorption and scattering coefficients, respectively. Seeing that the result of absorption and scattering are the loss of visibility, we can combine the two coefficients as β_{ex}, *i.e.*, the extinction coefficient which is related to the visibility range of the atmosphere, V [15]:

$$\beta_{ex} = \frac{3.912}{V} \quad (m^{-1}) \ . \tag{2}$$

Then, in terms of Beer-Lambert law of attenuation [16], the intensity of a beam with the original radiance I_o after traveling a distance d through the particle layer can be expressed as:

$$I = I_o exp(-\beta_{ex} d) \ , \tag{3}$$

also, the opacity can be defined as:

$$O = 1 - exp(-\beta_{ex} d) \ . \tag{4}$$

From Eq. 3 we can notice that when the extinction coefficient is constant, the loss of the radiance intensity is directly related to the distance d. All the previous image-based methods we mentioned in Sect. 1.1 took depth values of the scene as the distances d. Supposing in a plan-parallel atmospheric layer with

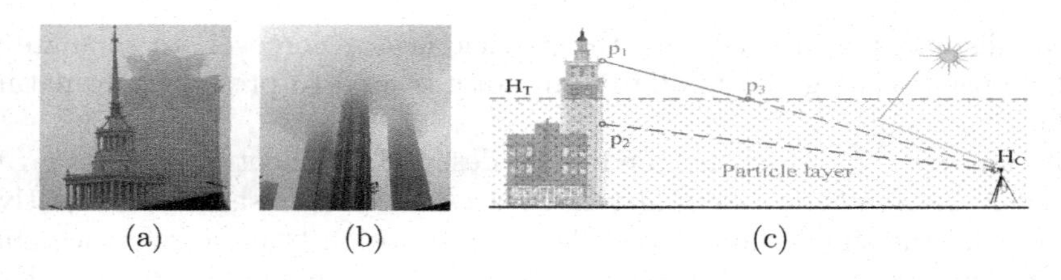

(a) (b) (c)

Fig. 1. (a) and (b) are images taken in reality. (a): Different elevations will result in different visibilities. (b): Fog/haze may occur at particular elevation. (c): Suppose the green dotted line denotes the top of the haze. Although p_1 and p_2 have equal depth value to the camera, the distances through the haze and reaching the camera (blue/red dotted line) are not equal. (Color figure online)

thickness H_T, and the elevation of the camera is H_C. As shown in Fig. 1(c), obviously, some pixels in images have large differences in elevations but share the same depth value. In order to obtain more accurate distances from objects through particle layer to camera, we first estimate pixels' elevations in images.

The field of view (FOV) of a camera is a pyramid, and the camera is located on the top of the cone. The pyramid is truncated by front clipping plane (FCP) and back clipping plane (BCP), thus forming a frustum which is called view frustum. Only those objects inside the view frustum are visible. 3D objects can be displayed onto 2D images by perspective projection transformation [17].

Given an RGB image, we segment main objects using GrabCut algorithm [18] and their depth values are set manually. With the depth map and vertical FOV of the camera, we can estimate all the elevations of pixels in the image.

Two terms must be considered if we want to decide the intensity of pixel p in image: 1) I_{ex}, the light reflected by p toward the direction of camera, and attenuated by particles. 2) I_{al}, the light scattered toward the direction of camera by particles. The former term can be expressed as:

$$I_{ex} = I_p exp(-\beta_{ex} r_p), \tag{5}$$

where I_p denotes the color of p in RGB channels, and r_p is the distance through the particle layer and reaching the camera. Since we have acquired the elevations of all the pixels, r_p can be easily derived following the properties of similar triangles. The latter term I_{al} is the color of fog or haze. We combine them as:

$$I = I_{ex} + O_p * I_{al}, \tag{6}$$

where O_p denotes the opacity of p calculated by Eq. 4. We can handle each pixel in three color channels of fog-free/haze-free images by Eq. 6 to simulate fog/haze effects.

2.2 Heterogeneous Fog

Perlin noise [19] is usually used to simulate inhomogeneous fog caused by atmospheric turbulence. [8,10,20] all used precomputed 2D Perlin noise to simulate the inhomogeneous foggy scenes. Due to the different depth values of pixels, it's well-founded to adopt 3D Perlin noise to achieve more natural effect. Given a 2D image and its corresponding depth map, we generate three 3D perlin noise with different amplitude and frequency of parameters, and combine them by:

$$noise = \frac{1}{3}\sum_{i=1}^{3}\frac{noise_i}{2^{i-1}} \ . \tag{7}$$

The computed noise coefficient is multiplied by β_{ex} in Eq. 5 when simulating heterogeneous fog.

3 AuthESI: Authenticity Evaluator for Synthetic Foggy/Hazy Images

In this section, we give an elaborated description of AuthESI, which is designed to objectively measure the authenticity of simulated foggy/hazy images. AuthESI is based on constructing a collection of typical natural scene statistic (NSS) features of fog/haze and fitting them to a multivariate Gaussian(MVG) model. The overall flowchart of AuthESI is shown in Fig. 2.

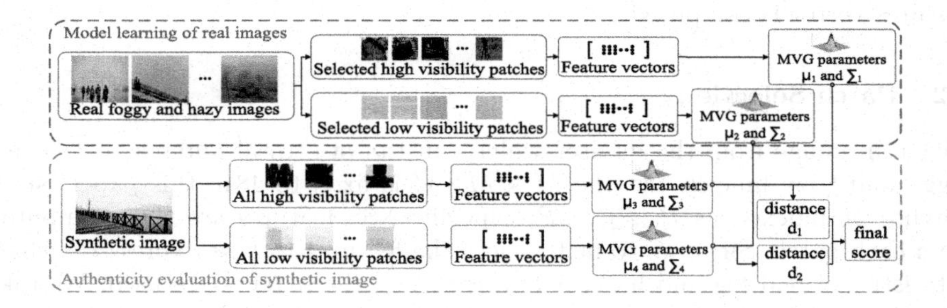

Fig. 2. Process flow of AuthESI.

3.1 NSS Features

Ruderman [21] has pointed out that the luminance of an input gray-scale image I conforms to a Gaussian distribution. The mean subtraction and divisive normalization operators (MSCN) can be computed as:

$$\hat{I} = \frac{I(i,j) - \mu(i,j)}{\sigma(i,j) + 1}, \tag{8}$$

where i and j are spatial coordinates, and

$$\mu(i,j) = \sum_{k=-K}^{K}\sum_{l=-L}^{L}\omega_{k,l}I(i+k, j+l), \tag{9}$$

$$\sigma(i,j) = \sqrt{\sum_{k=-K}^{K}\sum_{l=-L}^{L} \omega_{k,l}[l(i+k,j+l) - \mu(i,j)]^2} \quad (10)$$

are the local image mean and contrast, where $\omega = \{\omega_{k,l}|k = -K,...,K, l = -L,...,L\}$ defines a unit-volume Gaussian window.

Zhang et al. [22] have found that the log-derivative statistics are effective for analyzing natural images. After logarithmically converting the MSCN, we get:

$$J(i,j) = log[\hat{I}(i,j) + C], \quad (11)$$

where C is a small constant added to avoid numerical instabilities. Then, we compute the seven types of neighbours: $J(i,j+1) - J(i,j)$, $J(i+1,j) - J(i,j)$, $J(i+1,j+1) - J(i,j)$, $J(i+1,j-1) - J(i,j)$, $J(i-1,j) + J(i+1,j) - J(i,j-1) - J(i,j+1)$, $J(i,j) + J(i+1,j+1) - J(i,j+1) - J(i+1,j)$ and $J(i-1,j-1) + J(i+1,j+1) - J(i-1,j+1) - J(i+1,j-1)$.

Both MSCN values and log-derivative values can be modeled using a Generalized Gaussian Distribution (GGD), which is presented by: $f(x:\alpha,\beta) = \frac{\alpha}{2\beta\Gamma(1/\alpha)} exp(-(\frac{|x|}{\beta})^{\alpha})$, where $\Gamma(x) = \int_0^{\infty} t^{(x-1)} e^{-t}\, dt, x > 0$ denotes the gamma function. The variables α and β are shape and scale parameters which can be effectively used to describe the authenticity of fog and haze. Thereby, a 16-features vector is computed.

3.2 Patch Selection

Given an image from the dataset which consists of a total number of 180 real foggy and hazy images. Let the $M \times M$ (M is fixed to 48 in this paper) sized patches be indexed as $P_1, P_2, ..., P_n$. Specific NSS features are then computed from each patch. Since we want to select the features to best express the characteristics of real fog and haze, only a subset of the patches are used. We take the strategy of collecting two sets of patches, S_1 has the highest visibilities and S_2 has the lowest visibilities.

The dark channel prior proposed by He et al. [1] can be defined as $I_{dark}(i,j) = \min_{c \in R,G,B} [I_c(i,j)]$, where $c \in R, G, B$ represents the RGB color channels. Sky, fog or haze regions in photos usually have high value in I_{dark}, On the contrary, regions with high visibilities represent low value in I_{dark}. For each patch P, We first compute the average of dark channel values:

$$A_p = \frac{\sum_{i_p=1}^{M}\sum_{j_p=1}^{M} I_{dark}(i_p, j_p)}{M \times M}, \quad (12)$$

where $I_{dark}(i_p, j_p)$ is the dark channel values in patch P. Next, the binarization is performed in the patch as:

$$B(i,j) = \begin{cases} 0 & G(i,j) < \delta \\ 255 & G(i,j) \geq \delta \end{cases}, \quad (13)$$

where $G(i, j)$ is the gradient magnitude (GM) of $P(i, j)$ computed by *Sobel* operator, and δ is a threshold, which is fixed to 20 in this paper. For each patch P, we record the richness R_p by counting the numbers of 255 in P. If A_p is less than 30 and R_p is more than 400, this patch will be put into S_1. And if A_p is more than 100 and R_p is less than 80, this patch will be put into S_2.

3.3 Evaluate the Authenticity of Synthetic Images

The average of 16-features vector for all patches in S_1 and S_2 are computed separately. Then we use the MVG probability density to fit them:

$$MVG(f) = \frac{1}{(2\pi)^{d/2}|\sum|^{1/2}} exp[-\frac{1}{2}(f-\mu)^T \sum{}^{-1}(f-\mu)], \quad (14)$$

where f is the set of NSS features, μ and $\sum$ denotes the mean and covariance matrix of the MVG model.

Similarly, any synthetic image with fog or haze effect is treated like this and fit their features with two MVG model. But this time, all patches are used and if A_p of patch in Eq. 12 is less than 30, the patch will be put into the set S_3, otherwise, we put them into S_4. The authenticity of the synthetic image is expressed as the sum of modified Bhattacharyya distances:

$$D = \sqrt{(\mu_1 - \mu_3)^T (\frac{\sum_1 + \sum_3}{2})^{-1} (\mu_1 - \mu_3)} + \sqrt{(\mu_2 - \mu_4)^T (\frac{\sum_2 + \sum_4}{2})^{-1} (\mu_2 - \mu_4)}, \quad (15)$$

where $\mu_1, \mu_3, \mu_2, \mu_4$ and $\sum_1, \sum_3, \sum_2, \sum_4$ are the mean vectors and covariance matrices of the natural MVG model and the simulated image's MVG model.

4 Experiments and Discussions

The performance of our proposed FoHIS and AuthESI are evaluated in this section. The experiments were conducted on a PC equipped with an Intel(R) Core(TM) i7-4790 3.60 GHz CPU, 16 GB RAM.

Dataset: We select 18 fog-free/haze-free images with different estimated maximum depth, and each image is used to simulate fog or haze with FoHIS, Adobe Photoshop [23] and Guo's method [10]. FoHIS and Guo's method are implemented with Python. Since Guo's method cannot simulate fog/haze at particular elevations, two of the foggy images were only simulated by FoHIS and Photoshop. Thus, our dataset consists of 52 simulated images. Only 11 of these simulations are displayed in this paper, and their details are listed in Table 1. Please visit our source code page to check the complete dataset.

4.1 Evaluation of FoHIS

In this experiment, we organized the subjective assessment to evaluate the performance of FoHIS. A total number of 20 subjects who were 20 to 25 years old

Table 1. Details about the selected 11 simulated foggy/hazy images.

Input/Output	Maximum depth	Effect	Homogeneous	Particular elevation
(A)/(b), (c), (d)	150 m	Haze	Yes	No
(E)/(f), (g), (h)	800 m	Haze	Yes	No
(I)/(j), (k), (l)	30 m	Fog	No	No
(M)/(n), (o)	150 m	Fog	No	Yes

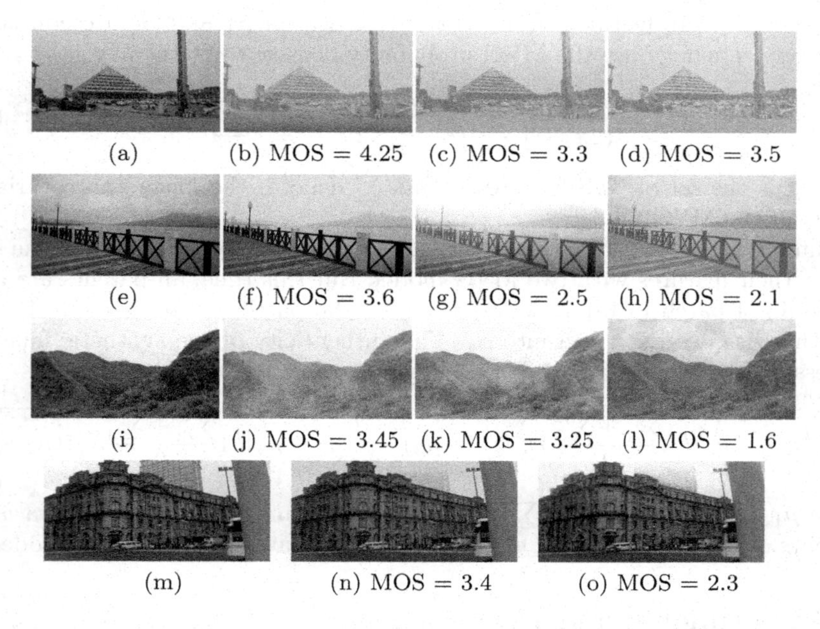

Fig. 3. For each row, from left to right: source image, FoHIS, Photoshop method and Guo's method [10]. The last row only contains ours and Photoshop method.

were involved in this experiment. According to the recommendations of [24], a LCD monitor with the resolution of 1920 × 1080 pixel was used to display one source image and corresponding synthetic images at the same time. We told the subjects that the first image was taken in real world, and the others (out of order) were simulated by different algorithms. The subjective score ranges from 1 to 5. 5 denotes the highest level of authenticity while 1 denotes the lowest. Each complete scoring operation (52 images) was limited to one hour so as to minimize the impact of fatigue. The benefits of doing so are mainly in three aspects. First, it enables the subjects to evaluate all images by using the same scoring strategy. Second, the relative ranking of the images simulated by different methods can be easily obtained. At last, authenticity comparison across different methods is meaningful in the evaluation of our AuthESI.

The mean opinion score (MOS) is computed by averaging all subjects' subjective scores. Pearson linear correlation coefficient (PLCC) and Spearman's rank-order correlation coefficient (SRCC) are computed as the evaluation criteria. Both coefficients range from [0, 1] with a higher value standing for better performance. The average values of PLCC and SRCC between the subjective scores of individual subjects and MOSs across the dataset are **0.7236** and **0.7014**, respectively. We can discover that the subjects reach a consensus on the authenticity of the synthetic images in the dataset. The MOSs of simulated images are shown in Fig. 3, which demonstrate that our FoHIS outperforms another two methods. Although Photoshop can generate reasonable effects, the results are not stable by reason of depending on user's skill level. Guo's method couldn't accurately estimate the transmission map all the time, hence their algorithm may fail in some images. Table 2 lists time cost of three simulation methods, from which we can see that although FoHIS needs preprocessing step, it runs fastest. When using the same source image to simulate dozens of foggy/hazy images with different extinction coefficients, FoHIS will show its high efficiency.

Table 2. Time cost of different methods for simulating an image. (640 × 480)

Method	Preprocessing	Homogeneous fog/haze	Heterogeneous fog
Photoshop	—	5 min	20 min
Guo's method	—	50 s	55 s
FoHIS	5 min	1.5 s	6 s

4.2 Evaluation of AuthESI

In this experiment, the performance of AuthESI was evaluated. We first computed the authenticity values (AV) of each simulated image in the dataset using Eq. 15. Then, we computed the PLCC and SRCC values of MOSs against AV across the dataset, which are **0.8124** and **0.8414**, respectively. The high values of these correlations indicate high consistency between prediction on simulated image's authenticity made by AuthESI and subjective judgements. In other words, our proposed innovative AuthESI can effectively evaluate the authenticity of simulated foggy/hazy images.

5 Conclusion

In this paper, a simulation framework FoHIS which can simulate natural-looking fog/haze effects in images is proposed. Meanwhile, an innovative objective evaluator AuthESI is designed and shows effectiveness in evaluating the authenticity of synthetic foggy/hazy images. Our future work may focus on constructing a large-scale dataset of foggy/hazy images and applying it to fog/haze removal.

Acknowledgments. This work was supported in part by the Natural Science Foundation of China under grant no. 61672380 and in part by the ZTE Industry-Academia-Research Cooperation Funds under grant no. CON1608310007.

References

1. He, K., Sun, J., Tang, X.: Single image haze removal using dark channel prior. IEEE Trans. PAMI **33**(12), 2341–2353 (2011)
2. Li, Y., Tan, R., Brown, M.S.: Nighttime haze removal with glow and multiple light colors. In: IEEE International Conference on Computer Vision, pp. 226–234 (2015)
3. Zhu, Q., Mai, J., Shao, L.: A fast single image haze removal algorithm using color attenuation prior. IEEE Trans. IP **24**(11), 3522–3533 (2015)
4. Zhang, D., Zhao, R., Shen, L., et al.: Action recognition in surveillance videos with combined deep network models. ZTE Commun. **14**(S1), 48–54 (2016)
5. Cai, B., Xu, X., Jia, K., Qing, C., Tao, D.: Dehazenet: an end-to-end system for single image haze removal. IEEE Trans. IP **25**(11), 5187–5198 (2016)
6. Kazufumi, K., Takashi, O., Eihachiro, N., Tomoyuki, N.: Photo realistic image synthesis for outdoor scenery under various atmospheric conditions. Vis. Comput. **7**(5), 247–258 (1991)
7. Zhou, K., Hou, Q., Gong, M., Snyder, J., Guo, B., Shum, H.Y.: Fogshop: real-time design and rendering of inhomogeneous, single-scattering media. In: Pacific Conference on Computer Graphics and Applications, pp. 116–125 (2007)
8. Anthony, G., Venceslas, B.: Modeling and rendering heterogeneous fog in real-time using B-spline wavelets. WSCG **2010**, 145–152 (2010)
9. Zhao, F., Zeng, M., Jiang, B., Liu, X.: Render synthetic fog into interior and exterior photographs. In: 12th ACM SIGGRAPH International Conference on Virtual-Reality Continuum and Its Applications in Industry, pp. 157–166 (2013)
10. Guo, F., Tang, J., Xiao, X.: Foggy scene rendering based on transmission map estimation. IJCGT **2014**, 10 (2014)
11. Liu, D., Klette, R.: Fog effect for photography using stereo vision. Vis. Comput. **32**(1), 1–11 (2016)
12. Zhang, L., Zhang, L., Bovik, A.C.: A feature-enriched completely blind image quality evaluator. IEEE Trans. IP **24**(8), 2579–2591 (2015)
13. Mittal, A., Soundararajan, R., Bovik, A.C.: Making a completely blind image quality analyzer. IEEE SPL **20**(3), 209–212 (2013)
14. Cox, L.J.: Optics of the atmosphere-scattering by molecules and particles. J. Mod. Optics **28**(7), 521–521 (1977)
15. Mahalati, R.N., Kahn, J.M.: Effect of fog on free-space optical links employing imaging receivers. Optics Express **20**(2), 1649 (2012)
16. Lykos, P.: The beer-lambert law revisited: a development without calculus. J. Chem. Educ. **69**(9), 730 (1992)
17. Carlbom, I., Paciorek, J.: Planar geometric projections and viewing transformations. ACM Comput. Surv. **10**(10), 465–502 (1978)
18. Rother, C., Kolmogorov, V., Blake, A.: "GrabCut": interactive foreground extraction using iterated graph cuts. ACM TOG **23**(3), 309–314 (2004)
19. Ken, P.: An image synthesizer. ACM SIGGRAPH Comput. Graph. **19**(3), 287–296 (1985)
20. Zdrojewska, D.: Real time rendering of heterogeneous fog based on the graphics hardware acceleration. Proc. CESCG **4**, 95–101 (2004)

21. Ruderman, D.L.: The statistics of natural images. Netw. Comput. Neural Syst. **5**(4), 517–548 (2009)
22. Zhang, Y., Chandler, D.M.: No-reference image quality assessment based on log-derivative statistics of natural scenes. J. Electron. Imaging **22**(4), 451–459 (2013)
23. TrickyPhotoshop.: Trickyphotoshop - How to Create Mist (Fog) using Photoshop CS6.http://www.youtube.com/watch?v=F9NTddMLSvM
24. ITU Radiocommunication Assembly: Methodology for the Subjective Assessment of the Quality of Television Pictures. International Telecommunication Union (2003)

Active Contours Driven by Saliency Detection for Image Segmentation

Guoqi Liu and Chenjing Li(✉)

Engineering Lab of Intelligence Business and Internet of Things, School of Computer and Information Engineering, Henan Normal University, Xinxiang 453007, Henan Province, China
gqliu@htu.edu.cn, 15249714369@163.com

Abstract. Aming at the over-segmentation problem of the active contour models, a new model based on the LBF (Local Binary Fitting) model driven by saliency detection is proposed. The proposed method consists of two main innovations: (1) The target object is located quickly and the initial contour is generated automatically by saliency detection method, which solves the problem that the LBF model is sensitive to the initial position, and the different targets can be segmented by selecting different initial contours. (2) The saliency detection results are transformed into priori energy functions, which are added to the energy model to prevent over-segmentation during the iterative process. We applied the proposed method to some gray images and real images, the simulation results show better segmentation accuracy.

Keywords: Active contour · Saliency detection · Prior region · Over segmentation · Initial contour

1 Introduction

The active contour model [1] is an effective method based on level set and curve evolution theory, which is widely used in image segmentation. Active contour model is mainly divided into edge-based [2,3] and region-based [4–8] models. The former relies on the edge information of the target object, which is difficult to deal with the weak boundary. The latter depends on the statistical information of the image, which is sensitive to the intensity inhomogeneity of images.

In order to deal with the image of intensity inhomogeneity [9,10], some region-based models are proposed, such as LBF (Local Binary Fitting) [6] model, LIF (Local Image Fitting) [7] model, LSACM (Local statistic active contour model) [8], etc. However, due to the nonconvexity problem of the energy function, these models are sensitive to the initial position and the energy function is easy to fall into the local minimum. It may leads to unexpected segmentation results, such as the over-segmentation problem. Especially for some real images, because the image intensity distribution is complex, we can not get the exact results.

Image saliency detection [11–13] is a technique for analyzing image features to generate image for saliency map. Hou et al. [11] proposed a spectral residual

D. Liu et al. (Eds.): ICONIP 2017, Part III, LNCS 10636, pp. 416–424, 2017.
https://doi.org/10.1007/978-3-319-70090-8_43

method, which is different from the traditional method of focusing on the whole image of the significant part, but to observe the changes of the background region of the image. Finally remove the background, and the rest is the prominent part of the image. Compared with the images of the log spectrum, the author found that the average value of log spectrum image is presented directly proportional to the frequency. Thus, the log amplitude spectrum of an image minus the mean log amplitude spectrum is obtained as a significant part. The method is characterized by high speed, high precision, and high robustness to different images, but the results have a certain distance to accurate segmentation.

In this paper, a new model is proposed, which combines the saliency detection with the LBF model. Firstly, by thresholding the detection results, the initial level set function is automatically generated. So that the initial contour can be close to the main target area, which avoids the problem that the traditional model needs to initialize the level set function for different images manually. Secondly, the saliency detection results are transformed into the prior region energy function and embedded into the LBF model. To prevent the contour far from the target object in the iterative process, while avoiding the over-segmentation of the target and preventing the energy function from being minimized locally.

2 Background

2.1 The Classical Active Contour Models

CV [4] model is a classic region-based model, the basic idea can be seen as a simplified Munford-shah function, whose energy function is expressed as:

$$E(C, c_1, c_2) = \mu \int_C \mathrm{d}s + \int_{\Omega_1} |I - c_1|^2 \mathrm{d}x + \int_{\Omega_2} |I - c_2|^2 \mathrm{d}x. \tag{1}$$

where Ω_1 and Ω_2 are the region of inside and outside the contour C respectively, whose intensity means are c_1 and c_2. I is image intensity. Moreover, the first term is the curve length to regularization with a weight μ and the last two terms are data fitting terms.

The CV model effectively solves the segmentation problem of the weak boundary image and the images without gradient. But because of its assumption that the background and foreground of the image intensity distribution is evenly, and in practical application, there is a large number of images with uneven intensity distribution, and the segmentation results are not satisfactory.

For the images of intensity inhomogeneity, the global region can not be reasonably fitted. But in the local area, the target object and the background can be binary fitted. So Li et al. [6] proposed the LBF model. The model joins the Gaussian kernel function to control the size of the region, thus extracting the local intensity information of the image and constructing a new energy function:

$$E^{LBF} = \sum_{i=1}^{2} \lambda_i \int e_i(x) \mathrm{d}x. \tag{2}$$

where $e_1(x)$ and $e_2(x)$ are defined as follows:

$$\begin{cases} e_1(x) = \int_\Omega K_\sigma(y-x)|I(y) - f_1(x)|^2 H(\phi)\mathrm{d}y, \\ e_2(x) = \int_\Omega K_\sigma(y-x)|I(y) - f_2(x)|^2 (1 - H(\phi))\mathrm{d}y. \end{cases} \quad (3)$$

K_σ is a kernel function, which is obtained by a truncated Gaussian function. $f_1(x)$ and $f_2(x)$ are computed as follows:

$$\begin{cases} f_1(x) = \frac{K_\sigma(x)*I(x)H(\phi(x))}{K_\sigma(x)*H(\phi(x))}, \\ f_2(x) = \frac{K_\sigma(x)*I(x)(1-H(\phi(x)))}{K_\sigma(x)*(1-H(\phi(x)))}. \end{cases} \quad (4)$$

The energy functional E^{LBF} is obtained by integrating the length regularization term in the above energy. Length regularization term is defined as follows:

$$L(\phi) = \int_C \mathrm{d}s = \int |\nabla \phi| \mathrm{d}x. \quad (5)$$

Then the evolution equation of level set function ϕ is computed as follows:

$$\frac{\partial \phi}{\partial t} = -\frac{\partial E}{\partial \phi} = -\delta_\epsilon(\phi)(e_1 - e_2) + \lambda \delta_\epsilon(\phi) div(\frac{\nabla \phi}{|\nabla \phi|}). \quad (6)$$

where ∇ is the gradient operator, $div(.)$ is the divergence operator.

The model can deal with the problem of image segmentation of intensity inhomogeneity. But due to the nonconvexity of energy function, it may fall into the local minimization, resulting in over-segmentation and sensitive to intialization.

Recently the Locally Statistical Active Contour Model (LSACM) [8] is proposed for images of intensity inhomogeneity. The model uses the local image statistical information to construct the level set energy function, which is not sensitive to initialization of the level set. But the model relies on a specific probabilistic model with high computational complexity, and the initialization of level set is still need to set manually.

2.2 Saliency Detection

In [11], by analyzing the logarithmic spectrum of the input image, the spectral residual of the image in the spectral domain is extracted, and a fast method for constructing the corresponding saliency map in the spatial domain is proposed. For a given an image I(x):

$$A(f) = \Re(F[I(x)]), \quad (7)$$

$$P(f) = \Im(F[I(x)]), \quad (8)$$

$$L(f) = log(A(f)), \quad (9)$$

$$R(f) = L(f) - h_n(f) * L(f), \quad (10)$$

$$S(x) = g(x) * F^{-1}[exp(R(f) + P(f))]^2. \quad (11)$$

where $A(f)$ denotes the general shape of log spectra. F and F^{-1} denote the Fourier Transform and Inverse Fourier Transform, respectively. $P(f)$ denotes the phase spectrum of the image, which is preserved during the process. The h is a convolution kernel of the mean filter with n*n, n is generally set to 3. $R(f)$ denotes the statistical singularities that is particular to the input image. Spectral Residual is added to the phase spectrum to obtain natural exponential result, and we can get the final saliency map.

3 The Proposed Model

3.1 Initialization

Initialization is the first step in the curve evolution of the active contour model. Inappropriate initial contours will keep the curve away from the real boundary, increase the number of iterations, and even get the correct segmentation result. Traditional active contours usually require artificial set of initial contours to achieve a reasonable initial position, but there are many uncertainties in manual initialization, the low efficiency of manual operation can affect the speed of curve evolution and the final segmentation effect. So it is necessary to replace the manual initialization with a simple and fast automation method.

The saliency detection results are usually probabilistic functions between [0, 1], In order to better combine with the active contour model, this paper first binarizes the results, and the thresholds are generally the mean values of the significance test results:

$$O(x) = \{-2, S(x) \geq mean(S(x)), 2, S(x) < mean(S(x)). \quad (12)$$

As shown in the Fig. 1, the first column is the result of the significance test,the second column is the image after the threshold. Then delete the small area, filling the hole and other morphological operations, we can get the third column of the level set function. Its 0 level set is the initial outline of the fourth column in the figure.

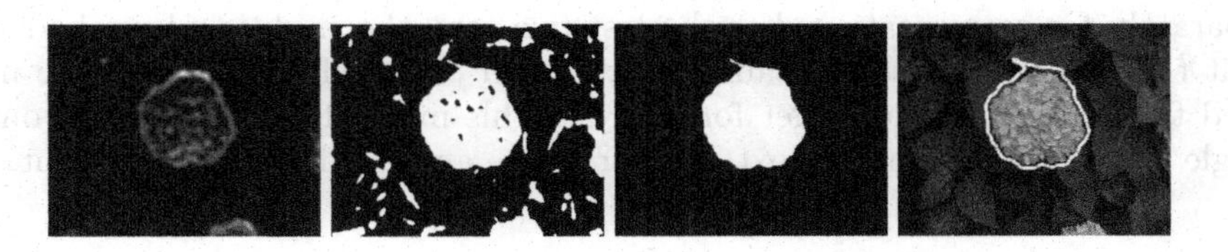

Fig. 1. The initial contour formation process. The column from left to right is the saliency detection result, the thresholding result, the initial level set function, the initial contour.

3.2 The Proposed Energy Model

The traditional region-based active contour model, such as CV, LBF. Due to the complexity of the image prospects or the large difference in intensity distribution, it usually results in over-segmentation of the target object. In this paper, the target object is positioned by the saliency detection method, and the result of the detection is embedded into the LBF energy function as a priori energy function [14,15]. So with the Formula (6), we get a new energy function:

$$\frac{\partial \phi}{\partial t} = -\delta_\epsilon(\phi)(e_1 - e_2) + \lambda_1 \delta_\epsilon(\phi) div(\frac{\nabla \phi}{|\nabla \phi|}) + \lambda_2 G_\sigma * O(x). \tag{13}$$

where λ_1, $\lambda_2$is the weight coefficient, G_σ is gaussian filter. Although the saliency detection method can not achieve the exact segmentation result, it can be used as a constraint term to prevent the energy function from falling into local minimization. In the area away from the boundary the priori energy plays the major role in preventing the occurrence of over-segmentation, and in the area close to the boundary, the LBF energy function plays the major role in achieving accurate segmentation results.

4 Experimental Results

4.1 Target Selection

In the practical application of image segmentation, the number of the targets is uncertain. Sometimes the need to split out all the targets, and sometimes need to select a specific single target for segmentation. In this paper, different splitting targets can be obtained by selecting different initial contours.

Figure 2 shows two representative multi-target images for illustration. Where the first column is the initial contour selected for the different targets, the second column is the corresponding a priori energy region, and the third column is the segmentation result. The traditional method usually ignores the half flower below the picture, but this paper can separate all the flower areas by selecting the initial contour and the priori region, and you can also select one of the flowers separately for segmentation. For the last two row, the traditional method will split the two aircraft at the same time. But in practical applications, we may need to extract a single target for analysis, this method can easily split out a single Of the target. So the next step for image analysis is more convenient.

4.2 Over-Segmentation Problem

For the image that the target foreground and background intensity contrast is not obvious, or the foreground intensity distribution is complex. It is often result in over-segmentation. Figure 3 shows the comparison of over-segmentation problem. The first row for the flower pattern, LBF in the image of the over segmentation problem is serious, CV model has a little over-segmentation. The second row

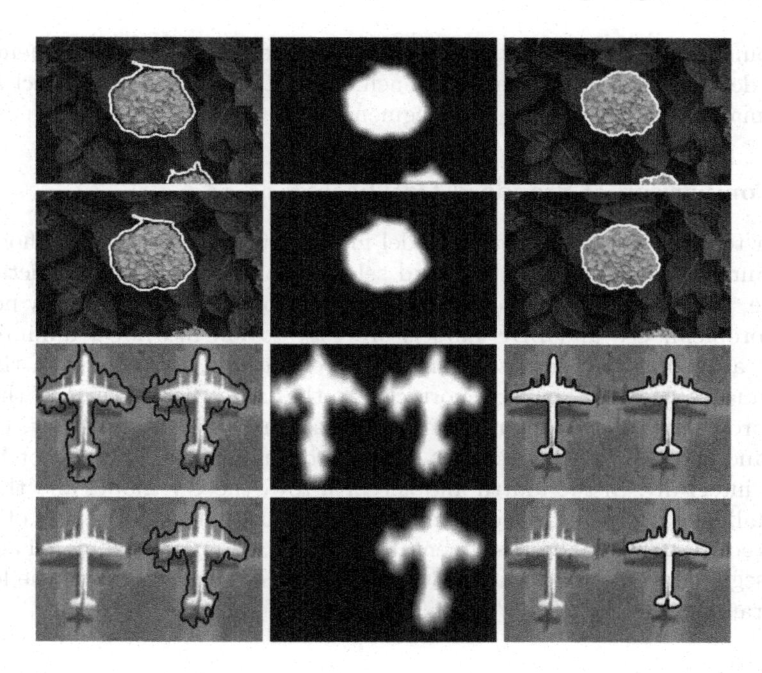

Fig. 2. Segmentation for different objects. The column from left to right is the initial contour, the priori energy function, the segmentation results.

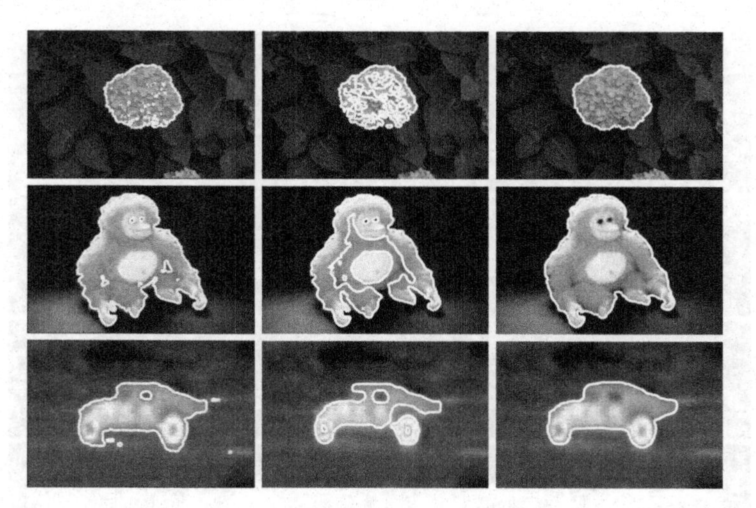

Fig. 3. Image segmentation results compared for over-segmentation. The column from left to right is the CV model, the LBF model and the proposed model.

for the toy pattern, CV and LBF model is over-segmentation. The eye part, in particular, is similar to the background and is easily segmented by individual. The third row is the car pattern. The image is more vague, the foreground and background intensity distribution is not obvious, resulting in over-segmentation. The model proposed in this paper can achieve better segmentation effect for the above images, and there is no over segmentation.

4.3 Comparison of Standard Databases

In order to measure the proposed model more accurately, this paper choose the Weizmann segmentation database and select several representative pictures to compare. Figure 4 shows the results of segmentation. Due to the energy nonconvexity problem, CV and LBF models are easy to fall into local minimization without achieving the correct segmentation effect. The CV model in the first row of images does not converge correctly to the target boundary, and the LBF model produces a leak segmentation. In the second and the third row, the CV model and the LBF model are over segmented, and the target can not be segmented integrally. In the fourth and the fifth row, the CV model and the LBF model fall into the local region, and the segmentation result is out of order. Compared with the standard segmentation results, the proposed model achieves higher segmentation accuracy and avoids the over segmentation and the leakage segmentation problem.

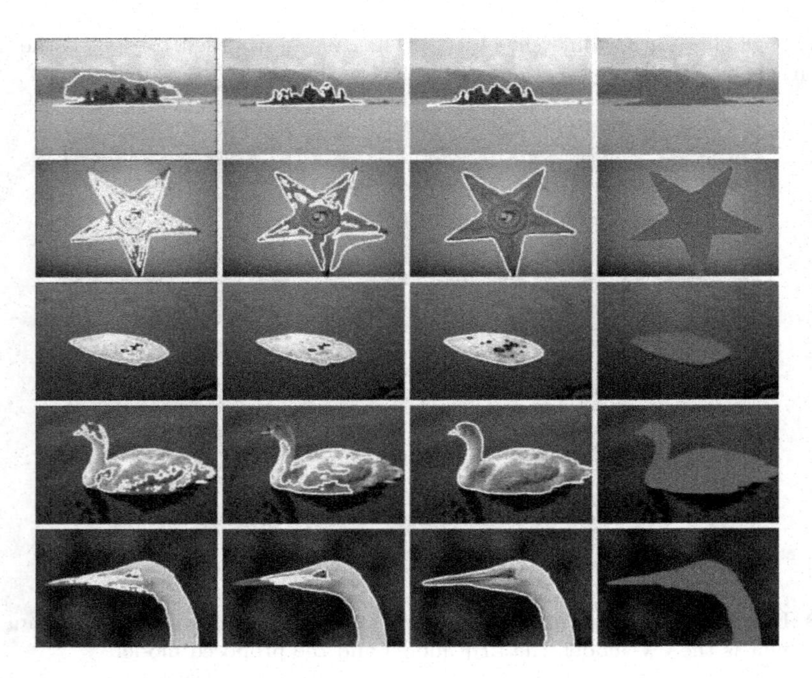

Fig. 4. Image segmentation results compared for Weizmann database. The column from left to right is the CV model, the LBF model, the proposed model and the standard segmentation results.

The Weizmann segmentation database adopts the F score evaluation method, and the formula is:

$$P = \frac{TP}{TP + FP}, \tag{14}$$

$$R = \frac{TP}{TP + FN}, \tag{15}$$

$$F = \frac{2 * PR}{P + R}. \tag{16}$$

where TP is True Positive, it means the correct target foreground segmentation sample. FP is False Positive, it means the sample which divides the target background into the foreground. FN is False Negative, it means the sample which divides the target foreground into the background wrongly. P is the precision rate and R is the recall rate. F means the accuracy rate, it can be interpreted as a weighted average of the precision and recall, where F reaches its best value at 1 and worst at 0.

Table 1. The F score value of the segmentation results in Fig. 4

Image	CV	LBF	The proposed model
1	0.6640	0.8833	0.9551
2	0.6504	0.7648	0.9896
3	0.9086	0.9087	0.9859
4	0.7711	0.5544	0.9770
5	0.8829	0.9098	0.9808

Table 1 shows the corresponding F score values of the segmentation results in Fig. 4. As can be seen from the table, the accuracy of CV and LBF models for multiple images is low, and the reason is that the segmentation results appear the over segmentation or the leakage segmentation problems. The proposed method in this paper shows the higher accuracy.

5 Conclusions

In this paper, we introduce the saliency detection method to the LBF model and a new model is constructed. Firstly, the initial contour is automatically generated by saliency detection results, and through the choice of initial contour to achieve the separation of different targets. Secondly, the significance detection results are added into the energy function as a priori region information, which effectively solves the problem of over-segmentation. Experimental results show that the proposed model has higher segmentation accuracy than the traditional models.

Acknowledgement. This work is jointly supported by the National Natural Science Foundation of China (No. U1404603).

References

1. Kass, M., Witkin, A., Terzopoulus, D.: Snakes: active contour model. Int. J. Comput. Vis. **1**(4), 321–331 (1988)
2. Caselles, V., Kimmel, R., Sapiro, G.: Geodesic active contours. Int. J. Comput. Vis. **22**(1), 61–79 (1997)
3. Xie, X., et al.: Active contouring based on gradient vector interaction and constrained level set diffusion. IEEE Trans. Image Process. **19**(1), 154–164 (2010)
4. Chan, T., Vese, L.: Active contours without edges. IEEE Trans. Image Process. **10**(2), 266–327 (2001)
5. Li, C.M., Kao, C., Gore, J., et al.: Minimization of region-scalable fitting energy for image segmentation. IEEE Trans. Image Process. **17**(10), 1940–1949 (2008)
6. Li, C.M., Kao, C., Gore, J., Ding, Z.: Implicit active contours driven by local binary fitting energy. In: Proceedings of IEEE conference on Computer Vision and Patter Recognition (2007)
7. Zhang, K., Song, H., Zhang, L.: Active contours driven by local image fitting energy. Pattern Recogn. **43**(4), 1199–1206 (2010)
8. Zhang, K., Zhang, L., Lam, K.M., et al.: A level set approach to image segmentation with intensity inhomogeneity. IEEE Trans. Cybern. **46**(2), 546–557 (2016)
9. Dong, F., Chen, Z., Wang, J.: A new level set method for inhomogeneous image segmentation. Image Vis. Comput. **31**(10), 809–822 (2013)
10. Li, C., Wang, X., Eberl, S., et al.: Robust model for segmenting images with/without intensity inhomogeneities. IEEE Trans. Image Process. **22**(8), 3296–3309 (2013)
11. Hou, X., Zhang, L.: Saliency detection: a spectral residual approach. In: IEEE Conference on Computer Vision and Pattern Recognition, pp. 1–8 (2007)
12. Borji, A., Cheng, M.M., Jiang, H., et al.: Salient object detection: a benchmark. IEEE Trans. Image Process. **24**(12), 5706–5722 (2015)
13. Tong, N., Lu, H., Ruan, X., et al.: Salient object detection via bootstrap learning. In: Proceedings of the IEEE Conference on Computer Vision and Pattern Recognition, pp. 1884–1892 (2015)
14. Tran, T., Pham, V.T., Shyu, K.K.: Moment-based alignment for shape prior with variational B-spline level set. Mach. Vis. Appl. **24**(5), 1075–1091 (2013)
15. Song, Q., Bai, J., Garvin, M.K., et al.: Optimal multiple surface segmentation with shape and context priors. IEEE Trans. Med. Imaging **32**(2), 376–386 (2013)

Robust Visual Tracking by Hierarchical Convolutional Features and Historical Context

Zexi Hu, Xuhong Tian, and Yuefang Gao(✉)

South China Agricultural University, Guangzhou, China
huzexi@outlook.com,
{tianxuhong,gaoyuefang}@scau.edu.cn

Abstract. In this paper, we present a visual tracking method to address the problem of model drift, which usually occurs because of drastic change on target appearance, such as motion blur, illumination, out-of-view and rotation. It has been proved that the hierarchical convolutional features of deep neural networks learned by huge classification datasets are generic for other task and can aid the tracker's power of discrimination. Ensemble-based trackers have been studied also to offer historical context for drift correction. We combine these two advantages into our proposed tracker, in which correlation filters are learned by hierarchical convolutional features and preserved as snapshots in an ensemble in certain occasion. Such an ensemble is capable of encoding the target appearance as well as provide historical context to prevent drift. Such context is considered to be complementary to correlation filters and convolutional features. The experimental results demonstrate the competitive performance against state-of-the-art trackers.

Keywords: Visual tracking · Correlation filters · Ensemble · Convolutional neural networks

1 Introduction

Visual Tracking plays a fundamental and important role over the past decades among topics in computer vision for its wide applications, such as video surveillance, automatic driving, human-computer interaction and so on. The problem of model drifting in tracking still remains challenging due to factors like illumination changes, partial or fully occlusion, fast motion, motion blur.

Previous methods attempt to handle these problems, ones of which proposed effective discriminative models like MIL [1], TLD [2], OAB [3]. Other studies focus on the representation abilities of features, which develop several hand-crafted features such as HOG [4, 5] and Color Name [6]. In [7], it has been studied that the discriminative model and the feature extraction play complementary roles in the tracker, which means a sophisticated model might not obtain a gain of performance with a well-crafted feature since they are both offering discriminative information. Therefore the bottleneck comes if improvement is solely made on these two components.

Recent years, as the development of deep learning achieves great progress, Convolutional Neural Networks (CNN) brings outstanding performance in several fields of

D. Liu et al. (Eds.): ICONIP 2017, Part III, LNCS 10636, pp. 425–434, 2017.
https://doi.org/10.1007/978-3-319-70090-8_44

computer vision, starting with object recognition, object detection. CNNs training requires a large amount of training data such as ImageNet [8] dataset and further studies prove these convolutional layers learned for recognition can extract generic features for other applications. Such a great success has spread to edge detection [9], scene classification [10] and so on. Several applications of CNNs has been made on visual tracking and most of the performance gain owe to convolutional layers' rich features. A convention Correlation Filter (CF) with the feature extracted from only one single layer can even outperform ones with an combination of hand-crafted features [11].

In spite of achieving state-of-the-art performance by CF and CNNs, another limitation still remains that a single-model tracker is prone to model drifting while error information accumulates during model updating and drastic change of target appearance accounts for this phenomenon such as motion blur, partial or fully occlusion, illumination variations. Strategies are proposed to tackle this model drift problem, one of which is to employ one or more components to detect and correct the tracker's error [2, 12–14], while the other one is to maintain an ensemble containing multiple trackers and making decision to each frame based on members of the ensemble [15–19].

In this paper, we propose a CNN-based tracker with multiple snapshots, which takes advantage of the rich hierarchical features offered by CNN to discriminate target from the background, while exploiting historical context offered by snapshots serving knowledge in past environment of the tracking processing to prevent the tracker from drifting. On a standard benchmark of 50 and 100 image sequences [20], our method achieves competitive performance in comparison with state-of-the-art trackers, especially CNN-based trackers.

2 Related Work

Correlation Filter Tracker. Correlation filters have been widely studied on visual tracking due to its computational efficiency. It is first proposed as a tracker in [21] learning a minimum output sum of squared error (MOSSE) filter encoding target appearance. Henriques et al. generalize correlation filter to kernel space and multiple channels in CSK [22] and KCF [4]. Danelljan et al. exploit the color attributes and a Gaussian kernel space in [6], introduce adaptive multi-scale correlation filters using HOG features in [23], and propose Spatially Regularized Correlation Filters (SRDCF) to address the boundary effects problem in [24]. For more details, we refer readers to a comprehensive survey of CF trackers [25].

Deep Learning Tracker. Several methods have been proposed to introduce deep learning into visual tracking. Numerous of them treat the deep neural networks as feature extractor [11, 26, 27]. In [11], Spatially Regularized Correlation Filters is adopted to learn CNN features. In [26, 27], multiple layers of CNN are exploited to calculate a response map by fixed weights and adaptive Hedge correspondingly. In [28], Wang et al. pick output of VGG-Net's two convolutional layers to train a long-term and short-term network correspondingly.

Ensemble-Based Tracker. Ensemble is usually regarded as a post-processor since generally it makes decision based on its containing member's decisions. In [15, 17], trackers making decisions with preserved snapshots is capable to recover after severe change of target appearance. The main insight is to exploiting historical context offered by snapshots preserved in the past of the tracking process and re-detect the target as soon as it reappear after drastic changes. In [17], entropy minimization is used as a metric for snapshot selection and in [15] trajectory consistency is adopted for CF due to its unambiguity. In [18, 19], trajectory consistency is also employed but both compare forward and backward trajectories of each tracker, while in [19] the ensemble is formed by different types of trackers and in [18] the trackers in the ensemble are generated from a tracker but the trajectories varies due to different update behaviors.

3 Proposed Method

In this section, we introduce the basic components, Correlation Filter and Hierarchical Features Extraction, and then they are employed under the framework of Historical Snapshot Selection in order to track with historical context. Finally, we present the whole pipeline of the proposed tracker.

3.1 Correlation Filter

Typical correlation filters tracker are adopted as a base tracker in our method. A CF tracker aims to find a classifier

$$f(\mathbf{x}) = \langle \mathrm{w}, \phi(\mathrm{x}) \rangle \tag{1}$$

to make prediction for the probability of image patch. Instead of sampling image patch through stepping, correlation filters model the appearance of a target object through training weights $\mathbf{w}$ on an image patch $\mathbf{x}$ of $M \times N$ pixels, which is usually an extended region centered at target with a padding, and all the training samples are extracted by the circular shift of $\mathbf{x}_i$, where $i \in \{0, 1, \ldots, M-1\} \times \{0, 1, \ldots, N-1\}$. These samples are assigned training label generated by Gaussian function y_i. The training process is to minimize the regression error:

$$\min_{\mathbf{w}} \sum\nolimits_i (f(\mathbf{x}) - y_i)^2 + \lambda \|\mathbf{w}\|^2, \tag{2}$$

where λ is a regularization parameter for preventing overfitting, and in $f(\mathbf{x})$, the function $\phi(\mathrm{x})$ maps x to the kernel space. Optimizing w by exhausting all the windows of the region of interest is considered to be computationally expensive, however, using the fast Fourier transformation (FFT), the filter w can be obtained as

$$\mathbf{w} = \sum_i \alpha_i \phi(\mathrm{x}_i), \tag{3}$$

where the coefficient $\boldsymbol{\alpha}$ is defined by

$$A = \mathcal{F}(\boldsymbol{\alpha}) = \mathcal{F}(\mathbf{y})/(\mathcal{F}(\phi(\mathbf{x}) \cdot \phi(\mathbf{x})) + \lambda) \tag{4}$$

In (4), $\mathcal{F}$ denotes the discrete Fourier operator and $\mathbf{y} = \{y_i\}$. In the new frame, the detection task is carried out on an image patch $\mathbf{z}$ cropped from the search window of $M \times N$ pixels and a response map is evaluated as

$$\mathrm{f}(\mathbf{z}) = \mathcal{F}^{-1}(A \odot \mathcal{F}(\phi(\mathbf{z}) \cdot \phi(\widehat{\mathbf{x}}))) \tag{5}$$

where $\widehat{\mathbf{x}}$ denotes the learned target appearance model.

To alleviate the problem of boundary discontinuities, input channels are weighted by a cosine window [21]. To keep the filters adaptive when target appearance changes, the update process is conducted as

$$\widehat{\mathbf{x}}^t = (1-\gamma)\widehat{\mathbf{x}}^{t-1} + \gamma \mathbf{x}^t \tag{6}$$

$$\widehat{A}^t = (1-\gamma)\widehat{A}^{t-1} + \gamma A^t \tag{7}$$

where t is the index of the current frame and γ is the learning rate.

3.2 Hierarchical Features Extraction

Several studied has investigated the application of the convolutional features on visual tracking. In [27], features are extracted from hierarchical layers of CNNs. Features of three layers are obtained and learned on three correlation filters separately to build a coarse-to-fine response map. Such a strategy can capture information from a high level, which is effective in discriminating the target from background against drastic changes with its semantic information, and low level, which offers spatial details to localize precisely with its larger resolution.

In this work, VGG-Net-19 [29] is chosen to extract the response map. In every frame, image patch of size $M \times N$ cropped from the search region centered at the target is resized to 224×224 and forward propagates through the network. The output of the *conv3-4*, *conv4-4* and *conv5-4* convolutional layers are taken as features encoding the search region. Each of the layer outputs will be learned a response map $f_i(\mathbf{x})$, where i indicates the index of the convolutional layer, following Eq. 2 and be updated following Eqs. 6 and 7.

Obtaining three response maps, a necessary process is conducted to resize these maps to the same size of $\frac{M}{4} \times \frac{N}{4}$ in order to operate the following composition process, which is defined as

$$\hat{f}(\mathbf{x}) = \frac{1}{n}\sum\nolimits_i \beta_i \mathrm{f}_i(\mathbf{x}) \tag{8}$$

where n is the number of selected layers and β_i is the weight of layer i. We follow the weight setting in [27] assigning 1, 0.5 and 0.02 for *conv5-4*, *conv4-4* and *conv3-4*.

3.3 Historical Snapshot Selection

Assuming a tracker $\mathcal{T}$, which is a classifier, in essence, updates its model while tracking in every incoming frame. Letting $\mathcal{T}$ save a snapshot at every interval τ and maintaining a snapshot ensemble $\mathbf{E}$ of size n which abandons oldest snapshot when new one comes, at frame t we have the ensemble $\mathbf{E} = \{\mathcal{S}_t, \mathcal{S}_{t-\tau}, \ldots, \mathcal{S}_{t-n\times\tau}\}$. The saved snapshots preserve the learned parameters of correlation filters from changing, as a result the ensemble serves as a role offering historical knowledge.

To make a prediction in every frame by ensemble $\mathbf{E}$, it is very critical to design a criterion for selecting the best snapshot $\mathcal{S}^*$ that generates the most robust results. The selection criterion is formally defined as follow,

$$\mathcal{S}^* = \mathrm{argmax}_{\mathcal{S}\in\mathbf{E}} \sum\nolimits_{k\in[t-\Delta]} \mathfrak{L}_{\mathcal{S}}^k \tag{9}$$

where $\mathfrak{L}_{\mathcal{S}}^k$ is the robust score of $\mathcal{S}_i$ and Δ is the size of the temporal window. This formula aims to measure the snapshots' robustness by their cumulative scores within a temporal window of fixed size.

The robust score $\mathfrak{L}_{\mathcal{S}}^k$ becomes the key of the measurement in the criterion. In [17], Zhang et al. apply entropy based loss function that is working well with a base tracker of online SVM method which shows ambiguity when a severe change occurs on target appearance. However, regarding of the less ambiguity of correlation filters, which are trained by a regression model, this function doesn't perform well with correlation filter trackers. Hence, in [15], Li et al. proposed to consider the trajectory consistency into the robust score, resulting in

$$\mathfrak{L}_{\mathcal{S}} = \ln C_{\mathcal{S}} - \eta H_{\mathcal{S}} \tag{10}$$

where the superscript of time t is omitted, η is the scalar controlling the tradeoff between the two terms. In the first term, trajectory consistency score $C_{\mathcal{S}}$ is defined as

$$C_{\mathcal{S}} = \frac{1}{n-1} \sum\nolimits_{(\mathcal{S}_i\in\mathbf{E})\cap(\mathcal{S}\neq\mathcal{S}_i)} \exp\left(-\frac{\|x_{\mathcal{S}} - x_{\mathcal{S}_i}\|^2}{\sigma^2}\right) \tag{11}$$

where $x_{\mathcal{S}_i}$ denotes the predicted bounding box of $\mathcal{S}_i$ in a specific frame. Hence, in this term, a snapshot with a trajectory more consistent with other snapshots is favored. In the second term, the entropy regularization is computed as

$$H_{\mathcal{S}} = -\sum_{Y\in Z} P(Y|X;\mathcal{S}) \log P(Y|X;\mathcal{S})$$

where X represents the target candidates proposed by the ensemble $\mathbf{E}$ and Y denotes the label indicating whether X is foreground or background while Z denotes the label set. Therefore, in this term, a snapshot with less ambiguity is favored.

We follow the setting of parameters in [15], η is 15 and σ is third of template size.

3.4 Pipeline

Given the above components, our tracker is proposed as HCNN (Historical CNN-based tracker). The pipeline of the tracker is described in this section.

At the beginning of the tracking process, a base tracker with correlation filters learned from the feature of the three layers in hierarchical features extraction is initialized and in the every incoming frame the tracker will predict the target location according to the response map composited by hierarchical features following Eq. 8 and update its model according to Eqs. 6 and 7. Meanwhile, following the description in Sect. 3.3, an ensemble **E** is maintained with n snapshots, which are generated at every interval, and the oldest one will be cast out when the number of snapshots exceeds n.

In every frame, the final decision is made by historical snapshot selection, calculating the robust scores of the snapshots according to their response maps, according to Eq. 9. Once the best snapshot $\mathcal{S}^*$ in the ensemble obtains a higher robust score than the current one, which keeps updating during the tracking process, we say a disagreement occurs in the ensemble that means the current tracker has updated unnecessary information in its model. Therefore, a rescue measure is taken to replace the current tracker with the best snapshot $\mathcal{S}^*$, which has preserved the uncorrupted model, as well the predicted result.

4 Experiments

4.1 Experiment Setup

We evaluate our proposed method on a public benchmark [20] containing 50 and 100 image sequences with state-of-the-art methods. Participating trackers comes mainly comes from three categories: (1) deep learning tracker: DeepSRDCF [11], HDT [26], CF2 [27], FCNT [28]; (2) multiple-snapshot tracker: MEEM [17], SME [15], (3) correlation filter trackers: SRDCF [24], Staple [13], DSST [23].

The experiment is performed in One Pass Evaluation (OPE) and two plots are generated as precision plot and success plot, the former of which shows the percentage of frames where the center location error of bounding boxes is within a threshold and the latter show the percentage of frames where the overlap ratio of bounding boxes exceeds a threshold. Trackers are ranked by precision scores which are the accuracy at 20-pixel threshold in the precision and success scores which are the Area Under Curve (AUC) score in the success plot.

In both datasets, our proposed tracker are evaluated with same parameters which are $\tau = 10$, $n = 7$ and $\Delta = 4$.

4.2 Experiment Results

Overall Performance. In Figs. 1 and 2, precision plots show our tracker HCNN performs favorably against state-of-the-art methods on both datasets, including trackers with deep learning features. With historical context, HCNN outperforms CF2 which is the non-historical version of HCNN, improving from 0.836 to 0.850 and from 0.803 to

0.817 in OTB100 and OTB50 correspondingly. In other hand, success plots show HCNN are still competitive but fall behind some trackers which have the capacity of handling the problem of target scale since success plots consider the overlap area between predicted bounding boxes and ground truth.

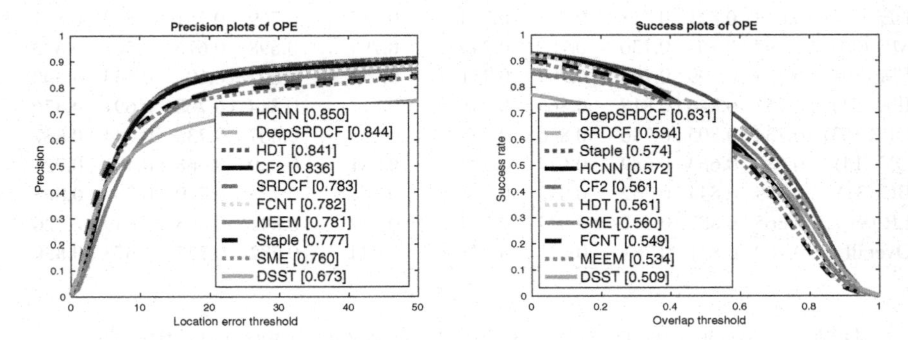

Fig. 1. Overall precision and success plots on OTB100 dataset.

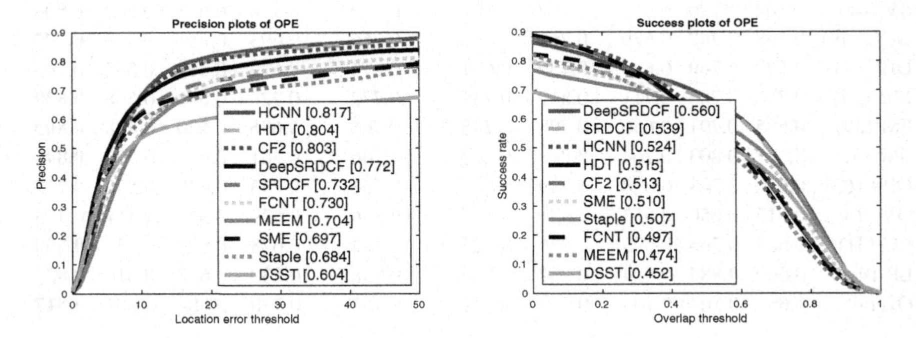

Fig. 2. Overall precision and success plots on OTB50 dataset.

Qualitative Evaluation. To gain further insight into the advantage of HCNN, we report results for 11 challenging attributes on OTB100 and OTB50 in Tables 1 and 2. These attributes are IV (illumination variation), SV (scale variation), OCC (occlusion), DEF (deformation), MB (motion blur), FM (fast motion), IPR (in-plane-rotation), OPR (out-of-plane rotation), OV (out-of-view), BC (background clutters) and LR (low resolution). The highest score in every attribute is set bold and the numbers following the abbreviations of attribute names indicate the number of image sequences.

HCNN ranks first in 6 attributes, especially in IV, MB, BC, outperforming the second one by about 2.5%, 3.0%, 2.5% in OTB100 and 4.6%, 3.3%, 4.4% in OTB50. In MB and OV, HCNN improves CF2 by 5.5% and 6.5% in OTB100 and 3.3% and 8.0% in OTB50, which is very significant in spite of not being the best one. It is noteworthy that due to typical correlation filter tracker's nature of the only searching limited area around the previous target location, it is prone to failure when MB or

Table 1. Precision scores of trackers for 11 challenging factors on OTB100 dataset.

	SME	HDT	MEEM	CF2	SRDCF	DeepSRDCF	FCNT	Staple	DSST	HCNN
IV (38)	0.753	0.820	0.771	0.831	0.792	0.791	0.748	0.782	0.721	**0.852**
SV (64)	0.725	0.808	0.737	0.799	0.745	**0.819**	0.755	0.727	0.638	0.818
OCC (49)	0.710	0.774	0.754	0.778	0.735	**0.825**	0.731	0.728	0.597	0.799
DEF (44)	0.673	**0.821**	0.718	0.791	0.734	0.783	0.769	0.751	0.542	0.811
MB (29)	0.697	0.766	0.730	0.780	0.743	0.799	0.698	0.676	0.543	**0.823**
FM (39)	0.739	0.799	0.709	0.797	0.751	0.797	0.697	0.693	0.534	**0.800**
IPR (51)	0.756	0.844	0.818	0.864	0.745	0.818	0.824	0.768	0.691	**0.876**
OPR (63)	0.734	0.805	0.786	0.816	0.742	0.835	0.807	0.738	0.644	**0.837**
OV (14)	0.680	0.663	0.689	0.677	0.597	**0.781**	0.620	0.668	0.481	0.721
BC (31)	0.759	0.844	0.752	0.843	0.775	0.841	0.723	0.749	0.704	**0.865**
LR (9)	0.665	**0.887**	0.842	0.847	0.765	0.847	0.841	0.695	0.649	0.820
Overall	0.760	0.841	0.781	0.836	0.783	0.844	0.782	0.777	0.673	**0.850**

Table 2. Precision scores of trackers for 11 challenging factors on OTB50 dataset.

	SME	HDT	MEEM	CF2	SRDCF	DeepSRDCF	FCNT	Staple	DSST	HCNN
IV (38)	0.702	0.802	0.697	0.806	0.744	0.701	0.686	0.693	0.684	**0.843**
SV (64)	0.704	0.790	0.678	0.797	0.704	0.767	0.710	0.635	0.567	**0.808**
OCC (49)	0.668	0.748	0.679	0.761	0.702	0.766	0.693	0.681	0.597	**0.772**
DEF (44)	0.597	**0.760**	0.603	0.739	0.671	0.674	0.679	0.691	0.542	0.745
MB (29)	0.720	0.783	0.711	0.812	0.737	0.772	0.723	0.646	0.508	**0.839**
FM (39)	0.695	0.791	0.650	0.798	0.745	0.766	0.676	0.640	0.448	**0.803**
IPR (51)	0.677	0.803	0.764	0.819	0.663	0.746	0.761	0.635	0.593	**0.841**
OPR (63)	0.630	0.745	0.698	0.749	0.666	0.751	0.743	0.630	0.552	**0.767**
OV (14)	0.615	0.650	0.657	0.671	0.573	**0.746**	0.629	0.669	0.411	0.725
BC (31)	0.684	0.763	0.693	0.766	0.723	0.763	0.678	0.624	0.659	**0.800**
LR (9)	0.669	**0.884**	0.824	0.852	0.736	0.828	0.830	0.667	0.616	0.822
Overall	0.697	0.804	0.704	0.803	0.732	0.772	0.730	0.684	0.604	**0.817**

OV occurs. This phenomenon has been discussed in [14]. Nevertheless, HCNN overcomes this drawback through its historical context offered by snapshots, correcting the drift model, while sustaining the discriminative power of hierarchical convolutional features, which is an asset in DEF, IPR and OPR.

5 Conclusion

In this paper, we presented a tracking method based on hierarchical convolutional features and historical context, which exploits the rich information of CNN features and recovery ability preventing model drift through preserved snapshots. The experimental results demonstrated such a combination can achieve favorable performance comparing with state-of-the-art trackers.

Acknowledgement. This research is supported by Science and Technology Planning Project of Guangdong Province, China (No. 2016A020210086, No. 2017A020208041).

References

1. Babenko, B., Yang, M.-H., Belongie, S.: Robust object tracking with online multiple instance learning. IEEE Trans. Pattern Anal. Mach. Intell. **33**, 1619–1632 (2011)
2. Kalal, Z., Mikolajczyk, K., Matas, J.: Tracking-learning-detection. IEEE Trans. Pattern Anal. Mach. Intell. **34**, 1409–1422 (2012)
3. Grabner, H., Grabner, M., Bischof, H.: Real-time tracking via on-line boosting. Proc. Br. Mach. Vis. Conf. **1**, 1–10 (2006)
4. Henriques, J.F., Caseiro, R., Martins, P., Batista, J.: High-speed tracking with kernelized correlation filters. IEEE Trans. Pattern Anal. Mach. Intell. **37**, 583–596 (2015)
5. Dalal, N., Triggs, B.: Histograms of oriented gradients for human detection. In: Proceedings - 2005 IEEE Computer Society Conference on Computer Vision and Pattern Recognition, CVPR 2005. pp. 886–893 (2005)
6. Danelljan, M., Khan, F.S., Felsberg, M., Van De Weijer, J.: Adaptive color attributes for real-time visual tracking. In: Proceedings of the IEEE Computer Society Conference on Computer Vision and Pattern Recognition, pp. 1090–1097 (2014)
7. Wang, N., Shi, J., Yeung, D.Y., Jia, J.: Understanding and diagnosing visual tracking systems. In: Proceedings of the IEEE International Conference on Computer Vision, pp. 3101–3109 (2015)
8. Deng, J., Dong, W., Socher, R., Li, L.J., Li, K., Fei-Fei, L.: ImageNet: a large-scale hierarchical image database. In: 2009 IEEE Conference on Computer Vision and Pattern Recognition, pp. 248–255 (2009)
9. Liu, Y., Cheng, M.-M., Hu, X., Wang, K., Bai, X.: Richer convolutional features for edge detection. (2016)
10. Zhou, B., Lapedriza, A., Xiao, J., Torralba, A., Oliva, A.: Learning deep features for scene recognition using places database. Adv. Neural. Inf. Process. Syst. **27**, 487–495 (2014)
11. Danelljan, M., Hager, G., Khan, F.S., Felsberg, M.: Convolutional features for correlation filter based visual tracking. In: Proceedings of the IEEE International Conference on Computer Vision, pp. 621–629 (2015)
12. Wang, N., Li, S., Gupta, A., Yeung, D.-Y.: Transferring rich feature hierarchies for robust visual tracking. http://arxiv.org/abs/1501.04587
13. Bertinetto, L., Valmadre, J., Golodetz, S., Miksik, O., Torr, P.: Staple: complementary learners for real-time tracking. In: Proceedings of the IEEE Conference on Computer Vision and Pattern Recognition (2016)
14. Ma, C., Yang, X., Zhang, C., Yang, M.H.: Long-term correlation tracking. In: Proceedings of the IEEE Computer Society Conference on Computer Vision and Pattern Recognition, pp. 5388–5396 (2015)
15. Li, J., Hong, Z., Zhao, B.: Robust visual tracking by exploiting the historical tracker snapshots. In: Proceedings of the IEEE International Conference on Computer Vision, pp. 604–612 (2016)
16. Kwon, J., Lee, K.M.: Tracking by sampling and integrating multiple trackers. IEEE Trans. Pattern Anal. Mach. Intell. **36**, 1428–1441 (2013)
17. Zhang, J., Ma, S., Sclaroff, S.: MEEM: robust tracking via multiple experts using entropy minimization. In: European Conference on Computer Vision, pp. 188–203 (2014)

18. Hu, Z., Gao, Y., Wang, D., Tian, X.: A universal update-pacing framework for visual tracking. In: Proceedings - International Conference on Image Processing, ICIP, pp. 1704–1708 (2016)
19. Lee, D., Sim, J., Kim, C.: Multihypothesis trajectory analysis for robust visual tracking. In: IEEE Computer Society Conference on Computer Vision and Pattern Recognition (CVPR), pp. 5008–5096 (2015)
20. Wu, Y., Lim, J., Yang, M.-H: Object tracking benchmark. IEEE Trans. Pattern Anal. Mach. Intell. **37**(9), 1834–1848 (2015). IEEE
21. Bolme, D., Beveridge, J.R., Draper, B. a., Lui, Y.M.: Visual object tracking using adaptive correlation filters. In: Proceedings of the IEEE Computer Society Conference on Computer Vision and Pattern Recognition, pp. 2544–2550 (2010)
22. Henriques, J., Caseiro, R., Martins, P., Batista, J.: Exploiting the circulant structure of tracking-by-detection with kernels. In: Comput. Vision–ECCV (2012)
23. Danelljan, M., Häger, G., Felsberg, M.: Accurate scale estimation for robust visual tracking. In: Proceedings of the British Machine Vision Conference (2014)
24. Danelljan, M., Gustav, H., Khan, F.S., Felsberg, M.: Learning spatially regularized correlation filters for visual tracking. In: IEEE International Conference on Computer Vision, pp. 4310–4318 (2015)
25. Chen, Z., Hong, Z., Tao, D.: An experimental survey on correlation filter-based tracking. http://arxiv.org/abs/1509.05520
26. Qi, Y., Zhang, S., Qin, L., Yao, H., Huang, Q., Lim, J., Yang, M.-H.: Hedged deep tracking. In: 2016 IEEE Conference on Computer Vision and Pattern Recognition (CVPR), pp. 4303–4311 (2016)
27. Ma, C., Huang, J. Bin, Yang, X., Yang, M.H.: Hierarchical convolutional features for visual tracking. In: Proceedings of the IEEE International Conference on Computer Vision, pp. 3074–3082 (2016)
28. Wang, L., Ouyang, W., Wang, X., Lu, H.: Visual tracking with fully convolutional networks. In: Proceedings of the IEEE International Conference on Computer Vision, pp. 3119–3127 (2015)
29. Simonyan, K., Zisserman, A.: Very deep convolutional networks for large-scale image recognition. In: International Conference on Learning Representations, pp. 1–14 (2015)

Learning Spatiotemporal and Geometric Features with ISA for Video-Based Facial Expression Recognition

Chenhan Lin[1], Fei Long[1(✉)], Junfeng Yao[1,2], Ming-Ting Sun[2], and Jinsong Su[1]

[1] Center for Digital Media Computing, Software School, Xiamen University, Xiamen 361005, China
linchenhan@stu.xmu.edu.cn, {flong,yao0010,jssu}@xmu.edu.cn

[2] Department of Electrical Engineering, University of Washington, Seattle, WA 98195-5852, USA
mts@uw.edu

Abstract. Many appearance-based and geometry-based approaches have been proposed in facial expression recognition. In this paper, we propose a method of learning and combining spatiotemporal features and geometric features for video-based expression recognition. Specifically, we first adopt a multi-layer independent subspace analysis (ISA) network to learn spatiotemporal features directly from videos, and then use another single layer ISA network to learn geometric features from the trajectories of the facial landmark points. The learned spatiotemporal features and geometric features are concatenated to be the final representation for the input video. We use a linear SVM in classification. Experiments on CK+ and MMI facial expression databases show that recognition performance can be improved effectively by incorporating geometric features into spatiotemporal features. Furthermore, comparison results with other related methods demonstrate that the overall accuracy of our method is comparable to some deep learning based methods and the learned features outperform popular hand-crafted features.

Keywords: Facial expression recognition · Independent subspace analysis · Spatiotemporal feature learning

1 Introduction

As a classic problem in computer vision, video-based facial expression recognition has attracted increasing attention in recent years for its broad application prospects, e.g. human-computer interface (HCI), perceptual user interfaces and social robotics [1, 2]. From the view of representation, the mainstream approaches for video-based facial expression recognition can be classified into appearance-based and geometry-based methods [3–5].

For the appearance-based methods, spatiotemporal hand-crafted descriptors such as HOG3D [6], SIFT3D [7], LBP-TOP [5] and GME [3] have been widely studied in

D. Liu et al. (Eds.): ICONIP 2017, Part III, LNCS 10636, pp. 435–444, 2017.
https://doi.org/10.1007/978-3-319-70090-8_45

video-based expression recognition. Sanin et al. [8] extend covariance descriptors from 2D to 3D, called Cov3D, and apply it to the recognition of CK+ facial expression. Experiments indicate the superior performance of Cov3D compared to other hand-crafted descriptors. As the shape of face can be captured by its landmarks, geometry-based methods can represent facial expressions by the movements of the facial landmark points. Jain et al. [9] propose a framework for video-based expression recognition by modeling temporal variations within shapes using Latent-Dynamic Conditional Random Fields. Kaya et al. [10] define 23 geometric features based on 49 facial landmark points and use them as one of the visual descriptors for emotion recognition.

Because of the importances of appearance and geometric features for facial expression recognition, some approaches based on the combination of these two kinds of features have been proposed. Yu et al. [11] combine covariance descriptors and landmark points for image-based expression recognition. In [12], Afshar et al. propose a system for expression recognition in videos by concatenating LGBP-TOP, improved dense trajectories and geometric features. For extracting temporal variations of facial expression automatically, Jung et al. [13] use two deep models to extract appearance and geometric variations in image sequences respectively and integrate these two models by joint fine tuning.

Recently, for the significant success in other related tasks, especially in face recognition, deep learning based models are also studied in facial expression recognition for its ability to learn features from raw data automatically. Kahou et al. [14] combine multiple deep neural networks for different data modalities to deal with the problem of emotion recognition in the wild. Ali et al. [15] propose a deep model for basic expression recognition by stacking Inception unit and test the model on six public expression databases to evaluate its generation ability. However, deep models with many parameters may suffer from over-fitting when applying them directly to facial expression databases because of the limitation of the amount of training data, especially for video-based expression recognition.

Compared to deep learning models, mid-level feature learning has better adaptive ability for small datasets nowadays. Liu et al. [16] model the local variations of faces in each emotion video as a manifold of mid-level features which are called as expression lets. In [17], Le et al. propose a multi-layer network to learn hierarchical invariant features from video data by integrating Independent Subspace Analysis (ISA) with deep learning techniques. ISA is a nonlinear extension of ICA with more robustness to local translations. Their method, combined with Bag-of-Feature (BoF) model, achieves excellent performance in action recognition.

In this paper, to leverage the strength of mid-level feature learning and the combination of appearance and geometric features, we propose a framework to learn and combine spatiotemporal and geometric features with ISA network for video-based expression recognition. For spatiotemporal feature learning, we train a multi-layer ISA network on a large number of 3D cuboids sampled from facial expression videos. For geometric feature learning, we first localize and normalize the facial landmarks. Then a single layer ISA network is trained to learn temporal geometric features from the trajectories of the facial landmarks. To reduce the computation cost, we conduct 3D pooling on the spatiotemporal features and temporal mean pooling on the geometric features respectively. Finally, the two kinds of features are concatenated to be the final representation for the input video. The proposed method is evaluated on two public facial expression databases: CK+ and MMI. Better recognition performance is obtained by combining spatiotemporal features and geometric features with respect to spatiotemporal features only. Furthermore, the classification accuracy of our method is higher than popular hand-crafted features and comparable to deep learning based methods.

The rest of this paper is organized as follows. We first describe our method in Sect. 2. Then the experimental results and discussions will be given in Sect. 3. Finally, we make a conclusion in Sect. 4.

2 Our Method

In this section, we first introduce the multi-layer ISA network which we adopt to learn spatiotemporal features, and then describe how to learn and combine spatiotemporal and geometric features for video-based facial expression recognition in our method.

2.1 Multi-layer ISA Network

Multi-layer ISA network can be viewed as a two-layered network stacked by ISA. The first layer is composed of multiple single layer ISAs which convolve with the local 3D cuboids in videos. Treating ISA on layer 1 (ISA1) as a feature learning kernel, the combined outputs of ISA1 are then given as input to ISA on the second layer (ISA2) with PCA as a preprocessing step. ISA1 and ISA2 are both trained on a set of local 3D cuboids sampled from videos in training set. The size of cuboids will be discussed in Sect. 4. The architecture of multi-layer ISA network is shown in Fig. 1, here the size of cuboids for ISA2 is $M_2 \times M_2$ (spatial) $\times$ T_2 (temporal).

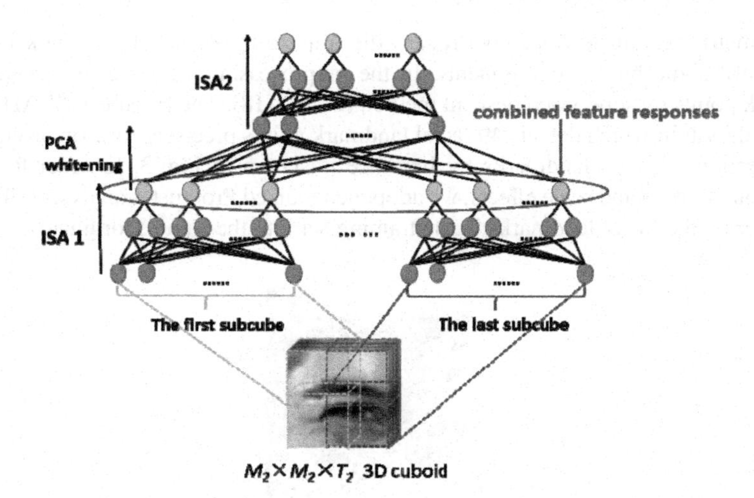

Fig. 1. Architecture of multi-layer ISA network [17]. (Color figure online)

2.2 Spatiotemporal Feature Learning

For spatiotemporal feature learning, we convolutionally sample local 3D cuboids from the input video and infer their features by a multi-layer ISA network (ST-ISA) which is trained by a large set of local 3D cuboids sampled from videos in the training set. Suppose we have N neurons on top of the second layer (yellow bubbles in Fig. 1), the local feature can be seen as a vector containing N channels. We put all the local features of the same channel together by the coordinates of their corresponding local cuboids in the video. In this way, we get N spatiotemporal feature maps denoted by $\{C^k|k = 1...N\}$ whose element $C^k_{x,y,t}$ represents the activation of *k*-th channel for the local cuboid at location (*x*, *y*, *t*). To reduce the calculation cost, we conduct 3D pooling on each feature map. The whole process is illustrated in Fig. 2.

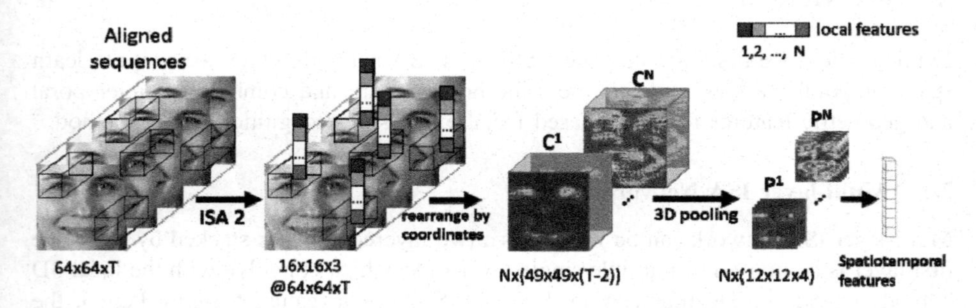

Fig. 2. Illustration of extracting spatiotemporal features by ST-ISA. Here **C** stands for channel while **P** represents channel after pooling. For easy to visualize, feature maps are visualized by color, here red color and blue color represent high and low activation respectively. (Color figure online)

2.3 Geometric Feature Learning

For geometric feature learning, we regard the trajectories of facial landmark points as mixtures of some hidden components. In the preprocessing step, we first detect facial landmark points by supervised descent method (SDM) [18]. The Intraface (SDM) method is a fast algorithm which detects 49 facial landmark points precisely for frontal-view face. The detection of facial landmarks by Intraface is shown in Fig. 3. In order to remove translation, rotation and scale effects, we adopt generalized Procrustes analysis (GPA) [19] to register all the facial landmarks in the training set into the same coordinate.

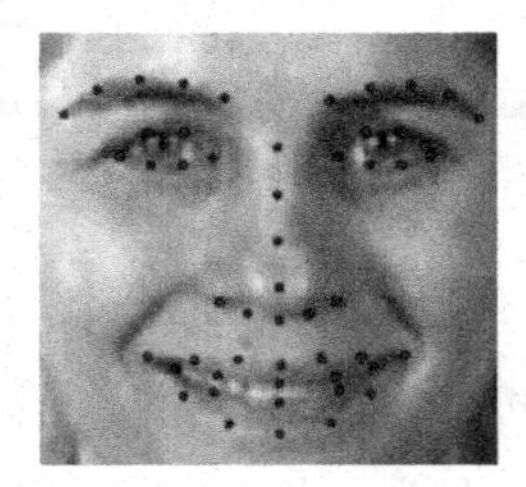

Fig. 3. The 49 facial landmark points detected by Intraface. Best visualization in color. (Color figure online)

After registration, a sliding window with a fixed length L is used to sample sequences of facial landmark points with a fixed stride so that each sequence contains trajectories of facial landmarks from L consecutive frames in the input video. It is natural to consider the trajectories of landmarks in each sequence as a one-dimension signal which is defined as follow.

$$X^t = \left\{x^t_{1,1}, y^t_{1,1}, x^t_{1,2}, y^t_{1,2} \ldots, x^t_{1,49}, y^t_{1,49}, \ldots, x^t_{i,j}, y^t_{i,j}, \ldots, x^t_{L,49}, y^t_{L,49}\right\}, \tag{1}$$

where $(x^t_{i,j}, y^t_{i,j})$ is the j-th point of i-th frame in t-th sequence of a given video after GPA registration.

We apply PCA to each frame of X, and then concatenate the results into one vector which is denoted as X_{pca}. In training phase, a single layer ISA network (G-ISA) is trained by all X_{pca} in the training set. In testing phase, the trained G-ISA receives each X_{pca} of testing video as input and outputs its geometric features. We concatenate these local features in chronological order to generate the geometric features for the whole video. In order to deal with the problem of variable length of videos in the database, we divide the concatenated features into a fixed number of groups along time and conduct temporal mean pooling on each group such that the final geometric representation is of the same length. The whole process of geometric feature extraction is shown in Fig. 4.

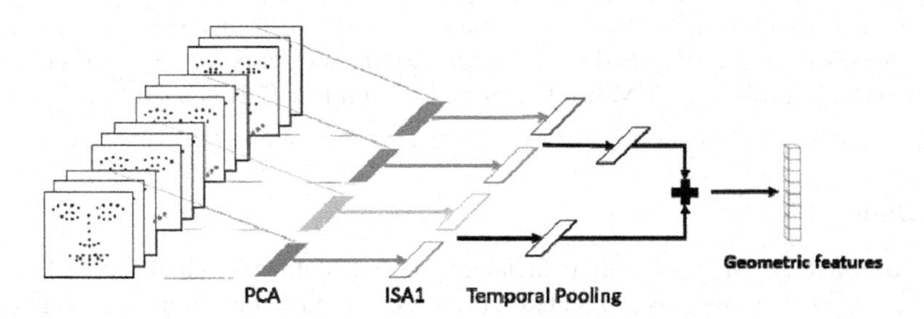

Fig. 4. Illustration of extracting geometric features by ISA network.

2.4 Feature Construction

The pooled spatiotemporal features and geometric features are concatenated to generate the final representations. For classification, multi-class classifier receives the concatenated features after L_2 normalization as input. The framework of the proposed method is shown in Fig. 5.

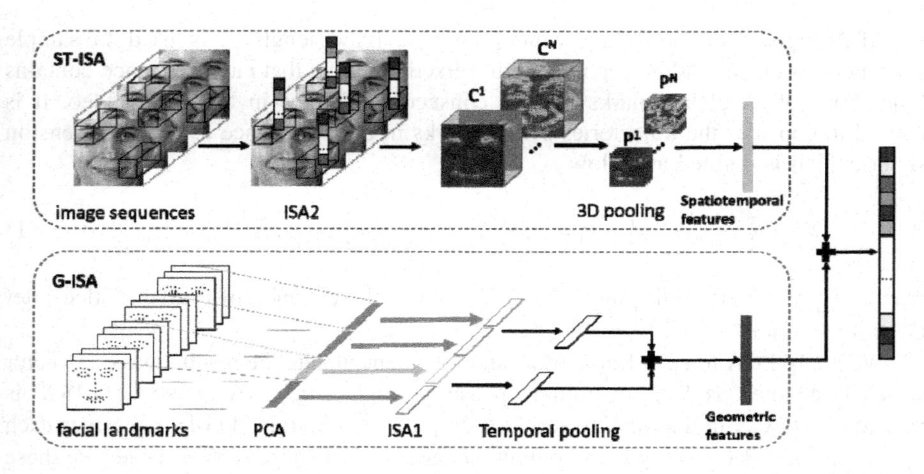

Fig. 5. Framework of our method.

3 Experiments

In this section, we evaluate our method by carrying out experiments on two widely used benchmark databases: CK+ [20] and MMI [21]. Based on the subject ID in the database, we construct 10 person-independent subsets to adopt 10-fold cross-validation. Nine subsets are used for training model, and the remaining subset is used for validation as in [16]. A multi-class linear SVM implemented by Liblinear [22] is used in classification stage.

3.1 Data

CK+ database is a representative facial expression database which consists of 327 sequences from 118 subjects annotated with the seven basic emotions (i.e. Anger (An), Contempt (Co), Disgust (Di), Fear (Fe), Happy (Ha), Sadness (Sa), and Surprise (Su)) according to FACS. The duration of each expression sequence varies from 10 to 60 frames, starting with a neutral face and ending with the peak of expression.

MMI database is a more challenging publicly available facial expression database with continually growing online resources. Our experiments are conducted on a total of 209 frontal view sequences selected from 31 subjects annotated with the six basic expressions: anger, contempt, disgust, fear, happy, sadness or surprise. Different from CK+, each sequence begins with a neutral facial expression, and has the peak of each emotion in the middle of the sequence, and ends with the neutral facial expression.

The sequence counts in each database are listed in Table 1.

Table 1. Sequence counts for each expression.

Expression	An	Co	Di	Fe	Ha	Sa	Su
CK+	45	18	59	25	69	28	83
MMI	33	N/A	32	29	42	32	41

3.2 Parameter Settings

For preprocessing, faces in each image sequence are normalized to 64 × 64 based on the eye locations. The detailed parameters are discussed below.

ST-ISA: For the first layer, we train the ISA1 on 3D cuboids of size 10 × 10 × 3. These 3D cuboids are convolutionally sampled with the stride of 4 pixels in spatial dimension and 3 frame in temporal dimension from the aligned image sequence. The algorithm learns 200 features with the subspace size of 1. For the second layer, we convolutionally sample 16 × 16 × 3 3D cuboids with the same stride as layer 1. The convolution step of ISA1 in the receptive field of ISA2 is 6 in spatial dimension, so that the dimension of the input of ISA2 is 800 (4 × 200 as shown in Fig. 1). ISA2 learns 200 features with the subspace size of 2 which yields the output with dimension of 100. In summary, each local appearance feature is a 100-dimensional vector. In 3D pooling, we divide the 3D feature map of each channel into 12 × 12 × 4 3D blocks.

G-ISA: The length of each sequence L is the hyper-parameter which is determined by specific database. The settings are $L = 3$ in CK+ and $L = 5$ in MMI, because the average length of video in MMI is greater than that of CK+. For training G-ISA, the algorithm learns 200 features with the subspace size of 1. In temporal pooling, we equally divide the sequences into 4 groups by their order in the temporal dimension, and apply mean pooling to each group.

STG-ISA: The final representation is formed by concatenating the two features above, which is called STG-ISA.

Except for the size of cuboid and the stride of sampling, all other parameters are chosen by cross validation on the training set.

3.3 The Role of Geometric Features

In this section, we compare the recognition performances of ST-ISA and STG-ISA to investigate the effectiveness of the combination of spatiotemporal and geometric features learned by ISA networks. Both expression recognition rates and the average accuracies are reported in Table 2 for CK+ and Table 3 for MMI.

Table 2. The expression recognition rates on CK+

Method	An	Co	Di	Fe	Ha	Sa	Su	Average
ST-ISA	91.11	88.89	100	60	98.55	82.14	98.8	88.5
STG-ISA	91.11	83.33	98.31	76	100	89.29	98.8	90.9

Table 3. The expression recognition rates on MMI

Method	An	Di	Fe	Ha	Sa	Su	Average
ST-ISA	69.7	75	10.71	83.33	56.25	80.49	62.58
STG-ISA	72.73	71.88	35.71	80.95	59.38	73.17	65.64

In the two databases, the average accuracy of STG-ISA is better than ST-ISA. Among the basic expressions, the recognition rates of Fear and Sadness are greatly improved, which may indicate that geometric feature is important for the recognition of the two expressions.

3.4 Comparison with Other Methods

We demonstrate the performance of our method by comparing it with other related hand-crafted features and deep learning based methods. The overall classification accuracies for the two databases are reported in Tables 4 and 5.

Table 4. The comparison results on CK+

Type	Method	Accuracy (%)
Hand-crafted	3D SIFT [7]	81.35
	LBP-TOP [5]	88.99
	HOG 3D [6]	91.4
	Cov3D [8]	92.3
Deep-learning	3DCNN + DAP [24] (15-fold)	92.4*
	DTAGN [13]	97.25*
Our method	**ST-ISA**	**92.97**
	STG-ISA	**94.5**

Table 5. The comparison results on MMI

Type	Method	Accuracy (%)
Hand-crafted	LBP-TOP [5]	57.2
	HOG 3D [6]	60.89
	3D SIFT [7]	64.4
Deep-learning	3D CNN [23]	53.2*
	3DCNN + DAP [24] (20-fold)	63.4*
Our method	**ST-ISA**	**65.38**
	STG-ISA	**67.31**

We separate all methods into 3 blocks by the type of features they used. The first block shows the experimental results of hand-crafted features. For LBP-TOP, we divide the spatial image into 8×8 blocks with overlapping of 3 pixels. For other local features, the settings are the same as [16]. The second block shows the results of some deep-learning based methods. The results with '*' are directly cited from their papers. Specifically, the results of 3DCNN + DAP adopt a 15-fold cross validation in CK+ and a 20-fold cross validation in MMI. DTAGN adopts a 10-fold subject independent cross validation. The results of our method are shown in the third block.

In the two databases, ST-ISA outperforms most of the hand-crafted features. By incorporating the geometric features leaned by G-ISA into spatiotemporal features,

STG-ISA achieves better performance (94.5% on CK+, 67.31% on MMI) compared to some deep learning based methods, such as 3D CNN and 3DCNN + DAP.

4 Conclusion

In this paper, we present a video-based expression recognition method by learning and combination of spatiotemporal features and geometric features with ISA network. Spatiotemporal features are learned directly from videos, while geometric features are learned from the trajectories of the facial landmarks. Experimental results on CK+ and MMI databases show the validity of incorporating the geometric features into the spatiotemporal features. Furthermore, the recognition performance obtained by our method is superior to traditional hand-crafted descriptors and comparative to some deep learning based methods. In our future work, we will try to apply our method to expression recognition in more challenging environments.

Acknowledgment. This work is supported by the Fundamental Research Funds for the Central Universities in China (No. 20720170056), the open funding project of State Key Laboratory of Virtual Reality Technology and Systems, Beihang University (Grant No. BUAAVR-14KF-01), and the Science and Technology Project of Quanzhou City (No. 2015G62).

References

1. De la Torre, F., Cohn, J.F.: Facial expression analysis. In: Moeslund, T.B., Hilton, A., Krüger, V., Sigal, L. (eds.) Visual Analysis of Humans, pp. 377–409. Springer, London (2011). doi:10.1007/978-0-85729-997-0_19
2. Fasel, B., Luettin, J.: Automatic facial expression analysis: a survey. Pattern Recogn. **36**(1), 259–275 (2003)
3. Wu, T., Bartlett, M.S., Movellan, J.R.: Facial expression recognition using Gabor motion energy filters. In: IEEE Computer Society Conference on Computer Vision and Pattern Recognition Workshops (CVPRW), pp. 42–47 (2010)
4. Ji, Y., Idrissi, K.: Automatic facial expression recognition based on spatio-temporal descriptors. Pattern Recogn. Lett. **33**(10), 1373–1380 (2012)
5. Zhao, G., Pietikainen, M.: Dynamic texture recognition using local binary patterns with an application to facial expressions. IEEE T PAMI **29**(6), 915–928 (2007)
6. Klaser, A., Marszalek, M.: A spatio-temporal descriptor based on 3d-gradients. In: BMVC 2008 (2008)
7. Scovanner, P., Ali, S., Shah, M.: A 3-dimensional sift descriptor and its application to action recognition. In: Proceedings of the 15th International Conference on Multimedia, pp. 357–360. ACM (2007)
8. Sanin, A., Sanderson, C., Harandi, M.T., Lovell, B.C.: Spatiotemporal covariance descriptors for action and gesture recognition. In: WACV 2013, pp. 103–110 (2013)
9. Jain, S., Hu, C., Aggarwal, J.K.: Facial expression recognition with temporal modeling of shapes. In: Computer Vision Workshops (ICCV Workshops), pp. 1642–1649 (2011)
10. Kaya, H., Gürpinar, F., Afshar, S., Salah, A.A.: Contrasting and combining least squares based learners for emotion recognition in the wild. In: Proceedings of the 2015 ACM, pp. 459–466. ACM (2015)

11. Yu, H., Liu, H.: Combining appearance and geometric features for facial expression recognition. In: Sixth International Conference on Graphic and Image Processing (ICGIP 2014), p. 944308. International Society for Optics and Photonics (2015)
12. Afshar, S., Ali Salah, A.: Facial expression recognition in the wild using improved dense trajectories and Fisher vector encoding. In: IEEE CVPR 2016, pp. 66–74 (2016)
13. Jung, H., Lee, S., Yim, J., Park, S., Kim, J.: Joint fine-tuning in deep neural networks for facial expression recognition. In: IEEE ICCV 2015, pp. 2983–2991 (2015)
14. Kahou, S.E., Pal, C., Bouthillier, X., Froumenty, P., Gülçehre, Ç., Memisevic, R., Vincent, P., Courville, A., Bengio, Y.: Combining modality specific deep neural networks for emotion recognition in video. In: Proceedings of the 15th ACM on International Conference on Multimodal Interaction (ICML), pp. 543–550 (2013)
15. Mollahosseini, A., Chan, D., Mohammad H.M.: Going deeper in facial expression recognition using deep neural networks. In: WACV 2016, pp. 1–10 (2016)
16. Liu, M., Shan, S., Wang, R., Chen, X.: Learning expression lets on spatiotemporal manifold for dynamic facial expression recognition. In: IEEE CVPR 2014, pp. 1749–1756 (2014)
17. Le, Q.V., Zou, W.Y., Yeung, S.Y., Ng, A.Y.: Learning hierarchical invariant spatio-temporal features for action recognition with independent subspace analysis. In: IEEE CVPR, pp. 3361–3368 (2011)
18. Xiong, X., De la Torre, F.: Supervised descent method and its applications to face alignment. In: IEEE CVPR 2013, pp. 532–539 (2013)
19. Gower, J.C.: Generalized procrustes analysis. Psychometrika **40**(1), 33–51 (1975)
20. Lucey, P., Cohn, J.F., Kanade, T., Saragih, J., Ambadar, Z., Matthews, I.: The extended Cohn-Kanade Dataset (CK+): a complete dataset for action unit and emotion-specified expression. In: IEEE Computer Society Conference on CVPR Workshops (CVPRW 2010), pp. 94–101. IEEE (2010)
21. Valstar, M., Pantic, M.: Induced disgust, happiness and surprise: an addition to the MMI facial expression database. In: LRECW (2010)
22. Fan, R.E., Chang, K.W., Hsieh, C.J., Wang, X.R., Lin, C.J.: LIBLINEAR: a library for large linear classification. J. Mach. Learn. Res. **9**, 1871–1874 (2008). http://www.csie.ntu.edu.tw/cjlinlliblinear
23. Ji, S., Xu, W., Yang, M., Yu, K.: 3D convolutional neural networks for human action recognition. IEEE Trans. Pattern Anal. Mach. Intell. **35**(1), 221–231 (2013)
24. Liu, M., Li, S., Shan, S., Wang, R., Chen, X.: Deeply learning deformable facial action parts model for dynamic expression analysis. In: Cremers, D., Reid, I., Saito, H., Yang, M.-H. (eds.) ACCV 2014. LNCS, vol. 9006, pp. 143–157. Springer, Cham (2015). doi:10.1007/978-3-319-16817-3_10

Robust Edge-Based Model with Sparsity Representation for Object Segmentation

Guoqi Liu[1], Haifeng Li[2(✉)], and Chenjing Li[1]

[1] School of Computer and Information Engineering, Henan Normal University, Xinxiang 453007, Henan, China
gqliu@htu.edu.cn, 15249714369@163.com

[2] College of Mathematics and Information Science, Henan Normal University, Xinxiang 453007, Henan, China
lihaifengxx@126.com

Abstract. Active contour models (ACM) based on level set method (LSM) are widely used in image segmentation. However, the classical edge-based models always extract some unnecessary objects or noise, and they lack robustness in segmenting weak boundary. In this paper, a method to constrain the evolution of contour is proposed. Firstly, extracting objects with a known topology (such as k connected objects) is viewed a sparse representation problem under a set of basis functions. According to sparse representation, a set of basis function is obtained with label operator to represent every connected region. Then, the corresponding energy functional model which views noise and non-objects as redundancy is defined based on basis functions. Furthermore, through the defined basis functions, a novel edge-stop term is designed and integrated into geometric active contour models. Experiments demonstrate that the proposed method improves the robust performances of ACM. On the other hand, the proposed method does not introduce any extra parameter.

Keywords: Active contour models · Image segmentation · Sparse representation · Object extraction · Level set

1 Introduction

Object segmentation [1] is an important topic in computer vision. Active contour models (ACMs) based on level set method (LSM) have advantages in topological flexibility and evolution robustness, which are researched in the past two decades [2,3]. The existing active contour models are always categorized into two classes: edge-based models and region-based models.

Region-based active contour models utilize the regional information to describe the foreground and background, and these models have good performance for the image with weak object boundaries [4–7]. Furthermore, they are significantly less sensitive to the location of initial contours. One of the most

D. Liu et al. (Eds.): ICONIP 2017, Part III, LNCS 10636, pp. 445–456, 2017.
https://doi.org/10.1007/978-3-319-70090-8_46

popular region-based active contour models is Chan-Vese (C-V) model [4]. C-V model has been successful for images with two regions, each having a distinct mean of image intensity. However, region-based active contour models usually have difficulty in handling images with intensity inhomogeneity or images having an intensity overlap between the foreground and background intensity distributions [8–10]. On the other hand, they always extract some non-objects and noise.

Edge-based models utilize image gradient to stop the evolving contours on the object boundaries [11–13]. Typical edge-based active contour models, such as geodesic active contour model (GAC) [11,12] and a distance regularized level set evolution (DRLSE) [13], define an edge-stop term to control the motion of the contour. The edge-based stop term serves to stop the contour on the desired object boundary. Generally, edge-based models are not sensitive to inhomogeneity of image intensities, and the evolution with edge-based models is more regularized and robust compared with utilizing region-based models. However, they are sensitive to poorly defined boundaries (such as weak boundary) and noise [14].

In this paper, in order to avoid extracting non-objects and noise, we contribute with a framework for segmenting objects in geometric active contour model based on sparsity constraint [15]. Firstly, every connected region is labeled and viewed as a basis function. Secondly, a set of basis function is integrated into GAC to replace the original indicator function and the energy functional equation is obtained with respect to basis function. Finally, the solution of evolution equation is obtained by gradient descent method. On the other hand, in order to improve the robustness in handling weak edge with edge-based models, a novel stopping term is proposed based on the analysis for the drawbacks of traditional edge stopping function. In the proposed method, the edge stopping function is computed the inner product of two defined terms, one is the regional stopping term, and the other is the traditional edge-stop term, which is viewed as an amplitude stopping term.

2 Related Work

2.1 Front Evolution Framework Based on Level Set Method

The level set technique developed by Osher and Sethian [3] represents the contour implicitly as the zero level set of a smooth, Lipschitz-continuous scalar function. The boundaries of the region of interest (ROI) are represented by the zero level of $\phi : C(t) = \{X|\phi(X,t) = 0\}$, where $X \in R^2$ in $2D$. A level set function is usually used to partition the image domain into several disjoint regions.

In order to represent the region of foreground and background, a regularized Heaviside function [4] with respect to ϕ is used to label the image regions, which is given as follows:

$$H_\epsilon(\phi) = \frac{1}{2}[1 + \frac{2}{\pi}arctan(\frac{\phi}{\epsilon})] \quad (1)$$

where ϵ is a regularized parameter. In the foreground, the values of Heaviside function approach to 1; whereas Heaviside function approaches to 0 in the background. According to the knowledge of LSM, the corresponding evolution equation based on level set is written as follows:

$$\begin{cases} \frac{\partial \phi(X,t)}{\partial t} = F(X,t)|\nabla \phi(X,t)| \\ \phi(X,0)) = \phi_0(X) \end{cases} \tag{2}$$

where gradient operator $\nabla(\cdot) = (\frac{\partial(\cdot)}{\partial x}, \frac{\partial(\cdot)}{\partial y})$ in R^2. F is the force pushing contour evolving to object. The above equation is the LSE equation of PDE-based LSMs. As pointed out in [16], the LSE equation in variational LSMs [17,18] can be similarly obtained. In order to obtain better performances, various energy functional [19–22] are proposed and integrated in geometric active contour models. Recently, some methods with sparse representation are presented to segment objects [15,23,24].

Generally, an evolution equation of edge-based model is calculated by minimizing the following energy functional $E(\phi)$:

$$Min_{\phi} E(\phi) = \int_{\Omega} \omega_b(X)|\nabla H(\phi(X))|dX \tag{3}$$

where ω_b is an edge stopping term. For classical GAC [11,12], the corresponding energy functional $\omega_b^t = \frac{1}{1+|\nabla G_\sigma \otimes I|^2}$ is a traditional edge stopping term, where G_σ is a Gaussian kernel with a standard deviation σ. The convolution $\otimes$ is used to smooth the image I to reduce the noise. This function usually takes smaller values at object boundaries than at other locations. In DRLSE, the traditional edge-stop term is also used though DRLSE integrated a regularized term for evolution. However, both GAC and DRLSE fail to stop the contour at object boundary because of the traditional edge-stop term. As shown in the first row of Fig. 1, a medical image is shown and two initial contours are set to extract two corresponding objects. Since the gray intensity between the two objects are

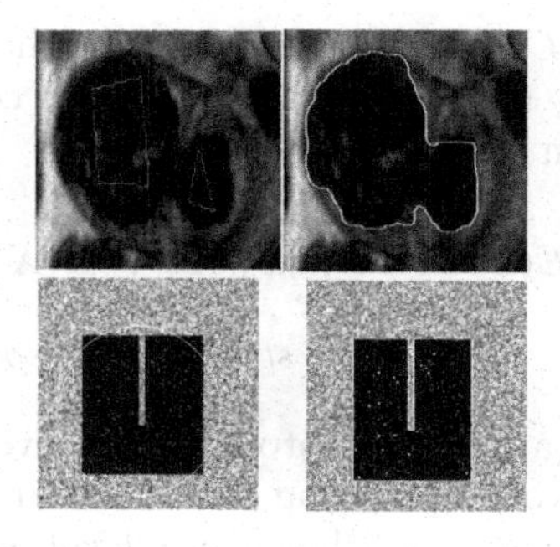

Fig. 1. Topology flexibility and traditional edge-stop term cause the undesired results with edge-based models. The first row shows that DRLSE extracts undesired object, and the second row shows that GAC extracts objects but including some noise.

similar and the edge gradient of both objects is weak, the two evolving contours merged into one contour and finally failed to extract objects. In the second row of Fig. 1, GAC extracts some non-objects.

3 Proposed Method

In this section, a model with sparsity constraint is proposed and stated. Based on the sparse basis functions, a novel edge-stop term is obtained by performing inner product between the defined regional stopping term and the traditional edge stopping term.

3.1 Energy Functional Model

According to the knowledge of LSM, contours are always represented by a Heaviside function [16]. In the proposed method, based on sparse decomposition [23], a set of basis function D is obtained to represent contours. Assuming the extracting objects are the union of several disjoint areas Ω_i, i.e., $\Omega = \sum_{i=1}^{n} \Omega_i$. Every connected area Ω_i is labeled and represented by a basis function D_i. For $D_i(x)$, the value is 1 when x is inside the connected region Ω_i, whereas the value is 0. Thus, with basis function, the Heaviside function H in LSM is represented as follows:

$$H = \sum_{i=1}^{n} s_i D_i, \quad i.e., \quad H = Ds, \qquad where \quad D = [D_1, \cdots, D_n], \quad s = [s_1, \cdots, s_n]^T \tag{4}$$

where s^T represents the transpose of a matrix s.

With $H(\phi) = Ds$, the above energy functional Eq. (3) is written as follows:

$$Min_{D,s} E(D, s) = \int_{\Omega} \omega_b(X) |\nabla (D(X)s)| dX \tag{5}$$

With initial contours C of known topology (such as k connected objects), basis functions $D_1, \cdots, D_k$ are used to describe these contours. Then, the above energy functional equation is represented as follows:

$$Min_{D,s} E(D, s) = \int_{\Omega} \omega_b(X) |\nabla (D(X)s)| dX \qquad st. \quad s \ is \ k\text{-}sparsity. \tag{6}$$

s is *k-sparsity*, then new appeared contours in the evolution are viewed redundant, which is used to avoid appearing new contours or splitting of evolving contours. From the above equation, the general level set energy functional equation is a special case of the proposed equation. Without the constraint condition, the proposed model becomes to the typical geometric active contour. After integrating the constraint in active contour model, our goal is to evolve contours to extract objects. With a steepest descent method, $D_i (1 \le i \le k)$ is computed as follows:

$$\frac{\partial D_i}{\partial t} = \omega_b^i(X)|\nabla D_i| div(\frac{\nabla D_i}{|\nabla D_i|}) \tag{7}$$

3.2 The Proposed Edge-Stop Term

Based on the above section, an energy functional is defined through basis functions. In this section, an edge-stop term will be computed, which is integrated into region information compared with traditional edge-stop term. Utilizing the above defined basis functions, a regional stopping term ω_r^i is computed as follows:

$$\omega_r^i = 1 - \sum_{j=1,\ j \neq i}^{n} D_j \tag{8}$$

The above regional stopping term is 0 inside the region of $\sum_{j=1,j\neq i}^{n} \Omega_j$, which can avoid merging evolving contours. When contour C_i evolves to the region $\sum_{j=1,j\neq i}^{n} \Omega_j$, C_i will stop evolving.

Then, a novel edge-stop term is obtained by computing the inner product between the above regional stopping term and the traditional edge-stop term.

$$\omega_b^i = \omega_r^i \omega_b^t \tag{9}$$

where ω_b^t is the traditional edge-stop term. The term ω_r^i mainly decides to the region where contour continues to evolve or not.

Based on the Eq. (7) and the proposed edge-stop term, $D_i (1 \le i \le k)$ is computed as follows:

$$\frac{\partial D_i}{\partial t} = \omega_r^i \omega_b^t |\nabla D_i| div(\frac{\nabla D_i}{|\nabla D_i|}) \tag{10}$$

An example is shown in Fig. 2, initial contour is set to converge to the target object.

After several iterations, the result with the proposed method is shown in the middle column of Fig. 2, the contour stops at object boundary and does not cross the weak edge because of the proposed edge-stop term. While the result with DRLSE is shown in the right hand of Fig. 2, the weak boundary leakage is shown. Traditional edge-stop term makes the contour cross the weak edge and fails to stop the contour at object boundary.

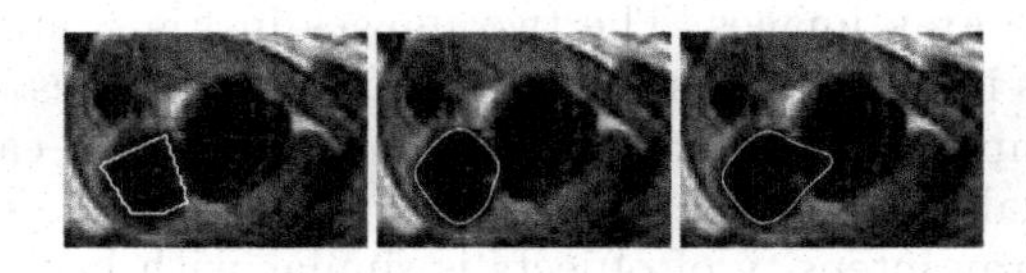

Fig. 2. A medical image is tested. The first image is the original image with initial contour, the second image is the result with the proposed method and the third image is the result with DRLSE model.

3.3 The Algorithm for the Proposed Model

In the following, an algorithm of solving the Eq. (6) is stated. The Eq. (6) has two block variables D and s, and the *k-sparsity* is a constrained condition. Thus, Eq. (6) is a normal optimization problem with sparsity constraint condition. Generally, Eq. (6) can be solved by convex relaxation method or greedy algorithm. In this paper, the sparsity condition must be satisfied. Therefore, the orthogonal matching pursuit (OMP) [25] is improved to make sure the sparsity constraint. In general, Eq. (6) is solved by two steps.

Firstly, fixing s, the Eq. (6) is minimized with respect to D. Secondly, utilizing label operator for D to label every connected region, then the proposed OMP algorithm is used to determine the k basis functions. To be specific, with a steepest descent method, $D_i (1 \leq i \leq k)$ is computed as follows:

$$\frac{\partial D_i}{\partial t} = \omega_r^i \omega_b^t(X) |\nabla D_i| div(\frac{\nabla D_i}{|\nabla D_i|}) \tag{11}$$

Then, for the updated D, the topology of D may increase since the Eq. (7) does not impose sparsity constraint. With label operator, a set of basis function to describe every connected area is obtained. In order to make sure the constraint condition, OMP algorithm [25] is improved to search the most relevant basis functions with respect to the previous basis function. Therefore, sparsity constraint is used to make sure robustness of the contour evolution and new contours by splitting of level set function are viewed redundant.

4 Experiments and Analysis

To demonstrate the advantages of the proposed method, some simulations are given for the comparisons of the tested methods. The DRLSE model is robust to noise and parameters. Therefore, DRLSE is selected as a basic model. The proposed constraint is integrated into DRLSE model, which is tested and compared with DRLSE. Some region based methods are also compared.

4.1 Simulations and Quantitative Evaluation for Comparisons with DRLSE

In this section, several simulations are given to compare the performances of tested methods. We compare our method with the DRLSE model for objects extraction on some gray images. The two images in Fig. 3 are from Weizmann database [26]. This image database includes many real images, and some images are selected to compare the tested methods. Which includes cases with ground-truth images to enable a fair comparison.

In Fig. 3, the gray intensity of objects is similar with background. The task is to extract the corresponding objects in all tested images. As shown in Fig. 3, initial contours are set near the objects. In the first row of Fig. 3, there are two objects to segment. Both the proposed method and DRLSE model success to extract the objects. However, the DRLSE model extracts some non-target objects, which is shown in the second column of Fig. 3. As comparison, the region of the extracted non-target object are clearly shown in Fig. 3. The proposed method integrates sparsity constraint, some unnecessary objects or noise are viewed redundant. Therefore, the proposed method could extract object with ideal result, as shown in the third column of Fig. 3.

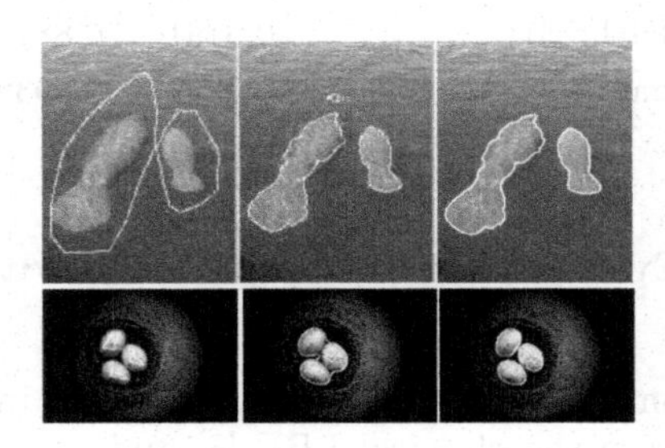

Fig. 3. The simulations are shown with the proposed method and DRLSE model. The first column is the tested images with initial contours, the second column is the result with DRLSE and the third column is the result with our proposed method.

A simulation for another image is shown in the second row of Fig. 3. The analysis for this example is given in this section. As shown in the second row of Fig. 3, three objects are need to extract, and the corresponding closed contours are initialed. The result with DRLSE model is shown in the second column. Evolving contours merge into one contour. While the proposed method extracts the corresponding three objects, which is shown in the final image of Fig. 3. The reason lies in the proposed edge-stop term. In the proposed method, the edge-stop term integrates into region information and evolving contours will stop near some defined region, which could avoid merging evolving contours or boundary leakage.

In order to make quantitative evaluation for tested method, the F-score is considered [26], which is the measure of extracted accuracy, which considers both the precision P and the recall R. It is the weighted, harmonic mean of precision and recall values, and F-score is given as follows:

$$F = \frac{2PR}{P + R} \tag{12}$$

The precision rate P is the number of correct results divided by the number of all returned results, recall rate (R) is the number of correct results divided by the number of results that should have been returned.

From the Table 1, the proposed method obtains better F values compared with DRLSE. The Fig. 3(i) ($i = 1, 2$) represents the result of i-th image with the tested methods in Table 1. Some non-objects are not extracted because of the proposed sparsity constraint, and the proposed edge-stop term makes contour precisely stop at target object compared with the traditional edge-stop term. Therefore, the proposed method obtains better segmented results compared with the DRLSE model, which can be seen from Table 1.

Table 1. Quantitative evaluation of results

Quantification (Method)	Fig. 3(1)	Fig. 3(2)
F (DRLSE)	0.9191	0.8815
F (ours)	0.9321	0.8932

4.2 Quantitative Evaluation for Comparisons with Some Region Based Models

In this section, more comparisons with some region based methods are given. The region based methods are the classical LBF [8] and the bi-convex fuzzy variational image segmentation method [27]. The quantitative evaluation is given and analyzed for the tested algorithms. The tested methods are also from the Weizmann segmentation evaluation database. The Fig. 4(i) ($i = 1, 2, 3$) represents the result of *i-th* image with the tested methods in Table 2.

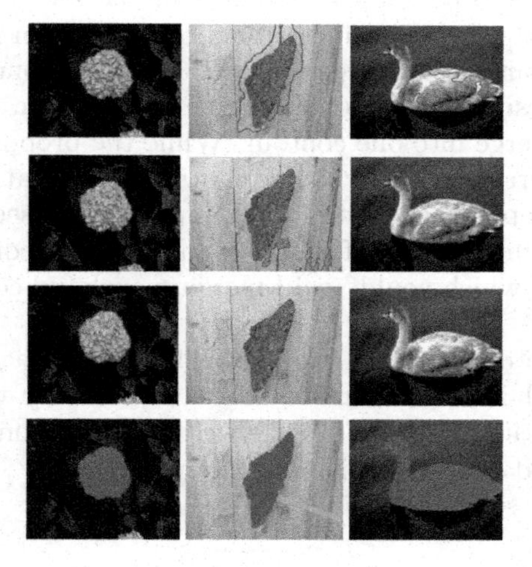

Fig. 4. The results with tested methods. The first row is the result with the LBF, the second row is the result with [27], the third row is the result with the proposed method, and the final row is the ground truth.

The region based methods are always less precise in segmenting objects since over-segmentation problem appears with these models. As shown in Fig. 4, the first row is the result with LBF and the second row is the result with [27].

Over-segmentation problem appears with these two methods and the precision is undesired. According to the Table 2, LBF obtains the lowest scores in quantitative scores(F). [27] seems obtain better results compared with LBF. However, our proposed method obtains the highest average scores in quantitative evaluation according to Table 2. On one hand, the proposed model integrated sparsity constraint to decrease non-objects. On the other hand, the proposed method includes edge information, it can make the contour stop at the objects. Therefore, the proposed method obtains the highest quantitative scores compared with LBF and [27].

Table 2. Quantitative evaluation of results

Q (%)	Fig. 4(1)	Fig. 4(2)	Fig. 4(3)
F (LBF)	85.85	48.25	56.26
F (Gong)	94.13	82.41	87.31
F (ours)	98.76	98.25	98.08

4.3 Analysis and Application on Medical Images

The proposed method is used to segment uterine fibroids in MR scans acquired by a 1.5T whole-body system (Genesis Signa; GE Medical Systems, Milwaukee, Wis.) [20]. The contrast of gray intensity in the image is weak and the boundaries of these objects are very close. Initial contours are set near the objects, which is shown in the first row of Fig. 5. DRLSE failed to extract objects, as shown in the second row of Fig. 5. Subsequent phases of the segmentation process, such as an expert-supervised segmentation method, is needed for these problem. While in the proposed method, without subsequent phases of the segmentation process, the extracted results are shown in the third row of Fig. 5.

The proposed method successes to extract objects, the analysis is given as follows. According to the proposed method, the updated equation with respect to $D = [D_1, D_2]$ are computed as follows:

$$\begin{aligned} \frac{\partial D_1}{\partial t} &= \omega_b^1(X)|\nabla D_1|div(\frac{\nabla D_1}{|\nabla D_1|}) \\ \frac{\partial D_2}{\partial t} &= \omega_b^2(X)|\nabla D_2|div(\frac{\nabla D_2}{|\nabla D_2|}) \end{aligned} \tag{13}$$

For contour C_1, $\omega_r^1 = 1 - D_2$, $\omega_b^1 = (1 - D_2)\omega_b^t$. $\omega_b^1 = (1 - D_2)\omega_b^t$ is 0 when C_1 is close to C_2, thus contour C_1 stop to evolve and could not cross with contour C_2. The same mechanism is also fit for contour C_2. Therefore, the proposed method could obtain robustness result compared with DRLSE.

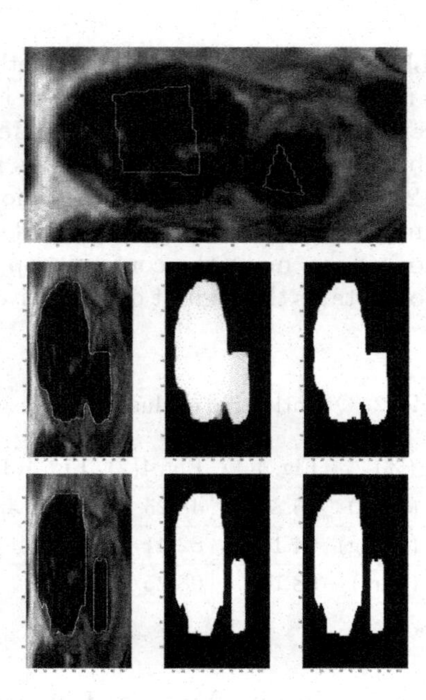

Fig. 5. A medical image is tested for the proposed method and DRLSE. The first row is the original images with initial contours, the second row is the result with DRLSE and the third row is the result with the proposed method.

5 Conclusion

In this paper, a novel frame for active contour model with sparse constraint is proposed. A set of basis function replaces the original indicator function of geometric active contour model to represent the region of extracted objects. Then, a novel energy functional model with these basis functions is defined. Besides, the corresponding edge-stop term is also designed through basis function. Simulations show the robust performances in segmenting target objects.

Acknowledgement. This work is jointly supported by the National Natural Science Foundation of China (No. U1404603).

References

1. Zhu, S., Yuille, A.: Region competition: "unifying snakes, region growing, and Bayes/MDL for multiband image segmentation". IEEE Trans. Pattern Anal. Mach. Intell. **18**, 884–900 (1996)
2. Kass, M., Witkin, A., Terzopoulus, D.: Snakes: active contour model. Int. J. Comput. Vis. **1**, 321–331 (1988)
3. Osher, S., Sethian, J.: Fronts propagating with curvature-dependent speed: algorithms based on Hamilton-Jacobi formulations. J. Comput. Phys. **79**, 12–49 (1988)

4. Chan, T., Vese, L.: Active contours without edges. IEEE Trans. Image Process. **10**, 266–27 (2001)
5. Vese, L., Chan, T.: A Multiphase Level Set Framework for Image Segmentation Partial Differential Equations and the Calculus of Variations. Springer, New York (2006)
6. Bresson, X., Esedoglu, S., Vandergheynst, P., et al.: Fast global minimization of the active contour/snake model. J. Math. Imaging Vis. **28**, 151–167 (2007)
7. Brown, E.S., Chan, T.F., Bresson, X.: Completely convex formulation of the Chan-Vese image segmentation model. Int. J. Comput. Vis. **98**, 103–121 (2012)
8. Li, C.M., Kao, C., Gore, J., Ding, Z.: Implicit active contours driven by local binary fitting energy. In: Computer Vision and Patter Recognition, pp. 1–7. IEEE Press, Minneapolis (2007)
9. Dong, F., Chen, Z., Wang, J.: A new level set method for inhomogeneous image segmentation. Image Vis. Comput. **31**, 809–822 (2013)
10. Zhang, K., Zhang, L., Lam, K.M., Zhang, D.: A level set approach to image segmentation with intensity inhomogeneity. IEEE Trans. Cybern. **46**, 546–557 (2016)
11. Caselles, V., Kimmel, R., Sapiro, G.: Geodesic active contour. Int. J. Comput. Vis. **22**, 61–79 (1997)
12. Paragios, N., Deriche, R.: Geodesic active contours and level sets for the detection and tracking of moving objects. IEEE Trans. Pattern Anal. Mach. Intell. **22**, 266–280 (2000)
13. Li, C., Xu, C., Gui, C., Fox, M.D.: Distance regularized level set evolution and its application to image segmentation. IEEE Trans. Image Process. **19**, 3243–3254 (2010)
14. Pratondo, A., Chui, C.K., Ong, S.H.: Robust edge-stop functions for edge-based active contour models in medical image segmentation. IEEE Signal Process. Lett. **23**, 222–226 (2016)
15. Zhang, S., Zhan, Y., Metaxas, D.N.: Deformable segmentation via sparse representation and dictionary learning. Med. Image Anal. **16**, 1385–1396 (2012)
16. Zhang, K., Zhang, L., Song, H., Zhang, D.: Re-initialization free level set evolution via reaction diffusion. IEEE Trans. Image Process. **22**, 258–271 (2013)
17. Mumford, D., Shah, J.: Optimal approximations by piecewise smooth functions and associated variational problems. Commun. Pure Appl. Math. **42**, 577–685 (1989)
18. Amini, A., Weymouth, T.E., Jain, R.: Using dynamic programming for solving variational problems in vision. IEEE Trans. Pattern Anal. Mach. Intell. **12**, 855–867 (1990)
19. Han, X., Xu, C., Prince, J.L.: A topology preserving level set method for geometric deformable models. IEEE Trans. Pattern Anal. Mach. Intell. **25**, 755–768 (2003)
20. Ben-Zadok, N., Riklin-Raviv, T., Kiryati, N.: Interactive level set segmentation for image-guided therapy: examples. http://www.eng.tau.ac.il/nk/ISBI09/
21. Li, C.M., Xu, C.Y., Gui, C.F., Fox, M.D.: Level set evolution without re-initialization: a new variational formulation. In: Computer Vision and Pattern Recognition, pp. 430–436. IEEE Press, San Diego (2005)
22. Ning, J., Zhang, L., Zhang, D., Yu, W.: Joint registration and active contour segmentation for object tracking. IEEE Trans. Circ. Syst. Video Technol. **23**, 1589–1597 (2013)
23. Zhang, S., Zhan, Y., Dewan, M., et al.: Towards robust and effective shape modeling: sparse shape composition. Med. Image Anal. **16**, 265–277 (2012)
24. Wang, G., Zhang, S., Xie, H., et al.: A homotopy-based sparse representation for fast and accurate shape prior modeling in liver surgical planning. Med. Image Anal. **19**, 176–186 (2015)

25. Tropp, J., Gilbert, A.: Signal recovery from random measurements via orthogonal matching pursuit. IEEE Trans. Inform. Theory **53**, 4655–4666 (2007)
26. Alpert, S., Galun, M., Basri, R., et al.: Image segmentation by probabilistic bottom-up aggregation and cue integration. In: Computer Vision and Pattern Recognition, pp. 1–8. IEEE Press, Minneapolis (2007)
27. Gong, M., Tian, D., Su, L., et al.: An efficient bi-convex fuzzy variational image segmentation method. Inf. Sci. **293**, 351–369 (2015)

Salient Object Detection Based on Amplitude Spectrum Optimization

Ce Li(✉), Yuqi Wan, and Hao Liu

College of Electrical and Information Engineering,
Lanzhou University of Technology, Lanzhou 730050, China
xjtulice@gmail.com

Abstract. Saliency detection is prerequisite for many computer vision tasks. The existing frequency domain models can not always detect a complete object. We propose a novel salient object detection model based on an optimized amplitude spectrum. This model computes saliency map in two steps. Firstly, we optimize amplitude spectrum by smoothing the peaks in log amplitude spectrum. The raw saliency maps are computed by combining the optimized amplitude spectrum and the original phase spectrum according to different thresholds. Secondly, we compute the entropy of raw saliency maps and select the raw saliency map with the smallest value of entropy as the final saliency map. Our model detects more complete object region. By testing on the databases ASD, MSRA10K, DUT-OMRON and SED2, experiments demonstrate that the proposed model outperforms the state-of-the-art models.

Keywords: Salient object detection · Optimized amplitude · Peaks

1 Introduction

Visual saliency has already got extensive studied by computer vision researchers and cognitive scientists. Saliency detection is a very important step to solve visual tasks, such as image segmentation [1], visual tracking [2], image and video compression [3]. Existing saliency detection models can be divided into spatial domain [4–6] and frequency domain [7–12] according to different computing domains.

In the frequency domain, the information of image is reflected in amplitude spectrum and phase spectrum. The existing frequency domain saliency detection models compute saliency map by using the information of amplitude spectrum or phase spectrum or combination of both. According to the different methods of using the information of frequency domain, we divide existing frequency domain saliency detection models into four groups: (i) the model uses original phase spectrum [7]; (ii) the model uses optimized amplitude spectrum and original phase spectrum [8,10,12,13]; (iii) the model uses optimized amplitude spectrum and optimized phase spectrum [11,14,15]; (iv) the model uses wavelet transform [16,17]. We propose model which can detect salient object by using optimized

D. Liu et al. (Eds.): ICONIP 2017, Part III, LNCS 10636, pp. 457–466, 2017.
https://doi.org/10.1007/978-3-319-70090-8_47

amplitude spectrum and original phase spectrum, so the proposed model belongs to the third group.

Hou and Zhang [8] introduced the frequency domain calculation into saliency detection. They combined spectral residual (SR) and phase spectrum to compute saliency map. In the first group, Guo et al. [7] found that only using phase spectrum from Fourier Transform (PFT) can achieve similar detection results to SR. On the basis of PFT, Guo extended the algorithm PFT to saliency detection of color image, and used phase spectrum from quaternion Fourier Transform (PQFT) to improve the overall performance of saliency map. In the second group, Li et al. [10] inspired by [7] and proposed the saliency model about Hypercomplex Fourier Transform (HFT), and the detection results were better. In the third group, Li et al. [15] proposed Hypercomplex Spectral Contrast (HSC), which utilized amplitude spectral contrast and phase spectral contrast to calculate saliency map and enhanced detection accuracy by average of multiscale saliency map. Li et al. [11] computed saliency map by combination of effective amplitude spectrum and phase spectrum, which were obtained by designed amplitude spectrum filters and phase spectrum filters, respectively. In the fourth group, Nevrez [16] proposed a saliency detection model, and the final saliency map was fusion of local saliency map and global saliency map both calculated by wavelet transform.

In the above algorithms, some algorithms directly abandon the amplitude spectrum, and others ignore the effect of different amplitudes to the salient object. In our work, we find that smoothing the peaks (above our selected threshold) in log amplitude spectrum can obtain better detection results. Thus, we propose a new saliency detection model which is based on the optimized amplitude spectrum by smoothing the peaks in log amplitude spectrum. **Compared with the state-of-the-act models, the proposed model makes full use of the amplitude spectral information and detects a more complete object region.**

2 Amplitude Spectrum Analysis

In the frequency domain, the peaks in log amplitude correspond to the repetitive patterns [10]. In this paper, we define the peaks as the amplitude higher than the selected threshold. The log amplitude spectrum is smoothed to suppress the repetitive background regions, which filter out non-salient regions. But, many salient regions are also repetitive patterns which correspond to some local peaks in log amplitude spectrum. These local peaks are also suppressed when smoothing the entire log amplitude spectrum, so the information of corresponding salient regions is also naturally filtered out. If we only smooth the peaks in log amplitude spectrum and don't change the original local peaks, which will keep more salient information and have a better detection results. Therefore, we propose a novel model which smooth the peaks in log amplitude spectrum, and the saliency map is obtained by inverse transform combining smoothed amplitude spectrum and original phase spectrum.

To demonstrate the effectiveness of our method, we construct one dimensional signals $f_1(t)$ and $f_2(t)$. The signal $f_1(t)$ is a periodic signal, called the first signal. And the signal $f_2(t)$ is generated by the frequency doubling of the 701–800 sampling segment in $f_1(t)$, called second signal. In Fig. 1, row 1st and 2nd show the original waveforms and the corresponding log amplitude spectrum of the first signal. Row 3rd and 4th show the original waveforms and the corresponding log amplitude spectrum of $f_2(t)$. In this paper, the high frequency part in $f_2(t)$ is called salient segment (701–800 sampling segment), and the low frequency part is called non-salient segment.

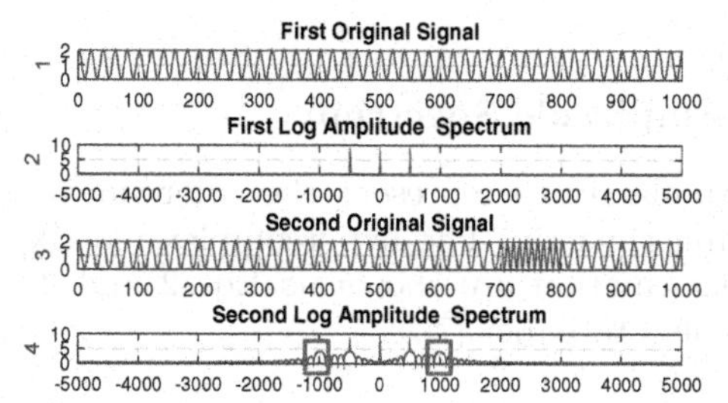

Fig. 1. The non-salient segment leads to the three peaks, the salient segment leads to the local peaks (in red boxes). (Color figure online)

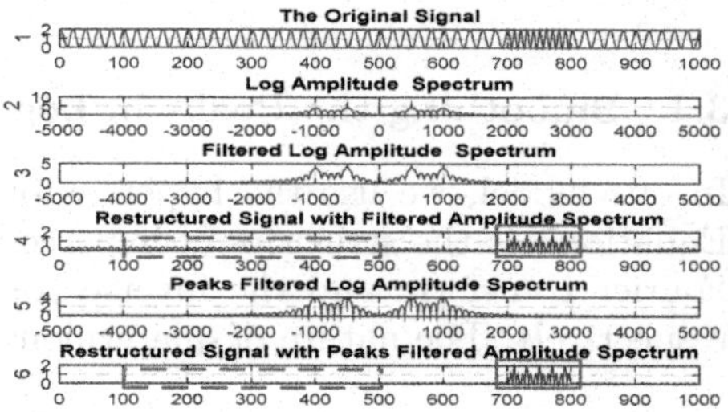

Fig. 2. Detection results by using the two models. Smoothing the peaks in log amplitude can get better detection result. (Color figure online)

According to [10], the higher amplitude in amplitude spectrum correspond to the repetitive patterns. So the three peaks of row 2nd in Fig. 1 are caused by repetitive patterns in $f_1(t)$. In fact, the three peaks of row 4th in Fig. 1 are also caused by the repetitive non-salient segment and the two local peaks (in red boxes) are caused by the repetitive salient segment in $f_2(t)$. If the entire log amplitude spectrum in row 2nd is smoothed, the three peaks are suppressed, so as the two local peaks are also suppressed, which weaken the non-salient segment and also weaken the salient segment. If only the peaks in log amplitude spectrum are smoothed, the obtained results only weaken the non-salient segment and get better saliency map. So we propose the model that only needs to smooth the peaks in log amplitude spectrum and only weakens the non-salient segment, and the proposed model obtains better saliency detection results.

The detection results of the models smoothing the entire log amplitude and the peaks are shown in Fig. 2. The detected non-salient segment has a lower amplitude (red dotted boxes) and the detected salient segment has a higher amplitude (red solid boxes) by the proposed model. Thus, the proposed model can obtain better detection results.

3 The Methodology

According to the analysis in Sect. 2, the peaks in log amplitude spectrum are caused by background regions and the local peaks are caused by salient regions. Thus, we propose the saliency detection model which is based on the optimized amplitude spectrum and the original phase spectrum. Firstly, we optimize the amplitude spectrum by smoothing the peaks in log amplitude spectrum according to different thresholds. Then, we obtain raw saliency maps by the inverse transform combining the optimized amplitude spectrum and original phase spectrum. Lastly, we select the raw saliency map with smallest entropy as the final saliency map.

3.1 Smoothing the Peaks in Log Amplitude Spectrum

In our model, we use the hypercomplex matrix to represent the color image. Equation 1 is the structure of hypercomplex matrix. The hypercomplex matrix Fourier transform and the inverse transformation are shown as Eqs. 2 and 3, respectively. The nature of quaternion refers to paper [7].

$$f(n,m) = a + bi + cj + dk \tag{1}$$

$$F(u,v) = \frac{1}{\sqrt{MN}} \sum_{m=0}^{M-1} \sum_{n=0}^{N-1} e^{-u2\pi((\frac{mv}{M})+(\frac{nu}{N}))} f(n,m) \tag{2}$$

$$f(n,m) = \frac{1}{\sqrt{MN}} \sum_{v=0}^{M-1} \sum_{u=0}^{N-1} e^{u2\pi((\frac{mv}{M})+(\frac{nu}{N}))} F(u,v) \tag{3}$$

RGB color system is the most commonly used color system, but it is device-dependent color system. *Lab* color system is based on physiological characteristics, and has a larger color space than *RGB* color system. *L* channel represents brightness, *a* channel represents the range from red to green and *b* channel represents the range from yellow to blue. *Lab* color system is more in line with human visual perception system. Thus, we use *Lab* color system to represent images in our work. Equation 4 is a pure imaginary matrix, where L, a and b are three imaginary parts. Equation 2 can be rewritten after Fourier Transform, see Eq. 5. $A(u,v) = \|F(u,v)\|$ and $p(u,v) = angle(F(u,v))$ are amplitude spectrum and phase spectrum, respectively. $L(u,v) = log(A(u,v))$ is the log amplitude spectrum.

$$f(n,m) = Li + aj + bk \tag{4}$$

$$F(u,v) = \|A(u,v)\| e^{uP(u,v)} \tag{5}$$

The original amplitude spectrum has a large spectral drop. When optimizing amplitude spectrum, therefore, we choose to smooth the peaks in log amplitude spectrum rather than original amplitude spectrum. How to determine the peaks in log amplitude spectrum? We choose a suitable threshold. The peaks are the amplitude over than the threshold. We use matrix $\Gamma(u, v)$ to record the positions of peaks in the amplitude spectrum in Eq. 6. $L_S(u, v)$ is the log amplitude spectrum after smoothing the peaks and $g(\sigma)$ is a Gaussian kernel (Eq. 7).

$$\Gamma(u, v) = \begin{cases} 1, & A(u, v) > threshold, \\ 0, & otherwise. \end{cases} \quad (6)$$

$$L_S(u, v) = [g(\sigma) * L(u, v)] \cdot \Gamma(u, v) + L(u, v) \cdot [1 - \Gamma(u, v)] \quad (7)$$

3.2 Choose the Thresholds and Smoothing Scale

Choose the appropriate thresholds. In our work, the threshold is an important parameter. Due to the diversity of images, it is difficult to find a suitable threshold for all images, so we decide to choose a number of thresholds. We have done a statistic of the images in the database ASD [18], after normalized the size of images to $128 * 128$. In Table 1, The $num1$ is the number of mean peaks higher than the corresponding threshold. The $key = mean(A(u, v))$ is the average amplitude of the amplitude spectrum.

If the number of peaks is less than one, smoothing the peaks in log amplitude spectrum is similar to what has nothing changed in the amplitude spectrum. If the number of peaks is too large, smoothing the peaks in log amplitude spectrum is similar to smoothing the entire log amplitude spectrum. Thus, when the number of peaks is too much nor too little, the detection results will be poor. In Table 1, when threshold is equal to key or $1.25key$, there are too many peaks; when threshold is equal to $2.75key$, the number of peaks is too small. And they are not suitable as thresholds. Finally, we choose the best threshold from experimental result in Eq. 8. And we assume that the optimal threshold for determining the peaks must appear in Eq. 8.

Table 1. $num1$ are the number of mean peaks in different thresholds.

$threshold$	key	$1.25key$	$1.5key$	$1.75key$	$2key$	$2.25key$	$2.5key$	$2.75key$
$num1$	7490.39	2031.65	383.97	61.79	9.79	1.83	0.58	0.05

Table 2. $num2$ are the number of images in different best thresholds

$threshold$	key	$1.25key$	$1.5key$	$1.75key$	$2key$	$2.25key$	$2.5key$	$2.75key$
$num2$	28	34	78	171	229	286	142	32

According to the eight thresholds in Table 1, we count the number which each threshold is selected as the optimal threshold, in the database ASD [18]. We obtain eight raw saliency maps by eight thresholds, and then we calculate the entropy of each raw saliency map. The threshold of the salient map with smallest entropy is the optimal threshold of the image. The statistical results are shown in Table 2, and $num2$ is the number of images which the corresponding threshold is the optimal threshold. The probability of the optimal threshold appears within our chosen best threshold (Eq. 8) is $(78+171+229+286+142)/1000 = 90.6\%$. So the chosen best thresholds are reasonable.

Choose the appropriate smoothing scale. In saliency map, the brighter the salient regions, the darker the non-salient regions, the better of detection results. So we define the Salient Contrast Ratio (SCR) as measure the detection effect in Eq. 9. The greater of SCR, the better of detection results.

$$threshold = u \cdot key, (u = 1.5, 1.75, 2, 2.25, 2.5) \tag{8}$$

$$SCR = S \cdot \Theta(*)/sum(\Theta(*)) - S \cdot (1 - \Theta(*))/sum(1 - \Theta(*)) \tag{9}$$

where, S is the raw saliency map. According to the ground truth, $\Theta(*)$ are equal to one in the salient positions and are equal to zero in the non-salient positions. We determine the optimal smoothing scale σ according to the images in the database ASD [18]. We calculate the saliency map to each smoothing scale in [0.4, 10] at 0.1 step length, and then compute the average SCR of each scale about all images in ASD. The curve of SCR is shown in Fig. 3 with increasing of σ. When $\sigma = 1.6$, the average SCR reaches the maximum. Thus, we select the best smoothing scale $\sigma = 1.6$.

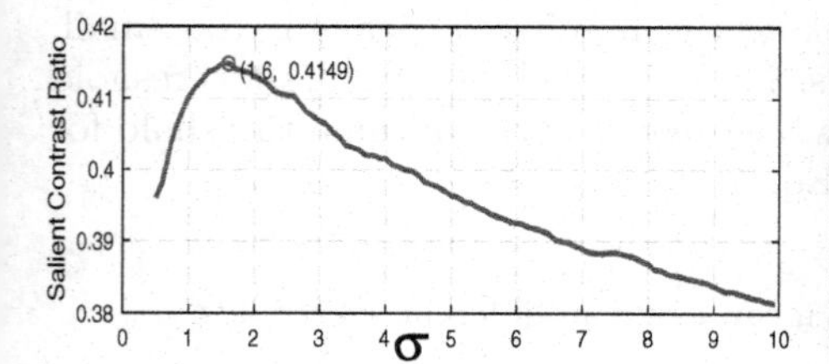

Fig. 3. SCR is increased firstly and then reduced. When $\sigma = 1.6$, SCR reaches the maximum.

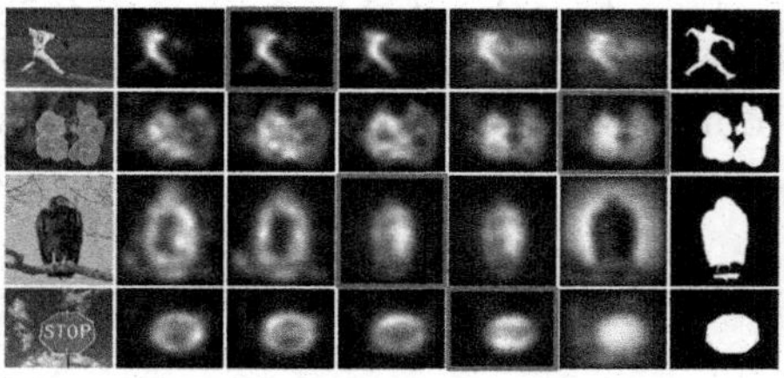

Fig. 4. Raw saliency maps shown in 2nd–6th columns. The final saliency map is shown in the red box. (Color figure online)

3.3 Optimized Amplitude and Final Saliency Map

After selecting the suitable thresholds in Eq. 8 and smoothing scale $\sigma = 1.6$, we can obtain five optimized log amplitude $L_S(u, v; k)$ according to Eqs. 7 and 8. The raw saliency maps are calculated (Eq. 10) by the inverse Fourier Transform combining optimized amplitude and original phase spectrum. In Eq. 10, $h(\sigma_0)$ is low-pass filter, $(\sigma_0 = 5)$.

The entropy of image represents aggregation characteristic of the gray distribution. The smaller the entropy, the more concentrated the gray distribution.

Thus, we can determine the final salient map by the entropy of image. We calculate the entropy of each raw saliency map, and the final saliency map is the raw saliency map with smallest entropy (Eq. 11).

$$S(x, y; k) = h(\sigma_0) * \|F^{-1}[exp(L_S(u, v; k) + i \cdot P(u, v))]\|^2 \quad (10)$$

$$S_f \Leftarrow argmin(entropy(S(x, y; k))) \quad (11)$$

In Fig. 4, the raw saliency maps are calculated by optimized amplitude spectrum and original phase spectrum according to different thresholds in Eq. 8, shown in 2nd–6th columns. The final saliency map is chosen by Eq. 11, and it is the raw saliency map with smallest entropy, shown in the red box in Fig. 4. Our model which computes the saliency map by using the optimized Amplitude spectrum by Hypercomplex Fourier Transform is referred as AHFT in this paper. The AHFT model is summarized in Algorithm 1.

Algorithm 1. AHFT saliency model

1: Adjust the resolution of the input color image to 128*128;
2: Use hypercomplex matrix representing input image, according to Eq.4;
3: Calculate amplitude spectrum and phase spectrum, according to Eq.2 and Eq.5;
4: Optimize amplitude spectrum, according to Eq.6 and Eq.7;
5: Calculate the raw saliency maps $S(x, y; k)$, according to Eq.10;
6: Choose the final saliency map S_f by the entropy, according to Eq.11;

4 Experimental Results

We tested AHFT on four open authoritative databases, including: ASD [18], MSRA10K [5], DUT-OMRON [19] and SED2 [1]. These four databases contain a sea of images and have manually marked Ground Truth. ASD database is the most cited database. MSRA10K database contains 10k images which are selected from MSRA database. The images in DUT-OMRON database have more complex background than the images in MSRA10K database. In SED2 database, images contain two or more salient objects. We compared the detection results among the following methods, including our model, the highly cited model of IT [4], the frequency domain models of SR [8], PFT [7], PQFT [7], DCT [9], QDCT [12], FT [18], WAVE [16], HFT [10], and the latest model of UHM [6].

Figure 5 shows the saliency maps. Most algorithms can only detect salient object boundary. The algorithms of FT [18] and WAVE [16] can highlight the entire salient region, but the brightness are not enough. UHM [6] can also enhance the entire salient region, but the background region cannot be suppressed effectively, thus the precision is poor. Only the proposed model can detect the whole salient object region and suppress the background region effectively.

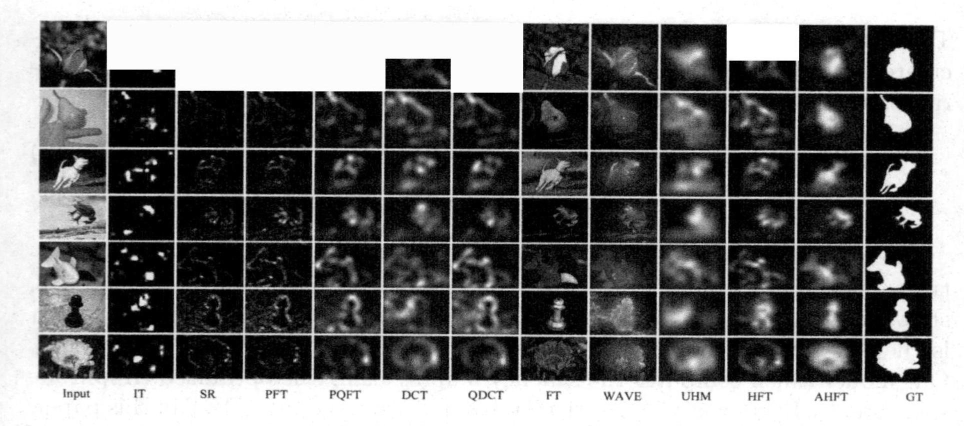

Fig. 5. Comparison saliency maps between our model and other models.

In order to evaluate the performance of our model comprehensively, we compare the detection results of various algorithms with three objective evaluation criteria AUC [11], F_β [5] and MAE [20]. AUC score is the area under the receiver operating characteristic (ROC) curve. The best AUC score is equal to 1. F_β is the comprehensive *Precision* and *Recall* evaluation criteria and is more objective to reflect the performance of saliency detection models. In Eq. 12, when $\beta^2 < 1$, *Recall* has a greater impact; when $\beta^2 > 1$, *Precision* has a greater impact. According to [5,16], we also choose $\beta^2 = 0.3$ in our experiment. Mean absolute error (MAE) is a good measure of detection accuracy. The smaller of MAE, the better of detection results.

$$F_\beta = \frac{(1 + \beta^2) \cdot Precision \cdot Recall}{\beta^2 \cdot Precision + Recall} \tag{12}$$

Table 3. Comparison scores of AUC, F_β and MAE between our model and other models.

Model	ASD			MASR10K			DUT-OMRON			SED2		
	AUC	F_β	MAE	AUC	F_β	MAE	AUC	F_β	MAE	AUC	F_β	MAE
IT [4]	0.648	0.383	0.194	0.640	0.378	0.213	0.636	0.349	0.198	0.682	0.476	0.245
SR [8]	0.715	0.376	0.214	0.750	0.431	0.239	0.702	0.304	**0.195**	0.795	0.490	0.226
PFT [7]	0.730	0.404	0.224	0.745	0.431	0.241	0.701	0.307	0.198	0.790	0.492	0.227
PQFT [7]	0.785	0.492	0.245	0.799	0.534	0.251	0.747	0.341	0.245	0.833	0.532	0.233
DCT [9]	0.839	0.569	0.258	0.823	0.576	0.267	0.784	0.374	0.280	0.830	0.514	0.262
QDCT [12]	0.827	0.545	0.242	0.825	0.571	0.252	0.779	0.367	0.255	0.839	0.538	0.239
FT [18]	0.866	0.724	0.206	0.790	0.638	0.245	0.682	0.365	0.250	0.819	**0.722**	0.206
WAVE [16]	0.911	0.652	0.249	0.876	0.641	0.262	0.821	0.390	0.256	0.879	0.660	0.248
UHM [6]	0.897	0.604	0.265	0.884	0.632	0.267	0.814	0.368	0.321	0.830	0.490	0.299
HFT [10]	0.931	0.716	0.178	0.906	0.696	0.202	0.858	0.492	0.198	0.871	0.595	0.208
AHFT [Ours]	**0.953**	**0.794**	**0.169**	**0.920**	**0.751**	**0.198**	**0.867**	**0.528**	0.206	**0.890**	0.678	**0.201**

The objective evaluation scores are shown in Table 3, and the bold numbers are the best values. Our method is obviously better than others in AUC. On the whole, our model outperforms the state-of-the-art models.

5 Conclusions

By analyzing amplitude spectrums of signal, we find that the repeated background in signal always corresponds to the peaks in log amplitude spectrum. Similarly, the salient object is also a repetitive pattern, which corresponds to the local peaks in log amplitude spectrum. If the entire log amplitude spectrum is smoothed, it will not only suppress the peaks but also suppress the local peaks in log amplitude spectrum. If only the peaks in log amplitude spectrum are smoothed, the background information is suppressed and the salient information is preserved. Obviously, the latter model achieves better detection result than the former model. Therefore, we propose a novel saliency model which combines the optimized amplitude spectrum by smoothing the peaks and the original phase spectrum to compute the saliency map. The proposed model can detect more complete saliency objects and obtain better detection results.

Acknowledgments. The paper was supported in part by the National Natural Science Foundation (NSFC) of China under Grant No. 61365003 and Gansu Province Basic Research Innovation Group Project No. 1506RJIA031.

References

1. Alpert, S., Galun, M., Brandt, A., Basri, R.: Image segmentation by probabilistic bottom-up aggregation and cue integration. TPAMI **34**(2), 315–327 (2012)
2. Borji, A., Frintrop, S., Sihite, D.N., Itti, L.: Adaptive object tracking by learning background context. In: CVPR, pp. 23–30 (2012)
3. Xue, J.R., Li, C., Zheng, N.N.: Proto-object based rate control for JPEG2000: an approach to content-based scalability. TIP **20**(4), 1177–1184 (2011)
4. Itti, L., Koch, C., Niebur, E.: A model of saliency-based visual attention for rapid scene analysis. TPAMI **20**(11), 1254–1259 (1998)
5. Cheng, M.M., Mitra, N.J., Huang, X.L., Philip, H.S., Hu, S.M.: Global contrast based salient region detection. TPAMI **37**(3), 569–582 (2015)
6. Tavakoli, H.R., Laaksonen, J.: Bottom-up fixation prediction using unsupervised hierarchical models. In: Chen, C.-S., Lu, J., Ma, K.-K. (eds.) ACCV 2016. LNCS, vol. 10116, pp. 287–302. Springer, Cham (2017). doi:10.1007/978-3-319-54407-6_19
7. Guo, C.L., Ma, Q., Zhang, L.M.: Spatio-temporal saliency detection using phase spectrum of quaternion Fourier transform. In: CVPR, pp. 1–8 (2008)
8. Hou, X.D., Zhang, L.Q.: Saliency detection: a spectral residual approach. In: CVPR, pp. 1–8 (2007)
9. Hou, X.D., Harel, J., Koch, C.: Image signature: highlighting sparse salient regions. TPAMI **34**(1), 194–201 (2012)
10. Li, J., Levine, M.D., An, X.J., Xu, X., He, H.E.: Visual saliency based on scalespace analysis in the frequency domain. TPAMI **35**(4), 996–1010 (2013)

11. Li, J., Duan, L.Y., Chen, X., Huang, T., Tian, Y.: Finding the secret of image saliency in the frequency domain. TPAMI **37**(12), 2428–2440 (2015)
12. Schauerte, B., Stiefelhagen, R.: Predicting human gaze using quaternion DCT image signature saliency and face detection. In: WACV, pp. 137–144 (2012)
13. Fang, Y.M., Lin, W.S., Lee, B.S., Lau, C.T., Chen, Z.Z., Lin, C.W.: Bottom-up saliency detection model based on human visual sensitivity and amplitude spectrum. Trans. Multimedia **14**(1), 187–198 (2012)
14. Li, C., Xue, J.R., Tian, Z.Q., Li, L., Zheng, N.N.: Saliency detection based on biological plausibility of hypercomplex Fourier spectrum contrast. Opt. Lett. **37**(17), 3609–3611 (2012)
15. Li, C., Xue, J.R., Zheng, N.N., Lan, X.G., Tian, Z.Q.: Spatio-temporal saliency perception via hypercomplex frequency spectral contrast. Sensors **13**(3), 3409–3431 (2013)
16. Imamoglu, N., Lin, W.S., Fang, Y.M.: A saliency detection model using low-level features based on wavelet transform. Trans. Multimedia **15**(1), 96–105 (2013)
17. Murray, N., Vanrell, M., Otazu, X., Parraga, C.A.: Low-level spatiochromatic grouping for saliency estimation. TPAMI **35**(11), 2810–2816 (2013)
18. Achanta, R., Hemami, S., Estrada, F., Susstrunk, S.: Frequency-tuned salient region detection. In: CVPR, pp. 1597–1604 (2009)
19. Yang, C., Zhang, L.H., Lu, H.C., Ruan, X., Yang, M.H.: Saliency detection via graph based manifold ranking. In: CVPR, pp. 3166–3173 (2013)
20. Zhu, W.J., Liang, S., Wei, Y.C., Sun, J.: Saliency optimization from robust background detection. In: CVPR, pp. 2814–2821 (2014)

Optimized Image Up-Scaling from Learning Selective Similarity

He Jiang and Jie Yang(✉)

Shanghai Jiaotong University, Shanghai, China
{jianghe2012, jieyang}@sjtu.edu.cn

Abstract. Remarkable up-scaling results are obtained from local self-examples (LSE) [1] at low cost. However, fine-detailed and cluttered regions are not reproduced realistically due to inappropriate addition of high frequency, and thus appear somewhat faceted and slightly distorted at edges. In this paper, an optimized algorithm that reduces these artifacts is proposed. A selective search is applied in the original restricted searching area. Mismatches can be avoided when patches are with extremely random details. Meanwhile we extend the local self-similarity on natural images to sub-pixel level which makes the assumption of local self-similarity more suitable especially for fine edge areas. It corrects the slightly misalignments when scaled with small factor. The proposed algorithm remains simple, and generates clear and believable textures compared with the original LSE and other mainstream methods.

Keywords: Selective similarity · Facet improvement · Sub-pixel self-similarity · Super-resolution

1 Introduction and Related Work

With the rapid increase of HD screen resolution and the growing demand of web video data transmission, the problem of creating artifact-free up-sampling image becomes an increasingly hot topic. In order to save network bandwidth and reduce complexity of codec, videos are often down sampled and transmitted at low resolution in the form of stream media and then up-sampled at terminals like digital TV, monitor, smart phone. Those image enlargements require not only clear and stable display results to human feelings but suitable hardware architecture for implementation.

In most image processing algorithms and commercial applications, classic bilinear, bi-cubic and their improved versions are widely applied. Despite robustness and effectiveness, the magnified results are beyond satisfactory for causing blur and jaggies. Many polynomial-based and edge-based interpolation methods [3, 7, 10] are also widely researched. Image super-resolution [4, 5, 11, 12, 14] is a rising family of methods to construct a high-resolution image from one or multiple low-resolution initial frames with promising results.

Reconstruction-based and example-based strategies are its two categories. The former one acquires sufficient number of down sampled, warped and blurred versions of the desired high-resolution image. And the reconstruction process is an inverse problem according to certain assumed prior knowledge of an observation model [6].

D. Liu et al. (Eds.): ICONIP 2017, Part III, LNCS 10636, pp. 467–475, 2017.
https://doi.org/10.1007/978-3-319-70090-8_48

However, this approach are numerically limited to a scaling factor of two [13] and the effect need to be improved. Example-based super-resolution uses a database of training images to create plausible high-frequency details in zoomed images. Pre-processing steps allow the use of details from regions of the training images which may look quite different from the image to be proposed.

Freedman and Fattal propose a new high-quality and efficient single-image up-scaling technique that extends the existing example-based super-resolution framework [1]. The method performs considerably less computations and presents state-of-the-art results. For simple edge structures, it can generate sharp and clean high resolution image with a decreased amount of jaggies and ringings. However, fine detailed and cluttered regions are not reproduced realistically, and appear somewhat faceted and slightly distorted at edges.

In this paper, an optimized algorithm that reduces the artifact is proposed. Similarities in different directions in original image are assigned with different probabilities when searching across different scales. Meanwhile we extend the local self-similarity on natural images to a selective similarity on sub-pixel level which makes the assumption of local self-similarity more suitable especially for fine edges. The rest of this paper is organized as follows. In Sect. 2 we go deep into some details of LSE method and analyze its drawbacks. Then we propose an improved one to overcome the drawbacks. In Sect. 4 we present examples and compare our results with classical methods. Finally concluding remarks are provided in Sect. 5.

2 Local Self-example and Its Limitations

The LSE exploits a local self-similarity assumption in natural images where small patches are very similar to themselves upon small scaling factors (normally less than 1.25 according to Fig. 3 in [1]). In other words, for every patch in an image, very similar one can be found in its slightly up-scaled version at localized regions around the relative point. As shown in Fig. 1, a fine detailed up-sampled image is obtained by iteratively combining a low frequency band and a high frequency band which is absorbed in H0 through patch matching. Proper high frequency band plays a key part in the effect of super-resolution result. The method does not rely on external or prior knowledge as database like the traditional super resolution algorithm. And searches for similar patches are not performed in the whole image either, but against restricted local windows, which ensures the validity of results. LSE produces good effect especially for natural images. The self-similarity assumption holds for various image singularities such as edges and corners. As shown in Fig. 2, patches in lowercase letters are slightly enlarged to get the patches with uppercase letters, and then they are cropped as shown in the white boxes. Patch (a), (A) and patch (b), (B) are very similar respectively. In patch (c) (dense fur), pixel values are distributed randomly that no obvious texture can be traced. A small magnification does not maintain such details that their resemblance is strongly weakened compared with the first two sets of patches. Such cluttered regions can be easily matched by mistake and appear somewhat faceted (some uncomfortable fake contour).

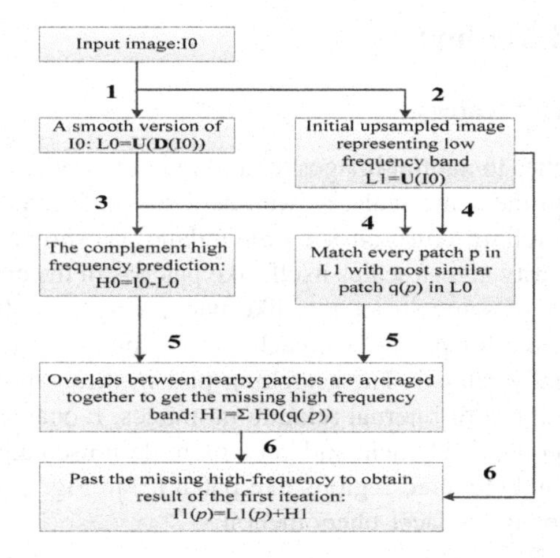

Fig. 1. Flow chart of local self-example based up-scaling for a small up-scaling factor. The interpolation operator U and D are implemented with new dedicated non-dyadic filter banks.

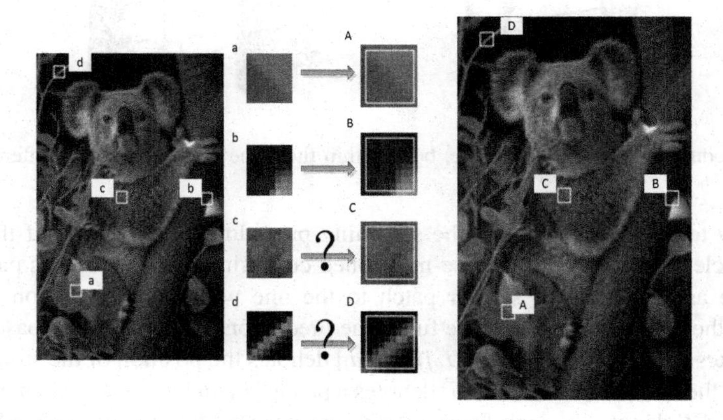

Fig. 2. From right to the left: original image koala, original patches, enlarged patches (and their cropped ones in the white box), the enlarged image.

By contrast, patch (d) (vine) demonstrates strong texture compared with patch (c), but it is difficult to match finely. How to present a fine structure with a small up-scaling factor? For example, in patch (d), the vine is 2-pixel wide and in patch (D), the vine is 2.5-pixel wide theoretically if the up-scaling factor is 1.25. Now it brings out a question of the mapping between an integer precision image and a non-integer precision image. Resultantly in patch (D), it is hard to find a theoretically same one in patch (d). Misregistration is inevitable as there is no theoretically right matching pair. As iteration goes on, mistakes accumulate and these fine details cannot be reproduced realistically and cause slight distortions along edges.

3 The Proposed Method

3.1 Selective Search Strategy

According to [2], patches in natural images tend to redundantly recur many times inside the image, both within the same scale, as well as across different scales. The key that we insist on patch searching relies on such redundancy as shown in Fig. 3. When an image possesses similarity across scales itself, like patches in the circular cone, or when an image has a bunch of same structures, like patches in the cylinder, the searching process makes it possible for patches on each side of the arrow a matching pair even they are not theoretically related. The searching part in these above mentioned situations helps to make full use of internal relation in images. It enables each pixel to grab more possible high frequency element and get a more de-noised and smoothing output. However, in random and cluttered regions like patch (c) in Fig. 2, searching can cause mismatch and accompany by facet phenomenon.

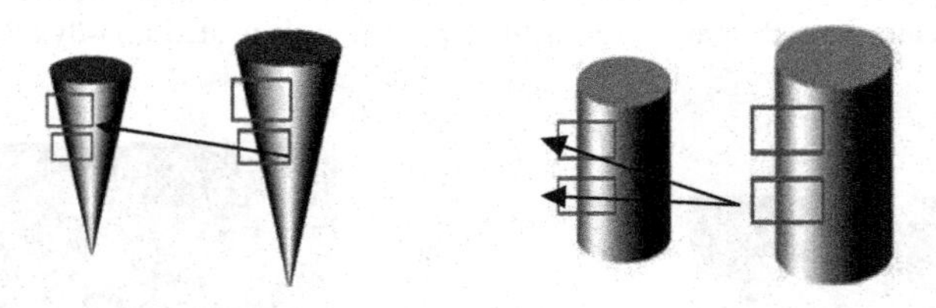

Fig. 3. Redundancy inside the image, both within the same scale and across different scales

We try to selectively turn on the searching procedure by predicting if the current patch has clear texture. In the above-mentioned confusing patches such as patch (c) in Fig. 2, we assign the most similar patch to the one with closest position to ensure validity rather than search for it. We fulfill the prediction with a direction-based search. Let I denotes an image indexed by (i, j). $[d_i, d_j]$ denotes the position of the similar patch. W denotes the searching window. LP denotes a patch from a larger version of an original image and OP denotes a patch from the original image. $[d_i, d_j]$ is determined by:

$$[d_i, d_j] = \begin{cases} \arg\min\limits_{[d_i,d_j]\in W}(|LP_{[d_i,d_j]} - OP_{[d_i,d_j]}| + c|LP_{[0,0]} - LP_{[d_i,d_j]}|) & min\_value \leq threshold \\ [d_0, d_0] & min\_value > threshold \end{cases}, \quad (1)$$

where c is the weight controlling similarity inside single image. When the *min_value* is larger than a *threshold* we set, we consider that the patch contains no texture, and thus $[d_i, d_j]$ is assigned to $[d_0, d_0]$ according to (1).

3.2 Sub-pixel Based Strategy

We extend the local self-similarity on natural images to sub-pixel level to address the problem of patch (d) in Fig. 2. It improves the prediction accuracy by creating a subpixel displacement of the original image. Namely, the regular part of an up-scaled image may find its matching counterpart in its original version of image, while singularity patches may be matched in its sub-pixel version. As Fig. 4 shows, sub-pixel images contain as much possible structures as small up-scaling process can get. Patches in (b) can find its most truthful downscaled version in sub-pixel ones (c)–(f). Resultantly appropriate high frequency band is pasted.

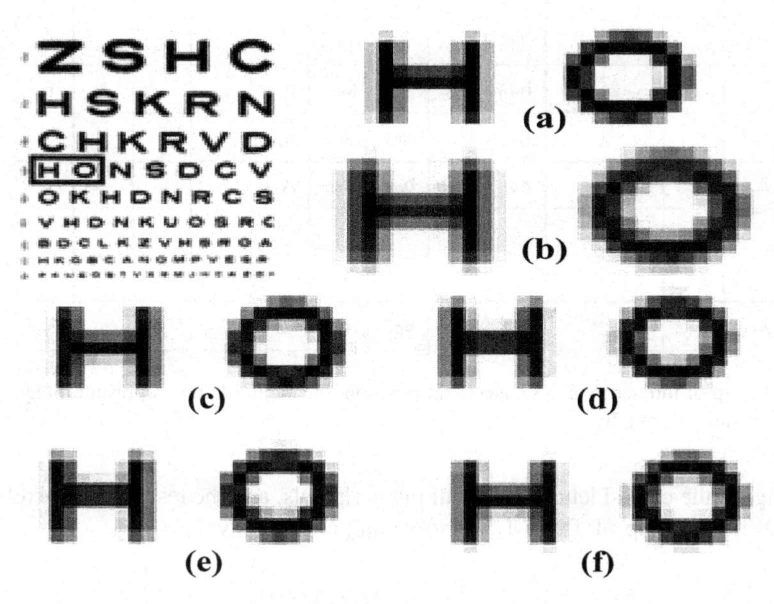

Fig. 4. Sub-pixel misalignments. (a) is a patch from original image, (b) is the up-scaled version with a factor of 1.25, (c)–(f) are sub-pixel misaligned versions.

In our research, sub-pixels are obtained using a filter derived from discrete Fourier transform (DCT) and inverse Fourier transform. An 8-tap separable DCT-based interpolation filter is used for 1/2 precision, and a 7-tap separable DCT-based interpolation filter is used for 1/4 precision (if necessary) as shown in Table 1.

Table 1. Coefficients of half pixel and quarter-pixel

Index	-3 1	-2 2	-1	0	3	4
Half-filter	-1 40	4 -11	-11	40	4	1
Quarter-filter	-1 17	4 -5	-10	58	1	

$A_{-1,-1}$				$A_{0,-1}$	$a_{0,-1}$	$b_{0,-1}$	$c_{0,-1}$	$A_{1,-1}$				$A_{2,-1}$
$A_{1,0}$				$A_{0,0}$	$a_{0,0}$	$b_{0.0}$	$c_{0.0}$	$A_{1,0}$				$A_{2,0}$
$d_{-1,0}$				$d_{0,0}$	$e_{0,0}$	$f_{0,0}$	$g_{0,0}$	$d_{1,0}$				$d_{2,0}$
$h_{-1,0}$				$h_{0,0}$	$i_{0,0}$	$j_{0,0}$	$k_{0,0}$	$h_{1,0}$				$h_{2,0}$
$n_{-1,0}$				$n_{0,0}$	$p_{0,0}$	$q_{0,0}$	$r_{0,0}$	$n_{1,0}$				$n_{2,0}$
$A_{-1,1}$				$A_{0,1}$	$a_{0,1}$	$b_{0,1}$	$c_{0,1}$	$A_{1,1}$				$A_{2,1}$
$A_{-1,2}$				$A_{0,2}$	$a_{0,2}$	$b_{0,2}$	$c_{0,2}$	$A_{1,2}$				$A_{2,2}$

Fig. 5. A map of integer pixel and sub-pixel position, the capital letters represent integer pixels, and the rest are sub-pixels.

In Fig. 5, the capital letters represent integer pixels, and the rest are sub-pixels to be solved. With the help of Table 1, we solve $a_{0,0}$ and $b_{0,0}$ by

$$a_{0,0} = \sum\nolimits_{i=-3}^{3} A_{I,j}\ Qfilter(i) \tag{2}$$

$$b_{0,0} = \sum\nolimits_{i=-3}^{3} A_{I,j}\ Hfilter(i) \tag{3}$$

$c_{0,0}$ and $d_{0,0}$ are calculated in a similar way as shown in (2) and (3). $e_{0,0}$, $f_{0,0}$, $g_{0,0}$, $i_{0,0}$, $j_{0,0}$, $k_{0,0}$, $p_{0,0}$, $q_{0,0}$ and $r_{0,0}$ are then calculated with their vertical neighbors $a_{0,0}$, $b_{0,0}$ and $c_{0,0}$:

$$e_{0,0} = \sum\nolimits_{i=-3}^{3} a_{0,i}\ Qfilter(i) \tag{4}$$

$$k_{0,0} = \sum\nolimits_{i=-3}^{4} c_{0,j}\ Hfilter(i) \tag{5}$$

As we can discover, vertical and horizontal interpolation can be separable. Half-pixel and quarter-pixel can be separable as well. This makes it easier for parallel realization and has already been used in High Efficiency Video Coding (HEVC) standard [15].

4 Experiments and Results

In this section, we show experimental results of our proposed algorithm. We select 300 images from the internet randomly, including different quality, resolution and different scenes such like landscape, person and cartoon. In order to demonstrate the effectiveness, we compare our algorithm with the current state-of-the-art upscale algorithms, NEDI [7], Shan [9], ICBI [8] and LSE [1]. Since the effect of image upscale largely depends on the subjective visual experience, we make a questionnaire to get the subjective evaluation of different people. Some examples are shown in Fig. 6. As the

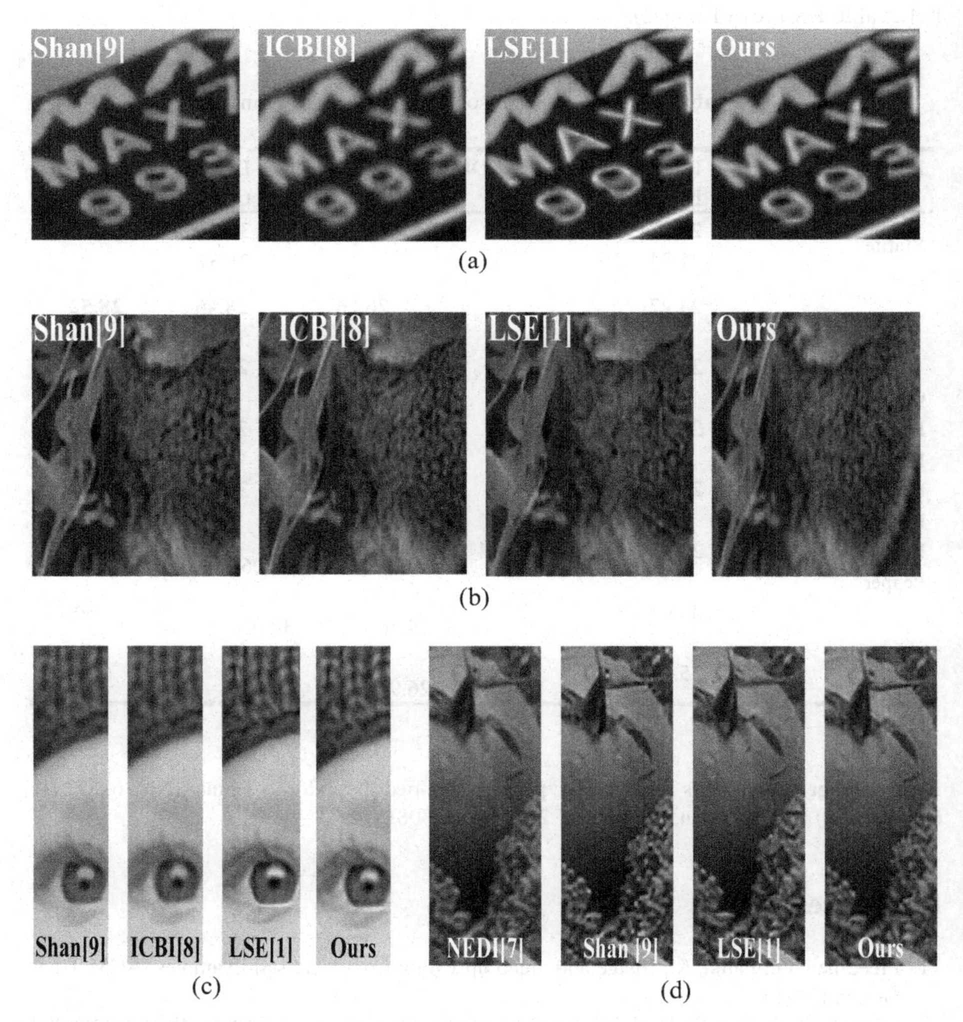

Fig. 6. (a) ×2, (b), (c), (d) ×4 up-scaled results of the proposed method and some mainstream methods.

comparison demonstrated, generally, the LSE and the proposed algorithm enjoy a better performance compared with the other mainstream ones in terms of subjective observation. For patches with general structures, the difference between the LSE and the proposed method can be vaguely discerned. However for fine detail and random cluttered regions that we devote to improve, artifacts are effectively suppressed. In Fig. 6, the traditional LSE produces some fake border along edges and cause slight distortion. This is particularly obvious in the characters 'A' and 'X', as well as rigid edges along the black eyes of the kid, while the proposed algorithm not. And the proposed method produces more natural structures in dense details like the broccoli and the fur, while the LSE exhibit somewhat distortion for details. Generally, our method remains simple and easy for hardware implementation, and generates clear and believable textures (Table 2).

Table 2. PSNR comparison of different algorithms

Images	Bicubic (dB)	NEDI(dB)	ICBI(dB)	LSE(dB)	Ours(dB)
Statue	29.22	30.73	31.24	31.44	**31.57**
Child	24.27	24.93	26.26	28.46	**28.52**
Resistance	26.06	28.92	28.91	29.35	**29.46**
Dog	22.62	24.12	24.88	25.16	**25.24**
Koala	27.56	29.33	29.87	30.15	**30.23**
Room	22.11	22.84	22.79	23.06	**23.12**
Lena	28.03	29.72	29.56	30.03	**30.11**
Pepper	25.84	26.23	26.91	27.07	**27.10**

Acknowledgements. This research is partly supported by NSFC, China (No: 61572315, 6151101179) and 863 Plan, China (No. 2015AA042308).

References

1. Freedman, G., Fattal, R.: Image and video upscaling from local self-examples. ACM Trans. Graph. (2011)
2. Glasner, D., Bagon, S., Irani, M.: Super-resolution from a single image. In: IEEE Conference on Computer Vision (2009)

3. Han, J.-W., et al.: A novel image interpolation method using the bilateral filter. IEEE Trans. Consum. Electron. (2010)
4. Yang, C.-Y., Huang, J.-B., Yang, M.-H.: Exploiting self-similarities for single frame super-resolution. In: Asian Conference on Computer Vision (2011)
5. Freeman, W.T., Jones, T.R., Pasztor, E.C.: Example-based super-resolution. Comput. Graph. Appl. **22**, 56–65 (2002)
6. Park, S.C., Park, M.K., Kang, M.G.: Super-resolution image reconstruction: a technical overview. Sign. Process. Mag. **20**, 21–36 (2003)
7. Li, X., Orchard, M.T.: New edge-directed interpolation. IEEE Trans. Image Process. **10**, 1521–1527 (2001)
8. Giachetti, A., Asuni, N.: Real-time artifact-free image upscaling. IEEE Trans. Image Process. **20**, 2760–2768 (2011)
9. Shan, Q., et al.: Fast image and video up-sampling. ACM Trans. Graph. (2008)
10. Fattal, R.: Image upsampling via imposed edge statistics. ACM Trans. Graph. (2007)
11. Kwatra, V., et al.: Texture optimization for example-based synthesis. ACM Trans. Graph. **24**, 795–802 (2005)
12. Chang, H., Yeung, D.-Y., Xiong, Y.: Super-resolution through neighbor embedding. In: IEEE Conference on Computer Vision and Pattern Recognition (2004)
13. Lin, Z., Shum, H.-Y.: Fundamental limits of reconstruction-based super resolution algorithms under local translation. IEEE Trans. Pattern Anal. Mach. Intell. **26**, 83–97 (2004)
14. Yang, J., et al.: Image super-resolution as sparse representation of raw image patches. In: IEEE Conference on Computer Vision and Pattern Recognition (2008)
15. JCTVC: High Efficiency Video Coding (HEVC) Test Model 10 Encoder Description. In JCTVC-L1002_v3 (2013)

The Camouflage Color Target Detection with Deep Networks

Ce Li(✉), Xinyu Zhao, and Yuqi Wan

College of Electrical and Information Engineering,
Lanzhou University of Technology, Lanzhou 730050, China
xjtulice@gmail.com

Abstract. The camouflage color target is similar to the background, so the detection is very difficult. How to identify the camouflage color target is still a challenging visual task. In order to solve the problem, we propose a camouflage color target detection algorithm based on image enhancement. Firstly, the image enhancement algorithm is used to realize the difference between the target and the background feature. Secondly, the region proposal network (RPN) is used to realize the accurate positioning of the specific target, and the extraction area ROI is identified by the classification layer in the deep neural network. Finally, we realize the detection with a camouflage color target. In this paper, the detection algorithm received better detection results in the leaves of butterfly and chameleon data collection.

Keywords: Target detection · Camouflage color · CLAHE · Faster R-CNN

1 Introduction

Target detection is an important branch of image processing and computer vision. The color and texture of the camouflage color target are very similar to the background. Therefore, the detection of the camouflage color target is a difficult problem in the field of computer vision. At present, the related research work on camouflage color target detection is less.

Pillai et al. [1] proposed a fusion algorithm for real-time detection of camouflage color targets. In [1], the targets are detected by using a novel target detection method by applying conventional image threshold methods in the wavelet domain. Huang and Jiang [2] proposed a method which based on weighted area consolidation to achieve the same color and background of the animal detection. Assefa et al. [3] proposed a detection algorithm by analysing the image texture and using quaternary Fourier Transform, according to the texture feature detection camouflage color target. The latest developments in target detection are driven by the regional proposal method [4] and the success of the region-based convolution neural network [5]. Although the region-based CNN was initially developed in terms of computations [5], the cost was significantly reduced by

D. Liu et al. (Eds.): ICONIP 2017, Part III, LNCS 10636, pp. 476–485, 2017.
https://doi.org/10.1007/978-3-319-70090-8_49

sharing convolution throughout the proposal [6,7]. The latest incarnation, Fast R-CNN [7] and Faster R-CNN [8], uses very deep networks to achieve near real-time rates.

The existing target detection algorithms rely on key points or characteristics to achieve the goal of positioning and identification, which greatly depends on the extraction of features in the characterization of the integrity of the target. In the case of similar backgrounds, the characterization of features is limited. Therefore, aiming at the camouflage color target, this paper proposes a method to realize the target detection by using the deep neural network. The experimental results show that our method has better detection results. An outline of the approach is presented in Fig. 1.

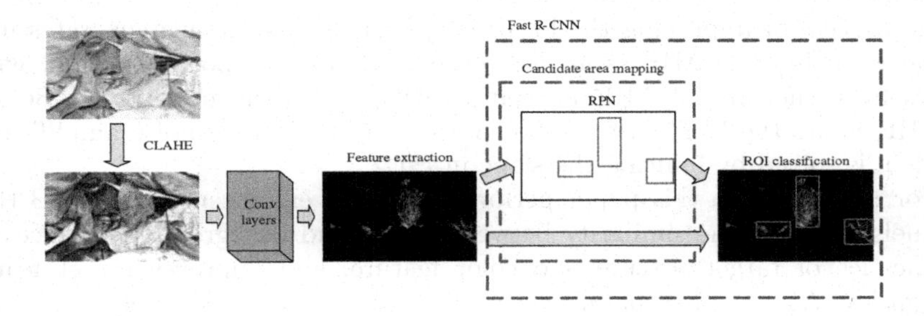

Fig. 1. Camouflage color target detection algorithm framework.

2 Contrast Limited Adaptive Histogram Equalization

As the camouflage color target and the background there are similar characteristics, the primary target is to achieve the object and the background elements of the distinction. In the natural environment, in order to adapt to the environment, some organisms evolved out of the camouflage color that can be fused in the environment. Due to the small difference between the camouflage color and the environment, the visual inspection method is difficult.

If we can increase the difference between the target and the background, the target detection will be easier and more accurate. So we can use the image enhancement algorithm to enlarge the difference between the target area and the background area. In this paper, we use the contrast limited adaptive histogram equalization proposed by Zuiderveld et al. [8,9] to enhance the pretreatment of the target region.

Histogram equalization is obtained by image gray scale histogram distribution to obtain the mapping curve, according to the gray probability distribution to achieve the spatial mapping of the image. The cumulative distribution is satisfied in Eq. 1, where $k \in \{1, 2, \cdots, L-1\}$ is the k-th grayscale value in the image represents, $P_k(x_k)$ is the probability of the k-th grayscale distribution of the image, and N and n_k are the number of pixel points and the number of k-th pixels in the image, respectively. The converted image satisfies Eq. 2, where S_j is the j-th gradation of the transformed image, r_{min} indicates the minimum gray scale value of the enhanced image, and $R = r_{max} - r_{min}$ indicates the maximum range of the gray value in the original image.

$$CDF(k) = \sum_{i=0}^{k} \frac{n_k}{N} = \sum_{i=0}^{k} P_k(x_k) \quad (1)$$

$$S_j = R * CDF(k) + r_{min} \quad (2)$$

The histogram equalization is the global transformation of the image, and the dynamic transformation of the local gray scale is not taken into account. The adaptive histogram equalization (AHE) is proposed to ensure the local dynamic information influence. AHE has more attention to detail information in the process of improving the local image contrast, and thus generates high-frequency noises. CLAHE is an improved AHE for the noise problem, which differs from the normal AHE in that CLAHE limits image contrast. Because CLAHE limits the local area of the contrast, and the local noise amplification problem is solved by limiting the slope in Eq. 2.

For color images, this paper performs CLAHE enhancement in RGB three channels, enlarges the similarity between target and background, enhances the significance of target texture and color feature, and improves target feature recognition. As the three channels are enhanced, the target color shift will occur. But this offset is floating within a certain range, and the target color shift contributes to the distinction from the background interference element. So the offset does not affect the follow-up target learning process. Figure 2 shows the enhanced result by CLAHE.

Fig. 2. The 1st row are the original images, and the 2nd row are the enhancement results by CLAHE.

3 Deep Neural Network Model

In our work, an camouflage color target detection algorithm based on deep learning is proposed. In order to ensure that the feature-aware network effectively extracts the target feature, the image enhancement algorithm is used to enhance the image to enlarge the difference between the target and the background in the image. Then, the Faster R-CNN is used to detect the target in order to improve the accuracy of the detection area. Figure 1 is the algorithmic diagram of the algorithm.

RPN [8] is the whole convolution neural network used in the detection to extract the target candidate box. The previous target detection network is basically determined by the regional partitioning algorithm target location, such as SSPnet [6] and Fast R-CNN [7], which reduces the real-time performance of the target detection algorithm in practical application. To solve this bottleneck, He et al. proposed the model by using CNN network to extract target region, that is, RPN target detection area extraction. In this paper, we use RPN to directly generate candidate regions, and use Fast R-CNN for border regression and attribute classification.

3.1 Region Proposal Network and Anchor

RPN is a fully connected convolution network, achieve end-to-end mapping in the detection process of the target area, and complete the target boundary and attribute prediction. RPN uses convolution network to generate target candidate region extraction, and use the multi-scale anchor to determine the candidate area. In the candidate region extraction process, using the SPP mapping mechanism, from the Conv5 [10] on the sliding window search. Refer to [10] for detailed process.

Anchor [8] is proposed in the Faster R-CNN. In the feature graph of the Conv5 output, Anchor is defined as an associated node that gets the center of the target area through the sliding window mapping for the RPN network to extract the candidate area. In the training RPN process, the initial stage cannot accurately determine the location of the target, and it is proposed to use Anchor for regional regression extraction.

In the process of using Anchors, each feature point is mapped to the center of the original experience field as the reference point, and the anchor box of different scale and different proportion is matched around the center point. In the training process, for each calibration of the target area, and its IoU overlap the largest proportion of Anchor marked as positive samples. In the remaining Anchors, the mark with the scale overlap ratio greater than 0.7 is a positive sample, that is, the foreground target; the overlap ratio is less than 0.3 as the negative sample, as the background area. For those who do not meet the conditions and cross the border Anchors directly discarded.

3.2 Loss Function

In the deep network based on RPN, in order to ensure the detection accurate of the target of the network, the training network needs to follow the rules according to certain rules, as shown in Fig. 3. In order to minimize the objective function

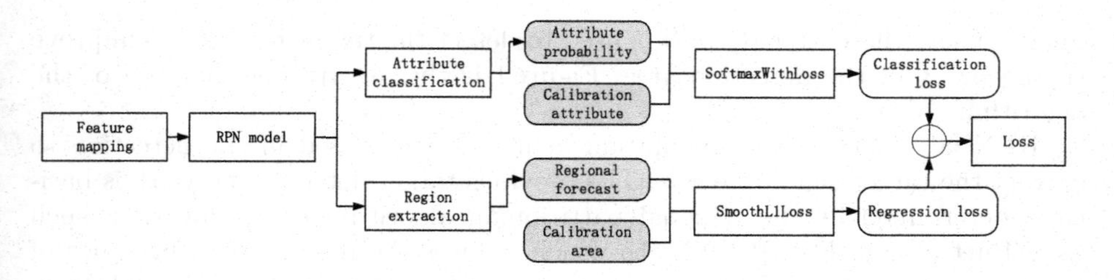

Fig. 3. Network loss structure diagram

on the regression classification of the target, we define the loss function for each image:

$$L(p_i, t_i) = \frac{1}{N_{cls}} \sum_i L_{cls}(p_i, p_i^*) + \lambda \frac{1}{N_{reg}} \sum_{i=0} P_i^* L_{reg}(t_i, t_i^*) \quad (3)$$

where i is the label of the anchor in the smallest batch image and p_i is the probability that the i-th anchor predicted as the target, conforms to the Softmax classification model, p_i^* is the standard quantity of the i-th anchor, satisfies:

$$p_i^* = \begin{cases} 0, & negativelabel, \\ 1, & positivelabel. \end{cases} \quad (4)$$

$t_i = \{t_x, t_y, t_w, t_h\}$ is a 4-parameter vector representing the predictive candidate box, t_x and t_y indicate that the target candidate region scale invariant translation parameter. t_w and t_h are the length and width of the relative candidate region in the logarithmic space. t_i^* represents t_i corresponding calibration target area. $L(p_i, t_i)$ consists of two parts, namely, attribute classification loss $L_{cls}(p_i, p_i^*)$ and border return loss $L_{reg}(t_i, t_i^*)$. Where $L_{cls}(p_i, p_i^*)$ satisfies:

$$L_{cls}(p_i, p_i^*) = -log[p_i^* p_i + (1 - p_i^*)(1 - p_i)] \quad (5)$$

The softmax classification model is a generalization of the logistic regression model in the multi-classification task, which guarantees that the proposed algorithm is suitable for multi-objective detection. However, the softmax model has a phenomenon of gradient disappearance at both ends of the mapping curve. In order to solve this problem, we can logarithmic spatial transformation, that is, the attribute classification loss function $L_{cls(p_i, p_i^*)}$ guarantees the characteristic descending gradient in the training process, and effectively achieves the classification network optimization.

$L_{reg}(t_i, t_i^*)$ is the border return loss, used to adjust the target area to meet:

$$L_{reg}(t_i, t_i^*) = R(t_i, t_i^*) \quad (6)$$

R is $SmoothL_1$ function:

$$Smooth_{L_1}(x) = \begin{cases} 0.5x^2, & \mid x \mid < 1, \\ \mid x \mid -0.5, & otherwise. \end{cases} \quad (7)$$

Relative loss function L_2, $Smooth_{L_1}$ is more robustness, for the deviation from the continuous area of the abnormal point, $Smooth_{L_1}$ passivation of abnormal points on the neural network, reduce the sensitivity of the deviation point, control the degree of gradient training to protect the stability of the neural network.

In Eq. 3, N_{cls} and N_{reg} are the minimum batch size and Anchors number are respectively expressed, and the parameters are normalized, $\lambda \approx N_{reg}/N_{cls}$ balance the weight between the attribute classification loss item and the border return loss item.

3.3 Regional Regression and Network Training

In the target detection task, the high-order features in the same node correspond to a number of Anchors in the target location process, and the network to determine the target area cannot be completely the same with the calibration target area in a variety of Anchors Boxes. So the training process needs which the candidate region performs a regression adjustment so that the candidate region contains the target completely, as shown in Fig. 4.

Non-maximum suppression essentially is the search for local maxima, suppressing non-maximal elements. In the detection task, we determine the best detection position in the repeated cross window. In the detection process, the specific target will appear a number of related overlapping areas. The probability is greater than the threshold p and IoU overlap rate greater than the threshold t of the region of the fusion.

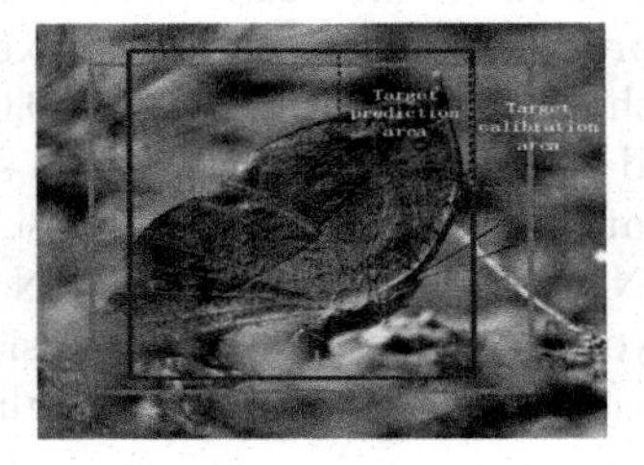

Fig. 4. Border regression schematic.

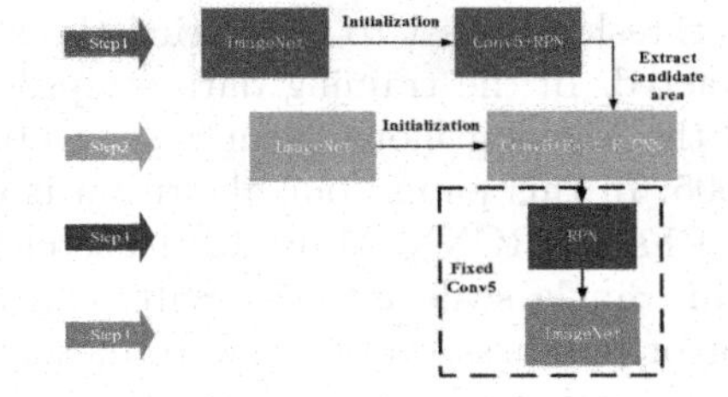

Fig. 5. Network training flow chart.

In this paper, the training model is divided into two modules, the RPN candidate area extraction module and the Fast R-CNN detection module. And the Faster R-CNN network composed of two modules, which realizes the sharing of the area extraction layer and the detection layer. The drawbacks of training. After the RPN extracts the candidate area, the final target detection and recognition are completed using Fast R-CNN, and the sharing of the five layers of ZF volume is realized by the alternate training method, as shown in Fig. 5. Specific steps are as follows:

Step 1: Use the ImageNet pre-training model to initialize the RPN network and use RPN to generate the candidate area of the target;

Step 2: Use the ImageNet model to initialize the Fast R-CNN network, take the candidate area which is generated by RPN in step 1 as input, and then obtain the Fast R-CNN network parameters;

Step 3: Fix the shared volume base layer and train the RPN network parameters to realize the fine tuning of the candidate area;

Step 4: Fix the volume base layer, train the Fast R-CNN network parameters, fine-tune the target attribute discrimination probability and the candidate box position parameter.

---

4 Experimental Results

In order to verify the effectiveness of our algorithm in detecting the camouflage color target, we used 300 images of the leaves and 100 images of chameleon which are selected from the Internet. Some of the training samples have similar elements to the background in order to effectively study the weak features in the experiment.

In the process of network training, the input image is normalized and the minimum edge has 600 pixel, and the long edge is scaled. In the feature detection network, the input feature graph is transformed into $224 * 224 * 3$. The IoU threshold is set to 0.7 during the NMS process and 300 Anchor Boxes are extracted. In the training data set process, the learning rate is set to 0.00005 and the learning momentum is set to 0.9 and the weight attenuation is set to 0.0005. In this paper, our algorithm is compared with Faster R-CNN [8], SSR [12] + Faster RCNN, MSR [12] + Faster RCNN, ACE [13] + Faster RCNN algorithm. Single scale retinex (SSR) can achieve dynamic range compression in color or lightness. Multi scale retinex (MSR) can compress color and lightness in a dynamic range at the same time.

The objective experiment is compared by the following two aspects: (1) IoU-Recall curve comparison, compared with the target detection accuracy; (2) comparative sample detection mAP. We assume as following: TP indicates which positive sample detection is true, FP indicates which negative sample detection is true, FN indicates which negative sample is false, TN indicates which positive sample is false, then recall rate can be expressed as:

$$R = \frac{TP}{TP + FN}, P = \frac{TP}{TP + FP} \tag{8}$$

$$mAP = \int P(R)dR \tag{9}$$

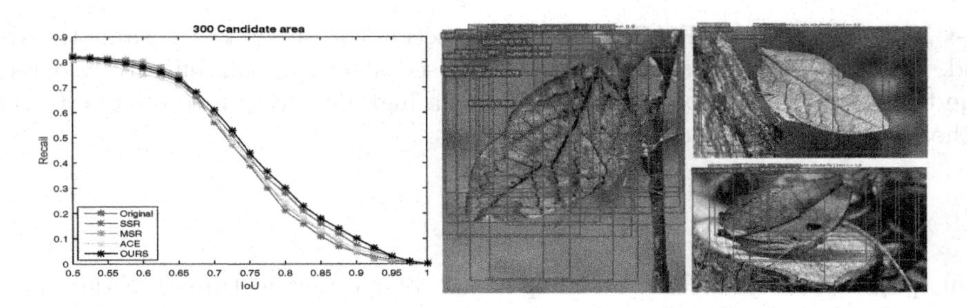

Fig. 6. The IoU-Recall curve comparison.

In the candidate region extracted in the sample, the candidate region is subjected to a different threshold value NMS, thereby calculating the recall rate of the target candidate frame to obtain the IoU-Recall curve, as shown in Fig. 6. It can be seen from the curve and Table 1 that the proposed algorithm has a better recall rate than Faster R-CNN, SSR + Faster RCNN, MSR + Faster RCNN, ACE + Faster RCNN. From the experimental results, the proposed method can more effectively amplify the difference between the target and the background in the detection of the disguised color target, extract the significant feature of the target, and combine the subsequent RPN-based detection network to achieve the accurate positioning of the target.

Table 1. Comparison of the proposed algorithm.

Performance	Original + Faster R-CNN	SSR + Faster R-CNN	MSR + Faster R-CNN	ACE + Faster R-CNN	OURS algorithm
mAP	0.6269	0.5833	0.5910	0.6083	**0.6462**

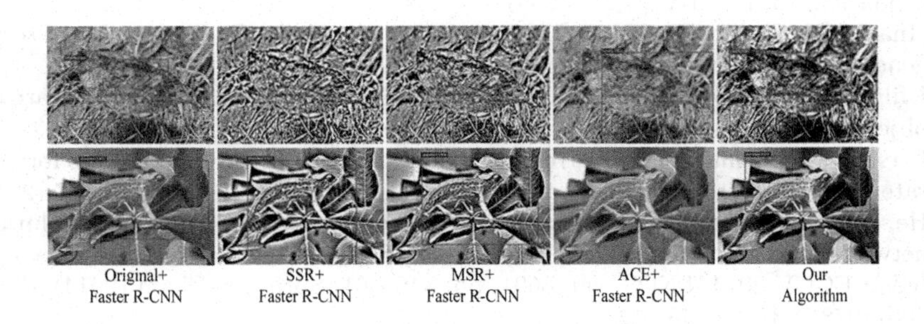

Fig. 7. The detection results compare between our algorithm with others.

Figure 7 shows the comparison of the detection results. It can be seen from the experiment that the detection algorithms SSR + Faster R-CNN, MSR + Faster

R-CNN, ACE + Faster R-CNN have no detection the object at the same threshold. From the whole observation, the proposed algorithm can effectively detect the target area in the experiment, and has a high detection rate relative to the other several enhanced detection algorithms.

5 Conclusions

This paper presents a RPN-based spurious color target feature detection algorithm. First of all, this algorithm through the image of the CLAHE enhanced processing to achieve the image of the target and background differences in amplification. Then, the deep neural network based on RPN is constructed, and the deep neural network is used to study the target characteristics, so as to complete the detection task. Based on the RPN deep neural network, our algorithm adds CLAHE enhancement processing to solve the interference of similar elements in the background to the target, and improves the perceived and weak features of the network. Compared with the experimental results of Faster R-CNN algorithm, our algorithm has some advantages in disguised color target detection. However, the detection performance of our algorithm depends on the results of enhanced stretching.

Acknowledgments. The paper was supported in part by the National Natural Science Foundation (NSFC) of China under Grant No. 61365003 and Gansu Province Basic Research Innovation Group Project No. 1506RJIA031.

References

1. Pillai, S.S., Swamy, M.N.S.: Camouflaged target detection using real-time video fusion algorithm based on multiscale transforms. In: CCECE, vol. 27, no. 1, pp. 1–5 (2014)
2. Huang, Z.Q., Jiang, Z.H.: Tracking camouflaged objects with weighted region consolidation. In: DICTA, p. 24 (2005)
3. Hnaidi, H., Guerin, E., Akkouche, S., Peytavie, A., Galin, E.: Feature based terrain generation using diffusion equation. CGF **29**(7), 2179–2186 (2010)
4. Uijlings, J.R.R., Sande, K.E., Gevers, T., Smeulders, A.W.M.: Selective search for object recognition. IJCV **104**(2), 154–171 (2013)
5. Girshick, R., Donahue, J., Darrell, T., Malik, J.: Rich feature hierarchies for accurate object detection and semantic segmentation. In: CVPR, pp. 580–587 (2014)
6. He, K., Zhang, X., Ren, S., Sun, J.: Spatial pyramid pooling in deep convolutional networks for visual recognition. In: Fleet, D., Pajdla, T., Schiele, B., Tuytelaars, T. (eds.) ECCV 2014. LNCS, vol. 8691, pp. 346–361. Springer, Cham (2014). doi:10.1007/978-3-319-10578-9_23
7. Girshick, R.: Fast R-CNN. In: ICCV, pp. 1440–1448 (2015)
8. Ren, S.Q., He, K.M., Girshick, R., Sun, J.: Faster R-CNN: towards real-time object detection with region proposal networks. In: arXiv e-prints arXiv:1506.01497v3 [cs.CV] (2016)
9. Zuiderveld, K.: Contrast limited adaptive histogram equalization. In: Graphics Gems, pp. 474–485 (1994)

10. Zeiler, M.D., Fergus, R.: Visualizing and understanding convolutional networks. In: Fleet, D., Pajdla, T., Schiele, B., Tuytelaars, T. (eds.) ECCV 2014. LNCS, vol. 8689, pp. 818–833. Springer, Cham (2014). doi:10.1007/978-3-319-10590-1_53
11. Lecun, Y., Jackel, L.D., Cortes, C., Denker, J.S., et al.: Learning algorithms for classification: a comparison on handwritten digit recognition. In: NNSMP, pp. 261–276 (2000)
12. Rahman, Z., Jobson, D.J., Woodell, G.A.: Multi-scale retinex for color image enhancement. In: ICIP, pp. 1003–1006 (1996)
13. Lee, E., Kim, S., Kang, W., Seo, D., Paik, J.: Contrast enhancement using dominant brightness level analysis and adaptive intensity transformation for remote sensing images. GRSL **10**(1), 62–66 (2013)

Action Prediction Using Unsupervised Semantic Reasoning

Cuiwei Liu[1(✉)], Yaguang Lu[2], Xiangbin Shi[1,2], Zhaokui Li[1], and Liang Zhao[1]

[1] School of Computer Science, Shenyang Aerospace University, Shenyang, Liaoning, People's Republic of China
liucuiwei@sau.edu.cn

[2] School of Information, Liaoning University, Shenyang, Liaoning, People's Republic of China

Abstract. This paper aims to address the problem of predicting the category of an ongoing action in a video, which enables us to react as quickly as possible. Action prediction is a challenge problem since neither the complete semantic information nor the definite temporal progress can be obtained from a partially observed video. In this paper, we propose to predict action categories of unfinished videos by using semantic reasoning. For the purpose of exploiting mid-level semantics from videos, we present an unsupervised semantic mining approach which expresses an observed video as a sequence of semantic concepts and learns the context relationship of various concepts by using a General Mixture Transform Distribution model (GMTD). Then the invisible future semantic concepts can be automatically estimated from the observed semantic concept sequence. Finally, we develop a discriminative structural model that integrates video observations, observed semantic concepts, and inferred semantic concepts for early recognition of incomplete videos. Experimental results on the UT-Interaction dataset show that the proposed method is able to effectively predict the action category of an unfinished video.

Keywords: Action prediction · Discriminative structural model · Semantic concept · General Mixture Transform Distribution model

1 Introduction

Recognizing human actions in videos is an important intelligent video analysis technique, and has attracted extensive research interest in the past decade. Recently, a variety of methods [3,5,13,21,22,24] have been proposed for after-the-fact recognition of human actions, which focuses on classifying a fully observed video into a certain action category. Some existing methods [3,21,24] represent an action video with low-level features including global features extracted from the entire video and local spatiotemporal features calculated around interest points, and discriminate the current action by using the mapping from low-level features to high-level action category. Another group of methods [7,17] adopt deep learning technique to learn discriminative features directly

D. Liu et al. (Eds.): ICONIP 2017, Part III, LNCS 10636, pp. 486–496, 2017.
https://doi.org/10.1007/978-3-319-70090-8_50

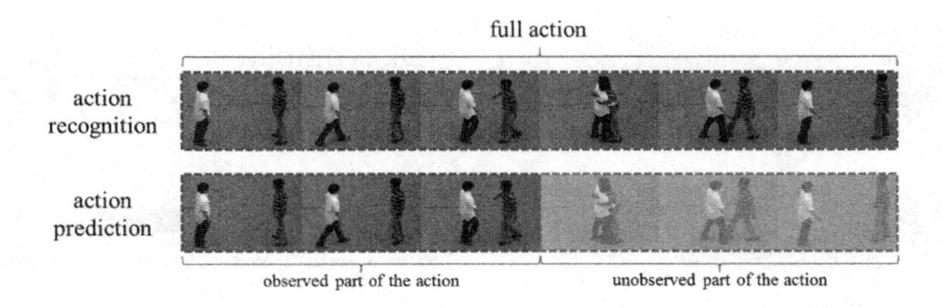

Fig. 1. Comparison between action recognition and action prediction.

from the raw data, and several neural networks have been developed in the recent years. Nevertheless, both of low-level features and deep learned features have innate limitations in semantic description of complex actions. In fact, human cognition of action videos is not based on visual features, but is dependent on high-level semantics. Consequently, some recent literatures [5,13,14,18,19,22,23] are devoted to exploring implicit semantic concepts in videos for recognizing real-world actions. Among them, some methods [5,13,18] explicitly define a set of semantic concepts and learn action models by taking advantages of semantic concept annotations. In contrast, another strategy [14,19,22,23] is to exploit the latent semantic information in action videos by introducing a group of data driven generated states.

However, in many real-world scenarios, it is desired that the intelligent system can distinguish a human action before it is fully executed. In fact, the ability of predicting intended actions greatly promotes the system performance in some applications such as intelligent video surveillance and health care assistance. For example, an intelligent surveillance system is desired to detect ongoing criminal activities as early as possible so as to avoid serious consequences. Similarly, it is not so useful if a nurse assistant robot alarms after a patient falls down. A desirable autonomous robot should predict the fall activity when it observes that a person is losing balance, which makes it be able to prevent this activity before any damage is caused.

We focus on predicting ongoing actions by using temporally incomplete observations. As is shown in Fig. 1, different from traditional action recognition methods which takes complete videos including full details as input, action prediction methods have to make decisions with partial semantic information extracted from the beginning of an action. It is common that the unobserved contents may include some important concepts for discriminating a human action. Therefore, the deficiency of subsequent semantic concepts poses great challenges for accurate action prediction. In addition, the temporal progress of an unfinished video is not available during testing. Therefore, how to effectively infer unknown semantic information of future observations is critical in action prediction.

In this paper, we present a novel action prediction method based on semantic reasoning. Our method starts with dividing an input incomplete video into a series of fixed-length segments, and then the convolutional neural network is

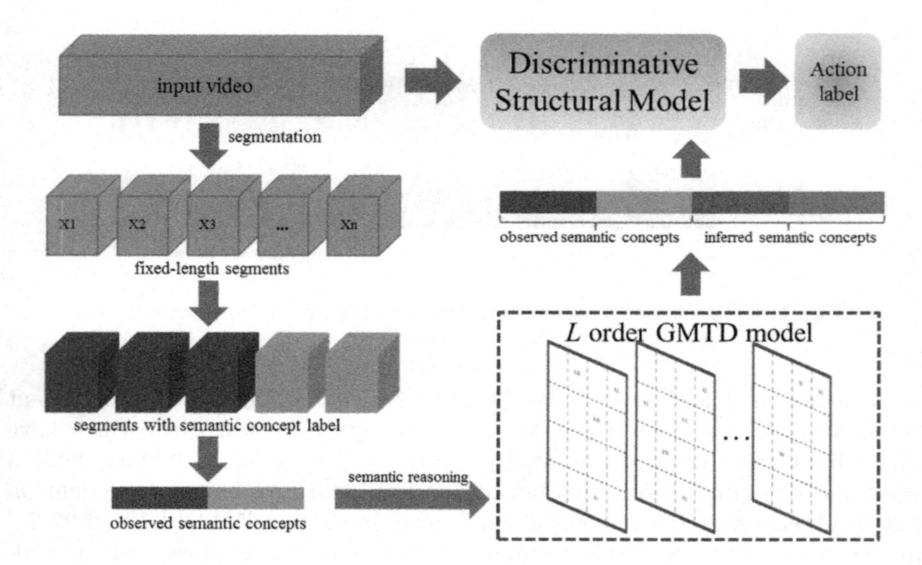

Fig. 2. Overview of the proposed method.

employed to extract features from each segment. In order to exploit video semantics, we propose an unsupervised semantic mining approach, which associates each observed video segment with a mid-level semantic concept and automatically infers the future semantic information. Particularly, the semantic mining approach represents the observed video as a sequence of semantic concepts, and captures the contextual relationship of these semantic concepts by using a General Mixture Transform Distribution (GMTD) model. It is difficult to determine how many semantic concepts need to be inferred, since we cannot explicitly obtain the temporal progress of a partially observed video. To this end, a specific concept is introduced for achieving automatic termination of concept inference. Then a discriminative structural model is developed to capture the relationship among actions, observed semantic concepts, inferred semantic concepts, and video observations. In the proposed model, the inferred concepts are treated differently from the semantic concepts extracted from observed videos so that they have different contributions to the final classification. Figure 2 depicts the framework of our method.

The main contribution of this paper includes: (1) a novel unsupervised semantic mining approach which is able to extract semantic concepts from an incomplete video as well as to infer the following concepts in the near future, and (2) a discriminative structural model that jointly learns the context relationship among actions, observed semantic concepts, inferred semantic concepts, and video observations.

2 Related Work

Action prediction is a challenging problem in computer vision domain. A formal definition of the concept of action prediction was first proposed by Ryoo [16]. In [16], action prediction is defined as an inference of the ongoing action given temporally incomplete observations, and the integral bag-of-words (IBoW) and dynamic bag-of-words (DBoW) methodologies are designed to cope with this problem. More generally, Cao et al. [2] aimed to recognize partially observed videos, in which the unobserved subsequence may occur at any time. They employed sparse coding to learn bases of video segments, and calculated a global posterior of actions by integrating action likelihoods of video segments. Xu et al. [25] analogized action prediction to the problem of query auto-completion in information retrieval. They developed an activity auto-completion model, which treats a partially observed video and a full video as a prefix and a query, respectively. Wang et al. [20] proposed a temporally-weighted generalized time warping algorithm to align an incomplete action video with a full action video. Their method is derived from the original generalized time warping algorithm, and can deal with action prediction problem by encouraging alignment in the early part of an action video. Hu et al. [4] developed a soft regression-based model to predict unfinished actions recorded in RGB-D sequences. Different from [2,16] that infer action category from low-level features with probabilistic frameworks and [4,20,25] which directly map a test unfinished action video to an appropriate full action video, our method explores the latent semantic concepts in unfinished action videos for action prediction.

Some recent literatures also achieve action prediction by exploiting hierarchical semantic information. Lan et al. [10] achieved action prediction by constructing a coarse-to-fine hierarchical description for a short video clip or a still image containing incomplete actions. In their method, different layers represent actions at different semantics, which is achieved by introducing extra annotations of training videos, such as human bounding boxes in each frame and motion tracks across frames. Li et al. [11,12] decomposed a incomplete long-duration activity into a set of action units according to the motion velocity curve which is not robust to noise in complex outdoor scene. Kong et al. [8,9] developed a multiple temporal scale support vector machine which integrates global progress model and local progress model to learn the evolution of actions. Their method divides every full action video into a fixed number of segments, each of which represents a temporal progress of the action and is associated with a local progress model. For an unfinished test video, their method takes advantage of the status of its temporal progress which is not available, actually.

3 Our Method

The goal of this work is to predict the action category y of an input video X containing the early part of a human action. Given an input video, we firstly divide it into a sequence of fixed-length video segments, each of which is described with a deep-learned representation. Concretely, we apply Convolutional Neural Network [6] to all frames, and utilize the output of the second fully connected layer (fc7) to represent them. Segment representation $\mathbf{x}$ is achieved by averaging CNN features of all the frames within it. Similarly, the mean feature of all frames constitutes a global video representation $\mathbf{x}^{global}$. Then an input video X is represented as $X = (\mathbf{x}^{global}, \{\mathbf{x}_t\}_{t=1:T})$, where T indicates the number of segments in X and is proportional to the video length.

We propose an unsupervised semantic mining approach, which associates each video segment $\mathbf{x}$ with a semantic concept to obtain the observed semantic concept sequence $\mathbf{S}^O$, and automatically infers an unobserved future semantic concept sequence $\mathbf{S}^U$. A novel discriminative structural model is developed to achieve the final action prediction by integrating the global video representation, the observed semantic concepts, and the inferred unobserved semantic concepts.

3.1 Unsupervised Semantic Mining Approach

The proposed unsupervised semantic mining approach firstly exploits a set of semantic concepts shared among different actions by using a group of fully observed training videos $D_{full} = \{X_n^{full} = \{\mathbf{x}_{n,t}^{full}\}_{t=1:T_n}\}_{n=1:N}$. As mentioned above, each full action video X_n^{full} is split into T_n fixed-length segments described with deep-learned features. And then we run k-means algorithm to cluster all training video segments $\{\mathbf{x}_{n,t}^{full}\}_{n=1:N,t=1:T_n}$ into K clusters, with each cluster center corresponding to a semantic concept. We can associate segments of an input video X with semantic concepts according to their cluster assignments. Then the action video X can be expressed as a concise sequence of observed semantic concepts $\mathbf{S}^O = [s_1^O, s_2^O, ..., s_p^O]$ by merging identical semantic concepts associated with neighboring segments, where p is the number of concepts in $\mathbf{S}^O$.

The proposed unsupervised semantic mining approach can automatically infer the unobserved semantic concepts for an incomplete action video with a General Mixture Transform Distribution (GMTD) model [1], which is a variation of the high-order Markov model. Suppose that there are K semantic concepts in total, the L-order Markov model requires $K^L(K-1)$ parameters to capture the context relationship among concepts, while the GMTD model needs $LK(K-1)+L-1$ parameters by assuming that the effects of nodes with different lags are considered separately.

The transform matrices of the GMTD model $\{\mathbf{g}^1, \mathbf{g}^2, \ldots, \mathbf{g}^L\}$ are obtained with statistics on all the semantic concept sequences $\{\mathbf{S}_{n,1}^O\}_{n=1:N}$ of full action videos in D_{full}. And the weights of different orders $\boldsymbol{\lambda} = (\lambda^1, \lambda^2, \ldots, \lambda^L)$ are computed through Eq. 1.

$$\lambda^l = \delta^{l-1}(1-\delta), 0 < \delta < 1, \tag{1}$$

where δ is set to a constant value. In the GMTD model, $\mathbf{g}_{i,j}^l$ indicates the possibility of semantic concept i transforming to semantic concept j in order l.

Given an incomplete action video X associated with an observed semantic concept sequence $\mathbf{S}^O = [s_1^O, s_2^O, ..., s_p^O]$, the GMTD model is able to infer the unobserved semantic concept sequence $\mathbf{S}^U = [s_{p+1}^U, s_{p+2}^U, ..., s_{p+q}^U]$ by Eq. 2.

$$
\begin{aligned}
s_{p+\tau}^U &= \arg\max_{j=1,2,...,K} \Big(\sum_{k=1}^{\tau-1} \lambda^{\tau-k} g_{s_{p+k}^U,j}^{\tau-k} + \sum_{k=1}^{\ell-\tau+1} \lambda^{\ell-k+1} g_{s_{p+k+\tau-\ell-1}^O,j}^{\ell-k+1}\Big), \\
\ell &= \min(L, p).
\end{aligned} \tag{2}
$$

The proposed method achieves automatic termination of concept inference by introducing an extra semantic concept s^{end}. The future semantic concept is inferred one by one until either of the following two conditions is satisfied: (1) the latest inferred concept is s^{end}, and (2) the length of $\mathbf{S}^U$ goes beyond the limit $L_{\max}$.

3.2 Discriminative Structural Model

Model Formulation. We develop a discriminative structural model for predicting the action category y of an incomplete video expressed as a triple $(\mathbf{x}^{global}, \mathbf{S}^O, \mathbf{S}^U)$, where $\mathbf{x}^{global}$ is the global representation of this video, while $\mathbf{S}^O = [s_1^O, s_2^O, ..., s_p^O]$ and $\mathbf{S}^U = [s_{p+1}^U, s_{p+2}^U, ..., s_{p+q}^U]$ indicate the observed semantic concept sequence and the inferred unobserved semantic concept sequence, respectively. Our model aims to learn a discriminative compatibility function $f(\mathbf{x}^{global}, \mathbf{S}^O, \mathbf{S}^U, y)$ which measures how compatible the action label y is suited to an input video $(\mathbf{x}^{global}, \mathbf{S}^O, \mathbf{S}^U)$.

$$
f(\mathbf{x}^{global}, \mathbf{S}^O, \mathbf{S}^U, y) = \mathbf{w}^{\mathrm{T}} \boldsymbol{\Phi}(\mathbf{x}^{global}, \mathbf{S}^O, \mathbf{S}^U, y). \tag{3}
$$

The model parameter $\mathbf{w}$ includes three parts $\{\alpha, \beta, \gamma\}$, and the joint feature vector $\boldsymbol{\Phi}(\mathbf{x}^{global}, \mathbf{S}^O, \mathbf{S}^U, y)$ formulates the relationship among the global video representation, the observed semantic concept sequence, the inferred unobserved semantic concept sequence, and the action label as

$$
\begin{aligned}
\mathbf{w}^{\mathrm{T}} \boldsymbol{\Phi}(\mathbf{x}^{global}, \mathbf{S}^O, \mathbf{S}^U, y) &= \alpha^{\mathrm{T}} \phi(\mathbf{x}^{global}, y) + \beta^{\mathrm{T}} \varphi(\mathbf{S}^O, y) + \gamma^{\mathrm{T}} \psi(s_p^O, \mathbf{S}^U, y), \\
\alpha^{\mathrm{T}} \phi(\mathbf{x}^{global}, y) &= \alpha_y^{\mathrm{T}} \cdot \mathbf{x}^{global}, \\
\beta^{\mathrm{T}} \varphi(\mathbf{S}^O, y) &= \sum_{k=2}^{p} \beta_{s_{k-1}^O, s_k^O, y}^{\mathrm{T}}, \\
\gamma^{\mathrm{T}} \psi(s_p^O, \mathbf{S}^U, y) &= \gamma_{s_p^O, s_{p+1}^U, y}^{\mathrm{T}} + \sum_{k=2}^{q} \gamma_{s_{p+k-1}^U, s_{p+k}^U, y}^{\mathrm{T}},
\end{aligned} \tag{4}
$$

where s_p^O indicates the last semantic concept in $\mathbf{S}^O$.

In Eq. 4, $\alpha^{\mathrm{T}} \phi(\mathbf{x}^{global}, y)$is a standard linear potential function for predicting the action label y with the global video feature $\mathbf{x}^{global}$. Since the inferred semantic information and the ground-truth may not perfectly coincide, two potential functions are employed to discriminate the contributions of $\mathbf{S}^O$ and $\mathbf{S}^U$. The potential function $\beta^{\mathrm{T}}\varphi(\mathbf{S}^O, y)$ measures the compatibility between an observed semantic concept sequence $\mathbf{S}^O$ and action y by modeling the co-occurrence relationships of two adjacent semantic concepts. Similarly, correlation between an inferred unobserved semantic concept sequence $\mathbf{S}^U$ and action y is evaluated by the potential function $\gamma^{\mathrm{T}}\psi(s_p^O, \mathbf{S}^U, y)$.

Learning Framework. Our training dataset $D = \{(X_{n,\rho}, \mathbf{S}^O, \mathbf{S}^U, y_n)\}_{n=1:N,\rho\in\{0.1,\ldots,0.9\}}$ is derived from the full action dataset $D_{full} = \{(X_n^{full}, y_n)\}_{n=1:N}$ by varying the observation ratio ρ from 0.1 to 0.9 by interval of 0.1. Nine incomplete videos $\{X_{n,\rho}\}_{\rho\in\{0.1,\ldots0.9\}}$ share the action label y_n with the full video X_n^{full}. As mentioned above, we extract a global representation $\mathbf{x}_{n,\rho}^{global}$ from video $X_{n,\rho}$, and associate it with an observed semantic concept sequence $\mathbf{S}^O$ and an inferred semantic concept sequence $\mathbf{S}^U$ by utilizing the proposed unsupervised semantic mining approach. The model parameter $\mathbf{w}$ is learned with a Structural SVM framework by solving the following optimization problem:

$$\min_{\mathbf{w}} \frac{1}{2}\|\mathbf{w}\|^2 + C_1 \cdot \sum_{n,\rho} \xi_{n,\rho} + C_2 \cdot \sum_{n,\rho} \zeta_{n,\rho}$$

$$s.t. \mathbf{w}^{\mathrm{T}}\boldsymbol{\Phi}(\mathbf{x}_{n,\rho}^{global}, \mathbf{S}_{n,\rho}^O, \mathbf{S}_{n,\rho}^U, y_n) - \mathbf{w}^{\mathrm{T}}\boldsymbol{\Phi}(\mathbf{x}_{n,\rho}^{global}, \mathbf{S}_{n,\rho}^O, \mathbf{S}_{n,\rho}^U, y) \geq \rho \cdot \Delta(y, y_n) - \xi_{n,\rho}, \forall n, \forall \rho, \forall y,$$
$$\alpha^{\mathrm{T}}\phi(\mathbf{x}_{n,(\rho+0.1)}^{global}, y_n) - \alpha^{\mathrm{T}}\phi(\mathbf{x}_{n,\rho}^{global}, y) \geq 0.1 \cdot \Delta(y, y_n) - \zeta_{n,\rho}, \forall n, \forall \rho, \forall y, \quad (5)$$

where $\xi_{n,\rho}$ and $\zeta_{n,\rho}$ are slack variables, and $\Delta(y, y_n)$ is a loss function measuring the cost of predicting the ground-truth y_n as y. Here $\Delta(y, y_n) = 1$ if $y \neq y_n$ and 0 otherwise.

The first constraint in Eq. 5 is a usual SVM max margin constraint which optimize $\mathbf{w}$ by classifying each training video correctly. Different from traditional classification framework, the loss function is scaled by the corresponding observation ratio ρ for the purpose of enforcing the progress levels of videos with different observation ratios. The second constraint in Eq. 5 enforces a monotonically increasing potential function $\alpha^{\mathrm{T}}\phi(\mathbf{x}_{n,\rho}^{global}, y)$ for sequentially arriving data in action prediction.

4 Experiments

4.1 Datasets and Experimental Setting

The proposed method is evaluated on the public available UT-Interaction dataset [15]. The whole dataset is further divided into two sets: the UT-Interaction Set 1 (UT-I #1) and UT-Interaction Set 2 (UT-I #2), each of which contains six human actions, including shake-hands, hug, kick, point, punch, and push, with 10 videos per class. UT-Interaction Set 1 is taken on a parking lot while UT-Interaction Set 2 is captured on a lawn in a windy day. Generally, UT-Interaction Set 2 contains a lot of camera jitters and is more difficult for action prediction than UT-Interaction Set 1. Examples of the two datasets are shown in Fig. 3.

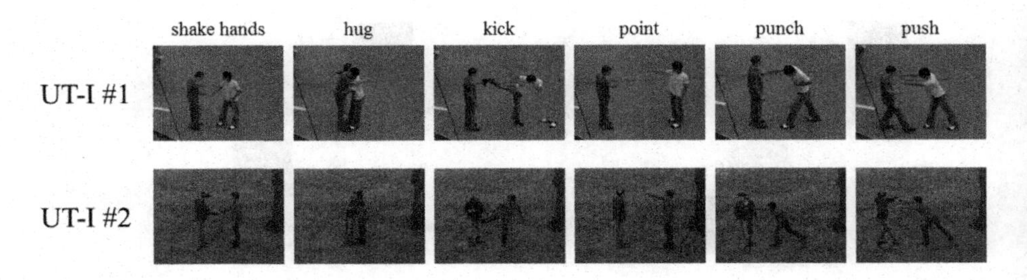

Fig. 3. Examples of the UT-Interaction dataset.

In our implementation, an action video is partitioned into a sequence of segments, each of which is composed of 10 adjacent frames. The proposed unsupervised semantic mining approach associates each video segment with a semantic concept by using k-means algorithm, and the number of clusters is set to 55. We set the order of GMTD model to 3, and stipulate that length of an inferred semantic concept sequence cannot go beyond 3. In order to compute the weights of different orders of the GMTD model, the parameter δ in Eq. 1 is set to 0.1.

4.2 Experimental Results

In our experiments, we adopt the 10-fold leave-one-out cross validation evaluation strategy in [16] for fair comparison. The proposed method is compared with state-of-the-art methods including Integral BoW and Dynamic BoW [16], SC and MSSC [2], TGTW-D and TGTW [20].

Figure 4(a) reports the prediction results on the UT-I #1 dataset corresponding to observation ratios from 0.1 to 0.9. Our method achieves 65% recognition accuracy at the beginning of actions when only the first 30% of testing videos are observed, and an impressive 82% recognition accuracy is accomplished at 0.5 observation ratio. The experimental results demonstrate that our method

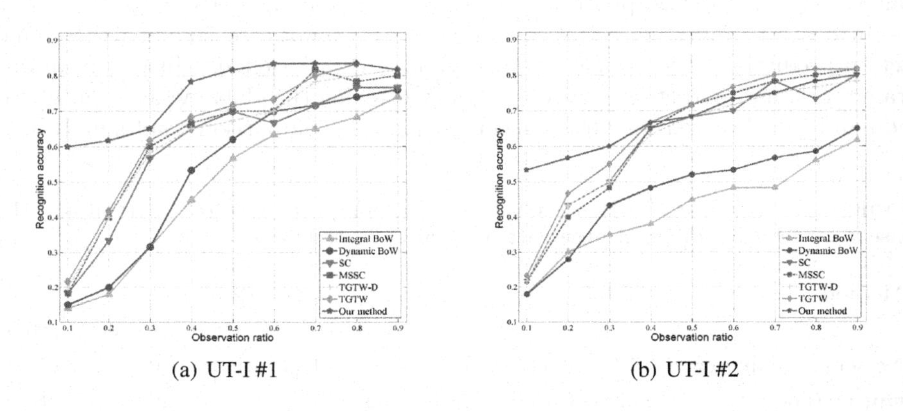

(a) UT-I #1 (b) UT-I #2

Fig. 4. Prediction results on the UT-Interaction dataset.

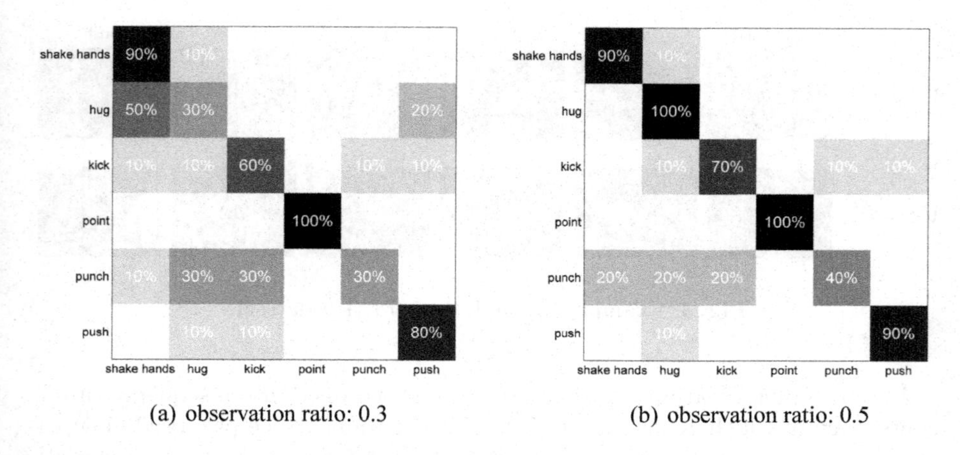

(a) observation ratio: 0.3

(b) observation ratio: 0.5

Fig. 5. Confusion matrices of our method on the UT-I #1 dataset.

is able to discriminate actions according to the beginning parts of them. More encouragingly, it is observable that our method outperforms all the other methods at different observation ratios. Confusion matrices of our prediction results at 0.3 and 0.5 observation ratios are depicted in Fig. 5(a) and (b), respectively. As is shown in Fig. 5(a), "shake hands", "point", and "push" actions can be recognized at their early stage, while "hug", "kick", and "punch" actions are more difficult for the prediction task. When more observations are available (see Fig. 5(b)), our method is able to achieve better performance.

Comparison results on the UT-I #2 dataset are depicted in Fig. 4(b). Our action prediction results on this dataset are slightly lower than the results on the UT-I #1 dataset at the same observation ratio. One reasonable explanation is that the UT-I #2 dataset is more challenging than UT-I #1 due to the camera jitters. Nevertheless, our method achieves better results than other methods when the observation ratio is less than 0.5, which strongly confirms the advantages of our method on predicting ongoing actions at the early stage.

In order to investigate the contribution of semantic reasoning to the final action prediction, we compared our method with a baseline, which is from our framework and predicts the action category utilizing the global video representation and the observed semantic concept sequence. Comparison results in Table 1

Table 1. Comparison of action prediction performance between our method and the baseline method at different observation ratios indicated by ρ.

Method	UT-I #1				UT-I #2			
	$\rho=0.2$	$\rho=0.3$	$\rho=0.4$	$\rho=0.5$	$\rho=0.2$	$\rho=0.3$	$\rho=0.4$	$\rho=0.5$
No semantic inference	0.57	0.62	0.72	0.8	0.53	0.55	0.6	0.65
Our method	0.62	0.65	0.78	0.82	0.57	0.6	0.67	0.68

illustrate that the our method performs better than the baseline method by incorporating the inferred unobserved semantic concepts, which is able to verify the effectiveness of the proposed unsupervised semantic mining approach.

5 Conclusion

In this paper, we have presented to predict ongoing actions using semantic reasoning. For the purpose of exploiting semantic information implicit in a video, an unsupervised semantic mining approach is proposed to express it with an observed semantic concept sequence. We employ a GMTD model to capture the context among different concepts, and further infer the future semantic information. The action prediction task is accomplished by a novel discriminative structural model which captures the intrinsic relationship among actions, observed semantic concepts, inferred semantic concepts, and video observations. Experiments on the UT-Interaction dataset have verified that our method can effectively recognize ongoing actions, especially at their early stage.

Acknowledgments. This work was supported in part by the Natural Science Foundation of China (NSFC) under Grant No. 61602320 and No. 61170185, Liaoning Doctoral Startup Project under Grant No. 201601172 and No. 201601180, Foundation of Liaoning Education al Committee under Grant No. L201607 and No. L2015403, and the Young Scholars Research Fund of SAU under Grants No. 15YB37.

References

1. Berchtold, A., Raftery, A.E.: The mixture transition distribution model for high-order Markov chains and non-Gaussian time series. Stat. Sci. **17**(3), 328–356 (2002)
2. Cao, Y., Wang, S., Barrett, D., Barbu, A., Narayanaswamy, S., Yu, H., Michaux, A., Lin, Y., Dickinson, S., Siskind, J.M.: Recognize human activities from partially observed videos. In: CVPR, pp. 2658–2665 (2013)
3. Efros, A.A., Berg, A.C., Mori, G., Malik, J.: Recognizing action at a distance. In: ICCV, vol. 2, pp. 726–733 (2003)
4. Hu, J.-F., Zheng, W.-S., Ma, L., Wang, G., Lai, J.: Real-time RGB-D activity prediction by soft regression. In: Leibe, B., Matas, J., Sebe, N., Welling, M. (eds.) ECCV 2016. LNCS, vol. 9905, pp. 280–296. Springer, Cham (2016). doi:10.1007/978-3-319-46448-0_17
5. Izadinia, H., Shah, M.: Recognizing complex events using large margin joint low-level event model. In: Fitzgibbon, A., Lazebnik, S., Perona, P., Sato, Y., Schmid, C. (eds.) ECCV 2012. LNCS, vol. 7575, pp. 430–444. Springer, Heidelberg (2012). doi:10.1007/978-3-642-33765-9_31
6. Jia, Y., Shelhamer, E., Donahue, J., Karayev, S., Long, J., Girshick, R., Guadarrama, S., Darrell, T.: Caffe: convolutional architecture for fast feature embedding. In: ACM International Conference on Multimedia, pp. 675–678 (2014)
7. Karpathy, A., Toderici, G., Shetty, S., Leung, T., Sukthankar, R., Fei-Fei, L.: Large-scale video classification with convolutional neural networks. In: CVPR, pp. 1725–1732 (2014)

8. Kong, Y., Fu, Y.: Max-margin action prediction machine. T-PAMI **38**(9), 1844–1858 (2015)
9. Kong, Y., Kit, D., Fu, Y.: A discriminative model with multiple temporal scales for action prediction. In: Fleet, D., Pajdla, T., Schiele, B., Tuytelaars, T. (eds.) ECCV 2014. LNCS, vol. 8693, pp. 596–611. Springer, Cham (2014). doi:10.1007/978-3-319-10602-1_39
10. Lan, T., Chen, T.-C., Savarese, S.: A hierarchical representation for future action prediction. In: Fleet, D., Pajdla, T., Schiele, B., Tuytelaars, T. (eds.) ECCV 2014. LNCS, vol. 8691, pp. 689–704. Springer, Cham (2014). doi:10.1007/978-3-319-10578-9_45
11. Li, K., Fu, Y.: Prediction of human activity by discovering temporal sequence patterns. T-PAMI **36**(8), 1644–1657 (2014)
12. Li, K., Hu, J., Fu, Y.: Modeling complex temporal composition of actionlets for activity prediction. In: Fitzgibbon, A., Lazebnik, S., Perona, P., Sato, Y., Schmid, C. (eds.) ECCV 2012. LNCS, vol. 7572, pp. 286–299. Springer, Heidelberg (2012). doi:10.1007/978-3-642-33718-5_21
13. Liu, C., Wu, X., Jia, Y.: A hierarchical video description for complex activity understanding. IJCV **118**(2), 240–255 (2016)
14. Pirsiavash, H., Ramanan, D.: Parsing videos of actions with segmental grammars. In: CVPR, pp. 612–619 (2014)
15. Ryoo, M.S., Aggarwal, J.K.: UT-interaction dataset, ICPR contest on Semantic Description of Human Activities (SDHA) (2010). http://cvrc.ece.utexas.edu/SDHA2010/Human_Interaction.html
16. Ryoo, M.S.: Human activity prediction: early recognition of ongoing activities from streaming videos. In: ICCV, pp. 1036–1043 (2011)
17. Simonyan, K., Zisserman, A.: Two-stream convolutional networks for action recognition in videos. In: NIPS, pp. 568–576 (2014)
18. Sun, C., Nevatia, R.: Active: activity concept transitions in video event classification. In: ICCV, pp. 913–920 (2013)
19. Tang, K., Li, F.F., Koller, D.: Learning latent temporal structure for complex event detection. In: CVPR, pp. 1250–1257 (2012)
20. Wang, H., Yang, W., Yuan, C., Ling, H., Hu, W.: Human activity prediction using temporally-weighted generalized time warping. Neurocomputing **225**, 139–147 (2017)
21. Wang, H., Kläser, A., Schmid, C., Liu, C.L.: Dense trajectories and motion boundary descriptors for action recognition. IJCV **103**(1), 60–79 (2013)
22. Wang, L., Qiao, Y., Tang, X.: Latent hierarchical model of temporal structure for complex activity classification. T-IP **23**(2), 810–22 (2014)
23. Wang, L., Qiao, Y., Tang, X.: Mining motion atoms and phrases for complex action recognition. In: ICCV, pp. 2680–2687 (2013)
24. Wu, X., Xu, D., Duan, L., Luo, J.: Action recognition using context and appearance distribution features. In: CVPR, pp. 489–496 (2011)
25. Xu, Z., Qing, L., Miao, J.: Activity auto-completion: predicting human activities from partial videos. In: ICCV, pp. 3191–3199 (2015)

Robust Facial Alignment for Face Recognition

Kuan-Pen Chou[1], Dong-Lin Li[4], Mukesh Prasad[2](✉), Mahardhika Pratama[3], Sheng-Yao Su[4], Haiyan Lu[2], Chin-Teng Lin[2], and Wen-Chieh Lin[1]

[1] Department of Computer Science, National Chiao Tung University, Hsinchu, Taiwan
kpchou.cs00g@g2.nctu.edu.tw, wcline@cs.nctu.edu.tw

[2] Centre for Artificial Intelligence, School of Software, FEIT, University of Technology Sydney, Sydney, Australia
{mukesh.prasad,haiyan.lu,chin-teng.lin}@uts.edu.au

[3] School of Computer Science and Engineering, Nanyang Technological University, Singapore, Singapore
mpratama@ntu.edu.sg

[4] Department of Electrical Engineering, National Chiao Tung University, Hsinchu, Taiwan
lazybones@g2.nctu.edu.tw, mikea1d2c3@gmail.com

Abstract. This paper proposes a robust real-time face recognition system that utilizes regression tree based method to locate the facial feature points. The proposed system finds the face region which is suitable to perform the recognition task by geometrically analyses of the facial expression of the target face image. In real-world facial recognition systems, the face is often cropped based on the face detection techniques. The misalignment is inevitably occurred due to facial pose, noise, occlusion, and so on. However misalignment affects the recognition rate due to sensitive nature of the face classifier. The performance of the proposed approach is evaluated with four benchmark databases. The experiment results show the robustness of the proposed approach with significant improvement in the facial recognition system on the various size and resolution of given face images.

Keywords: Face recognition · Face alignment · Facial feature localization and sparse representation classifier

1 Introduction

Face detection is a challenging and important issue to the facial recognition system In the real world application for security and safety purposes. The accurate detection of the face image increases the performance of the facial recognition system. Viola and Jones [1] introduced a fast object detection method based on a boosted cascade of haar-like features. This technique is widely used in the area of face recognition and detection research due to its computational capabilities and considerable performance. However, the variations of human faces based on different environment and conditions are still a challenging task in the real-world

D. Liu et al. (Eds.): ICONIP 2017, Part III, LNCS 10636, pp. 497–504, 2017.
https://doi.org/10.1007/978-3-319-70090-8_51

facial recognition system. The changes of facial expression, and pose, which are caused by uncontrolled environments deteriorate the accuracy of facial recognition system. Numerous approaches have been developed to handle the following face variation: (1) appearance-based methods [2–4], (2) train multiple detectors for each poses of human face [5–8], (3) nonlinear regression method [9–14], (4) manifold embedding method [12–17], and (5) geometrical analysis [18,19]. This paper uses a very simple geometric information to evaluate the human facial pose. For geometric approach, the accuracy of face feature point is very important. To find an accurate location searching for the face feature point and achieving a computational compatibility, an ensemble of regression tree based on gradient boosting technique [20] is applied in the proposed approach. The proposed approach precisely finds the face feature point of the test image and perform the facial correction by a geometrical transform technique.

The reminder of the proposed approach is organized as follows: Sect. 2 provides detail descriptions of the proposed approach which include architecture of the proposed approach, face alignment system, and face correction algorithm. Section 3 shows the experimental results on various facial benchmark databases and finally the conclusions are covered in Sect. 4.

2 Proposed System

2.1 Face Alignment System

The architecture of the proposed system is shown in Fig. 1. At first, the proposed system localizes the landmark of input face obtained from the face detector [1] via a regression tree based method. According to the position of landmarks, the facial orientation can be easily estimated from two-dimensional geometric analysis. As the face orientation is determined, the input face is normalized to the frontal pose by compensating the face misalignment on each axis.

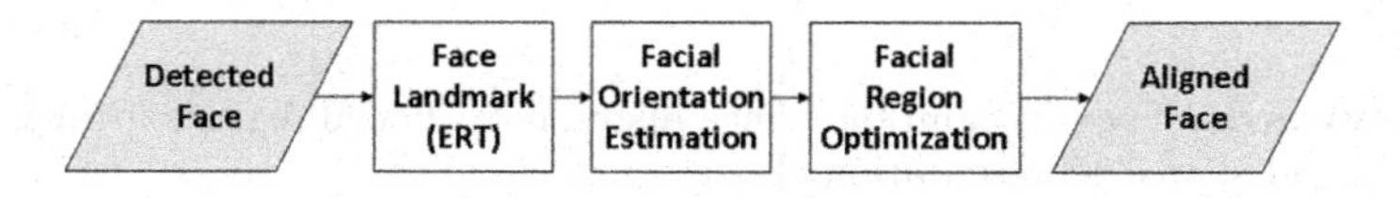

Fig. 1. The flowchart of our face correction system

2.2 Face Correction Algorithm

Compensation of Pitch: The rotation motion of face can be represented as pitch (rotation about the x-axis), yaw (rotation about the y-axis), and roll (rotation about the z-axis). In order to explicitly explain the proposed approach, we suppose that the z-axis is the frontal orientation of human face; that is, the frontal pose can be described in x-y plane, as shown in Fig. 2.

Figure 3 shows a geometric perspective that represents the projection from 3-D human face onto 2-D image plane. Let f is the focus distance of camera,

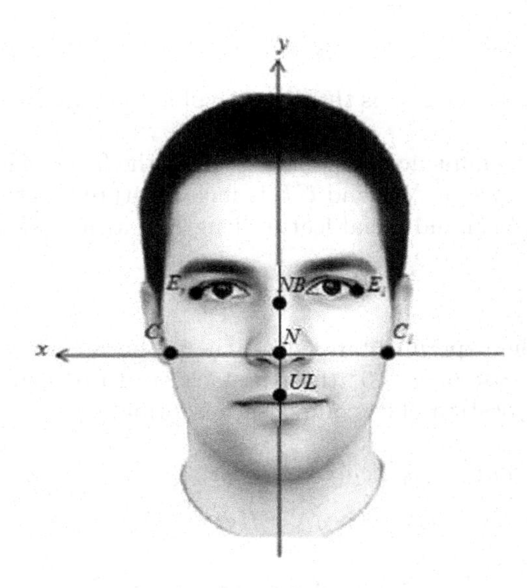

Fig. 2. The position of facial feature points

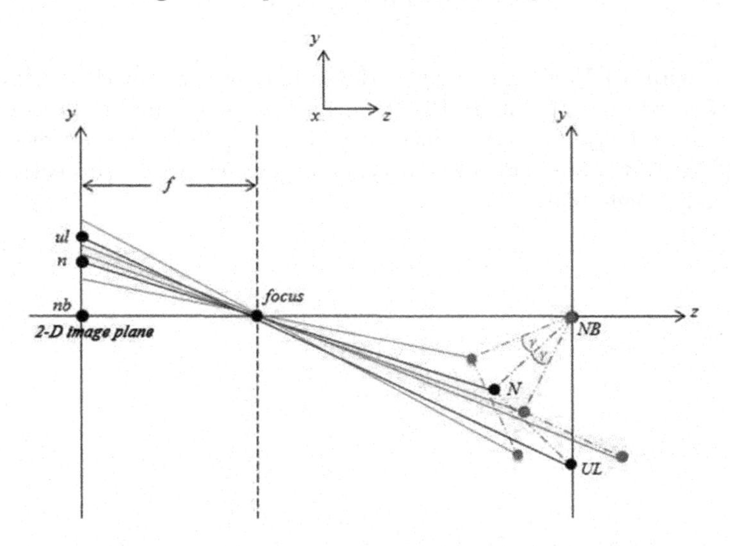

Fig. 3. The position of facial feature points

γ is the angle of pitch, N is tip of the nose, NB is the location of nose bridge, UL is the central point of upper lip, and n, nb, and ul are the projection of each points in 2-D image plane. In $\gamma = 0°$ case, UL should locate on the $z = 0$ plane if we set NB as the origin,

For the convenience of computation, we define two parameters as following:

$$y_u = nb - n \tag{1}$$

$$y_d = n - ul \tag{2}$$

where y_u is the nose length, y_d is the distance of nose and upper lip in 2-D image plane.

Next, observe the influence of pitch angle γ on the facial feature in 2-D image plane. Let the distance of NB and UL is facial feature length, and now define the ratio of nose length and facial feature length $\hat{P}$ as follows:

$$\hat{P} = \frac{y_u}{y_u + y_d} \tag{3}$$

According to the experimental results, the proposed system only aligns the face images in the case of $\gamma > 0°$ that the facial feature length is shortened by the effect of pitch rotation. The compensation formula is shown below:

$$Center_{pitch}(x, y') = Center_{origin}(x, y + \Delta y) \tag{4}$$

$$Height_{pitch}(x, y') = Height_{origin} - \Delta y \tag{5}$$

$$\Delta y = \begin{cases} y_u & , if \gamma > 0° \\ 0 & , if \gamma \leq 0° \end{cases} \tag{6}$$

Compensation of Yaw: Similarly as the case of pitch angle, the frontal human face is described in x-y plane. In Fig. 2, C_r and C_l are left and right side point of the face, respectively. E_r and E_l are outer corner of right and left eyes, respectively. In Fig. 4, β is the angle of yaw, c_r, c_l, e_r, and e_l are the projection of each points in 2-D image plane.

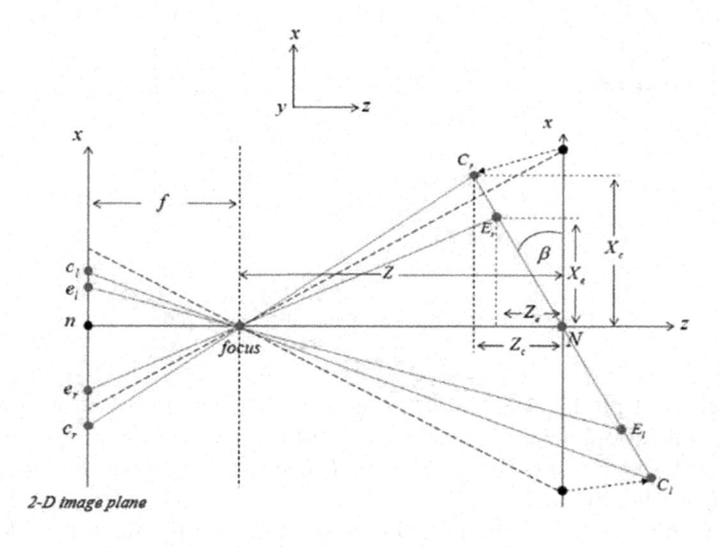

Fig. 4. Geometric perspective of the yaw rotation.

According to the geometric perspective for Fig. 4, the projection terms c_r and c_l can be described as:

$$c_r = \frac{fX_c}{Z - Z_c} = \frac{f\bar{C_r E_r}\cos\beta}{Z - Z_c} \tag{7}$$

$$c_l = \frac{fX_c}{Z + Z_c} = \frac{f\bar{C_l E_l}\cos\beta}{Z + Z_c} \tag{8}$$

and the yaw angle β can be represented in terms of c_r and c_l as follows:

$$\beta = \arccos(\frac{c_l(Z - Z_c)}{f\bar{C_r E_r}}) \tag{9}$$

$$\beta = \arccos(\frac{c_l(Z + Z_c)}{f\bar{C_l E_l}}) \tag{10}$$

Clearly, c_r and c_l have inverse relationship while the yaw angle β is small enough. The length of $\bar{c_r n}$ increases and $\bar{c_l n}$ decreases if we increase the yaw angle β. By taking advantage of this property, the compensation formula for y-axis can be designed as follows:

$$\Delta x = \begin{cases} \left(1 - \frac{c_r}{c_l}\right)\left(c_l - e_l\right), & if\beta < 0 \\ \left(1 - \frac{c_l}{c_r}\right)\left(c_r - e_r\right), & if\beta \geq 0 \end{cases} \tag{11}$$

$$Center_{yaw}(x', y) = Center_{origin}(x + \Delta x, y) \tag{12}$$

where $Center_{origin}$ is the original location of the central point of face and $Center_{yaw}$ is the compensated location of the central point of face.

Compensation of Roll: Outer corner of eyes are very robust facial feature points for adjusting roll rotation. Therefore, we only select the outer corner of eyes as the facial feature points in the compensation task, as shown in Fig. 2. For roll rotation, the coordinates of Er and El are only change in x and y axis, respectively, and the distance between the camera focus and the central point of Er and El are fixed. It means that the roll angle has same value with the rotation angle of human face displayed in Fig. 5. From Fig. 5(b), the roll angle α can be written as follows:

$$\alpha = \arctan(\frac{Y_e}{X_e}) \tag{13}$$

where $X_e = e_r^x - e_l^x$ and $Y_e = e_r^y - e_l^y$. In order to adjust the face to the frontal direction, we directly rotate the face image by performing the affine transform as follows:

$$M = \begin{bmatrix} \cos\alpha & -\sin\alpha \\ \sin\alpha & \cos\alpha \end{bmatrix} \tag{14}$$

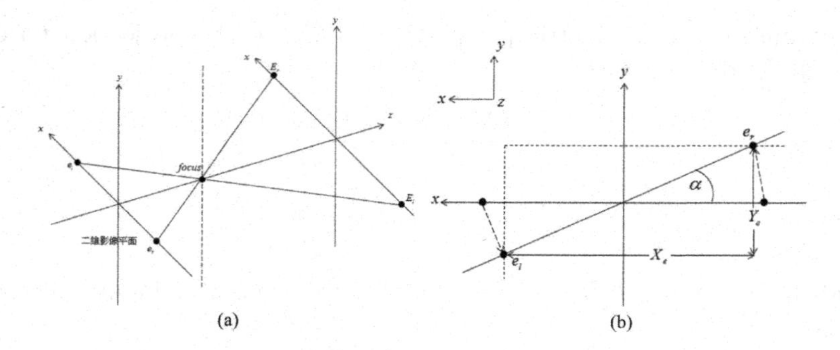

Fig. 5. Geometric perspective of the roll rotation in (a) 3-D space (b) 2-D space.

3 Experiment Results

This section demonstrates alignment method, which can significantly enhance the performance of facial recognition system. The FEI database [22] that includes 14 images for each of 200 individuals are selected to evaluate the effect of face alignment. For each individual, only 9 images without extreme illumination and yaw rotation are considered in the evaluation process. Figure 6 describes the classification result of SRC [21] with our proposed face alignment processes. It outperforms SRC without alignment by 10% under all input dimension. That is, face alignment plays a vital roles to the classification work of SRC.

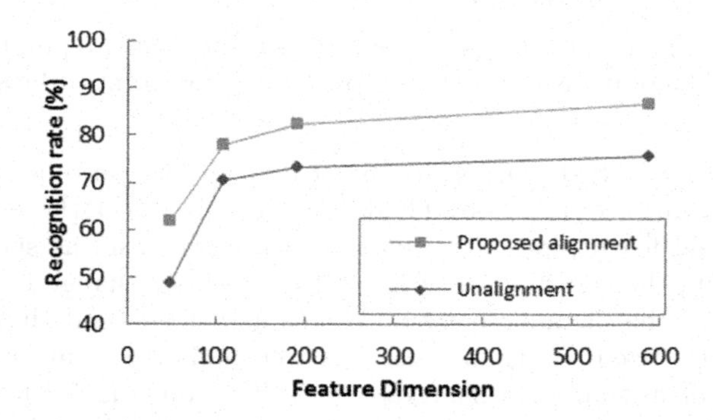

Fig. 6. FEI database recognition result of SRC under various input.

4 Conclusions

This paper proposes a facial recognition system with sparse representation classifier, which has been proven to be an effective classifier for illumination, occlusion, and corruption problems. The proposed system significantly improves the performance of facial recognition system with the help of face alignment method.

In the proposed system, the pose-correction process is only applies to the z-axis. Further, the proposed system will apply to address the pose-correction to x-axis and y-axis, because 3D human face model helps to understand the geometric relationship between different human poses and based on this knowledge, every poor-aligned face image can be correct to frontal pose.

References

1. Viola, P., Jones, M.J.: Robust real-time face detection. Int. J. Comput. Vis. **57**(2), 137–154 (2004)
2. Niyogi, S., Freeman, W.T.: Example-based head tracking. In: 2nd IEEE International Conference on Automatic Face and Gesture Recognition, pp. 374–378. IEEE (1996)
3. De Vel, O., Aeberhard, S.: Line-based face recognition under varying pose. IEEE Trans. Pattern Anal. Mach. Intell. **21**(10), 1081–1088 (1999)
4. Sherrah, J., Gong, S., Ong, E.J.: Face distributions in similarity space under varying head pose. Image Vis. Comput. **19**(12), 807–819 (2001)
5. Huang, J., Shao, X., Wechsler, H.: Face pose discrimination using support vector machines (SVM). In: 14th International Conference on Pattern Recognition, vol. 1, pp. 154–156 (1998)
6. Jones, M., Viola, P.: Fast Multi-view Face Detection. Mitsubishi Electric Research Lab TR2000396, 3, 14 (2003)
7. Rowley, H.A., Baluja, S., Kanade, T.: Rotation invariant neural network-based face detection. In: IEEE Computer Society Conference on Computer Vision and Pattern Recognition, pp. 38–44. IEEE (1998)
8. Liu, Q., Deng, J., Tao, D.: Dual sparse constrained cascade regression for robust face alignment. IEEE Trans. Image Proc. **25**(2), 700–712 (2016)
9. Li, Y., Gong, S., Sherrah, J., Liddell, H.: Support vector machine based multi-view face detection and recognition. Image Vis. Comput. **22**(5), 413–427 (2004)
10. Moon, H., Miller, M.L.: Estimating facial pose from a sparse representation. In: International Conference on Image Processing, vol. 1, pp. 75–78 (2004)
11. Seemann, E., Nickel, K., Stiefelhagen, R.: Head pose estimation using stereo vision for human-robot interaction. In: 6th IEEE International Conference on Automatic Face and Gesture Recognition, pp. 626–631. IEEE (2004)
12. Fu, Y., Huang, T.S.: Graph embedded analysis for head pose estimation. In: 7th IEEE International Conference on Automatic Face and Gesture Recognition, p. 68. IEEE (2006)
13. Ren, S., Cao, X., Wei, Y., Sun, J.: Face alignment via regressing local binary features. IEEE Trans. Image Proc. **25**(3), 1233–1245 (2016)
14. Tai, Y., Yang, J., Zhang, Y., Luo, L., Qian, J., Chen, Y.: Face recognition with pose variations and misalignment via orthogonal procrustes regression. IEEE Trans. Image Proc. **25**(6), 2673–2683 (2016)
15. Wu, J., Trivedi, M.M.: A two-stage head pose estimation framework and evaluation. Pattern Recogn. **41**(3), 1138–1158 (2008)
16. Balasubramanian, V.N., Ye, J., Panchanathan, S.: Biased manifold embedding: a framework for person-independent head pose estimation. In: IEEE Conference on Computer Vision and Pattern Recognition, pp. 1–7. IEEE (2007)
17. Asthana, A., Zafeiriou, S., Tzimiropoulos, G., Cheng, S., Pantic, M.: From pixels to response maps: discriminative image filtering for face alignment in the wild. IEEE Trans. Pattern Anal. Mach. Intell. **35**, 1312–1320 (2015)

18. Horprasert, T., Yacoob, Y., Davis, L.S.: Computing 3-D head orientation from a monocular image sequence. In: 2nd IEEE International Conference Automatic Face and Gesture Recognition, pp. 244–252. IEEE (1996)
19. Wang, J.G., Sung, E.: EM enhancement of 3D head pose estimated by point at infinity. Image Vis. Comput. **25**(12), 1864–1874 (2007)
20. Kazemi, V., Sullivan, J.: One millisecond face alignment with an ensemble of regression trees. In: IEEE Conference on Computer Vision and Pattern Recognition, pp. 1867–1874. IEEE (2014)
21. Wright, J., Yang, A.Y., Ganesh, A., Sastry, S.S., Ma, Y.: Robust face recognition via sparse representation. IEEE Trans. Pattern Anal. Mach. Intell. **31**(2), 210–227 (2009)
22. http://fei.edu.br/~cet/facedatabase.html

Automatic Leaf Recognition Based on Deep Convolutional Networks

Huisi Wu, Yongkui Xiang, Jingjing Liu, and Zhenkun Wen(✉)

College of Computer Science and Software Engineering,
Shenzhen University, Shenzhen, China
wenzk@szu.edu.cn

Abstract. Leaf recognition remains a hot research topic receiving intensive attention in computer vision. In this paper, we propose deep convolutional networks with deep learning framework on the large scale of leaf databases. Different from the existing leaf recognition algorithms that mainly depend on traditional feature extractions and pattern matching operations, our method can achieve automatic leaf recognition based on deep convolutional networks without any explicit feature extraction or matching. Because it does not require any feature detection and selection, the advantages of our framework are obvious, especially for the large scale leaf databases. Specifically, we design deep convolutional networks structure and adopt fine-tuning strategy for our network initialization. In addition, we also develop a visualization-guided parameter tuning scheme to guarantee the accuracy of our deep learning framework. Our method is evaluated on several different databases with different scales. Comparison experiments are performed and demonstrate that the accuracy of our method outperforms traditional methods.

Keywords: Leaf recognition · Deep convolutional networks · Learning feature visualization

1 Introduction

Plants are the most widely distributed biological on the earth. The existence of the plant is very important to human's life activities. Botanists typically study different kinds of plants according to their leaves, flowers, or vein structures. Due to the limited special knowledge on the plants for average person, it is essential to develop an automatic plant recognition system. In this paper, we select leaf as the key to identify a plant, because it can be easily obtained and has abundant texture details, which is also one of the important factors of plants.

In the past years, researchers have proposed many methods to identify plants. Researchers usually are tending to recognize plants by extracting leaf features or machine learning manner. On account of extracting feature, Wang et al. [1] proposed a method for recognizing leaf images based on shape features and hyper-sphere classifier. Wu et al. [2] proposed a rotation invariant shape context (RISC) for automatic leaf recognition. Ojala and Pietikainen [3] proposed the Local Binary Pattern (LBP) method, which utilized a set of binary patterns to represent the image. Zhang et al. [4] proposed

D. Liu et al. (Eds.): ICONIP 2017, Part III, LNCS 10636, pp. 505–515, 2017.
https://doi.org/10.1007/978-3-319-70090-8_52

Binary Gabor Pattern (BGP) method to extract texture feature. Relying on machine learning, back propagation network (BP), radial basis function network (RBF) and support vector machine (SVM) are all applied in leaf recognition. The improved version for the SVM is developed in LIBSVM [5]. For neural network classifier, extreme learning machine (ELM) [6] which is a simple learning algorithm for single-hidden-layer feed-forward neural networks (SLFNs). In the past few years, many papers showed significant advantages by deep learning in large database for the field of classification. Angie [7] utilize fine tune model for plant classification, but only for small sample library. Deep learning, which is widely applied to voice recognition, face identification, images recognition and so on. Despite this encouraging progress, there is still limited in plant leaf recognition with large database. Recently, deep learning framework provides a better way to understand internal leaf feature and recognition, which may help us to achieve a better leaf recognition performance.

In this paper, we employ deep learning network structure based on CAFFE [8] and convolutional neural networks (CNNs). Compared to the traditional feature-based methods or matching-based methods, our method emphasizes learning deep characteristics including edge, color, texture, and the overall feature in an implicit way (without any feature detection and selection operations). Furthermore, we also perform a visualization of network layers to adjust our network parameters. Convincing experiments and comparisons with state-of-the-art methods are conducted on different scales of databases to demonstrate the effectiveness of our deep learning framework.

2 Methodology

To achieve a more efficient and robust performance, our leaf recognition framework combines deep feature extraction and deep layer visualization on CNNs. As a result, high accuracy leaf recognition is obtained through a visualization-guided [9] designing and fine-tuning network. The framework of our system is as shown in Fig. 1

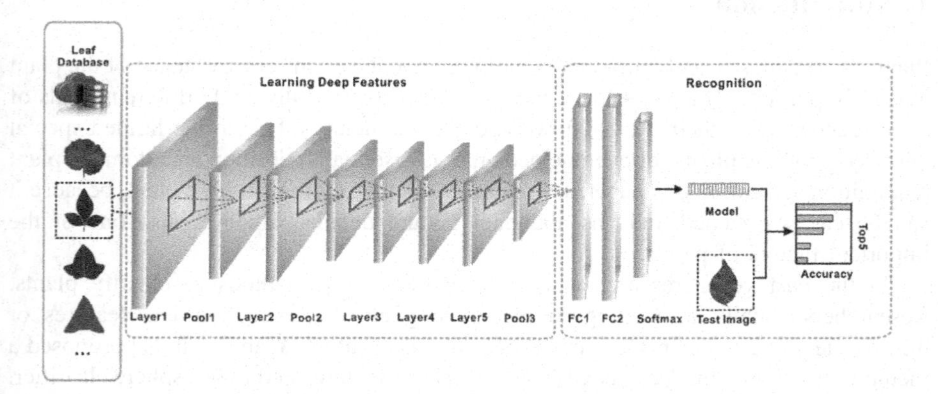

Fig. 1. Framework of our leaf recognition system.

Specifically, we first perform an automatic cropping as a preprocessing for learning deep features. We then divide leaves in database into two different kinds of sets, including training set and validation set. To verify the accuracy of our leaf recognition, we need to add labels to the two data sets. In addition, we also transform image data into formatted data, which can be handled with deep learning CAFFE before learning deep features in convolutional neural networks. Given an input leaf image for recognition, we normalize the image resolution by resampling it to 256 × 256 and crop it to 227 × 227 before input to the network. We also transform labeled image to specific suffix data format and compute a mean of the leaf image. Finally, we can predict the unknown leaf image through learning model and display an accuracy of the top 5 most similar images.

2.1 Convolutional Neural Networks

Deep learning is developed in recent years with the emergence of numerous deep learning models. Convolutional neural networks have attracted extensive attention especially in the field of pattern classification [10–15]. The basic structure of the convolutional neural networks contains both feature extraction and feature mapping layer, which are implemented by operators of convolution, activation function and pooling. Finally, the feature map through the full connected layer into the softmax classifier to classify.

Firstly, feature extraction process is calculated by convolution. The convolution can be formulated as follows,

$$x_j^\ell = \sum\nolimits_{i \in M_j} x_i^{\ell-1} * k_{ij}^\ell + b_j^\ell \tag{1}$$

where x_j^ℓ denotes the j-th output feature map in ℓ layer, M_j is the number of input map for this layer, $x_i^{\ell-1}$ represent i-th ouput map in layer $\ell - 1$, k_{ij}^ℓ is the convolutional kernel and b_j^ℓ is bias parameter for the j-th output map in layer ℓ.

Feature mapping process is computing by activation function. The activation function can be formulated as follows,

$$x_j^\ell = F\left(\sum\nolimits_{i \in M_j} x_i^{\ell-1} * k_{ij}^\ell + b_j^\ell\right) \tag{2}$$

Actually, feature map is to use activation function to map characteristics of convolutional to the same plane, as formula (2) shows. There are several types of the activation function such as sigmoid function, tanh function or Rectified Linear Units (ReLU) function $f(x) = \max(x, 0)$. We applied ReLU function in our network [16].

With feature map, our classifier can be implemented. However, network still requires a huge amount of calculation. In order to solve this problem, the next important operation is the pooling. Its function is to reduce the amount of parameters. During the pooling, a maximum or an average is computed.

The last process performed by a fully connected layer is the classification of the feature maps. The fully connected layer will compute the class and accuracy by the softmax function.

2.2 Deep Network Structure

Our deep network structure is shown in Fig. 2. Based on our hardware conditions and database, we set the network structure to 8 layers, including 5 layer convolution, 2 full connected layers and one layer classification.

Given a normalized input image (256 $\times$ 256 with 3 color channels), we randomly crop it into 227 $\times$ 227 to fit the first layer. Then, we can filter the 227 $\times$ 227 $\times$ 3 input leaf images with 96 kernels of size 9 $\times$ 9 with stride of 4 pixels on the first convolutional layer. The size of the generated feature maps is 60 $\times$ 60. The first convolutional layer is followed by a max pooling operation with a size of 3 $\times$ 3. The size of the generated feature maps is 29 $\times$ 29. Then, the second convolutional layer takes the pooled feature maps from the first layer and convolved with 256 filters on a kernel size of 7 $\times$ 7. In addition, the third and fourth convolutional layers which have 384 kernels of size 5 $\times$ 5 respectively perform only convolution without pooling. The fifth convolutional layer has 384 kernels of size 3 $\times$ 3. The size of the generated feature maps is 13 $\times$ 13. After performing convolution and pooling in the fifth layer, the final result is put into two full connected layers which have 4096 neurons.

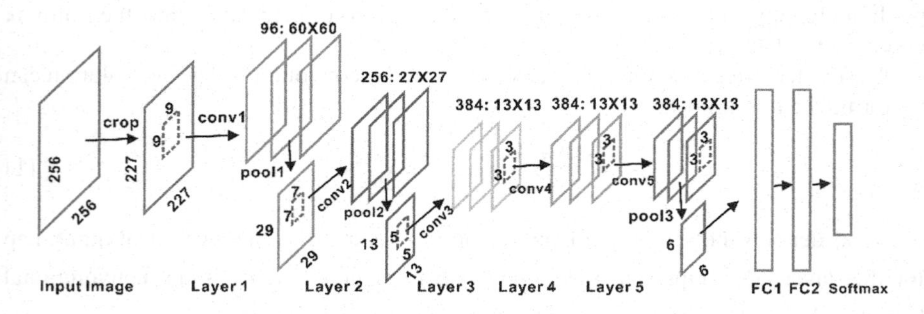

Fig. 2. The architecture of convolutional network

2.3 Network Visualization and Verification

For the above deep network structure, we use deconvnet [17] to visualize each layer and verify its effect. For visualization, the purpose is to help us intuitively understand what property of extracted characteristics. We refer to the previous methods for visual processing. A deconvnet can be thought of as a convolution model that uses the same operations (filtering, pooling) but in reverse.

In this paper, the deconvnet process does not have the ability to learn, but is used to visualize a trained convolution network model without learning training. The main visualization step is to add deconvnet layer to each convolution layer and the output of the convolution layer as the input for deconvnet layer. Then the deconvnet layer is followed by unpooling, ReLU and deconvnet. For unpooling, a table is used to record the location of each maximum, and then the position is unpoolled to fill in the maximum value to maintain the positions on the zero. For ReLU, deconvnet is directly performed with a ReLU function. For deconvnet, the same convolution kernel transpose is used as the core. For our network structure, we use the deconvnet to complete the visualization of convolution filter for each layer.

3 Experimental Results

In this paper, our system is developed based on the deep learning framework CAFFE. We have implemented our novel method using Ubuntu 14.04 on server with GPU k40. Our database has 18,000 leaves, of which 3,000 are test leaves. Typical leaf image examples in our database are shown in Fig. 3.

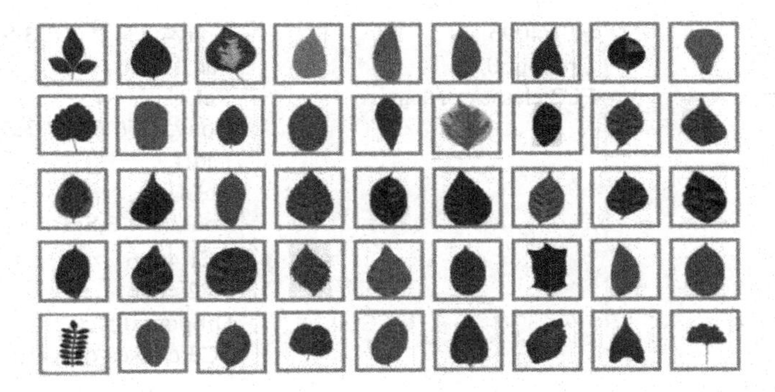

Fig. 3. Typical leaf image examples in our database

3.1 Deep Network Evaluation

We have 300 leaf classes database where each class includes 60 images. So we have 18000 leaves in total. We find that fine-tuning caffenet with some special tracks can get good accuracy on our dataset. To improve the accuracy, we need to design different networks to test accuracy and loss function and set the configuration file according to the experience of modifying learning rate and weight decay. Table 1 shows the parameters of three typical networks in our experiments. The last network is modified based on caffenet with smaller convolution kernels.

Through the experiments, we found that the accuracy can be significantly improved for extracting deep feature with the increase of network layers. The learning deep features of each layer are verified in our visualization experiments (Fig. 6). The experimental results for our database (training from scratch) are shown in Fig. 4. From the results, we can easily observe that our method can clearly demonstrate the effect of our deep learning framework for leaf recognition.

Table 1. Comparison of three kinds of network parameters

Network 1	Network 2	Network 3
Convl:96 * 9 * 9,S:2, Pad:0 ReLU Pool1:3 * 3,S:2	Convl:96 * 9 * 9,S:2,Pad:0 ReLU Pool1:3 * 3,S:2	Convl:96 * 9 * 9,S:2,Pad:0 ReLU Pool1:3 * 3,S:2
FC1:4096 Dropout:0.5 FC2:classes of leaves	Conv2:256 * 7 * 7,S:1,G:2, Pad:2 ReLU Pool2:3 * 3,S:2	Conv2:256 * 7 * 7,S:1,G:2, Pad:2 ReLU Pool2:3 * 3,S:2
	Conv3:384 * 5 * 5,S:1,G:2, Pad:2 ReLU Pool3:3 * 3,S:2	Conv3:384 * 5 * 5,S:1,G:2, Pad:2 ReLU
	FC1:4096 Dropout:0.5 FC2:classes of leaves	Conv4:384 * 3 * 3,S:1,G:2, Pad:1 ReLU Conv5:256 * 3 * 3,S:1,G:2, Pad:1 ReLU Pool3:3 * 3,S:2
		FC1:4096 FC2:4096 Dropout:0.5 FC3:classes of leaves

S:stride G:group

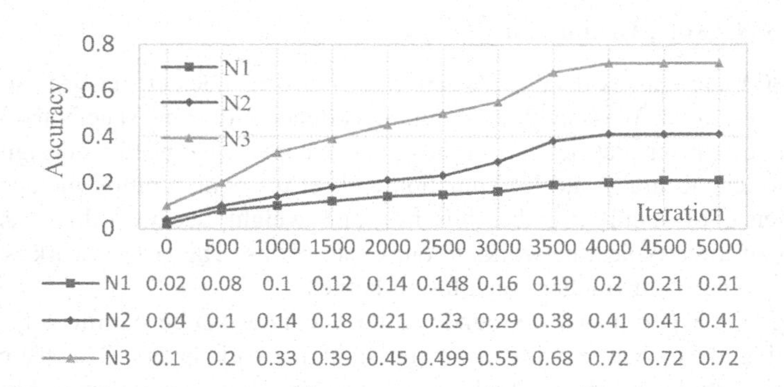

Fig. 4. Accuracy of different networks on our database (300 species)

3.2 Statistical Evaluation

In our experiments, we use the network with 5 convolutional layers (2 full connected layers). Compared with other two kinds of networks, our method has a significantly higher accuracy. To further improve our accuracy, we find that our database should be extended to cover more deep features. Therefore, we choose Imagenet [18] database to train from scratch. The trained model is used for the parameter initialization of our network. Based on our extended databases and a better initialization for the deep learning parameters, our method obtains a much better accuracy that our average accuracy rate reached 96% in different database. Statistical experimental results are shown in the following Fig. 5.

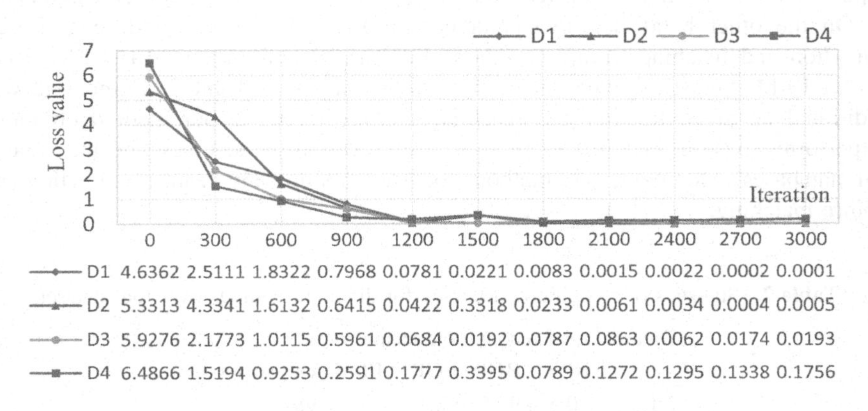

	0	300	600	900	1200	1500	1800	2100	2400	2700	3000
D1	4.6362	2.5111	1.8322	0.7968	0.0781	0.0221	0.0083	0.0015	0.0022	0.0002	0.0001
D2	5.3313	4.3341	1.6132	0.6415	0.0422	0.3318	0.0233	0.0061	0.0034	0.0004	0.0005
D3	5.9276	2.1773	1.0115	0.5961	0.0684	0.0192	0.0787	0.0863	0.0062	0.0174	0.0193
D4	6.4866	1.5194	0.9253	0.2591	0.1777	0.3395	0.0789	0.1272	0.1295	0.1338	0.1756

(a) Loss value of our network in different database

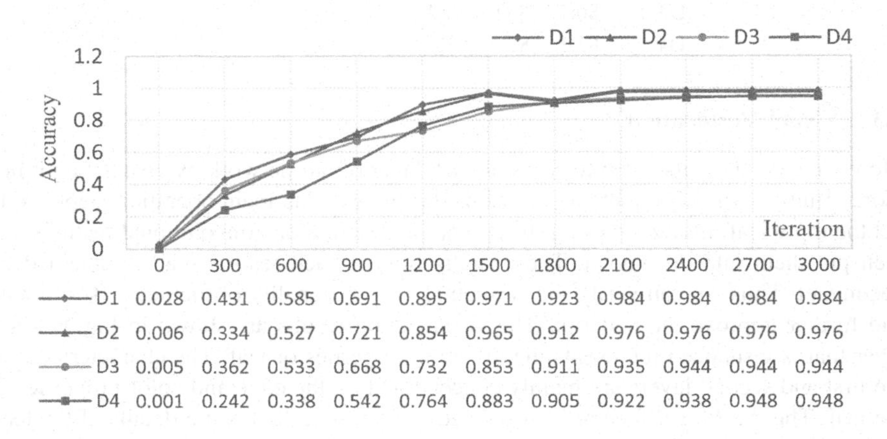

	0	300	600	900	1200	1500	1800	2100	2400	2700	3000
D1	0.028	0.431	0.585	0.691	0.895	0.971	0.923	0.984	0.984	0.984	0.984
D2	0.006	0.334	0.527	0.721	0.854	0.965	0.912	0.976	0.976	0.976	0.976
D3	0.005	0.362	0.533	0.668	0.732	0.853	0.911	0.935	0.944	0.944	0.944
D4	0.001	0.242	0.338	0.542	0.764	0.883	0.905	0.922	0.938	0.948	0.948

(b) Accuracy of our network in different database

Fig. 5. Loss changes and accuracy in different databases using model initialization

To evaluate the effect of our deep learning framework, we use four different databases with different scales to test the leaf recognition accuracy. Database 1 (D1), Database 2 (D2), Database 3 (D3), and Database 4 (D4) have 50, 100, 200, and 300 species, respectively. From Fig. 5, we can find that D1 (50 species), D2 (100 species), and D3 (200 species) database accuracy are converge after training 3000 times. When accuracy reaches between 94%–98%, the loss function value is halt. When the database becomes larger as D4 (300 species), the final accuracy rate still can reach 96.7%, but it require much more number of iterations (iterating 20000 times before convergence).

On the other hand, we also compare our method with the traditional feature extraction and machine learning methods for leaf recognition, such as RISC, BGP, BGP + ELM. Statistical experimental results are shown in Table 2. Comparing with traditional feature extraction and machine learning methods, our method obviously outperforms existing methods in terms of both recognition accuracy and the scale of leaf databases. Moreover, our method does not require any feature extraction and feature matching.

Table 2. Statistics of recognition accuracy for different methods over four datasets

Dataset	Leaf recognition accuracy (%)			
	RISC	BGP	BGP + ELM	Our method
D1	80.3	90.5	92.1	98.4
D2	81.8	82.4	90.5	97.6
D3	80.7	83.6	84.4	98.2
D4	83.1	82.1	88.4	96.7

3.3 Visual Verification

Moreover, we show the deep features of each layer of our network by visualizing it in a more intuitive way. The method we use is deconvnet. The main operation steps are to get the top 9 leaf images by using the activation function on convolutional features and then put them into deconvnet. The visualization of each feature map is obtained by deconvnet. Thus we can verify the correctness and rationality of the network structure and further improve the network. The visualization results are shown in Fig. 6. Each layer that is visualized represents the different properties of leaf. The characteristics of the first and second layers are mainly concentrated on the edge and color properties of the leaf. The third layer feature is visualized to represent the texture details of the leaf. The fourth layer shows the distinct features. The fifth layer shows the characteristics of the integral leaf. From the selected top 9 activation images, we can find that learning deep features is a process from low level features to the higher level features. From this, we can verify that our five convolutional layers network is reasonable and the advantages of our method are also clearly visualized and demonstrated.

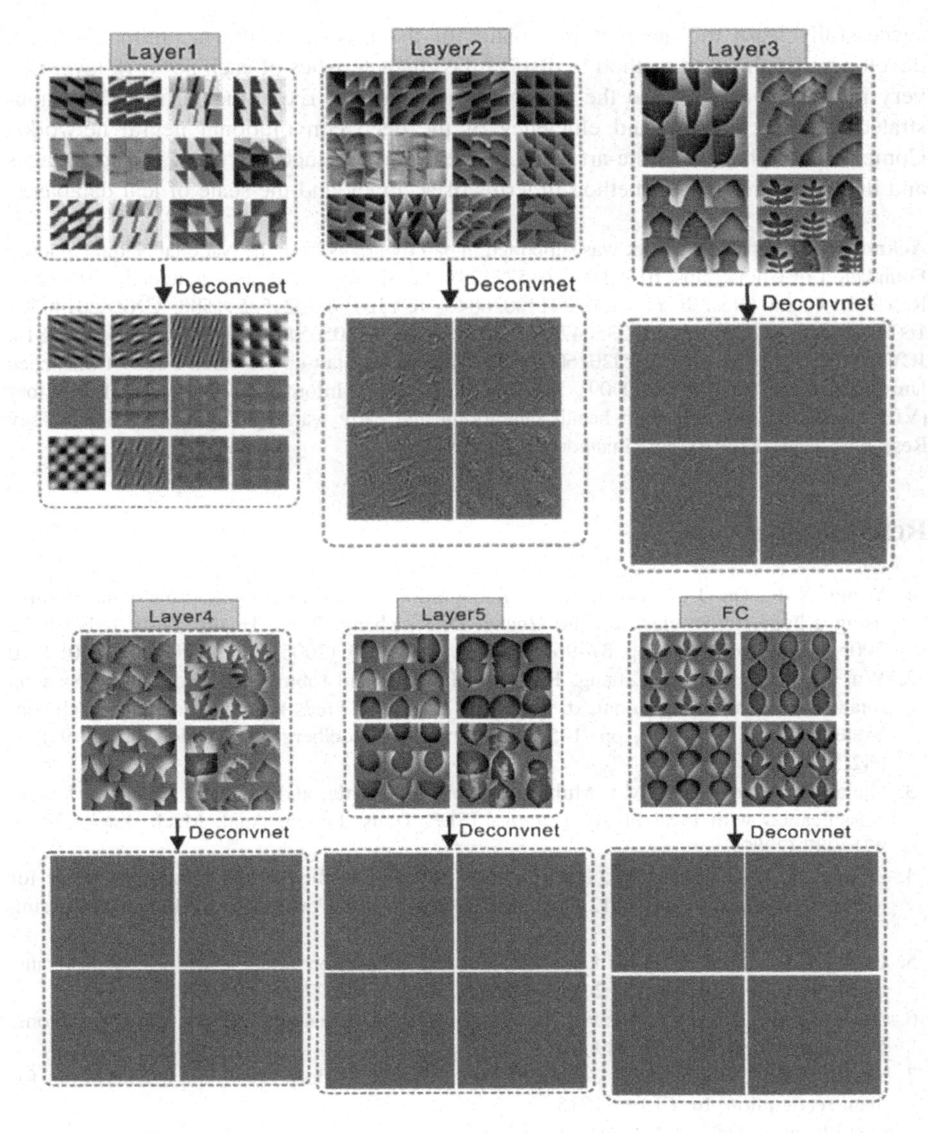

Fig. 6. Visualization of features in each layer of the trained model

4 Conclusion

In this paper, we present a novel deep learning framework for leaf recognition. By relying on the deep learning framework CAFFE, we successfully achieve automatic leaf recognition by designing and fine-tuning CNNs. To the best of our knowledge, our method is one of the early attempts to employ deep learning technique for leaf recognition. By designing our own deep convolutional neural networks, we also successfully learn the deep characteristics of the leaves. At the meantime, we also develop a visualization method to display the characteristics of each network layer in a very intuitive way to guide the parameter adjustments. Experimental results demonstrated the effectiveness and efficiency of the deep convolutional neural networks. Comparison with state-of-the-art methods were also conducted to show the advantages and improvements of our method in terms of accuracy and the scale of leaf databases.

Acknowledgments. This work was supported in part by grants from the National Natural Science Foundation of China (Nos. 61303101, 61572328), the Shenzhen Research Foundation for Basic Research, China (Nos. JCYJ20150324140036846, JCYJ20170302153551588, CXZZ20140902 160818443, CXZZ20140902102350474, CXZZ20150813151056544, JCYJ20150630105452814, JCYJ20160331114551175, JCYJ20160608173051207), the Start-up Research Fund of Shenzhen University (Nos. 2013-827-000009), the China-UK Visual Information Processing Laboratory (VIPL) and Maternal and child health monitoring and early warning Engineering Technology Research Center (METRC) of Guangdong Province.

References

1. Wang, X.-F., Du, J.-X., Zhang, G.-J.: Recognition of leaf images based on shape features using a hypersphere classifier. In: Huang, D.-S., Zhang, X.-P., Huang, G.-B. (eds.) ICIC 2005. LNCS, vol. 3644, pp. 87–96. Springer, Heidelberg (2005). doi:10.1007/11538059_10
2. Wu, H., Pu, P., He, G., Zhang, B., Yuan, L.: Fast and robust leaf recognition based on rotation invariant shape context. In: Wen, Z., Li, T. (eds.) Foundations of Intelligent Systems. AISC, vol. 277, pp. 145–153. Springer, Heidelberg (2014). doi:10.1007/978-3-642-54924-3_14
3. Ojala, T., Pietikainen, M.: Multi-resolution gray-scale and rotation invariant texture classification with local binary patterns. IEEE Trans. Pattern Anal. Mach. Intell. **24**(7), 971–987 (2002)
4. Zhang, L., Zhou, Z., Li, H.: Binary gabor pattern: an efficient and robust descriptor for texture classification. In: 2012 19th IEEE International Conference on Image Processing (ICIP), IEEE (2012)
5. Chang, C.C., Lin, C.J., et al.: LIBSVM: a library for support vector machines. ACM Trans. Intell. Syst. Technol. (TIST) **2**(3), 27 (2011). Article No. 27
6. Huang, G.B., Zhu, Q.Y., Siew, C.K.: Extreme learning machine: theory and applications. Neurocomputing **70**(1–3), 489–501 (2006)
7. Reyes, A.K., Caicedo, J.C., Camargo, J.E.: Fine-tuning deep convolutional networks for plant recognition. In: CLEF (2015)
8. CAFFE. http://caffe.berkeleyvision.org/
9. Zeiler, M.D., Fergus, R.: Visualizing and understanding convolutional networks. In: Fleet, D., Pajdla, T., Schiele, B., Tuytelaars, T. (eds.) ECCV 2014. LNCS, vol. 8689, pp. 818–833. Springer, Cham (2014). doi:10.1007/978-3-319-10590-1_53
10. Sainath, T.N., Kingsbury, B., Saon, G., Soltau, H., Mohamed, A.R., Dahl, G., Ramabhadran, B.: Deep convolutional neural networks for large-scale speech tasks. Neural Netw. **64**, 39–48 (2015)
11. Zagoruyko, S., Komodakis, N.: Learning to compare image patches via convolutional neural networks. In: Proceedings of the IEEE Conference on Computer Vision and Pattern Recognition (CVPR), pp. 4353–4361 (2015)

12. Jaderberg, M., Simonyan, K., Vedaldi, A., Zisserman, A.: Reading text in the wild with convolutional neural networks. Int. J. Comput. Vis. **116**(1), 1–20 (2016)
13. Tymoshenko, K., Bonadiman, D., Moschitti, A.: Convolutional neural networks vs. convolution kernels: feature engineering for answer sentence reranking. In: Proceedings of NAACL-HLT, pp. 1268–1278 (2016)
14. Dosovitskiy, A., Tobias Springenberg, J., Brox, T.: Learning to generate chairs with convolutional neural networks. In: Proceedings of the IEEE Conference on Computer Vision and Pattern Recognition, pp. 1538–1546 (2015)
15. Tuama, A., Comby, F., Chaumont, M.: Camera model identification with the use of deep convolutional neural networks. In: IEEE International Workshop on Information Forensics and Security, IEEE, Abu Dhabi, United Arab Emirates (2016)
16. ReLU. https://www.quora.com/Deep-Learning/What-is-special-about-rectifier-neural-units-used-in-NN-learning
17. Zeiler, M., Taylor, G., Fergus, R.: Adaptive deconvolutional networks for mid and high level feature learning. In: ICCV, IEEE, Barcelona, Spain (2011)
18. Imagenet. http://www.image-net.org/download-images

Evaluation of Deep Models for Real-Time Small Object Detection

Phuoc Pham, Duy Nguyen, Tien Do(✉), Thanh Duc Ngo, and Duy-Dinh Le

Multimedia Communications Laboratory at University of Information Technology, Vietnam National University, Ho Chi Minh City, Vietnam
{13520653,13520154}@gm.uit.edu.vn, {tiendv,thanhnd,duyld}@uit.edu.vn

Abstract. Real-time object detection is crucial for many applications. Approaches based on Deep Learning have achieved state-of-the-art performance on challenging datasets. Although several evaluations of the models have been conducted, there is no extensive evaluation with specific focuses on real-time small object detection. In this work, we present an in-depth evaluation of existing deep learning models in detecting small objects. We evaluate three state-of-the-art models including You Only Look Once (YOLO), Single Shot MultiBox Detector (SSD), and Faster R-CNN with related trade-off factors i.e. accuracy, execution time and resource constraints. Experiments were conducted on benchmark datasets and a newly generated dataset for small object detection. All analyses and findings are then presented.

Keywords: Real-time object detection · Small object detection

1 Introduction

In this paper, we research the task of detecting small objects in real time and refer to common objects that have small sizes. It means the objects located on a few small parts of an image. Small object detection has been applied to many existing systems, e.g. Automotive, Smart Transportation, Army Projects, etc. as Fig. 1. To do this, we encounter a lot of challenges. For example, an image is in high-dimensional space, come with various sizes or resolutions. The detected objects are too small compared to the size of an image. They also can be deformable or be viewed from different viewpoints. Generally, visual information of small objects is less than medium or big objects, it is hard to exploit these information for detecting small objects.

Approaches based on Deep Learning have achieved state-of-the-art performance on challenging datasets. Although several evaluations of the models have been conducted, there is no extensive evaluation with specific focuses on real-time small object detection. Thus, an evaluation of current deep models is really necessary including accuracy, resource usage, especially the execution time. We perform our evaluation on the current state-of-the-art approaches based on Deep Learning as YOLO [11] and SSD [9] models to point out the limitations

D. Liu et al. (Eds.): ICONIP 2017, Part III, LNCS 10636, pp. 516–526, 2017.
https://doi.org/10.1007/978-3-319-70090-8_53

Fig. 1. The illustration of real applications about detecting small objects. (a), a robot is detecting small objects on the floor as a mouse, a stick and a clock. (b), there is a rock as a barrier on the street and need to detect to alert the driver.

of the models. We then show how well-performed the detection models are when applying them to detect small objects. Furthermore, we conduct an assessment on Faster R-CNN [12] as a bridge between previous and novel approaches in object detection.

So far, almost detection methods all perform and evaluate their performance on common datasets as COCO, PASCAL VOC or ILSVRC. These datasets commonly contain objects filling medium or big parts on an image that do not concentrate on small objects. In addition, the number of classes of current small object datasets are less than common datasets. In this work, besides using the standard small object dataset [3], we also rely on the definitions of small objects from [3,13,15] so as to propose an appropriate dataset consisting of subsets taken from PASCAL VOC 2007. The dataset serves in the small object detection problem and is a reference for research group to research or develop from it. In summary, the main contributions in this paper are as follows:

- We evaluate state-of-the-art real-time object detectors based on Deep Learning on standard small object datasets from different aspects including the execution time, accuracy, resource usage, along with the trade-off among a variety of different inputs and its base network. Finally, we outline an overall picture of trade-off among these aspects.
- We propose a dataset consisting of subsets taken from PASCAL VOC 2007 based on the definitions of famous papers with many different ratios to let our evaluation fairly and find the relations among models with the mAP on different datasets.
- Following the results we achieved, we make our analyses how to choose a suitable models for detecting small objects.

2 Related Works

2.1 Small Object Definitions

So far, there are no any explicit and standard definitions about small objects. However, to perform the task of detecting small objects, researchers release different definitions for different datasets instead of using the size of bounding boxes containing objects to consider if the objects are small or not. For example, Zhu et al. [15] mentions that small objects are objects whose sizes filling 20% of an image when releasing his dataset about traffic signs. If the traffic sign has its square size, it is a small object when the width of bounding box less than 20% of an image and the height of bounding box less than the height of an image. In [13], Torralba et al. supposed small objects are less than or equal to 32×32 pixels. In Small Object Dataset [3], objects are small when they have mean relative overlap (the overlap area between bounding box area and the image are) from 0.08% to 0.58%, respectively, 16×16 to 42×42 pixel in a VGA image. In this work, we reuse the above definitions, especially the definitions from [3,15] as the main references because they are reliable resources and are widely accepted by other researchers.

2.2 Deep Models for Object Detection

Recently, in widespread developments of Deep Learning, it is known that Convolutional Neural Networks (CNN) approaches have showed lots of improvements and achieved good results in various tasks. Therefore, it is commonly applied to well-known works. Most of the works all have showed significant improvements in detecting objects filling medium or big parts on an image.

RCNN [5] is one of the pioneer. The following methods are an improvement form R-CNN such as [4,6,12]. Especially, Faster R-CNN [12] is considered as a state-of-the-art-approach. Although these works use a lot of different and breakthrough ideas from Sliding Window to Object Proposals or Deep Learning and mostly achieve the best results as state-of-the-art methods on standard datasets as COCO, PASCAL VOC, ILSVRC. However, their representations take time to run on an image completely and it makes the detectors cannot meet real-time conditions.

In terms of real-time detection, the methods, instead of using object proposal to get Region-of-Interesting (ROI) before moving to classifier like Faster R-CNN, use local information to predict objects as YOLO and SSD. Both methods process images in real-time and detect objects correctly and still have a high point of mAP. Nevertheless, both of two papers just mention that the models can detect small objects and have good results but they do not show any evidences to prove that.

2.3 Datasets

In context of small object detection, there are just a few works that completely conduct research on the problem and also deal with a small fraction of the overall

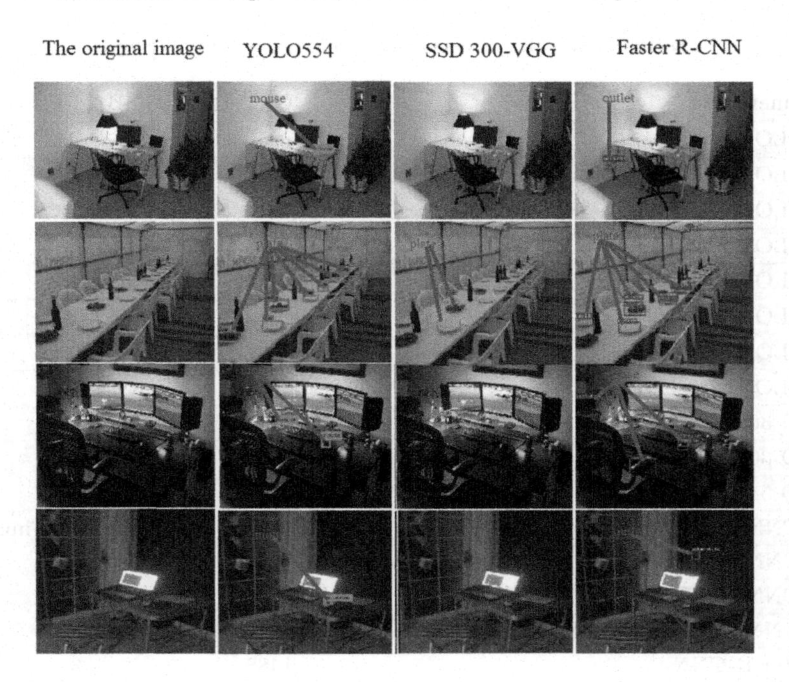

Fig. 2. The visualization of detector results on Small Object Dataset.

picture of small object detection. So far, most of these works are just about to detect sign traffic [15] or vehicles [2,7,10] or pedestrians [1] that do not detect common objects in real world as bottles, mouse, jars, clocks, telephones, etc. Fortunately, recent researches begin focusing on common small objects; particularly, Chen et al. [3] proposed Small Object Dataset by combining the Microsoft COCO [8] and SUN datasets [14] that consists of common objects such as mouse, telephone, switch, outlet, clock, tissue box, faucet, plate, and jar. Chen also augments the R-CNN algorithm with some modifications to improve performance of detecting small objects. Therefore, in this paper, we choose Small object Dataset and our proposed dataset to make our evaluation because these datasets contain common objects and the number of images are large so the evaluations are objective (Fig. 2).

2.4 Evaluated Deep Models

Faster R-CNN is one of a pioneer which is open for the trend of object detection based on Deep Leaning. In this work, R. Shaoqing et al. showed the progress to create hypotheses before taking them into classifiers is a crucial step in detection and it takes most of the time of data processing of the entire progress. The authors indicate that this is a bottleneck so they have proposed a new method called Region Proposal Network (RPN) that shares convolutional features of the whole image with the network used for detection, hence it enables mostly cost-free region proposals. By using the RPN, Faster R-CNN is speeded up.

Table 1. Results of execution time and resource usage.

Frameworks	GPU RAM (MiB)	FPS
YOLO 416 × 416	943	24
YOLO 448 × 448	989	21
YOLO 480 × 480	1044	19
YOLO 512 × 512	1094	20
YOLO 554 × 554	1174	18
YOLO 640 × 640	1383	13
YOLO 800 × 800	1839	9
YOLO 1024 × 1024	2621	6
SSD 300-Resnet	1177	6
SSD 300-VGG	962	6
SSD 512 × 512	2434	4
R-CNN + RPN prop.+ VGG16	1453	14 s/image
R-CNN + Alexnet, 7x, modified RPN proposals		
R-CNN + VGG16, 7x, modified RPN proposals		
R-CNN ContextNet(AlexNet, 7x)		
Faster R-CNN	1488	3

YOLO is inspired from Faster R-CNN and is a first approach that processes data in real time, which has several improvements including batch normalization, anchor boxes, dimension clusters. These lead to high mAP in standard datasets like PASCAL VOC, Microsoft COCO. Specially, thank to make the use of a novel, multi-scale training method the same YOLO model can deal with different sizes and offer an suitable trade-off between speed and accuracy.

SSD is YOLO's a competitor which outperforms mAP in common standard datasets as PASCAL VOC, COCO. Although the SSD approach is similar to YOLO, the difference is that uses multi-scale feature maps by adding many feature maps decreased progressively in late layers of the base network and make object predictions based on these feature maps instead of using only a feature map in the last layers.

3 Experimental Setup

3.1 Evaluation Criterias

In this paper, to evaluate efficiencies of models on the whole datasets used we use mAP measurement for accuracy following PASCAL VOC criteria, FPS for excution time and RAM GPU for resource usage.

3.2 Evaluation

We also make our evaluation following development toolkit of PASCAL VOC as a standard framework to take advantage from it.

As mentioned above, we use two models that SSD and YOLO to make our evaluation. Nonetheless, to get a general assessment of small object detection, we expand our judgements on YOLO and SSD versions that do not meet the condition of real time detection including YOLO 640, YOLO 800, YOLO 1024, SSD 512, SSD Resnet 512. For the entire models mentioned in our evaluation, we do not change the architecture of networks.

YOLO and SSD Configuration We have to retrain the entire models of YOLO and SSD based on the directions of the paper's authors when evaluating them on Small Object Dataset.

We still use the parameters that the authors have provided when training with other datasets including momentum, learning rate, stepsize,weight decay, etc. However, in order to initialize good default bounding boxes in testing phase when testing with our selected datasets, we have to change the anchors value (for YOLO models), or scales and aspect ratios (which have the same meaning) for SSD models. In order to get these values, the K-means clustering algorithm is used to find five 'anchors' of default bounding boxes. We use the five anchors directly for YOLO models. For each of these anchors, we calculate the ratio and then use it for SSD models. The following are ve anchors after running K-means algorithm: [3.5 4.1] [0.7 1.1] [0.3 0.4] [1.6 1.6] [1.0 0.6]

The early time, we set up YOLO and SSD like as Table 3 and we find that almost models get converge at 40k iterations after the processing time of models' converge. However, in order to unify these models for training, we force the iteration to stop at 60k times for all models. When we train both models on our proposed dataset, to ensure accuracy, all of models that we choose to use are the original models provided by the authors of the papers.

Faster R-CNN We have to retrain Faster R-CNN on Small Object Dataset. However, to be more objective, we reuse directly the anchor scales and aspect ratios following the paper [3] such as anchor scales = 16×16, 40×40, 100×100 pixels and aspect ratio = 0.5, 1, 2, instead of having to cluster a set of default bounding boxes similarly to SSD and YOLO. The remaining parameters are in Table 3.

3.3 Our Proposed Dataset

To find out the effects of object sizes among factors including models, excution time, accuracy, resource usage, we proposed a dataset consisting of common objects having many different ratios. The dataset was proposed by reusing the common definitions about small objects and then filtering images taken from PASCAL VOC 2007.

Table 2. Detection Results on Small Object Dataset. YOLO versions get the top results and are less than R-CNN about 6%.

Fameworks	Clock	Faucet	Jar	Mouse	Outlet	Plate	Switch	Tel	t. box	t. paper	mAP
YOLO416	22.8	30.8	4	52	20.4	13.1	13	6.1	0	35.3	19.39
YOLO448	23	36.9	9	52.5	18.4	13.6	17.5	4.2	0	34.3	20.13
YOLO480	34.2	37.3	**9.1**	53.3	21.4	13.6	15.8	9.1	9.1	34.2	23.71
YOLO512	23.1	36.6	6.1	59.8	24.6	14.2	15.7	9.1	4.5	32.4	22.61
YOLO554	23.4	37.2	**9.1**	60.1	27.2	13.4	19.9	9.1	4.5	34.5	23.84
YOLO640	20.2	36.2	3.2	59.8	27.8	11.7	18.1	8.2	4.5	35.6	22.53
YOLO800	27.6	36	2.3	60.2	32.8	13.1	23.3	9.1	9.1	26.7	24.02
YOLO1024	21.7	29.3	1.4	58.3	26.4	11.8	17.5	9.1	9.1	15.7	20.03
SSD300-Resnet	5.5	9.1	0	25.5	6.1	4.5	0	4.5	9.1	18.2	8.25
SSD300-VGG	9.1	17.1	0	26.1	9.1	9.1	0	4.5	0	16.7	9.16
SSD512	9.1	17.1	0	43	9.1	9.1	9.1	9.1	0	7.6	11.32
R-CNN(RPN prop. + VGG16)[3]	31.9	31.3	4.2	56.8	31.1	9.3	14.2	**16.4**	**23.4**	29.4	24.8
R-CNN(Alexet, 7x, 300 pro.)[3]	32.4	27.2	5.1	56.9	28	9.8	13.6	12.4	17.9	35.6	23.9
R-CNN(VGG16, 7x, 300 pro.)[3]	**37.3**	30.3	7.2	**60.6**	**41.5**	**15.8**	**21.5**	13.7	22	33.3	**28.4**
R-CNN(ContextNet (Alexnet, 7x))[3]	32.7	26.8	4.6	56.4	26.3	9.9	12.9	12.2	18.7	34	23.5
Faster R-CNN	23.76	**37.65**	8.03	54	16.16	11.88	15.12	9.1	6.25	**37.29**	21.92

Table 3. The parameters of models

Fameworks	Momentum	Decay	Gamma	Leaning_rate	Base_lr	Max_iteration	Batch_size	Training days	Stepsize
YOLO	0.9	0.0005		0.001		120000	8	5	
SSD300	0.9	0.0005	0.1		0.000004	120000	12	9	40000, 80000, 100000, 120000
SSD512	0.9	0.0005	0.1		0.000004	120000	12	14	
Faster R-CNN	0.9	0.0005	0.1		0.001		2	2	50000, 100000

- Based on the definition from [15], we choose objects whose width and height are less than 20% of an image's width and height and make our evaluation on it. Finally, we simply call the subset as VOC2007_WH_0.2. The subset has fewer than PASCAL VOC 2007 two classes such as dinning table, sofa because there are no objects meet the definition.
- Based on the definition on [3] and combining with a small modification to let our evaluation fairly and find the correlation among models with the mAP on different datasets, we expand the definition of Small Object Dataset that the maximum mean relative area of the original is just 0.58%, we totally add three levels to evaluate that are 0.58% (following the original definition), 10% and 20% (based on the definition of traffic sign) and filter images on VOC 2007 test

Table 4. The information of the subsets

Datasets	The number of classes	The number of images	The number of small object instances
VOC_WH_20	18	1070	2313
VOC_MRA_0.58	16	329	529
VOC_MRA_10	20	2231	5893
VOC_MRA_20	20	2970	7867

set. We call the subsets as VOC_MRA_0.58, VOC_MRA_10, VOC_MRA_20, respectively. The subsets are have the same number of PASCAL VOC 2007 classes except for VOC_MRA_0.58 has fewer than PASCAL VOC 2007 four classes because there are no objects meet the definition and these are dinning table, dog, sofa, train.

The following are the details of the number of small objects and images containing them for subsets (Table 4):

4 Results and Analyses

In this section, we present results that we achieved through experiments. All of models mentioned in this section except models cited from other papers are trained on environment: Ubuntu 14.04 64 bits, Intel(R) Xeon(R) CPU E5-2620 v3 @ 2.40 GHz, 65 GB RAM DDR3, GPU Tesla K20c. Following the setup of the original papers, running at Titan X or GPU K40 almost models all meet real time detection, but under our modest configuration when conducting experiments there are some model cannot meet real time detection. However, our experiments and results we made and achieved are a reliable preference for researchers. We separate our results into 3 parts as follows:

Accuracy:
In Fig. 3, we percieve that SSD recieve the top results and more than the versions of YOLO with objects have mean relative average 10%–20%. On the other hand, the versions of YOLO get top results with objects have mean relative average less than 10%.

SSD versions are not good at the VOC_PASCAL_0.58 and Small Object Dataset because of its architecture. Unlike YOLO, which uses the last feature map of the base network for prediction, SSD combines its results of prediction from many added feature maps to make prediction for objects as well as increasing mAP. It is a good idea to take advantage of prediction results from different layers as these layers contain a lot of information about objects of interest, these layers are also used for detecting big objects, whereas the early layers are used for detecting small objects. Therefore, we use SSD for detecting small objects, these layers are inadvertently lost information about small objects, especially the late layers, this leads to failure during the combination of results.

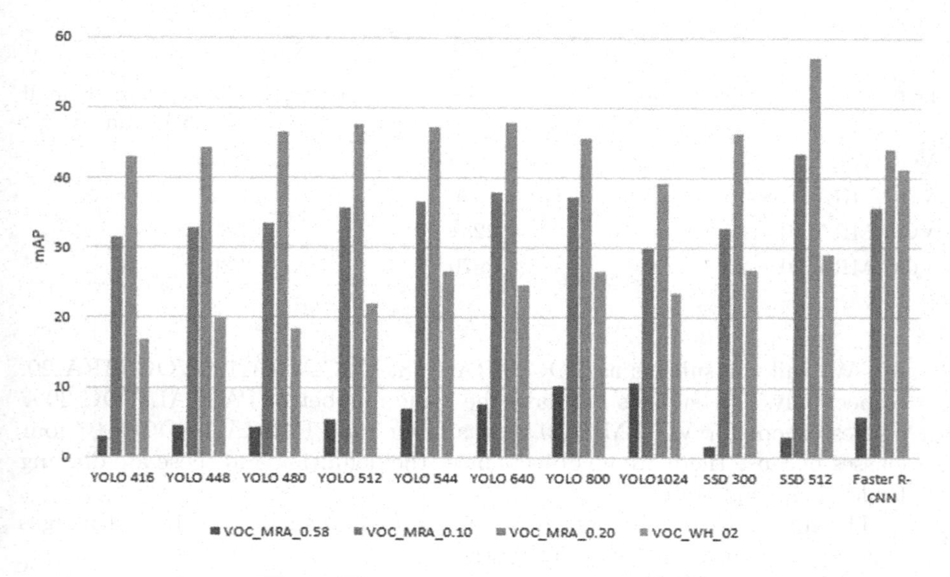

Fig. 3. Detection Results on the Subsets

Unlike the previous works, both YOLO and SSD models all scale input images to a fixed size before taking them into models, this helps models to process data effectively. On the other hand, this leads to a weak point that it is difficult to apply the models to datasets having multi-scale objects or small objects like our experiments. Because most of the objects in images all are deformed, especially small objects are very vulnerable when input images have to resize into a fixed size. In particular, we find that when we test on Small Object Dataset. The average size of images in the dataset is about to 915 width x 788 height, if the size of images is resized far from this, the results are bad and vice versa. These changes show at Table 2. For YOLO, YOLO800 achieves the top results compared to the others, whereas, for SSD, SSD512 (the highest resolution of SSD) also gets the top results. Increasing the size of input images is one of best ways to increase mAP following YOLO and SSD. However, increasing the size of input images is good when the size is close to the real size of images to let objects in the images be less affected.

Execution time and resource usage:

The size of an image is a proportional function to the usage of GPU RAM memory and is inversely proportional to the number of frame handled per second - FPS. The larger the size of an input image is, the higher the amount of RAM used is as showed Table 1, however almost models run with batch size = 1, the RAM memory used is from 1G to 2G. With this modest resource, models can be deployed on low-end computers or even embedded systems.

The analyses of the trade-off among detectors:
The ideas, which are single shot or unified detection models like SSD or YOLO, are a completely novel approach compared to previous approaches. The biggest advantage of this method has achieved is that it ensures the execution time of an image which is nearly a constant number, and it is very easy to estimate or optimize so as to move to and achieve the processing of data in real time. As to accuracy and execution time, that have been demonstrated and reported on both papers as YOLO and SSD, which outperformed in common datasets when comparing them to other models on the same datasets including R-CNN and Faster R-CNN.

Both models just get marginally lower results than the current methods when applying them to detecting small objects. Having fixed input sizes to predict objects in constant time proves very effectively on VOC or COCO. Howerver it leads to the lack of information of small objects, even object distortion or infomation loss if we resize an image to the sizes far from the original size. But with fixed sizes, we can choose the input reasonably to suit to our configuration.

Until now, the use of object proposals combined with classifier seems the most efficient method compared to the others about accuracy when it is applied for many datasets and this is also a natural method that human usually use to conduct a search on objects in a image.

5 Conclusion

In this paper, we evaluate state-of-the-art real-time detectors based on Deep Learning such as YOLO and SSD models on small object datasets about effects of different factors objectively including execution time, resource usage. Besides, we compare them to state-of-the-art methods cannot run in real time such as R-CNN and Faster R-CNN. We also propose a small object dataset consisting of subsets taken from PASCAL VOC 2007 that have common objects based on the definitions of famous papers about small objects to let our evaluation fairly and find the relation among models with the mAP on different datasets.

Through our experiments, if we focus on detecting small objects in real time and trade off a little about accuracy, YOLO554 is the best choice. On the other hand, if we concentrate on accuracy and do not care about the other aspects we can choose models whose sizes of an input are close to the average size of the real image. If objects are in 10%–20% of an image, SSD get better results than versions of YOLO, especially SSD512. If objects are less than 10% of an image, the YOLO get better than SSD.

Acknowledgement. This research is funded by Vietnam National University HoChiMinh City (VNU-HCM) under grant number B2017-26-01.

References

1. Alahi, A., Goel, K., Ramanathan, V., Robicquet, A., Fei-Fei, L., Savarese, S.: Social LSTM: Human trajectory prediction in crowded spaces. In: Proceedings of the IEEE Conference on Computer Vision and Pattern Recognition, pp. 961–971 (2016)
2. Geiger, A., Lenz, P., Urtasun, R.: Are we ready for autonomous driving? The kitti vision benchmark suite. In: Proceedings of the IEEE Conference on Computer Vision and Pattern Recognition, pp. 3354–3361 (2012)
3. Chen, C., Liu, M.-Y., Tuzel, O., Xiao, J.: R-CNN for small object detection. In: Lai, S.-H., Lepetit, V., Nishino, K., Sato, Y. (eds.) ACCV 2016. LNCS, vol. 10115, pp. 214–230. Springer, Cham (2017). doi:10.1007/978-3-319-54193-8_14
4. Girshick, R.: Fast R-CNN. In: Proceedings of the IEEE International Conference On Computer Vision, pp. 1440–1448 (2015)
5. Girshick, R., Donahue, J., Darrell, T., Malik, J.: Rich feature hierarchies for accurate object detection and semantic segmentation. In: Proceedings of the IEEE Conference on Computer Vision and Pattern Recognition, pp. 580–587 (2014)
6. He, K., Zhang, X., Ren, S., Sun, J.: Spatial pyramid pooling in deep convolutional networks for visual recognition. In: Fleet, D., Pajdla, T., Schiele, B., Tuytelaars, T. (eds.) ECCV 2014. LNCS, vol. 8691, pp. 346–361. Springer, Cham (2014). doi:10.1007/978-3-319-10578-9_23
7. Kembhavi, A., Harwood, D., Davis, L.S.: Vehicle detection using partial least squares. IEEE Trans. Pattern Anal. Mach. Intell. **33**(6), 1250–1265 (2011)
8. Lin, T.-Y., Maire, M., Belongie, S., Hays, J., Perona, P., Ramanan, D., Dollár, P., Zitnick, C.L.: Microsoft COCO: common objects in context. In: Fleet, D., Pajdla, T., Schiele, B., Tuytelaars, T. (eds.) ECCV 2014. LNCS, vol. 8693, pp. 740–755. Springer, Cham (2014). doi:10.1007/978-3-319-10602-1_48
9. Liu, W., Anguelov, D., Erhan, D., Szegedy, C., Reed, S., Fu, C.-Y., Berg, A.C.: SSD: single shot multibox detector. In: Leibe, B., Matas, J., Sebe, N., Welling, M. (eds.) ECCV 2016. LNCS, vol. 9905, pp. 21–37. Springer, Cham (2016). doi:10.1007/978-3-319-46448-0_2
10. Morariu, V.I., Ahmed, E., Santhanam, V., Harwood, D., Davis, L.S.: Composite discriminant factor analysis. In: 2014 IEEE Winter Conference on Applications of Computer Vision (WACV), pp. 564–571. IEEE (2014)
11. Redmon, J., Farhadi, A.: YOLO9000: better, faster, stronger. arXiv preprint arXiv:1612.08242 (2016)
12. Ren, S., He, K., Girshick, R., Sun, J.: Faster R-CNN: towards real-time object detection with region proposal networks. In: Advances in Neural Information Processing Systems, pp. 91–99 (2015)
13. Torralba, A., Fergus, R., Freeman, W.T.: 80 million tiny images: a large data set for nonparametric object and scene recognition. IEEE Trans. Pattern Anal. Mach. Intell. **30**(11), 1958–1970 (2008)
14. Xiao, J., Ehinger, K.A., Hays, J., Torralba, A., Oliva, A.: Sun database: exploring a large collection of scene categories. Int. J. Comput. Vis. **119**(1), 3–22 (2016)
15. Zhu, Z., Liang, D., Zhang, S., Huang, X., Li, B., Hu, S.: Traffic-sign detection and classification in the wild. In: Proceedings of the IEEE Conference on Computer Vision and Pattern Recognition, pp. 2110–2118 (2016)

SPMVP: Spatial PatchMatch Stereo with Virtual Pixel Aggregation

Peng Yao[1,2], Hua Zhang[1,2(✉)], Yanbing Xue[1,2], and Shengyong Chen[1,2]

[1] Key Laboratory of Computer Vision and System (Ministry of Education), Tianjin University of Technology, Tianjin, China
yp19880120@sina.com,
{hzhang, Xueyb0718, csy}@tjut.edu.cn
[2] Tianjin Key Laboratory of Intelligence Computing and Novel Software Technology, Tianjin University of Technology, Tianjin, China

Abstract. Stereo matching is one of the critical problems in the field of computer vision and it has been widely applied to *3D Reconstruction*, *Image Refocusing* and *etc*. Recently proposed PatchMatch (PM) stereo algorithm effectively overcomes the limitation of *integer-value* within the support window but it is still inferior in twofold: (1) *view propagation* of PM stereo algorithm generally yields underwhelming particle propagation; (2) it still suffers from a coarse performance in textureless regions. To mitigate these weaknesses, a Spatial-PM stereo algorithm without *view propagation* is proposed for improving the original one at first. Then a virtual pixel based cost aggregation framework with *two* sped-up strategies is proposed for tackling the problem of textureless mismatching. Jointing the *two* incremental improvements, we name the novel one as *Spatial PatchMatch Stereo with Virtual Pixel Aggregation* (SPMVP). Experiments show that SPMVP achieves superior results than other *four* challenging PM based stereo algorithms both in *integer & subpixel* level accuracy on all *31* Middlebury stereo pairs; and also performs better on Microsoft *i2i* stereo videos.

Keywords: PatchMatch stereo · Particle propagation · Virtual pixel · Sped-up strategy · Subpixel accuracy

1 Introduction

According to the researches of [1–3], stereo matching algorithms can be divided into *two* categories: global and local ones. Both of them are implemented with *one* or all of the following *four* steps: matching cost computation; cost aggregation; disparity computation; and disparity refinement.

Dynamic Programming (DP) [4], Graph Cut (GC) [5] and Belief Propagation (BP) [6] based stereo matching algorithms are the typical global ones. They usually omit the cost aggregation step, and seek a disparity solution by optimizing a predefined energy function. Despite the reliable results generated they still time-consumed.

D. Liu et al. (Eds.): ICONIP 2017, Part III, LNCS 10636, pp. 527–542, 2017.
https://doi.org/10.1007/978-3-319-70090-8_54

Local window based stereo algorithms, such as Adaptive Support Weight (ASW) [7], Geodesic Stereo (GS) [8], Virtual Support Window ASW (VSW-ASW) [9] and Histogram Aggregation (HA) [10, 11] are enforced matching cost computation at first and then followed with cost aggregation within a local window. The basic idea behind these algorithms is the local disparity consistency assumption [12] which demonstrates that all the pixels within a support window have constant disparity. However, this assumption is unlikely to hold when the window contains pixels that lie on a different surface than the center pixel and captures a surface that is non-fronto-parallel. After cost aggregation, a Winner-Takes-All (WTA) strategy is employed for disparity computation which selects the corresponding disparity value of the minimum aggregated cost for each pixel. Like most local stereo matching algorithms, they reflect higher time efficiency and approximate accuracy than global ones.

For both categories of stereo matching algorithms some disparity refinement strategies such as Left-Right Crosscheck (LRC), Weighted Median Filter (WMF) and Occlusion Filling (OF) are widely used for obtaining final results.PM algorithm [13] was firstly proposed for computing a *Nearest Neighbor Field* (NNF) between *two* images. It was reported for *Image Editing* and so on. Due to its computational efficiency, the similar idea was employed into PM stereo algorithm [14]. It integrates particle propagation and resampling into local window based aggregation framework. The most praiseworthy is that it estimates a *subpixel* disparity at each pixel by searching for an over-parameterized *3D* plane.

PatchMatch Belief Propagation (PMBP) [15] drew a new connection between BP and PM stereo for estimating correspondence fields of *two* images; and provided an explicit *pairwise* term to handle textureless regions. P. Heise et al. proposed a PM-Huber algorithm [16] by introducing a *Huber Regularization* into smooth term of an energy formulation for solving the perturbations of noises and outliers. C. Zhang et al. imposed the *Laplacian* operator as second-order smoothness and the prior of SLIC superpixels [17] to improve the original PM stereo algorithm (named as ARAP) [18]. More recently, a Cross-Scale framework [19, 20] was proposed which has improved aggregated algorithms by introducing *Multi-Scale* opinion and *Generalized Tikhonov Regularizer*. Nevertheless, these improved PM stereo algorithms still face prohibitively high computational challenges when the candidate disparities reside in a huge or very densely sampled label space.

Some other various PM based algorithms, such as PatchMatch Filter (PMF) [21] and Sped-up PMBP (SPM-BP) [22] both have provided better and faster performances. They utilized the advantages of SLIC superpixel and advanced efficient Edge Aware Filtering (EAF) method [22] to speed up PM and PMBP stereo algorithms respectively. Unfortunately, they remain performing weak in textureless regions than former ones.

Different from previous researches, our contributions mainly focus on solving the problems as follows:

(1) In the procedure of *view propagation* for original PM stereo algorithm, the propagated particle (plane) is assigned with a random value at the same horizontal location in another view. Actually, these corresponding particles of *two* views generally have different disparities due to the displacement of multiple cameras for one scene. So any propagated particle which usually provides a minimum but

error aggregated cost would lead mess performance in local horizontal regions. In order to mitigate this weakness, we have to abandon the *view propagation*. However, without *view propagation* would perform insufficient particle propagation. No matter PM or its various improvements, they were all enforced by *spatial propagation* in an *interleave* order. During *even* iterations they proceeded from *top-left* to *bottom-right* of the image with the particle propagation directions from *top* and *left* for each pixel, and *vice versa*. The workflow of *spatial propagation* inspires us an excellent way to offset the lost propagated information by enforcing these *two* (both *even* and *odd*) situations during each iteration. Then after every center-biased *resampling* (*randomly search*), the labeled particles are reassigned with new random disparities. Based on the modified workflow of original PM algorithm, an enhanced one is performed.

(2) Inspired by the work of VSW-ASW [9], a *color-based virtual pixel* aggregation framework is proposed to tackle textureless mismatching problem while implementing *spatial propagation* and *resampling*. According to the observation we found that in most of the textureless regions the neighbored pixels generally share with same or similar color. These textureless pixels may produce negative influence (*zero* or similar matching cost) into cost aggregation. So we combine the pixels with same color and their nearest *horizontal* neighbors as block-based *virtual pixels* within a local window to alleviate this weakness. Such strategy yields much better results while employing into the former enhanced PM stereo framework.

Based on the analyses we have mentioned above and the original workflow of original PM algorithm (details are shown in Sect. 2), the scheme of the proposed SPMVP stereo matching algorithm can be easily illustrated in Fig. 1. Generally speaking, the crux of SPMVP is on jointing the enhanced PM stereo algorithm with *virtual pixel aggregation* strategy (see the last *two* procedures in green rectangle with dash lines of Fig. 1). Though the same order of magnitude as original PM stereo algorithm (only *0.54* s slower on average), our SPMVP achieves superior results when compared with other *four* most challenging state-of-the-arts PM based stereo algorithms.

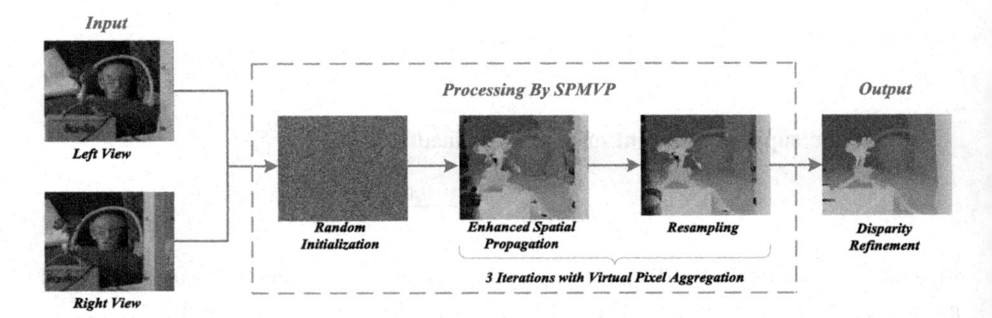

Fig. 1. Scheme of the proposed SPMVP stereo matching algorithm from input to output (with disparity refinement step).

Regarding the organization of this paper we firstly have a brief review of the original PM stereo algorithm in Sect. 2. Then a Spatial-PM (SPM) stereo algorithm without *view propagation* is proposed in Sect. 3.1. Based on the SPM, we innovatively exploit our SPMVP algorithm with *two* sped-up strategies for improving the textureless mismatching problem in Sect. 3.2. More performance evaluations will be given in Sect. 4. At last we draw the conclusions and plan the future work in Sect. 5.

2 PatchMatch Stereo Algorithm

We start by reviewing the initial algorithmic steps of original PM stereo [14] firstly. All the following steps have been enforced the same procedures into *two* corresponding views.

2.1 Random Initialization

PM stereo algorithm has assigned a random disparity $z_0 \in [0, maxdisp]$ from infinite label space for each pixel $p(x_0, y_0) \in I$, which can be represented as $\bar{\boldsymbol{p}}(x_0, y_0, z_0)$. Corresponding to p, an over-parameterized *3D* slanted plane f_p can be formulated as:

$$f_p = a_f x_0 + b_f y_0 + c_f \tag{1}$$

Where

$$(a_f, b_f, c_f) = (-n_x, -n_y, \bar{\boldsymbol{n}} \cdot \bar{\boldsymbol{p}})/n_z \tag{2}$$

and $\bar{\boldsymbol{n}} = (n_x, n_y, n_z)$ denotes the plane's normal as a randomly assigned unit vector.

2.2 Iteration

During each iteration, PM algorithm has included *three* stages: *spatial* & *view propagation* and *resampling*. At each stage, the particle propagation runs through for each pixel. The propagated information can be represented as the aggregated cost for any pixel p by using a *3D* slanted plane f within a local window W_p:

$$C_A(p,f) = \sum_{q \in W_p} \omega(p,q) \cdot C(q, q - (a_f q_x + b_f q_y + c_f)) \tag{3}$$

Where the supported weight $\omega(p,q)$ is defined as:

$$\omega(p,q) = e^{-\frac{\|I_p - I_q\|}{\gamma}} \tag{4}$$

$\|\mathrm{I}_p - I_q\|$ computes the L_1-distance of p and q in color space; and γ is a user-specified parameter. The matching cost $C(q, q')$ is generated by using the method in [24, 25].

Both *spatial* & *view* propagations' implementing order of original PM stereo algorithm performs from *top-left* to *bottom-right* at even iteration, and *vice versa*.

2.2.1 Spatial Propagation

Checking the situation of $C_A(p,f_q) < C_A(p,f_p)$ in a local window, where q is one of the spatial neighbors of current pixel p. If this is the case, f_q is accepted. In *even* iterations q is searching from *top* and *left* neighbors, whereas in *odd* iterations the *bottom* and *right* neighbors are required.

2.2.2 View Propagation

A randomly assigned plane $f_{p'}$ denotes f_p's plane transformation of current view and its parameters are directly generated from former steps. Specifically, a random *3D* plane ($f_{p'}$) of the same location in another view is assigned for computing the horizontal propagated information (aggregated cost) of the current view. If $C_A(p,f_{p'}) < C_A(p,f_p)$, the slanted plane f_p is updated by $f_{p'}$. The same proceeding order is enforced as *spatial propagation* too.

2.2.3 Resampling (Randomly Search)

For further refining the plane f_p, a center-biased random sampling is performed to prevent the solution from being trapped in local minima. This strategy is employed to produce $z_0' = z_0 + \Delta_{z_0}$ and $\bar{\boldsymbol{n}}' = u(\bar{\boldsymbol{n}} + \Delta_n)$($u(\cdot)$ computes the unit vector). Here $\Delta_{z_0} \in [-\Delta_{z_0}^{\max}, \Delta_{z_0}^{\max}]$ and $\Delta_{\mathrm{n}} \in [-\Delta_n^{\max}, \Delta_n^{\max}]$. It begins with $\Delta_{z_0}^{\max} = maxdisp/2$, $\Delta_n^{\max} = 1.0$ and stops at $\Delta_{z_0}^{\max} < 0.1$. The procedure from initial settings to closure is enforced by $\Delta_{z_0}^{\max} := \Delta_{z_0}^{\max}/2$ and $\Delta_n^{\max} := \Delta_n^{\max}/2$ during each iteration. With $\bar{\boldsymbol{p}}'(x_0, y_0, z_0')$ and $\bar{\boldsymbol{n}}'$, an updated plane $f_{p'}$ can be generated. In the situation of $C_A\left(p,f_p'\right) < C_A(p,f_p)$, f_p should be replaced by $f_{p'}$.

2.3 Disparity Refinement

The same as most local and global stereo matching algorithms, LRC, WMF and OF strategies are also applied to produce final refined disparity maps for original PM stereo algorithm.

3 Proposed Algorithm

In this part, we propose *two* incremental improvements based on the original PM stereo algorithm. At first, we simplify the original structure by removing the *view propagation* and modify the original *spatial propagation*, which makes it more accurate in horizontal regions. Based on the initial improvement, we innovatively combine the pixels with same color and their nearest neighbors into a pixel block, called *virtual pixel* and utilize them for recalculating cost aggregation in *spatial propagation & resampling*.

3.1 Spatial-PM Stereo Algorithm Without View Propagation

For *view propagation*, a random *3D* plane of the same location in another view is assigned for computing the horizontal propagated information (aggregated cost) of current view. In fact, the same locations of *two* views usually have different disparities (because of horizontal displacement of multiple cameras for one scene), so these obtained slanted planes generally perform incorrect. In one case but usually occurred that these planes generate minimum but error aggregated costs. That would lead mess performance in local horizontal regions. For this reason, we decide to remove the *view propagation*.

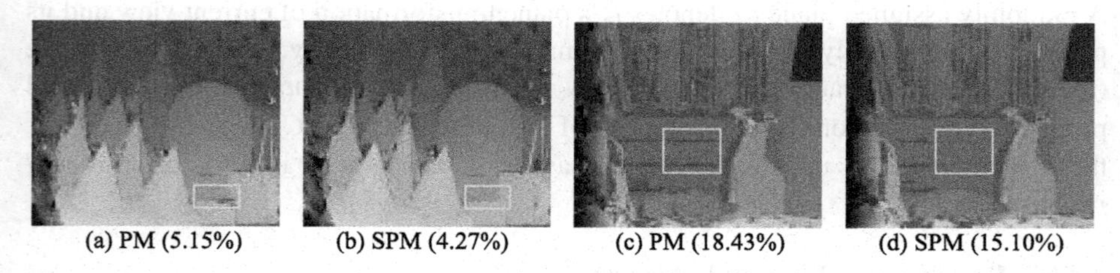

(a) PM (5.15%) (b) SPM (4.27%) (c) PM (18.43%) (d) SPM (15.10%)

Fig. 2. Results of *Cones* and *Laundry* by *two* PM based algorithms from top to bottom rows. (a), (c) Original PM. (b), (d) Spatial-PM (SPM). There are no disparity refinement strategies employed and for all the bad pixels are marked with red dots. The error rates in *non-occluded* regions are indicated below for each disparity map and error threshold is *1.0* pixel. (Color figure online)

However, without *view propagation* the PM stereo algorithm provides insufficient particle propagation. In order to offset the lost propagated information, we enforce the *two* (both *even* and *odd*) situations of *spatial propagation* at each iteration. Although the stable minimum aggregated cost obtained from *four* neighbored *spatial propagation* for each particle at the first iteration, it still can be randomly updated by the following *resampling* step. So after the first iteration, an enhanced and efficient PM stereo algorithm can be performed.

Figure 2. illustrates some results of *two* PM based algorithms without disparity refinement. SPM denotes the Spatial-PM stereo algorithm without *view propagation* and modified *spatial propagation*. We have experimentally implemented SPM with the same *3* iterations and a *35 * 35* window size as original PM. It can be easily observe that SPM outperforms the original one in matching accuracy. Pay attention to the regions in yellow rectangles, the negative influence of *view propagation* (horizontal error regions) has been greatly improved by SPM.

3.2 Spatial-PM Stereo with Virtual Pixel Aggregation

Based on the former proposed SPM stereo algorithm, we further propose a *color-based virtual pixel aggregation* framework while implementing *spatial propagation & resampling*.

This idea primarily stems from the assumption that local pixels with same or similar color should have similar disparities, especially in textureless regions. However, they may provide error matching costs (equal or close to *zero*) into window based cost aggregation and inevitably introduce noises. One effective way to alleviate this weakness is to joint them as one textureless pixel with some rules. Here we combine them as block-based *virtual pixels* for recalculating particle propagation.

First of all, we combine the pixels into *virtual pixels* which perform with same color. In our implementation, one *color-based virtual pixel* $q^{<v>}$ can be represented as the *union* of textureless pixels $q(x, y)$ in a local window:

$$q^{<v>}(x, y) = \bigcup_{q \in W_p} q(x, y) \tag{5}$$

Where for each combined pixel $q(x, y)$ performs color identical is satisfied with one of the following conditions from *top-left* to *bottom-right* within a local window:

$$I_q(x_0, y_0) = \begin{cases} I_q(x_0 + 1, y_0) \\ I_q(x_0, y_0 + 1) \end{cases} \tag{6}$$

And $-Z \leq x_0, y_0 < Z$, $S = 2Z + 1$ denotes the window length. Through formula (6), some cases of *virtual pixel* can be exhibited as in Fig. 3, where the *cerulean* pixel in each *pixel-combined* block which firstly starts to perform color identical with neighbored pixels represents one *virtual pixel*. That means there should be generated massive *virtual pixels* within a fixed size local window, such as in textureless regions.

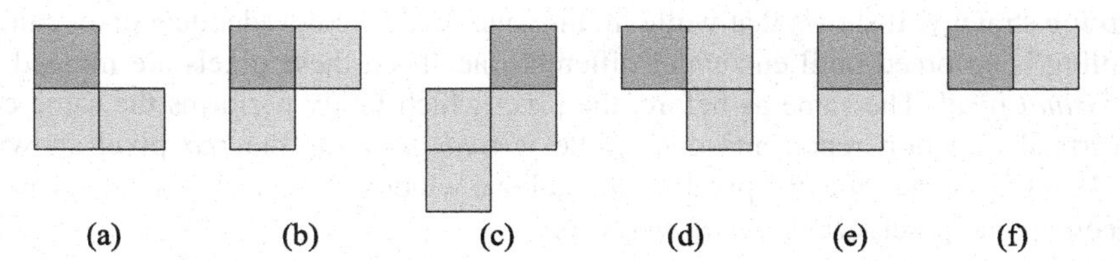

Fig. 3. Some cases of *virtual pixel*. Any *pixel-combined* block (as shown from (a) to (f)) can be seen as a *virtual pixel* and the *cerulean* pixel in each block represents it. (Color figure online)

However, comparing with one's *vertical* and *horizontal* neighbors in RGB color space consumes too much time. Besides, these cases of *virtual pixel* still produce significant noises. To speed up the implementing time and further alleviate the negative influence of noises, we innovatively exploited *two* strategies into SPMVP as mentioned below.

3.2.1 First Sped-up: Vertical Direction Virtual Pixel

To overcome the computational bottleneck in measuring color difference, here we propose a *vertical direction virtual pixel*. In this strategy, any combined pixel q just performs color identical on *vertical direction*:

$$I_q(x_0, y_0) = I_q(x_0, y_0 + 1) \tag{7}$$

Through formula (7) we can know that this category of *virtual pixel* can be shaped as vertical line segments within a local window. With this strategy, only one direction's pixels are estimated in RGB color space. More importantly, with this new structure a strategy for faster generating block-based *virtual pixel* can be easily and efficiently achieved.

3.2.2 Second Sped-up: Spatial Sampling on Vertical Direction Virtual Pixel

The cases of *virtual pixel* (as shown in Fig. 3) strictly combine the pixels which perform color identical but ignore the situation of similar ones. In fact, they also introduce some noises and error aggregated costs as we have mentioned above. One effective solution of this problem is further combing nearest neighbored pixels. Moreover, estimating the similarity of each pixel pair on *vertical direction* inevitably omit horizontal textureless pixels. Hence we propose a *spatial sampling* strategy on *vertical direction virtual pixel*, which can be defined as:

$$S_x(q) = \begin{cases} 1, & I_q(x, y) = I_q(x, y + 1) \\ 0, & Otherwise \end{cases} \tag{8}$$

Where $S_x(q)$ is a binary function capturing the regular sampled pixels on *horizontal direction*. This strategy cleverly utilizes the characteristic of local color consistency which neighbored pixels always share with same or similar colors in regular images. Figure 4 visualizes an example of *virtual pixel* with the proposed *horizontal spatial sampling* strategy. It shows that while finding one vertical color identical pixel pair, the sampling is enforced until encounter different one. Then these pixels are merged into one *virtual pixel*. The same as before, the pixel which firstly performs the same color on vertical direction represents one whole *virtual pixel* (as the *red* pixel shown in Fig. 4). At the same time, the pixels within a local window which are excluded can also be seen as one-pixel-sized *virtual pixels* too.

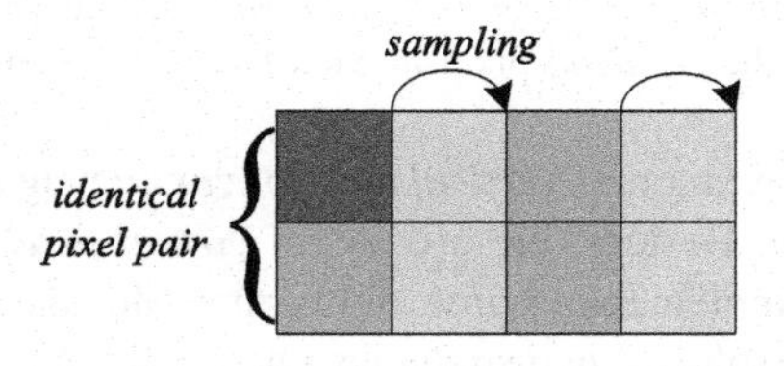

Fig. 4. An example of *virtual pixel* which is generated by using *two* proposed sped-up strategies. (Color figure online)

Finally, the updated model of propagated information (cost aggregation) with *virtual pixel* can be reformulated as follows:

$$C_A(p,f) = \sum_{q^{<v>} \in W_p} \omega(p, q^{<v>}) \cdot C\left(q^{<v>}, q^{<v>} - \left(a_f q_x^{<v>} + b_f q_y^{<v>} + c_f\right)\right) \quad (9)$$

In fact, the implementation of SPMVP with *two* sped-up strategies performs an average *23%* faster than without them and shows an approximate implementing time as original PM stereo algorithm. Figure 5 illustrates the results of PM, SPM and SPMVP without disparity refinement on *Bowling1*. It shows definitely that the SPMVP performs the lowest error rates in *non-occluded* regions and the best textureless handling (especially for the bottom-right textureless regions of the disparity maps). The same as in Sect. 3.1, we also choose the window size as *35 * 35* pixels and *3* iteration times (about how to choose the window size will be discussed in Sect. 4).

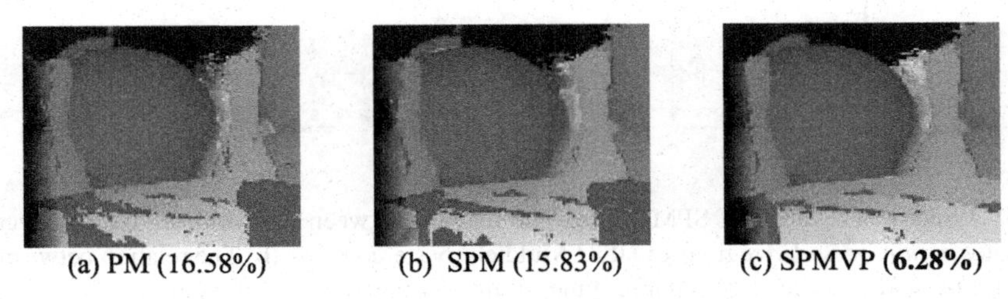

(a) PM (16.58%) (b) SPM (15.83%) (c) SPMVP (**6.28%**)

Fig. 5. Stereo results of *Bowling1* by PM, SPM and SPMVP without disparity refinement. All the error rates in *non-occluded* regions are indicated below and for all the bad pixels are marked with red dots. Error threshold is *1.0* pixel. (Color figure online)

4 Experimental Results

In this section, our proposed SPMVP is estimated with original PM [14] and other *four* most recently challenging PM based algorithms: ARAP [18], SPM-BP [22] and CS-PM [20]. All the implementations were performed on a same PC platform with a *3.60 GHz Intel Core i7* CPU, *16 GB* RAM and *64-bits* OS. And, they were completely done in *C++* code. That means we could fairly evaluate them both in accuracy & time efficiency. To proof the effectiveness we not only estimate the *integer* level but also *subpixel* accuracy on all *31* Middlebury stereo pairs from Sects. 4.1 to 4.4. We set $\gamma = 10$ in (4), $S_x(q) = 1$ in (8) and we also experimentally found that the iteration time should be set as *3* for SPMVP. For PM, ARAP, SPM-BP and CS-PM we hold the original parameters. We use the raw matching cost computation method as follows:

$$C(q) = \alpha \cdot \min\left(\left\|I_q - I_{q'}\right\|, \tau_{col}\right) + (1 - \alpha) \cdot \min\left(\left\|\nabla I_q - \nabla I_{q'}\right\|, \tau_{grad}\right) \quad (10)$$

The parameters are set as $\{\alpha, \tau_{col}, \tau_{grad}\} = \{0.1, 10, 2\}$.

4.1 Parameter Evaluation

Like other local window based cost aggregation stereo matching algorithms, the size of support window affects the performance of SPMVP significantly. So we plot the *window-error* (Fig. 6(a)) and *window-time* (Fig. 6(b)) performances after *3* iterations by SPMVP. Figure 6(a) illustrates the relation between window size and average error rates (*1.0* pixel) in *non-occluded* regions on all *31* Middlebury stereo pairs [26]. It can be easily concluded from the trend that the error rates becomes lower and convergent with the increasing of window size. It shows definitely that for the window size with *35 * 35* pixels, SPMVP achieves the lowest error rate and larger ones would not provide better results.

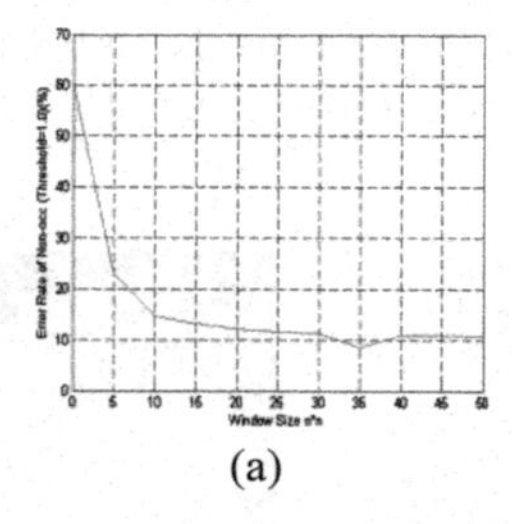

(a)

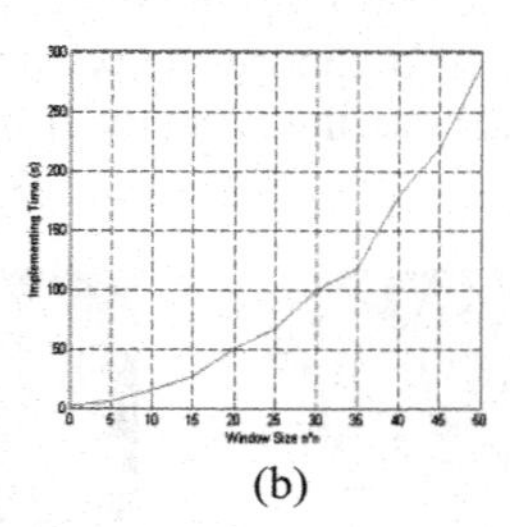

(b)

Fig. 6. Window size study of SPMVP. (a) The relation between the window size and average error rates in *non-occluded* regions of all *31* Middlebury stereo pairs. (b) The relation between the window size and average implementing time of all *31* Middlebury stereo pairs.

Moreover, the computational complexity of local window based algorithms is *O (NDW)* where *N* is the number of pixels in the image, *D* denotes the max disparity and *W* represents the numbers of pixels within a window [12]. That means for the same stereo pairs, the window size also determines the running time significantly. The *vertical axis* of Fig. 6(b) denotes the average implementing time of all *31* Middlebury stereo pairs with various window sizes (*horizontal axis*). We can observe that the running time rockets much more when the window size performs larger than *35 * 35* pixels but with non-optimal results (as shown in Fig. 6(a)). Based on the analyses we choose the best suited window size as *35 * 35* pixels for SPMVP.

4.2 Quantitive Evaluations of PM and SPM

Here we evaluate the original PM and our proposed SPM. To ensure fairness, there are no disparity refinement strategies employed and only the *non-occluded* regions are estimated. Table 1 simply shows the quantitive evaluations on all *31* Middlebury stereo pairs. It illustrates that the SPM outperforms in accuracy & time efficiency while comparing with original PM. For time efficiency, the PM performs *7%* slower than the proposed SPM.

Table 1. Quantitive evaluations on *31* Middlebury stereo pairs.

Algorithms	PM	SPM
Avg. Error (1.0 pixel)	11.50	**10.18**
Avg. Error (0.5 pixel)	31.46	**30.50**
Avg. Time (s)	117.79	**110.33**

4.3 Evaluations Without Disparity Refinement

In this part, we measure our SPMVP with original PM, ARAP, SPM-BP and CS-PM stereo algorithms. The same as in Sect. 4.2, there are no disparity refinement strategies employed and only *non-occluded* regions are measured. In Table 2, the first *two* rows with normal numbers show the average error rates both in *integer* & *subpixel* level accuracy. We can conclude from the table that SPMVP perform the best overall accuracy (marked with bold fonts). Although SPM-BP performs the fastest implementing time as shown in the last row, but it demonstrates a lower-ranking accuracy performance among the *five* estimated algorithms.

Table 2. Quantitive evaluations on all *31* Middlebury stereo pairs.

Algorithms	PM	ARAP	SPM-BP	CS-PM	SPMVP
Avg. Error (1.0)	11.50	20.48	12.21	9.40	**8.44**
Avg. Error (0.5)	31.46	36.59	30.74	30.04	**28.53**
Avg. Time (s)	117.79	1229.43	**33.81**	528.07	118.33

In fact, the running time of SPMVP is just *0.54 s* slower than PM, but performs more accurate than the slowest ARAP and the fastest SPM-BP. That means there is an alternative PM algorithm which could perform the same time efficiency as original one but provides superior results.

Figure 7 illustrates the results of *Aloe, Flowerpots* and *Lampshade2* from top to bottom rows by *five* PM based algorithms in *integer* level accuracy. Especially for *Flowerpots* and *Lampshade2*, SPMVP performs the lowest error rates and the most textureless sensitive. Then in Fig. 8 we have shown the comparisons with *subpixel* level accuracy. For the estimated *Bowling1* and *Monopoly* disparity maps, SPMVP shows more robust to disparity consistency assumption.

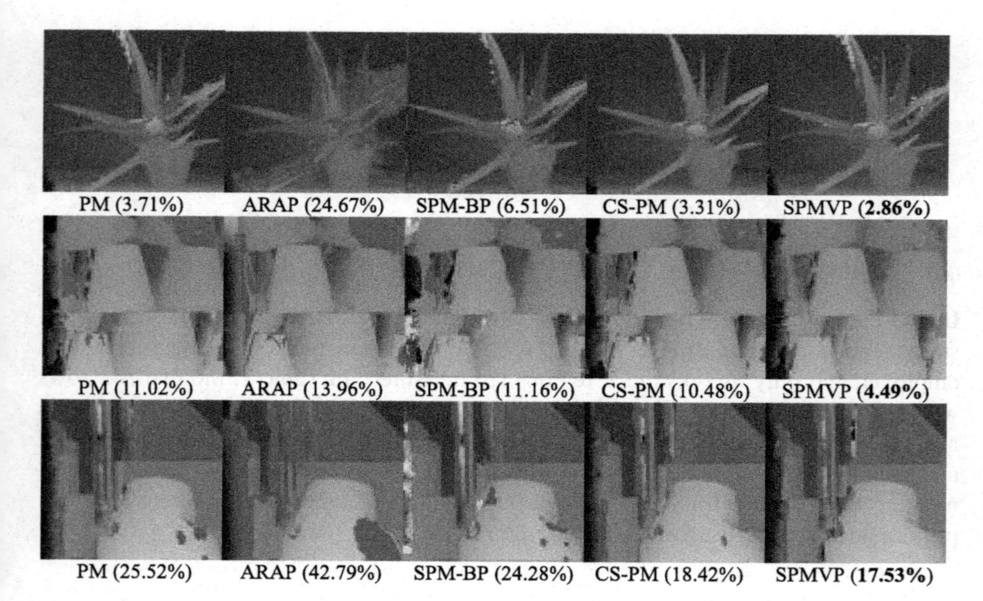

Fig. 7. Results of *Aloe*, *Flowerpots* and *Lampshade2* from top to bottom rows by *five* PM based algorithms: PM, ARAP, SPM-BP, CS-PM and SPMVP. For all the disparity maps are implemented without disparity refinement and error threshold is *1.0* pixel. All the error rates in *non-occluded* regions are indicated below and the bad pixels are marked with red dots. (Color figure online)

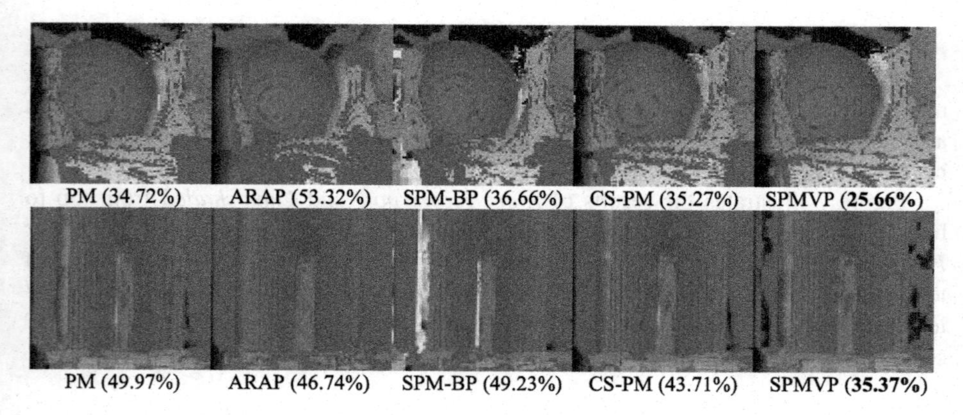

Fig. 8. Results of *Bowling1* and *Monopoly* from top to bottom rows by *five* PM based algorithms: PM, ARAP, SPM-BP, CS-PM and SPMVP. For all the disparity maps are implemented without disparity refinement and error threshold is *0.5* pixel. All the error rates in *non-occluded* regions are indicated below and the bad pixels are marked with red dots. (Color figure online)

4.4 Evaluations with Disparity Refinement

Table 3 simply shows the quantitive evaluations of *five* algorithms in *integer* level accuracy with disparity refinement. Different from former contents, we also evaluate the error rates in *all* pixels. For employing disparity refinement strategies, all the implementations have increased the running time slightly and decreased the error rates (in *non-occluded* regions) respectively. Even in this circumstance, SPMVP still reaches the least error rates in *non-occluded* regions and *all* pixels. Then we also measure the conditions with *subpixel* level accuracy as in Sect. 4.3. We can still easily conclude from Table 4 that the proposed SPMVP robustly keeps the top accuracies both in *non-occluded* regions and *all* pixels.

Table 3. Quantitive evaluations in *integer* level accuracy.

Algorithms	PM	ARAP	SPM-BP	CS-PM	SPMVP
Avg. Error (non-occ)	8.80	20.48	9.85	7.98	7.08
Avg. Error (all)	16.65	25.35	18.61	15.61	14.74
Avg. Time (s)	120.06	1289.63	36.36	532.10	120.53

Table 4. Quantitive evaluations in *subpixel* level accuracy.

Algorithms	PM	ARAP	SPM-BP	CS-PM	SPMVP
Avg. Error(non-occ)	29.54	36.59	29.35	29.03	27.87
Avg. Error(all)	25.40	33.79	29.18	24.77	23.35
Avg. Time(s)	120.06	1289.63	36.36	532.10	120.53

Figure 9 illustrates the final results with same disparity refinement strategies of *Baby3*, *Teddy* and *Wood1* from Middlebury stereo data set. For fair evaluating, we enforce the same strategies for refining the disparities, and only the error rates in *non-occluded* regions are measured here. In the comparisons of the *five* algorithms in Fig. 9, we set the error threshold as *0.5* pixel instead. From the measurements we can observe that SPMVP shows competitive results with other *four* PM based algorithms. Notice that for SPM-BP, it overuses the piecewise constancy assumption and generates massive erroneous fronto-parallel planes, as in *Teddy*.

4.5 Stereo Videos

Additionally, the results of *two* random snapshots of the *Ilkay* stereo video from Microsoft *i2i* data set [27] are shown in Fig. 10. We evaluate the same *five* PM based stereo algorithms with disparity refinement too. Our SPMVP establishes the most smooth disparity maps and has the best performance near foregrounds' depth boundaries.

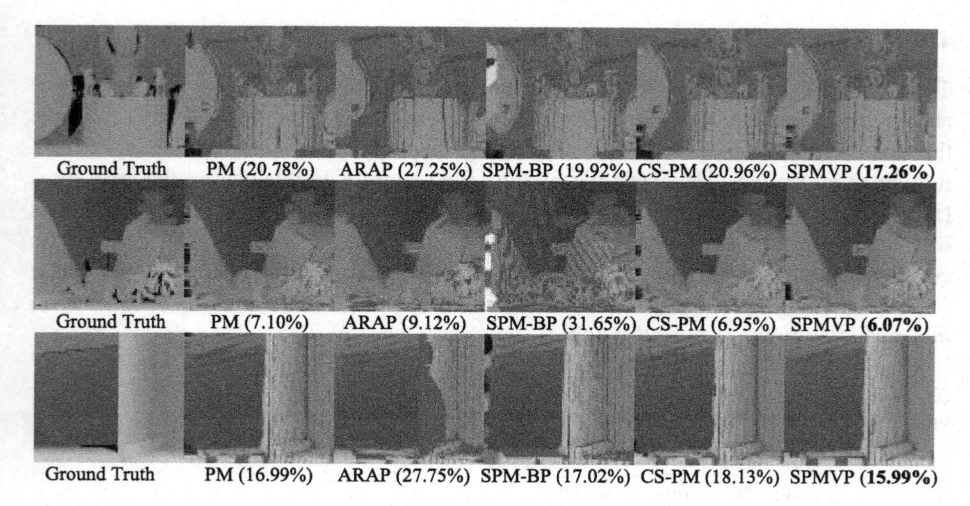

Fig. 9. Final results with same disparity refinement strategies of *Baby3*, *Teddy* and *Wood1* from top to bottom rows by *five* PM based algorithms: PM, ARAP, SPM-BP, CS-PM and proposed SPMVP. For all the error rates in *non-occluded* regions are indicated below and the bad pixels are marked with red dots. Error threshold is *0.5* pixel. (Color figure online)

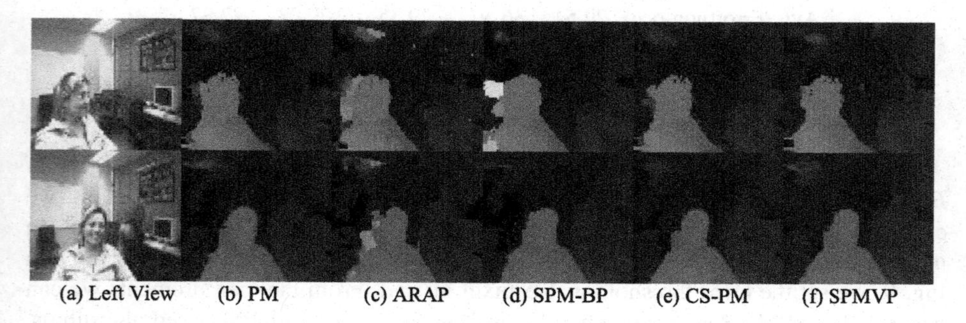

Fig. 10. Disparity maps of *two* random snapshots are established by using the PM, ARAP, SPM-BP, CS-PM and SPMVP. SPMVP performs better than the other *four* algorithms in the foregrounds. Disparity refinement technique is employed for all the algorithms. (a) Left view. (b) Results by PM algorithm. (c) Results by ARAP algorithm. (d) Results by SPM-BP algorithm. (e) Results by the proposed CS-PM algorithm. (f) Results by the proposed SPMVP algorithm.

5 Conclusions

In this paper, we proposed SPMVP algorithm to effectively solve the matching accuracy problem for stereo matching. At first, a SPM stereo algorithm without *view propagation* is proposed and performs more accurate & fast than the original one. Then based on it we innovatively utilize a *virtual pixel aggregation* framework with *two* sped-up strategies for alleviating the weakness for textureless regions' handling and disparity inconsistency, which leads the approximate running time of original PM stereo algorithm but powerfully performs more accurate results. Performance evaluations demonstrate that SPMVP robustly shows superior results not only in *integer* but also in *subpixel* level accuracy than other *four* challenging PM based stereo algorithms on all *31* Middlebury stereo pairs; and also performs better on Microsoft *i2i* stereo video data set. Our future works will focus on how to further optimize the SPMVP and employ the proposed *color-based virtual pixel* aggregation framework to improve other local aggregated based stereo algorithms.

Acknowledgements. This research has been supported by National Natural Science Foundation of China (U1509207, 61325019, 61472278, 61403281 and 61572357), Key project of Natural Science Foundation of Tianjin (14JCZDJC31700).

References

1. Scharstein, D., Szeliski, R.: A taxonomy and evaluation of dense two-frame stereo correspondence algorithms. Int. J. Comput. Vis. **47**, 7–42 (2002). Springer
2. Gong, M., Yang, R., Wang, L., et al.: A performance study on different cost aggregation approaches used in real-time stereo matching. Int. J. Comput. Vis. **75**, 283–296 (2007). Springer
3. Tombari, F., Mattoccia, S., Stefano, L., et al.: Classification and evaluation of cost aggregation methods for stereo correspondence. In: IEEE International Conference on Computer Vision Pattern Recognition, pp. 1–8 (2008)
4. Wang, L., Yang, R., Gong, M., et al.: Real-time stereo using approximated joint bilateral filtering and dynamic programming. J. Real-Time Image Proc. **9**, 447–461 (2014). Springer
5. Taniai, T., Matsushita, Y., Naemura, T.: Graph cut based continuous stereo matching using locally shared labels. In: IEEE International Conference on Computer Vision Pattern Recognition, pp. 1613–1620 (2014)
6. Yang, Q., Wang, L., Ahuja, N.: A constant space belief propagation algorithm for stereo matching. In: IEEE International Conference on Computer Vision Pattern Recognition, pp. 1458–1465 (2010)
7. Yoon, K.J., Kweon, I.S.: Adaptive support-weight approach for correspondence search. IEEE Trans. Pattern Anal. **28**, 650–656 (2006)
8. Hosni, A., Bleyer, M., Gelautz, M., et al.: Local stereo matching using geodesic support weights. In: IEEE International Conference Image Processing, pp. 2093–2096 (2009)
9. Hu, W., Zhang, K., Sun, L., et al.: Virtual support window for adaptive-weight stereo matching. In: IEEE Visual Communications and Image Processing, pp. 1–4 (2011)
10. Min, D., Lu, J., Do, M.N.: A revisit to cost aggregation in stereo matching: how far can we reduce its computational redundancy? In: IEEE International Conference on Computer Vision, pp. 1567–1574 (2011)
11. Min, D., Lu, J., Do, M.N.: Joint histogram based cost aggregation for stereo matching. IEEE Trans. Pattern Anal. **35**, 2539–2545 (2013). IEEE
12. Bleyer, M., Breiteneder, C.: Stereo matching—state-of-the-art and research challenges. In: Farinella, G., Battiato, S., Cipolla, R. (eds.) Advanced Topics in Computer Vision. Advances in Computer Vision and Pattern Recognition. Springer, London (2013). doi:10.1007/978-1-4471-5520-1_6

13. Barnes, C., Shechtman, E., Finkelstein, A., Goldman, D.B.: PatchMatch: a randomized correspondence algorithm for structural image editing. ACM Trans. Graph. **28**(3), 24:1–24:11 (2009)
14. Bleyer, M., Rhemann, C., Rother, C.: PatchMatch stereo - stereo matching with slanted support windows. In: British Machine Vision Conference (2011)
15. Besse, F., Rother, C., Fitzgibbon A., et al.: PMBP: PatchMatch belief propagation for correspondence field estimation. In: British Machine Vision Conference (2012)
16. Heise, P., Klose, S., Jensen, B., et al.: PM-Huber: PatchMatch with huber regularization for stereo matching. In: IEEE International Conference on Computer Vision, pp. 2360–2367 (2013)
17. Achanta, R., Shaji, A., Smith, K., et al.: SLIC Superpixels Compared to State-of-The-Art Superpixels Methods. IEEE Trans. Pattern Anal. Mach. Intell. **34**, 2274–2282 (2012). IEEE
18. Zhang, C., Li, Z., Cai, R., et al.: As-rigid-as-possible stereo under second order smoothness priors. In: European Conference on Computer Vision, pp. 112–126 (2014)
19. Zhang, K., Fang, Y., Min, D., et al.: Cross-scale cost aggregation for stereo matching. In: IEEE International Conference on Computer Vision and Pattern Recognition, pp. 1590–1597 (2014)
20. Zhang, K., Fang, Y., Min, D., et al.: Cross-scale cost aggregation for stereo matching. In: IEEE Transactions on Circuits and Systems for Video Technology (2016)
21. Lu, J., Yang, H., Min D., et al.: PatchMatch filter: efficient edge aware filtering meets randomized search for fast correspondence field estimation. In: IEEE International Conference on Computer Vision and Pattern Recognition, pp. 1854–1861 (2013)
22. Li, Y., Min, D., Brown, M.S., et al.: SPM-BP: sped-up PatchMatch belief propagation for continuous MRFs. In: IEEE International Conference on Computer Vision, pp. 4006–4014 (2015)
23. Lu, L., Shi, K., Min, D., et al.: Cross-based local multipoint filtering. In: IEEE International Conference on Computer Vision and Pattern Recognition, pp. 430–437 (2012)
24. Christo, R., Hosni, A., Bleyer, M., et al.: Fast cost-volume filtering for visual correspondence and beyond. In: IEEE International Conference on Computer Vision and Pattern Recognition, pp. 3017–3024 (2011)
25. Hosni, A., Christo, R., Bleyer, M., et al.: Fast cost-volume filtering for visual correspondence and beyond. IEEE Trans. Pattern Anal. **32**, 504–511 (2013)
26. Middlebury Online Stereo Evaluation (2016). http://vision.middlebury.edu/stereo/data/
27. Microsoft *i2i* Stereo Videos (2016). http://research.microsoft.com/enus/projects/i2i/data.aspx

Uncalibrated Trinocular-Microscope Visual Servo Control Strategy

Xuewei Wang[1(✉)], Qun Gao[2], and Fucheng You[1]

[1] School of Information Engineering,
Beijing Institute of Graphic Communication, Beijing 100026, China
wangxuewei@bigc.edu.cn
[2] Changchun Institute of Optics, Fine Mechanics and Physics,
Chinese Academy of Sciences, Changchun 130033, China

Abstract. Considering that both calibtating accurate the camera parameters and establishing a precise robot kinematics model are hardly, the uncalibrated trinocular-microscope visual servoing control strategy used for achieving precise positioning of the cylindrical target is proposed in this paper. Firstly, using Canny-algorithm, polar coordinate scanning and Ransac Least-Square fitting to extract the features of the target image and Moving-edge algorithm is used to realize real-time tracking of the target. Secendly, dynamic Quasi-Newton algorithm is adopted to estimate the image Jacobian matrix of the trinocular-microscope visual system. Thirdly, use the variance minimization strategy of the target pose error function to control the end movement of positioning robot, and use the strategy of iterative least-squares to improve the stability of the whole system. Furthermore, obtain the initial value of image Jacobian matrix according to the target moving which is in form of discrete linear independence movement near the desired pose. Finally, the dynamic residuals are adjusted in order to achieve precise positioning of the target under the condition that the basic platform of the positioning robot is in disturbance. Experimental results demonstrate the effectiveness of the proposed strategy.

Keywords: Quasi-Newton algorithm · Image jacobian · Dynamic residuals · Visual servoing control

1 Introduction

The 3-D scene can be built by using the visual servo control system, through which images can be captured, detected and processed. With these information, the target can be tracked and positioned in real-time [1, 2]. In order to achieve micron grade positioning accuracy of the target, microscope visual system is adopted [3]. However, visual the monocular microscopic visual system has the disadvantages of short depth, small view field and needing high real-time image algorithm, hence it is necessary to establish multi-camera microscopic visual to realize high-precision tracking and positioning of the target [4–6]. As to realize the need of high-precision positioning of the irregular parts for the progress of micro-assembly, four-camera microscopic visual system was built by Minnesota University of the United States [7].

D. Liu et al. (Eds.): ICONIP 2017, Part III, LNCS 10636, pp. 543–553, 2017.
https://doi.org/10.1007/978-3-319-70090-8_55

There are two kinds of servo control system. One is calibration-based and the other one is uncalibrated-based. In the first kind, parameters of the camera and robot are needed to accurately calibrated, and the accuracy of these parameters determine the control system's precision [8, 9]. If the configuration of the visual system is very complex, the calibration will become difficult. Meanwhile, tiny change of the calibration's parameters can cause terrible influence on control results. In the second kind, it doesn't need calibrating the parameters of the visual control system, but takes the advantage of image Jacobian matrix to transform the error of image feature into the error of Cartesian space. The robot is driven by the designed control law and it makes sure the system error is converged to the allowable range.

Kim et al. from Seoul National University proposed the uncalibrated visual servo control strategy for the static target [10]. In his system, when the sampling period is small enough, the positioning robot can track the still or moving slowly target. J.A. Piepmeier from the US Naval Academy designed the positioning robot's visual servo control strategy, which can track the dynamic target by the fixed camera, and it did not limited on the system's sampling cycle and the target's speed [11]. The author has proved the validity of the method by simulations and experiments, and this paper established the theoretical basis for the control of the uncalibrated visual servo. However, the strategy neglected the influence of dynamic residuals. According to the influence of dynamic residuals in the visual servo control system, Zhao jie and Li Mu from Harbin Industrial University presented the way to minimize the nonlinear variance to track the dynamic target for robot, and the authors have verified the correctness of the method by simulation [12, 13].

Based on the above research results, the trinocular-microscope servo control strategy has proposed in this paper, and the target achieved high-precision positioning driven by the positioning robot. Firstly, Jacobian matrix of the trinocular-microscope visual has been obtained by the dynamic Quasi-Newton algorithm., The robot's terminal motion is controlled by the target's pose minimization equation, and the target's position and gesture are adjusted in real-time. For improving the system's stability, the iterative least squares method is adopted to eliminate the divergence of the Jacobi matrix. Secondly, the initial value of the Jacobian matrix is obtained through the target's movement near the expected position off-line. Finally, by adding the adjustment of the dynamic residuals, the target high precision positioning is completed when the robot's basic is disturbed. Then, the validity and correctness of the strategy are verified by experiments.

2 Uncalibrated Trinocular-Microscope Microscopic Visual Servo Control Algorithm

Trinocular-microscope visual system is used to monitor the target's position in real time and sends the position error to the robot's control system. The error is gained by the image Jacobian matrix which is measured through dynamic Quasi-Newton algorithm. Because this is an iterative algorithm, so the initial value of the Jacobian matrix is very important. It is feasible to estimate the initial value off line, because the microscopic visual has the feature of the short depth and target's movement is also small in the descartes space. At the meanwhile, considering the effect of dynamic residuals on the accuracy of the target's position, it should be properly estimated.

2.1 Dynamic Quasi-Newton Algorithm

In the camera plane, $y^*(t)$ is the position of the target at the time t. $y(q)$ is the target's expected position, and it is just the function of the robot's joint vector q, $q \in R^n$. Because the time changes, the joint angle can be arbitrary, so q and t are irrelevant. The deviation between target's expected position and motion position record as $f(q, t)$,

$$f(q,t) = y(q) - y*(t) \tag{1}$$

$f(q, t)$ is the nonlinear function, and its variance is defined as the function's minimization, and follows:

$$F(q,t) = \frac{1}{2} f^T(q,t) f(q,t) \tag{2}$$

$F(q, t)$ is the sum of the functions and it can be written as the sum of the functions:

$$F(q+h_q, t+h_t) = F(q,t) + F_q h_q + F_t h_t + \ldots \tag{3}$$

where F_q and F_t are the partial differential of the q and t for function F respectively, h_q and h_t are the increment of the q and t. If F_q is set to zero, the least value of the $F(q, t)$ is obtained.

$$\frac{\partial F(q+h_q, t+h_t)}{\partial q} = F_q + F_{qq} h_q + F_{tq} h_t + o(h^2) = 0 \tag{4}$$

$o(h^2)$ is the second order series of the h_q and h_t, and these terms can be ignored, so the form is changed

$$\begin{gathered} J \equiv \frac{\partial f}{\partial p}, S \equiv \frac{\partial J^T}{\partial p} f \\ F_q = J^T f, F_{qq} = J^T J + S, F_{tq} = J^T \frac{\partial f}{\partial t} \end{gathered} \tag{5}$$

where, S is the dynamic residuals, J is the image Jacobian matrix, which describes the relationship of the variation between the target's position and image feature. The formula of (4) can be altered as,

$$q + h_q = q - (J^T J + S)^{-1} J^T (f + \frac{\partial f}{\partial t} h_t) \tag{6}$$

Assuming the estimated Jacobian matrix is $\hat{J}$, and it is taken into the Eq. (6), the joint angle q_{k+1} of the robot at the moment k is written as:

$$q_{k+1} = q_k - (\widehat{J}_k^T \widehat{J}_k + S)^{-1} \widehat{J}_k^T (f_k + \frac{\partial f_k}{\partial t} h_t) \tag{7}$$

and $h_q = q_{k+1} + q_k$. If the target's expected position has been set before, in other words, the product between velocity deviation and time increment is zero. Equation (7) is rewritten as:

$$q_{k+1} = q_k - (\widehat{J}_k^T \widehat{J}_k + S)^{-1} \widehat{J}_k^T f_k \tag{8}$$

Ignoring high orders of the derivative terms, we define the affine model $m(q, t)$ of the deviation function $f(q, t)$ as the following,

$$m_k(q, t) = f_k + \widehat{J}_k (q - q_k) + \frac{\partial f_k}{\partial t}(t - t_k) \tag{9}$$

The affine model at the step of k is the iteration of the step of k-1, therefore,

$$m_k(q_{k-1}, t_{k-1}) = f_{k-1} = f(q_{k-1}, t_{k-1}) \tag{10}$$

Then, merging the equation **Error! Reference source not found.**) and **Error! Reference source not found.**), the result is as following,

$$\Delta f = \widehat{J}_k h_q + \frac{\partial f_k}{\partial t} h_t \tag{11}$$

where $\Delta f = f_k - f_{k-1}$. Both side of the Eq. (11) are subtracted $\widehat{J}_{k-1} h_q$ at the same time, then transposed the equation,

$$h_q^T \Delta \widehat{J}^T = (\Delta f - \frac{\partial f_k}{\partial t} h_t - \widehat{J}_{k-1} h_q)^T \tag{12}$$

and $\Delta \widehat{J} = \widehat{J}_k - \widehat{J}_{k-1}$, the estimation expression of Jacobi matrix at the step of k is

$$\widehat{J}_k = \widehat{J}_{k-1} + \frac{(\Delta f - \widehat{J}_{k-1} h_q) h_q^T}{h_q^T h_q} \tag{13}$$

For the divergence of the image Jacobi matrix, iteration least square method is used to improve the stability of the system in this paper. Therefore, we define the cost function G_k of the affine model as

$$G_k = \sum_{i=1}^{k} \lambda^{k-i} \| m_k(q_{k-1}, t_{k-1}) - m_{k-1}(q_{k-1}, t_{k-1}) \|^2 \tag{14}$$

After minimizing G_k, the Eq. (15) is the result as following,

$$\begin{aligned} \widehat{J}_k &= \widehat{J}_{k-1} + \frac{(\Delta f - \widehat{J}_{k-1} h_q) h_q^T P_{k-1}}{\lambda + h_q^T P_{k-1} h_q} \\ P_k &= \frac{1}{\lambda} (P_{k-1} - \frac{P_{k-1} h_q h_q^T P_{k-1}}{\lambda + h_q^T P_{k-1} h_q}) \end{aligned} \tag{15}$$

where, the range of the weighted λ coefficient is (0,1].

2.2 Initial Value of the Image Jacobian Matrix

In the trinocular-microscope visual system, each sub-microscopic visual system contains different image features, and these features form the sub- Jacobi matrix. In order to reduce the complexity of the algorithm, stacking matrixs are adopted to gain the initial value of the Jacobi matrix.

Image features of the target include two ellipses at the both sides of the target, two straight lines and irregular edges, and so on. These features are extracted by Canny operator, polar coordinate scanning and Ransac least squares fitting method. Then tracking algorithm based on Moving-edge is implemented to track the ellipses and straight lines in real-time.

Ellipse features are expressed by central point (x_e, y_e), major-minor axis (a, b) and rotation angle θ. The Jacobi matrix can be showed as $J_e = [J_{xe}^T \quad J_{ye}^T \quad J_a^T \quad J_b^T \quad J_\theta^T]^T$, where, J_{xe}^T, J_{ye}^T, J_a^T and J_θ^T are the vectors of the Jacobian matrix based on the elliptic features, respectively. The linear features are represented by the distance ρ from the origin of the coordinate system to the lines, angle α between the direction of the normal line and x axis. Using these features, the Jacobian matrix is expressed as $J_l = [J_\rho^T \quad J_\alpha^T]^T$, where J_ρ^T and J_α^T are the vectors of the Jacobian matrix of the image features (ρ, α), which are the features vectors of the lines. Edge point feature (x_d, y_d) is written as the Jacobian matrix $J_d = [J_{xd}^T \quad J_{yd}^T]^T$, where, J_{xd}^T and J_{yd}^T are the vectors of the Jacobian matrix of the image features (x_d, y_d), which are the feature vectors of the edge point.

Stacking image Jacobian matrixs of the visual sub-system

$$\begin{aligned} J &= [J_{e1}^T J_{e2}^T J_{l1}^T J_{l2}^T J_d^T]^T \\ &= [J_{ex1}^T J_{ey1}^T J_{a1}^T J_{b1}^T J_{\theta 1}^T J_{ex2}^T J_{ey2}^T J_{a2}^T J_{b2}^T J_{\theta 2}^T J_{\rho 1}^T J_{\alpha 1}^T J_{\rho 2}^T J_{\alpha 2}^T J_{xd}^T J_{yd}^T]^T \end{aligned} \tag{16}$$

The features variation Δr_n in image space and target movement Δs_n are obtained by the target's moving tentatively near the expected position, and the initial value of the Jacobi matrix is expressed as the Eq. (17).

Where Δx_{ij}, Δy_{ij}, Δa_{ij}, Δb_{ij} and $\Delta \theta_{ij}$ (i = 1, 2; j = 1, 2, …, n) are the features variations of the target's left and right sides. $\Delta \rho_{ij}$ and Δa_{ij} (i = 1, 2; j = 1, 2, …, n) are the features variations of the line. Δx_{dj} and Δy_{dj} (i = 1, 2) are the central variations of the target's edge. $\Delta t_{xi,}$, $\Delta t_{yi,}$, $\Delta t_{zi,}$, $\Delta\theta_{xi}$, $\Delta\theta_{yi}$ and $\Delta\theta_{zi}$ ($i = 1, 2$) are the target's position variations in Cartesian space. Δs_n and Δr_n are the matrixs of $16 \times n$ and $6 \times n$, respectively. In order to ensure the stability of the control system, the dimension in feature space must be equal or greater than this in the motion space of the target. Δr_n is prone to singularity when target moves near the expected position, which makes target to deviate from the anticipated place and causes the control to fail. Therefore, the threshold is set at the beginning of the experiment. When the conditional number of the matrix is greater than the threshold, image feature variations are extremely small and the image Jacobian matrix exist the singularity.

2.3 The Estimation of Dynamic Residual Error

In the formula (8), If S is ignored when the dynamic residuals S is greater than $\widehat{J}_k^T\widehat{J}_k$, it will not only affect the response of the positioning robot, but also make the target not converge to the desired position. So the method estimating the value of S is proposed. By the definition of the S, the equation is gained as:

$$S_k = \frac{\partial J^T}{\partial p} f = \sum_{i=1}^{m} \frac{\partial^2 f(q_k, t_k)^T}{\partial q^2} f(q_k, t_k) \tag{17}$$

With the formula (19),

$$\sum_{i=1}^{m} \frac{\partial J_k^T}{\partial p} = \sum_{i=1}^{m} \frac{\frac{\partial f(q_k, t_k)^T}{\partial q} - \frac{\partial f(q_{k-1}, t_{k-1})^T}{\partial q}}{q_k - q_{k-1}} \tag{18}$$

We can get the formula as,

$$S_k = \sum_{i=1}^{m} \frac{\frac{\partial f(q_k, t_k)^T}{\partial q} - \frac{\partial f(q_{k-1}, t_{k-1})^T}{\partial q}}{q_k - q_{k-1}} f(q_k, t_k) \tag{19}$$

Multiplied the variation of Δq and transposed the formula (20), we can obtained the solution of the minimum norm.

$$S_k = \frac{(J_k - J_{k-1})^T f_k \Delta q^T}{\Delta q^T \Delta q} \tag{20}$$

The whole steps of the uncalibrated visual servo control strategy in the trinocular-microscope are as following:

Firstly, the value of q_0, q_1, p_0 and $\widehat{J}_0$ are acquired, where, q_0, q_1,$\in R^n$; $\widehat{J}_0 \in R^{m\times n}$; $P_0 \in R^{n\times m}$ and the initial value of the λ is set up, and its range is $(0, 1]$

Secondly, through the iterative Eq. (22), motion control of the positioning robot's end is got and the position is adjusted in real time by the iterative equations .

$$\begin{aligned}
&\Delta f = f_k - f_{k-1}, k = 1, 2, \ldots \\
&\frac{\partial f_k}{\partial t} h_t = -(y_k^* - y_{k-1}^*) \\
&h_q = q_{k+1} - q_k \\
&\widehat{J}_k = \widehat{J}_{k-1} + \frac{(\Delta f - \widehat{J}_{k-1} h_q) h_q^T P_{k-1}}{\lambda + h_q^T P_{k-1} h_q} \\
&P_k = \frac{1}{\lambda}(P_{k-1} - \frac{P_{k-1} h_q h_q^T P_{k-1}}{\lambda + h_q^T P_{k-1} h_q}) \\
&S = \frac{(\widehat{J}_k - \widehat{J}_{k-1})^T f_k h_q^T}{h_q^T h_q} \\
&q_{k+1} = q_k - (\widehat{J}_k^T \widehat{J}_k + S)^{-1} \widehat{J}_k^T (f_k + \frac{\partial f_k}{\partial t} h_t)
\end{aligned} \tag{21}$$

3 Experimental Results and Analysis

The experimental platform of the uncalibrated visual servo control strategy in the trinocular-microscope is shown in Fig. 1, where #1 and #3 microscopic visual are taken to detect and track the elliptical features of the target, and #2 microscopic visual is used to detect and track image features of lines and edge of target's sides. Because the robots have the repeated positioning accuracy of 2 μm, 2 μm and 2 μm along the x, y and z axis, and each has the repeated accuracy of 3μrad, 3μrad and 15μrad revolving round the x, y and z axis. Hence, the positioning robot has the high accuracy of the position, so it is viable to use these dates which gained from the robot's controller to evaluate the advantages and disadvantages of the algorithm. Because positioning robot's basic has the characteristics of the large moving range and low frequency. Therefore, in the experimental system, except supporting the robot and target, the positioning robot's basic can also be used as the interference source, which produces the disturbance to disturb the robot's positioning.

Using errors between the current image feature and excepted feature and pseudo-inverse matrix of the Jacobian, the target's pose errors ($\Delta t_{x,}$, Δt_y, Δt_z, $\Delta t_{u,}$, Δt_v, Δt_w) can be got. The control strategy of discrete incremental PI is adopted to gain the robot's outputs, which can control the movement of the target. Figure 2 shows the diagram of the uncalibrated visual servo control system.

Considering the factors such as image acquisition, feature extraction, robot's calculation of controlled variations and system delay, the cycle of sampling and calculation is set as 400 ms.

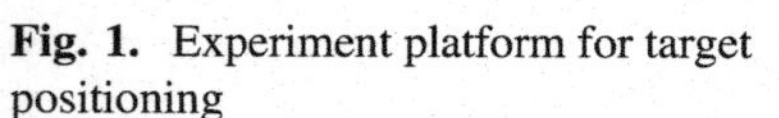

Fig. 1. Experiment platform for target positioning

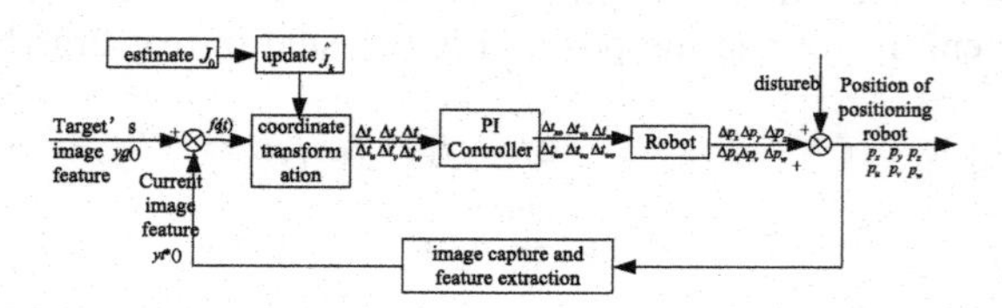

Fig. 2. Diagram of the visual servo control

3.1 Positioning Results and Analysis of the Target

The target's initial position is set at (0.7 0.3 0.1 0° 0° 0°). Since the target's both sides are elliptical and the ellipses have no-observed features, it is difficult to detect its rotation round the axis. Therefore, we set the target's expected position as (−0.6 −0.1 0.3 0.3° 0° −0.6°), which means the target's moving at five free degree freedoms in the Cartesian space, and the point to point movement is taken as the positioning robot's mode of motion.

Drive the positioning robot to move, we can make the target to move nearly the desired position by tentative way off line and obtain the initial value of Jacobian matrix. When the error between current position and desired position is less than 10 μm and the angle error is less than 0.02°, the movement of the positioning robot is finished and the initial value of Jacobian matrix is showed by equation. However, the initial value of the Jacobian matrix needs to be recalculated when the beginning pose of the positioning robot is changed.

When the positioning robot's basic is stationary, the robot's PI parameters are set as {0.07, 0.50}, {0.65, 0.46}, {0.06, 0.42}, {0.04, 0.45} and {0.08, 0.33}. After the experimental test, the value of λ is 0.03 when the position error is more than 0.01 mm and angle error is greater than 0.02°. At the same time, the value of λ is 0.99 when the position error is less than 0.01 mm and angle error is under 0.02°. The target converges to the desired position without adjustment of the dynamic residuals and its image frame is showed in Fig. 3.

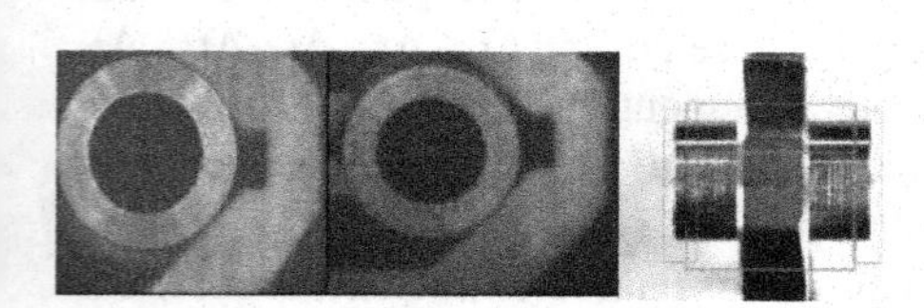

Fig. 3. Image frame of the target expected position without dynamic residual error

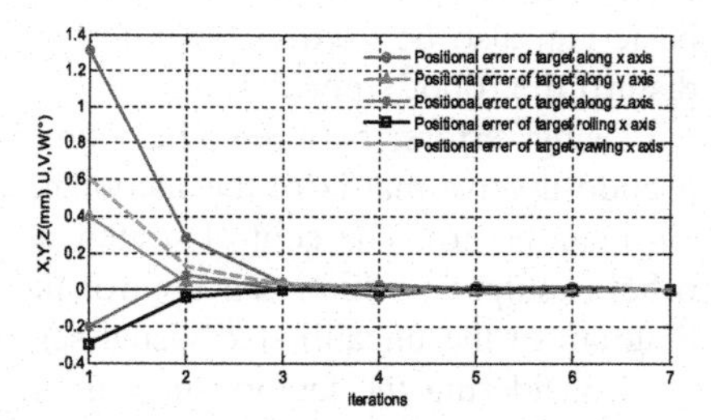

Fig. 4. Variety curve of the target's position error without distraction

Figure 4 shows the variety curve of the target's pose error. In the figure, it can be seen that the target needs 6 steps to reach the excepted pose, and the actual position read can be gained from the controller and the value of which is (−0.581 −0.092 0.285 0.295° 0° −0.6020°)

The positioning robot is disturbed when the basic of the positioning robot is moving along the x, y and z axis. During the process of the target moving toward to the anticipated location, the variety curve of the target's pose error is shown in Fig. 5 and the curve is not converged. From the figure, it is concluded that the positioning robot cannot complete the task of precise positioning when the positioning robot is disturbed without the adjustment of the dynamic residuals.

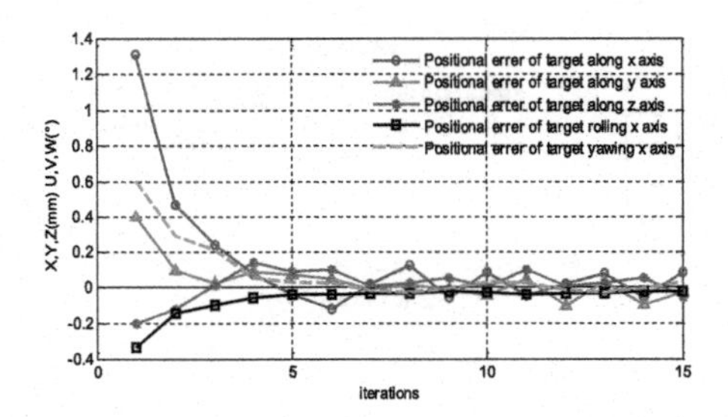

Fig. 5. Variety curve of the target's position error with the distraction

Therefore, the adjustment of dynamic residual error is needed when the basic of positioning robot moved. Then the robot's PI parameters are set as {0.45, 0.06}, {0.33, 0.10}, {0.40, 0.06}, {0.41, 0.08}, {0.42, 0.04}. After several tests, the value of λ is 0.25 when the position error is more than 0.01 mm and angle error is greater than 0.02°. Meanwhile, the value of λ is 0.97 when the position error is less than 0.01 mm and angle error is under 0.02°. In the Fig. 6, the image frame of target converges to the desired position is shown.

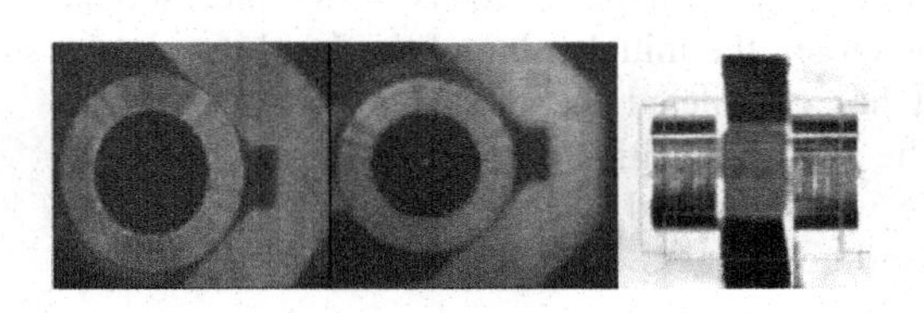

Fig. 6. Image frame of the target expected position the dynamic residual error position error

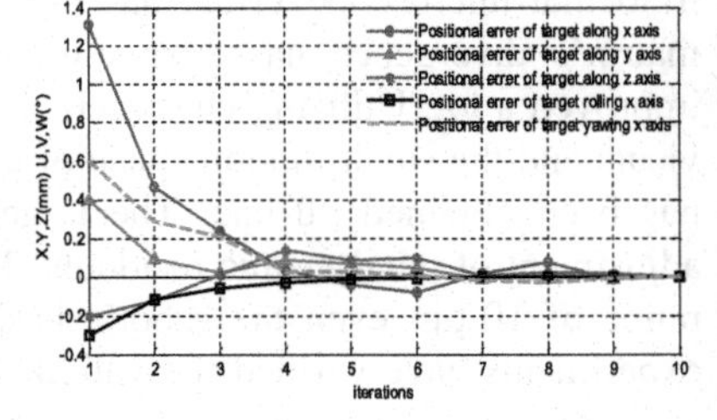

Fig. 7. Variety curve of the target's with the dynamic residual error

Figure 7 shows the variety curve of the target's pose error. From the figure, it can be seen that the target needs 10 steps to reach the excepted pose, and gets the actual position which is read from the controller and the value is (−0.600 −0.083 0.298 0.289° 0° −0.594°). In addition, it also shows that the target can converge to desired position quickly by adjusting the dynamic residuals when the basic of the robot is disturbed.

The comparison of target's trajectory between the poisoning robot with disturbance and without noise is shown in Fig. 8. From the figure, it is possible that the trinocular-microscope visual servo algorithm based on the Dynamic Quasi-Newton can realize the target's positioning precise when the robot is disturbed.

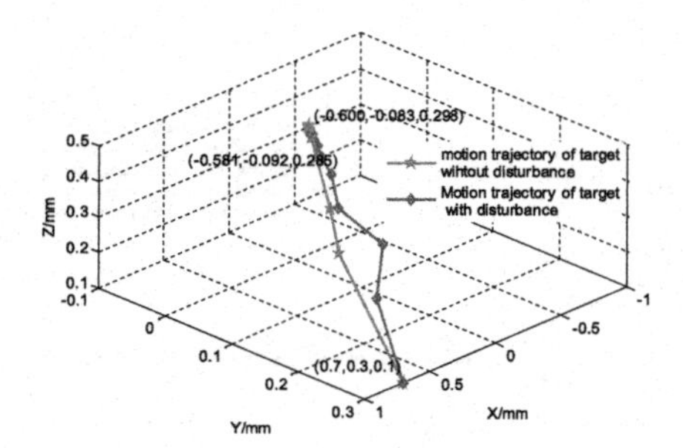

Fig. 8. Comparison of the target's motion track between the poisoning robot with disturbance and without noise

4 Conclusion

Considering the parameters of camera could not be calibrated accurately in the trinocular-microscope visual and kinematics model of the positioning robot is difficult to establish, the high-precision positioning method based on uncalibrated visual servo control strategy has been proposed, which has achieved the goal of the target's exact position. Through the dynamic Quasi-Newton algorithm, Jacobian matrix of the trinocular-microscope visual has been acquired. And with the least squares method, matrix's divergence has been eliminated so the stability of the system has been improved a lot. Utilizing characteristics of the short depth of focus in the microscopic visual and the small motion space of the target, the initial value of the Jacobian matrix has been obtained off-line. The target has completed the precision position by the adjustment of the dynamic residuals. The target's position can also achieve less in the range of 10 μm even the robot's basic was disturbed. At the end of the paper, the experiments have verified the validity and correctness of the method.

Acknowledgment. This work was supported by a grant from the Joint Funding Project of Beijing Municipal Commission of Education and Beijing Natural Science Fund Committee (201710015010).

References

1. Chaumette, F., Hutchinson, S.: Visual servo control i: basic approaches. J. IEEE Robot. Autom. Mag. **13**(4), 82–90 (2016)
2. Chaumette, F., Hutchinson, S.: Visual servo control ii: advanced approaches. J. IEEE Robot. Autom. Mag. **14**(1), 109–118 (2017)
3. Wang, L., Mills, J.K.: Automatic microassembly using visual servo control. J IEEE Trans. Electron. Packag. Manuf. **31**(4), 316–325 (2008)
4. Zhang, Y.N., Chen, X.: Visual measurement of micro-target pose in ICF experiment. J. Opto-Electron. Eng. **39**(5), 18–24 (2012)
5. Tamadazte, B., Thomas, A.: Automatic micromanipulation using multiscale visual servoing. In: Proceedings of IEEE Conference of Advanced Intelligent Mechatronics, pp. 88–93 (2009)
6. Wang, H., Jiang, M., et al.: Visual servoing of robots with uncalibrated robot and camera parameters. J. Mechatron. **22**(6), 661–668 (2012)
7. Kosmopoulos, D.I.: Robust Jacobian matrix estimation for image-based visual servoing. J. Robot. Comput. Integr. Manuf. **27**(1), 82–87 (2011)
8. Liu, M.: Research of uncalibrated visaual servoing robot. D. Master dissertion of Yanshan University, pp. 30–39 (2011)
9. Kim, G.W., Lee, B.H.: Uncalibrated visual vervoing technique using large residual. In: International Conference on Robotics and Automation, pp. 3315–3320 (2003)
10. Piepmeier, J.A., McMurray, G.V.: Uncalibrated dynamic visual servoing. J. IEEE Trans. Robot. Autom. **20**, 143–147 (2004)
11. Zhao, J., Li, M., Li, G.: Study on uncalibrated visual servoing technique. J. Control Decis. **21**(9), 1015–1020 (2006)
12. Li, M.: Research on robot uncalibrated visual servoing key technique. D. Ph.D. dissertion of Harbin Institute of Technology, p. 59-9 (2008)
13. Malis, E.: Visual servoing invarant to changes in camera-intrinsic parameters. J. IEEE Trans. Robot. Autom. **20**(1), 72–81 (2004)

Automatic Multi-view Action Recognition with Robust Features

Kuang-Pen Chou[1], Mukesh Prasad[2(✉)], Dong-Lin Li[4], Neha Bharill[3], Yu-Feng Lin[1], Farookh Hussain[2], Chin-Teng Lin[2], and Wen-Chieh Lin[1]

[1] Department of Computer Science, National Chiao Tung University, Hsinchu, Taiwan
kpchou.cs00g@g2.nctu.edu.tw, linyf915@gmail.com, wclin@cs.nctu.edu.tw

[2] Centre for Artificial Intelligence, School of Software, FEIT, University of Technology, Sydney, Australia
{mukesh.prasad, Farookh.Hussain, Chin-Teng.Lin}@uts.edu.au

[3] Department of Computer Science and Engineering, Birla Institute of Technology and Science, Pilani, Hyderabad, India
bharillneha@gmail.com

[4] Department of Electrical Engineering, National Chiao Tung University, Hsinchu, Taiwan
lazybones000@yahoo.com.tw

Abstract. This paper proposes view-invariant features to address multi-view action recognition for different actions performed in different views. The view-invariant features are obtained from clouds of varying temporal scale by extracting holistic features, which are modeled to explicitly take advantage of the global, spatial and temporal distribution of interest points. The proposed view-invariant features are highly discriminative and robust for recognizing actions as the view changes. This paper proposes a mechanism for real world application which can follow the actions of a person in a video based on image sequences and can separate these actions according to given training data. Using the proposed mechanism, the beginning and ending of an action sequence can be labeled automatically without the need for manual setting. It is not necessary in the proposed approach to re-train the system if there are changes in scenario, which means the trained database can be applied in a wide variety of environments. The experiment results show that the proposed approach outperforms existing methods on KTH and WEIZMANN datasets.

Keywords: Action recognition · Feature extraction · Background subtraction · Classification · Tracking

1 Introduction

Recently, human action recognition research brings many challenges in the areas of sports, securities and personal health care system. There is a need to develop an automatic video analysis, which can recognize the events related to human actions.

D. Liu et al. (Eds.): ICONIP 2017, Part III, LNCS 10636, pp. 554–563, 2017.
https://doi.org/10.1007/978-3-319-70090-8_56

Human action recognition is one of the most active research areas in computer vision and has many real-world applications, such as searching for the structure of large video archives, gesture recognition, video indexing, and video surveillance. Human computer interaction, in particular, is one of the most significant applications for action recognition. Visual cues are the most important mode of nonverbal communication, and the effective utilization of this mode to identify gestures and actions holds the promise of creating computers that can better interact with humans. Most of the current action recognition methods resort to sampling an action sequence manually before it is recognized in a film. However, it is not practical that setting the beginning and ending of an action sequence of the film previously. Therefore, a practical recognition system needs to be able to automatically separate many actions in an image sequence.

The action recognition based on optical flow [1, 2], tracking [3–6] and spatiotemporal shape template [7–9] has been proposed previously. The computation of optical flow helps to construct action templates for flow and tracking based approaches. However, at the boundary of the segmented human body, the features are more sensitive to noise, which are extracted from the flow templates. The action recognition problem treated as 3D object recognition by spatiotemporal shape template approaches [7–9]. These approaches require the extraction of highly detailed silhouettes, which may not be possible when there is real-world noisy video input. Further, recognition rate with 100% has been shown on the WEIZMAN dataset [7], however, these approaches does not work properly on a noisy dataset such as the KTH dataset [10]. The KTH dataset contains features like as low resolution, strong shadows, and camera movement, which renders clean silhouette extraction impossible. Space-time interest point-based approaches have become increasingly popular for addressing this problem. Further, the 2D SIFT descriptors [12] is extended into 3D with addition of dimension to the histogram orientation by Scovanner et al. [11]. Due to encoded temporal information, the extended 3D descriptors performs better than 2D descriptors in action recognition. Furthermore, Willems et al. [13] proposed the spatiotemporal domain which is extension of the SURF descriptor. Schuldt et al. [10] and Dollar et al. [14] described sparse spatiotemporal features to deal with the complexity of human actions recognition [16, 17]. Schuldt et al. [10] proposed the representation of action using 3D spacetime interest points detected from video. Schuldt also adopted the codebook and bag-of-words approach, which is often applied to object recognition, to produce a histogram of informative words for each action. The detected points are clustered to form a dictionary of prototypes or video-words. Similarly, Dollar et al. [14] introduced a multidimensional linear filter detector, which results in the detection of denser interest points. Dollar took the bag-of-words approach but argued for an even sparser sampling of the interest points. Niebles and Fei-Fei [15] introduced hierarchical modeling, which can be characterized as a constellation of bags of words and provides better performance. The approaches [10, 14, 15] represents Bag of Words (BOW) feature, which are adopted successfully for 2D object categorization and recognition. Compared with object tracking and spatiotemporal shape based approaches, which are robust to handle noise, moving camera, and very low quality resolution input. Moreover, these approaches mainly focused on individual local spacetime descriptors rather than global space time descriptors. The proposed approach does not requires any specific parameter tuning for data processing and it explicitly exploits spatiotemporal information at multiple temporal scales. Therefore, the proposed approach is able to capture local and global temporal information as well, for interesting points of distribution.

The proposed approach labels the beginning and end of the action sequence automatically. In addition, the proposed approach exploits only the global spatiotemporal information about where the interest points are and when they are detected. Therefore it is able to capture smooth motions and occlusions at low computational cost. In particular, the proposed approach uses view-invariant features to address multi-view action recognition from a range of perspectives.

2 Feature Description

This section introduces the feature vectors of the proposed system, which are described by the location of moving object and interest points. Sections 2.1 and 2.2 illustrate the boxfeature and cloud feature respectively, and Sect. 2.3 describes the quantization for reducing the dimension of the feature vectors of the proposed system.

2.1 Box Feature

The first set of features concerning the shape and speed of the foreground object are global and holistic. The Prewitt edge detector [19] is used for foreground detection to segment the object. After segmentation, mainly two features consider for further process; B_t^r represents the ratio of the height and width of the object, and B_t^{Sp} measures the absolute speed of the object, which is normalized by the height of the object for scale invariance. There is one B_t^r feature and one B_t^{Sp} feature for each image frame I_t.

2.2 Cloud Feature

The special information about gesture of human pose is preserved by the interest points. To make use of the temporal dependency between frames, a point cloud is formed by accumulating the points of interest extracted from a set of consecutive frames [18]. The spatiotemporal representation for human actions is provided by interest point cloud.

Let us consider an action video sequence A consists of T image frames, $A = [I_1, \ldots I_t, \ldots, I_T]$, where I_t is the t-th image frame. Now consider I_t denotes the current image frame and N_s denotes the size of a temporal scale. The sets of the previous K cumulative scales are defined as $[I_{t-N_s}, \ldots, I_t]$, $[I_{t-2\times N_s}, \ldots, I_t]$, $\ldots, [I_{t-K\times N_s}, \ldots, I_t]$. The image frame I_t, a total of K interest point clouds of different temporal scales are formed. The point clouds are denoted as $\left[C_t^1, \ldots, C_t^S, \ldots, C_t^K\right]$ as shown in Fig. 1. The cloud of the S-th scale is constructed by accumulating the interest points detected over the previous $S \times Ns$ frames.

Other sets of features are called cloud features, which are extracted from the interest point clouds of different scales. The cloud features are scale dependent. There are total

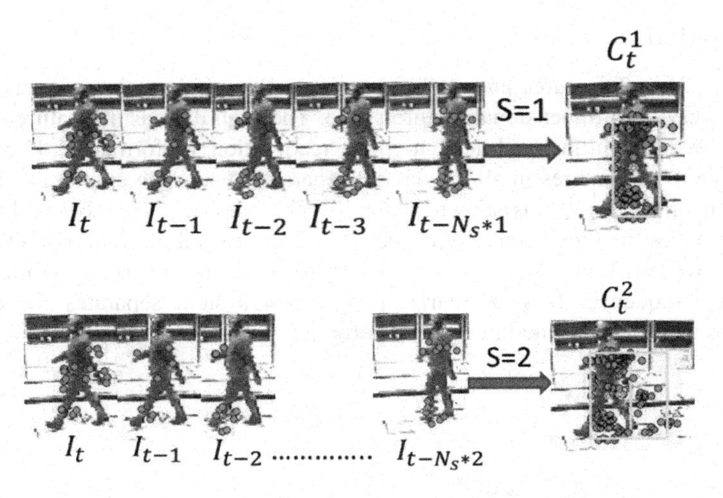

Fig. 1. Cloud for different temporal scale S

eight features computed from the s-th scale cloud. The representation of s-th scale cloud is as follows:

$$\left[C_s^r, C_s^{Sp}, C_s^{Vd}, C_s^{Hd}, C_s^{Hr}, C_s^{Wr}\right]$$

Since each image frame has temporal scales (i.e. interest point clouds in a frame), there are 6 features from the interest point clouds in total. There are also two other features from the foreground area. As a result, now the representation of each frame becomes with $6S + 2$ features, where S is the total number of scales (i.e. 6 features for each scale along with 2 scale-independent features B_t^r and B_t^{Sp}). The overview of the features of the proposed approach are shown in Fig. 2.

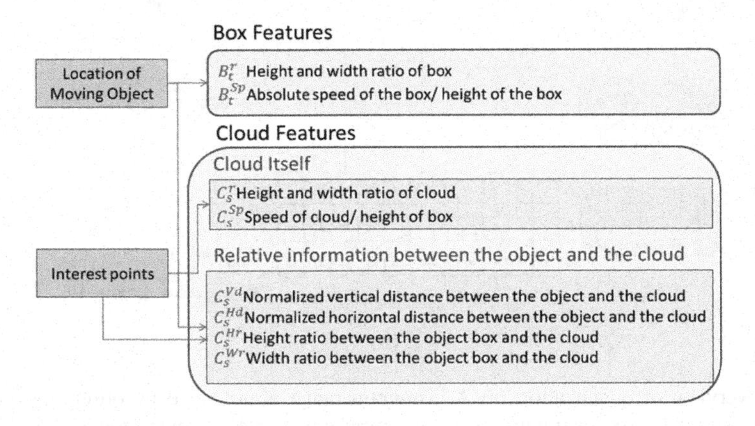

Fig. 2. Overview of the features of the proposed approach

2.3 Quantization

Total $(6S + 2) \times T$ features are used to represent the whole action sequence, which leads to a very high-dimensional feature space. The high dimension feature space can be caused for overfitting and lead to poor recognition performance. If $S = 6$, we observe one of all features in all the dataset separately using the empirical cumulative distribution function [20], as shown in Fig. 3. The empirical cumulative distribution function reduces the dimensionality of the feature space, and more importantly, makes the system representation less sensitive to feature noise and invariant to duration T of each action sequence. In particular, the proposed system separates the empirical cumulative distribution function into N_b portions.

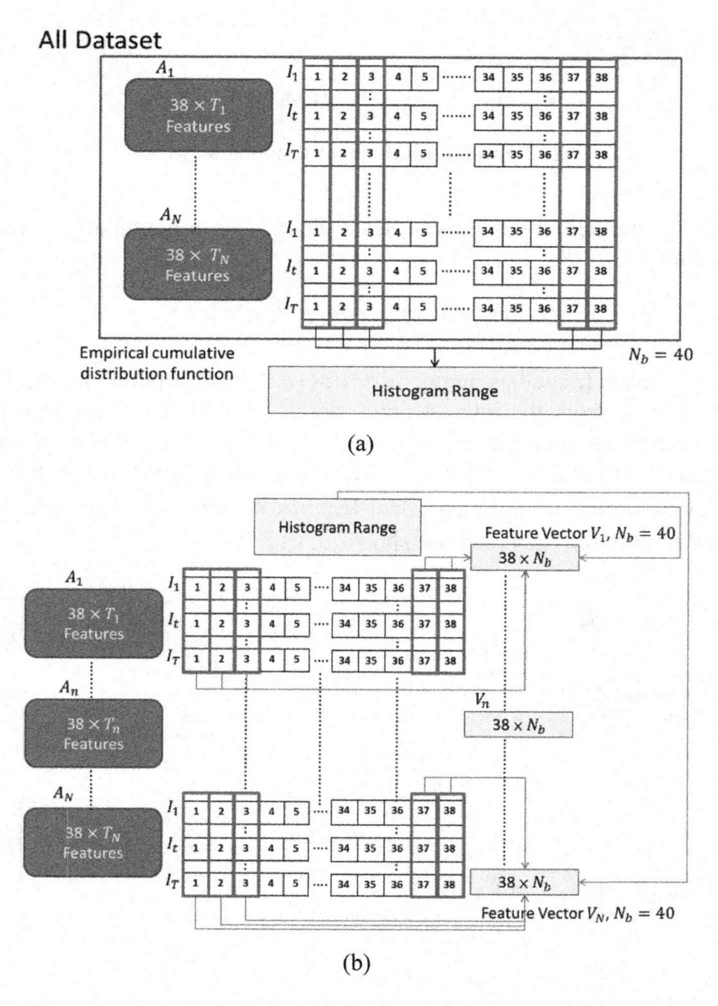

Fig. 3. Overview of quantization (a) A histogram range is produced by observing one of all features in all the dataset separately. (b) Each action sequence A is represented as $(6S + 2) \times N_b$ features.

3 Feature Reduction and Classification

In offline training, the proposed system stores quantized feature vectors, which are described in Sect. 2.3. In online testing, the proposed system uses the histogram range of the training database and transforms the testing data A_{test} to a feature vector V_{test}. Three classifiers are separately used to recognize the testing data for different recognition rates. Figure 4 shows the overview of feature reduction and classification.

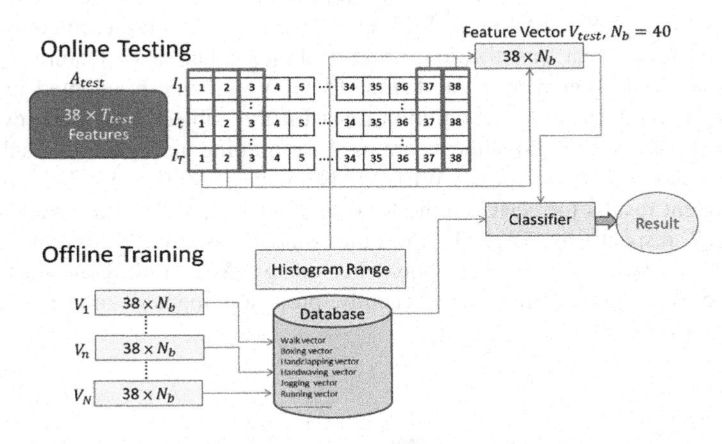

Fig. 4. Overview of feature reduction and classification

The performance of action recognition rate is calculated by using Nearest Mean Classifier (NMC). NMC helps to obtain the result using the minimum distance between the testing vector and training vector, which is the mean value of the feature vectors of the same action and the same view. Lastly, an absolute distance is chosen for the recognition decision, as shown in Fig. 5. We find that NMC is more suitable for the proposed system for real-time recognition and has a better recognition rate. Moreover, the dimension of the subject is reduced to one, which improves performance and makes recognition more efficient.

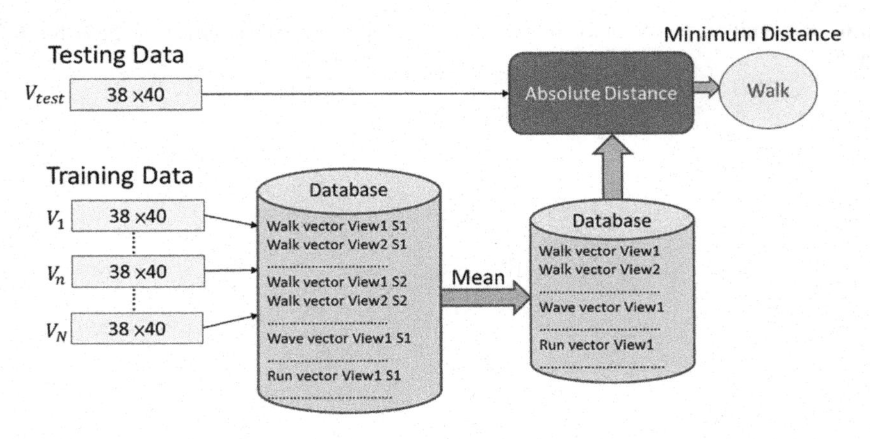

Fig. 5. Overview of NMC for our work

4 Experiment Result

This section presents the performance of the proposed system on two publicly available databases, KTH and WEIZMANN. The proposed system is implemented on Matlab 2010 in a Windows 7 environment with Intel Core *i5* 3.3 GHz and 8 GB RAM. All the testing inputs for the experiments are uncompressed AVI video files. The video frame resolution is based on the testing datasets. To construct the multi-scale interest point clouds, *Ns* and the total number of scales are chosen as 5 and 6 respectively. This parameter setting gives a total of 38 features, each of which is represented as a 40-bin histogram through linear quantization, i.e.1520 dimensional space feature representation. The Leave-One-Out Cross-Validation (LOOCV) scheme is adopted in the proposed system to compute the recognition rates for the subject invariance evaluation. For the KTH dataset, total 24 subjects are used for training and remaining subjects for validation. The training sets for the WEIZMANN dataset contains total eight subjects. The experiment results for NMC on the KTH and WEIZMANN datasets are shown in Figs. 6 and 7, respectively. Table 1 shows the comparison of the recognition accuracy of the proposed system with other approaches using NMC on different datasets. The proposed system outperforms most recently proposed methods on the KTH and WEIZMANN datasets.

BINS = 40 RATE = 90.5797%

	boxing	handclapping	handwaving	jogging	running	walking
boxing	.95	.02	.01	.01	.00	.01
handclapping	.02	.92	.04	.00	.01	.00
handwaving	.02	.01	.97	.00	.00	.00
jogging	.00	.00	.01	.85	.08	.07
running	.00	.00	.00	.13	.87	.00
walking	.00	.00	.00	.10	.02	.88

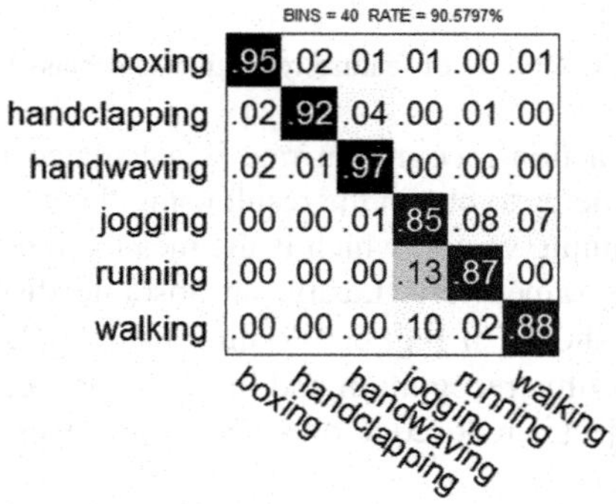

Fig. 6. Recognition performance of the proposed approach with accuracy 90.58% for KTH dataset

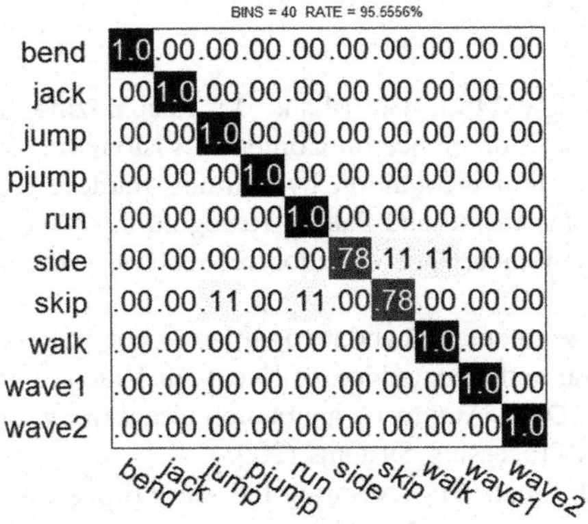

BINS = 40 RATE = 95.5556%

	bend	jack	jump	pjump	run	side	skip	walk	wave1	wave2
bend	1.0	.00	.00	.00	.00	.00	.00	.00	.00	.00
jack	.00	1.0	.00	.00	.00	.00	.00	.00	.00	.00
jump	.00	.00	1.0	.00	.00	.00	.00	.00	.00	.00
pjump	.00	.00	.00	1.0	.00	.00	.00	.00	.00	.00
run	.00	.00	.00	.00	1.0	.00	.00	.00	.00	.00
side	.00	.00	.00	.00	.00	.78	.11	.11	.00	.00
skip	.00	.00	.11	.00	.11	.00	.78	.00	.00	.00
walk	.00	.00	.00	.00	.00	.00	.00	1.0	.00	.00
wave1	.00	.00	.00	.00	.00	.00	.00	.00	1.0	.00
wave2	.00	.00	.00	.00	.00	.00	.00	.00	.00	1.0

Fig. 7. Recognition performance of the proposed approach with accuracy 95.56% for WEIZMANN dataset

Table 1. Comparison of the proposed approach with other methods on the KTH, WEIZMANN, MuHAVi datasets for subject invariance

Method	Dataset	
	KTH	WEIZMANN
Proposed Approach (NMC)	90.58%	95.56%
S. Gong et al. [18]	93.17%	96.66%
Niebles et al. [15]	83.30%	90.00%
Dollar et al. [14]	81.17%	85.20%
Zhang et al. [23]	91.33%	92.89%
Gilbert et al. [21]	89.92%	–
Savarese et al. [22]	86.83%	–
Nowozin et al. [24]	84.72%	–

5 Conclusion

This paper presents an approach for real world application which automatically labels the beginning and ending of the action sequence. The system uses the proposed view-invariant features to address multi-view action recognition from different perspectives for accurate and robust action recognition. The view-invariant features are obtained by extracting holistic features from different temporal scale clouds, which are modeled on the explicit global, spatial and temporal distribution of interest points. The experiments on the KTH and WEIZMANN datasets demonstrate that using view-invariant features obtained by extracting holistic features from clouds of interest points is highly discriminative and more robust for recognizing actions under different view changes. The experiments also show the proposed approach performs well with cross tested datasets using previously training data, that means there is no need to retrain the system if the scenario changes.

References

1. Efros, A.A., Berg, A.C., Mori, G., Malik, J.: Recognizing action at a distance. In: Ninth IEEE International Conference on Computer Vision, vol. 2, pp. 726–733 (2003)
2. Fathi, A., Mori, G.: Action recognition by learning midlevel motion features. In: IEEE Conference on Computer Vision and Pattern Recognition (2008)
3. Rao, C., Shah, M.: View-invariance in action recognition. Comput. Vis. Pattern Recognit. **2**, 316–322 (2001)
4. Ali, A., Aggarwal, J.: Segmentation and recognition of continuous human activity. In: IEEE Workshop on Detection and Recognition of Events in Video, p. 28 (2001)
5. Ramanan, D., Forsyth, D.A.: Automatic annotation of everyday movements. In: Conference on Neural Information Processing Systems (2003)
6. Sheikh, Y., Sheikh, M., Shah, M.: Exploring the space of a human action. In: International Conference on Computer Vision (2005)
7. Gorelick, L., Blank, M., Shechtman, E., Irani, M., Basri, R.: Actions as space-time shapes. Pattern Anal. Mach. Intell. **29**(12), 2247–2253 (2007)
8. Ke, Y., Sukthankar, R., Hebert, M.: Efficient visual event detection using volumetric features. In: IEEE Computer Society, Los Alamitos, CA, USA, vol. 1, pp. 166–173 (2005)
9. Yilmaz, A., Shah, M.: Actions sketch: a novel action representation. In: Computer Vision and Pattern Recognition, pp. 984–989 (2005)
10. Schuldt, C., Laptev, I., Caputo, B.: Recognizing human actions: a local SVM approach. In: International Conference on Pattern Recognition, pp. 32–36 (2004)
11. Scovanner, P., Ali, S., Shah, M.: A 3-Dimensional sift descriptor and its application to action recognition. In: International conference on Multimedia, pp. 357–360 (2007)
12. Lowe, D.: Distinctive image features from scale-invariant key-points. Int. J. Comput. Vision **20**, 91–110 (2003)
13. Willems, G., Tuytelaars, T., Van Gool, L.: An efficient dense and scale-invariant spatio-temporal interest point detector. In: European Conference on Computer Vision, vol. 2, pp. 650–663 (2008)
14. Dollar, P., Rabaud, V., Cottrell, G., Belongie, S.: Behavior recognition via sparse spatio-temporal features. In: International Conference on Computer Communications and Networks, pp. 65–72 (2005)
15. Niebles, J.C., Fei-Fei, L.: A hierarchical model of shape and appearance for human action classification. In: Computer Vision and Pattern Recognition (2007)
16. Singh, S., Velastin, S.A., Ragheb, H.: MuHAVi: a multicamera human action video dataset for the evaluation of action recognition methods. In: 2nd Workshop on Activity Monitoring by Multi-Camera Surveillance Systems (AMMCSS), 29 August, Boston, USA (2010)
17. Eweiwi, A., Cheema, S., Thurau, C., Bauckhage, C.: Temporal key poses for human action recognition. In: International Conference on Computer Vision Workshops (2011)
18. Bregonzio, M., Gong, S., Xiang, T.: Recognising action as clouds of space-time interest points. In: Computer Vision and Pattern Recognition, pp. 1948–1955 (2009)
19. Parker, J.: Algorithms for Image Processing and Computer Vision. Wiley Computer Publishing, New York (1997)
20. Shorack, G.R., Wellner, J.A.: Empirical Processes with Applications to Statistics. Wiley, New York (1986)
21. Gilbert, A., Illingworth, J., Bowden, R.: Scale invariant action recognition using compound features mined from dense spatio-temporal corners. In: Forsyth, D., Torr, P., Zisserman, A. (eds.) ECCV 2008. LNCS, vol. 5302, pp. 222–233. Springer, Heidelberg (2008). doi:10.1007/978-3-540-88682-2_18

22. Savarese, S., Pozo, A.D., Niebles, J., Fei-Fei, L.: Spatial temporal correlations for unsupervised action classification. In: IEEE Workshop on Motion and Video Computing (2008)
23. Zhang, Z., Hu, Y., Chan, S., Chia, L.-T.: Motion context: a new representation for human action recognition. In: Forsyth, D., Torr, P., Zisserman, A. (eds.) ECCV 2008. LNCS, vol. 5305, pp. 817–829. Springer, Heidelberg (2008). doi:10.1007/978-3-540-88693-8_60
24. Nowozin, S., Bakir, G.H., Tsuda, K.: Discriminative sub-sequence mining for action classification. In: International Conference on Computer Vision, pp. 1–8 (2007)

Learning Discriminative Convolutional Features for Skeletal Action Recognition

Jinhua Xu(✉), Yang Xiang, and Lizhang Hu

Shanghai Key Laboratory of Multidimensional Information Processing,
Department of Computer Science and Technology, East China Normal University,
3663 North Zhongshan Rd, Shanghai, China
jhxu@cs.ecnu.edu.cn

Abstract. Human action recognition is an important yet challenging computer vision task. With the introduction of RGB-D sensors, human body joints can be extracted with high accuracy, and skeleton-based action recognition has been investigated and gained some success. Convolutional Neural Networks (ConvNets) have been proved to be the most effective representation learning method for visual recognition tasks, but have not been applied to skeletal action recognition due to the lack of a big dataset. In this paper, we propose a convolutional network for skeletal action recognition. Different from the supervised training of ConvNets using backpropagation, we learn the convolutional features using projective dictionary pair learning. The advantages of our model include: First, the learned convolutional features are discriminative; Second, no big dataset is needed for training the ConvNet. Experimental results on three benchmark datasets demonstrate the effectiveness of our approach.

Keywords: Convolutional Neural Networks · Skeletal action recognition · Deep learning · Supervised dictionary learning

1 Introduction

Action recognition is an active research field in computer vision. It has many applications, including video surveillance, human-computer interaction and health care. Traditional research mainly concentrates on action recognition from video sequences of 2D frames with RGB channels [1], which only capture projective information of the real world and are sensitive to lighting conditions and occlusion. Recently, with the introduction of the low-cost RGB-D sensors such as Microsoft Kinect [2], depth information has been employed for action recognition [3,4].

There are two kinds of approaches for human action recognition based on the depth information, depth map-based approaches and skeleton-based approaches [4]. The former relies mainly on features that extracted from the space time volume (STV). The latter has been explored since the early work by Johansson [5] and attracted an increasing attention due to their robustness to variations of

D. Liu et al. (Eds.): ICONIP 2017, Part III, LNCS 10636, pp. 564–574, 2017.
https://doi.org/10.1007/978-3-319-70090-8_57

viewpoint, human body scale and motion speed as well as the realtime performance. (see [6] for a review). Our proposed approach is skeleton-based.

Deep learning, especially Convolutional Neural Networks (ConvNets or CNNs) has gained great success in all vision tasks including image recognition, segmentation, detection and retrieval. ConvNets have also been applied to action recognition for RGB video sequences [7–10]. Usually supervised method is used to train ConvNets and a big labelled dataset is needed, for example, Sports-1M was used to train the 3D ConvNets in [8], which consists of 1 million YouTube videos belonging to a taxonomy of 487 classes of sports. Due to lack of a big skeleton action dataset, ConvNets have not been applied to skeletal action recognition. To address this problem, we propose a new discriminative convolutional feature learning approach in this work.

We make the following contributions in this paper. A convolutional network model is proposed for skeletal action recognition and the features are learned through projective dictionary pair learning (PDPL), instead of the traditional back propagation. In this way, no big dataset is needed. Meanwhile, the convolutional features are discriminative since they are learned using supervised dictionary learning. Experiments on three benchmark datasets demonstrate the effectiveness of our approach and our results are better than or comparable to the prior models.

The remainder of this paper is organized as follows. In Sect. 2, related work is introduced. In Sect. 3, we describe the model architecture and the projective dictionary pair learning for convolutional feature learning. In Sect. 4, the experimental results of our method on three datasets are reported and compared to other methods. In Sect. 5, we briefly summarize the paper.

2 Related Work

Some hand-crafted mid-level features have been proposed for skeletal action recognition. In [11], a moving pose descriptor was proposed that considers both pose information as well as differential quantities (speed and acceleration) of the human body joints within a short time window around the current frame. In [12], an EigenJoints descriptor was proposed based on position differences of joints, which combines action information including static posture in the current frame, motion between the current frame and the preceding frame, and offset between the current frame and the initial frame. In [13], an actionlet ensemble model was proposed. Two types of frame-level features were utilized including the 3D joint position feature and depth appearance feature. In [14], data mining techniques were applied to obtain a representation for the spatial-temporal structures of human actions. This representation captures the spatial configurations of body parts in one frame by spatial-part-sets as well as the body part movements by temporal-part-sets. In [15], a feature called Sequence of the most informative joints (SMIJ) was proposed to represent an action. SMIJ is the ordered set of the most informative joints in each temporal window. The informative joints were selected based on highly interpretable measures such as the mean or variance of joint angles, maximum angular velocity of joints etc. In [16], a spatio-temporal hierarchy of skeletal configurations was proposed, where each configuration represents the motion of a set of joints at a particular temporal scale. Each of these skeletal configurations was modeled as a Linear Dynamical System (LDS) and the entire human activity was represented as a hierarchical set of LDSs.

Data-driven feature learning techniques such as ConvNets have been applied to RGB video-based action recognition. Among them, two different kinds of ConvNets are used. One is the frame based 2D ConvNets [9,17–19], where ConvNets are employed for feature learning at each frame and the temporal dynamics are modelled by using Temporal Pyramids, Hidden Markov Models (HMMs) or recurrent neural networks (RNNs). The other kind is the 3D ConvNets where temporal dimension is processed in the same way as the spatial dimensions and 3D convolutions are performed, thereby capturing the motion information encoded in multiple adjacent frames [7,8,10].

Some skeletal feature learning methods have also been proposed for action recognition [20–22]. In [20], a hierarchical compositional model was proposed that operates at three semantic levels. In [21], a hierarchical RNN model was proposed for skeleton based action recognition. The whole skeleton is divided into five parts, which are fed into five bidirectional recurrent neural networks (BRNNs). As the number of layers increases, the representations extracted by the subnets are hierarchically fused to be the inputs of higher layers. A fully connected layer and a softmax layer are performed on the final representation to classify the actions. In [22], a body-part motion based feature called Moving Poselet was proposed, which corresponds to a specific body part configuration undergoing a specific movement. Temporal pyramid pooling was then performed over the response maps of each body parts, and the pooled features from all body parts were concatenated to form the global feature of the action and fed into a classifier for recognition. Moving Poselets and action classifiers were jointly learned using the stochastic gradient descent algorithm.

Our work is most related to Moving Poselets [22]. The main difference is the training strategy. Instead of jointly learning the features and the classifier using back propagation, we learn them separately. We learn the discriminative convolutional features using projective dictionary pair learning [23], a supervised dictionary learning approach. In this way, the convolutional features can be learned layer by layer. Since the number of parameters in one layer is much less than that of the whole network, no big training dataset is needed. And the convolutional features are more discriminative due to the supervised dictionary learning.

3 Learning Discriminative Convolutional Features

Our proposed model is an 1D ConvNet in which convolutions are performed along the time dimension, as shown in Fig. 1. The kernel size represents the length of time slot and the channel number is equivalent to the dimension of input data. The abs activation function is used after the convolutional layer instead of the rectified linear unit (ReLU) because the positive response and negative response play similar roles in our dictionary learning. We have a temporal pooling layer after the convolutional layer to implement the temporal invariance. Since the number of frames (T) is various in videos, the pooling size is not fixed. We split a video into a fixed number of segments and pool in each segment. In this way, the output of the pooling layer is a fixed length vector, which will be fed to SVM for classification.

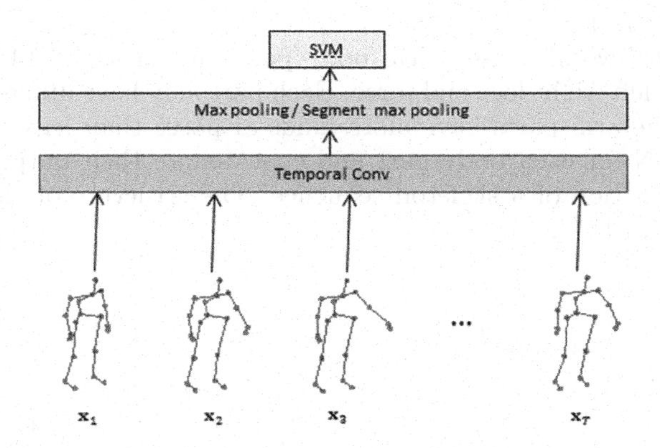

Fig. 1. The proposed CNN model

Convolutional networks are generally trained in a supervised way using the stochastic gradient descent (SGD) algorithm. The drawbacks of SGD include local minimum problem and slow convergence. Here we learn the convolutional features and the classifier separately. Specifically, we train the convolutional filters using projective dictionary pair learning (PDPL) method. Then linear SVM is trained for classification.

Let L be the kernel size of the convolutional filters. We sample the joints' 3D positions (x, y, z) from L consecutive frames densely in the training skeleton sequences. For a skeleton sequence with T frames, we will get $T - L + 1$ samples. If the number of joints is J, each sample is a $p(= 3 * J * L)$ dimensional vector, $\mathbf{x} \in \mathbf{R}^{p}$. We denote all samples from the k action class as X_k, and all training samples as $X = [X_1 \cdots X_k \cdots X_C]$, where C is the number of classes. Now we introduce how to learn the convolutional filters using PDPL method.

In [23], a projective dictionary pair learning method was proposed, which learns a synthesis dictionary and an analysis dictionary jointly. The analysis dictionary is composed of C sub-dictionaries, i.e., $P = [P_1^T, \cdots, P_k^T, \cdots, P_C^T]^T$. Similarly the synthesis dictionary $D = [D_1, \cdots, D_k, \cdots, D_C]$. $D_k \in \mathrm{R}^{p \times m_k}$, $P_k \in \mathrm{R}^{m_k \times p}$ form a sub-dictionary pair corresponding to the kth class, and m_k is the sub-dictionary size. The PDPL model was formulated as below:

$$< P^*, D^* >= arg \min_{P,D} \|X - DPX\|_F^2 + \lambda \sum_k \|P_k \bar{X}_k\|_F^2 \quad (1)$$

where $\bar{X}_k$ denotes the complementary data matrix of X_k in the whole training set X. The regularization term in (1) is to make the analysis sub-dictionary P_k generate significant coding coefficients for samples from class k and small coefficients for samples from other classes. Since each sample is extracted from L consecutive frames, the analysis dictionary P can be used as our convolutional filters. The convolutional features are obtained by PX. From the above analysis, the convolutional features are discriminative, that is, they are non-zeros for the filters from their own class, and are close to zeros for the filters from other classes, as will be shown in Fig. 3.

In this paper, we also use a five body parts model as in [14,21] including left/right arms, left/right legs and torso. Each part may have different dictionary size m_k since the arm parts have more kinds of poses than legs and torso. We construct a CNN for each body part and concatenate their outputs to get the feature representation of a skeleton sequence. The structure of the body parts model is shown in Fig. 2.

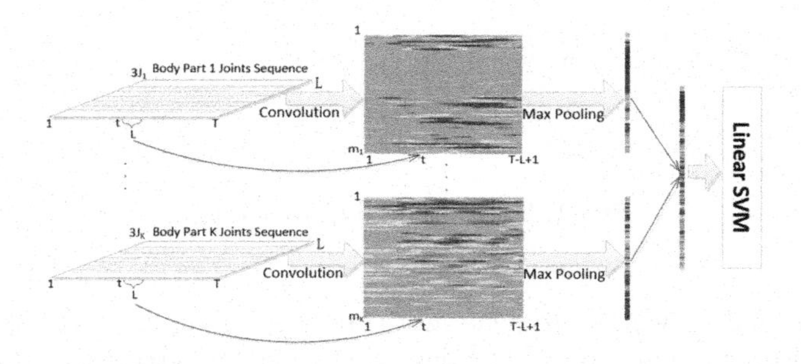

Fig. 2. The architecture of the body parts model

4 Experimental Results

We tested our method on three publicly available datasets: MSR Action3D dataset [24], MSR Daily Activity3D dataset [13] and Berkeley MHAD dataset [25]. To compare our training strategy with that in [22], we also used ConvNets with only one convolutional layer in all experiments.

4.1 MSR Action3D Dataset

The MSR Action3D dataset consists of skeleton data sequences of 20 actions such as hand waving and clapping. Each action is performed 2–3 times by 10 subjects, and the 3D body joint positions of 20 joints are extracted from RGB-D videos. These action sequences are relatively short sequences with 30–50 frames, and the frame rate is 15 frames per second. The dataset is divided into three action sets, AS1, AS2 and AS3 [24]. The AS1 and AS2 group actions with similar movement, while AS3 groups complex actions together. We followed the setting in [13]. All sequences from subjects 1, 3, 5, 7 and 9 were used for training and the remaining ones for testing.

Since the videos in this dataset are short, we did not split the videos into segments during temporal pooling. This means the pooling is performed on a whole sequence. The number of filters m was set as 240, and the filter size L was set as 5. The 20 joints were grouped into 5 body parts. We found the torso part was less discriminative, and used only four body parts including left/right arms and left/right legs on this dataset. The experimental results are shown in Table 1. Our average accuracy of three subsets is 92.6%, which is better than 78.97% of HOJ3D joints in [26], 83.3% of EigenJoints in [12], 92.46% of Lie Group [27]. Our result is also better than the deep RNNs in [21], but is slightly worse than the hierarchical RNNs in [21] and the moving poselets model in [22]. It may be because ten body parts were used in moving poselets model, rather than the four body parts in our model. Meanwhile, joints' position and velocity were used as the joint feature in moving poselets, while we used only the joints' position.

Table 1. Experimental results on MSR Action3D dataset.

Method	HOJ3D [26]	EigenJoints [12]	Lie Group [27]	DURNN [21]	DBRNN [21]	HURNN [21]	HBRNN [21]	Moving poselets [22]	Ours
Accuracy	78.97	83.3	92.46	77.11	83.52	93.57	94.49	93.50	92.60

To demonstrate the discriminativeness of the convolutional features, we show the convolutional feature maps of the right arm for the actions in AS3 subset in Fig. 3. There are eight action categories in this subset including high throw, forward kick, side kick, jogging, tennis swing, tennis serve, golf swing, pickup & throw. The filters from each category are concatenated sequentially. Each row in Fig. 3 corresponds to the response of a filter. It can be seen that the responses of high throw (the 1st class) are stronger on the top part of the response map which correspond to the filters from the first class. Similarly the responses of jogging (the 4th class), tennis swing (the 5th class) and golf swing (the 7th class) are stronger to the filters of their own classes respectively.

4.2 MSR Daily Activity3D Dataset

The MSR DailyActivity3D dataset consists of 16 daily activities such as drinking and reading books. Each action is performed twice by 10 subjects, making up 320 sequences in total. This dataset has longer sequences, with 100–300 frames. The skeleton data also contains 3D positions of the same 20 joints extracted from RGB-D videos. It is more challenging than MSR Action3D, since the actions are more complex and contain human-object interactions. Also following [13], we used the sequences from subject 1, 3, 5, 7 and 9 for training, and remaining ones for testing.

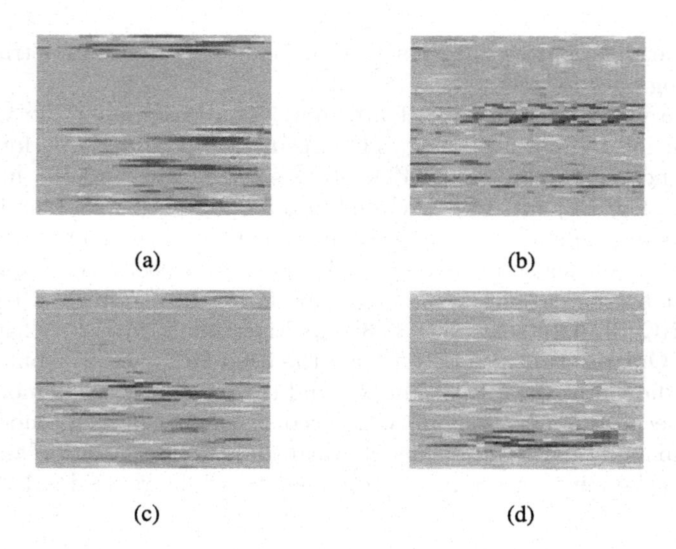

Fig. 3. Convolutional maps of actions from different classes in AS3 of MSR Action3D dataset

Since the videos in the MSR DailyActivity3D dataset are relatively long, we split the videos into six segments during temporal pooling. The number of filters m was set as 400, and the filter size L was set as 5. The 20 joints were organised into 5 body parts. We also only used four body parts excluding the torso. The accuracy of our algorithm is 75.0%, outperforming other state-of-the-art skeleton-based methods, as shown in Table 2. The best results in [13] that simultaneously used the depth information on this dataset were not compared to our method. HBRNN model was not tested on this dataset in [21] and we did not re-implement it, therefore no comparison was made with HBRNN.

Table 2. Results of our method and other methods on the MSR Daily Activity 3D dataset.

Method	Accuracy (%)
Only joint position features [13]	68.0
Actionlet ensemble on joint features [13]	74.0
Moving pose [11]	73.8
Moving poselets [22]	74.5
Ours	75.0

4.3 Berkeley MHAD Dataset

This dataset contains 11 actions performed by 12 subjects with 5 repetitions of each action, yielding a total of 659 action sequences (after excluding an erroneous sequence). The motion capture data was recorded at 480 fps and the action lengths vary from 773 to 14565 frames. Since the data has a high frame rate, we subsampled each sequence at every 16 frames so that the frame rate is reduced to 30 frames per second. Following [15,16,21,22], we used sequences from the first 7 subjects for training, and the remaining ones for testing.

In MHAD dataset, 3D positions of active LED markers were captured using optical motion capture system Impulse with the frequency of 480 Hz. Therefore, the joint positions are more accurate. We used only full body model on this less challenging dataset. The experimental results are shown in Table 3. Our model achieves 100% accuracy with a single convolutional layer. It can be seen that our model is competitive with hierarchical LDS [16], hierarchical RNN models [21] and moving poselets [22].

Table 3. Experimental results on Berkeley MHAD dataset.

Method	SMIJ [15]	Global LDS [16]	Hierarchical LDS [16]	HURNN [21]	HBRNN [21]	Moving poselets [22]	Ours
Accuracy (%)	95.37	99.27	100	99.64	100	100	100

4.4 Analysis

Since the Berkeley MHAD Dataset is less challenging and achieves 100% accuracy, we did parameter analysis only on MSR Action3D and MSR DailyActivity datasets. The two key parameters in our model are the filter kernel size and filter numbers. We selected different values for them.

Filter Kernel Size. To test the effect of the filter kernel size, we did experiments with different L on the MSR Action3D and MSR DailyActivity datasets. As shown in Table 4, the filter size has more effect on MSR DailyActivity dataset than on MSR Action3D. But we obtained the best results with L equal to 5 on both datasets.

Table 4. Recognition accuracies for different filter sizes.

L	4	5	6	7
MSR Action-3D	91.2	92.6	91.4	90.7
MSR DailyActivity	70.6	75.0	73.1	72.5

Filter Numbers. To evaluate the effect of the number of convolutional filters, we did experiments with different m on the MSR Action3D and MSR DailyActivity datasets. As shown in Table 5, the accuracies were improved when we increased the number of filters, but began to degrade after exceeding a threshold. We obtained the best result with m equal to 240 or 280 for MSR Action3D and with m equal to 400 for MSR DailyActivity dataset. Since the actions in MSR DailyActivity are more difficult to recognize than the ones in MSR Action3D, we need more convolutional filters to generate discriminative features.

In [22], totally 500 filters were used (50 filters for each body part) on all the three datasets. It can be seen that we need less filters to obtain similar results. This demonstrates that our convolutional features are more discriminative.

Table 5. Recognition accuracies for different number of filters.

MSR Action3D	m	160	200	240	280	320
	Accuracies	90.2	91.8	92.6	92.6	91.4
MSR DailyActivity	m	240	320	400	480	
	Accuracies	71.3	73.1	75.0	73.8	

Deep ConvNets. We tested our convolutional feature learning method on ConvNets with two convolutional layers, and no improvement was observed over single-layer networks. From the feature maps of the first convolutional layer shown in Fig. 3, we can see that the outputs of the filters from other classes except its own class are near zeros due to the regularization term in Eq. 1, therefore applying dictionary learning to the output of the first layer will get no more discriminability. We need to combine unsupervised dictionary learning on the lower layers with supervised dictionary learning on the top layer, but this is beyond the scope of this paper.

5 Conclusion

In this paper, we proposed a convolutional network for skeletal action recognition. Different from the supervised training of ConvNets using backpropagation, we learned the convolutional features using projective dictionary pair learning. Our model can be applied to recognition when no big training data is available. From the experimental results, it can be seen that our results were comparable to the moving Poselet [22], which was trained using back propagation.

In our future work, we will apply our method to deep ConvNets and employ different training method with different layer, for example, the lower convolutional layers can be trained using unsupervised dictionary learning such as Independent Component analysis (ICA) or sparse coding, and the higher convolutional layers using the proposed supervised dictionary training.

Acknowledgment. This work is supported by the National Natural Science Foundation of China under Project 61175116.

References

1. Aggarwal, J.K., Ryoo, M.S.: Human activity analysis: a review. ACM Comput. Surv. (CSUR) **43**(3), 16 (2011)
2. Han, J.: Enhanced computer vision with microsoft kinect sensor: a review. IEEE Trans. on Cybern. **43**(5), 1318–1334 (2013)
3. Chen, L., Wei, H., Ferryman, J.M.: A survey of human motion analysis using depth imagery. Pattern Recogn. Lett. **34**, 1995–2006 (2013)
4. Ye, M., Zhang, Q., Wang, L., Zhu, J., Yang, R., Gall, J.: A survey on human motion analysis from depth data. In: Grzegorzek, M., Theobalt, C., Koch, R., Kolb, A. (eds.) Time-of-Flight and Depth Imaging. Sensors, Algorithms, and Applications. LNCS, vol. 8200, pp. 149–187. Springer, Heidelberg (2013). doi:10.1007/978-3-642-44964-2_8
5. Johansson, G.: Visual motion perception. Sci. Am. **232**(6), 76–88 (1975)
6. Han, F., Reily, B., Hoff, W., Zhang, H.: Space-time representation of people based on 3D skeletal data: a review. Comput. Vis. Image Underst. **158**, 85–105 (2017)
7. Ji, S., Xu, W., Yang, M., Yu, K.: 3D convolutional neural networks for human action recognition. IEEE Trans. Pattern Anal. Mach. Intell. **35**(1), 221–231 (2013)
8. Karpathy, A., Toderici, G., Shetty, S., Leung, T., Sukthankar, R., Fei-Fei, L.: Large-scale video classification with convolutional neural networks. In: Proceedings of the IEEE conference on Computer Vision and Pattern Recognition, pp. 1725–1732 (2014)
9. Simonyan, K., Zisserman, A.: Two-stream convolutional networks for action recognition in videos. In: Advances in Neural Information Processing Systems, pp. 568–576 (2014)
10. Tran, D., Bourdev, L., Fergus, R., Torresani, L., Paluri, M.: Learning spatiotemporal features with 3D convolutional networks. In: Proceedings of the IEEE International Conference on Computer Vision, pp. 4489–4497 (2015)
11. Zanfir, M., Leordeanu, M., Sminchisescu, C.: The moving pose: an efficient 3D kinematics descriptor for low-latency action recognition and detection. In: IEEE International Conference on Computer Vision, pp. 2752–2759 (2013)
12. Yang, X., Tian, Y.L.: Effective 3D action recognition using eigenjoints. J. Vis. Commun. Image Represent. **25**(1), 2–11 (2014)
13. Wang, J., Liu, Z., Wu, Y., Yuan, J.: Mining actionlet ensemble for action recognition with depth cameras. In: IEEE Conference on Computer Vision and Pattern Recognition, pp. 1290–1297 (2012)
14. Wang, C., Wang, Y., Yuille, A.L.: An approach to pose-based action recognition. In: Proceedings of the IEEE Conference on Computer Vision and Pattern Recognition, pp. 915–922 (2013)
15. Ofli, F., Chaudhry, R., Kurillo, G., Vidal, R., Bajcsy, R.: Sequence of the Most Informative Joints (SMIJ): a new representation for human skeletal action recognition. J. Vis. Commun. Image Represent. **25**(1), 24–38 (2014)
16. Chaudhry, R., Ofli, F., Kurillo, G., Bajcsy, R., Vidal, R.: Bio-inspired dynamic 3D discriminative skeletal features for human action recognition. In: Proceedings of the IEEE Conference on Computer Vision and Pattern Recognition Workshops, pp. 471–478 (2013)
17. Yue-Hei Ng, J., Hausknecht, M., Vijayanarasimhan, S., Vinyals, O., Monga, R., Toderici, G.: Beyond short snippets: Deep networks for video classification. In: Proceedings of the IEEE Conference on Computer Vision and Pattern Recognition, pp. 4694–4702 (2015)

18. Mahasseni, B., Todorovic, S.: Regularizing long short term memory with 3D human-skeleton sequences for action recognition. In: Proceedings of the IEEE Conference on Computer Vision and Pattern Recognition, pp. 3054–3062 (2016)
19. Donahue, J., Anne Hendricks, L., Guadarrama, S., Rohrbach, M., Venugopalan, S., Saenko, K., Darrell, T.: Long-term recurrent convolutional networks for visual recognition and description. In: Proceedings of the IEEE Conference on Computer Vision and Pattern Recognition, pp. 2625–2634 (2015)
20. Lillo, I., Soto, A., Carlos Niebles, J.: Discriminative hierarchical modeling of spatio-temporally composable human activities. In: Proceedings of the IEEE Conference on Computer Vision and Pattern Recognition, pp. 812–819 (2014)
21. Du, Y., Wang, W., Wang, L.: Hierarchical recurrent neural network for skeleton based action recognition. In: IEEE Conference on Computer Vision and Pattern Recognition, pp. 1110–1118 (2015)
22. Tao, L., Vidal, R.: Moving poselets: a discriminative and interpretable skeletal motion representation for action recognition. In: Proceedings of the IEEE International Conference on Computer Vision Workshops, pp. 61–69 (2015)
23. Gu, S., Zhang, L., Zuo, W., Feng, X.: Projective dictionary pair learning for pattern classification. In: Advances in Neural Information Processing Systems, pp. 793–801 (2014)
24. Li, W., Zhang, Z., Liu, Z.: Action recognition based on a bag of 3D points. In: Workshop on Human Activity Understanding from 3D Data, pp. 9–14 (2010)
25. Ofli, F., Chaudhry, R., Kurillo, G., Vidal, R., Bajcsy, R.: Berkeley MHAD: a comprehensive multimodal human action database. In: 2013 IEEE Workshop on Applications of Computer Vision (WACV), pp. 53–60. IEEE (2013)
26. Xia, L., Chen, C.C., Aggarwal, J.K.: View invariant human action recognition using histograms of 3D joints. In: IEEE Conference on Computer Vision and Pattern Recognition Workshops (CVPRW), pp. 20–27 (2012)
27. Vemulapalli, R., Arrate, F., Chellappa, R.: Human action recognition by representing 3D skeletons as points in a lie group. In: Proceedings of the IEEE Conference on Computer Vision and Pattern Recognition, pp. 588–595 (2014)

Algorithm of Multi-camera Object Handoff Based on Object Mapping

Jianrong Cao(✉), Xuemei Sun, Zhenyu Li, and Yameng Wang

Shandong Jianzhu University, Jinan Shandong 250101, China
jrcao@sdjzu.edu.cn

Abstract. Aiming at the problem of multi-camera tracking and object matching in overlapped regions, an algorithm of multi-camera moving object handover based on object mapping is proposed in this paper. At First, the video image of the ground in the corridor is mapped to a building floor plan chart, and then the center point of the bottom line of minimum external object rectangle is defined as the foothold of moving object. The position of foothold is used to represent the moving object in order to achieve the mapping of moving object in the building floor plan. When the moving object is tracked and matched in the overlapping field of multi-camera, it is judged whether it is the same moving object according to the mapping position in the building floor plan, the moving direction and the color histogram of moving object. So an accurate tracking object handover matching can be obtained. The continuous tracking of the same moving object can be achieved between different cameras. The experimental results show that the proposed algorithm can accurately obtain the real time position of moving object and realize the continuous tracking of the same moving object by the multi-camera.

Keywords: Moving object · Multi-camera · Foothold · Object handoff

1 Introduction

The surveillance scope of a single camera is limited. The moving object tracking of multi-camera has to consider the matching problem of moving object in the multi-camera field. Actually it is a handover problem of moving object in multi-camera field. The handover of moving object is the necessary stage for subsequent processing of multi-camera. The characteristics of moving object is easy to change in the process of movement because of the non-rigid nature of the body. It is difficult to achieve the object handoff only using the feature fusion method of object. The main methods of multi-camera object handoff include object handoff algorithm based on field of view line [1], object handoff algorithm based on feature matching [2, 3] and object handoff algorithm based on projective invariants [4]. The object handoff algorithm based on field of view line is simple in principle and easy to implement, but there is a certain delay effect in the detection of new objects. Therefore, this method is easy to make errors when labeling an object. Because the size and shape of the same object are different from the different perspectives between different cameras, the object handoff algorithm based on feature matching will have a large deviation. The very important

D. Liu et al. (Eds.): ICONIP 2017, Part III, LNCS 10636, pp. 575–582, 2017.
https://doi.org/10.1007/978-3-319-70090-8_58

parameter in the object handoff is the positional relationship of object, and the position of the same moving object in the overlapping area is exactly the same under the ideal condition. Object handoff algorithm based on projective invariant is presented by establishing the positional correspondence of two cameras monitoring range. Related papers are summarized in [5, 6].

In this paper, the mapping relation of moving object on the floor plan of building is researched, and it can be used to mark the position of object on the floor plan of building. Then, according to the motion direction and color histogram of moving object, our method can determine whether the moving objects are the same one and achieve accurately handoff matching of moving object tracked between two adjacent cameras.

2 Mapping of Moving Object on Building Floor Plan

Two adjacent cameras are selected in the monitoring area as shown in Fig. 1. According to the position relation of the corridor, the doorway, the room and so on, the simplified floor plan chart of building can be obtained as shown in Fig. 2.

Fig. 1. Experimental scenario

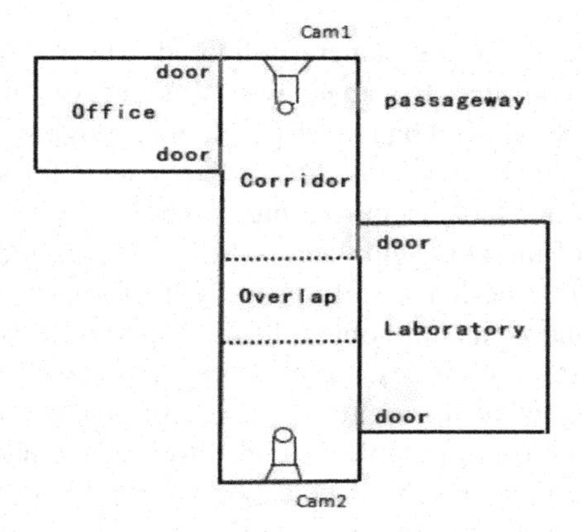

Fig. 2. Simplified floor plan chart of building

The floor plan chart of building is equivalent to projection of building in the horizontal plane. The floor of the corridor is the level surface of projection, so it is a two-dimensional image. The specific mapping is transformed into mapping the moving object in the corridor to the corridor floor on the building floor plan. Therefore, the first step is to map the image of corridor floor to the floor plan of building. Then moving object is mapped to the floor plan of building.

2.1 Mapping of Corridor Floor Image to Building Floor Plan

Because of the oblique installation, the shooting angle and the focus length of projection of camera, the images from the camera has the depth of focus. The closer the object is to the camera, the larger the image shows in the image, whereas the smaller the object shows in the image.

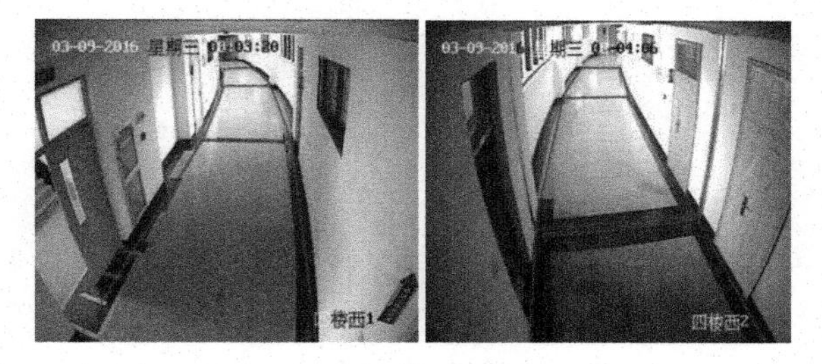

Fig. 3. Corridor image of experimental scene segmented by red lines. (Color figure online)

As shown in Fig. 3, the red line area is the corridor floor shot by the camera. It is assumed that every red line area is a near standard trapezoid. There is a certain degree of curvature on the edge of corridor floor in the floor plan of building, but it can be approximated as a rectangular in a small area. The mapping of the corridor image to the floor of building is essentially a mapping of trapezoidal region to a rectangular region in the image coordinate system, as shown in Fig. 4.

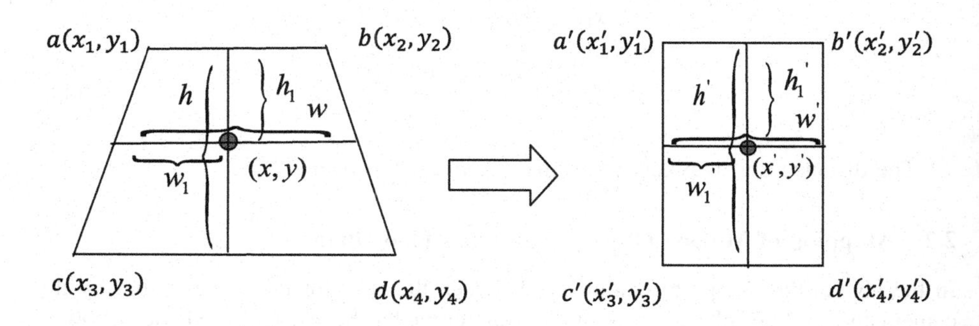

Fig. 4. Mapping of trapezoidal region to rectangular region

As shown in Fig. 4, the four vertices of trapezoid in the image coordinate system are $a(x_1, y_1)$, $b(x_2, y_2)$, $c(x_3, y_3)$, $d(x_4, y_4)$ and the four vertices of the rectangle are $a'(x_1', y_1')$, $b'(x_1', y_2')$, $c'(x_3', y_3')$, $d'(x_4', y_4')$. Assuming that (x, y) is a point in the trapezoidal region and the mapping point in the rectangular region is (x', y'). Our algorithm is based on the proportional mapping and the following relations are obtained.

$$\begin{cases} \frac{h_1}{h} = \frac{h_1'}{h'} \\ \frac{w_1}{w} = \frac{w_1'}{w'} \end{cases} \tag{1}$$

After deduction we can have:

$$h_1 = y - y_1 \tag{2}$$

$$h = y_3 - y_1 \tag{3}$$

$$w_1 = (x - x_1) + \frac{h_1}{h}(x_1 - x_3) \tag{4}$$

$$w = (x_2 - x_1) + \frac{h_1}{h}(x_1 + x_4 - x_2 - x_3) \tag{5}$$

$$h_1' = y' - y_1' \tag{6}$$

$$h' = y_3' - y_1' \tag{7}$$

$$w_1' = x' - x_1' \tag{8}$$

$$w' = x_2' - x_1' \tag{9}$$

In the end, we have:

$$x' = x_1' + \frac{(x - x_1) + \frac{h_1}{h}(x_1 - x_3)}{(x_2 - x_1) + \frac{h_1}{h}(x_1 + x_4 - x_2 - x_3)}(x_2' - x_1') \tag{10}$$

$$y' = y_1' + \frac{(y_3' - y_1')(y - y_1')}{y_3 - y_1} \tag{11}$$

The distance unit is pixel in the image coordinate system.

2.2 Mapping of Moving Object in Building Floor Plan

In the above, the mapping relationship between the corridor floor image shot by the camera and the floor plan chart of monitoring region has been established. Next, how to represent the position of moving object in the image should be considered. In the mapping relation, the reference object is the corridor ground. In order to study the mapping of moving object in the monitoring area, a clear relationship between the moving object and the corridor ground should be known. There is no doubt that the moving object is standing in the corridor floor, and it is in three-dimensional space represented in the world coordinate system. After the projection in the image plane, the moving object is in the 2D image coordinate. The moving object and corridor floor are on the same two-dimensional plane in the image and the moving object overlaps with the corridor floor. In many cases, the centroid of moving object is used to represent the position. However, the centroid is not on the corridor floor in the actual situation, so it is not accurate to represent the mapping position by the centroid of moving object. The feet of the pedestrian are in direct contact with the corridor floor in the movement process, so the central point of the bottom sideline in the minimum enclosing rectangle of moving object is used as a foothold of moving object. The foothold is mapped to the floor plan chart of monitoring region and used as the position of moving object.

In the image, the moving objects are in different positions and its sizes are different. The size of minimum enclosing rectangle of moving object is different in different position because of the depth of field. However, the foothold of pedestrian can be represented by the central point of the bottom sideline in the minimum enclosing rectangle of moving object and it is a stable mapping point. As shown in Fig. 5, the red box represents the enclosing rectangle of moving object and the blue point is a foothold.

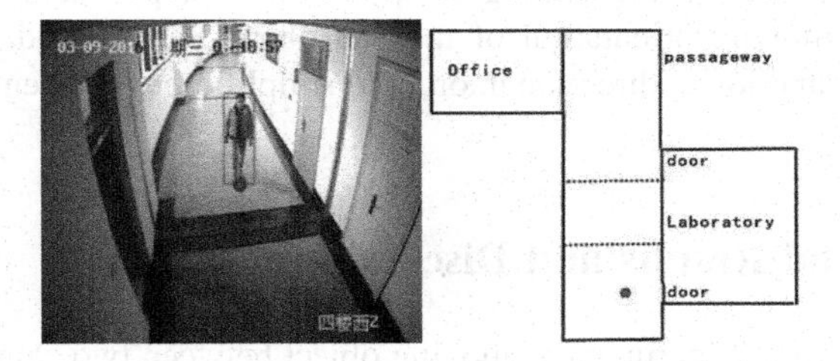

Fig. 5. Schematic diagram of pedestrian foothold. (Color figure online)

3 Multi-camera Monitoring Object Matching

In this paper, two adjacent cameras are represented by Cam1 and Cam2 and the monitoring area of Cam1 and Cam2 camera is the corridor in Fig. 1.

There is an overlapping region in the monitoring area of two cameras, which is the area covered by the two dashed lines in Fig. 2. Our algorithm will achieve the handover of moving object in overlapping areas. When the distance between multiple targets is very close in overlapping region, the mapping points will be very close on the building plan. If the distance between moving objects is only used to achieve the handover, sometimes there will produce some errors. Therefore, based on the judgment of moving object projection position, the moving object orientation is adopted to improve the accuracy of target handover. The moving orientation of object can be obtained by detection of moving object. The core idea of our algorithm is to determine whether the object is the same when the moving object enters overlapping region through the mapping position of object in the floor plan chart of monitoring region, the orientation and the color histogram of moving object. The algorithm is described as below.

① If a moving object enters the overlapping area of vision and there is only one moving object that the distance of moving object between Cam2 and Cam1 is less than the threshold, the moving object of two cameras is determined as the same object.

② If there are many moving objects that the distance between Cam2 and Cam1 is less than the threshold, it is necessary to compare the orientation of these objects in Cam2 and Cam1. If there is only one moving object in the Cam2 that is consistent with the motion direction of moving object in the Cam1, the two moving objects are judged to be the same one.

③ If there are a number of moving objects in Cam2 that the distance of moving object between Cam2 and Cam1 is less than the threshold and the motion direction of moving object is coincident, the moving objects are compared one by one by the histogram in Cam1 and Cam2. The most similar histogram of moving object is determined to be the same one.

The orientation judgment method is to record the position of moving object at the first time, then to record it again after the N frames. Finally, the orientation of moving object is judged by comparing two positions of moving object. The position of moving object at this time is used as the starting position of next step to determine the moving direction. So the moving orientation of moving object can be updated timely. It is divided into upper and lower direction in order to simplify the representation of moving orientation.

4 Experimental Results and Discussion

The real-time handover experiment of moving object between two cameras is shown in Figs. 6, 7, 8, 9 and 10, and the results are presented in time order. The image of Cam1 is on left and the image of Cam2 is on middle.

Fig. 6. Experimental results 1

Fig. 7. Experimental results 2

Fig. 8. Experimental results 3

Fig. 9. Experimental results 4

Fig. 10. Experimental results 5

In Figs. 6 and 7, the pedestrian is detected in the monitoring area of the Cam1, and it is tracked continuously as target 1. At some time, the real-time position of the target can be seen in the building floor plan. No pedestrians are detected in the monitoring area of the Cam2. In Fig. 8, the target 1 is tracked continuously in the monitoring area of the Cam1. In addition, a new pedestrian is detected by Cam1 as target 2 when the target 1 appears in the overlapping region. The Cam2 detects a pedestrian in the overlapping area, and the two cameras have detected the same target through the judgment using our algorithm. In Figs. 9 and 10, the target 2 is tracked continuously in the monitoring area of the Cam1, and the target 1 is tracked continuously in the monitoring area of the Cam2. Finally, the tracking results are mapped to the simplified building floor plan.

5 Conclusion

This paper presents a moving object handoff algorithm combining with building floor plan. A more accurate description of target position can be obtained, a good mapping between moving objects and a building floor plan are established. The multiple criterion such as the distance, the direction of motion and the histogram matching between targets of multi-camera in overlapping region is utilized to judge whether they are the same moving target. The results of discrimination are more accurate. In addition, the distance and direction of move object are firstly used in the judgment of same object, and then the color histogram matching of moving object is adopted. So the number of histogram matching is decreased largely, the computational complexity is simplified.

Acknowledgment. This work was supported by the University and College Independent Innovation Project of Jinan Science and Technology Bureau (201202002), Shandong Province Development Project of Science and Technology (2015GGX101024, 2013GGX10131) and Shandong Provincial Key Laboratory of Intelligent Building Technology.

References

1. Niu, E.G., Chen, Y.M., Pan, Q.: Fast algorithm for automatic generating field of view line of multiple cameras. Appl. Res. Comput. **24**(3), 297–299 (2007)
2. Yu, W.S., Tian, X.H., Hou, Z.Q.: Visual tracking algorithm based on feature matching of key regions. Acta Electron. Sin. **42**(11), 2150–2156 (2014)
3. Shah, J.H., Lin, M.Q., Chen, Z.H.: Multi-camera handoff for person re-identification. Neurocomputing **191**, 238–248 (2016)
4. Ji, H.Q., Li, N.: An improvement object handoff algorithm based on projective invariants. Comput. Digit. Eng. **40**(4), 78–80 (2012)
5. Wang, X.: Intelligent multi-camera video surveillance: a review. Pattern Recogn. Lett. **34**(1), 3–19 (2013)
6. Liem, M.C., Gavrila, D.M.: A comparative study on multi-person tracking using overlapping cameras. In: Chen, M., Leibe, B., Neumann, B. (eds.) ICVS 2013. LNCS, vol. 7963, pp. 203–212. Springer, Heidelberg (2013). doi:10.1007/978-3-642-39402-7_21

Illumination Quality Assessment for Face Images: A Benchmark and a Convolutional Neural Networks Based Model

Lijun Zhang[1], Lin Zhang[1](✉), and Lida Li[2]

[1] School of Software Engineering, Tongji University, Shanghai, China
{1632761,cslinzhang}@tongji.edu.cn
[2] Department of Computing, The Hong Kong Polytechnic University, Hong Kong, China
cslli@comp.polyu.edu.hk

Abstract. Many institutions, such as banks, usually require their customers to provide face images under proper illumination conditions. For some remote systems, a method that can automatically and objectively evaluate the illumination quality of a face image in a human-like manner is highly desired. However, few studies have been conducted in this area. To fill this research gap to some extent, we make two contributions in this paper. Firstly, in order to facilitate the study of illumination quality prediction for face images, a large-scale database, namely, *F*ace *I*mage *I*llumination *Q*uality *D*atabase (FIIQD), is established. FIIQD contains 224,733 face images with various illumination patterns and for each image there is an associated illumination quality score. Secondly, based on deep convolutional neural networks (DCNN), a novel highly accurate model for predicting the illumination quality of face images is proposed. To make our results reproducible, the database and the source codes have been made publicly available at https://github.com/zhanglijun95/FIIQA.

Keywords: Illumination quality assessment · Illumination transfer · Convolutional neural networks

1 Introduction

Over the past few decades, with the rapid development of e-commerce, more and more commercial institutions are going to provide remote services for customers to initiate their commercial activities anywhere [1]. Among the basic information required by those institutions, a face image under uniform and adequate lighting environment of the remote user is usually a must. For example, in a remote bank system, the success of account establishment process will be effected by the illumination condition on the input face image collected by the user's equipment. In such a case, if the system has a module that could dynamically evaluate the input face image's illumination quality and give the user some hints for adjusting

D. Liu et al. (Eds.): ICONIP 2017, Part III, LNCS 10636, pp. 583–593, 2017.
https://doi.org/10.1007/978-3-319-70090-8_59

the ambient light accordingly, it would be quite helpful. Therefore, a method that can automatically monitor the illumination quality of face images is desired.

Thus in this paper, we focus on addressing the face image illumination quality assessment (FIIQA) problem. Our goal is to design an algorithm that could automatically and efficiently evaluate the illumination quality of a given face image and the evaluation results should correlate well with human judgements. To demonstrate our goal more clearly, in Fig. 1, we show 4 face images along with their illumination quality scores predicted by our proposed approach $FIIQA_{DCNN}$ (see Sect. 3 for details). With $FIIQA_{DCNN}$, the predicted illumination quality scores of Fig. 1(a)–(d) are 0.0, 0.503, 0.670, and 1.0, respectively. It can be seen that the results are highly consistent with subjective evaluations.

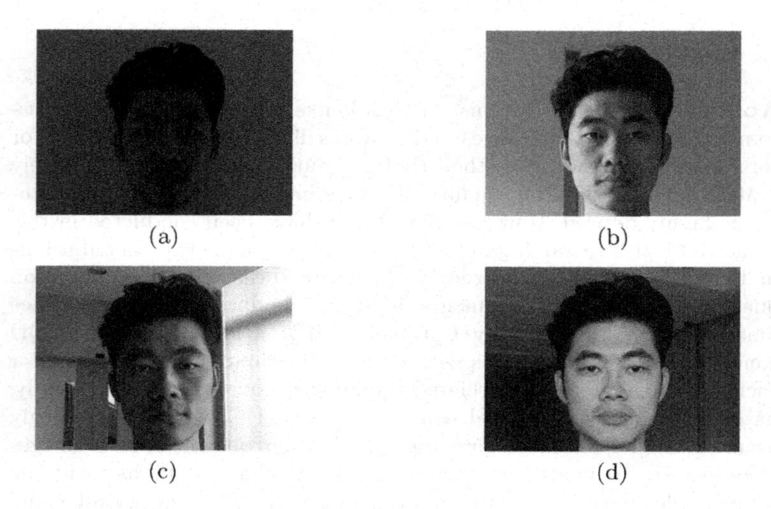

Fig. 1. (a)–(d) are four face images. Their illumination quality scores predicted by our approach $FIIQA_{DCNN}$ are 0.0, 0.503, 0.670, and 1.0, respectively.

1.1 Related Work

Studies particularly focusing on FIIQA are quite sporadic. In [2], Sellahewa and Jassim proposed an objective measure of face illumination quality to decide whether the face image should be preprocessed by illumination normalization. However, their quality index is adapted from a full-reference image quality assessment model UQI [3] and thus requires a reference image, which greatly reduces its practicability.

There are several studies focusing on face image quality assessment (FIQA). In these papers, usually various factors effecting quality are integrated together to induce a score as an overall quality index of the examined face image. These factors may include sharpness, noise, illumination, pose, etc. In [4], the authors considered several factors of face image quality and proposed an evaluation method for each factor separately. For illumination quality assessment, they partition the image into blocks and regard the weighted-average of blocks as the quality score. Such a simple model is not powerful enough to characterize the illumination quality of face images. In [5], a face selection technique using local patch-based probabilistic image quality assessment was proposed for video-based face recognition. However, this method is video-based and cannot be used for still images. In [6], Chen et al. proposed a learning-based approach for FIQA by fusing multiple features.

Actually, the FIIQA problem can also be considered as a special kind of NR-IQA problem. The aim of the NR-IQA research is to design an algorithm that can automatically evaluate the overall quality of a given image. In recent years, many eminent NR-IQA algorithms have emerged, such as BRISQUE [7], NIQE [8], SSEQ [9], LPSI [10], IL-NIQE [11], TCLT [12], OG-IQA [13], etc. Their performance for FIIQA have also been evaluated in our experiments (see Sect. 4 for details).

1.2 Our Motivations and Contributions

Having investigated the literature, we find that in the field of FIIQA, there is still large room for further improvement in at least two aspects. Firstly, though the problem of FIIQA is of paramount importance and has great demand for institutions providing remote services with customers face images, the studies in this area are quite rare. Hence, how to assess the illumination quality of a given image is still a challenging open issue. Secondly, for training and testing FIIQA algorithms, a public large-scale benchmark dataset, comprising face images with associated subjective scores and covering various real-world illumination patterns, is indispensable. Unfortunately, such a dataset is still lacking in this area.

In this work, we attempt to fill the aforementioned research gaps to some extent. Our contributions are summarized as follows:

(1) A large-scale database, namely, Face Image Illumination Quality Database (FIIQD) is constructed. This dataset comprises 224,733 face images with various real-world illumination patterns, each of which has an associated subjective score reflecting its illumination quality. To our knowledge, this is the first large-scale dataset established for the study of illumination quality assessment of face images.
(2) Recent years, the deep convolutional neural networks (DCNN) have gained researchers' much attention and achieved great success for numerous computer vision tasks [14,15]. In this paper, we make an attempt to adapt DCNN to solve the FIIQA problem. Consequently, a novel FIIQA model based on DCNN is proposed, namely FIIQA$_{DCNN}$. Experimental results have shown that FIIQA$_{DCNN}$ has an extremely strong capability in predicting the illumination quality of a given face image.

To make the results fully reproducible, the collected dataset and the source codes of FIIQA$_{DCNN}$ are publicly available at https://github.com/zhanglijun95/FIIQA.

The remainder of this paper is organized as follows. Section 2 presents steps for FIIQD construction. Section 3 describes the details of FIIQA_{DCNN}. Experimental results are reported in Sect. 4. Finally, Sect. 5 concludes the paper.

2 FIIQD: A Face Image Illumination Quality Database

In this section, the steps for establishing FIIQD are presented. To fulfill this task, we adopt a semi-automatic strategy. The construction of FIIQD mainly comprises four steps, the construction of the image set with source illumination patterns, the subjective evaluation of illumination pattern images, the construction of target face set, and illumination transfer. The pipeline of FIIQD construction is shown in Fig. 2.

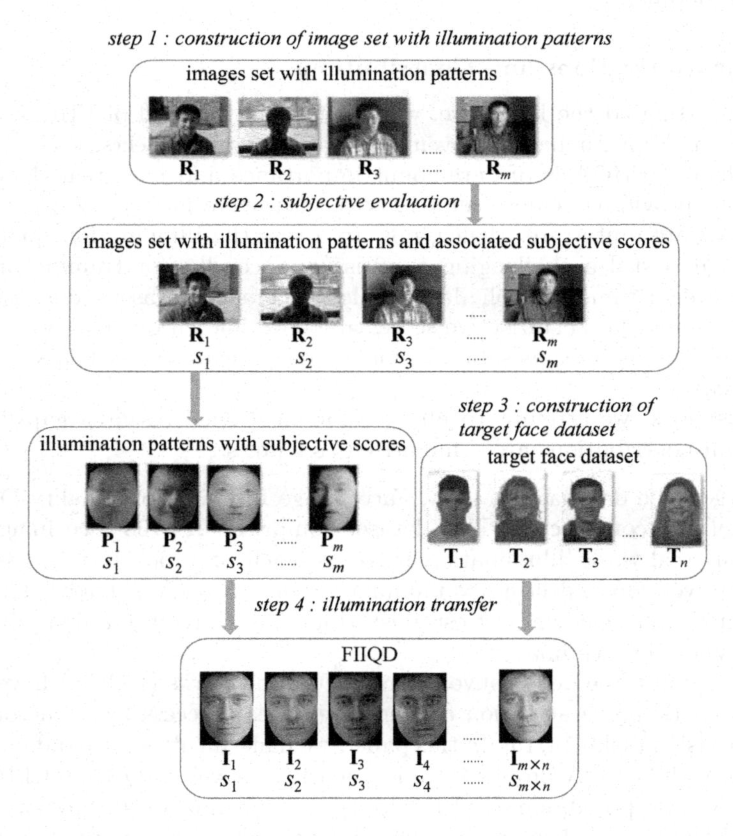

Fig. 2. The pipeline of FIIQD construction.

2.1 Step 1: The Construction of the Image Set with Source Illumination Patterns

In this step, our aim is to collect an image set containing face images with various real-world illumination patterns. These images are expected to provide source illumination patterns to be transferred to target face images so they should cover sufficient types of illumination conditions from real world.

Table 1. The plan for collecting images with source illumination patterns

Index	Scene	Illumination patterns	Time slots
1	Office	Toplight	Day/Night
		Lamp	
		Natural light	
2	Cafe	Window natural light	Day/Night
		Sunshade natural light	8 am/2 pm/5 pm
		Inside light	Sunny/Cloudy
3	Mall	Atrium light	Day/Night
		Shop light	
		Corridor light	
4	Library	Window natural light	Day/Night
		Toplight	3F/8F/14F
5	Home	Toplight	Day/Night
		Lamp	
6	Outside	Natural light	Day/Night
			8 am/2 pm/5 pm
			Sunny/Cloudy

Taking the abovementioned requirements and the typical application scenarios into consideration, we worked out a plan for collecting images with source illumination patterns as showed in Table 1. We selected 6 types of scenes to cover the most common application scenarios. For each scene, we define several different illumination situations by combining different illumination patterns with different time points of a day. And we captured at least 16 photos, 2 every 45° for each illumination situation. At last, we collected 499 images with various source illumination patterns and selected 200 from them, whose qualities are good enough, to form the image set $\mathcal{R}$ with illumination patterns. 6 samples from this set are showed in Fig. 3.

2.2 Step 2: Subjective Evaluation of Images in $\mathcal{R}$

In this step, the illumination quality of images in $\mathcal{R}$ is evaluated by subjective judgements. To achieve this goal, a single-stimulus continuous quality evaluation [16] was conducted. Then, we performed some postprocessing steps to the

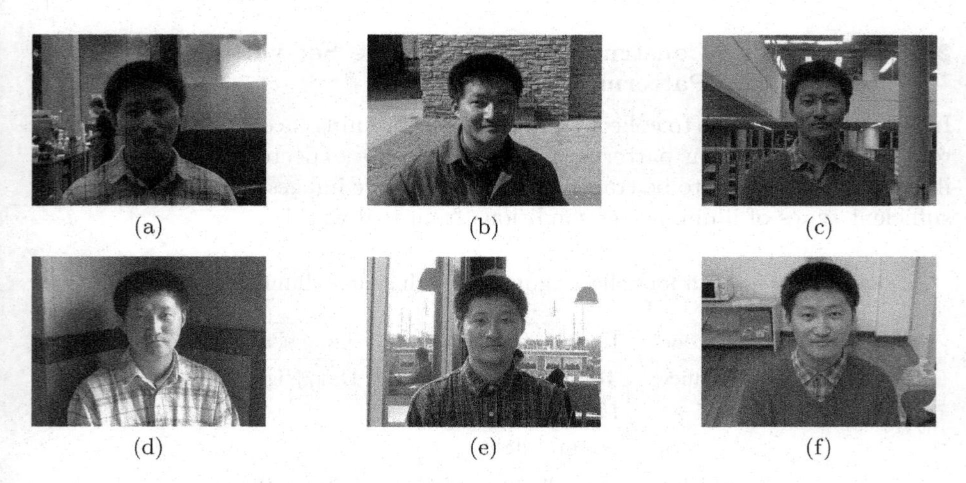

Fig. 3. (a)–(f) are six sample images with illumination patterns. Their associate subjective scores of illumination quality are 0.083, 0.175, 0.341, 0.667, 0.833 and 1.0, respectively.

raw scores. At first, we filtered out those heavily biased subjective scores that satisfy

$$d_{ij} - \overline{d_j} > T \cdot \sigma_j \tag{1}$$

where d_{ij} is the illumination quality score of the image $\mathbf{R}_j \in \mathcal{R}$ given by the ith evaluator, $\overline{d_j}$ is the mean score of $\mathbf{R}_j$, T is the threshold constant and σ_j is the standard deviation of $\mathbf{R}_j$'s scores. Then, to eliminate the effect of different subjective evaluation standards of evaluators, the raw scores d_{ij} were converted as,

$$z_{ij} = \frac{d_{ij} - \overline{d_i}}{\sigma_i} \tag{2}$$

where $\overline{d_i}$ is the mean score of the ith evaluator and σ_i is the standard deviation of his scores for all images in $\mathcal{R}$. We regard the mean evaluation score of $\mathbf{R}_j$ as its final subjective illumination quality score,

$$s_j = \frac{1}{N_j} \sum z_{ij} \tag{3}$$

where N_j is the number of the subjective scores for $\mathbf{R}_j$.

Now, for each image $\mathbf{R}_j \in \mathcal{R}$, an associated subjective score s_j reflecting its illumination quality is obtained.

2.3 Step 3: Target Face Set Construction

In this step, we built a target face image set including 1134 face images under uniform illumination from 1014 unique subjects. We established this set by selecting suitable face images from existing face datasets, such as YaleB [17], PIE [18],

FERET [19], etc. In consideration of diversity, the subjects in target face image set have wide distributions over several attributes, such as face shape, race, skin color, gender, generation, etc.

2.4 Step 4: Illumination Transfer

In this step, we transfer the illumination patterns from images in $\mathcal{R}$ to the images in the target face set by implementing the illumination transfer algorithm proposed in [20]. This step results images in the final FIIQD. Sample images in FIIQD are showed in Fig. 4. The first row are images with source illumination patterns; the first column are target images; the others are the corresponding transferred images in FIIQD. We obtain a database with 224,733 face images with various illumination patterns at last.

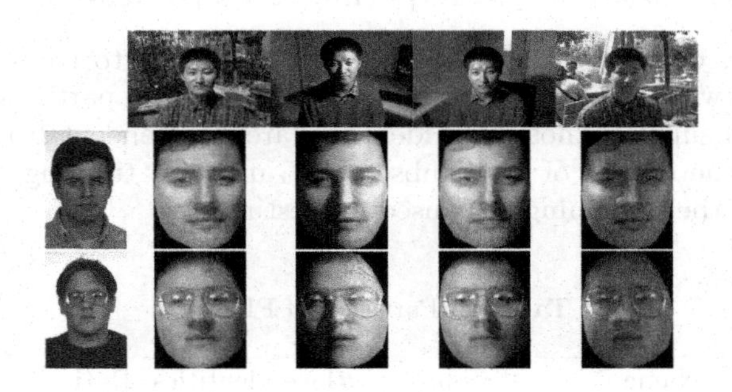

Fig. 4. Samples of the FIIQD.

Suppose that $\mathbf{R}_j \in \mathcal{R}$ is an image with a source illumination pattern and its subjective score is s_i. For the images whose illumination patterns are transferred from $\mathbf{R}_i$, their illumination quality scores are assigned as s_i.

3 FIIQA$_{DCNN}$: A Face Illumination Quality Assessment Model Based on DCNN

In this paper, we propose an FIIQA method, FIIQA$_{DCNN}$, based on DCNN and to our knowledge, our work is the first one to introduce DCNN to FIQA, not alone FIIQA.

In FIIQA$_{DCNN}$, we adopt the Deep Residual Networks in [15], which won the 1st places in ImageNet classification, ImageNet detection, ImageNet localization, COCO detection, and COCO segmentation in ILSVRC and COCO 2015 competitions. The key idea of [15] is to take a standard feed-forward ConvNet and add skip connections that bypass (or shortcut) a few convolution layers at a time. Each bypass gives rise to a residual block in which the convolution layers predict a residual that is added to the block's input tensor.

In FIIQA$_{DCNN}$, we select ResNet-50 as our model for its suitable depth and complexity to solve our problem. ResNet-50 has 53 convolution layers, 2 pooling layers and 1 fully connected layer. And the output number of the last fully connected layer is the number of classes, which is 200 in our case since we have 200 different illumination patterns in FIIQD.

In our implementation, we resize the input images into the size 224×224 and train the ResNet-50 model on the FIIQD training set. The learning rate starts from 0.01 and is divided by 10 when the error reaches plateaus, and the models are trained for up to 500K iterations. And we keep the weight decay and momentum the same with the original network settings.

4 Experimental Results and Discussion

4.1 Database Partition and Experimental Protocol

In experiments, we used FIIQD, constructed in Sect. 2, to train FIIQA$_{DCNN}$ and compare it with the other state-of-art algorithms. We partitioned the whole database into 3 subsets whose face identities are independent from each other. Table 2 shows the details of each subset, 71% used for training, 14% used for validation, and the remaining 15% used for testing.

Table 2. Partition of FIIQD

Name	#images	#face identities	Ratio
Training set	159159	709	71%
Validation set	30930	141	14%
Testing set	34644	164	15%

To evaluate the performance of our method, we adopted two correlation coefficients to measure the monotonic coherency between the prediction results and the subjective scores: Spearman rank-order correlation coefficient (SROCC) and Kendall rank-order correlation coefficient (KROCC). A value closer to 1 indicates a better result of quality estimation for both indices.

4.2 Comparisons with FIQA Methods

In this experiment, the performance of two state-of-the-art competing FIQA methods were evaluated. The first one is the RQS [6], which evaluates the overall quality of a face image. The other one is the method proposed in [4], which can evaluate the illumination quality of a face image. In Table 3, we list the two correlation coefficients, SROCC and KROCC, achieved by each method on the testing set of FIIQD.

Table 3. Performance comparison with FIQA methods

Method	SROCC	KROCC
RQS	0.125	0.086
Method in [4]	0.6873	0.5031
FIIQA$_{DCNN}$	**0.9477**	**0.8915**

4.3 Comparisons with NR-IQA Methods

In this experiment, we compared FIIQA$_{DCNN}$ with some state-of-the-art NR-IQA methods, including BRISQUE [7], NIQE [8], SSEQ [9], LPSI [10], IL-NIQE [11], TCLT [12], OG-IQA [13]. As no source codes of LPSI has been released yet, we implemented it by ourselves and tuned all parameters to achieve its best results. For the rest of the competing methods, we used the source codes provided by their authors. In Table 4, we list SROCC and KROCC achieved by each method on the testing set of FIIQA.

Table 4. Performance comparison with NR-IQA methods

Method	SROCC	KROCC
BRISQUE	0.0487	0.0333
NIQE	0.0260	0.0173
IL-NIQE	0.0459	0.0314
SSEQ	0.1185	0.0811
LPSI	0.1255	0.0847
TCLT	0.1600	0.1094
OG-IQA	0.1757	0.1209
FIIQA$_{DCNN}$	**0.9477**	**0.8915**

4.4 Discussion

Based on the results listed in Table 3, we could have the following findings. At first, FIIQA$_{DCNN}$ performs the best among all of the methods and achieves a high SROCC around 0.95. Secondly, the method in [4] performs much better than RQS. The major difference between them is that ROS evaluates the face image quality as a whole and does not consider the illumination factor separately while the method in [4] is specially designed for measuring illumination quality. This fact indicates that it is better to solve the FIIQA problem separately from the FIQA problem.

The superiority of our method FIIQA$_{DCNN}$ over the other competitors in NR-IQA can be clearly observed from Table 4. And it indicates that the predictions of NR-IQA can reflect the general quality of an image, but cannot reflect the illumination quality of a face image. It is more suitable to solve the problem using concrete analysis of face illumination.

5 Conclusions and Future Work

In this paper, we focus on addressing the problem of face image illumination quality assessment. Our contributions are twofold. First, we have constructed a large-scale database, namely FIIQD. It contains over $224K$ face images with various illumination patterns and each one is assigned an illumination quality score as ground truth. Second, we are the first to employ DCNN models to predict illumination quality of face images. Experiments conducted on FIIQD show that the proposed FIIQA model FIIQA_{DCNN} outperforms all its competitors by a large margin, making it quite attractive for real applications.

Acknowledgments. This work was supported in part by the Natural Science Foundation of China under grant no. 61672380 and in part by the ZTE Industry-Academia-Research Cooperation Funds under grant no. CON1608310007.

References

1. Xia, D., Cui, D., Wang, J., Wang, Y.: A novel data schema integration framework for the human-centric services in smart city. ZTE Commun. **13**(4), 25–33 (2015)
2. Sellahewa, H., Jassim, S.A.: Image-quality-based adaptive face recognition. IEEE Trans. IM **59**(4), 805–813 (2010)
3. Wang, Z., Bovik, A.C., Sheikh, H.R., Simoncelli, E.P.: Image quality assessment: from error visibility to structural similarity. IEEE Trans. IP **13**(4), 600–612 (2004)
4. Truong, Q.C., Dang, T.K., Ha, T.: Face quality measure for face authentication. In: Dang, T.K., Wagner, R., Küng, J., Thoai, N., Takizawa, M., Neuhold, E. (eds.) FDSE 2016. LNCS, vol. 10018, pp. 189–198. Springer, Cham (2016). doi:10.1007/978-3-319-48057-2_13
5. Wong, Y., Chen, S., Mau, S., Sanderson, C., Lovell, B.C.: Patch-based probabilistic image quality assessment for face selection and improved video-based face recognition. In: IEEE CVPR Workshops, pp. 74–81 (2011)
6. Chen, J., Deng, Y., Bai, G., Su, G.: Face image quality assessment based on learning to rank. IEEE SPL **22**(1), 90–94 (2015)
7. Mittal, A., Moorthy, A.K., Bovik, A.C.: No-reference image quality assessment in the spatial domain. IEEE Trans. IP **21**(12), 4695–4708 (2012)
8. Mittal, A., Soundararajan, R., Bovik, A.C.: Making a 'completely blind' image quality analyzer. IEEE SPL **20**(3), 209–212 (2013)
9. Liu, L., Liu, B., Huang, H., Bovik, A.C.: No-reference image quality assessment based on spatial and spectral entropies. Sig. Process. Image Commun. **29**(8), 856–863 (2014)
10. Wu, Q., Wang, Z., Li, H.: A highly efficient method for blind image quality assessment. In: IEEE ICIP, pp. 339–343 (2015)
11. Zhang, L., Zhang, L., Bovik, A.C.: A feature-enriched completely blind image quality evaluator. IEEE Trans. IP **24**(8), 2579–2591 (2015)
12. Wu, Q., Li, H., Meng, F., Ngan, K.N., Luo, B., Huang, C., Zeng, B.: Blind image quality assessment based on multichannel feature fusion and label transfer. IEEE Trans. CSVT **26**(3), 425–440 (2016)
13. Liu, L., Hua, Y., Zhao, Q., Huang, H., Bovik, A.C.: Blind image quality assessment by relative gradient statistics and adaboosting neural network. Sig. Process. Image Commun. **40**, 1–15 (2016)

14. Krizhevsky, A., Sutskever, I., Hinton, G.E.: ImageNet classification with deep convolutional neural networks. In: NIPS, pp. 1097–1105 (2012)
15. He, K., Zhang, X., Ren, S., Sun, J.: Deep residual learning for image recognition. In: IEEE CVPR, pp. 770–778 (2016)
16. Assembly, ITU Radiocommunication: Methodology for the Subjective Assessment of the Quality of Television Pictures. International Telecommunication Union (2003)
17. Georghiades, A.S., Belhumeur, P.N., Kriegman, D.J.: From few to many: Illumination cone models for face recognition under variable lighting and pose. IEEE Trans. PAMI **23**(6), 643–660 (2001)
18. Sim, T., Baker, S., Bsat, M.: The CMU pose, illumination and expression database. IEEE Trans. PAMI **25**(12), 1615–1618 (2003)
19. Phillips, P.J., Moon, H., Rizvi, S.A., Rauss, P.J.: The FERET evaluation methodology for face-recognition algorithms. IEEE Trans. PAMI **22**(10), 1090–1104 (2000)
20. Chen, X., Chen, M., Jin, X., Zhao, Q.: Face illumination transfer through edge-preserving filters. In: IEEE CVPR, pp. 281–287 (2011)

Graph Embedding Learning for Cross-Modal Information Retrieval

Youcai Zhang and Xiaodong Gu(✉)

Department of Electronic Engineering, Fudan University, Shanghai 200433, China
{yczhang12, xdgu}@fudan.edu.cn

Abstract. The aim of cross-modal retrieval is to learn mappings that project samples from different modalities into a common space where the similarity among instances can be measured. To pursuit common subspace, traditional approaches tend to solve the exact projection matrices while it is unrealistic to fully model multimodal data only by linear projection. In this paper, we propose a novel graph embedding learning framework that directly approximates the projected manifold and utilizes both the labeled information and local geometric structures. It avoids explicit eigenvector decomposition by iterating random walk on graph. Sampling strategies are adopted to generate training pairs to fully explore inter and intra modality among the data cloud. Moreover, graph embedding is learned in a semi-supervised learning manner which helps to discriminate the underlying representation over different classes. Experimental results on Wikipedia datasets show that the proposed framework is effective and outperforms other state-of-the-art methods on cross-modal retrieval.

Keywords: Graph embedding learning · Cross-modal retrieval · Semi-supervised learning

1 Introduction

In the information-overloaded age, co-occurred data from different modalities such as text, image, video and audio, usually delivers the same underlying semantic information. The rapid growth of multimodal data makes it difficult for traditional unimodal information retrieval system to acquire precise personal customization. There is a growing demand for cross-modal information retrieval that means the query and retrieved items are not required to share a common modality. For example, users can submit a text query to retrieve the relevant images and videos that best illustrate the query. However, there is no natural correspondence between representations of different modalities.

Canonical correlation analysis (CCA) [1] is a possible solution to analyze the correlation between two multivariate random vectors. Through CCA, we can find linear combinations between two sets of heterogeneous data which have maximum correlation with each other. Besides CCA, Partial Least Squares (PLS) [2] and cross-modal factor analysis (CFA) [3] are also used for cross-modal retrieval.

D. Liu et al. (Eds.): ICONIP 2017, Part III, LNCS 10636, pp. 594–601, 2017.
https://doi.org/10.1007/978-3-319-70090-8_60

Considering that traditional unsupervised approaches including CCA, PLS, CFA ignore the semantic information that may be effective to improve the retrieval accuracy, many supervised methods are proposed to exploit label information and obtain significant improvements. Rasiwasia et al. [1] designed semantic matching (SM) for different modalities. SM corresponds to using multiclass Logistic regression to classify both text and images. They further extended it to Semantic Correlation Matching (SCM), which aimed to pursue semantic spaces using the feature learned by CCA as low-level representation.

In this paper, we propose a novel **G**raph **E**mbedding **L**earning **(GEL)** framework to solve the problem of common subspace pursuit on cross-modal retrieval. Since the embedding are learned based on the graph structure, we propose a multimodal graph constructed via one-to-one pairwise constraints and K-nearest neighbors. Then we use sampling strategies to generate training pairs from random walk sequences, fully exploiting inter-modality and intra-modality similarity relationships among both labeled and unlabeled data. Further, we learn embedding based on the proposed graph structure and label information. During implementation, the black box model of the parameterized embedding can be expressed as a standard building block of neural network.

2 Related Work

Graph-based Semi-supervised learning has been proved to be effective in many fields. It is based on the assumption that nearby nodes tend to share the same semantics. Each data denotes a node and the edge weights indicate the affinities between both labeled and unlabeled multimodal data. Thus, we can acquire a similarity graph and construct graph regularization for optimization. Inspired by the idea of graph-based semi-supervised learning, Zhai et al. [4] incorporate graph regularization into a unified framework for cross-model retrieval named Joint Representation Learning (JRL). Specifically, JRL constructs a separate graph for each modal type and jointly models correlation and semantic information in a unified framework. Wang et al. [5] further takes the intra-modality categorical information into account and constructs a generalized multimodal graph to better model the similarity relationships among data from different modalities.

Embedding learning is a technique widely implemented in natural language processing and social network application. An embedding is a representation of instances in a new space such that the properties of, and the relationships between, the instances are preserved. It is closely related to other graph regularization methods and encodes graph structure directly into the embedding. According to [6], the central idea behind graph embedding is that distributed representation of each nodes is formulated as a probabilistic relationship between the contexts around it. In [7], the author designed skip-gram and CBOW model to learn the underlying embedding efficiently and capture meaningful semantic information among the data cloud. Planetoid [8] adopted embedding learning into semi-supervised learning process to leverage the labeled data and graph structure.

3 Graph Embedding Learning

3.1 Problem Formulation

Suppose we have a document corpus $L^{(i)} = \{x_i^1, x_i^2 \ldots, x_i^M\}$ containing data from M different modalities and representing the same underlying content. $x_i^p \in R^{1\times d^p}$ is the ith feature vectors from the p-th modality and is embedded in the d^p dimensional space. $X^p \in R^{N\times d^p}$ represents the p-th modality data with N instances. N_l and N_u denote the number of labeled instances and unlabeled instances respectively. $N = N_l + N_u$. Each corpus has a unique category, and we use the one-hot indicator vector $y^{(i)} \in R^{1\times C}$ to denote the labeled category information of training corpus, where C is the number of categories. Our purpose is to learn embedding $\varphi(x_i)$ where the similarity among different modalities can be measured.

3.2 Overall Architecture

In this article, we mainly focus on leveraging the labeled information and graph structure resided in the data cloud. In particular, we combine a supervised learner with a graph context learning process to perform cross-modal retrieval tasks (Fig. 1). Contrary to subspace pursuit method, we parameterized the projection of all the instances and directly optimize them with a unified loss function. We hope that under the manifold space to be solved, graph embedding with similar semantic information or strong cross-modal interconnection should be clustered closely to each other. At meanwhile, supervised classifier should still efficiently discriminate between instances according to different class distribution as well.

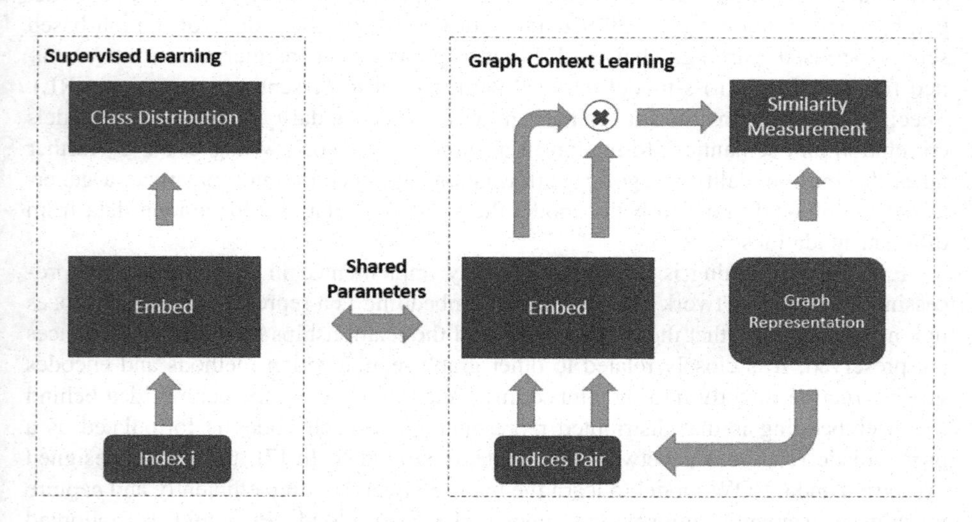

Fig. 1. Overview of proposed graph embedding learning framework

The loss function of the proposed framework can be expressed as Eq. (1).

$$L = L_s + \lambda L_u \tag{1}$$

To be more specific, in the supervised learning part, after obtaining the embedded representation of the input instance, we formulate a logistic regression to indicate its class distribution. We use cross entropy Eq. (2) as loss function for class distribution prediction.

$$L_s = \frac{1}{N_s}\sum_{i=1}^{N_s}\sum_{k=1}^{C} y_i \log(\tilde{y}_{ik}), \tag{2}$$

where

$$\tilde{y}_{ic} = \frac{e^{z_{ic}}}{\sum_{k=1}^{C} e^{z_{ik}}},\ z_i = W^T\varphi(x_i) + b \tag{3}$$

During graph context learning, we randomly choose a pair of instance $\{x_i, x_j, \gamma\}$ by some certain sampling strategies on the graph and then merge these two input embedding to acquire a similarity measurement on manifold space. Logistic loss function Eq. (4) is adopted.

$$L_u = \frac{1}{N_g}\sum_{i=1}^{N_g} \log(1 + exp(-\gamma \text{sim}(\varphi(x_i), \varphi(x_j)))) \tag{4}$$

And we adopt cosine distance Eq. (5) to compute similarity.

$$sim(\varphi(x_i), \varphi(x_j)) = cos(\varphi(x_i), \varphi(x_j)) = \frac{\varphi(x_i)^T \varphi(x_j)}{||\varphi(x_i)||||\varphi(x_j)||} \tag{5}$$

3.3 Multimodal Graph

Since the embedding is learned based on the graph structure, we propose a multi-modal graph via one-to-one pairwise constraints and K-nearest neighbors to preserve the inter and intra modality similarity relationship.

- Inter-modality connection

We use the one-to-one pairwise constraints among different modalities as the inter-modality connection. One-to-one pairwise constraints mean items from different modalities represent the same underlying content. The semantic affinity metric of two data from p-th modality and q-th modality are defined as Eq. (6).

$$a_{ij} = \begin{cases} 1 & if\ \mathbf{x}_i^p, \mathbf{x}_j^q\ in\, pairs \\ 0 & otherwise \end{cases} \tag{6}$$

- Intra-modality connection

We use the feature vectors of both labeled and unlabeled instances to establish the intra-modality connection. Distributional affinity metric of two data from the same modality is defined as Eq. (7).

$$a_{ij} = \begin{cases} 1 & \text{if } \mathbf{x}_i^p \in N_k\left(\mathbf{x}_j^p\right) \text{ or } \mathbf{x}_j^p \in N_k(\mathbf{x}_i^p) \\ 0 & \text{otherwise} \end{cases}, \quad (7)$$

where $N_k(\bullet)$ denotes the set of k-nearest neighbors.

3.4 Sampling Strategies on Graph

From Eq. (3), we can see that the loss function for graph context requires triplet input containing two distinct instances and a target signal. Judging only by the labeled cross-modal link from the training set, we could not efficiently induce the correlation among instances merely from the test set. In order to fully explore the inter and intra relationship among various modalities, we sample training pairs from iterating random walk on the multimodal graph and connect vertices according to some window size: $\{(S_j, S_k : |j-k| < d)\}$. Started by some arbitrary vertex, random walk sequence produces visiting path to other possible instances from intra and inter modality. It could be viewed as a diffusion process that propagates graph context information among similar vertices, which is ideal for cross-modal information retrieval tasks. Nevertheless, this sampling strategy is less computational-costly compared to other context based sampling method and feasible to be implemented on huge graph.

4 Experiment

4.1 Datasets and Evaluation Metrics

To demonstrate the effectiveness of our proposed GEL method, we conduct experiments on Wikipedia datasets for cross-modal retrieval tasks. Wikipedia Datasets[1] is generated from Wikipedia's "featured article". It contains 2866 image-text pairs, annotated with 10 semantic classes. To make this a fair comparison, we strictly follow the partition scheme on dataset described in [1]. 2173 document pairs are randomly selected for training and the remaining 693 pairs are used for test. The representation of the text is 10-dimension vector derived from a latent Dirichlet allocation model. The image features are 4096-dimension vector extracted from the last fully connected layer of VGG-19 model [9] pretrained on Imagenet classification tasks.

We adopt mean average precision (MAP) [1] and the normalized discount cumulative gain (NDCG) as evaluation metrics, which are widely used in the cross-modal retrieval literature. NDCG is defined as Eq. (8).

[1] http://www.svcl.ucsd.edu/projects/crossmodal/.

$$NDCG@k = \frac{1}{N^k}\sum_{j=1}^{k}\frac{2^{rel_j}-1}{\log(1+j)}, \quad (8)$$

where N^k is a normalization constant to ensure that the optimal top k ranking of the query is 1. rel_j is given by {Excellent = 3; Good = 1; Bad = 0}, Excellent refers to the retrieved instance represents the same underlying content to the query. Good denotes the retrieved instance belongs to the class of the query. Further, the average of the NDCGs on all the queries is the final evaluation result.

We also use precision-recall curve to evaluate the effectiveness of different methods.

4.2 Implementation Details

In experiments, we use Keras[2] to implement the proposed graph embedding learning framework. The dimension of embedding is set to 64. As a consequence, size of learnable parameters in the overall architecture is only proportional to the number of instances in the dataset. RMSprop method is utilized to fine-tune networks and its initial learning rate is given by 0.0001. RMSprop is an unpublished, adaptive learning rate method proposed by Geoff Hinton. During constructing the multimodal graph, we set the number of nearest neighbors to 10. We choose different batch size as 512 and 128 respectively for supervised and graph context learning. Our GEL algorithm is not very sensitive to these parameters. Settings on hyper-parameters of the model are shared across all the datasets, which indicates that our framework could be optimized easily and generalize well among different graph structure.

4.3 Experimental Results

The experimental results of cross-modal retrieval in terms of MAP and NDCG@25 are shown in Table 1.

Table 1. MAP and NDCG@25 comparison of different methods.

Experiment	MAP			NDCG@25		
	Img2txt	Txt2img	Average	Img2txt	Txt2img	Average
SM [1]	0.4187	0.3427	0.3807	0.2292	0.2338	0.2315
PCA+CM [1]	0.3836	0.3669	0.3753	0.2301	0.2827	0.2564
CFA [3]	0.3947	0.3729	0.3838	0.2453	0.2661	0.2557
PLS [2]	0.3940	0.3661	0.3801	0.2424	0.2849	0.2637
PCA+SCM [1]	0.4207	0.3789	0.3998	0.2509	0.2933	0.2721
JFSSL[a] [5]	0.4279	0.3957	0.4118			
GEL	**0.4403**	**0.4174**	**0.4289**	**0.2697**	**0.3253**	**0.2975**

[a]We site the publicly reported results of JFSSL

[2] https://github.com/fchollet/keras.

For fair comparison, we use all the retrieved results to calculate MAP scores and top25 retrieved results to calculate NDCG scores. PCA is performed on the original input to remove the redundancy in features. We can see that our method (GEL) slightly outperforms JFSSL, as GEL fully explores the inter and intra relationship among various modality via random walk. The learned embedding could be viewed as a form of manifold representation that is more efficient and compact compared to traditional subspace learning methods. SCM and JFSSL perform better than CM, CFA, PLS, which ignore the category information during super-vised learning process.

In addition, Fig. 2 gives the precision-recall curves of the image query and text query for the above approaches. At most recall levels, GEL method achieves comparable and higher precision compared with other state-of-the-art methods.

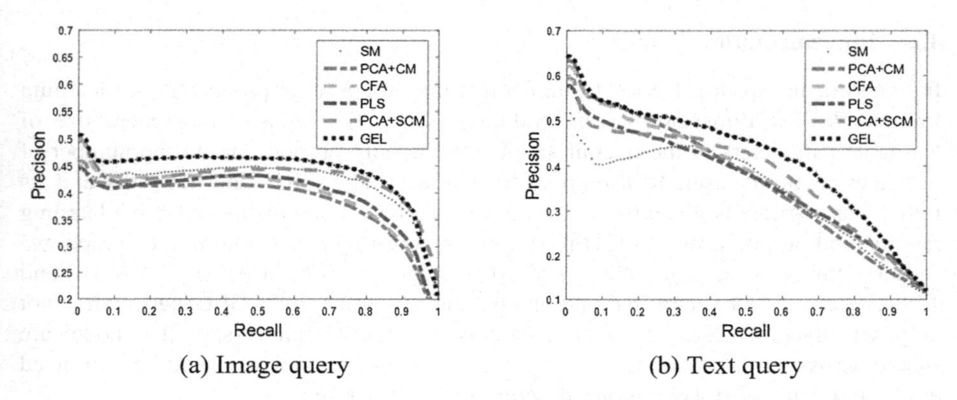

Fig. 2. Precision-recall curves of different methods on Wikipedia datasets.

Finally, we present two cross-modal retrieval examples using our GEL method in Fig. 3. The query text and the query image are shown at the left, and the top5 matched results are shown at the right. Double red triangles represent the one-to-one

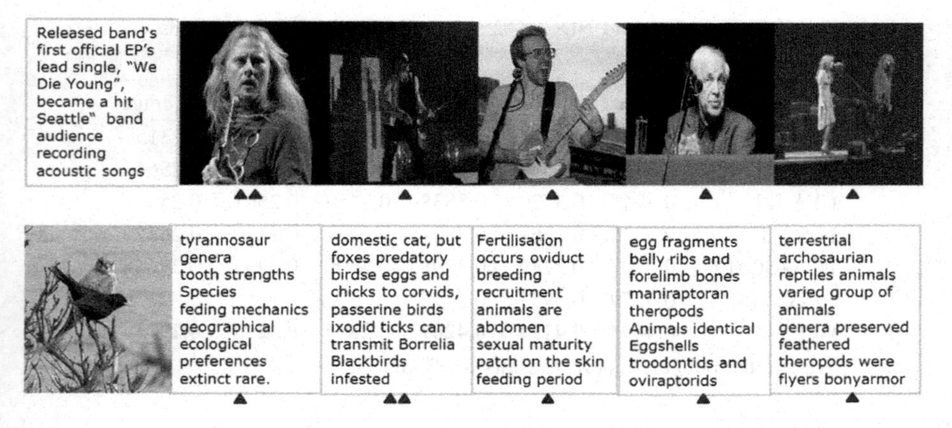

Fig. 3. Examples of cross-modal retrieval using text and image query respectively. (Color figure online)

correspondence, and single red triangle represents belonging to the same category label as the query. Note that our method finds the closest matches at semantic level. The top5 retrieved results share the same label with the queries. Moreover, our method retrieves the one-to-one correspondence.

5 Conclusion

This paper mainly addresses the problem of common subspace pursuit on cross-modal retrieval via graph embedding learning. Under the proposed framework, embedding learned by graph structure and label information directly approximates the projected manifold. During implementation, the black box model of the parameterized embedding is expressed as a standard building block of neural network. Furthermore, the inter and intra modality similarity are well preserved through random walk on the multimodal graph. Experimental results on Wikipedia datasets have demonstrated that proposed method outperforms other relevant state-of-the-art approaches.

Acknowledgments. This work was supported in part by National Natural Science Foundation of China under grants 61371148 and 61771145. The authors would like to greatly thank Jiayan Cao for his help in architecture modeling and implementation.

References

1. Rasiwasia, N., Costa Pereira, J., Coviello, E., et al.: A new approach to cross-modal multimedia retrieval. In: Proceedings of the 18th ACM International Conference on Multimedia, pp. 251–260 (2010)
2. Sharma, A., Jacobs, D.W.: Bypassing synthesis: PLS for face recognition with pose, low-resolution and sketch. In: Proceedings of the IEEE Conference on Computer Vision and Pattern Recognition, pp. 593–600 (2011)
3. Li, D., Dimitrova, N., Li, M., et al.: Multimedia content processing through cross-modal association. In: Eleventh ACM International Conference on Multimedia, Berkeley, CA, USA, November, pp. 604–611 (2003)
4. Zhai, X., Peng, Y., Xiao, J.: Learning cross-media joint representation with sparse and semisupervised regularization. IEEE Trans. Circ. Syst. Video Technol. **24**, 965–978 (2014)
5. Wang, K., He, R., Wang, L., et al.: Joint feature selection and subspace learning for cross-modal retrieval. IEEE Trans. Pattern Anal. Mach. Intell. **38**, 2010–2023 (2016)
6. Bengio, Y., Ducharme, R., Vincent, P., et al.: A neural probabilistic language model. J. Mach. Learn. Res. **3**, 1137–1155 (2003)
7. Mikolov, T., Chen, K., Corrado, G., et al.: Efficient estimation of word representations in vector space. Computer Science (2013)
8. Yang, Z., Cohen, W.W., Salakhutdinov, R.: Revisiting semi-supervised learning with graph embeddings. In: Proceedings of the International Conference on Machine Learning (2016)
9. Simonyan, K., Zisserman, A.: Very deep convolutional networks for large-scale image recognition. Computer Science (2014)

Particle Swarm Optimization Based Salient Object Detection for Low Contrast Images

Nan Mu[1], Xin Xu[1,2(✉)], Xiaolong Zhang[1,2], and Li Chen[1,2]

[1] School of Computer Science and Technology,
Wuhan University of Science and Technology, Wuhan 430065, China
xuxin@wust.edu.cn
[2] Hubei Province Key Laboratory of Intelligent Information Processing
and Real-Time Industrial System, Wuhan University of Science and Technology,
Wuhan 430065, China

Abstract. Saliency detection has attracted increasing attentions in computer vision. Although most traditional saliency models can effectively detect the salient objects in natural images, it is still a burning problem in low contrast images, for low lightness and few color information limit the applicability of these models. Different from conventional models, which are not robust on weak light environments, the proposed method uses the particle swarm optimization (PSO) algorithm to estimate the image saliency. First, the covariance feature is used to compute the local saliency of each superpixel region. Then, the PSO search is executed to measure the image saliency in a global perspective. Finally, the graph model is constructed to optimize the saliency value. As the proposed model incorporates both local and global cues, the generated salient objects have well-defined boundaries and uniform inner regions. Experimental results show that the proposed salient object detection model yields better results than eleven state-of-the-art saliency models on low contrast images.

Keywords: Salient object detection · Particle swarm optimization · Region covariance · Low contrast image

1 Introduction

Human visual system (HVS) has an intelligent mechanism, which can rapidly select the most important parts (so called salient objects) from a complex environment for further perception. Saliency detection aims to extract the most important visual information in an image, while the worthless information will be ignored in the subsequent processing. Thus, the detection of salient object can be widely used in image segmentation, object recognition, image retrieval, and so on.

Image data is increasing enormously quickly due to continuous growth of portable multimedia devices. How to effectively acquire the valuable information in image is the key to process big image data. Saliency detection, which focus on identify the most valuable regions, extremely suitable to deal with this problem. Currently, most

D. Liu et al. (Eds.): ICONIP 2017, Part III, LNCS 10636, pp. 602–612, 2017.
https://doi.org/10.1007/978-3-319-70090-8_61

computational models for saliency detection are designed for visible light images. If the scenes are under a poor light condition, these traditional saliency models will not perform well. The low contrast property in weak light scenes makes visible radiation of object reduce to a very low level. In such scenes, there is little difference between foreground and background, as a result, real salient objects are difficult to recognize.

Aiming to address the above problem, the proposed model evaluates salient objects mainly by four steps, which are shown in Fig. 1. Firstly, the input image is segmented into superpixels. Secondly, the covariance matrices of low level visual features, namely lightness, orientation, sharpness, and spectrum, are constructed, the nonlinear integration of different features make our model robust and suitable for the nighttime images. Thirdly, the local and global saliency of each superpixel patch is computed by measuring the region covariance difference between its surrounding superpixels and performing the PSO search algorithm. At last, the saliency value of salient superpixel patches can be strengthened by constructing manifold ranking through the graph. Experimental results on MSRA dataset [1] and the nighttime image dataset show that our model achieves better performance than the state-of-the-art saliency models.

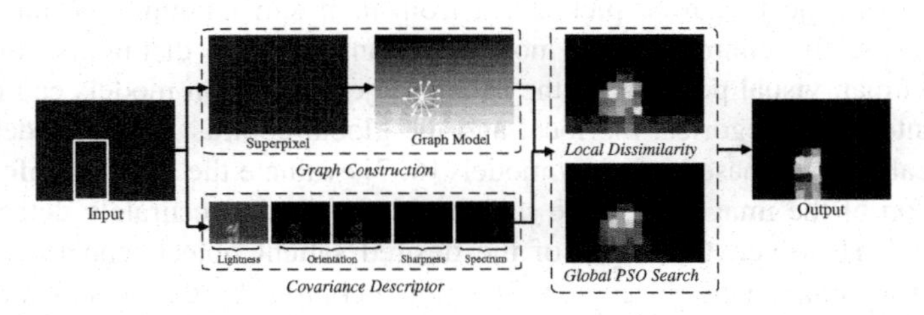

Fig. 1. Overview of the proposed model for salient object detection on nighttime image.

2 Related Works

2.1 Particle Swarm Optimization Algorithm

As a heuristic global optimization algorithm, particle swarm optimization algorithm [2] arouses much attention of contemporary researchers for its feasibility and effectiveness. The algorithm relies on the exchange of information between individuals to achieve the common evolution of the whole group. A set of random solutions, which called particles, are produced by the initialization process. All the particles have a fitness value by the optimized function. The flighting directions and distances of these particles are determined by a speed. The particles are optimized by tracking the personal extremum (pbest) and the global extremum (gbest) in the search space.

The initial population (swarm) of m particles in n dimension is denoted as $X = \{x_1, x_2, \cdots, x_m\}$, the position and the velocity of each particle are denoted as $x_i = \{x_{i1}, x_{i2}, \cdots, x_{in}\}$ and $v_i = \{v_{i1}, v_{i2}, \cdots, v_{in}\}$, respectively. Let $p_i = \{p_{i1}, p_{i2}, \cdots, p_{in}\}$

denotes the pbest of individual, and $p_g = \{p_{g1}, p_{g2}, \cdots, p_{gn}\}$ denotes the gbest of population, the speed and the position of the particles are updated as follow:

$$v_{ij}^{t+1} = w \times v_{ij}^t + c_1 \times r_1 \times (p_{ij}^t - x_{ij}^t) + c_2 \times r_2 \times (p_{gj}^t - x_{ij}^t), \tag{1}$$

$$x_{ij}^{t+1} = x_{ij}^t + v_{ij}^{t+1}, \tag{2}$$

where c_1 and c_2 are non-negative acceleration constants; r_1 and r_2 are the random numbers between [0, 1]. t is the current evolution algebra.

For the PSO search has less parameters, which is fast convergence, simple and easy to operate, it has been widely applied to the field of object optimization. In the area of image processing, the PSO algorithm is also successfully used in image segmentation [3], edge detection [4], image retrial [5], etc.

2.2 Saliency Detection Algorithm

The goal of saliency detection is to efficiently identify the *regions of interest* (ROI), thus the salient objects can be picked out from their surrounding, and attract human attention. Typically, contrast is the most important factor to distinguish the saliency region by human visual perception, the salient object detection models can be broadly classified into two categories: the local and the global contrast based models.

The local-contrast based saliency models [6, 7] evaluate the saliency information in a local region of the image, the large salient objects can be accurately detected by the local method. However, the edges of the desired salient object tend to have higher saliency values than the interior region. The global-contrast based saliency models [8, 9] estimate the salient information over the whole image, thus the salient objects can be uniformly highlighted. However, these models are susceptible to noise interference at nighttime images with low signal-to-noise ratio. The saliency models [10, 11] integrate both local and global contrast can achieve satisfactory performance in visible light environment, but they may face difficulties in weak light scenario.

Weak lighting condition makes the salient object detection of nighttime image becomes a challenging problem. There only exist a few papers considered the object and the contour detection for night images [12, 13], these models are not designed for salient object and are not robust enough to find object in real time. The main objective of the proposed model is to execute salient object detection for nighttime images. This model is applicable to a nighttime image dataset created by us via a standard camera.

3 Saliency Model for Low Contrast Images

3.1 Covariance-Based Local Saliency Estimation

In the proposed model, the input image is first divided into superpixels $SP(i)$, $i = 1, \cdots, N$ by using the *simple linear iterative clustering* (SLIC) algorithm [14]. Then, a graph $G = (V, E)$ of N nodes is generated to represent the image, where V is the nodes set and E is the edges set. Given G and the saliency seeds $s = [s_1, s_2, \cdots, s_N]^T$, the saliency diffusions $S' = [S'_1, S'_2, \cdots, S'_N]^T$ are calculated via:

$$S' = (I - \beta\Lambda)^{-1}s, \tag{3}$$

where I denotes the identity matrix of G, β controls the balance of unary and pairwise potentials in manifold ranking, and Λ is the normalized Laplacian matrix.

The covariance matrices of image superpixels are utilized as meta-features for local saliency estimation. The region covariance has a strong adaptability to rotation, scaling, and brightness changes, thus can be effective in complex low contrast image. It can also reflect the relationship between different features and integrate them in a nonlinearly way, thus the structure information of image can be effectively expressed.

After exploring the attributes of nighttime images, several visual features which can better represent the low contrast properties are first extracted, then the image is converted into a five-dimensional feature vector:

$$F(x,y) = \left[Ligh(x,y) \left|\frac{\partial I(x,y)}{\partial x}\right| \left|\frac{\partial I(x,y)}{\partial y}\right| Shar(x,y)\ Spec(x,y) \right]^T, \tag{4}$$

where $Ligh(x,y)$ denotes the lightness component in Lab color space, the glow amount is an important indicator to measure the saliency of the objects. $|\partial I/\partial x|$ and $|\partial I/\partial y|$ are the norm of first order derivatives of the intensity image, which can represent the edge orientation information. The sharpness feature $Shar(x,y)$ is computed by the convolution of grayscale image and the first-order derivatives of the Gaussian in vertical and horizontal directions [15]. The spectrum feature $Spec(x,y)$ is measured by computing the difference between log spectrum and amplitude [16], which is less affected by the image contrast and more robust to noise.

For each superpixel, it can be represented as a 5×5 covariance matrix [17] via:

$$C_i = \frac{1}{n-1}\sum_{k=1}^{n}(f_k - \mu^*)(f_k - \mu^*)^T, \tag{5}$$

where $\{f_k\}_{k=1,\cdots,n}$ are the five-dimensional feature points inside $SP(i)$ and μ^* is the mean of these points. The multiple features which might be correlated can be naturally fused by covariance matrix. The dissimilarity (denoted as $\rho(C_i, C_j)$) between different covariance matrices is measured by [18]. The smaller the value, the similar the two covariance matrixes.

$$\rho(C_i, C_j) = \sqrt{\sum\nolimits_{k=1}^{5} ln^2 E_k(C_i, C_j)}, \tag{6}$$

where $\{E_k(C_i, C_j)\}_{k=1,\cdots,5}$ are the generalized eigenvalues of C_i and C_j.

3.2 PSO-Based Global Saliency Optimization

In order to improve the robustness of the proposed model against background noise, global search algorithm is introduced to optimize the saliency results. Considering the strong global optimization properties of PSO algorithm, therefore, the PSO search is used to estimate the most different areas of each superpixel. Based on the central-surround principle [6], the greater the difference between an area and its surrounding areas, the more obvious the area is, and it is more likely in the salient object. Here, the saliency of a superpixel $SP(i)$ is defined by measuring the covariance matrix similarity with its adjacent superpixel regions.

Since the covariance matrix is a real symmetric positive definite matrix with a unique factorial decomposition property, its elements can be constructed into a group of points (known as Sigma Points) in the Euclidean space. In this way, the mean vector of the feature can be directly incorporated, the first and second order statistical codes can be integrated to obtain a rich feature representation [19]. For the covariance matrix C_i of superpixel $SP(i)$, the Sigma Point (denotes as $Sig = \{p_k\}$) can be obtained by conducting the Cholesky decomposition $C_i = LL^T$.

$$p_k = \begin{cases} a\sqrt{5}L_k & if\ 1 \leq k \leq 5 \\ -a\sqrt{5}L_k & if\ 5+1 \leq k \leq 2 \times 5 \end{cases}, \tag{7}$$

where L_k is the k column of the lower triangular matrix L, and a is usually set to 0.2. Through a simple connection, the 1×55 eigenvector $\Psi(C_i)$ can be achieved to describe the 5×5 covariance matrix C_i as:

$$\Psi(C_i) = (\mu^*,\ p_1, \cdots, p_5,\ p_6, \cdots, p_{10})^T. \tag{8}$$

Next, the local saliency of $SP(i)$ is calculated by measuring the similarity with its adjacent area. For each $SP(i)$, $i = 1, \cdots, N$, let R_i denotes the superpixel region, its adjacent superpixel regions (denoted as $\{R_j\}_{j=1,\cdots,num(i)}$, $num(i)$ is the number of affinity matrix of R_i) can be obtained according to the adjacent matrix of graph model. Then, the saliency of R_i is defined by computing the weighted average of covariance dissimilarities with its surrounding regions as:

$$s(R_i) = \frac{1}{num(i)} \sum\nolimits_{j=1}^{num(i)} d(R_i, R_j), \tag{9}$$

where $d(R_i, R_j)$ is the dissimilarity between R_i and R_j, which is given by:

$$d(R_i, R_j) = \frac{\|\Psi(C_i) - \Psi(C_j)\|}{1 + \|c^*(i) - c^*(j)\|}, \tag{10}$$

where $\Psi(C_i)$ and $\Psi(C_j)$ denote the vector representation of the covariance matrix, $c^*(i)$ and $c^*(j)$ denote the center of R_i and R_j, respectively.

In order to optimize the saliency results, a search algorithm need to be performed in the adjacent superpixel region to select the most similar areas. If there is a high difference between R_i and its most similar region, then it will have a higher difference with other adjacent superpixel regions. Therefore, the task of our model is to find the most similar adjacent region. For PSO algorithm has less control parameters, excellent search path and strong global optimization ability, it has shown a more efficient search effect in solving optimization problems. Thus, the PSO based algorithm is utilized to search the most similarity region of R_i, then the optimal saliency value of R_i can be obtained. The main steps of PSO search algorithm are shown as follows:

Algorithm 1: PSO based global search algorithm

Step 1: Initialize the randomly position x_{ij}^k and the velocity v_{ij}^k (i and j denote the particle and dimension, respectively). Initialize the parameters (iteration number: $k=1$; inertia weight: $w_{max}=0.9$, $w_{min}=0.4$; acceleration factor: $c_1=2$, $c_2=2$; random numbers: r_1 and r_2). Each $\Psi(C_i)$ of $SP(i)$ corresponds to a particle, its search space is $\{\Psi(C_j)\}_{j=1,\cdots,N}$.

Step 2: Compute the fitness $F_i^k = f(x_i^k)$ of particles and find the best particle.

Step 3: Select the personal and the global extremum $Pbest_i^k$ and $Gbest^k$.

Step 4: Update the velocity and the position of particles:
$v_{ij}^{k+1} = w \times v_{ij}^k + c_1 \times r_1 \times (Pbest_i^k - x_{ij}^k) + c_2 \times r_2 \times (Gbest^k - x_{ij}^k)$,
$x_{ij}^{k+1} = x_{ij}^k + v_{ij}^{k+1}$.

Step 5: Update the fitness of particles: $F_i^{k+1} = f(x_i^{k+1})$; the personal and the global extremum: $Pbest_i^{k+1}$ and $Gbest^{k+1}$.

Step 6: Goto step 4, and $k=k+1$ until k is greater than the set value.

Step 7: Output the optimum solution $Gbest^k$ (which is the covariance matrix vector $\Psi'(C_j)$ of the optimal adjacency region).
Compute the optimal saliency value of $SP(i)$ by Eq. (9) and (10).

By iteration, the convergence process of the objective function becomes the process of finding the optimal covariance matrix vector $\Psi'(C_j)$. After calculating the saliency value of all superpixels, a seed vector s is obtained. Then the diffusion process is utilized to optimize the saliency results by Eq. (3). The generated saliency maps are validated to have outstanding performance on the nighttime image datasets.

4 Experimental Results

Experiments have been conducted to evaluate the performance of the proposed approach on the visible light condition and poor illumination condition datasets include: (1) the MSRA dataset [1], in which the principle salient objects are labeled by different human subjects, and (2) the nighttime image (NI) dataset created by us, it contains various low lighting images in the evening, the resolution of these images is 640×480, and 300 typical nighttime images are selected for testing, the human-segmented ground truths for the principle salient objects are also provided.

The performance of the proposed approach is compared with other eleven state-of-the-art models including NP [20], IS [21], LR [22], CA [23], PD [24], GBMR [25], SO [26], BL [27], BSCA [28], GL [10], and GP model [29]. These saliency models are tested on the mentioned two datasets; the experimental data is the average of all the testing images in different datasets respectively.

For the objective comparison, the *True Positive Rate* (TPR) and the *False Positive Rate* (FPR) are computed to evaluate the accuracy. Given the obtained saliency map $S_I(x, y)$ and the ground-truth $G_T(x, y)$, a fixed threshold t is used to generate the binary masks $B_t(x, y)$. The TPR and FPR can be computed via:

$$\mathrm{TPR} = E(\prod_t B_t(x,y) \cdot G_T(x,y)), \mathrm{FPR} = E(\prod_t (1 - B_t(x,y)) \cdot G_T(x,y)). \tag{12}$$

The performance comparisons of TPR and FPR results of various saliency models are presented in Fig. 2, which are tested on the mentioned two datasets, respectively. It shows that the proposed model has superior performance.

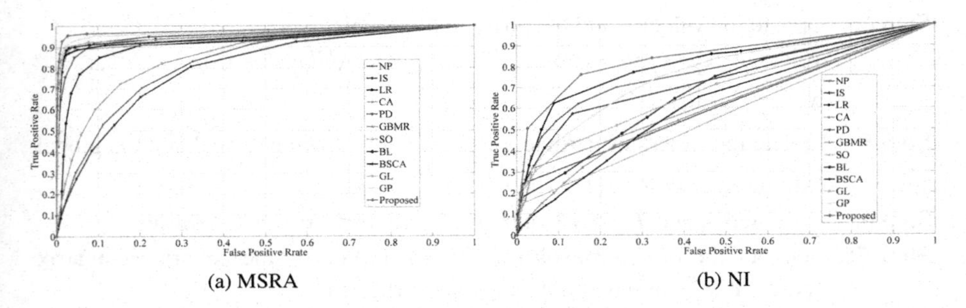

(a) MSRA (b) NI

Fig. 2. The TPR and FPR performance comparisons.

The precision, recall, and F-measure criteria are also used for performance evaluation. Let $R(\cdot)$ represents the salient region, given the ground-truth (denoted as G_T) and binary mask (denoted as B_m) of the obtained final saliency map, the precision and recall can be calculated via:

$$Precision = \frac{R(B_m \cap G_T)}{R(B_m)}, Recall = \frac{R(B_m \cap G_T)}{R(G_T)}. \tag{13}$$

The comprehensive evaluation index F-measure is got via

$$F{-}measure = \frac{(1 + \beta^2)\, precision \cdot recall}{\beta^2 \cdot precision + recall}. \tag{14}$$

where $\beta^2 = 0.3$ is to weigh the precision and recall. The performance comparisons of various models are shown in Fig. 3. The F-measure of our model is relatively higher than others on two datasets, which indicates a better performance.

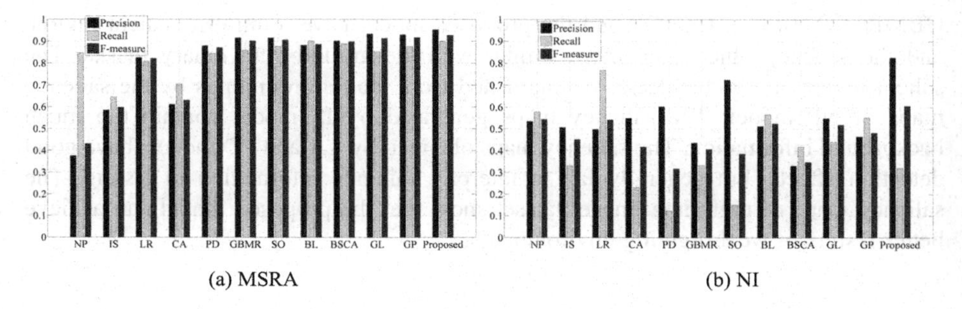

(a) MSRA (b) NI

Fig. 3. The precision, recall, and F-measure performance comparisons.

The *area under the curve* (AUC) is calculated to give an intuitive comparison. It indicates how well the generated saliency map predicts the human interesting region. The *mean absolute error* (MAE) is also introduced as another evaluation criterion, which is obtained by computing the average difference between the generated saliency map $S_I(x, y)$ and the ground-truth $G_T(x, y)$. The experiments are carried out using MATLAB on G2020 CPU PC with 12 GB RAM.

The AUC, MAE and the execution time performance comparisons of various saliency models are shown in Table 1, which demonstrates that our saliency maps achieve the best detection results and have a higher similarity with G-T. The computational complexity of the proposed model is slightly higher than the IS, GBMR, SO, BSCA and GP model, whereas our model can get more accurate estimations.

Table 1. The AUC, MAE and run-time (in second) performance comparisons.

	AUC (MSRA)	AUC (NI)	MAE (MSRA)	MAE (NI)	Time (MSRA)	Time (NI)
NP	0.8267	0.7734	0.3934	0.1941	2.055	18.316
IS	0.7990	0.7330	0.2668	0.1933	0.442	4.954
LR	0.9107	**0.8146**	0.1889	0.2437	45.339	216.16
CA	0.8535	0.6408	0.2550	0.1928	47.506	135.16
PD	0.9525	0.6156	0.1512	**0.1745**	15.885	41.066
GBMR	0.9368	0.6028	0.0846	0.2643	1.116	2.775
SO	0.9465	0.5679	**0.0753**	0.1997	0.127	0.204
BL	0.9503	0.6844	0.1215	0.3234	42.340	156.30
BSCA	0.9423	0.6199	0.0813	0.2855	1.622	2.965
GL	0.9551	0.6884	0.1124	0.2431	7.855	25.302
GP	**0.9667**	0.7676	0.0784	0.2681	1.710	7.555
Ours	***0.9846***	***0.8577***	***0.0591***	***0.1483***	3.203	5.912

The subjective performance comparisons on MSRA and NI dataset are presented in Fig. 4, which shows that the saliency maps obtained by the proposed model has good detection performance. As can be seen from Fig. 4, the superpixel-based model

GBMR, SO, BSCA, GL, GP and the proposed model have a uniform salient region, and the saliency objects are more similar with the ground-truth binary masks. The salient region cannot be clearly distinguished from their surroundings by the saliency maps of NP model. The saliency maps generated by IS model contain too much background information. The saliency maps obtained by CA and PD models have good detection effects, but generally fail to interpret saliency information in testing. The saliency maps of nighttime image dataset show that the proposed model can achieve better results in weak lighting environments.

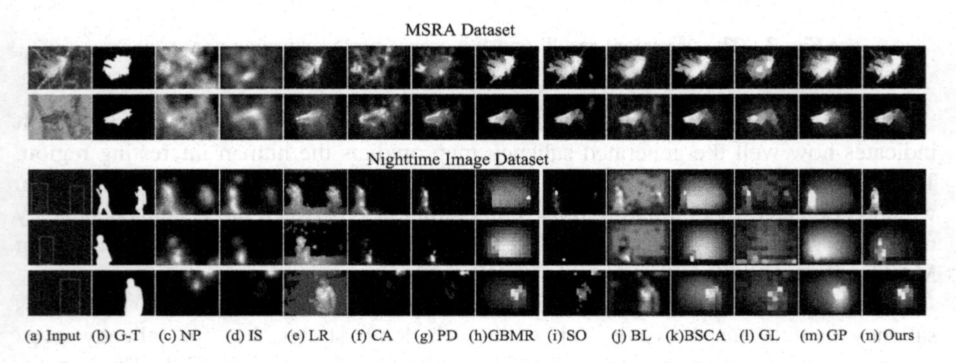

Fig. 4. Visual comparisons of saliency maps on the MSRA and NI dataset.

5 Conclusions

In this paper, the local covariance dissimilarity and the global PSO search are utilized to detect the salient object in low contrast images. The superpixel segmentation is adopted to simplify the calculation and the graph diffusion is utilized to optimize the results. Experiments have been carried out on the public available MSRA dataset and our nighttime image dataset for salient object detection, which indicate that the proposed method can achieve the best performance against eleven state-of-the-art salient object detection models in nighttime images. This study aims to solve the problem of saliency detection under weak lighting condition, which also provides a new thought for the research of saliency applications in big image data.

Acknowledgements. This work was supported by the Natural Science Foundation of China (61602349, 61440016, and 61273225), Hubei Chengguang Talented Youth Development Foundation (2015B22), and the Educational Research Project from the Educational Commission of Hubei Province (2016234).

References

1. Liu, T., Sun, J., Zheng, N.-N., Tang, X., Shum, H.-Y.: Learning to detect a salient object. In: CVPR, pp. 1–8 (2007)
2. Kennedy, J., Eberhart, R.: Particle swarm optimization. In: ICNN, pp. 1942–1948 (1995)
3. Gao, H., Pun, C.-M., Kwong, S.: An efficient image segmentation method based on a hybrid particle swarm algorithm with learning strategy. Inf. Sci. **369**, 500–521 (2016)
4. Uguz, S., Sahin, U., Sahin, F.: Edge detection with fuzzy cellular automata transition function optimized by PSO. Comput. Electr. Eng. **43**, 180–192 (2015)
5. Jiji, G.W., DuraiRaj, P.J.: Content-based image retrieval techniques for the analysis of dermatological lesions using particle swarm optimization technique. Appl. Soft Comput. **30**, 650–662 (2015)
6. Itti, L., Koch, C., Niebur, E.: A model of saliency-based visual attention for rapid scene analysis. TPAMI **20**(11), 1254–1259 (1998)
7. Zhang, J., Wang, M., Zhang, S., Li, X., Wu, X.: Spatiochromatic context modeling for color saliency analysis. TNNLS **27**(6), 1177–1189 (2016)
8. Xia, C., Qi, F., Shi, G., Wang, P.: Nonlocal center-surround reconstruction-based bottom-up saliency estimation. Pattern Recognit. **48**(4), 1337–1348 (2015)
9. Xu, X., Mu, N., Chen, L., Zhang, X.: Hierarchical salient object detection model using contrast based saliency and color spatial distribution. MTA **75**(5), 2667–2679 (2015)
10. Liu, J., Wang, S.: Salient region detection via simple local and global contrast representation. Neurocomputing **147**, 435–443 (2015)
11. Tong, N., Lu, H., Zhang, Y., Ruan, X.: Salient object detection via global and local cues. Pattern Recognit. **48**(10), 3258–3267 (2015)
12. Huang, K., Wang, L., Tan, T., Maybank, S.: A real-time object detecting and tracking system for outdoor night surveillance. Pattern Recognit. **41**(1), 432–444 (2008)
13. Han, J., Yue, J., Zhang, Y., Bai, L.-F.: Salient contour extraction from complex natural scene in night vision image. Infrared Phys. Technol. **63**, 165–177 (2014)
14. Achanta, R., Shaji, A., Smith, K., Lucchi, A., Fua, P., Susstrunk, S.: SLIC superpixels compared to state-of-the-art superpixel methods. TPAMI **34**(11), 2274–2282 (2012)
15. Xu, X., Mu, N., Zhang, H., Fu, X.: Salient object detection from distinctive features in low contrast images. In: ICIP, pp. 3126–3130 (2015)
16. Hou, X., Zhang, L.: Saliency detection: a spectral residual approach. In: CVPR, pp. 1–8 (2007)
17. Tuzel, O., Porikli, F., Meer, P.: Region covariance: a fast descriptor for detection and classification. In: Leonardis, A., Bischof, H., Pinz, A. (eds.) ECCV 2006. LNCS, vol. 3952, pp. 589–600. Springer, Heidelberg (2006). doi:10.1007/11744047_45
18. Forstner, W., Moonen, B.: A metric for covariance matrices. In: Geodesy-The Challenge of the 3rd Millennium, pp. 299–309 (2003)
19. Erdem, E., Erdem, A.: Visual saliency estimation by nonlinearly integrating features using region covariances. J. Vis. **13**(4), 11, 1–20 (2013)
20. Murray, N., Vanrell, M., Otazu, X., Parraga, C.A.: Saliency estimation using a non-parametric low-level vision model. In: CVPR, pp. 433–440 (2011)
21. Hou, X., Harel, J., Koch, C.: Image signature: highlighting sparse salient regions. TPAMI **34**(1), 194–201 (2012)
22. Shen, X., Wu, Y.: A unified approach to salient object detection via low rank matrix recovery. In: CVPR, pp. 853–860 (2012)
23. Goferman, S., Zelnik-Manor, L., Tal, A.: Context-aware saliency detection. TPAMI **34**(10), 1915–1926 (2012)

24. Margolin, R., Tal, A., Zelnik-Manor, L.: What makes a patch distinct? In: CVPR, pp. 1139–1146 (2013)
25. Yang, C., Zhang, L., Lu, H., Ruan, X., Yang, M.-H.: Saliency detection via graph-based manifold ranking. In: CVPR, pp. 3166–3137 (2013)
26. Zhu, W., Liang, S., Wei, Y., Sun, J.: Saliency optimization from robust background detection. In: CVPR, pp. 2814–2821 (2014)
27. Tong, N., Lu, H., Yang, M.: Salient object detection via bootstrap learning. In: CVPR, pp. 1884–1892 (2015)
28. Qin, Y., Lu, H., Xu, Y., Wang, H.: Saliency detection via cellular automata. In: CVPR, pp. 110–119 (2015)
29. Jiang, P., Vasconcelos, N., Peng, J.: Generic promotion of diffusion-based salient object detection. In: ICCV, pp. 217–225 (2015)

Locality-Constrained Iterative Matrix Regression for Robust Face Hallucination

Guangwei Gao[1,3,4(✉)], Huijuan Pang[2], Cailing Wang[2], Zuoyong Li[3], and Dong Yue[1]

[1] Institute of Advanced Technology, Nanjing University of Posts and Telecommunications, Nanjing 210023, China
csggao@gmail.com, medongy@vip.163.com

[2] School of Automation, Nanjing University of Posts and Telecommunications, Nanjing 210023, China
hjpangnupt@163.com, wangcl@njupt.edu.cn

[3] Fujian Provincial Key Laboratory of Information Processing and Intelligent Control, Minjiang University, Fuzhou 350121, China
fzulzydq@126.com

[4] Key Laboratory of Intelligent Perception and Systems for High-Dimensional Information of Ministry of Education, Nanjing University of Science and Technology, Nanjing 210094, China

Abstract. The performance of traditional face recognition approaches is sharply reduced when encountered with a low-resolution (LR) probe face image. The basic idea of a face super-resolution (SR) is to desire a high-resolution (HR) face image from an observed LR one with the help of a set of training examples. In this paper, we propose a locality-constrained iterative matrix regression (LCIMR) model for face hallucination task and use the alternating direction method of multipliers to solve it. LCIMR attempts to directly use the image matrix to compute the representation coefficients to maintain the essential structural information. A locality constraint is also enforced to preserve the locality and the sparsity simultaneously. Moreover, LCIMR iteratively updates the locality similarities and reconstruction weights based on the result (the hallucinated HR patch) from previous iteration, giving rise to improved performance. Experimental results on the benchmark FEI face database show the superiority of the proposed method over some state-of-the-art algorithms.

Keywords: Face hallucination · Locality constrained · Nuclear norm · Manifold

1 Introduction

In the past twenty years, face recognition has been an active topic in various applications [1–4]. However, due to the restrictions of surveillance system, such as long distance to the interest object, it is sometimes difficult to gain high-definition face images. Face hallucination, or face super-resolution, is a technology to obtain high-resolution (HR) face images

D. Liu et al. (Eds.): ICONIP 2017, Part III, LNCS 10636, pp. 613–621, 2017.
https://doi.org/10.1007/978-3-319-70090-8_62

from observed low-resolution (LR) inputs, thus providing more facial details for the following recognition process.

Compared with the interpolation-based methods, learning-based methods have gained more attention since they can significantly improve the visual quality for super-resolution reconstruction with large magnification factor. Inspired by the pioneering work of Barker [5], various learning-based face hallucination methods have been presented in the past few decades. Wang et al. [6] represented the input LR image as a linear combination over the LR training samples by principal component analysis (PCA). An [7] used canonical correlation analysis (CCA) to find a coherent subspace that maximizes the correlation between the PCA coefficients of corresponding LR and HR images. Chang et al. [8] presented a neighbor embedding (NE) based super-resolution method. Different from those approaches using a fixed number of neighbors for reconstruction, Yang [9] and Gao [10] introduced sparse coding technique that adaptively selects the most relevant neighbors for reconstruction.

Recognizing that human face is a highly structured object, the position prior can be fully incorporated into face hallucination procedure. Ma et al. [11] proposed a position-patch based face hallucination method using all patches from the same position in a dictionary, and used a least square representation (LSR) to obtain the optimal reconstruction weights. To tackle the unstable solution of least square problem in [11], Jung et al. [12] employed sparsity prior to improve the reconstruction result with several principal training patches. Recently, Wang et al. [13] further proposed a weighted adaptive sparse regularization (WASR) method to super-resolve face images. Jiang et al. [14] incorporated a locality constraint into the least square inverse problem to maintain locality and sparsity simultaneously.

In the aforementioned patch based methods, before calculating the representation coefficients, we must convert the patch matrices into vectors beforehand. In addition, they obtain the weights and the target HR patches separately and actually considers only one manifold (LR patch manifold), while ignoring the geometrical information of the HR patch manifold. In this paper, we propose to use the nuclear norm of the representation residual image to compute the representation coefficients straightforward to preserve the structure of error image. We also incorporate a locality constraint into the objective function to reach sparsity and locality simultaneously. In particular, an iteration approach is used to search the similarities in the HR patch manifold and calculate the reconstruction weights in the LR patch manifold. Experimental results on the benchmark FEI face database shows the superiority of the proposed method over some state-of-the-art algorithms.

The reminder of the paper is organized as follows. We present our face hallucination method in Sect. 2. Section 3 evaluates the performance of the proposed methods. Section 4 concludes this paper.

2 Locality-Constrained Iterative Matrix Regression

2.1 Problem Formulation

Unlike traditional methods, which aim to seek a package solution to construct the HR patch once and for all, we propose in this paper to iteratively update the HR patch.

In our method, all the face image patches are denoted in the matrix form. In other words, the input patch and training face image patches can be denoted as $y(i,j) \in \Re^{d\times d}$ and $A^m(i,j) \in \Re^{d\times d}(m = 1,\ldots,N)$, respectively. For the convenience of expression, we omit the position index (i, j) in the following text. Therefore, the HR patch guided locality-constraint is applied to construct the following objective function:

$$< \hat{w}, \hat{x} > \; = \arg\min_{w,x} \|A(w) - y\|_* + \lambda \|d \otimes w\|_2^2 \tag{1}$$

where $\otimes$ denotes point-wise vector product, $A(w) = w_1 A^1 + w_2 A^2 + \ldots + w_N A^N$, $\|\cdot\|_*$ is the nuclear norm (i.e. the sum of the singular values) of a matrix, d is locality similarities between the estimated HR patch x and bases in the HR patch dictionary:

$$d_j = \|x - B_j\|_2^2, j = 1,\ldots,N \tag{2}$$

The objective function (1) with respect to w and to x can be solved respectively and iteratively. For the p-th block, first set the $x(p)$ to its Bicubic interpolation version, then update the weight vector $w(p)$ with $x(p)$ by minimizing Eq. (1).

2.2 Optimization via ADMM

In the following text, we describe how to obtain w when getting the locality similarities d in the HR patch space.

For the convenience of expression, we rewrite the optimization problem of (1) as:

$$\begin{aligned} \min_{w,E} \; & \|E\|_* + \lambda \|d \otimes w\|_2^2 \\ s.t. \; & A(w) - y = E \end{aligned} \tag{3}$$

The above problem can be solved via the alternating direction method of multipliers (ADMM) [15, 16] with the following augmented Lagrangian function:

$$\begin{aligned} L_\mu(w, E) = \; & \|E\|_* + \lambda \|d \otimes w\|_2^2 \\ & + Tr\left(Z^T(A(w) - y - E)\right) + \frac{\mu}{2}\|A(w) - y - E\|_F^2 \end{aligned} \tag{4}$$

where Z is the Lagrange multiplier, μ is a penalty parameter.

Updating w: Given E, the optimization problem can be reformulated as

$$L_{\mu}(w) = \|d \otimes w\|_2^2 + \frac{\mu}{2\lambda}\left\| y + E - \frac{1}{\mu}Z - A(w) \right\|_F^2 \tag{5}$$

Following [14], the solution of problem (5) can be derived analytically as

$$w^{k+1} = \left(G + \tau D^2\right) \backslash ones(N, 1), \tag{6}$$

Where $ones(N, 1)$ is a $N \times 1$ column vector of ones, the operator "\" denotes the left matrix division operation, $\tau = 2\lambda/\mu$, D is a $N \times N$ diagonal matrix with entries $D_{mm} = d_m$, and G is the covariance matrix $G = C^T C$ with

$$C = \left(y + E - \frac{1}{\mu}Z \right) ones(N, 1)^T - H, \tag{7}$$

where $H = [Vec(A^1), Vec(A^2), \ldots, Vec(A^N)]$ and $Vec(\cdot)$ denotes the vectorization operator. The final optimal solution is obtained by rescaling to satisfy $\sum_{m=1}^{N} w_m = 1$.

Updating E: Given w, the optimization problem can be rewritten as

$$E^{k+1} = \arg\min_{E} \left(\frac{1}{\mu}\|E\|_* + \frac{1}{2}\left\| E - \left(A(w) - y + \frac{1}{\mu}Z \right) \right\|_F^2 \right) \tag{8}$$

The optimal solution is [17]:

$$E^{k+1} = U T_{\frac{1}{\mu}}[S] V \tag{9}$$

where $(U, S, V^T) = svd(A(w) - y + (1/\mu)Z)$. For a given $\tau > 0$, the singular value thresholding operator $T_{\tau}(\cdot)$ is defined as

$$T_{\tau}(Q) = U_{p \times r} \mathrm{diag}\left(\{\max(0, \sigma_j - \tau)\}_{1 \le j \le r} \right) V_{q \times r}^T \tag{10}$$

where r is the rank of S.

Here, we apply the following termination conditions:

$$\|A(w) - y - E\|_{\infty} \le \varepsilon \tag{11}$$

Algorithm 1. Face hallucination via LCIMR

Input: HR training images $A_H^1,\ldots,A_H^N$, corresponding LR training images $A_L^1,\ldots,A_L^N$, input LR images y, iterative number *maxIter*.

Initialize: $s = 0$, $x_0 = \text{Bicubic}(y)$;

1: **For** each input patch matrix in y:

repeat

a) Compute the locality similarities:

$$d_{m(s)}(i,j) = \left\| y_s(i,j) - A_H^m(i,j) \right\|_2^2, m = 1,\ldots,N$$

b) Calculate the optimal weights $w_{(s)}^*(i,j)$ according to Eq. (3) at a specific iteration number s;

c) Construct the desired HR patch by $x_{s+1}(i,j) = \sum_{m=1}^{N} A_H^m(i,j) w_{m(s)}^*(i,j)$

until $s > maxIter$

2: **End for**

3: The target HR image X can be obtained by integrating all the reconstructed HR patch matrices.

Output: The hallucinated HR face image X.

2.3 Face Hallucination via LCIMR

As for face hallucination, the training set consists of HR and LR face image pairs. A_H^m denote the HR face images, while A_L^m ($m = 1,\ldots, N$) denote their LR counterparts. The face hallucination task aims to acquire the HR face image X from its LR observation y. We summarize the whole face hallucination algorithm in Algorithm 1.

3 Experimental Results and Discussions

3.1 Datasets Description

Experiments are conducted on the FEI database [18] to demonstrate the effectiveness of the proposed algorithm. All the face images are manually aligned using the locations of three points: centers of left and right eyeballs and center of the mouth (some examples are shown in Fig. 1). We crop the region of the faces and normalize the HR images to the size of 120×100. The LR images are obtained by smoothing (an averaging filter of size 4×4) and down-sampling (the down-sampling factor is 4) the corresponding HR images, thus the LR images have size 30×25. 400 images from 200 subjects are selected and each subject has two frontal images, one with a neutral expression and the other with a smiling facial expression.

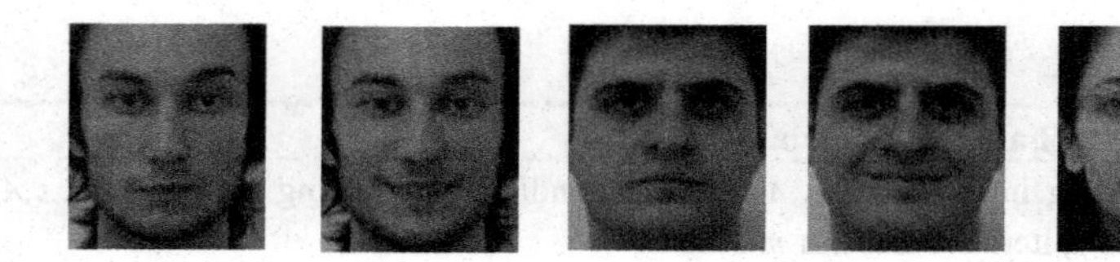

Fig. 1. Some sample images from the FEI face database.

3.2 Parameter Analysis

In this subsection, we investigate the effect of the locality regularization parameter λ of the proposed method. Figure 2 plots the average PSNR and SSIM values according to different values of λ. We can see that the regularization parameter plays an important role on the performance of the proposed method: as λ increases, more benefits on performance can be gained. However, it should be noted that the value of λ could not be set too high. With a proper regularization parameter λ, our proposed method will gain good results. We recommend setting λ between 0.005 and 0.05.

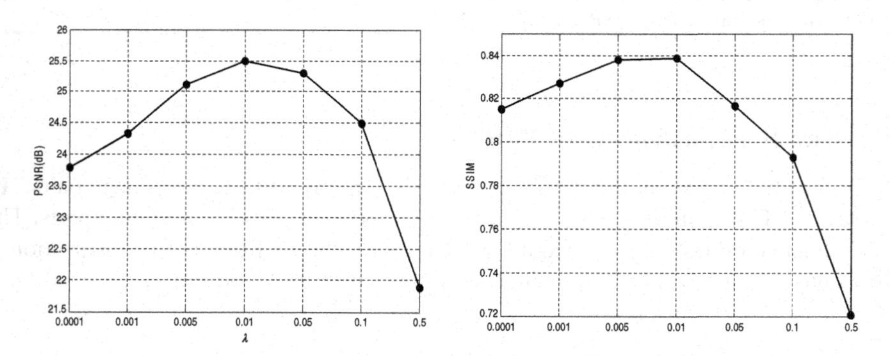

Fig. 2. The average PSNR and SSIM values of the proposed LCIMR method with different λ.

3.3 Hallucination Comparisons on FEI Database

In this part, we compare our proposed methods with Chang's neighbor embedding (NE) method [8], Ma's least square representation (LSR) method [11], Jung's sparse representation (SR) method [12] and Jiang's locality-constrained representation (LcR) method [14]. We randomly choose 250 images for training, and 40 images for testing. The iteration number is set to 3 in our method. As for other patch-based methods, we suggest using the size 3×3 pixels for LR patch, while the corresponding HR patch size is 12×12 pixels. As usual, we adopt the same assessment methods (PSNR and SSIM) to measure the qualities of the reconstructed images with other algorithms.

Each input is corrupted by a randomly located square block whose elements are random numbers between 0 and 255. We list some hallucinated face images in Fig. 3 to show the subjective reconstruction image equality. From Fig. 3, we can observe that outputs of NE, LSR and SR look similar to input noise images. Results from LCR seem more reliable than previous methods with less noise. However, block effects are still evident due to biased solution from input noise. Compared with other methods, our approach render smoother images with less block effect and more details with the help of structure noise depiction of nuclear norm. The average evaluation measures of different methods are tabulated in Table 1. The reconstruction qualities of all the super-resolution methods reduce in noise situation. While our method can still obtains the best performance in terms of PSNR and SSIM.

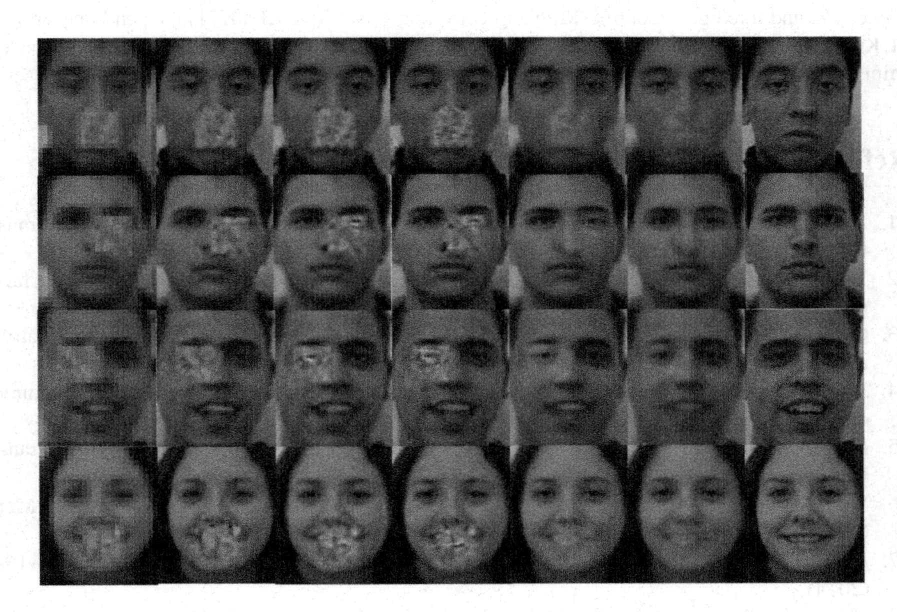

Fig. 3. Some sample images from the FEI face database.

Table 1. The average PSNR and SSIM values on the FEI database.

Methods	NE [7]	LSR [10]	SR [11]	LcR [13]	LCIMR
PSNR(dB)	23.6105	23.6790	23.7462	25.3105	**25.5942**
SSIM	0.7946	0.7995	0.8042	0.8121	**0.8396**

4 Conclusion

In this paper, we present a novel locality-constrained iterative matrix regression (LCIMR) model for robust face hallucination. LCIMR can directly compute the combination coefficients of input patch matrix without matrix-to-vector conversion by taking advantage of the structure characteristics of the representation error. LCIMR also enforces the locality constraint which can capture fundamental similarities between neighbor patches and derives an analytical solution to the constrained problem. Furthermore, LCIMR iteratively searches the similarities in the HR patch manifold and calculate the reconstruction weights in the LR patch manifold. Experiments conducted on the benchmark FEI face database clearly validate the advantages of our proposed LCIMR over the state-of-the-art methods in face hallucination both objectively and subjectively.

Acknowledgement. This work was partially supported by the National Natural Science Foundation of China under Grant nos. 61502245 and 61772254, the China Postdoctoral Science Foundation under Grant no. 2016M600433, the Natural Science Foundation of Jiangsu Province under Grant no. BK20150849, Open Fund Project of Fujian Provincial Key Laboratory of Information Processing and Intelligent Control (Minjiang University) (No. MJUKF201717), Open Fund Project of Key Laboratory of Intelligent Perception and Systems for High-Dimensional Information of Ministry of Education (Nanjing University of Science and Technology) (No. JYB201709).

References

1. Jain, A.K., Ross, A., Prabhakar, S.: An introduction to biometric recognition. IEEE Trans. Circ. Syst. Video Technol. **14**(1), 4–20 (2004)
2. Yang, M., Zhu, P., Liu, F., Shen, L.: Joint representation and pattern learning for robust face recognition. Neurocomputing **168**, 70–80 (2015)
3. Gao, G., Yang, J., Wu, S., Jing, X., Yue, D.: Bayesian sample steered discriminative regression for biometric image classification. Appl. Soft Comput. **37**, 48–59 (2015)
4. Jing, X.Y., Wu, F., Zhu, X., Dong, X., Ma, F., Li, Z.: Multi-spectral low-rank structured dictionary learning for face recognition. Pattern Recognit. **59**, 14–25 (2016)
5. Baker, S., Kanade, T.: Limits on super-resolution and how to break them. IEEE Trans. Pattern Anal. Mach. Intell. **24**(9), 1167–1183 (2002)
6. Wang, X., Tang, X.: Hallucinating face by eigentransformation. IEEE Trans. Syst. Man Cybern. C Appl. Rev. **35**(3), 425–434 (2005)
7. An, L., Bhanu, B.: Face image super-resolution using 2D CCA. Sig. Process. **103**, 184–194 (2014)
8. Chang, H., Yeung, D.Y., Xiong, Y.: Super-resolution through neighbor embedding. In: Proceedings of IEEE Computer Society Conference on Computer Vision and Pattern Recognition, pp. 275–282. IEEE, Washington (2004)
9. Yang, J., Wright, J., Huang, T.S., Ma, Y.: Image super-resolution via sparse representation. IEEE Trans. Image Process. **19**(11), 2861–2873 (2010)
10. Gao, G., Yang, J.: A novel sparse representation based framework for face image super-resolution. Neurocomputing **134**, 92–99 (2014)
11. Ma, X., Zhang, J., Qi, C.: Hallucinating face by position-patch. Pattern Recognit. **43**(6), 2224–2236 (2010)
12. Jung, C., Jiao, L., Liu, B., Gong, M.: Position-patch based face hallucination using convex optimization. IEEE Sig. Process. Lett. **18**(6), 367–370 (2011)
13. Wang, Z., Hu, R., Wang, S., Jiang, J.: Face hallucination via weighted adaptive sparse regularization. IEEE Trans. Circ. Syst. Video Technol. **24**(5), 802–813 (2014)
14. Jiang, J., Hu, R., Wang, Z., Han, Z.: Noise robust face hallucination via locality-constrained representation. IEEE Trans. Multimedia **16**(5), 1268–1281 (2014)
15. Yang, J., Luo, L., Qian, J., Tai, Y., Zhang, F., Xu, Y.: Nuclear norm based matrix regression with applications to face recognition with occlusion and illumination changes. IEEE Trans. Pattern Anal. Mach. Intell. **39**(1), 156–171 (2017)

16. Gao, G., Yang, J., Jing, X.Y., Shen, F., Yang, W., Yue, D.: Learning robust and discriminative low-rank representations for face recognition with occlusion. Pattern Recognit. **66**, 129–143 (2017)
17. Cai, J.F., Candes, E.J., Shen, Z.W.: A singular value thresholding algorithm for matrix completion. SIAM J. Optim. **20**(4), 1956–1982 (2010)
18. Thomaz, C.E., Giraldi, G.A.: A new ranking method for principal components analysis and its application to face image analysis. Image Vis. Comput. **28**(6), 902–913 (2010)

Structure-Preserved Face Cartoonization

Chenhao Gao, Bin Sheng(✉), and Ruimin Shen

Department of Computer Science and Engineering, Shanghai Jiao Tong University, No. 800 Dongchuan Road, Minhang District, Shanghai, China
gaochyz@163.com, shengbin@cs.sjtu.edu.cn, rmshen@sjtu.edu.cn

Abstract. Face cartoon synthesis has been proved to show wide range of uses in lots of fields, for example, instant message communication, suspects identity and online entertainment. In this paper, we propose a new dense descriptor based model to synthesize a face cartoon from a face photo which displays a great outcome. We generate two kinds of stylized face cartoons, one is called cartoon portraits and the other is called cartoon sketch. By integrating the two kinds of cartoons using guided filter, our results preserve the detail information of the input photo and generate good cartoon artistic style.

Keywords: Face cartoon · Texture synthesis · Local PatchMatch · Guided filter · Dense descriptor

1 Introduction

Human face displays wide range of uses in tasks of pattern recognition and computer vision. In recent years, a number of face sketch synthesize approaches [1–3] have been achieved and their applications in digital entertainment and law enforcement also proved to be a great success. However, most of the prior approaches focus on grayscaling sketch synthesizing with no color information, which makes the synthesized face sketch dull and unattractive. To synthesize a more vivid style of face portrait, several models have been established [4–7]. All of the above methods could be classified into two types, one is called component-based method [6,7] and the other is exemplar-based method [4,5]. But almost all the related works [4–6] generate a cartoon face with stiff or unnatural style, some of the organs on face are even distorted and hard to distinguish. Only semantic features on the those faces are deformed into a kind of cartoon style. How to generate a natural, undistorted and vivid face cartoon arouse our interest.

In our framework, we also find that patch matching between input photo and cartoon database also gives another style similar to xylograph. The xylograph style gives more details on edges of face which inspires our another contribution:using xylograph portraits as guidance to sharpen original cartoon sketches. In this paper, we refer xylograph style face to cartoon portrait because it portray the face features more accurately and the original exemplar-based generated face to cartoon sketch because it resembles the cartoon faces drawn by artist more. Our work achieves great visual result on the CUHK student database (Fig. 1) at a resolution of 630 × 500.

D. Liu et al. (Eds.): ICONIP 2017, Part III, LNCS 10636, pp. 622–631, 2017.
https://doi.org/10.1007/978-3-319-70090-8_63

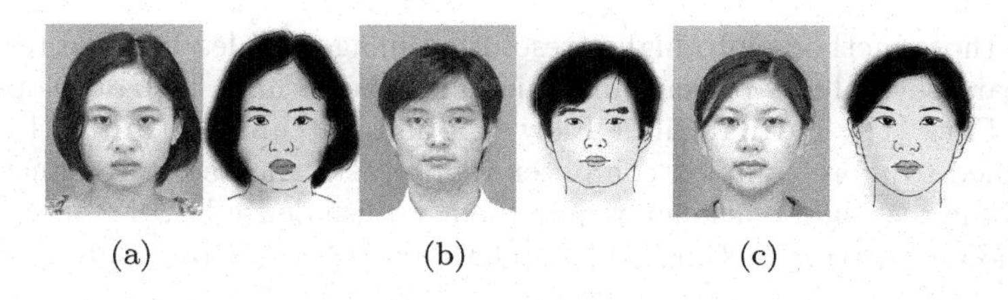

Fig. 1. Examples of CUHK face photo-cartoon pairs. Left photos in (a)–(c) Photos in face database. Right photos in (a)–(c) Corresponding cartoons drawn by an artist.

2 Related Work

2.1 Component-Based and Exemplar-Based Synthesis

Component-based synthesis focus on highlighting of face components and contours. In [6] an Active Appearance Model (AAM) are used to acquire feature points. Based on such feature points a face is decomposed into different organs, such as hair, mouth, eyes and nose. Then those organs are enhanced on a base image which is averaged throughout the whole database. A fetal drawback in this model lies in that components on face are coming from different training faces so that they might be totally unsymmetrical and weird (Fig. 2). As for [7] more complicated components are extracted from a wider range of database but the cartoon face organs are totally dissimilar to the original face organs. To summarize, component-based synthesis may result in obvious unlikeness between target input and output so we decide to use exemplar-based synthesis method.

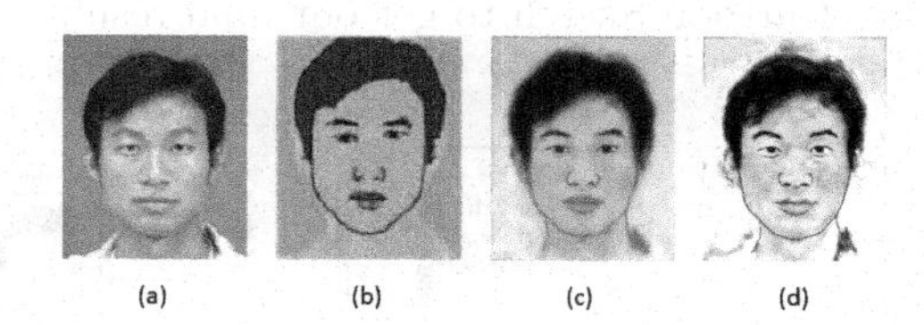

Fig. 2. Comparison between component-based approach in [6] and the exemplar-based approach. (a) The input face photo, (b) face cartoon generated in [6], (c)–(d) face cartoon generated using proposed approach, (c) is the cartoon sketch and (d) is the cartoon portrait. An unsymmetrical and weird effect appeared in eyes' region of (b) while (c) and (d) restore the organs perfectly.

2.2 Descriptor-Based Synthesis

Former exemplar-based face cartoon or face sketch synthesis are mainly based on image intensity. It might work well on low resolution cartoon or sketch synthesis, for example, the synthesized sketch in [1–3] has a resolution of 200×250. In our framework all the training images have a resolution of 630×500. Directly applying those methods into higher resolution image will lead to bad results. In recent years, several new package matching approaches have been developed such as SIFT Flow [11], DAISY Filter Flow [12] and PatchMatch Filter [13]. They are qualified to more complex computer vision tasks such as stereo and optical flow. For the reason of efficiency requirement we use dense DAISY descriptor as information extracting method and PatchMatch [14] to be our match method.

3 Our Approach

3.1 Overview

The algorithm of our descriptor-based cartoon synthesis is presented in this section. Figure 3 shows the overview of our algorithm. To save the computation time, we first compute all of the DAISY descriptor for each pixel in each image in database offline and store them for the matching process. In this system, a colorized face photo is grayed and then serves as input. We get an corresponding face cartoon as an output. First we calculate the standard upright DAISY descriptor for the graying input. Then we use the PatchMatch algorithm to find the best match for a test patch in a local area in one training photo. After we have matched all patches throughout the database we got offset vector and cost vector. Then we calculate K-NN search to find the closest k patches as candidates. We use those k candidates to formulate our raw synthesis result. The above process are execute twice, one for photo database and one for cartoon database. Then Spatial Sketch Denoising (SSD) [3] is used to denoise the raw synthesize result from photo database, we call it **Cartoon Sketch**. The raw result from cartoon database is named **Cartoon Portrait**. Finally we use Cartoon Portrait as a guidance to filter Cartoon Sketch to get our final result.

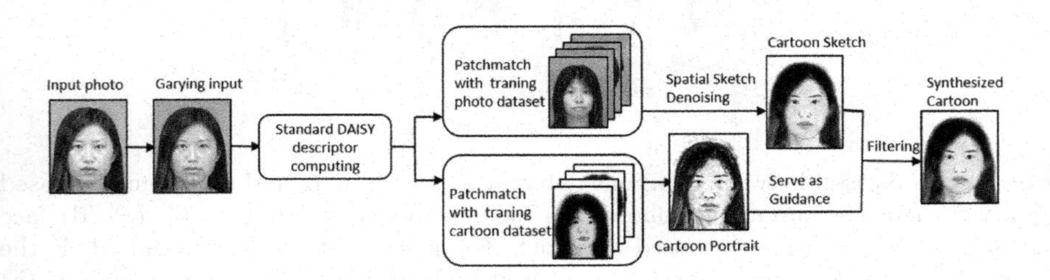

Fig. 3. Baseline approach of our methods.

3.2 Descriptor Formulation

Suppose I be the input photo and D_p represent the DAISY descriptor at pixel p in image I, our main goal is to find the most correspondent pixel q in a training image of a K elements training database $\{I_q^l\}_{l=1}^K$, then execute K-NN search to select k candidates for each pixel p in I. Here since the image size is large we choose the standard DAISY coefficient in [9], which means $Q = 3, T = 8, H = 8, R = 15$, then for each pixel in each training photo we have a 200 dimension vector to represent the descriptor, written as $\{D_q^l\}_{l=1}^K$. First we need to select candidate with closest Euclidean distance. For the lth database image, we choose the best match q for p, which means:

$$q = \arg\min_{q \subset \Phi_p} (D_p - D_q^l)^2 \tag{1}$$

where Φ_p represents the searching range, which will be detailed in Sect. 3.3.

3.3 Local PatchMatch Algorithm

PatchMatch [14] algorithm is a randomized searching algorithm for correspondence problems. In our task we detect the global PatchMatch and local PatchMatch differences. Global PatchMatch means searching around the whole image region through the whole training database. Figure 5 gives the explanation of such two methods. Theoretically global PatchMatch could find optimal correspondence point through training photo set but in our experiments global PatchMatch method is seriously constrained to the illumination conditions and in our cases leaves terrible halos on the foreheads, as shown in Fig. 5. The reason lies in that DAISY descriptor precisely capture the illumination changes on the forehead so as to restored such information that we exactly do not want. To overcome such unsatisfactory problem we use local PatchMatch algorithm instead. Local PatchMatch search only a small fixed region in the neighbourhood of central pixel which not only eliminate the above annoying effects but also improved the speed performance. Detailed PatchMatch process could be found in [14] (Fig. 4).

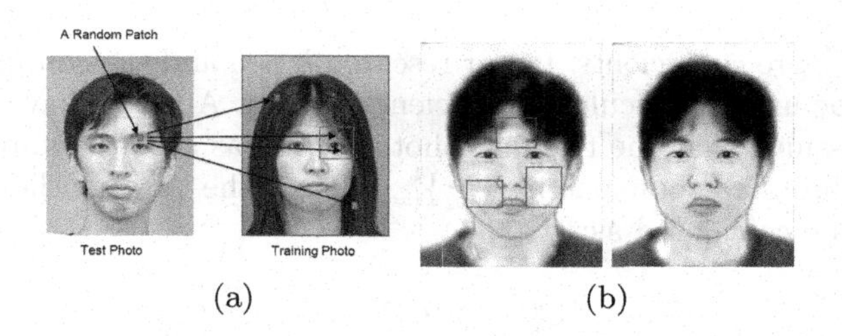

Fig. 4. Illustration of Local PatchMatch and Global PatchMatch. (a) Search range of Local PatchMatch and Global PatchMatch. Red square implicates the search range by Global PatchMatch while green square for Local PatchMatch. (b) Comparison between synthesized cartoon using Local PatchMatch and Global PatchMatch. The left result are generated using Global PatchMatch and obvious halos appear in red squares due to illumination changes. (Color figure online)

For a 200 dimension descriptor the brutal pixel-by-pixel scanline match is unbearable. Table 1 gives comparison of time consuming between different matching methods for a test photo.

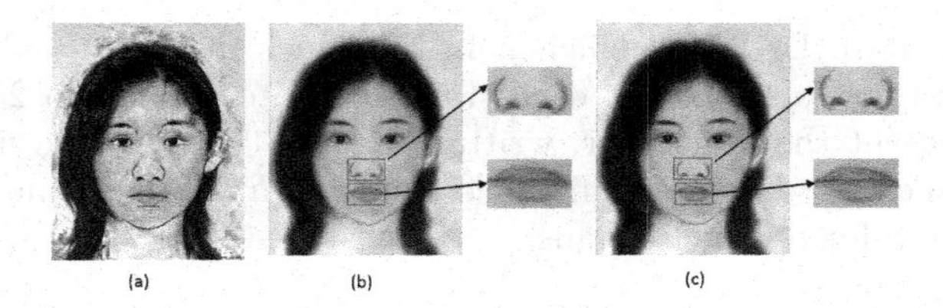

Fig. 5. Result of using Cartoon Portrait as guidance to filter Cartoon Sketch. (a) Cartoon Portrait C^P. (b) Filtered result F, in our experiment $|\omega| = 25$ and $\epsilon = 0.2$. (c) Original Cartoon Sketch C^S. We could find that due to (a)'s detail preservation effect (b) have better details in nose and mouse than (c).

Table 1. Time consuming comparison between different matching methods

Matching methods	Time (seconds)
Pixel-by-pixel scanline match	>3600
Local PatchMatch	257
Global PatchMatch	504

3.4 K-NN Search and Spatial Sketch Denoising (SSD)

After we have obtained the closest pixel q^l for p in lth training photo, we get totally K candidates. Then we execute K-NN search from those candidates and pick out k candidates as calculating component. Suppose I_p be the test patch center, and $I_{q_i}^{l_i} (i = 1, 2, \cdots, k)$ be the k corresponding matched patches, we have

$$x_p^1 I_{q_1}^{l_1} + x_p^2 I_{q_2}^{l_2} + \cdots + x_p^k I_{q_k}^{l_k} = I_p \tag{2}$$

where x_p^1 to x_p^k are coefficients. In our research $k = 5$ and the conjugate gradient solver could be used to calculate coefficients quickly. After we have got the coefficients, we assume that the training photo patch has a linear mapping relation [15] with training cartoon patch $\{C_{q_i}^{l_i}\}_{i=1}^k$. Then the raw synthesized cartoon sketch could be calculated as:

$$C_p = x_p^1 C_{q_1}^{l_1} + x_p^2 C_{q_2}^{l_2} + \cdots + x_p^k C_{q_k}^{l_k} \tag{3}$$

Spatial Sketch Denoising (SSD) is the main contribution of [3] and outperforms other denoising methods much. Ψ_p represents the denoising size around pixel p. In our research when denoising radius equals 11 and $\Psi_p = 121$ the result performs best.

3.5 Cartoon Portrait as Guidance

Former synthesizing approaches never consider there is any relationship between test photo and training cartoon because they share no stylistic similarity. But in our research since we use DAISY descriptor there do exist a map from test photo

Algorithm 1. Guided filtering by C^P for C^S.

Require: The Cartoon Portrait C^P and Cartoon Sketch C^S.
Ensure: The new synthesized cartoon result F.
1: ω_k is a square window in C^P and ϵ is the regularization parameter.
2: **for all** pixel k in C^P and C^S **do**
3: calculate $\mu_k(C^P)$ via boxfilter in ω_k for C^P
4: calculate $\mu_k(C^S)$ via boxfilter in ω_k for C^S
5: $cov_k = \mu_k(C^P \cdot C^S) - \mu_k(C^P) \cdot \mu_k(C^S)$
6: $\sigma_k^2 = \mu_k(C^P \cdot C^P) - \mu_k(C^P) \cdot \mu_k(C^P)$
7: $a_k = \frac{cov_k}{\sigma_k^2 + \epsilon}$
8: $b_k = \mu_k(C^S) - a_k \cdot \mu_k(C^P)$
9: **end for**
10: **for** each pixel F_i in F **do**
11: $\overline{a_i} = \frac{1}{|\omega|} \sum_{k \in \omega_i} a_k$
12: $\overline{b_i} = \frac{1}{|\omega|} \sum_{k \in \omega_i} b_k$
13: $F_i = \overline{a_i} \cdot C_i^P + \overline{b_i}$
14: **end for**

to training cartoon. For example, pixel at the center of eyes in test photo may have same gradient and orientation information as in training cartoon because human eyes approximately share a ellipse contour. This inspired us a new idea that matching the training cartoon rather than training photo may result in reasonable performance. From Fig. 5 we find that synthesized result with training cartoon is more similar to original face photo in detail and we decide to call it Cartoon Portrait. Compared with Cartoon Sketch, Cartoon Portrait displays a kind of xylograph style and preserve the sharp edges of chin and nose well. Besides some background noise and mosaic in overlapping area Cartoon Portrait restore the input face perfectly. This inspired us to use edge-aware filter to extract detail information from cartoon portrait to enhance edges in Cartoon Sketch. There are several kinds of edge-aware filter. The most widely used might be bilateral filter [16,17] and guided filter [18]. Here we use guided filter to process images. Suppose a new Cartoon Portrait C^P is generated via above workflow by cartoon database. Compared with Cartoon Sketch C^S generated in Sect. 3.4 we do not do any denoising on C^P because the details in C^P are exactly what we want. The whole algorithm could be summarized in Algorithm 1.

4 Experiments

In this section, we detect the performance of proposed face cartoon synthesis method with other methods. We set the search radius to 5 in the K-NN search process. The size of Ψ_p in SSD equals 121 which displays best visual result. In PatchMatch process the local search radius are set to 10 pixel. First we evaluate the proposed method against previous work qualitatively. Second we evaluate the proposed method quantitatively.

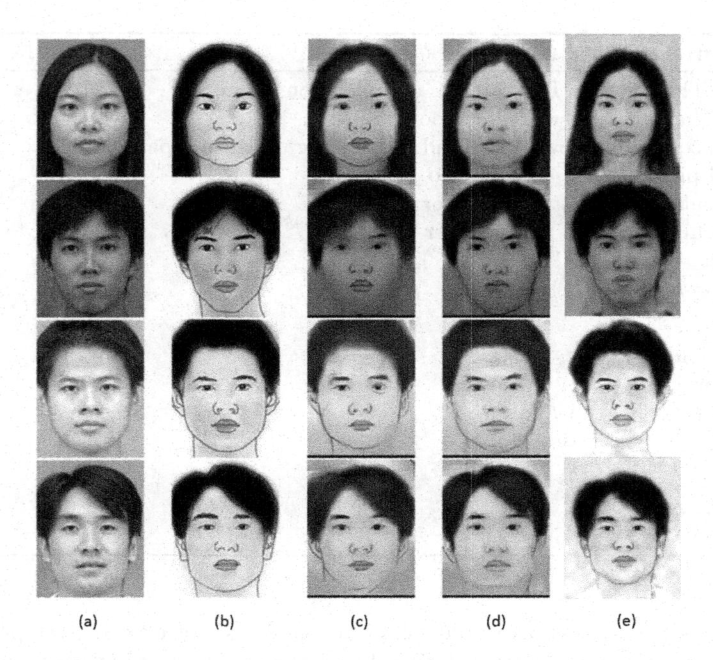

Fig. 6. Qualitative comparison with previous work. (a) Input test photos. (b) Face cartoons drawn by artist. (c) Results by the Markov network in [5]. (d) Results by guided synthesis in [4]. (e) Results by proposed integrated synthesis. (Color figure online)

4.1 Qualitative Comparison with Previous Work

Previous work are based on component-based synthesis method [6] and exemplar-based synthesis method [4,5]. Since we have compared our result with [6] in Sect. 2, now we focus on the comparison with [4,5]. Compared with our local PatchMatch method, the Markov Random Field based method [5] is a global matching approach, which means may result in the drawback brought by illumination changes. Another fetal drawback lies in that since we use descriptor-based method rather than traditional intensity-based method the time consume of MRF model may be intolerable.

As for the guided face cartoon synthesis in [4], it uses the average training photo patch I_p as a base then adds the filtered corresponding training cartoon patch C_p. In the filtered step four weights are calculated to form a filter kernel. Those weights are photo-photo weight, photo-cartoon weight, smoothness constraint weight and the spatial measure weight. Since this model are established in the average training photo, the overview of the synthesized result are inevitably resemble face photo as in Fig. 8. Another drawback of this model is the colorization problem. This model estimates the face cartoon with luminance channel and color channel. Luminace channel is responsible for the appearance reconstruction while color channel for colorization. However the model colorize the synthesized cartoon simply via two chrominance components C_b and C_r from input photo. Usually the background of a face photo is blue, which result in the blue background in the synthesized face cartoon (Fig. 6). But we all know that a more adaptive background color for a face cartoon is white because an artist usually will not consider the background when he is creating a cartoon work. In contrast, our approach synthesizes a face cartoon totally based on cartoon patches throughout the whole 3 chrominance channels and the background resemble the true cartoon by artist.

Table 2. Recognition rates of Rank-1 for different tests

Test method	Test 1	Test 2	Test 3	Test 4
Markov network [4]	100%	62%	98%	86%
Guided synthesis [4]	100%	82%	100%	86%
Proposed	100%	92%	100%	95%

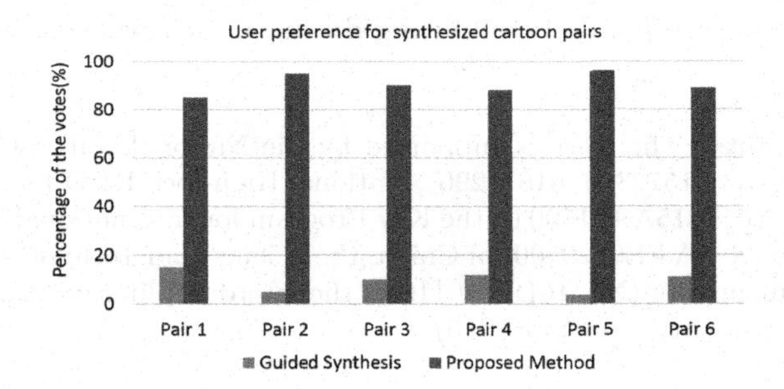

Fig. 7. User preference for synthesized cartoon pairs. There are totally 100 users are asked to choose their preferable cartoon in the totally 6 face cartoon pairs. The proposed cartoon synthesizing approach performs favorably against the guided synthesis approach across all the pairs.

4.2 Quantitative Comparison

Cartoon face is mainly rely on subjective evaluation and hard to be evaluated by quantitative criterion. We use recognition rate as an objective measurement to test our result. As in [4], we compute the rank-1 recognition rate and divide the recognition test into 4 part. Test 1 use face photo as query image, 50 synthesized cartoons by proposed method as database images. Test 2 use cartoon drawn by the artist as query image and 50 synthesized cartoons by proposed method as database images. Test 3 and Test 4 we use synthesized cartoon as query image and 50 test face photos or 50 test cartoons drawn by the artist as database images respectively. The recognition rates are measured as the quantitative criterion to evaluate our model's performance, which are given in Table 2. From Table 2 we could find that in the Test 2 and Test 4 the proposed method has a 10% advancement in contrast to previous work.

In order to make our research outcome more convincing we execute a user-subjective study. We use 6 set face cartoon generated using guided cartoon synthesis and proposed method and then ask 100 users to evaluate which cartoon is preferable for them. The testing result are shown in Fig. 7.

5 Conclusions and Future Works

This paper investigates a new descriptor-based algorithm to generate a cartoon face based on photo-cartoon database. By using DAISY descriptor and Local PatchMatch methodguided filter a satisfying structure-preserving performance has been achieved. From our work a new xylograph style cartoon face called Cartoon Portrait is generated and it displays face detail more vividly. Our possible future detection might focus on this Cartoon Portrait and generate more non-photorealistic style. Another improvement we need to focus on is to speed up the process. The whole rendering of a cartoon face in our research is about 8s, which can not be done real-time and constrains it's actual use. One possible improvement is to use parallel computing during match process, which requires further study.

Acknowledgments. The work is supported by the National Natural Science Foundation of China(No. 61572316, 61671290), National High-tech R&D Program of China (863 Program) (No. 2015AA015904), the Key Program for International S&T Cooperation Project (No. 2016YFE0129500) of China, the Science and Technology Commission of Shanghai Municipality (No. 16DZ0501100), the interdisciplinary Program of Shanghai Jiao Tong University (No. 14JCY10).

References

1. Wang, X., Tang, X.: Face photo-sketch synthesis and recognition. IEEE Trans. Pattern Anal. Mach. Intell. **31**(11), 1955–1967 (2009)
2. Zhou, H., Kuang, Z., Wong, K.Y.K.: Markov weight fields for face sketch synthesis. In: 2012 IEEE Conference on Computer Vision and Pattern Recognition (CVPR), pp. 1091–1097. IEEE (2012)
3. Song, Y., Bao, L., Yang, Q., Yang, M.-H.: Real-time exemplar-based face sketch synthesis. In: Fleet, D., Pajdla, T., Schiele, B., Tuytelaars, T. (eds.) ECCV 2014. LNCS, vol. 8694, pp. 800–813. Springer, Cham (2014). doi:10.1007/978-3-319-10599-4_51
4. Li, H., Liu, G., Ngan, K.N.: Guided face cartoon synthesis. IEEE Trans. Multimedia **13**(6), 1230–1239 (2011)
5. Zhang, C., Liu, G., Wang, Z.: Cartoon face synthesis based on Markov network. In: 2010 International Symposium on Intelligent Signal Processing and Communication Systems (ISPACS), p. 14. IEEE (2010)
6. Liu, S., Li, H., Xu, L.: Face cartoon synthesis based on the active appearance model. In: 2012 IEEE 12th International Conference on Computer and Information Technology (CIT), pp. 793–797. IEEE (2012)

7. Zhang, Y., Dong, W., Deussen, O., Huang, F., Li, K., Hu, B.G.: Data-driven face cartoon stylization. In: SIGGRAPH Asia 2014 Technical Briefs, pp. 1–4. ACM (2014)
8. Lowe, D.G.: Distinctive image features from scale- invariant keypoints. Int. J. Comput. Vis. **60**(2), 91–110 (2004)
9. Tola, E., Lepetit, V., Fua, P.: Daisy: an efficient dense descriptor applied to wide-baseline stereo. IEEE Trans. Pattern Anal. Mach. Intell. **32**(5), 815–830 (2010)
10. Mikolajczyk, K., Schmid, C.: A performance evaluation of local descriptors. IEEE Trans. Pattern Anal. Mach. Intell. **27**(10), 1615–1630 (2005)
11. Liu, C., Yuen, J., Torralba, A.: Sift flow: dense correspondence across scenes and its applications. IEEE Trans. Pattern Anal. Mach. Intell. **33**(5), 978–994 (2011)
12. Yang, H., Lin, W.Y., Lu, J.: Daisy filter flow: a generalized discrete approach to dense correspondences. In: 2014 IEEE Conference on Computer Vision and Pattern Recognition (CVPR), pp. 3406–3413 (2014)
13. Lu, J., Yang, H., Min, D., Do, M.N.: Patch match filter: efficient edge-aware filtering meets randomized search for fast correspondence field estimation. In: IEEE Conference on Computer Vision and Pattern Recognition, pp. 1854–1861 (2013)
14. Barnes, C., Shechtman, E., Finkelstein, A., Goldman, D.: Patchmatch: a randomized correspondence algorithm for structural image editing. ACM Trans. Graph. TOG **28**(3), 24 (2009)
15. Liu, Q., Tang, X., Jin, H., Lu, H., Ma, S.: A nonlinear approach for face sketch synthesis and recognition. In: IEEE Computer Society Conference on Computer Vision and Pattern Recognition, CVPR 2005, vol. 1, pp. 1005–1010. IEEE (2005)
16. Tomasi, C., Manduchi, R.: Bilateral filtering for gray and color images. In: Sixth International Conference on Computer Vision, pp. 839–846. IEEE (1998)
17. Petschnigg, G., Szeliski, R., Agrawala, M., Cohen, M., Hoppe, H., Toyama, K.: Digital photography with flash and no-flash image pairs. ACM Trans. Graph. (TOG) **23**, 664–672. ACM (2004)
18. He, K., Sun, J., Tang, X.: Guided image filtering. In: Daniilidis, K., Maragos, P., Paragios, N. (eds.) ECCV 2010. LNCS, vol. 6311, pp. 1–14. Springer, Heidelberg (2010). doi:10.1007/978-3-642-15549-9_1

Deep Metric Learning with Symmetric Triplet Constraint for Person Re-identification

Sen Li[1], Xiao-Yuan Jing[1,2(✉)], Xiaoke Zhu[1,3], Xinyu Zhang[1], and Fei Ma[1]

[1] State Key Laboratory of Software Engineering, School of Computer, Wuhan University, Wuhan, China
jingxy_2000@126.com

[2] College of Automation, Nanjing University of Posts and Telecommunications, Nanjing, China

[3] School of Computer and Information Engineering, Henan University, Kaifeng, China

Abstract. Deep metric learning is an effective method for person re-identification (PR-ID). In practice, impostor samples generally own more discriminative information than other well separable negative samples (WSN-samples). Specifically, existing triplet-based deep learning methods with asymmetric triplet constraint can not effectively remove impostors because they treat all different types of negative samples equally. To utilize discriminative information in negative samples more efficiently, we build a symmetric triplet constraint based deep metric learning network (STDML). STDML designs a symmetric triplet margin maximized objective function, which requires impostors to leave more than a margin from positive pair and requires the distance between WSN-samples maximized, simultaneously. Experiments on three benchmark datasets (CUHK03, CUHK01 and VIPeR) achieve better performance than existing methods and significantly improved the effectiveness of exploiting discriminative information from WSN-samples.

Keywords: Deep metric learning · Person re-identification · Impostor · WSN-sample · Symmetric triplet constraint

1 Introduction

Person re-identification is a great significant research in video surveillance which purpose is to match non-overlapping pedestrian captured in different cameras with different views. In practice, PR-ID meets great challenges due to the influence of large variations in illumination, viewpoint, background clutter, occlusion and image resolution. To solve the problem, a large number of metric learning (ML) methods are proposed in recent years by learning a distance metric to eliminate the influence of these variations. The major ML-based methods include the keep it simple and straightforward metric (KISSME) [1] and relative distance comparison (RDC) [2], etc. Those ML-based methods usually learn a distance metric based on hand-crafted features.

D. Liu et al. (Eds.): ICONIP 2017, Part III, LNCS 10636, pp. 632–641, 2017.
https://doi.org/10.1007/978-3-319-70090-8_64

To learn a distance metric with better discriminative capability, some deep metric learning (DML) methods for PR-ID have been presented with the development of deep learning, e.g., deep metric learning for practical person re-identification (DML) [3] and Embedding Deep Metric for Person Re-identification (EDML) [4]. The basic idea of both DML and EDML methods is to learn a deep metric network by maximizing similarity between inter-class in neighbors and minimizing similarity between intra-class simultaneously. In practice, DML-based methods have drawbacks on exploiting discriminant information in case of treating all different types of negative samples equally.

The impostor, whose clothes, stature and posture are similar to a another person, always owns more identical information than the other negative samples in reality. However, existing triplet-based methods [5–7] do not treat impostors differently and they make the similarity of negative pair larger and positive pair smaller in same triplet with asymmetric triplet constraint. Therefore, they can not effectively remove impostors because asymmetric triplet constraint always lead to situations as Fig. 1(a), (b) and (c). As illustrated in [8], there has much more discriminative information which has not been exploited in impostors and WSN-samples. So how to utilize these information effectively is a challenging problem.

To address this problem, this paper focuses on exploiting discriminative information more effectively from impostors and WSN-samples by deep metric leaning. We propose a novel deep metric learning method with symmetric triplet constraint and optimize the network by using symmetric triplet margin maximized objective function and a stochastic sub-gradient algorithm [9]. Our major contributions are summarized below:

(1) We build a deep metric learning framework with symmetric triplet constraint and design the symmetric triplet margin maximized objective function by treating impostors and WSN-samples differently, which makes impostors leave more than a margin distance between each sample in positive pair and maximizes the distance between the WSN-pair, simultaneously.
(2) We propose an optimization method for STDML network. In the procedure, we derive an update and back propagation rules for connecting weights, biases and the deep metric with a stochastic sub-gradient algorithm.
(3) We conduct experiments on three benchmark datasets (CUHK03, CUHK01 and VIPeR) to evaluate the effectiveness of STDML and outperform compared with state-of-the-art methods at matching rate.

2 Division of Negative Samples

According to relationships in negative samples illustrated in [8], negative samples can be divided into three categories, including symmetric correlated impostors, asymmetric correlated impostors and well separable negative samples.

Symmetric correlated impostors. Given a training set as $X = \{x_1, \ldots, x_i, \ldots, x_N\}$, x_i, x_j and x_k denote feature vector of i-th, j-th and k-th images in X, constituted a triplet $< i, j, k >$, where x_j is a positive sample

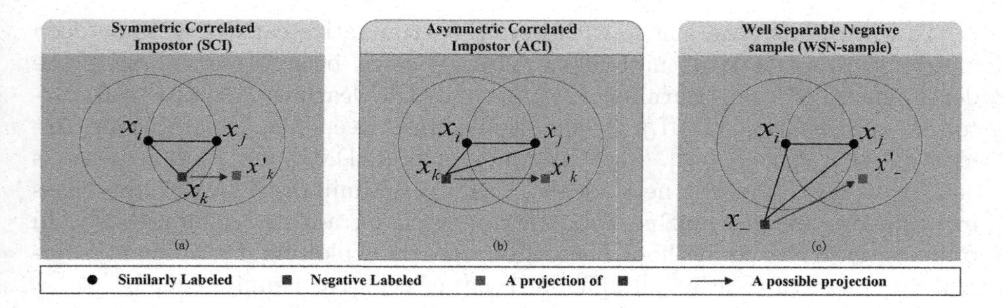

Fig. 1. Relationships in negative samples and probable impostor projection results of existing impostor-based deep learning methods with asymmetric triplet constraint.

and x_k is a negative sample correlated with x_i. If x_k satisfies the condition that $\|x_i - x_k\|^2 \leqslant \|x_i - x_j\|^2$ and $\|x_j - x_k\|^2 \leqslant \|x_i - x_j\|^2$ simultaneously, we call the x_k symmetric correlated impostor (SCI), referred in Fig. 1(a).

Asymmetric correlated impostors. If the triplet $< i, j, k >$ satisfies the condition that $\|x_i - x_k\|^2 \leqslant \|x_i - x_j\|^2$ but $\|x_j - x_k\|^2 > \|x_i - x_j\|^2$, we call x_k asymmetric correlated impostor (ACI), referred in Fig. 1(b).

Well separable negative samples. Existing x_i and x_- in X, where x_- is a negative sample correlated with x_i, if there is not any positive sample x_j with x_i satisfying SCI or ACI condition, we call x_- are well separable negative sample (WSN-sample) and constitute WSN-pair $< i, - >$ with x_i, referred in Fig. 1(c).

3 Proposed Approach

We propose our STDML for all impostors and WSN-samples. Then, we give the optimization and identification methods. Our basic idea of STDML is described in Fig. 2.

3.1 Pre-computing of Quadruplets Collection

Given pedestrian image set $X = \{x_1, \ldots, x_i, \ldots, x_N\}$, we extract their deep features by using AlexNet [10] for computing distance between two samples. For each x_i, we compute all quadruplets $< i, j, k, - >$ contained x_i, its positive sample x_j, impostor x_k and WSN-sample x_- by iterative combination of triplets $< i, j, k >$ and WSN-pair $< i, - >$ using rules in Sect. 2. Then, we obtain a collection $T = T_1, \ldots, T_i, \ldots, T_N$ where T_i contains all quadruplets of x_i, where $K(i)$ denotes the number of quadruplets in T_i.

3.2 STDML

With a quadruplet $< i, j, k, - >$, we use STDML with $L + 1$ layers. W^l and b^l denote l-th projection matrices and biases respectively, d^l denotes the number

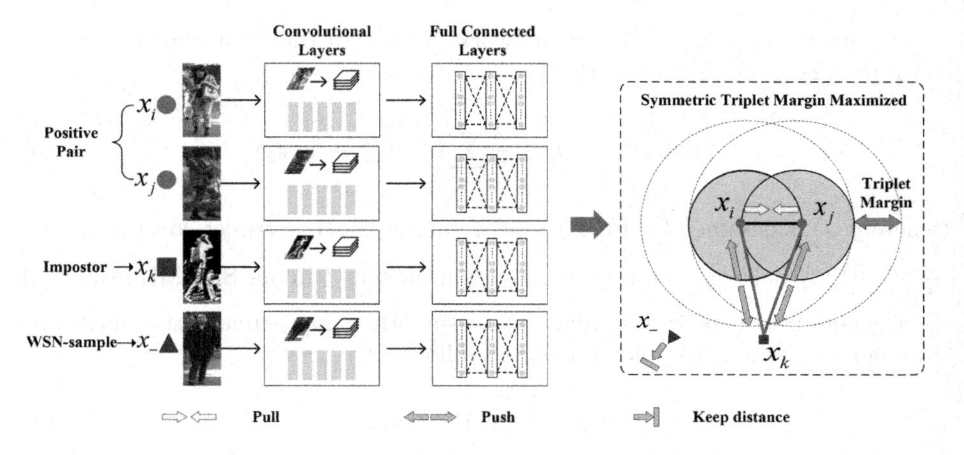

Fig. 2. Basic idea of our STDML.

of l-th nodes and σ is a nonlinear active function where $1 \leqslant l \leqslant L$. As illustrated in Fig. 2, a quadruplet $< i, j, k, - >$ is the input of first layer. Specifically, taking x_i for example, we compute the first output $h_i^1 = \sigma(W^1 x_i + b^1)$. For l-th layer, its output is $h_i^l = \sigma(W^l h_i^{l-1} + b^l)$ and the output of top layer is $h_i^L = \sigma(W^L h_i^{L-1} + b^L)$. For the $L+1$ layer, we design a symmetric triplet margin maximized objective function for L-th outputs of four samples in a quadruplet simultaneously.

In our approach, we compute the distance between outputs of L-th layer from x_i and x_j as below:

$$D(i,j) = \|h_i^L - h_j^L\|^2 \tag{1}$$

where H_i^L and H_j^L are outputs in L-th layer from image x_i and x_j.

To exploit discriminative information more effectively in negative samples, we require each quadruplet $\{i, j, k, -\}$ satisfying the condition that both $(\alpha+1)\|h_i^L - h_j^L\|^2 - \|h_i^L - h_k^L\|^2$ and $(\alpha+1)\|h_i^L - h_j^L\|^2 - \|h_j^L - h_k^L\|^2$ minimized and $\|h_i^L - h_-^L\|^2$ maximized simultaneously, where α is a margin factor. This idea is illustrated in Fig. 2. With the learned parameter $f = \{W^1, W^2, \ldots, W^L, b^1, b^2, \ldots, b^L\}$ in our network, the outputs of L-th layer should meet this objective function:

$$\begin{aligned} \min_f J = & \frac{1}{\|N\|} \sum_{i=1, T_i \in T}^{N} \frac{1}{K(i)} \sum_{<i,j,k,-> \in T_i} \{((\alpha+1)D(i,j) - D(i,k)) \\ & + ((\alpha+1)D(i,j) - D(j,k)) - \gamma D(i,-)\} \end{aligned} \tag{2}$$

where γ is a balance parameter between triplet term and WSN term and α is a margin ratio factor to improve learning effectiveness.

By introducing Eq.2 into our network, we design the optimization function with parameters of our network f:

$$\min_f H = g(J) + \frac{\lambda}{2}\sum_{l=1}^{L}(\|W^l\|_F^2 + \|b^l\|_2^2) \tag{3}$$

where $g(J)$ is designed for Eq. 2 to satisfy our symmetric triplet constraint, and $\frac{\lambda}{2}\sum_{l=1}^{L}(\|W^l\|_F^2 + \|b^l\|_2^2)$ is for regularizing parameters f in our STDML. And $g(a)$ is a generalized logistic loss function to smoothly approximate the hinge loss function $a = \max(a, 0)$, and is define as follows [9]:

$$g(a) = \frac{1}{\rho}\log(1 + exp(\rho a)) \tag{4}$$

3.3 Optimization

To obtain the best solution f, we propose an optimization method with iterative computing in our network. First, give a sequence suitable matrices to f for initialization and compute the quadruplets collection for training set, and then update our network by Eq. 3 until satisfying convergence condition.

We introduce the stochastic sub-gradient descent algorithm [9] to get the optimal value of $f = \{W^1, W^2, \ldots, W^L, b^1, b^2, \ldots, b^L\}$ in Eq. 3. The back propagation procedure of H in Eq. 3 is illustrated below:

$$\begin{aligned}\frac{\partial H}{\partial W^l} = &\sum_{<i,j,k,->\in T_i} (\delta_i^l(h_i^{l-1})' + \delta_j^l(h_j^{l-1})' + \delta_k^l(h_k^{l-1})' + \delta_-^l(h_-^{l-1})') \\ &+ \lambda W^l\end{aligned} \tag{5}$$

$$\frac{\partial H}{\partial b^l} = \sum_{<i,j,k,->\in T_i} (\delta_i^l + \delta_j^l + \delta_k^l + \delta_-^l) + \lambda b^l \tag{6}$$

where δ_i^l, δ_j^l, δ_k^l and δ_-^l denote updating functions. For the L-th layer, we compute them by:

$$\delta_i^l = g'(J)(4(\alpha+1)R_1 - 2R_2 - 2\gamma R_4) \odot \sigma'(y_i^L) \tag{7}$$

$$\delta_j^l = g'(J)(-4(\alpha+1)R_1 - 2R_3) \odot \sigma'(y_j^L) \tag{8}$$

$$\delta_k^l = g'(J)(2R_2 + 2R_3) \odot \sigma'(y_k^L) \tag{9}$$

$$\delta_-^l = g'(J)(2\gamma R_4) \odot \sigma'(y_-^L) \tag{10}$$

where

$$\begin{aligned}J \triangleq &\frac{1}{\|N\|}\sum_{i=1,T_i\in T}^{N}\frac{1}{K(i)}\sum_{<i,j,k,->\in T_i}\{((\alpha+1)D(i,j) - D^j(i,k)) \\ &+ ((\alpha+1)D(i,j) - D(j,k)) - \gamma D(i,-)\}\end{aligned} \tag{11}$$

$$R_1 \triangleq \frac{1}{K(i)} \sum_{<i,j,k,->\in T_i} h_i^l - h_j^l \tag{12}$$

$$R_2 \triangleq \frac{1}{K(i)} \sum_{<i,j,k,->\in T_i} h_i^l - h_k^l \tag{13}$$

$$R_3 \triangleq \frac{1}{K(i)} \sum_{<i,j,k,->\in T_i} h_j^l - h_k^l \tag{14}$$

$$R_4 \triangleq \frac{1}{K(i)} \sum_{<i,j,k,->\in T_i} h_i^l - h_-^l \tag{15}$$

$$y_i^l \triangleq W^l h_i^l + b^l \tag{16}$$

$$y_j^l \triangleq W^l h_j^l + b^l \tag{17}$$

$$y_k^l \triangleq W^l h_k^l + b^l \tag{18}$$

$$y_-^l \triangleq W^l h_-^l + b^l \tag{19}$$

For l-th layers, $1 \leqslant l \leqslant L-1$, we compute δ_i^l, δ_j^l, δ_k^l by:

$$\delta_i^l = (W^{l+1})^T \delta_i^{l+1} \odot \sigma'(y_i^l) \tag{20}$$

$$\delta_j^l = (W^{l+1})^T \delta_j^{l+1} \odot \sigma'(y_j^l) \tag{21}$$

$$\delta_k^l = (W^{l+1})^T \delta_k^{l+1} \odot \sigma'(y_k^l) \tag{22}$$

$$\delta_-^l = (W^{l+1})^T \delta_-^{l+1} \odot \sigma'(y_-^l) \tag{23}$$

where operation "$\odot$" is the element-wise multiplication.

Next, we update $f = \{W^1, W^2, \ldots, W^L, b^1, b^2, \ldots, b^L\}$ by:

$$W^l = W^l - \mu \frac{\partial H}{\partial W^l} \tag{24}$$

$$b^l = b^l - \mu \frac{\partial H}{\partial b^l} \tag{25}$$

where μ is the learning rate, $1 \leqslant l \leqslant L$.

The procedure of our optimization method can be seen in Algorithm 1.

3.4 Person Re-identification with Learned Network

For a pedestrian image y in testing set, we make image y as input of network with learned parameter f and obtain the deep feature h_y^L. Then, we compute distances between h_y^L and each image in training set by Eq. 1. Finally, we choose the smallest distance which is between h_y^L and the output of L-th layer from x_c and obtain the c as label of y. Denoted as follows:

$$L_y = arg \min_c (y_i, x_c).1 \leqslant c \leqslant C \tag{26}$$

where c is class of x_c, and C is total number of classes in X.

Algorithm 1. STDML

Input: Training set X, network layer number $L+1$, learning rate μ, iterative number K, parameter α, γ, λ, and convergence error ε;
Output: Parameters W^l and b^l, $1 \leqslant l \leqslant L$.
Initialization: Initialize W^l, b^l with appropriate values
for $k = 1, 2, \cdots, K$ **do**
    Compute the quadruplets collection T.
    for $l = 1, 2, \cdots, L$ **do**
        Compute h_i^l, h_j^l, h_k^l and h_-^l using the deep network.
    end
    for $l = L, L-1, \cdots, 1$ **do**
        Obtain the gradients according to (5)-(6).
    end
    for $l = 1, 2, \cdots, L$ **do**
        Update W^l and b^l according to (24)-(25).
    end
    Calculate H_k using (3).
    If $k > 1$ and $\|H_k - H_{k-1}\| < \varepsilon$, go to **Return.**
end
Return: W^l and b^l, where $1 \leqslant l \leqslant L$.

4 Experiments

To evaluate the proposed STDML, We conduct extensive experiments on three widely used datasets. Then, we show the comparable results compared with state of the arts.

4.1 Datasets

The **CUHK03** dataset [11] contains 13164 images from 1360 persons. In our experiment, we follow exactly the same setting as [11]. The **CUHK01** dataset [12] has 971 identities with 2 images per person in each view. We evaluate our STDML following [13]. The **VIPeR** dataset [14] contains 1264 outdoor images of 632 individuals from two non-overlapping cameras with varying illumination. The experiment setting is followed as [13].

4.2 Implementation Details

For evaluating our STDML, Our approach is implemented based on TensorFlow framework [15]. We constitute five convolutional layers to extract deep features and three fully connected layers for deep metric learning with parameter f, kernel and nodes initialized as AlexNet [10] for convolutional layers.

Input data organisation. Each image is resized to 224×224 before fed into network, and the output of the last FC layer is a 1024-dimensional feature vector. We compute quadruplets collections for three datasets, which most of samples are constituted at least five folds. At last, we set the batch size as 128.

Training Setting. Parameters of our method, including learning rate μ, parameter λ, margin factor α, are set as 10^{-3}, 10^{-5}, 0.15 for all three datasets, and balance parameter γ is set as 0.20, 0.30, 0.15 for CUHK03, CUHK01, and VIPeR respectively.

4.3 Results of STDML

We conduct random 10 trials to show average results with different Rank-k. The experiments on compared methods exactly use the same setting as their paper illustrated.

We evaluate our STDML with 10 representative person re-identification methods which report results on at least two datasets. Results with rank-1, rank-5 and rank-10 accuracies are reported in Table 1. Compared with existing representative methods, our method can significantly improve the effectiveness of exploiting discriminative information in WSN-samples and it perform a competitive performance than other methods.

Table 1. The matching rates of the state-of-the-art methods and our methods on three datasets. The bold indicates the best performance.

Method	CUHK03			CUHK01			VIPeR		
	r = 1	r = 5	r = 10	r = 1	r = 50	r = 10	r = 1	r = 5	r = 10
KISSME [1]	14.17	48.54	52.57	29.40	57.67	62.43	19.60	48.00	62.20
kLFDA [16]	48.20	59.34	66.38	42.76	69.01	79.63	32.33	65.78	79.72
SIRCIR [17]	52.17	85.00	92.00	72.50	91.00	95.50	35.76	67.00	82.50
NullReid [18]	58.90	85.60	92.45	64.98	84.96	89.92	42.28	71.46	82.94
Ensembles [19]	62.10	89.10	94.30	53.40	76.30	84.40	45.90	77.50	**88.90**
IDLA [20]	54.74	86.50	94.00	65.00	89.50	93.00	34.81	63.32	74.79
DeepRanking [21]	-	-	-	70.94	92.30	96.90	38.37	69.22	81.33
GatedSiamese [22]	68.10	88.10	94.60	-	-	-	37.80	66.90	77.40
ImpTrpLoss [6]	-	-	-	53.70	84.30	91.00	**47.80**	74.70	84.80
MDTnet [5]	74.68	**95.99**	97.47	77.50	95.00	**97.50**	45.89	71.84	83.23
STDML	**75.32**	93.67	**97.58**	**78.13**	**95.34**	96.81	45.75	**77.53**	87.65

5 Conclusion

In this paper, we propose a novel deep metric learning framework with symmetric triplet constraint for person re-identification, which exploits discriminative information more effectively from different type negative samples. With five convolutional layers and three fully connected layers, our STDML can extract deep features and learn a sequence deep matrices for removing impostors and remaining the independence of WSN-samples simultaneously. It can be seen on experimental results, our STDML outperform compared with state-of-the-art methods at matching rate.

Acknowledgments. The authors would like to thank the editors and anonymous reviewers for their constructive comments and suggestions. This work was supported by the National Key Research and Development Program of China under Grant No. 2017YFB0202001, by the National Nature Science Foundation of China under Grant Nos. 61272273, 61373038, 61672392, 61472178, 61672208, U1404618, the National Basic Research 973 Program of China under Project No. 2014CB340702, the Program of State Key Laboratory of Software Engineering under Grant No. SKLSE-1216-14, the Scientific Research Staring Foundation for Introduced Talents in NJUPT under NUPTSF No. NY217009, the Science and Technology Program in Henan province under Grant No. 1721102410064, the Science and Technique Development Program of Henan under Grant No. 172102210186, and the Province-School-Region Project of Henan University under Grant No.2016S11.

References

1. Kstinger, M., Hirzer, M., Wohlhart, P., Roth, P.M., Bischof, H.: Large scale metric learning from equivalence constraints. In: 25th IEEE Conference on Computer Vision and Pattern Recognition, pp. 2288–2295. IEEE Press, New York (2012)
2. Zheng, W.S., Gong, S., Xiang, T.: Reidentification by relative distance comparison. IEEE Trans. Pattern Anal. Mach. Intell. **35**, 653–668 (2013)
3. Yi, D., Lei, Z., Li, S.Z.: Deep metric learning for practical person re-identification. Comput. Sci. 34–39 (2014)
4. Shi, H., Yang, Y., Zhu, X., Liao, S., Lei, Z., Zheng, W., Li, S.Z.: Embedding deep metric for person re-identification: a study against large variations. In: Leibe, B., Matas, J., Sebe, N., Welling, M. (eds.) ECCV 2016. LNCS, vol. 9905, pp. 732–748. Springer, Cham (2016). doi:10.1007/978-3-319-46448-0_44
5. Chen, W., Chen, X., Zhang, J., Huang, K.: A multi-task deep network for person re-identification. In: 31th AAAI Conference on Artificial Intelligence, pp. 3988–3994. AAAI Press, New York (2017)
6. Cheng, D., Gong, Y., Zhou, S., Wang, J., Zheng, N.: Person re-identification by multi-channel parts-based CNN with improved triplet loss function. In: 29th IEEE Conference on Computer Vision and Pattern Recognition, pp. 1335–1344. IEEE Press, New York (2016)
7. Vijay Kumar, B.G., Carneiro, G., Ian, D.: Learning local image descriptors with deep Siamese and triplet convolutional networks by minimizing global loss functions. In: 29th IEEE Conference on Computer Vision and Pattern Recognition, pp. 5385–5394. IEEE Press, New York (2016)
8. Zhu, X., Jing, X.Y., Wu, F., Zheng, W., Hu, R., Xiao, C., Liang, C.: Distance learning by treating negative samples differently and exploiting impostors with symmetric triplet constraint for person re-identification. In: IEEE International Conference on Multimedia and Expo, pp. 1–6. IEEE Press, New York (2016)
9. Lu, J., Wang, G., Deng, W., Moulin, P., Zhou, J.: Multimanifold deep metric learning for image set classification. In: 28th IEEE Conference on Computer Vision and Pattern Recognition, pp. 1137–1145. IEEE Press, New York (2015)
10. Krizhevsky, A., Sutskever, I., Hinton, G.E.: Imagenet classification with deep convolutional neural networks. Commun. ACM **60**, 84–90 (2017)

11. Li, W., Zhao, R., Xiao, T., Wang, X.: Deepreid: deep filter pairing neural network for person re-identification. In: 27th IEEE Conference on Computer Vision and Pattern Recognition, pp. 152–159. IEEE Press, New York (2014)
12. Li, W., Zhao, R., Wang, X.: Human reidentification with transferred metric learning. In: 11th Asian Conference on Computer Vision, pp. 31–44. IEEE Press, New York (2012)
13. Lin, W., Shen, C., Van Den Hengel, A.: Deep linear discriminant analysis on fisher networks: a hybrid architecture for person re-identification. Pattern Recognit. **65**, 238–250 (2017)
14. Gray, D., Brennan, S., Tao, H.: Evaluating appearance models for recognition, reacquisition, and tracking. In: 9th Proceedings of the IEEE International Workshop on Performance Evaluation for Tracking and Surveillance, pp. 1–7. IEEE Press, New York (2007)
15. Abadi, M., Barham, P., Chen, J., Chen, Z., Davis, A., Dean, J., Devin, M., Ghemawat, S., Irving G., Isard, M., Kudlur, M., Levenberg, J., Monga, R., Moore, S., Murray, D.G., Steiner, B., Tucker, P., Vasudevan, V., Warden, P., Wicke, M., Yu, Y., Zheng, X.: Tensorflow: a system for large-scale machine learning. In: 12th USENIX Symposium on Operating Systems Design and Implementation, pp. 265–283. USENIX Association, Berkeley (2016)
16. Xiong, F., Gou, M., Camps, O., Sznaier, M.: Person re-identification using kernel-based metric learning methods. In: Fleet, D., Pajdla, T., Schiele, B., Tuytelaars, T. (eds.) ECCV 2014. LNCS, vol. 8695, pp. 1–16. Springer, Cham (2014). doi:10.1007/978-3-319-10584-0_1
17. Wang, F., Zuo, W., Lin, L., Zhang, D., Zhang, L.: Joint learning of single-image and cross-image representations for person reidentification. In: 29th IEEE Conference on Computer Vision and Pattern Recognition, pp. 1288–1296. IEEE Press, New York (2016)
18. Zhang, L., Xiang, T., Gong, S.: Learning a discriminative null space for person re-identification. In: 29th IEEE Conference on Computer Vision and Pattern Recognition, pp. 1239–1248. IEEE Press, New York (2016)
19. Paisitkriangkrai, S., Shen, C., Van Den Hengel, A.: Learning to rank in person re-identification with metric ensembles. In: 28th IEEE Conference on Computer Vision and Pattern Recognition, pp. 1846–1855. IEEE Press, New York (2015)
20. Ahmed, E., Jones, M., Marks, T.K.: An improved deep learning architecture for person re-identification. In: 28th IEEE Conference on Computer Vision and Pattern Recognition, pp. 3908–3916. IEEE Press, New York (2015)
21. Chen, S.-Z., Guo, C.-C., Lai, J.-H.: Deep ranking for person re-identification via joint representation learning. IEEE Trans. Image Process. **25**, 2353–2367 (2016)
22. Varior, R.R., Haloi, M., Wang, G.: Gated Siamese convolutional neural network architecture for human re-identification. In: Leibe, B., Matas, J., Sebe, N., Welling, M. (eds.) ECCV 2016. LNCS, vol. 9912, pp. 791–808. Springer, Cham (2016). doi:10.1007/978-3-319-46484-8_48

EPI-Patch Based Convolutional Neural Network for Depth Estimation on 4D Light Field

Yaoxiang Luo, Wenhui Zhou(✉), Junpeng Fang, Linkai Liang, Hua Zhang, and Guojun Dai

School of Computer Science and Technology,
Hangzhou Dianzi University, Hangzhou, China
{151050045,zhouwenhui,151050039,161050039,zhangh,daigj}@hdu.edu.cn

Abstract. Depth recovery from light field is an essential part of many light field applications. However, conventional methods usually suffers from two challenges: sub-pixel displacements and occlusions. In this paper, we propose an effective convolutional neural network (CNN) framework to perform the depth estimation on 4-dimensional (4D) light field. Based on the orientation-depth relationship of epipolar images (EPIs), we firstly build a training set by extracting a group of valid EPI-patch pairs with balanced depth distribution, and then an EPI-patch based CNN architecture is designed and trained to estimate the disparity of each pixel. Finally, a post-processing with global constrains is applied to the whole images to refine the output of CNN. Experimental results demonstrate the effectiveness and robustness of our method.

Keywords: Light field · Depth estimation · Deep learning · EPI · CNN

1 Introduction

Light-field cameras, also known as Plenoptic cameras, capture both spatial and angular information of the incoming light with a single shot. These 4-dimensional spatio-angular information can provide more helpful multiple viewpoints (sub-aperture images) for scene structure analysis and understanding.

Depth estimation from light-field camera has attracted extensive attention in recent years. Numerous algorithms have been developed using various depth cues from angular patches [1–4]. However, these methods actually suffer from two inevitable challenges: sub-pixel displacements and occlusions, that may lead to a loss of precision, especially in depth discontinuity regions. In order to evaluate the performance of depth estimation, several light-field datasets with ground truth have been developed [5–7].

In this paper, we take advantage of the recent success of deep learning to perform depth estimation of the 4D light-field. To the best of our knowledge, very little paper has been published on similar work. There are two key problems. First, existing light-field datasets not only are not large enough for the training of

D. Liu et al. (Eds.): ICONIP 2017, Part III, LNCS 10636, pp. 642–652, 2017.
https://doi.org/10.1007/978-3-319-70090-8_65

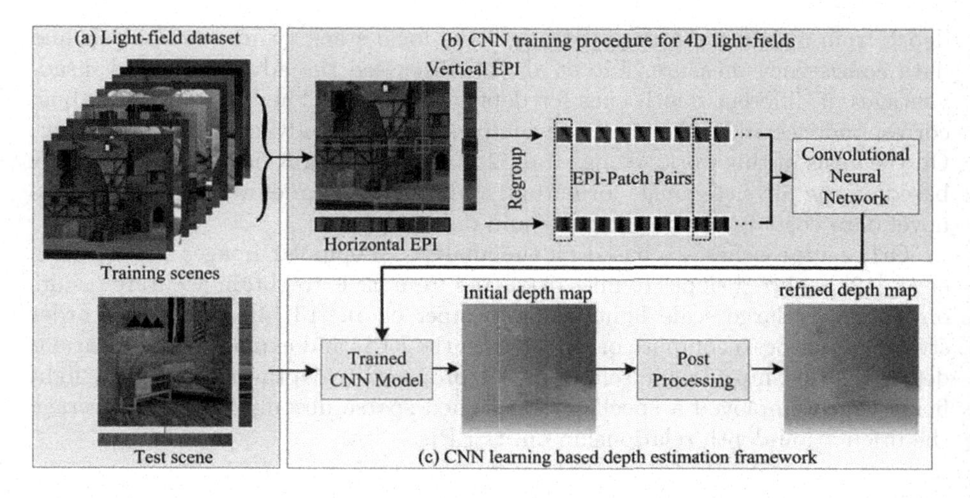

Fig. 1. Overview of our method. (a) We use the light-field dataset published by Honauer and Johannsen et al. [5,6]. We use their 16 additional scenes with ground truth as our training scenes, and the others as our test scenes. (b) A CNN architecture is proposed to take in a set of valid EPI-patch pairs extracted from the training scenes. (c) Finally, a post-processing with global constraints is applied to refine the output of the CNN.

deep learning, but also are imbalanced as far as depth distribution is concerned. Second, most of CNN architectures can not be directly adapted for 4D light-field. Our main contributions are shown in Fig. 1 and summarized below:

(1) An EPI-patch based CNN architecture is designed for 4D light-field. We firstly build a training set from the light-field benchmark dataset [5], which is composed of a set of valid EPI-patch pairs with balanced depth distribution extracted from the horizontal and vertical EPIs of each pixel. Then an EPI-patch based CNN architecture is trained to estimate the disparity of each pixel.

(2) A learning based depth estimation framework is developed. Since our CNN solution is based on the local EPI-patch pairs, it lacks global constraints and is unstable in occlusion or textureless regions, thus a post-processing with global constrains is applied to the whole images afterwards.

2 Related Work

Depth Estimation for Light-Field: There has been a substantial amount of research on depth estimation for light fields. Most of them focused on the solutions from different depth cues, such as correspondence, defocus, shading, and occlusion, etc. Georgiev and Lumsdaine [8] estimated disparity maps by computing a normalized cross correlation between microlens images. Yu et al. [9] analyzed the 3D geometry of lines in a light field image and computed the disparity maps through line matching between the sub-aperture images. Jeon et al. [1] estimated the multi-view stereo correspondences with sub-pixel accuracy using phase shift theorem. Lin et al. [10] described a technique to recover depth from a light field image based on the focal stack symmetry analysis and data consistency measure. Tao et al. [11] discussed the advantages and disadvantages of different depth cues for depth estimation. They combined shading, correspondence and defocus cues to complement the disadvantages of each other. On the basis of this work, Wang et al. [2] developed a light-field occlusion model based on the physical image formation, and then Williem et al. [3] proposed two novel data costs for correspondence and defocus cues.

Other approaches are based on the analysis of epipolar images (EPIs). Kim et al. [12] analyzed slopes of lines extracted from EPIs to obtain accurate results on extremely large scale light fields. Wanner et al. [13,14] used a first order structure tensor to compute orientation on the EPIs and exploit the orientation-depth relationship. Johannsen et al. [4] proposed a depth estimation in light fields which employed a specifically designed sparse decomposition to leverage the orientation-depth relationship on its EPIs.

Light-Field Datasets: Existing light field datasets can be classified into four types: synthetic light fields, real scene light fields captured with a plenoptic camera, a camera array or a gantry. One of the most well-known is the Stanford Light-field Archive [15], which provides about 20 light-fields sampled using a camera array, a gantry and a light-field microscope. The synthetic light-field archive by Marwah et al. [16] contains 5 camera light-fields and 13 display light-fields. Some real-world datasets were also introduced [17–19], that were usually captured by Lytro or Lytro Illum cameras. However, all these datasets are not suitable for machine learning because no ground truth is available.

Wanner et al. [7] introduced the first light field dataset with ground truth depth and an open benchmark: the HCI Light Field Benchmark. However, it is no longer satisfying the needs of the current light field community [5]. Therefore, Honauer and Johannsen et al. proposed a novel light field benchmark [5,6]. They provided 28 carefully designed synthetic, densely sampled 4D light fields with highly accurate disparity ground truth.

Convolutional Neural Networks: Convolutional neural networks have been proven to perform very well in computer vision tasks such as classification, object detection and segmentation. More recently, CNNs have been applied to multi-image tasks, such as optical flow prediction [20] and stereo match [21], etc. Since the work by Krizhevsky et al. [15] (a.k.a AlexNet), many advanced architectures have been introduced, including GoogleNet and VGG. However, most of existing CNN solutions are proposed on 2D or 3D data, and can not be directly applied to 4D light field data.

3 EPI-Patch Based CNN Architecture

Considering limited light field data, we propose an EPI-patch based CNN for depth estimation. We chose the EPI patches for several reasons. First, EPI based depth estimation has been proven effectively and successfully. Second, we can map 4D light field data onto a 2D CNN architecture by EPI patches. Finally, it is important that the number of EPI patches is large enough for learning.

3.1 Training EPI-Patch Pairs

We extract the horizontal and vertical EPI patches of each pixel and group them as a EPI-patch pair. After removing the invalid EPI-patch pairs and performing oversampling, the rest valid EPI-patch pairs in the training scenes compose our training set.

Let $\mathcal{L}_{\mathcal{F}}(x, y, s, t)$ is the 4D light-field data represented by the two plane parametrization, (x, y) and (s, t) are the spatial and angular coordinates, respectively. $N_x \times N_y$ is the spatial resolution of the light field, and $N_s \times N_t$ is the angular resolution. The central sub-aperture image (central viewpoint) is formed by the rays passed through the main lens optical center $(s = s_0, t = t_0)$.

As for a pixel $p(x_i, y_i)$ in the central sub-aperture image, the horizontal EPI of the image line y_i can be formulated as $\mathcal{L}_{\mathcal{F}}(x, y_i, s, t_0)$, and its size is $N_s \times N_x$. The horizontal EPI patch of the pixel p is a window patch with the center at (x_i, s_0) and the size of $N_s \times W_x$. Similarly, the vertical EPI of the image row x_i can be written as $\mathcal{L}_{\mathcal{F}}(x_i, y, s_0, t)$, and its size is $N_t \times N_y$. The vertical EPI patches of pixel p is a window patch with the center at (y_i, t_0) and the size of $N_t \times W_y$. As shown in Fig. 2(a).

However, not all the EPI-patch pairs are valid for depth estimation. According to the orientation-depth relationship of EPIs, the slope of straight lines on the EPIs is related to local depth [5]. Therefore we have to remove the invalid EPI-patch pairs in which we failed to extract lines, such as textureless patches. In this paper, we use an efficient and simple methods to detect the invalid EPI

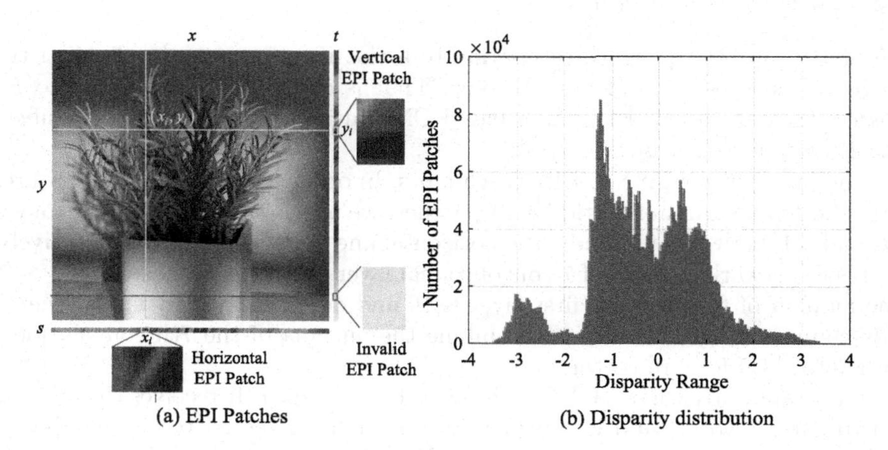

Fig. 2. EPI-patch pair extraction. (a) The horizontal and vertical EPI patches of one pixel. Invalid EPI patches are often those textureless patches. (b) The disparity distribution of all pixels of 16 training scenes.

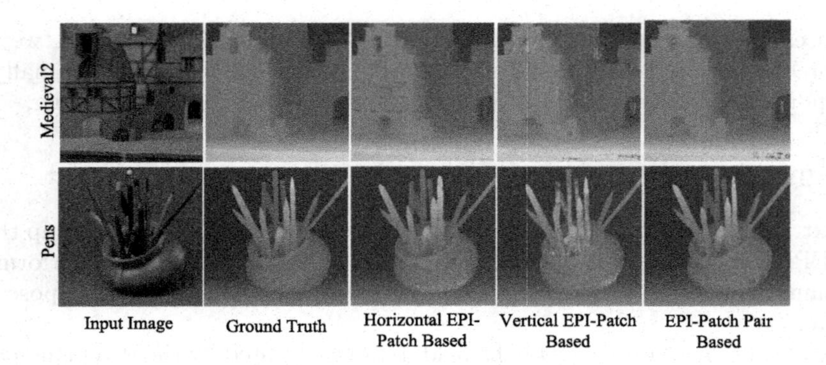

Fig. 3. The contribution of the horizontal and vertical EPI patches.From left to right are the input light field image, ground truth, horizontal EPI patch based CNN outputs, vertical EPI patch based CNN outputs, and EPI-patch pair based CNN outputs.

patches. A Canny operator is used to detect the edge information in the EPI patches, and the patches without edge information are marked as invalid patches.

Moreover, it should be noted that our training data are imbalanced data. The disparity distribution of all pixels is imbalance, as shown in Fig. 2(b). Dealing with imbalanced data is a well known challenge in machine learning [22]. The simplest method is oversampling, duplicating the EPI-patch pairs with small number until a balanced training set is created. Our experiment demonstrates using oversampling can achieve about 7% boost, increasing the depth estimation accuracy of the network from 75% to 82%. Figure 3 shows the contribution of the horizontal and vertical EPI patches.

3.2 Our CNN Architecture

We firstly convert the depth estimation to a classification issue by dividing the range of depth estimation into N_c bins. That is classifying the depth into N_c classes (bins). Then an EPI-patch based CNN architecture for depth classification is designed, as shown in Fig. 4.

The input of our CNN is EPI-patch pairs. In order to learn different features from the horizontal and vertical EPI patches, we adopt two independent convolutional sub-networks to handle the horizontal and vertical patches, respectively. Each sub-network consists of 7 convolutional layers with the kernel size of 2×2. The number of filters in the first layer is 16 and doubled layer by layer. Then a fully-connected layer is added to combine the outputs of the 7th convolutional layer into a 1D feature vector.

The output layer of our CNN is the classification layer. It receives the outputs of two sub-networks, and its output dimensionality is equal to the number of classes (bins). The probability of each class (bin) is computed by applying a Softmax function to the output vector.

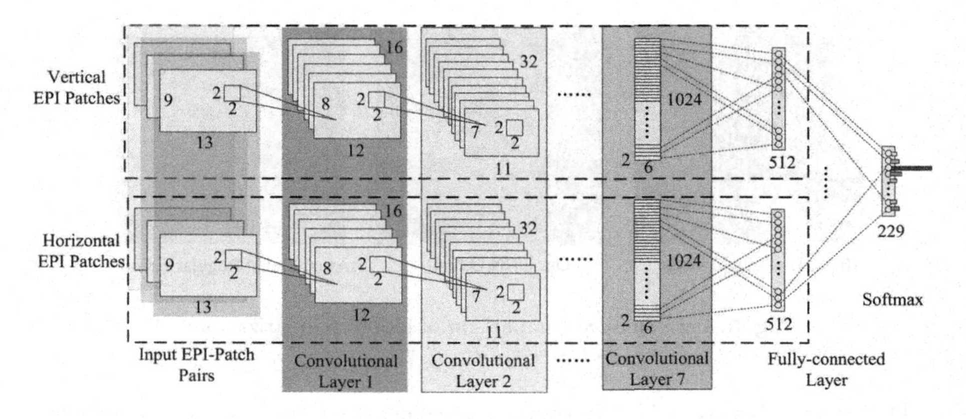

Fig. 4. Our EPI-patch based CNN architecture.

In the training procedure, our loss function is defined as the cross-entropy in multi-class problem.

$$\mathcal{E}\left(y, y^{gt}\right) = -\sum_{y_i} p\left(y_i, y^{gt}\right) * log\left(\frac{e^{-y_i}}{\sum_j e^{y_j}}\right) \tag{1}$$

$$p\left(y_i, y^{gt}\right) = \begin{cases} \lambda_1 & y_i = y^{gt} \\ \lambda_2 & \left|y_i - y^{gt}\right| = 1 \\ 0 & otherwise \end{cases} \tag{2}$$

where y is the network output, and y^{gt} is the label. In our experiments, $\lambda_1 = 0.5$ and $\lambda_2 = 0.25$.

4 Global Regularization for CNN Outputs

Since our CNN solution is based on the local EPI-patch pairs, it estimates the disparity of each pixel independently. It is obvious that our CNN outputs lack global constraints and are unstable in occlusion or textureless regions. As shown in Fig. 4.

To find an optimal estimation, we propose a global regularization for our CNN outputs. It is based on the energy minimization method with label costs [24]. The energy function combines data costs, smooth costs and label costs:

$$E\left(\alpha\right) = \overbrace{\sum_p E_{data}\left(\alpha_p\right)}^{\text{data costs}} + \overbrace{\sum_{\{p,q\}\in\mathcal{N}} E_{smooth}\left(\alpha_p, \alpha_q\right)}^{\text{smooth costs}} + \overbrace{\sum_{\alpha,\beta\in\mathcal{L}} E_{label}\left(\alpha, \beta\right)}^{\text{label costs}} \tag{3}$$

where $\mathcal{N}$ is neighborhood.

Fig. 5. Performance comparison of global regularization.

Data term $E_{data}(\alpha_p)$ is a unary term and measures the cost of assigning the label α_p to pixel p. Our network computes a N-dimensional feature representation for every pixel, and we reuse these values as initial costs in E_{data}.

Smooth term $E_{smooth}(\alpha_p, \alpha_q)$ is a pairwise term, and it penalizes assigning different labels α_p and α_q to the adjacent pixels p and q. It is used to impose piecewise smoothness constraints on the depth surface. Our smooth term used here is a weighted Potts model:

$$E_{smooth}(\alpha_p, \alpha_q) = w(p, q) \cdot T(\alpha_p \neq \alpha_q) \tag{4}$$

where $T(\cdot)$ is the generalized Potts metric. $w(p, q)$ are weights based on Canny edge e to preserve the edge information.

$$w(p, q) = \max(1 - |e(p) - e(q)|, c_e) \tag{5}$$

where c_e is a constant.

Label term $E_{label}(\alpha, \beta)$ is a high-order term and expresses the likelihood of assigning label β if current label is α. Our label term can be defined as follows,

$$E_{label}(\alpha, \beta) = \min\left(1 - e^{-\frac{(\alpha-\beta)^2}{2\sigma^2}}, c_{\mathcal{L}}\right) \tag{6}$$

where $c_{\mathcal{L}}$ is a constant.

5 Experimental Results and Analysis

Our experiments are conducted on a PC equipped with an Intel Core i7-4720HQ 2.6 GHz with 128 G memory and an NVIDIA GTX Titan X (pascal) GPU.

For the data term, the latest light field benchmark [5,6] is selected as our experimental dataset, because it provides the highly accurate disparity ground truth and the performance evaluation metrics. This dataset consists of 16 additional, 4 training, 4 test and 4 stratified scenes. We use the 16 additional scenes as our training scenes because the other scenes are all used for the performance evaluation in the benchmark website [6].

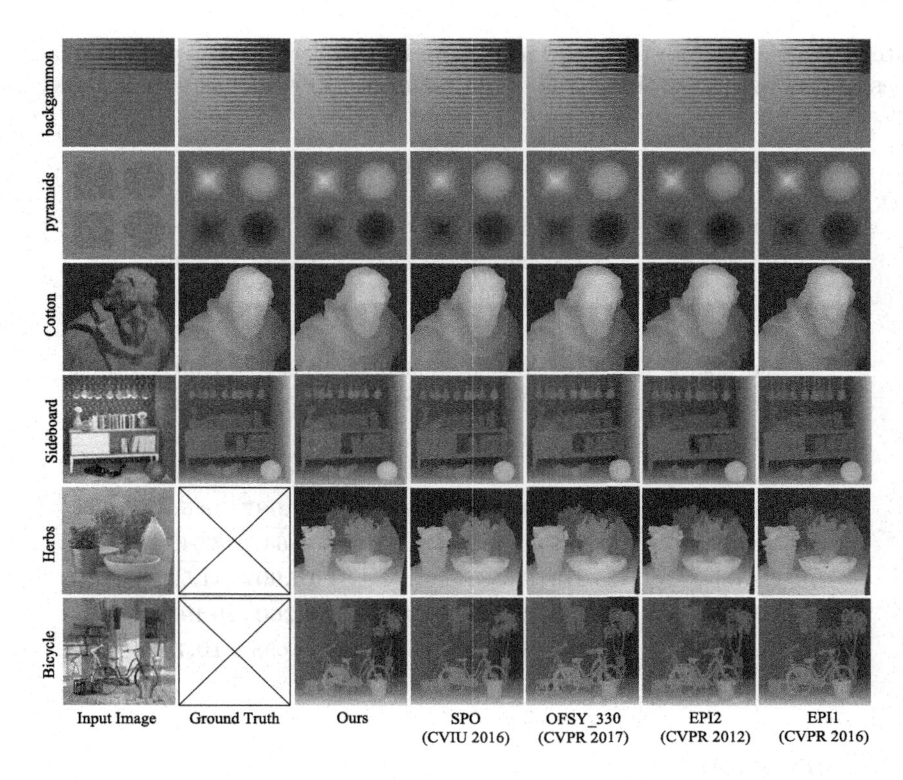

Fig. 6. Visual comparison results between our method and the SPO [25], OFSY_330 [26], EPI2 [14] and EPI1 [4] methods. The first two input images are selected from the stratified scenes. The next two ones are selected from the training scenes. The last two ones are selected from the test scenes and no ground truth is available.

Our CNN is implemented using the Tensorflow in python. The range of depth estimation is from -4 to 4, and the range of each class (bin) is 0.035, therefore the number of classes in the output layer is $N_c = 229$. The horizontal and vertical EPI patches have the same size: 9×13. Therefore, the numbers of feature maps from the first convolutional layer to 7th convolutional layer are $8 \times 12 \times 16$, $7 \times 11 \times 32$, ..., $2 \times 6 \times 1024$, as shown in Fig. 3.

The evaluation results of our method have been submitted on the website [6], named as EPN+OS+GC. Our camera-setup strategy is crosshair, that is only horizontal and vertical EPIs are used in our method. In order to better evaluate our method, here we compare our method to four state-of-the-art methods with the same camera-setup strategy. Figure 6 shows the visual comparison results between our method and the four state-of-the-art methods. There are only minor differences in visual comparison.

We use the BadPix (0.07) metric [5] for quantitative comparison, as shown in Table 1. Lower scores are better. It is obvious that our method has satisfactory performance, except two test scenes: dots and stripes. By analyzing the EPI patches extracted from these two scenes, we found there are lots of false straight lines caused by noise points. One of our future work will focus on this issue.

Table 1. Quantitative comparison of the BadPix (0.07) metric.

Test set	EPI1	EPI2	OFSY_330	SPO	Ours
Boxes	24.451	29.795	19.246	15.889	**15.304**
Cotton	13.928	16.694	3.306	2.594	**2.060**
Dino	10.350	15.667	3.434	**2.184**	2.877
Sideboard	18.376	18.953	10.355	9.297	**7.997**
Backgammon	21.331	22.076	4.828	3.781	**3.328**
Dots	62.001	46.534	37.670	**16.274**	39.248
Pyramids	0.858	1.083	0.861	0.356	**0.242**
Stripes	25.809	23.808	18.640	**14.987**	18.545
Bedroom	13.588	13.557	**4.743**	4.864	7.543
Bicycle	25.213	25.585	16.134	**10.907**	11.599
Herbs	47.082	30.944	12.304	**8.260**	9.190
Origaml	28.901	27.121	13.683	11.698	**10.749**

6 Conclusion

This paper develops an EPI-patch based CNN for depth estimation on 4D light field. The main idea is mapping 4D light field data onto convolutional CNN by extracting a group of valid horizontal and vertical EPI patches, and then designing an EPI-patch based CNN to learn the orientation-depth relationship from EPI patches. Finally, a global regularization is introduced to refine the output of CNN. Experimental results demonstrate our method has satisfactory performance.

Acknowledgments. This work is supported in part by the National High-tech R&D Program of China (863 Program, 2015AA015901), Key Program of Zhejiang Provincial Natural Science Foundation of China (No. LZ14F020003), and International Cooperation and Exchange of the National Natural Science Foundation of China (No. 2014DFA12040).

References

1. Jeon, H., Park, J., Choe, G., Park, J., Bok, Y., Tai, Y., Kweon, I.: Accurate depth map estimation from a lenslet light field camera. In: CVPR (2015)
2. Wang, T., Efros, A., Ramamoorthi, R.: Depth estimation with occlusion modeling using light-field cameras. IEEE TPAMI **38**(11), 2170–2181 (2016)

3. Williem and Park, I.: Robust light field depth estimation for noisy scene with occlusion. In: CVPR (2016)
4. Johannsen, O., Sulc, A., Goldluecke, B.: What sparse light field coding reveals about scene structure. In: CVPR, pp. 3262–3270 (2016)
5. Honauer, K., Johannsen, O., Kondermann, D., Goldluecke, B.: A dataset and evaluation methodology for depth estimation on 4D light fields. In: ACCV (2016)
6. 4D Light Field Benchmark Dataset and Evaluation. http://hci-lightfield.iwr.uni-heidelberg.de/
7. Wanner, S., Meister, S., Goldluecke, B.: Datasets and benchmarks for densely sampled 4D light fields. In: Vision, Modelling and Visualization (2013)
8. Georgiev, T., Lumsdaine, A.: Reducing plenoptic camera artifacts. Comput. Graph. Forum **29**(6), 1955–1968 (2010)
9. Yu, Z., Guo, X., Ling, H., Lumsdaine, A., Yu, J.: Line assisted light field triangulation and stereo matching. In: ICCV (2013)
10. Lin, H., Chen, C., Kang, S.B., Yu, J.: Depth recovery from light field using focal stack symmetry. In: ICCV (2015)
11. Tao, M., Srinivasan, P., Hadap, S., Rusinkiewicz, S., Malik, J., Ramamoorthi, R.: Shape estimation from shading, defocus, and correspondence using light-field angular coherence. In: IEEE TPAMI (2015)
12. Kim, C., Zimmer, H., Pritch, Y., Sorkine-Hornung, A., Gross, M.: Scene reconstruction from high spatio-angular resolution light fields. ACM Trans. Graph. (Proc. SIGGRAPH) **32**(4) (2013)
13. Wanner, S., Goldluecke, B.: Variational light field analysis for disparity estimation and super-resolution. IEEE TPAMI **36**(3), 606–619 (2014)
14. Wanner, S., Goldluecke, B.: Globally consistent depth labeling of 4D light fields. In: CVPR, pp. 41–48 (2012)
15. Wilburn, B., Joshi, N., Vaish, V., Talvala, E.V., Antunez, E., Barth, A., Adams, A., Horowitz, M., Levoy, M.: High performance imaging using large camera arrays. ACM Trans. Graph. (TOG) **24**, 765–776 (2005)
16. Marwah, K., Wetzstein, G., Bando, Y., Raskar, R.: Compressive light field photography using overcomplete dictionaries and optimized projections. ACM Trans. Graph. (Proc. SIGGRAPH) **32**, 1–11 (2013)
17. Mousnier, A., Vural, E., Guillemot, C.: Partial light field tomographic reconstruction from a fixed-camera focal stack. arXiv preprint (2015). arXiv:1503.01903
18. Rerabek, M., Ebrahimi, T.: New light field image dataset. In: 8th International Conference on Quality of Multimedia Experience (QoMEX) (2016)
19. Wang, T.-C., Zhu, J.-Y., Hiroaki, E., Chandraker, M., Efros, A.A., Ramamoorthi, R.: A 4D light-field dataset and CNN architectures for material recognition. In: Leibe, B., Matas, J., Sebe, N., Welling, M. (eds.) ECCV 2016. LNCS, vol. 9907, pp. 121–138. Springer, Cham (2016). doi:10.1007/978-3-319-46487-9_8
20. Fischer, P., Dosovitskiy, A., Ilg, E., Hausser, P., Hazirbas, C., Golkov, V.: FlowNet: Learning optical flow with convolutional networks. In: ICCV (2015)
21. Luo, W., Schwing, A.G., Urtasun, R.: Efficient deep learning for stereo matching,. In: CVPR, pp. 5695–5703 (2016)
22. Pulgar, F., Riveragar, A., Charte, F., del Jesus, M.: On the impact of imbalanced data in convolutional neural networks performance. In: de Pisón, F.M., Urraca, R., Quintián, H., Corchado, E. (eds.) HAIS 2017. LNCS, vol. 10334, pp. 220–232. Springer, Cham (2017). doi:10.1007/978-3-319-59650-1_19
23. Ioffe, S., Szegedy, C.: Batch normalization: Accelerating deep network training by reducing internal covariate shift. arXiv preprint (2015). arXiv:1502.03167

24. Delong, A., Osokin, A., Isack, H., Boykov, Y.: Fast approximate energy minimization with label costs. IJCV **96**(1), 1–27 (2012)
25. Zhang, S., Sheng, H., Li, C., Zhang, J., Xiong, Z.: Robust depth estimation for light field via spinning parallelogram operator. Comput. Vis. Image Underst. **145**, 148–159 (2016)
26. Strecke, M., Alperovich, A., Goldluecke, B.: Accurate depth and normal maps from occlusion-aware focal stack symmetry. In: CVPR (2017)

Deep Metric Learning with False Positive Probability

Trade Off Hard Levels in a Weighted Way

Jia-Xing Zhong[1,2], Ge Li[1(✉)], and Nannan Li[1]

[1] School of Electronic and Computer Engineering, Peking University, Shenzhen, China
jxzhong@pku.edu.cn, geli@ece.pku.edu.cn, linn@pkusz.edu.cn
[2] School of Data and Computer Science, Sun Yat-sen University, Guangzhou, China

Abstract. In recent years, deep metric learning has been an end-to-end fashion in computer vision community due to the great success of deep learning. However, existing deep metric learning frameworks are faced with a dilemma about the hard level trade-off for training examples. Namely, the “harder” examples we feed to neural networks, the more likely we attain highly discriminative models, but the more easily neural networks get stuck into poor local minimal in practice. To fight against this dilemma, we propose a deep metric learning method with **FA**lse **P**ositive **P**robabilit**Y** (FAPPY) to gradually incorporate different hard levels. Unlike mainstream deep metric learning schemes, the presented approach optimizes similarity probability distribution among training samples, instead of the similarity itself. Experimental results on *CUB-200-2011*, *Stanford Online Products* and *VehicleID* datasets show that our FAPPY method achieves or outperforms state-of-the-art metric learning methods on fine-grained image retrieval and vehicle re-identification tasks. Besides, the presented method has relatively low sensitivity of hyper-parameters and it requires minor changes on traditional classification networks.

Keywords: Convolutional neural network · Deep metric learning · Hard example mining · Fine-grained image retrieval · Vehicle re-identification

1 Introduction

Deep metric learning aims to learn a metric space that pushes away dissimilar data points farther and pulls similar ones closer by deep neural networks. As an end-to-end feature embedding approach, deep metric learning is widely used on computer vision area, such as fine-grained image recognition [1, 2], face identification/verification [3, 4], vehicle/person re-identification [5, 6], zero-shot learning [7, 8], *etc*.

However, deep metric learning suffers from a dilemma about mining hard training examples, where “hard” describes the distinguishable degree as shown in Fig. 1. On one hand, if we treat all the possible training samples without distinction, it is impossible or

D. Liu et al. (Eds.): ICONIP 2017, Part III, LNCS 10636, pp. 653–664, 2017.
https://doi.org/10.1007/978-3-319-70090-8_66

time-infeasible for the effective convergence of discriminative neural networks; on the other hand, if we force neural networks to pay excessive attention to hard examples, neural networks tend to get stuck in the local minimal in practice.

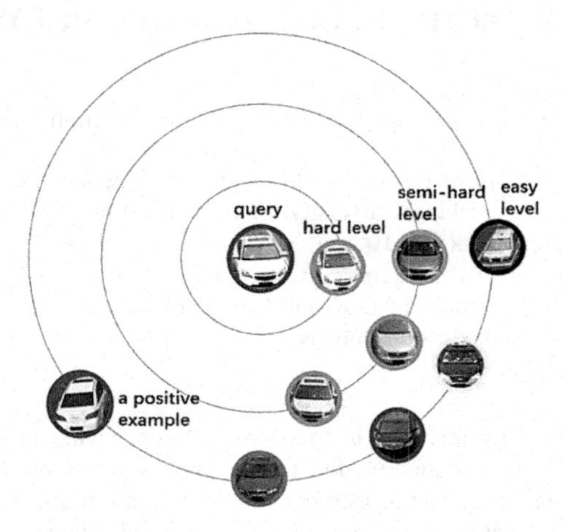

Fig. 1. Illustration of negative examples with different hard levels: In the metric space, shorter distance refers to higher similarity. The circles of the same color correspond to objects of the same label. Therefore, harder negative examples are closer to the query so that it is more difficult to tell the difference between them. (Color figure online)

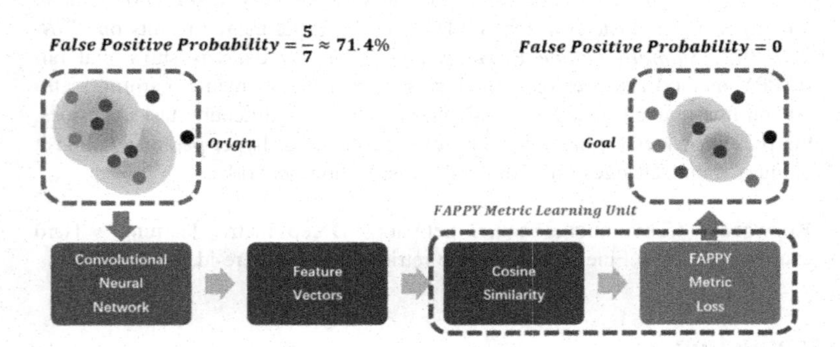

Fig. 2. Framework of the proposed approach: The neural network is designed to project the input data points into a cosine space at the lowest false positive rate. In the left-side original metric space, for a positive pair with the same label (two blue data points), there are 7 negative examples totally, 5 of which are closer to one positive example than the other one. Thus, the false positive probability is $\frac{5}{7} \approx 71.4\%$. Our goal is to minimize the false positive rate. (Color figure online)

In this paper, we propose a metric learning approach to deal with the above dilemma in a weighted way. Our FAPPY method is an end-to-end framework as shown in Fig. 2, which aims to minimize the false positive probability. Our implementation just needs to replace the last loss layer with the FAPPY metric learning unit on traditional classification networks.

Specifically, our FAPPY method optimizes probability distribution of relative similarity among training samples, rather than directly adjusts similarity as other methods. Compared with other deep metric learning algorithms, the proposed method is robust to hyper-parameter settings and do not require troublesome "hard example mining" [3] schemes.

Extensive experiments show that the performance of our approach is equal or better than state-of-the-art metric learning methods, without introducing any sensitive hyper-parameter. Our source code with *Caffe* and the trained models are available at: https://github.com/jx-zhong-for-academic-purpose/fappy_metric_learning.

2 Related Work

The first deep metric learning framework is siamese network [9], which strives to maximize the distance of positive examples and minimize the distance of negative ones. Sharing a similar goal, histogram loss [10] takes effort to separate similarity distributions of positive pairs from negative pairs. These two algorithms are based on absolute similarity/dissimilarity, however, the relative constraint seems of great importance on a majority of problems.

FaceNet [3] introduces triplet loss in the seminal work on deep metric learning with relative similarity, which optimizes relative distances from the anchor point. Later, Song et al. [11] proposes lifted structured embedding, which makes full use of the whole training batch. In recent works done by Yuan et al. [13] and Song et al. [12], relative similarity is adopted, as well as our FAPPY approach.

The above methods attempt to solve the problem on hard level trade-off in various ways. For instance, the lifted structured embedding [11] utilizes "log-sum-exp" function as a smooth upper bound, which may cause arithmetic overflow/underflow problems. FaceNet [3] applies the "semi-hard mining" strategy, so the batch size dramatically reaches up to 1800 for the selection of enough hard negative examples. To avoid these drawbacks, we fuse the false positive probability on various hard levels to settle this problem.

3 Deep Metric Learning with False Positive Probability

There are two stages in our FAPPY method, false positive probability estimation and weighted fusion. For clarity, we detail these two stages respectively in this section.

3.1 False Positive Probability Estimation

First of all, we estimate the false positive rate with different hard ranges respectively, where ***false positive*** means that a negative pair (two images with different labels) is determined as the positive one, as in Fig. 2.

Inspired by the histogram loss [10], we estimate the probability distribution with histograms. For each positive pair $\langle i,j \rangle$ belonging to a training batch set S, with a specific bin width Δ, we build a histogram to figure out its false positive rate.

We denote the cosine similarity between the training example $x, y \in S$ as $S_{xy} = cos\langle \overrightarrow{f_x}, \overrightarrow{f_y} \rangle$, where $\overrightarrow{f_x}$ and $\overrightarrow{f_y}$ are respective feature vectors. Since the cosine similarity ranges from -1 to +1, we define the bin boundaries of histograms as $b_1 = -1, b_2, \ldots, b_{n-1}, b_n = +1$ with n bins, where $n = \frac{2}{\Delta} + 1$ is determined by the bin width Δ.

Assuming $s_{ij} \in [b_t, b_{t+1}]$, we denote the heights of bins b_t and b_{t+1} as:

$$h_{ij}^{left} = w_{ij}^{t}, \quad (1)$$

$$h_{ij}^{right} = w_{ij}^{t+1}, \quad (2)$$

where w_{ij}^t indicates the probability weight for training pair $\langle i,j \rangle$ on the tth bin. The value of w_{ij}^t is determined by s_{ij} with triangular smoothing kernel [14]:

$$w_{ij}^{t} = \begin{cases} \frac{(s_{ij} - b_{t-1})}{\Delta}, & s_{ij} \in [b_{t-1}, b_t] \\ \frac{(b_{t+1} - s_{ij})}{\Delta}, & s_{ij} \in [b_t, b_{t+1}] \\ 0, & otherwise. \end{cases} \quad (3)$$

In the same way, for this positive pair $\langle i,j \rangle$, the similarity density of negative examples on the uth bin, are estimated respectively as follows, illustrated by the green and blue bins in Fig. 3:

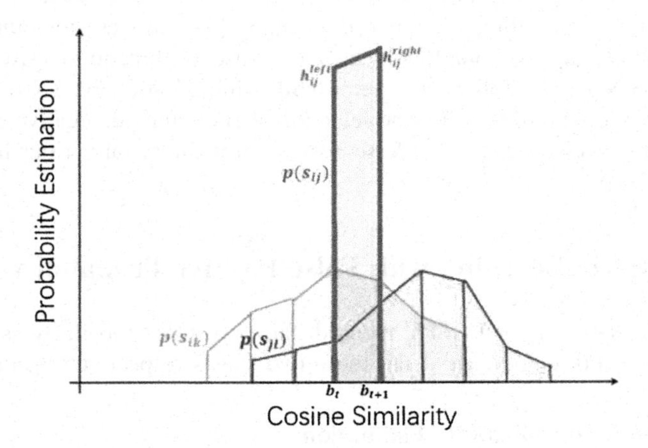

Fig. 3. For a specific positive pair and the bin width Δ, we build a histogram to estimate the false positive rate. The probability density function of s_{ij} is approximated by the red bins, and the green and blue portions indicate the similarity distribution of negative examples upon i and j in the same manner. (Color figure online)

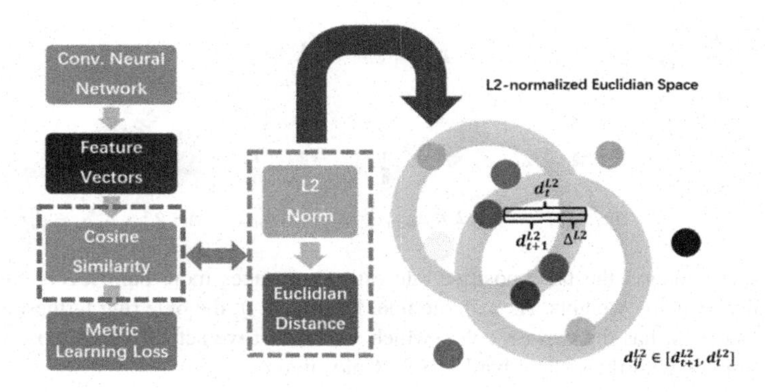

Fig. 4. In the equivalent L2-normalized Euclidian space, given the Δ^{L2} and a positive pair $\langle i,j \rangle$ with the distance $d_{ij}^{L2} \in \left[d_{t+1}^{L2}, d_t^{L2}\right]$, where $d_t^{L2} - d_{t+1}^{L2} = \Delta^{L2}, d_{ij}^{L2}, d_t^{L2}, d_{t+1}^{L2}$ and Δ^{L2} correspond to s_{ij}, b_{t+1}, b_t and Δ in the cosine space. During the backward propagation to minimize P_{ij}^{Δ}, ***only the blue-shaded area is taken into consideration***. The radius of blue shade circular ring locates in the range of $\left[d_{t+1}^{L2}, d_t^{L2}\right]$ from the data points i and j. The proof is given in Appendix A. (Color figure online)

$$h_i^u = \frac{1}{|N_{ij}|}\sum\nolimits_k w_{ik}^u, k \in N_{ij}, \tag{4}$$

$$h_j^u = \frac{1}{|N_{ij}|}\sum\nolimits_l w_{jl}^u, l \in N_{ij}, \tag{5}$$

where the negative set N_{ij} is composed of training examples whose labels are inconsistent with the example i (or j).

In Fig. 2, we have given an instance for the false positive probability, and its mathematical definition for the specific positive pair $\langle i,j \rangle$ is as follows:

$$\begin{aligned} P_{ij} &= P\left(s_{ik} \geq s_{ij}\right) + P\left(s_{jl} \geq s_{ij}\right) \\ &= \textstyle\int_{-1}^{1} p\left(s_{ij}\right)\left[\int_{s_{ij}}^{1} p(s_{ik})ds_{ik}\right]ds_{ij} + \int_{-1}^{1} p\left(s_{ij}\right)\left[\int_{s_{ij}}^{1} p(s_{jl})ds_{jl}\right]ds_{ij}, \forall k, l \in N_{ij}. \end{aligned} \tag{6}$$

As shown in Fig. 3, the estimation of $P\left(s_{ik} \geq s_{ij}\right)$ is obtained by the multiplication of the shade red area and the shade green area. Similarly, we approximate $P\left(s_{jl} \geq s_{ij}\right)$ with the shade red area timed by blue ones. Thus, the false positive probability P_{ij} for this positive pair $\langle i,j \rangle$ can be estimated as:

$$P_{ij}^{\Delta} = \left(h_{ij}^{left}\sum\nolimits_{u_1} h_i^{u_1} + h_{ij}^{right}\sum\nolimits_{u_2} h_i^{u_2}\right) + \left(h_{ij}^{left}\sum\nolimits_{u_3} h_j^{u_3} + h_{ij}^{right}\sum\nolimits_{u_4} h_j^{u_4}\right), \tag{7}$$

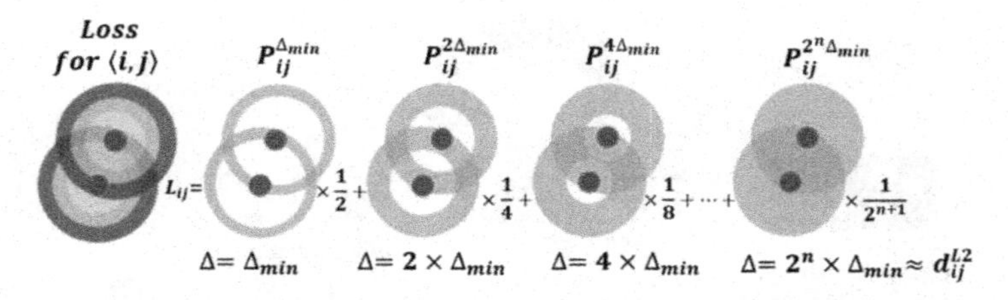

Fig. 5. Fusion strategy: the false positive rate estimation takes more hard levels into account with gradually reducing weights. As with the final loss for $\langle i,j\rangle$, the blue ring farthest away from a positive data point has the darkest color, which means that we actually give more weights to negative examples with the easier "hard level" (Color figure online)

where Δ is the bin width, and $u_1, u_1, u_1,\ u_4$ satisfy that $b_{u_1}, b_{u_3} \geq b_t$ while $b_{u_2}, b_{u_4} \geq b_{t+1}$ since $s_{ij} \in [b_t, b_{t+1}]$.

The corresponding relationship between Eqs. (6) and (7) is: $P_{ij} \approx P_{ij}^{\Delta}, P(s_{ik} \geq s_{ij}) \approx \left(h_{ij}^{left} \sum_{u_1} h_i^{u_1} + h_{ij}^{right} \sum_{u_2} h_i^{u_2} \right)$ and $P(s_{jl} \geq s_{ij}) \approx \left(h_{ij}^{left} \sum_{u_3} h_j^{u_3} + h_{ij}^{right} \sum_{u_4} h_j^{u_4} \right)$.

3.2 Weighted Fusion

Lemma 1. Given the bin width Δ and a positive pair $\langle i,j\rangle$, assuming $s_{ij} \in [b_t, b_{t+1}], \forall k, l \in N_{ij}$, the backward propagation gradients of false positive probability estimation P_{ij}^{Δ} Satisfy: $\frac{\partial P_{ij}^{\Delta}}{\partial s_{ik}} \neq 0 \Leftrightarrow s_{ik} \in [b_t, b_{t+1}], \frac{\partial P_{ij}^{\Delta}}{\partial s_{jl}} \neq 0 \Leftrightarrow s_{jl} \in [b_t, b_{t+1}]$.

In other words, the bin width Δ strongly constrains the range of hard levels. To make it more understandable, we illustrate Lemma 1 in Fig. 4 with the L2-normalized Euclidian space, an equivalent form of the cosine metric space. The proof of Lemma 1 and this equivalency is given in Appendix A.

The larger Δ value we set, the bigger metric space around positive examples we focus on, so that the more hard levels we choose. Therefore, we can merge different hard levels together and compute the loss function for $\langle i,j\rangle$: $L_{ij} = \alpha_1 P_{ij}^{\Delta_1} + \alpha_2 P_{ij}^{\Delta_2} + \ldots + \alpha_n P_{ij}^{\Delta_n}$ where α is the fusion weight and $\Delta_1, \Delta_2, \ldots, \Delta_n$ are different bin widths.

As is shown in Fig. 5, we gradually increase the delta value by 2x, while decrease its weight by 0.5x, and compute L_{ij} with the summation over these results. By this means, we attach greater importance to the easier "hard level". In the meanwhile, we also take harder levels into account. As for our FAPPY method, the only hyper-parameter we need to set is the minimum bin width Δ_{min}. It is worth mentioning that experimental results in Sect. 4 show that Δ_{min} is remarkably robust.

Algorithm 1. Fusion to Compute the Loss Function

Input: A training batch set S and the minimum bin width Δ_{min}
Output: The loss function L
1: $L \leftarrow 0$, $count_{positive\_pairs} \leftarrow |\{\langle i,j \rangle \in S | label_i = label_j\}|$, $\Delta_{cur} \leftarrow 2.0$
2: **repeat**
3: $L_{\Delta_{cur}} \leftarrow 0$
4: **for all** positive pair $\langle i,j \rangle$ such that $i, j \in S$ **do**
5: **if** $1.0 - s_{ij} \geq \Delta_{cur}$ **then**
6: compute $P_{ij}^{\Delta_{cur}}$ as **Eq. (7)**
7: $L_{\Delta_{cur}} \leftarrow L_{\Delta_{cur}} + P_{ij}^{\Delta_{cur}}$
8: **end if**
9: **end for**
10: $L \leftarrow L + L_{\Delta_{cur}} / count_{positive\_pairs}$
11: $L \leftarrow L/2.0$
12: $\Delta_{cur} \leftarrow \Delta_{cur}/2.0$
13: **until** $\Delta_{cur} < \Delta_{min}$

The average loss on each positive pair is the final loss function. The pseudocode of loss function computation is shown in Algorithm 1. To minimize the false positive error rate, the neural network is updated with $\frac{\partial L}{\partial \vec{f_i}}$ during the backward propagation process, where $\vec{f_i}$ is the feature vector of the training example i.

4 Experiments

We evaluate our FAPPY method on 3 datasets, *i.e. CUB-200-2011* [16], *Stanford Online Products* [11] and *VehicleID* [5]. The experiments consist of two typical tasks, fined-grained image retrieval and similar vehicle re-identification, implemented with

Table 1. Recall@K (%) of fine-grained image retrieval

Dataset	CUB-200-2011						Stanford online products			
K	1	2	4	8	16	32	1	10	100	1000
Contrastive [9]	26.4	37.7	49.8	62.3	76.4	85.3	42.0	58.2	73.8	89.1
Triplet [3]	36.1	48.6	59.3	70.0	80.2	88.4	42.1	63.5	82.5	94.8
LiftStruct [11]	47.2	58.9	70.2	80.2	89.3	93.2	62.1	79.8	91.3	97.4
Histogram [10]	50.3	61.9	72.6	82.4	88.8	93.7	**63.9**	**81.7**	**92.2**	**97.7**
Ours$^{0.01}$	50.4	**62.6**	74.8	**84.0**	90.5	94.9	63.3	**80.8**	**91.7**	**97.4**
Ours$^{0.001}$	50.5	**62.6**	**74.9**	**84.0**	**90.7**	**95.0**	**63.5**	80.7	91.6	97.2
Ours$^{0.0001}$	**50.6**	62.4	**74.9**	83.8	90.3	**95.0**	63.4	**80.8**	**91.7**	**97.3**

the *Caffe* [15] framework. Furthermore, we conduct a series of experimental studies about the sensitivity of hyper-parameters for both tasks, at various minimum bin widths $\Delta_{min} = \{0.01, 0.001, 0.0001\}$.

4.1 Datasets

The retrieval experiments are conducted on the *CUB-200-2011* and *Stanford Online Products* datasets, while the re-identification experiments are performed upon the benchmark *VehicleID* dataset. To ensure the apples-to-apples comparison to existing methods, we follow the standard train/test data split.

- *CUB-200-2011* dataset consists of 200 bird species with 11,788 images, where the first 100 species with 5,864 images make up the training set and the rest 100 species with 5,924 images constitute the test set.
- *Stanford Online Products* dataset consists of 22,634 classes with 120,053 online products images from eBay, where 59,551 images of 11,318 classes are for training and 60,502 images of 11,316 classes are for testing.
- *VehicleID* dataset consists of 26,267 vehicles with 221,763 images, where the training set includes 110,178 images of 13,134 vehicles and the test set comprises 111,585 images of 13,133 vehicles. In accordance with [5], we choose 3 test data splits with different scales. The small test set is composed of 800 vehicles with 7,332 images. The medium test set is composed of 1,600 vehicles with 12,995 images. The large test set is composed of 2,400 vehicles with 20,038 images.

4.2 Fine-Grained Image Retrieval

Fine-grained image recognition is such a challenging problem that researchers can just improve its performance by around 1% every year despite of their great efforts.

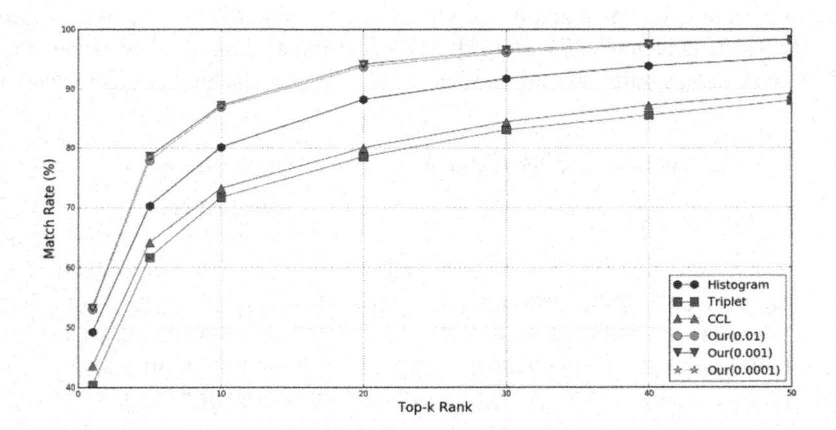

Fig. 6. CMC on the small gallery set

Table 2. Match rate (%) of vehicle re-identification task

Match rate	Small		Medium		Large	
Top K	Top 1	Top 5	Top 1	Top 5	Top 1	Top 5
Triplet [3]	40.4	61.7	35.4	54.6	31.9	50.3
CCL [5]	43.6	64.2	37.0	57.1	32.9	53.3
LiftStruct[a] [11]	/	/	/	/	/	/
Histogram [10]	49.3	70.9	43.7	65.1	39.5	59.9
Ours$^{0.01}$	53.1	77.9	46.9	71.9	42.4	66.2
Ours$^{0.001}$	53.3	**78.6**	**47.5**	**72.6**	**43.0**	**66.7**
Ours$^{0.0001}$	**53.5**	78.5	47.3	72.2	42.8	66.6

[a]The LiftStruct metric learning gets no results for it failed to converge in the experiment.

Typically, fined-grained image retrieval is a recognition problem on unseen categories, which means the object classes are entirely disjoint between train and test sets.

On fined-grained image retrieval tasks, we compare our FAPPY method with state-of-the-art deep metric learning approaches on two frequently-used datasets, *CUB-200-2011* and *Stanford Online Products*. We apply the standard Recall@K metric[1] [17] on the experiment result.

Following [10, 11], our basic network architecture is the *GoogLeNet* (up to "pool5") [18] pre-trained on *ImageNet ILSVRC* [19], appended by a fully connected product layer initialized with random weights. In addition, we keep the implementation details closely with above two papers.

As is shown in Table 1, the performance of FAPPY approach is marginally better than that of state-of-the-art on *CUB-200-2011* and just slightly inferior to the histogram loss on *Stanford Online Products*. Besides, the results of FAPPY by Recall@K metric remain stable (change less than 0.5%) even though the minimum bin width Δ_{min} varies across several orders of magnitude.

4.3 Vehicle Re-identification

For deep metric learning, vehicle re-identification is the touchstone of discernment, because numerous vehicles with the same vehicle model are almost exactly the same in appearance.

We make experience on the *VehicleID* dataset with different scales of gallery sets (*i.e.* small, medium and large). Following the paper [5], we adopt the pre-trained network *VGG_CNN_M_1024* [20] with the batch size of 150, and repeat the query 10 times in test phase.

[1] Recall@K is the average recall scores over all the query images in testing set. For each query image, the recall score is 1 if at least one positive image in the nearest K returned images and 0 otherwise.

The evaluation metric is the Top-k match rate and we draw the cumulative match characteristic (CMC) curve [21]. Our FAPPY method significantly outperforms the existing approaches and the result is pretty stable as shown in Table 2 and Fig. 6.

5 Conclusion

In this paper, we propose a deep metric learning framework with false positive probability (FAPPY), which can gradually incorporate training example with different hard levels. Neither complex "hard example mining" nor intricate hyper-parameters tuning is required in the presented method. The experimental results show that our FAPPY method is comparable to existing metric learning approaches on fine-grained image retrieval, and outperforms state-of-the-art on vehicle re-identification tasks.

Acknowledgements. This work is partially supported by National Science Foundation of China (No. U1611461), Shenzhen Peacock Plan (20130408-183003656), Science, Technology Planning Project of Guangdong Province (No. 2014B090910001) and National Natural Science Foundation of China (61602014). In addition, we would like to thank Guangzhou Supercomputer Center for providing us with Tianhe-2 system to conduct the experiment and giving us technical supports.

Appendix A

Lemma 1. *Proof.*
We compute the gradients of these pairs:

$$\frac{\partial P_{ij}^{\Delta}}{\partial s_{ij}} = h_i^t + h_j^t, \quad \frac{\partial P_{ij}^{\Delta}}{\partial s_{ik}} = \begin{cases} -h_{ij}^{right}, & s_{ik} \in [b_t, b_{t+1}] \\ 0, & otherwise \end{cases},$$
$$\frac{\partial P_{ij}^{\Delta}}{\partial s_{jl}} = \begin{cases} -h_{ij}^{right}, & s_{jl} \in [b_t, b_{t+1}] \\ 0, & otherwise \end{cases}$$

Strictly speaking, the necessary and sufficient condition of $\frac{\partial P_{ij}^{\Delta}}{\partial s_{ik}} \neq 0$ is $s_{ik}, s_{ij} \in (b_t, b_{t+1}]$. Likewise, $\frac{\partial P_{ij}^{\Delta}}{\partial s_{jl}} \neq 0$ if and only if $s_{jl}, s_{ij} \in (b_t, b_{t+1}]$.

The Equivalency *Proof.*

In the L2-normalized space, the feature vectors are defined as: $\overrightarrow{f_x^{L2}} = \frac{\overrightarrow{f_x}}{\|\overrightarrow{f_x}\|}$.

The Euclidian distance d_{xy}^{L2} and Δ^{L2} are as follows:

$$\mathrm{d}_{xy}^{L2} = \left\| \overrightarrow{f_x^{L2}} - \overrightarrow{f_y^{L2}} \right\| = \left\| \frac{\overrightarrow{f_x}}{\left\|\overrightarrow{f_x}\right\|} - \frac{\overrightarrow{f_y}}{\left\|\overrightarrow{f_y}\right\|} \right\| = \sqrt{1 - 2cos\left\langle \overrightarrow{f_x}, \overrightarrow{f_y} \right\rangle + 1} = \sqrt{2 - 2s_{xy}}$$

$$\Delta^{L2} = d_t^{L2} - d_{t+1}^{L2} = \sqrt{2 - 2b_t} - \sqrt{2 - 2b_{t+1}}$$

Therefore, Δ^{L2} is positively related to Δ while d_{xy}^{L2} is negatively related to s_{xy}, and the explanation in Fig. 4 is reasonable.

References

1. Cui, Y., Zhou, F., Lin, Y., Belongie, S.: Fine-grained categorization and dataset bootstrapping using deep metric learning with humans in the loop. In: 2016 IEEE Conference on Computer Vision and Pattern Recognition, pp. 1153–1162 (2016)
2. Zhang, X., Zhou, F., Lin, Y., Zhang, S.: Embedding label structures for fine-grained feature representation. In: 2016 IEEE Conference on Computer Vision and Pattern Recognition, pp. 1114–1123 (2016)
3. Schroff, F., Kalenichenko, D., Philbin, J.: Facenet: a unified embedding for face recognition and clustering. In: 2015 IEEE Conference on Computer Vision and Pattern Recognition, pp. 815–823 (2015)
4. Wen, Y., Zhang, K., Qiao, Y.: A discriminative feature learning approach for deep face recognition. In: 2016 European Conference on Computer Vision, pp. 499–515 (2016)
5. Liu, H., Tian, Y., Yang, Y., Pang, L., Huang, T.: Deep relative distance learning: tell the difference between similar vehicles. In: 2016 IEEE Conference on Computer Vision and Pattern Recognition, pp. 2167–2175 (2016)
6. You, J., Wu, A., Zheng, W.S.: Top-push video-based person re-identification. In: 2016 IEEE Conference on Computer Vision and Pattern Recognition, pp. 1345–1353 (2016)
7. Zhang, Z., Saligrama, V.: Zero-shot learning via joint latent similarity embedding. In: 2016 IEEE Conference on Computer Vision and Pattern Recognition, pp. 634–642 (2016)
8. Bucher, M., Herbin, S., Jurie, F.: Improving semantic embedding consistency by metric learning for zero-shot classification. In: 2016 European Conference on Computer Vision, pp. 730–746 (2016)
9. Chopra, S., Hadsell, R., LeCun, Y.: Learning a similarity metric discriminatively, with application to face verification. In: 2016 European Conference on Computer Vision, pp. 539–546 (2016)
10. Ustinova, E., Lempitsky, V.: Learning deep embeddings with histogram loss. In: Advances in Neural Information Processing Systems, pp. 4170–4178 (2016)
11. Song, H.O., Xiang, Y., Jegelka, S., Savarese, S.: Deep metric learning via lifted structured feature embedding. In: 2016 IEEE Conference on Computer Vision and Pattern Recognition, pp. 4004–4012 (2016)
12. Song, H.O., Jegelka, S., Rathod, V., Murphy, K.: Deep metric learning via facility location (2017)
13. Yuan, Y., Yang, K., Zhang, C.: Hard-aware deeply cascaded embedding. arXiv preprint:1611.05720 (2016)
14. Zucchini, W., Berzel, A., Nenadic, O.: Applied Smoothing Techniques. Part I: Kernel Density Estimation, pp. 5 (2003)

15. Jia, Y., Shelhamer, E., Donahue, J., Karayev, S., Long, J., Girshick, R., Guadarrama, S., Darrell, T.: Caffe: convolutional architecture for fast feature embedding. In: 22nd ACM International Conference on Multimedia, pp. 675–678 (2014)
16. Welinder, P., Branson, S., Mita, T., Wah, C., Schroff, F., Belongie, S., Perona, P.: Caltech-UCSD birds 200 (2011)
17. Jegou, H., Douze, M., Schmid, C.: Product quantization for nearest neighbor search. IEEE Trans. Pattern Anal. Mach. Intell. **33**(1), 117–128 (2011)
18. Szegedy, C., Liu, W., Jia, Y., Sermanet, P., Reed, S., Anguelov, D., Erhan, D., Vanhoucke, V., Rabinovich, A.: Going deeper with convolutions. In: 2015 IEEE Conference on Computer Vision and Pattern Recognition, pp. 1–9 (2015)
19. Russakovsky, O., Deng, J., Su, H., Krause, J., Satheesh, S., Ma, S., Huang, Z., Karpathy, A., Khosla, A., Bernstein, M., Berg, A.C., Fei-Fei, L.: Imagenet large scale visual recognition challenge. Int. J. Comput. Vis. **115**(3), 211–252 (2015)
20. Chatfield, K., Simonyan, K., Vedaldi, A., Zisserman, A.: Return of the devil in the details: delving deep into convolutional nets. arXiv preprint arXiv:1405.3531 (2014)
21. Gray, D., Brennan, S., Tao, H.: Evaluating appearance models for recognition, reacquisition, and tracking. In: IEEE International Workshop on Performance Evaluation for Tracking and Surveillance (PETS), vol. 3, no. 5 (2007)

Deep Learning Based Face Recognition with Sparse Representation Classification

Eric-Juwei Cheng[1], Mukesh Prasad[2(✉)], Deepak Puthal[3], Nabin Sharma[2], Om Kumar Prasad[4], Po-Hao Chin[1], Chin-Teng Lin[2], and Michael Blumenstein[2]

[1] Department of Electrical Engineering, National Chiao Tung University, Hsinchu, Taiwan
fred7018@gmail.com, jbh7749@gmail.com

[2] Centre for Artificial Intelligence, School of Software, FEIT, University of Technology Sydney, Sydney, Australia
{mukesh.prasad,Nabin.Sharma,Chin-Teng.Lin, Michael.Blumenstein}@uts.edu.au

[3] School of Electrical and Data Engineering, FEIT, University of Technology Sydney, Sydney, Australia
Deepak.Puthal@uts.edu.au

[4] International College of Semiconductor Technology, National Chiao Tung University, Hsinchu, Taiwan
omguptanctu@gmail.com

Abstract. Feature extraction is an essential step in solving real-world pattern recognition and classification problems. The accuracy of face recognition highly depends on the extracted features to represent a face. The traditional algorithms uses geometric techniques, comprising feature values including distance and angle between geometric points (eyes corners, mouth extremities, and nostrils). These features are sensitive to the elements such as illumination, variation of poses, various expressions, to mention a few. Recently, deep learning techniques have been very effective for feature extraction, and deep features have considerable tolerance for various conditions and unconstrained environment. This paper proposes a two layer deep convolutional neural network (CNN) for face feature extraction and applied sparse representation for face identification. The sparsity and selectivity of deep features can strengthen sparseness for the solution of sparse representation, which generally improves the recognition rate. The proposed method outperforms other feature extraction and classification methods in terms of recognition accuracy.

Keywords: Feature extraction · Face recognition · Deep learning · Convolutional neural network · Classification

1 Introduction

Among the most practical applications of image analysis and understanding, human face recognition have received significant attention in the recent years [1]. Some factors influencing the growing interest in face recognition include public security, identity

D. Liu et al. (Eds.): ICONIP 2017, Part III, LNCS 10636, pp. 665–674, 2017.
https://doi.org/10.1007/978-3-319-70090-8_67

verification, and demand for face analysis and modeling in multimedia data management. A significant progress have been made in this research area, and a number of face recognition and modeling systems have been developed and deployed. However, face recognition is still challenging problem to computer vision and pattern recognition researchers, especially under unconstrained conditions (e.g. arbitrary illumination and large pose variations and etc.). Zhao and Chellappa [2] provides a solution to the problem involves face detection from a scene image/video frame, feature extraction from face regions, and recognition. The precision of a face recognition system highly depends on features, whichis extracted to represent a face.

The traditional feature extraction methods can be categorized as model and appearance based technique. The model based techniques, also known as geometric techniques, use feature values such as the distance and angle between geometric points [3]. The appearance based techniques consider a face image as a high dimensional vector space and most of them use statistical approaches, mainly based on traditional Eigenfaces [4]. However, features extracted by traditional methods are sensitive to the elements, e.g. illumination, variation of poses, various expressions and etc. Recently, deep learning models have been used to extract the highlevel features, which have considerable tolerance to various conditions, such as illumination change and expression. Taigman, Yang, Ranzato and Wolf [5] derived face representations from a nine layer deep neural network with 97.35% accuracy on the Labeled Faces in the Wild (LFW) dataset. Huang, Lee and Learned-Miller [6] applied convolutional deep belief networks to learn a generative deep model without supervision. Sun, Wang and Tang [7, 8] learned high level face identity features with deep models for face identification, i.e. classifying a training image into one of 10,000 identities. In this paper, a two-layer deep convolutional neural network (CNN) for feature extraction and a sparse representation classification for face recognition is proposed. The combination is more robust to varying illumination and expression. The sparse property of deep neural activations is able to strengthen the sparseness as a solution of sparse representation, which generally improves the recognition rate.

2 Sparse Representation Algorithm

Representation of sparse signal has proven to be a very powerful tool for acquiring, representing, and compressing high-dimensional signals [9]. In object recognition, the basic problem is to use labeled training samples from different object classes to determine the correct class of an unknown/test sample. n_i training samples from the i-th class are given as columns of a matrix $A_i = \left[v_{i,1}, v_{i,2}, \ldots, v_{i,n_i}\right] \in IR^{m \times n_i}$. In face recognition, various statistical, generative, or discriminative models have been proposed for exploiting the structure of the A_i for recognition. An effective method is to model the samples from a single class as lying on a linear subspace. Additionally, it has been established that the images of faces with different illumination and expression lie on a special low dimensional subspace [10, 11]. Giving sufficient training samples of the i-th class, $A_i = \left[v_{i,1}, v_{i,2}, \ldots, v_{i,n_i}\right] \in IR^{m \times n_i}$, any unknown/test sample $y \in IR^m$ of the same class will lie in the linear span of the training samples:

$$y = \alpha_{i,1}v_{i,1} + \alpha_{i,2}v_{i,2} + \ldots + \alpha_{i,n_i}v_{i,n_i} \tag{1}$$

for $\alpha_{i,j} \in IR, j = 1, 2, \ldots, n_i$. The matrix A for the complete training set of all K object classes is:

$$A = [A_1, A_2, \ldots, A_k] \tag{2}$$

In terms of all training samples, the linear representation of y can be rewritten as:

$$y = Ax_0 \in IR^m \tag{3}$$

Where $x_0 = [0, \ldots, 0, \alpha_{i,1}, \alpha_{i,2}, \ldots, \alpha_{i,n_i}, 0, \ldots, 0]^T \in IR^n$ is the coefficient vector, whose entries are zero except those associated with the i-th class.

The equation $y = Ax$ has a unique solution if $m > n$. However, the solution of $y = Ax$ in robust face recognition is not unique. The difficulty can be resolved by using the minimum l^2-norm solution:

$$\hat{x}_2 = \arg\min\|x\|_2 \; subject\ to\ Ax = y \tag{4}$$

However, the solution $\hat{x}_2$ is not especially informative for recognizing y. The more sparse the recovered x is, the easier will it be to determine the accurate identity of y. As the theory of sparse representation and compressed sensing develops, the problem can be solved by finding the solution of l^1-minimization problem:

$$\hat{x}_2 = \arg\min\|x\|_1 \; subject\ to\ Ax = y \tag{5}$$

In a best case scenario, the nonzero entries in the $\hat{x}_1$ associate with the columns of A from a single object class. However, a few nonzero entries associated with multiple object classes may be created by noise and modeling error. The designing modeling of classifier depends on how accurate the coefficients defined with respect to the training samples of each object reproduce y. The characteristic function which selects the coefficients associated with the i-th class can be defined as $\delta_i : IR^n \rightarrow IR^n$. For $x \in IR^n, \delta_i(x) \in IR^n$ is the new one whose only nonzero entries are the entries in x which are related to the i-th class. The test sample y can be approximated as $\hat{y}_i = A\delta_i(\hat{x}_1)$ by using the coefficients associated with the i-th class. Soy can be classified by assign the approximation to the object class and minimizes the residual between y and $\hat{y}_i$. The sparse representation classification algorithm can be conclude in the Table 1.

Table 1. Classification algorithm using the sparse representation

1. A matrix of training samples $A = [A_1, A_2, \ldots, A_k] \in IR^{m \times n}$ for k classes, a test sample $\in IR^m$.
2. Normalize the columns of A to have unit l^2-norm.
3. Find the solution of the l^1-minimization problem
$$\hat{x}_1 = \arg\min\|x\|_1 \; subject\ to\ Ax = y$$
4. Compute the residuals $r_i(y) = \|y - A\delta_i(\hat{x}_1)\|_2$ for i=1,2,…,k
5. 5. The output is identity $(y) = \arg\min_i r_i(y)$

3 Proposed CNN Architecture

A CNN architecture is composed of one or more convolution layers and followed by one or more fully connected layers. A CNN is designed to take advantage of the 2D structure of an input image and can be achieved with local connections and tied weights followed by some form of pooling which results in translation invariant features [13]. Specifically, the architecture of CNN used in this paper is described in Fig. 1. The architecture contains two convolution layers with max-pooling, followed by a fully connected layer, and a Softmax output layer indicating identity of the classes. The convolution layers' operation is expressed as:

$$y^{j(r)} = \max\left(0, b^{j(r)} + \sum_i k^{ij(r)} * x^{i(r)}\right) \tag{6}$$

Where, x^i and y^i are the i-th input and the j-th output map, respectively, K^{ij} is the kernel used for convolution between the i-th input and the j-th output map. The convolution operation is represented as the symbol of *.

The max-pooling is overlapping in the first pooling layer and non-overlapping in the second pooling layer. The last hidden layer of the architecture is fully connected to the second pooling layer. The output is a Softmax classifier which predicts the probability distribution over n different identities. The probability of each identity is:

$$y_i = \frac{\exp\left(y_i^{'}\right)}{\sum_{j=1}^{n} \exp\left(y_j^{'}\right)} \tag{7}$$

Where $y_j^{'} = \sum_{i=1}^{fc} x_i.w_{i,j} + b_j$ linearly combines the fully connect layer features x_i as the input of neuron j, and y_j is the output.

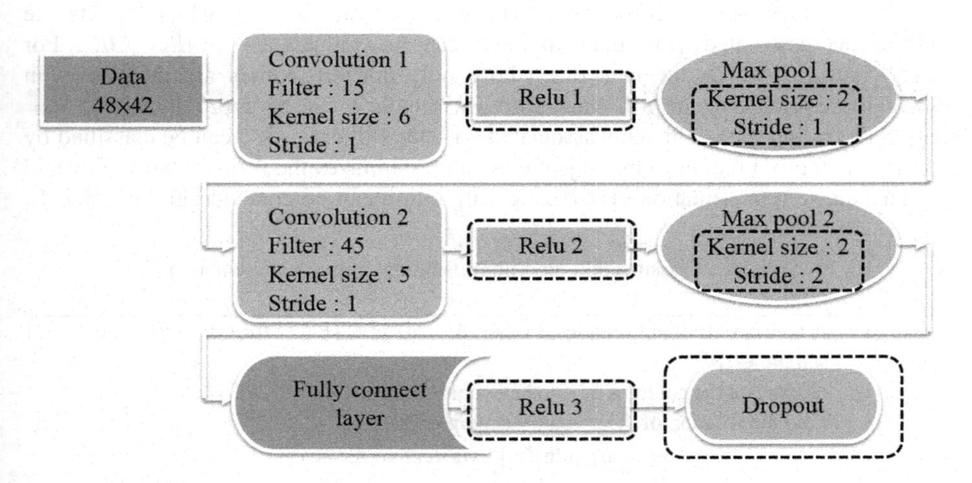

Fig. 1. Proposed CNN architecture with parameters

Some details of training will be introduced as follows. The model is trained by using Stochastic Gradient Descent (SGD), with a batch size of 32 samples, momentum of 0.9, and weight decay of 0.0005. The update rule for weight w is:

$$v_{i+1} = 0.9 \times v_i - 0.0005 \times \alpha \times w_i - \alpha \times \left\langle \frac{\partial L}{\partial w}\Big|_{w_i} \right\rangle_{D_i} \tag{8}$$

$$w_{i+1} = w_i + v_{i+1} \tag{9}$$

where, i is the iteration index, v is the momentum, α is the learning rate and $\left\langle \frac{\partial L}{\partial w}|_{w_i} \right\rangle_{D_i}$ is the average over the i-th batch D_i of the derivative of the derivative of the objective with respect to w, evaluated at w_i.

The weights in each layer are initialized by the Xavier algorithm that automatically determines the scale of initialization based on the number of input and output neurons. The neuron biases are initialized as constant, with the default filling value 0. The basic learning rate is set as 0.01. The initialized learning rate of weights is the same as the basic learning rate. Initialized learning rate of biases are twice as large as the basic learning rate. The settings usually lead to better convergence rates. By choosing the learning rate policy whose formula is $\alpha \times (1 + gamma \times i)^{-power}$ with $gamma = 0.0001$ and $wer = 0.75$, the learning rates of weights and biases can be updated during training period.

Yi Sun et al. [12] have founded some features of the deep neural activations, which play the vital role in the performance measurement, such as sparsity and selectivity. The phenomenon observed is that the neural activations are moderately sparse. Such sparsity distributions can maximize the discriminative power of the deep net as well as the distance between images. The various identities have different subsets of active neurons and two identical images have similar activation patterns.

By using the approach of visualization the properties of sparsity and selectivity can also be observed in Fig. 2 and can be used to distinguish different identities. On the other side, sparse representation will be chosen as the classifier for face recognition in this paper. The popular classifiers; nearest neighbor (NN) [14] and nearest subspace (NS) [15] are considered as a generalized classifiers. The approach considers all possible supports and adaptively chooses the minimal number of training samples needed to represent each test sample. Sparse representation has achieved good performance in face recognition with various expression and different illumination, as well as occlusion and disguise scenarios [16]. In the previous section, the algorithm of sparse representation has been introduced. The sparser the recovered coefficient vector is, the easier it can determine the correct identity of y. The sparsity and selectivity of deep neural activations will strengthen the sparseness of the coefficient vector to a certain extent, which will be helpful for final recognition. The combination has been proved effective and details of experiments will be introduced in the next section.

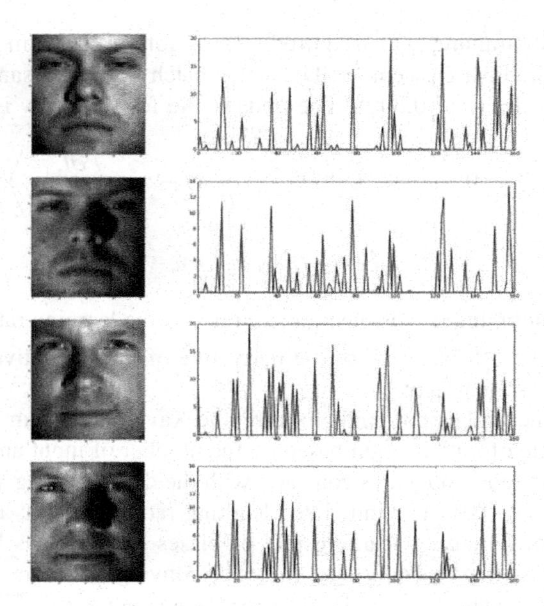

Fig. 2. Sparsity and selectivity of deep neural activations

4 Experiment Result

The proposed CNN is implemented on Ubuntu 14.04 OS, and integrated development environment is Caffe, which is a deep learning framework developed by the Berkeley Vision and Learning Center (BVLC) and by community contributors. Meanwhile, the sparse representation classification algorithm is implemented on Windows 7 OS and the integrated development environment is Microsoft Visual C++. The Extended Yale B database is used for experiments to analyze the performance of the proposed method.

4.1 Dataset

The Extended Yale B database is used to show the performance of the proposed method. This database contains 2,414 frontal face images of 38 individuals [17]. The cropped and normalized 192 $\times$ 168 face images are captured under different scenarios with various lighting conditions [18]. For each subject, half of the images are selected randomly for training and the rest for testing. The database is mainly used to examine the proposed method's resistance to extreme illumination change. Figure 3 shows some example images of one subject that represents various illumination conditions.

Fig. 3. Samples images from the Extended Yale B database.

4.2 Comparison

Experiment results are presented on the Extended Yale B database for face recognition, which demonstrates the efficacy of the proposed method. The proposed CNN architecture is trained with the Softmax classifier and implemented using Caffe library. The goals of experiments are as follows: (1) Compare the recognition rates of different feature dimensions; (2) Compare the recognition rates of the Softmax classifier and the sparse representation classifier with deep features; (3) Compare the recognition rates of different feature extraction methods with the sparse representation classifier. In this section, the 192 × 168 face images with various laboratory-controlled light conditions are resized to 48 × 42, which are the input of the CNN architecture. For each subject, 32 images are randomly selected for training and rest for testing. Total number of subjects are 38. Hence, the training set has 1216 images and the testing set has 1198 images. The training and testing set are randomly chosen five times and the average recognition rates are calculated as the final results. In each time, the feature space dimensions are 30, 56, 120, 160 and 504.

The final average results are presented in Table 2. In Table 2, the results of different dimensions with the same method and the results of different classifiers with the same features are compared. The recognition rate is increasing with the increase in the feature dimensions. As shown in Table 2, the average results of the deep features with spares representation classifier is better than with the Softmax classifier in each dimension, respectively. Deep features with spares representation classifier achieve a maximum accuracy of 99.2% with 504 dimension. The experimental results demonstrates that the sparse representation classifier has an advantage over the Softmax classifier, and deep features with different classifiers perform well in extreme illumination condition. In addition, a comparison is performed with the experiment results made by John Wright [16]. Here, both experiments considers 38 individuals of the Extended Yale B database and feature space dimension {30, 56, 120, 504}, and same configurations with the training and testing sets. In Fig. 4, the results of other different features including Eigenfaces [4], Laplacianfaces [19], Fisherfaces [10], randomfaces [20], and downsampled images with sparse representation classifier in different dimensions are presented. As is shown in Fig. 4, the method of deep learning feature extraction is superior to the other feature extraction methods in each dimension. In the 504 dimension, the maximum accuracy without deep features is 98.1% which is achieved by randomfaces [20] and deep features achieved the maximum accuracy of 99.2%. Table 3 provides a comparative study of different combinations. The best recognition rate and the corresponding dimension of each combination is shown. The proposed method achieves the highest accuracy, which confirms that the proposed combination works well for face recognition.

Table 2. Recognition accuracy comparison of Softmax and SRC with deep features on the Extended Yale B database

Methods	Dimension				
	30	56	120	160	504
	Accuracy				
Deep features+Softmax	96.68%	97.33%	97.78%	98.10%	98.85%
Deep features+SRC	97.00%	97.80%	98.33%	98.35%	99.17%

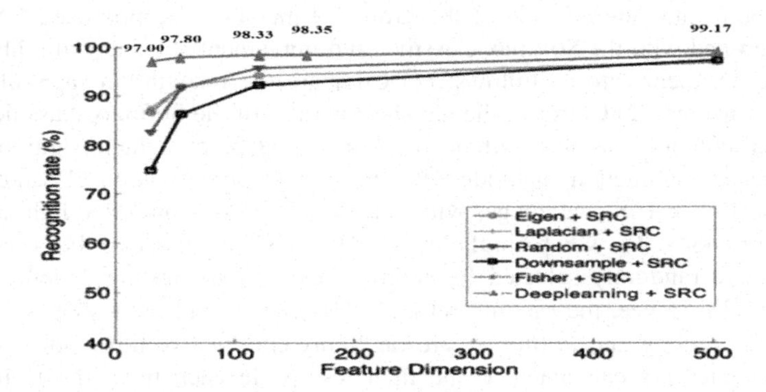

Fig. 4. Recognition accuracy comparison by using different feature extraction methods with SRC on the Extended Yale B database

Table 3. Comparative study using different combinations on the Extended Yale B database

Method	Feature dimension	Recognition accuracy
Eigen+SRC	504	96.77%
Laplacian+SRC	504	96.52%
Random+SRC	504	98.09%
Down sample+SRC	504	97.10%
Fisher+SRC	504	86.91%
Deep learning+Soft-max	504	98.92%
Deep learning+SRC	504	99.17%

5 Conclusion

This paper proposed a face recognition method composed of a two-layer deep CNN for feature extraction and sparse representation classification. Performance of the method is examined on two publicly available face databases. The proposed method combines strengths of deep features and sparse representation classifier. Features extracted from the CNN architecture are moderately sparse, highly selective to the identities, which maximize the discriminative power among different persons. Properties of deep features can be combined with the spares representation and have considerable tolerance to illumination change and expression. From the experimental results, with the same deep features, sparse representation classifier is superior to Softmax classifier, which is mainly used with CNN architectures. In addition, experiments also demonstrate that the method of deep learning feature extraction has an advantage over other feature extraction methods, including Eigenfaces, Laplacianfaces, Fisherfaces, randomfaces, and down-sampled images. The combination of deep features and sparse representation classifier achieved a better performance in the condition of varying illumination and expression.

References

1. Zhao, W., Chellappa, R., Phillips, P.J., Rosenfeld, A.: Face recognition: a literature survey. ACM Comput. Surv. **35**(4), 399–458 (2003)
2. Zhao, W., Chellappa, R.: Image-based face recognition: issues and methods. Opt. Eng. **78**, 375–402 (2002). Marcel Dekker Incorporated, New York
3. Ebied, R.M.: Feature extraction using PCA and kernel-PCA for face recognition. In: 8th International Conference on Informatics and Systems, vol. 8, pp. 72–77 (2012)
4. Turk, M.A., Pentland, A.P.: Face recognition using eigenfaces. In: IEEE Computer Society Conference on Computer Vision and Pattern Recognition (1991)
5. Taigman, Y., Yang, M., Ranzato, M., Wolf, L.: DeepFace: closing the gap to human-level performance in face verification. In: IEEE Conference on Computer Vision and Pattern Recognition, pp. 1701–1708 (2014)
6. Huang, G.B., Lee, H., Miller, L.E.: Learning hierarchical representations for face verification with convolutional deep belief networks. In: IEEE Conference on Computer Vision and Pattern Recognition, pp. 2518–2525 (2012)
7. Sun, Y., Wang, X., Tang, X.: Deep learning face representation from predicting 10,000 classes. In: IEEE Conference on Computer Vision and Pattern Recognition, pp. 1891–1898 (2014)
8. Sun, Y., Wang, X., Tang, X.: Deep learning face representation by joint identification-verification. In: Advances in Neural Information Processing Systems, pp. 1988–1996 (2014)
9. Wright, J., Ma, Y., Mairal, J., Sapiro, G.: Sparse representation for computer vision and pattern recognition. Proc. IEEE **98**(6), 1031–1044 (2009)
10. Belhumeur, P.N., Hespanha, J.P., Kriegman, D.J.: Eigenfaces vs. fisherfaces: recognition using class specific linear projection. IEEE Trans. Pattern Anal. Mach. Intell. **19**(7), 711–720 (1997)
11. Basri, R., Jacobs, D.: Lambertian reflectances and linear subspaces. IEEE Trans. Pattern Anal. Mach. Intell. **25**(2), 383–390 (2001)

12. Sun, Y., Wang, X., Tang, X.: Deeply learned face representations are sparse, selective, and robust. In: Proceedings of the IEEE Conference on Computer Vision and Pattern Recognition (2015)
13. Liang, M., Hu, X.: Recurrent convolutional neural network for object recognition. In: Proceedings of the IEEE Conference on Computer Vision and Pattern Recognition, pp. 3367–3375 (2015)
14. Duda, R., Hart, P.: Pattern Classification and Sciene Analysis. Wiley, New York (1973)
15. Ho, J., Yang, M., Lim, J., Lee, K., and Kriegman, D.: Clustering appearances of objects under varying illumination conditions. In: IEEE Computer Society Conference on Computer Vision and Pattern Recognition, vol. 1 (2003)
16. Wright, J., Yang, A.Y., Ganesh, A., Sastry, S.S., Ma, Y.: Robust face recognition via sparse representation. IEEE Trans. Pattern Anal. Mach. Intell. **31**(2), 210–227 (2009)
17. Belhumeur, P.N., Haven, N., Kriegman, D.J.: From few to many: illumination cone models for face recognition under variable lighting and pose. IEEE Trans. Pattern Anal. Mach. Intell. **23**(6), 643–660 (2001)
18. Lee, K.C., Ho, J., Kriegman, D.J.: Acquiring linear subspaces for face recognition under variable lighting. IEEE Trans. Pattern Anal. Mach. Intell. **27**(5), 684–698 (2005)
19. He, X., Yan, S., Hu, Y., Niyogi, P., Zhang, H.J.: Face recognition using Laplacian faces. IEEE Trans. Pattern Anal. Mach. Intell. **27**(3), 328–340 (2005)
20. Kaski, S.: Dimensionality reduction by random mapping: fast similarity computation for clustering. In: IEEE International Joint Conference on Neural Networks Proceedings, vol. 1, pp. 4–9 (1998)

Class-Wised Image Enhancement for Moving Object Detection at Maritime Boat Ramps

Jing Zhao[1], Shaoning Pang[1(✉)], Bruce Hartill[2], and Abdolhossein Sarrafzadeh[1]

[1] Department of Computing, Unitec Institute of Technology, 139 Carrington Road, Mount Albert, Auckland 1025, New Zealand
{jzhao,ppang}@unitec.ac.nz

[2] The National Institute of Water and Atmospheric Research, 41 Market Place, Viaduct Harbour, Auckland, New Zealand
http://www.dmli.info

Abstract. In the context of marine boat ramps traffic surveillance, we propose in this paper a novel image enhancement method for interpreting the traffic of boats passing across the boat ramps. As the background dynamics of land and water scenes differ markedly, this new approach classifies areas in each image as either land or water, so that different strategies can be adopted to enhance image on land and on the water, respectively. In particular, the use of the dynamic sharpening size and adaptive sharpening strength significantly increases the robustness of this enhancement method. Experimental results demonstrate that our method is much more able to cope with the highly dynamic land and water composition scenes compared with the state-of-the-art methods.

Keywords: Land and water separation · Image enhancement · Moving object detection · Adaptive sharpening

1 Introduction

Traffic surveillance at maritime boat ramps is a highly challenging problem in the computer vision community, since object detection and tracking in this scenario involves dynamic land water composition scene and severe unpredictability of the object. The region of interest (ROI) for a boat ramp is covers both land and water areas. On land, the background is usually static, with little or no change in topography. However water scenes are intrinsically dynamic, as water is a reflective surface that moves continuously, often to varying degrees. The reflection of the sun on water, coupled with the unpredictability of waves caused by the wind, moving vessels (wakes) and tidal flows in the maritime environment creates situations where object detection is far more challenging. In particular, the borderline between the water and land varies over time with the rise and fall of the tide and with the sea state. As a result, the spatial extent of the water and land changes over time in the ROI.

D. Liu et al. (Eds.): ICONIP 2017, Part III, LNCS 10636, pp. 675–685, 2017.
https://doi.org/10.1007/978-3-319-70090-8_68

The image capture rate used in this study is only one frame per minute (1-fpm), which is much lower than standard video frequency (30 frames per second). This low image capture rate is pragmatic when the images are interpreted manually and it minimizes the number of images that have to be transmitted and stored. Consequently, we have a sparse image series. Objects move abruptly from one image to another at such low capturing rate, and exit or enter the scene frequently (3–5 images of object lifespan in average). Unpredictable variations of an object, such as appearance, scale, and rate of movement, are particularly frequent in this situation. Conventional tracking algorithms used in this context will be unreliable as they are heavily dependent on motion and appearance continuity.

To cope with the dynamic marine environment, we have developed the automatic boat counting system to provide real-time traffic surveillance at maritime boat ramps in [1]. To improve performance in this scenario, we take advantage of the image enhancement techniques as the preprocessing step before moving object detection. This work focuses on the design of a new enhancing method for improving the interpretability of objects in the land water composition scene. Built upon the traditional unsharp masking techniques, the proposed approach performs enhancement on land and on the water, respectively, since the background dynamics of land and water scenes differ markedly. The impact of sunrise and sunset is also specifically considered by the proposed method, to allow for changes in outdoor luminance. Moreover, the use of the dynamic sharpening size and adaptive sharpening strength significantly increases the performance of this enhancing method. We apply the proposed algorithm to interpret a real-world boat-flow analysis and counting system, and compare the results against those generated by existing enhancing methods. The empirical results show that the performance of the proposed method is superior in this context.

2 The Computer Boat Counting System

The solution for computer boat counting is to develop specialized computer vision techniques for moving object detection and object of interest tracking/counting in a maritime boat ramp environment. As discussed in Sect. 1, the main challenges for developing such a computer vision based boat counting system are (1) the dynamically shifting background composition, which can easily cause object misdetection; and (2) the motion and appearance discontinuity seen in LFR videos, which can make object unpredictable and untraceable. To better cope with such challenges, we incorporate image enhancing techniques into the implemented boat counting system in [1].

Figure 1 gives a diagrammatic representation of the overall design of the proposed computer boat counting system. Identifying moving objects from a sequence of images is a fundamental step, which ensures that the obtained foreground pixels accurately correspond to the moving objects of interest [2]. These foreground pixels are further processed for object localization and tracking. Given the regions of objects in the frame, it is then the task of tracking

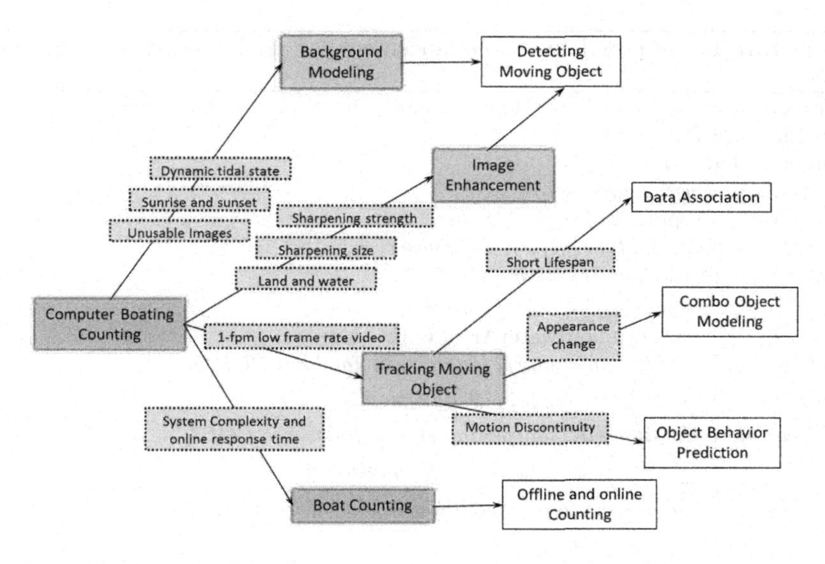

Fig. 1. Overall scheme of proposed computer boat counting system

algorithms to conduct object association from one image to the next to produce the tracks. An object tracker can generate the trajectory of an object over time by locating its place in each frame of the image sequence [3]. With the results of object detection and tracking, object of interest (boat) can be accurately counted by applying some strategies. Based on [1], we conduct image enhancement before moving object detection.

3 Proposed Image Enhancing Method

Image enhancement addresses the issue of improving the interpretability or perception of information in images for human viewers, or providing better input for other automated image processing techniques [4]. Our proposed image enhancing method is based on the conventional unsharp masking techniques, which is used when interpreting the traffic of boats passing across boat ramps. As the background dynamics of land and water scenes differ markedly, this new method classifies areas in each image as either land or water, given ancillary model data on predicted tidal height, so that different strategies can be used to enhance image on land and on the water, respectively. In particular, the use of the dynamic sharpening size and adaptive sharpening strength significantly increases the robustness of this enhancement method. The influence of sunrise and sunset is also considered by the new approach to account for changes in outdoor luminance.

The following subsections describe the proposed image enhancing method for land and water composition scenes in more detail. For an overview of the proposed approach, Algorithm 1 gives pseudocode of the core structure.

Algorithm 1. Proposed image enhancement for land and water composition scenes

Require: Current frame $\boldsymbol{I}_t$, ROI coordinates, shoreline area coordinates $\mathcal{S}$, and tidal height data H.

Ensure: Enhanced image $\boldsymbol{I}_{te}$.

/* *land and water area separation* */

1: $b^* \leftarrow$tide_interpolation(H); # *Subsec.* 3.1

2: pixel_classification($\boldsymbol{I}_t(x, y), \mathcal{S}$); # *Subsec.* 3.1

/* *search* k^* *in shoreline area* */

3: $k^* \leftarrow_{k' \in \mathcal{S}} (\arctan \frac{|k'-g|}{1+k'g})$; # *Equation* 4

4: $(\boldsymbol{I}_{lt}, \boldsymbol{I}_{wt}) \leftarrow$compute_landwaterArea($\boldsymbol{I}_t$, k^*, b^*);

/* *calculate the dynamic sharpening size* δ_l *for land ROI* */

5: $\delta_l = \left(\frac{\delta_{lmax} - \delta_{lmin}}{\theta_{lmax}}\right) \theta_l + \delta_{lmin}$; # *Equation* 5

/* *calculate the dynamic sharpening size* δ_l *for water ROI* */

6: $\delta_w = \left(\frac{\delta_{wmax} - \delta_{wmin}}{\theta_{wmax}}\right) \theta_w + \delta_{wmin}$; # *Equation* 6

/* *calculate the average gray-level intensity* $\tilde{\phi}_l$ *of land ROI* */

7: $\tilde{\phi}_l = \frac{\sum_{i=1}^{n_l} \phi_l(i)}{n_l}$; # *Equation* 7

/* *calculate the sharpening strength* ζ_l *for land ROI* */

8: $\zeta_l \leftarrow$compute_sharpenStrength($\tilde{\phi}_l$); # *Equation* 8

/* *calculate the average gray-level intensity* $\tilde{\phi}_w$ *of water ROI* */

9: $\tilde{\phi}_w = \frac{\sum_{i=1}^{n_w} \phi_w(i)}{n_w}$; # *Equation* 9

/* *calculate the sharpening strength* ζ_w *for water ROI* */

10: $\zeta_w \leftarrow$compute_sharpenStrength($\tilde{\phi}_w$); # *Equation* 10

/* *handle the influence of sunrise and sunset* */

11: **if** t falls in the sunrise/sunset zone **then**

12: $(\delta_l, \zeta_l, \delta_w, \zeta_w) \leftarrow$ sunrisesunset_handle(t);

13: **end if**

14: $\boldsymbol{I}_{lte} \leftarrow$ *enhance_image*$(\boldsymbol{I}_{lt}, \delta_l, \zeta_l)$;

15: $\boldsymbol{I}_{wte} \leftarrow$ *enhance_image*$(\boldsymbol{I}_{wt}, \delta_w, \zeta_w)$;

16: $\boldsymbol{I}_{te} \leftarrow \boldsymbol{I}_{lte} \cup \boldsymbol{I}_{wte}$

3.1 Land and Water Area Separation

The goal of separation is to segment the ROI into areas of land and water. In the image coordinate system shown in Fig. 2, the distinction between areas of land and water can be interpolated as a geometric problem, which is to find/fix a straight line as,

$$n = km + b \tag{1}$$

where k and b refer to the slope and intercept of straight line, respectively. Thus to determine the shoreline, the task is to search for optimal values of k and b.

Physically, we look the sea as a large container, with the amount of water determined the position of tidal boundary. In this sense, for a specific ramp, we are able to determine the optimal b^* given tidal height data H provided

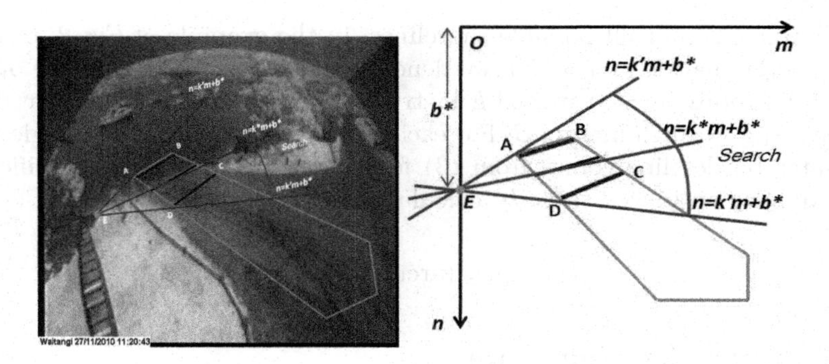

Fig. 2. Left: an example of maritime boat ramp. The region-of-interest (ROI) includes areas of land and water as seen inside the red polygon. Right: an illustration of searching optimal boundary between water and land. (Color figure online)

by an ancillary model using interpolation methods such as linear interpolation, polynomial interpolation, or spline interpolation, etc. As a result, we have the revised shoreline function as,

$$n = km + b^* \tag{2}$$

However, the slope k varies over time, as the direction of the boundary between the land and the water is not only determined by the shape of container, but also by the prevailing weather conditions such as the wind direction. Here, the proposed solution is to classify all ROI pixels into land and water area, then we seek the optimal slope k^* by a searching process described below.

Let $\boldsymbol{D}_t$ denote a land-water distribution matrix of current image $\boldsymbol{I}_t$, we can find the land water border line by accurately classifying every pixel as covering either land or water. $\boldsymbol{D}_t$ can be obtained by a binary pixel classification, which can be formulated as a convex optimization problem, i.e. the task of finding a minimizer of a convex function f that depends on a variable vector $\boldsymbol{w}$. Formally, we formulate this as an optimization problem, where the objective function is of the form

$$f(\boldsymbol{\omega}) = \frac{1}{2}\boldsymbol{\omega}^T\boldsymbol{\omega} + C\sum_{i=1}^{l}\max(1 - y_i\boldsymbol{\omega}^T\boldsymbol{x}_i, 0), \tag{3}$$

Here the vectors $x_i \in \mathcal{R}^d$ are the training data examples, for $1 \leq i \leq r$, and $y_i \in [-1, 1]$ are their corresponding labels, which we want to predict. Consequently with the SVM trained, every pixel in $\mathcal{S}$ is classified as either land or water. It is not difficult to model a line $n = gm + l$ that gives a pixel level shoreline approximation regardless of tide change.

Consider shoreline approximation in Fig. 2, by (2) we have point E that the actual land/water boundary should have gone through, and its distance to O is b^*. Without loss of generality, we can define for every ramp in surveillance a maximum margin for all possible shorelines. In the example of Fig. 2, rectangle $ABCD$ is the margin area which we denote hereafter as $\mathcal{S}$. To find the optimal slope k^*, we rotate line (2) around E by trying every possible slope k' that directs the line going through margin $\mathcal{S}$. For each test line, we calculate its angle to the land-water border line comes from (3) for land and water pixel classification. Thus, we have optimized slope k^* calculated as,

$$k^* \leftarrow_{k' \in \mathcal{S}} (\arctan \frac{|k' - g|}{1 + k'g}) \tag{4}$$

3.2 Dynamic Sharpening Size

Standard deviation of the Gaussian low-pass filter can be defined as a numeric value. This value controls the size of the region around the edge pixels that is affected by sharpening approach. A small value sharpens narrower regions around the edges, whereas a large value sharpens wider regions around edges. The sharpening size should be large enough so that image enhancing methods can increase the contrast along the edges, but small enough so that they are not misled by noise.

Consider boat ramps surveillance, we attempt to detect the moving objects at the interface between the land and the sea, at high traffic boat ramps on New Zealand's north-eastern coast. The background composition of the scene at a boat ramp can change substantially as tidal heights, levels of illumination, and reflectance vary throughout the day, and throughout the year. We have therefore adopted a dynamic sharpening size in the proposed method.

In the image data used for this work, the dimension of the objects of interest (i.e., boats and vehicles) shrinks from land to water area, since they are moving far away from the web cameras. Based on this observation, we design a scheme of dynamic sharpening size for enhancing images, in which the sharpening size is dynamically changing with the location of the moving objects. A larger value is used in the locations closer to the web cameras, whereas a smaller one is used in the locations farther away from the web cameras. This operation described above can be fitted by a linear model as shown in (5) and (6).

Let us denote δ_l as the sharpening size of the land ROI, which can be calculated as

$$\delta_l = \left(\frac{\delta_{lmax} - \delta_{lmin}}{\theta_{lmax}} \right) \theta_l + \delta_{lmin}, \tag{5}$$

where θ_l refers to the distance between the objects and the closest side to water in the land ROI, θ_{lmax} represents the length of the land ROI, and δ_{lmax} and δ_{lmin} are the sharpening size for the locations at the closet and farthest to the web cameras, respectively. δ_{lmax} and δ_{lmin} can be easily determined by cross validation.

Similarly, the sharpening size δ_w for the water ROI can be computed by (6).

$$\delta_w = \left(\frac{\delta_{wmax} - \delta_{wmin}}{\theta_{wmax}} \right) \theta_w + \delta_{wmin}, \tag{6}$$

3.3 Adaptive Sharpening Strength

The strength of the sharpening effect can be specified as a numeric value. A higher value of the sharpening strength leads to greater increase in the contrast of the sharpened pixels. Very large values for this parameter may produce undesirable effects in the output image. In contrast, lower values of this parameter may achieve smaller sharpening effect. Typical values for the sharpening strength are within the range of 0 to 2, although values larger than 2 are allowed.

In practice, the sharpening strength should be big enough so that image enhancing algorithms can adapt rapidly to changes in the background, but small enough so that they are not sensitive to noise. Because the background is influenced by changing luminance, different weather conditions, etc., it is essential for image enhancing methods to adopt an adaptive sharpening strength to optimize performance.

The image data used in this study is provided by web cameras overlooking boat ramps. In such scenes, a major proportion of each image is composed of the background, unless a significant amount of noise is present in the image. In other words, the binary object mask for a usable image would consist of a large number of pixels having the value 0 (i.e., background pixel), and a small number of pixels having the value 1 (i.e., object pixel). Based on this observation, we calculate the average gray-level intensity of the ROI. This computed value is used as a gating function that decides which sharpening strength to be used for enhancing the image. At those locations where the intensity is larger than the average (corresponding to the bright pixels), a lower sharpening strength is used. In contrast, at those locations where the intensity is below the average (corresponding to the dark pixels), a higher sharpening strength is employed. Such design is reasonable, because the dark objects are easily overlooked in our scenarios, and they need to be highlighted by a larger value of the sharpening strength.

Let us denote $\tilde{\phi}_l$ as the average gray-level intensity of the land ROI, which can be calculated by

$$\tilde{\phi}_l = \frac{\sum_{i=1}^{n_l} \phi_l(i)}{n_l}, \tag{7}$$

where $\phi_l(i)$ is the ith pixel in the land ROI, and n_l refers to the number of pixels in the land ROI.

Then the sharpening strength ζ_l for the land ROI can be computed as

$$\zeta_l = \begin{cases} \zeta_{hl} & \text{if } \phi_l(i) > \tilde{\phi}_l, \\ \zeta_{ll} & \text{otherwise.} \end{cases} \tag{8}$$

where ζ_{hl} and ζ_{ll} represent the sharpening strength for bright and dark pixels in land ROI, respectively. These two values can be easily determined by cross validation.

Similarly, the average gray-level intensity $\tilde{\phi}_w$ and sharpening strength ζ_w for the water ROI can be calculated by (9) and (10), respectively.

$$\tilde{\phi}_w = \frac{\sum_{i=1}^{n_w} \phi_w(i)}{n_w}, \quad (9)$$

$$\zeta_w = \begin{cases} \zeta_{hw} & \text{if } \phi_w(i) > \tilde{\phi}_w, \\ \zeta_{lw} & \text{otherwise.} \end{cases} \quad (10)$$

4 Experimental Results

The image data we used for our experiments was collected from a network of web cameras overlooking key boat ramps in New Zealand. The experimental data includes 2010–2012 image sequences captured at Waitangi, Takapuna and Raglan boat ramp. The frame size of the video is 720 × 576 pixels, and the frame rate is 1 frame/minute. The total number of frames tested for Waitangi, Takapuna and Raglan is 230,400, 241,920 and 249,120, respectively. In our experiment, the proposed image enhancing approach was compared with the state-of-the-art approaches. The same initializations were set to all algorithms for fair comparison. The parameters of all methods were tuned to achieve the best performance, and the ground truth was manually labeled in advance for comparisons.

4.1 Image Enhancement for Object Detection

To evaluate the performance of proposed image enhancing approach, we compare the proposed method with the following four different methods on the ability to detect moving objects at maritime boat ramps: (1) stretching the intensity values (Stretch) [5]; (2) histogram equalization (HisEqu) [6]; (3) contrast-limited adaptive histogram equalization (AdaHis) [7] and (4) unsharp masking method (Sharpen) [8].

Figure 3 presents the object detection results of the different image enhancing approaches in three frames, in which a rectangle represents a single object detected. The color of rectangle indicates different objects. Selected frames include scenarios where there are no moving objects (Left), a single object (Middle) and multiple objects (Right), at different tidal states. Figure 3(a) illustrates the truth objects extracted manually from three frames, and Fig. 3(b)–(f) display the object detection results from method (1)–(4) and the proposed method.

As seen in Fig. 3(a), the water area (top left of each frame) of the Waitangi boat ramp varies over time, with the rise and fall of the tide. The results produced by HisEqu and AdaHis are shown in Fig. 3(c) and (d), respectively. As shown in the Middle and Right frames, HisEqu and AdaHis can still extract the moving objects (i.e., cars) on the ground. However, they struggle to overcome dynamic illumination changes in the area of water, and objects (i.e., boats) in this area are missed. Moreover, many false alarms can be observed in the area of boat land and water as shown in the Left, Middle and Right frames. Figure 3(b) presents the results produced by Stretch, which contain fewer false alarms than those detected by HisEqu and AdaHis. However, this approach still struggles to interpret dynamic water background, and it tends to misidentify pixels with large variations of intensity values as shown in the Left frame. The results produced using Sharpen are demonstrated in Fig. 3(e). Sharpen extracts the foreground in a more reliable fashion than the other three existing methods. However, in the Right frame, the moving object (i.e., boat) cannot be identified in the border area of land and water. Results generated by the proposed method are displayed in Fig. 3(f). This algorithm obtains much clear foregrounds and fewer false positives in land and water areas under all scenarios.

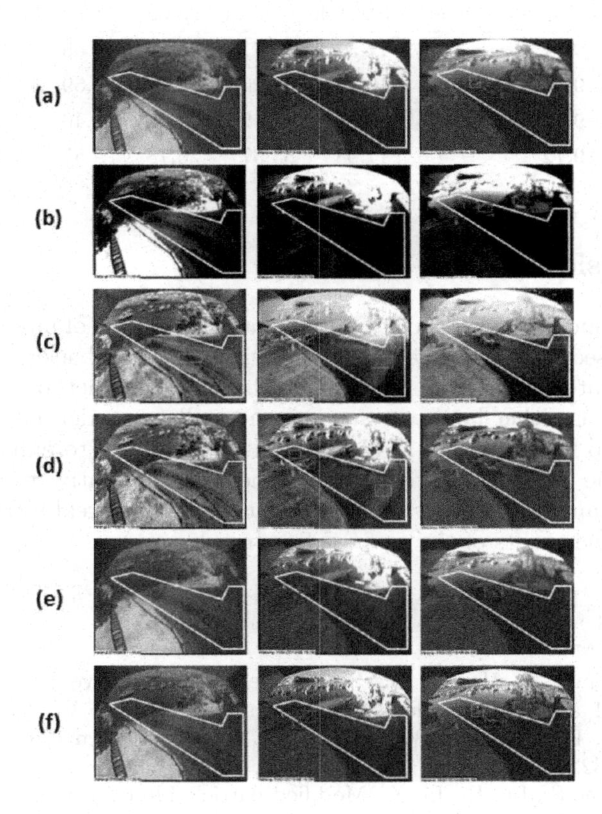

Fig. 3. Image enhancement for moving Object detection results. (a) Original frame and moving objects (ground truth), (b) Stretch, (c) HisEqu, (d) AdaHis, (e) Sharpen and (f) proposed method.

4.2 Boat Counting Performance

For boat counting, the proposed method was evaluated and compared with four conventional image enhancing methods, which include Stretch [5], HisEqu [6],

AdaHis [7] and Sharpen [8]. For performance evaluation, we calculate the differences between the ground truth (i.e., daily boat number from manual count), and daily boat number provided by different methods by NRMSE, normalized root mean squared error. Table 1 presents the NRMSE in percentage terms for five image enhancing methods at Waitangi, Raglan, and Takapuna, respectively. As seen from the table, our approach is much more able to cope with highly dynamic background and LFR compared with the state-of-the-art methods.

Table 1. Boat counting evaluation in NRMSE percentage

Method	No enhancement	Stretch	HisEqu	AdaHis	Sharpen	Proposed
Waitangi	9.71	17.32	35.71	26.12	9.56	**9.22**
Takapuna	9.83	19.79	37.29	28.73	9.69	**9.36**
Raglan	9.62	16.83	33.12	25.49	9.46	**9.20**
Average	9.72	17.98	35.37	26.78	9.57	**9.26**

5 Conclusion

This paper proposes a new algorithm for enhancing objects of interest (i.e., boats and vehicles) seen in one frame per minute videos. As the background dynamics of land and water scenes differ markedly, the proposed method classifies areas in each image as either land or water, given ancillary model data on predicted tidal height, so that different strategies can be employed to enhance image on land and on the water, respectively. Experimental comparative tests and quantitative performance evaluations on boat counting for three real-world boat ramps demonstrate the merits of the proposed approach.

References

1. Zhao, J., Pang, S., Hartill, B., Sarrafzadeh, A.: Adaptive background modeling for land and water composition scenes. In: Murino, V., Puppo, E. (eds.) ICIAP 2015. LNCS, vol. 9280, pp. 97–107. Springer, Cham (2015). doi:10.1007/978-3-319-23234-8_10
2. Zhang, T., Fei, S., Lu, H., Li, X.: Modified particle filter for object tracking in low frame rate video. In: Proceedings of the 48th IEEE Conference on Decision and Control, Held Jointly with the 28th Chinese Control Conference, pp. 2552–2557 (2009)
3. Yilmaz, A., Javed, O., Shah, M.: Object tracking: a survey. ACM Comput. Surv. **38**(4), 13 (2006)
4. Bedi, S.S., Khandelwa, R.: Various image enhancement techniques- a critical review. Int. J. Adv. Res. Comput. Commun. Eng. **2**(3), 1605–1609 (2013)
5. Lee, S.: An efficient content-based image enhancement in the compressed domain using retinex theory. IEEE Trans. Circ. Syst. Video Technol. **17**(2), 199–213. IEEE (2007)

6. Abdullah-Al-Wadud, M., and Kabir, M.H., Dewan, M.A.A., Chae, O.: A dynamic histogram equalization for image contrast enhancement. IEEE Trans. Consum. Electron **53**(2). IEEE (2007)
7. Reza, A.M.: Realization of the contrast limited adaptive histogram equalization (CLAHE) for real-time image enhancement. J. VLSI Sig. Process. **38**(4), 35–44 (2004)
8. Guang, D.: A generalized unsharp masking algorithm. IEEE Trans. Image Process. **20**(5), 1249–1261 (2011). IEEE

A Lagrange Programming Neural Network Approach for Robust Ellipse Fitting

Hao Wang, Ruibin Feng, Chi-Sing Leung(✉), and Hing Cheung So

Department of Electronic Engineering, City University of Hong Kong, Kowloon Tong, Hong Kong
wanghaocityu@gmail.com, rfeng4-c@my.cityu.edu.hk, {eeleungc,h.c.so}@cityu.edu.hk

Abstract. Ellipse fitting aims at constructing an elliptical equation that best fits the scattering points collected from an edge detection process. However, the edge detection process may introduce some noisy scattering points. This paper proposes a robust ellipse fitting model based on the Lagrange programming neural network (LPNN) framework. We formulate the ellipse fitting problem as a constrained optimization problem. The objective function contains an ℓ_1-norm term which can effectively suppress the effect of outliers. Since the LPNN framework cannot handle non-differentiable objective functions, we introduce an approximation for the ℓ_1-norm term. Besides, the local stability of the proposed LPNN method is discussed. Simulation results show that the proposed ellipse fitting algorithm can effectively reduce the influence of outliers.

Keywords: Ellipse fitting · LPNN · Outliers

1 Introduction

Ellipse fitting plays an important role in shape detection. Various ellipse fitting algorithms have been developed in the last two decades. Roughly, there are two categories of ellipse fitting algorithms. The first one is Hough transform (HT) and its variants [1]. However, the computational cost of this category is high.

The second category is based on the concept of least squares (LS) fitting. The algorithms [2] in this category estimate the elliptic parameters by minimizing the differences between the estimated ellipse and the scattering points obtained from the edge detection process. This category can be further divided into geometric approach and algebraic approach. The former one computes the sum of the orthogonal distances between the scattering points and the estimated ellipse [3]. In the algebraic approach, we compute the sum of algebraic distances. Since the algebraic approach is more computationally simple and attractive [4], many numerical algebraic distance based algorithms were proposed. Among them, the constrained least squares (CLS) algorithm [5] is a typical example. However, the CLS algorithm cannot handle the case that the scattering points contain non-Gaussian noise, or that some scattering points are outliers.

D. Liu et al. (Eds.): ICONIP 2017, Part III, LNCS 10636, pp. 686–696, 2017.
https://doi.org/10.1007/978-3-319-70090-8_69

Using neural circuits for solving constrained optimization problems has been studied for a long time [6]. Some projection models [7] were developed in the last decade. The Lagrange programming neural network (LPNN) approach [8–10] provides a general framework for solving various constrained optimization problems. Recently, some new applications of the LPNN approach [9,10], such as waveform design in radar systems, were developed. However, the LPNN framework can handle differentiable objective functions only.

This paper proposes a robust ellipse fitting model based on the LPNN framework. In order to improve the robustness of the algorithm, we use an ℓ_1-norm in our objective function. To solve the non-differentiable problem, we introduce an approximation for the ℓ_1-norm. The local stability of the proposed algorithm is also discussed.

The rest of this paper is organized as follows. Backgrounds on ellipse fitting and LPNN are given in Sect. 2. The proposed ellipse fitting algorithm is developed in Sect. 3. The local stability property is also discussed. Results of several experiments are provided in Sect. 4. Finally, conclusions are drawn in Sect. 5.

2 Background

Ellipse Fitting
The equation of an axis-aligned ellipse in the 2D space is given by $x^2/a^2+y^2/b=0$. It describes an ellipse with center at the origin. The widths along the two axes are a and b, respectively. For a general ellipse, the equation is given by

$$\frac{((x-c_x)\cos\phi+(y-c_y)\sin\phi)^2}{a^2}+\frac{(-(x-c_x)\sin\phi+(y-c_y)\cos\phi)^2}{b^2}=1, \quad (1)$$

where (c_x, c_y) is the center of the ellipse, and ϕ is the rotation angle of the ellipse.

Given n scattering points, $\{(x_i, y_i), i = 1, \cdots, n\}$, ellipse fitting aims at estimating the five parameters, $\{a, b, c_x, c_y, \phi\}$. However, it is cumbersome to estimate them directly, because Eq. (1) is highly nonlinear. Hence many methods [4,11] consider an alternative form, given by

$$Ax^2+Bxy+Cy^2+Dx+Ey+F=0. \quad (2)$$

The relationship between $\{A, B, C, D, E, F\}$ and $\{a, b, c_x, c_y, \phi\}$ is given by

$$A=\frac{\cos^2\phi}{a^2}+\frac{\sin^2\phi}{b^2},\ B=2\cos\phi\sin\phi\left(\frac{1}{a^2}-\frac{1}{b^2}\right),\ C=\frac{\sin^2\phi}{a^2}+\frac{\cos^2\phi}{b^2}$$
$$D=\frac{-2c_x\cos^2\phi-2c_y\sin\phi\cos\phi}{a^2}+\frac{-2c_x\sin^2\phi+2c_y\sin\theta\cos\phi}{b^2}$$
$$E=\frac{-2c_y\sin^2\phi-2c_x\sin\phi\cos\phi}{a^2}+\frac{-2c_y\cos^2\phi+2c_x\sin\phi\cos\phi}{b^2}$$
$$F=\frac{(c_x\cos\phi+c_y\sin\phi)^2}{a^2}+\frac{(c_x\sin\phi-c_y\cos\phi)^2}{b^2}-1.$$

Let $\mathcal{P} = \{(x_i, y_i) : i = 1, \cdots, n\}$ be the n scattering points. Define

$$\boldsymbol{\alpha} = [A, B, C, D, E, F]^{\mathrm{T}}, \psi_i = [x_i^2, x_i y_i, y_i^2, x_i, y_i, 1]^{\mathrm{T}}, \text{ and } \boldsymbol{\Psi} = [\psi_1, \cdots, \psi_n].$$

The "algebraic distance" [11] from a point (x_i, y_i) to an ellipse is given by

$$\text{algebraic distance} = \psi_i^{\mathrm{T}} \boldsymbol{\alpha}. \tag{3}$$

It is frequently used as an objective function for ellipse fitting. Under noise-free environment, each scattering point is on the ellipse, i.e.,

$$\psi_i^{\mathrm{T}} \boldsymbol{\alpha} = 0, \forall i = 1, \cdots, n, \text{ or saying } \boldsymbol{\Psi}^{\mathrm{T}} \boldsymbol{\alpha} = \mathbf{0}, \tag{4}$$

where $\mathbf{0}$ is a zero vector.

When the scattering points contain some noise, fitting ellipse can be formulated as the following constrained optimization problem, given by

$$\min_{\boldsymbol{\alpha}} \left\| \boldsymbol{\Psi}^{\mathrm{T}} \boldsymbol{\alpha} \right\|_2^2 \quad \text{s.t. } \boldsymbol{\alpha}^{\mathrm{T}} \boldsymbol{\alpha} = 1. \tag{5}$$

In (5), the unit-norm constraint is introduced to avoid the redundant solutions and the trivial solution ($\boldsymbol{\alpha} = \mathbf{0}$). The CLS method was proposed to solve the problem stated in (5). When the scattering points contain Gaussian-like noise, the CLS method works very well. However, when the scattering points contain impulsive noise or some of scattering points are outliers, the performance of CLS is very poor.

LPNN

The LPNN approach can be used to solve a general nonlinear constrained optimization problem, given by

$$\min_{\boldsymbol{z}} f(\boldsymbol{z}) \quad \text{s.t. } \boldsymbol{h}(\boldsymbol{z}) = 0, \tag{6}$$

where $\boldsymbol{z} = [z_1, \cdots, z_n]^{\mathrm{T}}$ denotes decision variable vector, $f : \mathbb{R}^n \to \mathbb{R}$ is the objective function, and $\boldsymbol{h} : \mathbb{R}^n \to \mathbb{R}^m$ $(m < n)$ is a vector valued function to define the m constraints. The LPNN approach requires that f and $\boldsymbol{h}$ should be twice differentiable. The first step in the LPNN approach is to set up its Lagrangian, given by

$$L(\boldsymbol{z}, \boldsymbol{\kappa}) = f(\boldsymbol{z}) + \boldsymbol{\kappa}^{\mathrm{T}} \boldsymbol{h}(\boldsymbol{z}), \tag{7}$$

where $\boldsymbol{\kappa} = [\kappa_1, \cdots, \kappa_m]^{\mathrm{T}}$ is the Lagrange multiplier vector. A LPNN has n variable neurons, which hold the decision variable vector $\boldsymbol{z}$, and m Lagrangian neurons, which hold the Lagrange multiplier vector $\boldsymbol{\kappa}$. In the second step, we construct its dynamics according to the Lagrangian, given by

$$\frac{d\boldsymbol{z}}{dt} = -\frac{\partial L(\boldsymbol{z}, \boldsymbol{\kappa})}{\partial \boldsymbol{z}}, \quad \frac{d\boldsymbol{\kappa}}{dt} = \frac{\partial L(\boldsymbol{z}, \boldsymbol{\kappa})}{\partial \boldsymbol{\kappa}}. \tag{8}$$

In (8), $\frac{d\boldsymbol{z}}{dt}$ is used to seek for the minimum objective value, while $\frac{d\boldsymbol{\kappa}}{dt}$ is used to restrict the decision variable vector within the feasible region. The network will settle down at a stable state if several conditions are satisfied [8,9]. After the neurons settle down at an equilibrium point, the solution is obtained by measuring the neuron outputs. From (8), we can see that f and $\boldsymbol{h}$ should be differentiable.

3 Development of Proposed Algorithm

To improve the robustness, we replace the l_2-norm in (5) with the l_1-norm. The constrained optimization becomes

$$\min_{\boldsymbol{\alpha}} \left\| \boldsymbol{\Psi}^{\mathrm{T}} \boldsymbol{\alpha} \right\|_1 \quad \text{s.t. } \boldsymbol{\alpha}^{\mathrm{T}} \boldsymbol{\alpha} = 1, \; B^2 - 4AC < 0, \tag{9}$$

where the second constraint is introduced to guarantee the solution corresponding to an ellipse only. Before we apply the LPNN approach, we need to consider two issues. First, we should transform the inequality constraint in (9) to an equality, because LPNN can handle equality constraints only. Another issue is that the objective function of (9) is not differentiable.

To solve the first issue, we introduce a new variable G into the inequality constraint, and modify the constraint as $B^2 - 4AC + G^2 = 0$. After the modification, we have

$$\min_{\tilde{\boldsymbol{\alpha}}} \left\| \tilde{\boldsymbol{\Psi}}^{\mathrm{T}} \tilde{\boldsymbol{\alpha}} \right\|_1 \qquad \text{s.t. } \tilde{\boldsymbol{\alpha}}^{\mathrm{T}} \boldsymbol{\Phi} \tilde{\boldsymbol{\alpha}} = 1, \; \tilde{\boldsymbol{\alpha}}^{\mathrm{T}} \boldsymbol{\Theta} \tilde{\boldsymbol{\alpha}} = 0, \tag{10}$$

where $\tilde{\boldsymbol{\alpha}} = [A, B, C, D, E, F, G]^{\mathrm{T}}$, $\tilde{\psi}_i = [x_i^2, x_i y_i, y_i^2, x_i, y_i, 1, 0]^{\mathrm{T}}$, $\tilde{\boldsymbol{\Psi}} = [\tilde{\psi}_1, \cdots, \tilde{\psi}_n]$,

$$\boldsymbol{\Phi} = \begin{bmatrix} \mathbf{I}_{6\times 6} & \mathbf{0}_{6\times 1} \\ \mathbf{0}_{1\times 6} & 0 \end{bmatrix}, \boldsymbol{\Theta} = \begin{bmatrix} \boldsymbol{\Lambda} & \mathbf{0}_{3\times 3} & \mathbf{0}_{3\times 1} \\ \mathbf{0}_{3\times 3} & \mathbf{0}_{3\times 3} & \mathbf{0}_{3\times 1} \\ \mathbf{0}_{1\times 3} & \mathbf{0}_{1\times 3} & 1 \end{bmatrix}, \text{ and } \boldsymbol{\Lambda} = \begin{bmatrix} 0 & 0 & -2 \\ 0 & 1 & 0 \\ -2 & 0 & 0 \end{bmatrix}.$$

Now, in (10), the constraints are all equality.[1] To address the second issue, we introduce an approximation for the absolute operation, given by

$$|x| \approx \frac{1}{\rho} \ln(\cosh(\rho x)), \tag{11}$$

where ρ is large is a positive constant. The approximation has similar shape with the ℓ_1-norm but differentiable at everywhere, as shown in Fig. 1. **One may**

[1] In this formulation, the constructed shape may be an ellipse or two parallel-lines. We can strictly restrict the estimated parameters to form an ellipse by changing the constraint "$\tilde{\boldsymbol{\alpha}}^{\mathrm{T}} \boldsymbol{\Theta} \tilde{\boldsymbol{\alpha}} = 0$" to "$\tilde{\boldsymbol{\alpha}}^{\mathrm{T}} \boldsymbol{\Theta} \tilde{\boldsymbol{\alpha}} = \delta$", where δ is a small positive constant.

argue that the approximation is computational complicated. However, we do not need to implement this approximation, and we only need to implement the derivative of the approximation, which is equal to $\tanh(\rho x)$.**This function is a commonly used activation function in the neural network community.** We set $\rho = 100$ in our experiment.

Using this approximation, (10) can be rewritten as,

$$\min_{\tilde{\boldsymbol{\alpha}}} \frac{1}{b} \sum_{i=1}^{n} \ln(\cosh(b\tilde{\psi}_i^{\mathrm{T}} \tilde{\boldsymbol{\alpha}})) \quad \text{s.t. } \tilde{\boldsymbol{\alpha}}^{\mathrm{T}} \boldsymbol{\Phi} \tilde{\boldsymbol{\alpha}} = 1, \; \tilde{\boldsymbol{\alpha}}^{\mathrm{T}} \boldsymbol{\Theta} \tilde{\boldsymbol{\alpha}} = 0. \tag{12}$$

Theoretically, we can directly use the concept of LPNN to solve the problem given in (12). According to our preliminary experiment, the dynamics constructed by (12), may not be stable. Hence, we further introduce a dummy variable vector $\boldsymbol{z}$, and three augmented functions into (12). After the modification we have

$$\min_{\tilde{\boldsymbol{\alpha}}, \boldsymbol{z}} \frac{1}{b} \sum_{i=1}^{n} \ln(\cosh(bz_i)) + \frac{c_1}{2} \|\boldsymbol{z} - \tilde{\boldsymbol{\Psi}}^{\mathrm{T}} \tilde{\boldsymbol{\alpha}}\|_2^2 + \frac{c_2}{2} \|\tilde{\boldsymbol{\alpha}}^{\mathrm{T}} \boldsymbol{\Phi} \tilde{\boldsymbol{\alpha}} - 1\|_2^2 + \frac{c_2}{2} \|\tilde{\boldsymbol{\alpha}}^{\mathrm{T}} \boldsymbol{\Theta} \tilde{\boldsymbol{\alpha}}\|_2^2$$

$$\text{s.t. } \boldsymbol{z} = \tilde{\boldsymbol{\Psi}}^{\mathrm{T}} \tilde{\boldsymbol{\alpha}}, \quad \tilde{\boldsymbol{\alpha}}^{\mathrm{T}} \boldsymbol{\Phi} \tilde{\boldsymbol{\alpha}} = 1, \quad \tilde{\boldsymbol{\alpha}}^{\mathrm{T}} \boldsymbol{\Theta} \tilde{\boldsymbol{\alpha}} = 0, \tag{13}$$

where $\boldsymbol{z} = [z_1, \ldots, z_n]^{\mathrm{T}}$. Parameters c_1 and c_2 are used to modulate the effect of the augmented terms. When they are large enough, the augmented terms can further improve the convexity and stability [8–10]. At an equilibrium point, the three augmented terms are equal to 0. In other words, they do not influence the objective values at equilibrium points.

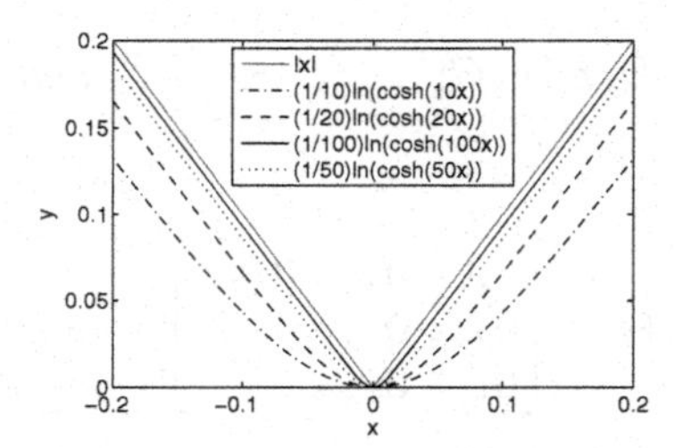

Fig. 1. Approximation of the l_1-norm.

The problem given in (13) can be solved by LPNN. First, we establish its Lagrangian, given by

$$L(\tilde{\boldsymbol{\alpha}}, \boldsymbol{z}, \boldsymbol{\zeta}, \beta, \gamma) = \frac{1}{\rho} \sum_{i=1}^{n} \ln(\cosh(\rho z_i)) + \boldsymbol{\zeta}^{\mathrm{T}} (\boldsymbol{z} - \tilde{\boldsymbol{\Psi}}^{\mathrm{T}} \tilde{\boldsymbol{\alpha}}) + \beta(\tilde{\boldsymbol{\alpha}}^{\mathrm{T}} \boldsymbol{\Phi} \tilde{\boldsymbol{\alpha}} - 1)$$

$$+ \gamma(\tilde{\boldsymbol{\alpha}}^{\mathrm{T}} \boldsymbol{\Theta} \tilde{\boldsymbol{\alpha}}) + \frac{c_1}{2} \|\boldsymbol{z} - \tilde{\boldsymbol{\Psi}}^{\mathrm{T}} \tilde{\boldsymbol{\alpha}}\|_2^2 + \frac{c_2}{2} \|\tilde{\boldsymbol{\alpha}}^{\mathrm{T}} \boldsymbol{\Phi} \tilde{\boldsymbol{\alpha}} - 1\|_2^2 + \frac{c_2}{2} \|\tilde{\boldsymbol{\alpha}}^{\mathrm{T}} \boldsymbol{\Theta} \tilde{\boldsymbol{\alpha}}\|_2^2. \tag{14}$$

In (14), $\tilde{\boldsymbol{\alpha}}$ and $\boldsymbol{z}$ are decision variable vectors, $\boldsymbol{\zeta} \in \mathbb{R}^n$, β and γ are the Lagrange multipliers. After we obtain the Lagrangian, we can deduce the neural dynamics as

$$\frac{d\boldsymbol{z}}{dt} = -\frac{\partial L(\tilde{\boldsymbol{\alpha}}, \boldsymbol{z}, \boldsymbol{\zeta}, \beta, \gamma)}{\partial \boldsymbol{z}} = -\tanh(\rho \boldsymbol{z}) - \boldsymbol{\zeta} - c_1 \left(\boldsymbol{z} - \tilde{\boldsymbol{\Psi}}^{\mathrm{T}} \tilde{\boldsymbol{\alpha}}\right), \tag{15}$$

$$\frac{d\tilde{\boldsymbol{\alpha}}}{dt} = -\frac{\partial L(\tilde{\boldsymbol{\alpha}}, \boldsymbol{z}, \boldsymbol{\zeta}, \beta, \gamma)}{\partial \tilde{\boldsymbol{\alpha}}} = \tilde{\boldsymbol{\Psi}}\boldsymbol{\zeta} - 2\beta\boldsymbol{\Phi}\tilde{\boldsymbol{\alpha}} - 2\gamma\boldsymbol{\Theta}\tilde{\boldsymbol{\alpha}} + c_1\tilde{\boldsymbol{\Psi}}\left(\boldsymbol{z} - \tilde{\boldsymbol{\Psi}}^{\mathrm{T}} \tilde{\boldsymbol{\alpha}}\right) \\ -2c_2\left(\tilde{\boldsymbol{\alpha}}^{\mathrm{T}}\boldsymbol{\Phi}\tilde{\boldsymbol{\alpha}} - 1\right)\boldsymbol{\Phi}\tilde{\boldsymbol{\alpha}} - 2c_2\left(\tilde{\boldsymbol{\alpha}}^{\mathrm{T}}\boldsymbol{\Theta}\tilde{\boldsymbol{\alpha}}\right)\boldsymbol{\Theta}\tilde{\boldsymbol{\alpha}}, \tag{16}$$

$$\frac{d\boldsymbol{\zeta}}{dt} = \frac{\partial L(\tilde{\boldsymbol{\alpha}}, \boldsymbol{z}, \boldsymbol{\zeta}, \beta, \gamma)}{\partial \boldsymbol{\zeta}} = \boldsymbol{z} - \tilde{\boldsymbol{\Psi}}^{\mathrm{T}} \tilde{\boldsymbol{\alpha}}, \tag{17}$$

$$\frac{d\beta}{dt} = \frac{\partial L(\tilde{\boldsymbol{\alpha}}, \boldsymbol{z}, \boldsymbol{\zeta}, \beta, \gamma)}{\partial \beta} = \tilde{\boldsymbol{\alpha}}^{\mathrm{T}}\boldsymbol{\Phi}\tilde{\boldsymbol{\alpha}} - 1, \tag{18}$$

$$\frac{d\gamma}{dt} = \frac{\partial L(\tilde{\boldsymbol{\alpha}}, \boldsymbol{z}, \boldsymbol{\zeta}, \beta, \gamma)}{\partial \gamma} = \tilde{\boldsymbol{\alpha}}^{\mathrm{T}}\boldsymbol{\Theta}\tilde{\boldsymbol{\alpha}}. \tag{19}$$

Here (15) and (16) are used to minimize the objective function, and (17)–(19) are used to restrict the solution into the feasible region. After the state of the network settles down at an equilibrium point, we can obtain an estimated $\tilde{\boldsymbol{\alpha}}$, denoted by $\tilde{\boldsymbol{\alpha}}^*$. According to $\tilde{\boldsymbol{\alpha}}^*$, the corresponding ellipse parameters $\{a^*, b^*, c_x^*, c_y^*, \theta^*\}$ can be calculated.

Figure 2 shows typical dynamics of the estimated parameters. The details of experimental settings are given in Sect. 4. It can be seen that the network settles down within 50 characteristic times.

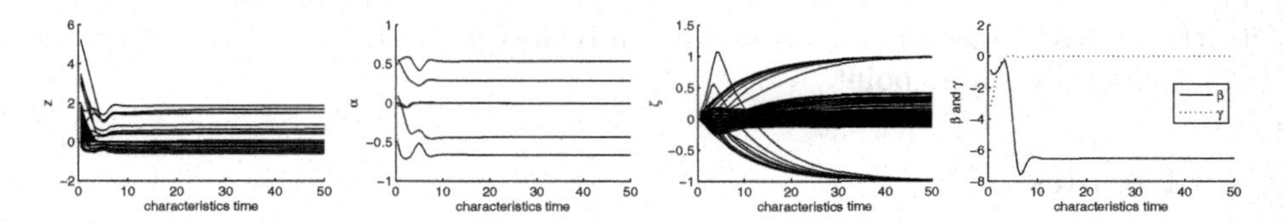

Fig. 2. Dynamics of the estimated parameters under impulse noise where the standard deviation of noise is equal to 0.6. (a) $\boldsymbol{z}$; (b) $\tilde{\boldsymbol{\alpha}}$; (c) $\boldsymbol{\zeta}$; (d) β and γ.

In the rest of this section, we discuss the asymptotical convergence (or saying local stability) of our proposed algorithm at a stationary point $(\boldsymbol{\alpha}^*, \boldsymbol{z}^*, \boldsymbol{\zeta}^*, \beta^*, \gamma^*)$ of Lagrangian function (14). Based on Theorem 1 of [8], there are two sufficient conditions for asymptotic convergence of the LPNN approach. The first one is convexity, i.e., the Hessian matrix of (14) at a stationary point should be positive definite. This condition can be proved by Lemma 1.25 of [8] under mild conditions.

For the second condition, at the stationary point $(\boldsymbol{\alpha}^*, \boldsymbol{z}^*, \boldsymbol{\zeta}^*, \beta^*, \gamma^*)$, the gradient vectors of the constraints with respect to the decision variables should be linearly independent, i.e., $\{\tilde{\boldsymbol{\alpha}}^*, \mathbf{z}^*\}$ should be a regular point. In our case, we have $n+2$ constraints:

$$h_1(\tilde{\boldsymbol{\alpha}}, \boldsymbol{z}) = \tilde{\boldsymbol{\alpha}}^{\mathrm{T}}\boldsymbol{\Phi}\tilde{\boldsymbol{\alpha}} - 1, h_2(\tilde{\boldsymbol{\alpha}}, \boldsymbol{z}) = \tilde{\boldsymbol{\alpha}}^{\mathrm{T}}\boldsymbol{\Theta}\tilde{\boldsymbol{\alpha}}, \text{ and } h_{i+2}(\tilde{\boldsymbol{\alpha}}, \boldsymbol{z}) = z_i - \tilde{\boldsymbol{\alpha}}^{\mathrm{T}}\tilde{\psi}_i, \forall i = 1, \cdots, n. \quad (20)$$

At the stationary point, the gradient vectors are given by

$$\left\{\begin{bmatrix}\frac{\partial h_1(\tilde{\boldsymbol{\alpha}}^*, \boldsymbol{z}^*)}{\partial \tilde{\boldsymbol{\alpha}}} \\ \frac{\partial h_1(\tilde{\boldsymbol{\alpha}}^*, \boldsymbol{z}^*)}{\partial \boldsymbol{z}}\end{bmatrix}, \cdots, \begin{bmatrix}\frac{\partial h_{n+2}(\tilde{\boldsymbol{\alpha}}^*, \boldsymbol{z}^*)}{\partial \tilde{\boldsymbol{\alpha}}} \\ \frac{\partial h_{n+2}(\tilde{\boldsymbol{\alpha}}^*, \boldsymbol{z}^*)}{\partial \boldsymbol{z}}\end{bmatrix}\right\} = \left\{\begin{bmatrix}2A \\ 2B \\ 2C \\ 2D \\ 2E \\ 2F \\ 0 \\ 0 \\ 0 \\ \vdots \\ 0\end{bmatrix}, \begin{bmatrix}-2C \\ B \\ -2A \\ 0 \\ 0 \\ 0 \\ G \\ 0 \\ 0 \\ \vdots \\ 0\end{bmatrix}, \begin{bmatrix}-x_1^2 \\ -x_1 y_1 \\ -y_1^2 \\ -x_1 \\ -y_1 \\ -1 \\ 0 \\ 1 \\ 0 \\ \vdots \\ 0\end{bmatrix}, \cdots, \begin{bmatrix}-x_n^2 \\ -x_n y_n \\ -y_n^2 \\ -x_n \\ -y_n \\ -1 \\ 0 \\ 0 \\ 0 \\ \vdots \\ 1\end{bmatrix}\right\} \quad (21)$$

There are $n+2$ gradient vectors. Each of them has $n+7$ elements. Firstly, it is easy to note that the last n vectors are linearly independent. Besides, they are all linearly independent with the first two vectors, providing that $G \neq 0$. Secondly, the first two vectors are linearly independent with each other. To make sure the fitting result is an ellipse, $B^2 - 4AC < 0$ must be satisfied, i.e., for $B^2 - 4AC + G^2 = 0$. Therefore, the $n+2$ gradient vectors in (21) are linearly independent, as long as at the stationary point $G \neq 0$. Furthermore, according to the first and second conditions we can deduce that $(\boldsymbol{\alpha}^*, \boldsymbol{z}^*, \boldsymbol{\zeta}^*, \beta^*, \gamma^*)$ is an asymptotically stable point.

4 Results

In this section, we conduct several experiments to evaluate the performance of the proposed algorithm. We need to initialize several parameters before we implement our proposed algorithm. First, for c_1 and c_2, we use a trial-and-error method to select them, and set them to $c_1 = 10$ and $c_2 = 20$. Second, we need to initialize the variables $(\tilde{\boldsymbol{\alpha}}, \boldsymbol{z}, \boldsymbol{\zeta}, \beta, \gamma)$. The initialization of $\tilde{\boldsymbol{\alpha}}$ represents a circle centered at the midpoint of the data set, and its radius is a small positive random value. Afterwards, $\boldsymbol{z}$ can be directly initialize as $\boldsymbol{z} = \tilde{\boldsymbol{\Psi}}^{\mathrm{T}}\tilde{\boldsymbol{\alpha}}$. Finally the Lagrangian variables $\boldsymbol{\zeta}, \beta$ and γ are initialized as small random values.

Several ellipse fitting algorithms are considered for performance comparison. They are the traditional CLS method [5], direct least squares fitting (DLSF) [11], and the l_2-norm LPNN. The l_2-norm LPNN is similar to the proposed algorithm but it uses an l_2-norm objective function.

In the following two subsections, we test the performance of our proposed algorithm under different noise environments. Before that we need to generate

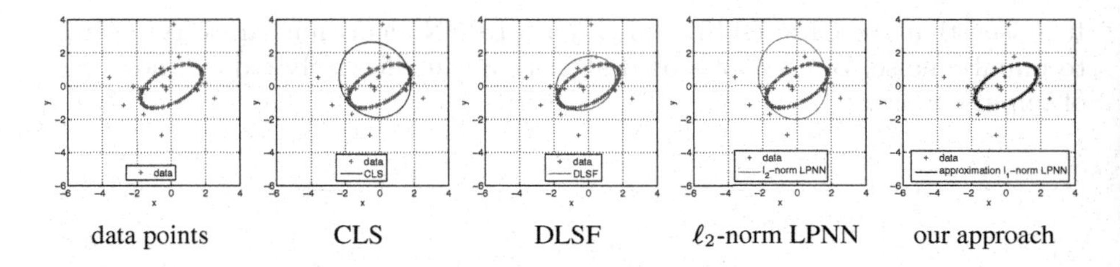

Fig. 3. Fitting results under Laplacian noise.

the data set which are used to do ellipse fitting. First, we set $c_x = 0$, $c_y = 0$, $a = 2$, $b = 1$, $\theta = 30°$ to obtain an ellipse, and select 100 data points from this ellipse. Then, we add different kinds of noise into the data set.

Experiment 1: Ellipse Fitting in Impulse Noise

In this experiment, we test the performance of the algorithms under impulse noise. Hence, we randomly choose 20 points from the data set and add some Laplacian noise into them. An example of the data set is given by Fig. 3. In our experiment, we change the standard deviation of the Laplacian noise from 0 to 1 and repeat the experiment 100 times at each noise level. Then we compute the mean absolute deviation (MAD) of the estimated parameters (c_x^*, c_y^*, a^*, b^*, ρ^*). The results are given in Fig. 4. From the figure, the proposed approach is superior to other comparison algorithms. A typical example of the fitting results is shown in by Fig. 3, where the standard deviation of the Laplacian noise is 0.8.

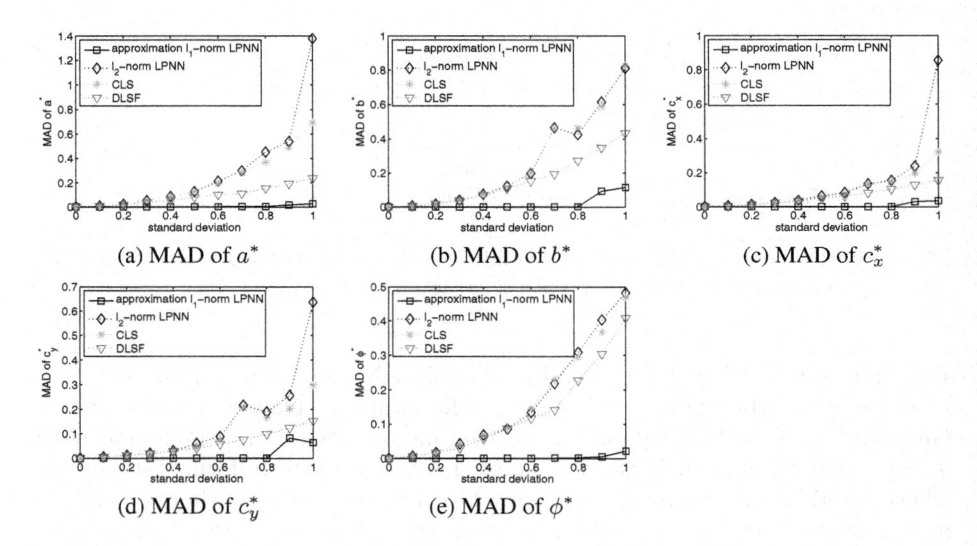

Fig. 4. MAD under Laplacian noise. The Laplacian noise level is varied from 0 to 1. We repeat the experiment 100 times at each noise level.

It is seen that the CLS, DLSF and l_2-norm LPNN algorithms are very sensitive to impulse noise. But our proposed algorithm can effectively reduce the impact of impulse noise.

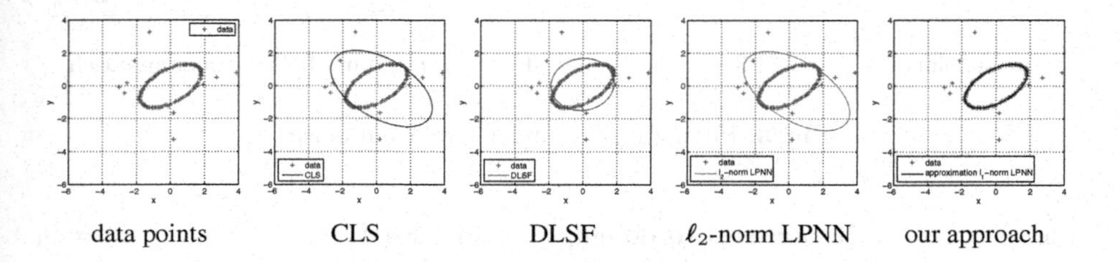

data points CLS DLSF ℓ_2-norm LPNN our approach

Fig. 5. Fitting results under uniform noise.

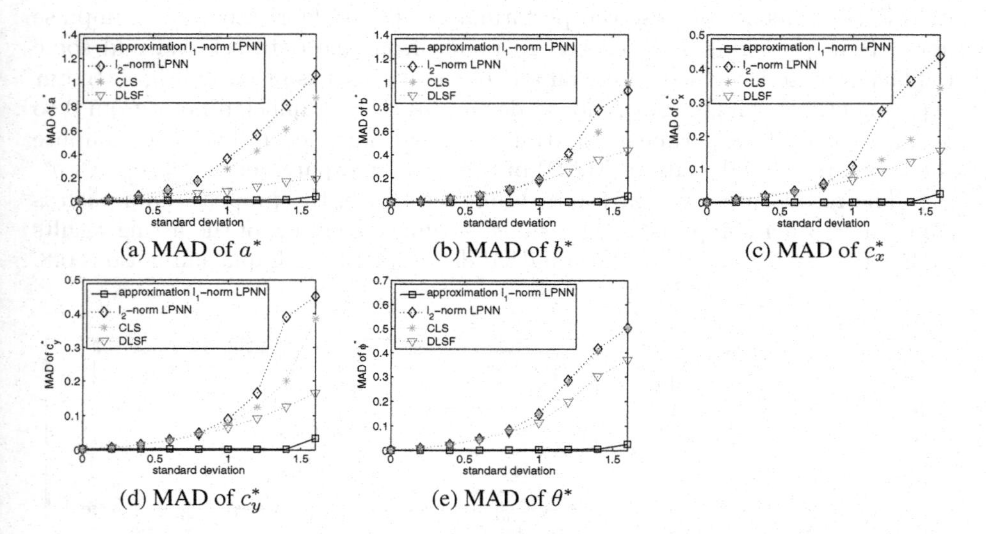

(a) MAD of a^* (b) MAD of b^* (c) MAD of c_x^*

(d) MAD of c_y^* (e) MAD of θ^*

Fig. 6. MAD under uniform noise. The uniform noise level is varied from 0 to 2.4. We repeat the experiment 100 times at each noise level.

Experiment 2: Ellipse Fitting in Gaussian Noise with Outliers
In the second experiment, we test the performance of different methods under Gaussian Noise with a few outliers. The standard deviation of Gaussian noise is fixed at 0.02. The outliers are generated by zero mean uniform distribution, whose standard deviation are range form 0 to 1.6. We randomly choose 10 points in the data set and add outliers into them. After that, an example of the geometric distribution of the data set is shown in Fig. 5. We repeat the experiment 100 times at each noise level and compute the MAD of the estimated parameters. The results are shown in Fig. 6. From the figure, the proposed approach is

superior to other comparison algorithms. A typical example is given by Fig. 5, where the standard deviation of outliers is 1.4. From the figures, the CLS, DLSF, and l_2-norm LPNN algorithms are very sensitive to impulse noise. Our proposed algorithm can effectively reduce the impact of noise.

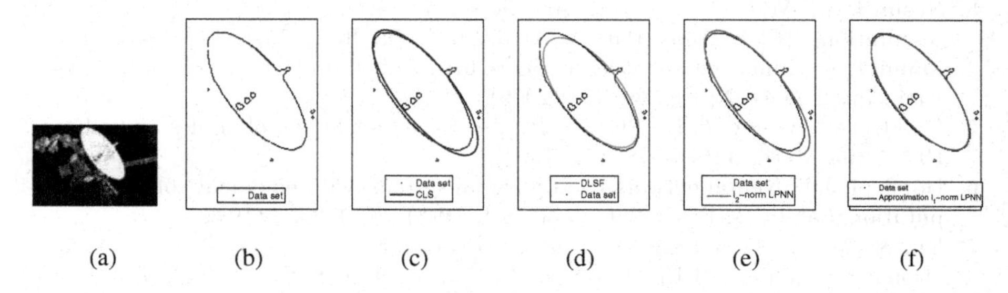

Fig. 7. Fitting results of a real image. (a) Original image. (b) Data points after edge extraction. (c)–(f) Fitting results of various algorithms.

Experiment 3: Real Image
In the third experiment, we test the performance of the proposed algorithm by a real image. Figure 7(a) is an image of a space probe. After edge detection, Fig. 7(b) is obtained. There are some outliers data points. The fitting results are shown in Fig. 7(c)–(f). From the figure, the fitting result of the proposed algorithm is much better than those of other algorithms.

5 Conclusion

This paper studies the ellipse fitting problem under complicated noise environment. The major purpose is to estimate an appropriate elliptic equation which can best match the coplanar scattering points and minimize the influence of noise. In order to achieve this goal, we utilize the robustness of l_1-norm. We use the LPNN framework to solve the problem. Since the LPNN framework can handle the equality constrained problems only, we modify the original constraints such that the problem is with the equality constraints only. In addition, the LPNN framework requires that its objective function and constraints are all twice differentiable. The paper introduces a differentiable approximation for the l_1-norm function. Moreover, to improve the system stability and accelerate its convergence, several augmented terms are used in the modification optimal function. The local stability of the proposed approach is also discussed. Simulation results show that the proposed algorithm is an effective approach to handle impulse noise and outliers.

Acknowledgment. The work was supported by a research grant from the Government of the Hong Kong Special Administrative Region (CityU 11259516).

References

1. Ballard, D.H.: Generalizing the hough transform to detect arbitrary shapes. Pattern Recogn. **13**(2), 111–122 (1981)
2. Barwick, D.: Very fast best-fit circular and elliptical boundaries by chord data. IEEE Trans. Pattern Anal. Mach. Intell. **31**(6), 1147–1152 (2009)
3. Rosin, P.L., West, G.A.: Nonparametric segmentation of curves into various representations. IEEE Trans. Pattern Anal. Mach. Intell. **17**(12), 1140–1153 (1995)
4. Maini, E.S.: Enhanced direct least square fitting of ellipses. Int. J. Pattern Recogn. Artif. Intell.gence **20**(6), 939–953 (2006)
5. Gander, W., Golub, G.H., Strebel, R.: Least-squares fitting of circles and ellipses. BIT Num. Math. **34**(4), 558–578 (1994)
6. Hopfield, J.J.: Neural networks and physical systems with emergent collective computational abilities. Proc. Nat. Acad. Sci. **79**(8), 2554–2558 (1982)
7. Hu, X., Wang, J.: A recurrent neural network for solving a class of general variational inequalities. IEEE Trans. Syst. Man Cybern. B Cybern. **37**(3), 528–539 (2007)
8. Zhang, S., Constantinides, A.G.: Lagrange programming neural networks. IEEE Trans. Circ. Syst. II, Analog Digit. Sig. Process. **39**(7), 441–452 (1992)
9. Liang, J., So, H.C., Leung, C.S., Li, J., Farina, A.: Waveform design with unit modulus and spectral shape constraints via Lagrange programming neural network. IEEE J. Sel. Top. Sig. Process. **9**(8), 1377–1386 (2015)
10. Liang, J., Leung, C.S., So, H.C.: Lagrange programming neural network approach for target localization in distributed MIMO radar. IEEE Trans. Sig. Process. **64**(6), 1574–1585 (2016)
11. Fitzgibbon, A., Pilu, M., Fisher, R.: Direct least square fitting of ellipses. IEEE Trans. Pattern Anal. Mach. Intell. **21**(5), 476–480 (1999)

Neurodynamics

Synchronization of Memristor-Based Time-Delayed Neural Networks via Pinning Control

Zhanyu Yang[1], Biao Luo[2(✉)], and Derong Liu[3]

[1] School of Automation and Electrical Engineering, University of Science and Technology Beijing, Beijing 100083, China

[2] The State Key Laboratory of Management and Control for Complex Systems, Institute of Automation, Chinese Academy of Sciences, Beijing 100190, China

biao.luo@hotmail.com

[3] School of Automation, Guangdong University of Technology, Guangzhou 510006, China

Abstract. As the realization of memristor by HP Lab, more and more researchers pay attention to the memristor-based neural networks (MNNs). In this paper, a pinning control method is applied to drive two MNNs to achieve synchronization. Conditions about the pinning controllers are given to guarantee the asymptotic synchronization of MNNs with time-varying delays. Furthermore, MNNs are nonlinear state-dependent systems with discontinuous right-hand sides such that all the dynamic analyses are under the framework of Filippov's solutions and the theory of differential inclusions. The effectiveness of the proposed pinning method is verified by a numerical example.

Keywords: Memristor-based neural networks · Asymptotic synchronization · Pinning control · Time-delay

1 Introduction

In 1971, Chua proposed a theory that there may exist the fourth fundamental circuit element besides resistor, inductor and capacitor, which was named memristor (short for memory and resistor) [1]. The memristor is an adjustable two-terminal resistive devices which can describe the relationship between electric charge and magnetic flux. Furthermore, its information cannot disappear when it loses the power. In 2008, Hewlett-Packard Lab realized the practical memristor [2]. After that, more and more attentions have been attracted because of its nonlinear and memory characteristics.

The conventional neural networks can be implemented by circuits, and the connection between neurons is usually resistor. By replacing the resistor with memristor in neural networks, memristor-based neural networks (MNNs) is proposed. Although conventional electrical circuit can emulate some operations of a

D. Liu et al. (Eds.): ICONIP 2017, Part III, LNCS 10636, pp. 699–708, 2017.
https://doi.org/10.1007/978-3-319-70090-8_70

neuron, the functionality of the synapse which plays an important role in biological neural networks is more difficult to mimic and memristor can easily realize synaptic behaviour [3]. Thus, the MNNs are more realistic models to emulate neural systems in human brains.

Synchronization of drive-response systems in neural networks draws many attentions from scholars because of the potential application in secure communication and information science [4–6]. Due to the memristors' advantages, many researchers focus on the synchronization of MNNs. In the circuit implementation of MNNs, the time delays often occur with transmitting the signals and may cause some undesired influences such as oscillation, instability and bifurcation [7]. Therefore, time delays are very important for analysing dynamic behaviors of MNNs.

MNNs are nonlinear systems with discontinuous right-hand side and their dynamic behaviors cannot be obtained by classical theory of differential equation. To solve the situation, the nonsmooth analysis and the framework of Filippov's solution are necessary. Up to now, there are already some considerable results for synchronization of MNNs under the framework of Filippov's solution [8–11]. In some situations, the networks cannot synchronize by themselves and the controllers are necessary for synchronization. However, most of the control methods need to add controllers to every node of the networks. In some large-scale networks or specific networks, it is impossible to control every node for the desired goal. To deal with this situation, pinning control which applies controllers to a fraction of nodes in the networks is introduced. Some results on synchronization of MNNs via pinning control have already been presented. In [12], Guo et al. presented some results on global synchronization of multiple memristive neural networks in the presence of external noise by pinning impulsive control law and a pinning adaptive control law. However, most papers focus on the multiple MNNs and coupled MNNs [12,13] and few researchers investigate the synchronization of MNNs via pinning control.

Motivated by above discussions, this paper focuses on the synchronization of MNNs with time-varying delays via pinning control. The rest of this paper is organized as follows. In Sect. 2, the basic mathematic model of MNNs and some problem descriptions are introduced. In Sect. 3, the designed pinning controllers are introduced and some sufficient criteria for asymptotical synchronization of MNNs are discussed. The numerical example is presented to ensure the effectiveness of the proposed pinning control in Sect. 4. The conclusions are given in Sect. 5.

2 Preliminaries

In this paper, a class of memristor-based neural networks are considered as following differential equation:

$$\dot{x}_i(t) = -d_i(x_i(t))x_i(t) + \sum_{j=1}^{n} a_{ij}(x_j(t))f_j(x_j(t)) + \sum_{j=1}^{n} b_{ij}(x_j(t-\tau_j(t))) \times g_j(x_j(t-\tau_j(t))) + I_i, \qquad t \geq 0, i = 1,2,...,n, \tag{1}$$

where $x_i(t)$ is the state of the ith neuron; I_i denotes the external input to the ith neuron; $\tau_j(t)$ corresponds to the transmission delay; $f_j(x_j(t))$ and $g_j(x_j(t-\tau_j(t)))$ denote the feedback functions without and with time-varying delays, respectively; $d_i(x_i(t))$, $a_{ij}(x_j(t))$ and $b_{ij}(x_j(t-\tau_j(t)))$ represent memristor-based weights. Based on the current-voltage characteristic of memristors, the memristive weights can be described as follows:

$$d_i(x_i(t)) = \begin{cases} d_i^{'}, & |x_i(t)| < \chi_i, \\ d_i^{''}, & |x_i(t)| \geq \chi_i, \end{cases} \qquad a_{ij}(x_j(t)) = \begin{cases} a_{ij}^{'}, & |x_j(t)| < \chi_j, \\ a_{ij}^{''}, & |x_j(t)| \geq \chi_j, \end{cases}$$

$$b_{ij}(x_j(t-\tau_j(t))) = \begin{cases} b_{ij}^{'}, & |x_j(t-\tau_j(t))| < \chi_j, \\ b_{ij}^{''}, & |x_j(t-\tau_j(t))| \geq \chi_j, \end{cases}$$

where the switching jump $\chi_j > 0$, and $d_i^{'}, d_i^{''}, a_{ij}^{'}, a_{ij}^{''}, b_{ij}^{'}, b_{ij}^{''}$ are all constant numbers.

Note that the memristor-based weights d_i, a_{ij}, b_{ij} are discontinuous and dependent on the states of the system. The MNN (1) is a nonlinear system with discontinuous right-hand side and the conventional solutions and dynamic analyses are invalid. In this situation, the solutions in Filippove's sense [14] are introduced for investigating dynamic behaviors of MNNs. By applying the theories about differential inclusions with Filippove-framework and nonsmooth analysis, the drive system can be written as follows:

$$\dot{x}_i(t) \in -co[d_i(x_i(t))]x_i(t) + \sum_{j=1}^{n} co[a_{ij}(x_j(t))]f_j(x_j(t)) + \sum_{j=1}^{n} co[b_{ij}(x_j(t-\tau_j(t)))] \times g_j(x_j(t-\tau_j(t))) + I_i, \qquad for \quad t \geq 0, \quad i = 1,2,...,n, \tag{2}$$

where

$$co[d_i(x_i(t))] = \begin{cases} d_i^{'}, & |x_i(t)| < \chi_i, \\ co\{d_i^{'}, d_i^{''}\}, & |x_i(t)| = \chi_i, \\ d_i^{''}, & |x_i(t)| > \chi_i, \end{cases}$$

$$co[a_{ij}(x_j(t))] = \begin{cases} a_{ij}^{'}, & |x_j(t)| < \chi_j, \\ co\{a_{ij}^{'}, a_{ij}^{''}\}, & |x_j(t)| = \chi_j, \\ a_{ij}^{''}, & |x_j(t)| > \chi_j, \end{cases}$$

$$co[b_{ij}(x_j(t-\tau_j(t)))] = \begin{cases} b_{ij}^{'}, & |x_j(t-\tau_j(t))| < \chi_j, \\ co\{b_{ij}^{'}, b_{ij}^{''}\}, & |x_j(t-\tau_j(t))| = \chi_j, \\ b_{ij}^{''}, & |x_j(t-\tau_j(t))| > \chi_j, \end{cases}$$

where $co\{d_i^{'}, d_i^{''}\}$ denotes closure of the convex hull generated by constant $d_i^{'}$ and $d_i^{''}$. If $d_i \in co\{d_i^{'}, d_i^{''}\}$, we have $\min\{d_i^{'}, d_i^{''}\} \leq d_i \leq \max\{d_i^{'}, d_i^{''}\}$.

Based on the concept of drive-response synchronization, the MNN (2) is regarded as drive system and the corresponding response system can be considered as following differential inclusions:

$$\begin{aligned}\dot{y}_i(t) \in &- co[d_i(y_i(t))]y_i(t) + \sum_{j=1}^{n} co[a_{ij}(y_j(t))]f_j(y_j(t)) + \sum_{j=1}^{n} co[b_{ij}(y_j(t-\tau_j(t)))] \\ &\times g_j(y_j(t-\tau_j(t))) + I_i + u_i(t), \qquad t \geq 0, \quad i = 1, 2, ..., n, \end{aligned} \tag{3}$$

which have the same parameters as drive system (2), and $u_i(t)$ is appropriate control input for the desired goal. The synchronization between drive system and response system is defined as $e_i(t) = y_i(t) - x_i(t), i = 1, 2, ..., n$, and the synchronization error system of MNNs can be considered as follows:

$$\begin{aligned}\dot{e}_i(t) \in &- \{co[d_i(y_i(t))]y_i(t) - co[d_i(x_i(t))]x_i(t)\} + \sum_{j=1}^{n} \{co[a_{ij}(y_j(t))]f_j(y_j(t)) \\ &- co[a_{ij}(x_j(t))]f_j(x_j(t))\} + \sum_{j=1}^{n} \{co[b_{ij}(y_j(t-\tau_j(t)))]g_j(y_j(t-\tau_j(t))) \\ &- co[b_{ij}(x_j(t-\tau_j(t))]g_j(x_j(t-\tau_j(t)))\} + u_i(t). \end{aligned} \tag{4}$$

There are some assumptions and lemmas in the following for synchronizing drive system and response system.

Assumption 1. *There exist constants $F_j > 0, G_j > 0$ to ensure that the neuron activation functions f_j and g_j, $j = 1, 2, ..., n$ satisfy the following conditions:*

$$\begin{aligned}|f_j(s_1) - f_j(s_2)| &\leq F_j|s_1 - s_2|, \\ |g_j(s_1) - g_j(s_2)| &\leq G_j|s_1 - s_2|. \end{aligned} \tag{5}$$

Assumption 2. *The transmission delay $\tau_j(t)$ is a differential function and there exists a constant $\mu > 0$, such that*

$$\dot{\tau}_j(t) \leq \mu < 1, \quad 0 < \tau_j(t) \leq \tau.$$

Lemma 1. *Assume that $V(x)$ is a positive definite continuous differential and radially unbounded function. If $V(x)$ satisfies the following conditions, the systems are globally asymptotically stable [15].*

1. *$V(x) \geq 0$, $W(x) \geq 0$, and $W(x) = 0$, if and only if $x = 0$.*
2. *$W(x)$ is strictly increasing on R_+.*
3. *The derivative of $V(x)$ with respect to time satisfies $\dot{V}(x) \leq -W(x)$.*

Lemma 2. *Under Assumption 1 and $f_j(\pm\chi_j) = g_j(\pm\chi_j) = 0$, we can have*

$$
\begin{aligned}
&|co[d_i(y_i(t))]y_i(t) - co[d_i(x_i(t))]x_i(t)| \le D_i|y_i(t) - x_i(t)| + \chi_i|d_i^{'} - d_i^{''}|, \\
&|co[a_{ij}(y_j(t))]f_j(y_j(t)) - co[a_{ij}(x_j(t))]f_j(x_j(t)| \le A_{ij}F_j|y_j(t) - x_j(t)|, \\
&|co[b_{ij}(y_j(t-\tau_j(t)))]g_j(y_j(t-\tau_j(t))) - co[b_{ij}(x_j(t-\tau_j(t))]g_j(x_j(t-\tau_j(t)))| \\
&\qquad \le B_{ij}G_j|y_j(t-\tau_j(t)) - x_j(t-\tau_j(t))|,
\end{aligned}
$$

where $A_{ij} = \max\{|a_{ij}^{'}|, |a_{ij}^{''}|\}$, $B_{ij} = \max\{|b_{ij}^{'}|, |b_{ij}^{''}|\}$ *and* $D_i = \max\{|d_i^{'}|, |d_i^{''}|\}$.

Proof. Here, we prove the inequality in this lemma by the following cases.

1. If $|y_i(t)| < \chi_i$ and $|x_i(t)| < \chi_i$, by defining $\Theta_i(t) = co[d_i(y_i(t))]y_i(t) - co[d_i(x_i(t))]x_i(t)$, we have
$$|\Theta_i(t)| = |d_i^{'}y_i(t) - d_i^{'}x_i(t)| \le D_i|y_i(t) - x_i(t)|.$$
2. If $|y_i(t)| > \chi_i$ and $|x_i(t)| > \chi_i$, we have
$$|\Theta_i(t)| = |d_i^{''}y_i(t) - d_i^{''}x_i(t)| \le D_i|y_i(t) - x_i(t)|.$$
3. If $|y_i(t)| < \chi_i$ and $x_i(t) > \chi_i$, we have
$$
\begin{aligned}
|\Theta_i(t)| \le& |d_i^{'}||\chi_i - y_i(t)| + |d_i^{''}||x_i(t) - \chi_i| + |\chi_i||d_i^{'} - d_i^{''}| \\
\le& D_i|y_i(t) - x_i(t)| + \chi_i|d_i^{'} - d_i^{''}|.
\end{aligned}
$$
4. If $|y_i(t)| < \chi_i$ and $x_i(t) < -\chi_i$, we have
$$
\begin{aligned}
|\Theta_i(t)| \le& |d_i^{'}||y_i(t) + \chi_i| + |d_i^{''}||x_i(t) + \chi_i| + \chi_i|d_i^{'} - d_i^{''}| \\
\le& D_i|y_i(t) - x_i(t)| + \chi_i|d_i^{'} - d_i^{''}|.
\end{aligned}
$$
5. Similarly, if $|x_i(t)| \ge \chi_i$ and $y_i(t) < -\chi_i$ or $|x_i(t)| \ge \chi_i$ and $y_i(t) \ge \chi_i$, the same conclusion can be obtained.

Above all, we can prove that the first inequality can be true at all time. The proof of the other two inequalities in the lemma can be referred into [10]. The proof of Lemma 2 is completed. □

Remark 1. In [10], a similar lemma was proposed and the following inequality was given.

$$|co[d_i(y_i(t))]y_i(t) - co[d_i(x_i(t))]x_i(t)| \le D_i|y_i(t) - x_i(t)|.$$

Actually, the inequality can not be true at all time. For example, the following condition is satisfied.

$$
co[d_i(x_i(t))] = \begin{cases} 1, & |x_i| < 1, \\ co\{1, 3\}, & |x_i| = 1, \\ 3, & |x_i| > 1. \end{cases}
$$

If $y_i(t) = 3$ and $x_i(t) = 0.5$ at time t, the aforementioned inequality can not be true. In this paper, we adopt a new condition which is always true to deal with the synchronization error systems and the proof is also given in Lemma 1.

3 Main Results

Pinning control aims to achieve the desired goal for entire network by controlling a fraction of nodes in networks. In this section, a new pinning control method is designed to synchronize drive system and response system and some sufficient conditions for synchronization of MNNs are also given.

Throughout this paper, the pinning controller in response system (3) can be considered as follows:

$$u_i(t) = \begin{cases} -sgn(e_i(t))\beta(t) \sum_{j=1}^{n} (\alpha |e_j(t)| + \gamma), & 1 \leq i \leq m, \\ 0, & m < i \leq n, \end{cases} \tag{6}$$

where

$$sgn(e_i(t)) = \begin{cases} 1, & e_i(t) \geq 0, \\ -1, & e_i(t) < 0, \end{cases} \qquad \beta(t) = \frac{\sum_{j=1}^{n} |e_j(t)| + \zeta}{\sum_{j=1}^{m} |e_j(t)| + \zeta},$$

and α, γ are both positive constants. m is the number of controlled nodes for MNNs. ζ is a small enough positive constant.

To obtain desired performance of synchronization, the number of controlled nodes can be appropriate, satisfying $m \geq 2$. Furthermore, the initial condition of errors in controlled nodes cannot be zero.

Theorem 1. *If Assumptions 1 and 2 are satisfied and $f_j(\pm\chi_j) = g_j(\pm\chi_j) = 0$, systems* (2) *and* (3) *can be synchronized asymptotically with the following parameters of the controllers* (6)*:*

$$\alpha = A + \delta, \quad \gamma \geq \chi_i |d_i^{'} - d_i^{''}|, \tag{7}$$

where

$$\delta \geq D_i + \frac{1}{2}\sum_{j=1}^{n} B_{ij}G_j + \frac{1}{2(1-\mu)}\sum_{j=1}^{n} B_{ji}G_i + 1,$$
$$A = max\{A_{ij}F_j\}, \quad i,j = 1,2,...,n.$$

Proof. Consider the following Lyapunov functional candidate for error systems (4):

$$V(t) = \frac{1}{2}\sum_{i=1}^{n} e_i^2(t) + \frac{1}{2(1-\mu)}\sum_{i=1}^{n}\sum_{j=1}^{n}\int_{t-\tau_j(t)}^{t} B_{ij}G_j e_j^2(s)ds. \tag{8}$$

The derivative of $V(t)$ along the trajectories of error system can be calculated as follows:

$$\dot{V}(t) = \sum_{i=1}^{n} e_i(t)\dot{e}_i(t) + \frac{1}{2(1-\mu)}\sum_{i=1}^{n}\sum_{j=1}^{n} B_{ij}G_j \left(e_j^2(t) - (1-\dot{\tau}_j(t))e_j^2(t-\tau_j(t))\right). \tag{9}$$

Under Assumption 2, the following inequalities are always true.

$$-\frac{1-\dot{\tau}_j(t)}{1-\mu} \leq -1, \qquad (\sum_{j=1}^{n}|e_j(t)|)^2 = \sum_{i=1}^{n}\sum_{j=1}^{n}|e_i(t)||e_j(t)| \geq \sum_{i=1}^{n}|e_i^2(t)|. \quad (10)$$

By using Lemma 2, the following results are obtained.

$$\sum_{i=1}^{n} e_i(t)\dot{e}_i(t) \leq \sum_{i=1}^{n} D_i e_i^2(t) + \sum_{i=1}^{n}\sum_{j=1}^{n} A_{ij}F_j|e_i(t)||e_j(t)| + \sum_{i=1}^{n}\sum_{j=1}^{n} B_{ij}G_j \times |e_j(t-\tau_j(t))||e_i(t)| + \sum_{i=1}^{n}\chi_i\hat{D}_i|e_i(t)| + \sum_{i=1}^{m} u_i(t)e_i(t), \quad (11)$$

where $\hat{D}_i = |d_i^{'} - d_i^{''}|$. ζ is a small enough positive constant such that $\sum_{j=1}^{m}|e_j(t)| + \zeta$ can be considered as $\sum_{j=1}^{m}|e_j(t)|$, approximately. Thus, we can obtain the following inequality.

$$\sum_{i=1}^{m} u_i(t)e_i(t) \approx -\sum_{j=1}^{n}|e_j(t)|(\alpha\sum_{j=1}^{n}|e_j(t)| + \gamma). \quad (12)$$

Based on the above inequalities, the derivative of $V(t)$ satisfies the following inequality:

$$\begin{aligned}\dot{V}(t) \leq & \sum_{i=1}^{n} D_i e_i^2(t) + \frac{1}{2}\sum_{i=1}^{n}\sum_{j=1}^{n} B_{ij}G_j e_i^2(t) + \sum_{i=1}^{n}\chi_i\hat{D}_i|e_i(t)| - \gamma\sum_{j=1}^{n}|e_j(t)| \\ & + \frac{1}{2(1-\mu)}\sum_{i=1}^{n}\sum_{j=1}^{n} B_{ij}G_j e_j^2(t) + \sum_{i=1}^{n}\sum_{j=1}^{n} A_{ij}F_j|e_i(t)||e_j(t)| \\ & - \alpha\left(\sum_{j=1}^{n}|e_j(t)|\right)^2 \leq \sum_{i=1}^{n} e_i^2(t)\left\{D_i + \frac{1}{2}\sum_{j=1}^{n} B_{ij}G_j + \frac{1}{2(1-\mu)}\sum_{j=1}^{n} B_{ji}G_i - \delta\right\} \\ & + \sum_{i=1}^{n}|e_i(t)|(\chi_i\hat{D}_i - \gamma) + \sum_{i=1}^{n}\sum_{j=1}^{n}(A_{ij}F_j - A)|e_i(t)||e_j(t)|. \end{aligned} \quad (13)$$

When the condition (7) is satisfied, we have $\dot{V}(t) \leq -\sum_{i=1}^{n} e_i^2(t) \leq 0$. From Lemma 1, the error systems are globally asymptotically stable. Thus, the synchronization errors $e(t)$ converge to zero asymptotically. It means that drive system and response system are synchronized asymptotically. □

4 Numerical Example

In this section, the numerical example is given to verify the effectiveness of the proposed pinning control. Consider a four-dimensional memristor-based neural network with time-varying delays as drive system:

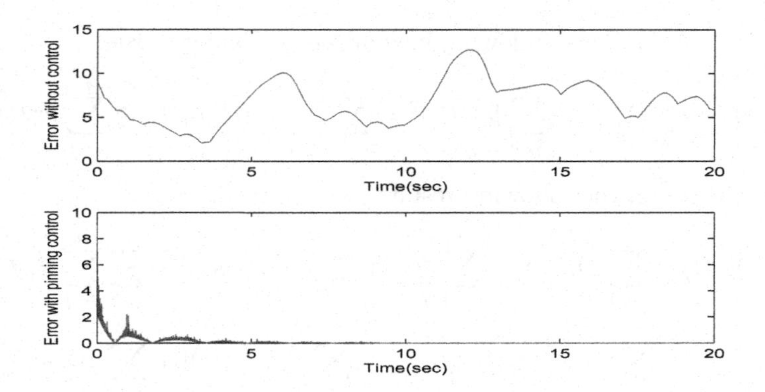

Fig. 1. The total error without and with pinning control

$$\dot{x}_i(t) = -d_i(x_i(t))x_i(t) + \sum_{j=1}^{4} a_{ij}(x_j(t))f_j(x_j(t)) + \sum_{j=1}^{4} b_{ij}(x_j(t-\tau_j(t))) \times g_j(x_j(t-\tau_j(t))) + I_i, \qquad t \geq 0, \quad i = 1,2,3,4, \tag{14}$$

where the parameters are shown as follows:

$$A^{'} = \begin{pmatrix} -0.8 & 1.2 & 0.8 & 1.6 \\ 0.6 & -1.2 & 1.8 & 1.4 \\ 0.4 & 0.9 & -1.0 & 0.8 \\ 1.0 & 1.4 & 0.9 & -1.8 \end{pmatrix}, \quad A^{''} = \begin{pmatrix} -1.0 & 1.4 & 1.2 & 1.2 \\ 0.8 & -1.6 & 2.0 & 1.6 \\ 0.8 & 1.2 & -1.8 & 1.2 \\ 0.8 & 1.0 & 1.2 & -2.0 \end{pmatrix},$$

$$B^{'} = \begin{pmatrix} -2.0 & 1.2 & 1.0 & 1.4 \\ 0.8 & -1.4 & 0.6 & 0.5 \\ 1.6 & 0.8 & -0.8 & 1.3 \\ 1.2 & 1.4 & 0.8 & -2.6 \end{pmatrix}, \quad B^{''} = \begin{pmatrix} -2.5 & 0.8 & 1.6 & 1.1 \\ 1.0 & -1.8 & 0.8 & 0.7 \\ 1.2 & 0.6 & -1.2 & 1.1 \\ 0.8 & 1.0 & 1.0 & -2.0 \end{pmatrix},$$

$$D^{'} = [d_1^{'}, d_2^{'}, d_3^{'}, d_4^{'}]^{\mathsf{T}} = [1.4, 0.9, 1.1, 1.0]^{\mathsf{T}},$$
$$D^{''} = [d_1^{''}, d_2^{''}, d_3^{''}, d_4^{''}]^{\mathsf{T}} = [1.0, 1.2, 1.3, 0.8]^{\mathsf{T}}.$$

The time-varying delay is $\tau_j(t) = e^t/(1+e^t)$, which satisfies Assumption 2 with $\mu = 0.25, \tau = 0.5$. The feedback activation functions are considered as $f_j(x) = g_j(x) = tanh(|x|-1)$. The switching rules are $\chi_j = 1, j = 1, 2, ..., n$. The external inputs are considered as $I = [I_1, I_2, I_3, I_4] = [0, 0, 0, 0]^{\mathsf{T}}$. The initial condition for drive system is $x(t) = [1.9, 2.4, -2.2, 2.5]^{\mathsf{T}}, \quad t \in [-\tau, 0)$. The response system is considered as follows:

$$\dot{y}_i(t) = -d_i(y_i(t))y_i(t) + \sum_{j=1}^{4} a_{ij}(y_j(t))f_j(y_j(t)) + \sum_{j=1}^{4} b_{ij}(y_j(t-\tau_j(t))) \times g_j(y_j(t-\tau_j(t))) + I_i + u_i(t), \qquad t \geq 0, \quad i = 1,2,3,4, \tag{15}$$

which shares the same parameters as drive system (14), i.e., they have the same matrixes $A^{'}, A^{''}, B^{'}, B^{''}, D^{'}, D^{''}$ and the same switching rules $\chi_j = 1$.

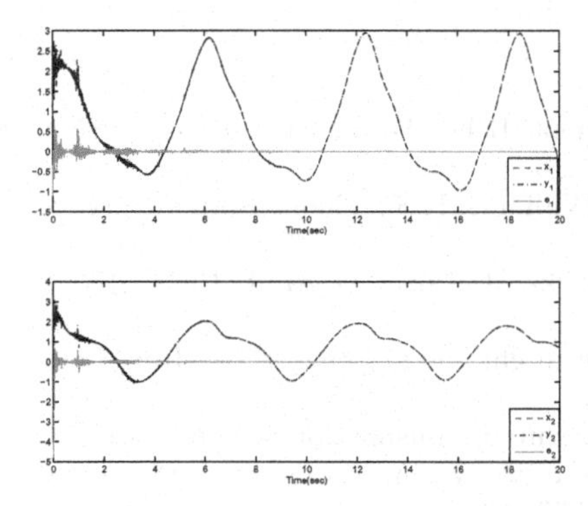

Fig. 2. The trajectories of states x_i and y_i for drive system and response system and their synchronization errors, $i = 1, 2$

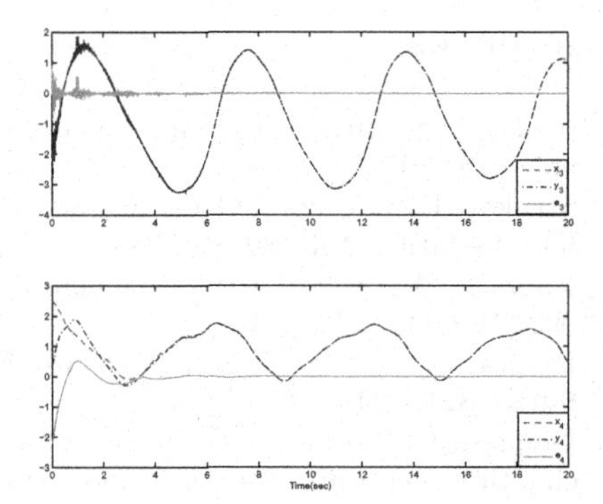

Fig. 3. The trajectories of states x_i and y_i for drive system and response system and their synchronization errors, $i = 3, 4$

The number of controlled nodes is $m = 3$, which means the controllers are added to i^{th} node $(i = 1, 2, 3)$. Select $\alpha = 12$, $\gamma = 0.4$ and $\zeta = 10^{-6}$ as parameters of pinning controller (6) for response system, satisfying Theorem 1. The external inputs are the same as drive system. The initial condition for response system is $y(t) = [0.8, -2.4, -1.3, 0.3]^{\mathsf{T}}, \quad t \in [-\tau, 0)$.

Figure 2 shows the trajectories of states in i^{th} nodes and their synchronization errors, $i = 1, 2$. The states of 3rd and 4th nodes for drive system and response system and their synchronization errors are shown in Fig. 3. The trajectories of total error $\sum_{i=1}^{n} |e_i(t)|$ without and with pinning control are given in Fig. 1. Based on the aforementioned figures, the effectiveness of the designed pinning control is proved.

5 Conclusion

In this paper, we focus on applying pinning control methods to memristor-based neural networks with time-varying delays. The dynamic behaviors and nonsmooth analysis are based on the theory of differential inclusions and the framework of Filippov's solutions. A novel pinning control policy is proposed to synchronize two MNNs with time-varying delays. Furthermore, some sufficient conditions of pinning controller (6) are introduced to achieve the asymptotical synchronization of MNNs with time-varying delays. The effectiveness of the pinning control is proved by numerical example in this paper.

Acknowledgments. This work was supported in part by the National Natural Science Foundation of China under Grants 61233001, 61273140, 61304086, 61374105, 61503377, 61533017, 61473011 and U1501251.

References

1. Chua, L.: Memristor-the missing circuit element. IEEE Trans. Circ. Theory **18**(5), 507–519 (1971)
2. Strukov, D.B., Snider, G.S., Stewart, D.R., Williams, R.S.: The missing memristor found. Nature **453**, 80–83 (2008)
3. Thomas, A.: Memristor-based neural networks. J. Phys. D Appl. Phys. **46**(9), 093001–093012 (2013)
4. Pecora, L.M., Carroll, T.L.: Synchronization in chaotic systems. Phys. Rev. Lett. **64**(8), 821–824 (1990)
5. Dimassi, H., Loria, A., Belghith, S.: A new secured transmission scheme based on chaotic synchronization via smooth adaptive unknown-input observers. Commun. Nonlinear Sci. Numer. Simul. **17**(9), 3727–3739 (2012)
6. Zhang, H., Xie, Y., Wang, Z., Zheng, C.: Adaptive synchronization between two different chaotic neural networks with time delay. IEEE Trans. Neural Netw. **18**(6), 1841–1845 (2007)
7. Ding, S., Wang, Z., Wang, J., Zhang, H.: H-infinity state estimation for memristive neural networks with time-varying delays: the discrete-time case. Neural Netw. **84**, 47–56 (2016)
8. Han, X., Wu, H., Fang, B.: Adaptive exponential synchronization of memristive neural networks with mixed time-varying delays. Neurocomputing **201**, 40–50 (2016)
9. Wen, S., Bao, G., Zeng, Z., Chen, Y., Huang, T.: Global exponential synchronization of memristor-based recurrent neural networks with time-varying delays. Neural Netw. **48**, 195–203 (2013)
10. Wang, L., Shen, Y., Yin, Q., Zhang, G.: Adaptive synchronization of memristor-based neural networks with time-varying delays. IEEE Trans. Neural Netw. Learn. Syst. **26**(9), 2033–2042 (2015)
11. Velmurugan, G., Rakkiyappan, R., Cao, J.: Finite-time synchronization of fractional-order memristor-based neural networks with time delays. Neural Netw. **73**, 36–46 (2016)
12. Guo, Z., Yang, S., Wang, J.: Global synchronization of memristive neural networks subject to random disturbances via distributed pinning control. Neural Netw. **84**, 67–79 (2016)
13. Yang, S., Guo, Z., Wang, J.: Robust synchronization of multiple memristive neural networks with uncertain parameters via nonlinear coupling. IEEE Trans. Syst. Man Cybern. **45**(7), 1077–1086 (2015)
14. Filippov, A.: Differential Equations with Discontinuous Right Hand Sides. Kluwer, Dordrecht (1988)
15. Clarke, F., Ledyaev, Y., Stern, R., Wolenski, P.: Nonsmooth Analysis and Control Theory. Springer-Verlag, New York (1998)

Identifying Intrinsic Phase Lag in EEG Signals from the Perspective of Wilcoxon Signed-Rank Test

Yunqiao Wu[1], John Q. Gan[2], and Haixian Wang[1](✉)

[1] Key Laboratory of Child Development and Learning Science of Ministry of Education, School of Biological Sciences and Medical Engineering, Southeast University, Nanjing 210096, Jiangsu, People's Republic of China
hxwang@seu.edu.cn

[2] School of Computer Science and Electronic Engineering, University of Essex, Colchester, UK

Abstract. In brain functional network connectivity analysis, phase synchronization has been effective in detecting regions demonstrating similar dynamics over time. The previously proposed connectivity indices such as phase locking value (PLV), phase lag index (PLI) and weighted phase lag index (WPLI) are widely used. They are, however, influenced by volume conduction or noise. In addition, appropriate thresholds have to be chosen in order to employ them successfully, which leads to uncertainty. In this paper, a novel connectivity index named phase lag based on the Wilcoxon signed-rank test (PLWT) is proposed under the framework of Wilcoxon signed-rank test, which avoids using thresholds to identify effective connections. We analyzed and compared PLWT with previous indices by simulating volume conduction and testing the scale-free character of brain networks constructed based on EEG signals. The experimental results demonstrated that PLWT can be utilized as a reliable and convincing measure to reveal true connections while effectively diminishing the influence of volume conduction.

Keywords: Functional connectivity · Phase synchronization · Volume conduction · Wilcoxon signed-rank test · Scale-free character

1 Introduction

The human brain is a complex network consisting of interacting subsystems. There has been a growing interest in studying functional connectivity, which is defined as the relationship between brain regions that share identical or similar functional properties. More specifically, it can be defined as the temporal correlation between spatially remote neurophysiological events [1]. Currently, there are several different imaging techniques which can be used to study brain connectivity such as EEG, MEG and fMRI. While fMRI obtains high spatial resolution, EEG/MEG has high temporal resolution. Based on the temporal information, phase synchronization is a manifestation of interaction between neuronal groups measurable from EEG or MEG signals [2]. To quantify the phase synchronization, some classical measures were developed.

D. Liu et al. (Eds.): ICONIP 2017, Part III, LNCS 10636, pp. 709–717, 2017.
https://doi.org/10.1007/978-3-319-70090-8_71

Specifically, phase locking value (PLV) is defined as the length of the averaged differences of instantaneous phases projected onto the unit circle [3]. Similar but modified indices were also proposed, for instance, phase lag index (PLI) [4], and weighted phase lag index (WPLI) [2].

A major problem that has always been the obstacle influencing the assessment of phase synchronization, is the so-called volume conduction (i.e. conductive characteristic of brain, soft tissue and skull). The volume conduction usually leads to spatial similarity between EEG signals [5, 6]. As proved, PLV suffers from the volume conduction between spatially adjacent electrodes while the other two indices can appropriately circumvent this problem [4]. However, they cannot avoid the influence of noise. More importantly, the thresholding scheme applied to these indices is recognized as an uncertain factor which would make great difference in the analysis of brain connectivity.

In this paper, we introduce a new phase synchronization index called the phase lag based on the Wilcoxon signed-rank test (PLWT). The utilization of the framework of Wilcoxon signed-rank test avoids the step of choosing threshold while verifying the existence of intrinsic neural interaction. Meanwhile, this index can effectively avoid the influence of the volume conduction. The experimental comparison of the PLWT approach with existing indexes confirms its effectiveness.

The rest of the paper is organized as follows. Section 2 briefly describes several classical phase synchronization indices and then proposes the procedure of the novel measure. Section 3 reports the experimental comparison between PLWT and other indices. Finally, the effectiveness and reliability of PLWT is discussed in Sect. 4.

2 Methods

2.1 Brief Review of Existing Neural Phase Synchronization Indices

Lachaux et al. proposed that phase-locking reflects underlying neural interaction, and presented the PLV index that directly measures frequency-specific synchronization by using only phase quantities for investigating long-range neural integration [3], given by

$$\mathrm{PLV_{xy}(f)} = \left|\left\langle \mathrm{e}^{\mathrm{i}\Delta\theta} \right\rangle\right| \tag{1}$$

where x and y are two signals recorded from two sensors, f represents a particular frequency, and $\Delta\theta$ is the difference of the instantaneous phases. The PLV equals to one if phase differences are constant while tends to zero if the phase differences are randomly distributed on the unity circle.

A useful index of measuring synchronization would be exclusively based on phase quantities while circumventing the effect of common sources (volume conduction). Accordingly, the phase lag index (PLI) is developed [4]. Its basic idea is to investigate the asymmetry of the distribution of phase differences around zero, given by

$$\mathrm{PLI_{xy}(f)} = |\langle \mathrm{sgn}(\sin(\Delta\theta)) \rangle| \tag{2}$$

where sgn(*) is the sign function. The value of PLI falls between zero, which corresponds to a centered phase differences around zero (mod π) (i.e., a symmetric distribution of phase differences), and one, which corresponds to completely consistent lag of phases between the two signals. The more consistent the lag, the more asymmetric the distribution of the phase differences is. The resulting non-zero value of PLI, associated with the effect of phase lag, cannot be caused by the volume conduction from a single strong source. Nevertheless, it would happen that a small perturbation of phases would change the signs of the phase differences with small magnitudes, which makes PLI sensitive to noise and volume conduction. The PLI is thus extended by weighting the phase differences according to their magnitudes, resulting in weighted phase lag index (WPLI) [2], given by

$$\mathrm{wPLI}_{xy}(f) = \frac{|\langle |L_1L_2 \sin(\Delta\theta)| \mathrm{sgn}(\sin(\Delta\theta))\rangle|}{\langle |L_1L_2 \sin(\Delta\theta)|\rangle} = \frac{|\langle L_1L_2 \sin(\Delta\theta)\rangle|}{|\langle L_1L_2 \sin(\Delta\theta)\rangle|} \tag{3}$$

where L_1 and L_2 are the amplitudes of the two signals after the Hilbert transform. WPLI aims to weaken the contribution of small phase differences.

2.2 A Neural Phase Synchronization Measure Based on Wilcoxon Signed-Rank Test

We propose a novel measure by utilizing the framework of Wilcoxon signed-rank test to verify the existence of neural interaction. The procedure is as follows:

(a) Sort the non-zero values of $|\sin(\Delta\theta_k)|$ (k = 1, …, N) in ascending order while leaving out zero values (θ_k is the phase difference at the time point t_k and N is the number of samples). We say that the rank of $|\sin(\Delta\theta_k)|$, denoted by $R(|\sin(\Delta\theta_k)|)$, is i if $|\sin(\Delta\theta_k)|$ stands at the ith position of the ordered series. If there exists a tie between two values, then their ranks are averaged.
(b) Compute two statistics

$$T^+ = \sum\nolimits_{\mathrm{sgn}(\sin(\Delta\theta_k)) > 0} R(|\sin(\Delta\theta_k)|), \tag{4}$$

$$T^- = \sum\nolimits_{\mathrm{sgn}(\sin(\Delta\theta_k)) < 0} R(|\sin(\Delta\theta_k)|), \tag{5}$$

(c) Let $T = \min\{T^+, T^-\}$.
(d) Given a significance level α, if T is smaller than the critical value found in the look-up table for the Wilcoxon signed-rank test, then we conclude that the distribution of phase differences is asymmetric significantly. Let n ($n \leq N$) be the number of non-zero values of $\sin(\Delta\theta_k)$. Alternatively, if $n > 25$, we can use the approximate critical value of the Gaussian distribution with the normalized statistic given by

$$Z = \frac{T - n(n+1)/4}{\sqrt{n(n+1)(2n+1)/24}} \tag{6}$$

(e) To quantify the phase synchronization, we propose the following index

$$W = \frac{|T^+ - T^-|}{|T^+ + T^-|} \tag{7}$$

We term the proposed method as phase lag based on the Wilcoxon signed-rank test (PLWT). PLWT is based solely on the sine function of phase differences. Its underlying thought is to identify the asymmetry of the phase difference distribution which indicates the occurrence of phase lag between the two signals. As expounded in PLI, the presence of phase lag, which cannot be resulted from volume conduction, suggests neural interaction. However, different from PLI, the phase differences are not equally treated in PLWT. Rather, as advocated by WPLI, the small phase differences, which are vulnerable to the impact of noise and volume conduction, should be assigned with small weights. We adopt the ranks of the phase differences instead of their magnitudes used in WPLI as weights. With the moderate weights of ranks, the proposed method is established with a solid framework of statistical hypothesis test. We therefore could reach a conclusion of the existence of intrinsic neural interaction in terms of statistical significance.

3 Experiments and Results

First, we performed a simulation experiment of volume conduction with different levels. Second, we compared the scale-free characteristic of brain functional networks under different thresholding schemes by considering the fitting of truncated power law which have been found as the best form of degree distribution [7].

3.1 Kuramoto Model

We employed the Kuramoto model [8, 9] to construct several simulated datasets according to the following equation

$$\dot{\theta}_i = \omega_i + \frac{K}{M}\sin\left(\theta_j - \theta_i\right) \tag{8}$$

where θ_i denotes the phase of the ith oscillator and ω_i is its natural frequency, and $K \geq 0$ is the coupling strength while M is the number of the oscillators. The frequencies ω_i are distributed according to a certain density $g(\omega)$, for example, Lorentzian density given by

$$g(\omega) = \frac{\gamma}{\pi\left[\gamma^2 + (\omega - \omega_0)^2\right]} \tag{9}$$

There is a phase transition from desynchronization to partially synchronization at a critical point. The critical coupling strength K_c can be denoted as

$$K_c = 2\gamma \tag{10}$$

We set the mean frequency of the oscillators as 10 Hz and the width of the Lorentz distribution γ as 1, so the theoretical critical value for K was 2. Here we simulated 10 trials of time series of 64 oscillators. The state of oscillator i can be described as A $\sin\theta_t$ in which A is the assumed constant amplitude for all the oscillators. Regarding the 64 oscillators as 64 EEG channels, we created three levels of overlap on the basis of EEG time series to simulate the existence of volume conduction [4]. The voltage of the ith channel at time t can be reached through the superstition of the state of the jth oscillator at time t as

$$V_i(t) = \frac{1}{2i_0 + 1} \sum_{j=i-i_0}^{j=i+i_0} O_j(t) \tag{11}$$

where i0 determines the degree of overlap. Here we set i_0 as 0, 4 or 8, and thus the corresponding numbers of common sources was 2 i_0 as 0, 8, or 16, respectively. The coupling strength K was set to be ranged from 0 to 8 with steps of 0.5. The resulting datasets were used for synchronization analysis including the computation of PLV, PLI, WPLI and PLWT. The experimental results were summarized in Fig. 1.

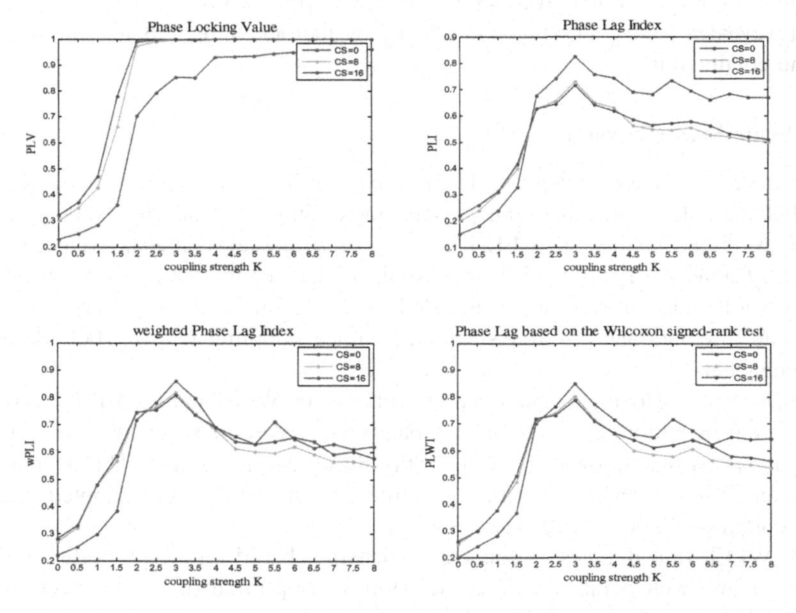

Fig. 1. Mean values of PLV, PLI, WPLI and PLWT as a function of coupling strength K (from 0 to 8) under three degrees of common sources (CS = 0, 8, 16). The results were averaged over all possible pairs of 64 EEG channels within one single trial, and then averaged across 10 trials.

It was observed that the mean PLV showed an upward trend with the increase of the coupling strength K under three levels of common sources, and an upward displacement of the curve with the increase of the level of overlap. After the critical point $K_c = 2$ (as demonstrated by the Kuramoto theory), the global PLV approximately reached the top (around 0.9 under CS = 0 and close to 1 under CS = 8 as well as CS = 16) and maintained stable, which meant that the oscillators were fully synchronized after the bifurcation. Thus, while PLV revealed the synchronization changes with changes in the connection strength K, it was also sensitive to volume conduction. Unlike PLV, the curves of PLI either started or ended at lower points. When $K < K_c$ ($K_c = 2$), as the number of common sources increased, the curve slightly moved upward across coupling strength. However, when $K > K_c$, it showed the opposite movement. Specifically the mean values of PLI under CS = 0 were higher than values under both CS = 8 and CS = 16, and there was no clear difference between common sources of 8 or 16. Overall, the range of the PLI values remained relatively conservative compared to PLV, which meant PLI underestimated the true levels of coupling. The general trend of the global WPLI was similar to that of the global PLI from the beginning to end, including the same upward displacement before the demarcation point ($K_c = 2$). Nevertheless, the most conspicuous distinction was that the three curves were interlaced and showed no significant difference after this point. Thus, while WPLI showed the expected trend with the increase of K, it was less sensitive to volume conduction than PLI or PLV. Finally, the line chart of PLWT was analogous to that of WPLI, either in the range of the magnitude or in the variation tendency. When the coupling strength K varied from 0 to 8, the PLWT values showed no significant difference under the three overlapping degrees, which came to the conclusion that PLWT revealed real coupling condition of oscillators without being influenced by volume conduction.

3.2 Scale-Free Character

Previous studies have revealed a scale-free character in brain functional networks based on fMRI [10]. Here we studied the same character on EEG data. The data used for this section were the resting state EEG datasets of adolescent subjects collected at the Research Center for Learning Science, Southeast University, China [11]. Twenty four healthy adolescent subjects were included. The EEG datasets were recorded by the neuroscan international 10–20 system [12] with a sampling rate of 1000 Hz and 60 EEG channels.

After constructing brain functional networks using WPLI and PLWT, we calculated the degree distribution in four frequency bands and tested the scale-free level. Here the parameter R^2 was employed to evaluate the fitting curves between the degree distribution and the power-law distribution. Different thresholds were adopted and their fitting conditions were summarized in Table 1.

For WPLI, four degrees of connection densities based on the proportional thresholding scheme were employed, and the adopted proportion of the strongest weights were p = 0.2, 0.14, 0.1 and 0.05. Here p = 0.14 was the generally accepted proportion concerning the number of EEG channels (N = 60). The results of WPLI showed that lower or smaller connection density led to higher R^2-values in each frequency band.

Moreover, under the commonly used proportion (p = 0.14), R^2-values systematically increased in higher frequency bands. As for PLWT, it was not necessary to select a parameter as threshold for the connectivity matrix, since we would directly obtain the adjacent matrix (binary) for one certain significance value, which was the biggest advantage of our algorithm for avoiding subjective factors when choosing the threshold. Thus we tested the fitting condition under five significance values ($\alpha = 0.05, 0.01$, 0.005, 0.001, and 0.0001. The upper half of the PLWT table showed the R^2-values under the five different significance values across frequency bands, and the lower half showed corresponding connection densities in four frequency bands. The R^2-values under any of the five significance values were systematically higher than that of WPLI under p = 0.14, and the corresponding densities showed favorable sparseness. In addition, the variation of these densities was very small, which meant that prominent and robust connections were already selected under $\alpha = 0.01$. Thus, for the analysis of functional connectivity based on PLWT, we can choose the wildly recognized significance values $\alpha = 0.01$ to define an adjacency matrix in a more conservative way.

Table 1. Comparison of fitting conditions between WPLI and PLWT under different thresholding schemes in terms of R^2-values and the corresponding connection densities of PLWT-based networks.

WPLI (R^2-values)	Connection density			
Frequency band	p=0.2	p=0.14	p=0.1	p=0.05
Theta	0.1262	0.4585	0.6932	0.8860
Alpha	0.1068	0.4935	0.7070	0.9148
Beta	0.1871	0.6573	0.8432	0.9365
Gamma	0.2276	0.7287	0.9073	0.9667

PLWT (R^2-values)	Significance level				
Frequency band	α=0.0001	α=0.001	α=0.005	α=0.01	α=0.05
Theta	0.6912	0.6625	0.6510	0.6464	0.6185
Alpha	0.6533	0.6185	0.5815	0.5614	0.5511
Beta	0.8222	0.8236	0.7901	0.7505	0.7097
Gamma	0.8776	0.8486	0.8385	0.8246	0.7829
(Connection densities)					
Theta	0.3326	0.3459	0.3556	0.3603	0.3714
Alpha	0.2657	0.2833	0.2967	0.3032	0.3186
Beta	0.1349	0.1565	0.1738	0.1825	0.2038
Gamma	0.1270	0.1492	0.1669	0.1758	0.1975

4 Discussion and Conclusion

In this paper, we have introduced a novel index PLWT in verifying the existence of connection between two signals under the framework of Wilcoxon signed-rank test. Before conducting experiments on real EEG data, the Kuramoto model of 64 coupled oscillators was employed to simulate the influence of volume conduction under three degrees on PLV, PLI, WPLI and PLWT as a function of the coupling strength K. Specifically, all the indices increased along with the increase of coupling strength K then remained approximately stable after one certain critical point, which was consistent with the prediction of this model. In addition, PLWT showed less sensibility to the influence of volume conduction than PLV or PLI. The result of PLWT was similar to that of WPLI in both the magnitude range and the variation trend, which implied that PLWT could be utilized as a reliable connectivity index.

After the preliminary experiment based on the simulated datasets, the index PLWT was basically recognized as convincible and reliable. Further, the real EEG datasets of children resting state were employed to analyze the scale-free character of different thresholding schemes across frequency bands. Under different connectivity densities, the R^2-values increased with the decreasing of connectivity density for WPLI. Considering PLWT, we found that the R^2-values slightly increased for lower significance levels across frequency bands and were generally higher than that of WPLI for any of the significance levels, indicating that PLWT was more robust under the generally accepted standard $\alpha = 0.01$. Also, the corresponding connectivity densities for all significance values in every frequency band showed good sparseness of constructed brain networks.

As a novel connectivity index under the framework of Wilcoxon signed-rank test, although PLWT contains slower calculation because of the process of hypothesis testing, it requires no carefully chosen thresholds before generating adjacency matrix. It can reveal robust connections while hardly influenced by volume conduction. Also, the special thresholding way of PLWT leads to scale-free character of brain functional networks and higher R^2-values across frequency bands, indicating better reliability.

Acknowledgment. This work was supported in part by the National Basic Research Program of China under Grant 2015CB351704, the Key Research and Development Plan (Industry Foresight and Common Key Technology) - Key Project of Jiangsu Province under Grant BE2017007-3, and the National Natural Science Foundation of China under Grants 61773114 and 61375118.

References

1. Biswal, B.B., Van, K.J., Hyde, J.S.: Simultaneous assessment of flow and BOLD signals in resting-state functional connectivity maps. NMR Biomed. **10**(4–5), 165–170 (1997)
2. Vinck, M., Oostenveld, R., Wingerden, M.V., et al.: An improved index of phase-synchronization for electrophysiological data in the presence of volume-conduction, noise and sample-size bias. Neuroimage **55**(4), 1548–1565 (2011)
3. Lachaux, J., Rodriguez, E., Martinerie, J., et al.: Measuring phase synchrony in brain signals. Hum. Brain Mapp. **8**(4), 194 (1999)

4. Stam, C.J., Nolte, G., Daffertshofer, A.: Phase lag index: assessment of functional connectivity from multi channel EEG and MEG with diminished bias from common sources. Hum. Brain Mapp. **28**(11), 1178–1193 (2007)
5. Nunez, P.L., Srinivasan, R.: Electric Fields of the Brain: The Neurophysics of EEG, 2nd edn. Oxford University Press, New York (2006)
6. Nunez, P.L., Williamson, S.J.: Neocortical dynamics and human EEG rhythms. Phys. Today **49**(4), 57 (2008)
7. Dimitriadis, S.I., Laskaris, N.A., Tsirka, V., et al.: Tracking brain dynamics via time-dependent network analysis. J. Neurosci. Methods **193**(1), 145–155 (2010)
8. Kuramoto, Y.: Self-entrainment of a population of coupled non-linear oscillators. In: Araki, H. (ed.) International Symposium on Mathematical Problems in Theoretical Physics, vol. 39, pp. 420–422. Springer, Heidelberg (1975). doi:10.1007/BFb0013365
9. Strogatz, S.H.: From Kuramoto to Crawford: exploring the onset of synchronization in populations of coupled oscillators. Physica D **143**(1–4), 1–20 (1975)
10. Eguiluz, V.M., Cecchi, G., Chialvo, D.R., et al.: Scale-free structure of brain functional networks. Phys. Rev. Lett. **94**(2), 018102 (2003)
11. Zhang, L., Gan, J.Q., Wang, H.: Mathematically gifted adolescents mobilize enhanced workspace configuration of theta cortical network during deductive reasoning. Neuroscience **289**, 334–348 (2015)
12. Homan, R.W., Herman, J., Purdy, P.: Cerebral location of international 10–20 system electrode placement. Electroencephalogr. Clin. Neurophysiol. **66**(4), 376–382 (1987)

Exponential Stability of Matrix-Valued BAM Neural Networks with Time-Varying Delays

Călin-Adrian Popa(✉)

Department of Computer and Software Engineering,
Polytechnic University Timişoara, Blvd. V. Pârvan, No. 2,
300223 Timişoara, Romania
calin.popa@cs.upt.ro

Abstract. Matrix-valued BAM neural networks are a generalization of real-valued BAM neural networks, for which the states, weights, and outputs are square matrices. This paper gives a sufficient criterion expressed in terms of linear matrix inequalities, for which the equilibrium point of these networks with time-varying delays is exponentially stable. A numerical example is provided to demonstrate the effectiveness of the proposed criterion.

Keywords: Matrix-valued BAM neural networks · Exponential stability · Linear matrix inequality · Time-varying delays

1 Introduction

The domain of multidimensional neural networks has received an increasing interest over the last few years. Complex-valued networks were the first such type of networks [1], presently having applications in telecommunications, complex-valued signal processing, and image processing [2]. Another type of multidimensional networks are the hyperbolic-valued neural networks [3]. Defined on the 4-dimensional algebra of quaternion numbers, quaternion-valued neural networks [4] have been applied to the 4-bit parity problem, chaotic time series prediction, and quaternion-valued signal processing. Complex, hyperbolic, and quaternion algebras are all Clifford algebras, thus Clifford-valued neural networks were also proposed [5].

The complex, hyperbolic, quaternion, and Clifford numbers all have a matrix representation [6]. Thus, matrix-valued neural networks that generalize all the above types of networks were introduced [6–10]. Because of their degree of generality, they have potential applications in high-dimensional data processing.

An extension of the unidirectional Hopfield neural networks, BAM neural networks [11] have many applications in pattern recognition and automatic control. Time delays appear unavoidably in real life implementations of neural networks, which can lead to oscillations and chaos. Complex-valued BAMs were introduced in [12], quaternion-valued BAMs in [13], and Clifford-valued BAMs in [14]. These

D. Liu et al. (Eds.): ICONIP 2017, Part III, LNCS 10636, pp. 718–727, 2017.
https://doi.org/10.1007/978-3-319-70090-8_72

facts into account, we study the exponential stability of the equilibrium point of matrix-valued BAM neural networks with time-varying delays.

The outline of the rest of the paper is as follows. Matrix-valued BAM neural networks are introduced, and an assumption and a useful lemma are given in Sect. 2. A sufficient criterion for the exponential stability of matrix-valued BAM neural networks with time-varying delays is established in Sect. 3. The effectiveness of the theoretical result is proved by a numerical example in Sect. 4. Section 5 is dedicated to the conclusions of the study.

Notations: $\mathbb{R}$ is the real number set, $\mathbb{R}^n$ is the n dimensional Euclidean space, and $\mathbb{R}^{n\times n}$ contains the real matrices of order n. A^T is the transpose of A and I_n is the identity matrix of order n. $A > 0$ ($A < 0$) means that matrix A is positive definite (negative definite). $||\cdot||$ is the vector Euclidean norm or the matrix Frobenius norm. $\lambda_{\min}(P)$ represents the smallest eigenvalue of matrix P.

2 Preliminaries

In this paper, we consider matrix-valued BAM neural networks, for which the states, weights, and outputs are square matrices from $\mathbb{R}^{n\times n}$. This type of network is defined by the following set of differential equations:

$$\begin{cases} \dot{X}_i(t) = -d_i^1 X_i(t) + \sum_{j=1}^{P} A_{ij}^1 f_j^1(Y_j(t)) + \sum_{j=1}^{P} B_{ij}^1 g_j^1(Y_j(t-\tau(t))) + U_i^1, \forall i \in \{1,\ldots,N\}, \\ \dot{Y}_j(t) = -d_j^2 Y_j(t) + \sum_{i=1}^{N} A_{ji}^2 f_i^2(X_i(t)) + \sum_{i=1}^{N} B_{ji}^2 g_i^2(X_i(t-\tau(t))) + U_j^2, \forall j \in \{1,\ldots,P\}, \end{cases} \tag{1}$$

where $X_i(t)$, $Y_j(t) \in \mathbb{R}^{n\times n}$ represent the states of the neurons at time t, d_i^1, $d_j^2 > 0$ represent the self-feedback weights, A_{ij}^1, $A_{ji}^2 \in \mathbb{R}^{n\times n}$ represent the weights without delay, B_{ij}^1, $B_{ji}^2 \in \mathbb{R}^{n\times n}$ represent the weights with delay, f_j^1, $f_i^2 : \mathbb{R}^{n\times n} \to \mathbb{R}^{n\times n}$ represent the nonlinear matrix-valued activation functions without delay, g_j^1, $g_i^2 : \mathbb{R}^{n\times n} \to \mathbb{R}^{n\times n}$ represent the nonlinear matrix-valued activation functions with delay, $\tau(t)$ represents the time-varying delay and we assume that $0 < \tau(t) \le \tau$, and $\dot{\tau}(t) \le \tau_d < 1$, $\forall t \ge 0$, and U_i^1, $U_j^2 \in \mathbb{R}^{n\times n}$ represent the external inputs, $\forall i \in \{1,\ldots,N\}$, $\forall j \in \{1,\ldots,P\}$.

The derivative $\dot{X}_i(t)$ is considered to be the matrix whose entries are the derivatives of the entries of $X_i(t)$: $\dot{X}_i(t) = \frac{dX_i(t)}{dt} := \left(\frac{d([X_i(t)]_{ab})}{dt}\right)_{1\le a,b\le n} = \left([\dot{X}_i(t)]_{ab}\right)_{1\le a,b\le n}$, and analogously for $\dot{Y}_j(t)$, $\forall i \in \{1,\ldots,N\}$, $\forall j \in \{1,\ldots,P\}$. Also, the activation functions $f_j^1, f_i^2, g_j^1, g_i^2$ are each formed of n^2 functions $[f_j^1]_{ab}$, $[f_i^2]_{ab}$, $[g_j^1]_{ab}$, $[g_i^2]_{ab} : \mathbb{R}^{n\times n} \to \mathbb{R}$, $1 \le a, b \le n$, $\forall i \in \{1,\ldots,N\}$, $\forall j \in \{1,\ldots,P\}$: $f_j^1(X) = \left([f_j^1]_{ab}(X)\right)_{1\le a,b\le n}$, and the analogous ones.

We need to make an assumption about the activation functions in order to study the stability of the above defined network.

Assumption 1. *The matrix-valued activation functions f_j^1, f_i^2, g_j^1, g_i^2 satisfy the following Lipschitz conditions, for any X, $X' \in \mathbb{R}^{n\times n}$:*

$$||f_j^1(X) - f_j^1(X')|| \leq l_j^{f^1} ||X - X'||, \quad ||f_i^2(X) - f_i^2(X')|| \leq l_i^{f^2} ||X - X'||,$$

$$||g_j^1(X) - g_j^1(X')|| \leq l_j^{g^1} ||X - X'||, \quad ||g_i^2(X) - g_i^2(X')|| \leq l_i^{g^2} ||X - X'||,$$

where $l_j^{f^1}$, $l_i^{f^2}$, $l_j^{g^1}$, $l_i^{g^2} > 0$ are the Lipschitz constants, $\forall i \in \{1,\ldots,N\}$, $\forall j \in \{1,\ldots,P\}$. Moreover, we denote $\overline{L_{f^1}} = diag\left(l_j^{f^1} I_{n^2}\right)_{1\leq j\leq P}$, $\overline{L_{f^2}} = diag\left(l_i^{f^2} I_{n^2}\right)_{1\leq i\leq N}$, $\overline{L_{g^1}} = diag\left(l_j^{g^1} I_{n^2}\right)_{1\leq j\leq P}$, $\overline{L_{g^2}} = diag\left(l_i^{g^2} I_{n^2}\right)_{1\leq i\leq N}$.

Now, we will transform the matrix-valued differential equations (1) into real-valued differential equations. We start by expanding each equation in (1) into n^2 real-valued equations, one corresponding to each entry in the original matrices:

$$\begin{aligned}
[\dot{X}_i(t)]_{ab} &= -d_i^1[X_i(t)]_{ab} + \sum_{j=1}^{P}\sum_{c=1}^{n}[A_{ij}^1]_{ac}[f_j^1]_{cb}(Y_j(t)) \\
&\quad + \sum_{j=1}^{P}\sum_{c=1}^{n}[B_{ij}^1]_{ac}[g_j^1]_{cb}(Y_j(t-\tau(t))) + [U_i^1]_{ab}, \\
[\dot{Y}_j(t)]_{ab} &= -d_j^2[Y_j(t)]_{ab} + \sum_{i=1}^{N}\sum_{c=1}^{n}[A_{ji}^2]_{ac}[f_i^2]_{cb}(X_i(t)) \\
&\quad + \sum_{i=1}^{N}\sum_{c=1}^{n}[B_{ji}^2]_{ac}[g_i^2]_{cb}(X_i(t-\tau(t))) + [U_j^2]_{ab},
\end{aligned} \tag{2}$$

for $1 \leq a, b \leq n$, $i \in \{1,\ldots,N\}$, $j \in \{1,\ldots,P\}$. By using the vectorization operation, the above differential equations can be written more compactly as:

$$\begin{aligned}
\text{vec}(\dot{X}_i(t)) &= -d_i^1 I_{n^2}\text{vec}(X_i(t)) + \sum_{j=1}^{P}(I_n \otimes A_{ij}^1)\text{vec}(f_j^1(Y_j(t))) \\
&\quad + \sum_{j=1}^{P}(I_n \otimes B_{ij}^1)\text{vec}(g_j^1(Y_j(t-\tau(t)))) + \text{vec}(U_i^1), \ \forall i \in \{1,\ldots,N\}, \\
\text{vec}(\dot{Y}_j(t)) &= -d_j^2 I_{n^2}\text{vec}(Y_j(t)) + \sum_{i=1}^{N}(I_n \otimes A_{ji}^2)\text{vec}(f_i^2(X_i(t))) \\
&\quad + \sum_{i=1}^{N}(I_n \otimes B_{ji}^2)\text{vec}(g_i^2(X_i(t-\tau(t)))) + \text{vec}(U_j^2), \ \forall j \in \{1,\ldots,P\},
\end{aligned} \tag{3}$$

where $A \otimes B$ denotes the Kronecker product of matrices A and B. If we denote $Z(t) = \left(\text{vec}(X_i(t))^T\right)^T_{1\leq i\leq N}$, $W(t) = \left(\text{vec}(Y_j(t))^T\right)^T_{1\leq j\leq P}$, $\overline{D^1} = \text{diag}\left(d_i^1 I_{n^2}\right)_{1\leq i\leq N}$, $\overline{A^1} = \left(I_n \otimes A_{ij}^1\right)_{\substack{1\leq i\leq N,\\ 1\leq j\leq P}}$, $\overline{B^1} = \left(I_n \otimes B_{ij}^1\right)_{\substack{1\leq i\leq N,\\ 1\leq j\leq P}}$, $\overline{f^1}(W(t)) =$

$\left(\text{vec}(f_j^1(Y_j(t)))^T\right)^T_{1\leq j\leq P}$, $\overline{g^1}(W(t-\tau(t))) = \left(\text{vec}(g_j^1(Y_j(t-\tau(t))))^T\right)^T_{1\leq j\leq P}$, $\overline{U^1} = \left(\text{vec}(U_i^1)^T\right)^T_{1\leq i\leq N}$, $\overline{D^2} = \text{diag}\left(d_j^2 I_{n^2}\right)_{1\leq j\leq P}$, $\overline{A^2} = \left(I_n \otimes A_{ji}^2\right)_{\substack{1\leq j\leq P,\\ 1\leq i\leq N}}$, $\overline{B^2} = \left(I_n \otimes B_{ji}^2\right)_{\substack{1\leq j\leq P,\\ 1\leq i\leq N}}$, $\overline{f^2}(Z(t)) = \left(\text{vec}(f_i^2(X_i(t)))^T\right)^T_{1\leq i\leq N}$, $\overline{g^2}(Z(t-\tau(t))) = \left(\text{vec}(g_i^2(X_i(t-\tau(t))))^T\right)^T_{1\leq i\leq N}$, $\overline{U^2} = \left(\text{vec}(U_j^2)^T\right)^T_{1\leq j\leq P}$, system (1) becomes:

$$\begin{cases} \dot{Z}(t) = -\overline{D^1}Z(t) + \overline{A^1}\,\overline{f^1}(W(t)) + \overline{B^1}\,\overline{g^1}(W(t-\tau(t))) + \overline{U^1} \\ \dot{W}(t) = -\overline{D^2}W(t) + \overline{A^2}\,\overline{f^2}(Z(t)) + \overline{B^2}\,\overline{g^2}(Z(t-\tau(t))) + \overline{U^2}. \end{cases} \tag{4}$$

If we assume that $\left(\hat{Z}^T, \hat{W}^T\right)^T$ is the equilibrium point of (4), we can shift it to the origin, to obtain

$$\begin{cases} \dot{\tilde{Z}}(t) = -\overline{D^1}\tilde{Z}(t) + \overline{A^1}\,\tilde{f^1}(\tilde{W}(t)) + \overline{B^1}\,\tilde{g^1}(\tilde{W}(t-\tau(t))) \\ \dot{\tilde{W}}(t) = -\overline{D^2}\tilde{W}(t) + \overline{A^2}\,\tilde{f^2}(\tilde{Z}(t)) + \overline{B^2}\,\tilde{g^2}(\tilde{Z}(t-\tau(t))), \end{cases} \tag{5}$$

where $\tilde{Z}(t) = Z(t) - \hat{Z}$, $\tilde{W}(t) = W(t) - \hat{W}$, $\tilde{f^1}(\tilde{W}(t)) = \overline{f^1}(W(t)) - \overline{f^1}(\hat{W})$, $\tilde{g^1}(\tilde{W}(t-\tau(t))) = \overline{g^1}(W(t-\tau(t))) - \overline{g^1}(\hat{W})$, $\tilde{f^2}(\tilde{Z}(t)) = \overline{f^2}(Z(t)) - \overline{f^2}(\hat{Z})$, $\tilde{g^2}(\tilde{Z}(t-\tau(t))) = \overline{g^2}(Z(t-\tau(t))) - \overline{g^2}(\hat{Z})$.

Remark 1. Because system (5) is equivalent with system (1), and so any property that holds for system (5), will also be true for system (1), we will only study the exponential stability of the origin of system (5).

Remark 2. A matrix-valued BAM neural network is not equivalent with a general $nN \times nP$-dimensional real-valued BAM neural network, because, for such a network, the matrices $\overline{A^1}$, $\overline{B^1}$, $\overline{A^2}$, $\overline{B^2}$ would be general unconstrained matrices, and wouldn't have the particular form given above.

We will also need the following lemma:

Lemma 1 ([15]). *The following inequality holds for any positive definite matrix $M \in \mathbb{R}^{n^2N\times n^2N}$ and any vector function $Z : [a,b] \to \mathbb{R}^{n^2N}$:*

$$\left(\int_a^b Z(s)ds\right)^T M \left(\int_a^b Z(s)ds\right) \leq (b-a)\int_a^b Z^T(s)MZ(s)ds,$$

in which the integrals are well defined.

3 Main Result

We give a sufficient criterion which assures the exponential stability of the origin of system (5).

Theorem 1. *If Assumption 1 holds, then the origin of system (5) is exponentially stable if there exist positive definite matrices* P_1^1, P_2^1, $P_1^2, \ldots, P_6^2$, P_1^3, P_2^3 *of appropriate dimensions, positive definite block-diagonal matrices* $R_1, \ldots, R_8$, *and* $\varepsilon > 0$, *which satisfy the following linear matrix inequality (LMI):*

$$(\Pi)_{14\times 14} < 0, \tag{6}$$

where $\Pi_{1,1} = 2\varepsilon P_1^1 - 2\overline{D^1}P_1^1 + P_1^2 + \tau\overline{D^1}P_1^3\overline{D^1} - \tau^{-1}e^{-2\varepsilon\tau}P_1^3 + \overline{L_{f^2}}^T R_1\overline{L_{f^2}} + \overline{L_{g^2}}^T R_3\overline{L_{g^2}}$, $\Pi_{1,2} = \tau^{-1}e^{-2\varepsilon\tau}P_1^3$, $\Pi_{1,11} = P_1^1\overline{A^1} - \tau\overline{D^1}P_1^3\overline{A^1}$, $\Pi_{1,14} = P_1^1\overline{B^1} - \tau\overline{D^1}P_1^3\overline{B^1}$, $\Pi_{2,2} = -\tau^{-1}e^{-2\varepsilon\tau}P_1^3$, $\Pi_{3,3} = -e^{-2\varepsilon\tau}(1-\tau_d)P_1^2 + \overline{L_{f^2}}^T R_2\overline{L_{f^2}} + \overline{L_{g^2}}^T R_4\overline{L_{g^2}}$, $\Pi_{4,4} = P_2^2 + \tau\overline{A^2}^T P_2^3\overline{A^2} - R_1$, $\Pi_{4,7} = \tau\overline{A^2}^T P_2^3\overline{B^2}$, $\Pi_{4,8} = \overline{A^2}^T P_2^1 - \tau\overline{A^2}^T P_2^3\overline{D^2}$, $\Pi_{5,5} = -e^{-2\varepsilon\tau}(1-\tau_d)P_2^2 - R_2$, $\Pi_{6,6} = P_3^2 - R_3$, $\Pi_{7,7} = -e^{-2\varepsilon\tau}(1-\tau_d)P_3^2 + \tau\overline{B^2}^T P_2^3\overline{B^2} - R_4$, $\Pi_{7,8} = \overline{B^2}^T P_2^1 - \tau\overline{B^2}^T P_2^3\overline{D^2}$, $\Pi_{8,8} = 2\varepsilon P_2^1 - 2\overline{D^2}P_2^1 + P_4^2 + \tau\overline{D^2}P_2^3\overline{D^2} - \tau^{-1}e^{-2\varepsilon\tau}P_2^3 + \overline{L_{f^1}}^T R_5\overline{L_{f^1}} + \overline{L_{g^1}}^T R_7\overline{L_{g^1}}$, $\Pi_{8,9} = \tau^{-1}e^{-2\varepsilon\tau}P_2^3$, $\Pi_{9,9} = -\tau^{-1}e^{-2\varepsilon\tau}P_2^3$, $\Pi_{10,10} = -e^{-2\varepsilon\tau}(1-\tau_d)P_4^2 + \overline{L_{f^1}}^T R_6\overline{L_{f^1}} + \overline{L_{g^1}}^T R_8\overline{L_{g^1}}$, $\Pi_{11,11} = P_5^2 + \tau\overline{A^1}^T P_1^3\overline{A^1} - R_5$, $\Pi_{11,14} = \tau\overline{A^1}^T P_1^3\overline{B^1}$, $\Pi_{12,12} = -e^{-2\varepsilon\tau}(1-\tau_d)P_5^2 - R_6$, $\Pi_{13,13} = P_6^2 - R_7$, $\Pi_{14,14} = -e^{-2\varepsilon\tau}(1-\tau_d)P_6^2 + \tau\overline{B^1}^T P_1^3\overline{B^1} - R_8$.

Proof. Consider the following Lyapunov-Krasovskii functional

$$V(t) = V_1(t) + V_2(t) + V_3(t),$$

where

$$V_1(t) = e^{2\varepsilon t}\begin{bmatrix}\tilde{Z}(t)\\ \tilde{W}(t)\end{bmatrix}^T P_1 \begin{bmatrix}\tilde{Z}(t)\\ \tilde{W}(t)\end{bmatrix}, \quad P_1 = \mathrm{diag}(P_1^1, P_2^1),$$

$$V_2(t) = \int_{t-\tau(t)}^{t} e^{2\varepsilon s}\xi^T(s)P_2\xi(s)ds, \quad P_2 = \mathrm{diag}(P_1^2, P_2^2, P_3^2, P_4^2, P_5^2, P_6^2),$$

$$\xi(s) = \left[\tilde{Z}^T(s)\ \tilde{f^2}^T(\tilde{Z}(s))\ \tilde{g^2}^T(\tilde{Z}(s))\ \tilde{W}^T(s)\ \tilde{f^1}^T(\tilde{W}(s))\ \tilde{g^1}^T(\tilde{W}(s))\right]^T,$$

$$V_3(t) = \int_{-\tau}^{0}\int_{t+\theta}^{t} e^{2\varepsilon s}\begin{bmatrix}\dot{\tilde{Z}}(s)\\ \dot{\tilde{W}}(s)\end{bmatrix}^T P_3 \begin{bmatrix}\dot{\tilde{Z}}(s)\\ \dot{\tilde{W}}(s)\end{bmatrix} dsd\theta, \quad P_3 = \mathrm{diag}(P_1^3, P_2^3).$$

The derivative of $V(t)$ along the trajectories of system (5) is

$$\dot{V}(t) = \dot{V}_1(t) + \dot{V}_2(t) + \dot{V}_3(t),$$

where

$$\dot{V}_1(t) = 2e^{2\varepsilon t}\left[\varepsilon\tilde{Z}^T(t)P_1^1\tilde{Z}(t) + \varepsilon\tilde{W}^T(t)P_2^1\tilde{W}(t) + \dot{\tilde{Z}}^T(t)P_1^1\tilde{Z}(t) + \dot{\tilde{W}}^T(t)P_2^1\tilde{W}(t)\right], \tag{7}$$

$$\begin{aligned}\dot{V}_2(t) &= e^{2\varepsilon t}\left[\xi^T(t)P_2\xi(t) - e^{-2\varepsilon\tau(t)}(1-\dot{\tau}(t))\xi^T(t-\tau(t))P_2\xi(t-\tau(t))\right]\\ &\le e^{2\varepsilon t}\left[\xi^T(t)P_2\xi(t) - e^{-2\varepsilon\tau}(1-\tau_d)\xi^T(t-\tau(t))P_2\xi(t-\tau(t))\right],\end{aligned} \tag{8}$$

$$\begin{aligned}\dot{V}_3(t) &= e^{2\varepsilon t}\Big[\tau\dot{\tilde{Z}}^T(t)P_1^3\dot{\tilde{Z}}(t) - \int_{t-\tau}^{t} e^{2\varepsilon(s-t)}\dot{\tilde{Z}}^T(s)P_1^3\dot{\tilde{Z}}(s)ds \\ &\quad +\tau\dot{\tilde{W}}^T(t)P_2^3\dot{\tilde{W}}(t) - \int_{t-\tau}^{t} e^{2\varepsilon(s-t)}\dot{\tilde{W}}^T(s)P_2^3\dot{\tilde{W}}(s)ds\Big] \\ &\leq e^{2\varepsilon t}\Big[\tau\dot{\tilde{Z}}^T(t)P_1^3\dot{\tilde{Z}}(t) - \tau^{-1}e^{-2\varepsilon\tau}\left(\int_{t-\tau}^{t}\dot{\tilde{Z}}(s)ds\right)^T P_1^3\left(\int_{t-\tau}^{t}\dot{\tilde{Z}}(s)ds\right) \\ &\quad +\tau\dot{\tilde{W}}^T(t)P_2^3\dot{\tilde{W}}(t) - \tau^{-1}e^{-2\varepsilon\tau}\left(\int_{t-\tau}^{t}\dot{\tilde{W}}(s)ds\right)^T P_2^3\left(\int_{t-\tau}^{t}\dot{\tilde{W}}(s)ds\right)\Big], \end{aligned} \quad (9)$$

where, for the last inequality, we used Lemma 1.

From Assumption 1, we have that

$$||f_j^1(X) - f_j^1(X')|| \leq l_j^{f^1}||X - X'||$$

$$\Leftrightarrow ||\text{vec}(f_j^1(X)) - \text{vec}(f_j^1(X'))|| \leq l_j^{f^1}||\text{vec}(X) - \text{vec}(X')||,$$

and the analogous ones for f_i^2, g_j^1, g_i^2, $\forall i \in \{1,\ldots,N\}$, $\forall j \in \{1,\ldots,P\}$, for any X, $X' \in \mathbb{R}^{n\times n}$, which allow us to deduce the existence of positive definite block-diagonal matrices $R_1,\ldots,R_8$, such that the following inequalities hold:

$$0 \leq \tilde{Z}^T(t)\overline{L_{f^2}}^T R_1\overline{L_{f^2}}\tilde{Z}(t) - \tilde{f^2}^T(\tilde{Z}(t))R_1\tilde{f^2}(\tilde{Z}(t)), \quad (10)$$

$$0 \leq \tilde{Z}^T(t-\tau(t))\overline{L_{f^2}}^T R_2\overline{L_{f^2}}\tilde{Z}(t-\tau(t)) - \tilde{f^2}^T(\tilde{Z}(t-\tau(t)))R_2\tilde{f^2}(\tilde{Z}(t-\tau(t))), \quad (11)$$

$$0 \leq \tilde{Z}^T(t)\overline{L_{g^2}}^T R_3\overline{L_{g^2}}\tilde{Z}(t) - \tilde{g^2}^T(\tilde{Z}(t))R_3\tilde{g^2}(\tilde{Z}(t)), \quad (12)$$

$$0 \leq \tilde{Z}^T(t-\tau(t))\overline{L_{g^2}}^T R_4\overline{L_{g^2}}\tilde{Z}(t-\tau(t)) - \tilde{g^2}^T(\tilde{Z}(t-\tau(t)))R_4\tilde{g^2}(\tilde{Z}(t-\tau(t))), \quad (13)$$

$$0 \leq \tilde{W}^T(t)\overline{L_{f^1}}^T R_5\overline{L_{f^1}}\tilde{W}(t) - \tilde{f^1}^T(\tilde{W}(t))R_5\tilde{f^1}(\tilde{W}(t)), \quad (14)$$

$$0 \leq \tilde{W}^T(t-\tau(t))\overline{L_{f^1}}^T R_6\overline{L_{f^1}}\tilde{W}(t-\tau(t)) - \tilde{f^1}^T(\tilde{W}(t-\tau(t)))R_6\tilde{f^1}(\tilde{W}(t-\tau(t))), \quad (15)$$

$$0 \leq \tilde{W}^T(t)\overline{L_{g^1}}^T R_7\overline{L_{g^1}}\tilde{W}(t) - \tilde{g^1}^T(\tilde{W}(t))R_7\tilde{g^1}(\tilde{W}(t)), \quad (16)$$

$$0 \leq \tilde{W}^T(t-\tau(t))\overline{L_{g^1}}^T R_8\overline{L_{g^1}}\tilde{W}(t-\tau(t)) - \tilde{g^1}^T(\tilde{W}(t-\tau(t)))R_8\tilde{g^1}(\tilde{W}(t-\tau(t))). \quad (17)$$

Multiplying inequalities (10)–(17) by $e^{2\varepsilon t}$, and adding them to (7)–(9), we obtain that

$$\dot{V}(t) \leq e^{2\varepsilon t}\zeta^T(t)\Pi\zeta(t),$$

where

$$\zeta(t) = \big[\,\tilde{Z}^T(t)\ \tilde{Z}^T(t-\tau)\ \tilde{Z}^T(t-\tau(t))\ f^2(\tilde{Z}(t))\ f^2(\tilde{Z}(t-\tau(t)))\ g^2(\tilde{Z}(t))\ g^2(\tilde{Z}(t-\tau(t)))$$

$$\tilde{W}^T(t)\ \tilde{W}^T(t-\tau)\ \tilde{W}^T(t-\tau(t))\ f^1(\tilde{W}(t))\ f^1(\tilde{W}(t-\tau(t)))\ g^1(\tilde{W}(t))\ g^1(\tilde{W}(t-\tau(t)))\big]^T,$$

and Π is given in (6). Also from (6), we have that $\Pi < 0$, thus $\dot{V}(t) < 0$, showing that $V(t)$ is strictly decreasing for $t \geq 0$. From the definition of $V(t)$, we can write the following inequalities:

$$e^{2\varepsilon t}\lambda_{\min}(P_1)\left\|\begin{bmatrix}\tilde{Z}(t)\\ \tilde{W}(t)\end{bmatrix}\right\|^2 \leq e^{2\varepsilon t}\begin{bmatrix}\tilde{Z}(t)\\ \tilde{W}(t)\end{bmatrix}^T P_1 \begin{bmatrix}\tilde{Z}(t)\\ \tilde{W}(t)\end{bmatrix} \leq V(t) \leq V_0,\ \forall t \geq T,\ T \geq 0,$$

where $V_0 = \max_{0 \leq t \leq T} V(t)$. Consequently, we have that

$$\left\|\begin{bmatrix}\tilde{Z}(t)\\ \tilde{W}(t)\end{bmatrix}\right\|^2 \leq \frac{V_0}{e^{2\varepsilon t}\lambda_{\min}(P_1)} \Leftrightarrow \left\|\begin{bmatrix}\tilde{Z}(t)\\ \tilde{W}(t)\end{bmatrix}\right\| \leq Me^{-\varepsilon t},\ \forall t \geq 0,$$

where $M = \sqrt{\frac{V_0}{\lambda_{\min}(P_1)}}$. Thus, we obtained the exponential stability for the origin of system (5), ending the proof of the theorem.

4 Numerical Example

In order to illustrate the correctness of the above result, we give a numerical example.

Example. Let us consider the following matrix-valued BAM neural network with time-varying delays:

$$\begin{cases}\dot{X}_1(t) = -d_1^1 X_1(t) + A_{11}^1 f_1^1(Y_1(t)) + B_{11}^1 g_1^1(Y_1(t-\tau(t))) + U_1^1,\\ \dot{X}_2(t) = -d_2^1 X_2(t) + A_{21}^1 f_1^1(Y_1(t)) + B_{21}^1 g_1^1(Y_1(t-\tau(t))) + U_2^1,\\ \dot{Y}_1(t) = -d_1^2 Y_1(t) + \sum_{i=1}^2 A_{1i}^2 f_i^2(X_i(t)) + \sum_{i=1}^2 B_{1i}^2 g_i^2(X_i(t-\tau(t))) + U_1^2,\end{cases} \quad (18)$$

where $N = 2$, $P = 1$, $d_1^1 = d_2^1 = d_1^2 = 20$,

$$A_{11}^1 = \begin{bmatrix}1 & 1\\ 2 & 2\end{bmatrix},\ A_{21}^1 = \begin{bmatrix}1 & 1\\ 2 & 2\end{bmatrix},\ A_{11}^2 = \begin{bmatrix}1 & 1\\ 1 & 1\end{bmatrix},\ A_{12}^2 = \begin{bmatrix}1 & 2\\ 2 & 1\end{bmatrix},\ B_{11}^1 = \begin{bmatrix}1 & 1\\ 2 & 2\end{bmatrix},\ B_{21}^1 = \begin{bmatrix}1 & 2\\ 3 & 2\end{bmatrix},$$

$$B_{11}^2 = \begin{bmatrix}1 & 2\\ 2 & 1\end{bmatrix},\ B_{12}^2 = \begin{bmatrix}1 & 3\\ 2 & 3\end{bmatrix},\ U_1^1 = \begin{bmatrix}-14 & -35\\ 23 & 44\end{bmatrix},\ U_2^1 = \begin{bmatrix}-43 & -23\\ 33 & 13\end{bmatrix},\ U_1^2 = \begin{bmatrix}-14 & -45\\ 23 & 34\end{bmatrix},$$

$$f_1^1\left(([X]_{ab})_{1\leq a,b\leq 2}\right) = f_1^2\left(([X]_{ab})_{1\leq a,b\leq 2}\right)$$
$$= f_2^2\left(([X]_{ab})_{1\leq a,b\leq 2}\right) = \left(\frac{1}{2(1+e^{-[X]_{ab}})}\right)_{1\leq a,b\leq 2},$$

$$g_1^1\left(([X]_{ab})_{1\leq a,b\leq 2}\right) = g_1^2\left(([X]_{ab})_{1\leq a,b\leq 2}\right)$$
$$= g_2^2\left(([X]_{ab})_{1\leq a,b\leq 2}\right) = \left(\frac{1-e^{-[X]_{ab}}}{1+e^{-[X]_{ab}}}\right)_{1\leq a,b\leq 2},$$

from which we get that $l_1^{f^1} = l_1^{f^2} = l_2^{f^2} = \frac{1}{4}$ and $l_1^{g^1} = l_1^{g^2} = l_2^{g^2} = 1$. The time-varying delay is taken to be $\tau(t) = 0.9|\sin t|$, which implies that $\tau = \tau_d = 0.9$.

From Theorem 1 we get that the equilibrium point of system (18) is exponentially stable, if LMI condition (6) is satisfied. By solving (6), we obtain $\varepsilon = 0.1$, $R_1 = \text{diag}(2.3413I_4,\ 2.8806I_4)$, $R_2 = \text{diag}(0.1380I_4, 0.0783I_4)$, $R_3 = \text{diag}(1.5670I_4, 1.6963I_4)$, $R_4 = \text{diag}(0.1849I_4, 0.2477I_4)$, $R_5 = 3.5840I_4$, $R_6 = 0.0426I_4$, $R_7 = 1.4531I_4$, $R_8 = 0.2159I_4$. (The values of the other matrices are not given due to space limitations.) The state trajectories of the elements of matrices X_1, X_2, and Y_1 are depicted in Fig. 1, for four initial values. It can be seen that the 4 elements of the matrices X_1, X_2, and Y_1, respectively, converge to the equilibrium point $(\hat{X}_1, \hat{X}_2, \hat{Y}_1)$ of (18), given by

$$\hat{X}_1 = \begin{bmatrix} -0.6568 & -1.7305 \\ 1.2364 & 2.2390 \end{bmatrix},\ \hat{X}_2 = \begin{bmatrix} -2.0795 & -1.0956 \\ 1.7237 & 0.6498 \end{bmatrix},\ \hat{Y}_1 = \begin{bmatrix} -0.5221 & -2.1166 \\ 1.2273 & 1.7230 \end{bmatrix}.$$

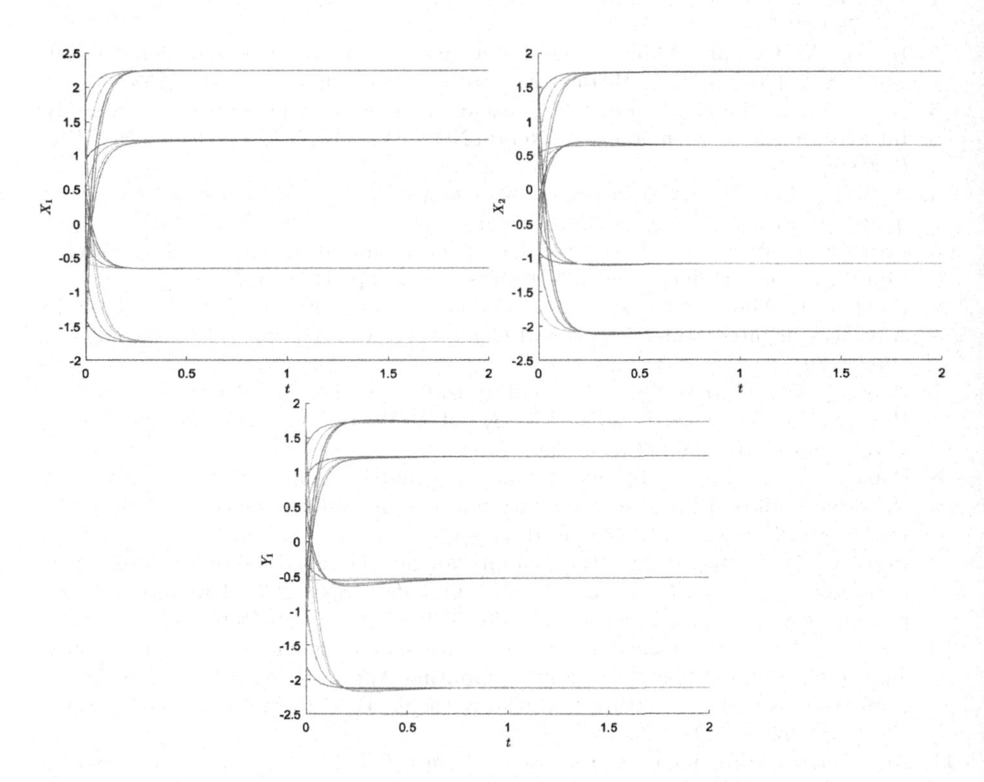

Fig. 1. State trajectories of the elements of X_1, X_2, and Y_1 in the Example

5 Conclusions

By making the assumption that the Lipschitz condition is satisfied by the nonlinear matrix-valued activation functions, a sufficient criterion expressed in terms of linear matrix inequalities was given for the exponential stability of matrix-valued BAM neural networks with time-varying delays. A numerical example was provided to verify the correctness of the theoretical result.

Matrix-valued neural networks offer potential for further study, both in terms of applications as well as in terms of dynamical behavior, especially due to their degree of generality, which encompasses complex-, hyperbolic-, quaternion-, and Clifford-valued neural networks.

References

1. Widrow, B., McCool, J., Ball, M.: The complex LMS algorithm. Proc. IEEE **63**(4), 719–720 (1975)
2. Hirose, A.: Complex-Valued Neural Networks. Studies in Computational Intelligence, vol. 400. Springer, Heidelberg (2012). doi:10.1007/978-3-642-27632-3
3. Nitta, T., Buchholz, S.: On the decision boundaries of hyperbolic neurons. In: International Joint Conference on Neural Networks (IJCNN), pp. 2974–2980. IEEE (2008)
4. Arena, P., Fortuna, L., Muscato, G., Xibilia, M.: Multilayer perceptrons to approximate quaternion valued functions. Neural Netw. **10**(2), 335–342 (1997)
5. Pearson, J., Bisset, D.: Back propagation in a Clifford algebra. In: International Conference on Artificial Neural Networks, vol. 2, pp. 413–416 (1992)
6. Popa, C.A.: Matrix-valued neural networks. In: Matoušek, R. (ed.) Mendel 2015. Advances in Intelligent Systems and Computing, vol. 378, pp. 245–255. Springer, Cham (2015)
7. Popa, C.-A.: Matrix-valued hopfield neural networks. In: Cheng, L., Liu, Q., Ronzhin, A. (eds.) ISNN 2016. LNCS, vol. 9719, pp. 127–134. Springer, Cham (2016). doi:10.1007/978-3-319-40663-3_15
8. Popa, C.A.: Global asymptotic stability for matrix-valued recurrent neural networks with time delays. In: 2017 International Joint Conference on Neural Networks (IJCNN), pp. 4474–4481 (2017)
9. Popa, C.-A.: Global exponential stability for matrix-valued neural networks with time delay. In: Cong, F., Leung, A., Wei, Q. (eds.) ISNN 2017. LNCS, vol. 10261, pp. 429–438. Springer, Cham (2017). doi:10.1007/978-3-319-59072-1_51
10. Popa, C.-A.: Matrix-valued bidirectional associative memories. In: Balas, V.E., Jain, L.C., Balas, M.M. (eds.) Soft Computing Applications. Advances in Intelligent Systems and Computing, vol. 634. Springer, Heidelberg (2018). doi:10.1007/978-3-319-62524-9_4
11. Kosko, B.: Bidirectional associative memories. IEEE Trans. Syst. Man Cybern. **18**(1), 49–60 (1988)
12. Lee, D., Wang, W.: A multivalued bidirectional associative memory operating on a complex domain. Neural Netw. **11**(9), 1623–1635 (1998)
13. Kuroe, Y.: Models of Clifford recurrent neural networks and their dynamics. In: International Joint Conference on Neural Networks (IJCNN), pp. 1035–1041. IEEE (2011)

14. Vallejo, J., Bayro-Corrochano, E.: Clifford hopfield neural networks. In: International Joint Conference on Neural Networks (IJCNN), pp. 3609–3612. IEEE, June 2008
15. Gu, K.: An integral inequality in the stability problem of time-delay systems. In: Proceedings of the 39th IEEE Conference on Decision and Control, pp. 2805–2810 (2000)

Asymptotic Stability of Delayed Octonion-Valued Neural Networks with Leakage Delay

Călin-Adrian Popa(✉)

Department of Computer and Software Engineering,
Polytechnic University Timişoara, Blvd. V. Pârvan, No. 2,
300223 Timişoara, Romania
calin.popa@cs.upt.ro

Abstract. This paper gives a sufficient criterion for the asymptotic stability of the equilibrium point of delayed octonion-valued neural networks with leakage delay. Defined over the normed division algebra of octonions, these networks represent a generalization of the complex- and quaternion-valued neural networks that have been intensely studied over the last few years, which doesn't fall into the Clifford-valued category. A numerical example is given to prove the effectiveness of the main result.

Keywords: Octonion-valued neural networks · Asymptotic stability · Linear matrix inequality · Time delay

1 Introduction

The last few years saw an increasing interest in the study of neural networks with values in multidimensional domains. Complex-valued neural networks [1] presently have numerous applications, ranging from those in complex-valued signal processing to those in telecommunications and image processing [2]. With applications in chaotic time series prediction, the 4-bit parity problem, and signal processing, are the quaternion-valued neural networks [3]. Clifford-valued neural networks represent a generalization of these two types of networks [4]. They have values in 2^n-dimensional Clifford algebras of which the complex and quaternion algebras are special cases.

A different generalization of the complex and quaternion algebras is the octonion normed division algebra of dimension 8. The easiest way to see this is by considering the fact that Clifford algebras are associative, whereas the octonion algebra is not. The complex, quaternion, and octonion algebras are the only normed division algebras that can be defined over the field of real numbers. Thus, octonion-valued neural networks were introduced [5–8], which have potential applications in signal processing and other areas related to higher-dimensional object processing.

D. Liu et al. (Eds.): ICONIP 2017, Part III, LNCS 10636, pp. 728–736, 2017.
https://doi.org/10.1007/978-3-319-70090-8_73

On the other hand, Hopfield neural networks [9] have been applied to the synthesis of associative memories, image processing, speech processing, control systems, signal processing, pattern matching, etc. In real life implementations of neural networks, time delays appear unavoidably and can lead to oscillations and chaos. Complex-valued Hopfield networks were discussed in [10], quaternion-valued Hopfield networks in [11], and Clifford-valued Hopfield networks in [12]. Taking these facts into account, in this paper, we discuss the asymptotic stability of delayed octonion-valued neural networks with leakage delay.

The rest of the paper is organized in the following manner: Sect. 2 provides the definition of delayed octonion-valued Hopfield neural networks with leakage delay, an assumption, and a useful lemma. A sufficient condition for the asymptotic stability of the equilibrium point of these networks is given in Sect. 3. The correctness of the main result is proved by a numerical example in Sect. 4. Section 5 concludes the paper.

Notations: $\mathbb{R}$ is the set of real numbers, $\mathbb{R}^n$ is the n dimensional Euclidean space, and $\mathbb{R}^{n\times n}$ the algebra of real square matrices of dimension $n \times n$. A^T is the transpose of matrix A and I_n is the order n identity matrix. $||\cdot||$ is the vector Euclidean norm or the matrix Frobenius norm. $A > 0$ $(A < 0)$ means that A is a positive definite (negative definite) matrix. $\lambda_{\min}(P)$ is the smallest eigenvalue of positive definite matrix P.

2 Preliminaries

We will first give the definition and some properties of the algebra of octonions.

The algebra of octonions is defined as

$$\mathbb{O} := \left\{ x = \sum_{p=0}^{7} [x]_p e_p \,\middle|\, [x]_0, [x]_1, \ldots, [x]_7 \in \mathbb{R} \right\},$$

in which e_p are the octonion units, $0 \leq p \leq 7$.

Octonion addition is defined by $x + y = \sum_{p=0}^{7}([x]_p + [y]_p)e_p$. Scalar multiplication is given by $\alpha x = \sum_{p=0}^{7}(\alpha[x]_p)e_p$, and octonion multiplication is given by the multiplication of the octonion units:

$\times$	e_0	e_1	e_2	e_3	e_4	e_5	e_6	e_7
e_0	e_0	e_1	e_2	e_3	e_4	e_5	e_6	e_7
e_1	e_1	$-e_0$	e_3	$-e_2$	e_5	$-e_4$	$-e_7$	e_6
e_2	e_2	$-e_3$	$-e_0$	e_1	e_6	e_7	$-e_4$	$-e_5$
e_3	e_3	e_2	$-e_1$	$-e_0$	e_7	$-e_6$	e_5	$-e_4$
e_4	e_4	$-e_5$	$-e_6$	$-e_7$	$-e_0$	e_1	e_2	e_3
e_5	e_5	e_4	$-e_7$	e_6	$-e_1$	$-e_0$	$-e_3$	e_2
e_6	e_6	e_7	e_4	$-e_5$	$-e_2$	e_3	$-e_0$	$-e_1$
e_7	e_7	$-e_6$	e_5	e_4	$-e_3$	$-e_2$	e_1	$-e_0$

These operations make $\mathbb{O}$ a real algebra. From the multiplication table, we have that $e_ie_j = -e_je_i \neq e_je_i$, $\forall i \neq j$, $0 < i,j \leq 7$, from which we deduce that $\mathbb{O}$ is not commutative, and that $(e_ie_j)e_k = -e_i(e_je_k) \neq e_i(e_je_k)$, for i,j,k distinct, $0 < i,j,k \leq 7$, or $e_ie_j \neq \pm e_k$, which allows us to see that $\mathbb{O}$ is not associative.

The conjugate of an octonion x is defined as $\overline{x} = [x]_0e_0 - \sum_{p=1}^{7}[x]_pe_p$, its norm as $||x|| = \sqrt{x\overline{x}} = \sqrt{\sum_{p=0}^{7}[x]_p^2}$, and its inverse as $x^{-1} = \frac{\overline{x}}{||x||^2}$. We can now see that $\mathbb{O}$ is a normed division algebra, and it can be proved that the only three division algebras that can be defined over the reals are the complex, quaternion, and octonion algebras.

Now, we can define delayed octonion-valued Hopfield neural networks with leakage delays, for which the states and weights are from $\mathbb{O}$. This type of network is described by the following set of differential equations:

$$\dot{x}_i(t) = -d_ix_i(t-\delta) + \sum_{j=1}^{N} a_{ij}f_j(x_j(t)) + \sum_{j=1}^{N} b_{ij}g_j(x_j(t-\tau)) + u_i, \quad (1)$$

for $i \in \{1,\ldots,N\}$, where $x_i(t) \in \mathbb{O}$ is the state of neuron i at time t, $d_i \in \mathbb{R}$, $d_i > 0$, is the self-feedback connection weight of neuron i, $a_{ij} \in \mathbb{O}$ is the weight connecting neuron j to neuron i without delay, $b_{ij} \in \mathbb{O}$ is the weight connecting neuron j to neuron i with delay, $f_j : \mathbb{O} \to \mathbb{O}$ is the nonlinear octonion-valued activation function without delay, $g_j : \mathbb{O} \to \mathbb{O}$ is the nonlinear octonion-valued activation function with delay, $\tau \in \mathbb{R}$, $\tau > 0$, is the time delay, $\delta \in \mathbb{R}$, $\delta > 0$ is the leakage delay, and $u_i \in \mathbb{O}$ is the external input of neuron i, $\forall i,j \in \{1,\ldots,N\}$.

The derivative is defined as the element-wise derivative of $x_i(t)$ with respect to t: $\dot{x}_i(t) = \sum_{p=0}^{7} \frac{d([x_i(t)]_p)}{dt} e_p$. The multiplication between the weights and the values of the activation functions in the set of differential equations (1) is the octonion multiplication, defined above.

In order to study the dynamic properties of (1), we need to make the following assumption about the activation functions:

Assumption 1. *The octonion-valued activation functions f_j and g_j satisfy the following Lipschitz conditions, $\forall x, x' \in \mathbb{O}$:*

$$||f_j(x) - f_j(x')|| \leq l_j^f ||x - x'||,$$

$$||g_j(x) - g_j(x')|| \leq l_j^g ||x - x'||,$$

where $l_j^f > 0$ and $l_j^g > 0$ are the Lipschitz constants, $\forall j \in \{1,\ldots,N\}$. We also denote $\overline{L_f} = diag\left(l_j^f I_8\right)_{1\leq j\leq N}$, $\overline{L_g} = diag\left(l_j^g I_8\right)_{1\leq j\leq N}$.

We will now transform the octonion-valued system (1) into a real-valued one. For this, each equation in (1) can be split into 8 real-valued equations:

$$[\dot{x}_i(t)]_p = -d_i[x_i(t-\delta)]_p + \sum_{j=1}^{N}\sum_{q=0}^{7}[a_{ij}]_{pq}[f_j(x_j(t))]_q + \sum_{j=1}^{N}\sum_{q=0}^{7}[b_{ij}]_{pq}[g_j(x_j(t-\tau))]_q + [u_i]_p, \tag{2}$$

for $0 \le p \le 7$, $i \in \{1,\ldots,N\}$, where $[x]_{pq}$ is an element of the matrix $\mathrm{mat}(x)$, defined by

$$\mathrm{mat}(x) = \begin{bmatrix} [x]_0 & -[x]_1 & -[x]_2 & -[x]_3 & -[x]_4 & -[x]_5 & -[x]_6 & -[x]_7 \\ [x]_1 & [x]_0 & -[x]_3 & [x]_2 & -[x]_5 & [x]_4 & [x]_7 & -[x]_6 \\ [x]_2 & [x]_3 & [x]_0 & -[x]_1 & -[x]_6 & -[x]_7 & [x]_4 & [x]_5 \\ [x]_3 & -[x]_2 & [x]_1 & [x]_0 & -[x]_7 & [x]_6 & -[x]_5 & -[x]_4 \\ [x]_4 & [x]_5 & [x]_6 & [x]_7 & [x]_0 & -[x]_1 & -[x]_2 & -[x]_3 \\ [x]_5 & -[x]_4 & [x]_7 & -[x]_6 & [x]_1 & [x]_0 & [x]_3 & -[x]_2 \\ [x]_6 & -[x]_7 & -[x]_4 & [x]_5 & [x]_2 & -[x]_3 & [x]_0 & [x]_1 \\ [x]_7 & [x]_6 & -[x]_5 & -[x]_4 & [x]_3 & [x]_2 & -[x]_1 & [x]_0 \end{bmatrix}.$$

By defining $\mathrm{vec}(x) = ([x]_0, [x]_1, \ldots, [x]_7)^T$, the equations in (2) can be written compactly as

$$\mathrm{vec}(\dot{x}_i(t)) = -d_i I_8 \mathrm{vec}(x_i(t-\delta)) + \sum_{j=1}^{N}\mathrm{mat}(a_{ij})\mathrm{vec}(f_j(x_j(t))) + \sum_{j=1}^{N}\mathrm{mat}(b_{ij})\mathrm{vec}(g_j(x_j(t-\tau))) + \mathrm{vec}(u_i), \tag{3}$$

for $i \in \{1,\ldots,N\}$. If we denote $y(t) = \left(\mathrm{vec}(x_i(t))^T\right)^T_{1\le i\le N}$, $\overline{D} = \mathrm{diag}\,(d_i I_8)_{1\le i\le N}$, $\overline{A} = (\mathrm{mat}(a_{ij}))_{1\le i,j\le N}$, $\overline{B} = (\mathrm{mat}(b_{ij}))_{1\le i,j\le N}$, $\overline{f}(y(t)) = \left(\mathrm{vec}(f_j(x_j(t)))^T\right)^T_{1\le j\le N}$, $\overline{g}(w(t-\tau)) = \left(\mathrm{vec}(g_j(x_j(t-\tau)))^T\right)^T_{1\le j\le N}$, system (3) becomes:

$$\dot{y}(t) = -\overline{D}y(t-\delta) + \overline{A}\,\overline{f}(y(t)) + \overline{B}\,\overline{g}(y(t-\tau)) + \overline{u}. \tag{4}$$

By shifting the equilibrium point $\hat{y}$ of (4) to the origin, we obtain

$$\dot{\tilde{y}}(t) = -\overline{D}\tilde{y}(t-\delta) + \overline{A}\,\tilde{f}(\tilde{y}(t)) + \overline{B}\,\tilde{g}(\tilde{y}(t-\tau)), \tag{5}$$

where $\tilde{y}(t) = y(t) - \hat{y}$, $\tilde{f}(\tilde{y}(t)) = \overline{f}(\tilde{y}(t)+\hat{y}) - \overline{f}(\hat{y})$, $\tilde{g}(\tilde{y}(t)) = \overline{g}(\tilde{y}(t)+\hat{y}) - \overline{g}(\hat{y})$.

Remark 1. Systems (1) and (5) are equivalent. This means that any property of system (5), will also be true for system (1). Thus, the asymptotic stability of the origin of system (5) will imply the asymptotic stability of the equilibrium point of (1).

We will also need the following lemma:

Lemma 1. ([13]) *For any vector function $y : [a, b] \to \mathbb{R}^{8N}$ and any positive definite matrix $M \in \mathbb{R}^{8N\times 8N}$, the following linear matrix inequality (LMI) holds:*

$$\left(\int_a^b y(s)ds\right)^T M \left(\int_a^b y(s)ds\right) \leq (b-a)\int_a^b y^T(s)My(s)ds,$$

where the integrals are well defined.

3 Main Result

We give an LMI-based sufficient condition for the asymptotic stability of the origin of (5).

Theorem 1. *If Assumption 1 holds, then the origin of system (5) is globally asymptotically stable if there exist positive definite matrices P_1, P_2, P_3, P_4, P_5^1, P_5^2, P_5^3, P_6, positive definite block-diagonal matrices R_1, R_2, R_3, R_4, all from $\mathbb{R}^{8N\times 8N}$, such that the following linear matrix inequality (LMI) holds*

$$(\Pi)_{8\times 8} < 0, \tag{6}$$

where $\Pi_{1,1} = -\overline{D}P_2 - P_2\overline{D} + P_3 + \delta P_4 + P_5^1 - \tau^{-1}P_6 + \overline{L_f}^T R_1\overline{L_f} + \overline{L_g}^T R_3\overline{L_g}$, $\Pi_{1,2} = -P_1\overline{D}$, $\Pi_{1,3} = \tau^{-1}P_6$, $\Pi_{1,4} = P_1\overline{A} + P_2\overline{A}$, $\Pi_{1,7} = P_1\overline{B} + P_2\overline{B}$, $\Pi_{1,8} = \overline{D}P_2\overline{D}$, $\Pi_{2,2} = -P_3 + \tau\overline{D}P_6\overline{D}$, $\Pi_{2,4} = -\tau\overline{D}P_6\overline{A}$, $\Pi_{2,7} = -\tau\overline{D}P_6\overline{B}$, $\Pi_{3,3} = -P_5^1 - \tau^{-1}P_6 + \overline{L_f}^T R_2\overline{L_f} + \overline{L_g}^T R_4\overline{L_g}$, $\Pi_{4,4} = P_5^2 + \tau\overline{A}^T P_6\overline{A} - R_1$, $\Pi_{4,7} = \tau\overline{A}^T P_6\overline{B}$, $\Pi_{4,8} = -\overline{A}^T P_2\overline{D}$, $\Pi_{5,5} = -P_5^2 - R_2$, $\Pi_{6,6} = P_5^3 - R_3$, $\Pi_{7,7} = -P_5^3 + \tau\overline{B}^T P_6\overline{B} - R_4$, $\Pi_{7,8} = -\overline{B}^T P_2\overline{D}$, $\Pi_{8,8} = -\delta^{-1}P_4$.

Proof. We define the following Lyapunov-Krasovskii functional

$$V(t) = V_1(t) + V_2(t) + V_3(t) + V_4(t) + V_5(t) + V_6(t),$$

where

$$V_1(t) = \tilde{y}^T(t)P_1\tilde{y}(t),$$

$$V_2(t) = \left(\tilde{y}(t) - \overline{D}\int_{t-\delta}^t \tilde{y}(s)ds\right)^T P_2\left(\tilde{y}(t) - \overline{D}\int_{t-\delta}^t \tilde{y}(s)ds\right),$$

$$V_3(t) = \int_{t-\delta}^t \tilde{y}^T(s)P_3\tilde{y}(s)ds,$$

$$V_4(t) = \int_{-\delta}^0\int_{t+\theta}^t \tilde{y}^T(s)P_4\tilde{y}(s)dsd\theta,$$

$$V_5(t) = \int_{t-\tau}^t \xi^T(s)P_5\xi(s)ds, \quad P_5 = \text{diag}(P_5^1, P_5^2, P_5^3),$$

$$\xi(s) = \left[\tilde{y}^T(s)\ \tilde{f}^T(\tilde{y}(s))\ \tilde{g}^T(\tilde{y}(s))\right]^T,$$

$$V_6(t) = \int_{-\tau}^{0}\int_{t+\theta}^{t} \dot{\tilde{y}}^T(s)P_6\dot{\tilde{y}}(s)dsd\theta.$$

The derivative of $V(t)$ along the trajectories of system (5) is

$$\dot{V}(t) = \dot{V}_1(t) + \dot{V}_2(t) + \dot{V}_3(t) + \dot{V}_4(t) + \dot{V}_5(t) + \dot{V}_6(t),$$

where

$$\dot{V}_1(t) = \dot{\tilde{y}}^T(t)P_1\tilde{y}(t) + \tilde{y}^T(t)P_1\dot{\tilde{y}}(t), \tag{7}$$

$$\begin{aligned}\dot{V}_2(t) &= \left(\dot{\tilde{y}}(t) - \overline{D}\tilde{y}(t) + \overline{D}\tilde{y}(t-\delta)\right)^T P_2\left(\tilde{y}(t) - \overline{D}\int_{t-\delta}^{t}\tilde{y}(s)ds\right)\\ &+ \left(\tilde{y}(t) - \overline{D}\int_{t-\delta}^{t}\tilde{y}(s)ds\right)^T P_2\left(\dot{\tilde{y}}(t) - \overline{D}\tilde{y}(t) + \overline{D}\tilde{y}(t-\delta)\right),\end{aligned} \tag{8}$$

$$\dot{V}_3(t) = \tilde{y}^T(t)P_3\tilde{y}(t) - \tilde{y}^T(t-\delta)P_3\tilde{y}(t-\delta), \tag{9}$$

$$\begin{aligned}\dot{V}_4(t) &= \delta\tilde{y}^T(t)P_4\tilde{y}(t) - \int_{t-\delta}^{t}\tilde{y}^T(s)P_4\tilde{y}(s)ds\\ &\le \delta\tilde{y}^T(t)P_4\tilde{y}(t) - \delta^{-1}\left(\int_{t-\delta}^{t}\tilde{y}(s)ds\right)^T P_4\left(\int_{t-\delta}^{t}\tilde{y}(s)ds\right),\end{aligned} \tag{10}$$

$$\dot{V}_5(t) = \xi^T(t)P_5\xi(t) - \xi^T(t-\tau)P_5\xi(t-\tau), \tag{11}$$

$$\begin{aligned}\dot{V}_6(t) &= \tau\dot{\tilde{y}}^T(t)P_6\dot{\tilde{y}}(t) - \int_{t-\tau}^{t}\dot{\tilde{y}}^T(s)P_6\dot{\tilde{y}}(s)ds\\ &\le \tau\dot{\tilde{y}}^T(t)P_6\dot{\tilde{y}}(t) - \tau^{-1}\left(\int_{t-\tau}^{t}\dot{\tilde{y}}(s)ds\right)^T P_6\left(\int_{t-\tau}^{t}\dot{\tilde{y}}(s)ds\right),\end{aligned} \tag{12}$$

where we have used Lemma 1 to obtain the inequalities in (10) and (12).

From Assumption 1 about the Lipschitz condition, we can deduce that

$$||f_j(x) - f_j(x')|| \le l_j^f||x - x'||$$

$$\Leftrightarrow ||\text{vec}(f_j(x)) - \text{vec}(f_j(x'))|| \le l_j^f||\text{vec}(x) - \text{vec}(x')||,$$

$\forall j \in \{1, \ldots N\}$, $\forall x, x' \in \mathbb{O}$, and analogously for the functions g_j. Thus, there exist positive definite block-diagonal matrices $R_1 = \text{diag}\left(r_j^1 I_8\right)_{1\le j\le N}$, $R_2 = \text{diag}\left(r_j^2 I_8\right)_{1\le j\le N}$, $R_3 = \text{diag}\left(r_j^3 I_8\right)_{1\le j\le N}$, $R_4 = \text{diag}\left(r_j^4 I_8\right)_{1\le j\le N}$, such that

$$0 \le \tilde{y}^T(t)\overline{L_f}^T R_1\overline{L_f}\tilde{y}(t) - \tilde{f}^T(\tilde{y}(t))R_1\tilde{f}(\tilde{y}(t)), \tag{13}$$

$$0 \leq \tilde{y}^T(t-\tau)\overline{L_f}^T R_2 \overline{L_f}\tilde{y}(t-\tau) - \tilde{f}^T(\tilde{y}(t-\tau))R_2\tilde{f}(\tilde{y}(t-\tau)), \quad (14)$$

$$0 \leq \tilde{y}^T(t)\overline{L_g}^T R_3 \overline{L_g}\tilde{y}(t) - \tilde{g}^T(\tilde{y}(t))R_3\tilde{g}(\tilde{y}(t)), \quad (15)$$

$$0 \leq \tilde{y}^T(t-\tau)\overline{L_g}^T R_4 \overline{L_g}\tilde{y}(t-\tau) - \tilde{g}^T(\tilde{y}(t-\tau))R_4\tilde{g}(\tilde{y}(t-\tau)). \quad (16)$$

Now, adding inequalities (13)–(16) to (7)–(12), we obtain

$$\dot{V}(t) \leq \zeta^T(t)\Pi\zeta(t),$$

where Π is defined by (6), and

$$\zeta(t) = \Big[\tilde{y}^T(t)\ \tilde{y}^T(t-\delta)\ \tilde{y}^T(t-\tau)\ \tilde{f}^T(\tilde{y}(t))\ \tilde{f}^T(\tilde{y}(t-\tau))\ \tilde{g}^T(\tilde{y}(t))\ \tilde{g}^T(\tilde{y}(t-\tau)) \\ \Big(\textstyle\int_{t-\delta}^{t}\tilde{y}(s)ds\Big)^T\Big]^T.$$

From (6) we have that $\Pi < 0$, which implies that $\dot{V}(t) < 0$, meaning that $V(t)$ is strictly decreasing for $t \geq 0$. It can be further deduced from the definition of $V(t)$ that

$$\lambda_{\min}(P_1)||\tilde{y}(t)||^2 \leq \tilde{y}^T(t)P_1\tilde{y}(t) \leq V(t) \leq V_0,\ \forall t \geq T,\ T \geq 0,$$

where $V_0 = \max\limits_{0 \leq t \leq T} V(t)$. Thus,

$$||\tilde{y}(t)||^2 \leq \frac{V_0}{\lambda_{\min}(P_1)} \Leftrightarrow ||\tilde{y}(t)|| \leq M,\ \forall t \geq 0,$$

where $M = \sqrt{\frac{V_0}{\lambda_{\min}(P_1)}}$. The above inequality proves the asymptotic stability of the origin of system (5), completing the proof of the theorem.

4 Numerical Example

To assess the effectiveness of the main result, we give a numerical example.

Example. Consider the following two-neuron delayed octonion-valued Hopfield neural network with leakage delay:

$$\begin{cases} \dot{x}_1(t) = & -d_1x_1(t-\delta) + \sum_{j=1}^{2} a_{1j}f_j(x_j(t)) + \sum_{j=1}^{2} b_{1j}g_j(x_j(t-\tau)) + u_1, \\ \dot{x}_2(t) = & -d_2x_2(t-\delta) + \sum_{j=1}^{2} a_{2j}f_j(x_j(t)) + \sum_{j=1}^{2} b_{2j}g_j(x_j(t-\tau)) + u_2, \end{cases} \quad (17)$$

where $d_1 = 30$, $d_2 = 20$, and

$$\mathrm{vec}(a_{11}) = (1,1,2,2,1,-1,-1,1)^T,\ \mathrm{vec}(a_{12}) = (2,1,1,-2,2,1,-2,2)^T,$$

$$\mathrm{vec}(a_{21}) = (2,-2,2,1,2,-2,1,2)^T,\ \mathrm{vec}(a_{22}) = (1,2,2,-2,1,1,2,-2)^T,$$

$$\mathrm{vec}(b_{11}) = (2,1,2,1,-2,2,-1,2)^T,\ \mathrm{vec}(b_{12}) = (-2,2,-2,2,1,2,-2,2)^T,$$

$$\mathrm{vec}(b_{21}) = (1,-2,2,-2,1,2,2,2)^T,\ \mathrm{vec}(b_{22}) = (1,2,2,1,2,-2,-2,1)^T,$$
$$\mathrm{vec}(u_1) = (10,-20,30,-40,50,-70,80,-90)^T,$$
$$\mathrm{vec}(u_2) = (90,-40,10,-60,30,-80,50,-20)^T,$$
$$f_j\left([x]_p\right) = \frac{1}{1+e^{-[x]_p}},\ g_j\left([x]_p\right) = \frac{1-e^{-[x]_p}}{1+e^{-[x]_p}},\ p \in \{0,1,\dots,7\},\ j \in \{1,2\},$$

from where we can deduce that $l_1^f = l_2^f = \frac{\sqrt{2}}{2}$ and $l_1^g = l_2^g = \sqrt{2}$. The leakage delay is $\delta = 0.02$ and the time delay is $\tau = 0.7$.

Applying Theorem 1, if condition (6) is satisfied, then the equilibrium point of system (17) is asymptotically stable. The LMI condition can be solved to yield the matrices $R_1 = \mathrm{diag}(0.0107I_8, 0.0080I_8)$, $R_2 = \mathrm{diag}(0.0005I_8, 0.0003I_8)$, $R_3 = \mathrm{diag}(0.0034I_8, 0.0026I_8)$, $R_4 = \mathrm{diag}(0.0023I_8, 0.0019I_8)$. (The values of the other matrices are not given due to space limitations.) The state trajectories of the elements of octonions x_1 and x_2 are given in Fig. 1, for four initial values. In the figure, it can be seen that the 8 elements of the octonions x_1 and x_2, respectively, converge to the equilibrium point $(\hat{x}_1, \hat{x}_2)$ of (17), given by

$$\mathrm{vec}(\hat{x}_1) = (0.7190, -0.3986, 1.3953, -1.4788, 1.6174, -1.9163, 2.5857, -2.9574)^T,$$

$$\mathrm{vec}(\hat{x}_2) = (4.2318, -1.9018, 1.2301, -3.0213, 1.5983, -4.0516, 2.9698, -1.0117)^T.$$

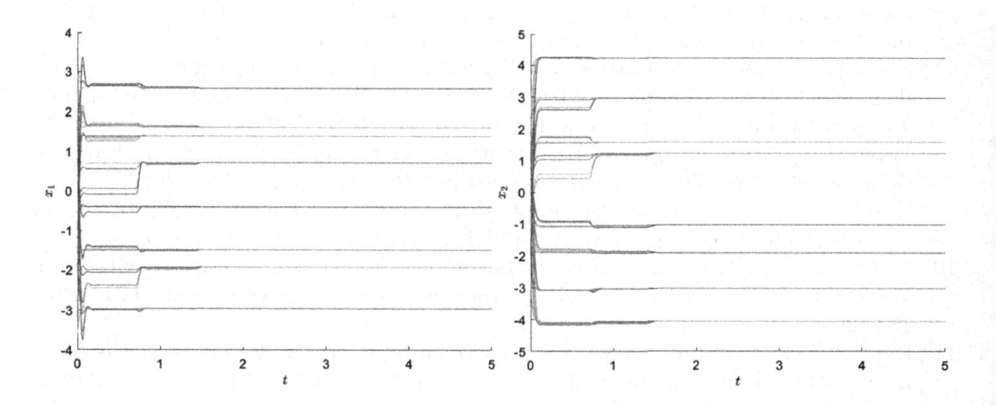

Fig. 1. State trajectories of the elements of x_1 and x_2 in the Example

5 Conclusions

A sufficient criterion, expressed in the form of a linear matrix inequality, for the asymptotic stability of the equilibrium point of delayed octonion-valued neural networks with leakage delay was derived, by making the assumption that the octonion-valued activation functions of such networks satisfy the Lipschitz condition. A numerical example was provided to assess the effectiveness of the proposed criterion.

Octonion-valued neural networks offer room for further development, especially due to the fact that the octonion algebra is the only normed division algebra that can be defined over the reals, except for the complex and quaternion algebras, which it generalizes. They can be used as an alternative to networks defined on Clifford algebras of dimension 8, because they don't fall into the Clifford-valued neural network category.

References

1. Widrow, B., McCool, J., Ball, M.: The complex LMS algorithm. Proc. IEEE **63**(4), 719–720 (1975)
2. Hirose, A.: Complex-Valued Neural Networks. SCI, vol. 400. Springer, Heidelberg (2012). doi:10.1007/978-3-642-27632-3
3. Arena, P., Fortuna, L., Occhipinti, L., Xibilia, M.: Neural networks for quaternion-valued function approximation. In: International Symposium on Circuits and Systems (ISCAS), vol. 6, pp. 307–310. IEEE (1994)
4. Pearson, J., Bisset, D.: Back propagation in a Clifford algebra. In: International Conference on Artificial Neural Networks, vol. 2, pp. 413–416 (1992)
5. Popa, C.-A.: Octonion-valued neural networks. In: Villa, A.E.P., Masulli, P., Pons Rivero, A.J. (eds.) ICANN 2016. LNCS, vol. 9886, pp. 435–443. Springer, Cham (2016). doi:10.1007/978-3-319-44778-0_51
6. Popa, C.-A.: Exponential stability for delayed octonion-valued recurrent neural networks. In: Rojas, I., Joya, G., Catala, A. (eds.) IWANN 2017. LNCS, vol. 10305, pp. 375–385. Springer, Cham (2017). doi:10.1007/978-3-319-59153-7_33
7. Popa, C.-A.: Global asymptotic stability for octonion-valued neural networks with delay. In: Cong, F., Leung, A., Wei, Q. (eds.) ISNN 2017. LNCS, vol. 10261, pp. 439–448. Springer, Cham (2017). doi:10.1007/978-3-319-59072-1_52
8. Popa, C.A.: Octonion-valued bidirectional associative memories. In: 2017 International Joint Conference on Neural Networks (IJCNN), pp. 783–787 (2017)
9. Hopfield, J.J.: Neural networks and physical systems with emergent collective computational abilities. Proc. Natl. Acad. Sci. U.S.A. **79**(8), 2554–2558 (1982)
10. Song, Q., Zhao, Z.: Stability criterion of complex-valued neural networks with both leakage delay and time-varying delays on time scales. Neurocomputing **171**, 179–184 (2016)
11. Liu, Y., Zhang, D., Lu, J., Cao, J.: Global μ-stability criteria for quaternion-valued neural networks with unbounded time-varying delays. Inf. Sci. **360**, 273–288 (2016). doi:10.1016/j.ins.2016.04.033
12. Liu, Y., Xu, P., Lu, J., Liang, J.: Global stability of Clifford-valued recurrent neural networks with time delays. Nonlinear Dyn. **84**(2), 767–777 (2016)
13. Gu, K.: An integral inequality in the stability problem of time-delay systems. In: Proceedings of the 39th IEEE Conference on Decision and Control, pp. 2805–2810 (2000)

Training the Hopfield Neural Network for Classification Using a STDP-Like Rule

Xiaolin Hu[1(✉)] and Tao Wang[2]

[1] Tsinghua National Laboratory for Information Science and Technology (TNLIST), Department of Computer Science and Technology, Center for Brain-Inspired Computing Research (CBICR), Tsinghua University, Beijing, China
xlhu@tsinghua.edu.cn
[2] Huawei Technology, Beijing, China

Abstract. The backpropagation algorithm has played a critical role in training deep neural networks. Many studies suggest that the brain may implement a similar algorithm. But most of them require symmetric weights between neurons, which makes the models less biologically plausible. Inspired by some recent works by Bengio et al., we show that the well-known Hopfield neural network (HNN) can be trained in a biologically plausible way. The network can take hierarchical architectures and the weights between neurons are not necessarily symmetric. The network runs in two alternating phases. The weight change is proportional to the firing rate of the presynaptic neuron and the state (or membrane potential) change of the postsynaptic neuron between the two phases, which approximates a classical spike-timing-dependent-plasticity (STDP) rule. Several HNNs with one or two hidden layers are trained on the MNIST dataset and all of them converge to low training errors. These results further push our understanding of the brain mechanism for supervised learning.

Keywords: Hopfield neural network · Spike-timing-dependent-plasticity · Backpropagation

1 Introduction

In recent years, supervised deep learning models equipped with the backpropagation (BP) algorithm have achieved remarkable success in many applications including image classification, object detection and speech recognition [10]. But it remains unclear how the brain performs supervised learning, especially in the hierarchical sensory systems such as the visual system and the auditory system.

In a series of papers [2,4,15], Bengio et al. argue that the brain may implement a similar algorithm to BP for supervised learning. A biologically plausible leaky integrator network is proposed for this purpose. The learning process is split into two phases. First, only the input is given and the network relaxes to a fixed point, which is called "negative phase". Second, the corresponding teaching signal appears and drives the network to evolve, which is called "positive phase".

D. Liu et al. (Eds.): ICONIP 2017, Part III, LNCS 10636, pp. 737–744, 2017.
https://doi.org/10.1007/978-3-319-70090-8_74

It is shown that in the beginning of the positive phase the neurons' states change in the direction of reducing the prediction error. A learning rule for the weights is proposed, which depends on the activities of presynaptic and postsynaptic neuron pairs in the two phases. This rule can be viewed as a firing rate version of the spike-timing-dependent-plasticity (STDP) rule [5], which depicts a basic form of synapse plasticity between real spiking neurons. With a slight modification the rule becomes a contrastive Hebbian learning (CHL) rule [1,14], which is able to train the model to recognize handwritten digits [15].

A limitation of the learning algorithm in Ref. [15], as well as in earlier works [14,16], is that it requires the weights between neurons to be symmetric, which is unrealistic in a biological system. A possible explanation for the discrepancy between the model and the brain is that a neuron in the model does not correspond to a single real neuron, but a population of real neurons, and the weight between two model neurons correspond to average synaptic strength between two populations of real neurons. However, if a learning algorithm could be devised without this restriction, it would push us closer to a more biologically detailed model for the brain.

Note that weight symmetry is also required in standard BP algorithm. In fact this algorithm requires symmetric feedforward and feedback weights between two layers, which is known as the "weight transport problem" [7]. This is one of the main reasons for BP being long criticized by neuroscientists about its biological plausibility [3,13]. However, two recent works [11,12] empirically showed that weight symmetry is unnecessary to train multi-layer Perceptrons. Good performance could be achieved by using random weights in backward computation. This result motivates the question whether weight symmetry is really needed in those more biologically plausible learning algorithms [14–16]. In the paper, we present some empirical results about the learning principle proposed by Bengio et al. [2,4,15] by removing the constraint of weight symmetry. The algorithm was successfully used to train the well-known Hopfield neural network (HNN) with asymmetric weights. Finally, we stress that the aim of this paper is not to provide a better training algorithm for deep learning, but to provide some insights for bridging deep learning and neuroscience.

2 Hopfield Neural Network

Suppose that each neuron i has a state $s_i(t)$ and its firing rate is delineated by a nonlinear activation function $f(s_i(t))$. Consider the following additive neural network

$$\tau \frac{ds_i}{dt} = -s_i + \sum_j w_{ij} f(s_j) + I_i \tag{1}$$

or in the vector form

$$\tau \frac{d\mathbf{s}}{dt} = -\mathbf{s} + \mathbf{W}\mathbf{f}(\mathbf{s}) + \mathbf{I} \tag{2}$$

where τ stands for a time constant, w_{ij} stands for the connection weight from the presynaptic neuron j to the postsynaptic neuron i, and I_i stands for the external input (from other brain areas) to neuron i. If the weight matrix $\mathbf{W}$ is symmetric, and sigmoid functions are used as activation functions, this model is called Hopfield neural network [8]. In this case, an energy function $E(\mathbf{s}(t))$ can be constructed which always decreases along time as the dynamic equation (2) evolves [8]. In this sense, the dynamic equation is said to be stable. But in this paper, we generalize the definition of HNN to any model governed by the dynamic equation (2) without imposing any restrictions on $\mathbf{W}$ and $f(\cdot)$. The results about the stability of HNN are abundant in the literature (e.g., [6,9]), but most are sufficient conditions only.

We consider a hierarchical architecture, which means that the weight matrix $\mathbf{W}$ is a sparse matrix when the model is deep. In Fig. 1, a model with two hidden layer and one output layer is depicted. It is assumed that the input neurons v_i are fully observed (or clamped), therefore other neurons cannot influence their states. In this case, only the hidden neurons and output neurons dynamically change their states according to (2) with

$$\mathbf{s} = \begin{pmatrix} \mathbf{h1} \\ \mathbf{h2} \\ \mathbf{y} \end{pmatrix}, \mathbf{W} = \begin{pmatrix} \mathbf{W}_{h1h1} & \mathbf{W}_{h1h2} & \mathbf{0} \\ \mathbf{W}_{h2h1} & \mathbf{W}_{h2h2} & \mathbf{W}_{h2y} \\ \mathbf{0} & \mathbf{W}_{yh2} & \mathbf{W}_{yy} \end{pmatrix}, \mathbf{I} = \begin{pmatrix} \mathbf{W}_{h1v}\mathbf{v} \\ \mathbf{0} \\ \mathbf{0} \end{pmatrix}, \tag{3}$$

where the input $\mathbf{v}$ is applied to the first hidden layer neurons via the weights $\mathbf{W}_{h1v}$. In supervised training setting, the output neurons accept teaching signals (e.g., class labels for classification) as external input, which will be discussed later.

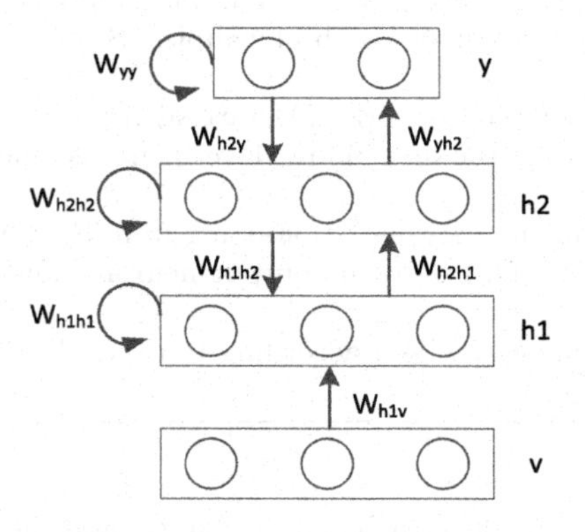

Fig. 1. An HNN with one input layer $\mathbf{v}$, two hidden layers $\mathbf{h}1$ and $\mathbf{h}2$ and one output layer $\mathbf{y}$.

3 Supervised Learning

In this section we present supervised learning scheme for HNN, which is inspired by Refs. [2,4,15]. The basic procedure is as follows. First, let the network (2) relax to its equilibrium point with an observation $\mathbf{v}$. This is called "negative phase" and denote the states by $\mathbf{s}^-$. Then inject a signal $-\beta(\mathbf{y}-\mathbf{d})$ to the output neurons in the dynamic Eq. (2), where $\beta > 0$ is a small constant and $\mathbf{d}$ is the teaching signal. The network evolves in reaction to the new input. This is called "positive phase". After a short period of evolution, record the new state as $\mathbf{s}^+$ and update the weights according to

$$w_{ij} \leftarrow w_{ij} + \eta \Delta w_{ij} \tag{4}$$

where

$$\Delta w_{ij} = (s_i^+ - s_i^-) f(s_j^+), \tag{5}$$

and $\eta > 0$ is the learning rate. According to this learning rule the synaptic change Δw_{ij} is proportional to the internal state change of the postsynpatic neuron i and the firing rate of the presynpatic neuron j. Note that if the presynaptic neuron denotes an input x_j, then $f(s_j^+)$ in (5) is changed to x_j. This rule approximates a typical type of STDP in biological systems [4].

The above two phases alternate with each other. Algorithm 1 presents the batch mode training details.

Algorithm 1. Training the Hopfield neural network.

Require: The current batch of samples, $(\mathbf{V}_n, \mathbf{D}_n) = \{(\mathbf{v}, \mathbf{d})\}^{batchsize}$; The weights learned on the previous batch, $\mathbf{W}_{n-1}$; The final state of the network (2) in the negative phase for the previous batch of samples, $\mathbf{S}_{n-1}^-$

Ensure:

1: *(Negative phase)* Run the network (2) starting from $\mathbf{S}_{n-1}^-$ till convergence with input $\mathbf{W}_{h1v,n-1}\mathbf{V}_n$ onto the first hidden layer neurons, and obtain final states $\mathbf{S}_n^-$;

2: *(Positive phase)* Run the network (2) starting from $\mathbf{S}_n^-$ for a few steps with additional input $-\beta(\mathbf{Y}_n - \mathbf{D}_n)$ onto the output neurons, and obtain final states $\mathbf{S}_n^+$;

3: Calculate new weight matrix $\mathbf{W}_n$ according to (4) and (5);

4: **return** $\mathbf{W}_n$ and $\mathbf{S}_n^-$;

Note that in step 1 of Algorithm 1, we have assumed that the input $\mathbf{v}$ is applied to the first hidden layer $\mathbf{h}_1$, which is the case in Fig. 1. But the input can be applied to any layer with skip-layer connections.

A critical condition for the success of the algorithm refers to the stability of the network during the negative phase, otherwise all analysis in what follows would be invalid. One possible solution is to restrict the weight matrix somewhere in its space which ensures the stability according to certain theoretical results (e.g. [6,9]) by a rectification step after step 3 in Algorithm 1. However, such an operation would be too artificial if we could not find biological evidence. Fortunately, in practice we found that with proper initialization and learning rate, instability could be avoided.

3.1 Implementation of the Model

Without loss of generality, the time constant τ in (2) is set to 1. Then the differential Eq. (2) can be approximated by the difference equation

$$\mathbf{s(t+1)} = \mathbf{s(t)} + \epsilon \mathbf{R}(\mathbf{s}(t)) \tag{6}$$

where $\mathbf{R}(\mathbf{s})$ stands for the right hand side of (2) and ϵ stands for the step size. Same as in [15], a modified version of the above equation is used

$$\mathbf{s(t+1)} = \mathbf{f}(\mathbf{s(t)} + \epsilon \mathbf{R}(\mathbf{s}(t))). \tag{7}$$

Empirical results suggested that this slight modification made learning more robust.

Simulating difference equations is time consuming. For saving computation, the final states of the negative phase for each input sample are stored, which are then used as the initial states for the same sample in the next epoch [15]. See Algorithm 1.

As claimed before, the stability of HNN is critical for the success of learning. The initial weights are chosen such that the network is stable at first. If the linear rectifier activation $f(x) = \max\{x, 0\}$ or the piecewise linear activation function $f(x) = \min\{\max\{x, 0\}, 1\}$ is used, according to Corollary 2 in Ref. [6], symmetric $\mathbf{W}$ and positive definiteness of $\mathbf{\Lambda} - \mathbf{W}$ ensures globally asymptotically stability of the network. This is easily achieved by setting small initial values for the weight matrix. To maintain stability, keeping the change of the magnitude of the weights small is important. Therefore small learning rates were used.

4 Experiments

We designed several hierarchical networks with one or two hidden layers for classifying the handwritten digits in the MNIST dataset. Each hidden layer had 256 neurons. The input layer and output layer had 784 and 10 neurons, respectively, since the images are of the size 28-by-28 and there are 10 categories. There are 60,000 training images and 10,000 test images.

4.1 Parameter Setting

For the negative phase, 20 steps were used for simulating the difference Eq. (7), and for the positive phase 4 steps were used. The step size $\epsilon = 0.5$. Consistent with [15], the results were insensitive to the step size if only it was between 0 and 1. The constant β in the algorithm was 1. The weight decay coefficient was

0.005. The weights were randomly initialized according to a normal distribution with zero mean and standard deviation 0.01. Batch size was set to 100. We found that an appropriate learning rate setting was important for the success of training. Inappropriate learning rates would lead to explosion of weights or slow convergence. Inspired by [15], in lower layers the learning rates were set larger and in higher layers, the learning rates were set smaller. See below for detailed settings. All results presented in the paper were obtained by using the activation function $f(x) = \min\{\max\{x, 0\}, 1\}$.

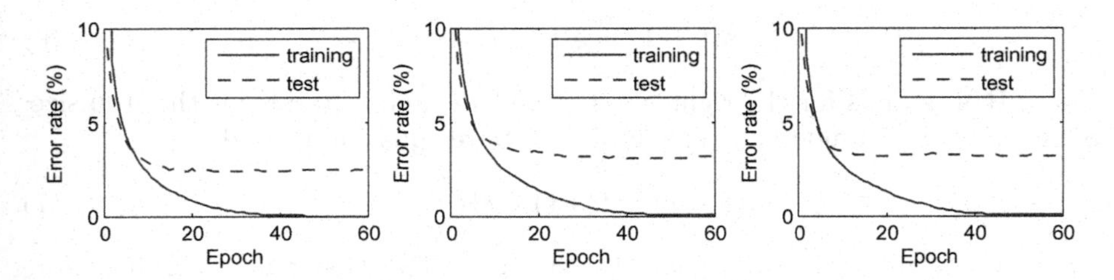

Fig. 2. Error curves of HNNs with a single hidden layer. *Left:* The baseline model. *Middle:* A network with recurrent connections between hidden neurons. *Right:* A network with symmetric weights between neurons

4.2 Results

We first designed a network with one hidden layer, where only feedforward and feedback connections between adjacent layers were present. The learning rates were 0.002 for $\mathbf{W}_{hv}$ and 0.001 for $\mathbf{W}_{yh}$ and $\mathbf{W}_{hy}$. Figure 2 (*left*) shows the error curves. After 45 epochs the training error reached exactly zero. This is the baseline setting.

Unlike the algorithm in [16] which is specifically designed to train a hierarchical network without intra-layer recurrent connections, our algorithm can be used to train any type of connections including intra-layer recurrent connections and skip-layer connections. To demonstrate this, we added recurrent connections in the hidden layer $\mathbf{W}_{hh}$. Self-connections are excluded, i.e., the diagonal elements of $\mathbf{W}_{hh}$ were all zeros. The learning rate for these weights was 0.001, and other parameters were the same as in the baseline setting. Figure 2 *middle* shows the error curves of the learning algorithm. After 60 epochs, the training error rate reached 0.02%.

Next we imposed a constraint of symmetric weights on the baseline network and trained the network with the learning rule (7). The learning rates were 0.002 for $\mathbf{W}_{hv}$ and 0.0005 for $\mathbf{W}_{yh}$ and $\mathbf{W}_{hy}$. The network achieved 0.01% training error rate within 60 epochs (Fig. 2, *right*).

We also tested the proposed learning algorithm on two HNNs with two hidden layers. The first one had only feedforward and feedback connections between adjacent layers. The learning rates for $\mathbf{W}_{h1v}$, $\mathbf{W}_{h2h1}$, $\mathbf{W}_{h1h2}$, $\mathbf{W}_{yh2}$ and $\mathbf{W}_{h2y}$

were 0.005, 0.001, 0.001, 0.0002 and 0.0002, respectively. The training error reached 1.28% in 200 epochs (Fig. 3, *left*). The second one was the same as the first one but had additional skip-layer connections between the first hidden layer and the output layer. The learning rates for the feedforward weights $\mathbf{W}_{yh1}$ and feedback weights $\mathbf{W}_{h1y}$ were both 0.0002. These additional connections helped the network to achieve lower training error, which was 0.03% after 100 epochs (Fig. 3, *right*).

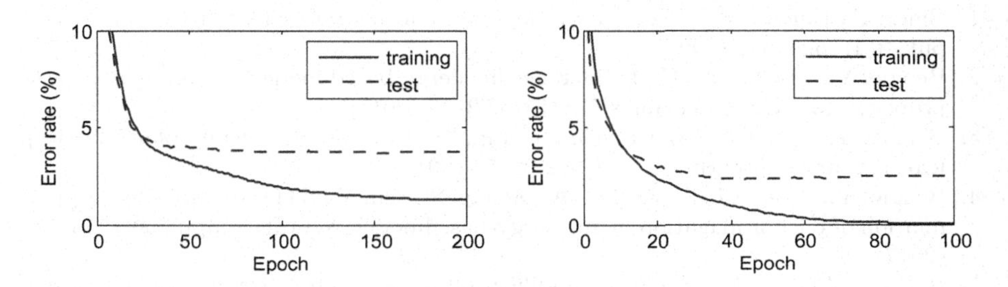

Fig. 3. Error curves of HNNs with two hidden layers, where the first model has only feedforward and feedback connections (*left*), and the second model has additional skip connections between the first hidden layer and the output layer (*right*).

5 Discussion

Understanding the brain mechanism for supervised learning is an old topic. Recently, Scellier et al. [15] showed that a biological plausible neural network equipped with CHL can achieve low training error on a pattern recognition task. But this learning algorithm requires symmetric weights between neurons which is incompatible with neurobiology. We have complemented this work by empirically showing that the well-known HNN can be trained using a STDP-like rule, which does not require symmetric weights between neurons.

The proposed approach has several limitations. First, the stability of HNN is not guaranteed during learning, though with proper initialization and small learning rates HNN was always stable in our experiments. But on the other hand, it is debatable that real neural circuits are theoretically guaranteed to be stable. Second, when the weight matrix is asymmetric, λ^* is not proportional to $(\mathbf{s}^- - \mathbf{s}^+)$. We could only show that their signs were largely congruent. This indicates that the proposed updating rule that depends on the difference of states between the two phases leads to "not so bad" path in searching for the optimal weights in the high dimensional space. But a rigorous proof about why this approximation can lead to near zero training loss is lacked. Third, we found that setting larger learning rates for lower layers was important for the convergence of the algorithm, but this makes the algorithm less biologically plausible. All of these are open questions.

Acknowledgments. This work was supported in part by the National Natural Science Foundation of China under Grant 91420201, Grant 61332007, Grant 61621136008 and Grant 61620106010, in part by the Beijing Municipal Science and Technology Commission under Grant Z161100000216126, and in part by Huawei Technology under Contract YB2015120018.

References

1. Baldi, P., Pineda, F.J.: Contrastive learning and neural oscillations. Neural Comput. **3**(4), 526–545 (1991)
2. Bengio, Y., Fischer, A.: Early inference in energy-based models approximates back-propagation. arXiv preprint arXiv:1510.02777 (2015)
3. Bengio, Y., Lee, D.H., Bornschein, J., Lin, Z.: Towards biologically plausible deep learning. arXiv preprint arXiv:1502.04156 (2015)
4. Bengio, Y., Mesnard, T., Fischer, A., Zhang, S., Wu, Y.: STDP-compatible approximation of backpropagation in an energy-basedmodel. Neural Comput. **29**, 555–577 (2017)
5. Bi, G.Q., Poo, M.M.: Synaptic modification by correlated activity: Hebb's postulate revisited. Annu. Rev. Neurosci. **24**(1), 139–166 (2001)
6. Forti, M., Tesi, A.: New conditions for global stability of neural networks with application to linear and quadratic programming problems. IEEE Trans. Circ. Syst. I: Fundam. Theory Appl. **42**(7), 354–366 (1995)
7. Grossberg, S.: Competitive learning: from interactive activation to adaptive resonance. Cogn. Sci. **11**(1), 23–63 (1987)
8. Hopfield, J.J.: Neurons with graded response have collective computational properties like those of two-state neurons. Proc. Natl. Acad. Sci. **81**(10), 3088–3092 (1984)
9. Hu, S., Wang, J.: Absolute exponential stability of a class of continuous-time recurrent neural networks. IEEE Trans. Neural Netw. **14**(1), 35–45 (2003)
10. LeCun, Y., Bengio, Y., Hinton, G.: Deep learning. Nature **521**(7553), 436–444 (2015)
11. Liao, Q., Leibo, J.Z., Poggio, T.: How important is weight symmetry in backpropagation? In: Proceedings of the Thirtieth AAAI Conference on Artificial Intelligence (AAAI), pp. 1837–1844 (2016)
12. Lillicrap, T.P., Cownden, D., Tweed, D.B., Akerman, C.J.: Random feedback weights support learning in deep neural networks. arXiv preprint arXiv:1411.0247 (2014)
13. Mazzoni, P., Andersen, R.A., Jordan, M.I.: A more biologically plausible learning rule for neural networks. Proc. Natl. Acad. Sci. **88**(10), 4433–4437 (1991)
14. Movellan, J.R.: Contrastive Hebbian learning in the continuous hopfield model. In: Connectionist Models: Proceedings of the 1990 Summer School, pp. 10–17 (1991)
15. Scellier, B., Bengio, Y.: Towards a biologically plausible backprop. arXiv preprint arXiv:1602.05179 (2016)
16. Xie, X., Seung, H.S.: Equivalence of backpropagation and contrastive Hebbian learning in a layered network. Neural Comput. **15**(2), 441–454 (2003)

Symbolic Solutions to Division by Zero Problem via Gradient Neurodynamics

Yunong Zhang[1,2,3,4(✉)], Huihui Gong[1,2,3,4], Jian Li[1,2,3,4], Huanchang Huang[1,2,3,4], and Ziyu Yin[1,2,3,4]

[1] School of Information Science and Technology, Sun Yat-sen University (SYSU), Guangzhou 510006, China
zhynong@mail.sysu.edu.cn
[2] Key Laboratory of Autonomous Systems and Networked Control, Ministry of Education, Guangzhou 510640, China
[3] SYSU-CMU Shunde International Joint Research Institute, Foshan 528300, China
[4] Key Laboratory of Machine Intelligence and Advanced Computing, Ministry of Education, Guangzhou 510006, China

Abstract. Division by zero (DBZ) problem, or say, division singularity problem, has perplexed scientists and engineers in many fields for centuries. How to solve DBZ problem has actually been discussed for more than 1200 years. Despondingly, plenty of efforts failed to solve DBZ problem effectively, and it is still considered as a formidable conundrum. This paper introduce an extension of the division operation from a time-varying perspective. Most problems in science and engineering fields are time-varying, and thus the extension is reasonable and practical. Furthermore, by employing the neat gradient-based neurodynamics (or say, gradient neurodynamics) equation different times, a series of different "symbolic" solutions are proposed. Note that the proposed symbolic solutions have the ability to conquer DBZ problem. The symbolic solutions to DBZ problem may promote the development of more complete singularity-conquering applications.

Keywords: Division by zero · Gradient neurodynamics · Singularity · Symbolic solutions

1 Introduction

People have been trying to solve division by zero (DBZ) problem for more than 1200 years [1,2]. Kaplan's book [1] presents an interesting outlook towards DBZ problem. In recent centuries, science and engineering have developed rapidly, and many predecessors have worked hard to solve DBZ problem. Howbeit their efforts have hitherto not paid off. Nowadays, DBZ problem remains to be an unimaginable, unfathomable and inconceivable problem. Analyzing the predecessors' work, the authors find that they spent too many endeavors on investigating DBZ problem in a time-invariant view and in a direct analytical manner.

D. Liu et al. (Eds.): ICONIP 2017, Part III, LNCS 10636, pp. 745–750, 2017.
https://doi.org/10.1007/978-3-319-70090-8_75

Contemporary studies [3–7] throw new light on DBZ problem from a time-varying perspective. In the last two decades, the authors have already studied the solutions of a rich repertoire of time-varying problems by neurodynamics approaches, such as time-varying reciprocal computation [4], time-varying linear system solving [5], time-varying matrix inversion [6] and time-varying inverse square root solving [7]. Note that the above problems involve the division operation, where the divisor is likely to be zero. Therefore, we can investigate DBZ problem from a time-varying perspective and in a numerical manner.

The rest of this paper is organized into three sections. Section 2 specifically presents the DBZ problem and presents one symbolic solution to DBZ problem. Section 3 demonstrates numerical experimental results of a representative example, which shows the efficacy and superiority of the symbolic solution. Besides, new more symbolic solutions to DBZ problem are proposed from a time-varying perspective in Sect. 4. Section 5 concludes this paper with final remarks.

2 DBZ Problem and Symbolic Solution

To specify DBZ problem, $n(t) \in \mathbb{R}$ is defined as the dividend (or say, numerator), $d(t) \in \mathbb{R}$ is the divisor (or say, denominator), and $q(t) \in \mathbb{R}$ is the quotient. Thus, the division operation with DBZ problem (i.e., standard division) can be depicted as

$$q(t) = \frac{n(t)}{d(t)}, \tag{1}$$

where $d(t)$ is sometimes equal to zero. Besides, $n(t)$, $d(t)$ and $q(t)$ are time-varying, including the time-invariant as a special case.

In order to find a symbolic solution to DBZ problem, we utilize gradient neurodynamics (GN) method [8–11] to overcome the singularity of DBZ. First of all, by applying neurodynamics method [12–15], the following function is transformed and obtained:

$$f(q(t), t) = q(t)d(t) - n(t) \to 0.$$

With GN method, we obtain the following energy function from the above function:

$$\xi_1(q(t), t) = \frac{(q(t)d(t) - n(t))^2}{2} \geq 0. \tag{2}$$

Then, we define the quotient's time derivative $\dot{q}(t)$ via the following GN design formula:

$$\dot{q}(t) = -\lambda_1 \frac{\partial \xi_1(q(t), t)}{\partial q(t)}, \tag{3}$$

where $\lambda_1 > 0$ is a design parameter and should be set as large as the hardware would permit, or appropriately large for simulative, experimental or practical purposes. Combining (2) and (3), we eventually get a symbolic solution in term of $\dot{q}(t)$ and $q(t)$ to DBZ problem:

$$\dot{q}(t) = -\lambda_1 d(t)(q(t)d(t) - n(t)). \tag{4}$$

Table 1. Numerical experimental results synthesized by symbolic solution (4) with increasingly larger λ_1 compared with standard division (1).

t	2	6	$9\pi/2$	22	24
$q^*(t)$ via (1)	−1.2358	−3.1425	$-\infty$	−11.7805	−12.0642
$q_{z1}(t)$ ($\lambda_1 = 1 \times 10^3$)	−1.2357	−3.1423	-294.3190	−11.7806	−12.0640
$q_{z2}(t)$ ($\lambda_1 = 5 \times 10^3$)	−1.2358	−3.1425	-569.4684	−11.7805	−12.0642
$q_{z3}(t)$ ($\lambda_1 = 1 \times 10^4$)	−1.2358	−3.1425	-755.6165	−11.7805	−12.0642
$q_{z4}(t)$ ($\lambda_1 = 5 \times 10^4$)	−1.2358	−3.1425	−1453.4050	−11.7805	−12.0642
$q_{z5}(t)$ ($\lambda_1 = 1 \times 10^5$)	−1.2358	−3.1425	−1924.5878	−11.7805	−12.0642

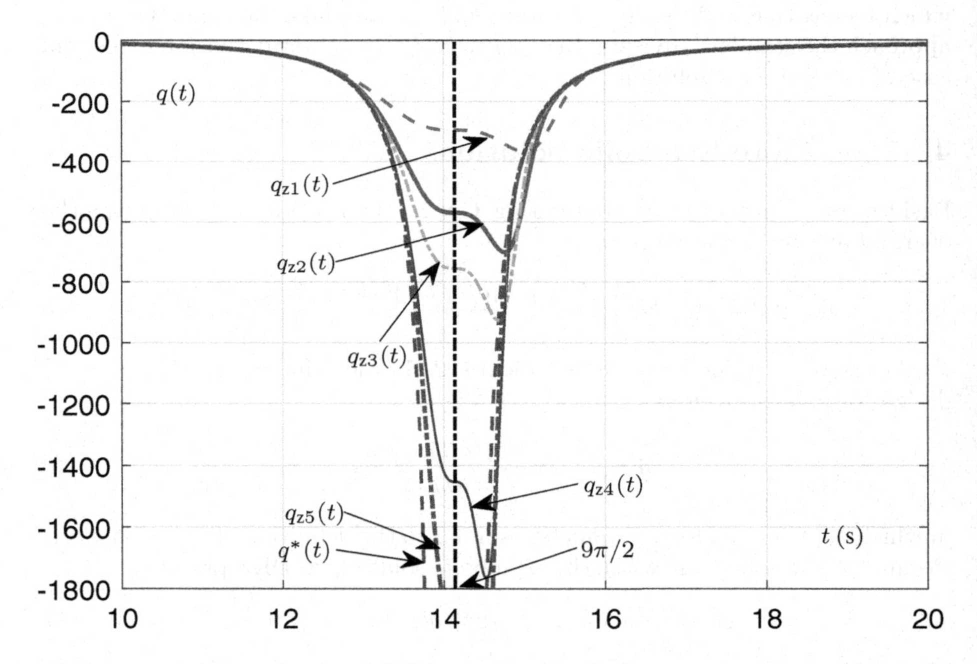

Fig. 1. Graphical results of symbolic solution (4) with increasingly larger λ_1 compared with standard division (1), where original DBZ singularity at time instant $t = 9\pi/2$ is avoided by (4).

That is, Eq. (4) is the symbolic solution to DBZ problem in the form of the 1st-order time derivative, which can primely solve DBZ problem by choosing appropriately large values of parameter λ_1 (though positive small-value λ_1 also works).

3 Numerical Experimental Results

For better illustration of symbolic solution (4), we conduct numerical experiments with an example

$$q(t) = \frac{-t}{(1 + \sin(t/3))},$$

which serves as a representative example with $-t$ being the dividend and $1 + \sin(t/3)$ being divisor. Table 1 and Fig. 1 intuitively show that, different from the standard division $q^*(t)$ at $t = 9\pi/2$ being infinite, the GN solutions q_{zi} (with $i = 1, 2, 3, 4, 5$) with five different values of λ_1 are all finite. Additionally, with the increase of λ_1 value, the symbolic solution achieves more exact precision, which means that sufficiently large values of λ_1 can make the symbolic solution approach the standard division. In other words, the standard division is the limit case of the symbolic solution.

4 New More Symbolic Solutions

Besides, more importantly, considering GN method again, we obtain another energy function as follows:

$$\xi_2(\dot{q}(t), q(t), t) = \frac{(\dot{q}(t) + \lambda_1 d(t)(q(t)d(t) - n(t)))^2}{2} \geq 0. \tag{5}$$

Afterwards, by defining the quotient's 2nd-order time derivative $\ddot{q}(t)$, GN design formula is given as

$$\ddot{q}(t) = -\lambda_2 \frac{\partial \xi_2(\dot{q}(t), q(t), t)}{\partial \dot{q}(t)}, \tag{6}$$

in which $\lambda_2 > 0$ is a design parameter (similar to the definition of λ_1). Combining (5) and (6), we obtain new another symbolic solution to DBZ problem:

$$\ddot{q}(t) = -\lambda_2(\dot{q}(t) + \lambda_1 d(t)(q(t)d(t) - n(t))). \tag{7}$$

That is, Eq. (7) is the symbolic solution to DBZ in the form of the 2nd-order derivative.

Likewise, defining the 3rd-order derivative of the quotient as $\dddot{q}(t)$, we utilize the energy function and GN method for three times, yielding

$$\xi_3(\ddot{q}(t), \dot{q}(t), q(t), t) = \frac{(\ddot{q}(t) + \lambda_2(\dot{q}(t) + \lambda_1 d(t)(q(t)d(t) - n(t))))^2}{2},$$

$$\dddot{q}(t) = -\lambda_3 \frac{\partial \xi_3(\ddot{q}(t), \dot{q}(t), q(t), t)}{\partial \ddot{q}(t)},$$

where, again, $\lambda_3 > 0$ is a design parameter. In the similar manner, we get another symbolic solution to DBZ problem:

$$\dddot{q}(t) = -\lambda_3(\ddot{q}(t) + \lambda_2(\dot{q}(t) + \lambda_1 d(t)(q(t)d(t) - n(t)))).$$

Moreover, repeatedly using GN method for n times, we obtain many other symbolic solutions to DBZ problem in the general form of

$$\begin{aligned} q^{(n)}(t) &= -\lambda_n(q^{(n-1)}(t) + \lambda_{n-1}(q^{(n-2)}(t) \\ &\quad + \lambda_{n-2}(\cdots + \lambda_2(\dot{q}(t) + \lambda_1 d(t)(q(t)d(t) - n(t)))\cdots))), \end{aligned}$$

in which $q^{(k)}$ with $k = 4, 5, 6, \cdots, n$ are the k-order time derivatives of $q(t)$, and $\lambda_k > 0$ with $k = 1, 2, 3, \cdots, n$ are the design parameters.

With the aid of GN method and its extensions, we (including interested readers) may even further propose many new more symbolic solutions to DBZ problem by obtaining and utilizing the 1st-order time derivative, the 2nd-order time derivative, the 3rd-order time derivative, $\cdots$, or the nth-order time derivative differently, being challenging future works.

5 Conclusion

In this paper, we have introduced an extension of the division operation from a time-varying perspective as most problems in science and engineering fields are time-varying. Furthermore, by employing the neat gradient neurodynamics equation different times, a series of different "symbolic" solutions have been proposed. Note that the proposed symbolic solutions have the ability to conquer DBZ problem. Some numerical experiments have been conducted to show the effectiveness to solve time-varying division with DBZ problem conquered.

At the end of this paper, it is worth summarizing its main contribution as follows: from a time-varying perspective and in an incremental manner, a series of different symbolic solutions in terms of different-order derivatives have been proposed in this paper to solve DBZ problem, while previous works such as [12–16] only provided one (or say, one kind of) specific symbolic solution in terms of the first-order derivative to solve DBZ problem.

Acknowledgments. The work is supported by the National Natural Science Foundation of China (with number 61473323), by the Foundation of Key Laboratory of Autonomous Systems and Networked Control, Ministry of Education, China (with number 2013A07), and also by the Laboratory Open Fund of Sun Yat-sen University (with number 20160209). Kindly note that all authors of the paper are jointly of the first authorship, with the following thoughts shared: (1) "By my understanding, science, including mathematics, is tree-like, web-like, city-like, $\cdots$, growing", (2) "For a scientifically new thing, there are many levels of contributions: the 1st one is creating; the 2nd one is proving; the 3rd one is applying; $\cdots$; the nth one is knowing; $\cdots$", (3) "It is better (and may be dangerously better) that a Ph.D. dissertation researches everything about something, absolutely new; it is better that a Master thesis researches some things about something, new; it is at least that a Bachelor thesis researches something about something, relatively new", (4) "Be a good person for oneself; be a good son, a good husband and a good father (or be a good daughter, a good wife and a good mother) for family; $\cdots$; be a good labor and a good boss for workunit; $\cdots$", (5) "You are what you have been thinking and doing on the basis of ground", and (6) "This ends, and that starts". Thanks a lot and best regards.

References

1. Kaplan, R.: The Nothing That Is: A Natural History of Zero. Oxford University Press, New York (2000)
2. Speijer, R.: Defining numbers in terms of their divisors. Nature **461**(7260), 37 (2009)
3. Zhang, Y., Yi, C.: Zhang Neural Networks and Neural-Dynamic Method. Nova Science Publishers, New York (2011)
4. Zhang, Y., Li, F., Yang, Y., Li, Z.: Different Zhang functions leading to different Zhang-dynamics models illustrated via time-varying reciprocal solving. Appl. Math. Model. **36**(9), 4502–4511 (2012)
5. Zhang, Y., Chen, Z., Chen, K.: Convergence properties analysis of gradient neural network for solving online linear equations. Acta Automatica Sinica **35**(8), 1136–1139 (2009)
6. Zhang, Y., Li, Z., Li, K.: Complex-valued Zhang neural network for online complex-valued time-varying matrix inversion. Appl. Math. Comput. **217**(24), 10066–10073 (2011)
7. Zhang, Y., Li, Z., Guo, D., Li, W., Chen, P.: Z-type and G-type models for time-varying inverse square root (TVISR) solving. Soft Comput. **17**(11), 2021–2032 (2013)
8. Carmesin, H.O.: Multilinear back-propagation convergence theorem. Phys. Lett. A **188**(1), 27–31 (1994)
9. Cai, B., Zhang, Y.: Equivalence of velocity-level and acceleration-level redundancy-resolution of manipulators. Phys. Lett. A **373**(38), 3450–3453 (2009)
10. Hajieghrary, H., Hsieh, M.A., Schwartz, I.B.: Multi-agent search for source localization in a turbulent medium. Phys. Lett. A **380**(20), 1698–1705 (2016)
11. Wang, G., Zhang, W., Lu, J., Zhao, H.: Dispersion and optical gradient force from high-order mode coupling between two hyperbolic metamaterial waveguides. Phys. Lett. A **380**(35), 2774–2780 (2016)
12. Zhang, Y., Zhang, Y., Chen, D., Xiao, Z., Yan, X.: Division by zero, pseudo-division by zero, Zhang dynamics method and Zhang-gradient method about control singularity conquering. Int. J. Syst. Sci. **48**(1), 1–12 (2017)
13. Zhang, Y., Xiao, Z., Guo, D., Mao, M.: Singularity-conquering tracking control of a class of chaotic systems using Zhang-gradient dynamics. IET Control Theor. Appl. **9**(6), 871–881 (2015)
14. Zhang, Y., Yu, X., Yin, Y., Peng, C., Fan, Z.: Singularity-conquering ZG controllers of z2g1 type for tracking control of the IPC system. Int. J. Control **87**(9), 1729–1746 (2014)
15. Zhang, Y., Yu, X., Yin, Y., Xiao, L., Fan, Z.: Using GD to conquer the singularity problem of conventional controller for output tracking of nonlinear system of a class. Phys. Lett. A **377**(25–27), 1611–1614 (2013)
16. Zhang, Y., Qiu, B., Ling, Y., Yang, Z., Peng, C.: What is 1/0 in the general sense of physics, applied computation, and/or electronics? In: The 34th Chinese Control Conference, pp. 1105–1110. IEEE (2015)

An Image Enhancement Algorithm Based on Fractional-Order Relaxation Oscillator

Xiaoran Lin, Shangbo Zhou(✉), Hongbin Tang, and Ying Qi

College of Computer Science, Chongqing University, Chongqing, China
shbzhou@cqu.edu.cn

Abstract. In this paper, a cortex rhythms mimicking in fractional-order Relaxation oscillator is implemented and the existence of the rhythm is proved. Furthermore, the Quasi Gamma Curve (QGC) model is established based on the fraction-order Relaxation oscillator in the rhythm oscillation and we certify that the property of QGC model is similar to that of Gamma Curve by curve fitting methods. The proposed model is utilized to enhance the low contrast images. Different quantity measures demonstrate that the proposed model is effective.

Keywords: Fractional-order · Relaxation oscillator · Image enhancement

1 Introduction

Nervous systems are information network made up of large number of neurons. In recent years, some scholars have considered to model visual systems [1, 2]. Simulating the mechanism of visual selection and shift is an important challenge for computational neuroscience. These works have been done by many researchers [3–5]. Chaotic neural networks are adopted to model scene segmentation. This category of algorithms mainly takes advantage of the principle, which the neurons representing the same object are synchronized and neurons representing different objects are desynchronized [6]. In Ref [7], a form of the Retinex algorithm, which is used for color image enchantment, is implement by the Wilson–Cowan equations representing large-scale activity in interacting neural populations.

The aforementioned systems are network composed by many neurons. There is few mention on a single fractional-order oscillator used to enhance image. An integer order Wilson–Cowan oscillator is used to enhance gray image [8]. In this paper we will construct a model which can used for image enchantment by single fractional order Relaxation oscillator. Because the fractional-order systems exhibit dynamic phenomena more complex than integer order ones and fractional-order systems can explain some unique properties which are unable to complete by integer order system [9]. We propose a model based on fractional-order Relaxation oscillator. The oscillation of biological systems was called physiological rhythms which are ubiquitous in living organisms [10]. Rhythms have also been found in brain activity, olfactory and visual system [11, 12]. The existence of the limit cycle ensures that the dynamic systems generate rhythm oscillation phenomenon. We propose a Quasi Gamma Curve

D. Liu et al. (Eds.): ICONIP 2017, Part III, LNCS 10636, pp. 751–759, 2017.
https://doi.org/10.1007/978-3-319-70090-8_76

(QGC) model based on the fractional-order relaxation system which is in limit cycle oscillation state.

The paper is organized as follows. Section 2 gives a brief review of the theories on fractional derivative calculation method and Gamma correction. In Sect. 3, the rhythm in the fractional and integer-order Relaxation oscillator is presented and the QGC model is established. The QGC model is used to enhance underexpose images in Sect. 4. The conclusions are drawn in Sect. 5.

2 Preliminary

2.1 Fractional Derivative and Its Approximation

The fractional-order integro-differential operator is an extending concept of inter-order operator. In this paper, we solve the fractional-order Relaxation system by the Grunwald-Letnikov (GL) definition [13, 14]. The definition of GL can be described as

$$ {}_aD_t^{q_1}f(t) = \lim_{h\to 0} h^{-q_1} \sum_{k=0}^{\left[\frac{t-a}{h}\right]} (-1)^k \binom{q_1}{k} f(t-kh), \quad (1) $$

where $\binom{q_1}{k} = \frac{q_1(q_1-1)\cdots(q_1-k+1)}{k!}$.

This formula can be reduced as

$$ {}_0D_t^{q_1}y(t_m) = h^{-q_1} \sum_{k=0}^{m} (-1)^k \varphi_k^{(q_1)} y_{m-k}, \quad (2) $$

where $\varphi_k^{(q_1)} = (-1)^k \binom{q_1}{k}, k = 0, 1, 2\cdots, \binom{q_1}{k} = \frac{q_1(q_1-1)\cdots(q_1-k+1)}{k!}$, q_1, h is the fractional-order of the systems and the time step, respectively.

2.2 Theory About Gamma Correction Function

Gamma correction function [15] is a simple but general function for image enhancement. It is defined by the following power law formula: $I_{out} = Q * I_{in}^{\alpha}$, where input I_{in} is raised to the power α and the constant Q determined by the maximum pixel intensity in the input image. Thus, Gamma correction function can be used to improve the image contrast by the smoothing curve. Intuitively, the function can be used to enhance underexposure images when $\alpha < 1$. The function with $\alpha > 1$ is exploited to improve overexposed images. It has no effect on images when $\alpha = 1$.

3 Model

3.1 Fractional-Order Relaxation Oscillator

Relaxation oscillator [16] has been studied in many fields, such as biology and engineering mechanics. But there are few studies on the fractional-order Relaxation oscillator. The fractional-order model defined by (3), is composed of an excitatory subpopulation whose activity is described by x and an inhibitory subpopulation whose activity is described by y. A relaxation oscillator l is typically defined by a reciprocally connection which are constituted by pair of excitatory unit x_l and inhibitory unit y_l.

$$\begin{cases} \frac{d^{q_1}x}{dt^{q_1}} = 3x_l - x_l^3 + 2 - y_l + I_l \\ \frac{d^{q_2}y}{dt^{q_2}} = \varepsilon\left(\sigma\left(1 + \tanh\left(\frac{x_l}{\rho}\right)\right) - y_l\right) \end{cases}, \quad (3)$$

where σ, ε and ρ are parameters. In addition, ρ determines the steep degree of the sigmoid function and ε is a positive number and determines the time scale for the evolution of y_l. I_l denotes the external stimulation to the oscillator. The parameters q_1 and q_2 are derivative orders. Due to space limitation, we only take the case of $q_1 = q_2 = q = 0.8$ and 1 to exhibit the dynamic characteristics about Relaxation oscillator. The other parameters σ, ρ and ε are set to be 4, 0.1, 0.02 respectively. The system's limit cycle caused by external stimulations I_l = 1, 0.5, 0.005 are presented as waveform diagram in Figs. 1(a), 2(a) and 3(a). The x-nullcline, y-nullcline and the limit cycles are shown in Figs. 1(b), 2(b) and 3(b). The x-nullcline and y-nullcline are cubic function, sigmoid function, respectively. If two nullclines intersect only at a point along the middle branch of the cubic, then the oscillator presents a stable limit cycle [17], such as being depicted in Figs. 1(b), 2(b) and 3(b). From Figs. 1(b), 2(b) and 3(b), we can see that the x-nullcline is decline as the out simulate I_l decreasing. When I_l = 0, the x-nullcline and y-nullcline has three intersection points. One of them is a fix point and cannot make any limit cycle. The rhythm phenomena cannot happen. Therefore, in this paper, the external stimulations $I_l = (I/255)$, (I = 2, 3, 4… 255, $I \in N$) are needed to be discussed.

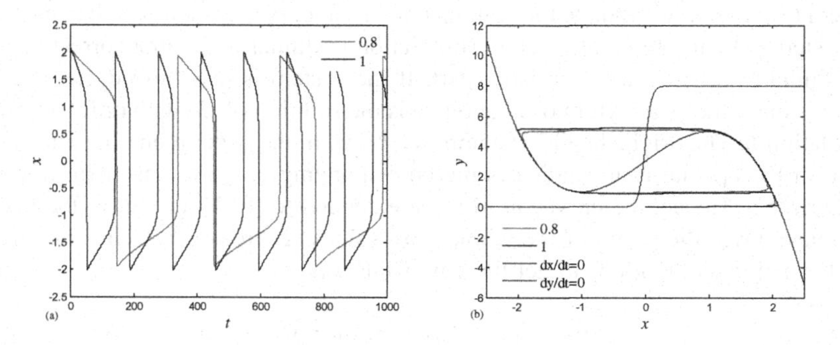

Fig. 1. System (3) with I_l = 1. (a) Waveform diagram, (b) limit cycles and nullclines.

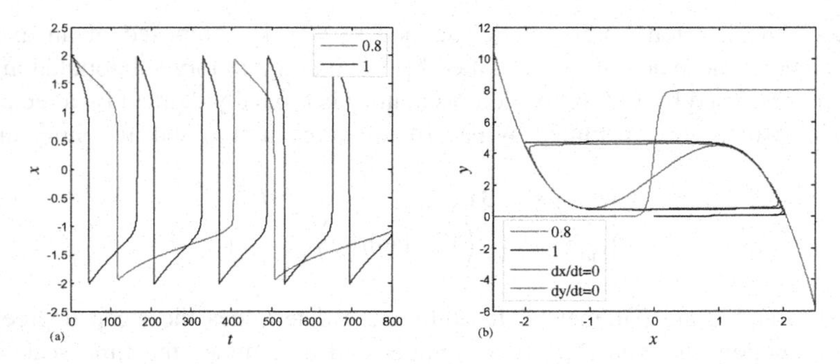

Fig. 2. System (3) with $I_l = 0.5$. (a) Waveform diagram, (b) limit cycles and nullclines.

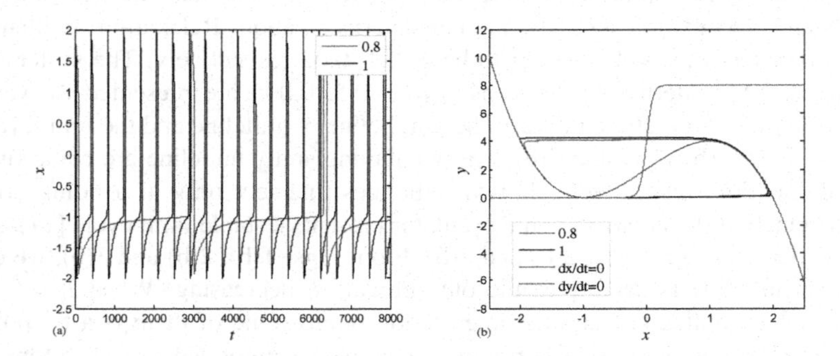

Fig. 3. System (3) with $I_l = 0.005$. (a) Waveform diagram, (b) limit cycles and nullclines.

3.2 The QGC in Fractional-Order Relaxation Oscillator

System (3) is used to simulate the gamma correction curve. The output characteristic of the system (3) with different external stimulation is similar to gamma correction curve. The similarity is proved in the latter part. If the system is in a limit cycle oscillation which is the same as a rhythm oscillation, x_l is the solution of (3) with different external stimulation I_l. The original QGC is composed by x_l. In the experiment, the interval of x_l is not in [0, 1]. The result indicates the corresponding curve is unsuitable for image enhancement. Hence, the curves need to be extended to [0, 1] and make the pixels in the image cover the scope of the whole gray values. Therefore, we construct a linear transformation as (4) for x_l to get the new value x_{ln}.

$$x_{ln} = (x_l - x_{lmin})/(x_{lmax} - x_{lmin}), \tag{4}$$

where x_{lmin} and x_{lmax} are respectively the max and min value of the solutions to responded for different stimulations I_l. The original QGC transforms into the QGC by (4). The results which are simulate Gamma correction curves with $\alpha = 0.25$ and 0.44 by QGC are shown in Fig. 4. From Fig. 4, we can appreciate the QGC is similar to Gamma correction curve. They are convex curves, which can expand the intermediate grayscale range. The QGC, with the characteristic of enlargement the contrast of the output and input value, is similar to Gamma curve with $\alpha < 0$.

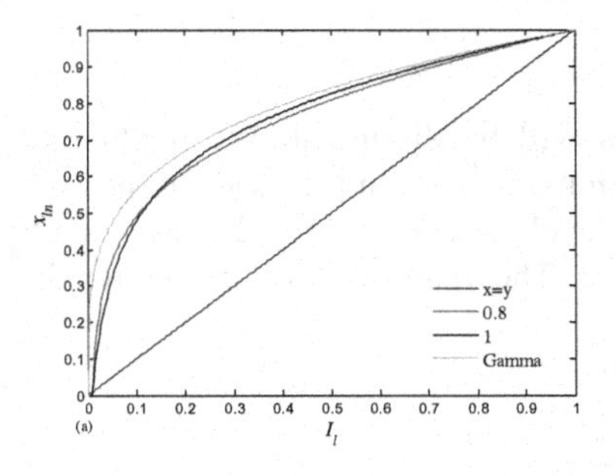

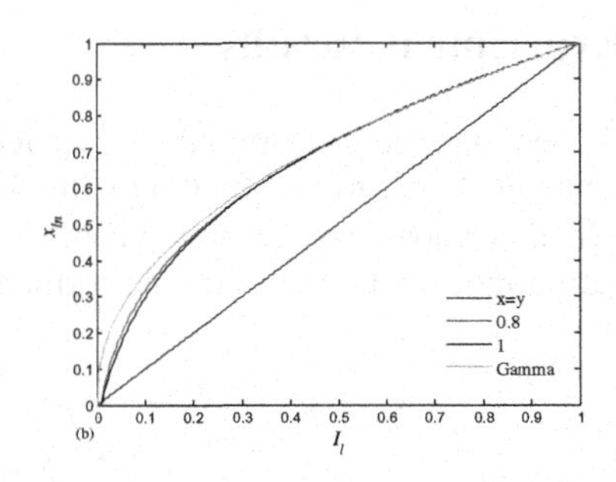

Fig. 4. The QGCs and the Gamma correction curve (a) the QGCs simulating $\alpha = 0.25$ (b) the QGCs simulating $\alpha = 0.44$.

From the Fig. 4, we can see that the QGC is an approximate Gamma curve. So the QGC almost matches the power law formula. To further demonstrate the effectiveness of QGC model for image enhancement, we certify the property of the QGC model is similar to that of Gamma Curve by curve fitting methods. We use the QGC simulation Gamma curve with different α shown in the following Table 1. The QGC model is $y = ax^b$.

Table 1. The fitting results of QGC model.

	$\alpha = 0.25$	$\alpha = 0.29$	$\alpha = 0.32$	$\alpha = 0.44$	$\alpha = 0.45$
Order b	0.3037	0.3152	0.3259	0.4607	0.4673
Coefficient a	0.9987	0.999	0.9994	1.004	1.004
SSE	0.00204	0.00188	0.00172	0.00436	0.00480
R-square	0.9998	0.9998	0.9998	0.9997	0.9997

From the Table 1, we can see that all of the coefficients a are approximate 1, and the orders b are approach to α. For every fitting curve, the sum of squares due to error (SSE) is small and the coefficient of determination (R-square) approximates to 1. The goodness of fit indexes are fairly good. The fitting results indicate that the QGC is approach to Gamma curve. As we all know, Gamma curve is effective for image enhancement. The QGC has been proved to approximate Gamma curve. So the QGC model is also effective for image enhancement.

Analogously, the Gamma correction curve with $\alpha > 1$, are effective for overexposed images and can be mimicked by the proposed system (3). Due to limited space, no more tautology here.

4 Experiment Results

In this section, the performance of QGCs will be discussed. We qualitatively and quantitatively compare our method with Gamma correction [15] and Retinex algorithm [18]. In this paper, we employ visibility level descriptor [19, 20] as our objective evaluation indexes to assess the performance. The visibility level descriptor described as (5)

$$\bar{r} = \exp\left(\frac{1}{N_r}\sum\nolimits_{r_i \in \wp_r} logr_i\right) \tag{5}$$

where N_r is the number of the set of visible edges in the enhanced image I_r; $\wp_r$ denotes the set of I_r; r_i is a ration of visibility level in the enhanced image to original image for every pixels. The $\bar{r}$ approximates the average visibility enhancement obtained by the image enhancement algorithm. Namely, it shows the contrast improvement quality of an enhanced image. Generally, a higher visibility level descriptor corresponds to more obvious details.

In the enhancing processes for different pixels, a pixel as an out stimulate, can get different enhanced value by system (3). All of $\boldsymbol{\alpha}$ used in the experiment can make the image get the best results by the Gamma curve method. Figures 5, 6, 7, 8 and 9 show low contrast images and enhancement results with proposed method and other comparison algorithms. As shown in the Figs. 5, 6, 7, 8 and 9, the enhancement effect of our proposed model is better than that of Gamma curve and Retinex. The improvement is also reflected by the objective measures shown in Table 2. In Table 2, the values use bold font for the best results. The results of the Retinex algorithm for every test image look unnatural due to over-enhancement. This is also obvious observed in the Figs. 5, 6, 7, 8 and 9. The visibility level descriptors got by our model, which is generated by fractional-order system, are the best. This means our model can make the low contrast image show more details.

Fig. 5. Results for image Housing. (a) Original image. Image processed by (b) QGC with q = 0.8, (c) QGC with $q = 1$, (d) Gamma curve with $\alpha = 0.25$, and (e) Retinex algorithm.

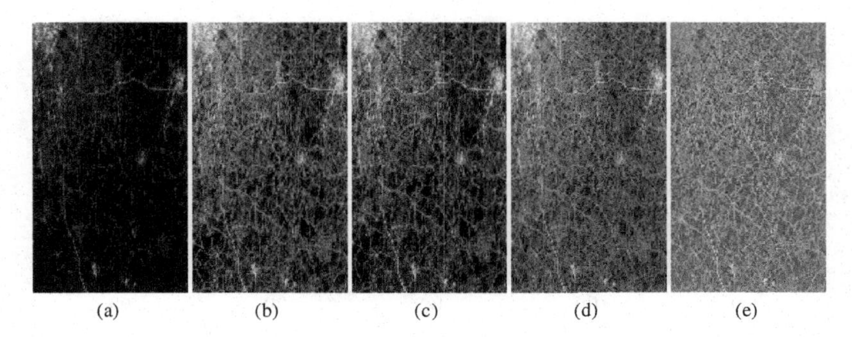

Fig. 6. Results for image *Ray*. (a) Original image. Image processed by (b) QGC with $q = 0.8$, (c) QGC with $q = 1$, (d) Gamma curve with $\alpha = 0.29$, and (e) Retinex algorithm.

Fig. 7. Results for image *Castle*. (a) Original image. Image processed by (b) QGC with $q = 0.8$, (c) QGC with $q = 1$, (d) Gamma curve with $\alpha = 0.32$, and (e) Retinex algorithm.

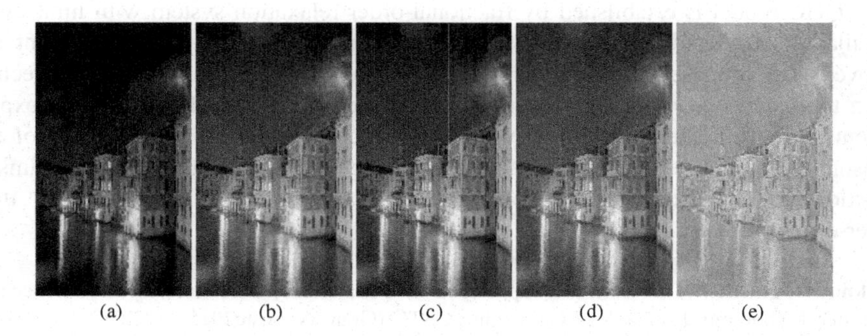

Fig. 8. Results for image *Thunder*. (a) Original image. Image processed by (b) QGC with $q = 0.8$, (c) QGC with $q = 1$, (d) Gamma curve with $\alpha = 0.44$, and (e) Retinex algorithm.

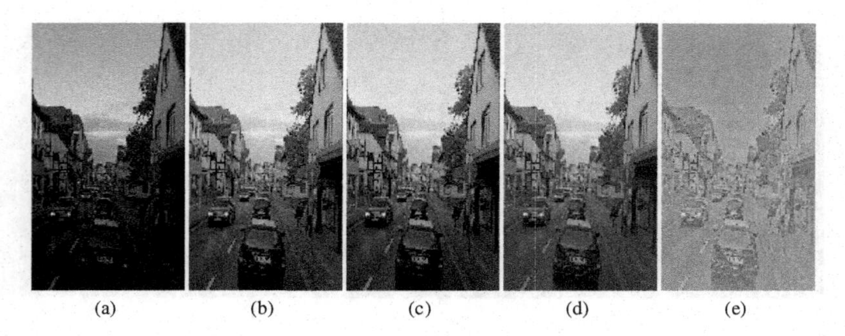
(a) (b) (c) (d) (e)

Fig. 9. Results for image *Car*. (a) Original image. Image processed by (b) QGC with $q = 0.8$, (c) QGC with $q = 1$, (d) Gamma curve with $\alpha = 0.45$, and (e) Retinex algorithm.

Table 2. Quantitative measurement results of visibility level descriptor.

Image	Methods				
	Orig.	$q = 0.8$	$q = 1$	Gam	Retinex
Housing	1.000	**7.0970**	6.0456	4.5337	4.6647
Ray	1.000	**4.9343**	4.6083	2.9733	3.3569
Castle	1.000	**5.8611**	4.8702	4.1358	4.7359
Thunder	1.000	**1.8341**	1.8295	1.4308	1.0218
Car	1.000	**2.6110**	2.5560	2.4336	2.3296
Average	1.000	**4.3939**	3.9044	3.0393	3.1613

5 Conclusions

In this paper, the rhythms in fractional-order relaxation oscillator are discussed. The QGC model is established by fractional-order relaxation system with limit cycle oscillation state and the effectiveness of QGC model for image enhancement are proved. The proposed model generated by fractional-order systems is more effective than that generated by integer-order systems, which is confirmed by abundant experiments. From another point of view, the existence of rhythms could be one of the reasons that human visual system can achieve image enhancement. And the rhythms in fraction-order Relaxation oscillator could be more objective in description than inter-order oscillator.

Acknowledgments. This work was supported by the Major Project of Fundamental Science and Frontier Technology Research of Chongqing CSTC (Grant No. cstc2015cyjBX0124).

References

1. Mcaulay, A.D.: Comparing artificial and biological dynamical neural networks. In: 15th International Conference on Signal Processing, Sensor Fusion, and Target Recognition XV, pp. 1–7 (2006)
2. Wagatsuma, N., von der Heydt, R., Niebur, E.: Modeling attention-induced reduction of spike synchrony in the visual cortex. In: Hirose, A., Ozawa, S., Doya, K., Ikeda, K., Lee, M., Liu, D. (eds.) ICONIP 2016. LNCS, vol. 9947, pp. 359–366. Springer, Cham (2016). doi:10.1007/978-3-319-46687-3_40
3. Qiao, Y., Liu, X., Miao, J., Duan, L.: A neural network model for visual selection and shifting. J. Integr. Neurosci. **15**, 321–335 (2016)
4. Qiao, Y., Meng, Y., Duan, L., Fang, F., Miao, J.: Qualitative analysis and application of locally coupled neural oscillator network. Neural Comput. Appl. **21**, 1551–1562 (2012)
5. Quiles, M., Wang, D., Zhal, L., Romero, R.: Selecting silent objects in real scenes: an oscillatory correlation model. Neural Netw. **24**, 54–64 (2011)
6. Zhao, L., Cupertino, T.H., Bertini Jr., J.R.: Chaotic synchronization in general network topology for scene segmentation. Neurocomputing **71**, 3360–3366 (2008)
7. Bertalmío, M., Cowan, J.D.: Implementing the Retinex algorithm with Wilson-Cowan equations. J. Physiol. Paris **103**, 69–72 (2009)
8. Wu, X., Li, Y., Kurths, J.: A new color image encryption scheme using CML and a fractional-order chaotic system. PLoS ONE **10**, e0119660 (2015)
9. Du, X., Yao, D.: An image enhancement method based on coupled Wilson-Cowan oscillators with double nodes. Acta Bioph. Sin. **28**(3), 242–253 (2012)
10. Zweifel, P.F.: From clocks to chaos: the rhythms of life. Phys. Today **42**, 72 (1989)
11. Buzsáki, G., Draguhn, A.: Neuronal oscillations in cortical networks. Science **304**, 1926–1929 (2004). (New York, N.Y.)
12. Eckhorn, R., Bauer, R., Jordan, W., Brosch, M., Kruse, W., Munk, M., Reitboeck, H.J.: Coherent oscillations: a mechanism of feature linking in the visual cortex? Biol. Cybern. **60**, 121–130 (1988)
13. Butzer, P.L., Westphal, U.: An introduction to fractional calculus. World Scientific, Singapore (2000)
14. Wang, D.: The time dimension for scene analysis. IEEE Trans. Neural Netw. **16**, 1401–1426 (2005)
15. Gonzalez, R.C., Woods, R.E.: Digital Image Processing, 2nd edn. Prentice Hall, Upper Saddle River (2002)
16. Van der Pol, B.: On "relaxation oscillation". Philos. Mag. **2**, 978–992 (1926)
17. Wang, D.: Relaxation oscillators and networks. Wiley Encycl. Electr. Electron. Eng. **18**, 396–405 (1999)
18. Jobson, D.J., Rahman, Z., Woodwell, G.A.: Properties and performance of center/surround Retinex. IEEE Trans. Image Process. **6**(3), 451–462 (1997)
19. Wang, S., Zheng, J., Hu, H.M., Li, B.: Naturalness preserved enhancement algorithm for non-uniform illumination images. IEEE Trans. Image Process. **22**, 3538–3548 (2013)
20. Hautière, N., Tarel, J.P., Aubert, D., Dumont, È.: Blind contrast enhancement assessment by gradient rationing at visible edges. Image Anal. Stereol. **27**, 87–95 (2008)

Bridging the Gap Between Probabilistic and Deterministic Models: A Simulation Study on a Variational Bayes Predictive Coding Recurrent Neural Network Model

Ahmadreza Ahmadi[1,2] and Jun Tani[1,2](✉)

[1] Department of Electrical Engineering, KAIST, Daejeon 305-701, Korea
ar.ahmadi62@gmail.com

[2] Okinawa Institute of Science and Technology, Okinawa 904-0495, Japan
tani1216jp@gmail.com

Abstract. The current paper proposes a novel variational Bayes predictive coding RNN model, which can learn to generate fluctuated temporal patterns from exemplars. The model learns to maximize the lower bound of the weighted sum of the regularization and reconstruction error terms. We examined how this weighting can affect development of different types of information processing while learning fluctuated temporal patterns. Simulation results show that strong weighting of the reconstruction term causes the development of deterministic chaos for imitating the randomness observed in target sequences, while strong weighting of the regularization term causes the development of stochastic dynamics imitating probabilistic processes observed in targets. Moreover, results indicate that the most generalized learning emerges between these two extremes. The paper concludes with implications in terms of the underlying neuronal mechanisms for autism spectrum disorder.

Keywords: Recurrent neural network · Variational Bayes · Predictive coding · Generative model

1 Introduction

Cognitive agents dealing with a changing environment need to develop internal models accounting for such fluctuations by extracting underlying structures through learning. Recently, many schemes have been proposed for learning fluctuated temporal patterns by extracting latent probabilistic structures. Those schemes include conventional dynamic Bayesian networks such as HMM and Kalman filters, and the recently developed variational Bayes recurrent neural network (RNN) models [2, 5]. At the same time, there have been alternative trials with a deterministic dynamics approach in which predictive models for probabilistically generated target sequences are learned by embedding extracted probabilistic structures into deterministic chaos self-organized in RNN models [10].

D. Liu et al. (Eds.): ICONIP 2017, Part III, LNCS 10636, pp. 760–769, 2017.
https://doi.org/10.1007/978-3-319-70090-8_77

Because these two approaches - one probabilistic and the other deterministic - have been developed relatively independently, research bridging the gap between them seems worthwhile.

In this direction, Murata et al. [8] developed a predictive coding-type stochastic RNN model inspired by the free energy minimization principle [4]. This model learns to predict the mean and variance of sensory input for each next time step of multiple perceptual sequences, mapping from its current latent state. The learning optimizes not only connectivity weights but also the latent state at the initial step for each training sequence by way of back-propagation through time [9]. Murata and colleagues experimented with the degree of the initial state dependency in learning to imitate probabilistic sequences, and examined how the internal dynamic structure develops differently in different cases. It turned out that initial state dependency arbitrated the development between deterministic and probabilistic dynamic structure. In the case of strong initial state dependency, deterministic chaos is dominantly developed, while stochastic dynamics develops by estimating larger variance in the case of weak initial state dependency.

However, the model in [8] suffers a considerable drawback. It cannot estimate variance for latent variables (the context units). This significantly constrains the capability of the model in learning latent probabilistic structures in target patterns. The current paper proposes a version of a variational Bayes RNN (VBRNN) model, which can estimate variance for each context unit at each time step (time varying variance). The proposed model is simpler than other VBRNN models [2,5] because it employs a predictive coding scheme rather than an autoencoder. Thus, the model is referred to as the variational Bayes predictive coding RNN (VBP-RNN). It assumes that learning aims to maximize the lower bound L [1], which is represented by the weighted sum of a negative regularization term for the posterior distribution of the latent variable and the likelihood term in the output generation.

The current study investigates how differently weighting these two terms in the summation during learning influences the development of different types of information processing in the model. To this end, we conducted a simulation experiment in which the VBP-RNN learned to predict/generate fluctuated temporal patterns containing probabilistic transitions between prototypical patterns. Consistent with Murata et al. [8], results showed that the different weighting arbitrates between two extremes at which the model develops either a deterministic dynamic structure or a probabilistic one. Analysis on simulation results clarifies how the degree of generalization in learning as well as the strength of the top-down intentionality in generating patterns changes from one extreme to another.

2 Model

This section introduces the lower bound equation. Then, the variational Bayes predictive coding RNN (VBP-RNN) is described.

2.1 Generative Model and the Lower Bound

A generative model can provide probabilistic prediction about fluctuating sensation. The joint probability of sensation x and latent variable z in a generative model can be written as a product of likelihood and a prior as:

$$P_\theta(x, z) = P_\theta(x|z)P(z) \tag{1}$$

where the likelihood $P_\theta(x|z)$ is parameterized by learning parameter θ. On the other hand, perception of x can be considered as a process of inferring posterior z as $P(z|x)$ which, however, becomes intractable when the likelihood is a nonlinear function of θ. Then, the problem is to maximize the joint probability $P_\theta(x, z)$ for a given sensory dataset $X = \{x_t\}_{t=1}^{N}$ by inferring both θ and the true posterior which is approximated using a recognition model $Q_\theta(z|x)$. In variational Bayes, it has been well known that this approximation by means of minimization of the KL-divergence between the model approximation Q and the true posterior P is equivalent to maximizing a value referred to as the lower bound [1]. The lower bound L to be maximized can be written as:

$$L = -KL(Q_\theta(Z|X)||P(Z)) + E_{Q_\theta(Z|X)}[log(P_\theta(X|Z)] \tag{2}$$

where the first term on the right hand side is the regularization term by which the posterior distribution of the latent variable is constrained to be similar to its prior, usually taken as a unit Gaussian distribution. The second term minimizes the reconstruction error.

2.2 Variational Bayes Predictive Coding RNN Models

Here, we describe the implementation of the aforementioned formulation in a continuous-time RNN (CTRNN) model as well as in a multiple timescale RNN (MTRNN) model [11]. If we take $X^l = \{x_t^l\}_{t=1}^{N}$ to be the lth teaching sensory sequence pattern used for training the VBP-RNN model, the regularization term in the lower bound for the all teaching sequences L_z can be written as:

$$L_z = \sum_{l=1}^{L} -KL(Q_\theta(z_{1:T}^l|x_{1:T}^l)||P(z_{1:T}^l)) \tag{3}$$

where the posterior z_t^l is approximated by the recognition model Q_θ as a conditional probability with a given sensory sequence pattern. The prior $P(z_t^l)$ can be given, for example as a normal distribution. Next, the reconstruction error term L_x can be described as:

$$L_x = \sum_{l=1}^{L}[E_{Q_\theta(z_{1:T}^l|x_{1:T}^l)} \sum_{t=1}^{T}[log(P_\theta(x_t^l|z_t^l)]] \tag{4}$$

The total lower bound is obtained as a weighted sum of the regularization term and the reconstruction error term, with a weighting parameter W as shown in Eq. 5.

$$L = W.L_z + (1 - W)L_x \tag{5}$$

Finally, the objective of learning is to maximize the total lower bound by optimizing both learning parameter θ and z_1^l as the latent state at the initial step for each latent state sequence, given a specific value for W. Following the Kingma and Welling's reparameterization trick [7], a random value is sampled from a standard normal distribution at each time step, i.e. $\varepsilon_t^l \sim N(0,1)$ which is used to sample $z_{1:T}^l$ from Q. In the current RNN implementation, the latent state is represented as the ensemble of the internal state values of all context units at each step, as $z_{t,1:C}^l$. Then, the internal state of the ith context unit at time step t in the lth sequence can be computed with the estimation of time varying mean $\mu_{t,i}^l$ and variance $\sigma_{t,i}^l$ in the following way:

$$z_{t,i}^l = \mu_{t,i}^l + \sigma_{t,i}^l \varepsilon_{t,i}^l \tag{6}$$

$$\mu_{t,i}^l = (1 - \frac{1}{\tau_i}) z_{t-1,i}^l + \frac{1}{\tau_i} (\sum_j w_{ij}^{\mu c} c_{t-1,j}^l + \sum_k w_{ik}^{\mu x} x_{t-1,k}^l + b_i^{\mu}) \tag{7}$$

$$\sigma_{t,i}^l = \exp(0.5 \times \sum_j (w_{ij}^{\sigma c} c_{t-1,j}^l + b_i^{\sigma})) \tag{8}$$

where τ_i is time constant of the ith context unit. Although the VBP-CTRNN uses the same time constant value for all context units, the VBP-MTRNN uses different time constant values for different units. Learning parameters $w_{ij}^{\mu|\sigma}$ and $b_i^{\mu|\sigma}$ are connectivity weights and biases, respectively. $c_{t,i}^l$ is the activation of the ith context unit the value of which is computed as $c_{t,i}^l = tanh(z_{t,i}^l)$. The ith dimension of the prediction output is computed in the form of probabilistic distribution by using the SoftMax function with M elements. Eventually, the distribution is computed by mapping from the current context unit activation patterns at each time step. The probability of the ith dimension of the prediction output is computed with w_{ij}^{xc} learnable connectivity weights and b_i^x biases as:

$$x_{t,i}^l = \frac{exp(\sum_j (w_{ij}^{xc} c_{t,j}^l + b_i^x))}{\sum_{i=1}^M exp(\sum_j (w_{ij}^{xc} c_{t,j}^l + b_i^x))} \tag{9}$$

Figure 1 outlines the information flow in the VBP-MTRNN. The learning process starts with random initialization for all learning parameters and the initial latent state for each latent sequence. The lower bound L with a given W can be obtained for each epoch by computing the latent state sequences as well as the output sequences using Eqs. 6–9. L is maximized by optimizing the learning parameters and the initial latent state for each latent sequence by using the back-propagation through time (BPTT) algorithm [9]. With an optimal W at the convergence of training, we expected the model's posterior sequence to approximate the true one. This model is simple compared to other variational Bayes RNN models [2,5]. Those models are built from separate functions, of the decoder RNN and the encoder RNN. By optimizing the connectivity weights and the initial latent states, in the current model the same RNN computes both the prediction output sequences by means of its forward dynamics and the posterior of the latent state sequences by means of BPTT.

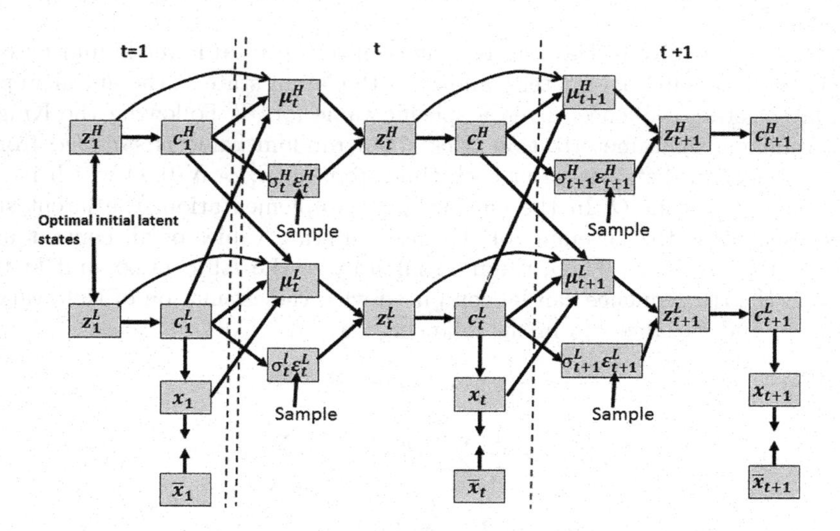

Fig. 1. The scheme of variational Bayes predictive MTRNN (VBP-MTRNN).

3 Simulation Experiments

We conducted simulation experiments to determine how learning in the proposed model depends on meta-prior W as well as on the number of training patterns. Figure 2 illustrates simulation and analysis performed in this study. First, a human generated patterns like ABB ABC ABC ABB ABC in which after AB, B is 30% probable C is 70%. These probability transitions are the same as those of a probabilistic finite state machine (pFSM) as shown in the bottom left of Fig. 2. Then, we trained two types of MTRNNs, a target generator MTRNN (Tar-Gen-MTRNN) and an output classifier MTRNN (Class-MTRNN) with those prototypical patterns. The Tar-Gen-MTRNN was prepared for autonomous generation of target temporal patterns (consisting of 100000 steps) which were then used for training the VBP-MTRNN. The Class-MTRNN was prepared for autonomous segmentation of temporal patterns into sequences of labels assigned to different prototypical patterns, which were used for the N-GRAM analysis. The patterns generated by the Tar-Gen-MTRNN were used as the target teaching patterns for the main experiment, training the VBP-MTRNN under different conditions. After training, the characteristics of output patterns generated by the VBP-MTRNN were quantitatively compared with those of the Tar-Gen-MTRNN. Using the trained Class-MTRNN, we computed the probabilistic distribution of N consecutive labels corresponding to different prototypical patterns classified from output patterns generated for 100000 steps by both the Tar-Gen-MTRNN and the VBP-MTRNN. Finally, the N-GRAM KL-divergence between these two probability distributions was computed in order to obtain a measure of similarity in the output generation between the Tar-Gen-MTRNN and the VBP-MTRNN trained in different conditions.

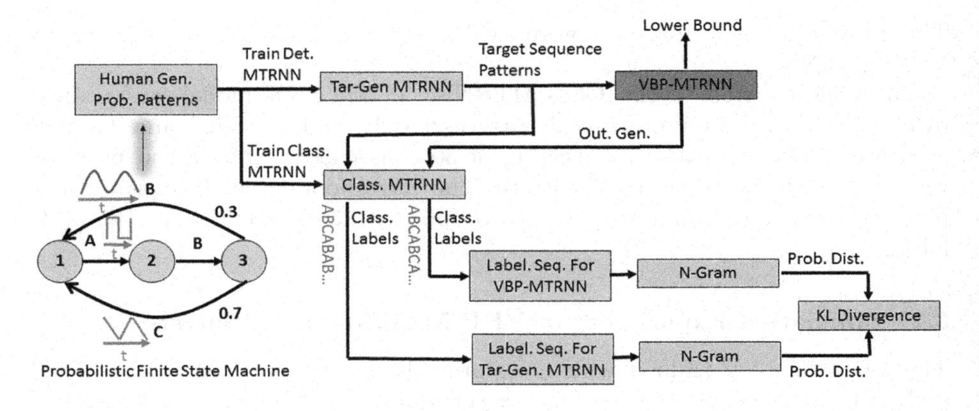

Fig. 2. Illustration of the simulation and analysis procedure.

All network models consisted of 7 context layers with (from lowest to highest layer) 121, 60, 30, 15, 10, 10, 10 context units, and all were trained for 100,000 epochs. The time constants of the 7 context layers were set to 2, 4, 8, 16, 32, 64, and 128 from lowest to highest layer. There were 121 input and output units in each network. All VBP-MTRNN models used a mini-batch size equal to 8, and the ADAM optimizer [6] was used to maximize the weighted lower bound.

3.1 Target Generator MTRNN and Classifier MTRNN

The Tar-Gen-MTRNN was prepared for autonomous generation of fluctuating target patterns of 100000 steps, which were used for the main experiment training the VBP-MTRNN. The 2-D temporal patterns for training the Tar-Gen-MTRNN were provided by a human generate pattern compositions from a set of different prototypical patterns using a tablet input device. The target sequence pattern was generated by concatenating 30 prototypical patterns. Each prototypical pattern is a different periodic pattern with 3 cycles, fluctuating in amplitude and periodicity at each appearance. The prototypical pattern "A" for example is generated 10 times in training patterns by a human. Because it is generated by a human on a tablet, each prototypical pattern is expressed with different amplitudes and periodicity in each trial. After training with these human expressed patterns, the Tar-Gen-MTRNN generated an output sequence pattern for 100,000 steps. This output sequence pattern was generated closed-loop (feeding next step inputs with current step prediction outputs) while adding Gaussian noise with zero mean and with constant σ of 0.1 into the internal state of each context unit at each time step. This was done in order to make the network output patterns stochastic while maintaining a certain probabilistic structure in transitions between the prototypical patterns. We sampled the sequence pattern generated from the 50,000th step to the 100,000th step. Then, two groups of target patterns were sampled, one consisting of 16 sequence patterns each with 400 step length and the other of 128 sequence patterns with the same step length.

These target sequence patterns were used in the main experiment, training the VBP-MTRNN.

In order to prepare the Class-MTRNN, an MTRNN model was trained to classify 3 different prototypical patterns (A, B, and C). The same human-generated sequence pattern consisting of 30 consecutive prototypical patterns was used as the teaching input pattern. The corresponding label sequence pattern was used as the target of SoftMax output with 3 elements for A, B, and C labels.

3.2 Simulation Experiment of VBP-MTRNN and Analysis

The VBP-MTRNN trained with meta-prior W set to 0.0, 0.01, 0.1, and 0.2 and with both 16 and 128 teaching target sequences that had been generated by the Tar-Gen-MTRNN. After training for 100,000 epochs under each condition, closed-loop output patterns were generated starting from all different initial latent states obtained for all target sequences after learning. Figure 3 compares one target sequence pattern and its corresponding closed-loop regeneration by the VBP-MTRNN trained with 16 target sequences. The first row shows the target pattern and the second, the third and the fourth rows show regenerated patterns with W set to 0.0, 0.1, and 0.2, respectively. Each pattern is associated with a sequence of labels classified by the Class-MTRNN. When $W = 0.0$, the target sequence pattern was completely regenerated for all steps. When $W = 0.1$, target and the regenerated patterns begin to significantly diverge at around 170 steps. Local deviation from each prototypical pattern arose soon after onset. When $W = 0.2$, the divergence starts earlier. These observations suggest that the VBP-MTRNN trained with $W = 0.0$ develops deterministic dynamics. Additional analysis on the output sequence generated for 100,000 steps revealed that there was no periodicity, suggesting that deterministic chaos or transient chaos developed in this learning condition. On the other hand, when W was increased to larger values during training, the model developed probabilistic processing in which more randomness was generated internally.

In order to quantify differences in network characteristics when trained under different meta-prior W settings and different numbers of training sequence patterns, the average divergence step (ADS) and the N-GRAM KL-divergence between the Tar-Gen-MTRNN and the VBP-MTRNN were computed for each condition. The ADS was computed for all target sequences by taking the average step at which the target sequence pattern and the regenerated one diverged. Starting from the initial step for both 16 and 128 target sequences, divergence was detected when the mean square error between the target and the generated pattern exceeded a threshold (0.025 in the current experiment). N-GRAM KL-divergence can be obtained as described previously. Setting N = 3, the Tri-gram was computed for the Tar-Gen-MTRNN. For this purpose, a sample label sequence was generated by feeding the Class-MTRNN with a sequence pattern generated for 100,000 steps by the Tar-Gen-MTRNN. In the same way, a Tri-gram for the VBP-MTRNN which had been trained with each different condition was computed by generating a sample sequence pattern with the same step length.

Table 1 shows the results of this analysis. As expected, for both 16 and 128 target sequences, ADS decreases as W increases.

Table 1. Average divergence step (ADS) and Tri-gram KL-divergence of Tar-Gen-MTRNN and VBP-MTRNN trained under various conditions.

	No. seq.	Weighting parameters			
		W = 0.0	W = 0.01	W = 0.1	W = 0.2
ADS	16 seq.	370	182	77	50
	128 seq.	207	169	75	41
Tri-gram KL-div.	16 seq.	0.0699	0.01	0.016	0.0817
	128 seq.	0.015	0.0151	0.0033	0.0575

The ADS obtained for the 128 target case decreases as the W value increases, but not as much overall as does the 16 target case. This suggests that the top-down intention for regenerating a learned sequence pattern decays as W and number of training sequences increases. The Tri-gram KL-divergence is minimized in between the two extreme W settings, 0.0 and 0.2. This value is smaller for 128 targets than for 16 for each W value (except for $W = 0.01$, which is very close). The training condition with 128 training target sequences and W set to 0.1 turned out to generate the minimum KL-divergence. Interestingly, we found that the probability distribution of the Tri-grams generated by the Tar-Gen-MTRNN and the VBP-MTRNN become quite similar in this condition. Good generalization in learning can be achieved in such a condition by extracting precise probabilistic structures. It should be taken into account that the average divergence step (ADS) was computed with a time step length of 400 for each target teaching sequence pattern, and that generated patterns exhibited the same time step length, so a higher ADS value does not signify better generalization capability. It is a proper measure only of exact similarity between the target teaching patterns and the generated output patterns. However, each Tri-gram KL-divergence value was computed by comparing long sequence patterns (100,000 steps) test-generated by the Tar-Gen-MTRNN and by the trained VBP-MTRNN. This value represents the capability for generalization in learning, as it shows how much each VBP-MTRNN model was able to extract of the probabilistic structure latent in the teaching target patterns.

4 Discussion and Summary

The current paper proposed a novel variational Bayes predictive coding RNN model, which can learn to predict/generate fluctuated temporal patterns from exemplars in order to shed new light on the gap between deterministic and probabilistic modeling. The model network is characterized by a meta-prior W which balances two cost functions, the regularization term and reconstruction

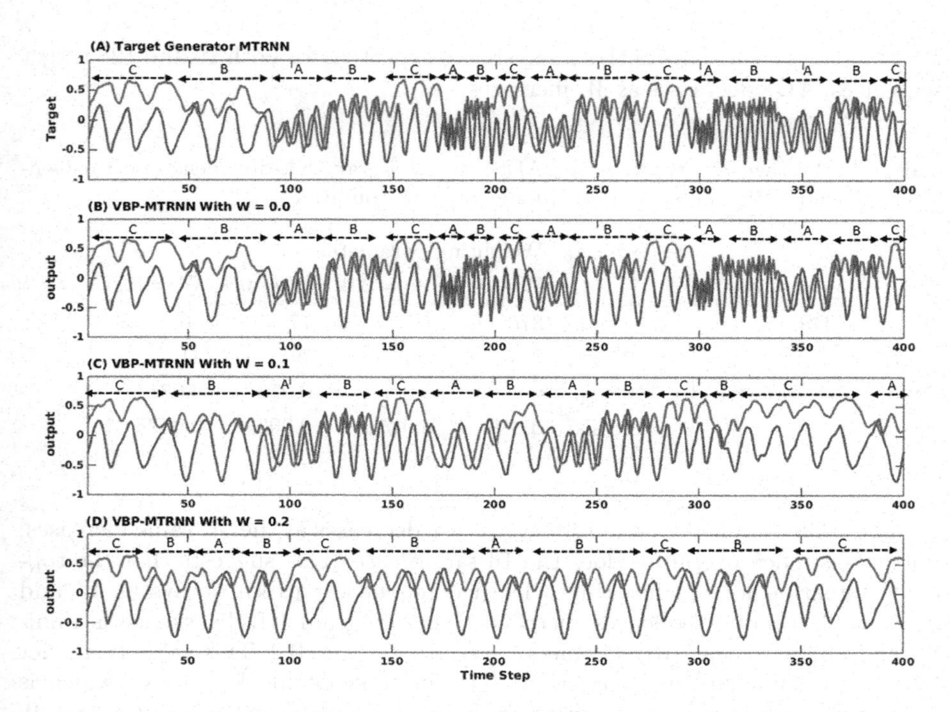

Fig. 3. A typical comparison between a teaching target sequence pattern and the closed-loop output generated by the VBP-MTRNN model trained with 16 target sequences and set with different values for W. The capital letters segmenting the sequence patterns indicate label sequences as classified by the Class-MTRNN.

error term. We investigated how this meta-prior parameter along with the number of taught sequence patterns affects model learning performance through simulation experiments which involved learning sequence patterns which exhibit probabilistic transitions between prototypical patterns. Results are summarized as follows. When the meta-prior W was set to 0.0, the model learned to imitate the probabilistic transitions observed in taught sequences through the development of deterministic chaos. It was able to repeat taught sequences exactly for long periods. However, with strong top-down intentionality developing in initial latent states, generalization in learning turned out to be poor. When W was set with larger values, the model exhibited stochastic dynamics as it adapted time varying sigma to noise sampling. With W set too high, reconstruction error increased with additional randomness in generated patterns as taught sequences could be repeated only for shorter periods. For every value of W, when the number of taught patterns increased, generalization in learning improved. Moreover, the highest degree of generalization in learning is exhibited in the middle between extreme W values.

The current results bear on the task of learning to extract latent structures in observed fluctuating temporal patterns, and as such may inform inquiry into the mechanism underlying autism spectrum disorders (ASD). Van de Cruys et al. [3] have suggested that ASD might be caused by overly strong top-down prior potentiation to minimize prediction error, which can enhance capacities for rote learning while losing the capacity to generalize what is learned, a pathology typical of ASD. With meta-prior W set below a threshold value, the proposed model naturally reflects such pathology. Future study is expected to validate this hypothesis. Future studies should concern scaling of the proposed model in various real world applications including robot learning, which will inevitably involve dealing with fluctuating temporal patterns. Moreover, studies should explore the organizing principles of cognitive brains both in normal and abnormal conditions by selectively extending the model, and comparing model with empirical data.

References

1. Bishop, C.M.: Pattern Recognition and Machine Learning. Springer, New York (2006)
2. Chung, J., Kastner, K., Dinh, L., Goel, K., Courville, A.C., Bengio, Y.: A recurrent latent variable model for sequential data. In: Advances in Neural Information Processing Systems, pp. 2980–2988 (2015)
3. Van de Cruys, S., Evers, K., Van der Hallen, R., Van Eylen, L., Boets, B., de Wit, L., Wagemans, J.: Precise minds in uncertain worlds: predictive coding in autism. Psychol. Rev. **121**(4), 649 (2014)
4. Friston, K.: A theory of cortical responses. Philos. Trans. R. Soc. Lond. B Biol. Sci. **360**(1456), 815–836 (2005)
5. Gregor, K., Danihelka, I., Graves, A., Rezende, D., Wierstra, D.: Draw: a recurrent neural network for image generation. In: International Conference on Machine Learning, pp. 1462–1471 (2015)
6. Kingma, D., Ba, J.: Adam: a method for stochastic optimization. arXiv preprint arXiv:1412.6980 (2014)
7. Kingma, D.P., Welling, M.: Auto-encoding variational bayes. arXiv preprint arXiv:1312.6114 (2013)
8. Murata, S., Yamashita, Y., Arie, H., Ogata, T., Sugano, S., Tani, J.: Learning to perceive the world as probabilistic or deterministic via interaction with others: a neuro-robotics experiment. IEEE Trans. Neural Netw. Learn. Syst. **28**(4), 830–848 (2017)
9. Rumelhart, D.E., Hinton, G.E., Williams, R.J.: Learning internal representations by error propagation. Technical report, California Univ San Diego La Jolla Inst for Cognitive Science (1985)
10. Tani, J., Fukumura, N.: Embedding a grammatical description in deterministic chaos: an experiment in recurrent neural learning. Biol. Cybern. **72**(4), 365–370 (1995)
11. Yamashita, Y., Tani, J.: Emergence of functional hierarchy in a multiple timescale neural network model: a humanoid robot experiment. PLoS Comput. Biol. **4**(11), e1000220 (2008)

Stability of Periodic Orbits and Fault Tolerance in Dynamic Binary Neural Networks

Shunsuke Aoki and Toshimichi Saito(✉)

Hosei University, Koganei, Tokyo 184-8584, Japan
tsaito@hosei.ac.jp

Abstract. The dynamic binary neural network is characterized by ternary connection parameters and can generate various binary periodic orbits. This paper considers two interesting problems based on typical examples. First, effect of connection sparsity on stability of target periodic orbits is considered: adding branches adequately to the most sparse network, stability of the periodic orbits can be reinforced. Second, fault tolerance of the network is considered: cutting one branch from the network, storage and stability of the periodic orbits are preserved in high probability.

Keywords: Dynamic binary neural networks · Fault tolerance · Stability

1 Introduction

The dynamic binary neural network (DBNN) is constructed by applying delayed feedback to layer-type artificial neural network. The DBNN is characterized by signum activation function and ternary connection parameters $\{-1, 0, 1\}$ [1–6]. Such systems are suitable for hardware implementation [7]. Depending on the parameters, the DBNN can generate various binary periodic orbits (BPOs). The dynamics can be integrated into the digital return map (Dmap) defined on a set of points. The Dmap can be regarded as a digital version of analog return one-dimensional maps [8] and is useful in visualization/consideration of the dynamics. Applications of the DBNN include control of switching circuits [9] and associative memories [10]. This paper considers relation between the connection parameters and stability of target binary periodic orbits (TBPOs). Such consideration is important in both fundamental study and engineering applications.

First, we introduce dynamics of the DBNN and derive the Dmap. Using the Dmap, we define stability of TBPOs. Second, effect of connection sparsity on stability of a TBPO is considered. In the consideration, we use TBPOs embedded in a DBNN with the most sparse connection. The most sparse DBNN is equivalent to the shift register [11] where all the BPOs are not stable. Adding branches adequately to the most sparse DBNN, the connection sparsity decreases and stability of the TBPO can be reinforced. Third, fault tolerance of the DBNN is considered. Performing fundamental numerical experiments, it is shown that storage

D. Liu et al. (Eds.): ICONIP 2017, Part III, LNCS 10636, pp. 770–778, 2017.
https://doi.org/10.1007/978-3-319-70090-8_78

and stability of TBPOs are preserved in high probability even if one branch of the DBNN is cut/removed. It should be noted that stability for branch addition and the fault tolerance have not been discussed in previous works.

2 Dynamic Binary Neural Networks

In this section, as preparations, we introduce the DBNN and Dmap presented in [1–3]. Applying a delayed feedback to the two-layer network, the DBNN is constructed as shown in Fig. 1(a). The dynamics is described by

$$x_i^{t+1} = F\left(\sum_{j=1}^{N} w_{ij} x_j^t - T_i\right),\ i = 1 \sim N,\quad F(x) = \begin{cases} +1 \text{ if } x \geq 0 \\ -1 \text{ if } x < 0 \end{cases} \tag{1}$$

The signum activation function F outputs +1 or −1 and $x_i^t \in \{-1, +1\} \equiv \boldsymbol{B}$ is the i-th binary state at discrete time t. The signum activation function is used in various artificial neural networks [4–6]. We abbreviate Eq. (1) by $\boldsymbol{x}^{t+1} = \boldsymbol{F}_D(\boldsymbol{x}^t)$ where $\boldsymbol{x}^t \equiv (x_1^t, \cdots, x_N^t)$. As an initial state vector $\boldsymbol{x}^1$ is applied, the DBNN generates a sequence of binary vectors $\{\boldsymbol{x}^t\}$. The connection parameters w_{ij} are ternary and the threshold parameters T_i are integer:

$$w_{ij} \in \{-1, 0, +1\},\ T_i \in \{0, \pm 1, \pm 2, \pm 3, \cdots\} \tag{2}$$

In order to visualize the dynamics DBNN, we introduce the Dmap. The domain of the DBNN is a set of binary vectors $\boldsymbol{B}^N$ that is equivalent to a set of points $L_D = (C_1, \cdots, C_{2^N})$, $C_i \equiv i/2^N$. Hence the dynamics of the DBNN can be integrated into the digital return map (Dmap) from L_D to itself:

$$\boldsymbol{x}^{t+1} = F_D(\boldsymbol{x}^t),\ \boldsymbol{x}^t \equiv (x_1^t, \cdots, x_N^t) \in \boldsymbol{B}^N \tag{3}$$

Figure 1(c) illustrates a Dmap for $N = 4$. Since the number of the points is finite, the steady state of the Dmap must be a periodic orbit defined as the following.

Here we give several basic definitions. A point $\theta_p \in L_D$ is said to be a periodic point with period p if $f^p(\theta_p) = \theta_p$ and $f(\theta_p)$ to $f^p(\theta_p)$ are all different where f^k is the k-fold composition of f. A sequence of the periodic points, $\{F(\theta_p), \cdots, F^p(\theta_p)\}$, is said to be a periodic orbit.

A point $\theta_q \in L_D$ is said to be an eventually periodic point (EPP) with step q if θ_q is not a periodic point but falls into some periodic orbit after q steps: $f^q(\theta_q) = \theta_p$ where θ_p is some PEP.

Since one BPO of the DBNN is equivalent to one periodic orbit of the Dmap, we use the term BPO instead of the periodic orbit hereafter.

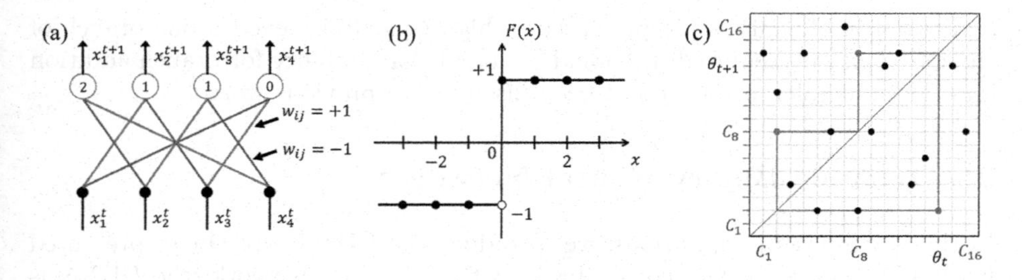

Fig. 1. DBNN. (a) Network configuration. Red and blue branches mean $w_{ij} = +1$ and $w_{ij} = -1$, respectively. $w_{ij} = 0$ means no connection. Threshold parameters T_i are shown in the green circle. (b) Signum activation function. (c) Dmap and periodic orbit with period 3. It corresponds to BPO with period 3 of the DBNN: $c_2 \equiv (-1, -1, -1, +1) \rightarrow c_8 \equiv (-1, +1, +1, +1) \rightarrow c_{14} \equiv (+1, +1, -1, +1)$. (Color figure online)

3 Feature Quantities and Stability of Periodic Orbits

First, we define a target binary periodic orbit (TBPO) with period p:

$$\boldsymbol{z}^1, \cdots, \boldsymbol{z}^p, \cdots; \ \boldsymbol{z}^i = \boldsymbol{z}^j \text{ for } |i - j| = np, \ \boldsymbol{z}^i \neq \boldsymbol{z}^j \text{ for } |i - j| \neq np \tag{4}$$

where n denotes positive integers. In associative memories, a TBPO is an object of storage. However, this paper does not consider storage of a TBPO but stability of a TBPO embedded in DBNN. Tables 1 and 2 show two examples of TBPOs (TBPO1 and TBPO2). The TBPO1 and TBPO2 are applicable to switching signal of AC/DC and DC/AC converters, respectively. They are fundamental and important switching circuits in power electronics [9]. In the AC/DC (respectively, DC/AC) converter, a 3-phase AC input (respectively, a DC input) is converted into a DC-like output (respectively, an AC-like output) via 6 switches. Replacing switch-on and switch-off with 1 and −1, respectively, the switching signals are represented by TBPO1 and TBPO2. The actual circuits can be found in [3].

Table 1. TBPO1 with period 6 from a control signal of AC/DC converter

z^1	+1	−1	−1	−1	−1	+1
z^2	+1	+1	−1	−1	−1	−1
z^3	−1	+1	+1	−1	−1	−1
z^4	−1	−1	+1	+1	−1	−1
z^5	−1	−1	−1	+1	+1	−1
z^6	−1	−1	−1	−1	+1	+1

Table 2. TBPO1 with period 6 from a control signal of AC/DC converter

z^1	+1	−1	−1	−1	+1	+1
z^2	+1	+1	−1	−1	−1	+1
z^3	+1	+1	+1	−1	−1	−1
z^4	−1	+1	+1	+1	−1	−1
z^5	−1	−1	+1	+1	+1	−1
z^6	−1	−1	−1	+1	+1	+1

In order to consider effects of connection sparsity on stability of TBPO, we define two feature quantities. The first quantity characterizes stability of a TBPO:

$$\alpha = \frac{\text{The number of EPPs falling into a TBPO}}{2^N - p}, \quad 0 \le \alpha \le 1 \tag{5}$$

As α increases, the number of EPPs increases and the stability of the TBPO becomes stronger. If $\alpha = 1$, the stability is the strongest. In this case, the TBPO is said to be completely stable. The second quantity characterizes the sparsity of the connection matrix $\boldsymbol{W} \equiv (w_{ij})$:

$$\text{SR} = \frac{\text{The number of zeros in } \boldsymbol{W}}{N^2 - N}, \quad 0 \le \text{SR} \le 1 \tag{6}$$

where each input neuron is assumed to connect to at least one output neuron. As SR increases, the sparsity increases.

If SR = 1, the connection is the most sparse as shown in Fig. 2(a) where the network connection between inputs and outputs is one to one. This DBNN is equivalent to the shift register [11] and the Dmap has no EPP: all the points are periodic points and all the BPOs are not stable. The TBPO1 and TBPO2 are included in the BPOs of the most sparse DBNN.

First, we consider the TBPO1. In the most sparse DBNN of SR = 1, stability quantity $\alpha = 0/58$ for TBPO1: it is unstable. Adding negative connection branches ($w_{ij} = -1$) to the DBNN, the stability of TBPO1 can be reinforced. Figure 2 shows typical examples: In the DBNN of SR = 28/30, stability of TBPO1 is not so strong ($\alpha = 33/58$). In the DBNNs of SR = 25/30 and SR = 10/30, TBPO1 is completely stable ($\alpha = 1$). Figure 2 suggests that stability of the TBPO can be reinforced by adding negative branches successively to the most sparse DBNN. In order to confirm this suggestion, we present the following algorithm.

Step 1: Select one TBPO (TBPO1 or TBPO2) in the most sparse DBNN.
Step 2: Add one negative connection branch ($w_{ij} = -1$) to either one position in the DBNN. If we can find some connection where the TBPO is stored and its stability is reinforced (α increases) or is saturated ($\alpha = 1$) then the added branch is saved and go to Step 4. Otherwise, remove the added branch and go to Step 3.

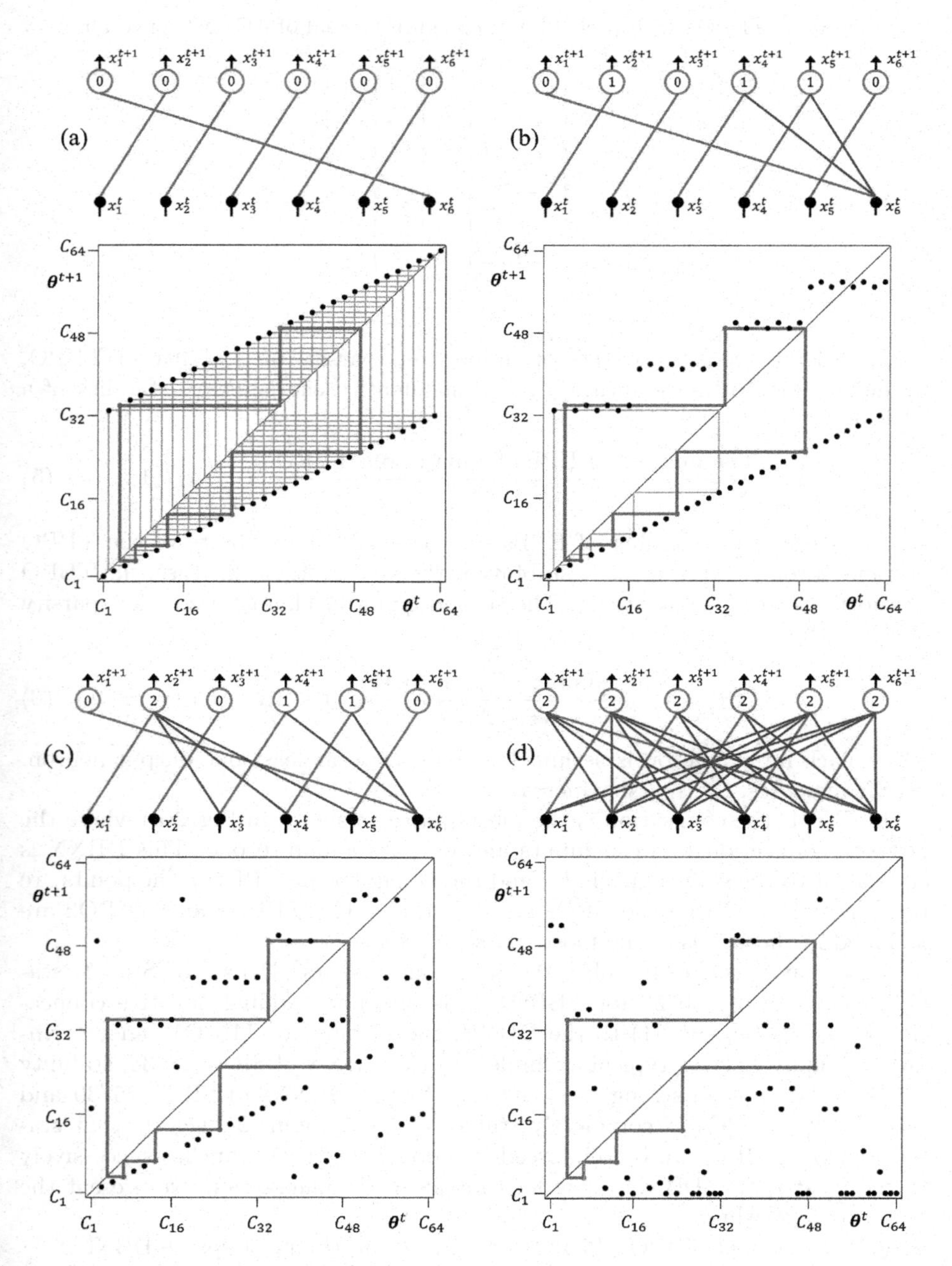

Fig. 2. DBNN and Dmap. Red orbit: TBPO1. Green orbits: spurious BPOs. (a) SR = 1. DBNN is equivalent to the shift register. $\alpha = 0/58$: TBPO1 is unstable. (b) SR = 28/30. $\alpha = 33/58$: TBPO1 is stable. (c) SR = 25/30. $\alpha = 1$: TBPO1 is completely stable. (d) SR = 10/30. $\alpha = 1$: TBPO1 is completely stable. (Color figure online)

Step 3: Add two negative connection branches to either two positions in the DBNN. If we can find some two connections where the TBPO is stored and its stability is reinforced (α increases) or is saturated ($\alpha = 1$) then the added branches are saved and go to Step 4. Otherwise, the algorithm is terminated.
Step 4: Go to Step 2 and repeat until SR $= 0$.

We have applied this algorithm to the TBPO1 and have confirmed that the TBPO1 can be completely stable ($\alpha = 1$) for $9/30 \leq \text{SR} \leq 26/30$. In a likewise manner, we have applied this algorithm to the TBPO2 and have confirmed that the TBPO2 can be completely stable ($\alpha = 1$) for $12/30 \leq \text{SR} \leq 26/30$. If SR decreases further, α decreases and stability is weakened. We have applied this algorithm to various TBPOs selected from BPOs of the most sparse DBNN, and have confirmed that the TBPOs can be completely stable. Mechanism of such stability reinforcement will be discussed elsewhere.

4 Fault Tolerance of the DBNN

In this section, we consider fault tolerance of the DBNN. First, we show two examples. Figure 3 shows that the DBNN of TBPO1 is robust for cutting one branch: even if one branch is cut (SR $= 20/30 \rightarrow 21/30$), storage and complete stability ($\alpha = 1$) of the TBPO1 are preserved. Figure 4 shows that the DBNN of TBPO2 is robust for cutting one branch: even if one branch is cut (SR $= 20/30 \rightarrow 21/30$), storage and complete stability ($\alpha = 1$) of the TBPO2 are preserved.

In order to confirm the fault tolerance of the DBNN, we have performed numerical experiments in the range $12/30 \leq \text{SR} \leq 25/30$. The results for TBPO1 and TBPO2 are summarized in Tables 3 and 4, respectively. In the experiments, we use a typical DBNN of each SR, cut either one branch, and investigate storage/stability of TBPO where 6 branches of the most sparse DBNN is survives. For each SR, the number of trials is the number of branches minus 6. The fault tolerance is evaluated by the following measures.

Storage rate (STR): Rate of trials where storage of TBPO is preserved after cutting one branch.
Stability rate (STA): Rate of trials where complete stability of TBPO is preserved after cutting one branch.
$\bar{\alpha}$: Average values of feature quantity α in the case where TBPO is stored. For convenience, the value is displayed by the closest fraction of denominator 58.

In the tables, we can see that storage of the TBPOs can be preserved after cutting one branch. Table 4 of TBPO1 shows higher probability of storage preservation than Table 3 of TBPO2. For the stored TBPO, the complete/strong stability can be preserved. In the measure STA, the stability of TBPO1 is stronger than TBPO2, however, average of stability quantity $\bar{\alpha}$ keeps large value even for the TBPO2. These results suggest that the DBNN has sufficient fault tolerance capability.

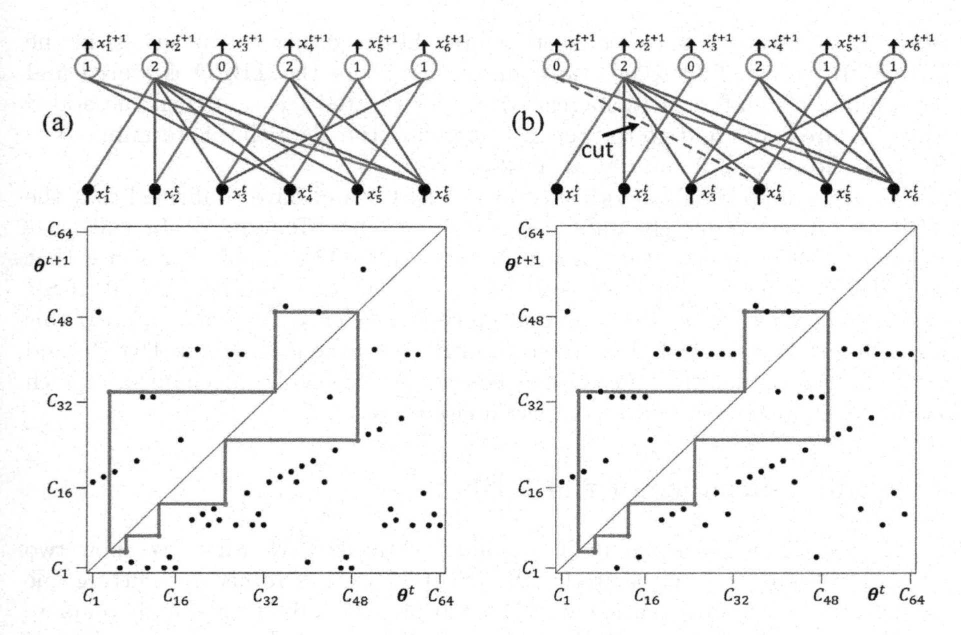

Fig. 3. Fault tolerance of DBNN for TBPO1. (a) SR = 20/30. α = 1: TBPO1 is completely stable. (b) SR = 21/30 (cutting $w_{14} = -1$). α = 1: TBPO1 is completely stable.

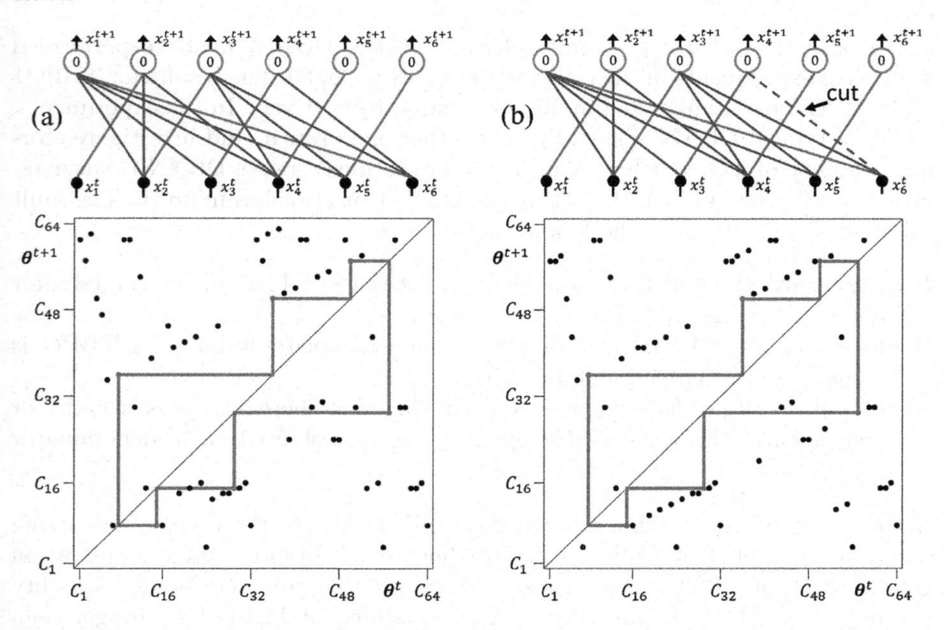

Fig. 4. Fault tolerance of DBNN for TBPO2. (a) SR = 20/30. α = 1: TBPO2 is completely stable. (b) SR = 21/30 (cutting $w_{46} = -1$). α = 1: TBPO2 is completely stable.

Table 3. Fault tolerance for TBPO1. SR: sparsity rate, STR: storage rate, STA: stability rate, $\bar{\alpha}$: average of α (displayed by the closest fraction of denominator 58).

SR	$\frac{12}{30}$	$\frac{13}{30}$	$\frac{14}{30}$	$\frac{15}{30}$	$\frac{16}{30}$	$\frac{17}{30}$	$\frac{18}{30}$	$\frac{19}{30}$	$\frac{20}{30}$	$\frac{21}{30}$	$\frac{22}{30}$	$\frac{23}{30}$	$\frac{24}{30}$	$\frac{25}{30}$
STR [%]	100	100	100	100	100	100	100	100	100	100	100	86	83	100
STA [%]	94	94	75	80	79	77	58	73	60	56	50	57	17	20
$\bar{\alpha}$	$\frac{44}{58}$	$\frac{55}{58}$	$\frac{55}{58}$	$\frac{57}{58}$	$\frac{51}{58}$	$\frac{56}{58}$	$\frac{47}{58}$	$\frac{44}{58}$	$\frac{41}{58}$	$\frac{40}{58}$	$\frac{42}{58}$	$\frac{46}{58}$	$\frac{38}{58}$	$\frac{34}{58}$

Table 4. Fault tolerance for TBPO2. SR, STR, STA, and $\bar{\alpha}$ are as in Table 3.

SR	$\frac{12}{30}$	$\frac{13}{30}$	$\frac{14}{30}$	$\frac{15}{30}$	$\frac{16}{30}$	$\frac{17}{30}$	$\frac{18}{30}$	$\frac{19}{30}$	$\frac{20}{30}$	$\frac{21}{30}$	$\frac{22}{30}$	$\frac{23}{30}$	$\frac{24}{30}$	$\frac{25}{30}$
STR [%]	89	76	81	80	64	62	67	64	70	67	88	71	83	40
STA [%]	61	65	69	67	43	46	50	27	50	44	38	43	33	20
$\bar{\alpha}$	$\frac{48}{58}$	$\frac{44}{58}$	$\frac{47}{58}$	$\frac{46}{58}$	$\frac{36}{58}$	$\frac{35}{58}$	$\frac{38}{58}$	$\frac{36}{58}$	$\frac{40}{58}$	$\frac{38}{58}$	$\frac{46}{58}$	$\frac{41}{58}$	$\frac{43}{58}$	$\frac{15}{58}$

5 Conclusions

Stability of TBPO for connection parameters is considered in this paper. Using typical examples of TBPO, two suggestions are given. First, the connection increases from the most sparse case, the stability of TBPO can be reinforced. Second, if the DBNN have TBPO whose stability is strong, the storage and stability of the TBPO can be preserved in high probability. Future problems include analysis of stability and fault tolerance for various DBNN/TBPOs, hardware implementation of DBNN, and engineering applications.

Acknowledgement. This work is supported in part by JSPS KAKENHI #15K00350.

References

1. Kouzuki, R., Saito, T.: Learning of simple dynamic binary neural networks. IEICE Trans. Fund. **E96-A**(8), 1775–1782 (2013)
2. Sato, R., Aoki, S., Saito, T.: Connection sparsity versus orbit stability in dynamic binary neural networks. In: Proceedings of IJCNN, pp. 4482–4487 (2017)
3. Sato, R., Saito, T.: Stabilization of desired periodic orbits in dynamic binary neural networks. Neurocomputing **248**, 19–27 (2017)
4. Gray, D.L., Michel, A.N.: A training algorithm for binary feed forward neural networks. IEEE Trans. Neural Netw. **3**(2), 176–194 (1992)
5. Kim, J.H., Park, S.K.: The geometrical learning of binary neural networks. IEEE Trans. Neural Netw. **6**(1), 237–247 (1995)
6. Chen, F., Chen, G., He, Q., He, G., Xu, X.: Universal perceptron and DNA-like learning algorithm for binary neural networks: non-LSBF implementation. IEEE Trans. Neural Netw. **20**(8), 1293–1301 (2009)
7. Courbariaux, M., Bengio, Y., David, J.-P.: Binary connect: training deep neural networks with binary weights during propagations. Adv. Neural Inf. Process. Syst. **28**, 3105–3113 (2015)

8. Ott, E.: Chaos in Dynamical Systems. Cambridge University Press, Cambridge (1993)
9. Vithayathil, J.: Power Electronics. McGraw-Hill, New York (1992)
10. Jiang, X., Gripon, V., Berrou, C., Rabbat, M.: Storing sequences in binary tournament-based neural networks. IEEE Trans. Neural Netw. **27**(5), 913–925 (2016)
11. Saravanan, S., Lavanya, M., Vijay Sai, R., Kumar, R.: Design and analysis of linear feedback shift register based on various tap connection. Procedia Eng. **38**, 640–646 (2012)

Basic Analysis of Cellular Dynamic Binary Neural Networks

Kazuma Makita, Takahiro Ozawa, and Toshimichi Saito(✉)

Hosei University, Koganei, Tokyo 184-8584, Japan
tsaito@hosei.ac.jp

Abstract. This paper studies cellular dynamic binary neural networks that can generate various periodic orbits. The networks is characterized by signum activation function and local connection parameters. In order to visualize/analyze the dynamics, we present a feature plane of present two simple feature quantities. We also we present normal form equations that can describe all dynamics of the networks. Using the normal form equation and feature plane, various phenomena are investigated.

Keywords: Dynamic binary neural networks · Cellular automata · Binary periodic orbits · Stability

1 Introduction

This paper studies basic dynamics of cellular dynamic binary neural networks (CDBNN), a simplified systems of the dynamics binary neural networks with the signum activation function [1,2]. The CBDNN is constructed by applying a delayed feedback to a feed forward neural network and is characterized by local connection parameters. Although the CDBNN has much less number of parameters than the dynamics binary neural networks [3–5], the CDBNN can generate a variety of binary periodic orbits (BPOs) in the steady state. The CDBNN a simple digital dynamical system the cellular automata [6,7]. Application of the digital dynamical systems includecontrol of switching circuits [3], signal processing [7], and ultla-wide-band communication [8]. Analysis of the CDBNN is basic to consider various digital dynamical systems and their applications.

First, we present a feature plane of two simple feature quantities. The first feature quantity characterizes balance between the steady states and transient phenomena. The second feature quantity characterizes the number of the steady state. Dynamics of a CDBNN is represented by a point on the feature plane. The feature plane is useful in visualized/analysis of the CDBNN dynamics. Derivation of the feature quantities is based on the digital return map (Dmap) into which the dynamics of the CDBNN is integrated. The Dmap is defined on a set of points and can be regarded as a digital version of the one dimensional maps such as the logistic map [9].

D. Liu et al. (Eds.): ICONIP 2017, Part III, LNCS 10636, pp. 779–786, 2017.
https://doi.org/10.1007/978-3-319-70090-8_79

Second, we present normal form equations that can describe all the CDBNNs. Using the normal form equations and feature planes, dynamics of all the CDBNN are investigated. Comparison of the CDBNN and cellular automata is also discussed.

2 Cellular Dynamic Binary Neural Networks

We present the cellular dynamic binary neural networks (CDBNN). The CDBNN is characterized by the signum activation function and local connection parameters. The CDBNN is constructed by applying delayed feedback to a feed-forward networks. For simplicity, we consider the case of the connection parameters in three closest neighbors. The dynamics of the CDBNN is described by

$$x_i^{t+1} = \text{sgn}\left(w_1 x_{i-1}^t + w_2 x_i^t + w_3 x_{i+1}^t - T\right) \tag{1}$$

$$\text{sgn}(x) = \begin{cases} +1 \text{ for } x \geq 0 \\ -1 \text{ for } x < 0 \end{cases} \quad i = 1 \sim N$$

$$\text{ab. } \boldsymbol{x}^{t+1} = F(\boldsymbol{x}^t),\ \boldsymbol{x}^t \equiv (x_1^t, \cdots, x_N^t) \in \boldsymbol{B}^N$$

where $\boldsymbol{x}^t$ is a binary state vector at discrete time t and $x_i^t \in \{-1, +1\} \equiv \boldsymbol{B}$ is the i-th element. $x_0^t = x_N^t$ and $x_{N+1}^t = x_1^t$ on the ring connection. T is threshold parameter. Let the connection parameters be $w_j \in \boldsymbol{R}$ $(j = 1 \sim 3)$ in consideration of basic dynamics. Note that the CDBNN is characterized by only four parameters: w_1, w_2, w_3, and T. It is much smaller than $N^2 + N$ parameters in dynamic binary neural networks (DBNN) [3–5].

In order to analyze the dynamics, we introduce the Dmap. Since the set of all the binary vectors $\boldsymbol{B}^N$ is equivalent to a set of 2^N points L_{2^N}.

$$x^{t+1} = F_D(x^t),\ x^t \in \{C_1, \cdots, C_{2^N}\} \equiv L_{2^N}$$

The 2^N points C_1 to C_{2^N} are expressed by decimal value of the binary code. x^t denotes a variable of Dmap corresponding to $\boldsymbol{x}^t$ of DBNN. Figure 1(b) illustrates Dmap where $L_{2^4} = \{C_1, \cdots, C_{16}\}$; $C_1 = 0 \equiv (-1, -1, -1, -1)$, $\cdots$, $C_{16} = 15 \equiv (+1, +1, +1, +1)$.

Since the number of the points is finite, the steady states are periodic orbits defined as the following. A point $x_p \in L_{2^N}$ is said to be a periodic point (PEP) with period p if $F_D^p(x_p) = x_p$ and $F_D(x_p)$ to $F_D^p(x_p)$ are all different where F_D^p is the p-fold composition of F_D. A periodic point with period 1 is referred to as a fixed point. A sequence of the PEPs, $\{F_D(x_p), \cdots, F_D^p(x_p)\}$, is said to be a periodic orbit (PEO). The PEO corresponds to a BPO of the CDBNN.

It should be noted that, since the number of points in L_{2^N} is 2^N, direct memory of all the inputs/outputs becomes hard as N increases. However, even if N is small, the CDBNN can exhibit a variety of periodic orbits some of which are applicable to engineering systems. For example, in the case of $N = 6$ as shown in Fig. 2(2), (3). The Dmap in Fig. 2 has a PEO with period 8 and three PEOs with period 6 and a fixed point. Figure 2(2) shows a periodic orbit with period 6 correspond to switching signal of DC/AC converters [3–5].

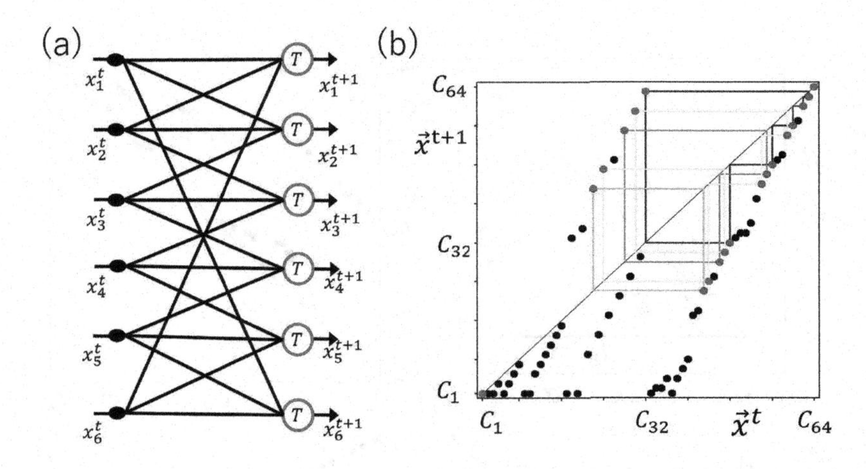

Fig. 1. Cellular dynamic binary neural networks (CDBNN) and Digital retern map (Dmap). (a) Network configuration. Black branches are connection parameters $w_j \in R$. Threshold parameters are shown in the green circle. $x_0^t = x_N^t$ and $x_{N+1}^t = x_1^t$ on the ring connection. (b) Dmap. Red points are PEPs. This Dmap has 4 PEOs classified by four colors. (Color figure online)

3 Feature Quantities and Feature Plane

In order to analyze the dynamic of the CDBNN, we present two simple feature quantities, and feature plane. The first quantity is the rate of PEPs. It characterizes balance between the steady states (PEPs) and transient phenomena.

$$\alpha = \frac{\text{The number of PEPs}}{2^N}, \quad \frac{1}{2^N} \le \alpha \le 1. \tag{2}$$

The second quantity is the rate of PEOs. It characterizes the number of the steady state.

$$\beta = \frac{\text{The number of PEOs}}{2^N}, \quad \frac{1}{2^N} \le \beta \le \alpha. \tag{3}$$

Using two feature quantities α and β, we construct a feature plane as shown in Fig. 2. The feature plane is useful in visualization and classification of the dynamic. We define three lines on the plane.

The first line is l_d: $\alpha = \beta$. All the PEOs are fixed points. For example, the Dmap in Fig. 2(1). This Dmap has 19 fixed points. The second line is l_r: $\alpha = 1$. The CDBNN has no transient phenomena. For example, the Dmap in Fig. 2(2). The third line is l_b: $\beta = 1/2^N$. The CDBNN has only one PEO.

Note that (α, β) is plotted in the triangle surrounded by these three lines. These lines can be criteria for the analysis/classification of CDBNN. Typical examples of Dmaps are shown in Fig. 2:

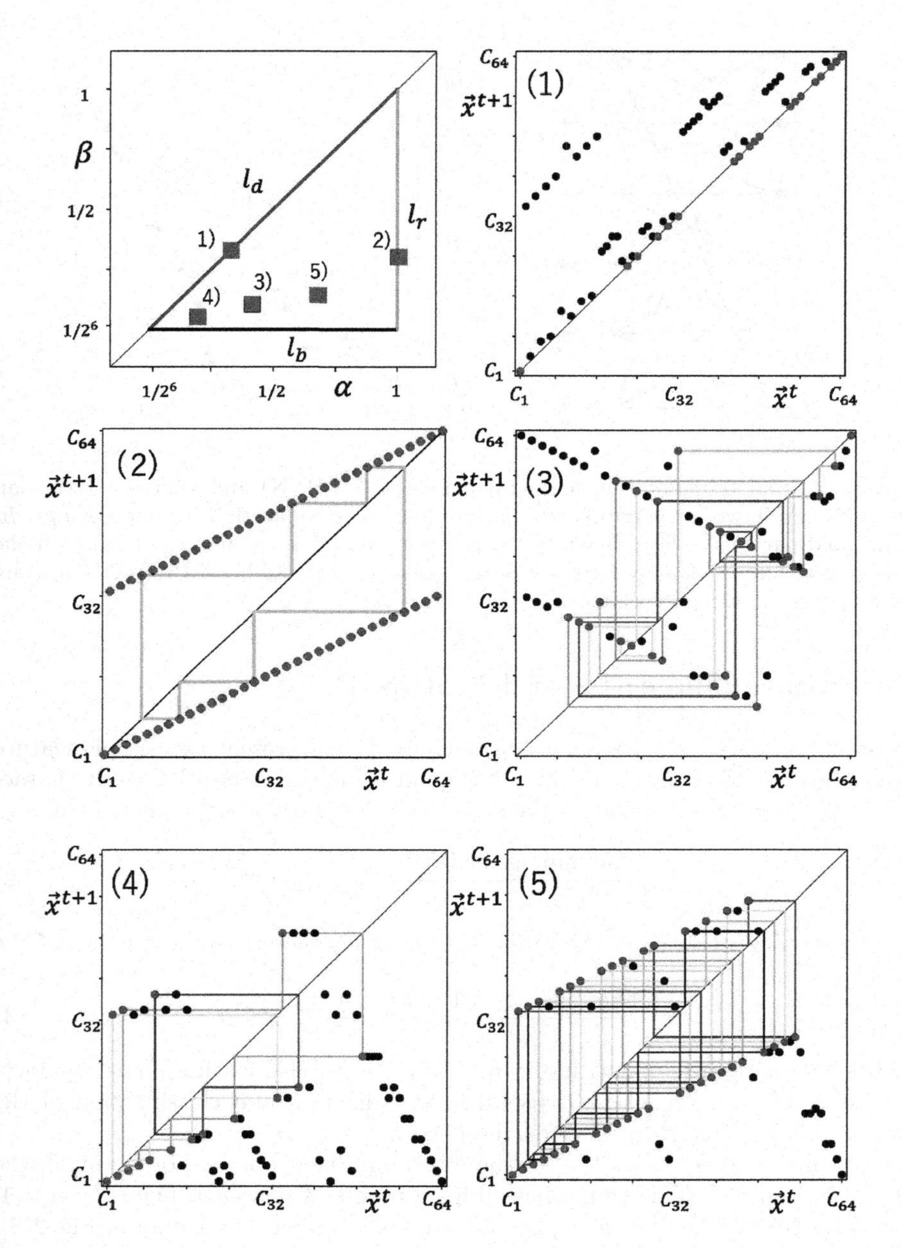

Fig. 2. Feature plane and Digital return map (Dmap). α versus β feature plane. (1) (α, β) = (19/64, 19/64) $\in l_d$. All PEPs are fixed points. (2) (α, β) = (64/64, 16/64).$\in l_r$. The CDBNN has no transient phenomena. A green orbit shows switching signal of DC/AC converters. (3) (α, β) = (25/64, 5/64). (4) (α, β) = (16/64, 3/64). (5) (α, β) = (39/64, 12/64). (Color figure online)

4 Normal Form Equations

As shown in Eq. (1), each binary output x_i^{t+1} is determined by three binary inputs $x_{i-1}^t, x_i^t, x_{i+1}^t$: It is equivalent to 3-bits linear separable Boolean function (LSBF). The number of 3-bit LSBF is 104. Although the CDBNN has four real

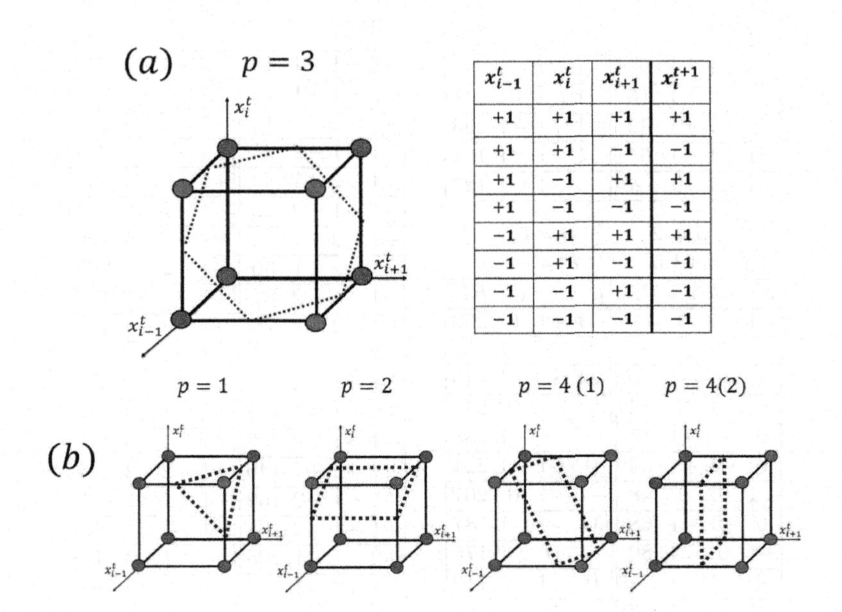

x_{i-1}^t	x_i^t	x_{i+1}^t	x_i^{t+1}
+1	+1	+1	+1
+1	+1	−1	−1
+1	−1	+1	+1
+1	−1	−1	−1
−1	+1	+1	+1
−1	+1	−1	−1
−1	−1	+1	−1
−1	−1	−1	−1

Fig. 3. Boolean cubes. (a) A Boolean cube ($p = 3$) and Truth table. $R = (1,0,1,0,1,0,0,0)$. RN = 168. A red point is +1. A blue point is −1. (b) Boolean cubes ($p = 1, 2, 4(1), 4(2)$). (Color figure online)

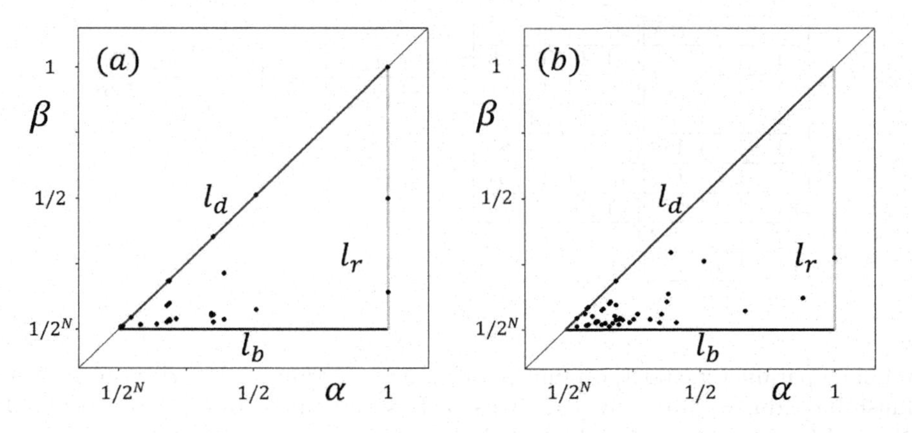

Fig. 4. The feature plane and scatter plots. (a) The feature plane for the 104 CDBNNs. (b) The feature plane for the 152 rules are non linear separable Boolean function.

Table 1. Ternary parameters and Rule numbers

$p=1$, $T=+2.5$				$p=7$, $T=-2.5$			
w_1	w_2	w_3	RN	w_1	w_2	w_3	RN
−1	−1	−1	1	+1	+1	+1	254
−1	−1	+1	2	+1	+1	−1	253
−1	+1	−1	4	+1	−1	+1	251
−1	+1	+1	8	+1	−1	−1	247
+1	−1	−1	16	−1	+1	+1	239
+1	−1	+1	32	−1	+1	−1	223
+1	+1	−1	64	−1	−1	+1	191
+1	+1	+1	128	−1	−1	−1	127

$p=2$, $T=+1.5$				$p=6$, $T=-1.5$			
w_1	w_2	w_3	RN	w_1	w_2	w_3	RN
−1	−1	0	3	+1	+1	0	252
−1	0	−1	5	+1	0	+1	250
−1	0	+1	10	+1	0	−1	245
−1	+1	0	12	+1	−1	0	243
0	−1	−1	17	0	+1	+1	238
0	−1	+1	34	0	+1	−1	221
+1	−1	0	48	−1	+1	0	207
0	+1	−1	68	0	−1	+1	187
+1	0	−1	80	−1	0	+1	175
0	+1	+1	136	0	−1	−1	119
+1	0	+1	160	−1	0	−1	95
+1	+1	0	192	−1	−1	0	63

$p=4(1)$, $T=+0.5$				$p=4(2)$, $T=+0.5$			
w_1	w_2	w_3	RN	w_1	w_2	w_3	RN
−1	−1	−1	23	+1	0	0	240
−1	−1	+1	43	0	+1	0	204
−1	+1	−1	77	0	0	+1	170
−1	+1	+1	142	0	0	−1	85
+1	−1	−1	113	0	−1	0	51
+1	−1	+1	178	−1	0	0	15
+1	+1	−1	212				
+1	+1	+1	232				

$p=3$, $T=+1.5$				$p=5$, $T=-1.5$			
w_1	w_2	w_3	RN	w_1	w_2	w_3	RN
−2	−1	−1	7	+2	+1	+1	248
−2	−1	+1	11	+2	+1	−1	244
−2	+1	−1	13	+2	−1	+1	242
−2	+1	+1	14	+2	−1	−1	241
−1	−2	−1	19	+1	+2	+1	236
−1	−1	−2	21	+1	+1	+2	234
−1	−2	+1	49	+1	+2	−1	206
−1	−1	+2	42	+1	+1	−2	213
+1	−2	−1	35	−1	+2	+1	220
+1	−2	+1	50	−1	+2	−1	205
−1	+1	−2	81	+1	−1	+2	174
−1	+2	−1	76	+1	−2	+1	179
+1	−1	−2	69	−1	+1	+2	186
+1	+1	−2	84	−1	−1	+2	171
+2	−1	−1	112	−2	+1	+1	143
−1	+1	+2	138	+1	−1	−2	117
−1	+2	+1	140	+1	−2	−1	115
+1	−1	+2	162	−1	+1	−2	93
+1	+1	+2	168	−1	−1	−2	87
+2	−1	+1	208	−2	+1	−1	47
+1	+2	−1	196	−1	−2	+1	59
+1	+2	+1	200	−1	−2	−1	55
+2	+1	−1	176	−2	−1	+1	79
+2	+1	+1	224	−2	−1	−1	31

$p=0$, $T=+3.5$				$p=8$, $T=-3.5$			
w_1	w_2	w_3	RN	w_1	w_2	w_3	RN
+1	+1	+1	0	−1	−1	−1	255

number parameter and the number of parameter values, the essential number must be reduced into only 104. This fact is very important. We present 104 Normal form equations for the CDBNN.

For convenience, we introduce the rule number: replacing −1 with 0 and applying the decimal expression to the binary code R, we obtain the rule number

RN. For example, RN = 168 for $R = (1, 0, 1, 0, 1, 0, 0, 0)$, the normal form equation is

$$x_i^{t+1} = \mathrm{sgn}\left(x_{i-1}^t + x_i^t + 2x_{i+1}^t - 1.5\right) \tag{4}$$

Figure 3 shows Boolean Cubes for $p = 1, 2, 3,$ and 4. Reversing output 1 by -1 in the Boolean cube of $p = 1, 2, and\ 3$, we obtain Boolean cube of $p = 7, 6, and\ 5$, respectively. 104 LSBF corresponds to 104 Boolean cubes. Based on the Boolean Cubes, we obtain 104 normal form equations whose parameters are summarized in Table 1. Note that, when $p = 3$ or 5, Ref. Table 1 describes LSBF by connection parameters in the range [−3,3]; whereas on connection parameters w_j in $\{-2, -1, 1, 2\}$.

Using the 104 normal form equations of CDBNN, we can calculate Dmap and feature quantities α and β. The 104 CDBNNs can be represented by points in the feature planes, Fig. 4(a) shows the feature plane for the 104 CDBNNs.

Elementary Cellular Automata (ECA) are defined by 3-bit Boolean functions and number of rules is 256 in which 104 rules are LSBF and 152 rules are non linear separable Boolean function. Figure 4(b) shows the feature plane for the 152 rules.

Finally, we compare the CDBNN with ECA from the three viewpoints.

1. 104 LSBF. 2-Layer CDBNN and 104 rule tables of ECA.
2. 152 non LSBF. Multi-Layer CDBNN and 152 rule tables of ECA. This theme will be discussed in the future.
3. Hardware implementation of CDBNN is simpler than ECA.

5 Conclusions

The CDBNN is presented and its various phenomena are considered in this paper. The dynamics is integrated into the Dmap and the feature plane is presented. The 104 normal form equations are derived. Using the feature plane and normal form equation, various phenomena are analyzed.

Future problems include more detailed analysis of the periodic orbits, comparison of wider class of CDBNN and ECA, and engineering application.

Acknowledgement. This work is supported in part by JSPS KAKENHI #15K00350.

References

1. Gray, D.L., Michel, A.N.: A training algorithm for binary feed forward neural networks. IEEE Trans. Neural Netw. **3**(2), 176–194 (1992)
2. Chen, F., Chen, G., He, Q., He, G., Xu, X.: Universal perceptron and DNA-like learning algorithm for binary neural networks: non-LSBF implementation. IEEE Trans. Neural Netw. **20**(8), 1293–1301 (2009)
3. Kouzuki, R., Saito, T.: Learning of simple dynamic binary neural networks. IEICE Trans. Fund. **E96–A**(8), 1775–1782 (2013)

4. Sato, R., Makita, K., Saito, T.: Analysis of various periodic orbits in simple dynamic binary neural networks. In: Proceedings of IJCNN, pp. 2031–2038 (2016)
5. Makita, K., Sato, R., Saito, T.: Stability of periodic orbits in dynamic binary neural networks with ternary connection. In: Hirose, A., Ozawa, S., Doya, K., Ikeda, K., Lee, M., Liu, D. (eds.) ICONIP 2016. LNCS, vol. 9947, pp. 421–429. Springer, Cham (2016). doi:10.1007/978-3-319-46687-3_47
6. Chua, L.O.: A Nonlinear Dynamics Perspective of Wolfram's New Kind of Science, I, II. World Scientific (2005)
7. Rosin, P.L.: Training cellular automata for image processing. In: Kalviainen, H., Parkkinen, J., Kaarna, A. (eds.) SCIA 2005. LNCS, vol. 3540, pp. 195–204. Springer, Heidelberg (2005). doi:10.1007/11499145_22
8. Iguchi, T., Hirata, A., Torikai, H.: Theoretical and heuristic synthesis of digital spiking neurons for spike-pattern-division multiplexing. IEICE Trans. Fund. **E93–A**(8), 1486–1496 (2010)
9. Ott, E.: Chaos in Dynamical Systems. Cambridge University Press, Cambridge (1993)

Robust Control of Uncertain Nonlinear Systems Based on Adaptive Dynamic Programming

Jing Na[1(✉)], Jun Zhao[1], Guanbin Gao[1], and Ding Wang[2]

[1] Faculty of Mechanical and Electrical Engineering, Kunming University of Science and Technology, Kunming, China
najing25@163.com, junzhao1993@163.com, gbgao@163.com

[2] Institute of Automation, Chinese Academy of Sciences, Beijing 100190, China
ding.wang@ia.ac.cn

Abstract. In this paper, we propose a new approach to address robust control problem of nonlinear systems with uncertainties based on an adaptive dynamic programming (ADP) algorithm. After reformulating the robust control problem as an optimal control problem, we propose a modified ADP method to solve the derived Hamilton-Jacobi-Bellman (HJB) equation, where the optimal cost function is approximated by online training a critic neural network (NN). Then the approximated optimal control action can be derived to guarantee the stability of the controlled system with uncertainties. The closed-loop system stability and convergence have been proved. A simulation example is provided to illustrate the effectiveness of the method.

Keywords: Robust control · Optimal control · Nonlinear uncertain systems · Adaptive dynamic programming

1 Introduction

In real life, the system to be controlled is always affected by external disturbances and/or model uncertainties. These factors must be taken into account in the controller design such that the closed-loop system has a good response even in the presence of such dynamics. This creates the problem of robust control design, which has been widely studied during the past decades [1–3]. However, it is well-known that the direct robust control design for nonlinear systems is not straightforward in comparison to its counterparts for linear systems. Hence, some effort has also been made to seek alternative solutions. In [3], it has been shown that robust control problems can be transformed into an equivalent optimal control problems for nominal systems, which can be easier to solve. However, the online solution of the derived optimal control problem is not addressed in [3].

Regarding to optimal control for linear systems, several techniques have been proposed to solve the associated Riccati equation, while for nonlinear systems, solving the HJB equation is quite challenging. Recently, the principle of dynamic programming has been extended to study optimal control problem in [4, 5]. The basic idea is to use

D. Liu et al. (Eds.): ICONIP 2017, Part III, LNCS 10636, pp. 787–796, 2017.
https://doi.org/10.1007/978-3-319-70090-8_80

neural networks to approximate the optimal cost function [6, 7], which leads to recent research work on adaptive/approximate dynamic programming (ADP) [7–11]. In [12–15], the idea of ADP has been further tailored to solve robust control problem, while the NN weights can be updated via the gradient algorithm. Inspired by the above fact, we revisit robust control design for nonlinear systems by using the methodology of ADP, while a modified NN weights updating law is adopted to online train the critic NN with guaranteed convergence [16–18]. The stability and convergence of the closed-loop system is analyzed. Finally, a simulation example is presented to show the efficacy.

2 Problem Formulation

We consider the following nonlinear system defined as [3]:

$$\dot{x} = A(x) + B(x)f(x) + B(x)u \tag{1}$$

where $x \in R^n, u \in R^m$ are the system state and control input, $A(x), B(x)$ are known system functions fulfilling $A(0) = 0$, and $f(x)$ is an unknown bounded function fulfilling $f(0) = 0$. Without loss of generality, we assume that $f(x)$ is bounded by a known function $f_{\max}(x)$, such that

$$\|f(x)\| \le f_{\max}(x) \tag{2}$$

The robust control problem to solve is to find a feedback control law $u = k(x)$ such that the closed-loop system

$$\dot{x} = A(x) + B(x)f(x) + B(x)k(x) \tag{3}$$

is asymptotically stable for all admissible disturbance $f(x)$.

It is well known that the above nonlinear robust control is difficult to solve directly. Hence, inspired by the result of [3], we will reformulate robust control as an optimal control problem first, and then provide an alternative solution by using a recently developed methodology named ADP [15].

3 Robust Control Solution via ADP

3.1 Reformulation of Robust Control

We first present that the above robust control is equivalent to the optimal control for the following nominal system without uncertainties

$$\dot{x} = A(x) + B(x)u \tag{4}$$

where a feedback control law $u = k(x)$ can be found to minimize the following cost function

$$V(x_0) = \int_0^\infty (f_{\max}^2(x) + x^T x + u^T u)dt \tag{5}$$

The above cost function (5) includes three elements: the first term $f_{\max}^2(x)$ denotes the effect of the uncertainties in the cost function; the second term $x^T x$ is devoted to the regulation of system state; and the final term $u^T u$ is the control action cost.

Hence, we have the following results:

Theorem 1 [3]. If the solution of optimal control for nominal system (4) with cost function (5) existed, then this solution can guarantee asymptotic stability of the original system (1) with uncertainties, i.e. robust control of system (1) can be solved by using the derived optimal control of nominal system (4).

Proof. To solve optimal control problem of (4) subject to cost function (5), we can define the Hamiltonian function as

$$H(x, u, V_x) = f_{\max}^2(x) + x^T x + u^T u + V_x^T(x)(A(x) + B(x)u) \tag{6}$$

where $V_x(x) = \partial V(x)/\partial x$ is the partial derivative of cost function $V(x)$ regarding to x.

Then the Hamilton-Jacobi-Bellman equation can be obtained as:

$$0 = \min_u H(x, u^*, V_x^*) = f_{\max}^2(x) + x^T x + u^{*T} u^* + V_x^{*T}(x)(A(x) + B(x)u^*) = 0 \tag{7}$$

and the optimal cost function can be given as

$$V^*(x) = \min_u \int_0^\infty (f_{\max}^2(x) + x^T x + u^T u)dt \tag{8}$$

Hence, from the optimal principle, the optimal control action can be calculated by using $\partial H/\partial u^* = 0$ as

$$u^* = k^*(x) = -\frac{1}{2} B^T(x) V_x^*(x) \tag{9}$$

From (7) and (9), we can have the following conditions

$$f_{\max}^2(x) + x^T x + k^{*T}(x)k^*(x) + V_x^{*T}(x)(A(x) + B(x)k^*(x)) = 0 \tag{10}$$

$$2k^*(x)^T + V_x^{*T}(x)B(x) = 0 \tag{11}$$

Then, we need to prove that the equilibrium $x = 0$ of the controlled system

$$\dot{x} = A(x) + B(x)f(x) + B(x)k^*(x) \tag{12}$$

is globally stable for all admissible uncertainties $f(x)$.

For this purpose, the Lyapunov theory will be used. One can easily verify that the function V^* in (8) fulfills

$$V^*(x) > 0, x \neq 0, \text{ and } V^*(x) = 0, x = 0 \tag{13}$$

Hence, $V(x)$ is a Lyapunov function. Moreover, we can calculate the derivative of $V(x)$ along the control system (12) as

$$\begin{aligned}\dot{V}^* &= \left(\frac{\partial V^*(x)}{\partial x}\right)^T \cdot \frac{dx}{dt} = V_x^{*T}(x)(A(x) + B(x)f(x) + B(x)k^*(x)) \\ &= V_x^{*T}(x)(A(x) + B(x)k^*(x)) + V_x^{*T}(x)B(x)f(x)\end{aligned} \tag{14}$$

Then from (10) and (11), we know that

$$\begin{aligned}\dot{V}^* &= -f_{\max}^2(x) - x^T x - k^{*T}(x)k^*(x) + V_x^{*T}(x)B(x)f(x) \\ &= -f_{\max}^2(x) - x^T x - k^{*T}(x)k^*(x) - 2k^{*T}(x)f(x)\end{aligned} \tag{15}$$

By applying the Young's inequality on the last term $2k^{*T}(x)f(x)$, we have

$$\begin{aligned}\dot{V}^* &\leq -f_{\max}^2(x) - x^T x - k^{*T}(x)k^*(x) + f(x)^T f(x) + k^{*T}(x)k^*(x) \\ &= -[f_{\max}^2(x) - f(x)^T f(x)] - x^T x \\ &\leq -x^T x < 0\end{aligned} \tag{16}$$

holds for all $x \neq 0$.

Therefore, the conditions of the Lyapunov stability analysis can be fulfilled. Moreover, similar to the analysis as shown in [3, 15], we can claim that the system states will eventually converge to zero, i.e. $\lim_{t \to \infty} x(t) = 0$. This completes the proof.

The above Theorem 1 provides an alternative solution to study robust control problem of uncertain system (1) by solving optimal control of nominal system (12), which can be solved in terms of a newly developed ADP technique. Hence, in the following section, we will discuss how to solve optimal control problem for (4) and (5).

3.2 Solving Optimal Control with ADP

In this subsection, we will further exploit the principle of a recent developed ADP scheme to derive the optimal solution. The basic idea is to train an NN to online estimate the optimal cost function in (5). For this purpose, we assume that the cost function $V(x)$ is continuously and differential [6]. Hence, it is possible to estimate $V(x)$ by the following single-layer NN over a compact set

$$V(x) = \omega_c^T \sigma_c(x) + \varepsilon_c(x) \tag{17}$$

where $\omega_c \in R^l$ is the ideal weight. $\sigma_c \in R^l$ is the regressor vector, and l is the number of neurons, $\varepsilon_c(x)$ is the NN approximate error. Hence, the partial derivative of cost function can be reformulated as

$$V_x(x) = (\nabla\sigma_c(x))^T\omega_c + \nabla\varepsilon_c(x) \tag{18}$$

where $\nabla\sigma_c(x) = \partial\sigma_c(x)/\partial x \in R^{l\times n}$ is a gradient matrix and $\nabla\varepsilon_c(x) = \partial\varepsilon_c(x)/\partial x \in R^n$ is the NN error.

Without loss of generality, we can assume that the NN weight ω_c, the gradient regressor matrix $\nabla\sigma_c(x)$ and the approximate errors $\varepsilon_c(x)$, $\nabla\varepsilon_c(x)$ are all bounded. In particular, from the perspective of theoretical analysis, we know the NN approximation can be arbitrarily small, i.e. $\varepsilon_c(x) \to 0$ when $l \to \infty$.

Since the ideal NN weights $\omega_c \in R^l$ are unknown, a practical critical NN can be used, such that the estimated cost function is given by

$$\hat{V}(x) = \hat{\omega}_c^T\sigma_c(x) \tag{19}$$

where $\hat{\omega}_c$ is the approximation of ω_c. Therefore, we can get the approximated cost function as

$$\hat{V}_x(x) = (\nabla\sigma_c(x))^T\hat{\omega}_c \tag{20}$$

With the NN approximation (17)–(20), we know that the ideal optimal control (9) can be reformulated as

$$u^*(x) = -\frac{1}{2}B^T(x)((\nabla\sigma_c(x))^T\omega_c + \nabla\varepsilon_c(x)) \tag{21}$$

and the approximated control control action can be given as

$$\hat{u}(x) = -\frac{1}{2}B^T(x)(\nabla\sigma_c(x))^T\hat{\omega}_c \tag{22}$$

The final problem to be solved is to determine the NN weights $\hat{\omega}_c$, which could guarantee the stability of the control system with (22).

From (7) and (18), the estimated Hamilton-Jacobi-Bellman equation is

$$0 = H(x, u, V_x) = f_{\max}^2(x) + x^Tx + u^Tu + \omega_c^T\nabla\sigma_c(x)(A(x) + B(x)u) + \epsilon_{HJB} \tag{23}$$

where $\varepsilon_{HJB} = \nabla\varepsilon_c^T(x)(A(x) + B(x)u)$ is a bounded residual HJB equation error due to the NN approximation error $\nabla\varepsilon_c(x)$. Hence, ε_{HJB} can be sufficiently small for $l \to \infty$.

To design an adaptive law to estimate ω_c, we define the known terms in (23) as $\Xi = \nabla\sigma_c(x)(A(x) + B(x)u)$ and $\Theta = f_{\max}^2(x) + x^Tx + u^Tu$, so that the approximated HJB Eq. (23) is rewritten as

$$\Theta = -\omega_c^T \Xi - \varepsilon_{HJB} \tag{24}$$

As shown in (24), the unknown critic NN weights ω_c^T has a linearly parameterized form, and thus can be '*directly*' estimated by using our recently proposed algorithm [18]. We define auxiliary regressor matrix $P \in \mathbb{R}^{l \times l}$ and vector $Q \in \mathbb{R}^l$ as

$$\begin{cases} \dot{P} = -\ell P + \Xi \Xi^T, \ P(0) = 0 \\ \dot{Q} = -\ell Q + \Xi \Theta, \ Q(0) = 0 \end{cases} \tag{25}$$

where $\ell > 0$ is a constant.

Then the adaptive law for the critic NN can be designed as

$$\dot{\hat{\omega}}_c = -\Gamma W \tag{26}$$

where $\Gamma > 0$ is the adaptive gain, and the auxiliary vector $W \in \mathbb{R}^l$ is given as

$$W = P\hat{\omega}_c + Q \tag{27}$$

Now, we have the following results

Lemma 1. For variables P, Q and W defined in (25)–(27), then W can be represented as $W = -P\tilde{\omega}_c + \upsilon$, where $\tilde{\omega}_c = \omega_c - \hat{\omega}_c$ is the NN weights error and $\upsilon = -\int_0^t e^{-\ell(t-r)} \varepsilon_{HJB}(r) \Xi^T(r) dr$ is bounded by $\|\upsilon\| \le \varepsilon_\upsilon$ for a positive constant ε_υ.

The proof of Lemma 1 can be conducted by solving the Eq. (25) with $Q = -PW + \upsilon$. A similar proof can be found in [16].

Lemma 2 [18]. If the regressor vector Ξ in (24) is persistently excited (PE), then the matrix P defined in (25) is positive definite, i.e., $\lambda_{\min}(P) > \sigma > 0$ for a constant $\sigma > 0$.

The main results of this subsection can be given as:

Theorem 2. For adaptive law (26) of critic NN with the regressor vector Ξ in (24) being PE, then the critic NN error $\tilde{\omega}_c$ converges to a compact set around zero. Moreover, in the idea case ($\varepsilon_{HJB} = 0$), $\tilde{\omega}_c$ converges to zero exponentially.

The proof of Theorem 2 can be conducted by using the fact $\dot{\tilde{\omega}}_c = -\dot{\hat{\omega}}_c = \Gamma W = -\Gamma(P\tilde{\omega}_c - \upsilon)$ from Lemma 1 and selecting a Lyapunov function $L = tr(\tilde{\omega}_c^T \Gamma^{-1} \tilde{\omega}_c)$. Here, we do not present the detailed proof due to the limited space. For those interested readers, please refer to [16] for a similar proof.

3.3 Stability Analysis

Theorem 3. For the nominal control system (4) with he approximated optimal control (22) and the adaptive law (26) for critical NN, for any initial state x_0, the system state $x(t)$ is uniformly ultimately bounded.

Proof. Substitute (22) in system (4), we get the closed-loop system dynamics as

$$\begin{aligned}\dot{x} &= A(x) + B(x)\hat{u}(x) = A(x) + B(x)u^* + B(x)(\hat{u} - u^*) \\ &= A(x) + B(x)u^* + \frac{1}{2}B(x)B^T(x)(\nabla\sigma_c^T\tilde{\omega}_c + \nabla\varepsilon_c)\end{aligned} \tag{28}$$

On the other hand, the HJB Eq. (7) can be given as

$$f_{\max}^2(x) + x^T x + u^{*T}u^* + V_x^{*T}(x)(A(x) + B(x)u^*) = 0 \tag{29}$$

Then we calculate the time derivative of $V^*(x)$ along (28) and (29) as

$$\begin{aligned}\dot{V}^* &= V_x^{*T}[A(x) + B(x)u^* + \frac{1}{2}B(x)B^T(x)(\nabla\sigma_c^T\tilde{\omega}_c + \nabla\varepsilon_c)] \\ &= -f_{\max}^2(x) - x^T x - u^{*T}u^* + \Delta\end{aligned} \tag{30}$$

where $\Delta = \frac{1}{2}(\omega_c^T\nabla\sigma_c + \varepsilon_c^T)B(x)B^T(x)(\nabla\sigma_c^T\tilde{\omega}_c + \nabla\varepsilon_c)$ is a bounded residual error due to the NN approximation error.

According to Theorem 2, we know that the residual error term Δ is bounded because the ideal NN weights ω_c, NN approximation errors ε_c, $\nabla\varepsilon_c$, regressor $\nabla\sigma_c$ and the weight error $\tilde{\omega}_c$ are all bounded. We denote $|\Delta| \leq \beta_M$ for a positive constant β_M, then

$$\dot{V}^* \leq -x^T x + \beta_M \tag{31}$$

Therefore, we can find that $\dot{V}^* < 0$ whenever $\|x\| > \sqrt{\beta_M}$. Hence, based on Lyapunov theory, we know that x will uniformly ultimately converge to a compact set defined as

$$\Omega_x = \{x : \|x\| \leq \sqrt{\beta_M}\} \tag{32}$$

This completes the proof.

Finally, we will prove that the approximated solution of the HJB equation is again indeed the solution of the optimal control solution, which can be summarized as

Theorem 4. If the solution of the optimal control problem based on the neural network existed, then the approximated optimal control (22) can ensure that the closed-loop control system (1) with uncertainties is asymptotically stable for any $\|f(x)\| \leq f_{\max}(x)$.

Proof. We denote $\hat{V}(x)$ as the solution of

$$0 = f_{\max}^2(x) + x^T x + \hat{u}^T\hat{u} + \hat{V}_x^T(A(x) + B(x)\hat{u}(x)) \tag{33}$$

where $\hat{u}(x)$ is the approximated optimal control given by (22). Then similar to (11), we can verify that

$$2\hat{u}^T(x) = -\hat{V}_x^T B(x) \tag{34}$$

We need to show based on the above condition that the closed-loop system remains asymptotically stable for all possible uncertainties $f(x)$. From (19), we can select $\sigma_c(x)$ so that $\hat{V}(0) = 0$ and $\hat{V}(x) > 0$ for all $x \neq 0$. Then following similar mathematical manipulations as those conducted in the proof of Theorem 1, we can easily obtain $\dot{\hat{V}}(x) \leq -x^T x < 0$ for all $x \neq 0$ as (16).

4 Simulations

Consider the following continuous-time nonlinear system:

$$\dot{x} = \begin{bmatrix} -x_1 + x_2 \\ -0.5x_1 - 0.5x_2(1 - (\cos(2x_1) + 2)^2) \end{bmatrix} + \begin{bmatrix} 0 \\ \cos(2x_1) + 2 \end{bmatrix}(u + 0.5px_1 \sin x_2) \tag{35}$$

where $x = [x_1, x_2]^T$ is the state variable and $u \in R$ is the control variable, p is an unknown parameter. The term $d(x) = 0.5px_1 \sin x_2$ reflects the uncertainties If $p \in [-1, 1]$, we can select $f_{\max}(x) = \|x\|$.

After reformulating this problem as an optimal control problem for nominal system [15], we can find a feedback control policy $u(x)$ for the following cost function

$$J(x) = \int_0^\infty \left\{ \|x\|^2 + x^T x + u^T u \right\} d\tau \tag{36}$$

We can derive the optimal cost function as $J^*(x) = x_1^2 + 2x_2^2$ and the optimal control action as $u^*(x) = -(\cos(2x_1^2) + 2)x_2$. We construct a critic NN to approximate the cost function with the weight vector $\omega_c = [\omega_1, \omega_2, \omega_3]^T$ and regressor $\sigma_c(x) = [x_1^2, x_1 x_2, x_2^2]^T$. Then, the ideal NN weight is $[1, 0, 2]^T$. In the simulation, we can select the initial weights of the critic NN within $[0, 2]$, the initial states of the controlled are $x_0 = [1, -0.5]^T$. To fulfill the PE condition, we introduce a noise in the system input. Simulation results are given in Fig. 1, where it is shown that the weights of the critic NN converge to a compact set around true values, which means the derived approximated control action converge to true values. With the obtained optimal control action, the evolution of the state trajectory is shown in Fig. 2, which implies that the system states converge to zero.

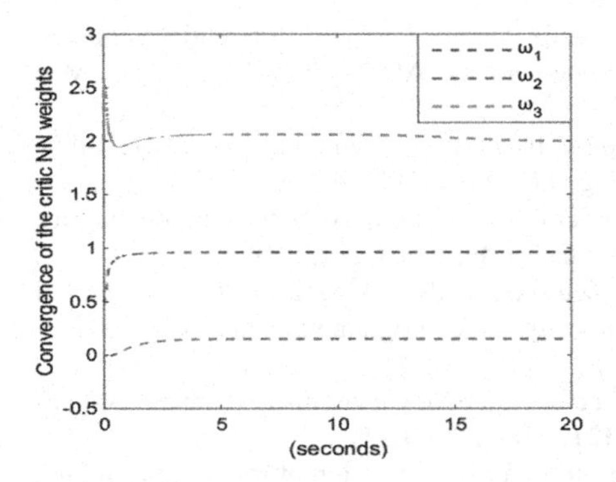

Fig. 1. Convergence of the critic NN weights.

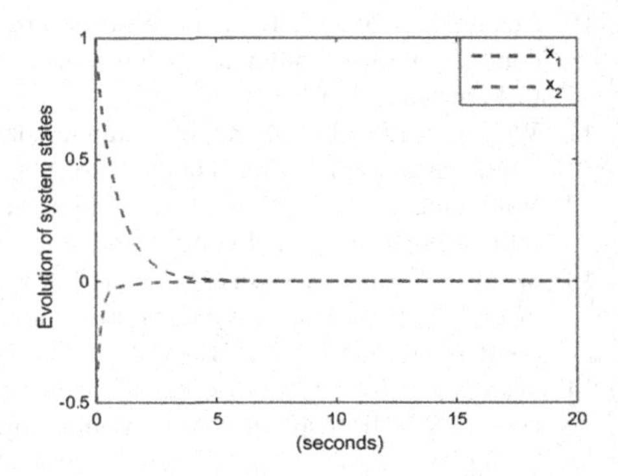

Fig. 2. Evolution of system states.

5 Conclusion

An alternative solution for robust control of nonlinear uncertain systems is addressed in this paper. The robust control problem can be transformed an optimal control problem by selecting an appropriate cost function. Then, to solve the derived optimal equation (e.g. HJB equation), a critic NN is online trained to learn the optimal cost function and to calculate the realistic control action. The stability and convergence of the closed-loop system is rigorously proved, and a simulation example is provided to show its feasibility.

References

1. Barmish, B.R., Shi, Z.: Robust stability of a class of polynomials with coefficients depending multilinearly on perturbations. IEEE Trans. Auto. Control **35**, 1040–1043 (1990)
2. Kravaris, C., Palanki, S.: A Lyapunov approach for robust nonlinear state feedback synthesis. IEEE Trans. Auto. Control **33**, 1188–1191 (1988)
3. Lin, F., Brandt, R.D., Sun, J.: Robust control of nonlinear systems: compensating for uncertainty. Int. J. Control **56**, 1453–1459 (1992)
4. Werbos, P.J.: Approximate dynamic programming for real-time control and neural modeling, Brentwood, U.K (1992)
5. Si, J., Barto, A., Powell, W., Wunsch, D.: Handbook of Learning and Approximate Dynamic Programming, Los Alamitos, U.S (2004)
6. Abu-Khalaf, M., Lewis, F.L.: Nearly optimal control laws for nonlinear systems with saturating actuators using a neural network HJB approach. Automatica. **41**, 779–791 (2005)
7. Lewis, F.L., Vrabie, D.: Reinforcement learning and adaptive dynamic programming for feedback control. IEEE Circuits Syst. **9**, 32–50 (2009)
8. He, H., Ni, Z., Fu, J.: A three-network architecture for on-line learning and optimization based on adaptive dynamic programming. Neurocomputing **78**, 3–13 (2012)
9. Wu, H.N., Luo, B.: Neural network based online simultaneous policy update algorithm for solving the HJI equation in nonlinear H_∞ control. IEEE Trans. Neural Networ. **23**, 1884 (2012)

10. Zhang, H., Cui, L., Luo, Y.: Near-optimal control for nonzero-sum differential games of continuous-time nonlinear systems using single-network ADP. IEEE Trans. Syst. Man Cybern. **43**, 206 (2013)
11. Wei, Q., Liu, D.: A novel iterative-adaptive dynamic programming for discrete-time nonlinear systems. IEEE Trans. Auto. Sci. Eng. **11**, 1176–1190 (2014)
12. Vamvoudakis, K.G., Lewis, F.L.: Online actor–critic algorithm to solve the continuous-time infinite horizon optimal control problem. Automatica **46**, 878–888 (2010)
13. Wang, D., Liu, D., Wei, Q.: Finite-horizon neuro-optimal tracking control for a class of discrete-time nonlinear systems using adaptive dynamic programming approach. Neurocomputing **78**, 14–22 (2012)
14. Wang, D., Liu, D.: Neuro-optimal control for a class of unknown nonlinear dynamic systems using SN-DHP technique. Neurocomputing **121**, 218–225 (2013)
15. Wang, D., Liu, D., Li, H.: Policy iteration algorithm for online design of robust control for a class of continuous-time nonlinear systems. IEEE Trans. Auto. Sci. Eng. **11**, 627–632 (2014)
16. Lv, Y., Na, J., Yang, Q., Wu, X., Guo, Y.: Online adaptive optimal control for continuous-time nonlinear systems with completely unknown dynamics. Int. J. Control **89**, 99–112 (2016)
17. Na, J., Herrmann, G.: Online adaptive approximate optimal tracking control with simplified dual approximation structure for continuous-time unknown nonlinear systems. IEEE/CAA J. Automatica Sinica **1**, 412–422 (2014)
18. Na, J., Mahyuddin, M., Herrmann, G., Ren, X., Barber, P.: Robust adaptive finite-time parameter estimation and control for robotic systems. Int. J. Robust Nonlinear Control **25**, 3045–3071 (2015)

A Fast Precise-Spike and Weight-Comparison Based Learning Approach for Evolving Spiking Neural Networks

Lin Zuo, Shan Chen, Hong Qu(✉), and Malu Zhang

School of Computer Science and Engineering, University of Electronic Science and Technology of China, Chengdu 610054, People's Republic of China
hongqu@uestc.edu.cn

Abstract. Evolving spiking neural networks (ESNNs) evolve the output neurons dynamically based on the information presented in the incoming samples and the information stored in the network. In order to improve the learning efficiency of the existing algorithms for ESNNs, this paper presents a fast precise-spike and weight-comparison based learning algorithm, called PSWC. PSWC can dynamically add a new neuron or update the parameters of existing neurons according to the precise time of the incoming spikes and the similarities of the weights. The proposed algorithm is demonstrated on several standard data sets. The experimental results demonstrate that PSWC has a significant advantage in terms of speed performance and provides competitive results in classification accuracy compared with SpikeTemp and rank-order-based approach.

Keywords: Evolving spiking neural networks · Precise spike · Weight comparison · Classification

1 Introduction

Spiking neurons, considered as the third generation of artificial neurons, are more biologically plausible than traditional artificial neurons [1]. Additionally, these biologically plausible neurons show powerful computing capabilities. However, the complexity of spike sequences might limit their use in spiking neural networks(SNNs), which inspires us to develop efficient learning algorithms.

A number of learning algorithms, SpikeProp [2], ReSuMe [3] and others [4–6], have been developed for SNNs with fixed network structures. The disadvantage of fixed structure is that the size of hidden and output layers have to be satisfied as a priori [7]. To overcome those issue, Kasabov presented evolving spiking neural networks(ESNNs), SNNs with a dynamically adaptive structure [8]. The point of ESNNs is that the output neurons are added dynamically. ESNNs use the principles of evolving connectionist systems [9] which use the way to increase neurons dynamically to captures clusters of input data. For ENNs, the rank-order approach which pays close attention to rank ordering of spikes has been proposed in [7, 10]. However, different spike sequences may have the same rank ordering.

D. Liu et al. (Eds.): ICONIP 2017, Part III, LNCS 10636, pp. 797–804, 2017.
https://doi.org/10.1007/978-3-319-70090-8_81

To circumvent the issue, SpikeTemp which adopts the precise time has been proposed in [11]. Although the classification efficiency significantly improves, the time cost increases compared with the rank-order-based approach.

The main contribution of this paper lies in the development of a learning method, called PSWC, which is more appropriate for online systems. PSWC follows the dynamic structure. And the fire output neuron is determined by similarity, which is defined by the inverse of the Euclidean distance among the weight vectors of neurons. Therefore, PSWC is faster than traditional learning algorithm in terms of calculating speed. And meanwhile providing the precise time to the weight vectors makes the algorithm be sufficiently effective.

The rest of this paper is structured as follows. Section 2 describes the information encoding schemes and the SNN structure which includes the learning procedure. Section 3 presents the experimental results on some data sets from the UCI machine learning repository. The results obtained are compared with the results obtained from SpikeTemp and the rank-order-based approach. Finally, Sect. 4 concludes this paper and presents future directions.

2 Information Encoding Schemes and Network Structure

In this section, the scheme of the information encoding is described, followed by the architecture of the ESNNs. During learning, the training samples and their class labels which are given as $\{(X_1, C_1), (X_2, C_2), ... (X_n, C_n), ...\}$ are presented to the network. Each sample, denoted by the m-dimensional input features and the corresponding class label, has been presented to the network at interval of time T only once. In this paper, T has been set to 9 ms.

2.1 Information Encoding

A real valued input is supposed to be transformed into a spike pattern in SNNs. Population Encoding using Gaussian receptive field provides a strategy to map from real value to a series of spike for input neurons [2]. Each feature will be encoded by p dimension Gaussian receptive fields. For m dimension features $(x_1, x_2, ..., x_m)$, Population Encoding can encode them into $m * p$ spiking time. The mean value and the standard deviation for the ith feature at j^{th} receptive field are defined as:

$$\mu_i^j = I_{\min}^i + \frac{(2j-3)(I_{\max}^i - I_{\min}^i)}{2(p-2)} \tag{1}$$

$$\sigma_i^j = \frac{1}{\beta}\frac{(I_{\max}^i - I_{\min}^i)}{(p-2)} \tag{2}$$

where: $I_{\max}^i$ and $I_{\min}^i$ are the maximum and minimum value of the ith feature; β effects the range of overlap in different Gaussian receptive fields, which is set as 1.5 in this paper. The encoding results are determined by the Gaussian function:

$$\varphi_i^j = \exp(-\frac{(x_i - u_i^j)^2}{2(\sigma_i^j)^2}) \tag{3}$$

with the response value of each Gaussian function, we can calculate the spiking time for each input neuron:

$$t_i^j = \begin{cases} [T(1-\varphi_i^j)], \varphi_i^j \neq 0 \\ -0.01, \varphi_i^j = 0 \end{cases} \tag{4}$$

where T is the length of time interval. If the spiking time is closed to T, the input neuron is seen as silent and set as −0.01.

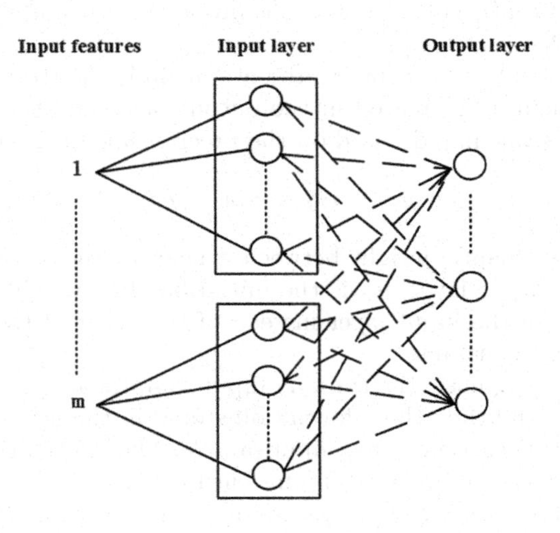

Fig. 1. Architecture of the evolving spiking neural network

2.2 Network Topology

Figure 1 shows the architecture of the Evolving Spiking Neural Network which consists of a layer of input neurons and a layer of output neurons. The input features are encoded into a set of spiking times fed to the neurons of the input layer using population encoding. Each neuron in the input layer represents a spike time and the number of neurons in this layer is $m * p$.

The neural network initially has no connections between the input layer and the output layer owing to the fact that the output layer has no neuron at the start of the learning procedure. When a sample is transmitted to the network, the output layer dynamically adds a neuron. The neurons in the input layer and the neurons in the output layer are fully connected. The neurons connections, i.e. the weight vectors, are established using the precise spike of the input layer. The weight vectors of each output neurons represent a pattern [8]. When a new output neuron is created, the weight vectors of the new one are compared with the already existing neurons using Euclidean distance. The closest output neuron in terms of weight vectors is the winner [8]. The class represented by the winner neuron and the class label of the training sample jointly determine the training methods. And after training, the predicted class for an incoming sample is the class of the winner output neuron.

2.3 Learning

The objective of the learning algorithm is to simulate the functional relationship between the input features and the class label. The weight vectors of the output neurons represent centres of clusters in the problem space and can be represented as fuzzy rules [12].

An individual output neuron represents an input pattern. For each input pattern, a new neuron is created in the output layer and the weight vector of the new neuron is calculated based on the precise time of the incoming spikes:

$$w_{ij} = w_0 + \exp(-t_i/\tau) \tag{5}$$

where: w_{ij} is the synaptic weight between a neuron i in the input layer and a neuron j in the output layer; w_0 is the initial baseline weight; t_i is the precise time of the spike in the input layer instead of the order of the first spike as in [13–16]; τ is a time constant.

During a training phase, the training input samples are presented to the network one by one. And then the information stored in the network are compared with the information carried by the input sample. The information represents the functional relationship between the input features and the class label of samples. The algorithm makes each sample choose one of the learning strategies:

(1) the strategy of adding a neuron: When the difference of the information stored in the network and the information carried by the input sample is relatively large, a new neuron in the output layer is added to record new information.
(2) the strategy of merging neurons: When the information of the input sample is sufficiently similar to the information stored in a already existing neuron in the network, the new neuron will be merged with the most similar one. The trained output neurons represent clusters of spatio-temporal spike patterns. Merging neurons and predicting the class label according to the similarities among neurons make it possible to achieve sufficiently fast learning, while providing the precise time to the weight vectors make the algorithm be sufficiently effective.

The detailed algorithm of the PSWC is presented in Algorithm 1.

3 Experiments and Results

In this section, PSWC is compared with the precise-spike based learning rule and the rank-order based learning rule. Four data sets were used for comparison in classification accuracy and running time performance.

Algorithm 1. Pseudo-code of the PSWC.

```
1  Set ESNN parameters: Th_sim, τ, w0 and Population Encoding parameters;
2  Establish neuron repository R;
3  for Each input sample (X_n, C_n) do
4      Encode input pattern into m * p spike time using population encoding;
5      Creat a new output neuron k for this input pattern;
6      Calculate the initial values of the weight vector w_k using the formula (5);
7      if R has no neuron representing the class of the input sample, i.e. C_n then
8          Add the neuron k to R;
9          set the weight vector as w_k;
10         set the initial number of trained sample as 0, i.e. N_k=0;
11     end
12     else if R has any neuron representing the class of the input sample then
13         Calculate the similarity of the neuron k and each neuron in R;
14         find the closest neuron q, q = arg max_j 1/(Σ_{j∈R} (w_k − w_j)^2);
15         record the similarity s(q) = 1/(w_k − w_q)^2;
16         record the class C_q;
17         if C_q ≠ C_n or (C_q = C_n and s(q) < Th_sim) then
18             Add the neuron k to R (including the setting of w_k and N_k);
19         end
20         else if C_q = C_n and s(q) > Th_sim then
21             merge the neuron k and q;
22             update the weight vector w_q = (w_q * N_q + w_k)/(N_q + 1);
23             increase the number of the trained sample by the neuron q;
24             N_q = N_q + 1;
25         end
26     end
27 end
```

3.1 Data Description and Simulation Results

The following benchmark data sets, downloaded from the UCI machine learning repository (http://www.ics.uci.edu/~mlearn/MLRepository.html), are used to test the performance of PSWC: (*i*) Iris; (*ii*) Wisconsin Breast Cancer (WBC); (*iii*) Abalone; (*iv*) Yeast. Each data set is divided into training sets and testing sets denoted by Tr and Ts respectively in Table 1. Besides, Table 1 lists some properties of each data set, including the size of data set (Instances), the number of features and the number of classes. And Accuracy_Tr/ Te (Ac_Tr/ Te) represents the training/testing accuracy. For example, the Iris data set consists of 150 instances with 50 samples in each class. There are 4 features in each sample. 10 random trials are conducted with 30 samples from each class, i.e. 90 samples are used for training and the remaining samples are used for testing. The results shown in Table 1 respectively are the mean and the best results obtained among the 10 random trials.

Table 1. Experiment and performance of chosen data sets

Database	Instance	Feature	Class	Tr	Te	Mean	Best
						Accuracy_Tr/Te(%)	Accuracy_Tr/Te
Iris	150	4	3	90	60	100/96.0	100/98.3
WBC	683	9	2	455	228	100/96.8	100/98.3
Abalone	4177	7	3	2000	2177	95.0/48.0	97.1/49.3
Yeast	1484	8	10	990	494	98.9/54.0	100/55.1

3.2 Analysis of Results

In the following table, Time is the running time in minutes which includes the time of training and testing. Demonstrated in Table 2, PSWC has high classification accuracies among all the training data sets and the testing data sets, such as Iris, WBC. The results in Table 2 demonstrate that, compared with the precise-spike based learning rule, classification accuracies of PSWC are higher on data sets, such as Abalone, Yeast, and comparable classification accuracies is achieved on the remaining data sets. Compared with the rank-order based learning rule, better accuracies are achieved for the Iris, Abalone, Yeast, and a comparable classification accuracy is achieved on the remaining data sets. In addition, the most important advantage is the running time performance. The simulation time of the algorithm is much less than the precise-spike based learning rule and the rank-order based learning rule on all the data sets.

Table 2. Performance comparison of chosen data sets

Database	PSWC		Precise spike approach		rank order approach	
	Ac_Tr/Te(%)	Time(min)	Ac_Tr/Te	Time	Ac_Tr/Te	Time
Iris	100/96.0	0.003	100/96.7	0.85	100/95.0	11.7
WBC	100/96.8	0.006	99.1/98.3	3.59	99.6/98.7	209.8
Abalone	95.0/48.0	1.07	45.7/47.8	6.56	44.5/44.8	66.0
Yeast	94.9/54.0	0.21	56.7/31.6	14.3	50.5/31.4	307.9

3.3 Evaluaton of Various Parameters Effect

In this section, Iris dataset was used to study the effect of various parameters on the performance of PSWC. These parameters consist of the number of Gaussian Fields (p), the threshold value of similarity (Th_sim), the time constant (τ) and the initial baseline weight (w_0).

In Fig. 2(a), the best testing performance can be reached when the number of Gaussian receptive fields is set to 25. And when the parameter p is set in the range of [25, 30], relatively high accuracies for both training and testing data sets are maintained. In Fig. 2(b), when the threshold value of similarity is set to 0.8, the best classification performance can be obtained. And the performance may be lower with the parameter increasing. In Fig. 2(c), the best classification performance can be obtained when the time constant is set to 45. And when the parameter is over 30, relatively high accuracies for both training and testing data sets will be maintained. And Fig. 2(d) shows that when the initial baseline weight is set to 0.01 or 0.6, the best classification performance can be obtained.

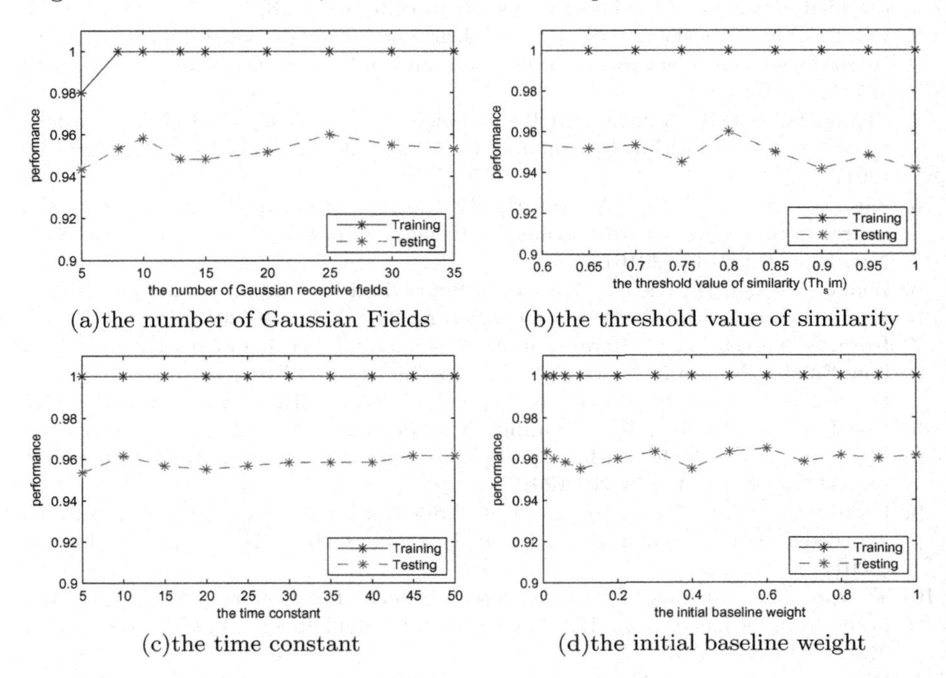

(a)the number of Gaussian Fields (b)the threshold value of similarity

(c)the time constant (d)the initial baseline weight

Fig. 2. Performance for different parameters

4 Conclusion and Future Works

In this paper, a new fast PSWC approach has been proposed to be used as an evolving spiking neural network classifier. The real valued input features are converted into a set of spiking sequences using the population encoding scheme. The spike patterns are fed to a two layered architecture and then the weight vectors are trained using the precise time and similarities of various neurons. A comparative study has been conducted between the PSWC and other two approachs. The results show the improvement of performance. PSWC provides competitive results in classification accuracy and speed performance. More experiments and comparisons are also another approach that can be applied. Moreover, the proposed approach might be attempted to be investigated as a clustering method.

Acknowledgments. This work was supported in part by the National Science Foundation of China under Grants 61573081, 61273308 and the Fundamental Research Funds for Central Universities under Grant ZYGX2015J062.

References

1. Maas, W.: Networks of spiking neurons: the third generation of neural network models. Neural Networks **10**, 1659–1671 (1997)
2. Bohte, S.M., Poutre, J.A.L., Kok, J.N.: Error-backpropagation in temporally encoded networks of spiking neurons. Neurocomputing **48**, 17–37 (2001)
3. Ponulak, F., Kasiski, A.: Supervised learning in spiking neural networks with ReSuMe: sequence learning, classification, and spike shifting. Neural Comput. **22**, 467–510 (2009)
4. Zhang, M., Qu, H., Xie, X.: EMPD: an efficient membrane potential driven supervised learning algorithm for spiking neurons. IEEE T. Cogn. Dev. Syst. **99**, 1–1 (2017)
5. Xie, X., Qu, H., Zhang, Y., Kurth, J.: Efficient training of supervised spiking neural network via accurate synaptic-efficiency adjustment method. IEEE Neur. Net. Lear. **99**, 1–14 (2016)
6. Zhang, M., Qu, H., Xie, X., Kurth, J.: Supervised learning in spiking neural networks with noise-threshold. Neurocomputing **219**, 333–349 (2017)
7. Dora, S., Suresh, S., Sundararajan, N.: A sequential learning algorithm for a Minimal Spiking Neural Network (MSNN) classifier. In: 2014 International Joint Conference on Neural Networks (IJCNN), pp. 2415–2421. IEEE Press, New York (2014)
8. Kasabov, N., Dhoble, K., Nuntalid, N., Giacomo, I.: Dynamic evolving spiking neural networks for on-line spatio- and spectro-temporal pattern recognition. Neural Networks **41**, 188–201 (2012)
9. Kasabov, N.: Evolving fuzzy neural networks for supervised/unsupervised online knowledge-based learning. IEEE Trans. Syst. Man Cybern. B Cybern. **31**, 902–918 (2001)
10. Thorpe, S., Gautrais, J.: Rank order coding. In: Bower, J.M. (ed.) Computer Neuroscience, pp. 113–118. Springer, Heidelberg (1998). doi:10.1007/978-1-4615-4831-7_19
11. Wang, J., Belatreche, A., Maguire, L.P., Thomas, M.M.: SpikeTemp: an enhanced rank-order-based learning approach for spiking neural networks with adaptive structure. IEEE Trans. Neural Networks Learn. **28**, 1–14 (2015)
12. Soltic, N.K.: Knowledge extraction from evolving spiking neural networks with rank order population coding. Int. J. Neural Syst. **20**, 437–445 (2010)
13. Wysoski, S.G., Benuskova, L., Kasabov, N.: On-line learning with structural adaptation in a network of spiking neurons for visual pattern recognition. In: Kollias, S.D., Stafylopatis, A., Duch, W., Oja, E. (eds.) ICANN 2006. LNCS, vol. 4131, pp. 61–70. Springer, Heidelberg (2006). doi:10.1007/11840817_7
14. Dhoble, K., Nuntalid, N., Indiveri, G., Kasabov, N.: Online spatio-temporal pattern recognition with evolving spiking neural networks utilising address event representation, rank order, and temporal spike learning. In: The 2012 International Joint Conference on Neural Networks (IJCNN), pp. 1–7. IEEE Press, New York (2012)
15. Dora, S., Subramanian, K., Suresh, S., Sundararajan, N.: Development of a self regulating evolving spiking neural network for classification problem. Neurocomputing **171**, 1216–1229 (2016)
16. Dora, S., Suresh, S., Sundararajan, N.: A sequential learning algorithm for a spiking neural classifier. Appl. Soft. Comput. **36**, 255–268 (2015)

An Energy-Aware Hybrid Particle Swarm Optimization Algorithm for Spiking Neural Network Mapping

Junxiu Liu[1], Xingyue Huang[1], Yuling Luo[1(✉)], and Yi Cao[2]

[1] Guangxi Key Lab of Multi-source Information Mining and Security, Faculty of Electronic Engineering, Guangxi Normal University, Guilin 541004, China
yuling0616@mailbox.gxnu.edu.cn

[2] Department of Business Transformation and Sustainable Enterprise, Surrey Business School, University of Surrey, Surrey GU2 7XH, UK

Abstract. Recent approaches to improving the scalability of Spiking Neural Networks (SNNs) have looked to use custom architectures to implement and interconnect the neurons in the hardware. The Networks-on-Chip (NoC) interconnection strategy has been used for the hardware SNNs and has achieved a good performance. However, the mapping between a SNN and the NoC system becomes one of the most urgent challenges. In this paper, an energy-aware hybrid Particle Swarm Optimization (PSO) algorithm for SNN mapping is proposed, which combines the basic PSO and Genetic Algorithm (GA). A Star-Subnet-Based-2D Mesh (2D-SSBM) NoC system is used for the testing. Results show that the proposed hybrid PSO algorithm can avoid the premature convergence to local optimum, and effectively reduce the energy consumption of the hardware NoC systems.

Keywords: Particle swarm algorithm · Genetic Algorithm · Spiking Neural Networks · Networks-on-Chip

1 Introduction

As an artificial neural network with the biological details, Spiking Neural Network (SNN) emulates information processing and communication capabilities of the mammalian brain [1]. Due to its biological properties and computational power, researchers aim to investigate custom hardware architectures to simulate information processing mechanisms of mammalian brain [2,3]. Recently, researchers used the Networks-on-Chip (NoC) strategy to interconnect the large-scale hardware SNNs [1,4,5]. A typical NoC architecture includes computing and communication subsystems. The computing subsystem consists of a number of processing elements (PEs). The communication subsystem is composed of the routers and the channels, which are responsible for the communication between the PEs [6]. For realizing SNNs, PEs are used to implement the basic functions

D. Liu et al. (Eds.): ICONIP 2017, Part III, LNCS 10636, pp. 805–815, 2017.
https://doi.org/10.1007/978-3-319-70090-8_82

of spike neurons. In order to achieve this, spike neurons should be assigned to the appropriate PEs. This process of assignment is defined as SNN mapping. The correspondence between the SNN neurons and the NoC PEs is a one-to-one correspondence, thus the mapping problem of the SNN belongs to a typical quadratic assignment problem. The heuristic algorithms, e.g. ant colony optimization [7], Genetic Algorithm (GA) [8] and simulated annealing [9] etc., can be used to solve the quadratic assignment problems. Compared with Ant Colony Optimization and GA, particle swarm optimization (PSO) [10,11] does not have complex parameters, which is easy for the implementation. Thus the PSO has been applied to the NoC mapping [12,13]. In this paper, the PSO is applied to solve the problem of SNN mapping. Meanwhile, in order to avoid the fast convergence and local optimum, this approach combines the mutation operations of GA. For the SNN hardware systems, energy consumption affects the system performance, e.g. high energy consumption reduces the lifetime of the system and also affect the reliability. Therefore, this paper proposes an energy-aware mapping algorithm for the hardware SNN. The rest is organized as follows. Section 2 presents the basic concepts and definitions of SNN mapping and Sect. 3 presents the proposed hybrid PSO algorithm. Section 4 reports the experimental results and performance analysis. Section 5 provides a summary.

2 SNN Mapping Problem

This section introduces the SNN mapping problem, which includes the basic concepts and definitions of SNN mapping. In this paper, the SNN mapping aims to assign the spiking neurons to the hardware NoC for specific applications under the optimization rules. The target is normally to minimize the cost, e.g., energy consumption or latency, to allow the SNN to achieve a good performance under the mapping. In this approach, the target is to minimize the energy consumption of the NoC system. It is assumed that one neuron in the SNN corresponds to one PE in the NoC, i.e. the function of each neuron can be designed using one PE. This process can be described by Fig. 1, where the neurons of the SNN are mapped to the hardware NoC system. A SNN communication graph (SNNCG) and a NoC architecture graph (NoCAG) [14] are employed for the mapping.

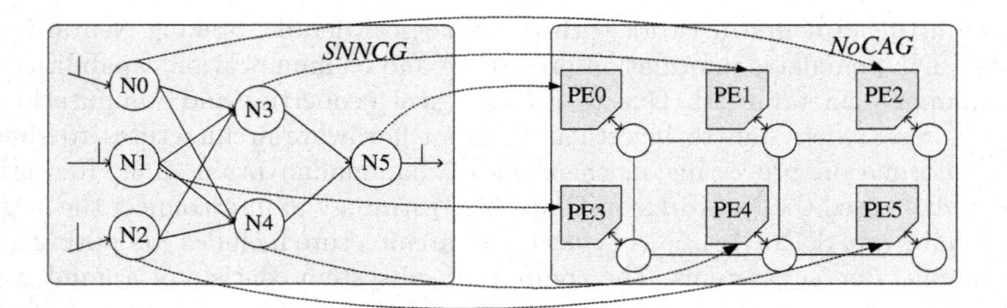

Fig. 1. SNN mapping process.

Definition 1: In the directed graph SNNCG (N, E), each vertex $n_i \in N$ represents a neuron, and each edge $e_{ij} \in E$ represents the communication path from the neuron n_i to n_j. The value t_{ij} of each edge e_{ij} represents the communication traffic between the neuron n_i to n_j, and the L_{ij} represents the maximum communication delay from the neuron n_i to n_j.

Equation (1) gives an example of communication matrix $T = [t_{ij}]$ between n neurons ($0 \le i \le n-1, 0 \le j \le n-1$). If there is no communication between the neuron n_i and n_j, then $t_{ij} = 0$.

$$T = \begin{bmatrix} 0 & t_{01} & t_{02} & \cdots t_{0,n-1} \\ t_{10} & 0 & t_{12} & \cdots t_{1,n-1} \\ t_{20} & t_{21} & 0 & \cdots t_{2,n-1} \\ \cdots & \cdots & \cdots & \cdots \; \cdots \\ t_{n-1,0} & t_{n-1,1} & t_{n-1,2} & \cdots \quad 0 \end{bmatrix} \quad (1)$$

Definition 2: In the directed graph NoCAG (V, P), each vertex $v_i \in V$ denotes a PE on the NoC. The edge $p_{ij} \in P$ denotes the communication path between the PE v_i and v_j. The f_{ij} on the edge p_{ij} represents the communication traffic from the PE v_i to v_j. The B_{ij} is the maximum bandwidth that the path p_{ij} can provide. The h_{ij} represents the distance from v_i to v_j, i.e. the number of hops in this approach.

In summary, given the SNNCG (N, E) and NoCAG (V, P), the SNN mapping problem is defined by

$$Mapping : N \to V \Rightarrow v_j = map(n_j) \in N, \exists v_j \in V. \quad (2)$$

The next section will introduce the details of the hybrid PSO algorithm for the SNN mapping problems.

3 Hybrid PSO Algorithm

3.1 Objective Function

According to the NoC communication energy model given in the approach of [15] and combined with the definition of Sect. 2, the energy consumption of 1 bit data from PE v_i to v_j is defined as

$$E_{ij} = (h_{ij} + 1) \times Es + h_{ij} \times E_l, \quad (3)$$

where E_s and E_l are the energy consumption of 1 bit transmission through the router and the adjacent channels, respectively. Then the NoC system communication energy E is given by

$$\begin{aligned} E &= \sum_{0 \le i \le n-1}^{0 \le j \le n-1} [(h_{ij}+1) \times E_s + h_{ij} \times E_l] \times t_{ij} \\ &= \sum_{0 \le i \le n-1}^{0 \le j \le n-1} t_{ij} \times h_{ij} \times (E_s + E_l) + \sum_{0 \le i \le n-1}^{0 \le j \le n-1} E_s \times t_{ij}. \end{aligned} \quad (4)$$

It can be seen that except $\sum_{0\leq i\leq n-1}^{0\leq j\leq n-1} t_{ij} \times h_{ij}$, the rest are constants. Thus the hybrid PSO algorithm objective function can be defined as

$$\min\{\sum_{0\leq i\leq n-1}^{0\leq j\leq n-1} t_{ij} \times h_{ij}\}, \tag{5}$$

s.t.

$$\forall n_i \in N \Rightarrow map(n_i) \in V \tag{6}$$

$$\forall n_i \neq n_j \Rightarrow map(n_i) \neq map(n_j) \tag{7}$$

$$\text{size(SNNCG)} \leq \text{size(NoCAG)} \tag{8}$$

$$\forall t_{ij} \leq B_{ij} \tag{9}$$

It can be seen that Eqs. (6) and (7) ensure that the neuron and the PE satisfy the one-to-one mapping requirement, and (8) and (9) ensure that the network size and bandwidth meet the requirements. The optimization goal of energy consumption is to minimize the sum of weighted distance between PEs.

3.2 Hybrid PSO

In the PSO, a solution for the optimization problem is a particle. Each particle has its own position, velocity, and a fitness value. A particle swarm consists of many particles. For each iteration, there is an optimal particle which has the best fitness. Other particles memorize and follow this optimal particle. Therefore, iteration process of the PSO is not completely random. According to the optimal particles at each iteration, the PSO can find the best solution using the algorithm update rule. In this paper, in order to apply PSO to the SNN mapping problem, several aspects should be considered, including (a) the particle position representation, (b) algorithm update rule, and (c) the velocity representation. They are discussed as follows.

(a) ***Particle position representation***. The particles are expressed by the D-dimensional position vector $X = (x_0, x_1, x_2, \cdots, x_{n-1})$, where x_i denotes the neuron number in the SNN and $0 \leq i \leq n-1$, n is the total number of neurons, the index i represents the PE where the neuron is to be placed in NoC. For example, the particle $X = (2, 1, 3, 4, 0)$ indicate that there is 5 neurons in the SNN, neuron #2 is placed on the first PE of NoC, neuron #1 is placed on the second PE, and same for the others.
(b) ***Algorithm update rule***. The traditional PSO is not suitable for optimizations in the discrete space, e.g. the mapping problems [16]. In order to overcome this drawback, this paper updates the positions of particles by "jump" operations. That is, for a multidimensional particle, each update

makes at least one dimension equal to one dimension of global optimal particle. Figure 2(a) shows an example. During the iterations of the hybrid PSO, the algorithm firstly selects the global optimal particle $X_g = (5, 1, 4, 0, 3, 2)$, then the particle $X_p = (4, 3, 2, 5, 1, 0)$ follows it to update. Each update makes at least one dimension to be the same as one dimension of global optimal particle. After four iterations, the current particle X_p is the same as the particle X_g.

(c) ***Velocity representation***. The velocity update rule changes the particle position in the traditional PSO where the velocity changes with the iterations. However, the hybrid PSO does not require the specific velocity update rule. Figure 2(a) shows that in this approach two dimensions will change after each jump. Therefore, the velocity is defined as 2.

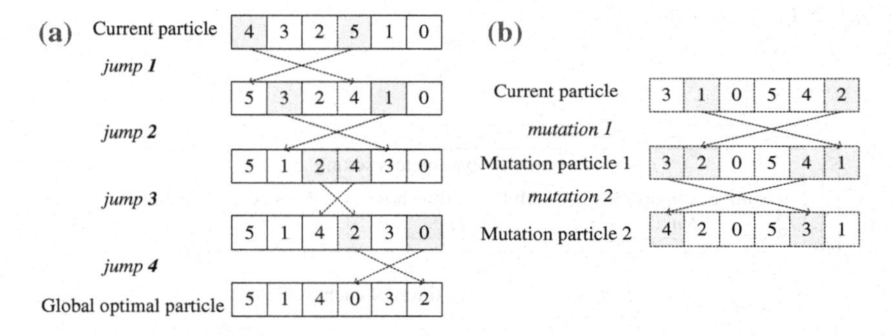

Fig. 2. Particle update process (a) and mutation operations (b).

The basic PSO has the disadvantages of being convergent too quickly and falling into local optimum easily. In order to overcome them, this paper defines the concept of similarity of particle swarm and makes the hybrid PSO combined with mutation operations of GA. When the similarity of particle swarm exceeds the threshold, all particles are mutated except the global optimal particle. The mutation operation can timely and effectively reduce the similarity of particle swarm, which enables the hybrid PSO to have the ability to search for global optimal solution.

The similarity of particle swarm (S_{ps}) can be calculated by

$$S_{ps} = \frac{\sum_{i=0}^{N-2} S_{p^i}}{N-1}, \tag{10}$$

where N denotes the total number of particles and S_{p^i} denotes the individual similarity of particle i which is given by

$$S_{p^i} = \frac{S_{dims}}{N_{dims}}, \tag{11}$$

where S_{dims} denotes the number of dimensions that the particle i is the same as the global optimal particle, N_{dims} denotes the number of total dimensions of particles.

When the similarity of particle swarm exceeds the threshold, each particle randomly swaps its own two dimensions until the similarity of particle swarm below the threshold. This process is the mutation process. Figure 2(b) is an example. When the similarity of particle swarm exceeds the threshold, the particle $X_c = (3, 1, 0, 5, 4, 2)$ has two mutation operations, and becomes $X_{m2} = (4, 2, 0, 5, 3, 1)$ eventually whose similarity is below the threshold. This process also applies to other particles in the particle swarm.

In the SNN mapping process, neurons and PEs are one-to-one correspondence. However, the number of neurons (N_n) and PEs (N_{pe}) is not always equal. If $N_n < N_{pe}$, some virtual neurons can be added to make them equal. But they are unused (i.e. only for solving the mapping) and can be removed after the hybrid PSO completes.

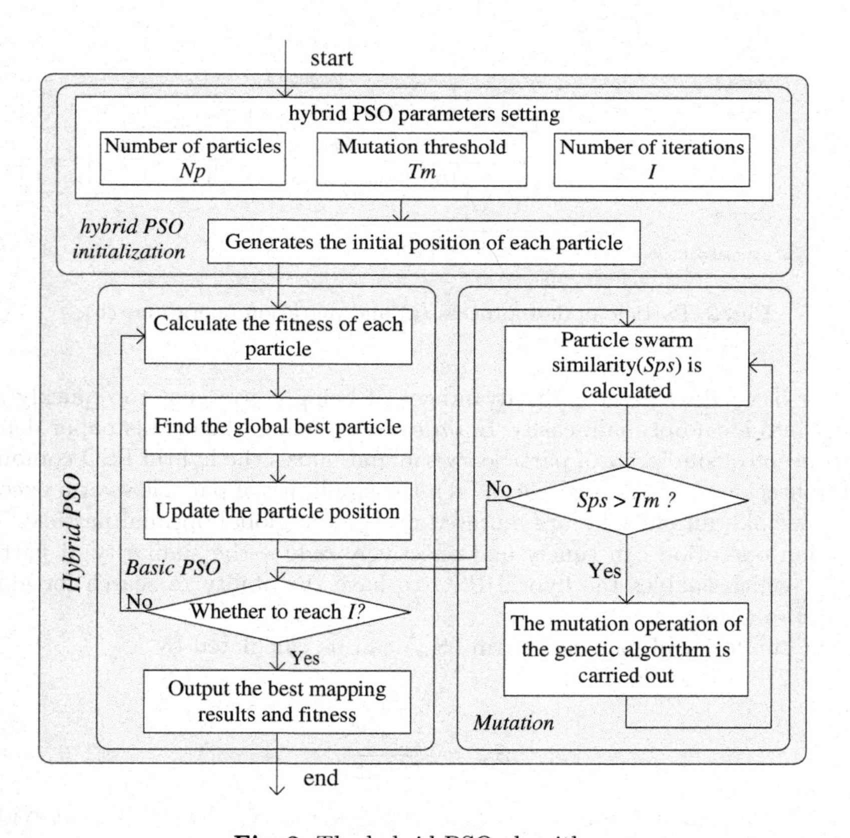

Fig. 3. The hybrid PSO algorithm.

3.3 The Hybrid PSO Running Process

Figure 3 describes the hybrid PSO running process, which includes three processes:

(a) ***The hybrid PSO initialization***. The main function of this process is to set the parameters of the hybrid PSO, and to add virtual neurons to let N_n equal to N_{pe} (if needed). The initial particle swarm is generated in this process.
(b) ***The basic PSO operation***. The basic operations of PSO are completed in this process. The PSO iterates and searches for the best mapping solution.
(c) ***The particle mutation operation***. The main function of this process is to avoid rapid convergence and reduce the possibility of falling into the local optimum, which enable the hybrid PSO to have the ability to search for global optimal solution.

4 Experimental Results

This section provides the experimental results for the hybrid PSO. In this paper, the SNN mapping algorithm is implemented in C++. The SNN used in this paper is shown in Fig. 4, which is a feedforward network and has an input layer, a hidden layer, and an output layer. The traffic of each communication path in the SNN is set to be the same in this experiment.

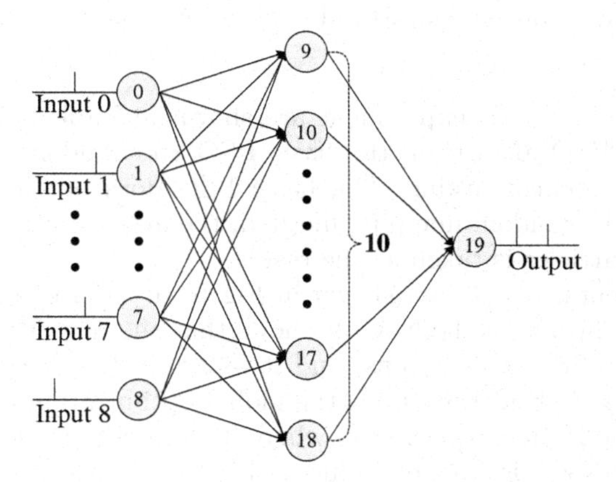

Fig. 4. Feedforward SNN architecture.

The hybrid PSO algorithm is validated on a Star-Subnet-Based-2D Mesh (2D-SSBM) NoC architecture which is based on our previous works [2,17], see Fig. 5. The 2D-SSBM NoC architecture has two hierarchical levels, i.e. the top layer is 2D mesh and the bottom is a star. For the local communication in the star topology, the packet can be transmitted via the node router, e.g. the path between the source node #14 to the destination node #8 is shown by the dash brown in Fig. 5. The tile router is employed to forward the global data transmission and it uses the XY routing algorithm, see the dash blue as an example.

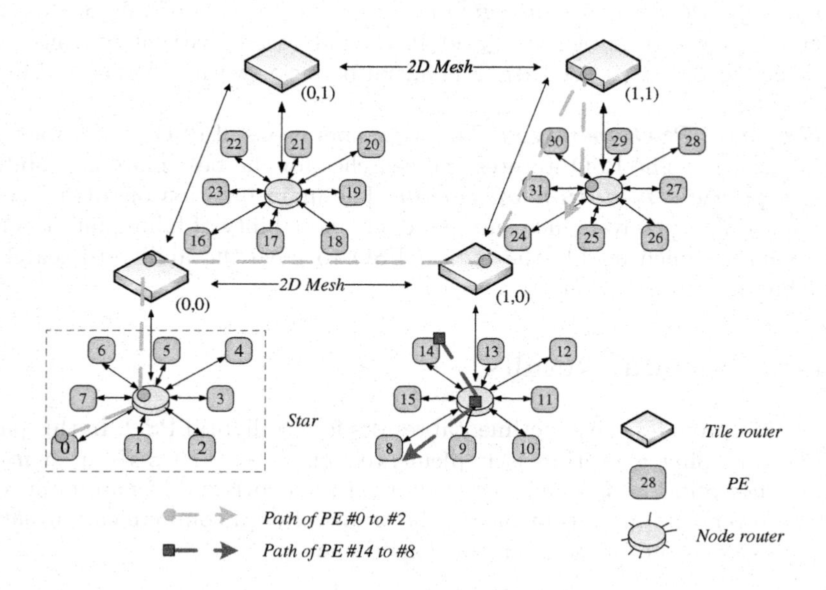

Fig. 5. Star-Subnet-Based-2D Mesh NoC architecture.

The parameters of this experiment are shown in Table 1. In order to evaluate the hybrid PSO algorithm, the basic PSO and random mapping method are employed for benchmarking. The basic PSO does not have the mutation operations. For the random mapping method, the average fitness of the random mapping of 30 particles is taken as the result.

The experimental results are shown in Fig. 6. In the Fig. 6(a), the fitness is the energy consumption of the NoC. It shows that the best fitness of basic PSO does not change after 24 iterations, and it falls into the local optimum. However, the proposed hybrid PSO finds the global optimization after a number of iterations (i.e. 130 in this experiment) where its fitness (i.e. energy consumption of the hardware NoC) is further reduced. Figure 6(b) shows that the particle swarm similarity of the basic PSO reaches 1 and particle swarm convergence completes early (i.e. at the iteration #46). However by using the hybrid PSO, the similarity of the particle swarm is always around the threshold which is due to the thresholding and the mutation operations. Therefore the hybrid PSO is capable to search the global optimal solution.

Table 1. Experiment parameter setting

	Parameters	Basic PSO	Hybrid PSO	Random
2D-SSBM NoC	2D mesh size		3×3	
	Number of nodes in the star subnetwork		4	
	Total number of nodes		36	
Feedforward SNN	Total number of neurons		20	
	Number of layers		3	
	Number of communication paths		100	
The hybrid PSO	Number of particles		30	
	Mutation threshold	N/A	0.5	N/A
	Total iterations	200	200	N/A

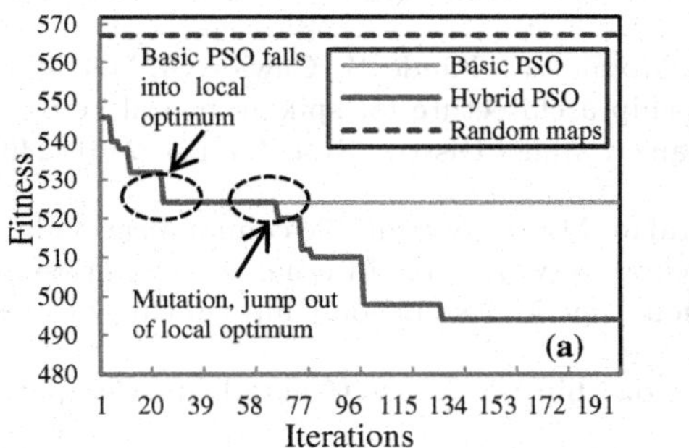

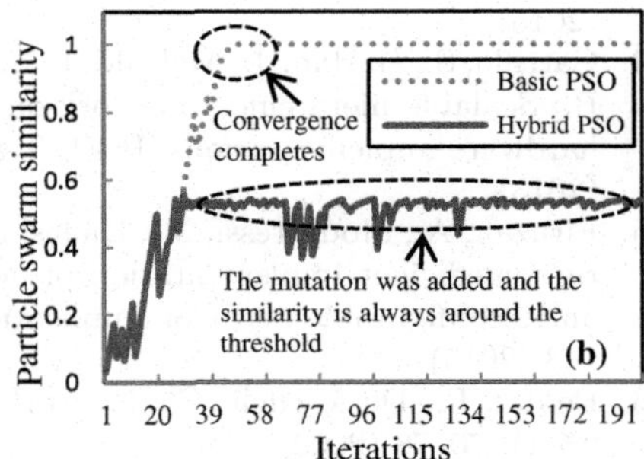

Fig. 6. The change of Fitness (a) and Particle swarm similarity (b) with iterations.

5 Conclusions

An energy-aware hybrid PSO algorithm for SNN mapping is proposed in this paper. The relationship between energy consumption and communication path is analysed. The proposed algorithm is able to search for the optimal solution through the basic PSO. In the meantime, the mutation operations via the GA is employed to avoid the premature convergence. Experimental results show that the proposed hybrid PSO algorithm can avoid the local optimum and achieve a lower energy consumption compared to the benchmarking algorithms. Future works include the optimization and further improvement of the proposed algorithm.

Acknowledgments. This research was supported by the National Natural Science Foundation of China under Grants 61603104 and 61661008, the Guangxi Natural Science Foundation under Grant 2015GXNSFBA139256 and 2016GXNSFCA380017, the funding of Overseas 100 Talents Program of Guangxi Higher Education, the

Research Project of Guangxi University of China under Grant KY2016YB059, Guangxi Key Lab of Multi-source Information Mining & Security under Grant MIMS15-07, and the Doctoral Research Foundation of Guangxi Normal University.

References

1. Liu, J., Harkin, J., Maguire, L.P., McDaid, L.J., Wade, J.J., Martin, G.: Scalable networks-on-chip interconnected architecture for astrocyte-neuron networks. IEEE Trans. Circuits Syst. I Regul. Pap. **63**(12), 2290–2303 (2016)
2. Liu, J., Harkin, J., Maguire, L.P., Mcdaid, L.J., Wade, J.J.: SPANNER: A self-repairing spiking neural network hardware architecture. IEEE Trans. Neural Netw. Learn. Syst. 1–14 (2017, inpress)
3. Akopyan, F., Sawada, J., Cassidy, A., Alvarez-Icaza, R., Arthur, J., Merolla, P., Imam, N., Nakamura, Y., Datta, P., Nam, G.J., Taba, B., Beakes, M., Brezzo, B., Kuang, J.B., Manohar, R., Risk, W.P., Jackson, B., Modha, D.S.: TrueNorth: design and tool flow of a 65 mW 1 million neuron programmable neurosynaptic chip. IEEE Trans. Comput. Aided Des. Integrated Circuits Syst. **34**(10), 1537–1557 (2015)
4. Carrillo, S., Harkin, J., McDaid, L.J., Morgan, F., Pande, S., Cawley, S., McGinley, B.: Scalable hierarchical network-on-chip architecture for spiking neural network hardware implementations. IEEE Trans. Parallel Distrib. Syst. **24**(12), 2451–2461 (2013)
5. Firuzan, A., Modarressi, M., Daneshtalab, M.: Reconfigurable communication fabric for efficient implementation of neural networks. In: International Symposium on Reconfigurable and Communication-Centric Systems-on-Chip (ReCoSoC), pp. 1–8 (2015)
6. Benini, L., De Micheli, G.: Networks on chips: a new SoC paradigm. Computer **35**(1), 70–78 (2002)
7. Gambardella, L.M., Taillard, É., Dorigo, M.: Ant colonies for the quadratic assignment problem. J. Oper. Res. Soc. **50**(2), 167–176 (1999)
8. Misevicius, A.: An improved hybrid genetic algorithm: new results for the quadratic assignment problem. Knowl. Based Syst. **17**(2–4), 65–73 (2004)
9. Paul, G.: Comparative performance of tabu search and simulated annealing heuristics for the quadratic assignment problem. Oper. Res. Lett. **38**(6), 577–581 (2010)
10. Eberhart, R., Kennedy, J.: A new optimizer using particle swarm theory. In: International Symposium on MICRO Machine and Human Science, pp. 39–43 (1995)
11. Kennedy, J., Eberhart, R.: Particle swarm optimization. In: Proceedings of IEEE International Conference on Neural Networks (ICNN), pp. 1942–1948 (1995)
12. Bahirat, S., Pasricha, S.: A particle swarm optimization approach for synthesizing application-specific hybrid photonic networks-on-chip. In: International Symposium on Quality Electronic Design (ISQED), pp. 78–83 (2012)
13. Sahu, P.K., Shah, T., Manna, K., Chattopadhyay, S.: Application mapping onto mesh-based network-on-chip using discrete particle swarm optimization. IEEE Trans. Very Large Scale Integr. (VLSI) Syst. **22**(2), 300–312 (2014)
14. Singh, A.K., Srikanthan, T., Kumar, A., Jigang, W.: Communication-aware heuristics for run-time task mapping on NoC-based MPSoC platforms. J. Syst. Architect. **56**(7), 242–255 (2010)
15. Hu, J., Marculescu, R.: Energy-aware mapping for tile-based NoC architectures under performance constraints. In: Proceedings of Asia South Pacific Design Automation Conference (ASP-DAC), pp. 233–239 (2003)

16. Kennedy, J., Eberhart, R.: A discrete binary version of the particle swarm algorithm. In: Proceedings of IEEE International Conference on Systems, Man, and Cybernetics. Computational Cybernetics and Simulation, pp. 4104–4108 (1997)
17. Liu, J., Harkin, J., McDaid, L., Martin, G.: Hierarchical networks-on-chip interconnect for astrocyte-neuron network hardware. In: Villa, A.E.P., Masulli, P., Pons Rivero, A.J. (eds.) ICANN 2016. LNCS, vol. 9886, pp. 382–390. Springer, Cham (2016). doi:10.1007/978-3-319-44778-0_45

A Dynamic Region Generation Algorithm for Image Segmentation Based on Spiking Neural Network

Lin Zuo, Linyao Ma, Yanqing Xiao, Malu Zhang, and Hong Qu(✉)

School of Computer Science and Engineering, University of Electronic Science and Technology of China, Chengdu 610054, People's Republic of China
hongqu@uestc.edu.com

Abstract. We propose a dynamic region generation algorithm for image segmentation based on spiking neural network inspired by human visual cortex that shows the tremendous capacity of processing image. The network structure generated by the proposed algorithm is automatically and dynamically. An image can be decomposed into several different shape and size of regions that look like superpixels. Merging these regions based on the color space similarity can extract contour. Dynamic network architecture brings stronger computing power. Dynamic generation method leads to more flexible network. Experimental results on BCDS300 dataset confirm that our approach achieves satisfactory segmentation results for different images compared with SLIC.

Keywords: Spiking neural network · Image segmentation · Pattern recognition

1 Introduction

Inspired by real neuron cell, Spiking Neuron Network (SNN) has unique way to transmit and calculate information. The biological neuronal cells transmit information by releasing and receiving pulses of charged ions. In order to simulate these pulses, input data is coded in spike timing sequence [1]. Constructing neuron models by mathematical method is also an important basis for realizing SNN [2]. Currently, Threshold-Fire Model and Conductance-Based Model are most popular neuron models in SNN field [3]. According to the Hebbian learning rule [4], neuron has the ability of learning. Therefore, the research of SNN learning algorithms based on Spike Timing Dependent Plasticity (STDP) [5] becomes an important hotspot in the study field. Given the resemblance of biological nervous system, SNN shows great potential in dealing with complex real-world problems [6] and accomplishing brain-like computation [7,8].

Medical classification and detection [9], human gesture recognition [10] and memory design [11] by SNN methods have been proposed recent years. A few works on application of membrane computing to image segmentation have been

D. Liu et al. (Eds.): ICONIP 2017, Part III, LNCS 10636, pp. 816–824, 2017.
https://doi.org/10.1007/978-3-319-70090-8_83

accomplished. Wu [12] developed a network with several ON/OFF color pathways for image segmentation inspired by the visual system. Lin [13] proposed a three-layer network using different ON/OFF pathways with genetic algorithm. Feature extraction were accomplished through retina-inspired SNN models [14]. [15] focused on the respective fields to extract progressively image features. In our research, we apply SNN to image preprocessing, which will reduce running time if SNN is implemented on hardware by the memristor [16] in the future.

The main contribution of this paper is to propose a dynamic region generation algorithm based on SNN. The dynamic network structure not only accelerates the calculation speed, but also gives SNN the ability to deal with multi-dimensional data. This algorithm showed a new perspective to generate regions like superpixels with a stronger bionic, creating the conditions for SNN feature extraction and classification. The rest of the paper is organized as follows. Section 2 describes the neural model we used, the architecture and segmentation algorithm we proposed. Section 3 shows the experimental results. Some conclusions are drawn in Sect. 4.

2 Network Structure and Segmentation Algorithm

2.1 Spiking Neuron Model

Our network uses a neuron model that mimics a visual cortex neuron, namely Spike Response Model (SRM) [17] which introduces a better nonlinear neuronal spike generation process, compared with Integrate-and-fire (IF). In the SRM model, the state of neuron i is only represented by its membrane potential. If there is no presynaptic spike, membrane potential will stay in resting potential. The arrival of presynaptic leads to a postsynaptic potential (PSP) which is the product of synaptic efficiency and the normalized kernel, described by

$$PSP_i(t) = \sum_j \omega_{ij}\varepsilon_{ij}(t - t_j), \tag{1}$$

where ω_{ij} is synaptic efficacy and the arriving time of neuron i is t_j. The spike-response function ε describes the response of u_i to presynaptic spike, given by

$$\varepsilon(t) = \frac{t}{\tau} * e^{1-\frac{t}{\tau}}(t > 0), \tag{2}$$

where τ determines the rise and decay time of the PSP.

When membrane potential exceeds the threshold ϑ under the sum of several presynaptic spikes, an output spike will be generated and membrane potential starts getting back to resting potential at the same time. This is the firing mechanism of SNN. Supposed that the last firing time of neuron i is $\hat{t}_j$. The membrane potential u_i of neuron i at time t can be expressed as

$$u_i(t) = \eta(t - \hat{t}_j) + \sum_j \omega_{ij}\varepsilon_{ij}(t - t_j). \tag{3}$$

Once membrane potential u_i starts dropping after exceeds the neuron threshold ϑ, the neuron can not fire again. This is called refractory period described by $\eta(t-\hat{t}_j)$. In order to save computational cost and simplify model, setting resting potential to zero while still having essential spiking neuron properties.

2.2 Architecture of Model

A dynamic region generation algorithm based on SNN performs well to image segmentation. Network modifies parameters and architecture by automatically searching the inherent laws and essential attributes in the samples. Uncertain network architecture brings stronger computing power. Dynamic generation makes the whole network more flexible. Inspired by the findings that there are multiple receptive fields existed in the visual cortex and each of them makes different response to varied patterns [18], the architecture for neural network is shown in Fig. 1(a).

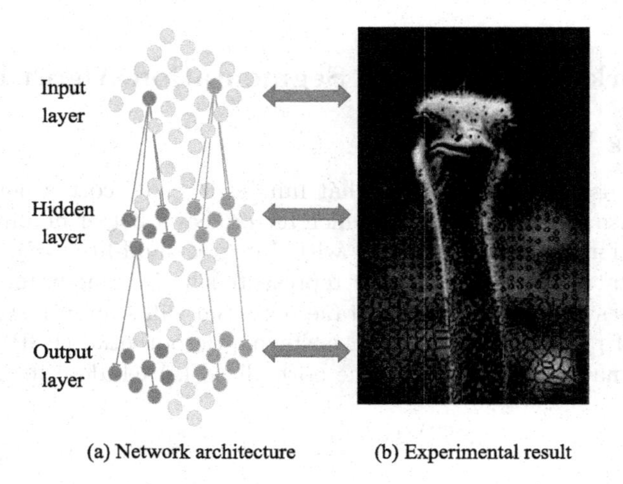

Fig. 1. The process of SNN-based segmentation algorithm. In the first layer, blue nodes and orange nodes are two initial fire-point corresponding to the first layer in (a). After several time steps, they fire some neighborhood nodes respectively shown in the second layer. This process repeats over and over again until time exceeds the specified range. After that, the image is divided into several regions showed in layer three of (b). (Color figure online)

Each input neuron is regarded as an independent network started with no hidden layer neurons. Each neuron in hidden layers represents a region that needs to be aggregated. Firing mechanism connects the hidden neurons to the input neurons. The output of upper layer is entered as the input in next layer. When time exceeds the specified range, this process stops and each input neuron forms different size of sub-networks. In essence, the input neurons resemble retinal neurons and these networks seem like receptive fields in the visual cortex.

2.3 SNN-Based Segmentation Algorithm

In this section we detail our SNN-based segmentation algorithm, designed to deal with variety of image conditions. It contains three critical strategies. First, selecting the initial fire points and using those pixels as the first layer of SNN. Second, calculating weights of lower layer neurons through the dissimilarity among the adjoining pixels. Third, dynamically building the tree-like network structure to get the final segmented regions.

Initial Fire-Points. In the image, each pixel is regarded as a neuron. The whole SNN structure will be dynamically generated during the algorithm implementation, so that each image has a different network structure. The ultimate goal is to form a forest of neurons, each tree representing a segmented region. Our premier task is to find a root node for each region tree, which is also called the initial firing spiking neuron. The image is firstly divided into equally-sized patches equaled to the number of input neurons in RGB color space. Then we calculate and compare the gradient between pixel i and its adjacent pixels j, select the points with the smallest gradient in each partition as the initial fire points. This is done to avoid selecting an initial node on an edge.

To calculate the gradient, we transfer the image from RGB color space to LAB color space and the gradient is defined by the following formula

$$GRA = \sqrt{(L_i - L_j)^2 + (A_i - A_j)^2 + (B_i - B_j)^2}, \quad (4)$$

where GRA is the gradient between pixel i and pixel j. L, A, B are pixel numbers in three channels of LAB color space respectively.

Weight Setting. The weight of each neuron is related to the value of corresponding pixels and the value of surrounding pixels. More precisely, it is associated with the difference between adjacent pixel values. It is expected that the more different the adjacent pixel values are, the more difficult the pulses pass between neurons, the greater the neuron weight should be set. Obviously, the weight and pixel difference are correlated negatively. A negative correlation function will do the job to convert the pixel difference to the neuron weight. Different functions will lead to different neural network structures and varied region segmentation. In this paper we propose two negative correlation functions performing well in image region segmentation and the comparison of two methods will be elaborated in Sect. 3.

We firstly use the Euclidean distance D_r in RGB color space to represent the dissimilarity between adjacent pixels, determined by the following equations

$$D_r = \sqrt{(R_i - R_j)^2 + (G_i - G_j)^2 + (B_i - B_j)^2}, \quad (5)$$

where i and j are the adjacent pixels, and R, G, B are pixel values in three channels of RGB color space respectively.

It is necessary to limit the range of distance so as to avoid the firing time too scattered and reduce computation cost. Thus, we propose the first method

$$D_1 = \alpha\sqrt{(1 - D_r)} - \gamma_1. \tag{6}$$

Another way to set weight is represented as follows

$$D_2 = \alpha\sqrt{(1 + e^{\beta * D_r})} - \gamma_2, \tag{7}$$

where $\alpha, \beta, \gamma_1, \gamma_2$ are all constant.

Building Network Structure. The output of SNN-based segmentation algorithm is segmentation regions, transformed from the final network structure. Since there is only root nodes at the beginning, leaf nodes will be dynamically added into the tree network in the specified time interval. When more than one different tree-like networks snatch the same node at the same time, a conflict may occur. The competitive method is that which fires this node first will aggregate this node. SRM model will be used to determine which tree will first get current leaf node. As time goes on, the tree-like network structure is also growing. When the segmentation areas become larger, the number of pixels that are not classified is gradually reduce.

The number of segmented regions is determined by the size and complexity of the image. Since spiking neurons transmit information in spiking time series, time window $[0, T_{max}]$ is given in order to build a spiking network. Starting fire from the root node, each moment we calculate the voltage value $h_{ij}(t)$ of the postsynaptic neuron node P_j. If $h_{ij}(t)$ exceeds the threshold ϑ, node P_j will immediately be added into region S_j. When t reaches T_{max}, the final set of regions S_j can be obtained.

2.4 Regions Merging

Selective search [19] is used in our superpixel-like segmented regions merging. Merging most regions will reduce the difficulty of distinguishing between the target and background, qualified for image preprocessing methods. In our research, we can visually intuitively compare the merged regions with the original image to verify the validity of the segmentation algorithm.

3 Experiment and Results

The algorithm is simulated in Matlab using different settings for the BCDS300 dataset [20,21] with time step of 1s. Considering the computation cost, we scale each image in BCDS300. This dataset contains 200 color level images along with ground truth segmentations and another 100 images for testing. The results below are based on the two stages of our algorithm. One compares the performance of segmenting small regions between our algorithm and superpixels

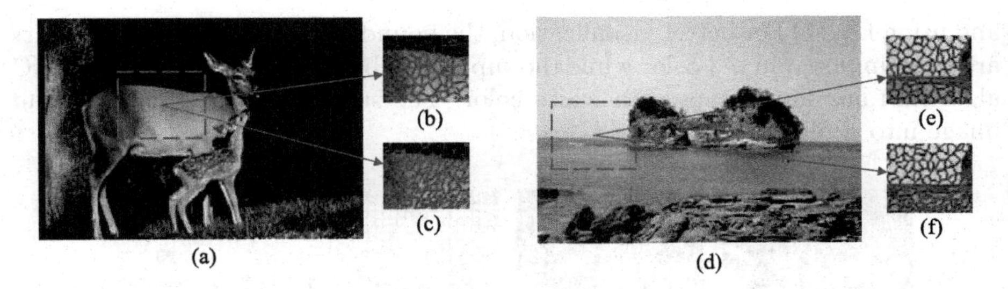

Fig. 2. Result of two weight-setting methods. The basic issues are the same, but each result is different. (b), (e) are segmentation result using Eq. (6) and (c), (f) are segmentation result using Eq. (7). (c) has a better performance than (b) on the deer's back.

generated by Simple Linear Iterative Cluster (SLIC) [22]. Another shows the merging result of selective search. The following parameters for SNN are used in our experiment: $\tau = 4$, $T_{max} = 10$. These parameters are determined by trail and error.

3.1 Effect of Different Weight Setting

In the first experiment, we analyze the effect and efficiency of the different weight setting methods in the same image. Since SNN-based segmentation algorithm is an unsupervised method, weight-setting plays a critical role in segmentation. Each methods has advantage in different parts of image. The detail result of each methods is shown in Fig. 2. Figure 2(b), (e) correspond to Eq. (6) and Fig. 2(c), (f) correspond to Eq. (7) where $\alpha = 1$, $\beta = 0.25$, $\gamma_1 = 5$, $\gamma_2 = 1.8$.

Comparing Fig. 2(b) with Fig. 2(c), although the regions of Fig. 2(b) are more irregular, the edge is more in line with the deers back in Fig. 2(c). As for the second input image, Fig. 2(e) has a better result than Fig. 2(f), considering the regions on the reef are neater. The experiment result shows that although the internal segmentation is a little messy, D_2 details the edge better. D_1 can lead to neat segmented regions, no matter which part of the image. We can make a choice based on the actual expectation. For example, in our research D_1 is used to compare with superpixels, because an excellent and clear segmentation is required in the whole image. When it comes to merging, the internal division is less important than the fitting of edge, so we use D_2 to adherent to the image boundaries.

3.2 Comparison with SLIC

In the second experiment, we evaluate the segmentation results visually. The segmentation results of the two images are shown in Fig. 3(a) and (c). Moreover, SLIC results are given by Fig. 3(b) and (d), respectively. To set initial fire-points, we divide the image into equal-size square patches and the length of square patch is 10, making a trade-off between efficiency and the quality of segmentation. In each square patches, choosing the minimum gradient node as initial fire-point and using Eq. (7) For better visualization, the boundaries of SNN-based segments are superimposed in red color while the superpixel boundaries recovered by SLIC algorithm are superimposed in white color. The superpixels tend to divide an image into similar-sized regions.

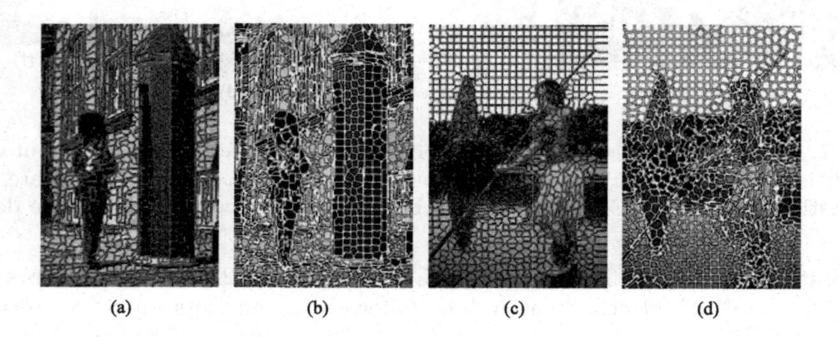

Fig. 3. Segmentation results compared with SLIC. (Color figure online)

These segmentation results indicate that the SNN-based segmentation algorithm achieves good segmentation results on these images. Compared to traditional method based on k-means, the proposed method has a stronger bionic. The input neurons resemble retinal cells and the dynamically generated hierarchy network may help us understand how the visual cortex extracts the information from image.

3.3 Merging Result

In the third experiment, we merge the superpixels together using selective search algorithm as regions can yield more information than superpixels. Figure 4 compares the original image to the result after processing with SNN-based segmentation algorithm and selective search. Figure 4(a) presents the original image. Figure 4(b) shows the segments produced by SNN-based algorithm. In the process, several parameters are set. We gradually reduced lots of small regions to just 40 larger regions or less. Figure 4(c) visually shows that segmented regions fit the edge of bird and branches perfectly, adequate for detection.

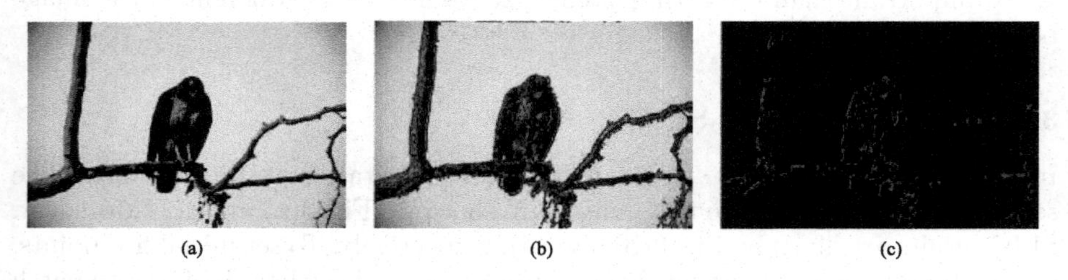

Fig. 4. The result of regions merging. (a) Original input data. (b) The intermediate results. (c) The final segmented object in black background.

4 Disucssion and Conclusion

This paper presented a dynamic region generation algorithm for image segmentation. A great performance has been achieved through SNN-based segmentation algorithm. Regions are generated by dynamically constructing network structures. Each pixel is regarded as a spiking neuron. Multiple input neurons produce varied size and shape of networks. Each network represents a segmented region. Unlike traditional segmentation methods, our method integrates with biological mechanism closely. Morphological diversity comes from the Multiple receptive fields in visual cortex. Firing mechanism comes from the way in which neurons transmit information. In order to give visual contrast, a merging algorithm successfully combines similar regions together. We tested our model ın the BCDS300 dataset, compared with superpixel, and received a satisfactory result.

Experiments show that SNN-based region segmentation algorithm we proposed has the ability to segment the image, as a pre-processing of image recognition. Also, the result greatly supports the argument that SNN has a strong computational potential and shows the value of SNN in the field of image processing application. Due to special biological characteristics, excellent effect has been achieved and no worse than superpixels. Although on the simulation platform, it is difficult to catch up with the efficiency of superpixels, SNN-based segmentation method will be much quicker if implemented in the hardware. Researches about using memristors to build SNN has made great progress. The next work is to combine all SNN-based image processing methods to build a complete SNN-based image recognition system, including segmentation, feature extraction and classification. The entire system will be transferred to the hardware to get better efficiency and less power consumption. Furthermore, we plan to combine SNN-based segmentation with deep learning. For example, using the segmentation results as the input to a new SNN combined with convolution neural network.

Acknowledgments. This work was supported in part by the National Science Foundation of China under Grants 61573081, 61273308 and the Fundamental Research Funds for Central Universities under Grant ZYGX2015J062.

References

1. Ghosh-Dastidar, S., Adeli, H.: Spiking neural networks. Int. J. Neural. Syst. **19**, 295–308 (2009)
2. Qu, H., Xie, X., Liu, Y., Zhang, M., Lu, L.: Improved perception-based spiking neuron learning rule for real-time user authentication. Neurocomputing **151**, 310–318 (2015)
3. De Berredo, R.C.: A review of spiking neuron models and applications. Doctoral dissertation, M.Sc. Dissertation, University of Minas Gerais (2005)
4. Wolters, A., Sandbrink, F., Schlottmann, A., Kunesch, E., Stefan, K., Cohen, L.G., Classen, J.: A temporally asymmetric Hebbian rule governing plasticity in the human motor cortex. J. Neurophysiol. **89**, 2339–2345 (2003)
5. Masquelier, T., Guyonneau, R., Thorpe, S.J.: Competitive STDP-based spike pattern learning. Neural. Comput. **21**, 1259–1276 (2009)

6. Qu, H., Yang, S.X., Willms, A.R., Yi, Z.: Real-time robot path planning based on a modified pulse-coupled neural network model. IEEE Trans. Neural. Netw. **20**, 1724–1739 (2009)
7. Xie, X., Qu, H., Yi, Z., Kurths, J.: Efficient training of supervised spiking neural network via accurate synaptic-efficiency adjustment method. IEEE Trans. Neural Netw. Learn. Syst. **28**, 1411–1424 (2017)
8. Zhang, M., Qu, H., Belatreche, A., Xie, X.: EMPD: an efficient membrane potential driven supervised learning algorithm for spiking neurons. IEEE Trans. Cogn. Dev. Syst (2017)
9. Ghosh-Dastidar, S., Adeli, H.: Improved spiking neural networks for EEG classification and epilepsy and seizure detection. Integr. Comput. Aided. Eng. **14**, 187–212 (2007)
10. Meng, Y., Jin, Y., Yin, J.: Modeling activity-dependent plasticity in BCM spiking neural networks with application to human behavior recognition. IEEE Trans. Neural. Netw. **22**, 1952–66 (2011)
11. Ang, C.H., Jin, C., Leong, P.H., van Schaik, A.: Spiking neural network-based auto-associative memory using FPGA interconnect delays. In: Field-Programmable Technology, pp. 1–4. IEEE Press (2011)
12. Wu, Q., McGinnity, T.M., Maguire, L., Cai, R., Chen, M.: A visual attention model based on hierarchical spiking neural networks. Neurocomputing **116**, 3–12 (2013)
13. Lin, X., Wang, X., Cui, W.: An automatic image segmentation algorithm based on spiking neural network model. In: Huang, D.-S., Bevilacqua, V., Premaratne, P. (eds.) ICIC 2014. LNCS, vol. 8588, pp. 248–258. Springer, Cham (2014). doi:10.1007/978-3-319-09333-8_27
14. Sun, Q.Y., Wu, Q.X., Wang, X., Hou, L.: Fruit Image Segmentation Based on a Colour Perception Neural Network Inspired by the Retina Structure. ATLANTIS Press (2015)
15. Kerr, D., McGinnity, T.M., Coleman, S., Clogenson, M.: A biologically inspired spiking model of visual processing for image feature detection. Neurocomputing **158**, 268–280 (2015)
16. Afifi, A., Ayatollahi, A., Raissi, F.: Implementation of biologically plausible spiking neural network models on the memristor crossbar-based CMOS/nano circuits. In: IEEE ECCTD 2009, pp. 563–566. IEEE Press (2009)
17. Gerstner, W., Kistler, M.: Spiking Neuron Models: Single Neurons, Populations, Plasticity. Cambridge University Press, Cambridge (2002)
18. Hosoya, T., Baccus, S.A., Meister, M.: Dynamic predictive coding by the retina. Nature **436**, 71 (2005)
19. Van de Sande, K.E., Uijlings, J.R., Gevers, T., Smeulders, A.W.: Segmentation as selective search for object recognition. IEEE ICCV 2011, pp. 1879–1886 (2011)
20. Segmentation evaluation database. http://www.wisdom.weizmann.ac.il/vision/Seg_Evaluation_DB/index.html
21. Alpert, S., Galun, M., Brandt, A., Basri, R.: Image segmentation by probabilistic bottom-up aggregation and cue integration. IEEE Trans. Pattern Anal. Mach. Intell. **34**, 315–327 (2012)
22. Achanta, R., Shaji, A., Smith, K., Lucchi, A., Fua, P., Ssstrunk, S.: SLIC superpixels compared to state-of-the-art superpixel methods. IEEE Trans. Pattern Anal. Mach. Intell. **34**, 2274–2282 (2012)

Global Stability of Complex-Valued Neural Networks with Time-Delays and Impulsive Effects

Dongwen Zhang, Haijun Jiang(✉), Cheng Hu, Zhiyong Yu, and Da Huang

College of Mathematics and System Sciences,
Xinjiang University, Urumqi 830046, Xinjiang, China
jianghai@xju.edu.cn

Abstract. The global exponential stability problem for a class of complex-valued recurrent neural networks with both asynchronous time-varying delays and impulse is concerned in this paper. By using Schur complement and Lyapunov functional, some new sufficient criteria to ascertain globally exponential stability of the equilibrium point are obtained in terms of linear matrix inequality. An example is given to illustrate the effectiveness of the results.

Keywords: Stability criterion · Neural networks · Time delays

1 Introduction

In the nearest decades, it has been proved that analyzing the recurrent neural networks have attracted more and more scientific research interesting due to their widespread applications and with it produce various classes of artificial neural networks, such as Hopfled neural network, recurrent neural network, cellular neural network. Moreover, their dynamic behaviors have also been extensively investigated and numerous excellent results related to the subject have been studied. As an extensive extension of real-valued recurrent neural networks, complex-valued recurrent neural networks with complex-valued external inputs, connection weight matrix, activation function and states become strongly desired due to their practical applications in various aspects, such as, face detection, motion control, associative memory design and control and optimization.

However, in electronic implementation of recurrent neural networks, because of the information communications between the different neurons, various kinds of time delay parameters are inevitable considered into the research equations that depict the recurrent neural network models. On the one hand, the appearance of time delays could lead to oscillation or unstable behaviors in the complex-valued neural networks. In [1], authors pointed out that the interconnections of different nodes or computers are generally asynchronous in practice, that is to say, it is inevitable for us to deal with different time delays between different nodes. In recent years, time delays have been extensively studied in the complex-valued recurrent neural networks by analyzing their dynamic behaviors because

D. Liu et al. (Eds.): ICONIP 2017, Part III, LNCS 10636, pp. 825–835, 2017.
https://doi.org/10.1007/978-3-319-70090-8_84

of their applications in engineering optimization, signal processing, bidirectional associative memory, etc (see [2]), and the references therein. On the other hand, the impulsive effect is also presented in the scientific research field that also lead to some unstable phenomena or others nature phenomena, in which some of these phenomena even can not be explained by means of we have already known. Recently, there has been increasing interesting attention on the stability analysis of impulsive complex-valued neural networks with time delays. For example, there are some literatures in the area. In [3], global stability of complex-valued neural networks on time scales has been researched to guarantee the ψ-global exponential stable based on delta differential operator and a special discontinuous conditions. In [4], the authors studied a array of generalized discrete complex-valued recurrent neural networks and discussed the existence of a unique equilibrium pattern and proposed a sufficient condition to guarantee the global exponential stability of equilibrium point. In [5], authors exhibited a complex-valued recurrent neural network and assumed that it connection weight matrix was Hermitian with nonnegative diagonal elements in order to preserve the asymptotically stability of the network. However, in [6], the condition of connection weight matrix was weakened by the authors in [5] in addition to all other things being equal. In order to study the stable problems, there are several methods to analyze the stability of neural networks, such as Lyapunov method, energy function method and synthesis method. However, compared with making great efforts of the real valued-neural recurrent neural networks, the complex-valued networks develops slowly and little progress because of too many complicated properties.

As one of hot topics in the field of artificial neural networks, Cohen-Grossberg neural networks have lots of development potential and obviously advantages. Motivated by the above discussions, the basic idea of this paper is to solve more extensive complex-valued Cohen-Grossberg neural network. In the literature, we will derive three classes of activation functions and systematically explore the global stability problem of complex-valued recurrent neural networks with both mixed time-delays and impulsive effect.

The remainder of this paper is organized as follows. In Sect. 2, we give some useful Lemmas, definitions and model for complex-valued neural network with time delays. We obtain some global stability results for complex-valued recurrent neural networks with three classes of activation in Sect. 3. An example is given to demonstrate the effectiveness of our results in Sect. 4. Conclusions are given in Sect. 5.

2 Preliminary

In this section, first of all, some basic concepts about inequality and nonsingular M-matrix are illustrated. We mainly consider the following complex-valued recurrent neural networks with both mixed time delays and impulsive effect.

$$\begin{cases} \dot{z}_i(t) = -e_i(z_i(t))[d_i z_i(t) - \sum_{j=1}^n a_{ij} f_j(z_j(t)) - \sum_{j=1}^n b_{ij} g_j(z_j(t-\tau_{ij}(t)) \\ \qquad - \sum_{j=1}^n c_{ij} \int_{-\infty}^t k_{ij}(t-s) h_j(z_j(s))ds + I_i], \qquad t \neq t_k, \\ z_i(t) = p_{ik}(z_1(t^-), \ldots, z_n(t^-)) + q_{ik}(z_1(t-\tau_{i1})^-, \ldots, z_n(t-\tau_{in})^-) + J_i, \; t = t_k \end{cases} \tag{1}$$

For $t \geq 0$, where $z_i(t) \in C$ is the state of the ith neuron at time t from the neuron field F_u; $e_i(z_i(t)) : C^n \longrightarrow R_+^n$ denotes matrix amplification function; $f_j(z_j(t)), g_j(z_j(t-\tau_{ij}(t)), h_j(z_j(t)) \in C$ are the activation functions; $\tau_{ij}(t)$ represents to the transmission delay; $D = diag\{d_1, \ldots, d_n\} \in R^n$ is the self-feedback connection weight matrix, where $d_i > 0$; $A, C, B \in C^{n\times n}$ are the connection weight matrices; $I = (I_1, \ldots, I_n) \in C^n$ is the external input vector; $K_{ij} : [0, +\infty) \longrightarrow [0, +\infty)$ is the delay kernel function. The second part is discrete part of model (1) where $z_j(t^-)$ and $z_j((t-\tau_{ij}(t))^-)$ denote the left limit of $z_j(t)$ and $z_j((t-\tau_{ij}(t)))$, respectively; $p_{ik}(z_1(t^-), \ldots, z_n(t^-))$ denotes impulsive perturbations of the ith unit at time t_k, and $q_{ik}(z_1(t-\tau_{i1})^-, \ldots, z_n(t-\tau_{in})^-)$ represents impulsive perturbations of the ith unit at time $t_k - \tau(t_k)$; The initial conditions of system (1) are in the form of $z_i(s) = \phi_i(s), s \in (-\infty, 0]$, where ϕ_i is bounded and continuous on $(-\infty, 0]$.

For a complex-valued recurrent neural network, the choice of activation functions is one of the fundamental challenges. Any regular analytic function cannot be bounded unless it reduces to a constant according to *Liouville's theorem*. In other words, activation functions in complex-valued neural networks cannot be both bounded and analytic. Throughout this literature, let $f_j(\cdot)$ be a set of complex-valued function, there classes of complex-valued activation function satisfying the following assumptions.

Assumption 1. For any i, there exist positive diagonal matrices.
$L = diag\{l_1, \ldots, l_n\}$, $K = diag\{k_1, \ldots, k_n\}$, and $M = diag\{m_1, \ldots, m_n\}$ such that

$$\begin{cases} |f_i(x_1) - f_i(x_2)| \leq l_i|x_1 - x_2|, x_1, x_2 \in C \\ |g_i(x_1) - g_i(x_2)| \leq k_i|x_1 - x_2|, x_1, x_2 \in C \\ |h_i(x_1) - h_i(x_2)| \leq m_i|x_1 - x_2|, x_1, x_2 \in C. \end{cases} \tag{2}$$

Assumption 2. The delay kernel $K_{ij} : [0, +\infty) \rightarrow [0, +\infty)$ is real valued nonnegative continuous function and satisfies

$$\int_0^{+\infty} e^{xs} K_{ij}(s)ds = G_{ij}(x) \tag{3}$$

where $G_{ij}(x)$ is continuous function in $[0, r), r > 0$, and $G_{ij}(0) = 1$.

Assumption 3. Model (1) has a unique equilibrium point.

Assumption 4. Let acquire two nonnegative matrices $P_k = (p_{ij}^k)_{n\times n}$ and $Q_k = (q_{ij}^k)_{n\times n}$ such that

$$|p_{ik}(u_1, \ldots, u_n) - p_{ik}(v_1, \ldots, v_n)| \leq \Sigma_{i=1}^n p_{ij}^k |u_i - v_i|,$$
$$|q_{ik}(u_1, \ldots, u_n) - q_{ik}(v_1, \ldots, v_n)| \leq \Sigma_{i=1}^n q_{ij}^k |u_i - v_i|,$$

for all $u_1, \ldots, u_n \in C$, $v_1, \ldots, v_n \in C, i = 1, 2, \ldots, n$; $k = 1, 2, \ldots$.

Assumption 5. For any $i \in 1, 2, \ldots, n$, there exist two positive constants $e_i^- < e_i^+$, such that $e_i^-|x_1 - x_2| \leq |e_i(x_1) - e_i(x_2)| \leq e_i^+|x_1 - x_2|, x_1, x_2 \in C$.

Remark 1. As an extension of the activation functions of real-valued neural networks satisfying the Lipschitz condition, complex-valued functions satisfying the *Assumption* 1 are actually change absolute value of real-valued functions into modules of the complex-valued in the literature. Compared with [7], we use a more extensively conditions.

Definition 1. The equilibrium point $\bar{z} = (\bar{z_1}, \cdots, \bar{z_n})^T$ of system (1) is said to be globally exponentially stable, if there exist constants $\varepsilon > 0$ and $M > 0$ such that

$$\|z(t) - \bar{z}\| \leq M\|\phi(s) - \bar{z}\|e^{-\varepsilon(t-t_0)},$$

for all $t > 0$, where $z(t) = (z_1(t), \ldots, z_n(t))^T$ is any solution of system (1) with initial value $\phi(s) = (\phi_1(s), \ldots, \phi_n(s))^T$, $s \in (-\infty, 0]$ and $\|\phi(s) - \bar{z}\| = \sup_{s\in[-\infty,0]}(\sum_{i=1}^n |\phi_i(s) - \bar{z_i}|^2)^{\frac{1}{2}}$.

Lemma 1. ([7]) Let Q be $n \times n$ real matrix with non-positive off-diagonal elements, then Q is a nonsingular M-matrix if there exists a vector $\xi > 0$ such that $\xi^T Q > 0$ or $Q\xi > 0$.

Lemma 2. ([8]) Let A be a nonnegative matrix, then $\rho(A)$ is a nonnegative eigenvalue of A and its corresponding eigenvectors have at least one to be positive. When A is a nonsingular M-matrix, B is a nonnegative matrix, we denote $\Omega(A) = \{\xi \in R^n | A\xi > 0, |\xi > 0\}, \Gamma(B) = \{\xi \in R^n | B\xi = \rho(B), |\xi > 0\}$.

3 Main Results

Theorem 1. Suppose that *Assumptions* (1–5) hold, then the equilibrium point of model (1) is globally exponentially stable if the following conditions are satisfied

(1) $\Phi = \begin{pmatrix} -|A|L - |B|K - |C|M & \sqrt{D} \\ \sqrt{D} & -I_n \end{pmatrix}$ can be seen as a nonsingular M-matrix, where $D = \sqrt{D}\sqrt{D}$.
(2) $E = \Omega(W) \bigcap \{\bigcap_{K=1}^{\infty}[\Gamma(P_k) \bigcap \Gamma(Q_k)]\}$ has at least a column vector with positive, where $W = D - |A|L - |B|K - |C|M$.
(3) There exists a constant $\alpha > 0$ such that $\frac{\ln \alpha_k}{t_k - t_{k-1}} \leq \alpha < \varepsilon, k = 1, 2, \ldots$. where scalar $\varepsilon > 0$ is determined by inequality

$$\begin{aligned} &\xi_i(\varepsilon - d_i)e_i^- + \{\textstyle\sum_{j=1}^n \xi_j |a_{ij}| l_j + e^{\varepsilon\tau} \sum_{j=1}^n \xi_j |b_{ij}| k_j \\ &+ \textstyle\sum_{j=1}^n \xi_j |c_{ij}| m_j G_{ij}(\varepsilon)\} e_i^+ < 0, i = 1, 2, \ldots, n, \end{aligned} \tag{4}$$

for a given $\xi = (\xi_1, \ldots, \xi_n)^T \in E$, and

$$\alpha_k \geq \max\{1, \rho(P_k) + e^{\varepsilon\tau}\rho(Q_k)\} \tag{5}$$

where $\tau = \max_{1\leq i,j\leq n}\{\tau_{ij}\}$, $k = 1, 2, \ldots$.

Proof. Let $\bar{z} = (\bar{z_1}, \dots, \bar{z_n})$ be an equilibrium point of the system (1) and $e_i(z_i) = 0$ if and only if $z_i = \bar{z}_i$. Denote $u_i(t) = z_i(t) - \bar{z}_i$, $\tilde{f}_j(u_j(t)) = f_j(u_j(t) + \bar{z_j}) - f_j(\bar{z_j})$, $\tilde{g}_j(u_j(t)) = g_j(u_j(t) + \bar{z_j}) - g_j(\bar{z_j})$, $\tilde{h_j}(u_j(t)) = h_j(u_j(t) + \bar{z_j}) - h_j(\bar{z_j})$, $\tilde{e_i}(u_i(t)) = e_i(u_i(t) + \bar{z_i}) - e_i(\bar{z_i})$, $\bar{p_{ik}}(u_1(t^-), \dots, u_n(t^-)) = p_{ik}(u_1(t^-) + \bar{z_1}, \dots, u_n(t^-) + \bar{z_n}) - p_{ik}(\bar{z_1}, \dots, \bar{z_n})$, $\bar{q_{ik}}(u_1((t-\tau_{i1})^-), \dots, u_n((t-\tau_{in})^-)) = q_{ik}(u_1((t-\tau_{i1})^-) + \bar{z_1}, \dots, u_n((t-\tau_{in})^-) + \bar{z_n}) - q_{ik}(\bar{z_1}, \dots, \bar{z_n})$.

From *Assumption* 2, we know that the model (1) can be rewritten as follows:

$$\begin{cases} \dot{u_i}(t) = -\tilde{e_i}(u_i(t))[d_i u_i(t) - \sum_{j=1}^n a_{ij}\tilde{f}_j(u_j(t)) - \sum_{j=1}^n b_{ij}\tilde{g}_j(u_j(t-\tau_{ij}(t))) \\ \qquad - \sum_{j=1}^n c_{ij}\int_{-\infty}^t k_{ij}(t-s)\tilde{h_j}(u_j(s))ds], & t \neq t_k, \\ u_i(t) = \bar{p_{ik}}(u_1(t^-), \dots, u_n(t^-)) + \bar{q_{ik}}(u_1((t-\tau_{i1})^-), \dots, u_n((t-\tau_{in})^-)), & t = t_k \end{cases} \tag{6}$$

Define

$$v_i(t) = e_i^- \int_{-\infty}^t \varepsilon |u_i(s)| ds + \ln |u_i(t)|.$$

Calculating the derivative of $v_i(t)$ along with (6), we can get from *Assumption* 1 that

$$\dot{v_i}(t) \leq (\varepsilon - d_i)e_i^- |u_i(t)| + [\sum_{j=1}^n |a_{ij}| l_j |u_j(t)| + \sum_{j=1}^n |b_{ij}| k_j |u_j(t-\tau_{ij}(t))|$$
$$+ \sum_{j=1}^n |c_{ij}| \int_{-\infty}^t k_{ij}(t-s) m_j |u_j(s)| ds] e^+,$$

for $i = 1, 2, \dots, n$; $t_{k-1} < t < t_k$, $k = 1, 2, \dots, n$.

Let $y = \frac{(1+r)\|\phi - z\|}{min_{1 \leq i \leq n}\{\xi_i\}}$ (r is a positive constant), then

$$e^{\varepsilon t}|u_i(t)| \leq |u_i(t)| = |\phi_i - z_i| \leq \|\phi - z\| < \xi_i y, t \in [-\infty, 0), i = 1, 2, \dots, n.$$

Next, we will prove the following equation

$$|u_i(t)| < \xi_i y \exp\{-\varepsilon t\}, t \in [-\infty, t_1), i = 1, 2, \dots, n \tag{7}$$

hold. In fact, if Eq. (7) is not true, then there exist some $\tilde{i}$ and $t' \in [-\infty, t_1)$, we can make these inequalities $|u_{\tilde{i}}(t')| = \xi_{\tilde{i}} y \exp\{-\varepsilon t'\}$, $\dot{v_{\tilde{i}}}(t') \geq 0$ and $u_j(t) \leq \xi_j y \exp\{-\varepsilon t\}, t \in (-\infty, t']$, $j = 1, 2, \dots, n$.

Therefore, we have

$$\dot{v_{\tilde{i}}}(t') \leq (\varepsilon - d_{\tilde{i}})e_i^- |u_{\tilde{i}}(t')| + \{\sum_{j=1}^n |a_{\tilde{i}j}| l_j |u_j(t')| + \sum_{j=1}^n |b_{\tilde{i}j}| k_j |u_j(t' - \tau_{\tilde{i}j}(t'))|$$
$$+ \sum_{j=1}^n |c_{\tilde{i}j}| \int_{-\infty}^{t'} k_{\tilde{i}j}(t'-s) m_j |u_j(s)| ds\} e_i^+$$

$$\leq [(\varepsilon - d_{\widehat{i}})e_i^- \xi_i + (\sum_{j=1}^{n} |a_{\widehat{i}j}|l_j\xi_j + \exp\{\varepsilon\tau\} \sum_{j=1}^{n} |b_{\widehat{i}j}|k_j\xi_j$$
$$+ \sum_{j=1}^{n} \xi_j |c_{\widehat{i}j}| G_{\widehat{i}j}(\varepsilon) m_j) e_i^+] y \exp\{-\varepsilon t'\}.$$

Since W is a nonsingular M-matrix and the set E is nonempty, from *Lemma* 1, Φ is equal to W according to *Liouville's theorem*, there exist $\xi = (\xi_1, \dots, \xi_n)^T \in E \subseteq \Omega(W)$ and $\xi > 0$ such that $W\xi < 0$ yield that

$$-\xi_i d_i + \sum_{j=1}^{n} \xi_j (|a_{ij}|l_j + |b_{ij}|k_j + |c_{ij}|m_j) < 0, i = 1, 2, \dots, n.$$

Furthermore, we can choose $\varepsilon = \min_{1 \leq i \leq n}\{\varepsilon_i\}$, then we have that

$$\xi_i(\varepsilon - d_i)e_i^- + \{\sum_{j=1}^{n} \xi_j |a_{ij}|l_j + e^{\varepsilon\tau} \sum_{j=1}^{n} \xi_j |b_{ij}|k_j + \sum_{j=1}^{n} \xi_j |c_{ij}| m_j G_{ij}(\varepsilon)\} e_i^+ < 0,$$

where $i = 1, 2, \dots, n$. Motivated by the above discussions, we can conclude that it is a contradiction.

Moreover, considering the impulsive effect, we will use the mathematical induction to prove that

$$|u_i(t)| < \alpha_0\alpha_1 \dots \alpha_k \xi_i y \exp\{-\varepsilon t\}, t \in [t_k, t_{k+1}), \tag{8}$$

where $\alpha_0 = 1$.

When $k = 0$, we know that (7) and (8) hold. Suppose that inequalities

$$|u_i(t)| \leq \alpha_0\alpha_1 \dots \alpha_k \xi_i y \exp\{-\varepsilon t\}, t \in [t_k, t_{k+1}), k = 1, 2, \dots, m. \tag{9}$$

By *Lemma* 2 and *Assumption* 4, the discrete part of model (6) satisfies that

$$|u_i(t_m)| \leq |p_{im}^-(u_1(t_m^-), \dots, u_n(t_m^-))| + |q_{im}^-(u_1(t_m - \tau_{i1})^-, \dots, u_n(t_m - \tau_{in})^-)|$$
$$\leq [\sum_{j=1}^{n} p_{ij}^m \xi_j + e^{\varepsilon\tau} \sum_{j=1}^{n} q_{ij}^m \xi_j] \alpha_0\alpha_1 \dots \alpha_{m-1} y \exp\{-\varepsilon t_m\}$$
$$\leq [\rho(P_m) + \exp\{\varepsilon\tau\}\rho(Q_m)]\xi_i \alpha_0\alpha_1 \dots \alpha_{m-1} y \exp\{-\varepsilon t_m\}$$
$$\leq \xi_i \alpha_0\alpha_1 \dots \alpha_{m-1}\alpha_m y \exp\{-\varepsilon t\}.$$

Next, we will prove that $|u_i(t)| \leq \xi_i\alpha_0\alpha_1 \dots \alpha_m y \exp\{-\varepsilon t\}$, $i = 1, 2, \dots, n$; $t \in (-\infty, t_{m+1})$ hold. In fact, if this is not true. Then there exist some $\widehat{i}$ and $t'' \in (-\infty, t_{m+1})$ such that $|u_{\widehat{i}}(t'')| = \alpha_0\alpha_1 \dots \alpha_{m-1}\alpha_m \xi_{\widehat{i}} y \exp -\varepsilon t''$, $\dot{v}_{\widehat{i}}(t'') \geq 0$, $u_j(t) \leq \alpha_0\alpha_1 \dots \alpha_{m-1}\alpha_m \xi_j y \exp\{-\varepsilon t\}$, $t \in (-\infty, t'']$, $j = 1, 2, \dots, n$. However, according to the last discussions, we can also get

$$\dot{v}_{\widehat{i}}(t'') \leq \xi_{\widehat{i}}(\varepsilon - d_{\widehat{i}})e_i^- + \{\sum_{j=1}^{n} \xi_j |a_{\widehat{i}j}|l_j + e^{\varepsilon\tau} \sum_{j=1}^{n} \xi_j |b_{\widehat{i}j}|k_j$$
$$+ \sum_{j=1}^{n} \xi_j |c_{\widehat{i}j}| m_j G_{\widehat{i}j}(\varepsilon)\} y \exp\{-\varepsilon t''\} e_i^+ < 0.$$

which this is a contradiction. So the conjecture holds. From (4), we have

$$|u_i(t)| \leq \frac{(1+r)\xi_i}{\min_{1\leq i\leq n}\{\xi_i\}}\|\phi - \bar{z}\|e^{-(\varepsilon-\alpha)t}$$

for any $t \in (-\infty, t_k)$, $k = 1, 2, \ldots$.

This means that the equilibrium point of model (1) is globally exponentially stable, and the exponential convergence rate equals $\varepsilon - \alpha > 0$.

Corollary 1. Under *Assumptions* (1–4) and $|u_i(t)| \leq Re(e_i(u_i(t)))$ hold, then the equilibrium point of model (1) is globally exponentially stable such that the conditions (1–3) hold in *Theorem* 1.

Remark 2. The condition $W = D - |A|L - |B|K - |C|M > 0$ is weaken than the condition $W = -D - |A|L - |B|K - |C|M > 0$ because we can use the former condition to induce the latter condition based on the external inputs matrix of D. When the external inputs are not considered and the amplification function $e_i(u_i(t)) : C \longrightarrow C$ are chosen as a class of special function $e_i(u_i(t)) : C \longrightarrow R^+$, such as $|u_i(t)|$, we can translate $z_i(t)$ into $u_i(t)$, the model (1) can be rewritten as follows:

$$\begin{cases} \dot{u}_i(t) = -|u_i(t)|[d_i u_i(t) - \sum_{j=1}^n a_{ij}\tilde{f}_j(u_j(t)) - \sum_{j=1}^n b_{ij}\tilde{g}_j(u_j(t-\tau_{ij}(t)) \\ \qquad -\sum_{j=1}^n c_{ij}\int_{-\infty}^t k_{ij}(t-s)\tilde{h}_j(u_j(t))ds] \qquad t \neq t_k, \\ u_i(t) = \bar{p_{ik}}(u_1(t^-), \ldots, u_n(t^-)) + \bar{q_{ik}}(u_1((t-\tau_{i1})^-), \ldots, u_n((t-\tau_{in})^-), t = t_k \end{cases} \tag{10}$$

Furthermore, when the amplification function are not considered in the system (10), the system turns into the following complex-valued neural networks

$$\begin{cases} \dot{u}_i(t) = -[d_i u_i(t) - \sum_{j=1}^n a_{ij}\tilde{f}_j(u_j(t)) - \sum_{j=1}^n b_{ij}\tilde{g}_j(u_j(t-\tau_{ij}(t)) \\ \qquad -\sum_{j=1}^n c_{ij}\int_{-\infty}^t k_{ij}(t-s)\tilde{h}_j(u_j(t))ds] \qquad t \neq t_k, \\ u_i(t) = \bar{p_{ik}}(u_1(t^-), \ldots, u_n(t^-)) + \bar{q_{ik}}(u_1((t-\tau_{i1})^-), \ldots, u_n((t-\tau_{in})^-), t = t_k. \end{cases} \tag{11}$$

Theorem 2. Suppose that *Assumptions* (1–4) hold, then the equilibrium point of model (10) is globally exponentially stable if the conditions (1–3) hold in *Theorem* 1.

Proof. Choose a Lyapunov functional as $v_i(t) = \int_0^t \varepsilon|u_i(s)|ds + \ln|u_i(t)|$.

Calculating the derivative of $v_i(t)$, one obtains that

$$\dot{v}_i(t) \leq (\varepsilon - d_i)|u_i(t)| + \sum_{j=1}^n |a_{ij}|l_j|u_j(t)| + \sum_{j=1}^n |b_{ij}|k_j|u_j(t-\tau_{ij}(t))|$$
$$+\sum_{j=1}^n |c_{ij}|\int_{-\infty}^t k_{ij}(t-s)m_j|u_j(s)|ds$$

for $i = 1, 2, \ldots, n$; $t_{k-1} < t < t_k, k = 1, 2, \ldots, n$.

Let $y = \frac{(1+r)\|\phi - z\|}{min\{\xi_i\}}$ ($r > 0$ is a positive constant), then, we will prove.

$$|u_i(t)| < \xi_i y \exp\{-\varepsilon t\}, t \in [-\infty, t_1), i = 1, 2, \ldots, n. \tag{12}$$

hold. In fact, if (12) is not the true, then there exist $\widetilde{i}$ and $t' \in [-\infty, t_1)$, we can obtain that $|u_{\widetilde{i}}(t')| = \xi_{\widetilde{i}} y \exp\{-\varepsilon t'\}, \dot{v}_{\widetilde{i}}(t') \geq 0$ and $u_j(t) \leq \xi_i y \exp\{-\varepsilon t\}, t \in (-\infty, t'], j = 1, 2, \ldots, n$.

However, it can be derived that this is a contradiction.

$$\dot{v}_{\widetilde{i}}(t') \leq [(\varepsilon - d_{\widetilde{i}})\xi_i + \sum_{j=1}^{n} |a_{\widetilde{i}j}| l_j \xi_j + \sum_{j=1}^{n} |b_{\widetilde{i}j}| k_j \xi_j$$
$$+ \sum_{j=1}^{n} \xi_j |c_{\widetilde{i}j}| G_{\widetilde{i}j}(\varepsilon) m_j] y \exp\{-\varepsilon t'\} < 0$$

We use the similar way to consider the effect of the impulsive effect

$$|u_i(t)| \leq \xi_i \alpha_0 \alpha_1 \ldots \alpha_{m-1} \alpha_m y \exp\{-\varepsilon t\}, \tag{13}$$

for $i = 1, 2, \ldots, n; t \in [t_{m-1}, t_m]$.

In the following, we will prove that

$|u_i(t)| \leq \xi_i \alpha_0 \alpha_1 \ldots \alpha_m y \exp\{-\varepsilon t\}$, $i = 1, 2, \ldots, n$; $t \in [t_m, t_{m+1})$ hold. In fact, if this is not true, then there exist some $\widehat{i}$ and $t'' \in [t_m, t_{m+1})$ such that

$$|u_{\widetilde{i}}(t'')| = \alpha_0 \alpha_1 \ldots \alpha_{m-1} \alpha_m \xi_{\widetilde{i}} y \exp -\varepsilon t'', \dot{v}_{\widetilde{i}}(t'') \geq 0$$

$$u_j(t) \leq \alpha_0 \alpha_1 \ldots \alpha_{m-1} \alpha_m \xi_i y \exp\{-\varepsilon t\}, t \in [t_{m-1,}, t''], j = 1, 2, \ldots, n,$$

for $t \in (-\infty, t'']$.

However, from the above, we can get

$$\dot{v}_{\widetilde{i}}(t'') \leq [(\varepsilon - d_{\widetilde{i}})\xi_{\widetilde{i}} + \sum_{j=1}^{n} |a_{\widetilde{i}j}| l_j \xi_j + \sum_{j=1}^{n} |b_{\widetilde{i}j}| k_j \xi_j$$
$$+ \sum_{j=1}^{n} \xi_j |c_{\widetilde{i}j}| \int_{-\infty}^{t''} k_{\widetilde{i}j}(t'' - s) m_j ds] \alpha_0 \alpha_1 \ldots \alpha_{m-1} \alpha_m \xi_{\widetilde{i}} y \exp\{-\varepsilon t''\} < 0,$$

this is a contradiction. So the conjecture holds. From (11), we have

$$\|z(t) - \bar{z}\| \leq M \|\phi - \bar{z}\| e^{-(\varepsilon - \alpha)(t - t_0)}$$

for $t \geq t_0$, where $M = \sqrt{\sum_{j=1}^{n} (\frac{(1+r)\xi_i}{\min_{1 \leq i \leq n}\{\xi_i\}})} \geq 1$. We complete the proof.

Theorem 3. Under *Assumptions* (1–4), then the equilibrium point of model (11) is globally exponentially stable if the conditions (1–3) hold in *Theorem* 1.

Proof. Define $v_i(t) = \frac{1}{2}e^{2\varepsilon t}u_i(t)\bar{u}_i(t)$,

Calculating the derivative of $v_i(t)$ along the solution of (11), one has

$$\dot{v}_i(t) \leq 2(\varepsilon - d_i)v_i(t) + 2\sum_{j=1}^{n}|a_{ij}|l_j v_j(t)$$
$$+2\sum_{j=1}^{n}|b_{ij}|k_j v_j(t-\tau_{ij}(t)) + 2\sum_{j=1}^{n}|c_{ij}|\int_{-\infty}^{t}k_{ij}(t-s)m_j v_j(s)ds$$

for $i = 1, 2, \ldots, n$; $t_{k-1} < t < t_k$, $k = 1, 2, \ldots, n$.

Let $y = \frac{(1+r)\|\phi - z\|^2}{2\min\{\xi_i\}}(r > 0)$, then $\frac{1}{2}exp\{2\varepsilon t\}|u_i(t)|^2 \leq |u_i(t)|^2 = |\phi_i - z_i|^2 \leq \|\phi - z\|^2 < \xi_i y$, which $t \in [-\infty, 0), i = 1, 2, \ldots, n$. That is:

$$|u_i(t)| < \sqrt{2\xi_i y}\exp\{-\varepsilon t\}, t \in (-\infty, 0), i = 1, 2, \ldots, n \tag{14}$$

Besides, we will prove the following equation

$$v_i(t) < \xi_i y, t \in (-\infty, t_1), i = 1, 2, \ldots, n \tag{15}$$

hold. In fact, if the Eq. (15) is not true, then there present some $\tilde{i}$ and $t' \in [-\infty, t_1)$, one can make these inequalities. $|v_{\tilde{i}}(t')| = \xi_{\tilde{i}}y, \dot{v}_{\tilde{i}}(t') \geq 0$ and $v_j(t) \leq \xi_{\tilde{i}}y, t \in (-\infty, t'], j = 1, 2, \ldots, n$. Therefore, one has

$$\dot{v}_{\tilde{i}}(t') \leq 2y[(\varepsilon - d_{\tilde{i}})\xi_i + \sum_{j=1}^{n}|a_{\tilde{i}j}|l_j\xi_j + e^{\varepsilon\tau}\sum_{j=1}^{n}|b_{\tilde{i}j}|k_j\xi_j$$
$$+\sum_{j=1}^{n}|c_{\tilde{i}j}|\int_{-\infty}^{t'}k_{\tilde{i}j}(t'-s)m_j\xi_j ds] < 0$$

this is a contradiction. Hence

$$v_i(t) < \xi_i y, t \in [-\infty, t_1), i = 1, 2, \ldots, n,$$
$$|u_i(t)| < \sqrt{2\xi_i y}\exp\{-\varepsilon t\}, t \in [-0, t_1), i = 1, 2, \ldots, n. \tag{16}$$

Next, we can use the similar way to prove the following process.

Remark 3. Based on the new criterion, we can guarantee the existence, uniqueness, and global exponential stability of the equilibrium point of the complex-valued recurrent neural networks. Compared with the before papers, our conditions are less conservative in this paper.

Corollary 2. Supposing that *Assumptions* 1 and 2 hold, then the first part of the model (6) has a unique equilibrium point if $W = D - |A|L - |B|K - |C|M$ assumed to be treated as a nonsingular M-matrix.

4 Numerical Examples

Example 1. Consider two dimensional impulsive complex-valued neural networks with mixed time delays as model (1). Assume that parameters are given as follows: $e(z(t)) = \begin{pmatrix} 1 & 0 \\ 0 & 1 \end{pmatrix}$, $D = \begin{pmatrix} 3.5 & 0 \\ 0 & 3.6 \end{pmatrix}$, $A = \begin{pmatrix} 0.3+0.2i & 0.8+1i \\ 0.9-0.7i & 2.2+0.8i \end{pmatrix}$,
$B = \begin{pmatrix} 0.2-0.3i & 0.1-0.3i \\ 0.8+0.2i & 1.3-0.7i \end{pmatrix}$, $C = \begin{pmatrix} -0.2-0.15i & -0.1-0.1i \\ -0.1-0.1i & -0.2-0.15i \end{pmatrix}$, $I = \begin{pmatrix} 0 \\ 0 \end{pmatrix}$,
$P = \begin{pmatrix} -0.02+0.37i & -0.4+0.08i \\ -0.07+0.5i & -0.06-0.9i \end{pmatrix}$, $Q = \begin{pmatrix} 0.07-0.2i & 0.07-0.2i \\ 0.04-0.02i & 0.03+0.2i \end{pmatrix}$.

$f(z(t)) = g(z(t)) = h(z(t)) = z(t)$. By simple calculation, we can see that all conditions are satisfied. Then the model can be stabilized by the impulsive effects as shown in Figs. 1 and 2.

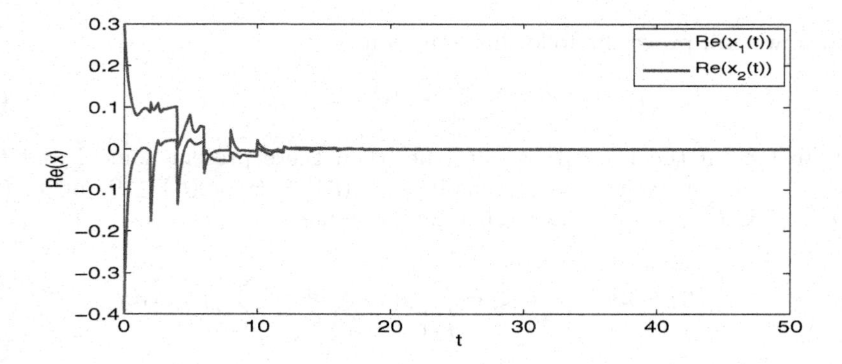

Fig. 1. Real part of state trajectories

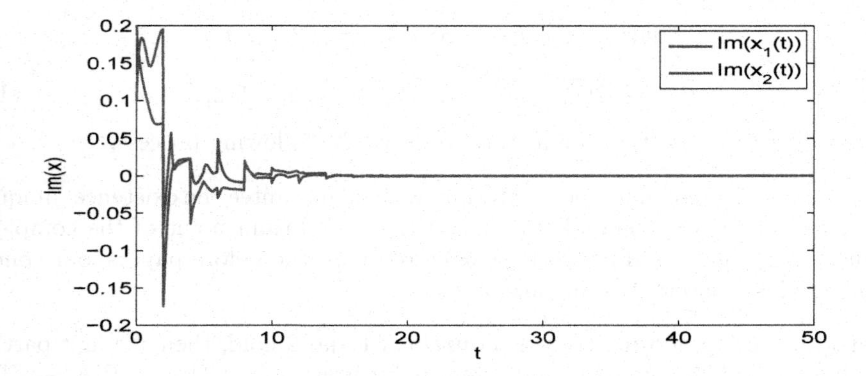

Fig. 2. Imaginary part of state trajectories.

5 Conclusion

In the literature, we have derived several new sufficient conditions for ascertaining the global stability of complex-valued impulsive recurrent neural networks with mixed time-varying delays. When impulsive effects are considered in the complex-valued neural networks, the new conditions can be used to guarantee the existence, uniqueness, and global exponential stability of the equilibrium point.

Acknowledgments. This work was supported by the Excellent Doctor Innovation Program of Xinjiang University (Grant No. XJUBSCX-2015006), the Excellent Doctor Innovation Program of Xinjiang Uyghur Autonomous Region (Grant No. XJGRI2016001), the National Natural Science Foundation of People's Republic of China (Grant No. 61164004).

References

1. Liu, X., Chen, T.: Global exponential stability for complex-valued recurrent neural networks with asynchronous time delays. IEEE Trans. Neural Netw. Learn. Syst. **27**(3), 593–606 (2016)
2. Lu, W., Chen, T.: New conditions on global stability of cohen-grossberg neural networks. Neural Comput. **15**, 1173 (1989)
3. Bohner, M., Rao, V.S.: Global stability of complex-valued neural networks on time scales. Differ. Equ. Dyn. Syst. **19**, 3–11 (2011)
4. Sree, H., Murthy, G.: Global dynamics of a class of complex-valued neural networks. Int. J. Neural Syst. **18**, 165–171 (2008)
5. Jankowski, S., Lozowski, A., Zurada, J.: Complex-valued multistate nueral associative memory. IEEE Trans. Neural Netw. Learn. Syst. **7**(6), 1491–1496 (1996)
6. Lee, D.: Relexation of stability condition of the complex-valued neural networks. IEEE Trans. Neural Netw. Learn. Syst. **12**(5), 1260–1262 (2001)
7. Song, Q., Cao, J.: Stability analysis of Cohen-Grossberg neural network with both time-varying and continuously distributed delays. J. Comput. Appl. Math. **197**(1), 188–203 (2006)
8. Fang, T., Sun, J.: Stability of complex-valued recurrent neural networks with time-delays. IEEE Trans. Neural Netw. Learn. Syst. **25**(9), 1709–1713 (2014)

Pinning Synchronization in Heterogeneous Networks of Harmonic Oscillators

Zhengxin Wang[1,2](✉), Jingbo Fan[1], He Jiang[2], and Haibo He[2]

[1] College of Science, Nanjing University of Posts and Telecommunications, Nanjing 210023, China
zwang@njupt.edu.cn, fan1025607790@163.com

[2] Department of Electrical, Computer, and Biomedical Engineering, University of Rhode Island, Kingston, RI 02881, USA
{hjiang,he}@ele.uri.edu

Abstract. In this paper, a networked heterogeneous system coupled of multiple nonidentical harmonic oscillators is investigated. The synchronization problem for the networked heterogeneous system is studied. To synchronize the heterogeneous network, a leader is introduced. Based on the pinning control, a distributed control input is proposed to synchronize the heterogeneous network to the leader. By Lyapunov functional method and matrix theory, sufficient conditions for guaranteeing quasi-synchronization between the heterogeneous network and the leader are obtained. The theoretical results show that all the heterogeneous oscillators can tend eventually to the leader oscillator within a bounded error. Finally, numerical simulations are provided to verify the effectiveness of the theoretical results.

Keywords: Pinning control · Synchronization · Quasi-synchronization · Heterogeneous networks · Harmonic oscillators

1 Introduction

Complex networks are ubiquitous in the real world, such as neural networks, multi-agent systems and social networks. Synchronization is a typical collective behaviour of complex networks and has been studied extensively in the past few years [1–5].

In the last decade, synchronization in coupled harmonic oscillators, which are the important coupled networks, has received particular attention. By means of continuous-time distributed protocols, synchronization of coupled harmonic oscillators was studied in [6, 7] and sufficient conditions for synchronization were presented. By proposing a distributed impulsive control protocol, Zhou *et al.* [8] studied synchronization of coupled harmonic oscillators over both fixed and switching topologies. In the case of sampled-data information and intermittent control, global synchronization in coupled harmonic oscillators was proved in [9]. Based on the distributed sampled-data control protocols, synchronization of coupled harmonic oscillators was studied in [10, 11] under various conditions.

D. Liu et al. (Eds.): ICONIP 2017, Part III, LNCS 10636, pp. 836–845, 2017.
https://doi.org/10.1007/978-3-319-70090-8_85

It should be noted that the aforementioned results concerning synchronization of coupled harmonic oscillators mainly focused on the homogeneous networks. Because of external disturbance, parameter uncertainty and individual difference, heterogeneous networks are more accordant with reality than homogeneous networks. Cooperative control of the heterogenous networks has gained a great attention in recent years [12–19]. However, synchronization of the heterogeneous harmonic oscillators received very little attention so far. The cases of synchronization of heterogeneous networked systems are various. In the case of impulsive coupling, both synchronization [12] and quasi-synchronization [13] can be reached. When relying only on the coupling of the networked system, quasi-synchronization can be guaranteed under the given assumptions [14,15]. However, an external control law is an effective way to synchronize the heterogeneous networked systems [16–18]. Even if the position interaction topology and velocity interaction topology are heterogeneous, containment tracking can be reached in multi-agent systems with multiple leader by adopting external controllers [19]. Since a complex network has mass nodes, pinning control is a desirable selection by pinning a small fraction of nodes [20–22].

Motivated by the above discussions, this paper focuses on the synchronization problem of a networked heterogeneous system coupled by multiple non-identical harmonic oscillators. Since the dynamics of heterogeneous oscillators are different, heterogeneous oscillators usually can not reach synchronization. By introducing a leader oscillator and employing a pinning distributed protocol, quasi-synchronization in leader-following heterogeneous oscillator systems is analyzed. Sufficient criteria for quasi-synchronization are derived.

The notations used throughout this paper is summarized as follows. Let $\mathbb{R}$ denote the real numbers, O_N and I_N be the $N \times N$ zero matrix and identity matrix, respectively. $\text{diag}\{\cdots\}$ stands for a diagonal matrix. For a real matrix A, let A^T be its transpose, $A_s = (A + A^T)/2$ be its symmetric part. A symmetric matrix $P > 0$ means that P is positive definite. $\|x\| = \sqrt{x^T x}$ for all $x \in \mathbb{R}^n$. Let $\lambda_{\min}(M)$ and $\lambda_{\max}(M)$ be the minimal and maximal eigenvalue of a symmetric square matrix M, respectively.

2 Preliminaries

2.1 Graph Theory

A digraph denoted by $\mathcal{G} = (\mathcal{V}, \mathcal{E}, \mathcal{A})$, where $\mathcal{V} = \{v_1, v_2, \cdots, v_N\}$ and $\mathcal{E} \subseteq \mathcal{V} \times \mathcal{V}$ are the sets of nodes and edges, respectively. An directed edge (v_i, v_j) from v_i to v_j is denoted by e_{ij}. A weighted adjacency matrix $\mathcal{A} = [a_{ij}]_{N\times N} \in \mathbb{R}^{N\times N}$ is defined as $a_{ii} = 0$ and $a_{ij} > 0$ if $e_{ji} \in \mathcal{E}$. A directed path from v_i to v_j is a sequence $v_i, v_{i_1}, v_{i_2}, \cdots, v_{i_j}, v_j$ of distinct nodes such that $(v_i, v_{i_1}), (v_{i_1}, v_{i_2}), \cdots, (v_{i_j}, v_j)$ are edges in $\mathcal{E}$. A directed graph is called strongly connected if there is a directed path between any pair of distinct nodes. A digraph $\mathcal{G}$ has a spanning tree if there exists at least one node which has a directed path to all the other nodes in digraph $\mathcal{G}$. The Laplacian matrix $L = [l_{ij}]_{N\times N}$ of graph $\mathcal{G}$ is defined as $l_{ii} = \sum_{j\neq i} a_{ij}$ and $l_{ij} = -a_{ij}$ for $i \neq j$.

2.2 Problem Formulations

Consider a heterogeneous network composed of N nonidentical harmonic oscillators:

$$\begin{cases} \dot{x}_i(t) = v_i(t), \\ \dot{v}_i(t) = -\omega_i^2 x_i(t) - u_i(t), \quad i = 1, 2, \cdots, N, \end{cases} \tag{1}$$

where $\omega_1, \omega_2, \cdots, \omega_N$ are the frequencies of N oscillators and are different with each other, $x_i(t), v_i(t) \in \mathbb{R}$ are the position state and velocity state of node i, respectively, $u_i \in \mathbb{R}$ is the control input for node i.

The dynamics of the leader is given by

$$\begin{cases} \dot{x}_0(t) = v_0(t), \\ \dot{v}_0(t) = -\omega_0^2 x_0(t), \end{cases} \tag{2}$$

where ω_0 is the frequency of the leader, $x_0(t), v_0(t) \in \mathbb{R}$ are the position state and velocity state of the leader, respectively.

The initial values of the follower oscillators and the leader are assumed to be $[x_i(0), v_i(0)]^T$ $(i = 1, 2, \cdots, N)$ and $[x_0(0), v_0(0)]^T$, respectively.

To realize synchronization between N heterogeneous coupled oscillators (1) and the leader (2), the following distributed protocol enlightened by [20] is employed:

$$\begin{aligned} u_i(t) = \alpha &\left(\sum_{j=1}^{N} a_{ij}(x_i(t) - x_j(t)) + b_i(x_i(t) - x_0(t)) \right) \\ +\beta &\left(\sum_{j=1}^{N} a_{ij}(v_i(t) - v_j(t)) + b_i(v_i(t) - v_0(t)) \right), \quad i = 1, 2, \cdots, N, \end{aligned} \tag{3}$$

where $\alpha, \beta > 0$ are the coupling strengths, b_i is the pinning gain for node i. $b_i > 0$ if the ith node is pinned, otherwise $b_i = 0$.

Let $\widehat{x}_i(t) = x_i(t) - x_0(t)$, $\widehat{v}_i(t) = v_i(t) - v_0(t)$ be the synchronization errors. In view of input protocol (3), one can derive the error systems as follows

$$\begin{cases} \dot{\widehat{x}}_i(t) = \widehat{v}_i(t), \\ \dot{\widehat{v}}_i(t) = -\alpha \sum_{j=1}^{N} l_{ij}\widehat{x}_j(t) - \alpha b_i \widehat{x}_i(t) - \beta \sum_{j=1}^{N} l_{ij}\widehat{v}_j(t) - \beta b_i \widehat{v}_i(t) \\ \qquad\quad -\omega_i^2 \widehat{x}_i(t) + \delta_i(t), \end{cases} \tag{4}$$

where $\delta_i(t) = (\omega_0^2 - \omega_i^2)x_0(t), i = 1, 2, \cdots, N$.

Assumption 1. The leader has a directed path to each follower oscillator.

Assumption 2. The leader starts from a bounded region. That is, there exists a compact $\mathcal{C}$ such that $[x_0(0), v_0(0)]^T \in \mathcal{C} \subset \mathbb{R}^2$.

Definition 1 ([13]). The heterogeneous networks (1) and (2) are said to reach quasi-synchronization (or bounded synchronization) with an error bound $\varepsilon > 0$, if there exists a compact set M such that, for any $[x_i(0), v_i(0)]^T \in \mathbb{R}^2$ and $[x_0(0), v_0(0)]^T \in \mathcal{C}$, the errors $\widehat{x}_i(t)$ and $\widehat{v}_i(t)$ both converge into the set $M = \{e \in \mathbb{R} : |e| \leq \varepsilon\}$ as $t \to \infty$.

Remark 1. If $\varepsilon = 0$, then the errors $\widehat{x}_i(t)$ and $\widehat{v}_i(t)$ both converge to zero. Therefore, synchronization is reached.

Lemma 1 (Schur Complement [23]). The following linear matrix inequality (LMI)

$$\begin{bmatrix} Q(x) & S(x) \\ S^T(x) & R(x) \end{bmatrix} > 0,$$

where $Q(x) = Q^T(x)$, $R(x) = R^T(x)$, is equivalent to either of the following conditions:

(1) $Q(x) > 0, \; R(x) - S^T(x)Q^{-1}(x)S(x) > 0;$
(2) $R(x) > 0, \; Q(x) - S(x)R^{-1}(x)S^T(x) > 0.$

Lemma 2 ([24]). Let A and B be n by n Hermitian matrices, $\alpha_1 \geq \alpha_2 \geq \cdots \geq \alpha_n$, $\beta_1 \geq \beta_2 \geq \cdots \geq \beta_n$ and $\gamma_1 \geq \gamma_2 \geq \cdots \geq \gamma_n$ be eigenvalues of matrices A, B and $A + B$, respectively. Then $\alpha_i + \beta_n \leq \gamma_i \leq \alpha_i + \beta_1$.

Lemma 3 ([25]). Let $A \in \mathbb{R}^{n\times n}$ be a symmetric matrix. One has

$$\lambda_{\min}(A)x^T x \leq x^T Ax \leq \lambda_{\max}(A)x^T x, \quad x \in \mathbb{R}^n.$$

3 Synchronization with Continuous-Time Communication

In this section, some sufficient conditions are derived for guaranteeing quasi-synchronization between N heterogeneous coupled harmonic oscillators (1) and the leader (2).

Let $\widehat{x}(t) = [\widehat{x}_1(t), \widehat{x}_2(t), \cdots, \widehat{x}_N(t)]^T$, $\widehat{v}(t) = [\widehat{v}_1(t), \widehat{v}_2(t), \cdots, \widehat{v}_N(t)]^T$. Rewrite (4) as a compact form:

$$\begin{cases} \dot{\widehat{x}}(t) = \widehat{v}(t), \\ \dot{\widehat{v}}(t) = -A\widehat{x}(t) - \alpha H\widehat{x}(t) - \beta H\widehat{v}(t) + \delta(t), \end{cases} \tag{5}$$

where $A = \mathrm{diag}\{\omega_1^2, \omega_2^2, \cdots, \omega_N^2\}, B = \mathrm{diag}\{b_1, b_2, \cdots, b_N\}$, $H = L + B, \delta(t) = [\delta_1(t), \delta_2(t), \cdots, \delta_N(t)]^T$.

Remark 2. Under Assumption 2 and by solving the Eq. (2) directly, one can obtain $\delta(t)$ is bounded. That is, there exists a constant $\overline{\delta}$ such that $|\delta(t)| \leq \overline{\delta}$.

Let $G = -L$ and $D(\rho) = \dfrac{G+G^T}{2} + \rho I_N = G_s + \rho I_N$.

Without loss of generality, suppose that the first $l(1 \leq l < N)$ oscillators in heterogenous network (1) are pinned. Then, $B = \mathrm{diag}\{\gamma_1, \gamma_2, \cdots, \gamma_l, 0, \cdots, 0\}$ with $\gamma_i > 0$, $i = 1, 2, \cdots, l$. Similar to [26], using matrix decomposition gets

$$D(\rho) - B = \begin{bmatrix} D_{11}(\rho) - \widetilde{B} & D_{12} \\ D_{12}^T & [D(\rho)]_l \end{bmatrix},$$

where $[D(\rho)]_l = [G_s + \rho I_N]_l$ is the minor matrix of $D(\rho)$ by removing its first $l(1 \leq l < N)$ row-column pairs, $\widetilde{B} = \mathrm{diag}\{\gamma_1, \gamma_2, \cdots, \gamma_l\}$, $D_{11}(\rho)$ and D_{12} are the corresponding matrix blocks of $D(\rho)$.

Theorem 1. Suppose that Assumptions 1 and 2 hold. Quasi-synchronization in directed network of heterogeneous harmonic oscillators (1) and (2) is achieved if $\beta \leq 2\alpha$, $\lambda_{\max}((G_s)_l) < -\rho_0$ and $b_i > \lambda_{\max}\left(D_{11}(\rho_0) - D_{12}[D(\rho_0)]_l^{-1} D_{12}^T\right)$, where $\rho_0 = \frac{2\alpha+\beta}{2\beta^2}$.

Proof. Let $\eta(t) = [\widehat{x}^T(t), \widehat{v}^T(t)]^T$. Consider the following Lyapunov functional candidate

$$V(t) = \frac{1}{2}\eta^T(t) P \eta(t), \tag{6}$$

where

$$P = \begin{bmatrix} 2\alpha H_s + A & \frac{\alpha}{\beta} I_N \\ \frac{\alpha}{\beta} I_N & I_N \end{bmatrix}. \tag{7}$$

By Lemma 1, $P > 0$ is equivalent to $2\alpha H_s + A - \frac{\alpha^2}{\beta^2} I_N > 0$, which is further equivalent to

$$D(\frac{\alpha}{2\beta^2}) - B - \frac{1}{2\alpha} A < 0.$$

It follows from the condition $\lambda_{\max}((G_s)_l) < -\rho_0$ and Lemma 2 that $[D(\rho_0)]_l < 0$. Therefore, according to the condition $b_i > \lambda_{\max}\left(D_{11}(\rho_0) - D_{12}[D(\rho_0)]_l^{-1} D_{12}^T\right)$ and Lemma 1, one has $D(\rho_0) - B < 0$. In view of $\frac{\alpha}{2\beta^2} < \rho_0$, one has

$$D(\frac{\alpha}{2\beta^2}) - B - \frac{1}{2\alpha} A < D(\frac{\alpha}{2\beta^2}) - B < D(\rho_0) - B < 0.$$

That is, $P > 0$.

Taking the derivative of $V(t)$ along the trajectory (5) yields

$$\begin{aligned} \frac{dV(t)}{dt}\Big|_{(5)} = & \left[\widehat{x}^T(t)(2\alpha H_s + A) + \frac{\alpha}{\beta}\widehat{v}^T(t)\right]\widehat{v}(t) \\ & + \left[\frac{\alpha}{\beta}\widehat{x}^T(t) + \widehat{v}^T(t)\right]\left[-A\widehat{x}(t) - \alpha H\widehat{x}(t) - \beta H\widehat{v}(t) + \delta(t)\right] \end{aligned}$$

$$= -\frac{\alpha}{\beta}\widehat{x}^T(t)(\alpha H + A)\widehat{x}(t) - \widehat{v}^T(t)\left(\beta H - \frac{\alpha}{\beta}I_N\right)\widehat{v}(t) + \left[\frac{\alpha}{\beta}\widehat{x}^T(t) + \widehat{v}^T(t)\right]\delta(t). \tag{8}$$

Note that

$$\widehat{x}^T(t)\delta(t) \le \frac{1}{2}\widehat{x}^T(t)\widehat{x}(t) + \frac{1}{2}\delta^T(t)\delta(t), \tag{9}$$

$$\widehat{v}^T(t)\delta(t) \le \frac{1}{2}\widehat{v}^T(t)\widehat{v}(t) + \frac{1}{2}\delta^T(t)\delta(t). \tag{10}$$

Combining (9) and (10) with (8) derives

$$\dot{V}(t) \le -\eta^T(t)Q\eta(t) + \left(\tfrac{1}{2} + \tfrac{\alpha}{2\beta}\right)\delta^T(t)\delta(t), \tag{11}$$

where

$$Q = \begin{bmatrix} \frac{\alpha}{\beta}\left(\alpha H_s + A - \frac{1}{2}I_N\right) & O_N \\ O_N & \beta H_s - (\frac{\alpha}{\beta} + \frac{1}{2})I_N \end{bmatrix}.$$

In fact, $\alpha H_s + A - \frac{1}{2}I_N > 0$ and $\beta H_s - (\frac{\alpha}{\beta} + \frac{1}{2})I_N > 0$ are equivalent to $D(\frac{1}{2\alpha}) - B - \frac{1}{\alpha}A < 0$ and $D(\rho_0) - B < 0$, respectively. It follows from the condition $\beta \le 2\alpha$ that $\frac{1}{2\alpha} < \rho_0 = \frac{2\alpha+\beta}{2\beta^2}$. That is,

$$D(\frac{1}{2\alpha}) - B - \frac{1}{\alpha}A < D(\rho_0) - B < 0.$$

Therefore, $Q > 0$. It follows from (11) and Lemma 3 that

$$V(t) \le V(0)e^{-2rt} + \frac{(\alpha+\beta)\overline{\delta}^2}{4\beta r}(1 - e^{-2rt}), \tag{12}$$

where $r = \frac{\lambda_{\min}(Q)}{\lambda_{\max}(P)} > 0$. Together with Lemma 3 again,

$$\lim_{t\to\infty} \|\eta(t)\|^2 \le \frac{(\alpha+\beta)\overline{\delta}^2}{2\beta r\lambda_{\min}(P)}. \tag{13}$$

According to Definition 1, quasi-synchronization in (1) and (2) is reached with the upper bound $\sqrt{\frac{(\alpha+\beta)\overline{\delta}^2}{2\beta r\lambda_{\min}(P)}}$ at a exponential rate r. ■

Theorem 1 is restricted to $\beta \le 2\alpha$. If $\beta > 2\alpha$, then the relation $D(\frac{1}{2\alpha}) - B - \frac{1}{\alpha}A < D(\rho_0) - B$ can not be guaranteed. It follows from

$$D(\frac{1}{2\alpha}) - B - \frac{1}{\alpha}A = D(\rho_0) - B + \frac{(\alpha+\beta)(\beta-2\alpha)}{2\alpha\beta^2}I_N - \frac{1}{\alpha}A,$$

and $D(\rho_0) - B < 0$ that $D(\frac{1}{2\alpha}) - B - \frac{1}{\alpha}A < 0$ is equivalent to $\frac{(\alpha+\beta)(\beta-2\alpha)}{2\alpha\beta^2}I_N - \frac{1}{\alpha}A \le 0$. Thus, Theorem 2 is a straightforward conclusion from Theorem 1.

Theorem 2. Suppose that Assumptions 1 and 2 hold. Quasi-synchronization in directed heterogeneous oscillator networks (1) and (2) is achieved if $\beta \geq 2\alpha$, $\lambda_{\max}((G_s)_l) < -\rho_0$, $b_i > \lambda_{\max}\left(D_{11}(\rho_0) - D_{12}[D(\rho_0)]_l^{-1}D_{12}^T\right)$, and $\min\limits_{1\leq i\leq N}\{\omega_i^2\} \geq \frac{(\alpha+\beta)(\beta-2\alpha)}{2\beta^2}$.

4 Numerical Results

Consider the networked heterogeneous systems (1) and (2) with the following parameters: $A = \text{diag}\{1, 1.5, 2, 2.5, 3, 3.5, 4, 4.5, 5, 5.5\}$, $\omega_0^2 = 0.5$. Select $\alpha = 4, \beta = 7$. The topology is shown in Fig. 1.

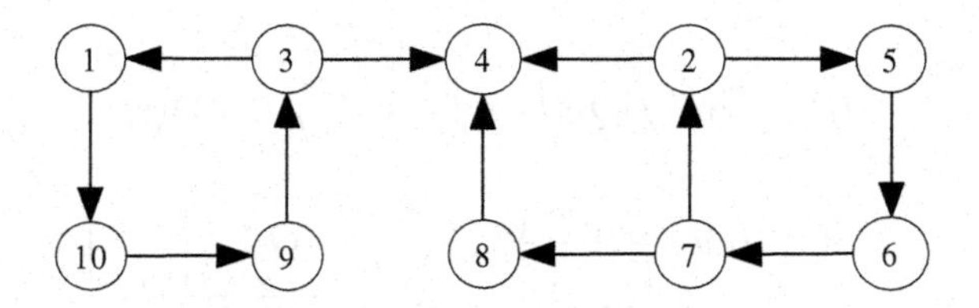

Fig. 1. Topology of the heterogeneous harmonic oscillators.

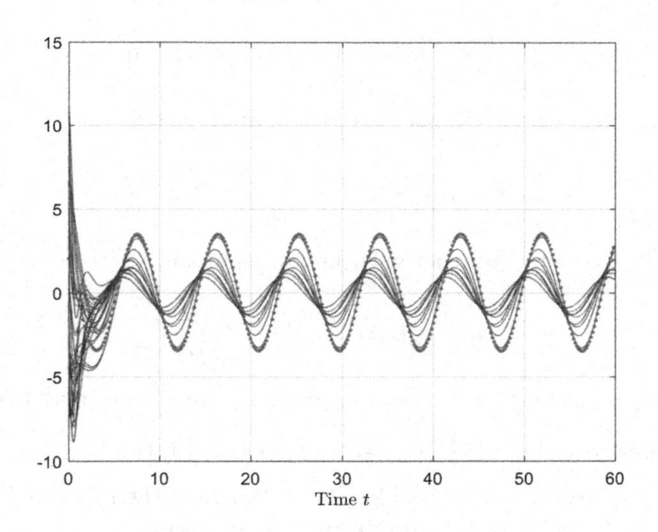

Fig. 2. Evolutions of position states of ten oscillators. (Color figure online)

By calculating, $\lambda_{\max}\left(D_{11}(\rho_0) - D_{12}[D(\rho_0)]_l^{-1}D_{12}^T\right) = 13.7325$. If one chooses $B = \text{diag}\{15, 14, 0, 0, 0, 0, 0, 0, 0, 0\}$, then the conditions of Theorem 1 are satisfied. Take initial value $[x_0(0), v_0(0)]^T = [2, -2]^T$ and random initial values for

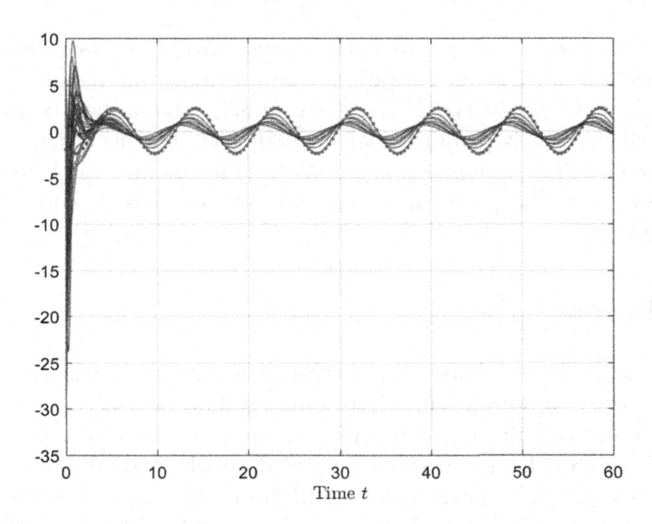

Fig. 3. Evolutions of velocity states of ten oscillators. (Color figure online)

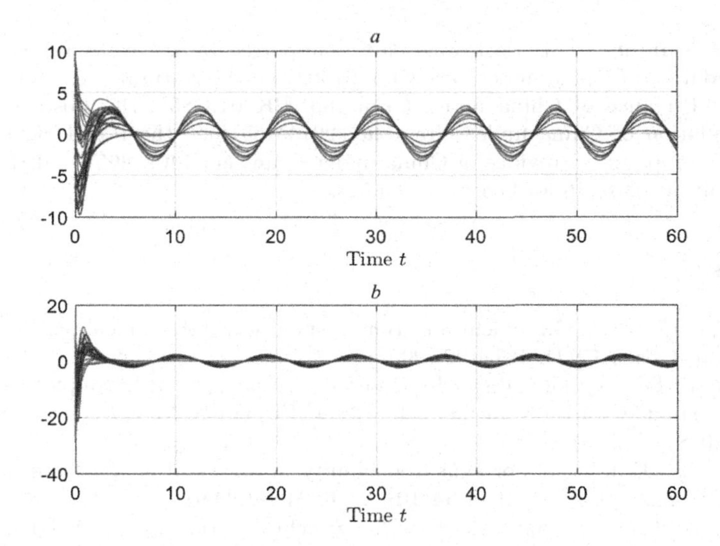

Fig. 4. Evolutions of position and velocity errors of ten oscillators.

the leader and the follower oscillators, respectively. The evolutions of position and velocity states of ten oscillators are drawn in Figs. 2 and 3, respectively. The red dotted lines in Figs. 2 and 3 are the trajectories of position state and velocity state of the leader, respectively.

The evolutions of position and velocity errors of ten oscillators are shown in Fig. 4(a) and (b), respectively.

Remark 3. According to Theorem 1, the upper bound of quasi-synchronization can be estimated. Under the condition that the network topology, the matrices P, Q and pinning matrix B are determined, the upper bound of quasi-synchronization errors is proportional to $\overline{\delta}$. In fact, $\overline{\delta}$ reflects the maximal difference between all heterogeneous harmonic oscillators and the leader oscillator. Figure 4 shows that the position errors and velocity errors tend to a bounded range eventually.

5 Conclusions

The quasi-synchronization problem is studied for the networked heterogeneous system coupled of N nonidentical harmonic oscillators. Based on a distributed pining control input, sufficient conditions for quasi-synchronization between the heterogeneous network and the leader are derived. According to the results, the errors between all heterogeneous oscillators and the leader will tend to be bounded ultimately. The upper bound can be estimated. Numerical simulations are given to verify the effectiveness of the theoretical results.

Acknowledgments. This work was jointly supported by the National Natural Science Foundation of China under Grant no. 61304169, the Natural Science Foundation of Jiangsu Province of China under Grant no. BK20130857, the Postdoctoral Science Foundation of China under Grant no. 2014M551629, the Postdoctoral Science Foundation of Jiangsu Province of China under Grant no. 1402086C, and the Jiangsu Government Scholarship for Overseas Studies.

References

1. Wang, X.F., Chen, G.: Synchronization in small-world dynamical networks. Int. J. Bifurcat. Chaos **12**(1), 187–192 (2002)
2. Wu, W., Chen, T.: Global synchronization criteria of linearly coupled neural network systems with time-varying coupling. IEEE Trans. Neural Netw. **19**(2), 319–332 (2008)
3. Tuna, S.E.: Conditions for synchronizability in arrays of coupled linear systems. IEEE Trans. Autom. Control **54**(10), 2416–2420 (2009)
4. Wang, Z., Cao, J., Duan, Z., Liu, X.: Synchronization of coupled Duffing-type oscillator dynamical networks. Neurocomputing **136**, 162–169 (2014)
5. Yang, S., Guo, Z., Wang, J.: Global synchronization of multiple recurrent neural networks with time delays via impulsive interactions. IEEE Trans. Neural Netw. Learn. Syst. (in press, 2017). doi:10.1109/TNNLS.2016.2549703
6. Ren, W.: Synchronization of coupled harmonic oscillators with local interaction. Automatica **44**(12), 3195–3200 (2008)
7. Su, H., Wang, X., Lin, Z.: Synchronization of coupled harmonic oscillators in a dynamic proximity network. Automatica **45**(10), 2286–2291 (2009)
8. Zhou, J., Zhang, H., Xiang, L., Wu, Q.: Synchronization of coupled harmonic oscillators with local instantaneous interaction. Automatica **48**(8), 1715–1721 (2012)
9. Zhang, H., Zhou, J.: Synchronization of sampled-data coupled harmonic oscillators with control inputs missing. Syst. Control Lett. **61**(12), 1277–1285 (2012)

10. Sun, W., Lü, J., Chen, S., Yu, X.: Synchronisation of directed coupled harmonic oscillators with sampled-data. IET Control Theory Appl. **8**(11), 937–947 (2014)
11. Song, Q., Liu, F., Wen, G., Cao, J., Tang, Y.: Synchronization of coupled harmonic oscillators via sampled position data control. IEEE Trans. Circuits Syst. I Regul. Pap. **63**(7), 1079–1088 (2016)
12. Wang, Z., Cao, J., Chen, G., Liu, X.: Synchronization in an array of nonidentical neural networks with leakage delays and impulsive coupling. Neurocomputing **111**, 177–183 (2013)
13. He, W., Qian, F., Lam, J., Chen, G., Han, Q.-L., Kurths, J.: Quasi-synchronization of heterogeneous dynamic networks via distributed impulsive control: error estimation, optimization and design. Automatica **62**, 249–262 (2015)
14. Zhao, J., Hill, D.J., Liu, T.: Global bounded synchronization of general dynamical networks with nonidentical nodes. IEEE Trans. Autom. Control **57**(10), 2656–2662 (2012)
15. Wang, Z., Jiang, G., Yu, W., He, W., Cao, J., Xiao, M.: Synchronization of coupled heterogeneous complex networks. J. Franklin Inst. **354**(10), 4102–4125 (2017)
16. Lunze, J.: Synchronization of heterogeneous agents. IEEE Trans. Autom. Control **57**(11), 2885–2890 (2012)
17. Liu, X.-K., Wang, Y.-W., Xiao, J.-W., Yang, W.: Distributed hierarchical control design of coupled heterogeneous linear systems under switching networks. Int. J. Robust Nonlinear Control **27**(8), 1242–1259 (2017)
18. Wang, Z., Duan, Z., Cao, J.: Impulsive synchronization of coupled dynamical networks with nonidentical duffing oscillators and coupling delays. Chaos **22**(1), 013140 (2012)
19. Qin, J., Zheng, W.X., Gao, H., Ma, Q., Fu, W.: Containment control for second-order multiagent systems communicating over heterogeneous networks. IEEE Trans. Neural Netw. Learn. Syst. (in press, 2017). doi:10.1109/TNNLS.2016.2574830
20. Song, Q., Cao, J., Yu, W.: Second-order leader-following consensus of nonlinear multi-agent systems via pinning control. Syst. Control Lett. **59**(9), 553–562 (2010)
21. Wen, G., Yu, W., Hu, G., Cao, J., Yu, X.: Pinning synchronization of directed networks with switching topologies: a multiple Lyapunov functions approach. IEEE Trans. Neural Netw. Learn. Syst. **26**(12), 3239–3250 (2015)
22. Liu, X., Chen, T.: Synchronization of linearly coupled networks with delays via aperiodically intermittent pinning control. IEEE Trans. Neural Netw. Learn. Syst. **26**(10), 2396–2407 (2015)
23. Boyd, S., Ghaoui, L., Feron, E., Balakrishnan, V.: Linear Matrix Inequalities in System and Control Theory. SIAM, Philadelphia (1994)
24. Wilkinson, J.H.: The Algebraic Eigenvalue Problem. Clarendon Press, Oxford (1965)
25. Horn, R.A., Johnson, C.R.: Matrix Analysis, 2nd edn. Cambridge University Press, New York (2013)
26. Song, Q., Cao, J.: On pinning synchronization of directed and undirected complex dynamical networks. IEEE Trans. Circ. Syst. I Regul. Pap. **57**(3), 672–680 (2010)

Prediction of Tropical Storms Using Self-organizing Incremental Neural Networks and Error Evaluation

Wonjik Kim(✉) and Osamu Hasegawa

Department of Systems and Control Engineering, Tokyo Institute of Technology, Yokohama, Japan
{kim.w.ab,hasegawa.o.aa}@m.titech.ac.jp

Abstract. In this paper, we propose a route prediction method that uses a self-organizing incremental neural network (SOINN). For the training and testing of the neural network, only the latitude and longitude of the tropical storm and atmospheric information around East Asia are required. Our proposed method can predict the movement of a tropical storm with only a short calculation time, and the prediction accuracy is close to the accuracy of the Japan Meteorological Agency. This paper describes the algorithm used for the neural network training, the process for handling the data sets and the method used to predict the storm trajectory. Additionally, experimental results that indicate the performance of our method are presented in the results section.

Keywords: Self-organizing incremental neural network · Meteorology · Natural disaster · Tropical storm

1 Introduction

Currently, natural disasters are being researched in a variety of ways. For example, there have been several investigations into the political [1], economic [2], and statistical aspects of natural disasters [3]. Since natural disasters have a large impact on society, many researchers have made substantial efforts to mitigate the damage that can arise from them. There are many phenomena that are classified as natural disasters such as floods, droughts, and earthquakes. In this paper, we only focus on tropical storms, particularly typhoons.

As technology has developed over the years, tropical storms have been analyzed in many ways including satellite imaging and weather radar [4]. Additionally, the prediction of the movement of tropical storms has also been previously considered [5, 6]. Since analyzing weather information is a big data problem, modern methods of weather prediction or simulation take a significant amount of time even if the calculations are done by a super computer.

To address this issue, we propose a novel method that uses a self-organizing incremental neural network (SOINN) that is trained on big data (i.e., the meteorological information). Since SOINN demonstrates good performance with respect to calculation times, the predicted route of a tropical storm can be calculated quickly. Moreover,

D. Liu et al. (Eds.): ICONIP 2017, Part III, LNCS 10636, pp. 846–855, 2017.
https://doi.org/10.1007/978-3-319-70090-8_86

accuracy is very important to evaluate the target; therefore, we will also calculate the accuracy of the predicted route. Finally, simulated tropical storms will be presented. Throughout this paper, we use an Intel® CoreTM i7 CPU 870 2.93 GHz with 16.0 GB RAM, and the results are calculated using MATLAB.

2 Training Procedure

2.1 Self-organizing Incremental Neural Network

A SOINN, which was originally proposed by Shen and Hasegawa [7], can be trained without initializing the number of nodes and locations. SOINNs have also previously been reported to display robust performances when handling noisy datasets. Figure 1 depicts the SOINN algorithm [8] that is used in this paper. By applying SOINN, it becomes possible to handle the problem even if there is only few data for training. We have a limitation of data, than, SOINN was decided to adapting for prediction model.

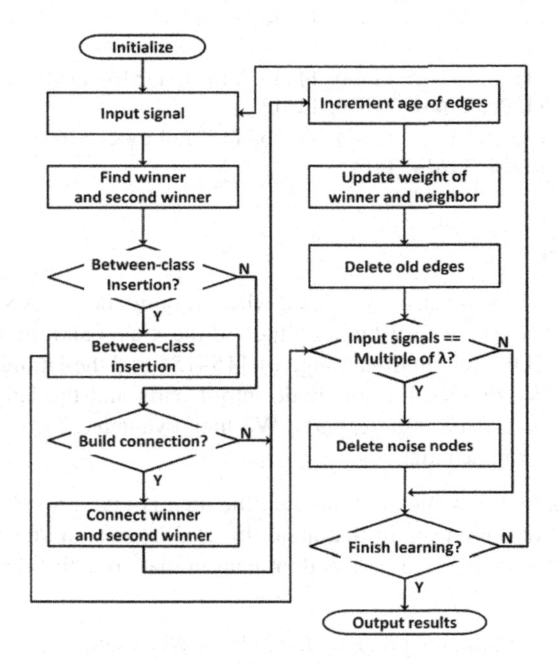

Fig. 1. Algorithm of SOINN

In this paper, we set $\lambda = 1000$, $age_{max} = 5000$. In SOINN, the nodes deleted if it doesn't updated until age_{max}. It was reported that if λ is low, SOINN becomes strong at noise and hard to make cluster, and if λ is high, it becomes weak at noise and easy to make cluster. We hope that data don't disappear by age_{max} and become easier to make cluster, the value of λ and age_{max} are considered.

2.2 Tropical Storms

Several names can be used to describe tropical storms, such as typhoons, cyclones, and hurricanes. The name that is applied to a tropical storm in question is selected based on where the storm is occurring. For example, if the storm occurs in the Atlantic of Northeast Pacific, the term "hurricane" is used. If the storm occurs in the Northwest Pacific, it is called a "typhoon". Lastly, the term "cyclone" is used if the storm occurs in the South Pacific or the Indian Ocean [9].

In this paper, we only consider data from typhoons, and the data for these typhoons are presented in Table 1. The location of the typhoon, starting from when it was born and ending when it disappeared and measured in 3-hour intervals, was acquired from the Japan Meteorological Agency (JMA) [10].

Table 1. Classification of typhoon used in learning (T1511 means 11th typhoon which came in 2015)

Classification	Numbering of typhoon
Landing to Japan	T1511, T1512, T1515, T1518, T1607, T1609, T1610, T1611, T1612, T1616
Approach to Japan	T1506, T1507, T1509, T1513, T1514, T1516, T1517, T1520, T1521, T1526, T1601, T1606, T1613
No affect to Japan	T1519, T1523, T1524, T1525, T1527, T1602, T1603, T1604, T1605, T1608, T1614, T1615

2.3 Atmosphere

Atmospheric conditions are also considered when training the SOINN on typhoon data. For atmospheric analysis, we set the longitude to the x-axis and the latitude to y-axis. We only considered the longitudinal range of 115–155 and the latitudinal range of 20–50 in this paper. We divided the longitude into 4 parts and the latitude into 3 parts, which divides the data area into 12 parts. We then evaluated each area's atmosphere based on the following scheme:

(1) If high pressure dominates in that area, the area is given a value of 1.
(2) If low pressure dominates in that area, the area is given a value of 0.
(3) If neither high nor low pressure is dominant in that area, the area is given a value of 0.5.

In this study, digitalizing process divided the study area in 12 parts. It ignores specific distribution, then, the error derived by this ignorance would be appeared. It is needed to consider the division which helps to reinforce the performance.

The atmospheric data was acquired from the Japan Weather Association [11] (Fig. 2).

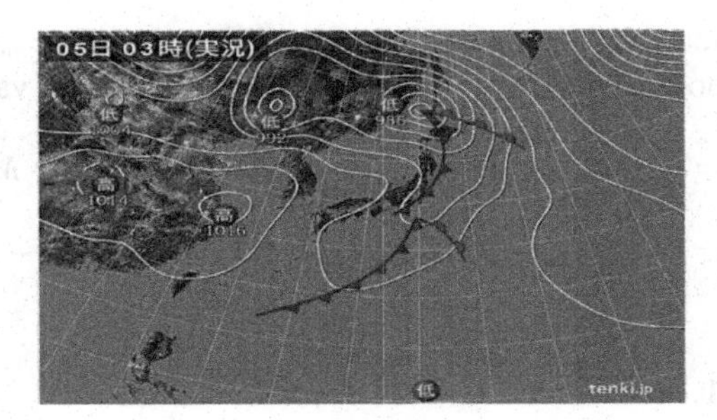

Fig. 2. Atmosphere map in 2015/05/05 3:00 Am, East Asia.

0	0.5	0	0.5
1	0.5	0.5	0.5
0.5	0.5	0.5	0.5

Fig. 3. Example of evaluating Fig. 2.

3 Experiment

To verify the effectiveness of the proposed route prediction method, we used the 5 typhoons listed in Table 2. We used all of the landing typhoons that occurred in 2015 and 2016 to train the SOINN, Thus, we selected typhoons T1408 and T1419, which occurred in the desired time interval. For the approaching typhoons, T1617 was selected because it is the most recent data point for the approaching category. Although T1618 is technically the most recent approaching typhoon, its route is similar to those of T1408 and T1419, and therefore we elected to use T1617 and not T1618. Most of the typhoons that fall into the "no effect" category disappeared very quickly. However, T1504 and T1626 occurred in the desired time interval and existed for more than one week. Therefore, they were selected for analysis.

Table 2. Classification of typhoon used for test. (T1408 means 8th typhoon which came in 2014)

Classification	Numbering of typhoon
Landing to Japan	T1408, T1419
Approach to Japan	T1617
No affect to Japan	T1504, T1626

When we use the typhoon's location data for learning and testing, the following normalization is applied. $T_{Latitude}$ denotes the latitude value of typhoon and

MAX (T_{La_all}) denotes the maximum value in all $T_{Latitude}$. $T_{Longitude}$ denotes the longitude value of typhoon and MIN(T_{Lo_all}) denotes the minimum value in all $T_{Longitude}$.

$$T_{NormalizedLatitude} = (T_{Latitude} - MIN(T_{La\_all}))/(MAX(T_{La\_all}) - MIN(T_{La\_all})) \quad (1)$$

$$T_{NormalizedLongitude} = (T_{Longitude} - MIN(T_{Lo\_all}))/(MAX(T_{Lo\_all}) - MIN(T_{Lo\_all})) \quad (2)$$

3.1 Prediction Method

To predict the route of the typhoon, the following algorithm is used.

Algorithm 1. Method of predicting route

(1) ***Parameters*** : $\overrightarrow{x_n}$: vector of location in time n, $\overrightarrow{x_m^*}$: vector of predicted location in time m, N : set of neurons.
(2) In time t, combine four vectors $\overrightarrow{x_t}$, $\overrightarrow{x_{t+1}}$, $\overrightarrow{x_{t+2}}$, $\overrightarrow{x_{t+3}}$ in one vector $\overrightarrow{X_t}$
(3) For every typhoons, in whole time T, create data set $\overrightarrow{X_T}$
(4) Get N from SOINN learned by $\overrightarrow{X_T}$
(5) In time c, Find the min Euclid-distance between ($\overrightarrow{x_c}$, $\overrightarrow{x_{c+1}}$, $\overrightarrow{x_{c+2}}$) and N.
(6) Define the nearest neuron between ($\overrightarrow{x_c}$, $\overrightarrow{x_{c+1}}$, $\overrightarrow{x_{c+2}}$) and N is $\overrightarrow{N_c}$
(7) Assuming that $\overrightarrow{x_{c+3}^*} = \overrightarrow{N_{c+3}} - \overrightarrow{N_{c+2}} + \overrightarrow{x_{c+2}}$.
(8) Includes $\overrightarrow{x_{c+3}^*}$ to $\overrightarrow{X_T}$
(9) Get N from SOINN learned by new $\overrightarrow{X_T}$

By repeating Algorithm 1 N times, we can predict the location of the typhoon after 3 N hours.

3.2 Location Adjustment

Because only 35 typhoons are used for training the algorithm, the amount of data is not enough. Therefore, we applied the Algorithm 2 after the Algorithm 1 before repeating.

Algorithm 2. Location adjustment

(1) ***Parameters*** : $\overrightarrow{x_n}$: vector of location in time n, $\overrightarrow{x_m^*}$: vector of predicted location in time m, N : set of neurons, $\overrightarrow{N_c}$: the nearest neuron between ($\overrightarrow{x_c}$, $\overrightarrow{x_{c+1}}$, $\overrightarrow{x_{c+2}}$) and N, r1 : coefficient of long-side, r2 : coefficient of short side.
(2) if $|\overrightarrow{N_{n+3}} - \overrightarrow{N_{n+2}}| > r1 \times average(|\overrightarrow{x_{n+2}} - \overrightarrow{x_{n+1}}| + |\overrightarrow{x_{n+1}} - \overrightarrow{x_n}|)$ then
(3) $\overrightarrow{x_{n+3}^*} = (\overrightarrow{x_{n+2}} - \overrightarrow{x_n})/2 + \overrightarrow{x_{n+2}}$
(4) else
(5) if $|\overrightarrow{N_{n+3}} - \overrightarrow{N_{n+2}}| < r2 \times average(|\overrightarrow{x_{n+2}} - \overrightarrow{x_{n+1}}| + |\overrightarrow{x_{n+1}} - \overrightarrow{x_n}|)$ then
(6) $\overrightarrow{x_{n+3}^*} = (\overrightarrow{x_{n+2}} - \overrightarrow{x_n})/2 + \overrightarrow{x_{n+2}}$
(7) end if

If the predicted location is evaluated too far by step (2) in Algorithm 2, $\overrightarrow{x_{n+3}^{*}}$ will be changed to follow their original movement. A similar thing happens when predicted location is evaluated too close. In this report, we set $r1 = 1.5$ and $r2 = 0.8$.

3.3 Result

By using Algorithms 1 and 2, we predicted the route of 5 typhoons for testing the performance. From the JMA homepage, we can check the average error for predicting 24 and 48 h ahead [12]. We also estimated the error for the 24- and 48-hour predictions from our algorithm. During this time, we predicted the movement of the storm until it changed to an extratropical low. The following table presents the average distance between the real storm location and the predicted location. According to Table 3, the error of prediction in landing typhoon is bigger than other error. One of the main reason is shortage of data in landing type. We expect that result might be improved with more data.

Table 3. Average error of location of proposed method

Typhoon number	24 h prediction	48 h prediction
T1408	163.5 km	383.0 km
T1419	120.8 km	257.9 km
T1504	55.9 km	89.0 km
T1617	85.3 km	221.2 km
T1626	102.1 km	278.7 km
Total average	105.5 km	250.0 km

We also present the average error for different years from the JMA. The JMA uses two super computers-a FUJITSU PRIMEHPC FX100 and a FUJITSU PRIMERGY CX2550M1-and professional datasets that are not available to the public for calculations [13] (Table 4).

The total average error of our predictions is 105.5 km for 24 h and 250.0 km for 48 h. The SOINN takes only 0.663 s to be trained on the dataset. The specific calculation times are shown in Table 5.

Figures 4, 5, 6, 7, 8, 9, 10 and 11 present the actual and predicted movement of the storm. The cross points in Figs. 7, 8, 9, 10 and 11 indicate the real location of the typhoon. The lines that originate from each cross point indicate the predicted route at that point. Due to space limitations, we only present the actual and predicted routes for T1408 at full size. The remaining figures are presented in a smaller size.

Table 4. Average error of location by Japan Meteorological Agency

Year	24 h prediction	48 h prediction
2016	79 km	142 km
2015	72 km	119 km
2014	101 km	177 km

Table 5. Calculation time for prediction

	Used time
Learning from original data set	0.663 s
T1408	15.190 s
T1419	15.510 s
T1504	12.591 s
T1617	10.945 s
T1626	10.991 s

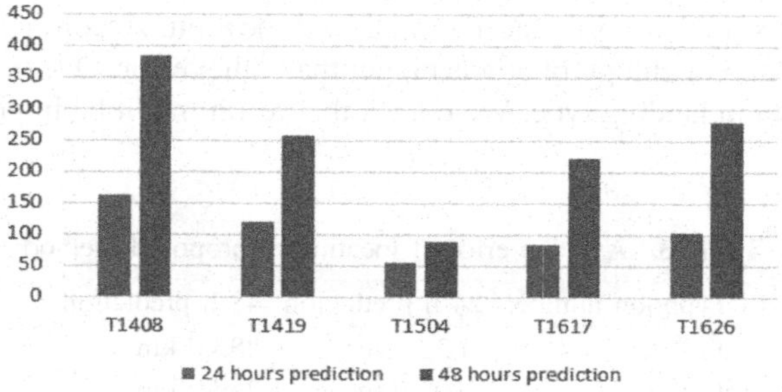

Fig. 4. Average error of location by our method

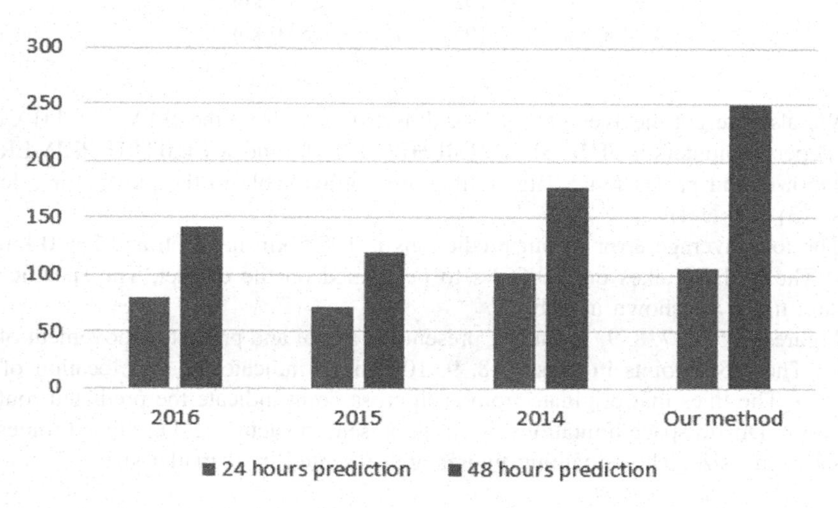

Fig. 5. Comparing error of ours with error of Japan Meteorological Agency

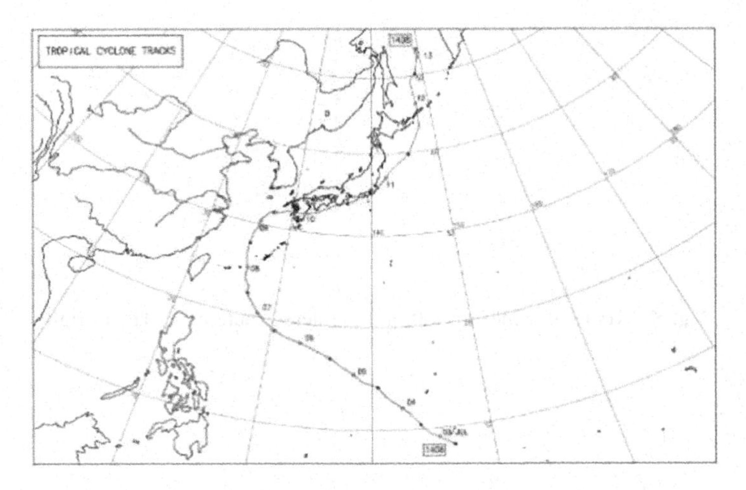

Fig. 6. Route of T1408

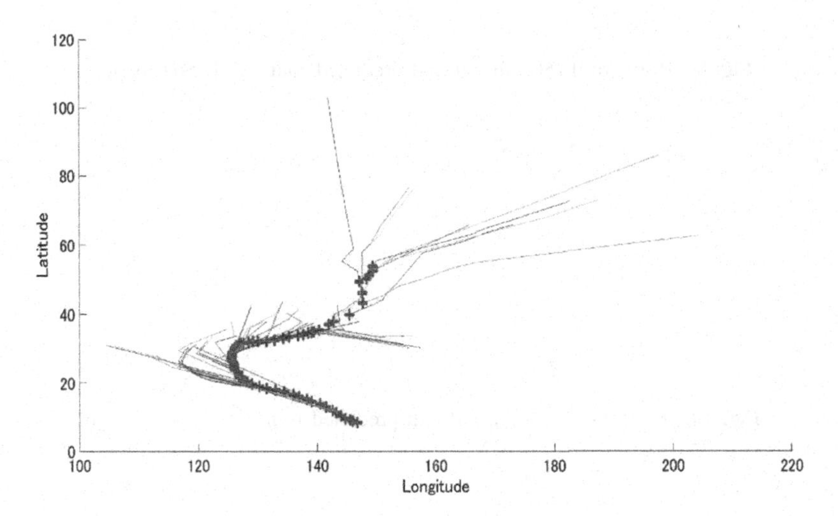

Fig. 7. Predicted route of T1408

In Figs. 7, 8, 9, 10 and 11, lines indicate the predicted movement four days into the future from each point. However, we did not calculate the error in the distance prediction two days into the future.

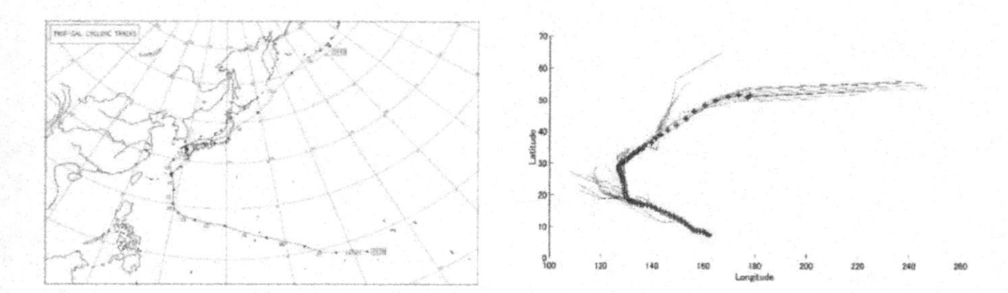

Fig. 8. Route of T1419 in left and predicted route of T1419 in right

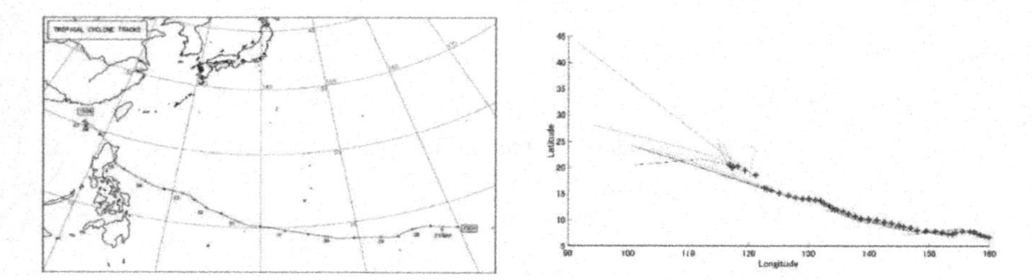

Fig. 9. Route of T1504 in left and predicted route of T1504 in right

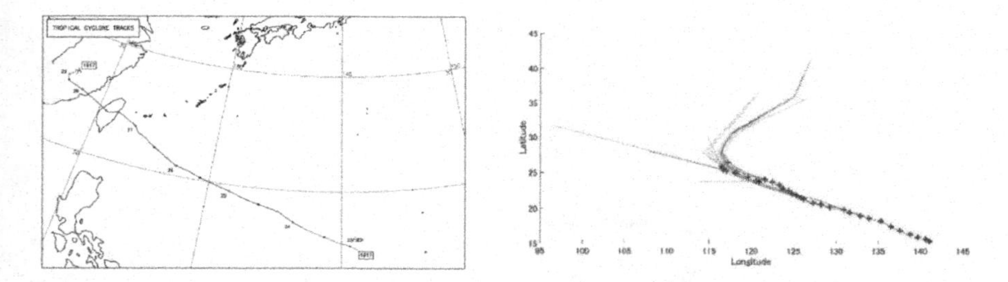

Fig. 10. Route of T1617 in left and predicted route of T1617 in right

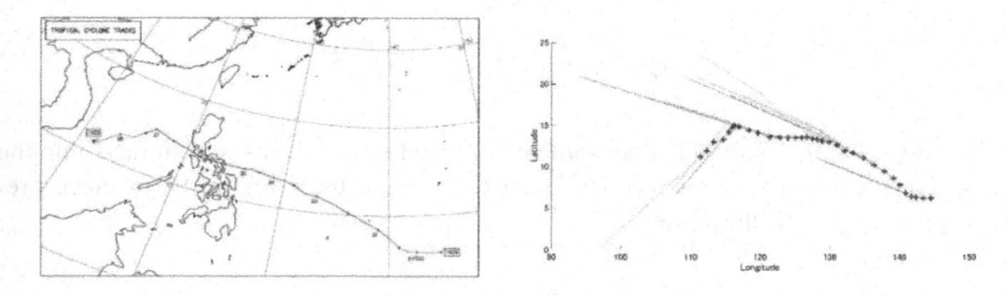

Fig. 11. Route of T1626 in left and predicted route of T1626 in right

4 Conclusion

In this paper, we used SOINN to predict the routes of tropical storms. Our proposed method shows a close accuracy to the JMA predictions, even though our method takes only a few seconds for the calculations. Furthermore, we only used open tropical storm data from 2015 and 2016. Though we used only open data that we refined slightly, our method indicates a high accuracy and fast calculation speed. This results based on one of the characteristic of SOINN: online learning. There is substantial potential for improvement if larger, more specific meteorological datasets are used. We, therefore, conclude that our method is effective and performs well in predicting the routes of tropical storms.

References

1. Fair, C.C., Kuhn, P.M., Malhotra, N., Shapiro, J.N.: Natural disasters and political engagement: evidence from the 2010-11 Pakistani Floods. Q. J. Political Sci. **12**(1), 99–141 (2017)
2. Toya, H., Skidmore, M.: Economic development and the impacts of natural disasters. Econ. Lett. **94**(1), 20–25 (2007)
3. Guha-Sapir, D., Vos, F., Below, R., Ponserre, S.: Annual disaster statistical review 2011: the number and trends. Centre for Research on the Epidemiology of Disasters (CRED) (2012)
4. Jochen, Z., Andreas, N.: Early Warning Systems for Natural Disaster Reduction, 1st edn. Springer, Heidelberg (2003)
5. Lin, T.C., Hamburg, S.P., Lin, K.C., Wang, L.J., Chang, C.T., Hsia, Y.J., Vadeboncoeur, M. A., McMullen, C.M., Liu, C.-P.: Typhoon disturbance and forest dynamics: lessons from a northwest Pacific subtropical forest. Ecosystems **14**(1), 127–143 (2011)
6. Bellingham, P.J., Takashi, K., Shin-ichiro, A.: The effects of a typhoon on Japanese warm temperate rainforests. Ecol. Res. **11**(3), 229–247 (1996)
7. Shen, F., Hasegawa, O.: An incremental network for on-line unsupervised classification and topology learning. Neural Netw. **19**(1), 90–106 (2006)
8. Yamasaki, K., Makibuchi, N., Shen, F., Hasegawa, O.: Self-organizing incremental neural Network-SOINN- and its usage. Brain Neural Netw. **17**(4), 187–196 (2010)
9. National Ocean Service Homepage. http://oceanservice.noaa.gov/facts/cyclone.html
10. Japan Meteorological Agency Homepage. http://www.data.jma.go.jp/fcd/yoho/typhoon/route_map/index.html
11. Japan Weather Association's information Homepage. http://www.tenki.jp/guide/chart/
12. Japan Meteorological Agency Homepage, Average error of the year. http://www.data.jma.go.jp/fcd/yoho/typ_kensho/table.html
13. Japan Meteorological Agency Homepage, Facility introduction. http://www.mri-jma.go.jp/Facility/supercomputer.html

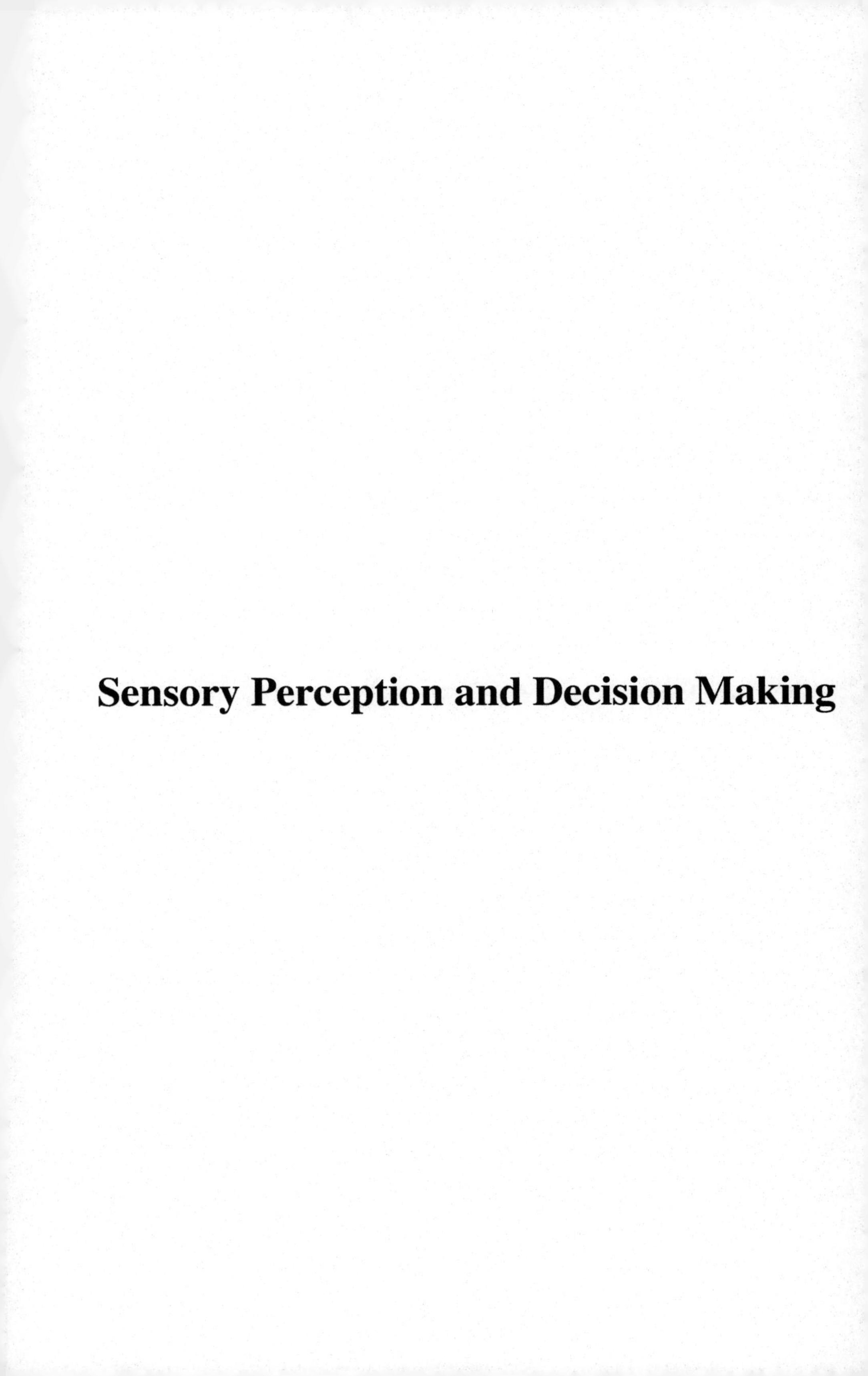

Sensory Perception and Decision Making

Formulation of Border-Ownership Assignment in Area V2 as an Optimization Problem

Zaem Arif Zainal(✉) and Shunji Satoh

Graduate School of Information Systems,
The University of Electro-Communications, Tokyo 182-8585, Japan
zaem@hi.is.uec.ac.jp, shunji@uec.ac.jp

Abstract. Border-ownership (BO) assignment, or the assignment of borders to an occluding object, is a primary step in visual perception. Physiological experiments have revealed the existence of neurons in area V2 that respond selectively to objects placed on a specific side of their response field. Although existing models can reproduce this phenomenon, they are not based on a clear computational theory. For this study, we formulated BO assignment as a well-defined optimization problem. We hypothesize that information related to BO assignment can be expressed as a conservative vector field. This conservative vector field is proposed as the gradient of a scalar field that carries information related to the depth order of the overlapping object. Conservative vector fields have zero curl (rotation). Using this theorem, we construct and solve an optimization problem. Numerical simulations demonstrate that a model based on our derived algorithm solves BO assignment for problems of perceived order and occlusion. Deduced neural networks provide insight into possible characteristics of lateral connections in area V2.

Keywords: Computational model · Perception · Depth order · Middle vision

1 Introduction

Humans can distinguish and identify mutually overlapping objects. When one object occludes another, the border separating the two objects is "owned" by the occluding object. Assignment of borders to image regions containing the occluding object, or border-ownership (BO) assignment, is believed to be a preliminary process of visual perception [1]. In the example of a shaded image in Fig. 1, a two-dimensional (2D) image that includes two shaded regions can be interpreted as one object overlapping another. Two interpretations might be made of the image: (a) a dark rectangle over a light-grey background or (b) a light-grey rectangular frame over a dark grey background. These distinct percepts of object depth order might be a function of BO assignment coding. In Fig. 1, arrows at the borders represent possible directions for BO assignment. The inner region is perceived as a protruding rectangle when BO is assigned inwards. Similarly, the outer region is perceived as a protruding rectangular frame when BO is assigned outwards. The likely interpretation by viewers is Fig. 1a, a protruding rectangular object, attributed to Gestalt factors such as closure.

D. Liu et al. (Eds.): ICONIP 2017, Part III, LNCS 10636, pp. 859–866, 2017.
https://doi.org/10.1007/978-3-319-70090-8_87

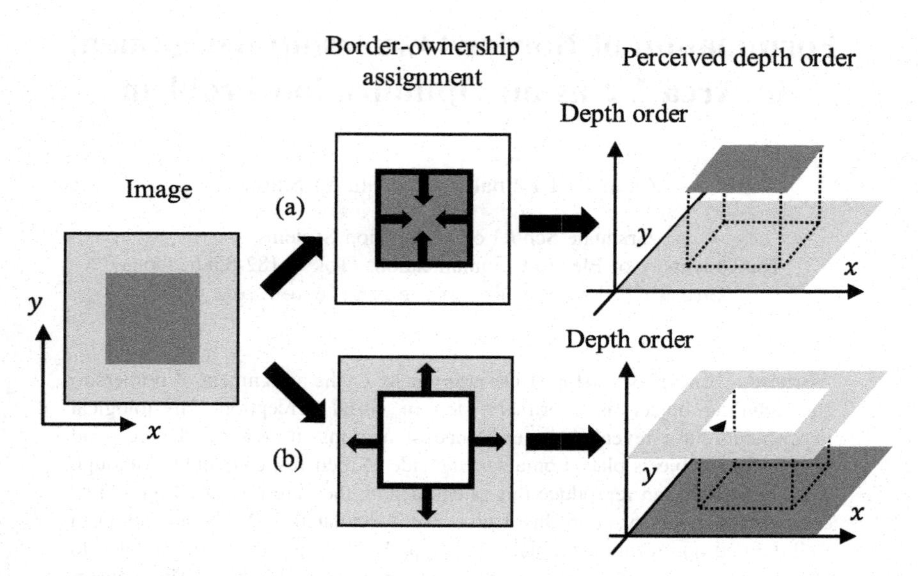

Fig. 1. A 2D image can be perceived in two different ways depending on how BO assignment is performed. Arrows at borders face the assigned region. A higher depth order value coincides with a region closer to the viewer. (a) "Inward" BO assignment results in perception of a rectangular object. (b) "Outward" BO assignment results in a rectangular frame.

Findings from physiological experiments support the assertion that the BO assignment is implemented in low/middle visual processing. Some neurons in area V2 display selectivity to objects on a specific side of the edge in their receptive fields [2]. Another set of neurons is influenced by binocular depth cues in addition to Gestalt cues [3]. These findings imply that V2 is involved in surface reconstruction.

Several proposed models can help to explain object-side selectivity of V2 neurons [4–6]. Weights representing the lateral connections in Li's model are based on the geometric relation between edge segments, considering cases in which objects a likely exist on a specific side [4]. The formula describing these neural weights includes 23 free parameters that must be determined appropriately. Sakai et al. discovered that randomly generated feedforward neural connections from V1 to V2 are capable of reproducing the diverse properties of object-side selective neurons [5]. Because of the random characteristics of the proposed neural connections, the underlying computational theory of their model is unclear. Craft et al. propose that responses of object-selective neurons can be modeled by the sum of responses to edges in their receptive field, outputs of neurons which respond selectively to T-junctions, and feedforward-feedback processing with "grouping cells" [6]. The neural components of their model are determined in a hypothetical manner. The primary objective of models proposed in their studies is to reproduce the phenomena of neurons with object-side selectivity. No report of the relevant literature describes a study approach using the subject in a deductive manner.

Defining a clear computational theory through formulation presents several advantages over existing approaches. Standard regularization theory has been proposed to formulate problems in early vision [7]. Algorithms and neural networks can be derived from this theory, which hypothesizes that the brain solves "ill-posed" problems. Similarly, we attempt to formulate BO assignment by defining a suitable energy function for minimization. Then we proceed to deduce the algorithm and neural networks that are involved.

2 Computational Theory and Model

Variables for BO and depth order must be defined to formulate the BO assignment problem. In Fig. 1, respective BO of the edge segments are expressed as arrows pointing to the object. If one considers each image pixel as a location for a neuron's receptive field, then the population output of these neurons can be expressed as a two-dimensional vector field in the Cartesian coordinate system, $\boldsymbol{E}(x,y) = \left(E_x(x,y), E_y(x,y)\right)^{\mathrm{T}}$, where $E_x(x,y)$ and $E_y(x,y)$ respectively represent vector components in the $\hat{\boldsymbol{x}}$ and $\hat{\boldsymbol{y}}$ direction. We define depth order as a scalar field $\phi(x,y)$ in which zero values of ϕ correspond to the background. Larger positive values of ϕ correspond to areas regarded as closer to the viewer. We propose that differentiation of a depth-order scalar field $\phi(x,y)$ is equal to BO vector field $\boldsymbol{E}(x,y)$. This relation can be described mathematically as

$$\boldsymbol{E}(x,y) = \boldsymbol{\nabla}\phi(x,y), \tag{1}$$

where $\boldsymbol{\nabla}$ is the partial differentiation operator in the $\hat{\boldsymbol{x}}$ and $\hat{\boldsymbol{y}}$ direction, $\nabla \stackrel{\text{def}}{=} (\partial/\partial x, \partial/\partial y)^{\mathrm{T}}$. Both $\boldsymbol{E}(x,y)$ and $\phi(x,y)$ are unknown values before the onset of a visual stimulus. Therefore, they must be determined appropriately

For this study, we assume that initial values for a two-dimensional vector field evolving over time $\boldsymbol{E}(x,y,t)$ are given for $t = 0$, and that must be updated accordingly until the relation in Eq. (1) is satisfied. By definition, the gradient of any scalar field is a conservative vector field. Conservative vector fields are curl-free, meaning that "curl" (rotation) at all spatial points is zero. The "curl" for a two-dimensional vector field $\boldsymbol{E}(x,y)$ is defined as the difference between partial differentiation on components $E_x(x,y)$ and $E_y(x,y)$ as

$$\mathrm{curl}(\boldsymbol{E}(x,y)) = \frac{\partial}{\partial x}E_y(x,y) - \frac{\partial}{\partial y}E_x(x,y). \tag{2}$$

We proceed to construct an energy function to minimize based on the theorem of zero "curl" as

$$J[\boldsymbol{E}(x,y)] = \frac{1}{2}\iint (\mathrm{curl}(\boldsymbol{E}(x,y)))^2 dxdy \to \min. \tag{3}$$

Integration is applied on the image region. Using the steepest descent method on energy function $J[\boldsymbol{E}(x,y)]$, we obtained an update rule for $\boldsymbol{E}(x,y,t)$, which can be expressed as individual update rules for $\hat{\boldsymbol{x}}$ and $\hat{\boldsymbol{y}}$ components, $E_x(x,y,t)$ and $E_y(x,y,t)$,

$$\frac{\partial}{\partial t}\boldsymbol{E}(x,y,t) = \begin{pmatrix} \frac{\partial}{\partial t}E_x(x,y,t) \\ \frac{\partial}{\partial t}E_y(x,y,t) \end{pmatrix} \propto \begin{pmatrix} \frac{\partial^2}{\partial y^2}E_x(x,y,t) - \frac{\partial^2}{\partial x \partial y}E_y(x,y,t) \\ -\frac{\partial^2}{\partial x \partial y}E_x(x,y,t) + \frac{\partial^2}{\partial x^2}E_y(x,y,t) \end{pmatrix}. \tag{4}$$

The above update rule is also expressible in terms of rotation $\mathrm{curl}(\boldsymbol{E}(x,y,t))$ and the perpendicular of the gradient operator $\nabla \stackrel{\mathrm{def}}{=} (-\partial/\partial y, \partial/\partial x)^{\mathrm{T}}$ to give the concise equation of

$$\frac{\partial}{\partial t}\boldsymbol{E}(x,y,t) \propto \boldsymbol{\nabla}^{\perp}\mathrm{curl}(\boldsymbol{E}(x,y,t)). \tag{5}$$

A model was constructed based on Eq. (6) by replacing partial differential operators with discrete differential filters, where x and y are of unit length one. For this study, we use a discrete two-dimensional Gaussian function of standard deviation σ,

$$G_{\sigma}(x,y) = \frac{1}{2\pi\sigma^2}\exp\left(-\frac{x^2+y^2}{2\sigma^2}\right), \tag{6}$$

and perform convolution $*$ of its derivatives with $E_x(x,y,t)$ and $E_y(x,y,t)$ as

$$\begin{pmatrix} \frac{\partial}{\partial t}E_x(x,y,t) \\ \frac{\partial}{\partial t}E_y(x,y,t) \end{pmatrix} \propto \begin{pmatrix} \frac{\partial^2}{\partial y^2}G_{\sigma}(x,y) * E_x(x,y,t) - \frac{\partial^2}{\partial x \partial y}G_{\sigma}(x,y) * E_y(x,y,t) \\ -\frac{\partial^2}{\partial x \partial y}G_{\sigma}(x,y) * E_x(x,y,t) + \frac{\partial^2}{\partial x^2}G_{\sigma}(x,y) * E_y(x,y,t) \end{pmatrix}. \tag{7}$$

The above update rules for horizontal and vertical vectors, $\frac{\partial}{\partial t}E_x(x,y,t)$ and $\frac{\partial}{\partial t}E_y(x,y,t)$ were conducted on border pixels until energy function $J[\boldsymbol{E}(x,y,t)]$ was minimized sufficiently. Depth order $\phi(x,y)$ is treated as a dependent variable, calculated by averaging results of line integration conducted on $\boldsymbol{E}(x,y,t)$ through horizontal and vertical paths through (x,y) after energy is minimized. It is noteworthy that, in theory, the result of line integration is independent of the integration path when Eq. (1) is satisfied and when $\boldsymbol{E}(x,y,t)$ is a conservative vector field.

3 Numerical Simulations

In our simulations, initial vectors at $t = 0$ were determined as unit vectors facing the inner side of L-junctions (right-angle contours). Vector updating was performed for $t = 1000$ iterations using Gaussian derivative filters of standard deviation $\sigma = 1$. Depth order $\phi(x,y)$ was rescaled to a maximum value of 1.

Numerical simulations were performed for problems of occlusion, as presented in Fig. 2a, which shows simulation results for an occlusion problem where the common percept of the figure is a stack of three rectangles. After vector updating, pixels along the borders where no vectors initially existed are assigned correctly to the occluding objects. Calculated depth order $\phi(x,y)$ agrees qualitatively with perception of three stacked-up rectangles. The three closed regions have different values of depth order ϕ. Vector magnitudes reflect the difference in values of depth order ϕ across object borders. Figure 2b presents a similar problem, but with the top-most rectangle removed. Calculated depth order $\phi(x, y)$ agrees with the percept of a C-shaped object occluding a rectangle. A difference of BO assignment is apparent around the concave region at $t = 1000$, shown in the insets of Fig. 2a and b. Despite similar local concavity, BO is assigned correctly according to depth-order perception.

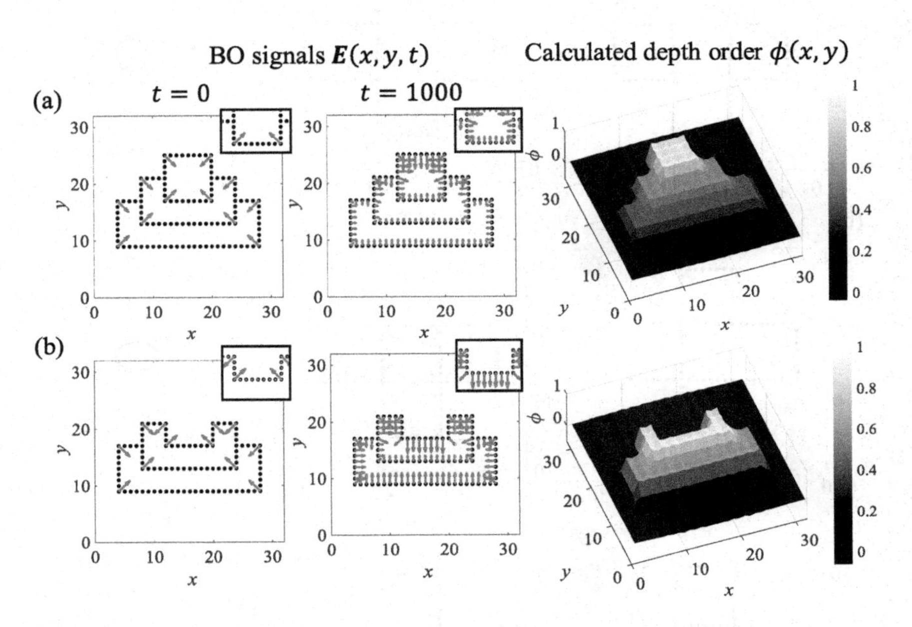

Fig. 2. Initial vectors, updated vectors corresponding to BO, and calculated depth order for (a) three overlapping rectangles and (b) a C-shaped figure occluding a rectangle.

We also performed simulations with a set of stimuli to demonstrate the effect of transparent overlay on BO assignment. Because our model uses no luminance contrast information, simulations are based on the edges of stimuli in the insets of Fig. 3. Consider the circled vector positioned at the vertical edge in the enlarged area of $\boldsymbol{E}(x, y, t)$ at $t = 1000$. The circled vectors in Fig. 3a points towards the left, where an opaque occluding object is presumed to exist. It is particularly interesting that the circled vector in Fig. 3b points to the left as well. A transparent occluding object is presumed to exist on the left. In contrast, when the image is separated into four squares, the circled vector points to the square on the right of its edge (Fig. 3c). Despite the absence of luminance, our model can reproduce depth order for a transparent overlay. Simulation results agree qualitatively with actual BO modulation for a neuron with a receptive field located at the same position of the circled vectors [8].

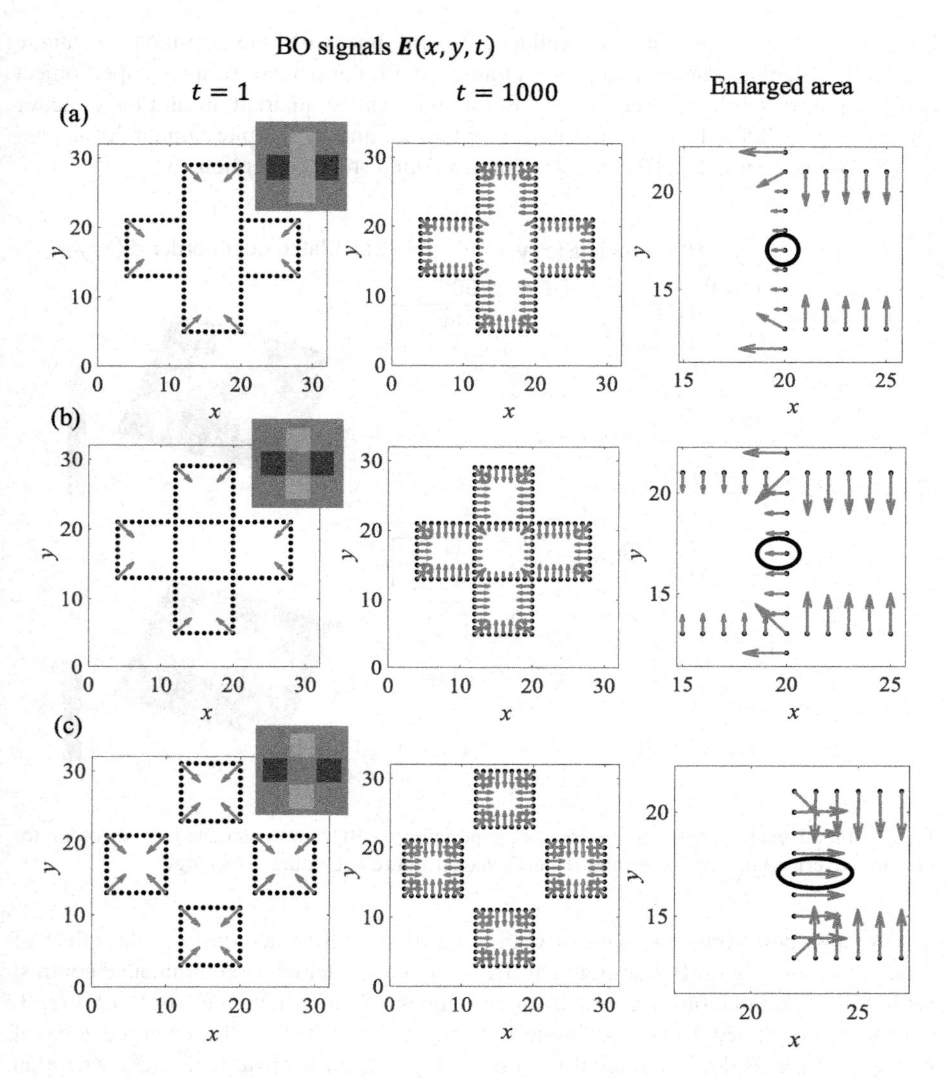

Fig. 3. Calculated BO assignment based on the contours of (a) overlapping rectangles, (b) a transparent overlay and (c) four separated squares. Directions of vectors circled in the enlarged area of (b) and (c) agree with the BO modulation of actual neurons [8].

4 Discussion

Results of numerical simulations demonstrate that our model can resolve BO assignment for occlusion problems without the use of an explicit T-junction detector. Furthermore, our model can distinguish between borders that are a result of occlusion and those which arise from the shape of the objects themselves (Fig. 2). It might also be adopted as the basis for a more complex model which, for example, incorporates luminance contrast for the quantitative reproduction of responses of object-side selective neurons to transparent overlays.

Neural connections weights can be inferred from the update rule in Eq. (7), derived from the formulation of BO assignment. Consider the update rule for $E_x(x, y, t)$. $E_x(x, y, t) > 0$ (<0) corresponds to a neuron at (x, y) with right-side (left-side) object preference. The spatial filters on the right side of the equation can be generalized as weights $w_x(x, y) \stackrel{\text{def}}{=} \frac{\partial^2}{\partial y^2} G_\sigma(-x, -y)$ and $w_y(x, y) \stackrel{\text{def}}{=} \frac{-\partial^2}{\partial xy} G_\sigma(-x, -y)$. Spatially discretized weights for a neuron at $(0, 0)$ can be expressed as

$$\frac{\partial}{\partial t} E_x(0, 0, t) = \sum_i \sum_j w_x(i, j) E_x(i, j, t) + \sum_i \sum_j w_y(i, j) E_y(i, j, t). \quad (8)$$

Weights can be expressed as vectors $\boldsymbol{w} = (w_x, w_y)^{\mathrm{T}}$ of magnitude $\|\boldsymbol{w}\| = \sqrt{w_x^2 + w_y^2}$. The weights of our model for $E_x(x, y, t)$ are visualized in Fig. 4a. Each vector illustrates the direction of $\boldsymbol{w}$, while pixel intensity represents vector magnitude $\|\boldsymbol{w}\|$. Vector magnitude was normalized to 1. Only vectors of magnitudes $\|\boldsymbol{w}\| > 0.0.57$ are displayed.

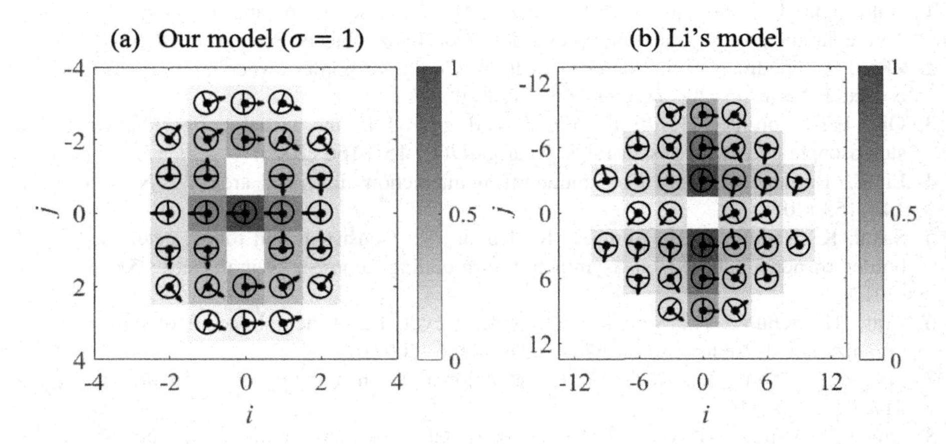

Fig. 4. Visualization of neural weights $\boldsymbol{w}$ for a neuron at $(0, 0)$ with a right-side object selectivity $E_x(0, 0, t)$. Image intensity corresponds to vector magnitude $\|\boldsymbol{w}\|$. Neural weights of (a) our model are qualitatively similar to that of (b) Li's model.

Characteristics of the deduced neural network in our study were found to be qualitatively similar to that of lateral connections proposed by Li, which was visualized similarly (Fig. 4b). It is noteworthy that neural weights in Li's model are not ground truths, and that the spatial characteristics of lateral connections in V2 are unknown now. BO signal propagation by lateral connections would imply that signal delay increases with figure size, which contradicts experimental findings [2]. However, our model requires only one parameter to express neural weights for spatial resolution σ: Li's model includes 23 parameters.

5 Conclusion

This report is the first of a study approaching BO assignment in V2 as an optimization problem. The assignment was achieved by minimizing the curl of a vector field representing BO. A simple model based on this computational theory can resolve BO assignment for simple occlusion problems. Unlike previous models, ours has a clear computational theory. Nevertheless, it has limitations. Initial vectors are provided deterministically. Factors other than the geometric configuration of edges, such as information related to contrast intensity, might be included in future models to construct more inclusive models of neurons in area V2. Characteristics of the derived neural networks might provide insight into how BO assignment is encoded. We hope that results of this study can serve as a foundation for research conducted to elucidate low/middle visual processing.

Acknowledgements. This work was partially supported by JSPS KAKENHI (16K00204).

References

1. Nakayama, K., Shimojo, S., Silverman, G.H.: Stereoscopic depth: Its relation to image segmentation, grouping, and the recognition of occluded objects. Perception **18**, 55–68 (1989)
2. Zhou, H., Friedman, H.S., von der Heydt, R.: Coding of border ownership in monkey visual cortex. J. Neurosci. **20**(17), 6594–6611 (2000)
3. Qiu, F.T., von der Heydt, R.: Figure and ground in the visual cortex: V2 combines stereoscopic cues with gestalt rules. Neuron **47**(1), 155–166 (2005)
4. Li, Z.: Border ownership from intracortical interactions in visual area V2. Neuron **47**(1), 143–153 (2005)
5. Sakai, K., Nishimura, H., Shimizu, R., Kondo, K.: Consistent and robust determination of border ownership based on asymmetric surrounding contrast. Neural Netw. **33**, 257–274 (2012)
6. Craft, E., Schütze, H., Niebur, E., von der Heydt, R.: A neural model of figure-ground organization. J. Neurophysiol. **97**(6), 4310–4326 (2007)
7. Poggio, T., Torre, V., Koch, C.: Computational vision and regularization theory. Nature **317**, 314–319 (1985)
8. Qiu, F.T., von der Heydt, R.: Neural representation of transparent overlay. Nat. Neurosci. **10**(3), 283–284 (2007)

A Joint Learning Framework of Visual Sensory Representation, Eye Movements and Depth Representation for Developmental Robotic Agents

Tanapol Prucksakorn, Sungmoon Jeong(✉), and Nak Young Chong

School of Information Science, Japan Advanced Institute of Science and Technology, Ishikawa, Japan
{tanapol.pr,jeongsm,nakyoung}@jaist.ac.jp

Abstract. In this paper, we propose a novel visual learning framework for developmental robotics agents which mimics the developmental learning concept from human infants. It can be applied to an agent to autonomously perceive depths by simultaneously developing its visual sensory representation, eye movement control, and depth representation knowledge through integrating multiple visual depth cues during self-induced lateral body movement. Based on the active efficient coding theory (AEC), a sparse coding and a reinforcement learning are tightly coupled with each other by sharing a unify cost function to update the performance of the sensory coding model and eye motor control. The generated multiple eye motor control signals for different visual depth cues are used together as inputs for the multi-layer neural networks for representing the given depth from simple human-robot interaction. We have shown that the proposed learning framework, which is implemented on the Hoap-3 humanoid robot simulator, can effectively learn to autonomously develop the sensory visual representation, eye motor control, and depth perception with self-calibrating ability at the same time.

1 Introduction

For living organisms such as humans and mammals, visual perception is one of the most important function. It gives the organism an ability to learn and interact with environments around them. However, when they were born, they do not instantly understand how to use the information they perceived. So, for their lifetime they continuously learn and improve their perception, while interact with the environments. In biological vision systems, the data that is collected by human or animals organs are very noisy and messy data. It is not self-explanatory meaningful information [10]. So, it is quite difficult for us to make use of these non-obvious data. In [17], they discussed that our brain is

T. Prucksakorn and S. Jeong—Contributed equally.

D. Liu et al. (Eds.): ICONIP 2017, Part III, LNCS 10636, pp. 867–876, 2017.
https://doi.org/10.1007/978-3-319-70090-8_88

not programmed to know how to use those data, but instead the brain is trained autonomously to understand how to translate those noisy unordered information into visual perception. In the same way, the developmental robotics agents which are not programmed with visual perception ability faced the same problem that they do not know how to utilize the data. Thus, to use those information, we must create a representation of the data that is packed with the vast information.

In the previous studies, an active efficient coding proposed in [1,2,4] is employed to encode and represent the information perceived by the robot by taking advantages of redundancies. Reinforcement learning algorithm is used as a learning scheme for the robot to generate eye movements based on the encoded information. It has been proven to be successful when learning of vergence and smooth pursuit eye movement are needed [7,12,15,18,19]. In [16], they have successfully demonstrated generating multiple eye movements, which are smooth pursuit and vergence to track a moving object, but depth perception is not included in the learning framework. Moreover, all of the generated eye movement information could not be used for depth perception because stationary observer cannot extract depth information from motion parallax or optic flow without a priori knowledge such as object size.

Especially, to actively perceive the depth information, the biological vision systems can autonomously generate multiple visual depth cues during the lateral body movement such as stereo disparity and motion parallax. When they keep the visual fixation during the body movement, both of the visual depth cues and eye movements are autonomously generated by the same intrinsically motivated learning principal to maximize the redundancy between sensory inputs in binocular and monocular viewing. In [12], they have proposed a developmental learning framework for active depth perception with self-induced lateral body movement, but they only considered a single depth cue as motion parallax. Interestingly, the organism does not use only one visual depth cue for their whole lifetime. They can integrate the information about multiple visual depth cues and analyze the eye movements to perceive the spatial information about the surrounding environment. Generally, in psychology, dominant eye is a concept that implies that one eye moves before another eye does. Recently there are studies that support the dominant eye hypothesis [6,13,14]. Also, according to [5], they reported that when a motion is self-induced by active observer, two visual depth cues (stereo vision and motion parallax) will be sequentially activated which is not observable in a static observer. Therefore, we may consider that two eye movements for different visual depth cues during the self-induced lateral body movement can be sequentially generated in an independent process to minimize the conflict of multiple cues and then finally multiple eye movements are used to analyze the depth information by integrating each of them. To the best of our knowledge, no one has attempted to propose such a learning framework for developmental robots under the efficient coding theory. This approach enables to autonomously learn not only sensory representation and eye movement controls for the multiple visual depth cue analysis but also active depth perception during self-induced body movements.

2 Methods

We combine the sparse coding and reinforcement learning algorithm together to achieve active efficient coding for learning of multiple cues from the dominant and non-dominant eyes, respectively. Sensory coding model learns how to encode and represent the two images which are generated by the dominant eye with self-induced lateral body movement for motion parallax and two eyes for stereo disparity. Reinforcement learner controls the motor based on the encoded information done by the sensory coding model to increase the efficiency of the coding model.

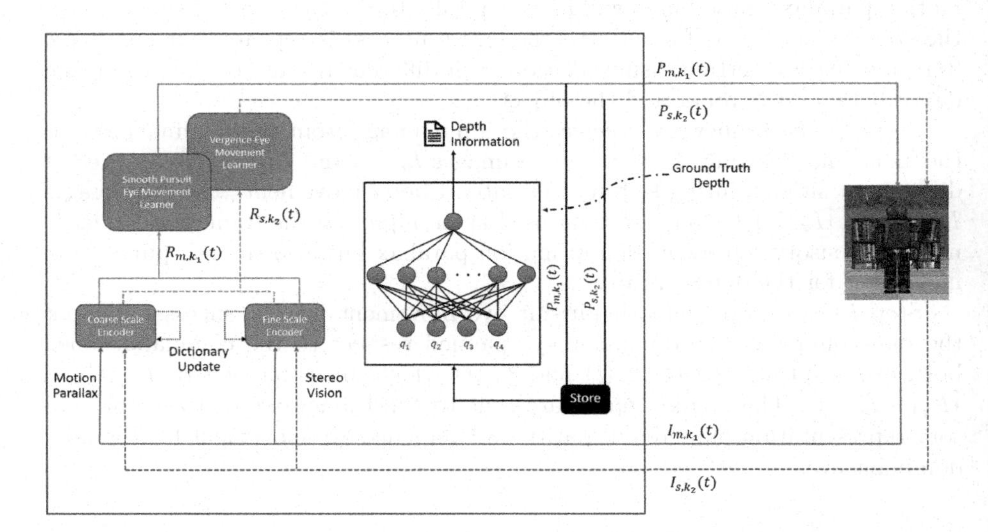

Fig. 1. Model architecture. (1) At the first step k_1, to perform the motion parallax, the robot captures the successive images $I_{m,k_1}(t)$ during the self-induced lateral body movement which are fed into the sensory encoders with multiple image scales. Later, an output reward signal, $R_{m,k_1}(t)$, is sent to the reinforcement learner to generate an appropriate eye movement to hold the fixation during the body movement. Finally, pan command $P_{m,k_1}(t)$ is sent to the robot and it generates the smooth pursuit eye movement for dominant eye camera to maximize the redundancy between the successive images. (2) At the second step k_2, stereo images $I_{s,k_2}(t)$ are captured from both two cameras and sent to the sensory encoders. An output reward signal, $R_{s,k_2}(t)$, is sent to the reinforcement learner to generate the vergence command $P_{s,k_2}(t)$ to maximize the redundancy between the stereo images. The visual dictionaries are then updated based on visual reconstruction errors for both of visual depth cues. Finally, the stored eye movements (q_1, q_2, q_3, and q_4) are used as an input for the neural network to represent the depth information which is given by human-robot interaction.

2.1 Model Architectures

Since the concept of integrating two cues with dominant eye requires that one eye should move before the another one, the framework is divided into 2 steps which are motion parallax for the dominant eye first and stereo vision for the non-dominant eye. The framework (Fig. 1) sequentially perform motion parallax, stereo vision and their integration to represent the depth information. In one iteration t, it is subdivided to 2 steps k_1 and k_2. At step k_1, the robot will perform motion parallax by moving laterally from original position to the *leftmost position*. Then at sub-iteration k_2, the robot will execute stereo vision after the motion parallax is done. After h iterations, the robot will perform the motion parallax again, but it will move laterally to the *rightmost position*. Then, the stereo vision is performed and the entire process is repeated for another h iterations with a certain visual fixation with different texture of the object and depth between the robot and the object.

Step k_1: the framework receives the input image from the dominant eye as the monocular viewing. Two successive images $I_{m_1}(t)$ and $I_{m_2}(t)$ are captured at different positions during self-induced lateral body movement. The two images $I_{m,k_1}(t) = \left[I_{m_1}(t)\ I_{m_2}(t)\right]$ are then used as an input for the framework to learn not only sensory representation of motion parallax but also smooth pursuit eye movement for the dominant eye.

Step k_2: after the smooth pursuit eye movement learner successfully sent the pan command, the dominant eye panned respect to the command. Then both eyes capture images $I_{s_1}(t)$ and $I_{s_2}(t)$ which are combined to $I_{s,k_2}(t) = \left[I_{s_1}(t)\ I_{s_2}(t)\right]$. The stereo images are sent to the framework to learn the sensory representation of stereo disparity and vergence eye movement for the non-dominant eye.

2.2 Sensory Coding Model

Two input images are then cropped by 128×128 pixels and 80×80 pixels from the center of the images. Two cropped images represent fine scale and coarse scale respectively. We use two scales of the images to represent the foveal system in human eyes. The fine scale image represents a foveal region ion our eyes which has more detail from the center of vision. While, coarse scale represents parafoveal area which has lower detail. Discussions and comparisons between using one scale and two scales have been done in [7]. They discussed how gaining the access of multi-scale images could improve the learning of the framework. While, having only one scale might prevent the system to learn.

After cropping, the cropped images are then convert to gray scale. 10 by 10 pixels patches are extracted from the gray scale images whose locations are generated by 1 pixel and 4 pixels shifts horizontally and vertically for coarse scale and fine scale, respectively. The image patches are then sub-sampled using Gaussian pyramid algorithm by a factor of 8 for coarse scale, and factor of 2 for fine scale. The patches are reshaped to be one-dimensional vectors which have zero mean and unit norm, $x_i^j(t)$. Where, i is the index of the patch, and $j \in \{C, F\}$. C is for coarse scale and F stands for fine scale.

For coarse scale and fine scale, the two one-dimensional vectors are then combined into a single vector $x^j(t)$. The first 100 elements of the vectors are from the first image and the remaining are from the second image. The combined vectors ($x^C(t)$ and $x^F(t)$) will consist of $P = 200$ elements.

Later, the patches are encoded by sparse coding algorithm in linear fashion. Each patch can be represented by a linear combination of basis functions picked from an over-complete dictionary $\phi^j(t) = \{\phi_n^j(t)\}_{n=1}^N$[11]. We use $N = 288$ basis functions. Two pairs of dictionaries are randomly initialized and normalized each pair contains coarse scale and fine scale dictionary for stereo vision ($d = s$) and motion parallax ($d = m$) as shown in Fig. 1. We use matching pursuit algorithm [8] to estimate and find the sparse representation of the input vector by the weighted sum as follows:

$$x_i^j(t) \approx \hat{x}_i^j(t) = \sum_{n=1}^{N} a_{i,n}^j(t)\phi_n^j(t) \; . \tag{1}$$

The matching pursuit algorithm suits to concept of sparse coding, which can estimate $x_i(t)$ by using a limited number of coefficients. In this research, the maximum number of non-zero scalar coefficients $a_{i,n}(t)$ is set to be 10 elements to ensure sparseness of the efficient coding. For later use in reinforcement learner part, pooled activity, $f_n(t)$, which represent the activity of each neuron cell is calculated from the coefficients from matching pursuit algorithm as follows:

$$f^j(t) = \begin{bmatrix} f_1^j(t) \\ f_2^j(t) \\ \vdots \\ f_P^j(t) \end{bmatrix} . \tag{2}$$

where, each element of the vector $f^j(t)$ is described as:

$$f_n(t) = \sum_{i=1}^{P} a_{i,n}(t)^2. \tag{3}$$

A reconstruction error is introduced as a cost function to be used in sensory coding model and reinforcement learner. It measures the estimation error of vector $x(t)$. The reconstruction error is defined as:

$$e(t) = \frac{1}{P}\sum_{i=1}^{P} \frac{\|x_i(t) - \sum_{n=1}^{N} a_{i,n}(t)\phi_n(t)\|^2}{\|x_i(t)^2\|}. \tag{4}$$

Gradient descent method is used to update the dictionaries with the reconstruction error as a cost function. After each update, the dictionaries are then normalized.

2.3 Reinforcement Learning

The state representation of the reinforcement learner can be described by combination of coarse scale and fine scale pooled activity, $f_n(t)$ as follows:

$$f(t) = \begin{bmatrix} f^C(t) \\ f^F(t) \end{bmatrix} . \tag{5}$$

The reward that is given to the learning agent is a negative of the summation of reconstruction error from both scales which is described as:

$$R_{d,k}(t) = -(e^C(t) + e^F(t)) . \tag{6}$$

where, $k \in \{k_1, k_2\}$ and $d \in \{m, s\}$. m is for motion parallax. s is for stereo vision. An actor-critic algorithm number 3 proposed in [3] is employed for the leaner agent. For action selection, we use Gibbs distribution (softmax) for probabilistically choosing an action as follows:

$$\pi(f(t), a_t) = \frac{e^{z_a}}{\sum_{a' \in A} e^{z_{a'}}} . \tag{7}$$

For each action, the activation value z_a is given by:

$$z_a = \sum_{n=1}^{N} w_a(t) f_n(t) , \tag{8}$$

where $w_a(t)$ is a weight vector from the state $f(t)$ to action a that is initially random. The action is pan angle of the cameras in degrees. Possible actions a are contained in a set of actions A. In this research we use $A = \{-0.2°, -0.1°, -0.05°, 0°, 0.05°, 0.1°, 0.2°\}$. Thus, the policy maps $f(t)$ to $a \in A$. The selected actions are $P_{m,k_1}(t)$ for motion parallax and $P_{s,k_2}(t)$ for stereo vision.

2.4 Depth Representation

A simple feed forward neural network with two layer is used to interpret between eye movements to the object's distance. In each iteration after stereo vision is executed, the eye movements are stored and accumulated for depth estimation. When the robot successfully performs motion parallax and stereo vision at both leftmost position and rightmost position, the amount of eye movements $\boldsymbol{q}$ are then used to train the neural network. $\boldsymbol{q}$ contains:

1. q_1, Left eye's pan movement at leftmost position
2. q_2, Vergence eye movement at leftmost position
3. q_3, Left eye's pan movement at rightmost position
4. q_4, Vergence eye movement at rightmost position

We use Levengerg-Marquardt method [9] for training the neural network. A sigmoid transfer function is used in the hidden layer which has 10 neurons. The input of the neural network is $\boldsymbol{q}$. While, the target is ground truth depth provided by supervisor.

3 Simulations and Results

3.1 Experimental Setup

We use V-REP, a robot simulator, as a 3D environment visualization for the framework. The framework is implemented and developed in MATLAB. The environment in the simulator comprises HOAP3 robot, an object with interchangeable texture, and a still background image. The lateral movement of the robot is simplified to be changing the position of the robot directly to cut out the travel time.

In this simulation, we test the multiple cues to estimate the depth between the robot and the object and it is from 1 m to 3 m with an 0.1 m interval, i.e. 1.0,1.1,1.2,...3.0 meters. The distance between the leftmost position and the rightmost position is 0.2 m, i.e. $\delta = 0.1$. The baseline, distance between two eyes, is 0.06 m. The number of iterations h is 30 iterations. We prepare 100 different images to learn the various visual textures of the environment. To evaluate the eye movement training, we define mean absolute error (MAE) for evaluating eye movements as follows:

$$\mathrm{MAE(t)} = \frac{1}{1000} \sum_{\mathrm{k}=0}^{999} \left| \theta(\mathrm{t}+29+30\mathrm{k}) - \theta^{*}(\mathrm{t}+29+30\mathrm{k}) \right| . \tag{9}$$

Where,

1. $\theta(t)$ represents the pan/vergence angle of the eye at time t
2. $\theta^{*}(t)$ represents the optimal pan/vergence angle at time t

3.2 Adaptive Visual Dictionary

The principal component analysis (PCA) is applied to visualize the distribution of the visual dictionaries. Because the visual dictionaries were randomly initialized, most of the elements are quite redundant between each other. The first and the second PCs are used to visualize the distribution of visual dictionary as shown in Figs. 2(a)–(c). We can see that the trained visual dictionaries are more sparsely distributed than the initial dictionary.

3.3 Joint Development of Active Depth Perception

The results of the training are shown in Fig. 3. Figure 3(a) shows the MAE of eye movements. The red line shows the MAE of the stereo vision, while the blue line shows the MAE of the smooth pursuit. To test the depth perception, all of the eye movements $\boldsymbol{q}$ with different experimental conditions are used as inputs for the neural networks. The outputs from the neural network are used to calculate the MAE at every time steps. We applied a moving average window with window size of 1,000 iterations to observe trend of the depth learning as shown in Fig. 3(b).

From the simulation results, we can see that the framework could jointly learn to improve the sensory encoding and represent the visual stimuli while learning to generate multiple eye motor control with depth perception.

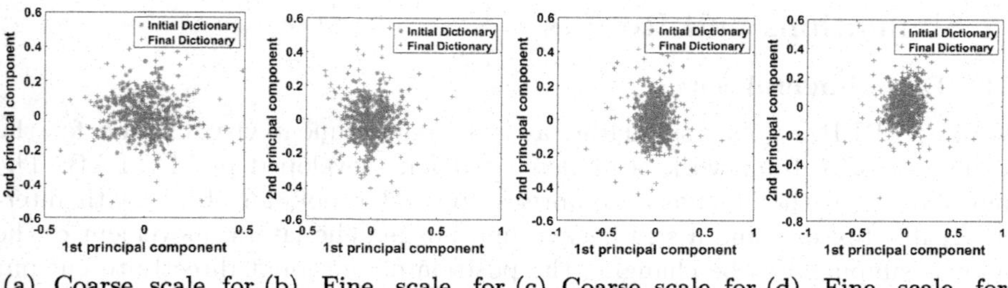

(a) Coarse scale for stereo disparity (b) Fine scale for stereo disparity (c) Coarse scale for motion parallax (d) Fine scale for motion parallax

Fig. 2. Visualization of development of the visual dictionaries. The distribution of the visual dictionaries using the first and second PCs at the initial time and the end of training, respectively.

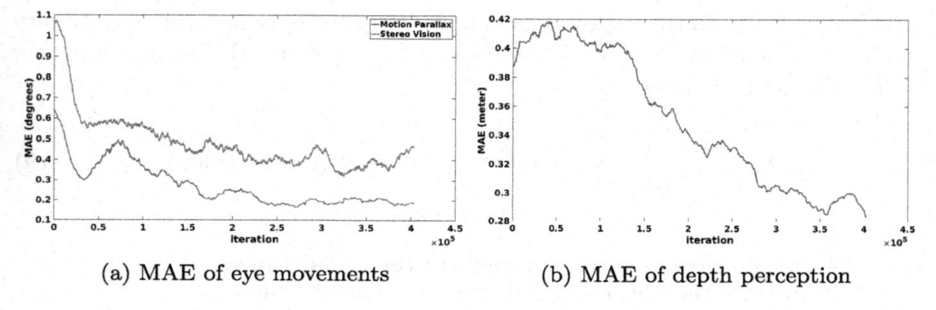

(a) MAE of eye movements (b) MAE of depth perception

Fig. 3. Development in each part of the system. The figures visualize the evolution of the visual representation (coding), eye movement and depth estimation. (a) represents the eye movement errors in form of mean absolute error. (b) shows how depth estimation develops through the learning.

4 Robustness Test

To verify the adaptation properties of the framework, perturbations are simulated by applying a constant in-plane roll rotation of each camera at a time as shown in Fig. 4. In Fig. 4(a), noticeable increases in eye movement errors are observed after inducing the disturbance which are presented by the gray dotted line in the figures. Smooth pursuit eye movements are not largely effected by the disturbance, while vergence eye movements are more susceptible to the interference. Because the vergence eye movements are dependent to the results of smooth pursuit eye movements. Even though the vergence control MAE is drastically increased as shown in Fig. 4(a), the MAE of depth perception is slowly increased at that time. Because, the depth perception is done by integrating both of eye movements and it could be recovered with the supports from both cues as shown in Fig. 4(b).

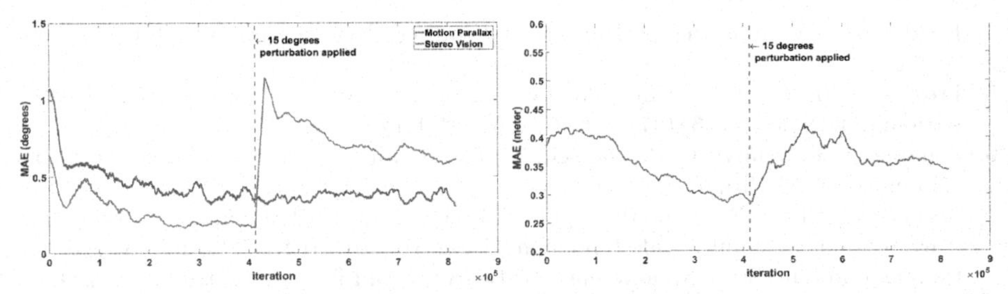

(a) MAE of eye movements with perturbation

(b) MAE of depth perception with perturbation

Fig. 4. Adaptation property from the perturbation. (a) MAE of the eye movements during execution of learning time with the perturbation. (b) MAE of the depth perception during execution of learning time with the perturbation

5 Conclusion

In this research, we proposed a novel developmental learning framework to actively the active depth perception during self-induced lateral body movements. The proposed framework can simultaneously develop the sensory representation, eye movement control and integration of the visual depth cues such as stereo disparity and motion parallax. In order to avoid the conflict of multiple eye movements, the two different eye movements are sequentially trained and generated, while they share the same learning architecture. Finally, the generated multiple eye movements are effectively used to represent the depth information. Also, the proposed learning framework can be seamlessly recovered from the external perturbations. To extend this to fully autonomous architecture, an unsupervised learning method will be employed instead. Moreover, the dominant eye may be competitively selected during the learning period.

Acknowledgement. This work was supported by Japan-Germany collaboration research project on computational neuroscience "Autonomous Learning of Active Depth Perception: from Neural Models to Humanoid Robots" from Japan Agency for Medical Research and Development (AMED) and was partially supported by EU-Japan coordinated R&D project on "Culture Aware Robots and Environmental Sensor Systems for Elderly Support" commissioned by the Ministry of Internal Affairs and Communications (MIC) of Japan and EC Horizon 2020.

References

1. Attneave, F.: Some informational aspects of visual perception. Psychol. Rev. **61**(3), 183–193 (1954)
2. Barlow, H.B.: Possible Principles Underlying the Transformation of Sensory Messages. MIT Press, Cambridge (1961)
3. Bhatnagar, S., Sutton, R.S., Ghavamzadeh, M., Lee, M.: Natural actor-critic algorithms. Automatica **45**(11), 2471–2482 (2009)

4. Field, D.J.: What is the goal of sensory coding? Neural Comput. **6**(4), 559–601 (1994)
5. Frey, J., Ringach, D.L.: Binocular eye movements evoked by self-induced motion parallax. J. Neurosci. **31**(47), 17069–17073 (2011)
6. Johansson, J., Seimyr, G.Ö., Pansell, T.: Eye dominance in binocular viewing conditions. J. vis. **15**(9), 21–21 (2015)
7. Lonini, L., Zhao, Y., Chandrashekhariah, P., Shi, B., Triesch, J.: Autonomous learning of active multi-scale binocular vision. In: 2013 IEEE Third Joint International Conference on Development and Learning and Epigenetic Robotics (ICDL), pp. 1–6, August 2013
8. Mallat, S.G., Zhang, Z.: Matching pursuits with time-frequency dictionaries. IEEE Trans. Signal Process. **41**(12), 3397–3415 (1993)
9. More, J.J.: The Levenberg-Marquardt algorithm: implementation and theory. In: Numerical Analysis, pp. 105–116. Springer, Berlin (1978)
10. Mugan, J., Kuipers, B.: Autonomous learning of high-level states and actions in continuous environments. IEEE Trans. Auton. Ment. Dev. **4**(1), 70–86 (2012)
11. Olshausen, B.A., Field, D.J.: Sparse coding with an overcomplete basis set: a strategy employed by v1? Vision. Res. **37**(23), 3311–3325 (1997)
12. Prucksakorn, T., Jeong, S., Triesch, J., Lee, H., Chong, N.Y.: Self-calibrating active depth perception via motion parallax. In: 2016 Joint IEEE International Conferences on Development and Learning and Epigenetic Robotics (ICDL-Epirob). IEEE (2016)
13. Shneor, E., Hochstein, S.: Eye dominance effects in feature search. Vision. Res. **46**(25), 4258–4269 (2006)
14. Shneor, E., Hochstein, S.: Eye dominance effects in conjunction search. Vision. Res. **48**(15), 1592–1602 (2008)
15. Teulière, C., Forestier, S., Lonini, L., Zhang, C., Zhao, Y., Shi, B., Triesch, J.: Self-calibrating smooth pursuit through active efficient coding. Robot. Auton. Syst. **71**, 3–12 (2015)
16. Vikram, T., Teuliere, C., Zhang, C., Shi, B., Triesch, J.: Autonomous learning of smooth pursuit and vergence through active efficient coding. In: 2014 Joint IEEE International Conferences on Development and Learning and Epigenetic Robotics (ICDL-Epirob), pp. 448–453. IEEE (2014)
17. Weng, J., Luciw, M.: Brain-like emergent spatial processing. IEEE Trans. Auton. Ment. Develop. **4**(2), 161–185 (2012)
18. Zhang, C., Zhao, Y., Triesch, J., Shi, B.E.: Intrinsically motivated learning of visual motion perception and smooth pursuit. In: 2014 IEEE International Conference on Robotics and Automation (ICRA), pp. 1902–1908. IEEE (2014)
19. Zhao, Y., Rothkopf, C., Triesch, J., Shi, B.: A unified model of the joint development of disparity selectivity and vergence control. In: 2012 IEEE International Conference on Development and Learning and Epigenetic Robotics (ICDL), pp. 1–6, November 2012

Neural Representation of Object's Shape at the Electroreceptor Afferents on Electrolocation

Kazuhisa Fujita[1,2](✉) and Yoshiki Kashimori[2]

[1] National Institute of Technology, Tsuyama Collage, 654-1 Numa, Tsuyama, Okayama 708-8506, Japan
[2] University of Electro-Communications, 1-5-1 Chofugaoka, Chofu, Tokyo 182-8585, Japan
k-z@nerve.pc.uec.ac.jp

Abstract. A weakly electric fish can recognize object's shape in the complete darkness. The ability to recognize object's shape is provided by the electrosensory system of the fish. The fish generates an electric field using its electric organs (EOD: electric organ discharge). An object around the fish modulates the self-generated EOD. Electroreceptor afferents on the fish's body surface convert the EOD amplitude modulation into firings. The fish can extract information about object's shape from the EOD amplitude modulation using its electrosensory system. In the present study, we calculated the EOD amplitude modulation evoked by objects that were various shapes and firing patterns of the electroreceptor afferents evoked by the EOD amplitude modulation using computer simulation. We found that the EOD amplitude modulation can be represented by firing patterns of the electroreceptor afferents. Furthermore, we demonstrated that the feature of object's shape appears in the variation of the peak of firing rate with the rotation of the object.

Keywords: Electrosensory system · Electrolocation · Object recognition

1 Introduction

A weakly electric fish can recognize various object's parameters such as size, distance, shape, electric properties, etc., in complete darkness [1]. The electrosensory system of the fish provides the ability to recognize these object's parameters, so called electrolocation. However, it is not well-known why the fish can recognize the object's parameters using the electrosensory system. To understand the neural mechanism for electrolocation, many researchers have studied the electric stimulus and neural activity on electrolocation using theoretical and experimental methods.

A weakly electric fish generates an electric field using its electric organ (Electric Organ Discharge: EOD). Objects with electric properties that differ from those of the surrounding water modulate the EOD around the fish's body [2].

D. Liu et al. (Eds.): ICONIP 2017, Part III, LNCS 10636, pp. 877–884, 2017.
https://doi.org/10.1007/978-3-319-70090-8_89

The fish can detect objects by sensing the EOD modulation [3,4]. The electroreceptors on fish's body surface always monitor the state of the EOD modulation. Furthermore, the fish can extract the object's parameters from the EOD modulation using the electrosensory system. Especially, the fish can recognize object's shape using the EOD modulation in complete darkness [1,5]. However, it is not well-known what features of the EOD modulation represent the object's parameters, especially object's shape, and how the nerve system extracts the object's features from the EOD modulation. To resolve these questions, we need to understand electrolocation at both stimulus level and neural activity level.

Spatial distribution of an EOD modulation generated by an object will depend on the size, lateral distance, shape, and electric properties of the object. The fish can extract information about object's shape from the EOD modulation. Many researchers has measured the EOD modulation using experimental methods [6–9]. However, it is difficult to accurately measure the spatial distribution of the EOD modulation in an experimental setting. Because of this problem, a theoretical approach has also been used to investigate the EOD modulation. Some modeling and theoretical studies have evaluated the EOD amplitude modulation induced by resistive objects [10–12]. These studies have focused on the amplitude of the EOD modulation caused by an object without considering object's shape.

The EOD modulation received by the fish is converted to neural activity by the electroreceptor afferents on fish's skin and is projected to the higher electrosensory nucleus. The activities of the electroreceptor afferents are projected to the electrosensory lobe of the hind brain and the torus semicircularis in the midbrain [13]. Many researchers have proposed the population coding of electrosensory information in the electroreceptors and the electrosensory lateral-line lobe for electrolocation [14–18]. It is important to clarify the representation of the EOD modulation generated by an object on a neural activity of the electroreceptor afferents because the ability of representation of the electroreceptor afferents affects the extraction of object's parameters on the higher electrosensory nucleus. In previous studies, we have simulated the activities of the electroreceptor afferents underlying amplitude coding without considering object's shape.

In this study, we calculated the EOD modulation generated by an object by physical simulation and investigated the dependence of the EOD amplitude modulation on object's shape. Furthermore, we simulated neural activities of the electroreceptor afferents evoked by the EOD amplitude modulation. To address this issue, we developed a physical model to accurately calculate the EOD modulation produced by objects that have various distances, sizes, and shapes. Furthermore, we developed the neural model of the electroreceptor afferents to calculate neural activity evoked by the EOD amplitude modulation.

2 Methods

2.1 Situation

The situation of the simulation performed in this study is shown in Fig. 1. The electric fish and an object are in the water tank. There is the object near the fish.

Three kinds of object's shape are square, triangle and circle. The objects have lower resistivity than water. We approximate the fish as a rectangle (200 mm × 8mm). The model fish consists of a body surrounded by skin with high resistivity and an electric dipole produced by the electric organ. The dipole (the open circles in Fig. 1) is used as the electric organ of the fish and generates an electric field around the fish. We set the magnitude of the dipole to 12.5 mV, as using in [10,12]. We set the resistivity of water to 3.8 kΩcm [10,12]. We set the resistivity of the interior tissue and the skin of the fish to 0.1 kΩcm and 1.2 kΩcm, respectively [10,12].

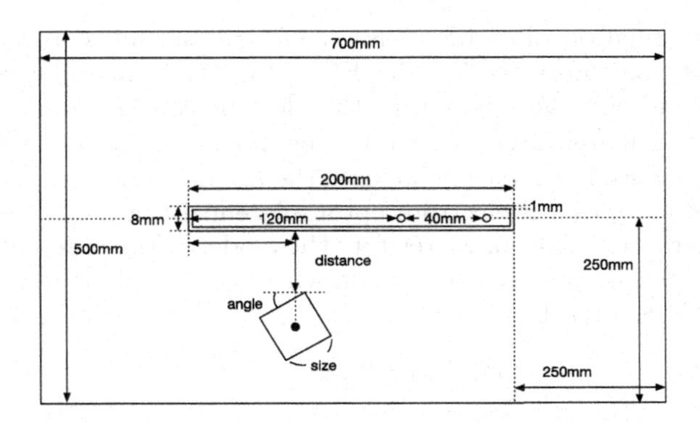

Fig. 1. An electric fish and a square object in the water tank.

2.2 A Model of an Electric Field

In this subsection, we denote the model for the calculation of an electrical field generated by a weakly electric fish and the EOD amplitude modulation produced by an object. In the real environment, it is difficult to measure the potential of the electric field generated by the fish at all locations and the EOD amplitude modulation received by the fish. Heiligenberg [11], Hoshimiya et al. [12], and Fujita and Kashimori [10] have developed the numerical model to calculate the electric field and have investigated the EOD amplitude modulation received by the weakly electric fish on its skin. Hoshimiya et al. [12] and Fujita and Kazhimori [10] have used the finite element method (FEM) to calculate the electric field.

The continum form of Ohm's Law is denoted by

$$\boldsymbol{E} = \rho \boldsymbol{j} \tag{1}$$

where $\boldsymbol{E}$ is electric field, $\boldsymbol{j}$ is current density, and ρ is resistivity. Because of the potential $\boldsymbol{E} = -\nabla V$, potential V is denoted by

$$\nabla V = -\rho \boldsymbol{j} \tag{2}$$

We obtain the potential at all locations of the water tank by solving this equation. The EOD amplitude modulation is calculated from the potential.

We also use the FEM to calculate the electric field and the EOD amplitude modulation generated by an object near the fish. Because this equation is similar to the heat spread equation, we could easily apply the FEM to the calculation of the electric field. The calculation of the electric field by the FEM consists of three calculation stages, preprocessing, processing, and postprocessing. We use Free FEM++, which is open source software to make mash and to solve the problem, to make a mesh in the preprocessing.

2.3 Neural Model

Figure 2 shows relation ship between an electric stimulus and electroreceptors. The electroreceptors receive the EOD amplitude modulation depending on object's shape and the location of the electroreceptor on the fish body. A P-electroreceptor afferent mainly encodes the magnitude of the EOD amplitude modulation as spikes. In this study, we used the simple model proposed by Brandman and Nelson [19] as a P-electroreceptor afferent. Because the original model represents difference equation, we rewrote the model to the differential equation. The voltage between inner and outer skin where the ith electroreceptor afferent locates, $v_i(t)$, is denoted by

$$v(t) = C \times \mathrm{AM}(t) + \xi, \tag{3}$$

where $\mathrm{AM}(t)$ is an intensity of the EOD amplitude modulation, ξ is a Gaussian random number with zero mean and the standard deviation σ, C is a constant value. The afferent nerve of the ith electroreceptor fires when the voltage, $v_i(t)$, is more than the threshold $\theta_i(t)$. Thus, the spike of the ith afferent nerve, $s_i(t)$, is indicated by

$$s_i(t) = \begin{cases} 1 & \text{if } v_i(t) > \theta_i(t) \\ 0 & \text{otherwise} \end{cases}. \tag{4}$$

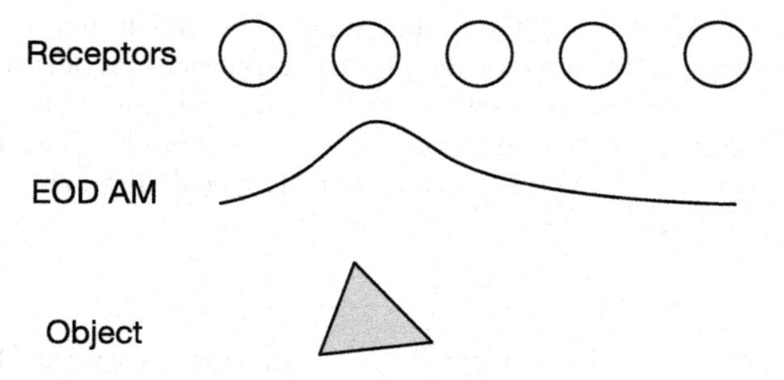

Fig. 2. Our neural model of electroreceptor afferents. The electroreceptors are on the fish's skin. The electroreceptors are placed on the one-dimensional array. The electroreceptors receive the EOD amplitude modulation induced by an object near the fish.

When the ith afferent nerve fires, $\theta_i(t + \Delta t) = \theta_i(t) + b$ to consider the adaptation of the threshold, where Δt is time interval. Because $\theta_i(t)$ recovers with time, $\theta_i(t)$ yields

$$\tau \frac{d\theta_i(t)}{dt} = -\theta_i(t) + \theta_0, \tag{5}$$

where τ is a time constant, and θ_0 is resting threshold.

We set the parameters: $C = 10$, $b = 3$, $\tau = 5$, $\theta_0 = 0.4$, and $\sigma = 0.1$. In this study, we did not consider mutual connections between the electroreceptors. In other words, the electroreceptors independently respond to electric stimuli.

3 Results

3.1 EOD Modulations

The left figure in Fig. 3 shows the spacial distributed EOD amplitude modulation on the skin of the fish induced by the 3 cm square placed 2 cm and 5 cm away from the fish at a point 4 cm posterior to the head of the fish. Spatial distribution of the EOD amplitude modulation resembled "Gaussian curve". The shape of the EOD amplitude modulation corresponds to experimental results. Thus, our simulation is experimentally valid.

The left, the middle, and the right figures in Fig. 4 respectively show the change in the peak values of the EOD amplitude modulations induced by the 2 cm square, the 3 cm square, and the 3 cm equilateral triangle varying the angle of the object and the lateral distance between the object and the fish. The peak values became smaller as the object was moved from the fish. Furthermore, the peak values became smaller as the angle of the object increased. However, the peak values increased at the largest angle. Moreover, we found the difference in variation of the peak values with object's angle. The form of reduction of the peak values induced by the square object was convex, but the form of reduction of the peak values induced by the equilateral triangle object was downward convex. These results suggest that the variation of the peak values of the EOD amplitude modulation with object's angle would represent a shape of an object.

3.2 Representation on Receptor Afferents

The right figure in Fig. 3 shows the firing rate of the electroreceptor afferents on the fish's skin evoked by the 3 cm square placed 2 cm and 5 cm away from the fish at a point 4 cm posterior to the head of the fish. We can see that the form of the spatial distribution of the firing rate of the electroreceptor afferents reflected the form of the EOD amplitude modulation. Furthermore, even if the intensity of the modulations is very small, the form of firing rate is almost same as the EOD amplitude modulation. This result suggests that the electroreceptor afferents would be able to extract features of EOD amplitude modulation related with object's shape.

The left, the middle, and the right figure in Fig. 5 respectively show the peak values of the firing rate of the electroreceptor afferents evoked by the 2 cm

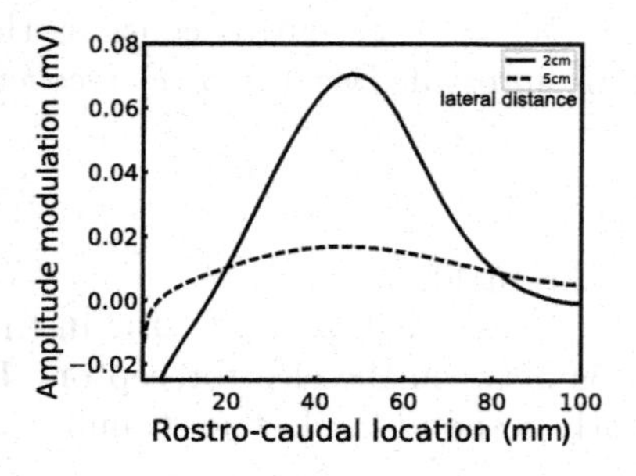

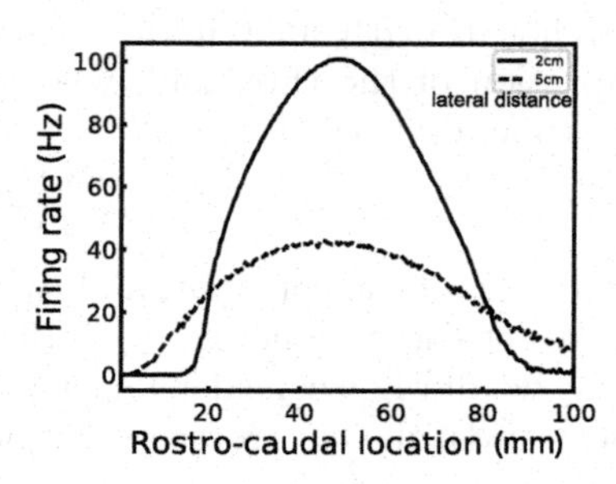

Fig. 3. The left figure shows the EOD amplitude modulations generated by the 3 cm square. The horizontal and the vertical line indicate the rostro-caudal location and the magnitude of the EOD amplitude modulation, respectively. The right figure shows the firing rate of the electroreceptor afferents evoked by the 3 cm square. The horizontal and the vertical line indicate the rostro-caudal location and the firing rate, respectively. The lateral distances of the object from the fish are 2 cm (solid line) and 5 cm (broken line).

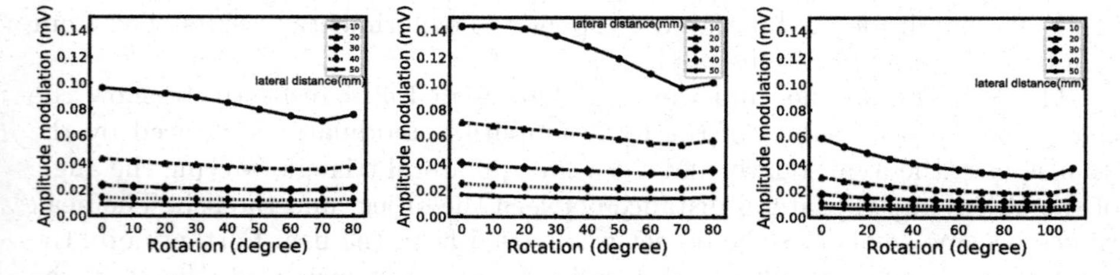

Fig. 4. These figures show the peak values of the EOD amplitude modulation evoked by 2 cm square (the left figure), 3 cm square (the middle figure), and 3 cm triangle (the right figure) with the angle. The labels of the lines indicate the lateral distance of the object from the fish. The horizontal and the vertical line indicate the angle of the object and the peak of the EOD modulation, respectively.

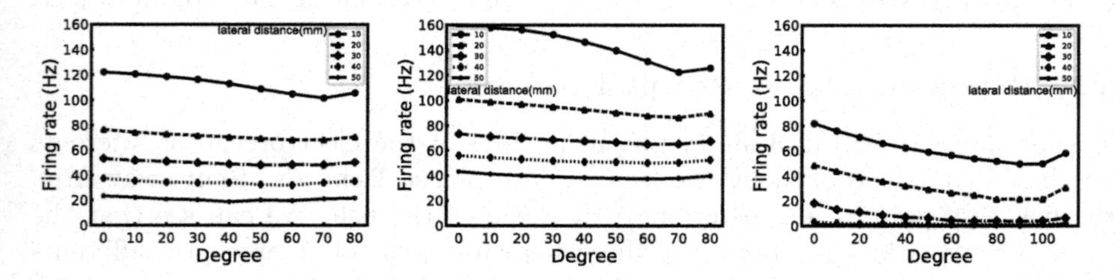

Fig. 5. These figures show the peak values of the firing rate of the electroreceptor afferents evoked by the EOD amplitude modulations induced by 2 cm square (the left figure), 3 cm square (the middle figure), and 3 cm triangle (the right figure) with the angle. The labels of the lines indicate the lateral distance of the object from the fish. The horizontal and the vertical line indicate the angle of the object and the peak of the firing rate of the electroreceptor afferents, respectively.

square, the 3 cm square, and the 3 cm equilateral triangle varying the angle of the objects and the lateral distance between the object and the fish. The EOD amplitude modulation is encoded into the firing rate of the electroreceptor afferents even if an object is far from the fish. Furthermore, the variation of the peak values of the firing rate reproduced the form of the variation of the peak values of the EOD amplitude modulations well. This result suggests that the electroreceptor afferents would be able to extract the variation of the EOD amplitude modulation depending on object's shape.

4 Conclusion

The purpose of the present study is the investigation of the dependence of the EOD amplitude modulation on object's shape and the neural activities of electroreceptor afferents. To address this issue, we calculated an electric field with modulation induced by an object and simulated the response of the electroreceptor afferents. We found that the shape of an object near the fish was represented by the variation of the peak value of the EOD amplitude modulation with object's angle. Furthermore, we showed that the electroreceptor afferents would be able to reproduce the variation of the peak value of the EOD amplitude modulation as the variation of the firing rate.

Acknowledgments. This work was supported by JSPS KAKENHI Grant Number 15K07146.

References

1. von der Emde, G., Behr, K., Bouton, B., Engelmann, J., Fetz, S., Folde, C.: 3-Dimensional scene perception during active electrolocation in a weakly electric pulse fish. Front. Behav. Neurosci. **4**, 26 (2010)
2. Lissman, H.W., Machin, K.E.: The mechanism of object location in gymnarchus niloticus. J. Exp. Biol. **35**, 451–486 (1958)
3. Bastian, J.: Electroloaction: behavior, anatomy, and physiology. In: Bullock, T.H., Heiligenberg, W. (eds.) Electroreception, pp. 577–612. Wiley, New York (1986)
4. Heiligenberg, W.: Neural Nets in Electric Fish. MIT press, Cambridge (1991)
5. Fechler, K., von der Emde, G.: Figure-ground separation during active electrolocation in the weakly electric fish, gnathonemus petersii. J. Physiol. Paris **107**(1–2), 72–83 (2013)
6. Assad, C., Rasnow, B., Stoddard, P.K., Bower, J.M.: The electric organ discharges of the gymnotiform fishes: Ii. eigenmannia. J. Comp. Physiol. A. **183**(4), 419–32 (1998)
7. Caputi, A.A., Budelli, R., Grant, K., Bell, C.C.: The electric image in weakly electric fish: physical images of resistive objects in gnathonemus petersii. J. Exp. Biol. **201**, 2115–2128 (1998)
8. von der Emde, G.: Capacitance detection in the wave-type electric fish eigenmannia during active electrolocation. J. Comp. Physiol. A. **182**, 217–224 (1998)

9. von der Emde, G., Ronacher, B.: Perception of electric properties of objects in electrolocating weakly electric fish: two-dimensional similarity scaling reveals city-block metric. J. Comp. Physiol. A. **175**, 801–812 (1994)
10. Fujita, K., Kashimori, Y.: Modeling the electric image produced by objects with complex impedance in weakly electric fish. Biol. Cybern. **103**(2), 105–118 (2010)
11. Heiligenberg, W.: Theoretical and experimental approaches to spatial aspects of electrolocation. J. Comp. Physiol. **103**(3), 247–272 (1975)
12. Hoshimiya, N., Shogen, K., Matsuo, T., Chichibu, S.: The apteronotus EOD field: waveform and EOD field simulation. J. Comp. Physiol. **135**, 283–290 (1980)
13. Bastian, J.: Electrolocation i. how the electroreceptors of apteronotus albifrons code for moving objects and other electrical stimuli. Ann. N. Y. Acad. Sci. **188**, 242–269 (1971)
14. Fujita, K., Kashimori, Y.: Population coding of electrosensory stimulus in receptor network. Neurocomputing **69**(10–12), 1206–1210 (2006)
15. Fujita, K., Kashimori, Y., Zheng, M., Kambara, T.: A role of synchronicity of neural activity based on dynamic plasticity of synapses in encoding spatiotemporal features of electrosensory stimuli. Math. Biosci. **201**, 113–124 (2006)
16. Fujita, K., Kashimori, Y., Kambara, T.: Spatiotemporal burst coding for extracting features of spatiotemporally varying stimuli. Biol. Cybern. **97**, 293–305 (2007)
17. Lewis, J.E., Maler, L.: Neuronal population codes and the perception of object distance in weakly electric fish. J. Neurosci. **21**(8), 2842–2850 (2001)
18. Nelson, M.E., Maclver, M.A.: Prey capture in the weakly electric fish apteronotus albifrons: sensory acquisition strategies and electrosensory consequences. J. Exp. Biol. **202**, 1195–1203 (1999)
19. Brandman, R., Nelson, M.E.: A simple model of long-term spike train regularization. Neural Comput. **14**, 1575–1597 (2002)

Performance Comparison of Motion Encoders: Hassenstein–Reichardt and Two-Detector Models

Hideaki Ikeda[1(✉)] and Toru Aonishi[2]

[1] Graduate School of Interdisciplinary Science and Engineering, Tokyo Institute of Technology, Nagatsuta-cho 4259–G5–17, Midori-ku, Yokohama, Kanagawa 226-8502, Japan
ikeda.h.al@m.titech.ac.jp
[2] School of Computing, Tokyo Institute of Technology, Nagatsuta-cho 4259–G5–17, Midori-ku, Yokohama, Kanagawa 226-8502, Japan
aonishi@c.titech.ac.jp

Abstract. Several motion-detection models have been proposed based on insect visual system studies. We specifically examine two models, the Hassenstein-Reichardt (HR) model and the two-detector (2D) model, before selecting model the more efficient motion encoders. We analytically obtained the mean and variance of stationary responses of the HR and the 2D models to white noise to evaluate performances of the two models. Especially when analyzing the 2D model, we calculated higher-order cumulants of a rectified Gaussian. Results show that the 2D model gives almost equal performance to that of the HR model in a biologically reasonable case.

Keywords: Motion detection · Neural coding · White noise analysis · Hassenstein–Reichardt model · Two–detector model

1 Introduction

Several motion detector models, classified as correlation-type elementary motion detectors (correlation-type EMDs), have been proposed based on insect visual system studies. In this type of model, speed and motion directions are extracted by calculating delayed cross-correlation of the two input signals. Reportedly, correlation-type EMD can be implemented in early visual systems of insects, especially in *Drosophila melanogaster* [1–3]. Additionally, the possibility has been suggested that the correlation-type EMD is a common algorithm for motion detection in many species including rabbit [4] and human models [5,6]. Furthermore, correlation-type EMD is expected to be extremely useful for artificial visual processing for micro-unmanned aerial vehicles, for reducing computational effort [7,8].

This paper presents specific examination of two correlation-type EMD models: the Hassenstein–Reichardt (HR) model and the two-detector (2D) model.

D. Liu et al. (Eds.): ICONIP 2017, Part III, LNCS 10636, pp. 885–893, 2017.
https://doi.org/10.1007/978-3-319-70090-8_90

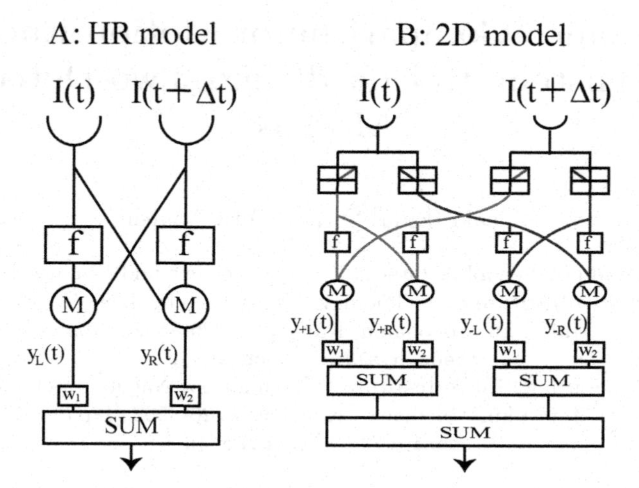

Fig. 1. Models are the following: A, the Hassenstein–Reichardt model (HR model), a motion encoder using delayed cross-correlation of the two input signals; B: the two-detector model (2D model). The 2D model has two parallel pathways, each of which functions similarly to the HR model and which processes only positive or negative components. f, M, and SUM respectively represent low-pass filter, multiplication, and summation. w_1 and w_2 are weight coefficients.

Since the HR model was first proposed (Fig. 1A), it has been extended to several models called the 2D model (Fig. 1B) and the four detector (4D) model, which has biologically more reasonable structures than the original HR model [1,9]. The extended models consist of a set of standard Reichardt detector (SRD) units corresponding to the original HR model. To satisfy Dale's rule, these separated SRD units respectively process various pairs of half-wave rectified signals. The whole input–output relation of the 4D model is mathematically equivalent to that of the HR model [9,10], whereas the whole input–output relation of the 2D model differs from that of the HR model.

Recently, some research groups have attempted mathematical evaluation of the performance of the two models in terms of motion encoders. The HR model was evaluated analytically by calculating its response to sine waves and white noise [5,11–15]. However, no analytical evaluation of the 2D model is available: only numerical evaluation by simulating its response to synthetic natural scene [16]. These earlier studies have mainly addressed amplitude response as a measure of encoding performance. However, to evaluate the robustness as an encoder, one must address the fluctuation of response of the models. It is difficult to treat the fluctuation of response of the correlation-type EMD analytically, especially the 2D model with the half-wave rectifiers.

For this study, we analytically obtained the mean and variance of responses of the HR and the 2D models to white noise; then we derived the signal-to-fluctuation-noise ratio (SFNR) to evaluate performances of the two models in terms of motion encoders. Especially, when analyzing the 2D model, we calculated the fourth-order cumulants of a rectified Gaussian distribution. Results showed that the 2D model gives almost identical performance to that of the HR model in a biologically reasonable case.

2 Methods

2.1 Formulation of Correlation-Type Elementary Motion Detectors

Hassenstein–Reichardt Model: We briefly explain the HR model (Fig. 1A).

Let I be the input signal. Here, x_L and x_R respectively denote a left-side signal and a right-side signal in the first processing stage, which are delayed through a low-pass filter. In the second processing stage, x_L and x_R are multiplied with each opposite input signal.

$$y_L(t) = \int_0^{\infty} dt' f(t') I(t - t') I(t + \Delta t) \tag{1}$$

$$y_R(t) = \int_0^{\infty} dt' f(t') I(t + \Delta t - t') I(t) \tag{2}$$

Therein, $f(t)$ is an impulse response function of the low-pass filter. For mathematical tractability, $f(t) = 0$ is satisfied if $t \leq 0$. Δt is the difference of detection time between the left-side and the right-side photoreceptors.

In the final stage, the multiplied signals y_L and y_R are summed up with the weight.

$$\mathcal{R}_{HR} = w_1 y_L + w_2 y_R \tag{3}$$

In that equation, w_1 and w_2 are weight coefficients with mutually different signs.

2D Model: We briefly explain the 2D model, which model consists of two SRD units equivalent to the HR model as shown in Fig. 1B. In the first processing stage of the left side and the right side of the 2D model, the input signal $I(t)$ is split into $I_+(t)$ and $I_-(t)$ through two half-wave rectifiers. The left SRD unit processes $I_+(t)$ and $I_+(t+\Delta t)$; the right SRD unit processes $I_-(t)$ and $I_-(t+\Delta t)$ (Fig. 1B). Additionally, x_{+L} and x_{+R} respectively denote a left-side signal and a right-side signal in the left SRD, which are delayed through the low-pass filter $f(t)$. Then, y_{+L} and y_{+R} are multiplied with each opposite rectified input signal as follows.

$$y_{+L}(t) = \int_0^{\infty} dt' f(t') I_+(t - t') I_+(t + \Delta t) \tag{4}$$

$$y_{+R}(t) = \int_0^{\infty} dt' f(t') I_+(t + \Delta t - t') I_+(t) \tag{5}$$

It is noteworthy that the right SRD unit is formulated similarly. Therefore, we omit the explanation of the right SRD unit.

In the final stage, the multiplied signals y_{+L} and y_{+R} in the left SRD unit and y_{-L} and y_{-R} in the right SRD unit are summed up with the weight.

$$\mathcal{R}_{2D} = w_1 y_{+L} + w_1 y_{-L} + w_2 y_{+R} + w_2 y_{-R} \tag{6}$$

In that equation, w_1 and w_2 are weight coefficients with mutually different signs.

2.2 Numerical Evaluation

We evaluate the analytical solutions by comparison with numerical simulation results. The conditions of numerical evaluations in Sects. 3.2 and 3.3 are given as shown below.

In the HR model and the 2D model SRD unit, the low-pass filter of the first processing stage was implemented with the first-order low-pass filter, as in earlier studies [9,15].

$$f(t) = \begin{cases} \frac{1}{\tau}\exp(-\frac{1}{\tau}t) & (0 < t) \\ 0 & (t \leq 0) \end{cases} \tag{7}$$

For mathematical tractability, $f(t) = 0$ is satisfied if $t \leq 0$.

In both models, the weight coefficients w_1 and w_2 of the final stage were given as the same as those of earlier studies [9,11].

$$w_1 = 1, \quad w_2 = -\alpha \tag{8}$$

where $0 < \alpha \leq 1$.

3 Results

3.1 Analytical Solutions of HR and 2D Models

Analysis of the 2D Model: We obtained the mean value of stationary response of the 2D model to the white Gaussian noise to rewrite the second joint moment derived from the above formulation as the sum of the products of the cumulants (we call this expression cumulant expansion).

$$\begin{aligned} < \mathcal{R}_{2D} > &= (\mathcal{A}_1 + \mathcal{A}_2)\sigma^2 \\ \mathcal{A}_1 &= \frac{\pi - 1}{\pi}(w_1 f(-\Delta t) + w_2 f(\Delta t)) \\ \mathcal{A}_2 &= \frac{1}{\pi}(w_1 + w_2)\int_0^\infty dt' f(t') \end{aligned} \tag{9}$$

Therein, brackets $< X >$ denote the mean of X.

Next, we obtained the variance of the stationary response of the 2D model to the white Gaussian noise to rewrite the fourth-order joint moment derived from the formulation as the cumulant expansion.

$$\begin{aligned} V[\mathcal{R}_{2D}] &= (\mathcal{B}_1 + \mathcal{B}_2 + \mathcal{B}_3 + \mathcal{B}_4 + \mathcal{B}_5)\sigma^4 \\ \mathcal{B}_1 &= \frac{3\pi^2 - 2\pi - 2}{2\pi^2}(w_1^2(f(-\Delta t))^2 + w_2^2(f(\Delta t))^2) \\ \mathcal{B}_2 &= \frac{\pi + 2}{\pi^2}(w_1^2 f(-\Delta t) + w_2^2 f(\Delta t)) \int_0^\infty dt' f(t') \\ \mathcal{B}_3 &= \frac{\pi - 2}{\pi^2} w_1 w_2 \int_0^\infty dt' f(t' + \Delta t) f(t') \\ \mathcal{B}_4 &= \frac{\pi - 1}{2\pi}(w_1^2 + w_2^2) \int_0^\infty dt' (f(t'))^2 \\ \mathcal{B}_5 &= \frac{\pi - 2}{2\pi^2}(w_1^2 + w_2^2)(\int_0^\infty dt' f(t'))^2 \end{aligned} \tag{10}$$

In those expressions, $V[X]$ represents the variance of X.

Analysis of the HR Model: We obtained the mean value of stationary response of the HR model to the white Gaussian noise as the following equation.

$$< \mathcal{R}_{HR} >= (w_1 f(-\Delta t) + w_2 f(\Delta t))\sigma^2 \tag{11}$$

This result is the same as the result obtained from an earlier study [5] for $w_1 = 1$ and $w_2 = -1$.

We obtained the variance of stationary response of the HR model as shown below.

$$\begin{aligned} V[\mathcal{R}_{HR}] &= (\mathcal{C}_1 + \mathcal{C}_2)\sigma^4 \\ \mathcal{C}_1 &= w_1^2(f(-\Delta t))^2 + w_2^2(f(\Delta t))^2 \\ \mathcal{C}_2 &= (w_1^2 + w_2^2) \int_0^\infty dt' (f(t'))^2 \end{aligned} \tag{12}$$

3.2 Numerical Evaluation of Analytical Solutions

We compared the analytical solutions with those obtained from numerical simulations. Figure 2 shows the mean and variance of responses of the two models as a function of τ and α. The analytical solutions matched those obtained by numerical simulations. When τ was small, however, a mismatch resulted from discretized approximation.

3.3 Signal-to-Fluctuation-Noise Ratios of the HR and 2D Models

SFNR of the HR and 2D Models: To evaluate the performance of the two models as an encoder, we calculated the signal-to-fluctuation-noise ratio (SFNR)[17]. The SFNR is defined as

$$\mathrm{SFNR}_x = \sqrt{\frac{< \mathcal{R}_x >^2}{V[\mathcal{R}_x]}}, \tag{13}$$

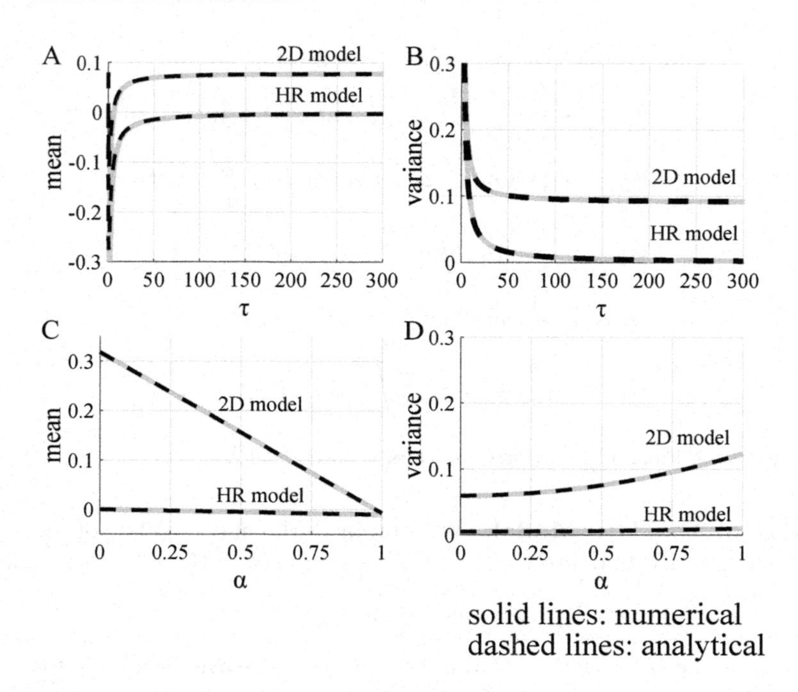

Fig. 2. Comparison between the analytical solutions of the stationary responses of the HR and the 2D models to white Gaussian noise and those obtained by numerical simulations. A and B respectively show the mean and variance as a function of τ, when α fixed to 0.75. C and D show them as a function of α, when τ fixed to 100 [ms]. Solid and dashed lines respectively show numerical solutions and analytical solutions. Here, the Gaussian noise variance is set to 1^2. (Color figure online)

where $x \in \{\text{HR, 2D}\}$ that indicates either the HR or 2D model. The larger SFNR stands for more precise motion detection. We analytically obtained the mean and variance of responses of the HR and the 2D models. Therefore, we can obtain the SFNR analytically.

Figures 3A and 3B respectively show SFNR as a function of α and τ in the 2D and the HR models. In the case of the 2D model, SFNR_{2D} became independent of τ if τ is larger than about 50 [ms]. In this region, SFNR_{2D} was larger as α approached zero. However, in the case of the HR model, SFNR_{HR} changed dependent on both of τ and α. The changes were smaller than those of the 2D model.

Figure 3C presents the difference of SFNRs between the two models. The white dashed lines represent differences of SFNRs of zero. Markers on the panel denote the values of parameters estimated in earlier studies [9,11,18]. They were near the dashed line (Fig. 3C).

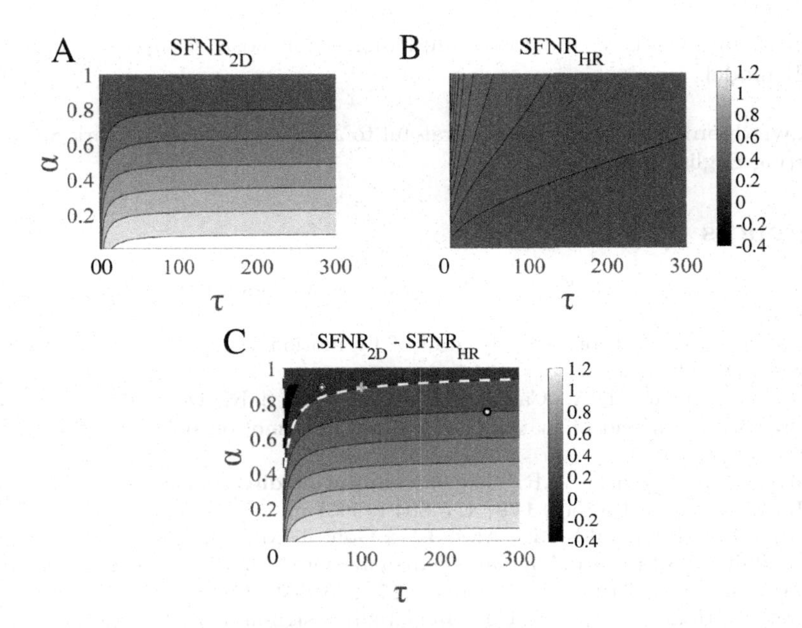

Fig. 3. Gray-scale plots of the SFNRs of the HR and the 2D models and their differences: A, SFNRs of the 2D model; B, SFNRs of the HR model; C, differences obtained when SFNRs of the HR model are subtracted from the corresponding SFNRs of the 2D model. The white dashed line represents the difference of SFNRs becoming zero. The markers denote the values of parameters estimated in the earlier works (diamond [9], circle [18], and cross [11]). Δt is set to 1 [ms]. (Color figure online)

4 Discussion

We analytically evaluated responses of two typical models of the correlation-type EMD to the white Gaussian noise. We analytically obtained the mean and variance of responses of the HR and the 2D models to the white noise. The HR model is fundamentally equivalent to the four detector models as described above. Therefore, we must deal with almost all correlation-type EMD models by analyzing the HR model. Here we have overcome the difficulty of analyzing the 2D model with half-wave rectifiers. We were able to obtain higher-order joint moments in the 2D model strictly by calculating higher-order cumulants of the rectified Gaussian.

Furthermore, to evaluate the performance of the two models as motion encoders, we calculated SFNR based on the analytical solutions. As shown in Fig. 3C, $SFNR_{2D}$ and $SFNR_{HR}$ were comparable at biologically reasonable parameter values. This result suggests that the 2D model can provide almost equivalent performance to the HR model in a biologically reasonable case.

The HR model must be implemented biologically with forms of the 4D model because of the restriction of the Dale's rule. The 2D model is structurally simpler than the 4D model. Therefore, we conclude that the 2D model has almost equivalent performance and lower implementation costs compared to those of the HR model.

Acknowledgement. We are deeply grateful to Japanese Neural Network Society for supporting English proofreading.

References

1. Borst, A., Haag, J., Reiff, D.F.: Fly motion vision. Annu. Rev. Neurosci. **33**, 49–70 (2010)
2. Borst, A.: In search of the holy grail of fly motion vision. Eur. J. Neurosci. **40**, 3285–3293 (2014)
3. Behnia, R., Clark, D.A., Carter, A.G., Clandinin, T.R., Desplan, C.: Processing properties of on and off pathways for drosophila motion detection. Nature **512**, 427–430 (2014)
4. Barlow, H.B., Levick, W.R.: The mechanism of directionally selective units in rabbit's retina. J. Physiol. **178**, 477–504 (1965)
5. Clark, D.A., Fitzgerald, J.E., Ales, J.M., Gohl, D.M., Silies, M.A., Norcia, A.M., Clandinin, T.R.: Flies and humans share a motion estimation strategy that exploits natural scene statistics. Nat. Neurosci. **17**, 296–303 (2014)
6. Borst, A., Helmstaedter, M.: Common circuit design in fly and mammalian motion vision. Nat. Neurosci. **18**, 1067–1076 (2015)
7. Franceschini, N.: Small brains, smart machines: from fly vision to robot vision and back again. Proc. IEEE **102**, 751–781 (2014)
8. Akima, H., Sato, S.: LSI implementation of neural network model for detecting local image motion in motion stereo vision [in japanese]. J. Jpn. Neural Netw. Soc. **22**, 152–161 (2015)
9. Eichner, H., Joesch, M., Schnell, B., Rei, D.F., Borst, A.: Internal structure of the fly elementary motion detector. Neuron **70**, 1155–1164 (2011)
10. Clark, D.A., Bursztyn, L., Horowitz, M.A., Schnitzer, M.J., Clandinin, T.R.: Defining the computational structure of the motion detector in drosophila. Neuron **70**, 1165–1177 (2011)
11. Egelhaaf, M., Borst, A., Reichardt, W.: Computational structure of a biological motion-detection system as revealed by local detector analysis in the flys nervous system. J. Opt. Soc. Am. A: **6**, 1070–1087 (1989)
12. Zanker, J.M., Srinivasan, M.V., Egelhaaf, M.: Speed tuning in elementary motion detectors of the correlation type. Biol. Cybern. **80**, 109–116 (1999)
13. Hidayat, E., Medvedev, A., Nordström, K.: Identification of the reichardt elementary motion detector model. In: Sun, C., Bednarz, T., Pham, T.D., Vallotton, P., Wang, D. (eds.) Signal and Image Analysis for Biomedical and Life Sciences. AEMB, vol. 823, pp. 83–105. Springer, Cham (2015). doi:10.1007/978-3-319-10984-8_5
14. Schnell, B., Weir, P.T., Roth, E., Fairhall, A.L., Dickinson, M.H.: Cellular mechanisms for integral feedback in visually guided behavior. Proc. Natl. Acad. Sci. **111**, 5700–5705 (2014)
15. Roy, S., van Steveninck, R.R.: Bilocal visual noise as a probe of wide field motion computation. J. Vis. **16**, 1–19 (2011)
16. Leonhardt, A., Ammer, G., Meier, M., Serbe, E., Bahl, A., Borst, A.: Asymmetry of drosophila on and off motion detectors enhances real-world velocity estimation. Nat. Neurosci. **19**, 706–715 (2016)

17. Friedman, L., Glover, G.H., Krenz, D., Magnotta, V.: BIRN, the first.: reducing inter-scanner variability of activation in a multicenter fMRI study: role of smoothness equalization. Neuroimage. **32**, 1656–1668 (2006)
18. Ikeda, H., Suzuki, Y., Morimoto, T., Aonishi, T.: Model selection of early vision system of drosophila melanogaster. IPSJ Trans. Math. Model. App. **9**, 24–31 (2016)

Wireless Network Gateway Placement by Evolutionary Graph Clustering

Maolin Tang(✉) and Chien-An Chen

Queensland University of Technology, 2 George Street,
Brisbane QLD 4001, Australia
m.tang@qut.edu.au

Abstract. Gateway placement is an important problem in the design of a backbone wireless network (BWN) as it directly affects the installation and ongoing running costs of the BWN. From a computational point of view, the gateway placement problem is a constrained combinatorial optimization problem. In this paper, we transform the gateway problem into a graph clustering problem and design a repairing genetic algorithm (RGA) to solve the graph clustering problem. Different from traditional GAs, this RGA embeds a procedure that can detect and repair those infeasible solutions generated by the crossover and mutation operators. Experimental results show that the infeasible solution detecting and repairing procedure can not only reduce the computation time of the RGA, but also improve the quality of the solutions generated by the RGA. In this paper, we also conduct an empirical study of the computational efficiency of the RGA. The analysis result shows that its computational efficiency is quadratic, which is computationally efficient.

Keywords: Backbone wireless network · Gateway placement · Graph clustering · Genetic algorithm · Evolutionary computation

1 Introduction

A wireless network is an ad hoc communication network that is made up of wireless communication nodes. It allows for continuous connections and reconfiguration around broken or blocked paths by hopping from node to node until the destination is reached. In a wireless network, communication nodes can connect to each other via multiple hops. The infrastructure that supports a wireless network is a backbone wireless network (BWN).

A BWN consists of a set of wireless routers. Each router forwards packages on behalf of other routers and clients. The wireless routers are generally not mobile, but their clients may be mobile. Clients are interconnected through a BWN in a wireless network. A client communicates with other networks through a wireless router, which is called *gateway*.

Gateway placement is an important problem in the design of a BWN. It is to determine which of the wireless routers should be selected as gateways.

D. Liu et al. (Eds.): ICONIP 2017, Part III, LNCS 10636, pp. 894–902, 2017.
https://doi.org/10.1007/978-3-319-70090-8_91

This is so called BWN gateway placement problem. The challenge in the BWN gateway problem is to find a minimal number of wireless routers from the BWN while guaranteeing the Quality-of-Service (QoS) of the BWN. There are many applications of the gateway placement problem. For example, the base placement problem in Internet of Things (IoT) is basically an instance of the BWN gateway placement problem.

There are two popular QoS constraints needed to be considered in the BWN gateway placement problem: *delay constraint* and *gateway capacity constraint*. A wireless network is a multi-hop network where significant delay may occur due to contention for wireless channel, pocket processing, and pocket queuing. The delay in a BWN can be measured by the number of communication hops between the router connected to a client and the gateway of the wireless router [1]. The delay constraint requires that the maximal number of hops from any router to its gateway is not greater than a given number R. In addition, a gateway has a capacity S, which is measured by the maximal number of clients that it can serve at the same time. The capacity constraint requires that the maximal number of wireless routers that share a gateway should not exceed the capacity of the gateway S.

In this paper we transform the gateway placement problem into a graph clustering problem and propose a genetic algorithm (GA) to solve the graph clustering problem. Since infeasible might occur in the encoding scheme of the GA, the GA uses a repairing technique to transform any infeasible solution that might be generated in the initial population and during the evolution process into a feasible one. Thus, the GA is called Repairing GA, or RGA in this paper. The proposed RGA has been implemented and evaluated by experiments, and the experimental results show that the proposed RGA has great performance and good scalability.

In the rest of the paper, firstly we discuss related work briefly in Sect. 2 and then transform the gateway placement problem into a graph clustering problem in Sect. 3. Then, the RGA for solving the graph clustering problem is presented in Sect. 4 and is evaluated in Sect. 5. This research is concluded in Sect. 6.

2 Related Work

From a computational point of view, the gateway placement problem is conjectured as an NP-hard problem as it can be transformed into the minimum dominating set problem [1], which has been proven to be NP-complete. Thus, brute-force algorithms are not suitable for solving the problem because they would lead to combinatorial explosion in their search space when the problem size becomes large. Due to the reason, many heuristic algorithms have been proposed to address various gateway placement problems.

In [11], Wong, *et al.* addressed two gateway placement problems: one aimed to optimize the communication delay of the BWMN; another was to optimize the communication cost of the BWN. However, neither of the algorithms can be immediately used to handle the gateway placement problem with QoS constraints. In [3] Bejerano adopted a clustering technique to tackle a gateway placement problem that considered some QoS constraints. The technique found a gateway placement solution in four consecutive stages: (1) select cluster heads; (2) assign each node to an identified cluster satisfying the delay constraint; (3) break down the clusters that do not satisfy relay load constraint or the gateway capacity constraint; (4) select gateways to reduce the maximum relay load. Aiming at optimizing the throughput of a wireless network, Li, *et al.* proposed an algorithm im [6]. In [1], Aoun, *et al.* transformed a gateway placement problem considering various QoS constraints in the design of BWN into the minimum dominating set problem and adopted a recursive dominating set algorithm to tackle the minimum dominating set problem. In [9], the same gateway placement problem was modelled as a graph clustering problem and an incremental clustering algorithm was proposed to solve the graph clustering problem.

There are other related research work. In [4], Chandra, et al. explored the placement problem of Internet Transit Access Points (ITAPs) in wireless neighborhood networks under three wireless link models, and for each of the wireless link models, they developed algorithms for the placement problem based on neighborhood layouts, user demands, and wireless link characteristics. The placement problem is similar to the gateway place of BWN. However, their algorithms consider only users bandwidth requirements. There are other gateway placement problems that are similar to the gateway placement problem in the design of BWMN, such as the gateway placement in hybrid manet-satellite networks [5], the gateway placement in multihop in-vehicle internet access [8], the energy-aware gateway placement in green wireless mesh networks [2], and the router node placement problem in wireless mesh network considering service priority in [7]. However, they are different from the gateway placement problem to be addressed in this paper as their objectives and constraints are different.

The RGA makes use of the problem modelling ideas proposed in [9]. But we refine the problem formulation in this paper. In addition, instead of using heuristics, we design a RGA to solve the graph clustering problem.

3 Problem Formulation

A BWN can be represented by a directed graph $G = (V, E)$, where $v \in V$ represents a router in the BWN and $< v_k, v_l > \in E$ stands for router v_l is within the transmission range of router v_k. The BWN gateway placement problem is to cluster the vertices in V (routers) into a number of clusters C_1, C_2, $\cdots$, C_n and to pick up a vertex (router) g_i from each of the clusters C_i such that the number of clusters n is minimized subject to the following constraints:

$$C_1 \cup C_2 \cup \cdots \cup C_n = V \tag{1}$$

$$C_i \cap C_j = \phi \tag{2}$$

$$|C_i| \leq S \tag{3}$$

where S is a given integer constant representing the capacity of the gateways.

$$\forall v \in C_i, dist(v, g_i) \leq D \tag{4}$$

where $1 \leq i, j \leq n$ and $i \neq j$, $dist(v, g_i)$ is the shortest distance from v to g_i in G, and D is another given integer constant standing for the maximal number of hops between any router to its designated gateway.

Equations 1 and 2 ensure that each of the wireless routers belongs to one and only one cluster, in Eq. 3 ensures the gateway capacity constraint is met, and in equation enures the delay constraint is satisfied in the design of BWN.

4 The RGA

The RGA embeds a procedure to repair those infeasible solutions generated by the crossover and mutation operators of the RGA. A chromosome in the RGA represents a gateway placement solution and is encoded as a binary string, which will be discussed later. Solutions in the initial generation are initially randomly generated. Since some of the randomly generated solutions may be infeasible, the infeasible solution repairing procedure is then used to fix up all the infeasible solutions in the initial population. The selection strategy used by the RGA is roulette selection. The crossover operators adopted in the RGA is a classical uniform crossover [10] and the mutation operator randomly picks up a gene and changes the value of the gene to its complement. The RGA terminates its evolution when there is no improvement in its best solution within a given number of consecutive generations.

Chromosomes randomly generated in the initial generation or by the genetic operators during the evolution may not be a feasible gateway placement solution. If this occurs, the GA uses a repairing procedure to convert the chromosome to a feasible chromosome, i.e. a chromosome representing a feasible solution.

This section discusses some key issues in the design of the RGA, including genetic encoding and decoding, fitness function construction, detecting and repairing an infeasible chromosome.

4.1 Encoding

The RGA uses a binary string $p = r_1 r_2 \cdots r_{|V|}$ to represent a chromosome, where r_i represents wireless router i, and $1 \leq i \leq |V|$.

$$r_i = \begin{cases} 1, \text{ if router } i \text{ is selected as a gateway;} \\ 0, \text{ otherwise.} \end{cases} \tag{5}$$

4.2 Decoding

Given a chromosome, $p = r_1 r_2 \cdots r_{|V|}$, the RGA uses the following procedure to transform the chromosome into a BWN gateway placement, which is represented in a matrix $C = [c_{ij}]_{m \times n}$, where $m = \sum_{i=1}^{|V|} r_i$ is the number of gateways, $n = |V|$, and $m \leq n$. Each row in the matrix stands for a selected gateway and each column in the matrix represents a wireless router in the BWN.

$$c_{ij} = \begin{cases} 1, \text{ if the number of hops from router j to gateway i} \\ \quad \text{is no more than D;} \\ 0, \text{ otherwise.} \end{cases} \tag{6}$$

For the BWN graph shown in Fig. 1, for example, chromosome $p = 1001001$ will be transformed into

$$\begin{pmatrix} 1\,1\,1\,1\,1\,0\,1 \\ 1\,1\,1\,1\,1\,1\,0 \\ 1\,1\,1\,0\,1\,1\,1 \end{pmatrix}$$

where the delay constraint is 2, the first row corresponds to wireless router 1, the second row corresponds to wireless router 4, and the third row corresponds to wireless router 7.

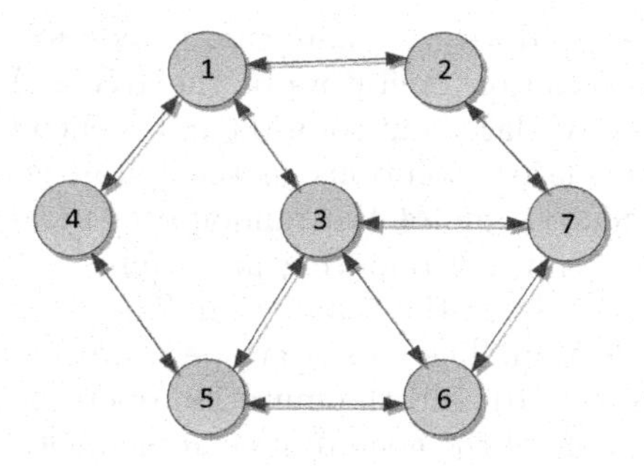

Fig. 1. A BWN graph

4.3 Detecting and Repairing an Infeasible Chromosome

A chromosome is infeasible, if its corresponding gateway placement solution is infeasible. There are two scenarios where a chromosome is infeasible. One is the scenario where there exists one wireless router that is not included into any cluster; another is the scenario where there exists a cluster that include more than S wireless routers (exceeding the capacity of the gateway). The former scenario covers the situation where there is no path from the router to any of the gateways and the situation where there exists a path from the router to a gateway, but the number of hops is more than D.

Given a chromosome $p = r_1 r_2 \cdots r_n$, the RGA uses the above decoding technique to transform it into a matrix representation C. Since it can be guaranteed by the decoding technique that all 1 s in matrix C represent that there exists a path from the corresponding router to its gateway with no more than D hops. Thus the RGA needs to check two situations: one is if the chromosome exceeds the gateway capacity constraint; another is if there exists a router that is not included into any cluster.

To check if the chromosome violates the gateway capacity constraint, the RGA checks if the total number of 1 s in each row of C exceeds S. If there exists any row in which the total number of 1 s is greater than S, the RGA uses a strategy to remove some 1 s in that row such that the total number of 1 s in each row is reduced to S. The strategy that the RGA uses is to remove those 1 s that have more 1 s in their corresponding column as by doing this we can minimize the chance to add new gateways.

To check if there exists a wireless router which has not been included into any existing cluster, the RGA checks if there exists any column in C where all the elements in that column are all 0s. If there exists any, then it indicates the chromosome is infeasible and therefore needs to be repaired.

To repair this kind of infeasible chromosome, the RGA finds all the wireless routers that have not been included into any gateway cluster in the chromosome. Then, the RGA randomly selects among those wireless routers as a gateway one by one until all the wireless routers are included into a new gateway cluster.

4.4 Fitness Function

The fitness value of a feasible chromosome $p = r_1 r_2 \cdots r_n$ is defined in the following equation:

$$fitness(p) = 1 - \frac{\sum_{i=1}^{n} r_i}{|V|} \tag{7}$$

5 Evaluation

The proposed RGA is evaluated by experiments in this section. The design of the experiments is explained, and the experimental results are presented and discussed in this section.

5.1 Experimental Design

The experiments are designed to evaluate the quality of solutions generated by the RGA, which is performed by comparing the quality of solutions generated by the incremental graph clustering algorithm in [9], which has the best performance among those existing algorithms for the BWN gateway placement problem, for the randomly generated test problems. Both the RGA and the incremental graph clustering algorithm are implemented in $C\#$ and all experiments are conducted on the same computer.

In the experiments, first of all we developed a program to randomly generate test problems of different size. To keep it simple, in all the randomly generated test problems, the coverage of all wireless routers is 100 m, and the minimal distance between any pair wireless routers is 30 m. It is assumed that the delay constraint is 3 hops and the gateway capacity for all gateways is 10.

For each of the randomly generated test problems, we used the incremental graph clustering algorithm to solve it and record its solution, and then we used the RGA to solve the same problem. Considering the stochastic nature of the RGA, we repeat the same experiment for 20 times and then calculate the average, best and worst solutions of the 20 runs.

The parameters of the RGA were both set as following: the probability for crossover was 0.95 and the probability for mutation was 0.05; the size of population was 100; and the termination condition was "no improvement in the best chromosome in 20 consecutive generations".

5.2 Experimental Results

Table 1 shows the experimental results. In the table, the solutions are measured by the number of gateways needed in the BWN design. Thus, the smaller the number, the better the solution. It can be seen from the table that the average number of gateways required in the solutions generated by the RGA is less than that in the solution created by the incremental graph clustering algorithm, which is a type of heuristic algorithm and therefore denoted by HA in the table, and that the larger the test problem size is, the more advantages the RGA has.

Table 1. Comparison of the performances of the HA and the RGA

Test problem	BWN nodes(#)	HA gateways(#)	RGA (gateways(#))		
			Ave	Best	Worst
1	10	3	2.00	2	2
2	20	4	4.00	4	4
3	30	6	5.05	5	6
4	40	8	8.15	7	9
5	50	11	7.85	7	8
6	60	12	11.55	10	12
7	70	16	11.05	10	11
8	80	18	16.85	16	18
9	90	21	19.95	18	22
10	100	26	21.70	19	24

5.3 Analysis of the Scalability of the RGA

To analyse the scalability of the RGA, we visualise how the computation time of the RGA varies with the size of the BWN gateway placement problem in Fig. 2.

We have conducted a non-linear regression analysis to find the correlation between the computation time and the number of nodes in the BWN graph using the experimental data. Here is the outcome of the non-linear regression analysis result:

$$T(n) = 0.01125n^2 + 0.0585n + 0.27 \tag{8}$$

From the above function, we know the computational efficiency class of the RGA belongs to $\Theta(n^2)$, where n is the number of nodes in the BWN graph. This indicates the RGA is scalable.

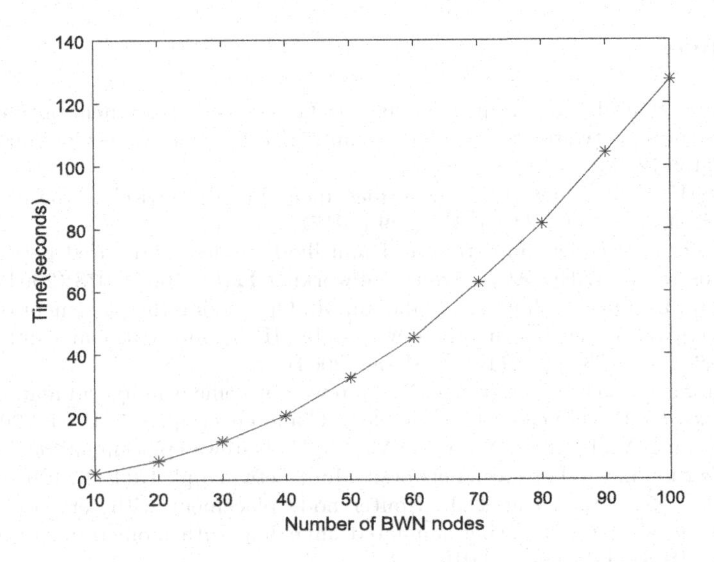

Fig. 2. Visualisation of the scalability of the GA

6 Conclusions

In this paper we have developed a RGA for solving the gateway placement problem in the design of BWN. The RGA has been evaluated by experiments and the experimental results have shown that the RGA has satisfactory performance. This paper has also analysed the scalability of the RGA. The major contributions and innovations of this paper include:

- This paper presents a new model for the BWN gateway placement problem. Using the new model, it is computationally cheaper and easier to calculate the quality of solution (the number of gateways needed in the solution) and to check if a solution violets the gateway capacity constraint and the delay constraint. The new model is not specific to the RGAs. It can also be suitable for developing other computational methods for the BWN gateway placement problem. The framework for the problem modelling can be used for other gateway placement problems, such as the base station placement problem in IoT.
- This paper finds a way of identifying and repairing infeasible solutions to the graph clustering problem.
- This paper proposes a method that uses the identifying and repairing infeasible solution procedure to generate better quality of solutions in shorter computation time.
- This paper shows the computational efficiency of the RGA is quadratic, which indicates the RGAs is scalable.

References

1. Aoun, B., Boutaba, R., Iraqi, Y., Kenward, G.: Gateway placement optimization in wireless mesh networks with qos constraints. IEEE J. Sel. Areas Commun. **24**(11), 2127–2136 (2006)
2. Ashraf, U.: Energy-aware gateway placement in green wireless mesh networks. IEEE Commun. Lett. **21**(1), 156–159 (2017)
3. Bejerano, Y.: Efficient integration of multihop wireless and wired networks with qos constraints. IEEE/ACM Trans. Networking **12**(6), 1064–1078 (2004)
4. Chandra, R., Qiu, L., Jain, K., Mahdian, M.: Optimizing the placement of internet taps in wireless neighborhood networks. In: IEEE International Conference on Network Protocols, pp. 271–282. IEEE (2004)
5. Hamdi, M., Franck, L., Lagrange, X.: Gateway placement in hybrid manet-satellite networks. In: IEEE Vehicular Technology Conference, pp. 1–5. IEEE (2012)
6. Li, F., Wang, Y., Li, X.Y., Nusairat, A., Wu, Y.: Gateway placement for throughput optimization in wireless mesh networks. Mob. Netw. Appl. **13**(1–2), 198–211 (2008)
7. Lin, C.C., Shu, L., Deng, D.J.: Router node placement with service priority in wireless mesh networks using simulated annealing with momentum terms. IEEE Syst. J. **10**(4), 1402–1411 (2016)
8. Omar, H.A., Zhuang, W., Li, L.: Gateway placement and packet routing for multihop in-vehicle internet access. IEEE Trans. Emerg. Top. Comput. **3**(3), 335–351 (2015)
9. Tang, M.: Gateways placement in backbone wireless mesh networks. Int. J. Commun. Netw. Syst. Sci. **2**(1), 44 (2009)
10. Tang, M.: A memetic algorithm for the location-based continuously operating reference stations placement problem in network real-time kinematic. IEEE Trans. Cybern. **45**(10), 2214–2223 (2015)
11. Wong, J.L., Jafari, R., Potkonjak, M.: Gateway placement for latency and energy efficient data aggregation [wireless sensor networks]. In: IEEE International Conference on Local Computer Networks, pp. 490–497. IEEE (2004)

Next Generation Hybrid Intelligent Medical Diagnosis Systems

Sabri Arik[1] and Laszlo Barna Iantovics[2(✉)]

[1] Istanbul University, Istanbul, Turkey
[2] Petru Maior University, Tirgu Mures, Romania
ibarna@science.upm.ro

Abstract. Many medical diagnosis problems (*MDPs*) are difficult to be solved by physicians. Frequently, the difficult *MDPs* solving by physicians require the assistance of the medical computing systems, which many times should be intelligent. Many papers presented in the specialized literature prove, that the intelligence of a system (frequently agent-based) can offer advantages in the *MDPs* solving versus a system that does not have such intelligence. Cooperative hybrid (human-machine) medical diagnosis systems seem to be well suited for the solving of many difficult *MDPs*. A difficult aspect in the design of such systems consists in the establishment of how to combine in an optimal way the humans and intelligent systems interoperation in order to solve the undertaken problems in the most efficient way. With this purpose, a novel hybrid medical system, called *Intelligent Medical Hybrid System* (*IntHybMediSys*) is proposed in this paper, a system which combines efficiently the humans and computing systems advantages in the problem-solving. We give a definition to the *Difficult Medical Diagnosis Problem Solving Intelligence*. *IntHybMediSys* is a highly complex hybrid system composed of physicians and intelligent agents that can interoperate intelligently in different points of decision in order to solve efficiently very difficult medical diagnosis problems. *IntHybMediSys* is able to handle emergent information that rise during the medical problems solving that allows the precise establishment of the most efficient contributor (a physician or an artificial agent) at each contribution during a problem-solving. This kind of problem-solving has as an effect the increase of accuracy of the elaborated diagnostic.

Keywords: Intelligent hybrid medical system · Machine intelligence · Difficult medical diagnosis problem

1 Introduction

In the healthcare [1, 2], there are many complex *MDPs* (comorbidities, rare illnesses, difficult clinical cases) that are difficult to be solved by clinicians and/or intelligent medical computational systems (medical expert systems for instance). As an example of a difficult problem, we mention the case of a patient that suffers by a comorbidity, more illnesses, between whose curing treatments there are some dependencies. For example, a very effective medicine for the curing of an illness has a negative effect to the curing of another illness.

D. Liu et al. (Eds.): ICONIP 2017, Part III, LNCS 10636, pp. 903–912, 2017.
https://doi.org/10.1007/978-3-319-70090-8_92

Agent-based systems represent one of the most important approaches that can be applied for the intelligent solving of many difficult problems [3, 4]. The main motivation consists in the properties of the agents that differentiate them from other systems capable to make computations. Agent-based systems can be endowed with capacities that allow to intelligently solve (could suppose cooperation) hard problems.

Difficulties in a *MDP* solving, the elaboration of a diagnostic consisting in the identification of the illness(es) and establishment of the corresponding treatment(s), appear based on considerations like: the solving requires a large amount of data (the distributed medical history of the patient for example); the solving requires heterogeneous problem-solving knowledge (detained by more clinicians and/or medical computational systems, for example, medical expert systems); the problem description contains missing information/data (the patient does not specify some of the symptoms of his/her illness, for example) or erroneous data (the patient erroneously specifies some of the symptoms of his/her illness for example); the solving of the problem is not completely known to the physicians; the solving of the problem is not completely known to existent computing medical systems etc.

The solving of many [5] *MDPs* or sub-problems (search for the answer to a medical issue, what is the most efficient medicine for curing an illness in a specific stage of evolution of that illness, for example) by medical systems capable of making computations requires machine intelligence. Most of the medical systems are agent- based systems, having applications like: Intelligent Monitoring of Intensive Care Units [6], different kind of Clinical Decision Support [7] and Decision Support System for Optimal Drug Prescription [8].

We would like to note that at the current stage of evolution of computing systems, the *MDPs* solving intelligence of humans and medical agent-based systems could not be directly compared because they are of completely different types. The humans attained their intelligence during a very long evolution. Agent-based systems have not attained such a high-level intelligence until now, and they will not attain a similar intelligence with the humans for a long time in the future, even if the evolution of hardware and software has itself been very fast.

In this paper, we propose a novel very complex hybrid medical system called *IntHybMediSys* capable to intelligently solve difficult medical diagnosis problems. *IntHybMediSys* combines efficiently the physicians and medical computing systems advantages in the *MDPs* solving, taking into consideration the complementarities in the human thinking and the systems intelligence to make computations.

The upcoming part of the paper is organized as follows: Sect. 2 presents the proposed *IntHybMediSys* hybrid system; in the Subsect. 2.1 there are presented some motivations for such a complex system development; Subsect. 2.2 presents the proposal; in the Subsect. 2.3 there is summarized a preliminary validation; Sect. 3 presents the conclusions of the research.

2 IntHybMediSys a Novel Hybrid Medical Complex System

2.1 Motivations for the Design of the IntHybMediSys System

Many medical diagnosis problems are inherently difficult. The difficulty can be seen from the physician's and from the computational point of view. A problem can be difficult for physicians but easily solvable by computing medical systems (the description contains a very large quantity of data, for example). A problem can be difficult for computing systems but easily solvable by physicians (the interpretation of some medical images for example). We will discuss the possibility to endow intelligent medical systems with: human kind of specialty knowledge, human kind of intuition, and human kind of common sense knowledge.

Medical Expert Systems (*MESs*) are well known based on their ability to solve *MDPs* similarly with the physicians [9, 10]. *MESs* detain similar knowledge as the physicians. Applications of *MESs* are by a large variety ranging from: bone scan interpretation [9], early detection of brain cancer from *CT* and *MR* images [10] and so on. Likewise, there are some novel agent-based systems that extend the *ESs* with other functionalities, able to solve similar problems with the *ESs* but exhibit more intelligence [11]. Extensions may consist in new abilities like: more autonomy in operation, capacity to perceive the environment via sensors, capacity to execute actions in the environment via effectors, cooperation capacity and learning capacity.

Physicians and intelligent agent-based medical systems have different advantages and limitations in the difficult *MDPs* solving. This motivates the necessity of physician-computer "hybridization" in some cases of difficult problem-solving or when this may reduce the human effort. Hybrid systems can outperform both physicians and computing medical systems that operate individually. By efficient cooperation, the human and artificial thinking advantages in the problem-solving are combined. Physicians can elaborate decisions using their medical knowledge, intuition and commonsense knowledge, what we call as the fundamental components of the "physicians' intelligence in solving difficult medical cases".

The positive influence of intuition to creative problem-solving by the humans is analyzed in the paper [12]. The importance of intuition in the clinical diagnoses establishment and different types of clinical decisions is studied in the paper [13]. We would like to conclude that the intuition is a very important component of the human intelligence in difficult problems solving, but unfortunately, it is almost totally missing even from the most intelligent artificial systems.

The importance of machines reasoning by common sense was analyzed first by McCarthy [14]. The paper [15] comments why the advances in the commonsense knowledge representation and reasoning are so slow and analyzes the impact of such an advance. The paper [16] analyzes the importance of specialty knowledge and common sense knowledge related to the diagnostics establishment in medicine. There are many computing systems that have some kind of specialty knowledge, able to solve problems or sub-problems like the physicians but there do not exist computing systems able to detain and process common sense knowledge comparable even to a young child.

We identified that, very difficult *MDPs* solving require what we call in our study *Medical Diagnosis Problem Solving Intelligence* (*MDPSI*), whose most important elementary components are the: diagnosing knowledge, intuition and common sense knowledge detained by the physicians and intelligent assistance offered by medical systems able to make computations (expert systems for instance). We consider that the analysis of the *MDPSI* is necessary for the design of hybrid systems able to solve difficult *MDPs* with a very high intelligence.

2.2 The Proposed IntHybMediSys Medical Complex System

In this section, we briefly describe the proposed highly complex next generation *IntHybMediSys* hybrid system's operation. We denote with *ICMS* such a hybrid system. *Glob_Medi* represents a memory/shared place by physicians and intelligent agents that allow the writing of different health data/information (symptoms, medical images, signals etc.) related to an ill person. $ICMS = HMD \cup IAS \cup IKS$.$HMD = \{HMD_1, HMD_2, \ldots HMD_n\}$ denote the physicians members of the *ICMS* system; they form a social network of medical specialists. $IKS = \{IKs_1, IKs_2, \ldots IKs_q\}$ represent the agents, which we call knowledge sources, who can contribute to the solving of problems written onto the *Glob_Medi*. The knowledge sources can have different specializations (medical image analysis, to give different suggestions to physicians, etc.) based on that they are able to make different type of analyzing/processing of the data/information from the *Glob_Medi*. Some knowledge sources could operate like agent-based medical expert systems or could be specialized in outlining and/or eliminating erroneous data/information from *Glob_Medi*. In the case of a patient, a high temperature that couldn't have a body can be recorded, for example, by mistake.

$IAS = \{IAS_1, IAS_2, \ldots IAS_n\}$ represent the intelligent assistant agents' members of the *ICMS* system. Each assistant agent is owned by a physician. For example, IAS_i is owned by the HMD_i physician. $IAS = \{IAS_1(HMD_1), IAS_2(HMD_2), \ldots IAS_n(HMD_n)\}$. An assistant agent's specialization contains the description of the help that it can offer to the owner physicians and other agents. A new assistant agent will be introduced when a new physician joins to *ICMS*. In the *IntHybMediSys* each physician interacts directly with a single assistant agent owned by him/her during a problem-solving. Each assistant agent knows how it must cooperate with some other agents in order to offer the necessary help to its owner physician.

We denote with *Pati* the patient whose illness/illnesses is diagnosed. *Pati* is the patient of the HMD_k physician. For the illustration of the *ICMS* system's operation, we present the *IHCPS* algorithm that describes a Prl_v difficult *MDP* solving (*Pati* illness diagnosis). $Solu_v$ denotes the identified illness(es) and the prescribed treatment(s). Figure 1 illustrates the cooperative solving of the Prl_v problem.

Step 1 describes the *Pati* patient's illness presented by HMD_k. *Glob_Medi* could include details, like: symptoms, syndromes, laboratory analysis results, etc. *ModM* is the moderator physician who will moderate the Prl_v solving. *COPM* is composed of physicians who jointly with the agents *IKS*, will contribute to the Prl_v solving.

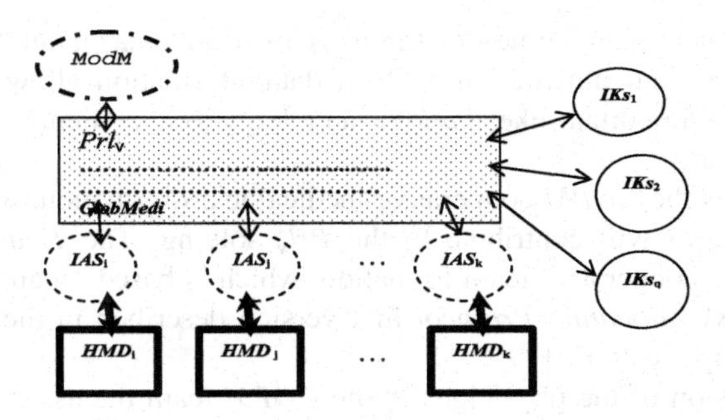

Fig. 1. Prl_v cooperative solving by *IntHybMediSys*

```
IHCPS: Intelligent Cooperative Problem-Solving Algorithm
IN: Prl_v; OUT: Solu_v;
Step 1: Data collection
@HMD_k consults Pati identifying the illness symptoms.
@HMD_k writes Prl_v onto the Glob_Medi.
Step 2: Preprocessing of the data from the Glob_Medi.
@Agents from the set IKS could make different processing
and analyzing of the data from the Glob_Medi.
Step 3: Formation of COPM the cooperative medical team
@Formation of a cooperative medical team denoted COPM.
@Establishment of a moderator physician denoted ModM.
Step 4: Cooperative diagnosis elaboration
While (the solution of Prl_v is not obtained) Do
  @Each member of the COPM∪IKS team estimates its
  capacity to contribute to the Prl_v solving.
  @Members that can contribute to the Prl_v solving report
  their contribution capacity to ModM.
  @ModM selects the best contributor Con_i (Con_i∈COPM∪IKS)
  and gives the writing right to the Glob_Medi.
  @Con_i contributes to the problem-solving by adding
  and/or retracting and/or modifying previously written
  data/information onto the Glob_Medi.
  @After Con_i finishes the writing onto the Glob_Medi,
  ModM retracts the writing right from Con_i.
EndWhile
Step 5:Verification of the solution correctness by agents
@If the solution Sol_v is obtained, the artificial agents
IKS will check for erroneous and missing data.
Step 6:Validation of the solution by physician(s)
@Sol_v validation by one or more physicians.
EndIntelligentProblemSolving
```

Step 2 presents the initial processing of data performed by knowledge sources, with the purpose to improve the description accuracy/completeness. Some knowledge sources may collect (many times distributed) medical data about the *Pati* medical history (past illnesses/treatments etc.). Such data/information allow the outlining of aspects that can be important like: *Pati* has an allergy to a medicine; *Pati* is frequently ill by an illness etc.

Step 3 outlines the *COPM* cooperative medical team formation, which jointly with the agents from *IKS* will contribute to the Prl_v solving. The *TeamForm* algorithm briefly describes a cooperative team formation, which is based on an adaptation of the *Contract Net Task Allocation Protocol* first version described in the papers [17, 18], having different applications [19].

For the inclusion of the physicians in the *COPM* team the assistant agent IAS_k of HMD_k establishes Prl_v initial description, and includes it in the announcement Ann_v (1): $Deadline_v$ denotes the problem-solving deadline (in case if there is established a deadline); $Emitted_v$ denotes the time when the announcement has been emitted; $Eligibility_v$ is the eligibility criteria for inclusion in the cooperative problem-solving team (could include constraints like: medical specialization, clinical practice and experience in the diagnosed-problem solving). Based on some considerations, like the specificity of the diagnosed illness, it could include some other details that could help the invited physicians to establish more precisely their offer.

```
TeamForm: Contract Net Based Team Formation Algorithm
IN: Prl_v; OUT: COPM;
@IAS_k helps HMD_k to formulate Prl_v announcement Ann_v.
@IAS_k establish PotMedSet the potential physicians able to
contribute to the Prl_v cooperative solving.
@IAS_k sends Ann_v to the assistant agents of the PotMedSet.
@IAS_k collects/filter the answers from the assistant
agents of PotMedSet and sends the result to HMD_k.
@Based on the received responses HMD_k will establish the
COPM cooperative team (COPM will include HMD_k).
EndContractNetTeamFormation
```

$$Ann_v = <Prl_v;\ Emitted_v;\ Deadline_v;\ Eligibility_v> \qquad (1)$$

Ann_v is sent by IAS_k to the assistant agents of a selected list of physicians that potentially can contribute to the Prl_v solving (they have at least the necessary medical specialization and based on the available information they could be available). Some of the assistant agents of the invited physicians will send an answer to the invitation. $Ans_{r,v}$ (2) presents the parameters of an answer: HMD_b denotes the physician who answered to the invitation; $Decision_{r,v}$ denotes the decision related to the invitation (acceptance or rejection); $Offer_{r,v}$ denotes the specification of how they can contribute to the Prl_v solving; $Experience_{r,v}$ could include indicators of the experience in the Prl_v's solving (number and difficulty of similar cases etc.); $Proposed\_Role_{r,v}$ is the proposed role by HMD_b (contributor to the Prl_v solving or acting as moderator); $Rem_{r,v}$ the answer related to local or remote participation to the Prl_v solving. Based on some considerations, like the specificity of the diagnosed illness, an answer could include some other parameters. For example, a medical student could assist to a problem solving, with learning purposes.

$$Ans_r = <Ann_v; HMD_b; Decision_{r,v}; Offer_{r,v}; Experience_{r,v}; Proposed\_Role_{r,v}; Rem_{r,v}> \quad (2)$$

IAS_k filters the responses and transmits the result to HMD_k who based on the received answers will form *COPM*. In the case of a very large network of physicians *HMD*, the cooperative team formation could be a recursive process. Assistant agents of the physicians who receive an invitation could forward that invitation to some other assistant agents of physicians that were not invited. This recursive search process can be continued until the cooperative team is formed.

The choosing from the *COPM* of the *ModM* is by joint agreement. The decision should be based on aspects like: the expertise in the Prl_v solving; the experience in cooperative *MDP* solving and the experience in moderating.

Step 4 describes the hybrid cooperative solving of Prl_v. The physicians and agents solve the Prl_v by: adding (a new supposed illness, so on), retracting (a previously written illness, so on) and modifying (an erroneous data/information, so on) information/data written onto the *Glob_Medi*. During the Prl_v solving the physicians can interoperate intelligently with the owned assistant agents.

ModM increases the coherence and efficiency of the Prl_v solving. For the illustration of the increased coherence and effectiveness in operation of the system, we consider the case when more physicians and agent(s) try to make different processing's onto the *Glob_Medi* at the same time. As a simple scenario, we mention when a knowledge source intends to write onto the *Glob_Medi* an important information, that the patient is allergic to the medicine denoted *med* (information extracted from the patient's medical history) and at the same time a physician wishes to recommend *med* considering it as very efficient in treating the patient's illness. Because the knowledge source writes an important detail related to the patient, *ModM* can allow to the agent to firstly write onto the *Glob_Medi*.

Step 5 presents Sol_v verification made by knowledge sources. They can verify, for example, in the *Pati* medical history details, if *Pati* was treated with the subscribed medicine in the past, and in the affirmative case, what the effect of the medicine was. A final diagnostic establishment and validation in the case of an illness(es) is the responsibility of physicians. Step 6 describes the final validation of the solution performed by *ModM* and HMD_k.

2.3 Experimental Evaluation of IntHybMediSys

There are some proposals of computing systems presented in the scientific literature that are somehow comparable with our proposal considering the *MDP* and different difficult clinical tasks solving ability. For example, the paper [2] presents a novel approach for a clinical problem considered difficult consisting in the correction of skin contour defects in leaking stomas by filler injection. The paper [5] presents an agent-based medical diagnosis system. In the papers [20–23] systems able of cooperative solving of different difficult problems are proposed. The most classical approaches of systems able to similarly elaborate diagnoses with the humans are the medical expert systems [9, 10].

As all complex systems, *IntHybMediSys* requires as well more consecutive versions until a fully functional system is obtained. As an illustrative motivation, we mention the rule based *MESs* whose endowment with knowledge could have the duration of more years. In our proposal just a single knowledge source can operate as an extended *MES*, however, our proposal is extremely complex.

For validation purposes, we implemented a first version of the *IntHybMediSys* considering the dermatology, were in some difficult cases (a very rare skin illness, a comorbidity) a dermatologist physician cannot elaborate a diagnostic. Any number of physicians assisted by assistant agents can contribute to a problem solving. There were implemented two knowledge sources specialized in the analysis of information/data posted by the physicians. The knowledge sources are able to: verify contradictions in the opinions of the physicians, helping the physicians in the formulation of a consensuses opinion by taking into consideration aspects like the experience of the physicians and expertise in the diagnosed illness and some other analyses that were considered useful.

The next research that we will perform will consist in the implementation of a new knowledge source able to make autonomously some skin image analysis having the following objectives: (1) to make different suggestions that could be taken into consideration by the physicians; (2) to outline mistakes in the physicians decisions by analyzing images of the skin, letting the physicians to make a decision on the suggestion (3) to make predictions related to the evolution of skin illness(es).

3 Conclusions

Even if the intelligence of medical diagnoses problems solving is different in the case of physicians and agent-based medical systems, we consider that could be designed efficient hybridizations based on complementarities in aspects of medical diagnosis problem-solving. For example, we mention the human intuition that is very specific to humans, necessary in many very difficult medical diagnoses, but is almost inexistent in intelligent medical systems, and the capacity of medical agent-based systems to make large amounts of verifications that could be missed by the physicians. In this paper, we give a definition of *Medical Diagnosis Problem Solving Intelligence* (*MDPSI*). Such intelligence, at a maximal level, could be attained only by very well designed hybrid human-medical systems. One of the primordial questions is how to design the architecture of such a hybrid system in order to have the most efficient operation and performance. We consider that, in order to design a hybrid system able to solve highly intelligently diagnosis problems, the *MDPSI* should be analyzed among others.

In this paper, a novel hybrid complex medical system called *IntHybMediSys* is proposed. The motivation of the research was the proposal of a solution for solving medical diagnosis problems of high difficulty, that cannot be solved (or cannot be solved accurately) by a single physician or a medical system individually. The proposal is general, and could be implemented in different individual medical fields (cardiology, dermatology, internal medicine and so on) or more medical fields combined (diagnosis of comorbidities). In the case study, we have considered the field of dermatology.

One of the most interesting definitions of machine intelligence was proposed by Alan Turing [24]. Turing's definition stemmed from the idea of an artificial cognitive system that was able to mimic the cognition of a human being. Over the years, many papers [25–27] have been published related to the Turing test. Based on the very high complexity of such a proposed *IntHybMediSys* system, it is very difficult to make comparisons with similar systems. Based on this aspect we consider a very tempting research direction the design of similar tests with the Turing test for measuring and comparing the intelligence of systems by such a high complexity and intelligence.

Acknowledgment. Laszlo Barna Iantovics acknowledge the support of the COROFLOW project PN-III-P2-2.1-BG-2016-0343, Contract: 114BG/2016.

References

1. Giovacchini, G., Giovannini, E., Leoncini, R., Riondato, M., Ciarmiello, A.: PET and PET/CT with radiolabeled choline in prostate cancer: a critical reappraisal of 20 years of clinical studies. Eur. J. Nuclear Med. Mol. Imaging **44**(10), 1751–1776 (2017)
2. Weidmann, A.K., Al-Niaimi, F., Lyon, C.C.: Correction of skin contour defects in leaking stomas by filler injection: a novel approach for a difficult clinical problem. Dermatol. Ther. **4** (2), 271–279 (2014)
3. Iantovics, L.B., Chira, C., Dumitrescu, D.: Principles of the Intelligent Agents. Casa Cartii de Stiinta Press, Cluj-Napoca (2007)
4. Imam, I.F.: Intelligent Adaptive Agents. Papers from the 1996 AAAI Workshop. Technical Report WS-96-04, AAAI Press, CA (1996)
5. Iantovics, L.B.: Agent-based medical diagnosis systems. Comput. Inform. **27**(4), 593–625 (2008)
6. Nouira, K., Trabelsi, A.: Intelligent monitoring system for intensive care units. J. Med. Syst. **36**(4), 2309–2318 (2012)
7. Shirabad, J.S., Wilk, S., Michalowski, W., Farion, K.: Implementing an integrative multi-agent clinical decision support system with open source software. J. Med. Syst. **36**(1), 123–137 (2012)
8. Miller, K., Mansingh, G.: OptiPres: a distributed mobile agent decision support system for optimal patient drug prescription. Inform. Syst. Front. **19**(1), 129–148 (2017)
9. Haupt, F., Berding, G., Namazian, A., Wilke, F., Böker, A., Merseburger, A., Geworski, L., Kuczyk, M.A., Bengel, F.M., Peters, I.: Expert system for bone scan interpretation improves progression assessment in bone metastatic prostate cancer. Adv. Ther. **34**(4), 986–994 (2017)
10. Kar, S., Majumder, D.D.: A mathematical theory of shape and neuro-fuzzy methodology-based diagnostic analysis: a comparative study on early detection and treatment planning of brain cancer. Int. J. Clin. Oncol. **22**(4), 667–681 (2017)
11. Al-Qaysi, I., Unland, R., Weihs, C., Branki, C.: Medical diagnosis decision support HMAS under uncertainty HMDS. Stud. Comput. Intell. **326**, 67–94 (2011)

12. Eubanks, D.L., Murphy, S.T., Mumford, M.D.: Intuition as an influence on creative problem-solving: the effects of intuition, positive affect, and training. Creativity Res. J. **22** (2), 170–184 (2010)
13. Pelaccia, T., Tardif, J., Triby, E., Charlin, B.: An analysis of clinical reasoning through a recent and comprehensive approach: the dual-process theory. Med. Educ. Online **16**(1), 5890 (2011)
14. McCarthy, J.: Programs with common sense, In. Proceedings of the Teddington Conference on the Mechanization of Thought Processes, Her Majesty's Stationery Oce, London (1959)
15. Davis, E., Marcus, G.: Commonsense reasoning and commonsense knowledge in artificial intelligence. Commun. ACM **58**(9), 92–103 (2015)
16. Keravnou, E.T., Dams, F., Washbrook, J., Dawood, R.M., Hall, C.M., Shaw, D.: Background knowledge in diagnosis. Artif. Intell. Med. **4**(4), 263–279 (1992)
17. Smith, R.G.: The contract net protocol: high-level communication and control in a distributed problem solver. IEEE Trans. Comput. **29**(12), 1104–1113 (1980)
18. Davis, R., Smith, R.G.: Negotiation as a metaphor for distributed problem solving. Artif. Intell. **20**, 63–109 (1983)
19. Alotaibi, S.J.: ICT classroom LMSs: examining the various components affecting the acceptance of college students in the use of blackboard systems. Adv. Intell. Syst. Comput. **498**, 523–532 (2016)
20. Zhao, H., Ma, W., Sun, B.: A novel decision making approach based on intuitionistic fuzzy soft sets. Int. J. Mach. Learn. Cybern. **8**(4), 1107–1117 (2017)
21. Jin, F., Ni, Z., Pei, L., Chen, H., Li, Y., Zhu, X., Ni, L.: A decision support model for group decision making with intuitionistic fuzzy linguistic preferences relations. Neural Comput. Appl., 1–22 (2017).
22. Iantovics, L.B.: Intelligent computations for complex problem solving. In: Hluchý, L., Kurdel, P., Sebestyénová, J. (eds.) Proceedings of the 7th International Workshop on Grid Computing for Complex Problems (GCCP 2011), pp. 27–36 (2011)
23. Iantovics, L.B., Kovacs, L., Fekete, G.L.: Next generation university library information systems based on cooperative learning. New Rev. Inform. Networking **21**(2), 101–116 (2016)
24. Turing, A.M.: Computing machinery and intelligence. Mind, New Series, vol. 59(236), pp. 433–460. Oxford University Press on behalf of the Mind Association (1950)
25. Kuppusamy, K.S., Aghila, G.: HuMan: an accessible, polymorphic and personalized CAPTCHA interface with preemption feature tailored for persons with visual impairments. Univ. Access Inform. Soc., 1–24 (2017).
26. Thwaites, A., Soltan, A., Wieser, E., Nimmo-Smith, I.: The difficult legacy of Turing's wager. J. Comput. Neurosci. **43**(1), 1–4 (2017)
27. Luger, G.F., Chakrabarti, C.: From Alan Turing to modern AI: practical solutions and an implicit epistemic stance. AI Soc. **32**(3), 321–338 (2017)

2-Tuple Prioritized Weighted Harmonic Operator and Its Use in Group Decision Making

Jin Han Park[1(✉)], Seung Bin Lee[1], Ja Hong Koo[2], and Young Chel Kwun[3]

[1] Department of Applied Math., Pukyong National University, Busan 48513, South Korea
jihpark@pknu.ac.kr, binii725@naver.com

[2] Department of Fiber Plastic Design, Dong-A University, Busan 604-714, South Korea
jhkoo@dau.ac.kr

[3] Department of Mathematics, Dong-A University, Busan 604-714, South Korea
yckwun@dau.ac.kr

Abstract. In this paper, we develop a 2-tuple prioritized weighted harmonic (2TPWH) operator and discuss its desirable properties such as idemtopency, boundedness and monotonicity. Then, we apply the 2TPWH operator to propose an approach to multiple attribute group decision making, with 2-tuple linguistic information, in which the decision makers and attributes are in different priority levels. Finally, an example is used to illustrate the applicability of the proposed approach.

Keywords: 2-Tuple variables · 2-Tuple prioritized weighted harmonic operator · Group decision making

1 Introduction

The goal of decision making problem is to find the most desirable alternative(s) from a given alternative set. Because of increasing complexity of the socio-economic environment, a single decision maker (DM) has not enough to consider all relevant parts of a problem, as a result, many decision making process take the form of group setting. Group decision making is a typical decision activity where utilizing some experts alleviate the decision difficulties due to the problem's complexity. In the real world, the uncertainty and unclear knowledge of the experts imply that they cannot provide exact values to express their preferences. Linguistic variable is an useful tool to express a MD's preference information over objects in process of decision making under uncertain environment [1–3]. The aggregation of linguistic variables is an important step to get a decision result. Over the past decades, lots of linguistic aggregation operators have been proposed. So we classify them into the following categories: (1) the semantic model based linguistic aggregation operators; (2) the symbolic model [4,9,10] based linguistic aggregation operators; (3) the linguistic aggregation operators, which compute with words directly; (4) the 2-tuple representation model [13,14] based

D. Liu et al. (Eds.): ICONIP 2017, Part III, LNCS 10636, pp. 913–923, 2017.
https://doi.org/10.1007/978-3-319-70090-8_93

linguistic aggregation operators, including 2-tuple weighted averaging (2TWA) operator [13], 2-tuple OWA (2TOWA) operator [13], 2-tuple weighted geometric (2TWG) operator [15], 2-tuple ordered weighted geometric (2TOWG) operator [15] and 2-tuple hybrid geometric (2THG) operator [15]. The operators in (1) and (2) describe approximation process to express the result in initial expression domain, which produce the loss of information and hence bring about the lack of precision, while the operators in (3) and (4) allow a continuous representation of the linguistic information on its domain, and thus they can represent any counting of information obtained in an aggregation process without any loss of information.

Most of linguistic aggregation operators are under the assumption that the attributes are at the same priority level. However, in practical multiple attribute group decision making (MAGDM) with linguistic information, the attributes generally have different priority levels. To overcome this problem, Zhang [16] developed the 2-tuple prioritized weighted averaging (2TPWA) operator and 2-tuple prioritized weighted geometric (2TPWG) operator for handling the MAGDM problems. Harmonic mean is used to aggregate central tendency data. In the existing literatures, harmonic mean is used as a fusion technique of numerical data. Thus, "how to aggregate 2-tuples by using the harmonic mean?" is worth paying attention to research. In this paper, we give our attention to develop a 2-tuple prioritized weighted harmonic (2TPWH) operator. To do so, the rest of this paper is organized as follows. Section 2 reviews the basic concept related to 2-tuples. Section 3 develops the 2TPWH operator and investigate its properties. Based on the developed operator, Sect. 4 presents an approach to MAGDM in which the attributes and decision makers are in different priority level. Section 5 illustrates the applicability of the developed approach with an example.

2 Basic Concepts

In this section, we first review some concepts of the 2-tuple.

Let $S = \{s_i : i = 0, 1, 2, \ldots, g\}$ be a finite and totally ordered discrete linguistic term set with odd cardinality, where s_i represents a possible value for a linguistic variable, and it satisfies the following conditions [8, 12]:

(1) The set is ordered: $s_i \geq s_j$ if $i \geq j$;
(2) There is the negation operator: $\text{neg}(s_i) = s_j$ such that $j = g - i$;
(3) Max operator: $\max(s_i, s_j) = s_i$ if $s_i \geq s_j$;
(4) Min operator: $\min(s_i, s_j) = s_i$ if $s_i \leq s_j$.

For example, S can be defined so as its elements are uniformly distributed on a scale on which a total order is defined [12]:

$$S = \{s_0 = \text{extremely poor}, s_1 = \text{very poor}, s_2 = \text{poor}, s_3 = \text{slightly poor}, \\ s_4 = \text{fair}, s_5 = \text{slightly good}, s_6 = \text{good}, S_7 = \text{very good}, \\ s_8 = \text{extremely good}\}.$$

Definition 1 **([13]).** Let β be the result of an aggregation of the indices of a set of labels assessed in linguistic term set S, i.e., the result of a symbolic aggregation operation. $\beta \in [0, g]$, being $g + 1$ the cardinality of S. Let $i = \text{round}(\beta)$ and $\alpha = \beta - i$ be two values such that $i \in [0, g]$ and $\alpha \in [0.5, -0.5)$ then α is called a *symbolic translation*.

From this concept, Herrera and Martínez [13] developed the linguistic representation model which represents the linguistic information by means of 2-tuple (s_i, α_i), $s_i \in S$ and $\alpha_i \in [-0.5, 0.5)$:

- s_i represents the linguistic label center of the information;
- α_i is a numerical value expressing the value of the translation from the original result β to the closest index label i in the linguistic term set S, i.e., the symbolic translation.

Definition 2 **([13]).** Let $S = \{s_0, s_1, \ldots, s_g\}$ be a linguistic term set and $\beta \in [0, g]$ be a value representing the result of a symbolic aggregation, then the 2-tuple that expresses the equivalent information to β is obtained with the function $\triangle : [0, g] \to S \times [-0.5, 0.5)$ defined by

$$\begin{aligned} &\triangle : [0, g] \to S \times [-0.5, 0.5) \\ &\triangle(\beta) = (s_i, \alpha_i), \text{ with } \begin{cases} s_i, & i = \text{round}(\beta), \\ \alpha_i = \beta - i, & \alpha_i \in [-0.5, 0.5), \end{cases} \end{aligned} \tag{1}$$

where round$(\cdot)$ is the usual round operation, s_i has the closest index label to β and α_i is the value of the symbolic translation.

Contrarily, let $S = \{s_0, s_1, \ldots, s_g\}$ be a linguistic term set and (s_i, α_i) be a 2-tuple. There is always a $\triangle^{-1}$ function:

$$\begin{aligned} &\triangle^{-1} : S \times [-0.5, 0.5) \to [0, g] \\ &\triangle^{-1}(s_i, \alpha_i) = i + \alpha_i = \beta \end{aligned} \tag{2}$$

such that from a 2-tuple it returns its equivalent numerical value $\beta \in [0, g]$.

It is obvious that the conversion of a linguistic term into a linguistic 2-tuple consists of adding a value zero as symbolic translation:

$$s_i \in S \Longrightarrow (s_i, 0). \tag{3}$$

Definition 3 **([13]).** Let (s_k, α_k) and (s_l, α_l) be two 2-tuples, with each one representing a counting of information, then:

- if $k < l$ then (s_k, α_k) is smaller than (s_l, α_l), denoted by $(s_k, \alpha_k) < (s_l, \alpha_l)$;
- if $k = l$ then

(1) if $\alpha_k = \alpha_l$ then (s_k, α_k) and (s_l, α_l) represent the same information, denoted by $(s_k, \alpha_k) = (s_l, \alpha_l)$;
(2) if $\alpha_k < \alpha_l$ then (s_k, α_k) is smaller than (s_l, α_l), denoted by $(s_k, \alpha_k) < (s_l, \alpha_l)$;
(3) if $\alpha_k > \alpha_l$ then (s_k, α_k) is bigger than (s_l, α_l), denoted by $(s_k, \alpha_k) > (s_l, \alpha_l)$.

3 2-Tuple Prioritized Weighted Harmonic Operator

The prioritized average (PA) operator was originally introduced by Yager [17], and has usually used in the situation where the input arguments are the exact values. However, in some situations, the input arguments take the form of 2-tuples rather than numerical ones because of lack of knowledge, time pressure, and the decision maker's limited information processing capabilities. Zhang [16] extended the PA operator to linguistic environment as follows:

Definition 4 ([16]). Let $\{(r_i, \alpha_i)\}$ $(r_i \in S,\ \alpha_i \in [-0.5, 0.5),\ i = 1, 2, \ldots, n)$ be a collection of 2-tuples, then the 2-tuple prioritized weighted average (2TPWA) and 2-tuple prioritized weighted geometric (2TPWG) operators is defined as:

$$\text{2TPWA}((r_1,\alpha_1),(r_2,\alpha_2),\ldots,(r_n,\alpha_n)) = \Delta\left(\sum_{i=1}^{n}\frac{T_i}{\sum_{j=1}^{n}T_j}\Delta^{-1}(r_i,\alpha_i)\right), \quad (4)$$

$$\text{2TPWG}((r_1,\alpha_1),(r_2,\alpha_2),\ldots,(r_n,\alpha_n)) = \Delta\left(\prod_{i=1}^{n}\Delta^{-1}(r_i,\alpha_i)^{\frac{T_i}{\sum_{j=1}^{n}T_j}}\right), \quad (5)$$

where $T_1 = 1$, $T_i = \prod_{k=1}^{i-1}\frac{1}{g}\left|\Delta^{-1}(r_i,\alpha_i)\right|$, $i = 2, 3, \ldots, n$ and $\frac{1}{g}\left|\Delta^{-1}(r_i,\alpha_i)\right|$ is the score values of (r_i,α_i) $(i = 2, \ldots, n)$.

Definition 5. Let $\{(r_i, \alpha_i)\}$ $(r_i \in S,\ \alpha_i \in [-0.5, 0.5),\ i = 1, 2, \ldots, n)$ be a collection of 2-tuples, then we define a 2-tuple prioritized weighted harmonic (2TPWH) operators as follows:

$$\text{2TPWH}((r_1,\alpha_1),(r_2,\alpha_2),\ldots,(r_n,\alpha_n)) = \Delta\left(\frac{1}{\sum_{i=1}^{n}\frac{T_i}{\sum_{j=1}^{n}T_j\Delta^{-1}(r_i,\alpha_i)}}\right), \quad (6)$$

where $T_1 = 1$, $T_i = \prod_{k=1}^{i-1}\frac{1}{g}\left|\Delta^{-1}(r_i,\alpha_i)\right|$, $i = 2, 3, \ldots, n$ and $\frac{1}{g}\left|\Delta^{-1}(r_i,\alpha_i)\right|$ is the score values of (r_i,α_i) $(i = 2, \ldots, n)$.

Clearly, the 2TPWH operator is a nonlinear weighted aggregation operator, and the weight $\frac{T_i}{\sum_{j=1}^{n}T_j}$ of the argument (r_i,α_i) depends on all the input arguments (r_j,α_j) $(j = 1, 2, \ldots, n)$ and allows the argument values to support each other in the prioritized aggregation process.

In the following we shall investigate some desirable properties of the 2TPWH operator.

Theorem 1. Let $\{(r_i, \alpha_i)\}$ $(r_i \in S,\ \alpha_i \in [-0.5, 0.5),\ i = 1, 2, \ldots, n)$ be a collection of 2-tuples, then we have the following properties:

(1) Idempotency: If $(r_i,\alpha_i) = (r,\alpha)$ for all i, then

$$\text{2TPWH}((r_1,\alpha_1),(r_2,\alpha_2),\ldots,(r_n,\alpha_n)) = (r,\alpha).$$

(2) Boundedness:

$$\min_i(r_i, \alpha_i) \leq 2\text{TPWH}((r_1, \alpha_1), (r_2, \alpha_2), \ldots, (r_n, \alpha_n)) \leq \max_i(r_i, \alpha_i).$$

(3) Monotonicity: Let $\{(r'_i, \alpha'_i)\}$ $(r'_i \in S,\ \alpha'_i \in [-0.5, 0.5),\ i = 1, 2, \ldots, n)$ be a collection of 2-tuples, if $(r_i, \alpha_i) \leq (r'_i, \alpha'_i)$, for all i, then

$$2\text{TPWH}((r_1, \alpha_1), (r_2, \alpha_2), \ldots, (r_n, \alpha_n)) \leq 2\text{TPWH}((r'_1, \alpha'_1), (r'_2, \alpha'_2), \ldots, (r'_n, \alpha'_n)).$$

4 An Approach to MAGDM with Linguistic Information

In this section, we shall utilize the 2TPWH operators to MAGDM with linguistic information. For a MAGDM problem with linguistic information, let $X = \{x_1, x_2, \ldots, x_n\}$ be a discrete set of n alternatives, $G = \{G_1, G_2, \ldots, G_m\}$ be a set of $m$attributes and that there is a prioritization between the attributes expressed by he linear ordering $G_1 \succ G_2 \succ \cdots \succ G_m$, indicate attribute G_j has a higher priority than G_s, if $j < s$. Let $D = \{d_1, d_2, \ldots, d_s\}$ be a set of s decision makers and there is a prioritization between the decision makers expressed by the linear ordering $d_1 \succ d_2 \succ \cdots \succ d_s$, indicate decision maker d_p has a high priority than d_q if $p < q$. Suppose that $A^{(k)} = (a_{ij}^{(k)})_{m\times n}$ $(k = 1, 2, \ldots, s)$ is the linguistic decision matrix, where $a_{ij}^{(k)} \in S$ is preference value, which takes the form of linguistic variables, given by the decision maker $d_k \in D$, for alternative $x_j \in X$ with respect to attribute $G_i \in G$.

Then, we utilize the 2TPWH operator to propose an approach to MAGDM with linguistic information, which involves the following steps:

Step 1: Transform the linguistic decision matrix $A^{(k)} = (a_{ij}^{(k)})_{m\times n}$ into 2-tuple linguistic decision matrix $\tilde{A}^{(k)} = ((a_{ij}^{(k)}, 0))_{m\times n}$, $k = 1, 2, \ldots, s$.

Step 2: Calculate the values of $T_{ij}^{(k)}$ $(k = 1, 2, \ldots, s)$ as follows:

$$T_{ij}^{(k)} = \prod_{l=1}^{k-1} \frac{1}{g} \left| \Delta^{-1}(a_{ij}^{(l)}, 0)\right|,\ i = 1, 2, \ldots, m, j = 1, 2, \ldots, n, k = 2, 3, \ldots, s,$$

$$T_{ij}^{(1)} = 1,\ i = 1, 2, \ldots, m, j = 1, 2, \ldots, n. \tag{7}$$

Step 3: Utilize the 2TLPWH operator (6):

$$(a_{ij}, \alpha_{ij}) = 2\text{TPWH}((a_{ij}^{(1)}, 0), (a_{ij}^{(2)}, 0), \ldots, (a_{ij}^{(s)}, 0))$$

$$= \Delta\left(\frac{1}{\sum_{l=1}^{s} \frac{T_{ij}^{(l)}}{\sum_{k=1}^{s} T_{ij}^{(k)} \Delta^{-1}(a_{ij}^{(l)}, 0)}}\right),\ i = 1, 2, \ldots, m, j = 1, 2, \ldots, n \tag{8}$$

to aggregate all the individual 2-tuple linguistic decision matrices $\tilde{A}^{(k)} = ((a_{ij}^{(k)}, 0))_{m\times n}$ $(k = 1, 2, \ldots, s)$ into the collective 2-tuple linguistic decision matrix $\tilde{A} = ((a_{ij}, \alpha_{ij}))_{m\times n}$.

Step 4: Calculate the values of T_{ij} $(i = 1, 2, \ldots, m,\ j = 1, 2, \ldots, n)$ as follows:

$$T_{ij} = \prod_{k=1}^{i-1} \frac{1}{g} \left| \Delta^{-1}(a_{kj}, \alpha_{kj}) \right|,\ i = 2, 3, \ldots, m, j = 1, 2, \ldots, n,$$
$$T_{1j} = 1,\ j = 1, 2, \ldots, n. \tag{9}$$

Step 5: To get the overall preference value (a_j, α_j) corresponding to the alternative x_j, we aggregate all the preference values (a_{ij}, α_{ij}) $(i = 1, 2, \ldots, m)$ in the jth column of $\tilde{A}$ by using the 2TPWH operator (6):

$$\begin{aligned}(a_j, \alpha_j) &= \text{2TPWH}((a_{1j}, \alpha_{1j}), (a_{2j}, \alpha_{2j}), \ldots, (a_{mj}, \alpha_{mj})) \\ &= \Delta\left(\frac{1}{\sum_{k=1}^{m} \frac{T_{kj}}{\sum_{i=1}^{m} T_{ij} \Delta^{-1}(a_{kj}, \alpha_{kj})}} \right),\ j = 1, 2, \ldots, n.\end{aligned} \tag{10}$$

Step 6: Utilize the collective overall attribute values (a_j, α_j) $(j = 1, 2, \ldots, n)$ and Definition 4 to rank the alternatives x_j $(j = 1, 2, \ldots, n)$, and then select the most desirable one.

Step 7: End.

5 An Illustrative Example

Let us suppose an investment company, which wants to invest a sum of money in the best option (adapted from [4]). There is a panel with five possible alternatives in which to invest the money: (1) x_1 is a car industry; (2) x_2 is a food company; (3) x_3 is a computer company; and (4) x_4 is an arms company. The investment company must take a decision according to the following four attributes: (1) G_1 is the risk analysis; (2) G_2 is the growth analysis; (3) G_3 is the social-political impact analysis; and (4) G_4 is the environmental impact analysis. The four possible alternatives x_j $(j = 1, 2, 3, 4)$ are evaluated using the linguistic term set

$$\begin{aligned} S = \{ & s_0 = \text{extremely poor}, s_1 = \text{very poor}, s_2 = \text{poor}, s_3 = \text{slightly poor}, \\ & s_4 = \text{fair}, s_5 = \text{slightly good}, s_6 = \text{good}, S_7 = \text{very good}, \\ & s_8 = \text{extremely good}\}. \end{aligned}$$

by three decision makers d_k $(k = 1, 2, 3)$ under the above four attributes G_i $(i = 1, 2, 3, 4)$, and construct, respectively, the linguistic decision matrix $A^{(k)} = (a_{ij}^{(k)})_{4\times 4}$ $(k = 1, 2, 3)$ as listed in Tables 1, 2 and 3. The decision maker d_1 has the absolute priority for decision making, the decision maker d_2 comes next. That is, there is a prioritization between three decision makers expressed by the linear ordering $d_1 \succ d_2 \succ d_3$. In three decision makers' opinion, there exists the prioritization relationship among these attributes, for example, the risk analysis of the candidate is the most important, but the environment impact analysis of the candidate is not so important comparing with other attributes. Therefore, the prioritization relationship can be denoted by $G_1 \succ G_2 \succ G_3 \succ G_4$.

Table 1. Linguistic decision matrix $A^{(1)}$

	x_1	x_2	x_3	x_4
G_1	s_4	s_3	s_3	s_8
G_2	s_4	s_6	s_2	s_1
G_3	s_1	s_5	s_7	s_3
G_4	s_5	s_8	s_5	s_6

Table 2. Linguistic decision matrix $A^{(2)}$

	x_1	x_2	x_3	x_4
G_1	s_5	s_7	s_7	s_8
G_2	s_2	s_4	s_8	s_6
G_3	s_7	s_8	s_6	s_5
G_4	s_3	s_8	s_6	s_2

Table 3. Linguistic decision matrix $A^{(3)}$

	x_1	x_2	x_3	x_4
G_1	s_2	s_7	s_5	s_6
G_2	s_1	s_8	s_6	s_8
G_3	s_2	s_6	s_4	s_5
G_4	s_8	s_8	s_4	s_7

To get the best alternative(s), we first utilize the 2TPWH operator to develop an approach to MAGDM problem with linguistic information, which can be described as following:

Step 1: Transform the linguistic decision matrix $A^{(k)} = (a_{ij}^{(k)})_{4\times 4}$ $(k = 1, 2, 3)$ into 2-tuple linguistic decision matrix $\tilde{A}^{(k)} = ((a_{ij}^{(k)}, 0))_{4\times 4}$ $(k = 1, 2, 3)$ (see Tables 4, 5 and 6):

Table 4. 2-tuple linguistic decision matrix $\tilde{A}^{(1)}$

	x_1	x_2	x_3	x_4
G_1	$(s_4, 0)$	$(s_3, 0)$	$(s_3, 0)$	$(s_8, 0)$
G_2	$(s_4, 0)$	$(s_6, 0)$	$(s_2, 0)$	$(s_1, 0)$
G_3	$(s_1, 0)$	$(s_5, 0)$	$(s_7, 0)$	$(s_3, 0)$
G_4	$(s_5, 0)$	$(s_8, 0)$	$(s_5, 0)$	$(s_6, 0)$

Table 5. 2-tuple linguistic decision matrix $\tilde{A}^{(2)}$

	x_1	x_2	x_3	x_4
G_1	$(s_5, 0)$	$(s_7, 0)$	$(s_7, 0)$	$(s_8, 0)$
G_2	$(s_2, 0)$	$(s_4, 0)$	$(s_8, 0)$	$(s_6, 0)$
G_3	$(s_7, 0)$	$(s_8, 0)$	$(s_6, 0)$	$(s_5, 0)$
G_4	$(s_3, 0)$	$(s_8, 0)$	$(s_6, 0)$	$(s_2, 0)$

Table 6. 2-tuple linguistic decision matrix $\tilde{A}^{(3)}$

	x_1	x_2	x_3	x_4
G_1	$(s_2, 0)$	$(s_7, 0)$	$(s_5, 0)$	$(s_6, 0)$
G_2	$(s_1, 0)$	$(s_8, 0)$	$(s_6, 0)$	$(s_8, 0)$
G_3	$(s_2, 0)$	$(s_6, 0)$	$(s_4, 0)$	$(s_5, 0)$
G_4	$(s_8, 0)$	$(s_8, 0)$	$(s_4, 0)$	$(s_7, 0)$

Step 2: Utilize (7) to calculate the values of $T_{ij}^{(k)}$ $(k = 1, 2, 3)$ which are contained in the matrices $T^{(k)} = (T_{ij}^{(k)})_{4\times 4}$, respectively

$$T^{(1)} = \begin{pmatrix} 1\,1\,1\,1 \\ 1\,1\,1\,1 \\ 1\,1\,1\,1 \\ 1\,1\,1\,1 \end{pmatrix}, \quad T^{(2)} = \begin{pmatrix} 0.500\ 0.375\ 0.375\ 1.000 \\ 0.500\ 0.750\ 0.250\ 0.125 \\ 0.125\ 0.625\ 0.875\ 0.375 \\ 0.625\ 1.000\ 0.625\ 0.750 \end{pmatrix},$$

$$T^{(3)} = \begin{pmatrix} 0.3125\ 0.3281\ 0.3281\ 1.0000 \\ 0.1250\ 0.3750\ 0.2500\ 0.0938 \\ 0.1094\ 0.6250\ 0.6563\ 0.2344 \\ 0.2344\ 1.0000\ 0.4688\ 0.1875 \end{pmatrix}.$$

Step 3: Utilize the 2TPWH operator (8) to aggregate all the individual 2-tuple linguistic decision matrices $\tilde{A}^{(k)} = ((a_{ij}^{(k)}, 0))_{4\times 5}$ $(k = 1, 2, 3)$ into the collective 2-tuple linguistic decision matrix $\tilde{A} = ((a_{ij}, \alpha_{ij}))_{4\times 5}$ (see Table 7):

Table 7. Collective 2-tuple linguistic decision matrix $\tilde{A}$

	x_1	x_2	x_3	x_4
G_1	$(s_4, -0.4198)$	$(s_4, -0.0738)$	$(s_4, -0.2364)$	$(s_7, 0.200)$
G_2	$(s_3, -0.4000)$	$(s_5, 0.2987)$	$(s_3, -0.3818)$	$(s_1, 0.1804)$
G_3	$(s_1, 0.1509)$	$(s_6, -0.1144)$	$(s_6, -0.4092)$	$(s_4, -0.4645)$
G_4	$(s_4, 0.2488)$	$(s_8, 0)$	$(s_5, -0.0309)$	$(s_3, 0.4048)$

Step 4: Utilize (9) to calculate the values of T_{ij} which are contained in the matrix $T = (T_{ij})_{4\times 4}$:

$$T = \begin{pmatrix} 1 & 1 & 1 & 1 \\ 0.4475 & 0.4908 & 0.4704 & 0.9000 \\ 0.1454 & 0.3251 & 0.1540 & 0.1330 \\ 0.0209 & 0.2391 & 0.1076 & 0.0590 \end{pmatrix}.$$

Step 5: Utilize the decision information given in matrix $\tilde{A}$ and the 2TPWH operator (10) to derive the collective overall preference value (a_j, α_j) of the alternative x_j $(j = 1, 2, 3, 4)$:

$$(a_1, \alpha_1) = (s_3, -0.2305),\ (a_2, \alpha_2) = (s_5, -0.2480),$$
$$(a_3, \alpha_3) = (s_4, -0.4980),\ (a_4, \alpha_4) = (s_2, 0.1870).$$

Step 6: Rank all the alternatives x_j $(j = 1, 2, 3, 4)$ in accordance with the collective overall attribute values (a_j, α_j) $(j = 1, 2, 3, 4)$ and Definition 2.3:

$$x_2 \succ x_3 \succ x_1 \succ x_4$$

and thus the best alternative is x_2.

In the following, we compare the proposed approach with other approaches proposed by Zhang [16] (i.e. utilize the 2TPWA operator and the 2TPWG operator in Steps 2 and 5). Each of approaches has its advantages and disadvantages and none of them can always perform better than the others in any situations. It perfectly depends on how we look at things, and not on how they are themselves. The relative comparison with the approaches are shown in Table 8. From Table 8, we know that the ranking results obtained by different aggregation operators are different. This is because these aggregation operators are based on different relationships and may produce different results.

Table 8. Comparison with other approaches

	Zhang [16]	Zhang [16]	Proposed approach
Problem type	MAGDM	MAGDM	MAGDM
Application area	Investment of money	Investment of money	Investment of money
Decision information	Linguistic decision matrix	Linguistic decision matrix	Linguistic decision matrix
Solution method			
Aggregation stage	2TPWA operator	2TPWG operator	2TPWH operator
Exploitation stage	2TPWA operator	2TPWG operator	2TPWH operator
Final decision	$x_2 \succ x_4 \succ x_3 \succ x_1$	$x_2 \succ x_3 \succ x_4 \succ x_1$	$x_2 \succ x_3 \succ x_1 \succ x_4$

6 Conclusions

In this paper, motivated by the idea of harmonic mean, we develop the 2TPWH operator for aggregating 2-tuple linguistic information. The prominent characteristic of the proposed operators is that it take into account prioritization among the attributes. Then, we utilize the 2TPWH operator to develop some approaches to solve the MAGDM problem, with linguistic information, in which the decision makers and attributes are in different priority levels. A practical example about investment selection is given to verify the developed approach and to demonstrate their practicality and effectiveness.

References

1. Zadeh, L.A.: The concept of a linguistic variable and its application to approximate reasoning. Part 1, 2, and 3, Inf. Sci. **8** 199–249, 301–357 (1975). **9**, 43–80 (1976)
2. Zadeh, L.A.: A computational approach to fuzzy quantifiers in natural languages. Comput. Math. Appl. **9**, 149–184 (1983)
3. Xu, Z.S.: Linguistic aggregation operators: an overview. In: Bustince, H., Herrera, F., Montero, J. (eds.) Fuzzy Sets and Their Extensions: Representation, Aggregation and Models. Studies in Fuzziness and Soft Computing, vol. 220, pp. 163–181. Springer, Heidelberg (2008). doi:10.1007/978-3-540-73723-0_9
4. Herrera, F., Herrera-Viedma, E.: Linguistic decision analysis: steps for solving decision problems under linguistic information. Fuzzy Sets Syst. **115**, 67–82 (2000)
5. Herrera, F., Herrera-Viedma, E., Verdegay, J.L.: A sequential selection process in group decision making with a linguistic assessment approach. Inf. Sci. **85**, 223–239 (1995)
6. Herrera, F., Herrera-Viedma, E., Verdegay, J.L.: Linguistic measures based on fuzzy coincidence for reaching consensus in group decision making. Int. J. Approx. Reason. **16**, 309–334 (1997)
7. Delgado, M., Herrera, F., Herrera-Viedma, E., Martínez, L.: Combining numerical and linguistic information in group decision making. Inf. Sci. **107**, 177–194 (1988)
8. Herrera, F., Herrera-Viedma, E.: Aggregation operators for linguistic weighted information. IEEE Trans. Syst. Man Cybern. Part A: Syst. Hum. **27**, 646–656 (1997)
9. Delgado, M., Verdegay, J.L., Vila, M.A.: On aggregation operators of linguistic labels. Int. J. Intell. Syst. **8**, 351–370 (1993)
10. Peláez, J.I., Doña, J.M.: LAMA: a linguistic aggregation of majority additive operator. Int. J. Intell. Syst. **18**, 809–820 (2003)
11. Xu, Z.S.: On generalized induced linguistic aggregation operators. Int. J. Gen. Syst. **35**, 17–28 (2006)
12. Xu, Z.S.: A method based on linguistic aggregation operators for group decision making with linguistic preference relations. Inf. Sci. **166**, 19–30 (2004)
13. Herrera, F., Martínez, L.: A 2-tuple fuzzy linguistic representation model for computing with words. IEEE Trans. Fuzzy Syst. **8**, 746–752 (2000)
14. Herrera, F., Martínez, L.: A model based on linguistic 2-tuples for dealing with multigranular hierarchical linguistic contexts in multi-expert decision-making. IEEE Trans. Syst. Man Cybern. Part B Cybern. **31**, 227–234 (2001)

15. Xu, Y.L., Huang, L.: An approach to group decision making problems based on 2-tuple linguistic aggregation operators. In: ISECS International Colloquium on Computing, Communication, Control, and Management, pp. 73–77. IEEE Computer Society, Guangzhou (2008)
16. Zhang, Z.: 2-tuple prioritized aggregation operators and their application to multiple attribute group decision making. British J. Math. Comput. Sci. **4**, 2278–2297 (2014)
17. Yager, R.R.: Prioritized aggregation operators. Int. J. Approx. Reason. **48**, 263–274 (2008)

Reasoning Under Conflicts in Smart Environment

Hela Sfar(✉), Badran Raddaoui, and Amel Bouzeghoub

CNRS Paris Saclay, Telecom SudParis, SAMOVAR, Paris, France
{hela.sfar,badran.raddaoui,amel.bouzeghoub}@telecom-sudparis.eu

Abstract. This paper addresses the problem of reasoning under conflicts within rule-based systems. Nowadays, using logical rules to infer knowledge, express restrictions and so on has shown a great interest in several domains which makes rule-based systems very popular. However, since rules may originate from heterogeneous sources and change over time, methods are required to maintain rules consistency by detecting and resolving the conflicts. The method we put forward in this work refines a form of non-monotonic reasoning called argumentation with new definitions. Through implementing and evaluating the framework to a smart home test-bed, the proposed framework dynamically detects and flexibly resolves conflicts. The result proves the efficiency and viability of our proposal.

Keywords: Smart environment · Conflict detection and resolution · Argumentation theory

1 Introduction

Rule-based systems differ from standard procedural or object-oriented programs in that there is no clear order in which code executes. Instead, the knowledge of the expert is captured in a set of rules, each of which encodes a small piece of the expert's knowledge. Rule-based systems are used in several domains including ubiquitous computing, sensors networks, and so on. In this paper we focus on rule-based systems in ubiquitous computing particularly for smart environment applications. As a matter of fact, as rules are defined by human experts, conflicts may occur during their update or management, when different human experts contribute to their definition or when there is a contradiction in the sources of information. As a result, a conflict decreases the system performance and changes the behavior of that system. To illustrate the problem, let us consider the following scenario:

Scenario 1: *Mary is an elderly woman living alone in her smart house. The house is equipped with multiple sensors and a rule-based framework is in charge of activity recognition depending on context data. During the set up of the system and according to the life routine of Mary, the expert defines a logic rule stating that it is possible to conclude that Mary is eating Bread ("EatBread") only*

D. Liu et al. (Eds.): ICONIP 2017, Part III, LNCS 10636, pp. 924–934, 2017.
https://doi.org/10.1007/978-3-319-70090-8_94

if Mary has fetched the Knife ("FetchKnife") and fetched the Bread ("FetchBread"). However, due to some health issues, the life routine of Mary changes: for eating bread she must be sitting also. Therefore, another expert updates the set of rules and adds another rule stating that the "EatBread" can be inferred from the three conditions "FetchKnife", "FetchBread" and "SitOn". This latter, nevertheless, does not remove the previous rule which gives rise to the creation of two conflicting rules in the system. As a result, erroneous activities are provided by the system which decreases its accuracy.

Scenario 1 illustrates possible conflicts that may often occur in rule-based systems for smart environments. To the best of our knowledge, this type of conflict which we call *contrariness* is not treated in the literature. Indeed, previous works on rule-based systems focus in general on handling conflict only based on *inconsistency* [11,12]. More precisely, existing approaches omit this form of conflicts that frequently occur in smart environment where additional preconditions are required for the conclusion of a rule to hold.

In this paper, we propose to address this problem by adopting argumentation theory to reason over conflicts in rule-based systems. Argumentation theory consists of a set of arguments, attacks between them and a semantics for identifying the acceptable arguments. An argument is a reason for believing a statement, doing an action, etc. One of the key aims of argumentation research is to provide principled techniques for handling conflicts. Various approaches have been proposed to formalize argumentation and there are two major families of structured argumentation systems: deductive argumentation [9] and defeasible argumentation [11,12]. Despite the popularity of these systems, the attack among arguments is mainly rooted by inconsistency. However, real-life (counter)-arguments do not necessarily appear mutually inconsistent. It is the exploration of this gap, then, that this paper is aimed at which is the advantage of this work. The paper presents an argumentation framework in the context of rule-based systems where the notion of counter-argumentation is introduced through two sorts of conflicts, namely, *inconsistency* and *contrariness* among two arguments.

To resume, two main contributions are presented in this work:

1. A proposal of a generic model for reasoning under conflicts in rule-based systems through the exploitation of argumentation theory. Particularly, for smart environments besides *inconsistency* of rules they also may be *contrariness*.
2. Usually, previous argumentation based methods for conflict resolution are generally validated or checked theoretically. To the best of our knowledge, this method has not been studied and applied in smart environment systems. In this paper, we first study the effect of conflicting rules into a rule-based system (AGACY Monitoring [8] for activity recognition in smart environments). Then, we show how our argumentation based method can be used to resolve all the conflicts among rules.

2 Related Work

In this work, we classify conflicts into two main categories: (1) Related rules conflicts [1,3] and (2) Related events conflicts [2,4]. The related rules conflicts category is about detecting and resolving conflicting logical rules within the system. Resolving conflicts coming from rules evolution and change is recently studied in [3]. The authors proposed a dynamic system for nutrition recommendation for elderly people based on the recognized activities within their smart home. The system is built based on ontologies and logical rules. In order to resolve the conflict that can occur during the rules evolution, the authors presented an algorithm that is triggered whenever any change of rules has been made. In the same vein, another work has been proposed by [1]. The authors designed a new method for detecting and resolving conflicting logical rules in wireless sensor networks. The proposed resolution method is based on the Feature Interaction (FI) techniques. In this work, different strategies for conflict detection and resolution according to the nature of conflict have been studied.

The category of related events conflict states that the source of conflict is not the rules format but the events in the environment that are the content of the rule. For example, this category of conflict can concern resolving conflicting activities of multiple users within a smart environment. To tackle this issue, the work in [2] introduced a method for resolving conflicts of multiple users who share context-aware applications within a smart home. Then, the authors proposed an ontology-based technique for conflict resolution. Furthermore, another work proposed a reminder system for elderly person living in a smart home and a method to resolve the conflict in the schedule coming from different events [4]. Conflicts in user contexts within ambient intelligent systems has been also studied in [5]. In fact, the authors introduced a formal model that enables the specification of an intelligent environment's components, whose interaction can lead to conflicts.

In this work, we are interested in recognition and resolution of rules related conflict. As we can see, this problematic has not been studied enough in the literature. Each of the studied work previously mentioned tackles the problem of conflicting rules and proposes a solution that can be applied only into its particular case of study that is for example the wireless sensors network in [3]. Therefore, the surveyed works suffer from the lack of a general model to recognize and resolve the conflicting rules in rule-based systems. To overcome this problem, we propose an argumentation based framework that is able to detect, explain, and resolve different kinds of conflicts among rules and it can be applied in any use case.

3 Argumentation Framework for Rules-Based Systems

Argumentation is a form of reasoning that studies different ways in which an argument for or against a proposition can be made, and then determine which arguments attack which other arguments in an effort to decide what can be reasonably concluded. In this section, we introduce a formalism, and investigate a natural extension of deductive argumentation's well-known model of argument systems [9].

3.1 Rule-Based Knowledge Representation

A logical rule is usually expressed in the form: "if the precondition ϕ holds then the conclusion ψ holds", where the precondition and the conclusion are logical properties. However, such a piece of knowledge cannot simply be formalized through the standard material implication of classical logic: in other words, it is not possible to capture the intended meaning of the above statement by an implication in classical first-order logic of the form $\phi \rightarrow \psi$. In this respect, a very relevant role is played by research in logic programming. In fact, logic program rules are implications with a non-standard semantics. In the sequel, we will consider a simple language $\mathcal{L}_R$ constructed from a set of rules where each rule is of the form $\alpha_1 \wedge \ldots \wedge \alpha_n \rightarrow \beta$ where α_1 to α_n and β are literals, and let $\vdash$ be the consequence relation for that language. The language $\mathcal{L}_R$ will be used throughout the paper in order to represent data in smart environment systems.

Since rules in rule-based systems for smart environments may originate from distinct sources and change over time, methods are required to maintain rule consistency. For this, we define the notion of *contrariness* between a pair of rules of $\mathcal{L}_R$ that is intended to encompass both standard inconsistency and a form of disagreement between them. The main idea is to replace classical inconsistency between conflict rules with a more general notion of conflict. Let us introduce the concept progressively before introducing the formal definition.

Let α and β be two rules of $\mathcal{L}_R$. α *is the contrary* of β in any of the following cases. First, α and β are mutually inconsistent in $\mathcal{L}_R$. Note that this also covers the standard-logic occurrences of conflict. Second, we need to address the case that requires α of the form $\phi \rightarrow \psi$ to contrary β of the form $\phi \wedge \gamma \rightarrow \psi$. First, α and β must be two rules about the same conclusion, hence $\psi \equiv \psi'$[1]. Second, we require the precondition of the first rule to entail the precondition of the second one but not conversely; formally, for $\alpha = \phi \rightarrow \psi$ and $\beta = \phi' \rightarrow \psi'$, we should have $\phi \vdash \phi'$ and $\phi' \nvdash \phi$.

Definition 1. *Let α and β be two rules of $\mathcal{L}_R$ such that $\alpha = \phi \rightarrow \psi$ and $\beta = \phi' \rightarrow \psi'$. Then, α is the contrary of β, denoted $\alpha \bowtie \beta$, iff:*

- *$\{\alpha, \beta\}$ is inconsistent, or*
- *$\phi \vdash \phi'$, $\phi' \nvdash \phi$, and $\psi \equiv \psi'$.*

Example 1. Let $\alpha = FetchKnife \wedge FetchBread \rightarrow EatBread$ and $\beta = FetchKnife \wedge FetchBread \wedge SitOn \rightarrow EatBread$ be two rules. Then, $\beta \bowtie \alpha$.

The concept of contrariness of a rule is naturally extended into a concept of contrariness of a set of rules.

Definition 2. *Let Φ and α be a subset and a rule of $\mathcal{L}_R$, respectively. α is the contrary of Φ, denoted $\alpha \bowtie \Phi$, iff there exists β in $\mathcal{L}_R$ such that $\Phi \vdash \beta$ and $\alpha \bowtie \beta$.*

[1] The symbol $\equiv$ denotes the logical equivalence, i.e. $\psi \vdash \psi'$ and $\psi' \vdash \psi$.

Obviously, $\bowtie$ is neither symmetric, nor antisymmetric, nor antireflexive. However, it is monotonic as shown by the following proposition.

Proposition 1. *Let Φ, Ψ, and α be two subsets and a rule of $\mathcal{L}_R$, respectively. If $\alpha \bowtie \Phi$, then $\alpha \bowtie \Phi \cup \Psi$.*

It is easy to show the following result.

Proposition 2. *Let Φ, Ψ, and α be two subsets and a rule of $\mathcal{L}_R$, respectively. If $\alpha \equiv \beta$, then $\alpha \bowtie \Phi$ iff $\beta \bowtie \Phi$.*

Next, we define our argumentation framework in the light of the notion of contrariness.

3.2 Arguments

In the literature, deductive argumentation adopts solely consistency to build arguments [9]. However, in the context of rule-based reasoning, we are interested in defining a stronger notion of argument and encompass a specific form of conflict that often occurs in real-life argumentation: i.e., claims that additional preconditions are required for the conclusion of a rule to hold. In what follows, the existence of a finite set of rules K from $\mathcal{L}_R$ is assumed.

Definition 3 (Argument). *An argument for a rule α with respect to K from $\mathcal{L}_R$ is a pair $\langle \Phi, \alpha \rangle$ s.t. the following conditions hold: (1) $\Phi \subseteq K$, (2) $\forall \beta$ s.t. $\Phi \vdash \beta$, $\beta \not\bowtie \Phi$ (3) $\Phi \vdash \alpha$, and (4) $\forall \Phi' \subset \Phi$, $\Phi' \nvdash \alpha$*

For an argument $\langle \Phi, \alpha \rangle$, we call Φ the *support*, α the *conclusion* of the argument. We also say that $\langle \Phi, \alpha \rangle$ is an argument for α. Notice that the usual non-contradiction condition in deductive argumentation expressed by $\Phi \nvdash \bot$ is naturally extended and replaced by a non-contrariness requirement (Condition (2)).

The following example shows that whilst it is clear that taking consistency of supports into account is a valuable development when generating arguments, it is not just consistency, per se, that is important.

Example 2. Let consider the set of rules $K = \{FetchBread \rightarrow EatBread, FetchKnife \wedge FetchBread \rightarrow EatBread, FetchKnife \wedge FetchBread \wedge SitOn \rightarrow EatBread, FetchKnife \rightarrow EatBread\}$. Then, some arguments of K are: $\langle \{FetchKnife \wedge FetchBread \rightarrow EatBread\}, FetchKnife \wedge FetchBread \rightarrow EatBread \rangle$ and $\langle \{FetchKnife \wedge FetchBread \wedge SitOn \rightarrow EatBread\}, FetchKnife \wedge FetchBread \wedge SitOn \rightarrow EatBread \rangle$. However, $\langle \{FetchBread \rightarrow EatBread, FetchKnife \rightarrow EatBread\}, FetchBread \wedge FetchKnife \rightarrow EatBread \rangle$ is not an argument.

A notion of *quasi-equivalent* arguments is now introduced as follows. It is intended to capture situations where two arguments can be said to make the same point on the same grounds. Not surprisingly, for an argument, its quasi-equivalent ones can be infinite.

Definition 4 (Quasi-equivalence). *Two arguments* $\langle \Phi, \alpha \rangle$ *and* $\langle \Phi, \beta \rangle$ *are quasi-equivalent iff* $\alpha \equiv \beta$.

Property 1. Let $\langle \Phi, \alpha \rangle$ be an argument. There is an infinite set of arguments of the form $\langle \Psi, \beta \rangle$ s.t. $\langle \Phi, \alpha \rangle$ and $\langle \Psi, \beta \rangle$ are quasi-equivalent.

Arguments are not necessarily independent. The definition of *more conservative arguments* captures a notion of subsumption between arguments, translating situations where an argument is in some sense contained within another one.

Definition 5 (Dependency). *An argument* $\langle \Phi, \alpha \rangle$ *is more conservative than an argument* $\langle \Psi, \beta \rangle$ *iff* $\Phi \subseteq \Psi$, $\beta \vdash \alpha$.

Then, a more conservative argument can be seen as more general in the sense that it is less demanding on the support and less specific with respect to the conclusion.

Actually, the concept of being more conservative induces the concept of quasi-equivalent arguments, and conversely.

Proposition 3. *Two arguments* $\langle \Phi, \alpha \rangle$ *and* $\langle \Psi, \beta \rangle$ *are quasi-equivalent iff each one is more conservative than the other.*

The notions of quasi-equivalence and of being more conservative will be used in the next subsection to avoid some redundancy when counter-arguments need to be generated.

3.3 Conflicts Among Arguments

Based on the notion of contrariness $\bowtie$ introduced in Subsect. 3.1, let us now explain how arguments can be challenged. To do so, we will distinguish tow kinds of counter-argument: *rebuttal* and *defeater*.

Definition 6 (rebuttal). *An argument* $\langle \Psi, \beta \rangle$ *is a rebuttal for an argument* $\langle \Phi, \alpha \rangle$ *iff* $\beta \bowtie \alpha$.

Clearly, the notion of rebuttal defined for arguments in $\mathcal{L}_R$ is thus asymmetric, since the relation $\bowtie$ is not symmetric.

Example 3. The argument $\langle \{FetchKnife \wedge FetchBread \wedge SitOn \rightarrow EatBread\}, FetchKnife \wedge FetchBread \wedge SitOn \rightarrow EatBread \rangle$ is a rebuttal for $\langle \{FetchKnife \wedge FetchBread \rightarrow EatBread\}, FetchKnife \wedge FetchBread \rightarrow EatBread \rangle$.

An argument attacks another if its conclusion is the contrary of some rules in the other. This leads to the notion of *defeater* defined as follows:

Definition 7 (Defeater). *An argument* $\langle \Psi, \beta \rangle$ *is a defeater for* $\langle \Phi, \alpha \rangle$ *iff* $\beta \bowtie \Phi$.

The next proposition shows a general relation between the two kinds of conflicting arguments, i.e. rebuttal and defeater.

Proposition 4. *If an argument* $\langle \Psi, \beta \rangle$ *is a rebuttal for* $\langle \Phi, \alpha \rangle$ *then* $\langle \Psi, \beta \rangle$ *is a defeater for* $\langle \Phi, \alpha \rangle$.

As intended, defeaters can exist even if there is no inconsistency involved, as shown by Example 3: the support of the defeater is consistent with the support of the argument A it attacks ; but, the attack relation is mainly based on the second form of contrariness among rules.

Since defeaters are arguments, then they can be ordered from more conservative to less conservative as shown by the following definition.

Definition 8 (Maximally conservative defeater). *An argument* $\langle \Psi, \beta \rangle$ *is a maximally conservative defeater for* $\langle \Phi, \alpha \rangle$ *iff* $\langle \Psi, \beta \rangle$ *is a defeater for* $\langle \Phi, \alpha \rangle$ *such that no defeaters for* $\langle \Phi, \alpha \rangle$ *are strictly more conservative than* $\langle \Psi, \beta \rangle$.

We assume that there exists an enumeration which we call canonical enumeration of all maximally conservative defeaters for $\langle \Phi, \alpha \rangle$.

Now, in order to avoid some redundancy among counter-arguments we can ignore the unnecessary variants of maximally conservative defeaters. To do so, we introduce the notion of *rational defeaters* as follows.

Definition 9 (rational defeater). *Let* $\langle \Psi_1, \beta_1 \rangle, \ldots, \langle \Psi_n, \beta_n \rangle, \ldots$ *be the canonical enumeration of all maximally conservative defeaters for* $\langle \Phi, \alpha \rangle$. *Then,* $\langle \Psi_i, \beta_i \rangle$ *is a rational defeater for* $\langle \Phi, \alpha \rangle$ *iff* $\forall\ j < i$, $\langle \Psi_i, \beta_i \rangle$ *and* $\langle \Psi_j, \beta_j \rangle$ *are not quasi-equivalent.*

Henceforth, the rational defeaters will gather all the possible attacks of a given argument in the same ones: the ones which are representative of all defeaters for that arguments.

3.4 Argument Graph

Given a rule-based knowledge base, the idea is to gather arguments and counter-arguments in the same structure in order to capture the way arguments can take place as a dispute develops. To do so, in this subsection, we can collect and organize all these arguments and rational defeaters in the same structure. If we construct an argument graph based on the arguments that can be constructed from a rule-based knowledge base, then we can be exhaustive in including all arguments and rational defeaters that can be constructed from that base. An argument graph is a directed graph where each node denotes an argument and each edge denotes an attack by one argument on another. Hence, when one argument is a counterargument to another argument, this is represented by an edge. More formally, given a rule-based knowledge base K, we can generate an argument graph $G = (\mathcal{A}, \mathcal{R})$ where $\mathcal{A}$ is the set of arguments obtained from K as mentioned previously in Definition 3 and $\mathcal{R}$ is a rational defeater relation.

Definition 10 (Argument graph). *Let K be a rule-based knowledge base. An argument graph for K is a graph $G = (\mathcal{A}, \mathcal{R})$ where $\mathcal{A}$ is the set of arguments generated from K and $\mathcal{R} = \{(A, B) | A, B \in \mathcal{A}$ and A is a rational defeater for $B\}$.*

Now, in order to evaluate the acceptability of arguments of a given argument graph we adopt Dung's argumentation system [10]. Given an argument graph for K and two arguments A and B from $\mathcal{A}$, we say that A *attacks* B if there is $(A, B) \in \mathcal{R}$. Moreover, a set $S \subseteq \mathcal{A}$ *attacks* an argument $B \in \mathcal{A}$ if there is $A \in S$ such that A attacks B. A set $S \subseteq \mathcal{A}$ is said to be *conflict-free* if there are no arguments $A, B \in S$ such that A attacks B. Moreover, an argument $A \in \mathcal{A}$ is *defended* by a set $S \subseteq \mathcal{A}$ iff $\forall B \in \mathcal{A}$ such that B attacks A, there is $c \in S$ such that c attacks b.

Using the notions of conflict-freeness and defense, we can define a number of argumentation semantics, each embodying a particular rationality criterion, in order to identify reasonable sets of arguments, called *extensions*.

Given a rule-based knowledge base K and an argument graph $G = (\mathcal{A}, \mathcal{R})$. A set $\mathcal{E} \subseteq \mathcal{A}$ of arguments is said to be:

- *admissible* iff $\mathcal{E}$ is conflict-free and all its arguments are defended by $\mathcal{E}$,
- a *complete* extension iff $\mathcal{E}$ is admissible and $\mathcal{E}$ contains all and only the arguments it defends,
- a *grounded* extension iff $\mathcal{E}$ is a minimal (w.r.t. set inclusion) complete set of arguments,
- a *preferred* extension iff $\mathcal{E}$ is a maximal (w.r.t. set inclusion) admissible set of arguments.

Basically, each of these semantics corresponds to some properties which certify whether a set of arguments can be profitably used to support a point of view in a dispute. In the sequel, we consider preferred semantics, which represents the main contribution in Dung's theory, as it allows multiple extensions (differently from grounded semantics), the existence of extensions is always guaranteed, and no extension is a proper subset of another extension (differently from complete semantics).

Example 4. Let the four arguments A, B, C and D constructed from the rule-based knowledge base $K = \{FetchKnife \wedge FetchBread \rightarrow EatBread, FetchBread \rightarrow EatBread, FetchBread \wedge SitOn \rightarrow EatBread, FetchKnife \wedge FetchBread \wedge SitOn \rightarrow EatBread\}$. The argument graph associated to K is depicted in Fig. 1 where $A = \langle\{FetchBread \wedge SitOn \rightarrow EatBread\}, FetchBread \wedge SitOn \rightarrow EatBread\rangle$, $B = \langle\{FetchBread \rightarrow EatBread\}, FetchBread \rightarrow EatBread\rangle$, $C = \langle\{FetchKnife \wedge FetchBread \wedge SitOn \rightarrow EatBread\}, FetchKnife \wedge FetchBread \wedge SitOn \rightarrow EatBread\rangle$, and $D = \langle\{FetchKnife \rightarrow EatBread\}, FetchKnife \rightarrow EatBread\rangle$. Let us determine the extensions of the argument graph of K under the preferred semantics. Clearly, $\mathcal{E} = \{C\}$ is the preferred extension of the argument graph.

4 Empirical Evaluation

In order to evaluate the proposed method for conflict resolution in rule-based systems, we tested our method into our previous work AGACY Monitoring system [8] for activity recognition in smart environments. The system returns the user activity for a defined time window. To achieve this aim it has two layers: (1) A semantic layer and (2) A data driven layer. The former is in charge of the inference of the events and actions that have been done by the user. This inference process requires the definition of logical rules from human experts. The second layer applies the Demspter Shafer Theory (DS) [6] or the Support Vector Machine (SVM) [7] for the activity recognition based on the inferred actions and events. The system was tested in [8] with Opportunity dataset[2]. The dataset holds (from AGACY Monitoring perspective) sensors data, events, actions, and activities that have been done by three persons with three different routines. According to the actions within the dataset, a set of 25 rules has been defined. Since the number of rules is not high and the rules have been defined by one expert, they do not contain any conflict. Consequently, for the same dataset, we have added in the system a number $nbCR$ of conflicting logical rules. Then, we have applied our argumentation-based method in order to resolve conflicts among rules. Firstly, we measured the precision of the system with 3 conflicting logical rules for different time windows with DS and SVM. The result of this experiment is depicted in Fig. 2. Then, with a time window equal to 180 s, we measured the precision of the system by varying the number of added conflicting rules as shown in Fig. 3. Notice that we choose the value 180 s for time window since it is the most suitable time window value to provide the best precision [8]. Finally, we applied the

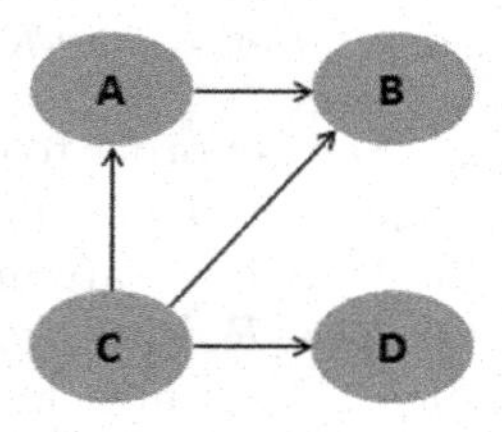

Fig. 1. An argument graph for K.

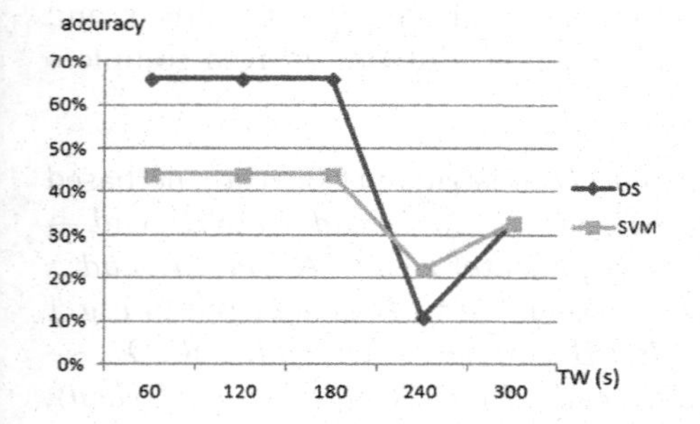

Fig. 2. Average recognition precision of AGACY Monitoring for all subjects over the three routines for three conflicting rules with different values of TW

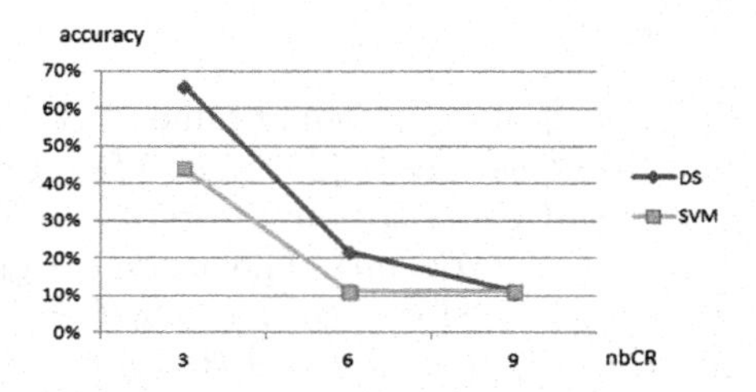

Fig. 3. Average recognition precision of AGACY Monitoring for all subjects over the three routines for TW = 180 s with different number of conflicting rules (nbCR)

[2] http://webmind.dico.unimi.it/care/annotations.zip.

proposed method for conflict resolution and we measured the precision of the system for the different time window values. The result is shown in Fig. 4.

On the one hand, we can observe in Fig. 2 that with 3 conflicting rules the system has an average precision level (66% for DS and 44% for SVM) for TW∈ [60 s..180 s]. Then, the precision value decreases with the increasing value of TW to be equal to 11% with TW = 240 s. On the other hand, by varying the number of conflicting rules with TW = 180 s we clearly see in Fig. 3 that the system has a poor precision level with 6 and 9 conflicting rules. Consequently, the presence of conflicting rules in such system has a great effect on its performance that decreases with the increasing number of conflicting rules. The resolution of the added 9 conflicting rules through our proposed argumentation based method (Sect. 3) has obviously increased the precision level of the system as depicted in Fig. 4. As we can see, the system with DS has a precision value of 91% with TW=180s instead of 10% of precision with 9 conflicting rules. The accuracy curves of AGACY Monitoring obtained after the conflict resolution are similar to that without conflict [8]. This result proves that our proposed method is perfectly able to resolve all conflicts in the system. Moreover, besides than decreasing the precision level, the conflicting rules have also changed the behavior of the system. As a matter of fact, the Fig. 2 shows that for TW <= 180 s, the DS and SVM have the same accuracy values. Nevertheless, for TW > 180 s the SVM works better. On the one hand, this behavior is different from the normal one (without conflict): The SVM has better precision than DS for TW < 180 s otherwise the DS works better than SVM.

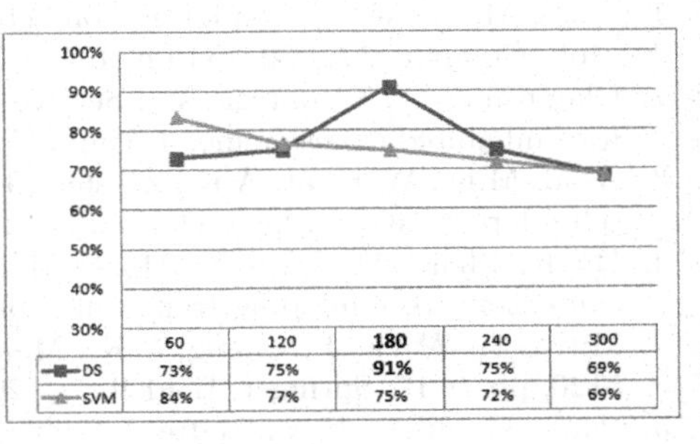

	60	120	180	240	300
DS	73%	75%	91%	75%	69%
SVM	84%	77%	75%	72%	69%

Fig. 4. Average recognition precision of AGACY Monitoring for all subjects over the three routines for different values of TW after conflict resolution

5 Conclusion

In this paper we proposed a new method for conflict handling in rule-based systems. The method is based on the exploitation of argumentation theory to resolve conflicts in smart environments. We evaluated the proposal into an activity recognition system for smart homes to resolve the added conflicting rules. The method has proved its ability to resolve all the added conflicts.

References

1. Evan, M., Jesse, B.: Exploring conflicts in rule-based sensor networks. Pervasive Mob. Comput. **27**, 133–154 (2016)
2. Choonsung, S., Woontack, W.: Service conflict management framework for multi-user inhabited smart home. J. Univ. Comput. Sci. **15**, 2330–2352 (2009)
3. Asad, M.K., Wajahat, A.K., Zeeshan, P., Farkhund, I., Sungyoung, L.: Towards a self adaptive system for social wellness. Sensors **16**, 531–548 (2016)
4. Du, K., Zhang, D., Zhou, X., Mokhtari, M., Hariz, M., Qin, W.: HYCARE: a hybrid context-aware reminding framework for elders with mild dementia. In: Helal, S., Mitra, S., Wong, J., Chang, C.K., Mokhtari, M. (eds.) ICOST 2008. LNCS, vol. 5120, pp. 9–17. Springer, Heidelberg (2008). doi:10.1007/978-3-540-69916-3_2
5. Paulo, C., Silvia, R.A., Andre, C.S.: Towards automatic conflict detection in home and building automation system. Pervasive Mob. Compu. **12**, 37–57 (2014)
6. Zadeh, L.A.: A simple view of the Dempster-Shafer Theory of evidence and its implication for the rule of combination. AI Mag. **7**, 85–90 (1986)
7. Hearst, M.A., Dumais, S.T., Osuna, E., Platt, J., Scholkopf, B.: Support vector machines. IEEE Intell. Syst. Appl. **13**, 18–28 (1998)
8. Sfar, H., Bouzeghoub, A., Ramoly, N., Boudy, J.: AGACY monitoring: a hybrid model for activity recognition and uncertainty handling. In: Blomqvist, E., Maynard, D., Gangemi, A., Hoekstra, R., Hitzler, P., Hartig, O. (eds.) ESWC 2017. LNCS, vol. 10249, pp. 254–269. Springer, Cham (2017). doi:10.1007/978-3-319-58068-5_16
9. Besnard, P., Hunter, A.: A logic-based theory of deductive arguments. Artif. Intell. **128**, 203–235 (2001)
10. Phan Minh, D.: On the acceptability of arguments and its fundamental role in nonmonotonic reasoning, logic programming and n-person games. Artif. Intell. **77**, 321–358 (1995)
11. Javier García, A., Ricardo Simari, G.: Defeasible logic programming: an argumentative approach. TPLP **4**, 95–138 (2004)
12. Amgoud, L., Besnard, P.: A formal characterization of the outcomes of rule-based argumentation systems. In: Liu, W., Subrahmanian, V.S., Wijsen, J. (eds.) SUM 2013. LNCS, vol. 8078, pp. 78–91. Springer, Heidelberg (2013). doi:10.1007/978-3-642-40381-1_7

Author Index

Printed in the United States
By Bookmasters

Printed in the United States
By Bookmasters